FRONTIERS
IN
CATECHOLAMINE RESEARCH

FRONTIERS
IN
CATECHOLAMINE RESEARCH

*Proceedings of the Third International Catecholamine Symposium
Held at the University of Strasbourg, Strasbourg, France
May 20–25,1973*

Edited by

EARL USDIN, Ph.D.
Psychopharmacology Research Branch,
National Institute of Mental Health,
Rockville, Maryland

and

SOLOMON H. SNYDER, M.D.
Department of Pharmacology and Experimental Therapeutics,
Johns Hopkins University School of Medicine,
Baltimore, Maryland

PERGAMON PRESS INC.

New York · Toronto · Oxford · Sydney · Braunschweig

PERGAMON PRESS INC.
Maxwell House, Fairview Park, Elmsford, N.Y. 10523

PERGAMON OF CANADA LTD.
207 Queen's Quay West, Toronto 11′, Ontario

PERGAMON PRESS LTD.
Headington Hill Hall, Oxford

PERGAMON PRESS (AUST.) PTY. LTD.
Rushcutters Bay, Sydney, N.S.W.

VIEWEG & SOHN GmbH
Burgplatz 1, Braunschweig

Copyright © 1973, Pergamon Press Inc.
Library of Congress Catalog Card No. 73-17046

Library of Congress Cataloging in Publication Data

International Catecholamine Symposium, 3rd, University of Strasbourg, 1973.
Frontiers in catecholamine research.

1. Catecholamines—Congresses.   I. Usdin, Earl, ed.   II. Snyder, Solomon H., 1938– ed.
III. Title [DNLM: 1. Catecholamines—Congresses. W3 IN1265 1973f/QV129 I601 1973f]
     QP801.C3315 1973   599′.01′9245   73-17046 ISBN 0–08–017922–3

Set by the Universities Press, Belfast. Printed by A. Wheaton & Co. Exeter

CONTENTS

v

III. Regulation

Chairmen: LESLIE L. IVERSEN and B. R. BELLEAU

Councillor: THEODORE SOURKES

Reporter: NORMAN WEINER

IV. Synaptic Dynamics of Receptors

Chairmen: ROBERT F. FURCHGOTT and NEIL C. MORAN
Councillor: SYDNEY SPECTOR
Reporter: G. ALAN ROBISON

V. Synaptic Dynamics

Chairmen: HERMANN K. F. BLASCHKO, ULF S. VON EULER, PARKHURST
            A. SHORE, ULLRICH TRENDELENBURG

Councillors: ANNICA DAHLSTROM and GEORGE HERTTING

Reporters: ROSS J. BALDESSARINI and BERTIL HAMBERGER

VI. Catecholamines in Central Nervous System

Chairmen: JOEL ELKES, MARTHE VOGT, ALBERT SJOERDSMA and HANS
             WEIL-MALHERBE

Councillors: DONALD REIS and CHRISTER OWMAN

Reporters: BARRY HOFFER and MERTON SANDLER

VIII. Catecholamines in Man

Frontiers in Catecholamine Research 1973, pp. 1 to 24. Pergamon Press. Printed in Great Britain.

INTRODUCTION

EARL USDIN and SOLOMON H. SNYDER

HISTORICAL

WHEREAS the origins of man are in dispute, the origins of the Third International Catecholamine Symposium are quite clear. At the ceremonies attendant to the 25th anniversary celebration of the founding of the National Institute of Mental Health in June of 1971, the seed was first brought forth. This seed was nurtured at lunch meetings held in conjunction with almost every major psychopharmacological conference held over the next year or so. During the course of these meetings, an Organising Committee was formed. The final composition of the Organising Committee is as follows:

SOLOMON H. SNYDER (Chairman), Johns Hopkins University
EARL USDIN (Executive Secretary), National Institute of Mental Health
PAUL MANDEL (Chairman, Strasbourg Organising Committee), University
of Strasbourg
JULIUS AXELROD, National Institute of Mental Health
FLOYD BLOOM, National Institute of Mental Health
WILLIAM E. BUNNEY, Jr., National Institute of Mental Health
ARVID CARLSSON, University of Göteborg
ERMINIO COSTA, National Institute of Mental Health
JACQUES DE CHAMPLAIN, University of Montreal
KJELL FUXE, Karolinska Institutet
SILVIO GARATTINI, Istituto di Ricerche Farmacologi 'Mario Negri'
JACQUES GLOWINSKI, College de France
MENEK GOLDSTEIN, New York University
LESLIE L. IVERSEN, Cambridge University
SEYMOUR KETY, Harvard University
NORMAN KIRSCHNER, Duke University
IRWIN J. KOPIN, National Institute of Mental Health
MORRIS LIPTON, University of North Carolina
WILLIAM MANGER, New York University
JOSÉ MUSACCHIO, New York University
ALFRED PLETSCHER, Hoffmann-LaRoche & Company
THEODORE RALL, Case-Western Reserve University
MERTON SANDLER, Queen Charlotte's Hospital
JOSEPH SCHILDKRAUT, Harvard University
ALBERT SJOERDSMA, Merrell International
THEODORE SOURKES, McGill University
SYDNEY SPECTOR, Roche Institute of Molecular Biology
HANS THOENEN, University of Basle
SIDNEY UDENFRIEND, Roche Institute of Molecular Biology

URBAN UNGERSTEDT, Karolinska Institutet
ULF S. VON EULER, Karolinska Institutet
NORMAN WEINER, University of Colorado
RICHARD WURTMAN, Massachusetts Institute of Technology

Among the crucial questions which had to be decided by the Organising Committee or one of its subcommittees were the site of the Symposium, dates, subject coverage, speakers, format, publication, financial support. Although the Israeli Center for Psychobiology offered to provide partial support for the Symposium if the meeting were held in Israel, the Organising Committee reluctantly decided to decline this offer since the additional travel costs were deemed too high. Professor Paul Mandel kindly offered to organise a meeting at the University of Strasbourg with the very generous support of Dr. Albert Sjoerdsma and the Merrell International Research Center. The extremely efficient Strasbourg Local Organising Committee consisted of the following: P. Mandel (Chairman), A. Sjoerdsma, J. C. Hermetet, I. G. Morgan and G. Vincedon.

Selection of the dates of May 20–May 25, 1973 for the Symposium allowed the meeting to be sandwiched in between a meeting in Stockholm on dynamics of degeneration and growth in neurons and a meeting in Sardinia on serotonin. In addition, the Scottish Rite scheduled a conference in Strasbourg on catecholamines and their enzymes in the neuropathology of schizophrenia. A Stockholm–Strasbourg–Sardinia circuit permitted one to be brought up to date at almost all levels of catecholaminology.

It rapidly became clear to the program subcommittee (headed by Dr. Floyd Bloom) that it would be impossible to include both adequate reviews as well as current research in the scope of the meeting. It was opted to concentrate on the latter and hence the title for this meeting: *Frontiers in Catecholamine Research*.

The problem of selection of speakers was never satisfactorily resolved. At the time of the first two International Catecholamine Symposia it was possible to schedule all speakers who wished to make presentations, but so much is now being accomplished in catecholamine research by so many that it just was not possible to schedule all of the excellent talks which were proposed, even though the Organising Committee reluctantly adopted a two simultaneous sessions format. It is recognised that the selection process, in spite of honest efforts to be fair, represents the bias of the organisers. To those who were not given adequate opportunity to present their results, apologies must be offered.

Several mechanisms were used to compensate for the fact that there were two simultaneous sessions. In addition to circulating abstracts of each of the scheduled talks, at the end of each day a Reporter from Session I summarised both the formal presentations and the open discussions in the Session II room while a Reporter from Session II summarised in the Session I room. In an attempt to maximise utilisation of the limited open discussion time, a Councillor was appointed for each session. The Councillor condensed questions and comments which were submitted in response to the formal presentations. Summaries of open discussions at individual sessions have been prepared by designated Reporters and are included at the end of each chapter.

To expedite publication, each author was required to submit his manuscript at the time of the meeting. It was extremely gratifying to receive 220 of the 222 manuscripts before the meeting closed. Primarily for archival purposes, the talks will be published not only in this volume but also in *Biochemical Pharmacology* or the *Journal of Neurochemistry*.

SUPPORT

This meeting would not have been possible without the generous sponsorship of the National Institute of Mental Health (Grant No. MH 22927-01) and the Merrell International Research Center.

We should also like to acknowledge the important contribution made by donations from the following organisations:

Abbott Laboratories, Ltd.
Astra Pharmaceutical
Beecham Research Labs.
Boots Company, Ltd.
Ciba Laboratories
Du Pont
Farbenwerke Hoechst A. G.
Geigy Pharmaceuticals
Hoffmann-LaRoche
Imperial Chemical Industries, Ltd.
Leitz-France

Eli Lilly & Company
May & Baker Ltd.
Merck Sharp & Dohme
New England Nuclear
Schering
Laboratoires Servier
Smith, Kline & French Laboratories, Ltd.
Ste. Parisienne d'Expansion Chimique
    SPECIA
The Wellcome Trust
Wyeth
Zeiss

Yeoman service in obtaining these funds was given by Dr. Richard Wurtman and Dr. Merton Sandler. Ms. Nancie Brownley masterfully processed the manuscripts and travel funds.

PARTICIPANTS

Twenty-nine countries were represented at the Third International Catecholamine Symposium:

Argentina—1
Australia—5
Austria—3
Belgium—8
Canada—10
China (People's Republic)—3
Czechoslovakia—1
Denmark—7
Finland—6
France—79
Germany—26
Hungary—2
Israel—5
Italy—19
Japan—3

Mexico—1
Netherlands—7
New Zealand—1
Norway—3
Poland—2
Portugal—3
South Africa—1
Sweden—48
Switzerland—16
United Kingdom—43
U.S.A.—169
U.S.S.R.—1
Venezuela—1
Yugoslavia—1

The addresses of the participants is as follows:

Dr. BENGT ÅBLAD
Hassle Ltd.
Kärragatan 5
Mölndal, Sweden

Dr. GEORGE K. AGHAJANIAN
Department of Psychiatry
Yale University
  School of Medicine
34 Park Street
New Haven, Connecticut 06520

Dr. SVEN AHLENIUS
Department of Pharmacology
University of Göteborg
Fack
S-400 33 Göteborg 33
Sweden

Dr. SERGIO ALGERI
Istituto di Ricerche Farm.
  'Mario Negri'
Via Eritrea, 62
20157 Milan, Italy

Dr. OLLE ALMGREN
Department of Pharmacology
University of Göteborg
Fack
S-400 33 Göteborg 33
Sweden

Dr. NILS-ERIK B. ANDEN
University of Göteborg
Fack
S-400 33 Göteborg 33
Sweden

Dr. EVANGELOS T. ANGELAKOS
Department of Physiology
  and Biophysics
Hahnemann Medical College
235 N. 15th Street
Philadelphia, Pennsylvania 19102

Dr. BURTON ANGRIST
Department of Psychiatry
New York University Medical Center
550 First Avenue
New York, New York 10016

Dr. PHILIPPE ASCHER
Laboratoire de Neurobiologie
Ecole Normale Superiure
46, Rue d'Ulm
Paris-5e, France

Mr. COLIN ATACK
Department of Pharmacology
University of Göteborg
Fack
S-400 33 Göteborg 33
Sweden

Dr. D. AUNIS
C.N.R.S., Centre de Neurochimie
67085 Strasbourg Cedex, France

Dr. MARIA BÄCKSTROM
Psychiatric Research Center
Univ. of Uppsala
Ulleraker Hospital
S-750 17 Uppsala, Sweden

Dr. ROSS J. BALDESSARINI
Department of Psychiatry
Harvard Medical School at
  Massachusetts General Hospital
Fruit Street
Boston, Massachusetts 02114

Dr. S. BANERJEE
Department of Pharmacology and
  Experimental Therapeutics
Johns Hopkins University
  School of Medicine
Baltimore, Maryland 21205

Dr. JACK BARCHAS
Department of Psychiatry
Stanford University
  School of Medicine
Palo Alto, California 94304

Dr. G. BARTHOLINI
Medical Research Department
F. Hoffmann-LaRoche and Company
CH-4002, Basel
Switzerland

Dr. P. BASSET
C.N.R.S., Centre de Neurochimie
67085 Strasbourg Cedex, France

Dr. NICOLE BAUMANN
Laboratoire de Neurochimie
Hôpital de la Salpêtrière
47, Boulevard de l'Hôpital
75013 Paris, France.

Dr. HANS GEORG BAUMAGARTEN
Department of Neuroanatomy, U.K.E.
University of Hamburg
Martinistrasse 52
2 Hamburg 20, Germany

Dr. MARCELLE BEAUVALLET
Directeur de Reserche au C.R.N.S.
Inst. de Pharm., Fac. de Med.
21 Rue de l'École de Médecine
Paris-6e, France

Dr. A. H. BECKETT
Department of Pharmacy
Chelsea College
Univesity of London
Manresa Road
London SW3 6LX, England

Dr. JORGE BELDAR
Heymans Institute
De Pintelaan Straat 135
Ghent, Belgium

Dr. B. R. BELLEAU
Department of Chemistry
McGill University
P.O. Box 6070
Montreal 101, Quebec
Canada

Dr. CLAUDE BENSCH
329 Boulevard President Wilson
33200 Bourdeaux-Cauderan, France

Dr. BRIGITTE BERGER
Lab. de Neuropathologie Charles Foix
Hôpital de la Salpêtrière
47, Boulevard de l'Hôpital
75013 Paris, France

Dr. BARRY BERKOWITZ
Roche Institute of Molecular Biology
Nutley, New Jersey 07110

Dr. BERNET
Dept. of General Physiology
University of Lille
Lille, France

Dr. M. J. BESSON
Lab. de Biol. Moléculaire
   College de France
11, Place Mercelin-Bertholet
Paris-5e, France

Dr. JOHN A. BEVAN
Department of Pharmacology
Univ. of California, Los Angeles
Los Angeles, California 90024

Dr. BUDH D. BHAGAT
Department of Physiology
St. Louis University
   School of Medicine
1402 South Grand Boulevard
St. Louis, Missouri 63104

Dr. JOHN H. BIEL
Pharmacology and Medicinal
   Chemistry Division
Abbott Laboratories
N. Chicago, Illinois 60064

Dr. PIER ANTONIO BIONDI
Istituto de Fisiologie della Nutrizione
Animale
Via Celoria 10
20133 Milano, Italy

Dr. IRA BLACK
Department of Neurology
Cornell Univ. School of Medicine
525 E. 68th Street
New York, New York 10021

Dr. HERMANN K. F. BLASCHKO
Department of Pharmacology
Oxford University
South Parks Road
Oxford, England

Dr. R. BLOCH
Laboratoire de Pharmacologie
Faculté de Médecine
67085 Strasbourg Cedex
France

Dr. SAUL BLOOMFIELD
Centre de Recherche International
16, rue d'Ankara
67000-Strasbourg, France

Dr. BOHUON
Institut Gustave-Roussy
16 Bis, Avenue
P.-Vaillant-Couturier
94-Villejuif (Val-de-Marne)
France

Dr. J. R. BOISSIER
Institut National de la Sante
   et de la Recherche Medicale
2, Rue d'Alésia
75014 Paris, France

Dr. WALTER M. BOOKER
Department of Pharmacology
Howard University
   College of Medicine
Washington, D.C. 20001

Dr. ALAN A. BOULTON
Psychiatric Research Unit
University Hospital of Saskatchewan
Saskatoon, Sask. 7N OW8,
Canada

Dr. PHILIP B. BRADLEY
Department of Pharmacology
The Medical School
Birmingham B15 2TJ, England

Dr. C. BRAESTRUP
Sct. Hans Hospital
4000 Roskilde
Denmark

Dr. KLAUS BRANDAU
Bayer Pharm Forschungszentrum
Postfach 130105
56 Wuppertal 1, Germany

Dr. GEORGE R. BREESE
Department of Psychiatry
School of Medicine
University of North Carolina
Chapel Hill, North Carolina 27514

Dr. WILLIAM S. BRIMIJOIN
Mayo Foundation
200 First Street Southwest
Rochester, Minnesota 55901

Dr. A. BRINI
Clinique Opthalmologique
Faculté de Médecine
67085 Strasbourg Cedex
France

Dr. J. BROEKHUYSEN
Avenue J. & P. Carsoel 118
B-1180 Brussels, Belgium

Dr. C. BROEKKAMP
Dept. of Pharmacology
Geert Grooteplein Noord 21
Nijmegen, The Netherlands

Dr. ROGER BROWN
Department of Pharmacology
University of Göteborg
Fack
S-400 33 Göteborg 33, Sweden

Ms. NANCIE BROWNLEY
Department of Pharmacology and
   Experimental Therapeutics
Johns Hopkins University
   School of Medicine
Baltimore, Maryland 21205

Dr. EDITH BULBRING
University Laboratory of Physiology
Parks Road
Oxford OX1 3PT, England

Dr. BENJAMIN S. BUNNEY
Department of Psychiatry
Yale University School of Medicine
34 Park Street
New Haven, Connecticut 06508

Dr. WILLIAM E. BUNNEY
National Institute of Mental Health
5600 Fishers Lane-Room 13–103
Rockville, Maryland 20852

Dr. JOHN BURKE
Centre de Recherche Merrell International
16, rue d'Anakara
67000 Strasbourg, France

Sister ELIZABETH M. BURNS
Department of Physiology
Saint Louis Univ. School of Medicine
1402 S. Grand Boulevard
St. Louis, Missouri 63104

Dr. G. BURNSTOCK
Department of Zoology
University of Melbourne
Parkville, Victoria 3052
Australia

Dr. LARRY L. BUTCHER
Department of Psychology
Univ. of California, Los Angeles
Los Angeles, California 90024

Dr. JEAN CADILHAC
Institut de Biologie
Faculté de Médecine
34000 Montpellier, France

Dr. MARIAGIULIA CAGNASSO
Istituto di Fisiologia della Nutrizione
Animale
Via Celoria 10
20133 Milano, Italy

Dr. J. CAHN
I.T.E.R.C.
6, rue Blanche
92129 Montrouge, France

Dr. JOHN CALDWELL
Biochemistry Department
St. Mary's Hospital Medical School
London W2 1PG, England

Dr. DONALD B. CALNE
Department of Medicine (Neurology)
Royal Postgraduate Medical School
London W.12, England

Dr. ARVID CARLSSON
Department of Pharmacology
University of Göteborg
S-400 33 Göteborg 33
Sweden

Dr. SUSAN M. BONHAM CARTER
Research Laboratories
Queen Charlotte's Maternity Hospital
London W6 OXG, England

Dr. F. CATTABENI
Department of Pharmacology
University of Milan
20129 Milan, Italy

Dr. M. N. CHALAZONITIS
C.N.R.S.
31 Chemin Joseph-Aiguier
13009 Marseille, France

Dr. JOHN P. CHALMERS
Department of Medicine
University of Sydney
New South Wales 2006
Australia

Dr. THOMAS N. CHASE
National Institutes of Health (NIMH)
Building 10—Room 2S-243
Bethesda, Maryland 20014

Lecturer H. Y. CHEN
Chinese Academy of Medical Science
Peking, Peoples Republic of China

Dr. A. CHERAMY
Lab. de biologie moléculaire
Collége de France
11, Place Marcelin-Bouthelot
Paris-5e, France

Professor Y. C. Chin
Chinese Acadamy of Medical Science
Peking, Peoples Republic of China

Dr. Roland Ciaranello
Laboratory of Clinical Science
National Institutes of Health (NIMH)
Building 10—Room 2D-45
Bethesda, Maryland 20014

Dr. J. Ciesielski-treska
C.N.R.S., Centre de Neurochimie
67085 Strasbourg Cedex, France

Dr. J. Clavet
Dept. of Embryology
Faculté de Médecine
67000 Strasbourg, France

Dr. Doris H. Clouet
New York State Narcotics Addiction
    Control Comm. Research Lab.
80 Hanson Place
Brooklyn, New York 11217

Dr. Gerald Cohen
College of Physicians and Surgeons
Columbia University
New York, New York 10032

Dr. Robert W. Colburn
Nat. Inst. of Mental Health
NIH, Bldg. 10, Room 3N 320
Bethesda, Maryland 20014

Dr. Erminio Costa
Laboratory of Preclinical Pharmacology
St. Elizabeth's Hospital
William A. White Building
Room 100
Washington D.C. 20032

Dr. Marcello Costa
Department of Anatomy
University of Helsinki
Siltavuorenpenger
Helsinki, Finland

Dr. L. Côté
College of Physicians and Surgeons
Columbia University
New York, New York 10032

Dr. Joseph T. Coyle, Jr,
National Institutes of Health (NIMH)
Building 10—Room 2D-45
Bethesda, Maryland 20014

Dr. H. Cramer
Neurologische Universitatsklinick
Abteilung fur Neurophysiologie
Hansastrasse 9a
78 Freiburg im Breisgan, Germany

Dr. Cyrus R. Creveling
National Institute of Arthritis
    and Metabolic Diseases
Building 4—Room 273
Bethesda, Maryland 20014

Dr. T. J. Crow
Department of Psychiatry
University Hospital of South
    Manchester
West Didsbury, Manchester M20 8LR
England

Dr. Annica Dahlström
Institute of Neurobiology
University of Göteborg
Medicinaregatan 5
Fack
S-400 33 Göteborg 33, Sweden

Dr. Wallace Dairman
Roche Institute of Molecular Biology
Nutley, New Jersey 07110

Dr. John W. Daly
National Institute of Arthritis
    and Metabolic Diseases
Building 4—Room 227
Bethesda, Maryland 20014

Dr. H. Daniel
C.H. Boehringer Sohn
6507 Ingelheim/Rhein, Germany

Dr. Donald S. Davies
Department of Clinical Pharmacology
Royal Postgraduate Medical School
Ducane Road
London W. 12, England

Dr. John M. Davis
Department of Pharmacology
Vanderbilt University
Nashville, Tennessee 37203

Dr. Virginia E. Davis
Neurochemistry and Addiction
    Research Laboratory
Veterans Administration Hospital
Houston,Texas 77031

Dr. Leonardo Angelis
Institute of Pharmacology & Pharmacognosy
University of Milano
Via A. del Sarto 21
20129 Milano, Italy

Dr. Jaques de Champlain
Department of Physiology
University of Montreal
Case Postale 6128
Montreal 101, Quebec
Canada

Dr. Takeo Deguchi
National Institutes of Health
Building 10—Room 2D-45
Bethesda, Maryland 20014

Dr. Wybren DeJong
Rudolf Magnus Institute for
  Pharmacology
University of Utrecht
Medical Faculty
Vondellaan 6
Utrecht, The Netherlands

Dr. W. P. de Potter
Department of Pharmacology
University of Ghent
De Pintelaan 135
B-9000 Ghent, Belgium

Dr. Vincent De Quattro
Department of Internal Medicine
White Memorial Center
1720 Brooklyn Avenue
Los Angeles, California 90033

Dr. M. Derome-Tremblay
Les Laboratoires Servier
22, rue Garnier
B.P. 110
92200 Neuilly-sur-Seine
France

Dr. Laurent Descarries
Department of Physiology
University of Montreal
Case postale 6128
Montreal 101, Quebec, Canada

Dr. Antoine D'Iorio
Department of Biochemistry
University of Ottawa
Ottawa, KIN 6N5 Ontario, Canada

Dr. J. G. Domenet
Medical Dept., Geigy Pharmaceuticals
Hurdsfield Industrial Estate
Macclesfield Cheshire SK10 2LY, U.K.

Dr. M. J. Dowdall
Max Planck Institute of
  Biophys. Chimie
Abt. f. Neurochemie
D-3400 Göttingen-Nikolausberg
Germany

Dr. A. Dresse
Institut de Therapeutique
32 Boulevard de la Constitution
B 4000 Liège, Belgium

Dr. Josée Dubreuil
Laboratoires Roger Bellon
90, rue Marcel Bourdarias
94140 Alfortville, France

Dr. Felix Ebef
Centre de Neurochimie du C.N.R.S.
11 rue Humann
67085 Strasbourg Cedex, France

Dr. Richard Ebstein
New York Medical Center
Department of Psychiatry
Room H 544
New York, New York 10016

Dr. Donald Eccleston
M.R.C. Brain Metabolism Unit
University Dept. of Pharmacology
1 George Square
Edinburgh EH8 9JZ, Scotland

Dr. Burr S. Eichelman, Jr.
Department of Psychiatry
Stanford University Medical Center
Stanford, California 94305

Dr. Joel Elkes
Department of Psychiatry and
  Behavioral Science
Johns Hopkins University
  School of Medicine
601 N. Broadway
Baltimore, Maryland 21205

Dr. Leon Ellenbogen
Dept. of Cardiovas.-Renal
  Pharmacology
Lederle Laboratories
Pearl River, New York 10965

Dr. Lynne G. Elliott
Department of Pharmacology
Howard University
  College of Medicine
Washington, D.C. 20001

Dr. Jörgen Engel
Department of Pharmacology
University of Göteborg
Fack
S-400 33 Göteborg 33
Sweden

Dr. Karl Engelman
Department of Medicine
University of Pennsylvania
  School of Medicine
3400 Spruce Street
Philadelphia, Pennsylvania 19104

Dr. S. J. Enna
Medical Research Department
Hoffman-LaRoche and Co. Ltd.
4002 Basle, Switzerland

Dr. Olavi Eränkö
Department of Anatomy
University of Helsinki
Siltavirorenpenger
Helsinki, Finland

Dr. Virgel Gene Erwin
School of Pharmacy
University of Colorado
Boulder, Colorado 80302

Mlle. CATHERINE EUVRARD
Centre de Recherche Delalande
30, rue Henri-Regnault
92402 Courbevoie, France

Dr. GUY M. EVERETT
Department of Pharmacology
University of California
   School of Medicine
San Francisco, California 94122

Dr. STANLEY FAHN
Dept. of Neurology
Univ. of Pennsylvania Hospital
Philadelphia, Pennsylvania 19104

Dr. LARS-OVE FARNEBO
Department of Histology
Karolinska Institutet
S-104 01 Stockholm 60
Sweden

Dr. ANNEMARIE FEDORCAK
Abteilung Biochemie
Chemisch-pharmazeutische Fabrik
Postfach 720
Birkendorfer Strasse 65
Biberach an der Riss, Germany

Dr. MARTON FEKETE
Department of Pharmacology
Research Institute for
   Pharmaceutical Chemistry
Szabadsagharcosok u 47–49
Budapest 4, Hungary

Dr. JACK H. FELLMAN
Department of Biochemistry
University of Oregon Medical School
3181 S.W. Sam Jackson Park Road
Portland, Oregon 97201

Dr. PAUL FELTZ
Laboratoire des Centres Nerveux
4, avenue Gordon-Bennett
75016-Paris, France

Dr. J. D. FERNSTROM
Dept. of Endocrinology & Metabolism
Mass. Inst. of Technology
Cambridge, Massachusetts 02139

Dr. ROBERT M. FERRIS
Department of Pharmacology
Burroughs Wellcome and Company
3030 Cornwallis Road
Research Triangle Park, N.C. 27709

Dr. HANS C. FIBIGER
Department of Psychiatry
Kinsmen Lab. of Neurological Research
University of British Colombia
Vancouver 8
British Columbia, Canada

Dr. MARIANNE FILLENZ
University Laboratory of Physiology
Oxford, England

Dr. ERNEST FOLLENIUS
Laboratoire de Cytologie Animale
Université Louis Pasteur
12, rue de l'Université
67000 Strasbourg, France

Dr. ANDREW G. FRANTZ
College of Physicians and Surgeons
Department of Medicine
Columbia University
New York, New York 10032

Dr. LEWIS S. FREEDMAN
New York University Medical Center
Department of Psychiatry
Room H 544
New York, New York 10016

Dr. RAY W. FULLER
Lilly Research Laboratories
Indianapolis, Indiana 46206

Dr. ROBERT F. FURCHGOTT
Department of Pharmacology
State University of New York
   Downstate Medical Center
450 Clarkson Avenue
Brooklyn, New York 11203

Dr. KJELL FUXE
Department of Histology
Karolinska Institutet
S-104 01 Stockholm 60, Sweden

Dr. E. MARTIN GÁL
Department of Psychiatry
University of Iowa
State Psychopathic Hospital
500 Newton Street
Iowa City, Iowa 52240

Dr. WILLIAM F. GANONG
Department of Physiology
University of California
   School of Medicine
San Francisco, California 94122

Dr. JOSE GARRETT
Laboratorio de Farmacologia
Faculdade de Medicina
Porto, Portugal

Dr. LAURIE BRUCE GEFFEN
School of Biological Sciences
Flinders University of South
   Australia
Bedford Park, S.A. 5042
Australia

Dr. JEAN GERARDY
Université de Liège
Institut de Thérapeutique
32, Bd de la Constitution
B-4000 Liège, Belgium

Dr. ELLIOT S. GERSHON
Jerusalem Mental Health Center
Ezrath Nashim
Givat Shaul
P.O. Box 140, Jerusalem, Israel

Dr. SAMUEL GERSHON
Department of Psychiatry and Neurology
New York Univ. School of Medicine
550 First Avenue
New York, New York 10016

Dr. GIAN LUIGI GESSA
Istituto de Farmacologia
University of Cagliari
Via Porcell, 4
09100 Cagliari, Italy

Dr. JAMES W. GIBB
Department of Pharmacology
University of Utah
Salt Lake City, Utah 84122

Dr. C. N. GILLIS
Department of Pharmacology
Yale University School of Medicine
New Haven, Connecticut 06510

Dr. ALFRED G. GILMAN
Department of Pharmacology
University of Virginia
  School of Medicine
Charlottesville, Virginia 22903

Dr LEIV GJESSING
Central Laboratory
Dikemark Hospital
Asker 1385, Solberg, Norway

Dr. JAQUES GLOWINSKI
Unite de Neuropharmacologie Biochimique
Lab de Neurophysiologie Generale
College de France
11, Place Mercelin-Berthelot
Paris, Veme, France

Dr. MENEK GOLDSTEIN
Department of Psychiatry and Neurology
New York University Medical Center
550 First Avenue
New York, New York 10015

Dr. G. GOMBOS
C.N.R.S., Centre de Neurochimie
67085 Strasbourg-Cedex, France

Dr. B. L. GOODWIN
Research Laboratories
Queen Charlotte's Maternity Hospital
London W6 OXG, England

Dr. FREDERICK GOODWIN
National Institutes of Health (NIMH)
Building 10–Room 4S-239
Bethesda, Maryland 20014

Dr. JAMES GOODWIN
Natl. Heart and Lung Institute
Building 10, Room 7N 242
Bethesda, Maryland 20014

Dr. JEAN GOODWIN
Department of Psychiatry
Georgetown University
Washington, D.C.

Dr. CHRISTO GORIDIS
C.N.R.S. Neurochem. Centre
11, rue Humann
F-67 Strasbourg, France

Dr. V. Z. GORKIN
Institute of Biological and
  Medicinal Chemistry
Academy of Medical Sciences
  of the U.S.S.R.
Pogodinskaya Street, 10
Moscow 119117, U.S.S.R.

Dr. JOHN W. GORROD
Department of Pharmacy
Chelsea College of Science
  and Technology
271/3 King Street, Hammersmith
London, W.6, England

Dr. Cl. GOTTESMANN
Université de Nice
Faculté des Sciences
Laboratoire de Psychophysiologie
Parc Valrose
06 Nice 02, France

Dr. PAUL GREENGARD
Department of Pharmacology
Yale University School of Medicine
333 Cedar Street
New Haven, Connecticut 06510

Dr. H. GROBECKER
Pharmacology Institute
University of Frankfurt
Ludwig Rehn Strasse
6000 Frankfurt/Main 70
Germany

Dr. A. GROPETTI
Department of Pharmacology
University of Milan
Via Vanvitelli, 32
20129 Milan, Italy

Mme. O. GRYNZPAN
Laboratoire de Cytologie
7 Quai St. Bernard
75005 Paris, France

Dr. ALESSANDRO GUIDOTTI
Laboratory of Preclinical
  Pharmacology
St. Elizabeth's Hospital (NIMH)
Wm. A. White Bldg.—Room 103
Washington, D.C. 20032

Dr. SERAFIN GUIMARAES
Department of Pharmacology
Faculty of Medicine
Porto, Portugal

Dr. LARS-MAGNUS GUNNE
Psychiatric Research Center
University of Uppsala
Ulleråker Hospital
S-75017 Uppsala, Sweden

Dr. YEHUDA GUTMAN
Department of Pharmacology
Hadassah Medical School
Hebrew University
P.O. Box 1172
Jerusalem, Israel

Dr. W. E. HAEFELY
Pharmacology Department
F. Hoffmann-LaRoche and Company
CH-4002 Basle, Switzerland

Dr. JAN HÄGGENDAL
Department of Pharmacology
University of Göteborg
Fack
S-400 33 Göteborg 33
Sweden

Dr. BERTIL HAMBERGER
Department of Histology
Karolinska Institutet
S-104 01 Stockholm 60
Sweden

Dr. BERND HAMPRECHT
Max Planck Institut fur Biochemie
8033 Martinsreid bei Munchen, Germany

Dr. INGEBORG HANBAUER
National Institutes of Health
  (NIMH)
Building 10–Room 2D-46
Bethesda, Maryland 20014

Dr. BOYD K. HARTMAN
Department of Psychiatry
Washington University
  School of Medicine
4940 Audubon Avenue
St. Louis, Missouri 63110

Dr. MATTI HÄRKÖNEN
Department of Clinical Chemistry
University of Helsinki
Meilahti Hospital
00290 Helsinki 29, Finland

Dr. JOHN A. HARVEY
Psychology Department
University of Iowa
Iowa City, Iowa 52240

Dr. KOROKU HASHIMOTO
Tohoku University School of Medicine
Department of Pharmacology
Sendai 980, Japan

Ms. ELIZABETH HASSELAGER
Institute of Biochemsitry
Odense University
Niels Bohr Alle
DK-5000 Odense, Denmark

Dr. G. HÄUSLER
F. Hoffmann-LaRoche and Company
CH-4002 Basle, Switzerland

Dr. PER HEDQVIST
Fysiologiska Institutionen I
Karolinska Institutet
S-104 01 Stockholm 60, Sweden

Dr. RICHARD HEIKKILA
Department of Neurology
Columbia University
  College of Physicians and Surgeons
New York, New York 10032

Dr. KAREN B. HELLE
Institute of Physiology
University of Bergen
Årstadveien 19
5000 Bergen, Norway

Dr. MATTS HENNING
Department of Pharmacology
University of Göteborg
Medicinaregatan
Göteborg SV, Sweden

Dr. STEVEN HENRIKSEN
Department of Psychiatry
Stanford Univ. School of Med.
Stanford, California 94305

Dr. WILLIAM F. HERBLIN
Experimental Station,
  E.I. du Pont
1007 Market Street
Wilmington, Delaware

Dr. J. C. HERMETET
Institut de Chemie Biologique
University of Strasbourg
Faculté de Médecine
F-67085 Strasbourg, Cedex, France

Dr. GEORGE HERTTING
Pharmacologisches Institut
  de Universität
Waehringerstrasse 13a
A-1090 Vienna IX, Austria

Dr. Hiroyoshi Hidaka
Department of Biochemistry
Institute for Dev. Res.
Aichi Prefectural Colony
Kasugai, Aichi 480-03, Japan

Dr. Barry J. Hoffer
St. Elizabeth's Hospital (NIMH)
Wm. A. White Bldg.—Room 44
Washington, D.C. 20032

Dr. Tomas Hökfelt
Karolinska Institutet
Department of Histology
S-104 01 Stockholm 60, Sweden

Dr. M. Holzbauer-Sharman
Agricultural Research Council
Institute of Animal Physiology
Babraham (Cambridge)
Great-Britain

Dr. A. S. Horn
M.R.C. Neurochemical Pharmacology Unit
Department of Pharmacology
Medical School, Hills Road
Cambridge CB2 2QD, England

Dr. Oleh Hornykiewicz
Department of Psychopharmacology
Clarke Institute of Psychiatry
250 College Street
Toronto 2B, Ontario, Canada

Dr Heide Hörtnagl
Department of Pharmacology
University of Innsbruck
Peter-Mayr-Strasse 1
A-6020 Innsbruck, Austria

Dr. M. D. Houslay
University of Cambridge
Department of Biochemistry
Tennis Court Road
Cambridge, England

Dr. Robert Howd
Dept. of Nutrition & Food Sci.
Massachusetts Institute of Technology
Cambridge, Mass. 02139

Dr. Susan Huszti
McGill University Department of Psychiatry
1033 Pine Avenue West
Montreal 112, Quebec
Canada

Dr. Matte O. Huttenen
Department of Medical Chemistry
University of Helsinki
Siltavuorenpenger 10
00170 Helsinki 17, Finland

Dr. F. Isch
Doyen, Faculté de Médecine
Université Louis Pasteur
rue Kirschleger
67085 Strasbourg Cedex, France

Dr. Leslie L. Iversen
Medical Research Council
  Neuropharmacology Unit
Department of Pharmacology,
  Medical School
Hills Road
Cambridge CB2 2QD, England

Dr. Susan D. Iversen
Department of Experimental Psychology
University of Cambridge
Downing Street
Cambridge CB2 3EB, England

Dr. David Jacobowitz
National Institutes of Health (NIMH)
Building 10—Room 2D-46
Bethesda, Maryland 20014

Dr. Christian Jacquot
Laboratoire de Pharmacodynamie
Centre Pharmaceutique
92290 Chatenay-Malabry
France

Dr. Paul A. J. Janssen
Janssen Pharmaceutica Research
  Laboratory
2340 Beerse, Belgium

Dr. S. Jard
Lab. de Physiologie Cellulaire
11 place Marcelin-Berthelot
College de France
75321 Paris, France

Dr. Bevyn Jarrott
Department of Physiology
University of Monash
Clayton, Victoria
Australia 3168

Dr. F. Javoy
Unite de Neuropharmacologie
  Biochimique
Lab de Neurophysiologie Generale
College de France
11, place Mercelin-Berthelot
Paris, Veme, France

Dr. Gunnar Johansson
Ylipiston Fysiologian Laitos
Institutum Physiologicum
Siltavuorenpenger 20 A.
Helsinki, Suomi-Finland

Dr. R. D. Johnson
Research Laboratories
Queen Charlotte's Maternity Hospital
London W6 OXG, England

Dr. Wulff Jonas
Jerusalem Mental Health Center
Ezrath Nashim
Givat Shaul
P.O. Box 140
Jerusalem
Israel

Dr. Jan Jonason
Department of Pharmacology
University of Göteborg
Fack
S-400 33 Göteborg 33, Sweden

Dr. Gösta Jonsson
Department of Histology
Karolinska Institutet
S-104 01 Stockholm 60
Sweden

Dr. A. Jori
Istituto de Ricerche Farm.
'Mario Negri'
Via Eritrea 62
20157 Milan, Italy

Dr. Michael Jouvet
Lab. de Pathologie Generale et Exp.
Faculté de Médecine
Université Claude-Bernard
8, avenue Rockefeller
69 Lyon 8ᵉ, France

Dr. A. V. Juorio
Department of Pharmacology
University College
Gower Street
London W.C.1., England

Dr. Michel Jung
Centre de Recherche
Merrell International
16, rue d'Ankara
6700 Strasbourg, France

Dr. Manifred E. Karobath
Psychiatric Clinic
University of Vienna
Lazarettgasse 14
A-1097 Vienna, Austria

Dr. H. Käser
Inst. f. Klin.-Exptl. Tumorforschung
3004 Bern, Switzerland

Dr. Seymour Kaufman
National Institutes of Health (NIMH)
Building 36—Room 3D-30
Bethesda, Maryland 20014

Dr. Tomislav Kažić
Medical Faculty
Department of Pharmacology
11.105 Belgrade, Yugoslavia

Dr. John W. Kebabian
Department of Pharmacology
Yale University School of Medicine
New Haven, Connecticut 06510

Dr. Wolfgang Kehr
Department of Pharmacology
University of Göteborg
Fack
S-400 33 Göteborg 33, Sweden

Dr. Carol Kellogg
Department of Psychology
University of Rochester
Rochester, New York 14642

Dr. E. Kempf
C.N.R.S., Centre de Neurochimie
67085 Strasbourg Cedex, France

Dr. Seymour S. Kety
Department of Psychiatry
Harvard Medical School at
    Massachusetts General Hospital
Boston, Massachusetts 02114

Dr. J. Killian
C.N.R.S., Centre de Neurochimie
67085 Strasbourg Cedex, France

Dr. Richard Kinsolving
Merrell-National Laboratories
Cincinnati, Ohio 45215

Dr. Norman Kirshner
Department of Biochemistry
Duke University Medical School
Durham, North Carolina 27706

Dr. David C. Klein
Laboratory of Biomedical Science
National Institute of Child Health
    and Development
National Naval Medical Center
Building 125—Room 35
Washington, D.C. 20014

Dr. Richard L. Klein and
    Dr. Asa Thureson-Klein
Department of Pharmacology and Toxicology
University of Mississippi Medical Center
Jackson, Mississippi 39216

Dr. Irwin J. Kopin
Laboratory of Clinical Science
National Institutes of Health (NIMH)
Building 10—Room 2D-46
Bethesda, Maryland 20014

Dr. Jakob Korf
Department of Biological Psychiatry
Ostersingel 59
Groningen, The Netherlands

Dr. Stephen H. Koslow
St. Elizabeth's Hospital (NIMH)
Wm. A. White Building—Room 116
Washington, D.C. 20032

Dr. Günther Kroneberg
Farbenfabriken Bayer A.G.
Fredrich-Ebert Strasse 217–219
Postfach 1301 05
56 Wuppertal 1, Germany

Dr. Hans Joerg Kruse
Farbwerke Hoechst A.G.
623 Frankfurt/Main 80
Postfach 800320
Germany

Dr. M. Kuhar
Department of Pharmacology and
  Experimental Therapeutics
Johns Hopkins University
  School of Medicine
Baltimore, Maryland 21205

Dr. Ronald G. Kuntzman
Biological Research
Hoffman-LaRoche, Inc.
Nutley, New Jersey 07110

Dr. Elwood H. LaBrosse
Department of Surgery
University of Maryland Medical School
29 S. Green Street
Baltimore, Maryland 21201

Dr. Pierre Laduron
Janssen Pharmaceutica Research
  Laboratory
Beerse, Belgium

Dr. Hugo Lagercrantz
Fysiologiska Institutionen I
Karolinska Institutet
S-104 01 Stockholm 60
Sweden

Dr. Friedhelm Lamprecht
National Institutes of Health (NIMH)
Building 10—Room 2D-46
Bethesda, Maryland 20014

Dr. Heinrich Langemann
Pharmakologisches Institut
University of Zürich
Gloriastrasse 32
8006 Zürich, Switzerland

Dr. Salomon Z. Langer
Institute of Pharmacological Research
Junin 956, 5°
Buenos Aires, Argentina

Dr. Richard Laverty
Department of Pharmacology
University of Otago Medical School
King Street
P.O. Box 913
Dunedin, New Zealand

Dr. Lazdunski
Service de Biochimie
Université de Nice
Faculté des Sciences
Parc Valrose
06034 Nice, France

Dr. Harold E. Lebovitz
Department of Medicine
Division of Endocrinology
Duke University
Durham, North Carolina 27706

Dr. Robert J. Lefkowitz
Massachusetts General Hospital
Cardiology Branch
Boston, Massachusetts 02114

Asst. Prof. H. P. Lei
Chinese Academy of Medical Science
Peking, Peoples Republic of China

Dr. Sarah Fryer Leibowitz
Department of Physiological Psychology
The Rockefeller University
York Avenue and East 66th Street
New York, New York 10221

Dr. Jean Leonardelli
Faculty of Medicine
Lab. d'Histologie et d'Embryologie
place de Verdun
59 Lille, France

Dr. Gerald Letertre
Centre de Recerche Merrell
  International
16, rue d'Ankara
67000-Strasbourg, France

Dr. Rita Levi-Montalcini
Laboratory of Cell Biology
Consiglio Nazionale dell Ricerche
via G. Romagnosi 18/A
00196 Rome, Italy

Dr. Robert H. Levin
Merrell-National Laboratories
Cincinnati, Ohio 45215

Dr. Tommy Lewander
Psychiatric Research Center,
University of Uppsala
Ulleråker Hospital
S-750 17 Uppsala, Sweden

Dr. W. Lichtensteiger
Pharmacology Institute
University of Zurich
Gloriastrasse 32
8006 Zurich, Switzerland

Dr. Peter Lidbrink
Department of Histology
Karolinska Institutet
S-104 01 Stockholm 60
Sweden

Dr. Ernst Lindner
Farbwerke Hoechst A.G.
Abteilung für Pharmakologie
D-6230 Frankfurt/M-80
Postfach 800320
Germany

Miss MARGIT LINDQVIST
Department of Pharmacology
University of Göteborg
Fack
S-400 33 Göteborg 33
Sweden

Dr. OLLE LINDVALL
Department of Histology
University of Lund
S-223 62 Lund, Sweden

Dr. MORRIS A. LIPTON
Department of Psychiatry
University of North Carolina
    School of Medicine
Chapel Hill, North Carolina 27514

Dr. ANTONIA LIUZZI
Laboratorio di Biologia
    Cellulare del C.N.R.
Via Romagnosi 18/A
00196-Rome, Italy

Dr. AKE LJUNGDAHL
Department of Histology
Karolinska Institutet
S-104 01 Stockholm 60
Sweden

Dr. KENNETH G. LLOYD
Department VI Med.
F. Hoffmann LaRoche Company
CH 4002 Basle, Switzerland

Dr. WALTER LOVENBERG
National Heart and Lung Institute
Building 10—Room 7N-242
Bethesda, Maryland 20014

Dr. PER LUNDBORG
Department of Pharmacology
University of Göteborg
Fack
S–400 33 Göteborg 33
Sweden

Dr. LOY D. LYTLE
Dept. of Nutrition and Food Sciences
Massachusetts Institute of Technology
Cambridge, Massachusetts 02139

Dr. JAMES W. MAAS
Department of Psychiatry
Yale University School of Medicine
333 Cedar Street
New Haven, Connecticut 06510

Dr G. MACK
C.N.R.S., Centre de Neurochimie
67085 Strasbourg Cedex, France

Dr. ANGUS V. P. MACKAY
University of Cambridge Medical School
Department of Pharmacology
Cambridge CB2 2QD
England

Mr. TOR MAGNUSSON
Department of Pharmacology
University of Göteborg
Fack
S-400 33 Göteborg 33
Sweden

Dr. LAURENT MAÎTRE
Ciba-Geigy Ltd.
Basle, Switzerland

Dr. J. MAJ
Institute of Pharmacology
Polish Academy of Science
52, Ojcowska-Street
31-344 Krakow, Poland

Dr. MANA
Laboratoires Hoechst
3 rue Clement Marot
Paris 8e, France

Dr. LUCIANO MANARA
Istituto di Ricerche Farmacologiche
    'Mario Negri'
Via Eritrea, 62
20157 Milano, Italy

Dr. PAUL MANDEL
C.N.R.S.
Centre de Neurochimie
11, rue Humann
67, Strasbourg, France

Dr. ARNOLD J. MANDELL
Department of Psychiatry
Univ. of California School
    of Medicine
P.O. Box 109
La Jolla, California 92037

Dr. WILLIAM M. MANGER
100 West 58th Street
New York, New York 10019

Dr. EMMANUEL MARKIANOS
Nervenklinik der Universität
Nussbaumstrasse 7
8 Munich 2, Germany

Dr. BERNARD H. MARKS
Department of Pharmacology
Ohio State University
    College of Medicine
Columbus, Ohio 43210

Dr. E. MARLEY
Department of Pharmacology,
University of London
Institute of Psychiatry
De Crespigny Park, Denmark Hill
London SE5 8AF, England

Dr. Per Martinson
Department of Pharmacology
University of Göteborg
Fack
S-400 33 Göteborg 33
Sweden

Dr. Ch. Marx
Institut de Physiologie
Faculté de Médecine
67085 Strasbourg Cedex
France

Dr. Karl Masek
Czechoslovak Academy of Sciences
Institute of Pharmacology
Albertov 4
Prague 2
Czechoslovakia

Dr. Aleksander A. Mathe
Division of Psychiatry
Boston University
   School of Medicine
80 East Concord Street
Boston, Massachusetts 02118

Dr. P. Mathieu
Laboratoire de Biologie
Hôpital Psychiatrique de Vinatier
95, Boulevard Pinel
69677 Bron, France

Dr. S. Matthysse
Department of Psychiatry
Massachusetts General Hospital
Boston, Massachusetts 02114

Mr. Robert Maxwell
Pergamon Press
Oxford, England

Dr. Robert A. Maxwell
The Wellcome Research Labs.
Burroughs Wellcome Company
3030 Cornwallis Road
Research Triangle Park, N.C. 27709

Dr. Steven E. Mayer
Department of Medicine
University of California, San Diego
La Jolla, California 92037

Dr. James L. Meek
St. Elizabeth's Hospital
Wm. A. White Building—Room 12,
Washington, D.C. 20032

Dr. J. J. Meisch
Netherland Central Institute for
   Brain Research
Ijdijk 28
Amsterdam-O, The Netherlands

Dr. A. Mentrup
C. H. Boehringer Sohn
6507 Ingelheim/Rhein, Germany

Dr. M. Miras-Portugal
Centre de Neurochimie du C.N.R.S.
67085 Strasbourg Cedex, France

Dr. Kjell Modigh
Department of Pharmacology
University of Göteborg
Fack
S-400 33 Göteborg 33
Sweden

Dr. Perry B. Molinoff
Department of Pharmacology
University of Colorado
4200 East 9th Avenue
Denver, Colorado 80220

Dr. Kenneth E. Moore
Department of Pharmacology
Michigan State University
B440 Life Sciences
East Lansing, Michigan 48823

Dr. Robert Y. Moore
Department of Pediatrics
University of Chicago
5825 Maryland Avenue
Chicago, Illinois 60637

Dr. Walter J. Moore
Department of Chemistry
Indiana University
Bloomington, Indiana 47401

Dr. Neil C. Moran
Department of Pharmacology
Emory University
Atlanta, Georgia 30322

Dr. Ian Morgan
C.N.R.S.
Centre de Neurochimie
11, rue Humann
67, Strasbourg, France

Dr. Robert A. Mueller
Department of Anesthesiology
University of North Carolina
   School of Medicine
Chapel Hill, North Carolina 27514

Dr. Eugenio E. Müller
Department of Pharmacology
University of Milan
Via Vanvitelli 32
20129 Milan, Italy

Dr. Dennis L. Murphy
National Institutes of Health (NIMH)
Building 10—Room 3S-229
Bethesda, Maryland 20014

Dr. José M. Musacchio
Department of Pharmacology
New York University Medical Center
550 First Avenue
New York, New York 10016

Dr. E. Muscholl
Pharmacology Institute
University of Mainz
Obere Zahlbacher Strasse 67
  (Hochhaus)
6500 Mainz, West Germany

Dr. E. Naftchi
Institute of Rehabilitation Medicine
New York University Medical Center
400 E. 34th Street
New York, New York 10016

Dr. Toshiharu Nagatsu
Department of Biochemistry
School of Dentistry
Aichi-Gakuin University
2-11 Suemori-dori
Chikusa-ku, Nagoya 464
Japan

Dr. O. A. Nedergaard
Department of Pharmacology
Odense University
Niels Bohrs Alle
DK-5000 Odense, Denmark

Dr. Norton H. Neff
Laboratory of Preclinical
  Pharmacology
St. Elizabeth's Hospital
Wm. A. White Building—Room 129
Washington, D.C. 20032

Dr. Anders Nobin
Department of Histology
University of Lund
S-223 62 Lund, Sweden

Dr. Henrik V. Nybäck
Department of Pharmacology
Karolinska Institutet
S-104 91 Stockholm 60
Sweden

Dr. J. Offermeier
Department of Pharmacology
Poschefstroom University
Potchefstroom, South Africa

Dr. Lars Olson
Department of Histology
Karolinska Institutet
S-104 01 Stockholm 60
Sweden

Dr. Lars Oreland
Department of Pharmacology
University of Umeå
S-901 87 Umeå, Sweden

Dr. W. Osswald
Department of Pharmacology
Faculty of Medicine
University of Porto
Porto, Portugal

Dr. J. Ostrowski
Cassella Farbwerke Mainkur
Medizinische Forschungsabteilung-
  Biochemie
6 Frankfurt am Main-Fechenheim
Germany

Dr. G. Ourisson
Président, Université Louis Pasteur
4, rue Blaise Pascal
B.P. 1032 F
67070 Strasbourg Cedex, France

Dr. Christer Owman
Department of Histology
University of Lund
Biskopsgatan 5
22362 Lund, Sweden

Dr. Michael G. Palfreyman
Beecham Research Labs.
Medicinal Research Centre
Coldharbor Road
The Pinnacles, Harlow, Essex CM195AD
England

Dr. Rodolfe Paoletti
Istituto di Farmacologia
Universitá di Milano
Via Andrea del Sarto 21
20129 Milano, Italy

Dr. R. Papeschi
Max Planck Inst. für Psychiatrie
Kraepelinstrasse 2-10
8 Munich 23, Germany

Dr. C. M. B. Pare
Department of Psychiatry
St. Bartholomew's Hospital
West Smithfield, London E.C.1.
England

Dr. Popat N. Patil
College of Pharmacy
Ohio State University
500 West 12th Avenue
Columbus, Ohio 43210

Dr. Robert L. Patrick
Department of Psychiatry
Stanford University School of Medicine
Stanford, California 94305

18   E. Usdin and S. H. Snyder

Dr. Miguel Perez de la Mora
Dept. of Biophysics
University College
Gower Street
London WC1E 6BT, England

Dr. John P. Perkins
Department of Pharmacology
University of Colorado
  School of Medicine
Denver, Colorado 80220

Dr. Marie-Lise Perrenoud
U.99 Hôpital Henri Mondor
9400 Creteil
179 rue de Tobriac
75013 Paris
France

Dr. Barbara Petrack
Department of Biochemistry
Ciba-Geigy Pharmaceutical Research
Ardsley, New York 10502

Dr. Arnaud-Philippart
Les Laboratories Servier
22, rue Gurnier
B.P. 110
92200 Neiully-sur-Seine
France

Dr. O. T. Phillipson
Research Laboratories
Queen Charlotte's Maternity Hospital
London W6 OXG, England

Dr. A. Philippu
Institut für Pharmakologie
  und Toxikologie
University of Würzburg
Koellikerstrasse 2
87 Würzburg, Germany

Dr. Alfred Pletscher
F. Hoffmann-LaRoche and Company
CH-4002 Basle, Switzerland

Dr. Salvatore Pluchino
School of Pharmacy
Central University of Venezuela
Apartado 40, 252-Neieva-Granada
Caracas, Venezuela

Dr. J. C. Poignant
Soc. Franc. Rech. Med. Sci. Union & Co.
Dept. de Pharm. Gen.
14, rue du Val d'Or
92150 Suresnes, France

Dr. Alan Poisner
Department of Pharmacology
University of Kansas Medical Center
Kansas City, Kansas 66103

Dr. Alfred Pope
McLean Hospital
Belmont, Massachusetts

Dr. Curt C. Porter
Merck Sharp and Dohme Research
  Laboratories
West Point, Pennsylvania 19486

Mme. Elena Portmann
Unité de Neuropharm. (INSERM)
2, rue d'Alésia
75014 Paris, France

Dr. J. F. Pujol
Department of Med. Exptl.
Université Claude-Bernard
Faculté de Médecine
8, avenue Rockefeller
69 Lyon 8e, France

Dr. Giorgio Racagni
Institute of Pharmacology & Pharmacognosy
University of Milano
Via A. del Sarto 21
20129 Milano
Italy

Dr. M. J. Rand
Department of Pharmacology
University of Melbourne
Parkville, Victoria 3032,
Australia

Dr. Axel Randrup
Sct. Hans Hospital
4000 Roskilde, Denmark

Dr. Jean Rapin
Laboratoire de Pharmacodynamie
Centre Pharmaceutique
92290 Chatenay-Malabry
France

Dr. Leena Rechardt
Department of Anatomy
University of Helsinki
Siltavuorenpenger 20
00170 Helsinki 17
Finland

Dr. John L. Reid
Department of Clinical Pharmacology
Royal Postgraduate Medical School
Ducane Road
London W12 OHS, England

Dr. Donald J. Reis
Department of Neurology
Cornell University Medical College
New York, New York 10021

Dr. Cecilia Reuter
Les Laboratoires Servier
22, rue Garnier
92200 Neuilly
France

Dr. J. G. RICHARDS
Department of Experimental Medicine
F. Hoffmann-LaRoche and Company
CH-4002 Basle, Switzerland

Dr. J. STEVEN RICHARDSON
National Institutes of Health (NIMH)
Building 10 Room 2D-46
Bethesda, Maryland 20014

Dr. ELLIOT RICHELSON
Department of Pharmacology
Johns Hopkins University
   School of Medicine
725 North Wolfe Street
Baltimore, Maryland 21205

Dr. A. PAULINE RIDGES
University of Liverpool
Department of Psychiatry
Life Sciences Building
Crown Street
Liverpool L 69 3BX
Great Britain

Dr. G. ALAN ROBISON
Department of Pharmacology
University of Texas Medical School
102 Jesse H. Jones Library Building
Houston, Texas 77025

Dr. MARK ROFFMAN
Department of Psychiatry
Room H 544
New York Univ. Medical Center
New York, New York 10016

Dr. BJÖRN-ERIK ROOS
Department of Pharmacology
University of Göteborg
Fack
S-400 33 Göteborg 33, Sweden

Dr. RICHARD K. RONDEL
CIBA Laboratories
Horsham Sussex RH12 4AB
England

Dr. KURT ROSENHECK
Weizmann Institute
Rehovat, Israel

Dr. SVANTE ROSS
Astra Läkemedel AB
S-151 85 Södertälje
Sweden

Dr. ROBERT ROTH
Department of Pharmacology
   (Psychiatry)
Yale Univ. School of Medicine
333 Cedar Street
New Haven, Connecticut 06510

Dr. COLIN R. J. RUTHVEN
Department of Chemical Pathology
Queen Charlotte's Maternity Hospital
Goldhawk Road
London W6 OXG, England

Dr. CHARLES O. RUTLEDGE
Department of Pharmacology
Univ. of Colorado Medical Center
Denver, Colorado 80220

Dr. JUAN SAAVEDRA
National Institute of Mental Health
Building 10—Room 2D-45
Bethesda, Maryland 20014

Dr. CHARLOTTE SACHS
Department of Histology
Karolinska Institutet
S-104 01 Stockholm 60
Sweden

Dr. MERTON SANDLER
Department of Chemical Pathology
Bernhard Baron Memorial Research
   Laboratories
Queen Charlotte's Maternity Hospital
Goldhawk Road
London W6 OXG, England

Dr. HENRY SARAU
Department of Pharmacology
Smith Kline and French Labs.
Philadelphia, Pennsylvania 19101

Dr. EVA SATORY
Department of Pharmacology
Medical University of Budapest
Ulloi ut 26
Budapest VIII, Hungary

Dr. U. SCAPAGNINI
Institute of Pharmacology
II Faculty of Medicine
Via Sergio Pansini 5
Naples, Italy

Dr. ULRICH SCHACHT
Farbwerke Hoechst A.G.
Pharma Biochemie H821
6230 Frankfurt/M-80
Postfach 800320
West Germany

Dr. J. SCHEEL-KRÜGER
Sct. Hans Hospital
4000 Roskilde, Denmark

Dr. JOSEPH J. SCHILDKRAUT
Department of Psychiatry
Massachusetts Mental Health Center
74 Fenwood Road
Boston, Massachusetts 02115

Dr. Klaus Schlossmann
Bayer-Pharma Forschungszentrum
Institut für Pharmakologie
Werk Elberfeld
Aprather Weg 18a
Postfach 130105
56 Wuppertal 1
Germany

Dr. Margret Schlumpf
Pharmakologisches Institut der
    Universität
Gloriastrasse 32
CH-8006 Zurich
Switzerland

Dr. Ronald I. Schoenfeld
Department of Psychiatry
Children's Hospital Medical Center
Basic Pediatric Science Research Bldg.
300 Longwood Avenue
Boston, Massachusetts 02115

Dr. Michael Schramm
Department of Biological Chemistry
Hebrew University
Jerusalem, Israel

Dr. M. Schreiner
Farbwerke Hoechst A.G.
Verkauf Arzneimittel
Ress. wiss. Med. Information
623 Frankfurt (Main) 80 Germany

Dr. J. Schrold
Department of Pharmacology
Odense University
Niels Bohr Alle
DK-500, Odense, Denmark

Dr. Hans J. Schumann
Pharmakologisches Institut
    Klinikum Essen
Hufelandstrasse 55
43 Essen, West Germany

Dr. J. Schwartz
Institut de Pharmacologie
    et de Med. Exptl.
Faculté de Médecine
11, rue Humann
67 Strasbourg, France

Dr. Goran C. Sedvall
Department of Pharmacolgy
Karolinska Institutet
10401 Stockholm 60
Sweden

Dr. Guldborg Serck-Hanssen
Institute of Pharmacology
University of Oslo
Blindern
Oslo 3, Norway

Dr. Pedro A. Serrano
Department of Endocrinology
Institute of Cardiology
Avenue Cuauhtemoc, No. 300
Mexico 7, D.F. Mexico

Dr. Paulette B. Setler
Department of Pharmacology
Smith Kline and French Labs.
1500 Spring Garden Street
Philadelphia, Pennsylvania 19101

Dr. Dennis F. Sharman
Agricultural Research Council
Institute of Animal Physiology
Babraham, Cambridge
England

Dr. Jean Chen Shih
Neuropsychiatric and Brain
    Research Institutes
UCLA Medical School
Los Angeles, California 90024

Dr. William J. Shoemaker
St. Elizabeth's Hospital
Wm. A. White Bldg.—Room 148
Washington, D.C. 20032

Dr. Baron Shopsin
Department of Psychiatry
New York Univ. Medical Center
550 First Avenue
New York, New York 10016

Dr. Parkhurst A. Shore
Department of Pharmacology
University of Texas
Southwestern Medical School
5323 Harry Hines Boulevard
Dallas, Texas 75235

Dr. Pierre Simon
Faculté de Médecine/Pitie Salpetriere
91 Bd. de l'Hôpital
75 Paris 13, France

Dr. Albert Sjoerdsma
Merrell International Research Center
16 rue d'Ankara
67-Strasbourg, France

Dr. Jef. L. Slangen
Rudolf Magnus Inst. f. Pharmacology
University of Utrecht Medical Fac.
Vondellaan 6
Utrecht, The Netherlands

Dr. Theodore A. Slotkin
Dept. of Physiology and Pharmacology
Duke University Medical Center
Durham, North Carolina 27710

Dr. John R. Smythies
Neurosciences Program
University of Alabama Medical Center
Birmingham, Alabama 35294

Dr. STUART SNIDER
Department of Pharmacology
University of Göteborg
Fack
S-400 33 Göteborg 33
Sweden

Dr. SOLOMON H. SNYDER
Department of Pharmacology
    and Experimental Therapeutics
Johns Hopkins University
    School of Medicine
725 North Wolfe Street
Baltimore, Maryland 21205

Dr. CONSTANTINO SOTELO
INSERM
Lab. d'Histologie Normale et
    Pathologique du Systéme Nerveux
128 Bd. de Port-Royal
Paris 14, France

Dr. THEODORE SOURKES
Department of Psychiatry
McGill University
1033 Pine Avenue West
Montreal 112, Quebec,
Canada

Dr. P. F. SPANO
Department of Pharmacology
University of Milan
Via A. Del Sarto, 21
I-202129 Milan, Italy

Dr. SHELDON B. SPARBER
Department of Pharmacology
University of Minnesota
Minneapolis, Minnesota 55455

Dr. SYDNEY SPECTOR
Department of Pharmacology
Roche Institute of Molecular Biology
Nutley, New Jersey 07110

Dr. RICHARD SQUIRES
A/S Ferrosan
Sydmarken 5
DK-2860 Søborg, Denmark

Dr. ROSALMA STANCHERIS
Istituto di Fisiologia della
Nutrizione Animale
Via Celoria 10
20133 Milano
Italy

Dr. KLAUS STARKE
Pharmakologischen Institut
Klinikum Essen
Hufelandstrasse 55
43-Essen-Holsterhausen
West Germany

Dr. JANINA STASZEWSKA-BARCZAK
Polish Academy of Sciences
Medical Research Centre
Laboratory of Neurophysiology
Dworkowa 3, Warsaw, Poland

Dr. LARRY STEIN
Department of Psychopharmacology
Wyeth Laboratories, Inc.
P.O. Box 8299
Philadelphia, Pennsylvania 19101

Dr. R. J. STEPHENS
Nicholas Research Institute
Neuropharmacology Lab.
225 Bath Road
P.O. Box 17
Slough, Bucks, SL1 41U
Great Britain

Dr. L. STINUS
Université de Bordeaux I
Laboratoire de Psychophysiologie
33 Talence, France

Dr. LENNART STJÄRNE
Department of Physiology
Karolinska Institutet
Stockholm 60, Sweden

Dr. GÜNTER STOCK
Department of Pharmacology
University of Göteborg
S-400 33 Göteborg 33
Sweden

Dr. JON M. STOLK
Department of Pharmacology
    and Toxicology
Dartmouth Medical School
Hanover, New Hampshire 03755

Dr. HARRY STRUYKER-BOUDIER
Institute of Pharmacology
University of Nijmegen
Nijmegen, The Netherlands

Dr. M. F. SUGRUE
Organon Laboratories Ltd.
Newhouse, Lanarkshire
ML1 5SH, Scotland

Dr. FRIDOLIN SULSER
Department of Pharmacology
Vanderbilt University
    School of Medicine
Nashville, Tennessee 37203

Dr. AININ SURIA
St. Elizabeth's Hospital
    (NIMH)
Washington, D.C. 20032

Dr. Torgny Svensson
Department of Pharmacology
University of Göteborg
Fack
S-400 33 Göteborg 33
Sweden

Dr. J. Szurssewski
Department of Physiology
  and Biophysics
Mayo Foundation
Rochester, Minnesota

Dr. Robert I. Taber
Department of Pharmacology
Schering Corporation
Bloomfield, New Jersey 07003

Dr. J. Taxi
Lab. de Biologie Animale
Faculté de Sciences
12 rue Cuvier
Paris 5ᵉ, France

Dr. A. M. Thierry
Lab. de Biologie Moleculaire
  College de France
11, place Marcelin-Bertholet
Paris 5ᵉ, France

Dr. Nguyen Bich Thoa
National Institutes of Health
  (NIMH)
Building 10—Room 2D-46
Bethesda, Maryland 20014

Dr. Hans Thoenen
Department of Pharmacology
Biozentrum der Universität
Klingelbergstrasse 70
CH-4056 Basle, Switzerland

Dr. James A. Thomas
Laboratory of Clin. Sci.
Natl. Inst. of Mental Health
Bethesda, Maryland 20014

Dr. K. F. Tipton
Department of Biochemistry
Cambridge University
Tennis Court Road
Cambridge, England

Dr. T. Budya Tjandramagu
Clinical Pharmacology Program
Emory University School of Medicine
Atlanta, Georgia 30303

Dr. J. P. Tranzer
Department of Experimental Medicine
F. Hoffmann-LaRoche and Company
CH-4002 Basle, Switzerland

Dr. Ullrich Trendelenburg
Institute for Pharmacology
  and Toxicology
University of Würzburg
D-8700 Würzburg, West Germany

Dr. José M. Trifaró
Department of Pharmacology
  and Therapeutics
McGill University
3655 Drummond Street
Montreal 109, Quebec, Canada

Dr. A. J. Turner
Istituto di Ricerche Farm.
  'Mario Negri'
Via Eritrea
Milan, Italy

Dr. Gertrude M. Tyce
Department of Biochemistry
Mayo Clinic and Foundation
Rochester, Minnesota

Dr. Sidney Udenfriend
Roche Institute of Molecular Biology
Nutley, New Jersey 07110

Dr. Guiliana Udeschini
Institute of Pharmacology & Pharmacognosy
University of Milano
Via A. del Sarto 21
20129 Milano, Italy

Dr. Urban Ungerstedt
Department of Histology
Karolinska Institutet
S-10401 Stockholm 60
Sweden

Dr. P. F. Urban
C.N.R.S., Centre de Neurochimie
67085 Strasbourg Cedex, France

Dr. Earl Usdin
Psychopharmacology Research Branch
National Institute of Mental Health
5600 Fishers Lane—Room 9-95
Rockville, Maryland 20852

Dr. Vera R. Usdin
Gillette Research Institute
1413 Research Boulevard
Rockville, Maryland 20850

Dr. Eric Vallee
Laboratoires Clin. Midy.
34034 Montpellier, France

Dr. William Van Dorsser
Université de Liège
Institut de Thérapeutique
32, Bd de la Constitution
B-4000 Liège, Belgium

Dr. Daniel P. van Kammen
Section on Psychiatry, NIMH
National Institutes of Health
Building 10—Room 3S229
Bethesda, Maryland 20014

Dr. Jacque Van Rossum
Department of Pharmacology
R.K. Universität
Kapittelweg 40
Nijmegen, The Netherlands

Dr. Bernardo B. Vargaftig
Centre de Recherhce
  Merrell International
16, rue d'Ankara
67000 Strasbourg, France

Dr. Guy Vincedon
C.N.R.S., Centre de Neurochimie
67085 Strasbourg, Cedex, France

Dr. Rene Viterbo
Centre de Recherche
  Merrell International
16, rue d'Ankara
67000 Strasbourg, France

Dr. H. Gerhard Vogel
Farbwerke Hoechst A.G.
Postfach 800320
6230 Frankfurt (Main) 80,
Germany

Dr. Marthe Vogt
Institute of Animal Physiology
Babraham, Cambridge
England

Dr. Ulf S. Von Euler
Department of Physiology
Karolinska Institutet
S-104 01 Stockholm 60
Sweden

Dr. Wilfred Von Studnitz
Medical Diagnostic Institute
Nussbaumstrasse 14
D 8 Munich, Germany

Mr. Bertil Waldeck
Department of Pharmacology
University of Göteborg
Fack
S-400 33 Göteborg 33
Sweden

Dr. P. Waldmeier
Ciba-Geigy S.A.
Basle, Switzerland

Dr. Marshall B. Wallach
Department of Experimental
  Psychiatry
Syntex Research
Palo Alto, California 94304

Dr. Judith Walters
Department of Psychiatry
Yale University School of Medicine
34 Park Street
New Haven,
Connecticut 06519

Dr. J. M. Warter
C.N.R.S., Centre de Neurochimie
67085 Strasbourg Cedex, France

Mme. J. Weil-Fugazza
Institut de Pharmacologie
21, rue de l'Ecole de Médecine
75005 Paris, France

Dr. Hans Weil-Malherbe
St. Elizabeth's Hospital
Wm. A. White Bldg.—Room 528
Washington, D.C. 20032

Dr. Murray Weiner
Merrell-National Laboratories
Cincinnati, Ohio 45215

Dr. Norman Weiner
Department of Pharmacology
University of Colorado Medical Center
4200 East 9th Avenue
Denver, Colorado 80220

Dr. Richard M. Weinshilboum
Department of Pharmacology
Mayo Graduate School of Medicine
Rochester, Minnesota 55901

Mrs. Virginia Weise
National Institutes of Health (NIMH)
Building 10—Room 2D-46
Bethesda, Maryland 20014

Dr. Benjamin Weiss
Department of Pharmacology
Medical College of Pennsylvania
3300 Henry Avenue
Philadelphia, Pennsylvania 19129

Dr. J. Wepiere
Faculté de Pharmacie
7 rue Baptist-Clement
92290 Chatenay Malabry
France

Dr. J. P. Werbenec
C.R.E.P.
Grenoble, France

Dr. Bèngt C. Werdínius
Department of Pharmacology
University of Göteborg
Fack
S-400 33 Göteborg 33
Sweden

Dr. G. Wermuth
Faculté de Pharmacie
Université Louis Pasteur
67083 Strasbourg Cedex, France

Dr. Erik Westermann
Institute for Pharmacology
Medizinische Hochshule Hannover
Roderbruchstrasse 101
3-Hannover-Kleefeld, Germany

Dr. Thomas C. Westfall
Department of Pharmacology
University of Virginia
   School of Medicine
Charlottesville, Virginia 22903

Dr. M. Wetzel
34, rue Principale
67300 Schiltigheim, France

Dr. Sherwin Wilk
Department of Pharmacology
Mt. Sinai School of Medicine
Fifth Avenue & 100th Street
New York, New York 10029

Dr. B. Will
Laboratoire de Psychobiologie
rue de l'Université
7000 Strasbourg, France

Dr. R. T. Williams
St. Mary's Hospital Medical School
London W2 1PG, England

Dr. Hans Winkler
Department of Medicine
University of Miami
P.O. Box 875, Biscayne A
Miami, Florida 33152

Dr. Anna Wirz-Justice
Psychiatrische Universitatsklinik Basel
Wilhelm-Klein-Strasse 27
4056 Basel, Switzerland

Dr. G. Frederick Wooten
Laboratory of Clinical Science
National Institutes of Health (NIMH)
Building 10—Room 2D-46
Bethesda, Maryland 20014

Dr. Richard J. Wurtman
Department of Endocrinology
   and Metabolism
Massachusetts Inst. of Technology
Cambridge, Massachusetts 02139

Dr. Melvin D. Yahr
Department of Neurology
College of Physicians and Surgeons
Columbia University
630 W. 168th Street
New York, New York 10032

Dr. Moussa B. H. Youdim
MRC Unit and Department of Clinical
   Pharmacology
University of Oxford
Radcliffe Infirmary
Oxford, England

Dr. Eleanor Zaimis
Department of Pharmacology
Royal Free Hospital
   School of Medicine
8 Hunter Street
London WC1N 1BP, England

Dr. Fernanda Zambotti
B. Baron Memorial Research Labs.
Queen Charlotte's Maternity Hospital
Goldhawk Road
London W6 OXG, England

Dr. E. Albert Zeller
Department of Biochemistry
Northwestern University
303 East Chicago Avenue
Chicago, Illinois 60611

Dr. Michael J. Zigmond
Department of Biology
University of Pittsburgh
Pittsburgh, Pennsylvania 15213

Dr. Earl A. Zimmerman
Box 100
New York Neurological Institute
710 W. 168th Street
New York, New York 10032

Dr. M. J. Zwiller
C.N.R.S., Centre de Neurochimie
67085 Strasbourg Cedex, France

Abbreviations

A = adrenaline
AADC = aromatic amino acid decarboxylase
ACh = acetylcholine
AD = analgesic dose
a-DBH = anti dopamine β-hydroxylase serum
ALB = Albany rat
AMP = adenosine monophosphate
AMPT = α-methyl-p-tyrosine
AMT = α-methyltyrosine
AN = arcuate nucleus
ANOVA = Analysis of Variance
APC = N,N'-bis-(ω-aminopentyl)-cystamine
Ar = aromatic
AS-NGF = antiserum to nerve growth factor
ATD = amantadine
ATPase = adenosine triphosphatase
ATV = area ventralis tegmenti

BA = bioassay
BBB = blood–brain barrier
biopterin-H_4 = tetrahydrobiopterin
BPRS = Brief Patient's Rating Scale
Br$_2$FA = 5-(3′,4′-dibromobutyl)picolinic acid

CA = catecholamine(s)
cAMP = cyclic adenosine monophosphate
CA-R$_\alpha$ = catecholamine α-receptor
CFA = complete Freund's adjuvant
CGI = Clinical Global Impressions
cGMP = cyclic guanosine monophosphate
ChAc = choline acetylase
CMC = carboxymethylcellulose
CNS = central nervous system
COMT = catechol O-methyltransferase
CPZ = chlorpromazine
CSF = C.S.F. = cerebrospinal fluid
cyclic AMP = cyclic adenosine monophosphate

D = dopamine; also, D = desynchronized sleep
DA = dopamine
d-A = D-amphetamine
d-AMP = D-amphetamine
DBcAMP = dBcAMP = dibutyryl cyclic adenosine monophosphate
DBH = DβH = dopamine β-hydroxylase
DC = decarboxylase (of aromatic amino acids)
DC-AMP = dibutyryl cyclic adenosine monophosphate
DDC = dopamine decarboxylase
DHAA = 3,4-dihydroxyphenylacetaldehyde
DHPAA = 3,4-dihydroxyphenylacetaldehyde
DHPE = 2-(3,4-dihydroxyphenyl)ethan-l-ol
DHPGA = 3,4-dihydroxyphenylglycolaldehyde
DHT = 5,6-DHT = 5,6-dihydroxytryptamine
DM = dopamine
DMI = desmethylimipramine
DMPH = $\Delta^{3,4}$-dimethylheptyltetrahydrocannabinol
DMPH$_4$ = 6,7-dimethyl-5,6,7,8-tetrahydropterin
DMPP = dimethylphenylpiperazinium
DOCA = de(s)oxcorticosterone acetate
DOMA = dihydroxymandelic acid

DOPA = dopa = Dopa = dihydroxyphenylalinine
DOPAC = dihydroxyphenylacetic acid
DOPAD = DOPA decarboxylase
DOPEG = 3,4-dihydroxyphenylglycol
DPI = diphosphatidyl inositol
D.S. = Down's syndrome

E = epinephrine
EB = early bleeding
ECD = electron capture detector
EDTA = ethylenediamine tetraacetic acid
EEDQ = 1-ethoxycarbonyl-2-ethoxy-1,2-dihydroquinoline
EGTA = ethylene glycol bis (β-aminoethyl ether)tetraacetic acid
EPI = epinephrine
EST = electroconvulsive shock treatment
ETA = 5,8,11,14-eicosatetraynoic acid

FA = fusaric acid
F.D. = familial dysautonomia
FFA = free fatty acids
FI = fixed interval
FR = fixed ratio
FRF = follicle stimulating hormone releasing factor
FSH = follicle stimulating hormone

GA = glyoxylic acid
GABA = γ-aminobutyric acid
GAD = glutamic acid decarboxylase
GBL = γ-butyrolactone
GC = gas chromatography
GEDTA = ethyleneglycol bis (β-aminoethyl ether)tetraacetic acid
GH = growth hormone
GHB = γ-hydroxybutyric acid
GIF = growth hormone inhibiting factor
GLC = gas–liquid chromatography
GMP = guanosine monophosphate
GRF = growth hormone releasing factor

HAL = haloperidol
H$_4$B = tetrahydrobiopterin
H.C. = Huntington's Chorea
HCG = Human Chorionic Gonadotrophin
6-HDA = 6-hydroxydopamine
HGH = human growth hormone
HHH = hypothalamus-hypophysiotropic hormone
5-HIAA = 5-hydroxyindoleacetic acid
HMPG = MHPG = 4-hydroxy-3-methoxyphenylethylene glycol
HRS = Hamilton Rating Scale
Ht = hypothalamus
5-HT = 5HT = 5-hydroxytryptamine = serotonin
5-HTP = 5-hydroxytryptophan
HVA = homovanillic acid

IA = indoleamine
i.c. = intracisternally
ICS = intracranial self-stimulation

25

IMAO = inhibitor of monoamine oxidase
IMO = immobilization
inc. FA = incomplete Freund's adjuvant
ISO = isoproterenol
IVT = intraventricularly

l-AMP = L-amphetamine
LB = late bleeding
LC = locus coeruleus
l.d.c. = large dense-core
LDH = lactate dehydrogenase
LDV = large dense-core vesicles
LGV = large granulated vesicles
LH = luteinizing hormone
L-PCS = lateral to pendunculus cerebellaris
superior
LRF = luteinizing hormone releasing factor

M = metanephrine
MAO = monoamine oxidase
MAOA = monoamine oxidase activity
MAOI = monoamine oxidase inhibitor
α-MD = α-methyldopa
α-MDA = α-methyldopamine
4-MDA = 4-methyldopamine
MFB = median forebrain bundle
MHPE = 2-(4-hydroxy-3-methoxyphenyl)ethan-
1-ol
MHPG = 3-methoxy-4-hydroxyphenyl-
ethyleneglycol
MID = multiple ion detection
MIF = melanocyte stimulating hormone
release inhibiting factor
MMT = α-methyl-m-tyrosine
MMTA = α-methyl-m-tyramine
α-m-Na = α-methylnoradrenaline
α-MNE = α-methylnorepinephrine
MOPEG = 3-methoxy-4-hydroxyphenyl glycol
6-MPH$_4$ = 6-methyl-pteridine
αMPT = αMpT = α-methyl-p-tyrosine
βMPT = β-methyl-p-tyrosine
MS = mass spectrometry
αMT = α-methyltyrosine
5-MTHF = 5-methyltetrahydrofolic acid

NA = noradrenaline = norepinephrine
(or, noradrenergic)
NCP = caudate nucleus + putamen
NE = norepinephrine
NEFA = non-esterified fatty acid
NEM = N-ethylmaleimide
NGF = nerve growth factor
NGS = normal goat serum
NIH = National Institutes of Health (U.S.)
NIMH = National Institute of Mental Health
(U.S.)
NM = normetanephrine
NMRI = National Medical Research Institute
(UK)
NTS = nucleus tractus solitarii

6-OHDA = 6-OH-DA = 6-hydroxydopamine
6-OHDM = 6-hydroxydopamine
6-OHDOPA = 6-hydroxydopa
OM = Osborne-Mendel rat
OMI = O-methylisoprenaline

PAPETA = p-aminophenethyltrimethylam-
monium

PAPS = phosphoadenosine phosphosulfate
PAS = Periodic Acid–Schiff
PBA = phenoxybenzamine
PBZ = phenoxybenzamine
PCA = p-chloroamphetamine
PCPA = pCPA = p-chlorophenylalanine
P.D. = Parkinson's disease
PDE III = high affinity phosphodiesterase
PEP = systolic pre-ejection period
PFP = pentafluoropropionic
PFPA = pentafluoropropionic anhydride
PG = prostaglandin
PGE = prostaglandin E
PGO = ponto-geniculo-occipital
PIF = prolactin inhibiting factor
PLP = pyridoxal 5′-phosphate
PMS = pregnant mare serum
PNMT = phenylethanolamine N-methyl-
transferase
POH = p-hydroxyamphetamine
pOHA = p-hydroxyamphetamine
pOHNE = p-hydroxynorephedrine
PPL = propranalol
PR-IH = prolactin release inhibiting hormone
PS = paradoxical sleep
PST = phenolsulfotransferase
PZ = pimozide

RBF = renal blood flow
REM = rapid eye movement (sleep)
RF = releasing-factor
RH = releasing hormone
RHA = Roman high avoidance rat
RIA = radioimmunological

S-AMe = S-adenosyl methionine
SCG = superior cervical ganglion
s.d.c. = small dense core
SDS = sodium dodecyl sulfate
SGV = small granular vesicles
SHR = spontaneous hypertensive rat
S-I = stimulation induced
SIF = small intensely fluorescent
SN = substantia nigra
SRIF = somatotropin release inhibiting factor
SWS = slow wave sleep

T3 = triiodothyronine
TH = tyrosine hydroxylase
THC = tetrahydrocannabinol
TIDA = tubero-infundibular dopamine
TIQ = tetrahydroisoquinoline
TLC = thin layer chromatography
TO = time out
T-OH = tyrosine hydroxylase
TRH = thyrotropin releasing hormone
TRP = tryptophan
TSH = thyroid stimulating hormone
TSMT = thiolmethyltransferase
TZ = thioproperazine

VI = variable interval
VMA = vanillylmandelic acid
VMT = ventromedial tegmental area
VT = ventral tegmental
W = Wistar rat
W/Ky = Wistar/Kyoto rat

ZC = zona compacta
ZR = zona reticulata

Frontiers in Catecholamine Research 1973, pp. 27 to 37. Pergamon Press. Printed in Great Britain.

THE IMPACT OF MONOAMINE RESEARCH ON DRUG DEVELOPMENT

A. PLETSCHER

Research Division, F. Hoffmann-La Roche and Co. Ltd, Basel, Switzerland

THE QUASI exponential course of progress since the middle of this century in the field of biogenic monoamines has been characterised by a close feedback interaction between basic research and drug developments. On the one hand, the advancement of fundamental knowledge on monoamines greatly stimulated the search for new drugs, and on the other hand, the use of drugs as tools was very helpful in the elucidation of basic mechanisms. For instance, two findings obtained with drugs were decisive for starting the "explosion" of monoamine research, namely that: (a) reserpine, a psychosedative drug, caused a long-lasting marked depletion of 5-hydroxytryptamine (5HT) in tissues including brain (SHORE et al., 1955; PLETSCHER et al., 1956) and, as shown later, also of noradrenaline (NA) and dopamine (DA) in animals, and (b) the monoamine oxidase (MAO) inhibitor iproniazid, which increased the monoamines in the brain and counteracted the amine-depleting and behavioural effects of reserpine in animals, improved mental depression in man (see PLETSCHER et al., 1965).

In this presentation, a short review will be given of the clinically and experimentally used drugs which have been developed or whose mechanism of action has been clarified through modern monoamine research. The drugs will be classified according to their presumable mechanism of action (Fig. 1).

1. BIOSYNTHESIS

1.1 Tyrosine hydroxylase

Hydroxylation of tyrosine is a rate-limiting step in the biosynthesis of catecholamines, and therefore in vivo inhibition of tyrosine hydroxylase decreases the formation of this amine. Two classes of inhibitors were developed several years ago, i.e. derivatives of tyrosine, e.g. α-methyl-tyrosine, 3-iodo-tyrosine, 3,4-diiodo-tyrosine, 3-iodo-α-methyl-tyrosine, and catechols, e.g. 3,4-dihydroxyphenylpropylacetamide (α-n-propyl-dopacetamide). Both types cause a reversible inhibition, the tyrosine derivatives compete with the substrate, the catechols probably interact with the pteridine cofactor (UDENFRIEND, 1966). As a consequence, the endogenous NA and DA considerably decrease after administration of the drugs, especially when repeated. The levels of other amines, e.g. 5HT, are not markedly affected. Recently, other compounds inhibiting tyrosine hydroxylase have been described, e.g. substances of microbial origin, such as oudenone (NAGATSU et al., 1971).

The tyrosine hydroxylase inhibitors do not yet have an established place in medical therapy, although α-methyl-p-tyrosine showed some clinical effects, e.g. lowering of elevated blood pressure in patients with pheochromocytoma and sedation (SJOERDSMA, 1966). However, the inhibitors have been used to elucidate the functional role of catecholamines and to measure the catecholamine turnover in the tissues.

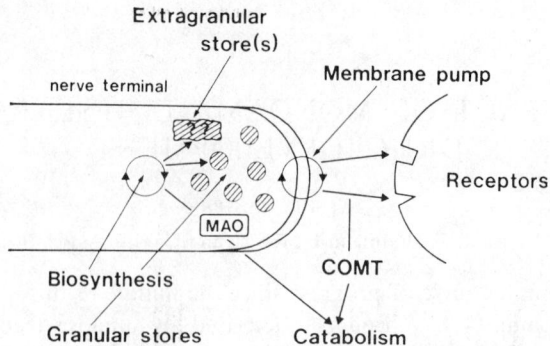

FIG. 1.—Various factors involved in monoaminergic neurohumoral transmission.

1.2 *Tryptophan hydroxylase*

Enzymes hydroxylating tryptophan in 5-position of the phenyl ring occur in various tissues, such as pineal gland, brain stem, mast cells, liver. Except for brain stem (rats), the enzymes also hydroxylate phenylalanine. Various inhibitors have been found *in vivo*, e.g. ring-substituted tryptophan and phenylalanine derivatives (inhibition competitive with substrate), catechols, e.g. α-n-propyl-dopacetamide (inhibition non-competitive with substrate and pteridine cofactor), or ion chelators (KOE, 1971).

Most of these inhibitors do not markedly alter 5HT tissue levels *in vivo*. An outstanding exception is *p*-chlorophenylalanine. This compound, owing to an "irreversible" inhibition of tryptophan hydroxylase, causes a marked, rather selective and long-lasting decrease of the 5HT content, e.g. in the brain of animals. In addition, *p*-chlorophenylalanine also markedly inhibits liver phenylalanine hydroxylase.

Several derivatives related to *p*-chlorophenylalanine, i.e. *p*-chloro-amphetamine, *p*-chloro-metamphetamine, phenfluramine, have also been reported to produce a marked, relatively elective decrease of 5HT and 5-hydroxyindoleacetic acid in the brain. Some of these compounds are considerably more potent, and their duration of action is longer compared to *p*-chlorophenylalanine. Various possibilities are being discussed concerning the mechanism of action, e.g. inhibition of tryptophan hydroxylase, 5HT release with concomitant MAO inhibition or destruction of 5-hydroxytryptaminergic nerve terminals (PLETSCHER *et al.*, 1966; OPITZ, 1967; SULSER and SANDERS-BUSH, 1971).

Neither *p*-chlorophenylalanine nor the halogenated phenylethylamines are being used in medical therapy. However, the compounds serve as valuable tools for research into the physiological and pathophysiological role of 5HT, e.g. regarding behaviour, pain, drug habituation, etc. Especially research on sleep and sexual behaviour has profited from these drugs. A derivative, *p*-chloro-α-methyl-phenylalanine, which strongly inhibits hepatic phenylalanine hydroxylase without interfering with brain tryptophan hydroxylase, may be useful in producing experimental phenylketonuria (KOE, 1971).

1.3 *Decarboxylase of aromatic amino acids (DC)*

Inhibitors of DC, an enzyme localised throughout the monoaminergic neurons, have been systematically searched for, because such compounds were thought to be

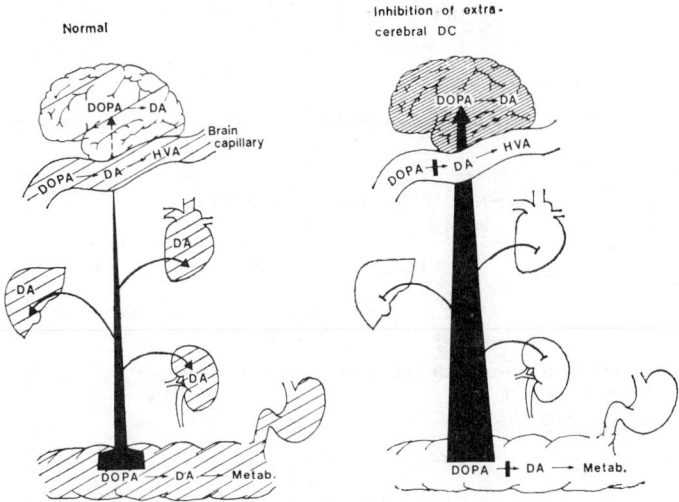

FIG. 2.—Penetration of L-dopa from the gastrointestinal tract into the brain in the absence and presence of an inhibitor of extracerebral decarboxylase. The shaded areas symbolize the concentration of dopamine.

useful in arterial hypertension and mental illness. Indeed, α-methyldopa, which inhibits the enzyme *in vitro* and *in vivo*, lowers elevated blood pressure, but it was soon realised that its hypotensive action did not depend on DC inhibition (PLETSCHER *et al.*, 1965).

Later, DC inhibitors belonging to the class of hydrazines or oxyamines have been developed. These compounds, although up to over 100 times more potent than α-methyldopa, have been shown not to markedly decrease blood pressure and to be devoid of psychotropic action in man, even in high doses. However, some of the inhibitors (e.g. Ro 4-4602* and MK 486†) markedly enhance the increase of cerebral DA following administration of L-3,4-dihydroxyphenylalanine (L-dopa). This action is due to a preferential inhibition by the drugs of the decarboxylation of L-dopa in extracerebral tissues. As a consequence, larger amounts of the amino acid penetrate into the brain parenchyma, where decarboxylation into DA occurs since cerebral DC has remained active (BARTHOLINI and PLETSCHER, 1968; PORTER, 1971) (Fig. 2).

Combination of L-dopa with DC inhibitors like Ro 4-4602 and MK 486 shows distinct advantages over L-dopa alone in the treatment of Parkinson's syndrome, especially with regard to occurrence and severity of peripheral side effects.

1.4 *Dopamine-β-hydroxylase*

This copper containing enzyme is localised in the catecholamine-storing granules of adrenal medulla and noradrenergic nerves. Various compounds, including benzyl-hydrazines, benzyloxyamines and derivatives of picolinic acid, have been found to be inhibitors *in vitro* and/or *in vivo*. Some of them, e.g. disulfiram and tropolone, probably act by copper chelation. Figure 3 indicates several DA-β-hydroxylase

---

* Ro 4-4602 = 1-DL-seryl-2-(2,3,4-trihydroxybenzyl) hydrazine hydrochloride.
† MK 486 = L-α-methyldopa-hydrazine.

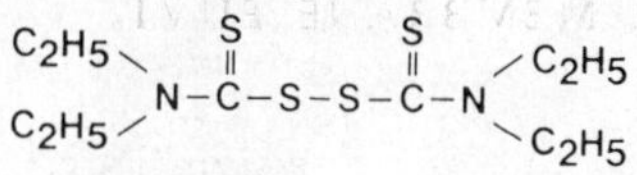

Disulfiram

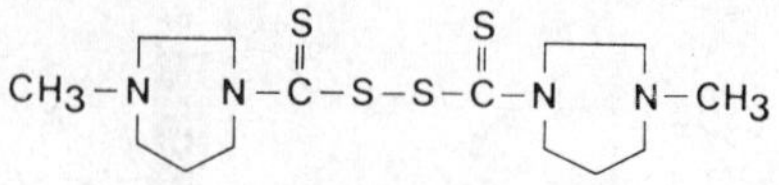

FLA 63

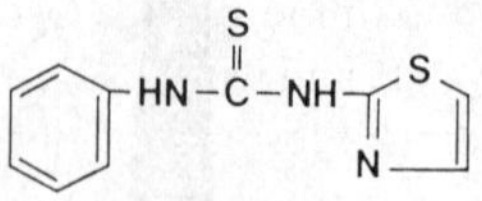

U–14624

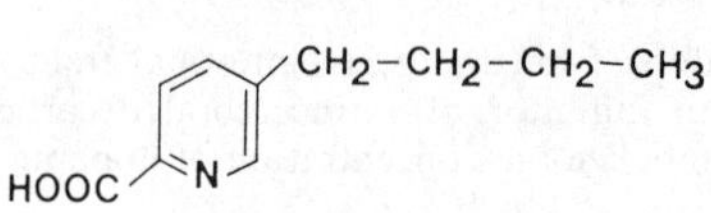

Fusaric acid

FIG. 3.—Some inhibitors of dopamine-β-hydroxylase.

inhibitors which decrease the NA and raise the DA levels in the tissues, especially in brain (GOLDSTEIN, 1966; SULSER and SANDERS-BUSH, 1971; HIDAKA, 1973).

DA-β-hydroxylase inhibitors are being used as tools for elucidating the physiological and pathophysiological role of NA and DA, e.g. regarding behaviour. Furthermore, it has been shown that some DA-β-hydroxylase inhibitors lower elevated blood pressure in animals and man and that fusaric acid possibly alleviates the involuntary movements due to L-dopa treatment in patients with Parkinson's syndrome (MENA et al., 1971).

1.5 Amine precursors

Catecholamines as well as 5HT are hardly able to pass the blood–brain barrier in adults. However, the immediate precursors of the amines, e.g. L-dopa, D-erythro-3,4-dihydroxyphenylserine and L-5-hydroxytryptophan (5HTP), probably because of the presence of specific transport systems, penetrate to some extent into the brain where decarboxylation to DA, NA and 5HT, respectively, takes place (Fig. 2). The amines seem to be mainly stored in the respective neurons, i.e. dopaminergic, noradrenergic and 5-hydroxytryptaminergic. With high doses of L-dopa, however, DA has also been observed to accumulate in 5-hydroxytryptaminergic neurons, and 5HT may also be stored by catecholaminergic neurons (CARLSSON, 1972; SHORE, 1972).

The experimental findings that monoamines are increased by application of their immediate precursors were the rational basis for the development of L-dopa as a therapeutic agent which is now widely used in Parkinson's syndrome.

The therapeutic usefulness of 5HTP and tryptophan (which also increases cerebral 5HT, though less than 5HTP, but specifically in 5-hydroxytryptaminergic neurons), e.g. in mental depression, is still under investigation.

2. MEMBRANE PUMP

The antidepressant action of imipramine, a non-MAO inhibitor, was an unexpected finding based on careful clinical observation (KUHN, 1957). Later, biochemical and histofluorimetric investigations indicated that imipramine probably inhibited the active uptake of monoamines ("amine pump") by the presynaptic membrane of the monoaminergic nerve terminal, leading to an increased concentration of amines in the synaptic cleft, i.e. in the vicinity of the amine receptors. The enhancement of some pharmacological actions of monoamines, e.g. NA, as well as the reversal of the effects of benzoquinolizines and reserpine by imipramine and related compounds are also in agreement with this hypothesis. An additional action of tricyclic antidepressants at the level of the intraneuronal monoamine storage organelles cannot, however, be excluded (SULSER et al., 1964; HIMWICH and ALPERS, 1970; SULSER, 1971).

Various pharmacological and biochemical test systems have been developed based on this concept and used in the search for further antidepressant drugs. Several new tricyclic compounds were found and are being widely used in the treatment of mental depression. These drugs differ in their effect on the uptake of catecholamines and 5HT. Thus, secondary amines (desmethylimipramine, nortriptyline, protriptyline) have been reported to be more potent than tertiary amines (imipramine, chloroimipramine and amitriptyline) in blocking the NA uptake by the neuronal membrane. In contrast, tertiary amines are more potent with regard to the inhibition of 5HT uptake (Lit. see SULSER and SANDERS-BUSH, 1971).

The uptake of DA by the membrane pump of the dopaminergic neurons of the striatum does not seem to be appreciably affected by tricyclic antidepressants like desmethylimipramine. However, the neuronal uptake of DA as well as of NA was found to be inhibited by certain anticholinergic and antihistamine drugs used in the treatment of Parkinson's syndrome (SHORE, 1972).

Lithium, a drug used in manic depressive illness, has been reported to enhance the uptake of NA into nerve terminals, but in addition this cation exerts many other metabolic effects (HIMWICH and ALPERS, 1970; SHORE, 1972).

3. INTRANEURONAL STORAGE

3.1 Granular membrane

3.1.1 Rauwolfia alkaloids. Good evidence exists that reserpine and similar drugs, e.g. deserpidine and rescinamine, act at the level of the amine storage organelles. They inhibit the uptake of 5HT, NA and DA which, when no longer protected from metabolic enzymes, are degraded by intraneuronal MAO. The uptake mechanism and the reserpine-induced behavioural changes, however, are restored before the amine content of the granules (SHORE, 1972).

Recent experiments indicate that the granular amine storage seems to involve at least two mechanisms, i.e. an active uptake at the granular membrane level and an intragranular, intermolecular bonding with nucleotides, bivalent cations and proteins decreasing the outflow of the amines from the granules. As shown with isolated membranes of adrenal chromaffin granules, reserpine inhibits the amine uptake at the membrane level. Whether or not the drug also interferes with the intragranular bonding remains to be investigated (PLETSCHER et al., 1973).

Reserpine has proven to be a most valuable tool for research, e.g. for the

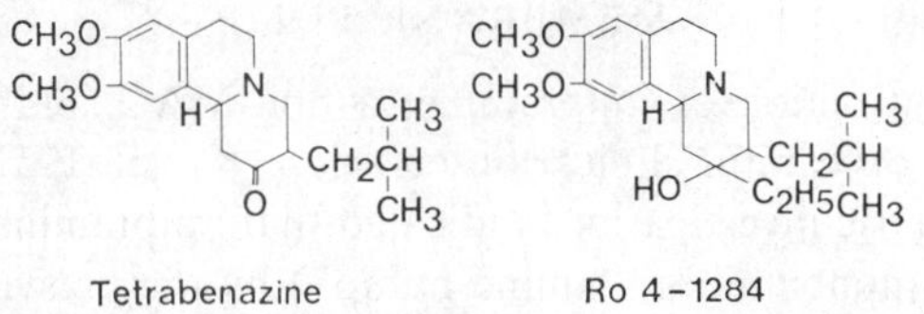

Tetrabenazine   Ro 4–1284

FIG. 4.—Structural formulas of two commonly known benzoquinolizine derivatives.

differentiation of various amine pools as well as for the elucidation of the intra-
and extraneuronal monoamine metabolism and of the functional role of monoamines
in various organs including brain. In addition, reserpine is also being used in medical
therapy, mainly in arterial hypertension and as a psychosedative agent.

3.1.2 *Other compounds.* After the discovery of the amine-lowering properties
of reserpine, a systematic research for compounds with a similar type of action was
initiated. Various derivatives of 1,2,3,4,6,7-hexahydro-11bH-benzo[a]quinolizine,
e.g. tetrabenazine and Ro 4-1284 (Fig. 4), were the first interesting compounds. They
act similarly to reserpine, but are less potent, of shorter duration of action and probably more selective for the brain (PLETSCHER *et al.*, 1962).

The "reversal" of the behavioural effects of tetrabenazine and Ro 4-1284 (SULSER
et al., 1964) serves as a basis for the screening of new antidepressants. In addition,
tetrabenazine is clinically used, especially in the treatment of abnormal extrapyramidal
movements, e.g. in Huntington's chorea.

Various phenylethylamine derivatives, e.g. bis(3,4-dichlorophenetyl)amine (PLET-
SCHER *et al.*, 1966) and prenylamine (JUORIO and VOGT, 1965) (which is used in
the treatment of coronary insufficiency) as well as compounds from other classes,
seem to decrease tissue monoamines by a reserpine-like mechanism. The amine-
depleting action of decaborane and ε-aminocaproic acid remains to be further elucidated (LIPPMANN and WISHNICK, 1965; MERITT and SULKOWSKI, 1966).

3.2 *Intragranular storage*

3.2.1 *Tyramine.* This compound belongs to the group of indirect sympathomimetic amines acting by displacement of intraneuronal NA from its storage sites. The
liberated amine not being degraded by intraneuronal MAO and therefore remaining
physiologically active, stimulates noradrenergic receptors. Tyramine probably acts
mainly on a relatively small pool of newly synthetised NA whose nature is still uncertain (KOPIN, 1964; WEINER, 1970).

3.2.2 *Amphetamine.* This drug, like tyramine, displaces endogenous NA, but
in addition probably DA. Stereotyped activity elicited by amphetamine is being
related to DA liberation, whereas the locomotor stimulation and increase in aggressive-
ness are thought to be also due to displaced NA (SULSER and SANDERS-BUSH, 1971).

By biochemical and combined histofluorimetric-biochemical studies, evidence
was obtained that the drug acts by liberation of catecholamines from a still unknown
extragranular, but intraneuronal pool resistant to reserpine. In addition, amphetamine has been shown to induce other changes, e.g. inhibition of catecholamine
uptake at the neuronal membrane, and to be transformed into a false neurohumoral
transmitter (see below) (FUXE and UNGERSTEDT, 1970; SULSER and SANDERS-BUSH,
1971).

3.2.3 *Amantadine*. The mechanism of action of this antiviral drug, which was found by chance to have a therapeutic effect in Parkinson's syndrome, is still controversial. It has been proposed that it acts in a similar, though less potent way than amphetamine, i.e. by liberating DA from an extragranular intraneuronal pool (FARNEBO *et al.*, 1971).

3.2.4 *False neurotransmitters*. A false neurotransmitter accumulates in the same storage sites as the physiological transmitter is released by nerve stimulation and depleted by agents that also deplete the physiological transmitter. Various amines fulfil these criteria, e.g. octopamine, α-methyloctopamine, *m*-octopamine, metaraminol, α-methylnoradrenaline, *p*-hydroxynorephedrine in the noradrenergic nerves, and α-methyl-*m*-tyramine and α-methyldopamine possibly in the dopaminergic system. These amines are metabolically formed from precursors, e.g. α-methyl-dopamine and α-methylnoradrenaline from α-methyldopa, α-methyl-*m*-tyramine and metaraminol from α-methyl-*m*-tyrosine and *p*-hydroxynorephedrine from amphetamine. They displace the physiological transmitter from the storage sites and are released together with the remaining physiological transmitter, e.g. on nerve stimulation.

The functional effect of a false transmitter that is present in the neuronal storage sites depends on various factors, e.g. on its action on the release and reuptake of the physiological transmitter, its interaction with the receptor including its direct receptor-stimulating properties. In many instances, the net functional effect of a false transmitter is inferior to that of the physiological transmitter. This may explain the blood pressure-lowering effect of some aromatic amino acids, e.g. of the widely used antihypertensive drug α-methyldopa as well as of α-methyl-*m*-tyrosine (KOPIN, 1969; SHORE, 1972).

The false transmitters have been serving as valuable tools, especially for differentiating between neuronal uptake and storage of amines as well as for assessing the role of MAO in monoaminergic neurons. In addition, formation of false transmitters may partly explain the mechanism of action of drugs like α-methyldopa and amphetamine and serve as a lead for the development of new antihypertensive and psychotropic drugs.

3.3 *Monoamine release*

Bretylium, bethanidine, guanethidine, guanisoquine and debrisoquine lower arterial blood pressure probably by a peripheral mechanism. The drugs are taken up into the noradrenergic neurons by the amine pump of the neuronal membrane and, probably because of a membrane stabilising action, inhibit the ability of a nerve impulse flow to release NA. In addition, some of the drugs (e.g. debrisoquine) may inhibit intraneuronal MAO which possibly plays a role in the antihypertensive effect. These compounds do not markedly decrease the total NA content of the tissues, although they may interfere with NA storage, e.g. with NA replenishment of a functional pool from the main store. In contrast, guanethidine has no MAO inhibitory effect and causes depletion of the neuronal NA (SHORE, 1972).

An interesting compound which induces a selective rise of DA in the basal ganglia of the brain, possibly by interfering with the release of the amine, is γ-hydroxybutyrate. Its mechanism of action, however, remains to be further elucidated (GESSA *et al.*, 1966).

3.4 *Destruction of monoamine nerve terminals*

6-Hydroxydopamine, in appropriate concentration, has been found to cause a relatively selective destruction of noradrenergic nerve terminals in extracerebral tissues and, if administered intracisternally, of catecholaminergic terminals in the brain. The compound is accumulated within the terminals by the amine pump of the neuronal membrane, and once a sufficient concentration has been reached the terminals are destroyed, probably by denaturation of the proteins (MALMFORS and THOENEN, 1971).

5,6-Dihydroxytryptamine preferentially destroys 5-hydroxytryptaminergic nerve terminals in the brain, probably by a mechanism similar to that described for 6-hydroxydopamine in noradrenergic neurons. In the extracerebral organs 5,6-dihydroxytryptamine also causes degeneration of noradrenergic nerve terminals.

Neither 6-hydroxydopamine nor 5,6-dihydroxytryptamine are used in medical therapy. However, both compounds, owing to their relative specificity, have proved to be valuable tools for differentiating the functional correlates of noradrenergic and 5-hydroxytryptaminergic transmission (BAUMGARTEN and LACHENMAYER, 1972; DA PRADA *et al.*, 1973).

3.5 *Monoamine catabolism*

3.5.1 *MAO inhibitors*. The compounds used today in medical treatment can be classified into three main groups, i.e. derivatives of phenylalkylhydrazines (e.g. iproniazid), of phenylcyclopropylamine (e.g. tranylcypromine), and of *N*-benzyl-propargylamine (e.g. pargyline). All these drugs have been developed on the basis of their marked and long-lasting inhibitory effect on MAO in animals *in vivo*. However, the inhibitors may also cause other changes, e.g. enhancement or inhibition of monoamine release (PLETSCHER *et al.*, 1965).

Good evidence exists for a causal connection between changes in monoamine metabolism and the clinical action of MAO inhibitors. The exact mechanism of the drugs and the functional role of the various monoamines remains, however, to be further clarified. An "overflow" of monoamines from maximally filled intraneuronal storage sites (psychostimulation), an inhibition of monoamine release as well as an accumulation of a false neurotransmitter (octopamine) in these sites (blood pressure-lowering action) are being discussed (PLETSCHER *et al.*, 1965; KOPIN, 1968).

Today, the MAO inhibitors have a limited use in a variety of clinical conditions, e.g. mental depression, arterial hypertension, angina pectoris. They also belong to the "armamentarium" of a laboratory which is involved in monoamine research. For instance, recently clorgyline and deprenyl, owing to their relatively specific inhibitory effect on type A and B MAO, respectively, were used for the characterisation of MAO isoenzymes (NEFF *et al.*, 1973).

3.5.2 *Catechol-O-methyltransferase (COMT) inhibitors*. Various substances have been found to be active *in vitro* and/or *in vivo* inhibitors of COMT, e.g. polyphenols such as pyrogallol, catechol derivatives such as dihydroxyphenylacetamides, tropolones, cycloheptimidazole derivatives, *S*-adenosyl homocystein, pyridoxal phosphate. They act either by competition with the catecholamine substrate (main mechanism of pyrogallol being a COMT substrate itself) or by non-competitive inhibition (tropolones) (UDENFRIEND *et al.*, 1959; BELLEAU and BURBA, 1961). The effect of the hitherto known COMT inhibitors on the endogenous catecholamines, e.g. of the

brain, seems to be rather weak. However, some pharmacological actions of exogenous catecholamines have been shown to be augmented or prolonged by the inhibitors (WYLIE *et al.*, 1960). Furthermore, in the brain of rats pyrogallol markedly
diminished the *O*-methylation of L-dopa and its catechol metabolites (BALESSANDRINI
and CHACE, 1972).

No COMT inhibitor has been introduced into medical therapy up to now, although
a potent, non-toxic derivative might be of use in the treatment of mental depression
or in enhancing the action of L-dopa in Parkinson's syndrome.

3.6 *Monoamine receptors*

Interference of drugs with monoamine receptors has been shown to influence the
monoamine metabolism via a feedback mechanism. In general receptor blocking
agents increase and receptor stimulant drugs decrease monoamine turnover. The
nature of the feedback mechanism is not yet clear in all respects.

In the following, some examples of drugs acting on monoamine receptors will be
briefly discussed.

3.6.1 *Receptor blockade. Neuroleptic drugs*, e.g. phenothiazines, thioxanthenes,
butyrophenones, methiothepin, clozapine, block DA and to various extents also NA
or 5HT (methiothepin) receptors, especially in the central nervous system. The DA
receptor blockade leads to catalepsy in animals and disturbances of extrapyramidal
functions in man. Thereby, the cerebral turnover of DA and NA is increased (Lit.
see BARTHOLINI *et al.*, 1972, 1973).

An interesting new type of "neuroleptic drug", represented by clozapine, is
claimed to have an antipsychotic action in schizophrenia, but to cause no major
disturbances of extrapyramidal functions in animals and man. The drug markedly
enhances the turnover of cerebral catecholamines in rats (BARTHOLINI *et al.*, 1972).

3.6.2 *Receptor stimulation. Apomorphine* is the prototype of a DA receptor
agonist. It has been shown to alleviate symptoms in Parkinson's syndrome and to
block haloperidol-induced dyskinesia in schizophrenic patients. Furthermore, the
drug produces stereotype behaviour as well as rotational movements towards the
unoperated site in rats with unilateral removal of one corpus striatum or with unilateral spreading depression. Concomitantly, apomorphine decreases the DA turnover in the striatum (Lit. see PLETSCHER, 1973).

Recently, another drug, ET 495 (2-(4-piperonyl-l-piperazinyl) pyrimidine), has
been described to stimulate DA receptors and to have a longer duration of action
than apomorphine in animals (CORRODI *et al.*, 1972).

Clonidine has been shown to be a centrally acting NA receptor agonist which
reduces the turnover of cerebral NA (ANDÉN *et al.*, 1970). Stimulation of central
NA receptors is probably the main reason for the blood pressure-lowering effect of
this drug which has been introduced into medical therapy for this indication. Receptor
agonists have also been found in the 5-hydroxytryptaminergic system, for instance,
lysergic acid diethylamide (LSD) and some other hallucinogenic compounds seem to
stimulate central 5HT receptors and to slow down the cerebral 5HT turnover
(ANDÉN *et al.*, 1969).

Some of the drugs interfering with monoamine receptors have proved to be of
value in neuropsychiatry and internal medicine (e.g. neuroleptics, clonidine), and the

TABLE 1. SUMMARY OF DRUGS ACTING ON MONOAMINERGIC TRANSMISSION

Action on	Point of attack	Drug
Biosynthesis	tyrosine | tryptophan | hydroxylase	α-methyltyrosine p-chlorphenylalanine
	decarboxylase	*Ro 4-4602**, *MK 486**
	DA-β-hydroxylase	disulfiram, FLA 63, fusaric acid
	amine precursors	*L-DOPA**, L-5HTP
Amine pump	presynaptic membrane	*imipramine, amitriptyline*
Intraneuronal storage	granular membrane	*reserpine, tetrabenazine**
	granular (and/or extra-granular) stores	tyramine, *amphetamine amantadine*
	false transmitters	*α-methyldopa** α-methyl-m-tyrosine
NA release	neuronal membrane etc.	*bretylium, guanethidine*
Structure of nerve terminals	catecholaminergic 5-hydroxytryptaminergic	6-hydroxydopamine 5,6-dihydroxytryptamine
Catabolism	MAO	*iproniazid, tranylcypromine*, pargyline**
	COMT	pyrogallol, tropolones
Receptors	catecholaminergic dopaminergic noradrenergic 5-hydroxytryptaminergic	*chlorpromazine, haloperidol* apomorphine *clonidine* LSD

Italics : in use in medical therapy.
Asterisk: discovered on the basis of a monoamine hypothesis.

basic knowledge gained with the presently available compounds will facilitate the future research for "receptor active" drugs.

A *summary* of the various types of drugs acting on monoamine metabolism is presented in Table 1.

REFERENCES

ANDÉN N. E., CARLSSON A. and HAGGENDAHL J. (1969) *Ann. Rev. Pharmac.* **9**, 119–134.
ANDÉN N. E., CORRODI H., FUXE K., HÖKFELT T., RYDIN C. and SVENSSON T. (1970) *Life Sci.* **9** (I), 513–523.
BALESSANDRINI R. J. and CHACE K. V. (1972) *Europ. J. Pharmac.* **17**, 163–166.
BARTHOLINI G. and PLETSCHER A. (1968) *J. Pharmac. exp. Ther.* **161**, 14–20.
BARTHOLINI G., HAEFELY W., JALFRE M., KELLER H. H. and PLETSCHER A. (1972) *Br. J. Pharmac.* **46**, 736–740.
BARTHOLINI G., KELLER H. H. and PLETSCHER A. (1973) *Neuropharmac.*, in press.
BAUMGARTEN H. G. and LACHENMAYER L. (1972) *Brain Res.* **38**, 228–232.
BELLEAU B. and BURBA J. (1961) *Biochem. biophys. Acta* **54**, 195–196.
CARLSSON A. (1972) *Acta Neurol. Scand.*, (Suppl.) **51**, 11–42.
CORRODI H., FARNEBO L. O., FUXE K., HAMBERGER B. and UNGERSTEDT U. (1972) *Europ. J. Pharmac.* **20**, 195–204.
DA PRADA M., O'BRIEN R. A., TRANZER J. P. and PLETSCHER A. (1973) J. *Pharmac. exp. Ther.*, in press.

FARNEBO L. O., FUXE K., GOLDSTEIN M., HAMBERGER B. and UNGERSTEDT U. (1971) *Europ. J. Pharmac.* **16,** 27–38.

FUXE K. and UNGERSTEDT U. (1970) In *Symposium on Amphetamines and Related Compounds* (COSTA E. and GARATTINI S., eds.) pp. 257–288. Raven Press, New York

GESSA G. L., VARGIU L., CRABAI F., BOERO G. C., CABONI F. and CAMBA R. (1966) *Life Sci.* **5** (II), 1921–1930.

GOLDSTEIN M. (1966) *Pharmac. Rev.* **18,** 77–94.

HIDAKA H. (1973) This Symposium.

HIMWICH H. E. and ALPERS H. S. (1970) *Ann. Rev. Pharmac.* **10,** 313–334.

JUORIO A. V. and VOGT M. (1965) *Br. J. Pharmac.* **24,** 566–573.

KOE B. K. (1971) *Fedn. Proc.* **30,** 886–896.

KOPIN I. J. (1964) *Pharmac. Rev.* **16,** 179–192.

KOPIN I. J. (1968) *Ann. Rev. Pharmac.* **8,** 377–394.

KUHN R. (1957) *Schweiz. med. Wschr.* **87,** 1135.

LIPPMANN W. and WISHNICK M. (1965) *J. Pharmac. exp. Ther.* **150,** 196–202.

MALMFORS T. and THOENEN H. (Eds.) (1971) *6-Hydroxydopamine and Catecholamine Neurons.* North-Holland, Amsterdam.

MERITT J. H. and SULKOWSKI T. S. (1966) *Pharmacologist* **8,** 196.

MENA I., CROUT J. and COTZIAS G. C. (1971) *J. Am. med. Ass.* **218,** 1829–1830.

NAGATSU T., NAGATSU I., UMEZAWA H. and TAKEUCHI T. (1971) *Biochem. Pharmac.* **20,** 2507.

NEFF N. H., YANG H. Y. T. and GORIDIS C. (1973) This Symposium.

OPITZ, K. (1967) *Naunyn-Schmiedeberg's Arch. exp. Path. Pharmakol.* **259,** 56–65.

PLETSCHER A. (1973) In: *Industrial Aspects of Biochemistry.* FEBS Special Meeting, Dublin, in press. North-Holland, Amsterdam.

PLETSCHER A., SHORE P. A. and BRODIE B. B. (1956) *J. Pharmac. exp. Ther.* **116,** 84–89.

PLETSCHER A., BROSSI A. and GEY K. F. (1962) *Int. Rev. Neurobiol.* **4,** 275–306.

PLETSCHER A., GEY K. F. and BURKARD W. P. (1965) In: *Handbook of Experimental Pharmacology.* (EICHLER O. and FARAH A., Eds.) Vol. **19,** pp. 593–735. Springer, Berlin.

PLETSCHER A., DA PRADA M., BURKARD W. P., BARTHOLINI G., STEINER F. A., BRUDERER H. and BIGLER F. (1966) *J. Pharmac. exp. Ther.* **154,** 64–72.

PLETSCHER A., DA PRADA M., STEFFEN H., LÜTOLD B. and BERNEIS K. H. (1973) *Brain Res.,* in press.

PORTER C. C. (1971) *Fedn. Proc.* **30,** 871–876.

SHORE P. A. (1972) *Ann. Rev. Pharmac.* **12,** 209–222.

SHORE P. A., SILVER S. L. and BRODIE B. B. (1955) *Science* **122,** 284.

SJOERDSMA A. (1966) *Pharmac. Rev.* **18,** 673–683.

SULSER F. and SANDERS-BUSH E. (1971) *Pharmac. Rev.* **11,** 209–230.

SULSER F., BICKEL M. H. and BRODIE B. B. (1964) *J. Pharmacol. exp. Ther.* **144,** 321.

UDENFRIEND S. (1966) *Pharmac. Rev.* **18,** 43–51.

UDENFRIEND S., CREVELING C. R., OZAKI M., DALY J. W. and WITKOP B. (1959) *Archs. Biochem. Biophys.* **84,** 249–251.

WEINER N. (1970) *Ann. Rev. Pharmac.* **10,** 273–290.

WYLIE D. W., ARCHER S. and ARNOLD A. (1960) *J. Pharmac. exp. Ther.* **130,** 239–244.

Enzymes

(May 21, 1973; 10:30 A.M.–6:30 P.M.)

Frontiers in Catecholamine Research 1973, pp. 41 to 45. Pergamon Press. Printed in Great Britain.

TWO MECHANISMS OF CATECHOLAMINE STORAGE IN ADRENAL CHROMAFFIN GRANULES

A. Pletscher, M. Da Prada, B. Lütold, K. H. Berneis,
H. Steffen and H. G. Weder*

Research Department, F. Hoffmann-La Roche and Co. Ltd., Basel, Switzerland

This presentation summarises some recent studies with isolated membranes as well as with the watersoluble contents of bovine adrenal chromaffin granules.

GRANULAR MEMBRANES

Membranes of high purity, obtained by osmotic shock of isolated granules (prepared by density gradient centrifugation), are thoroughly washed and then incubated in an artificial medium with various monoamines. The membrane preparations, which consist of electron transparent vesicular structures, contain an Mg^{2+}-dependent adenosine-5'-triphosphatase (ATP-ase). They take up monoamines, whereby noradrenaline (NA) has been shown to be transported into the interior of the vesicles (Lit. see Pletscher et al., 1973). The amine uptake exhibits several characteristics, the most important of which can be summarised as follows:

The uptake of dopamine (DA), NA and adrenaline (A) is activated by ATP and Mg^{2+} (Table 1) (Taugner, 1971a; Agostini and Taugner, 1973; Pletscher et al., 1973).

The uptake occurs against a considerable concentration gradient, since in the membrane vesicles a concentration more than 100 times higher than that of the medium builds up (Taugner, 1972; B. Lütold, M. Da Prada and A. Pletscher, in prep.).

When the quotient molar concentration of membrane-bound to free catecholamines (CA) is plotted against the molar concentration of bound CA (Scatchard plot) (Scatchard, 1949), two equilibrium constants are obtained. The K at low concentrations of the amine (K_1) shows rather high values, and in addition the initial course of the NA and A curves is typical for a cooperative effect. These features are characteristic for a process of relatively high specificity, e.g. an enzyme-dependent active transport. In contrast, the K at high amine concentrations is lower, indicating a non-specific absorption (e.g. to proteic and/or lipidic membrane components) (Fig. 1) (in preparation).

The stimulation of the uptake by Mg^{2+}/ATP is limited to some amines, e.g. CA and 5-hydroxytryptamine (5HT), whereas others, e.g. tryptamine, histamine and metaraminol, are not markedly influenced (Pletscher et al., 1973; Lütold et al., in prep.). Furthermore, the Mg^{2+}/ATP-stimulated uptake of NA shows some stereospecificity, (—)NA being taken up preferentially (Taugner, 1972).

The Mg^{2+}/ATP-stimulated CA uptake markedly decreases with diminishing temperature, whereas the CA uptake in the absence of Mg^{2+}/ATP as well as the

---

* Institute of Molecular Biology and Biophysics, ETH, Zürich

Table 1.   Uptake of dopamine (DA) by isolated membranes of bovine adrenal chromaffin granules incubated in an artificial medium (Taugner, 1971a) at 37°C for 15 min.

Incubation with	DA uptake
DA alone	$1·6 \pm 0·4$
DA + ATP + Mg^{2+}	$9·5 \pm 0·6$
DA + ATP + Mg^{2+} + NEM	$1·2 \pm 0·2$
DA + ATP + Mg^{2+} + reserpine	$1·9 \pm 0·2$

Initial concentration: amines 45 μM, ATP 5 mM; concentration of Mg^{2+} 5 mM. The figures show averages with s.e. of 3–7 experiments and are indicated in nmole per mg protein. NEM = N-ethylmaleimide.

uptake of tryptamine (which is considerable) are only slightly temperature-dependent (Lütold et al., in prep.).

N-Ethylmaleimide, an inihibitor of ATP-ase, as well as reserpine decrease the Mg^{2+}/ATP-dependent CA uptake (Table 1), but do not markedly influence the uptake of tryptamine (Taugner and Hasselbach, 1968; Lütold et al., in prep.).

A stoichiometric correlation between the ATP-ase activity of the vesicles and the influx of CA across the membrane has been demonstrated with various methods (Taugner and Hasselbach, 1968; Hasselbach and Taugner, 1970; Taugner, 1971b).

From these and other findings it may be concluded that the transport of CA through the granular membrane does not occur by mere passive diffusion, but that

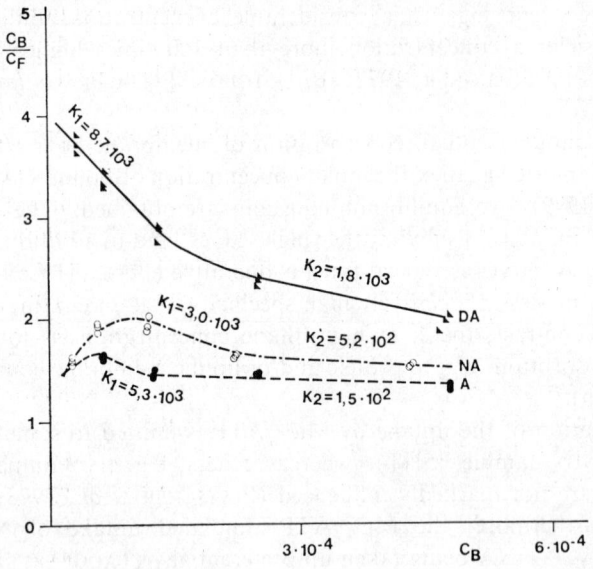

Fig. 1.—Scatchard plots and equilibrium constants of dopamine (DA), noradrenaline (NA) and adrenaline (A). Isolated membranes of bovine adrenal chromaffin granules were incubated in an artificial medium (Taugner, 1971a) containing 5 mM ATP and 5 mM Mg^{2+} with various concentrations of the amines at 37°C for 15 min. C_B = molar concentration of bound amine; C_F = molar concentration of free amine.

rather a specific mechanism is involved. The connection of the transport with ATP and ATP-ase probably indicates its dependence on energy which enables the amine to be pumped into the vesicular space.

INTRAGRANULAR INTERACTIONS

In monoamine-storing organelles, e.g. chromaffin granules and 5HT organelles of blood platelets, the amines are accumulated in very high concentration together with other constituents, e.g. nucleotides (ATP, etc.) and soluble proteins. Since the organelles could hardly be osmotically stable if amines and nucleotides were present in monomolecular form, a physico-chemical interaction between the constituents of the organelles is likely to occur (PLETSCHER et al., 1973). This interaction has been investigated by various physico-chemical methods, and the main results can be summarised as follows:

Nuclear magnetic resonance and infrared spectroscopy of solutions containing ATP and CA indicate that these two components form complexes in a molar ratio of 1:3–4 depending on the pH. Thereby, an ionic bond is formed between the positively charged nitrogen of CA and the negatively charged phosphate groups of the nucleotide. A hydrogen bridge stabilises this ionic bond (WEINER and JARDETZKY, 1964; PAI and MAYNERT, 1972).

Determinations of apparent average molecular weights by ultracentrifugation of aqueous solutions containing CA and ATP indicate that aggregation between amine and nucleotide occurs. Rising concentration, decreasing temperature and small amounts of bivalent cations (Ca^{2+}, Mg^{2+}) increase, whereas large amounts of these ions decrease the apparent average molecular weight. A aggregates less markedly with ATP than NA. Based on results obtained from sedimentation equilibrium and velocity experiments, it can be estimated that in aqueous solutions containing ATP, CA and Mg^{2+} in concentrations (17–18 % w/w) and molar ratios (4:1:0·25) similar to those occurring in vivo, more than 20 molecules of CA and ATP show mutual interaction at 37°C (PLETSCHER et al., 1973).

On analytical ultracentrifugation of a solution of NA/ATP/CA^{2+} (molar ratio 4:1:0·25) to which 2–3 per cent chromogranins are added, a single schlieren peak appears at various temperatures and concentrations. The sedimentation velocity of the observed schlieren peak is higher than that obtained with solutions of either NA/ATP/Ca^{2+} or of the chromogranins alone. After replacement of chromogranins by albumine, two distinctly different schlieren peaks are to be seen (BERNEIS et al., 1973).

Addition of CA to a solution of ATP results in a marked decrease of osmolality (measured by isothermic distillation) (Table 2) (BERNEIS et al., unpublished results).

On ultracentrifugation of isolated contents of adrenal chromaffin granules, NA, A and ATP sediment jointly at a rate much higher than that to be expected if the solutes were present as single molecules. Furthermore, the apparent average molecular weight of the granular contents markedly increases with rising concentration and diminishing temperature, whereas the concentration and temperature dependence of the molecular weights of the chromogranins is far less pronounced (DA PRADA et al., 1971).

Fluorescence spectroscopy shows that in suspensions of intact chromaffin granules (in which the local concentration of CA is high) fluorescence quenching is markedly higher and fluorescence life-time markedly lower than in preparations in which the

Table 2. Effect of added catecholamines (CA) on osmolality of an aqueous solution of 10% (w/w) ATP containing $CaCl_2$ at 37°C.
(Molar ratio $CaCl_2$:ATP = 0·24:1)

Catecholamine	Osmolality (mosm) Molar ratio CA:ATP		
	0:1	2:1	4:1
Adrenaline	400	165	250
Noradrenaline	400	100	200

same amount of chromaffin granules has been lysed so that the CA are equally distributed in the aqueous medium. Since—as demonstrated in artificial solutions—the quenching of fluorescence at higher CA concentrations is collision-induced, the above finding strongly suggests that within the chromaffin granules the CA retain a relatively good mobility. On the other hand, some interaction between CA and ATP is likely to occur. Thus, in aqueous CA solutions the quenching of the fluorescence is stronger and the fluorescence life time shorter in the presence than in the absence of ATP.

Comparisons of the p-values (indicating the degree of polarisation of the lowest singlet-singlet transition) of suspensions of intact chromaffin granules with those of aqueous solutions of A and A/ATP confirm that *in vivo* CA interact with ATP. The p-value has been shown by experiments with artificial solutions to be a measure for the mutual interaction between CA and ATP (H. Steffen, M. Da Prada and A. Pletscher, submitted for publication).

From these experiments it can be concluded that in the chromaffin granules in vivo the CA and ATP show mutual interaction which is influenced by bivalent cations and that CA/ATP aggregates probably also interact with chromogranins. The intramolecular bonding seems to be of dynamic nature. Preliminary experiments with microcalorimetry of CA/ATP solutions yielded an association constant indicative for dipole interactions ($K \sim 10^2$). This finding and the ultracentrifugation experiments therefore indicate that the intragranular aggregates probably have a superstructure which cannot, however, be expressed by a static, stoichiometric model.

CONCLUDING REMARKS

Two different processes seem to be involved in the storage of CA in adrenal chromaffin granules. The amines are apparently pumped into the organelles against a considerable concentration gradient by a Mg^{2+}/ATP-dependent mechanism operating at the level of the granular membrane and involving ATP-ase. Previous experiments have shown that the maximum NA concentration reached in isolated granular membranes amounts to 5 mM, but the CA concentration in intact chromaffin granules was calculated to be approximately 500 mM (Agostini and Taugner, 1973). The build-up of this high concentration is most likely due to the existence of the second mechanism, i.e. the intragranular interaction of the CA with nucleotides such as ATP and also with chromogranins. The storage of ATP in the adrenal chromaffin granules has been shown to precede that of CA in the course of the ontogenetic development of rabbits and rats (O'Brien *et al.*, 1972). Therefore, the amines transported into the organelles by the membrane pump* can undergo intramolecular bonding with the nucleotide. This interaction probably reduces the diffusion of CA out of the granules enabling the amine pump to operate against a much higher amine concentration gradient. Preliminary experiments with artificial lipid membranes confirm this view. Thus, addition of ATP and $CaCl_2$ to NA or A solutions decreases the velocity of membrane permeation of the amines probably as a result of a reduction of the concentration of non-aggregated amine. On the other hand, the binding of CA in the organelles is reversible, allowing the amines to be rapidly liberated, e.g. during the release process of CA.

REFERENCES

AGOSTINI B. and TAUGNER G. (1973) *Histochemie* **33**, 255–272.
BERNEIS K. H., GOETZ U., DA PRADA M. and PLETSCHER A. (1973) *Naunyn-Schmiedeberg's Arch. Pharmakol.*, **277**, 291–296.
DA PRADA M., BERNEIS K. H. and PLETSCHER A. (1971) *Life Sci.* **10** (I), 639–646
HASSELBACH W. and TAUGNER G. (1970) *Biochem. J.* **119**, 265–271.
O'BRIEN R. A., DA PRADA M. and PLETSCHER A. (1972) *Life Sci.* **11** (I), 749–759.
PAI V. S. and MAYNERT E. W. (1972) *Mol. Pharmac.* **8**, 82.
PLETSCHER A., DA PRADA M., STEFFEN H., LÜTOLD B. and BERNEIS K. H. (1973) *Brain Res.*, in press
SCATCHARD G. (1949) *Ann. N.Y. Acad. Sci.* **51**, 660.
TAUGNER G. (1971a) *Naunyn-Schmiedeberg's Arch. Pharmakol.* **270**, 392–406.
TAUGNER G. (1971b) *Biochem. J.* **123**, 219–225
TAUGNER G. (1972) *Naunyn-Schmiedeberg's Arch. Pharmakol.* **274**, 299–314.
TAUGNER G. and HASSELBACH W. (1968) *Naunyn-Schmiedeberg's Arch. Pharmakol. exp. Path.* **260**, 58–79.
WEINER N. and JARDETZKY O. (1964) *Naunyn-Schmiedeberg's Arch. exp. Path. Pharmakol.* **248**, 308.

---

* Granular membrane pump

Frontiers in Catecholamine Research 1973, pp. 47 to 52. Pergamon Press. Printed in Great Britain.

PROPERTIES OF TYROSINE HYDROXYLASE*

José M. Musacchio† and Gale L. Craviso

Department of Pharmacology, New York University School of Medicine,
550 First Avenue New York, New York 10016, U.S.A.

The conversion of tyrosine to dopa is catalyzed by tyrosine hydroxylase (E.C. 1.10.3.1.), and it is considered to be the rate-limiting step in catecholamine biosynthesis (Levitt et al., 1965). There is general agreement that the rate of catecholamine biosynthesis is adjusted to the physiological needs of the organism by at least two different mechanisms: feedback inhibition and induction of tyrosine hydroxylase. The catecholamine feedback inhibition of tyrosine hydroxylase has been repeatedly demonstrated in vitro and in vivo (Udenfriend et al., 1965; Costa and Neff, 1966; Alousi and Weiner, 1966; Spector et al., 1967), but there are still many aspects of this mechanism which are not fully understood. In order to understand this process fully, it is necessary to know the subcellular distribution and the kinetic characteristics of the enzyme. Therefore, we will review in this article some of the properties of tyrosine hydroxylase which will help in the understanding of the factors that regulate tyrosine hydroxylase activity.

SUBCELLULAR DISTRIBUTION OF TYROSINE HYDROXYLASE

Tyrosine hydroxylase was isolated from the adrenal gland and brain by Nagatsu et al. (1964a, b), and the enzyme was originally described as particle-bound; the fraction of tyrosine hydroxylase which was found in the supernatant was considered to have been solubilized by the prolonged homogenization. In the course of our efforts to purify the enzyme, we found that the enzyme from bovine adrenal glands has the tendency to precipitate; in contrast, the rat adrenal enzyme was found exclusively in soluble form (Musacchio, 1967, 1968), and it does not precipitate. These observations suggested that perhaps tyrosine hydroxylase was a soluble enzyme, and that in certain species it has the tendency to precipitate and to appear as particle-bound. In order to test this hypothesis, we studied the subcellular distribution of bovine tyrosine hydroxylase, using procedures specifically designed for subcellular fractionation of enzymes.

The subcellular distribution of tyrosine hydroxylase was studied using different homogenization media, differential centrifugation, and sucrose density gradient fractionation. Our results clearly indicated that tyrosine hydroxylase is a soluble enzyme (Wurzburger and Musacchio, 1971): most of the tyrosine hydroxylase (64%) remains at the top of the gradient, and only a very small fraction (10%) is associated with the chromaffin granules. About 20% of the enzyme, and 13% of the epinephrine, sediments at the bottom of the gradient with the unbroken cells and cell debris. The specific activity of the enzyme is highest in the soluble fraction. Our studies also indicated that the composition of the buffer used for homogenization

* This work was supported by Grant AM HE 13128
† Research Scientist Awardee of the Public Health Service Grant 5-K2-MH-17,885

has a marked effect on the distribution of the enzyme between the so-called "nuclear fraction" and the low-speed supernatant; about 36% of the total enzyme activity is found in the nuclear fraction when the glands are homogenized in isotonic sucrose, as compared with 13% when sucrose is replaced by isotonic KCl.

In addition to our studies, Laduron and Belpaire (1968) have independently reported that adrenal tyrosine hydroxylase is a soluble enzyme; more recently, Weiner et al. (1971) have also concluded that the enzyme is localized in the high-speed supernatant. Stjärne and Lishajko (1967) have clearly demonstrated that, in the sympathetic nerves, tyrosine hydroxylase is exclusively localized in the high-speed supernatant fractions.

A considerable fraction of rat brain tyrosine hydroxylase (60%) is contained in the synaptosomal fraction, but, when the synaptosomes are lysed, 82% of the enzyme is found in the soluble fraction (Coyle, 1972). Kuczenski and Mandell (1972) have studied the subcellular distribution of tyrosine hydroxylase in the rat midbrain and striatum: the midbrain enzyme is localized mainly in the soluble fraction, while the striatum tyrosine hydroxylase is found mainly in the synaptosomal fraction. This difference in the distribution is expected because the midbrain contains relatively more cell bodies than nerve endings, and the striatum contains dopaminergic nerve endings and no dopaminergic cell bodies. Between 60 and 70% of the enzyme contained in the synaptosomes can be solubilized by hypotonic shock in 2 mM phosphate buffer, while the rest remains bound, presumably to synaptosomal membranes (Kuczenski and Mandell, 1972). This fraction of tyrosine hydroxylase which remains adsorbed to synaptosomal membranes cannot be solubilized by repeated washing; in this respect, the rat striatal enzyme is similar to the aggregated bovine enzyme obtained from the high-speed adrenal gland supernatant and to the adrenal enzyme which adsorbs onto subcellular particles. We have observed that, when the rat caudate is homogenized in ten volumes of distilled water in a glass homogenizer, 98% of the tyrosine hydroxylase is obtained in soluble form; if 2 mM phosphate buffer is used instead of water, only 75–80 per cent of the enzyme is obtained in soluble form (Musacchio and Wurzburger, unpublished observation). The reason for this discrepancy is not known, but it is probable that a small concentration of certain salts may be necessary for the adsorption of tyrosine hydroxylase to some subcellular structures.

In conclusion, all the evidence indicates that tyrosine hydroxylase is a soluble enzyme in the adrenal glands and in the peripheral sympathetic nerves. The studies on the subcellular distribution of brain enzyme are consistent, but do not prove that tyrosine hydroxylase is a soluble enzyme. The association of a small percentage of the total tyrosine hydroxylase and the synaptosomal debris sedimented after osmotic shock perhaps could be explained by the known property of tyrosine hydroxylase to aggregate and adsorb to subcellular particles.

It is obvious that the question of the localization of tyrosine hydroxylase is quite important in order to understand the regulation of catecholamine biosynthesis and the mechanism of action of several drugs which modify this process. The catecholamine pool that controls the activity of tyrosine hydroxylase has to be in contact with the enzyme, and the localization of the enzyme will, in turn, indicate which pool regulates catecholamine biosynthesis. The localization of tyrosine hydroxylase in the cytosol is consistent with the pharmacological evidence which indicates that there is a small cytoplasmic pool of catecholamines which is an important regulatory factor in the biosynthesis of catecholamines (ALOUSI and WEINER, 1966); this view has been supported by a considerable number of investigators (WEINER and SELVARATNAM, 1968; KOPIN *et al.*, 1969; GOLDSTEIN *et al.*, 1970).

DIFFERENT MOLECULAR FORMS OF BOVINE ADRENAL TYROSINE HYDROXYLASE

In some of our early attempts to purify tyrosine hydroxylase, we observed that, when the enzyme was treated with trypsin, the patterns of elution from Sephadex G-200 indicated that the trypsin-treated enzyme was smaller than the native enzyme. This suggested that the trypsin-treated enzyme could have been partially degraded by the treatment; therefore, we decided to study some of the molecular parameters of both enzyme forms. We determined the sedimentation coefficients of both enzyme forms by sucrose density gradient centrifugation, and found that the native form of tyrosine hydroxylase has a sedimentation coefficient of 9·2 S; in contrast, the trypsin-treated enzyme has a sedimentation coefficient of 3·4 S. The molecular weight of the different enzyme forms can be estimated from their sedimentation coefficients, but a more accurate determination of the molecular weight can be obtained if, in addition to the sedimentation coefficient, the Stokes radius is used in the calculations (ACKERS, 1964). The molecular weight calculated from these two parameters is 34,000 daltons for the trypsin-treated tyrosine hydroxylase (MUSACCHIO *et al.*, 1971). The radius of the native form of tyrosine hydroxylase could not be measured, due to aggregation during gel filtration, but the results of gel filtration in 2 M urea and the determination of the sedimentation coefficient indicate that the native enzyme is about 4·4-fold larger than the trypsin-treated form; therefore, we calculate that the molecular weight of native tyrosine hydroxylase is about 150,000. It is obvious that the trypsin-treated fragment of tyrosine hydroxylase contains at least one catalytic site; the subunit composition of the native enzyme is not known.

KINETIC STUDIES ON TYROSINE HYDROXYLASE

In view of the large differences in the molecular weight between the native and the trypsin-treated tyrosine hydroxylase, we have studied the kinetic characteristics of both enzyme forms (MCQUEEN, 1972; MUSACCHIO *et al.*, 1973). The experiments were planned according to Cleland's method for three substrate systems: one substrate was varied at several fixed levels of the second, while the third substrate was held constant (CLELAND, 1970). Initial velocity studies were performed with both the native and the trypsin-treated forms of bovine adrenal tyrosine hydroxylase. In all the experiments using the trypsin-treated tyrosine hydroxylase, linear intersecting double reciprocal plots were obtained for all the substrates. When identical experiments were run with the native form of tyrosine hydroxylase, linear intersecting double reciprocal plots were also obtained for all the substrates. But, in our early experiments, when oxygen was taken as the variable substrate, the lines obtained were slightly curved, concave downward (MUSACCHIO *et al.*, 1973). In more recent experiments, in which the gassing with O_2 was improved and constantly monitored with an oxygen electrode, the lines are straight.

The Michaelis constants for both enzyme forms were determined for each substrate. The K_m for 6,7-dimethyl-5,6,7,8-tetrahydropterin (DMPH$_4$) was 5·65 for the native, and $6·24 \times 10^{-5}$ M for the trypsin-treated form; the difference was not significant. The tyrosine K_m for the native form ($4·76 \times 10^{-5}$ M) was almost twice that of the trypsin-treated form ($2·38 \times 10^{-5}$ M). The oxygen K_m of the native form (12·33 per cent) was about three times that of the trypsin-treated form (4·20 per cent) (McQueen, 1972; Musacchio et al., 1973).

The intersecting patterns of all the double reciprocal plots of both forms of tyrosine hydroxylase are indicative of a sequential mechanism. These findings are in agreement with the studies of Joh et al. (1969) on the trypsin-treated tyrosine hydroxylase, but they are at variance with those of Ikeda et al. (1966), which proposed a ping–pong mechanism for the native form of tyrosine hydroxylase.

Inhibition studies using both product and substrate inhibitors should be performed in order to determine the order of addition of substrates and to understand the regulatory properties of the enzyme. Native tyrosine hydroxylase is inhibited by catecholamines and dopa; this inhibition is competitive with DMPH$_4$ (Udenfriend et al., 1965; Ikeda et al., 1966) and also with tetrahydrobiopterin (Musacchio et al., 1971; Nagatsu et al., 1972). The inhibition produced by α-methyl-L-tyrosine is competitive with tyrosine, and non-competitive with DMPH$_4$ (Nagatsu et al., 1964b). Similar patterns of inhibition are obtained with the trypsin-treated form of the enzyme (McQueen, 1972). Unfortunately, the unavailability of certain inhibitors and some technical difficulties have not yet permitted us to complete the inhibition studies and to establish the order of substrate addition to the enzyme.

The relative effectiveness of dopamine, norepinephrine, and epinephrine to inhibit native tyrosine hydroxylase was also studied; it was found that dopamine ($K_i 2·1 \times 10^{-5}$ M) is twice as effective as norepinephrine and epinephrine ($K_i 4·6 \times 10^{-5}$ M) as feedback inhibitor of tyrosine hydroxylase (Musacchio et al., 1973). Since dopamine is formed in the cytosol, it is possible that dopamine may be the amine that controls the activity of tyrosine hydroxylase. This is consistent with the fact that more than 10% of the heart total catecholamines is dopamine (Costa et al., 1972), and that it is most likely located in the cytosol, since dopamine is very rapidly transformed into norepinephrine when it is taken up into the storage vesicles.

Since the preparation of bovine adrenal tyrosine hydroxylase used in our early studies has the tendency to aggregate, and since aggregation may, conceivably, change some of the kinetic characteristics of the enzyme, we decided to study the guinea pig adrenal tyrosine hydroxylase, which does not aggregate. We obtained linear intersecting double reciprocal plots for all the substrates. Two such experiments are shown in Fig. 1, and the K_m values for the different substrates are listed in Table 1. The results obtained with the guinea pig enzyme are in agreement with those previously obtained with the bovine adrenal enzyme; this confirms that the mechanism of substrate addition to tyrosine hydroxylase is sequential rather than ping-pong.

In the course of these experiments, which were performed using tetrahydrobiopterin (H$_4$B) instead of DMPH$_4$, we found that the atmospheric concentration of oxygen inhibits the enzyme, and that the inhibition is more pronounced at low levels of H$_4$B. It can also be observed in Table 1 that the K_m for oxygen for the

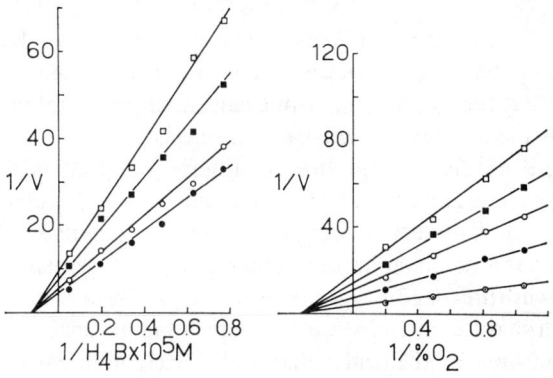

FIG. 1.

Left: Double reciprocal plot of tetrahydrobiopterin concentration against the activity of guinea pig adrenal tyrosine hydroxylase in the presence of several oxygen concentrations. The incubation mixture contained $0 \cdot 11$M Tris-acetate buffer, pH $6 \cdot 1$; $87 \cdot 5$ μM ferrous ammonium sulfate; $0 \cdot 1$M 2-mercaptoethanol; 30 μM 3,5-H^3-L-tyrosine; 1-tetrahydrobiopterin, 13-200 μM; enzyme and water to $0 \cdot 1$ ml. Oxygen concentrations were $0 \cdot 95 \%$ ($\square$), $1 \cdot 23 \%$ ($\blacksquare$), $1 \cdot 96 \%$ ($\bigcirc$), and $4 \cdot 99 \%$ ($\bullet$). Velocity was determined by measuring the amount of tritiated water released during the reaction and expressed as nmoles of product in 6 min at 37°C.

Right: Double reciprocal plot of oxygen concentration against the activity of guinea pig adrenal tyrosine hydroxylase in the presence of several tyrosine concentrations. Assay as described for previous figure. 1-Tetrahydrobiopterin was held constant at 200 μM. Oxygen concentrations were $0 \cdot 95 \%$–$4 \cdot 99 \%$, and tyrosine concentraions were $3 \cdot 66$ μM ($\square$); $4 \cdot 69$ μM ($\blacksquare$); $6 \cdot 52$ μM ($\bigcirc$); $10 \cdot 7$ μM ($\bullet$); and $30 \cdot 0$ μM ($\bigcirc$).

guinea pig adrenal enzyme depends on the structure of the pterin cofactor used in the reaction. The K_m for oxygen is five-fold larger when tetrahydrobiopterin is replaced by the artificial confactor DMPH$_4$. Similar findings have previously been reported by FISHER and KAUFMAN (1972) for the bovine brain and adrenal tyrosine hydroxylase. The native and the trypsin-treated bovine adrenal tyrosine hydroxylase can also be inhibited with high concentrations of O_2; similarly, the inhibition is more marked at low levels of DMPH$_4$ (McQUEEN and MUSACCHIO, unpublished observation).

TABLE 1. K_m VALUES FOR THE DIFFERENT SUBSTRATES OF GUINEA PIG ADRENAL TYROSINE HYDROXYLASE

Number of experiments	Constant substrate	H^{4}B K_m* ($\times 10^{-5}$M)	Tyrosine K_m ($\times 10^{-5}$M)	$O_2 K_m$ (Per cent)	DMPH^4 K_m ($\times 10^{-5}$M)
4	5% O$_2$	7·24	1·84		
1	20% O$_2$	9·09			
1	1·96% O$_2$	5·88	2·53		
2	30 μM Tyr	7·33		3·17	
2	200 μM H$_4$B		4·65	3·02	
1	20% O$_2$		8·66		13·33
1	200 μM DMPH$_4$		7·58	15·38	

* The K_m values were calculated by determining the reciprocal of the K_m directly from the double reciprocal plot in experiments similar to those described in Fig. 1.

SUMMARY

(1) The evidence indicating that tyrosine hydroxylase is a soluble enzyme located in the cell cytoplasm has been reviewed. This is in agreement with the pharmacological evidence that there is a cytoplasmic catecholamine pool which, by feedback inhibition, controls the activity of tyrosine hydroxylase.

(2) Some of the physical properties of different molecular forms of tyrosine hydroxylase have been examined. Bovine adrenal tyrosine hydroxylase subjected to trypsin digestion has a molecular weight of 34,000; the native form of the enzyme is about four-fold larger. In addition, it has a K_m for tyrosine about two-fold, and for oxygen about three-fold, larger than the K_m's of the trypsin-treated enzyme. The K_m values for the different substrates of guinea pig adrenal tyrosine hydroxylase were determined using the artificial cofactor $DMPH_4$ and tetrahydro-L-biopterin; the K_m's obtained with the artificial cofactor are two- to five-fold larger.

(3) All the initial velocity double reciprocal plots of tyrosine hydroxylase are intersecting in the absence of added products, indicating a sequential rather than a ping–pong mechanism.

REFERENCES

ACKERS G. K. (1964) *Biochemistry* 3, 723–730.
ALOUSI A. and WEINER N. (1966) *Proc. Natl. Acad. Sci. U.S.A.* 56, 1491–1496.
CLELAND W. W. (1970) In *The Enzymes*. (BOYER P., Ed.) Vol. II, pp. 1–65. Academic Press, New York.
COSTA E., GREEN A. R., KOSLOW S. H., LeFEVRE H. J., REVUELTA A. V. and WANG C. (1972) *Pharmacol. Rev.* 24, 167–190.
COSTA E. and NEFF N. H. (1966) In: *Biochemistry and Pharmacology of the Basal Ganglia*. (COSTA E., CÔTE L. T. and YAHR M. D., Eds.) pp. 141–155. Raven Press, New York.
COYLE J. T. (1972) *Biochem. Pharmacol.* 21, 1935–1944.
FISHER D. B. and KAUFMAN S. (1972) *J. Neurochem.* 19, 1359–1365.
GOLDSTEIN M., OHI Y. and BACKSTROM T. (1970) *J. Pharmacol. Exp. Ther.* 174, 77–82.
IKEDA M., FAHIEN L. A. and UDENFRIEND S. (1966) *J. Biol. Chem.* 241, 4452–4456.
JOH T. H., KAPIT R. and GOLDSTEIN M. (1969) *Biochim. Biophys. Acta* 171, 378–380.
KOPIN I. J., WEISE V. K. and SEDVALL G. C. (1969) *J. Pharmacol. Exp. Ther.* 170, 246–252.
KUCZENSKI R. T. and MANDELL A. J. (1972) *J. Biol. Chem.* 247, 3114–3122.
LEVITT M., SPECTOR S., SJOERDSMA A. and UDENFRIEND S. (1965) *J. Pharmacol. Exp. Ther.* 148, 1–8.
LADURON P. and BELPAIRE F. (1968) *Biochem. Pharmacol.* 17, 1127–1140.
McQUEEN C. A. (1972) M.S. Thesis, New York University Graduate School of Arts and Sciences.
MUSACCHIO J. M. (1967) *Pharmacologist* 9, 210.
MUSACCHIO J. M. (1968) *Biochem. Pharmacol.* 17, 1470–1473.
MUSACCHIO J. M., D'ANGELO G. L. and McQUEEN C. A. (1971) *Proc. Natl. Acad. Sci. U.S.A.* 68, 2087–2091.
MUSACCHIO J. M., McQUEEN C. A. and CRAVISO G. L. (1973) In: *New Concepts in Neurotransmitter Regulation*. (MANDELL A. J., Ed.) pp. 69–88. Plenum Press, New York.
MUSACCHIO J. M., WURZBURGER R. J. and D'ANGELO G. L. (1971) *Molec. Pharmacol.* 7, 136–146.
NAGATSU T., LEVITT M. and UDENFRIEND S. (1964a) *Biochem. Biophys. Res. Commun.* 14, 543–549.
NAGATSU T., LEVITT M. and UDENFRIEND S. (1964b) *J. Biol. Chem.* 239, 2910–2917.
NAGATSU T., MIZUTANI K., NAGATSU I., MATSUURA S. and SUGIMOTO T. (1972) *Biochem. Pharmacol.* 21, 1945–1953.
SPECTOR S., GORDON R., SJOERDSMA A. and UDENFRIEND S. (1967) *Mol. Pharmacol.* 3, 549–555.
STJÄRNE L. and LISHAJKO F. (1967) *Biochem. Pharmacol.* 16, 1719–1728.
UDENFRIEND S., ZALTZMAN-NIRENBERG P. and NAGATSU T. (1965) *Biochem. Pharmacol.* 14, 837–845.
WEINER N. and SELVARATNAM I. (1968) *J. Pharmacol. Exp. Ther.* 161, 21–33.
WEINER N., WAYMIRE J. C. and SCHNEIDER F. H. (1971) *Acta Cientifica Venezolana* 22, (Supl. 2), 179–183.
WURZBURGER R. J. and MUSACCHIO J. M. (1971) *J. Pharmacol. Exp. Ther.* 177, 155–167.

Frontiers in Catecholamine Research 1973, pp. 53 to 60. Pergamon Press. Printed in Great Britain.

COFACTORS OF TYROSINE HYDROXYLASE

Seymour Kaufman

Laboratory of Neurochemistry, National Institute of Mental Health, Bethesda,
Maryland 20014, U.S.A.

It would be a great convenience to scientists if the laws of nature adhered to the same table of contents as the one devised for this Symposium. Unfortunately, they do not. It is therefore not possible to discuss the cofactors of tyrosine hydroxylases without discussing the properties of tyrosine hydroxylase. Indeed, the single theme that I wish to develop is the following one: the kinetic and regulatory properties of tyrosine hydroxylase (as well as those of the other two pterin-dependent hydroxylases) are largely determined by the structure of the cofactor with which the enzyme is functioning.

The field of pterin-dependent enzymes was opened up with the discovery in 1958 that a non-protein cofactor isolated from rat liver extracts was an essential component of the hepatic phenylalanine hydroxylating system (Kaufman, 1958). In the following year, evidence was presented which indicated that the cofactor might be a pteridine. Furthermore it was shown that certain synthetic, unconjugated pterins (pterin is a 2-amino-4-hydroxy-pteridine), such as 6-methyl-(6-MPH$_4$) and 6,7-dimethyltetrahydropterin (DMPH$_4$), had high cofactor activity with the phenylalanine hydroxylating system (Kaufman, 1959; Kaufman and Levenberg, 1959). In 1963, the earlier indications that the naturally occurring cofactor was a pterin were substantiated with the report that the hydroxylation cofactor from rat liver was the reduced form of the unconjugated pterin, biopterin (Kaufman, 1963). The work on phenylalanine hydroxylase established the first role in metabolism for this class of compound.

Mainly through the use of the 6,7-dimethyl analogue of the cofactor, the general manner in which the pterin performs its cofactor role in hydroxylation reactions was elucidated. The conversion of phenylalanine to tyrosine was shown to proceed according to equations (1) and (2), where RH stands for phenylalanine, ROH for tyrosine, XH$_4$ for the tetrahydropterin and XH$_2$ for the quinonoid dihydropterin (Kaufman, 1964a)

$$RH + O_2 + XH_4 \rightarrow ROH + H_2O + XH_2 \tag{1}$$

$$XH_2 + TPNH + H^+ \rightarrow XH_4 + TPN^+ \tag{2}$$

Reaction (1) is catalysed by phenylalanine hydroxylase and reaction (2) is catalysed by a separate enzyme, dihydropteridine reductase. Although both TPNH and DPNH can function with the latter enzyme, the K_m is lower and the V_{max} is higher with DPNH (Nielsen et al., 1969; Scrimgeour and Cheema, 1971; Craine et al., 1972). In the absence of the reductase, chemical reducing agents such as mercaptans and ascorbate can also serve to reduce the quinonoid dihydropterin and allow the pterin to function catalytically (Kaufman, 1959).

In 1964, partially purified preparations of tyrosine hydroxylase from adrenal medulla were shown to have an absolute requirement for a tetrahydropterin (Nagatsu et al., 1964; Brenneman and Kaufman, 1964; Kaufman, 1964b), thus establishing this enzyme as the second pterin-dependent aromatic amino acid hydroxylase. It appeared likely that the role of the tetrahydropterin in this hydroxylating system would prove to be the same as the one previously established in the phenylalanine hydroxylating system, an expectation that was fufilled by subsequent work.

The criteria needed to establish the cofactor role of a pterin in a hydroxylation reaction (KAUFMAN, 1964b) were all met with tyrosine hydroxylase. These criteria are: (a) the pterin is specific (NAGATSU et al., 1964; BRENNEMAN and KAUFMAN, 1964; KAUFMAN, 1964b); (b) the pterin functions catalytically (BRENNEMAN and KAUFMAN, 1964; KAUFMAN, 1964b) ; (c) the pterin is utilised during the hydroxylation reaction and this utilisation is substrate dependent (BRENNEMAN and KAUFMAN, 1964; KAUFMAN, 1964b).

It was also shown that the hydroxylase can be coupled with dihydropteridine reductase (BRENNEMAN and KAUFMAN, 1964). This last finding strongly indicates that the tetrahydropterin is oxidised to the quinonoid dihydropterin in this hydroxylation reaction just as it is during the hydroxylation of phenylalanine.

The stoichiometry of the reaction catalysed by bovine adrenal tyrosine hydroxylase was only established fairly recently (SHIMAN et al., 1971). In the presence of tetrahydrobiopterin, the reaction proceeds as shown in equation (3) (where dopa stands for 3,4-dihydroxyphenylalanine)

L-tyrosine + O_2 + tetrahydrobiopterin →

$$L\text{-dopa} + H_2O + \text{quinonoid dihydrobiopterin} \quad (3)$$

A different aspect of the stoichiometry problem will be discussed in a later section.

Because soluble tyrosine hydroxylase has never been purified to any extent, almost all of our knowledge of the properties of the enzyme from bovine adrenal medulla have come from studies carried out with partially purified preparations that were solubilised by limited proteolysis with either tryspin (PETRACK et al., 1968) or chymotrypsin (SHIMAN et al., 1971). In the following discussion, the latter preparation will be referred to simply as the "solubilised" enzyme.

To return to the theme that I mentioned earlier, namely, that the behaviour of these hydroxylases varies dramatically with the pterin cofactor used, the first point worthy of note is that even the stoichiometry of the reaction catalysed by tyrosine hydroxylase varies with the cofactor. As already mentioned, in the presence of tetrahydrobiopterin, there is tight coupling between tetrahydropterin oxidation and dopa formed, i.e. the ratio of these two quantities is close to one. In the presence of $DMPH_4$, the commercially-available, commonly-used analogue of the naturally-occurring cofactor, however, the hydroxylation reaction is only loosely coupled to tetrahydropterin oxidation and the ratio of tetrahydropterin oxidised to dopa formed is about 1·75 (SHIMAN et al., 1971). This analogue-induced uncoupling phenomenon, which was first discovered with phenylalanine hydroxylase (KAUFMAN, 1961; STORM and KAUFMAN, 1968), leads to the formation of an abnormal product in the hydroxylation reaction, i.e. H_2O_2 (STORM and KAUFMAN, 1968), which could, in itself, alter the characteristics of the enzyme. A summary of the stoichiometry of the hydroxylation reaction in the presence of various model cofactors is shown in Table 1 (SHIMAN et al., 1971). It should be mentioned that loose coupling with the 7-methyltetrahydropterin is observed with both tyrosine and phenylalanine hydroxylases (STORM and

TABLE 1. STOICHIOMETRY OF TYROSINE HYDROXYLATION WITH MODEL
COFACTORS

Tetrahydropterin	Dihydropterin/dopa	Range	Number of determinations
6MPH$_4$	1·10 ± 0·05	(1·05–1·12)	6
7MPH$_4$	1·60 ± 0·05	(1·56–1·62)	4
PH$_4$	1·15 ± 0·05	(1·12–1·15)	4
DMPH$_4$	1·75 ± 0·15	(1·61–2·0)	10

PH$_4$ stands for 2-amino-4-hydroxytetrahydropteridine.

KAUFMAN, 1968), but that tight coupling is observed with phenylalanine hydroxylase in the presence of the dimethylpterin.

For both the solubilised and the particulate enzyme the rate of the hydroxylation reaction is about twice as fast in the presence of tetrahydrobiopterin as it is in the presence of DMPH$_4$ (SHIMAN et al., 1971). It should be recalled, however, that there is partial uncoupling in the presence of the latter compound; the activity of the enzyme, therefore, is the same in the presence of either pterin when this activity is measured as tyrosine-dependent oxidation of the tetrahydropterin.

As can be seen in Table 2 the K_m of both the solubilised and the particulate enzyme for tetrahydrobiopterin is lower than it is for DMPH$_4$ (SHIMAN and KAUFMAN, 1970). Although it is not surprising that the K_m for the naturally-occurring cofactor should be lower than that of the dimethyl compound, a much more interesting and somewhat unexpected finding is that for tyrosine hydroxylase, as well as for the other pterin-dependent hydroxylases, the apparent K_m values for its substrates (both the amino acid substrate and oxygen) also varies with the pterin used (KAUFMAN, 1970).

The summary of K_m values shown in Table 3 illustrates this point for tyrosine hydroxylase (SHIMAN and KAUFMAN, 1970). For both the purified, solubilised and the particulate enzyme, the K_m value for tyrosine is much lower in the presence of tetrahydrobiopterin than it is in the presence of DMPH$_4$.

The apparent K_m value for oxygen is also markedly lower in the presence of tetrahydrobiopterin than it is in the presence of DMPH$_4$. With the dimethyl compound, a value of 6 per cent has been reported (IKEDA et al., 1966), whereas in the presence of tetrahydrobiopterin the K_m value for oxygen is about 1 per cent (FISHER and KAUFMAN, 1972). With brain tyrosine hydroxylase, the K_m for oxygen is about 1 per cent in the presence of DMPH$_4$ and is less than 1 per cent in the presence of tetrahydrobiopterin (FISHER and KAUFMAN, 1972).

One of the most dramatic examples of the enzyme's properties being influenced by the nature of the pterin cofactor is that of the enzyme's amino acid substrate specificity. Although the adrenal enzyme was originally thought to be absolutely specific for L-tyrosine (NAGATSU et al., 1964), it was later shown that L-phenylalanine

TABLE 2. PTERIN MICHAELIS CONSTANTS FOR BOVINE ADRENAL TYROSINE
HYDROXYLASE

	Solubilised enzyme ($\times 10^{-5}$ M)	Particulate enzyme ($\times 10^{-5}$ M)
DMPH$_4$	30	30
6MPH$_4$	30	30
Tetrahydrobiopterin	10	10

TABLE 3. MICHAELIS CONSTANTS FOR TYROSINE WITH TYROSINE
HYDROXYLASE FROM BOVINE ADRENAL

Pterin used	Solubilised enzyme ($\times 10^{-5}$ M)	Particulate enzyme ($\times 10^{-5}$ M)
$DMPH_4$	20	10
$6MPH_4$	7	4
Tetrahydrobiopterin	1·5	0·4

could also serve as a substrate (IKEDA et al., 1965). The low rate of phenylalanine hydroxylation by tyrosine hydroxylase in the presence of $DMPH_4$ indicated that this reaction was probably of minor physiological significance.

A more detailed study of the phenylalanine hydroxylating activity of tyrosine hydroxylase, however, led to the surprising finding that in the presence of tetrahydrobiopterin, the rate of phenylalanine hydroxylation by both the highly purified, solubilised, enzyme and the particulate enzyme is equal to, or greater than, the rate of tyrosine hydroxylation (SHIMAN et al., 1971). Furthermore, as will be discussed in greater detail later, whereas under these conditions the enzyme is sensitive to inhibition by excess tyrosine, there is no evidence for inhibition by excess phenylalanine. These results made it highly likely that phenylalanine could serve as an important precursor of norepinephrine in vivo, a prediction that has been recently fulfilled (KAROBATH and BALDESSARINI, 1972).

A final pterin-dependent property of tyrosine hydroxylase that will be discussed is that of inhibition by its substrates and products. This enzyme shares with hepatic phenylalanine hydroxylase the property of being inhibited by excess concentrations of either of its substrates, tyrosine (SHIMAN et al., 1971) and oxygen (FISHER and KAUFMAN, 1972). With both enzymes, marked inhibition is only apparent in the presence of tetrahydrobiopterin. The inhibition of adrenal tyrosine hydroxylase by tyrosine has been observed with both the particulate and the solubilised enzyme (SHIMAN et al., 1971) (see Fig. 1). The brain enzyme is also inhibited by excess tyrosine in the presence of tetrahydrobiopterin, but it is less sensitive than the adrenal enzyme; whereas 50 per cent inhibition of the adrenal enzyme is apparent at about 0·1 mM tyrosine, 50 per cent inhibition of the brain enzyme is observed at about 0·35 mM tyrosine (LLOYD and KAUFMAN, unpublished).

Inhibition by the other substrate of the enzyme, oxygen, has been reported for both the solubilised adrenal enzyme and the brain enzyme (FISHER and KAUFMAN, 1972).

It has been reported that the product of the tyrosine hydroxylase-catalysed reaction, dopa, as well as a large variety of other catechols (IKEDA et al., 1966; UDENFRIEND et al., 1965; CARLSON et al., 1963), inhibit tyrosine hydroxylase, the inhibition being competitive with the tetrahydropterin (IKEDA et al., 1966).

SHIMAN and KAUFMAN (unpublished) found that the inhibition by dopa that is observed with both the solubilised and particulate adrenal enzyme is non-competitive with respect to both tyrosine and tetrahydrobiopterin. The data for the solubilised enzyme with tetrahydrobiopterin and the particulate enzyme with $DMPH_4$ are shown in Figs. 2 and 3. To forestall confusion, it should be noted that the extent of inhibition decreases as the concentration of tetrahydropterin increases, but that even at infinite concentrations of tetrahydropterin, inhibition is still manifest; in the accepted kinetic terminology, this pattern of inhibition is "non-competitive". The same conclusion applies to the effect of tyrosine on the extent of inhibition.

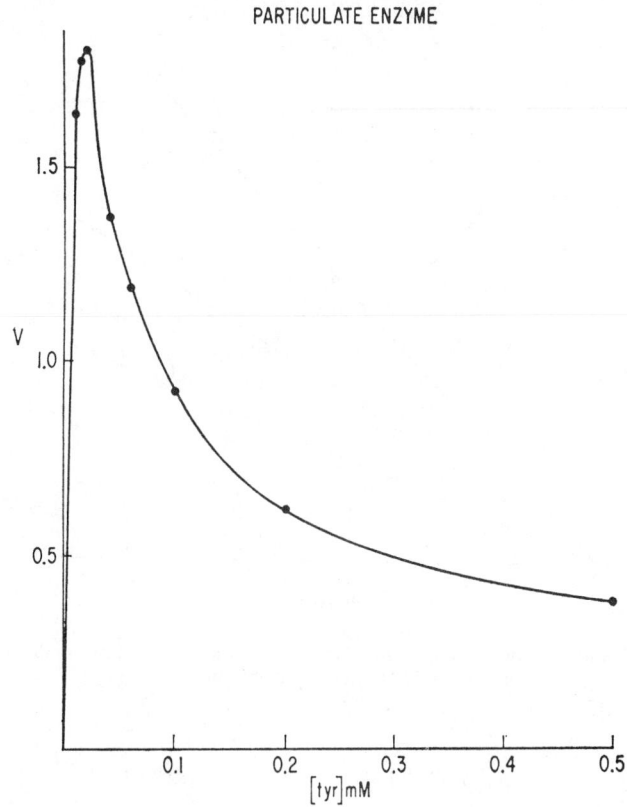

FIG. 1.—Substrate inhibition by excess tyrosine with particulate tyrosine hydroxylase from bovine adrenal medulla. The tetrahydrobiopterin concentration was 0·1 mM.

The data in Figs. 2 and 3 also show that the enzyme is much more sensitive to inhibition by dopa in the presence of tetrahydrobiopterin than in the presence of $DMPH_4$. This difference is shown by both the solubilised and the particulate enzyme. The latter form of the enzyme is even more sensitive to the dopa inhibition than is the former. Inhibition data for both forms of the enzyme in the presence of dopa are summarised in Table 4.

A large mass of data can be recapitulated by the following statement: the apparent affinity of the enzyme (as measured by K_m and K_i values) for its substrates and its product is higher in the presence of the naturally-occurring cofactor, tetrahydro-biopterin, than it is in the presence of the synthetic model cofactor, $DMPH_4$, i.e. substrates are better substrates and inhibitors are better inhibitors (including the substrates in high concentrations) in the presence of tetrahydrobiopterin than in the presence of $DMPH_4$.

It is apparent from these studies of the enzyme with different pterins that two important regulatory properties of tyrosine hydroxylase are lost or attenuated when $DMPH_4$ is substituted for tetrahydrobiopterin: inhibition by excess tyrosine and inhibition by dopa, the product of the reaction. It would be of interest to test the prediction that follows from these results, that is, that $DMPH_4$ *in vivo* might lead to norepinephrine synthesis that is relatively uncontrolled by these two restraints.

S. KAUFMAN

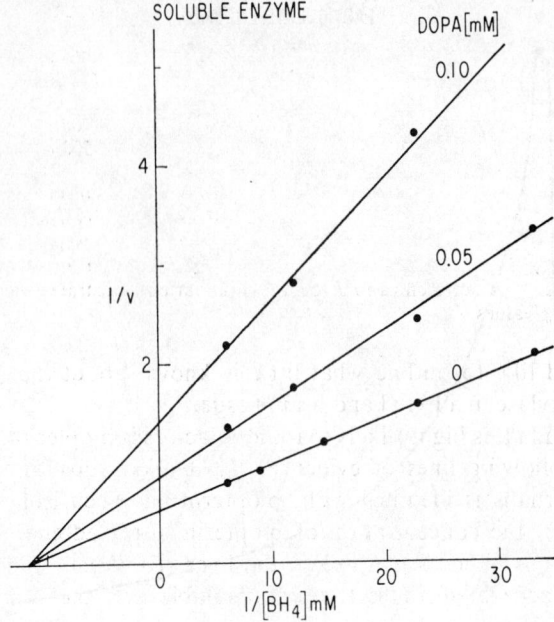

FIG. 2.—Inhibition of solubilised bovine adrenal tyrosine hydroxylase by dopa in the presence of BH_4 (tetrahydrobiopterin).

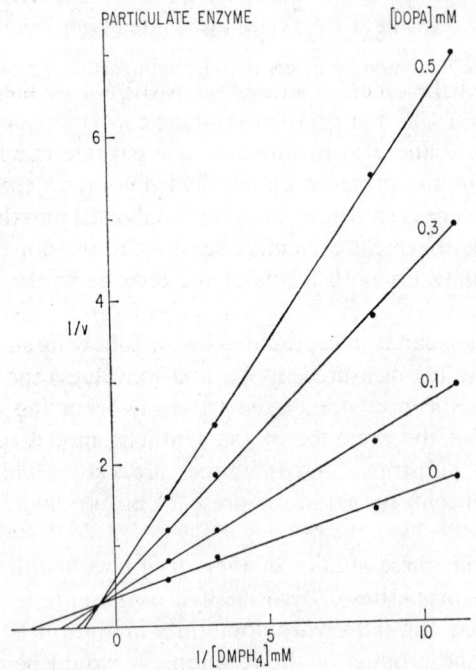

FIG. 3.—Inhibition of particulate bovine adrenal tyrosine hydroxylase by dopa in the presence of $DMPH_4$.

TABLE 4. INHIBITION OF TYROSINE HYDROXYLASE BY L-DOPA

Tetrahydropterin	Particulate enzyme K_i (slope)	Soluble enzyme K_i (slope)
Tetrahydrobiopterin	40 μM	60 μM
DMPH₄	150 μM	500 μM

Summary of K_i (slope) values obtained with two different pterins with dopa as the inhibitor. The K_i (slope) values were obtained from a replot of slope *versus* [L-dopa]. The K_i's obtained were identical within experimental error for inhibition against either pterin or L-tyrosine. The concentrations of the non-varied substrates were at their K_m values.

Finally I would like to outline what little is known about the natural cofactor for tyrosine hydroxylase in adrenal and brain tissue.

In adrenal medulla it is highly likely that the cofactors is biopterin. This conclusion is based on the following lines of evidence: (1) 6-alkyl substituted pterin that is similar to, and perhaps is identical with, biopterin has been isolated (LLOYD and WEINER, 1971); (2) the concentration of biopterin in rat adrenal glands is higher (4·8 μg/g) than in any other organ except the liver (REMBOLD, 1964); (3) tyrosine hydroxylase from adrenal medulla (both the soluble and the solubilised enzyme) requires an unconjugated pterin for activity and may be completely inactive with a conjugated pterin such as tetrahydrofolate (LLOYD, MORI and KAUFMAN, 1971); (4) at least one property of the enzyme in an intact tissue preparation resembles much more closely that of the isolated enzyme in the presence of tetrahydrobiopterin, than that of the isolated enzyme in the presence of DMPH₄, i.e. phenylalanine is an excellent substrate for tyrosine hydroxylase in isolated nerve endings (KAROBATH and BALDESSARINI, 1972).

It should be noted that the K_m value of 1×10^{-4} M of the adrenal enzyme for tetrahydrobiopterin is similar to the probable tissue concentration. Assuming uniform distribution within the tissue, a value of about 2×10^{-5} M can be calculated. It would thus appear that the activity of the adrenal enzyme may be somewhat limited by availability of the cofactor, but that the limitation is not nearly as severe as might have been estimated from the K_m of DMPH₄.

As far as brain is concerned, much less is known. Biopterin is present in brain, but its concentration is less than one-twentieth that of adrenal tissue (REMBOLD, 1964). Assuming uniform tissue distribution as before, this would give a molar concentration of about 1×10^{-6} M. Since the K_m of a partially purified preparation of the enzyme from brain is about $2·5 \times 10^{-4}$ M (LLOYD and KAUFMAN, 1973), there is some uncertainty about whether biopterin is the cofactor for tyrosine hydroxylase in this tissue.

REFERENCES

BRENNEMAN A. R. and KAUFMAN S. (1964) *Biochem. Biophys. Res. Commun.* **17,** 177–183.
CARLSON A., CORRODI H. and WALDECK B. (1963) *Helv. Chim. Acta* **46,** 2271–2285.
CRAINE J. E., HALL E. S. and KAUFMAN S. (1972) *J. Biol. Chem.* **247,** 6082–6091.
FISHER D. B. and KAUFMAN S. (1972) *J. Neurochem.* **19,** 1359–1365.
IKEDA M., FAHIEN L. A. and UDENFRIEND S. (1966) *J. Biol. Chem.* **241,** 4452–4456.
IKEDA M., LEVITT M. and UDENFRIEND S. (1965) *Biochem. Biophys. Res. Commun.* **18,** 482–488.
KAROBATH M. and BALDESSARINI R. J. (1972) *Nature (New Biol.)* **236,** 206–208.
KAUFMAN S. (1958) *J. Biol. Chem.* **230,** 931–939.
KAUFMAN S. (1959) *J. Biol. Chem.* **234,** 2677–2682.

KAUFMAN S. (1961) *Biochim. Biophys. Acta.* **51,** 619–621.
KAUFMAN S. (1963) *Proc. Nat. Acad. Sci. U.S.* **50,** 1085–1093.
KAUFMAN S. (1964a) *J. Biol. Chem.* **239,** 332–338.
KAUFMAN S. (1964b) *Trans. N.Y. Acad. Sci. Ser. II,* **26,** 977–983.
KAUFMAN S. (1970) *J. Biol. Chem.* **245,** 4751–4758.
KAUFMAN S. and LEVENBERG B. (1959). *J. Biol. Chem.* **234,** 2683–2688.
LLOYD T. and KAUFMAN S. (1973) *Mol. Pharmacol.* in press.
LLOYD T. and WEINER N. (1971) *Mol. Pharmacol.* **7,** 569–580.
NAGATSU T., LEVITT M. and UDENFRIEND S. (1964). *J. Biol. Chem.* **239,** 2910–2917.
NIELSEN K. M., SIMONSEN V. and LIND K. E. (1969) *Europ. J. Biochem.* **9,** 497–502.
PETRACK B., SHEPPY F. and FETZER V. (1968) *J. Biol. Chem.* **243,** 743–748.
REMBOLD H. (1964) In: *Pteridine Chemistry.* (PFLEIDERER W. and TAYLOR E. C., eds.), pp. 465–484.
  Pergamon Press, New York.
SCRIMGEOUR K. G. and CHEEMA S. (1971) *Ann. N.Y. Acad. Sci.* **186,** 115–118.
SHIMAN R., AKINO M. and KAUFMAN S. (1971) *J. Biol. Chem.* **246,** 1330–1340.
SHIMAN R and KAUFMAN S. (1970) *Methods in Enzymol.* **17A,** 609–615.
STORM C. B. and KAUFMAN S. (1968) *Biochem. Biophys. Res. Commun.* **32,** 788–793.
UDENFRIEND S., ZALTZMAN-NIRENBERG P. and NAGATSU T. (1965) *Biochem. Pharmacol.* **14,** 837–845.

Frontiers in Catecholamine Research 1973, pp. 61 to 67. Pergamon Press. Printed in Great Britain.

CHARACTERISATION OF DOPA DECARBOXYLASE

WALLACE DAIRMAN, JAMES CHRISTENSON
and SIDNEY UDENFRIEND
Roche Institute of Molecular Biology, Nutley, New Jersey 07110, U.S.A.

DOPA decarboxylase, which catalyses the conversion of L-dopa to dopamine was the first of the catecholamine biosynthetic enzymes to be discovered. This enzyme has been studied extensively in many laboratories ever since its enzymatic activity was demonstrated in 1938 (HOLTZ et al., 1938). In this report we will present our findings obtained over the past several years with regard to dopa decarboxylase.

PROPERTIES OF A HOMOGENOUS HOG
KIDNEY DOPA DECARBOXYLASE

In 1970 the dopa decarboxylase of hog kidney was purified to homogeneity (CHRISTENSON et al., 1970). The final preparations were approximately 300-fold purified compared to the initial 25,000 $\times$ g supernatant and were essentially homogeneous by polyacrylamide disc gel electrophoresis and analytical ultracentrifugation. Recently, similar purifications have been reported (VOLTATTORNI et al., 1971; LANCASTER and SOURKES, 1972; GOLDSTEIN et al., 1972).

The purified enzyme was found to require at least one free sulfhydryl group for activity (CHRISTENSON et al., 1970). Although the purified enzyme was associated with about 0·9 mole of pyridoxal 5'-phosphate (PLP) per mole of enzyme protein, maximal activity was attained only when exogenous cofactor was present. VOLTATTORNI et al., (1972) have reported a similar level of PLP, however LANCASTER and SOURKES (1972) found a somewhat lower and variable amount of PLP associated with their purified preparations. The activity of their enzyme preparations was not stimulated by exogenous cofactor, except when the substrate was o- or m-tyrosine.

We found that the purified enzyme decarboxylated dopa, 5-hydroxytryptophan (5-HTP), tryptophan, p-tyrosine and (very feebly) histidine. Thus we feel that the name aromatic-L-amino acid decarboxylase proposed by Lovenberg (LOVENBERG et al., 1962) accurately describes this enzyme.

We found the molecular weight of the enzyme to be 112,000 daltons by sedimentation equilibrium. Gel electrophoresis in the presence of sodium dodecyl sulfate (SDS) and 2-mercaptoethanol gave rise to three bands of protein corresponding to molecular weights of about 57,000; 40,000 and 21,000 daltons (CHRISTENSON et al., 1970; CHRISTENSON, 1972). LANCASTER and SOURKES (1972) found a molecular weight of 85–90,000 daltons for the native enzyme and two subunits of 48–50,000 and 40–44,000 daltons by SDS-gel electrophoresis. These authors suggested the possibility that the 21,000 dalton subunit in our preparations may have been separated from the enzyme by their purification procedure. It is tempting to suppose that the difference between the two preparations in PLP content and in the effect of PLP on the activity of the enzyme may be related to this factor but, of course, much more extensive and detailed chemical study of the enzyme will be required to resolve this point.

IMMUNOTITRATION OF DOPA AND 5HTP
DECARBOXYLATING ACTIVITY

An antiserum to hog kidney dopa decarboxylase was obtained, which was shown to be monospecific by double diffusion and immunoelectrophoresis experiments (Christenson et al., 1970, 1971). When a mixture of antibody and dopa decarboxylase from hog tissues was centrifuged at 100,000 × g for 60 min after a short incubation period, it was found that the enzyme activity of the supernatant could be completely abolished. Control sera had no effect. By this method, it was shown that the antiserum crossreacted with dopa decarboxylase from kidneys, adrenals, and brains of rats, rabbits, and guinea pigs (Christenson et al., 1971) and from human liver (Dairman and Christenson, in press).

It was clear that this antiserum provided an excellent tool for testing the "single enzyme" hypothesis; viz. that only a single protein is responsible for the decarboxylation of both dopa and 5HTP. The demonstration that a single, pure enzyme with such a capability does exist is a necessary, but not sufficient, condition for this hypothesis, since such a demonstration cannot exclude the possibility that there may also be a second enzyme, at least in some tissues.

Since the experimental approach demanded that only soluble activities be studied, it was necessary to show that the soluble activities were indeed representative of the overall enzyme content of the tissue. When hog and rat brains and rabbit pineals were homogenised in 5 mM phosphate buffer and centrifuged at 27,000 × g for 15 min, at least 85 per cent of each activity was found in the supernatant (Christenson et al., 1972). Moreover, washing of the sediment with the same buffer released most of the remainder of both activities into the supernatant. The decarboxylating activity of the hog and rat liver and kidney remained completely in the supernatant fraction following centrifugation. Neither 5HTP or dopa decarboxylating activity sedimented at 100,000 × g under the conditions used as controls in the immunological experiments, as discussed below.

The immunotitration experiments were designed such that data would be obtained under conditions of both antibody excess and antigen excess. The "single enzyme" hypothesis would predict that the loss of a given fraction of one activity should be correlated with the loss of the same fraction of the other activity. Such was found to be the case in all the tissues examined, including hog kidney, liver, whole brain, cerebellum, midbrain, and hypothalamus, as well as rat whole brain, and rabbit pineal glands (Christenson et al., 1972). These data are presented in Table 1.

The pineal gland was specifically chosen for study since the synthesis of serotonin and its derivative, melatonin, is a major pathway in this tissue. The levels of norepinephrine, serotonin, and melatonin in the pineal and in the nerve endings impinging upon it undergo large, rhythmic diurnal fluctuations which are out of phase (Snyder et al., 1965; Wurtman and Axelrod, 1966; Lynch 1971). Thus, one might expect the enzymes involved in the biosynthesis of these compounds to be under strict and independent control. Nevertheless, our evidence indicated that there was only a single decarboxylase, even in the pineal.

Recently, Sims, Davis, and Bloom (1973) have reported evidence that separate dopa and 5HTP decarboxylase may exist, at least in rat brain. This suggestion was based on the finding that the two activities had different subcellular and gross regional distributions, showed differences in reaction kinetics, and were affected differently

TABLE 1. SIMULTANEOUS IMMUNOLOGICAL TITRATION OF DOPA AND 5-HYDROXY-
TRYPTOPHAN DECARBOXYLASE ACTIVITIES

Tissue	Antiserum/ enzyme (μl/unit activity)*	Decarboxylase activity remaining in supernatant	
		Dopa (% of control)	5-Hydroxy-tryptophan (% of control)
Hog kidney	0·022	89	81
	0·041	64	62
	0·045	68	62
	0·082	36	39
	0·102	20	22
	0·134	7	7
	0·307	0	0
Hog liver	0·180–0·9	0	0
Hog brain (whole)	0·013	82	79
	0·025	63	65
	0·190	0	0
Cerebellum	0·016	78	83
	0·031	69	70
	0·078–0·23	0	0
Midbrain	0·007	92	84
	0·014	75	83
	0·016	79	78
	0·021	75	75
	0·039	47	59
	0·063	28	33
	0·106–0·24	0	0
Hypothalamus	0·014	85	83
	0·024	67	68
	0·048	37	40
	0·120	0	0
Rat brain (whole)	0·190	88	85
	0·635	56	50
	0·952	41	40
	1·90†	2	1
Rabbit pineal	0·028	93	94
	0·055	84	87
	0·138	61	62
	0·277	5	5

* The indicated amounts of antiserum and enzyme (1 unit $= 1\eta$ mole of dopa decarboxylated/min) were incubated for 1 hr at room temperature followed by centrifugation at 100,000 $\times$ g for 1 hr. The dopa and 5HTP decarboxylating activity was then assayed in the supernatant.

† An amount of rabbit anti-goat gammaglobulin equal to 20 times the volumes of anti-decarboxylase serum was added after the 1 hr incubation; the mixture was incubated for an additional 30 min before centrifugation.

Taken from CHRISTENSON et al., 1972.

by PLP and by a carbonyl reagent, aminooxyacetic acid. Furthermore, it was shown that after intracisternal administration of 6-hydroxydopamine, dopa decarboxylase activity was significantly decreased in a number of brain regions, while 5HTP decarboxylase activity was unchanged or slightly elevated (SIMS and BLOOM, 1973).

While these data appear to contradict our own results, it should be remembered that a *complete* separation of the two activities has never been demonstrated. In addition, it has been proposed (BENDER and COULSON 1972) that dopa decarboxylase has "separate affinity sites for the two substrates adjacent to a single catalytic site." Such a proposal could explain the observed kinetic differences. Finally, it is known that the catalytic properties of an enzyme may be modified by its association with structural components (Munkres and Woodward, 1966). Thus the finding that the decarboxylase activity which is associated with particulate cellular matter exhibits a different substrate specificity from the soluble activity (Sims et al., 1973) do not necessarily mean that the two activities are due to different proteins. The point may be relevant in the case of the profound structural changes attendant upon 6-hydroxy-dopamine treatment (Thoenen and Tranzer, 1968). These concepts should be taken into account in any future efforts to investigate the identity of dopa and 5HTP decarboxylase.

REGULATION OF DOPA DECARBOXYLASE

Data from a number of laboratories indicate that neuronal dopa decarboxylase is not subject to regulatory control under a variety of conditions during which catechol-amine synthesis is either accelerated or decreased in response to acute changes in sympathetic nerve activity (Gordon et al., 1966; Spector et al., 1967; Dairman et al., 1968). It has also been reported that conditions which cause an elevation of tyrosine hydroxylase and dopamine-β-hydroxylase (DBH) in sympathetic tissue fail to alter the activity of dopa decarboxylase (Thoenen et al., 1971). This latter finding has also been confirmed by results from our laboratory. Thus neuronal dopa decarboxylase appears not to be subject to regulatory influences. However, a number of observations on Parkinsonian patients treated with L-dopa suggested that upon chronic administration of this agent, a reduction in its metabolism might occur (Cotzias et al., 1971; Yahr et al., 1969; Whitsett et al., 1970). These clinical observations, plus the fact that we had observed reductions in the tissue activity of tyrosine hydroxylase following L-dopa administration to animals (Dairman and Udenfriend, 1972) prompted an investigation into the effect of L-dopa administration on the tissue levels of dopa decarboxylase in the rat.

Following 4 days of L-dopa administration to the rat (1000 mg/kg/day s.c.), a 50 per cent decrease in the activity of liver dopa decarboxylase was observed (Dairman et al., 1971). This effect appears to be specific for the liver in that the decarboxylase activity of kidney, heart, adrenal, brain, mesenteric artery, salivary gland, superior cervical ganglia and sciatic nerve remained unchanged. However, the liver may contain as much as 75 per cent of the total amount of this enzyme in an animal. Thus a decrease in the level of decarboxylase activity in the liver could have pharmacological significance in the treatment of Parkinson's disease with L-dopa. That is, if a similar phenomenon occurred in Parkinsonian patients taking L-dopa, less of the administered drug would be peripherally decarboxylated, making more of it available for conversion to dopamine in the brain.

When L-dopa was given to rats as previously described, a decrease in liver decar-boxylase activity could not be demonstrated before 3 days of drug administration. Doses of L-dopa ranging from 100 to 1000 mg/kg/day s.c. when given for 4 days resulted in a dose-dependent decrease in liver enzyme activity which was statisti-cally significant even at the lowest dose (Dairman et al., 1971). This dose (100 mg/kg/day) is similar to the daily intake of L-dopa in Parkinsonian patients. It was also possible to demonstrate a 40 per cent reduction in liver decarboxylase activity when L-dopa was administered orally to rats at a dose of 100 mg/kg/day for 7 days.

Two reports have appeared in the literature which indicate that the administration of dopa to rats resulted in a reduction of brain PLP levels (KURTZ and KANFER, 1971; EBADI and PFEIFFER, 1972). Since the decarboxylase is a PLP dependent enzyme, we investigated whether the observed reduction in liver activity might be related to a decreased availability of this cofactor. Over a wide range of PLP concentration (10^{-7}–10^{-4} M) the *in vitro* activation of liver decarboxylase activity was identical for enzyme preparations from dopa treated and control rats. The concomitant administration of pyridoxine (5 mg/kg/day) with L-dopa 500 mg/kg/day) did not alter the L-dopa mediated reduction in liver decarboxylase activity (DAIRMAN *et al.*, 1971). Finally, the PLP content of both liver and brain were assayed by two different analytical methods (MARUYAMA and COURSIN, 1968; ADAMS 1969) following a regimen of L-dopa administration which produced about a 50 per cent lowering in decarboxylase activity. In these experiments, the liver levels of PLP usually remained constant with an occasional reduction of 5–10 per cent. Reduction in the brain levels of the cofactor were not observable. The results of our studies indicate that the L-dopa mediated reduction in liver decarboxylase is not related to a deficiency of PLP or to a defect in the ability of the cofactor to activate the enzyme.

A titration of a given amount of monospecific antiserum to dopa decarboxylase with various amounts of liver supernatants obtained from L-dopa or saline treated animals showed that the L-dopa group had about one-half the amount of antigen as the controls (DAIRMAN *et al.*, 1971). This decrease in antigenic material was in excellent agreement with the reduction in decarboxylase activity which was obtained in these same livers following L-dopa administration. These data indicated that the reduction in activity observed is due to a net reduction in dopa decarboxylase enzyme protein.

It was of interest to determine if other substrate amino acids for the decarboxylase would also decrease the liver enzyme activity. Following 4 days of s.c. administration of either L-5HTP (500 mg/kg/day) or L-tyrosine (1000 mg/kg/day), a 35 per cent reduction in liver decarboxylase activity was observed.

The ability of a substrate to decrease the amount of an enzyme involved in its metabolism is rather unique. Generally the opposite effect has been noted; that is, the induction of metabolic enzymes by their substrates. The availability of potent inhibitors of monoamine oxidase, an enzyme which degrades the amine products of dopa decarboxylase allowed us to investigate whether these amine products might be responsible for the decreased liver activity. The administration of small amounts of either dopa, dopamine or 5-HTP, which by themselves were incapable of reducing liver decarboxylase levels caused significant reductions in enzyme activity when given in combination with the monoamine oxidase inhibitor pargyline (DAIRMAN *et al.*, 1972). These data indicate that the aromatic amino acids themselves are not directly responsible for the decreased levels of liver decarboxylase but that the amines formed via decarboxylation or the non-deaminated amine metabolites are probably the mediators of the effect.

AXOPLASMIC TRANSPORT OF DOPA DECARBOXYLASE

The subcellular distribution of tyrosine hydroxylase and dopa decarboxylase still remains somewhat controversial. Although considerable evidence indicates these enzymes are solubly distributed (CLARK *et al.*, 1954; BLASKO *et al.*, 1965; STJÄRNE and LISHAJKO, 1967; MUSACCHIO, 1968; LADURON and BELPAIRE, 1968 a,b), there are kinetic data which make it attractive to postulate a common vesicular location for these two enzymes and DBH (UDENFRIEND, 1966). If tyrosine hydroxylase and dopa decarboxylase are in, or firmly attached to, the DBH containing vesicle, then the axoplasmic transport rates of all three of these enzymes should be identical. Therefore the axoplasmic transport rates of dopa decarboxylase and DBH were determined in rat and guinea pig sciatic nerve. By measuring the accumulation of these two enzyme activities in a 0·5 cm segment of nerve proximal to a ligation at several time intervals following the ligation, proximal to distal rates of transport could be calculated. In rat sciatic nerve the transport rate for DBH was found to be 72 mm/day while that of dopa decarboxylase was 20 mm/day (DAIRMAN *et al.*, 1973). In the guinea pig sciatic nerve, the rates of transport were different by a greater extent, being 30 mm/day for the decarboxylase and 140 mm/day for DBH.

Despite the observation that the proximal accumulation of DBH was over 3 times as great as that of the decarboxylase, these two enzymes could be transported at identical rates if the decarboxylase was being degraded at a substantially faster rate than DBH within the sympathetic axons. In order to investigate this possibility, rat sciatic nerves were ligated at two points approximately 1·5 cm apart. Enzymes should be unable to move into or out of this segment of nerve. After 6 hr this segment was removed and it was found that the decarboxylase activity decreased by 15 per cent while that of DBH fell by 29 per cent in comparison to sham-operated controls (DAIRMAN *et al.*, 1973). Thus differences in the rates of inactivation of these two enzymes could not account for the three fold slower transport of dopa decarboxylase in relation to DBH.

The results of our findings with regard to axoplasmic transport of these two enzymes indicate that dopa decarboxylase cannot be contained within, or firmly attached to, the DBH and norepinephrine containing transport vesicle. However, one is still faced with providing some explanation to account for an efficient synthesis of norepinephrine in view of the fact that the K_m value of DBH for dopamine is $\sim 1 \times 10^{-3}$ M. One factor which may partially reconcile this difficulty is the existence of a vesicular uptake mechanism for dopamine (KIRSHNER, 1962; POTTER and AXELROD, 1964). Thus dopamine formed in the soluble phase of the neurone could be concentrated in the DBH containing vesicle for an efficient conversion to norepinephrine.

REFERENCES

ADAMS E. (1969) *Anal. Biochem.* **31**, 118–122.
BENDER D. A. and COULSON W. F. (1972) *J. Neurochem.* **19**, 2801–2810.
BLASCHKO H., HAGEN P. and WELCH A. D. (1955) *J. Physiol. (London)* **129**, 27–49.
CHRISTENSON J. G. (1972) Ph.D Thesis, City University of New York, New York, New York.
CHRISTENSON J. G., DAIRMAN W. D. and UDENFRIEND S. (1970) *Arch. Biochem. Biophys.* **141**, 356–367.
CHRISTENSON J. G., DAIRMAN W. D. and UDENFRIEND S. (1971) *Fedn. Proc.* **30**, 33 (Abstr).
CHRISTENSON J. G., DAIRMAN W. D. and UDENFRIEND S. (1972) *Proc. Nat. Acad. Sci. (U.S.A.)* **69**, 343–347.
CLARK C. T., WEISSBACH H. and UDENFRIEND S. (1954) *J. Biol. Chem.* **210**, 139–148.
COTZIAS G. C., PAPAVASILIOU P. S., GINOS J., STECK A. and DÜBY S. (1971) *Ann. Rev. Med.* **22**, 305–326.
DAIRMAN W., CHRISTENSON J. G. and UDENFRIEND S. (1971) *Proc. Nat. Acad. Sci.* **68**, 2117–2120.
DAIRMAN W., CHRISTENSON J. G. and UDENFRIEND S. (1972) *Pharmacol. Rev.* **24**, 269–289.
DAIRMAN W. and CHRISTENSON J. G. (1973) *Europ. J. Pharmacol.* **22**, 135–140.

DAIRMAN W., GEFFEN L. and MARCHELLE M. (1973) *J. Neurochem.* **20**, 1617–1623.
DAIRMAN W., GORDON R., SPECTOR S., SJOERDSMA A. and UDENFRIEND S. (1968) *Mol. Pharmacol.* **4**, 457–464.
DAIRMAN W. and UDENFRIEND S. (1972) *Mol. Pharmacol.* **8**, 293–299.
EBADI M. S. and PFEIFFER R. (1972) *Vth Int. Congr. Pharmacol.* **61**, (Abstr.)
GOLDSTEIN M., FUXE K., HÖKFELT T. (1972) *Pharmacol. Rev.* **24**, 293–309.
GORDON R., REID J. V. O., SJOERDSMA A. and UDENFRIEND S. (1966) *Mol. Pharmacol.* **2**, 606–613.
HOLTZ P., HEISE R. and LUDTKE K. (1938) *Arch. Exp. Pathol. Pharmacol.* **191**, 87–118.
KIRSHNER N. (1962) *J. Biol. Chem.* **237**, 2311–2317.
KURTZ D. J. and KANFER J. N. (1971) *J. Neurochem.* **18**, 2235–2236.
LADURON P. and BELPAIRE F. (1968) *Nature (Lond.)* **217**, 1155–1156.
LADURON P. and BELPAIRE F. (1968) *Biochem. Pharmacol.* **17**, 1127–1140.
LANCASTER G. A. and SOURKES T. L. (1972) *Canad. J. Biochem.* **50**, 791–797.
LOVENBERG W., WEISSBACH H. and UDENFRIEND S. (1962) *J. Biol. Chem.* **237**, 89–92.
LYNCH H. J. (1971) *Life Sci.* **10**, 791–795.
MARUYAMA H. and COURSON B. D. (1968) *Anal. Biochem.* **26**, 420–429.
MUNKRES K. D. and WOODWARD D. O. (1966) *Proc. Nat. Acad. Sci. (U.S.A.)* **55**, 1217–1224.
MUSACCHIO J. M. (1968) *Biochem. Pharmacol.* **17**, 1470–1473.
POTTER L. T. and AXELROD J. (1964) *J. Pharmacol. Exp. Ther.* **142**, 299–305.
SIMS K. L. and BLOOM F. E. (1973) *Brain Res.* **49**, 165–175.
SIMS K. L. DAVIS G. A. and BLOOM F. E. (1973) *J. Neurochem.* **20**, 449–464.
SYNDER S. H., ZWEIG M., AXELROD J. and FISCHER J. E. (1965) *Proc. Nat. Acad. Sci. (U.S.A.)* **53**, 301–305.
SPECTOR S., GORDON R., SJOERDSMA A. and UDENFRIEND S. (1967) *Mol. Pharmacol.* **3**, 549–555.
STJARNE L. and LISHAJKO F. (1967) *Biochem. Pharmacol.* **16**, 1719–1728.
THOENEN H., KETTLER R., BURKARD W. and SANER A. (1971) *Naunyn. Schmiedeberg. Arch. Pharmak.* **270**, 146–160.
THOENEN H. and TRANZER J. P. (1968) *Naunyn. Schmiedeberg Arch. Pharmakol. Exp. Path.* **261**, 271–288.
VOLTATTORNI C. B., MINELLI A. and TURANO C. (1971) *FEBS. Letters.* **17**, 231–235.
UDENFRIEND S. (1966) *Harvey Lecture* **60**, 57–83.
WHITSETT T. L., MCKINNEY A. S. and GOLDBERG L. I. (1970) *Clin. Res.* **18**, 28 (Abstr.).
WURTMAN R. J. and AXELROD J. (1966) *Life Sci.* **5**, 665–669.
YAHR M. D., DUVOISIN R. C., SCHEAR M. J., BARRETT R. E. and HOELN M. M. (1969) *Arch. Neurol.* **21**, 343.

Frontiers in Catecholamine Research 1973, pp. 69 to 78. Pergamon Press. Printed in Great Britain.

CHARACTERISATION, LOCALISATION AND REGULATION OF CATECHOLAMINE SYNTHESISING ENZYMES

Menek Goldstein[1], Berta Anagnoste[1], Lewis S. Freedman[1],
Mark Roffman[1], Richard P. Ebstein[1], Dong H. Park[1],
Kjell Fuxe[2] and Tomas Hökfelt[2]

[1]Department of Psychiatry, Neurochemistry Laboratories New York University
Medical Center, New York, N.Y. 10016, U.S.A.
and
[2]Department of Histology, Karolinska Institutet 104 01 Stockholm, Sweden

Immunochemical and immunohistochemical techniques provide a basis for new investigations in the field of neurotransmitters. Enzymes involved in the biosynthesis of catecholamines were purified and used as antigens to produce specific antibodies (Gibbs et al., 1967; Geffen et al., 1969; Hartman and Udenfriend, 1970; Goldstein et al., 1971a; Fuxe et al., 1971a). In this presentation we will describe some studies on the purification, characterisation and immunohistochemical localisation of three enzymes, namely aromatic L-amino acid decarboxylase (AADC),* dopamine-β-hydroxylase (DβH) and phenylethanolamine N-methyl-transferase (PNMT), involved in the biosynthesis of catecholamines.

Recent studies have implicated adenosine 3'5'-monophosphate (cyclic AMP) in the functions of the peripheral and central nervous system (McAffe and Greengard, 1972). We will present results of our studies which show that cyclic AMP may play an important role in the regulation of catecholamine biosynthesis.

AROMATIC L-AMINO ACID DECARBOXYLASE (EC 4.1.1.28)

AADC was purified from bovine adrenal glands (Goldstein, 1972). After immunization of rabbits with purified AADC (enzyme preparation obtained after polyacrylamide gel electrophoresis) specific antisera to AADC were obtained. Bovine AADC antiserum inhibits AADC activity from different tissues of various species. Immunochemical and immunohistochemical studies were applied to test whether the same enzyme catalyzes the decarboxylation of L-dopa and L-5-hydroxy-tryptophan (L-5HTP). We and others (Christenson et al., 1972) have shown that specific anti-serum directed against AADC inhibits L-Dopa and L-5HTP striatal decarboxylase activity proportionately. Furthermore, the immunohistochemical studies (see next section) and the findings that 6-hydroxydopamine induced degeneration of the nigro-striatal dopamine pathway causes a proportional reduction

---

Abbreviations used: AADC; aromatic L-amino acid decarboxylase (EC 4.1.1.28). DβH; dopamine-β-hydroxylase (EC 1.14.2.1.). PNMT; phenylethanolamine N-methytransferase (EC 2.2.1 −). Dopa; 3,4-Dihydroxyphenylalanine. 5-HTP; 5-Hydroxytryptophan. 5-HT; Serotonin. cyclic AMP; Adenosine 3',5'-monophosphate. dB-cAMP; N^6,O^2-dibutyryl adenosine 3',5'-cyclic monophosphate.

in striatal L-Dopa and L-5HTP decarboxylase activity (GOLDSTEIN and UNGERSTEDT, unpublished data; GOLDSTEIN et al., 1973a) support the hypothesis that a single enzyme catalyzes the decarboxylation of both aromatic L-amino acids in the striatum. However, our data does not exclude the possibility that in some regions of the brain an enzyme exists which catalyzes specifically either L-Dopa or L-5HTP decarboxylation (SIMS and BLOOM, 1971).

LOCALISATION OF AADC

Methodological improvements (HARTMAN et al., 1972; HÖKFELT et al., 1973a) have made it possible to study the localisation of catecholamine synthesising enzymes by immunofluorescence in peripheral and central monoaminergic systems (HÖKFELT et al., 1973b,c). The findings that catecholamine synthesising enzymes can be localised with immunofluorescence on unfrozen aldehyde fixed material opens up the possibility to carry out some studies on the ultrastructural level (HÖKFELT et al., 1973c).

A specific AADC immunofluorescence of strong intensity was observed in the cytoplasm of practically all medullary cells of the adrenal glands. No specific immunofluorescence was observed in the adrenal cortex. Immunofluorescence was observed in the endothelial cells of the tubuli and glomeruli of the kidneys, and in the norepinephrine cell bodies of the sympathetic ganglia (GOLDSTEIN et al., 1972).

The dopamine containing cell bodies and axons of the CNS exhibit immuno-fluorescence of strong intensity, while the dopamine containing terminals exhibit immunofluorescence of moderate intensity (i.e. the nigro-neostriatal dopamine system, the mesolimbic dopamine system and the tubero-infundibular dopamine system). Although the results obtained with the immunofluorescence technique confirm the previously reported findings obtained with the histochemical technique of Falck-Hillarp, the former technique is more sensitive. Only with the immuno-fluorescence technique does the dopamine pathway become strongly fluorescent and therefore it is possible to map out the dopamine fibers in detail in the intact brain. Strong AADC immunofluorescence was observed in the dopamine fibers of the globus pallidus. These findings could explain the results reported by BERNHEIMER and HORNYKIEWICZ (1966) that the levels of homovanilic acid are high in the globus pallidus.

The norepinephrine containing cell bodies and axons exhibit immunofluorescence of weak intensity and no immunofluorescence was observed in the norepinephrine containing terminals (the ascending norepinephrine system from the lower brain stem, e.g. from the locus coeruleus to the cortical areas of the hypothalamus).

The 5-HT containing axons and cell bodies exhibit immunofluorescence of strong intensity and very weak immunofluorescence was observed in the 5-HT containing terminals. These results demonstrate that L-Dopa and L-HTP are decarboxylated in their respective neurons by an antigenically indistinguishable enzyme.

Using the immunofluorescence technique some hitherto unknown 5-HT terminal systems could be demonstrated, e.g. in dorsal and median hypothalamic areas, in the cerebral cortex and the bulbus olfactorius. Many cell bodies in the preoptic area and in the hypothalamus show a moderate fluorescence. It was previously reported that these cells contain an amine-like substance (BUTCHER et al., 1972).

Figure 1 (a, b, c and d) depicts AADC immunofluorescence in the cell bodies of

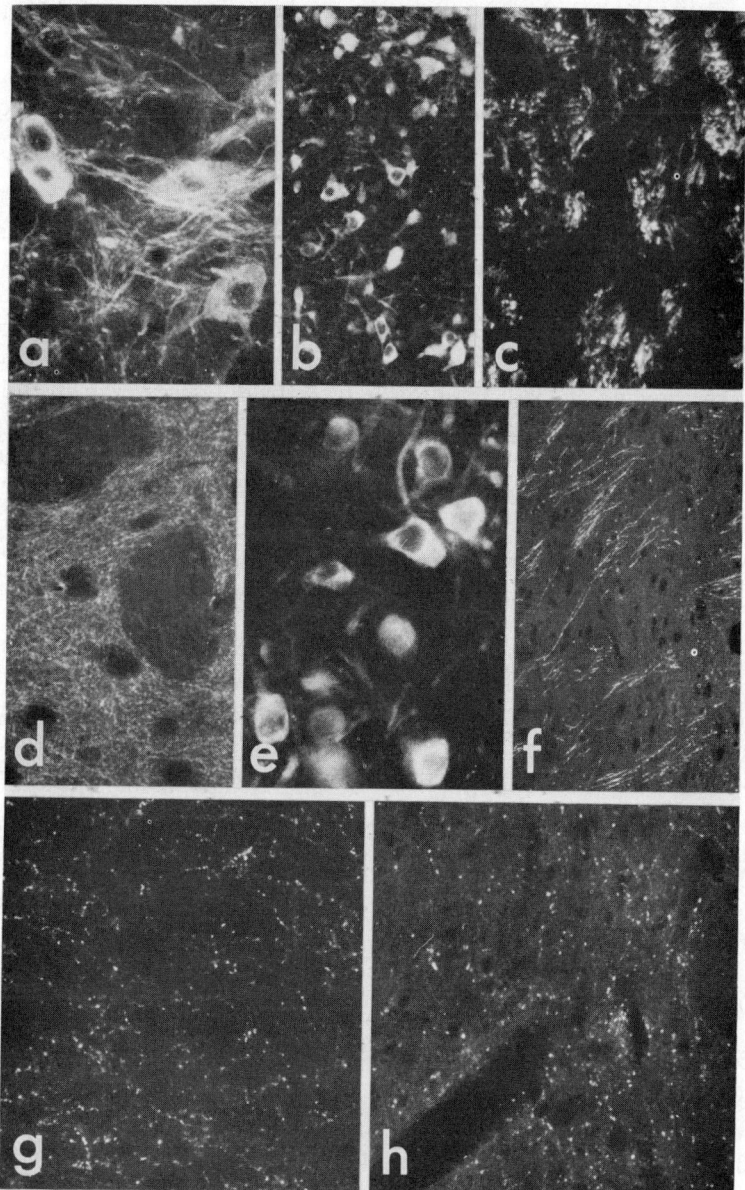

FIG. 1.(a–h)—Immunofluorescence micrographs of various brain regions after incubation with antibodies to a CA synthesising enzyme followed by incubation with FITC
conjugated antibodies (indirect technique). A strong immunofluorescence is seen in
cell bodies and cell processes of the substantia nigra (a), in cell bodies of the raphe
nuclei (b), in bundles of axons in the internal capsule (c) and in nerve terminals in the
neostriatum (d) after incubation with anti-AADC. After incubation with anti-DβH,
fluorescent cells are found in the locus coeruleus (e), fluorescent axons in the subthalamus
(Forel's field Hl) and (f) fluorescent nerve endings in the nucleus ventralis anterior
thalami (g). After incubation with anti-PNMT, immunofluorescence of a medium intensity is seen in the perifornical region of hypothalamus (h). Magnifications 300 × (a,c,e,)
120 × (b,f), 200 × (g,h).

the substantia nigra, in the raphe nucleus, in axons of the internal capsule and in the terminals of the striatum.

At the ultrastructural level ferritin molecules could be demonstrated in cell bodies and dendrites in the substantia nigra. It is therefore conceivable that the dopamine neurons could be identified in the electron microscope (HÖKFELT *et al.*, 1973c).

DOPAMINE-β-HYDROXYLASE (EC 1.14.2.1)

DβH is a copper protein and the valence of the copper undergoes cyclic changes during the enzymatic β-hydroxylation reaction (FRIEDMAN and KAUFMAN, 1965; BLUMBERG *et al.*, 1965). Increased oxygen concentration enhances the enzymatic activity and can replace fumarate in the stimulation of the enzymatic hydroxylation (GOLDSTEIN *et al.*, 1968). The initial velocity patterns are consistent with a mechanism in which the binding of the first substrate (ascorbate) to the enzyme is followed by the release of the product (dehydroascorbate) before a second substrate can react ("Ping–Pong" mechanism). The subsequent substrates (dopamine or O_2) add to the enzyme before either product is released and the interconversion of the central ternary complexes most likely represents the rate-limiting step in the overall β-hydroxylation reaction (GOLDSTEIN *et al.*, 1968).

Purification

Various procedures for the purification of DβH from bovine adrenal glands were described (LEVIN and KAUFMAN, 1961; GOLDSTEIN *et al.*, 1965; GOLDSTEIN, 1972). More recently we have purified the enzyme from human serum and from human pheochromocytoma tumors. The enzyme from human pheochromocytoma was purified by the previously described procedure (GOLDSTEIN, 1972) and as a final purification step the enzyme preparation was submitted to sucrose density gradient centrifugation (EBSTEIN, PARK and GOLDSTEIN, unpublished data). Following sucrose density gradient centrifugation two fractions with enzymatic activity were obtained; one fraction with high specific activity was associated with a high molecular form of the enzyme (approximately 300,000) and the other with a low specific activity was associated with a low molecular form of the enzyme (approximately 150,000) (Fig. 2).

Immunochemical studies

After immunisation of rabbits with purified bovine DβH or with human DβH (enzyme preparations purified on polyacrylamide disc gel electrophoresis) the corresponding antisera were obtained. Following immunoelectrophoresis bovine antiserum gives a single precipitin arc with bovine adrenal enzyme but does not give a precipitin arc with human enzyme. Human anti-serum gives a precipitin arc with human enzyme but does not give a precipitin arc with bovine adrenal enzyme. Neither human antiserum nor bovine antiserum gives a precipitin arc with rat adrenal enzyme. These studies indicate that DβH is heterogeneous among different species.

Assay of DβH activity

A sensitive and specific procedure was developed for assay of DβH activity in tissues and serum (GOLDSTEIN *et al.*, 1971b; MOLINOFF *et al.*, 1971). In the first of the coupled enzymatic reactions, tyramine (the substrate) is converted by DβH to

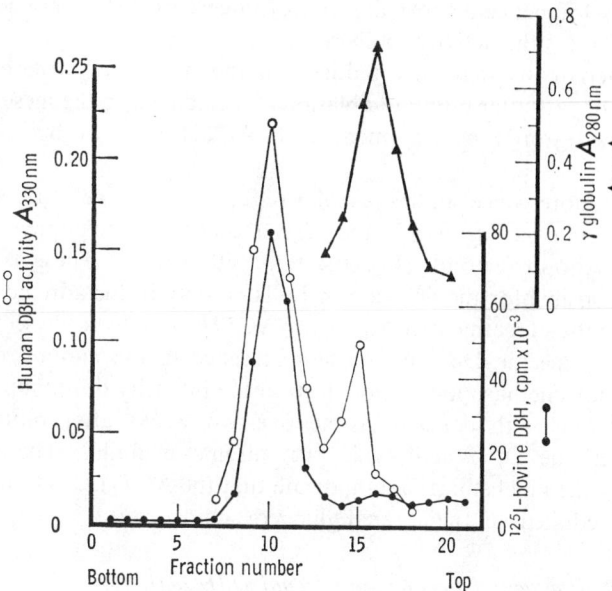

FIG. 2.—Sucrose density gradient centrifugation (linear gradient, 5–20 per cent sucrose) of DβH isolated from a human pheochromocytoma tumor. [125]I bovine DβH and human gamma gloubulin were used as marker proteins.

octopamine; in the second reaction, the octopamine is further converted by added PNMT to N-methyl octopamine. S-adenosyl-L-methionine-methyl-[14]C (SAM) serves as methyl donor and the N-methylated derivatives are separated by solvent extraction and their radioactivity is determined. Endogenous inhibitors of DβH are inactivated by the addition of ethylmaleimide (GOLDSTEIN et al., 1971b) or cupric ions (MOLINOFF et al., 1971).

The effect of stress

We have investigated the effects of a single swim stress (acute stress) and of repeated swim stress (chronic stress) on DβH activity in serum and in tissues containing either monoaminergic cell bodies or nerve terminals (ROFFMAN et al., 1973). The acute and chronic swim stress leads to an increase in serum DβH activity in rats. The acute but not the chronic swim stress resulted in a decrease in DβH activity in the mesenteric arteries. The results of our studies suggest that DβH in the arteries may serve as a source for circulatory DβH. The chronic but not the acute stress resulted in an increase in enzyme activity in the cervical ganglia and in adrenal glands. These findings are consistent with the idea that chronic stress induces de novo enzyme protein synthesis.

The effect of hypophysectomy

A study was undertaken to determine the effects of hypophysectomy on tissue and serum DβH (FUXE et al., 1971b; FUXE et al., 1972; FREEDMAN et al., 1973). Serum DβH activity is increased in rats after hypophysectomy. The maximal increase in the serum hydroxylase activity was reached 2–4 weeks after hypophysectomy and amounts to approximately 150 per cent above the control values. ACTH but not dexamethasone administration causes the activity to decrease.

The DβH activity is markedly reduced in the mesenteric arteries and in the cervical ganglia of rats after pituitary ablation. The activity in the mesenteric arteries can be partially restored by treatment with ACTH but not by treatment with dexamethasone.

The immunofluorescence studies reveal a reduction in the cell size and a decrease in the DβH protein concentration in the adrenal medulla of hypophysectomised rats (6 weeks after hypophysectomy (FREEDMAN *et al.*, 1973)). These two effects are most likely responsible for the decrease in DβH activity in the adrenals of hypophysectomised animals. Treatment of rats with ACTH after hypophysectomy resulted in an increase in specific DβH immunofluorescence in the adrenal medulla while treatment with dexamethasone had no effect on the intensity of immunofluorescence. However, treatment with dexamethasone one-two weeks after pituitary ablation partially restores the DβH activity in the adrenal medulla. The immunohisto-fluorescence and the biochemical data indicate that the ACTH effect on DβH activity is not entirely mediated via the adrenal glucocorticoids.

The effect of ascorbic acid deficiency on adrenal DβH activity

Ascorbate serves as a cofactor in the enzymatic dopamine-β-hydroxylation (LEVIN and KAUFMAN, 1961). It was therefore of interest to determine the effects of ascorbic acid deficiency on adrenal DβH activity. The results presented in Table 1 show that DβH activity is significantly reduced in the adrenals of scorbutic guinea pigs when compared with non-scorbutic animals. Since tyrosine hydroxylase activity was reported to also be reduced in the adrenals of the scorbutic guinea pigs (NAKASHIMA *et al.*, 1972) it will be of interest to determine whether the decrease in enzyme activities is specific for catecholamine synthesising enzymes or whether other adrenal enzymes are also altered in scurvy.

LOCALISATION OF DβH

DβH was localised in all medullary cells of the adrenal glands and in the norepinephrine containing cell bodies of the peripheral nervous system. In the central nervous system the enzyme could be localised in neurons which are known to contain norepinephrine. The cell bodies of the pons (e.g. in the locus coeruleus) and medulla oblongata exhibited specific immunofluorescence (FUXE *et al.*, 1971a). The intensity of the DβH immunofluorescence was high in the norepinephrine containing cell bodies, axons

TABLE 1. THE EFFECT OF ASCORBIC ACID DEFICIENCY ON ADRENAL DOPAMINE-β-HYDROXYLASE ACTIVITY IN THE GUINEA PIG

	Body weight (g)	Adrenal weight (mg/pair)	DβH Activity (nm/pair)
Nonscorbutic (8)	279 ± 12	102 ± 3	$72 \cdot 0 \pm 5 \cdot 6$
Scorbutic (6)	147 ± 8	144 ± 5	$40 \cdot 8 \pm 4 \cdot 0$
% Change	-47%	$+40\%$	-44%

The number of animals are given in parenthesis.
Ascorbic acid deficient diet fed *ad lib.* to weanling guinea pigs for 3 weeks.

and terminals. Figure 1(e, f, and g) depicts DβH immunofluorescence in the cell bodies of the locus coeruleus and in the nerve fibers of the subthalamus and in the nerve terminals of the nucleus ventralis anterior thalami.

Using the histochemical procedure of Falck-Hillarp it was difficult to demonstrate the presence of norepinephrine fibers in areas which contain high concentrations of dopamine. However, with the immunofluorescence procedure single DβH containing fibers were localised in the caudal part of the striatum. Also in the lateral parts of the external layer of the median eminence a thin strand of DβH containing terminals was localised at certain levels adjacent to the primary capillary plexus (HÖKFELT et al., 1973c). The DβH containing terminals in the median eminence have not been previously described and may have an important functional role in the control of the release of the hormones from the peptidergic neurons.

Preliminary studies at the ultrastructural level show heavy deposits of ferritin molecules around unmyelinated axons in the locus coeruleus area. Further studies are in progress to determine whether this represents an artefact or indicates that DβH is partially bound to the cell membrane.

The immunoperoxidase technique was also employed to localise DβH in the adrenals and in the CNS (ZIMMERMAN et al., 1973).

PHENYLETHANOLAMINE N-METHYLTRANSFERASE (EC 2.2.1.-)

PNMT was purified from bovine adrenal medulla (CONNET and KIRSHNER, 1970; JOH and GOLDSTEIN, 1973). The enzyme from the supernatant fraction was isolated in pure form and has a molecular weight of approximately 40,000. The enzyme also occurs in two higher molecular forms with molecular weights of approximately 80,000 and 160,000. Differently charged isozymes were separated from the low molecular weight form of the enzyme by DEAE-sephadex chromatography and polyacrylamide disc gel electrophoresis. The immunochemical analyses revealed heterogeneity of adrenal PNMT among different species. It appears that the corticoid-inducible, mammalian enzyme is immunologically distinguishable from the uninducible frog adrenal enzyme (JOH and GOLDSTEIN, 1973).

LOCALISATION OF PNMT

PNMT was localised only in specific populations of medullary cells of the adrenal glands. This finding supports the evidence that two types of medullary cells store norepinephrine and epinephrine (ERÄNKÖ, 1952; HILLARP and HÖKFELT, 1953). Preliminary results indicate the presence of weakly fluorescent terminals e.g. in the dorsomedial hypothalamic nuclei and in the perifornical area (Fig. 1h). It is conceivable that these terminals belong to central epinephrine containing neurons.

THE EFFECTS OF CYCLIC-AMP ON CATECHOLAMINE BIOSYNTHESIS

Evidence has been presented that agents which stimulate or block dopamine receptors affect the rate of dopamine synthesis in the striatum (GOLDSTEIN et al., 1970a; KEHR et al., 1972). Since it was reported that a dopamine-sensitive adenylate cyclase may be the receptor for dopamine in mammalian brain (KEBABIAN et al., 1972) we have investigated the effects of cyclic-AMP on ^{14}C-dopamine synthesis from ^{14}C-tyrosine in slices obtained from various regions of the brain (ANAGNOSTE et al., 1973; GOLDSTEIN et al., 1973). Results presented in Table 2 show that dB-cAMP

TABLE 2. THE EFFECT OF VARIOUS CONCENTRATIONS OF dB-cAMP ON
DOPAMINE CONCENTRATIONS AND ON ^{14}C-DOPAMINE SYNTHESIS FROM
^{14}C-TYROSINE IN STRIATAL SLICES

dB-cAMP (M)	Dopamine (μg/g)	% Change	^{14}C-Dopamine (ng/g/hr)	% Change
None	4·2 ± 0·6	—	248 ± 30	—
5 × 10^{-5}	N.E.	—	345 ± 45	+40
5 × 10^{-4}	9·5	+137	580 ± 68	+133

N.E., not estimated
In all incubations 5 × 10^{-5}M theophylline was added.
Results are the mean from at least three experiments ± S.E.M.

increases the concentration of dopamine in striatal slices and stimulates the synthesis of ^{14}C-dopamine from ^{14}C-tyrosine. The dB-cAMP-induced stimulation of dopamine synthesis is dose-dependent. In separate experiments it was established that the cyclic nucleotide and not the butyryl moiety of dB-cAMP is effective in the stimulation of ^{14}C-dopamine synthesis. dB-cAMP stimulates the synthesis of ^{14}C-catecholamines from ^{14}C-tyrosine in various dopamine containing regions of the CNS (i.e. median eminence, cortex). In order to determine whether the dB-cAMP effect occurs at the tyrosine hydroxylase step we have investigated the effects of end product inhibition on the dB-cAMP-induced stimulation of ^{14}C-dopamine synthesis. The addition of dopamine to the media results in an inhibition of ^{14}C-dopamine synthesis from ^{14}C-tyrosine in presence as well as in absence of dB-cAMP. However, the presence of dopamine in the medium does not abolish the stimulatory effects of dB-cAMP. Thus, the dB-cAMP-induced stimulation of dopamine synthesis occurs prior or at the tyrosine hydroxylase step and it is conceivable that dB-cAMP reduces the end product inhibition of catecholamine synthesis. It is generally assumed that the activation of tyrosine hydroxylase activity due to increased nerve impulses is associated with catecholamine release followed by a reduction in the end-product inhibition. However, it is conceivable that other factors such as activation of tyrosine hydroxylase by cyclic-AMP (or cyclic-AMP induced enhanced intraneuronal uptake of tyrosine) might contribute to the increased rate of catecholamine synthesis.

Since depolarising agents like K$^+$ or ouabain stimulate the synthesis of dopamine in striatal slices (GOLDSTEIN *et al.*, 1970b; HARRIS and ROTH, 1971) and stimulate the formation of cyclic-AMP (SHIMIZU and DALY, 1972) we have investigated whether these two effects are linked to each other. The results presented in Table 3 show that the effects of dB-cAMP and of K$^+$ on ^{14}C-dopamine synthesis are synergistic. These results indicate that the depolarisation of the cell membrane facilitates the dB-cAMP-induced stimulation of dopamine synthesis. We have previously shown that calcium deprivation stimulates the ^{14}C-dopamine synthesis from ^{14}C-tyrosine in striatal slices (GOLDSTEIN *et al.*, 1970c). These findings prompted us to investigate the effects of dB-cAMP on ^{14}C-dopamine in calcium-deprived medium. The dB-cCMP-induced stimulation of ^{14}C-dopamine synthesis is less effective in calcium deprived medium (Table 3). The effects of calcium-deprivation and of dB-cAMP on augmenting ^{14}C-dopamine synthesis are not additive, indicating that dB-cAMP and calcium deprivation stimulate dopamine synthesis by a similar mechanism. It is noteworthy that calcium deprivation results in an enhancement of adrenal adenylate cyclase (RUBIN *et al.*, 1972). It is therefore conceivable that calcium also inhibits

TABLE 3. THE STIMULATION OF ^{14}C-DOPAMINE SYNTHESIS BY dB-cAMP IN A K$^+$ ENRICHED MEDIUM OR IN A Ca^2 DEFICIENT MEDIUM

Medium	^{14}C-Dopamine (ng/g/hr)	% Change from controls
Regular (control)	250 ± 30	—
5 × 10^{-5}M dB-cAMP	245 ± 45	40
5 × 10^{-4}M dB-cAMP	580 ± 70	133
40 mM k$^+$	560 ± 65	124
40 mM k$^+$ + 5 × 10^{-5}M dB-cAMP	740 ± 75	195
40 mM k$^+$ + 5 × 10^{-4}M dB-cAMP	790 + 80	215
Ca^2 free	505 ± 65	100
Ca^2 free + 5 × 10^{-5}M dB-cAMP	535 + 70	115
Ca^2 free + 5 × 10^{-4}M dB-cAMP	650 ± 25	160

In all incubations 5 × 10^{-5}M theophylline was added.
Results are the mean from at least three experiments ± S.E.M.

striatal adenylate cyclase and that the stimulation of dopamine synthesis in calcium-deprived medium is due to an increase in cyclic-AMP concentration. The results of our study indicate that cyclic-AMP may play an important role in the short term regulation of the transmitter synthesis.

SUMMARY

(1) Three enzymes involved in catecholamine biosynthesis, namely DβH, AADC and PNMT were purified and characterised. The purified enzymes were used as antigens to produce specific antibodies in rabbits. The latter were utilised for immunochemical and immunohistochemical studies.

(2) The catecholamine synthesising enzymes were localised in the adrenal glands and in the peripheral and central nervous system. Using the immunofluorescence technique the dopaminergic, noradrenergic and serotoninergic systems were mapped out in the brain. Preliminary studies were carried out at the ultrastructural level.

(3) Changes in rat serum DβH activities in various physiological and pathological states were investigated. Acute swim stress causes an increase in serum DβH activity with a concomitant decrease in the enzyme activity in the mesenteric arteries. Chronic swim stress leads to an increase in serum DβH activity as well as to an increase in enzyme activity in the adrenals and cervical ganglia. Hypophysectomy causes an increase in serum DβH activity and a decrease in the enzyme activity in the mesenteric arteries.

(4) dB-cAMP induces a stimulation of ^{14}C-dopamine synthesis from ^{14}C-tyrosine in striatal slices. Evidence was obtained that dB-cAMP-induced stimulation occurs prior or at the tyrosine hydroxylase step. The results of our study indicate that cyclic-AMP may play an important role in the short term regulation of the transmitter synthesis.

Acknowledgements—This work was supported by USPHS Grant MH-02717, NSF Grant GB 27603 and by grants from the Swedish Medical Research Council (04X-2887, 14P-3262, 04X-715).

REFERENCES

ANAGNOSTE B., SHIRRON C. and GOLDSTEIN M. (1973) *Fedn. Proc.* **32**, No. 3, 3370 Abstr.
BERNHEIMER H. and HORNYKIEWICZ O. (1965) *Klin. Wscher.* **43**, 711–715.
BLUMBERG W. E., GOLDSTEIN M., LAUBER E. and PEISACH J. (1965) *Biochim. Biophys. Acta.* **99**, 188–190
BUTCHER L. L., ENGEL J. and FUXE K. (1972) *Brain Res.* **41**, 487–411.
CHRISTENSON J. G., DAIRMAN W. and UDENFRIEND S. (1972) *Proc. Nat. Acad. Sci.* **69**, 343–347.
CONNETT R. J. and KIRSHNER N. (1970) *J. Biol. Chem.* **245**, 329–334.
ERÄNKO O. (1952) *Acta Anat. (Basel)*, Suppl. 17.
FREEDMAN L. S., ROFFMAN M., GOLDSTEIN M., FUXE K. and HÖFELT T. (1973) *Europ. J. Pharmacol.* In press,
FRIEDMAN S. and KAUFMAN S. (1965) *J. Biol. Chem.* **240**, 552–554.
FUXE K., GOLDSTEIN M., HÖKFELT T. and JOH T. H. (1971a) *Prog. Brain Res.* **34**, 127–138.
FUXE K., GOLDSTEIN M. HÖKFELT T., FREEDMAN L. and ANAGNOSTE B. (1917b) *Acta Pharmacol. Toxicol.* **29**, Suppl. 4, 15.
FUXE K., GOLDSTEIN M. HÖKFELT T., JOH T. H., FREEDMAN L. S. and ANAGNOSTE B. (1972) *Fedn Proc.* **31**, 544.
GEFFEN L. B., LIVETT D. G. and RUSH R. A. (1972) *J. Physiol. Lond.* **204**, 593–605.
GIBBS J. W., SPECTOR S. and UDENFRIEND S. (1967) *Molec. Pharmacol.* **3**, 473–478.
GOLDSTEIN M., LAUBER E. and MCKEREGHAN M. R. (1965) *J. Biol. Chem.* **240**, 2066–2072.
GOLDSTIEN M., JOH T. H. and GARVEY T. Q. III. (1968) *Biochem.* **7**, 2724–2730.
GOLDSTEIN M., FREEDMAN L. S. and BACKSTROM T. (1970a) *J. Pharm. Pharmacol.* **22**, 715–717.
GOLDSTEIN M., OHI Y. and BACKSTROM T. (1970b) *J. Pharmacol. Exp. Therap.* **174**, 77–82.
GOLDSTEIN M., BACKSTROM T., OHI Y. and FRENKEL R. (1970c) *Life Sci.* **9**, 919–924.
GOLDSTEIN M., FUXE K. and HOKFELT T. (1972) *Pharmacol. Rev.* **24**, 293–309
GOLDSTEIN M., FUXE K., HÖKFELT T. and JOH T. H. (1971a) *Experientia (Basel)* **27**, 951–952.
GOLDSTEIN M., FREEDMAN L. S. and BONNAY M. (1971b) *Experientia (Basel)* **27**, 632–633.
GOLDSTEIN M. (1972) In: *Research Methods in Neurochemistry.* (MARKS N. and RODNIGHT, R., Eds.) Vol. 1, pp. 317–340. Plenum Press, New York.
GOLDSTEIN M., ANAGNOSTE B., FREEDMAN L. S., ROFFMAN M. and LELE K. P. (1973a) Wenner-Gren Meeting, Stockholm, To be published.
GOLDSTEIN M., ANAGNOSTE B. and SHIRRON C. (1973b) *J. Pharm. Pharmacol.* In press.
HARRIS J. E. and ROTH R. H. (1971) *Mol. Pharmacol.* **7**, 593—604.
HARTMAN B. K. and UDENFRIEND S. (1970) *Molec. Pharmacol.* **6**, 85–94.
HARTMAN B. K., ZIDE D. and UDENFRIEND S. (1972) *Proc. Nat. Acad. Sci.* **69**, 2722–2726.
HILLARP N-A. and HÖKFELT T. (1953) *Acta Physiol. Scand.* **30**, 55–68.
HÖKFELT T., FUXE K., GOLDSTEIN M. and JOH T. H. (1973a) *Histochemi.* **33**, 231–254.
HÖKFELT T., FUXE K. and GOLDSTEIN M. (1973b) *Brain Res.* **53**, 175–180.
HÖKFELT T., FUXE K. and GOLDSTEIN M. (1973c) *Brain Res.* In press.
JOH T. H. and GOLDSTEIN M. (1973) *Mol. Pharmacol.* **9**, 117–129.
KEBABIAN J. W., PETZOLD G. L. and GREENGARD P. (1972) *Proc. Nat. Acad. Sci.* **69**, No. 8, 2145–2149.
KEHR W., CARLSSON A., LINDQVIST M., MAGNUSSON T. and ATACK C. (1972) *J. Pharm. Pharmacol.* **24**, 744–747.
LEVIN E. Y. and KAUFMAN S. (1961) *J. Biol. Chem.* **236**, 2043–2049.
MCAFEE D. A. and GREENGARD P. (1972) *Science* **178**, 310–312.
MOLINOFF P. B., WEINSHILBOUM R. and AXELROD J. (1971) *J. Pharmacol. Exp. Ther.* **178**, 425–431.
NAKASHIMA Y., SUZUE R., SANADA H. and KOWADA S. (1972) *Arch. Biochem. Biophys.* **152**, 515–520.
ROFFMAN M., FREEDMAN L. S. and GOLDSTEIN M. (1973) *Life Sci.* **12**, 369–372.
RUBIN R. P., JOANUS S. D. and CARCHMAN R. A. (1972) *Nature (New Biol.)* **240**, 150–152.
SHIMIZU U. and DALY J. W. (1972) *Europ. J. Pharmacol.* **17**, 240–247.
SIMS K. and BLOOM F. (1971) *Trans. Am. Soc. Neurochem.* **2**, 109 Abstr.
ZIMMERMAN E. A., HSU K. C., COTE L., TANNENBAUM M., FREEDMAN L. S., ROFFMAN M. and GOLDSTEIN M. (1973) *Fedn. Proc.* **32**, No. 3, 457 Abstr.

Frontiers in Catecholamine Research 1973, pp. 79 to 81. Pergamon Press. Printed in Great Britain.

IMMUNOLOGICAL LOCALISATION OF DOPAMINE-β-HYDROXYLASE ON THE CHROMAFFIN GRANULE MEMBRANE

JAMES A. THOMAS[1], L. S. VAN ORDEN III[2], J. A. REDICK[2] and IRWIN J. KOPIN[1]

[1]Laboratory of Clinical Science National Institute of Mental Health Bethesda, Maryland 20014, U.S.A. and [2]Department of Pharmacology, The Toxicology Center, University of Iowa, Iowa City, Iowa 52240, U.S.A.

CATECHOLAMINE release is thought to occur by fusion of the vesicular membrane of a secretory granule with the plasma membrane of adrenal medullary chromaffin cells. Support for exocytosis has been derived from experiments in several laboratories. Less clear is the order of the process in which catecholamines are formed and packaged for release. The enzyme dopamine-β-hydroxylase (EC 1.14.2.1) (DBH) has been shown to be responsible for the conversion of dopamine to norepinephrine within the chromaffin cells of the adrenal medulla (KIRSHNER, 1959).

DBH is located on the vesicular membrane as well as in the contents of the vesicle. Using antibodies to DBH, this study was undertaken to locate DBH on the inner or outer surface of the vesicular membrane. Purified DBH from bovine adrenal medullae, prepared according to GOLDSTEIN et al. (1965), was injected into goats from which antibodies were subsequently harvested. For the study of antibody inhibition of enzyme activity, isolated chromaffin vesicles were prepared according to the method of KIRSHNER (1962). Purified chromaffin granules were prepared for electron microscopy by subcellular fractionation and sucrose density gradient centrifugation according to the method of WINKLER et al. (1972).

The effect of antibodies to bovine DBH on the activity of the enzyme in lysed and intact bovine medullary granules was examined as follows: Granules were lysed by homogenization in 0·005 M Tris HCl containing 0·1% Triton X-100, and aliquots of the homogenate were incubated with either normal goat serum (NGS) or anti-DBH serum (a-DBH) at 37°C. The solutions were then assayed for DBH activity by a modification of the method of FRIEDMAN and KAUFMAN (1965). Intact granules in isotonic sucrose medium were incubated with either NGS or a-DBH and then assayed for DBH activity. Comparable preparations of intact vesicles were incubated with either NGS or a-DBH, spun at 100,000 × g and the supernatant assayed for DBH activity.

Vesicles obtained from density gradients were fixed for electron microscopy by a modified Karnovsky method. After fixation, the vesicles were treated with a-DBH, linked to a peroxidase–anti-peroxidase complex by an anti-goat IgG, and peroxidase was then localised by osmium tetroxide precipitation using diaminobenzidine according to the method of REDICK, VAN ORDEN, THOMAS and KOPIN (manuscript in preparation).

Incubation of lysed granules with a-DBH resulted in about 80 per cent inhibition of the enzyme (Table 1). Intact granules had about 50 per cent of the DBH activity of lysed granules. Incubation of the intact granules with a-DBH reduced the activity to 22 per cent of the total DBH activity. This represents a 56 per cent inhibition of the enzyme activity of the intact granules. Since only 6 per cent of the DBH activity

TABLE 1

Incubation condition	DBH activity (counts/min)	% total DBH activity
Lysed granules + pre-immune serum	9241 ± 153 (4)*	100
Lysed granules + anti-DBH	1782 ± 47 (4)	19
Intact granules + pre-immune serum	4602 ± 118 (4)	50
Intact granules + anti-DBH	2000 ± 165 (4)	22
	100,000 × *g*	
Intact granules + pre-immune serum	545 ± 200 (4)	6
Intact granules + anti-serum	70 ± 2 (4)	1

* Number of determinations is indicated in parentheses.
DBH = Dopamine-β-hydroxylase.

in the intact vesicle preparation remained in the supernatant, the inhibition of DBH activity by the antibody must be the result of inhibition of vesicular DBH. It appears unlikely that the antibody penetrated the vesicle; it is thus probable that the DBH activity inhibited by the antibodies must lie at the surface of the vesicle.

Electron micrographs prepared according to the described method show a discrete black precipitate attached to and surrounding the unit membrane of the chromaffin vesicle in the preparation incubated with a-DBH. This is in contrast to preparations incubated with NGS in which either no precipitate is detected or a nonspecific amorphous adsorbed material is visible. The results of the present study indicate that a portion of the antigenic sites of DBH are located on the exterior of the vesicular membrane.

The DBH activity of intact vesicles is only about half that of the total activity. Of the 50 per cent DBH activity of intact vesicles, 28 per cent is blocked by exposure to a-DBH. Since a-DBH is only about 80 per cent effective in blocking DBH, it appears that about 35 per cent of the total DBH activity in vesicles is available for reaction with a-DBH. Presumably, this 35 per cent lies on the surface of the vesicular membrane.

Hortnagl et al. (1972) have shown that membrane-bound DBH is the major protein of the chromaffin granule membrane, indicating that the structural protein also has catalytic activity. DBH is a large molecule of molecular weight 300,000; it is probable that, as an integral structural component, both antigenic and enzymatic sites of the molecule extrude from both surfaces of the membrane. Both the inhibition by a-DBH of activity of the enzyme in intact vesicles and the electron microscopic demonstration of DBH on the membranes of intact vesicles support this view.

It has been suggested that dopamine is taken up into the chromaffin granule, hydroxylated to form norepinephrine, released into the cytoplasm where it is converted to epinephrine, which then reenters the storage granule and awaits release (Axelrod and Weinshilboum, 1972). This hypothetical process appears to be both inefficient and uneconomical. The DBH activity on the exterior of the vesicles suggests an alternate hypothesis for the synthesis of catecholamines within the chromaffin cell.

Dopamine could be hydroxylated on the exterior of the vesicle, thus forming nor-epinephrine, and then methylated by cytoplasmic phenylethanolamine-N-methyl-transferase to form epinephrine. The N-methylated catecholamine could then be taken up into vesicles for storage and/or release.

REFERENCES

AXELFORD J. and WEINSHILBOUM R. (1972) *New Eng. J. Med.* **287**, 237–242.
FRIEDMAN S. and KAUFMAN S. (1965) *J. Biol. Chem.* **240**, 4763–4773.
GOLDSTEIN M., LAUBER G. and MCKEREGHAN M. R. (1965) *J. Biol. Chem.* **240**, 2066–2072.
HORTNAGL H., WINKLER H. and LOCUS H. (1972) *Biochem. J.* **129**, 187–195.
KIRSHNER N. (1959) *Pharmacol. Rev.* **11**, 350–357.
KIRSHNER N. (1962) *J. Biol. Chem.* **237**, 2311–2317.
WINKLER H., SCHOPF J. A. L., HORTNAGL H. and HORTNAGL H. (1972) *Naunyn-Schmiedebergs Arch. Pharmakol.* **273**, 43–61.

Frontiers in Catecholamine Research 1973, pp. 83 to 86. Pergamon Press. Printed in Great Britain.

NEW INHIBITORS OF MICROBIAL ORIGIN FOR DOPAMINE-β-HYDROXYLASE

Toshiharu Nagatsu[1], Takeshi Kato[1], Hiroshi Kuzuya[1],
Hamao Umezawa[2] and Tomio Takeuchi[2]

[1]Department of Biochemistry, School of Dentistry, Aichi-Gakuin University, Chikusa-ku, Nagoya, Japan, and [2]Institute of Microbial Chemistry, Shinagawa-ku, Tokyo, Japan

SINCE 1965, Umezawa *et al.* initiated the screening of enzyme inhibitors found in culture filtrate of microorganisms (UMEZAWA, 1972). In the course of this project, several new inhibitors for enzymes of catecholamine biosynthesis and metabolism have been discovered, as shown in Table 1.

TABLE 1. INHIBITORS OF MICROBIAL ORIGIN FOR ENZYMES OF CATECHOL-AMINE BIOSYNTHESIS AND METABOLISM (UMEZAWA, 1972)

Inhibitor	Micro-organism	Enzyme	Hypotensive effect
OUDENONE	Oudemansiella	TH	+
FUSARIC ACID	Fusarium	DBH	+
DOPASTIN	Pseudomonas	DBH	+
AQUAYAMYCIN	Streptomyces	TH DBH	—
METHYLSPINAZARIN	Streptomyces	COMT (DBH)	+

TH: tyrosine hydroxylase; DBH: dopamine-β-hydroxylase; and COMT: catechol-*O*-methyltransferase.

The chemical structures of these natural enzyme inhibitors and the mechanism of inhibition may give a clue to understand the similarities and the phylogenetic relationship between these enzymes. Also, the inhibition *in vivo* in animals including man by these enzyme inhibitors may cause pharmacological effects such as hypotension which can be of pharmacological and clinical values.

Among these enzyme inhibitors, the inhibitors for tyrosine hydroxylase [L-tyrosine, tetrahydropteridine: oxygen oxidoreductase (3-hydroxylating)(EC 1.14.3.a.)], such as oudenone (Umezawa et al., 1970; Nagatsu, T. et al., 1971; Ohno et al., 1971), and the inhibitors for dopamine-β-hydroxylase (3, 4-dihydroxyphenylethylamine: oxygen oxidoreductase (hydroxylating) (EC 1.14.2.1.)), such as fusaric acid (Hidaka et al., 1969; Nagatsu, T. et al., 1970), were found to be potent hypotensive agents in various animals. These inhibitors showed a highly pronounced hypotensive action (Nagatsu, I. et al., 1971; Nagatsu, T. et al., 1972; Nagatsu, T. et al., 1973) on spontaneously hypertensive (SH) rats (Okamoto and Aoki, 1963).

Recently, another new inhibitor of dopamine-β-hydroxylase, dopastin, has been discovered by Iinuma et al. (1972), and many new derivatives of fusaric acid with higher inhibitor activities (50% inhibition at 10^{-9} M) have been synthesized by Umezawa et al. (1973). These new inhibitors also have hypotensive action.

Serum or plasma has dopamine-β-hydroxylase which is released from the sympathetic nerves (Weinshilboum and Axelrod, 1971; Goldstein et al., 1971; Nagatsu, T. and Udenfriend, 1972). We have found that the degree of inhibition of dopamine-β-hydroxylase in vivo by these new inhibitors in rats and humans can be examined by measuring the serum dopamine-β-hydroxylase activities after the administration of the inhibitors.

After the oral administration of 50–100 mg of fusaric acid to humans, serum dopamine-β-hydroxylase activity decreased rapidly at 3 hr(to 5–40%), then recovered nearly linearly with time, and almost completely recovered at 24 hr (Nagatsu, T. et al., 1972).

After the oral administration of 3·5 mg/kg of fusaric acid to rats, marked inhibition (50%) of the serum enzyme activity was observed at 1·5 hr, and then the activity gradually recovered. Twenty-four hr after the administration, the activity recovered to about 90% of the initial level. Norepinephrine levels in the heart also decreased up to 6 hr and then recovered gradually up to 24 hr nearly to the original level.

The inhibition of serum dopamine-β-hydroxylase activity after the oral administration (3·5 mg/kg) of the enzyme inhibitors including fusaric acid (5-butylpicolinic acid), 5-(4′-chlorobutyl)picolinic acid, 5-isopentylpicolinic acid, and dopastin, has been compared. As shown in Fig. 1, the degree of inhibition of the serum enzyme in vivo was nearly comparable to the degree of inhibition for the pure enzyme in vitro. Thus, dopastin, which was less active in vitro (the doses for 50% inhibition, ID 50; about 10^{-6} M) inhibited the serum enzyme in vivo by about 40%; fusaric acid which had a higher in vitro inhibition effect (ID 50; about 10^{-8} M), by about 50%; and the fusaric acid derivatives which were the most potent in vitro inhibitors (ID 50; about 10^{-9} M), by about 70%. The degrees of in vivo inhibition of the serum enzyme were also comparable to the hypotensive effect of these inhibitors; the fusaric acid derivatives were the most effective hypotensive agents.

Hidaka and Takeya (1972) also reported that 5-(3′,4′-dihalobutyl)picolinic acid and 5-(3′-halobutyl)picolinic acid, which are potent inhibitors of dopamine-β-hydroxylase, reduced the tissue catecholamine content of SH rat and rabbit and also reduced their blood pressure.

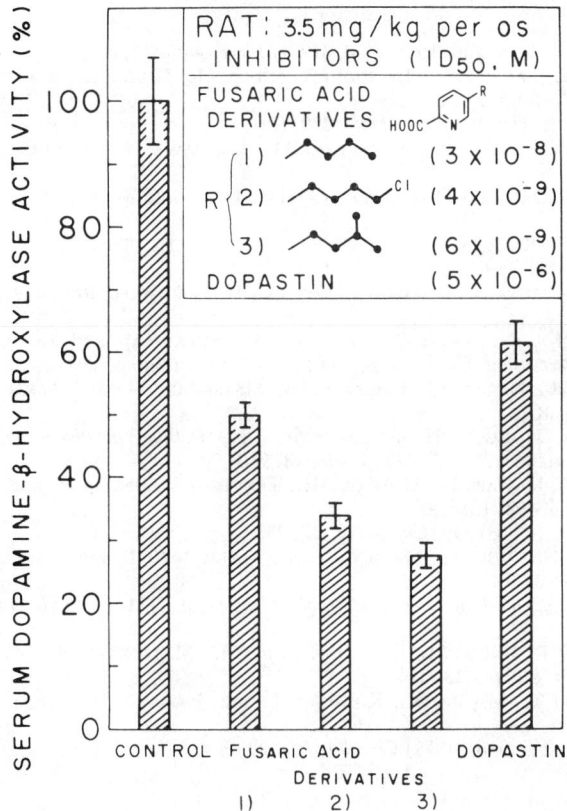

FIG. 1.—Inhibition of rat serum dopamine-β-hydroxylase activity after the oral admin-istration of new inhibitors of the enzyme. The values are the mean ± s.d. of 5 rats. Rat serum enzyme activity was measured by a modification of the method by NAGATSU, T. and UDENFRIEND (1972). The control (100%) activities were 0·74 ± 0·07 nmoles/min/ml serum (37°C).

Suda *et al.* (1969) reported that by increasing carbon numbers in 5-alkyl group of 5-alkyl-picolinic acid up to 4 or 5, the inhibitory effect for the enzyme increased in parallel with the hypotensive action, and that 5-butyl- or 5-pentyl-picloinic acid is the most effective inhibitor and hypotensive agent. As reported here, 5-(4'-chlorobutyl) picolinic acid and 5-isopentylpicolinic acid are more potent inhibitors and hypotensive agents than 5-butyl- or 5-pentyl-picolinic acid.

Oudenone is a specific inhibitor for tyrosine hydroxylase, but by modifying the structure of oudenone, the derivatives acquired *in vitro* inhibitory effects not only on tyrosine hydroxylase but also on dopamine-β-hydroxylase, and these oudenone derivatives showed more pronounced hypotensive effects than oudenone itself.

These results suggest that dopamine-β-hydroxylase among the catecholamine synthetic enzymes is highly susceptible to specific and potent enzyme inhibitors to produce *in vivo* inhibition and hypotensive effect. The measurement of the serum dopamine-β-hydroxylase activity would be a valuable screening test to find a potent enzyme inhibitor which is expected to have hypotensive action.

REFERENCES

GOLDSTEIN M., FREEDMAN L. S. and BONNAY M. (1971) *Experientia* **27**, 632.

HIDAKA H., NAGATSU T., TAKEYA K., TAKEUCHI T., SUDA H., KOJIRI K., MATSUZAKI M. and UMEZAWA H. (1969) *J. Antibiotics* **22**, 228.

HIDAKA H. and TAKEYA K. (1972) *Nature, Lond.* **239**, 334.

IINUMA H., TAKEUCHI T., KONDO S., MATSUZAKI M., UMEZAWA H. and OHNO M. (1972) *J. Antibiotics* **25**, 497.

NAGATSU I., NAGATSU T., MIZUTANI K., UMEZAWA H., MATSUZAKI M. and TAKEUCHI I. (1971) *Nature, Lond.* **230**, 381.

NAGATSU T., HIDAKA H., KUZUYA H., TAKEYA K., UMEZAWA H., TAKEUCHI T. and SUDA H. (1970) *Biochem. Pharmacol.* **19**, 35.

NAGATSU T., KATO T., KUZUYA H., OKADA T., UMEZAWA H. and TAKEUCHI T. (1972) *Experientia* **28**, 779.

NAGATSU T., MIZUTANI K., NAGATSU I., UMEZAWA H., MATSUZAKI M. and TAKEUCHI T. (1972) In: *Spontaneous Hypertension* (OKAMOTO K., *ed.*) 31–36. Igaku Shoin, Tokyo.

NAGATSU T., MIZUTANI K., NAGATSU I., UMEZAWA H., MATSUZAKI M. and TAKEUCHI T. (1973) *Mol. Cell. Biochem.* **1**, 107.

NAGATSU T., NAGATSU I., UMEZAWA H. and TAKEUCHI T. (1971) *Biochem. Pharmacol.* **20**, 2505.

NAGATSU T. and UDENFRIEND S. (1972) *Clin. Chem.* **18**, 980.

OHNO M., OKAMOTO M., KAWABE N., UMEZAWA H., TAKEUCHI T., IINUMA H. and TAKAHASHI S. (1971) *J. Am. Chem. Soc.* **93**, 1285.

OKAMOTO K. and AOKI K. (1963) *Jap. Circulat. J.* **27**, 282.

SUDA H., TAKEUCHI T., NAGATSU T., MATSUZAKI M., MATSUMOTO I. and UMEZAWA H. (1969) *Chem. Pharm. Bull.* **17**, 2377.

UMEZAWA H (1972) in *Enzyme Inhibitors of Microbial Origin.* pp. 1–114. Tokyo University Press, Tokyo.

UMEZAWA H., TAKEUCHI T., IINUMA H., SUZUKI K., ITO M., MATSUZAKI M., NAGATSU T. and TANABE, O. (1970) *J. Antibiotics* **23**, 514.

UMEZAWA H., TAKEUCHI T., MIYANO K., KOSHIGOE T. and HAMANO H. (1973) *J. Antibiotics* In press.

WEINSHILBOUM R. and AXELROD J. (1971) *Circulat. Res.* **28**, 307.

Frontiers in Catecholamine Research 1973, pp. 87 to 90. Pergamon Press. Printed in Great Britain.

PICOLINIC ACID DERIVATIVES AS INHIBITORS OF DOPAMINE β-HYDROXYLASE *IN VIVO*; THEIR EFFECTS ON BLOOD PRESSURE AND STRESS ULCER

HIROYOSHI HIDAKA

Department of Biochemistry, Institute for Developmental Research, Aichi Prefecture Colony,
Kasugai-city, Aichi, Japan

HIDAKA *et al.* (1969) first demonstrated fusaric (5-butylpicolinic) acid's ability to inhibit a purified bovine adrenal dopamine β-hydroxylase [3,4-dihydroxyphenyl-ethylamine, ascorbate: Oxygen oxidoreductase (hydroxylating), EC 1.14.2.1.] (DBH). Among derivatives of picolinic acids tested, 5-(3'4'-dihalobutyl)-picolinic acid and 5-(3'-halobutyl)picolinic acid are the most potent inhibitors of DBH which produced 50 per cent inhibition at the concentration of 1×10^{-8} M and 5-butyl picolinic acid (FA) and 5-(dimethyldithiocarbamyl) picolinic acid (YP-279) are the next potent inhibitors which produced 50 per cent inhibition at 8×10^{-8} M (HIDAKA *et al.*, 1973a).

A significant reduction of blood pressure in normotensive and hypertensive man was observed after a single administration (oral) of 5-butylpicolinic acid calcium salt at a dose of 6 mg/kg (HIDAKA *et al.*, 1973b; TERASAWA and KAMEYAMA (1971). The decrease in blood pressure was also demonstrable in spontaneously hypertensive rat (SHR) and normotensive rabbit after the administration of 50 mg/kg of 5-(3',4'-dibromobutyl) picolinic acid (Br$_2$FA) and FA. It has also been found (OSUMI *et al.*, 1973) that FA (100 mg/kg) prevents the gastric stress ulcer of restrained rat induced by soaking in water. Many mechanisms exist through which a substance may exert an antihypertensive and anti-stress ulcer effect. The role of the central nervous system in the depressor and the anti-ulcer response to picolinic acid derivatives as DBH inhibitors was evaluated in the present study. Data in Table 1 show that both FA, Br$_2$FA and YP-279 are effective antihypertensive agents at a dose of 50 mg/kg to SHR. Among these drugs, the most consistent and reproducible reduction of blood pressure was observed in the administration of YP-279 (50 mg/kg). When restrained rats were soaked in water ($23° \pm 1°$C) for 4 hr, all of the treated rats had the gastric ulcer (Fig. 1). Pretreatment of the rat with Br$_2$FA and FA 50 mg/kg prevented completely the gastric stress ulcer (erosion) induced by soaking the rat in water (Fig. 1). On the contrary, an administration of YP-279 failed to prevent the formation of this ulcer (Fig 1). The discrepancy between the effect of YP-279 on blood pressure and on the gastric stress ulcer was studied by determining both tissue distribution of YP-279 and *in vivo* inhibition of DBH in brain or heart by YP-279. Maximum concentration of Br$_2$FA, FA or YP-279 in rat plasma was detected 30 min after the drugs 75 mg/kg. Br$_2$FA, FA and YP-279 disappeared from the rat plasma by 6 hr. Although maximum concentration of Br$_2$FA or FA in rat brain was detected by about 1 hr, YP-279 could not be detected in rat brain at all at any time (0–6 hr).

TABLE 1. Effect of picolinic acid derivatives on blood pressure in SHR. Picolinic acid
derivatives (50 mg/kg) were given orally and systolic blood pressure in the tail was measured.

Time after drug injection (hr)	Blood pressure (% decrease)			
	Control	FA	Br$_2$FA	YP-279
1	186 ± 3·8 (100%)			151 ± 6·2 (−19%)
3	197 ± 3·1 (100%)	118 ± 10 (−40%)	133 ± 7·4 (−32%)	142 ± 8·1 (−28%)
6	193 ± 5·4 (100%)	132 ± 9·6 (−32%)	127 ± 6·9 (−34%)	135 ± 4·1 (−30%)
24	180 ± 6·0 (100%)			149 ± 8·2 (−17%)

YP-279 is thought be unable to pass blood brain barrier. This hypothesis was supported by the fact that the administration of YP-279 to the rat intraperitoneally
75 mg/kg did not change brain norepinephrine (NE) levels (Table 2). Maximum
depletion of brain NE by Br$_2$FA and FA was detected 3 and 2 hr respectively after
each drug 75 mg/kg. Brain NE returned to control level by 6 hr (FA) and by 9 hr
(Br$_2$FA) (Table 2). Maximum depletion of endogenous adrenal epinephrine (E), to
30 per cent (YP-279) and 50 per cent (Br$_2$FA, FA) of control levels, was detected 3 hrs
after these compounds 75 mg/kg. E levels returned to control level by 6 hrs after
YP-279, but was still lowered to approximately 40 per cent of control level 9 hr after
Br$_2$FA and FA. The fall in rat adrenal E by these three compounds was accompanied
by the increase in adrenal DA.

The effects of Br$_2$FA, FA and YP-279 on DBH in rat brain and heart were studied
in vivo by measuring the conversion of ^{3}H-DA into ^{3}H-NE in the brain or heart
after intracisternal or intravenous administration of ^{3}H-DA. Data in Fig. 2 show
the facts that the biosynthesis of ^{3}H-NE from ^{3}H-DA in rat brain was effectively
inhibited by Br$_2$FA (80 per cent) and by FA (66 per cent), but was not inhibited
significantly by YP-279. In contrast to the results obtained in the brain, Br$_2$FA,
FA and YP-279 inhibited effectively the conversion of ^{3}H-DA to ^{3}H-NE in the heart
by 81, 76 and 57 per cent respectively. These compounds did not have any effect on
the uptake of NE or DA (Data are not shown in this article).

Derivatives of picolinic acid tested (Br$_2$FA, FA and YP-279) which are the most
potent group of DBH inhibitors in vitro were found to be also effective inhibitors
of DBH in vivo. In contrast to Br$_2$FA and FA, YP-279 failed to inhibit brain DBH
because of its property unable to pass through blood brain barrier. Decrease in
blood pressure by these picolinic acid derivatives appeared to be related to their
inhibition of DBH in the peripheral tissues, whereas the anti-ulcer effect of picolinic
acid derivatives was due to the inhibition of brain DBH. Because YP-279 which

Fig. 1.—Effect of picolinic acid derivatives on the gastric stress ulcer of the rat. Picolinic
acid derivatives (Br$_2$FA, FA and YP-279) 50 mg/kg respectively suspended in 0·5%
carboxymethylcellulose (CMC) were administered i.p. just before soaking restrained
animals in the water (23 ± 1°C). Control rats were injected with 0·5% CMC. Rats
were sacrificed after soaking for 4 hr.

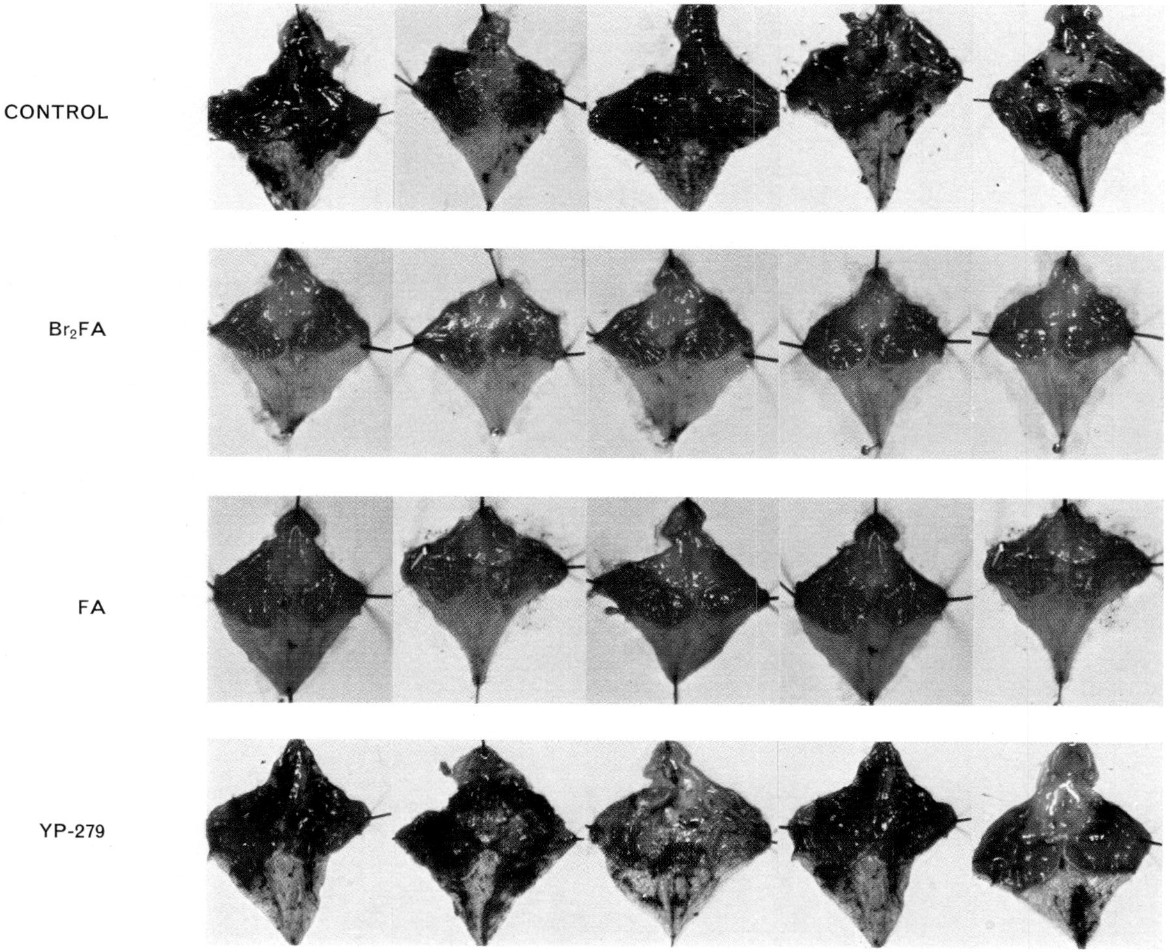

CONTROL

Br₂FA

FA

YP-279

Fig. 1.

TABLE 2. EFFECT OF PICOLINIC ACID DERIVATIVES ON BRAIN NE OF RATS. COMPOUNDS 75 mg/kg WERE INJECTED i.p. RATS (10 PER GROUP) WERE SACRIFICED BY DECAPITATION AT 1, 2, 3, 6 AND 9 hr AFTER EACH DRUG. CONTROL RATS RECEIVED 0·5% CMC WERE SACRIFICED

Time after drug (hr)	Norepinephrine (μg/g brain)			
Drug	Control	Br$_2$FA	FA	YP-279
1	0.37 ± 0.01	0.26 ± 0.01*	0.23 ± 0.01*	0.33 ± 0.01
2	0.35 ± 0.01	0.25 ± 0.02*	0.18 ± 0.02*	0.37 ± 0.02
3	0.35 ± 0.02	0.17 ± 0.01*	0.20 ± 0.01*	0.34 ± 0.01
6	0.34 ± 0.01	0.25 ± 0.01*	0.34 ± 0.01	0.36 ± 0.02
9	0.37 ± 0.01	0.32 ± 0.01	0.34 ± 0.01	0.37 ± 0.01

* Values are significantly different from the control ($p < 0.01$).

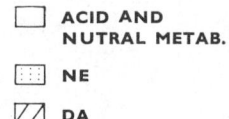

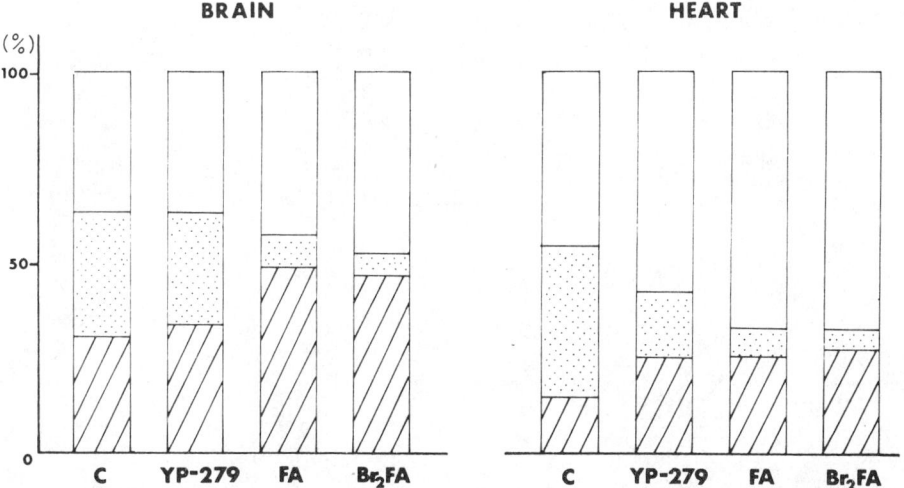

FIG. 2.—Effect of picolinic acid derivatives on the conversion of ³H–DA into ³H–NE in rat brain and heart. Br$_2$FA, FA and YP-279 were administered i.p. 60 min before injection of ³H–DA. Control rats were injected with 0·5% CMC. Animals (five per group) were sacrificed 20 min (brain) and 10 min (heart) after injection of ³H–DA (10 or 5 μCi). Experiments in rat brain and heart were made separately. Data were expressed as a percentage of total count (dis/min) in each tissue. 100 per cent corresponds to $144,447 \pm 10,021$ dis/min per g brain and $141,970 \pm 5,160$ dis/min per g heart.

did not pass through blood brain barrier was not an anti-ulcer agent but anti-hypertensive agent. In conclusion, DBH inhibitor reduced blood pressure and also prevented the gastric stress ulcer of the rat implying that the DBH in the central nervous system contributes importantly to the anti-stress ulcer action of the inhibitor and that the enzyme in the peripheral tissue contributes to the antihypertensive action of the inhibitor.

90 H. HIDAKA

REFERENCES

HIDAKA H., ASANO T. and TAKEMOTO N. (1973a) *Molecular Pharm.* **9,** 172–177.
HIDAKA H., NAGASAKA A. and TAKEDA A. (1973b) *J. Clin. Endocr. Metab.* in press.
HIDAKA H., NAGATSU T., TAKEYA K., TAKEUCHI T., SUDA H., KOJIRI K., MATSUZAKI M. and
  UMEZAWA H. (1969) *J. Antibiotics.* **22,** 228–230.
HIDAKA H. and TAKEYA K. (1972) *Nature, Lond.* **239,** 334–335.
OSUMI Y., TAKAORI S. and FUJIWARA M. (1973) *Life Sci.* in press.
TERASAWA F. and KAMEYAMA M. (1971) *Jap. Circul. J.* **35,** 339–357.

Frontiers in Catecholamine Research 1973, pp. 91 to 96. Pergamon Press. Printed in Great Britain.

THE INNERVATION OF CEREBRAL BLOOD VESSELS BY CENTRAL NORADRENERGIC NEURONS*

BOYD K. HARTMAN

Department of Psychiatry, Washington University School of Medicine, St. Louis, Missouri 63110, U.S.A.

DOPAMINE-β-hydroxylase (EC 14.2.1; DBH) is the enzyme that converts dopamine to norepinephrine (NE). This enzyme has been shown to be a specific and sensitive immunochemical marker for noradrenergic neurons both in the central nervous system (CNS) and in the peripheral sympathetic system (GEFFEN et al., 1969; HARTMAN and UDENFRIEND, 1970; FUXE et al., 1971; HARTMAN et al., 1972; HARTMAN, 1973). Improvements in methodology for the immunofluorescent localisation of this enzyme have permitted the resolution of the entire central noradrenergic nervous system, including neuron cell bodies, non-terminal axon fibres and terminal varicosities (HARTMAN et al., 1972; HARTMAN, 1973). In rat there is excellent agreement between the results obtained by DBH and NE localisation with regard to the location of NE-producing cell bodies, the course of major fibre pathways, and the major terminals areas (FUXE et al., 1972; HARTMAN et al., 1972; HARTMAN, 1973; HARTMAN and UDENFRIEND, 1972). The specificity of DBH for the noradrenergic system is shown by the fact that it is not seen in cell bodies in the substantia nigra or the midbrain raphe system, which have been shown to be dopaminergic and serotonergic respectively (FUXE et al., 1972; HARTMAN et al., 1972; HARTMAN, 1973).

An early observation made with the DBH immunofluorescence technique, that had not been described by investigators using NE histofluoresence, was the close association of DBH containing fibres and terminals with small blood vessels deep within the brain parenchyma (HARTMAN et al., 1972; HARTMAN, 1973; HARTMAN and UDENFRIEND, 1972). One possibility was that this innervation of small vessels originates in the sympathetic nerve plexus on pial arteries, and follows the arterial tree into the brain. Several anatomical characteristics of the small vessel innervation, however, made it appear unlikely that the peripheral sympathetic fibres could be the source of the terminals on small vessels. First, small vessels were observed to be associated with greater numbers of DBH-containing fibres and terminals than large vessels. Second, terminal fibres on some small vessels seemed to originate from pre-terminal fibres not associated with blood vessels, and presumably of central origin. Finally, areas known to receive large contributions of centrally-derived noradrenergic innervation, such as the hypothalamus, showed the greatest number of fibres associated with blood vessels; whereas, the caudate-putamen which is innervated by the central dopaminergic system showed no DBH-containing fibres associated with its blood supply. It was therefore proposed that the small artery and arteriolar innervation observed with DBH immunofluorescence was derived from noradrenergic neurons located within the brain itself. (HARTMAN AND UDENFRIEND, 1972; HARTMAN et al., 1972).

---

* This work was supported by the following grants from the United States Public Health Service: MH-21874, DA-00259, AA-00209 and a Research Scientist Development Award MH-70451 from the National Institutes of Mental Health.

This paper presents additional and more direct evidence to show that the DBH-containing fibres and terminals associated with small vessels in the brain are not derived from the peripheral sympathetic system and therefore must be of central origin. The observation of small cerebral vessel innervation by noradrenergic fibres is also extended to monkey and human brain.

The noradrenergic neuron cell bodies within the superior-cervical sympathetic ganglia are the source of the sympathetic nerve supply of pial blood vessels (NIELSEN and OWMAN, 1967; EDVINSSON et al., 1972). Thus surgical removal of the superior cervical ganglia represents a direct and unequivocal means of determining the source of the DBH-containing terminals associated with small blood vessels within the brain. If after ganglionectomy they disappear along with the sympathetic fibres on pial vessels their peripheral origin would be established, likewise if they remain their source must be the central noradrenergic system.

Bilateral superior cervical ganglionectomised rats and sham operated controls were examined using the immunofluoresence technique for DBH localisation as described in detail previously (HARTMAN et al., 1972; HARTMAN, 1973). The brains were examined 7 days, 14 days and 2 months after surgery. Sham operated controls showed the typical dense noradrenergic innervation of pial vessels (Fig. 1A). Bilateral ganglion-ectomy resulted, as expected, in the disappearance of DBH-containing fibres on pial blood vessels (Figs. 1B, C). The absence of the sympathetic innervation throughout the pial system was used as the criterion of complete removal of both superior cervical ganglia. By 7 days all traces of the pial vessel innervation had disappeared and in no case was there any apparent regeneration of sympathetic fibres even after 2 months.

Within the brain parenchyma, superior cervical ganglionectomy resulted in no detectable decrease in the presence of noradrenergic fibres or terminals associated with small arteries and arterioles. Figures 2A–E show small vessels as observed in the brains of superior cervical ganglionectomised rats.

Varicose DBH-containing fibres are seen in close association with these small vessels. The persistence of these terminals after the disappearance of the sympathetic innervation of the pial vessels establishes that they originate from DBH-containing neurons within the brain itself.

In the hypothalamus, vessels in certain areas invariably are associated with relatively large numbers of varicose terminals. The small vessels that supply the paraventricular nucleus have been shown to be densely innervated by DBH-containing fibres (HARTMAN et al., 1972). Many vessels throughout the hypothalamus show similarly dense innervation. In general, however, the regions supplied by the vessels are not as well defined. Figure 2A shows a small vessel from the lateral hypothalamus. Figures 2B, C show a vessel in the anterior ventral nucleus of the thalamus. In this nucleus most small vessels are associated with many DBH-containing varicosities.

Since there are no DBH-containing cell bodies rostral to the pons, it must be concluded that the vascular innervation in the rostral brain areas arrives via the ascending noradrenergic fibre pathways from the locus coeruleus and other clusters of noradrenergic cell bodies in the lower brain stem. These fibre pathways also give rise to neuronal contacts. Whether the same preterminal noradrenergic fibre may give rise to terminals both on neuronal elements and on vessels is not known.

In the medulla and pons the pattern of vascular innervation is also well defined.

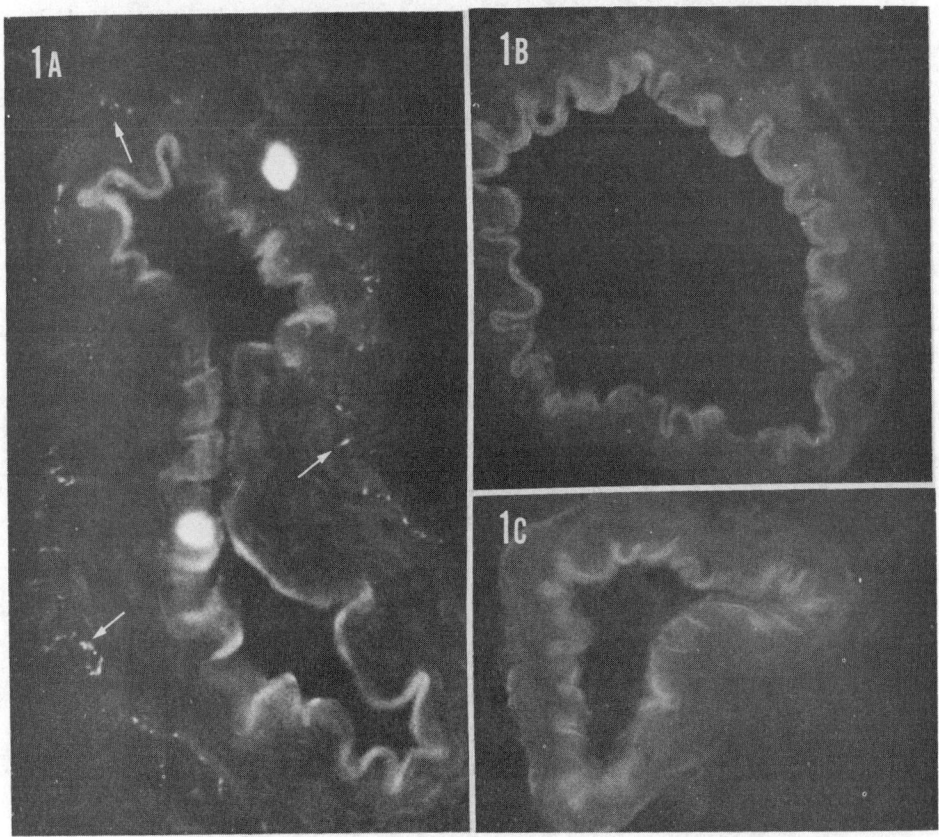

FIG. 1.—Pial Arteries: A. Pial artery from a sham operated superior cervical gang-lionectomised rat showing the presence of bright immunofluorescence of DBH in sympathetic fibres (↑); B and C. Pial arteries from superior cervical ganglionectomised rats 7 days (B) and 14 days (C) after ganglionectomy. There is complete absence of DBH-containing sympathetic innervation (×450).

The small penetrating branches of the basilar artery are contacted by DBH-containing fibres which appear to be derived from DBH-containing cell bodies in the ventral-lateral medulla. Noradrenergic fibres approach the midline from a lateral direction and turn dorsally and ventrally to contact these small penetrating branches (Figs. 2D, E). Vessels in the high spinal cord appear to be innervated by noradrenergic fibres derived from the descending noradrenergic fibre pathway.

In most areas of the brain (except those that are exclusively dopaminergic) some small vessels can be found with a few DBH-containing fibres near their surface. This type of association could be explained by chance and has not been considered to represent vessel innervation for the following reasons: (a) only small numbers of varicosities are associated with these vessels, (b) the same vessel in adjacent sections will not be associated with DBH-containing fibres, (c) the same pattern of association

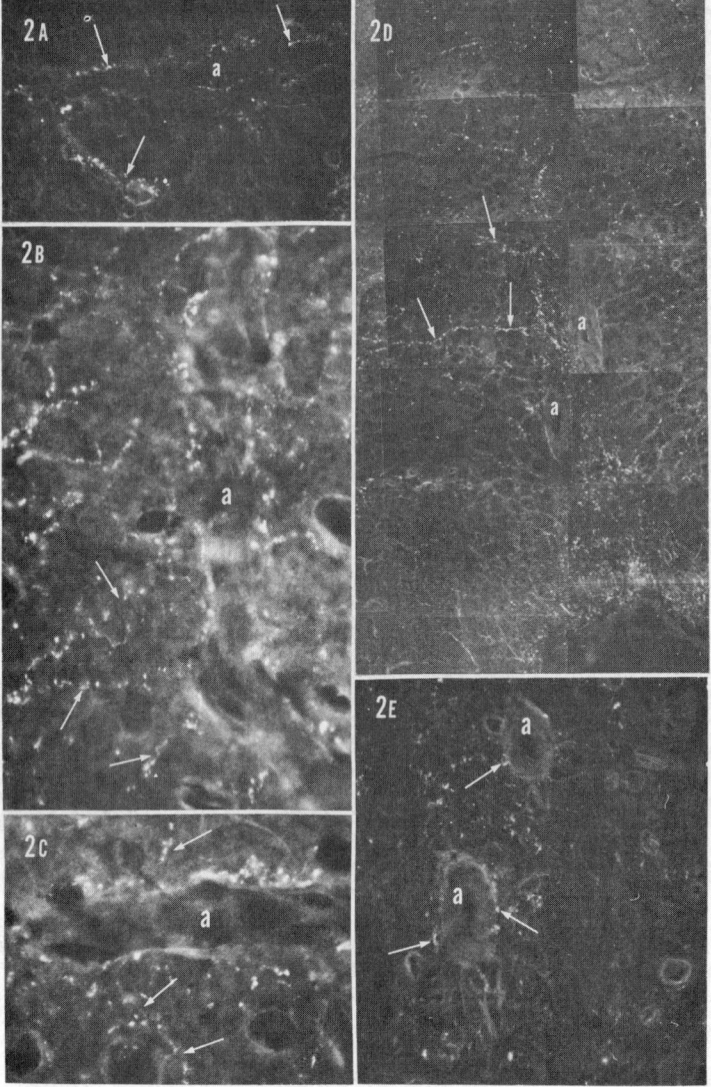

FIG. 2.—Small blood vessels in rat brain 14 days after bilateral superior cervical ganglionectomy. A. Small branching vessel (a) in the lateral hypothalamus with noradrenergic fibres showing DBH immunofluorescence associated with its wall (×350). B
and C. Small blood vessels (a) in the anterior ventral nucleus of the thalamus with a
dense coating of fluorescent DBH-containing varicose fibres. Nerve fibres are seen
approaching these vessels and frequently appear to give rise to varicose terminals on the
surface of the vessels (↑) (×850). D. Montage of coronal section through the ventral
medulla showing small penetrating branches (a) of the basilar artery. Fluorescent
DBH-containing nerve fibres approach these vessels from a lateral direction (↑) until
reaching the midline, where they appear to turn and course in a dorsal-ventral direction
parallel to the course of the vessels and in close association with them (×120). E.
Horizontal section through medulla at a level similar to that shown in Fig. 2D. Penetrating branches (a) of the basilar artery in cross section. These vessels are sparsely coated
with varicosities showing DBH immunofluorescence. Although few in number and of
low density, terminals such as these are seen along the entire length of these arteries
when followed in serial sections (×350).

is not reproducible from animal to animal. In the cerebral cortex for example, although in some areas there is a fairly dense network of DBH-containing fibres it has not been possible to recognise any consistent relationship to the vascular system.

Several generalisations concerning cerebral vascular innervation by noradrenergic nerves can be made. As has been repeatedly shown, the pial vessels receive noradrenergic innervation from the superior cervical ganglion. This innervation appears similar to the sympathetic innervation of vessels supplying other organs. Using the immuno-fluorescence method for localisation of DBH, it has been possible to establish that deep cerebral vessels *also* are innervated by noradrenergic fibres. These fibres, however, rather than originating from the peripheral sympathetic system, have been shown to take their source from the noradrenergic neurons located within the brain. The degree of this innervation and its distribution is not homogeneous throughout the brain. The pattern of innervation suggests that the blood flow in areas such as the hypothalamus—even within specific nuclei of the hypothalamus—and the pons-medulla may be markedly affected by the central noradrenergic system; whereas blood flow in other areas such as the cerebral cortex may be only minimally affected by this system.

Two other species were examined in order to show that innervation of small brain vessels was not unique to rats. Rhesus monkey and human brain obtained at autopsy were examined using the immunofluorescence technique for DBH localisation. The antibody to bovine adrenal DBH crossreacted well both with monkey and with human enzyme. As shown in Fig. 3A, B, DBH fibres and terminals were also associated with small blood vessels both in monkey and in human brain.

On the basis of these anatomical observations it would appear that one specific function of the central noradrenergic neurons is the regulation of cerebral (particularly brainstem) microcirculation. These findings may provide an anatomical basis for the fact that NE will reduce hypothalamic blood flow when injected locally into that area but does not effect cerebral blood flow when administered intravenously (ROSENDORFF, 1972).

The ultimate significance of these findings will depend upon the relative importance of the central noradrenergic system in the regulation of cerebral blood flow. This will require physiological as well as anatomical data.

Some recent observations made after spinal cord injury in cats relate directly to the significance of central noradrenergic control of blood flow. Following spinal injury, vasospasm occurs within small arteries in the cord. This vasospasm results in tissue necrosis which is frequently more damaging than the trauma itself. A rise in cord NE accompanies this phenomenon (OSTERHOLM and MATHEWS, 1972a). Using fluorescent histochemical localisation for NE, it has been found that prior to the trauma only a few faintly fluorescent terminals are seen associated with blood vessels. However, after truama the dense vascular innervation becomes visible (OSTERHOLM, 1973). The source of this vascular innervation has also been traced to central noradrenergic fibres that descend in the cord (OSTERHOLM, 1973). This central noradrenergic innervation has been implicated as the cause of the destructive vasospasm by the observation that a high cord section, cutting the descending NE fibre pathway, completely blocks the necrotic lesions. Of possible clinical importance is the finding that the tyrosine hydroxylase inhibitor α-methyl-*p*-tyrosine will also prevent the spinal cord lesions (OSTERHOLM and MATHEWS, 1972b).

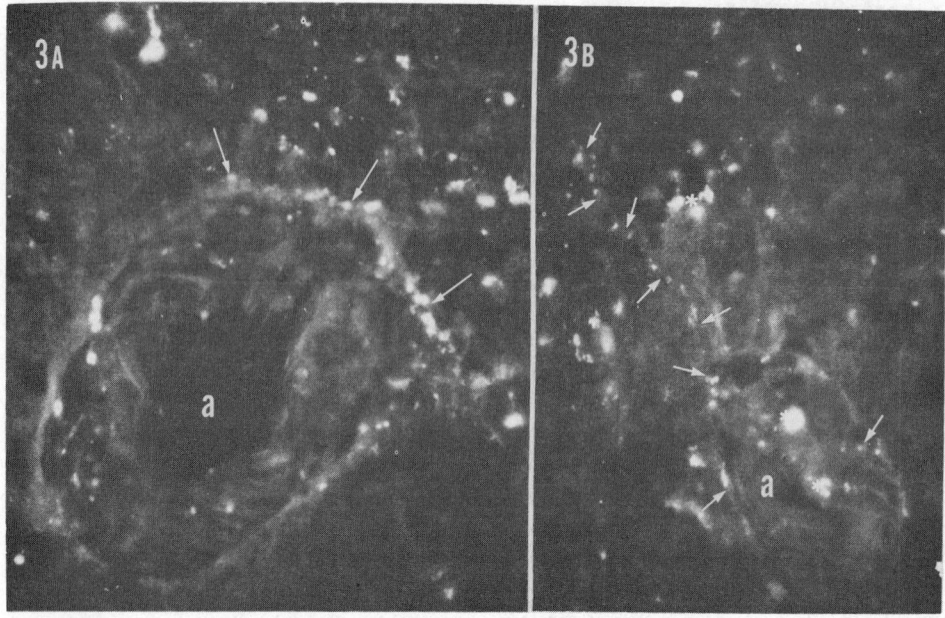

FIG. 3.—A. Normal monkey brain. Small artery (a) from horizontal section deep within the anterior-lateral hypothalamus. The outer wall of the artery is closely associated with many varicose terminals showing bright DBH immunofluorescence ($\times 850$). B. Human brain. Small artery (a) in lateral hypothalamus of a 74 year old man, with no cerebral disease, obtained at autopsy (3 h post mortem). Varicose fibres and terminals showing specific DBH fluorescence ($\uparrow$) are closely associated with the vessel. Lipofuscin pigment which exhibits bright yellow (rather than the green character-istics of fluoresceine) autofluorescence is also seen (*) ($\times 850$).

The fact that noradrenergic nerve fibres derived from cell bodies within the brain-stem have been implicated in the etiology of vasospasm in the spinal cord leads us to believe that fibres derived from the same central neurons but distributed to cerebral blood vessels may also play a role in cerebral vasospasm and infarction.

Acknowledgements—Grateful acknowledegment is extended to Dr. S. Udenfriend and Dr. E. Robins for their helpful suggestions and encouragement in this investigation and to D. Zide, A. Hartman and D. Blehm for their excellent technical assistance.

REFERENCES

EDVINSSON L., OWMAN C. H., ROSENGREN E. and WEST K. A. (1972) *Acta Physiol. Scand.* **85**, 201.
FUXE K., GOLDSTEIN M., HOKFELT T. and JOH T. H. (1971) In: *Progress in Brain Research* (Ed. ERÄNKO O. pp. 127–138. Elsevier, Amsterdam.
GEFFEN L. B. LEVETT B. G. and RUSH R. A. (1969) *J. Physiol.* **204**, 593.
HARTMAN B. K. (1973) *J. Histochem. Cytochem.* **21**, 312.
HARTMAN B. K. and UDENFRIEND S. (1970) *Mol. Pharmacol.* **6**, 85.
HARTMAN B. K. and UDENFRIEND S. (1972) *Pharmacol. Rev.* **24**, 311.
HARTMAN B. K., ZIDE D. and UDENFRIEND S. (1972) *Proc. Natn. Acad. Sci. U.S.A.* **69**, 2722.
NIELSEN K. C. and OWMAN C. H. (1967) *Brain Res.* **6**, 773.
OSTERHOLM J. L. (1973) *J. Neurosurg.* In press.
OSTERHOLM J. L. and MATHEWS G. T. (1972a) *J. Neurosurg.* **36**, 386.
OSTERHOLM J. L. and MATHEWS G. J. (1972b) *J. Neurosurg.* **36**, 395.
ROSENDORFF C. (1972) *Progr. Brain Res.* **35**, 115.

Frontiers in Catecholamine Research 1973, pp. 97 to 100. Pergamon Press. Printed in Great Britain.

COMPARATIVE STUDIES ON BOVINE AND GUINEA PIG TYROSINE HYDROXYLASE

BARBARA PETRACK, VALENTINA FETZER and RALPH ALTIERE

Research Department, Pharmaceuticals Division CIBA-GEIGY Corporation,
Ardsley, New York 10502, U.S.A.

THIS report compares some of the characteristics of bovine and guinea pig tyrosine hydroxylase as determined in experiments with 2-amino-4-hydroxy-6,7-dimethyl-tetrahydropteridine ($DMPH_4$) and tetrahydrobiopterin (biopterin-H_4) as cofactors.

Native and trypsinised bovine adrenal medulla tyrosine hydroxylases were prepared as described (NAGATSU *et al.*, 1964; PETRACK *et al.*, 1968). Tyrosine hydroxylase was partially purified from the supernatant of homgenised guinea pig adrenals, via ammonium sulfate fractionation and Sephadex-G-200 chromatography, followed by a second ammonium sulfate step. The enzyme was partially purified from the supernatant of homogenised guinea pig brains, via two fractionation steps with ammonium sulfate. L-Erythro-biopterin was generously supplied by Drs. P. Jacobi and E. C. Taylor (Princeton University). It was reduced by slight modification of the described procedure (KAUFMAN, 1967). It was quantitated via spectrophotometric measurements (KAUFMAN, 1967; NAGATSU *et al.*, 1972).

A primary difference between the adrenal tyrosine hydroxylase of the two species is the characteristic of the bovine enzyme to aggregate, either with itself, other proteins, or with particles. WURZBURGER and MUSACCHIO (1971) first reported this property of the enzyme, but others who have worked with it also know the virtual impossibility of keeping the enzyme in solution. In contrast, guinea pig adrenal tyrosine hydroxylase is soluble and can remain so, although it, too, has some tendency to aggregate. Thus guinea pig adrenal tyrosine hydroxylase may be purified, whereas the bovine enzyme cannot, until a method for keeping it deaggregated is discovered.

It was found that the activity of bovine adrenal medulla tyrosine hydroxylase, assayed at 38°C, was only slightly greater than at 25°C. This was observed with both native and trypsinised bovine adrenal enzyme, even when the latter was assayed after activation by preincubation with $FeSO_4$ and mercaptoethanol as described (PETRACK *et al.*, 1972). In contrast, guinea pig adrenal tyrosine hydroxylase is approximately fourfold more active at 38°C than it is at 25°C. This observation suggests an intrinsic difference between the enzymes from the two species.

The absence of a significant increase in activity at 38°C suggests that bovine tyrosine hydroxylase may be more labile than the guinea pig enzyme. A further indication for this possibility is the difference in the ease of rupturing the enzyme molecule with trypsin.

MUSACCHIO *et al.* (1971) have shown that the sedimentation coefficient of native bovine adrenal tyrosine hydroxylase is 9·2 S, whereas that of the trypsinised enzyme is only 3·7 S. Complete conversion to the low molecular weight enzyme is obtained by incubation of the bovine enzyme with as little as 1·5 μg trypsin/mg protein for 1 hr at 30°C. This treatment results in an increased activity, although higher concentrations of trypsin destroy activity (PETRACK *et al.*, 1968).

In contrast, the guinea pig enzyme remained a 9·2 S protein following the same treatment, although a small increase in activity was observed. Increasing the trypsin concentration to 9 μg/mg protein caused some inactivation. Nevertheless, as shown in Fig. 1, when analysed via sucrose density sedimentation, two discrete peaks were obtained, corresponding to 3·7 and 9·2 S proteins. Less than half was in the low molecular weight form.

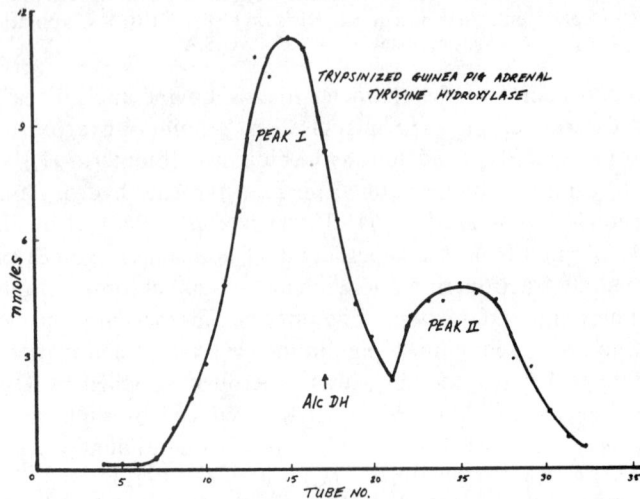

FIG. 1.—Sucrose density gradient pattern of trypsinised guinea pig adrenal tyrosine hydroxylase. The peak of marker enzyme, alcohol dehydrogenase (7·4 S), is indicated.

A further species difference was observed with respect to substrate inhibition of the enzyme when biopterin-H_4 was the cofactor. Shiman, Akino and Kaufman reported that the bovine enzyme was markedly inhibited at low concentrations of tyrosine when biopterin-H_4, but not when $DMPH_4$, served as cofactor (Shiman et al., 1971).

Figure 2 confirms that the bovine adrenal enzyme is indeed inhibited by substrate; more than 30 per cent inhibition was observed at 0·05 mM tyrosine. Similar effects were obtained with native and trypsinised enzyme. In contrast, both guinea pig

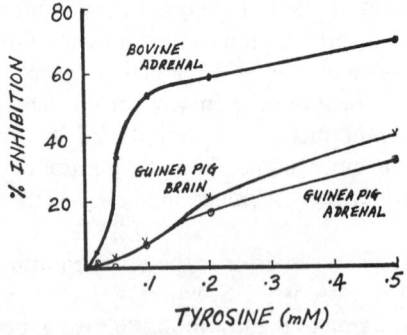

FIG. 2.—Substrate inhibition by excess tyrosine. Assays were carried out as described (1) except that biopterin-H_4 (0·2 mM) was used as cofactor, with 2-mercaptoethanol (50 mM) to regenerate the reduced form. The concentration of tyrosine-^{3}H was varied as shown; separate blanks were included for each concentration.

adrenal and brain tyrosine hydroxylases are much less sensitive to substrate inhibition. Significant inhibition required more than 0·2 mM tyrosine, concentrations which are unlikely to occur *in vivo*. No inhibition was observed with $DMPH_4$ as cofactor.

Table 1 summarises the K_m for tyrosine obtained in experiments similar to those in Fig. 2, in which biopterin-H_4 was used as cofactor. The K_m of trypsinised bovine adrenal enzyme agrees with that reported for the chymotrypsin-solubilised enzyme (SHIMAN and KAUFMAN, 1970). However, our K_m for the native bovine adrenal enzyme is considerably greater than the value of 7 μM reported (SHIMAN and KAUFMAN, 1970), but is similar to that obtained for both guinea pig adrenal and brain tyrosine hydroxylases.

Table 2 compares the K_m values for $DMPH_4$ and biopterin-H_4 with both guinea pig and bovine adrenal (native) tyrosine hydroxylase. It is seen that for both enzymes, the K_m for biopterin-H_4 is approximately one-fitth that of $DMPH_4$. In each case, the V_{max} is considerably greater with biopterin-H_4. Our data agree with that reported by NAGATSU *et al.* (1972) for native bovine adrenal enzyme. However, our K_m is approximately one-fifth of the K_m of 0·1 mM for biopterin-H_4 reported by SHIMAN *et al.* (1970); this latter value appears unexpectedly high for the naturally occurring cofactor.

In summary, bovine and guinea pig adrenal tyrosine hydroxylases appear to differ with respect to effect of temperature on reaction rate, ease of rupturing the

TABLE 1. TYROSINE k_m AT 0·2 mM
BIOPTERIN-H_4

Source of tyrosine hydroxylase	k_m (mM)
Bovine adrenal	
Trypsinised	0·020
Native	0·046
Guinea pig	
Adrenal	0·048
Brain	0·067

Specific activites, expressed as nmoles dopa formed/mg protein/10 min, at 38°C:
Bovine adrenal (native) = 34
Bovine adrenal (trypsinised) = 270
Guinea pig adrenal = 170
Guinea pig brain = 30

TABLE 2. k_m AND V_{max} VALUES WITH BIOPTERIN-H_4 AND $DMPH_4$ AS COFACTORS AT 0·02 mM TYROSINE

Source of tyrosine hydroxylase	k_m biopterin-H_4 (mM)	$DMPH_4$	V_{max} biopterin-H_4 (nmoles/10 min)	$DMPH_4$
Guinea pig adrenal	0·02	0·10	4·0	0·92
Bovine adrenal medulla (native enzyme)	0·01	0·05	4·8	2·08

enzyme molecule with trypsin, and sensitivity of the enzyme to substrate inhibition by excess tyrosine. Nevertheless, the tyrosine hydroxylase of both species exhibited similar K_m values for tyrosine and for the synthetic and naturally occurring pteridine cofactors. Biopterin-H_4 was demonstrated to be a more effective cofactor in both species, since the K_m values were considerably lower and the V_{max} values considerably higher than with $DMPH_4$.

Most studies concerned with elucidating mechanisms for the control of catecholamine biosynthesis, via an effect on tyrosine hydroxylase, have been carried out in rats. However, little attention has been directed toward characterisation of tyrosine hydroxylase in this species. This is also the case with the human enzyme.

Our findings that some intrinsic properties of tyrosine hydroxylase vary in different species suggest the need for additional studies with rat and human tyrosine hydroxylase.

REFERENCES

Kaufman S. (1967) *J. Biol. Chem.* **242,** 3934–3943.

Musacchio J. M., Wurzburger R. J. and D'Angelo G. (1971) *Mol. Pharmacol.* **7,** 136–146.

Nagatsu T., Levitt M. and Udenfriend S. (1964) *J. Biol. Chem.* **239,** 2910–2917.

Nagatsu T., Mizutani K., Nagatsu I., Matsuura S. and Sugimoto T. (1972) *Biochem. Pharmac.* **21,** 1945–1953.

Petrack B., Sheppy F. and Fetzer V. (1968) *J. Biol. Chem.* **243,** 743–748.

Petrack B., Sheppy F., Fetzer V., Manning T., Chertock H. and Ma D. (1972) *J. Biol. Chem.* **247,** 4872–4878.

Shiman R. and Kaufman S. (1970) In: *Methods in enzymology,* H. Tabor and C. W. Tabor (Eds.), Vol. 17A, Academic Press, New York, p. 609.

Shiman R., Akino M. and Kaufman S. (1971) *J. Biol. Chem.* **246,** 1330–1340.

Wurzburger R. J. and Musacchio J. M. (1971) *J. Pharmac. Exp. Ther.* **177,** 155–168.

Frontiers in Catecholamine Research 1973, pp. 101 to 105. Pergamon Press. Printed in Great Britain.

REGULATION OF PHENYLETHANOLAMINE N-METHYLTRANSFERASE

ROLAND D. CIARANELLO

National Institute of Mental Health, Laboratory of Clinical Science, Bethesda,
Maryland 20014, U.S.A.

THE LAST step in catecholamine biosynthesis, the conversion of noradrenaline to adrenaline, is a methylation reaction catalysed by phenylethanolamine N-methyltransferase (PNMT). The importance of ATP in this reaction was first demonstrated by BULBRING (1949) and further elucidated by KELLER et al. (1950), who established the importance of the methyl group of methionine in the transfer reaction. The previous discovery of S-adenosylmethionine (CANTONI, 1953) and its immediate implication in a number of biologic methylations suggested its importance in catecholamine metabolism in general and adrenaline synthesis in particular, a point which was confirmed in 1957 by KIRSHNER and GOODALL (1957).

ISOLATION AND CHARACTERISATION OF PHENYLETHANOLAMINE N-METHYLTRANSFERASE

Phenylethanolamine N-methyltransferase was partially purified by AXELROD (1962), who elucidated most of the properties of the enzyme. PNMT is localised almost exclusively to the adrenal medulla, although traces of activity have been reported in heart (AXELROD, 1962), brain (AXELROD, 1962; MCGEER and MCGEER, 1964; CIARANELLO et al., 1969a; POHORECKY et al.,1969), in extra-adrenal chromaffin tissue (ROFFI and MARGOLIS, 1966; CIARANELLO et al., 1973a) and in sympathetic ganglia (CIARANELLO et al., 1973a). PNMT catalyses the N-methylation of a number of primary and secondary β-hydroxyphenylethylamine derivatives. The requirement of the enzyme for β-hydroxylated phenylethylamine substrates appears to vary with species: PNMT from monkey (AXELROD, 1962), rabbit (FULLER et al., 1970), rat and human (AXELROD and VESELL, 1970) will N-methylate only phenylethylamine substrates bearing a β-hydroxyl group. PNMT purified from dog adrenals (AXELROD and VESELL, 1970) and from beef adrenals will N-methylate phenylethylamines to a limited extent (MOLINOFF et al., 1971; LADURON, 1972).

PNMT's from the rat and the frog share common substrate specificities, but virtually every other biochemical parameter of these enzymes differs (WURTMAN et al., 1968). These include: pH and thermal optima, thermal stability and starch-block migration. When these studies are compared with those of previous findings a striking uniformity of substrate specificity is observed, coupled with an equally striking heterogeneity of electrophoretic migration. These findings suggest that the evolution of PNMT has permitted amino acid substitutions which affect charge, molecular weight and stability of the enzyme, but has not tolerated alterations at the catalytic site of the molecule, which has remained relatively constant.

Purified bovine phenylethanolamine N-methyltransferase appears to be a protein of molecular weight in the range 37000–41000 daltons (CONNETT and KIRSHNER, 1970; JOH and GOLDSTEIN, 1973; CIARANELLO and AXELROD, 1973b). This measure has been complicated by the presence of multiple forms of the enzyme purified from beef adrenal. These "other" PNMT's probably represent a mixture of aggregated "native" enzyme molecules brought about by prolonged storage. However, compelling evidence for the existence of charge isozymes of beef adrenal PNMT has recently been presented (JOH and GOLDSTEIN, 1973). The amino acid composition of purified beef adrenal PNMT has been shown to consist of an unusually high concentration of glutamic acid.

REGULATION OF PHENYLETHANOLAMINE N-METHYLTRANSFERASE

The unique anatomic relationship between the adrenal cortex and the adrenal medulla posed an obvious question about regulatory relationships between these two embryologic and physiologically distinct structures. The biosynthetic and secretory activity of the adrenal cortex is probably under the exclusive control of ACTH elaborated by the pituitary gland, while the activity of the adrenal medulla is regulated by impulses from the splanchnic nerves to this modified sympathetic ganglion. A portal venous system supposedly drains the cortical hormones into the medulla, thus bathing medullary cells with extremely high concentrations of glucocorticoids. COUPLAND (1953) proposed that the adrenal glucocorticoids were important in the methylation of noradrenaline. This proposal was established by WURTMAN and AXELROD (1966), who showed that hypophysectomy markedly reduced PNMT levels in the adrenals of rats. Enzyme activity could be restored to normal values by the administration of ACTH or glucocorticoids. The enzyme response was specific to ACTH or glucocorticoids; a variety of trophic polypeptide hormones or other steroid hormones failed to restore PNMT in hypophysectomised rats (WURTMAN, 1966).

The glucocorticoid-mediated increase in PNMT activity is blocked by inhibitors of protein synthesis (WURTMAN and AXELROD, 1966). CIARANELLO and BLACK (1971) have proposed that the induction of PNMT by glucocorticoids involves stimulation of de novo enzyme synthesis rather than inhibition of the degradation of the existing enzyme molecules.

PNMT in the adrenals of intact rats did not respond as in hypophysectomised rats. Administration of ACTH or glucocorticoids in dosages and for times sufficient to induce the enzyme several-fold in hypophysectomised rats failed to elevate the enzyme in intact rats. Extraordinary conditions, such as implantation of an ACTH secreting tumor (VERNIKOS-DANELLIS et al., 1968), unilateral adrenalectomy (CIARANELLO et al., 1969b) or chronic, intermittent restraint stress (KVETNANSKY et al., 1970) produced modest to dramatic increases in PNMT activity but only over extremely long (1–6 week) periods of exposure.

Although the adrenal medulla is the principal site of adrenaline synthesis, PNMT has been reported in other tissues. It is of interest that the effect of glucocorticoids on extra–adrenal PNMT activity has been variable. In the olfactory bulb (POHORECKY et al., 1969) the enzyme has been reported to be induced by dexamethasone, although the enzyme elsewhere in the brain is unresponsive to glucocorticoid administration (CIARANELLO and BARCHAS, 1969, unpublished observations). Similarly, PNMT in the heart does not respond to glucocorticoid administration (CIARANELLO, JACOBOWITZ and AXELROD, 1971, unpublished observations).

Traces of PNMT activity have been reported in the organ of Zuckerkandl (ROFFI, 1968) and in the superior cervical ganglion (CIARANELLO et al., 1973). In both tissues the administration of hydrocortisone or dexamethasone to newborn rats results in an 8- to 15-fold increase in PNMT activity (CIARANELLO *et al.*, 1973a). This responsiveness is paralleled by proliferation of small, intensely fluorescent cells (ERANKO and ERANKO, 1972; CIARANELLO *et al.*, 1973a) in both tissues, and is exquisitely time-dependent. Administration of glucocorticoids to newborn rats between birth and five days of age resulted in a brisk PNMT induction; after five days of age hormone administration failed to induce the enzyme (CIARANELLO *et al.*, 1973a).

Chronic exposure to agents which result in a reflex increase in splanchnic nerve activity, such as reserpine (MOLINOFF *et al.*, 1970; CIARANELLO and BLACK, 1971) or 6-hydroxydopamine (THOENEN *et al.*, 1970) also result in increased PNMT activity. These data have led to the proposal of a minor regulatory role on PNMT activity by splanchnic nerve input to the adrenal medulla.

There is a striking consistency among mammalian species in the response of PNMT to hypophysectomy. In rats, dogs and lambs (WURTMAN *et al.*, 1968; COMLINE and SILVER, 1970) hypophysectomy is associated with a decline in PNMT activity. Hypophysectomy produces a decrease in urinary excretion in humans; (LUFT and VON EULER, 1956) presumably adrenal PNMT has declined as well. In the frog, however, hypophysectomy has no effect on PNMT activity, and the enzyme is unresponsive to exogenous hormone administration (WURTMAN *et al.*, 1968).

GENETIC CONTROL OF PNMT

In mice, the pituitary gland also plays an essential role in PNMT activity. Studies on inbred mouse strains have emphasised the importance of genetic determinants in the qualitative and quantitative response of adrenal PNMT to exogenous stimuli (CIARANELLO *et al.*, 1972b). These studies have revealed that, in three inbred strains studied, the pituitary gland was a common denominator in each strain. However, PNMT in one strain (DBA/2J) was regulated by both glucocorticoids and splanchnic nerve impulses, both stimuli producing extremely rapid enzyme responses. In another strain (CBA/J), only glucocorticoids were effective in increasing enzyme activity, while in a third strain (C57Bl/Ka), neither glucocorticoids nor nerve impulses controlled the enzyme. Instead pituitary ACTH appeared to be exerting a direct regulatory control over the enzyme independent of glucocorticoids. These data suggested that the determinants of which stimuli will be effective in inducing PNMT synthesis are probably under genetic control.

The role of genetic determinants in regulating steady-state levels of PNMT has recently been explored (CIARANELLO *et al.*, 1972; KESSLER *et al.*, 1972; CIARANELLO and AXELROD, 1973b). Studies in several different inbred mouse strains revealed wide variations in PNMT activity. When strains with extreme PNMT levels were interbred, hybrid progeny with intermediate enzyme values were obtained.

Recent studies have focused on the genetic control of steady-state enzyme activity (CIARANELLO and AXELROD, 1973b). In two sublines of the Balb/c strain, PNMT levels differed by nearly twofold. This difference was due to different levels of PNMT protein, as measured by antibody titration, and not to the presence of activators or inhibitors. F_1 heterozygotes were intermediate between the homozygous parents, suggesting an autosomal codominant mode of transmission. The F_2 progeny obtained by crossing members of the F_1 generation produced offspring whose PNMT levels did not differ from their parents, but which showed significantly greater population variance. These results suggested the F_2 progeny to be a heterogenous mixture of segregating genotypes. When the F_2 were subdivided into phenotype classes of high, intermediate and low PNMT activity, their distribution corresponded to that expected if PNMT levels were determined by an allele at a single gene locus.

Alterations in steady-state enzyme levels can be brought about by alterations in either the rate of synthesis or of degradation of PNMT. By precipitating with anti-phenylethanolamine N-methyltransferase antibody the PNMT from mice previously given radiolabelled amino acids we were able to establish that the rate of synthesis of PNMT in the two sublines was identical. However, the rate of degradation of PNMT in the low-PNMT Balb/cN subline was nearly twice as fast as in the Balb/cJ. Thus, it appeared that the Balb/cJ subline had twice as many PNMT molecules because they were being degraded at half the rate of the Balb/cN PNMT molecules, and that this trait was under the control of a single genetic locus. The rates of soluble adrenal protein synthesis and degradation were also measured; they were identical in the two sublines.

The gene controlling the altered rate of degradation could have been the structural gene for PNMT; mutation would produce a PNMT molecule whose properties as a substrate for degradation would be altered. Alternatively, the mutated gene could code for a product which was the degrading enzyme for PNMT. To test these possibilities, a number of biochemical and biophysical properties of PNMT in the two mouse sublines were tested: PNMT from the Balb/cJ and Balb/cN were identical with regard to their K_m's for phenylethanolamine and S-adenosylmethionine, their thermal stability at 50° C; their tryptic stability and their electrophoretic mobility on polyacrylamide. These results suggested that PNMT from the two sublines was similar, and may have been identical. However, other tests are currently being devised to establish if the mutation we are studying has occurred in the PNMT structural gene or in a control gene whose product degrades PNMT. Other mutant mice strains whose PNMT shows altered kinetic, thermal or electrophoretic properties are being crossbred with the Balb/c mice and tests of segregation of altered structural gene properties with altered rates of degradation will be performed.

The last decade has seen the history of PNMT proceed from its isolation and gross tissue and subcellular distribution to knowledge of the hormonal factors which regulate its activity to an analysis of the controls operant at the genome which dictate the level of steady-state activity. PNMT has surely been one of the most thoroughly studied proteins in the biogenic amine schema; the intervening years to the Fourth Catecholamine Symposium should be exciting ones indeed in terms of this fascinating enzyme.

Acknowledgement—I would like to express my deepest appreciation to Dr. Julius Axelrod, in whose laboratory the genetic studies on Balb/c mice were performed, for his sustained encouragement and support during the course of these studies.

REFERENCES

AXELROD J. (1962) *J. Biol. Chem.* **237**, 1657–1660.
AXELROD J. and VESELL E. S. (1970) *Mol. Pharmacol.* **6**, 78–84.
BÜLBRING E. (1949) *Brit. J. Pharmacol.* **4**, 234–244.
CANTONI G. L. (1953) *J. Biol. Chem.* **204**, 403–416.
CIARANELLO R. D., BARCHAS R. E., BYERS G. S., STEMMLE D. W. and BARCHAS J. D. (1969a) *Nature, Lond.* **221**, 368–369.
CIARANELLO R. D., BARCHAS J. D. and VERNIKOS-DANELLIS J. (1969b) *Life Sci.* **8**, 401–407.
CIARANELLO R. D. and BLACK I. B. (1971) *Biochem. Pharmacol.* **20**, 3529–3532.
CIARANELLO R. D., BARCHAS R. E., KESSLER S. and BARCHAS J. D. (1972a) *Life Sci.* **11**, 565–572.

CIARANELLO R. D., DORNBUSCH J. N. and BARCHAS J. D. (1972b) *Mol. Pharmacol.* **8**, 511–520.
CIARANELLO R. D., JACOBOWITZ D. and AXELROD J. (1973a) *J. Neurochem.* **20**, 799–805.
CIARANELLO R. D. and AXELROD J. (1973b) *J. Biol. Chem.*, in press.
COMLINE R. S. and SILVER M. (1970) *Nature, Lond.* **225**, 739–740.
CONNETT R. J. and KIRSHNER N. (1970) *J. Biol. Chem.* **245**, 329–334.
COUPLAND R. E. (1953) *J. Endocrinol.* **9**, 194–203.
ERANKO L. and ERANKO O. (1972) *Acta Physiol. Scand.* **84**, 125–133.
FULLER R. W., WARREN B. J. and MOLLOY B. B. (1970) *Biochem. Biophys. Acta* **222**, 210–212.
JOH T. H. and GOLDSTEIN M. (1973) *Mol. Pharmacol.* **9**, 117–129.
KELLER E. B., BOISSONNAS R. A. and DU VIGNEAUD V. (1950) *J. Biol. Chem.* **183**, 627–632.
KESSLER S., CIARANELLO R. D., SHIRE J. G. M. and BARCHAS J. D. (1972) *Proc. Nat. Acad. Sci.* **68**, 2448–2450.
KIRSHNER N. and GOODALL M. (1957) *Biochim. Biophys. Acta* **24**, 658–659.
KVETNANSKY R., WEISE V. K. and KOPIN I. J. (1970) *Endocrinology* **87**, 744–799.
LADURON P. (1972) *Arch. int. de Pharmacodyn. Ther.* **195**, 197–208.
LUFT R. and VON EULER U. S. (1956) *J. Clin. Endocrinol. Metab.* **16**, 1017.
MCGEER P. L. and MCGEER E. G. (1964) *Biochem. Biophys. Res. Commun.* **17**, 502–507.
MOLINOFF P. B., BRIMIJOIN S., WEINSHILBOUM R. M. and AXELROD J. (1970) *Proc. Nat. Acad. Sci. U.S.A.* **66**, 453–458.
MOLINOFF P. B., WEINSHILBOUM R. M. and AXELROD J. (1971) *J. Pharmacol. Exp. Ther.* **178**, 425–431.
POHORECKY L. A., ZIGMOND M., KARTEN H. and WURTMAN R. J. (1969) *J. Pharmacol. Exp. Ther.* **165**, 190–195.
POHORECKY L. A. and WURTMAN R. J. (1971) *Pharmacol. Rev.* **23**, 1–35.
ROFFI J. and MARGOLIS F. (1966) *C. R. hebd. Séance Acad. Sci., Paris* **263**, 1496–1499.
ROFFI J. (1968) *J. Physiol. Paris* **60**, 445–494.
THOENEN H., MUELLER R. A. and AXELROD J. (1970) *Biochem. Pharmacol.* **19**, 669–673.
VERNIKOS-DANELLIS J., CIARANELLO R. D. and BARCHAS J. D. (1968) *Endocrinology* **83**, 1357–1358.
WURTMAN R. J. and AXELROD J. (1966a) *J. Biol. Chem.* **241**, 2301–2305.
WURTMAN R. J. (1966b) *Endocrinology* **78**, 608–614.
WURTMAN R. J., CASPER A., POHORECKY L. A. and BARTTER F. C. (1968a) *Proc. Nat. Acad. Sci. U.S.A.* **61**, 522–528.
WURTMAN R. J., AXELROD J., VESELL E. S. and ROSS G. T. (1968b) *Endocrinology* **82**, 584–590.

Frontiers in Catecholamine Research 1973, pp. 107 to 112. Pergamon Press. Printed in Great Britain.

CATECHOL-*O*-METHYLTRANSFERASE

C. Bohuon and M. Assicot

Unité de Biologie Clinique et Experimentale, Institut Gustave Roussy,
9480-Villejuif, France

Catechol-*O*-methyltransferase (COMT)(EC 2-1-1-6) was discovered and described by Axelord and Tomchik (1958), following the finding of Armstrong, Mc Millan and Shaw (1957) that vanylmandelic acid is the major metabolic product of epinephrine in man. This enzyme is able to methylate all the compounds which have a catechol function.

In vivo catecholamines are catabolised by COMT and by mono-aminoxydase (MAO) but we know less about the former.

For the last 15 years the findings on COMT have been obtained essentially in the field of purification and the description of new inhibitors. Other papers have been published on the physiological role of COMT, the mechanism of transmethylation and its repartition in various tissues. It seems that an essential problem is the great lability of this enzyme, which complicates its isolation and makes accurate assays difficult.

HETEROGENEITY OR HOMOGENEITY OF THE PURIFIED SOLUBLE COMT FROM RAT LIVER?

During the purification of soluble COMT from rat liver two or three fractions with COMT activity have been observed. These fractions can be obtained by electrophoresis or chromatographic methods at various steps of the purification (Anderson and d'Iorio, 1968; Axelrod and Vesell, 1970; Flohe and Schwabe, 1970).

This problem of isozymes has been studied using isoelectrofocusing, a very high resolution system for the separation of proteins based on the differences between isoelectric pH (pH_i).

Using a method previously described (Assicot and Bohuon, 1969), it is possible to obtain an enzyme purified about four-hundred times (Table 1). At various steps of this purification, we have studied the homogeneity of COMT in a gradient of ampholines going from pH 4·10 to pH 5·00. After the Sephadex G 200 stage, which give a preparation where COMT is purified about fifty times, it is possible to find two bands with enzymatic activity (Fig. 1). One major band (80%) has an isoelectric pH at 4·66 and one minor band (20%) has an isoelectric pH at 4·54.

At the following stage (phosphate gel adsorption) the two bands were still present. But after the ultimate phase (DEAE cellulose chromatography) only the band with an isoelectric pH 4·66 remained. Unfortunately during this last phase a large amount enzyme activity is lost and it is possible that the minor band, simply cannot although present, be detected.

On the two bands collected after isoelectrofocusing, we have studied: K_m: 2·5 × 10^{-4} for epinephrine; pH, the optimum is between 7·8 and 8·2; inhibition by antibodies obtained after injection of the pure enzyme in rabbits; the action of inhibitors. In all cases, we have been unable to find any difference. This experiment was repeated with a new preparation of COMT with the same results.

Table 1a. Purification of rat liver catechol-*O*-methyltransferase

Fraction	Volume (ml)	Total protein (mg)	Total activity (units)	Specific activity (units/mg)	Yield (%)	Purification (-fold)
I Supernatant fraction of rat liver	480	14200	106	0·0075	(100)	
II Ammonium sulfate fractionation	22	1705	76	0·044	71·7	5·8
III Sephadex G-200 Chromatography	180	184	70	0·380	66	50·6
IV Hydroxyapatite treatment	215	61	59	0·969	55·6	129
V DEAE-Cellulose chromatography	33	3·9	13	3·30	12·2	440

Table 1b. Molecular weight of COMT (EC 2-1-1-6)

	Origin	Purification	Molecular weight	
Anderson-D'Iorio (1968)	Rat liver	186	29000	Sephadex G-100
Flohe-Schwabe (1970)	Rat liver	298		
Ball-Knuppen-Brener (1971)	Human liver	380	29000 25000	Ultracentrifugation Sephadex G-100
Assicot-Bohuon (1969) (1972)	Rat liver	440	20000 (19100–20800)	Sephadex G-100

At the present time, whilst it is not possible to draw any firm conclusions, we may infer that the minor band is perhaps a fraction corresponding to a primitive stage of the enzyme.

Some arguments in support of this hypothesis can be found in some recent papers: Marsden (1972) in the tissue of the snail, *Helix aspersa*, (Veriti and Bevan, 1972) in rabbit aorta.

The presence of COMT in the red blood cells was established by Axelrod and Cohn (1971); we began a detailed study of this COMT, using rat erythrocytes. In these cells, two types of COMT were found: one isozyme is soluble, the other is bound to the membrane. The erythrocyte-soluble enzyme has properties which are biochemically and immunologically similar to those described for the rat liver COMT (Assicot-Bohuon, 1971). On the other band, the *O*-methylating activity (10% of the total activity) found in the erythrocyte membrane preparation differs in pH optimum, heat stability K_m values and immunochemical reactivity. These isozymes were studied by an isoelectrofocusing method.

The soluble enzyme purified from rat erythrocytes was shown to have two bands; at pH_i 4·50 and pH_i 4·66 respectively. Before electrofocusing, the enzyme bound to the membrane was solubilised by Triton X100. Only one band was observed. This band had a pH_i of around 4·70.

In conclusion, it seems that in this particular tissue at least two well differentiated enzymes are present. The exact relation and function of these isozymes of COMT is still open to speculation.

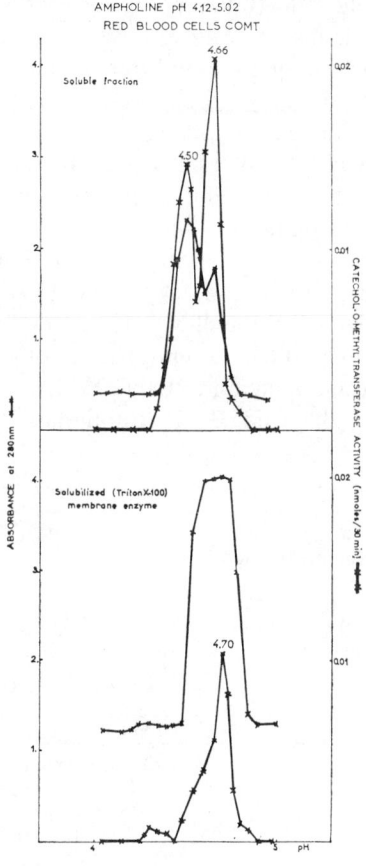

FIG. 1

INHIBITION OF COMT

Many inhibitors of COMT have been described, but almost all belong to the group of catechol compounds with an inhibition of a competitive type. Some of the more interesting compounds that we have studied have been gallic acid derivatives. *In vitro* propylgallate, and butylgallate have an inhibitor effect similar to pyrogallol, which is the reference compound (Table 2).

TABLE 2. INHIBITORS OF COMT

Inhibitors	% Inhibition	
	1×10^{-4}M	1×10^{-4}M
Gallic acid derivatives		
Propyl	83·5	49
Butyl	79	59
Methyl	79	44
Pyrogallol	80	47
4-methyltropolone	78	45
S-Adenosyl homocysteine	93	10

The assays were performed at pH 7·6; Epinephrine: $0·83 \times 10^{-3}$M and *S*-AMe: $6·6 \times 10^{-5}$M

Propylgallate and butylgallate (GPA 1714) are useful for *in vivo* inhibition of COMT and the latter tested in man by Ericson (1971). In association with L-DOPA, he noted an interesting effect in the release of tremor, but it seems that some secondary and toxic effects appeared later.

Furthermore, we have studied the analogs of S-AMe. This work was initiated recently in collaboration with Dr Hildesheim and Dr Lederer. With the exception of *S*-adenosylhomocystein, the high inhibition effect of which was previously reported by Coward *et al.* (1972), no significant inhibition at a concentration of 10^{-4} M was found. All these compounds show a competitive inhibition type.

In a different series of experiments (Wurtzbourger, Assicot and Bohuon, 1973, results not published), we studied the effect of α- and β-adrenergic inhibitor compounds on the two types of COMT in the erythrocytes of rats. This choice was made in an attempt to find a specific function of the COMT bound to the membrane.

The results are given in the Table 3. It is noticeable that propranolol does not

TABLE 3. INHIBITION OF RED BLOOD CELLS COMPT BY PROPRANOLOL

	Soluble fraction		Solubilized membrane enzyme	
	Propranolol		Propranolol	
	1.66×10^{-3}M	1.66×10^{-4}M	1.66×10^{-3}M	1.66×10^{-4}M
Epinephrine				
1.66×10^{-4}M	9	0	31	4
1.66×10^{-3}M	9	4	9	0

	Ghost fraction						
	Propranolol						
	1.66×10^{-3}M	1.66×10^{-4}M	1.66×10^{-5}M	1.66×10^{-6}M	1.66×10^{-8}M	1.66×10^{-10}M	1.66×10^{-12}M
Epinephrine							
1.66×10^{-6}M	50	28	—	14	14	17	0
1.66×10^{-4}M	36	13	2				
1.66×10^{-3}M	7	25					

have the same inhibitory effect on the two types of COMT. Membrane COMT is inhibited *in vitro* by a very low concentration of propranolol. We are unable to explain this fact. Under the same experimental conditions, phenoxybenzamine was without effect. Several attemps to obtain COMT from the membrane of rat hepatocytes, were unsuccessful, and consequently, it was not possible to test propranolol on this system. It is interesting to notice the remarkable effect of the association of propranolol with L-DOPA for the therapy of Parkinson's disease.

Recent studies have been made on the ratio of methylation in the meta or para position. Since Senoh's (1959) first observation, it has been accepted that some degree of para methylation can appear *in vitro*. It is clear now that the ratio of meta or para methylated compounds *in vitro* which are formed by COMT depends on pH, ionic strength, divalent cation and the nature and position of the substituents on the catechol ring (Creveling *et al.*, 1972; Katz and Jacobson, 1972). The results of studies of over 50 substituted catechols have been reported by Creveling (1972).

AMPHOLINE pH 4.12-5.02
RAT LIVER COMT

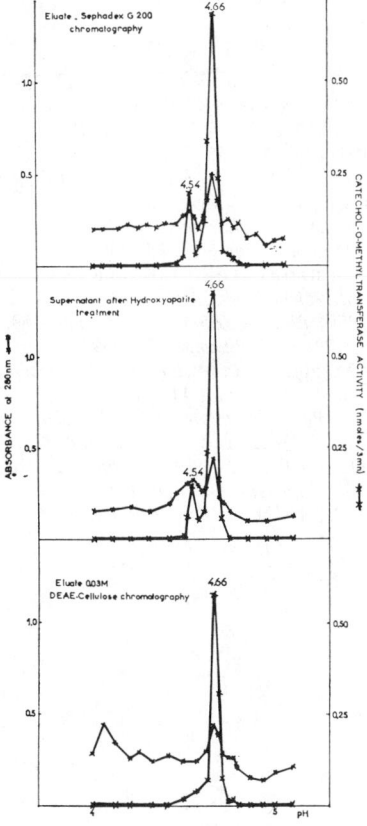

FIG. 2

The findings *in vivo*, on the ratio of isomeric *O*-methylated products are less exhaustive. THOMAS has published on the formation of vanillic and isovanillic acid after perfusion of an isolated rat liver with dihydroxybenzoic acid (THOMAS, 1972). We reported in 1971, that iso-HVA was frequently present in human urine, at very low concentration iso-HVA has also been found in spinal fluid (MATHIEU 1972). On the contrary no definitive evidence for the presence of iso-VMA in human urine has been obtained (COMOY-BOHUON, 1972).

The study of the paramethoxylated metabolite of dopamine in diseases will certainly prove useful. For example, recently we have found a very high excretion of iso-HVA in a patient treated with L-DOPA who has a hyper-sensitivity for this drug as well as dyskynesia. This biochemical observation indicates the potential interest of studies of paramethylation, especially in relation with psychiatric diseases.

CONCLUSION

Progress has certainly been made since the last Symposium on catecholamines, but nevertheless many problems remain to be solved. Perhaps the most interesting are the studies on the primary structure of COMT which would yield very important advances in the field of methyltransferases and consequently in methyl transfer-mechanism from S-AMe to a substrate.

REFERENCES

ANDERSON F. J. and D'IORIO A. (1968) *Biochem. Pharmacol.* **17,** 1943.
ARMSTRONG M. D., McMILLAN A. and SHAW K. N. (1957) *Biochim. Biophys. Acta* **25,** 422.
ASSICOT M. and BOHUON C. (1970). *Eur. J. Biochem.* **12,** 490.
ASSICOT M. and BOHUON C. (1971) *Biochim.* **53,** 871.
AXELROD J. and COHN C. K. (1971) *J. Pharmacol. Exp. Ther.* **170,** 650.
AXELROD J. and TOMCHIK R. (1958) *J. Biol. Chem.* **233,** 490.
AXELROD J. and VESELL E. S. (1970) *Mol. Pharmacol.* **6,** 78.
COMOY E. and BOHUON C. (1971) *Clin. Chim. Acta* **35,** 369.
CREVELING C. R., MORRIS N., SHIMIZU H., ONG H. H. and DALY J. (1972) *Mol. Pharmacol.* **8,** 398.
ERICSONN A. D. (1971) *J. Neurol. Sciences* **14,** L93.
FLOCHE L. and SCHWABE K. (1970) *Biochim. Biophys. Acta* **220,** 469.
KATZ R. and JACOBSON A. E. (1972) *Mol. Pharmacol.* **8,** 594.
LOWARD J. K., D'URSO-SCOTT M. and SWEET W. D. (1972) *Biochem. Pharmacol.* **21,** 1203.
MARSDEN C. A. (1973) *Comp. Biochem. Physiol.* **44,** 687.
MATHIEU P. and REVOL L. (1970) *Bull. Soc. Chim. Biol.* **52,** 1039.
SENOH S., DALY J., AXELDOR J. and WITKOP B. (1959) *J. Am. Chem. Soc.* **81,** 6240.
THOMAS H., MULLER-ENOCH D. and ROTH S. (1972) *Hoppe Seyler's Z. Physiol. Chem.* **353,** 1894.
VERITY A. and BEVAN J. A. (1972) *Biochem. Pharmacol.* **21,** 193.
VERITY A., SU C. and BEVAN J. A. (1972) *Biochem. Pharmacol.* **21,** 193.

Frontiers in Catecholamine Research 1973, pp. 113 to 115. Pergamon Press. Printed in Great Britain.

THE CELLULAR LOCALISATION AND PHYSIOLOGICAL ROLE OF CATECHOL-O-METHYL TRANSFERASE IN THE BODY

BEVYN JARROTT

Department of Physiology, Monash University,
Clayton, VIC. 3168, Australia

IN ORDER to understand the physiological role of catechol-O-methyltransferase (COMT, EC 2.1.1.6) in the body, it is necessary to know the cellular and subcellular distribution of the enzyme and the factors which govern the rate of O-methylation of catechol compounds in intact cells.

NEURONAL LOCALISATION

The initial studies of COMT activity in nervous tissue demonstrated enzyme activity in all regions of the central nervous system (AXELROD, ALBERS and CLEMENTE, 1959). Subcellular fractionation of brain homogenates shows that approximately 50 per cent of the activity is recovered in the supernatant fraction with the remainder in an occluded form within synaptosomes (ALBERICI, RODRIGUEZ DE LORES ARNAIZ, and DE ROBERTIS, 1965; BROCH and FONNUM, 1972). This suggestion that COMT is localised in nerve terminals has led the latter authors to suggest that COMT may be a better marker for synaptosomes than lactate dehydrogenase. However, the types of neurons in the CNS containing COMT are unknown.

In peripheral tissues, COMT activity is present in homogenates of sympathetic ganglia and postganglionic sympathetic axons (AXELROD et al., 1959; GIACOBINI and KERPEL-FRONIUS, 1969; JARROTT, 1971). Since COMT activity in tissues with a dense sympathetic innervation falls after surgical sympathectomy at a rate similar to the fall in endogenous noradrenaline (JARROTT and IVERSEN, 1971; JARROTT and LANGER, 1971), the presence of COMT in sympathetic nerve terminals may be postulated. In less densely innervated tissues, the fall in COMT activity after sympathectomy is less and in some tissues such as heart, is not significant (JARROTT, 1971). Other studies of the pattern of metabolism of noradrenaline released spontaneously or after sympathetic nerve stimulation of isolated tissues (LANGER, STEFANO and ENERO, 1972) further substantiate the postulate that COMT is present in sympathetic nerve terminals.

EXTRANEURONAL LOCALISATION

In densely innervated tissues, approximately 50–60 per cent of COMT activity remains after sympathectomy, suggesting an extraneuronal localisation of this remaining enzyme. In less densely innervated tissues such as heart, kidney and liver, denervation studies indicate that the bulk of the enzyme is localised in non-neuronal cells. After subcellular fractionation of homogenates of most peripheral tissues, more than 90 per cent of COMT activity is found in the supernatant fraction although in aorta homogenates, a significant proportion is also present in particulate fractions (VERITY, SU and BEVAN, 1972). This, together with pharmacological studies

(Eisenfeld, Landsberg and Axelrod, 1967) suggests that COMT is an intracellular enzyme.

COMT activity is present in homogenates of gliomas (Katz, Goodwin and Kopin, 1969) and cultured astrocytoma cells (Silberstein, Shein and Berv, 1972) and this may account for the almost even distribution of the enzyme throughout the CNS.

FACTORS AFFECTING O-METHYLATION IN INTACT TISSUES

Unfortunately, measurements of COMT activity in tissue homogenates using optimal concentrations of substrates and cofactor do not reflect the extent of O-methylation of noradrenaline in intact tissues. For example, the COMT activity in homogenates of rat and guinea-pig heart is comparable, however, noradrenaline is extensively O-methylated when perfused at high concentration through isolated rat hearts but not in perfused guinea-pig hearts (Jarrott, 1970). This observation led to a study of the factors controlling O-methylation in cardiac muscle. It was hypothesised (Jarrott, 1970) that guinea-pig hearts may have less S-adenosylmethionine available for O-methylation than rat hearts or that guinea-pig hearts lack a mechanism for transporting noradrenaline to COMT. A study of methionine adenosyltransferase (EC 2.5.1.6.), the enzyme that synthesises S-adenosylmethionine from methionine and ATP, shows that the activity in guinea-pig heart homogenates ($14 \cdot 2 \pm 2 \cdot 5$ nmol product formed/hr/g wet wt, $n = 5$) is approximately twice the activity in rat heart homogenates (Jarrott and Zelcer, unpublished) and therefore the first suggestion is unlikely. However, fluorescence histochemical studies show that guinea-pig cardiac muscle cells, in contrast to rat cardiac muscle cells, do not appreciably transport noradrenaline intracellularly (Jacobowitz and Brus, 1971; Jarrott and Zelcer, unpublished). Thus, as COMT is an intracellular enzyme, it is the presence or absence of this extraneuronal transport system rather than the amount of COMT protein or S-adenosylmethionine that is the major determinant of the extent of O-methylation of catecholamines in heart. Similarly, noradrenaline released by field stimulation is extensively O-methylated in the cat nictitating membrane but not at all in the rat vas deferens (Langer, 1970) and yet both tissue homogenates have comparable COMT activity. Fluorescence histochemical and biochemical studies show that the rat vas deferens does not significantly accumulate noradrenaline extraneuronally whereas the nictitating membrane does (Burnstock, McCullough, Story and Wright, 1972; Draskoczy and Trendelenburg, 1970).

PHYSIOLOGICAL ROLE OF COMT

The role of COMT in sympathetic nerve terminals is uncertain but the recent observation that the deaminated metabolites of noradrenaline as well as noradrenaline itself inhibit tyrosine hydroxylase (Rubio and Langer, 1973) suggests that intraneuronal COMT, by O-methylating these catechols, would increase noradrenaline synthesis by reducing feedback inhibition on tyrosine hydroxylase.

In tissues, the formation of normetanephrine from noradrenaline has been regarded as enzymic inactivation of noradrenaline since normetanephrine has approximately 1/1000th the potency of noradrenaline on the beta-adrenoceptors of cardiac muscle (Kukovetz, Hess, Shanfeld and Haugaard, 1959). However, on the alpha-adrenoceptors of nictitating membrane, normetanephrine has a similar potency to noradrenaline (Langer and Rubio, 1973) and thus O-methylation does not result in inactivation of noradrenaline. As the *O*-methylated, deaminated metabolite of noradrenaline (3-methoxy, 4-hydroxy-mandelate) has essentially no pharmacological activity on this latter preparation, it is apparent that both COMT and monoamine oxidase (EC1.4.3.4) are necessary for enzymic inactivation of noradrenaline. Since normetanephrine has a higher affinity for monoamine oxidase than noradrenaline (TIPTON, YOUDIM and SPIRES, 1972), *O*-methylation probably precedes deamination in the formation of 3-methoxy, 4-hydroxy-mandelate. The presence of COMT and MAO in glial cells suggests that one function of these cells in the CNS is to take up and inactivate catecholamines released at synaptic clefts (KATZ *et al.*, 1969).

REFERENCES

ALBERICI M., RODRIGUEZ DE LORES ARNAIZ G. and DE ROBERTIS E. (1965). *Life Sci.* **4**, 1951–1960.
AXELROD J., ALBERS W. and CLEMENTE C. D. (1959). *J. Neurochem.* **5**, 68–72.
BROCH O. J. and FONNUM F. (1972). *J. Neurochem.* **19**, 2049–2055.
BURNSTOCK G., MCCULLOUGH M. W., STORY D. F. and WRIGHT M. E. (1972). *Brit. J. Pharmac.* **46**, 243–253.
DRASKOCZY P. R. and TRENDELENBURG U. (1970). *J. Pharmac. exp. Ther.* **174**, 290–306.
EISENFELD A. J., LANDSBERG L. and AXELROD J. (1967). *J. Pharmac. exp. Ther.* **158**, 378–385.
GIACOBINI E. and KERPEL-FRONIUS S. (1969). *Acta. physiol. scand.* **75**, 523–529.
JACOBOWITZ D. and BRUS R. (1971). *Eur. J. Pharmac.* **15**, 274–284.
JARROTT B. (1970). *Br. J. Pharmac.* **38**, 810–821.
JARROTT B. (1971). *J. Neurochem.* **18**, 17–27.
JARROTT B. and IVERSEN L. L. (1971). *J. Neurochem.* **18**, 1–6.
JARROTT B. and LANGER S. Z. (1971). *J. Physiol. Lond.* **212**, 549–559.
KATZ R. I., GOODWIN J. S. and KOPIN I. J. (1969). *Life Sci.* **8**, 561–569.
KUKOVETZ W. R., HESS M. E., SHANFELD J. and HAUGAARD N. (1959). *J. Pharmac. exp. Ther.* **127**, 122–127.
LANGER S. Z. (1970). *J. Physiol., Lond.* **208**, 515–546.
LANGER S. Z. and RUBIO M. C. (1973). *Arch. Pharmacol.* **276**, 71–88.
LANGER S. Z., STEFANO F. J. E. and ENERO M. A. (1972). *J. Pharmac. exp. Ther.* **183**, 90–102.
RUBIO M. C. and LANGER S. Z. (1973). *J. Pharmac. exp. Ther.* (submitted for publication).
SILBERSTEIN S. D., SHEIN H. M. and BERV K. R. (1972). *Brain Res.* **41**, 245–248.
TIPTON K. F., YOUDIM M. B. H. and SPIRES I. P. C. (1972). *Biochem. Pharmac.* **21**, 2197–2204.
VERITY M. A., SU C. and BEVAN J. A. (1972). *Biochem. Pharmac.* **21**, 193–201.

Frontiers in Catecholamine Research 1973, pp. 117 to 119. Pergamon Press. Printed in Great Britain.

IMMUNOLOGICAL CHARACTERIZATION OF CATECHOL-O-METHYLTRANSFERASE

C. R. Creveling, R. T. Borchardt* and C. Isersky

National Institutes of Health, Bethesda, Maryland 20014, U.S.A.

Rabbits immunized with pure catechol-O-methyltransferase (COMT) EC. 2.1.1.6, responded by forming first one and later three distinct populations of enzyme-related antibodies. The antisera were shown to precipitate and neutralize the enzyme.

The enzyme protein used for the immunization was obtained from the soluble fraction of rat liver using the procedure of Nikodejevic et al. (1970) followed by affinity chromatography on Agarose 4B to which dopamine was attached through the amine group, gel filtration on Sephadex G-100 and filtration through an Amicon PM-30 membrane. The molecular weight of the protein was $23,000 \pm 700$ as determined by SDS-acrylamide disc gel electrophoresis, gel filtration on Sephadex G-50, 75 and 100, amino acid analysis and sedimentation equilibrium analysis. While this form of COMT was chosen as the antigen, present evidence suggests that at least two other forms of COMT with molecular weights of 11,250 and 37,000 are present in the soluble fraction from rat liver. Rabbits were immunised by intra-muscular injection of 0·4 mg antigen in CFA at day 0 followed by 0·4 mg in inc.FA at day 14, and thereafter repeated subcutaneous injections of 0·4 mg antigen in PBS at two-week intervals. Antisera obtained at day 14, 21 and 25 gave a single precipitin line on immunodiffusion and electrophoresis against COMT ranging in purity from crude liver supernatant (at high concentrations only) to pure 23,000-COMT. Antisera obtained subsequent to subcutaneous administration of the antigen show the gradual appearance of two additional COMT-related antibody populations as shown in Fig. 1. It is suggested that this heterogeneous response is due to the presence of three molecular species of the enzyme arising from the 23,000-COMT. These enzyme species would share some antigenic determinants and hence give partial antigenic identity. The antisera obtained prior to day 26, designated early bleeding antibody (EB), neutralised only 50 per cent of the COMT activity in a partially purified preparation of enzyme. The remaining enzyme activity was completely neutralised by addition of the "late bleeding antisera (LB)," the neutralisation of enzyme activity with the LB antisera followed the precipitin curve. Since complete neutralisation of the pure 23,000-COMT could be demonstrated with the EB antibody it appears that the higher molecular weight component has antigenic determinates absent in the 23,000-molecular weight component. At the present time the characterisation of the 11,250-COMT is incomplete. This molecular species has been determined by sedimentation equilibrium studies in the presence of 10^3 M mercapto-ethanol and by assuming the presence of a single cysteine residue per mole of monomer in amino acid analysis. Immunoelectrophoretic analysis also indicated presence of three enzyme related components. A recent report (White and Wu, 1973), however, has indicated the presence of such a low molecular species in the presence of high

---

* University of Kansas, Lawrence, Kansas 66044

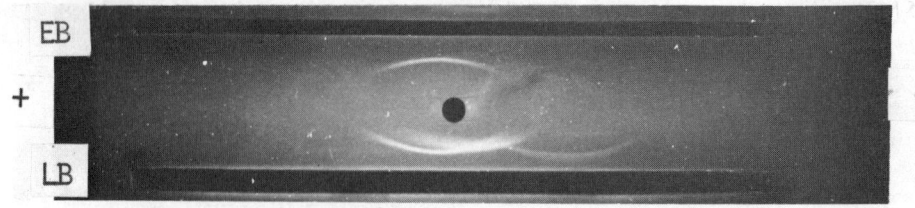

FIG. 1.—Immunoelectrophoresis of COMT against EB antisera (upper well) and against LB antisera (lower well). Noble Difco 1·5%; 0·2 m Tris buffer, pH 8·2; 30 mA for 45 min.

concentrations of dithiothreitol with preparations of COMT from rat brain and liver.

The 23,000-COMT contains two sulfhydryl groups, one of which is essential for enzymatic activity (MORRIS, MCNEAL and CREVELING, 1973). The majority of antigenic determinates are independent of these two sulfhydryl groups one of which is present in the active centre of the enzyme. COMT isotopically labelled by reaction of the two sulfhydryl groups with ^{14}C-N-ethylmaleimide was completely precipitated by the LB antisera and the resulting precipitation curve was identical to one obtained with native enzyme. This reaction provides a basis for a radio immunoassay for COMT.

Neutralisation and immunodiffusion studies have indicated rodent interspecies cross reactions in three strains of rat including the Aoki spontaneously hypertensive rat, mouse and guinea pig. Organ cross reaction was shown with COMT from rat liver, kidney, heart, blood vessels and brain. The present antisera did not show cross reaction with COMT from the liver of monkey, cat, dog and man. Of particular interest was the observation that liver microsomal COMT from rodents was neutralised by antibody to the soluble enzyme. Although it is known that the level of COMT activity associated with the microsomal fraction of liver varies considerably in different species and that certain biochemical parameters are common to the soluble and microsomal enzyme (CREVELING et al., 1972) the nature of this membrane-bound COMT remains obscure. While complete neutralisation of the COMT activity associated with rat liver microsomes was accomplished with the LB antibody no precipitin reaction could be demonstrated by immunodiffusion. Therefore it appeared that the molecular size of the membrane fragments of liver microsomes were greater than 10^6 since IgM with a molecular weight greater than 10^6 penetrates the 1·5% agar used for immunodiffusion. Purification of the microsomal COMT from rat liver microsomes and an associated membrane-bound thiolmethyltransferase (TSMT) was accomplished by classical differential centrifugation, repeated resuspension and sedimentation in isotonic sucrose until no COMT activity could be demonstrated in the 100,000 × g supernatant, centrifugation on a discontinuous sucrose gradient between 0·35 and 2·0 M sucrose, solubilization by treatment with 0·3% Triton-100 and chromatography on Sephadex G-200 yielding a preparation of solubilised membrane fragments bearing both enzymatic activities with molecular weights in excess of 200,000. The final preparation was chromatographed on Agarose 4B resulting in the separation of 90 per cent of the membrane-bound protein and TSMT activity from the membrane-bound COMT. The purified membrane-bound COMT fraction was neutralised by anti COMT and no neutralisation could be demonstrated against the

TSMT activity. Immunodiffusion studies with the purified membrane-bound COMT gave negative results suggesting that the molecular size of the membrane fragment was still in excess of 10^6. However when the preparation was incubated with 1% SDS for 30 min at pH 7·2 and then subjected to disc gel electrophoresis several high molecular weight proteins ($>$60,000) and two small proteins were demonstrated with

COMPARISON OF MEMBRANE BOUND COMT
WITH SOLUBLE COMT

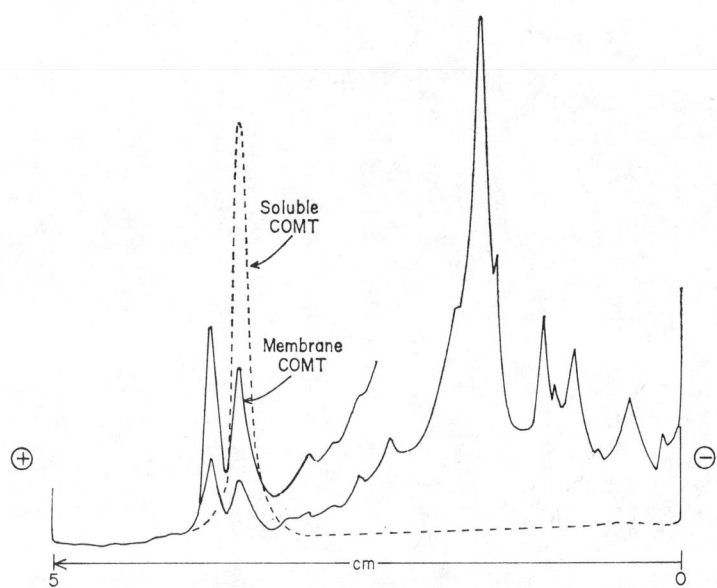

Fig. 2.—Electrophoretic mobility of proteins from purified microsomal membranes following treatment with 1 per cent SDS (7 per cent polyacrylamide, pH 8·3).

molecular weights of 23,000 and 15,000 respectively. The 23,000 molecular species obtained from purified microsomal membranes appeared to be isographic with authentic 23,000-COMT as illustrated in Fig. 2. While the immunological characterisation of the isographic protein obtained from purified liver microsomes is not complete the present evidence strongly suggests that when separated from the membrane it is similar to soluble COMT.

REFERENCES

NIKODEJEVIC B., SENOH S., DALY J. W. and CREVELING C. R. (1970). *J. Pharmacol. Exp. Ther.*, **174**, 83–93.
WHITE H. L. and WU J. C. (1973). *Fedn. Proc.* **32**, 797.
MORRIS N. D., McNEAL E. and CREVELING C. R. (1973) ACS Middle Atlantic Regional Meeting, Jan. 14, Washington, D.C., Abs. F-19 (1973).
CREVELING C. R., MORRIS N., SHIMIZU H., ONG H. H. and DALY J. (1972) *Mol. Pharmaocol.*, **8**, 398–409.

Frontiers in Catecholamine Research 1973, pp. 121 to 128. Pergamon Press. Printed in Great Britain.

NEW CONCEPTS ON THE *N*-METHYLATION REACTIONS OF BIOGENIC AMINES IN ADRENAL MEDULLA AND BRAIN

PIERRE LADURON

Department of Neurobiochemistry, Janssen Pharmaceutica, Beerse, Belgium

N-METHYLATION OF DOPAMINE AS A PREVIOUS STEP TO ADRENALINE FORMATION

IT IS generally accepted that adrenaline is enzymatically synthesised according to the sequence dopamine → noradrenaline → adrenaline. Up to recently, noradrenaline was thought to be the immediate precursor of adrenaline and nearly all the biochemical data seemed to converge to this view (AXELROD, 1966). Nevertheless this central dogma of the adrenergic neurotransmission has recently been questioned and the purpose of the present paper is to provide the main outlines allowing to consider the *N*-methylation of dopamine not as an alternative pathway but, more likely, as the major pathway in the biosynthesis of adrenaline.

In studying the intracellular localisation of enzymes involved in the biosynthesis of catecholamines in the adrenal medulla, we concluded (LADURON and BELPAIRE, 1968) that tyrosine hydroxylase (EC 1.10.3.1) is not contained within the catecholamine granules (Fig. 1) as it had previously been interpreted erroneously (NAGATSU *et al.*, 1964; UDENFRIEND 1966). The use of a more selective method of tissue fractionation, led us to propose a model which implies that the first two steps, i.e. hydroxylation of tyrosine and decarboxylation of dopa, occur outside the granules and that, after being taken up in the granules dopamine is converted into noradrenaline by the dopamine-β-hydroxylase (EC 1.14.2.14) which is certainly within the granules (Fig 1). Finally, since phenylethanolamine-*N*-methyltransferase (PNMT) (EC. 2.1. 1.X) is localised in the cytosol (Fig. 1) it was suggested that, in the adrenal medulla, a certain amount of noradrenaline, when coming out of the granules, is *N*-methylated into adrenaline which, afterwards, should return to the granules to be stored. It is precisely this point that raised difficulties in forwarding a valid interpretation for the intracellular pathway of catecholamines.

Why such a translocation? What is the physiological meaning of a translocation from an intragranular to an extragranular compartment and vice versa? How can the granular membrane possess a so highly elaborate screening mechanism, capable of recognising one specific kind of amine and allowing it to cross in a given direction? Many questions but no answers. In referring to the classical picture described in nearly all the textbooks and showing a diffusion phenomenon of catecholamines between an intra- and an (hypothetical) extragranular pool, it should be very easy to solve this problen. However, such a diffusion which is certainly no longer accepted for the process of catecholamine release, remains purely speculative and experimental evidence to support this view is still lacking. Nevertheless, this translocation mechanism for noradrenaline and adrenaline appeared to be the only possibility in 1968, since "*N*-methyltransferase, as AXELROD (1966) claimed, shows an absolute specificity towards phenylethanolamine derivates, none of the phenylethylamines being *N*-methylated".

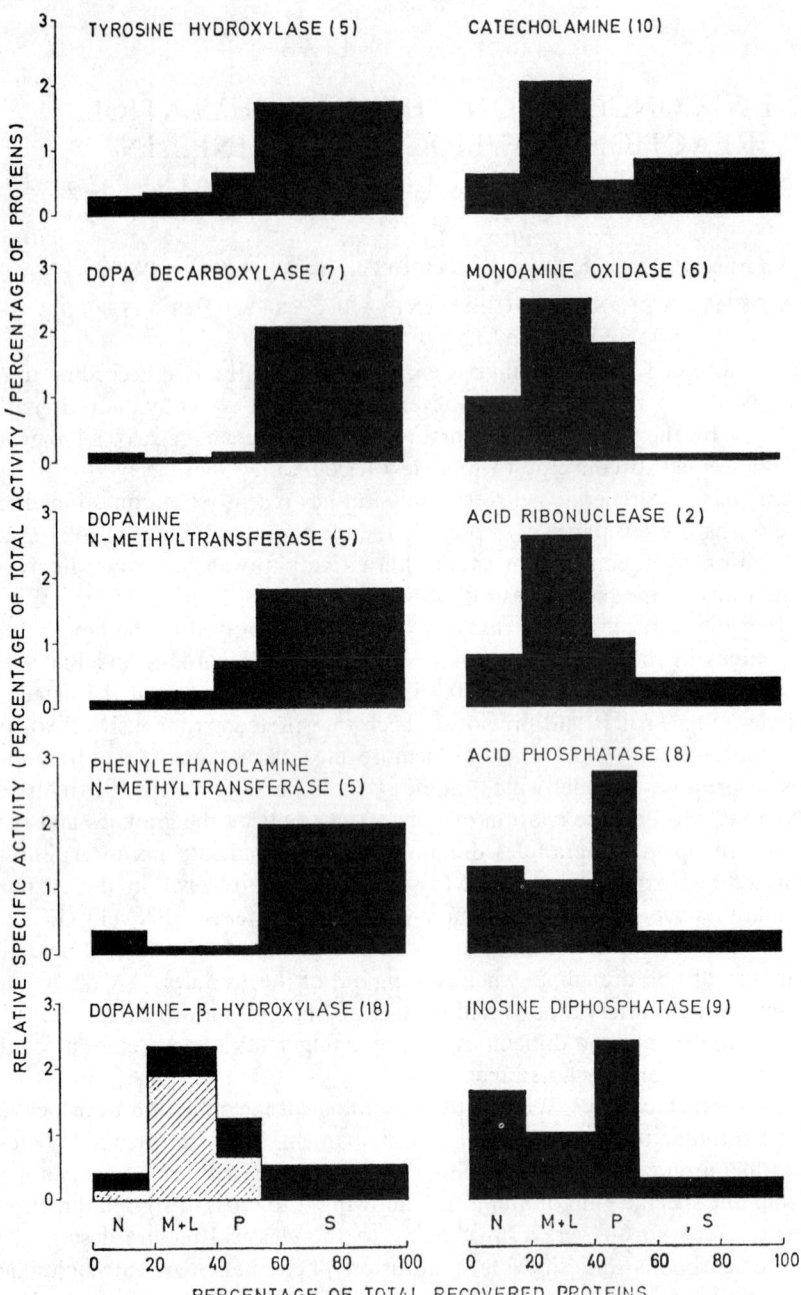

FIG. 1.—Distribution pattern of enzymes in various fractions obtained by differential centrifugation from bovine adrenal medulla. Homogenisation was carried out in 0·25 M sucrose, except in 0·16 M KCl for tyrosine hydroxylase. N = nuclear fraction; M + L = large mitochondrial fraction; P = microsomal fraction; S = supernatant. Each graph represents the mean of several experiments (number in parentheses). Shaded areas represent latent activities of dopamine β-hydroxylase (i.e. total activity, with Triton X-100, minus free activity).

Many considerations have prompted us to find out whether the formation of adrenaline could not be provided through an other biosynthetic pathway, apparently more logical and involving the *N*-methylation of dopamine. Firstly the concept of latency of dopamine-β-hydroxylase (BELPAIRE and LADURON, 1968) which must be attributed to the lack of permeability of granular membranes towards a given substrate, strengthened the idea of a compartmental structure in the chromaffin cells and therefore, was incompatible with a flow of amines from the granules into the cytosol and vice versa. Secondly, the concept of exocytosis, the introduction of which has radically changed the way of thinking about adrenergic neurotransmission, rendered highly improbable and is even opposed to the idea that a neurotransmitter must leave its storage site by diffusing through the cytoplasmic compartment (SMITH and WINKLER, 1972). Finally a third reason was that, contrary to the generally accepted opinion, we have never believed that exogenous noradrenaline can be taken up by the sites (synaptic vesicle or granule) which bind and store endogenous amines (LADURON *et al.*, 1966; LADURON 1969).

It has been recently demonstrated that dopamine can be converted to epinine *in vitro* by *N*-methyltransferase of bovine adrenal medulla (LADURON, 1972a). Although the properties and the identification of this reaction have already been described in detail, I would like, however, to draw the attention to some important cautions. Firstly, as it became rapidly evident that only quite irregular results could be obtained when dopamine replaced normetanephrine in the incubation mixture commonly used for the adrenal *N*-methyltransferase, some experimental conditions were then modified. The concentration of dopamine was first increased to reach 2 mM, which has allowed to observe the appearance of black colour in the tube throughout the incubation period. This oxidative process was then prevented by adding a mixture of metabisulphite and EDTA. This latter compound was found useful, and sometimes necessary, to inhibit remaining dopamine-β-hydroxylase in an incompletely purified enzyme preparation. However, both these compounds were not needed for incubation when the enzyme preparation was submitted to a more extensive purification through a Sephadex G-200 column In addition to this oxidation problem, special care must be taken of the dialysis period after precipitation with ammonium sulphate and purification on Sephadex, significant amounts of noradrenaline and adrenaline were still detectable in such preparations, which, therefore, could explain the high values generally observed for the blanks. Consequently, the occurrence of extremely high concentration of catecholamines in the adrenal medulla (about 5 mg per g tissue) represents a real drawback and, probably also, a source of errors in experiments dealing with the *in vitro* *N*-methylation of dopamine and much more especially in experiments using tissue slices. Figure 2 clearly illustrates this statement. For a 3×10^{-7} M concentration of noradrenaline, the *in vitro* conversion of dopamine into epinine was completely inhibited. Similarly, adrenaline also exerts an inhibitory effect but at lower concentrations.

Another point which has apparently given rise to difficulties to decide whether dopamine can be considered as a normal substrate for the *N*-methyltransferase, concerns the Michaelis–Menten constant for dopamine. At first sight, noradrenaline would seem to be a more appropriate substrate owing to its lower K_m (ranging between 10^{-4} and 10^{-6} M according to the data of the literature) as compared with 2.7×10^{-3} M for dopamine (LADURON, 1972a). Firstly, it must be emphasised that a

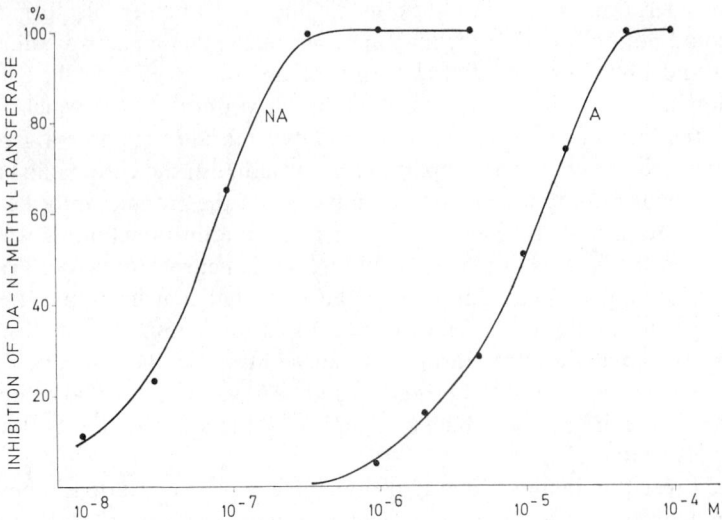

FIG. 2.—*In vitro* inhibition of dopamine *N*-methyltransferase with noradrenaline (NA) and adrenaline (A). Enzyme was partially purified by ammonium sulphate precipitation from a bovine adrenal medulla supernatant.

typical substrate inhibition curve has been reported for noradrenaline (FULLER *et al.*, 1970; CONNETT and KIRSHNER, 1970) so that saturation can never be obtained and the calculation of K_m values is rendered highly approximate. Let us recall too that the K_m which is defined as the substrate concentration at which half the maximum velocity is attained, is an index of the affinity of an enzyme for a substrate only under the Michaelis–Menten assumptions that K_m is equal to K_s, the enzyme–substrate dissociation constant. This assumption is certainly not justified for all reactions especially for those with two substrates or involving several enzyme–substrate intermediate complexes. Furthermore, contrary to an opinion often held, a higher affinity of an enzyme for a given substrate does not mean that this substrate should be the preferred substrate for the enzyme *in vivo*. In this regard, many synthetic substrates sometimes possess lower K_m than substrates commonly found in nature. For *N*-methyltransferase 3,4-dichlorophenethanolamine has been reported with a tenfold lower K_m than that of noradrenaline (FULLER *et al.*, 1970). Therefore the comparison of two K_m values is not necessarily a valid indication that one substrate rather than another is preferentially used *in vivo*. Consequently, it seems to us much more important to consider the intracellular distribution of a given substrate with respect to the localisation of the enzyme susceptible of binding it. In this regard, it is beyond doubt that dopamine is present in the cytosol since dopa-decarboxylase is certainly located in this compartment. (Fig. 1). However, there is no evidence supporting the view of an extragranular pool for noradrenaline. Even the concept of exocytosis makes this possibility quite unlikely (SMITH and WINKLER, 1972). For these reasons and in spite, of its high K_m, very similar to that for dopamine-β-hydroxylase, dopamine may be considered as a valid substrate for adrenal *N*-methyltransferase which is certainly much less rapidly saturated with it than with noradrenaline. The experiments performed *in vivo* and presented later on are quite consistent with this view.

Although the *in vitro* formation of epinine is now well established in the adrenal medulla, this does not mean that such a conversion can also occur *in vivo*. However, if epinine could be isolated from adrenal glands or *in vivo* synthesised from labelled precursors, one might put forward a general concept on the physiological role of this amine. Up to now, epinine has only been reported to occur normally in the parotid gland of *Bufo marinus* where an "unspecific" *N*-methyltransferase for dopamine was also described (MÄRKI *et al.*, 1962). As already mentioned (LADURON, 1973) the use of column and thin layer chromatography enabled us to isolate endogenous epinine from bovine adrenal medullae. For this, acid extracts of the medullae were submitted to three different runs of chromatography; firstly by adsorption on Al_2O_3 columns and then on Dowex to separate dopamine and epinine from the other catecholamines. Finally these two amines were separated and identified on thin layer using phenol HCl as solvent. This previous finding was now confirmed by means of other methods like gas chromatography and mass fragmentography (in preparation). Moreover, retrogradely perfusing bovine adrenal medullae with radioactive tyrosine enabled us to isolate labelled epinine. Therefore the formation of this catecholamine can also be demonstrated *in vivo*. Consequently, the foregoing data allow us to conclude that epinine is an important catecholamine in adrenal medulla and, most probably the immediate precursor of adrenaline. A new model for the intracellular biosynthesis of catecholamines has therefore been proposed presenting the great advantage of avoiding a translocation of noradrenaline and adrenaline from the granular to the extragranular space and vice versa (LADURON, 1972). In this model, two new steps are now well established, the *N*-methylation of dopamine and the β-hydroxylation of epinine. Indeed the conversion of epinine to adrenaline has been reported to occur much more easily than the β-hydroxylation of dopamine, since, with epinine as substrate, the reaction did not require ascorbic acid as cofactor (BRIDGERS and KAUFMAN, 1962). Recent confirmation of this has been obtained in our laboratory (unpublished results). This could perhaps explain the relatively low amount of endogenous epinine found in the adrenal medulla (very approximately one 100th of the amount of dopamine).

Concerning the uptake of epinine, preliminary experiments have indicated that this amine can be taken up by the granules of the adrenal medulla. However, as the radioactive epinine provided by a reaction mixture and then purified, still contained a little unlabelled dopamine, the real rate of uptake could not to be exactly measured. Further experiments must be performed to compare the rate at which this amine is taken up as compared with the other catecholamines such as dopamine, which has been found to be preferentially taken up (LADURON and BELPAIRE, 1968). It is noteworthy that good evidence for a similar uptake of noradrenaline or adrenaline is still lacking since the granular preparations used were generally contaminated by other subcellular organelles and since a distinction between an adsorption on the external membrane and a penetration within the granules has never been seriously considered. On the other hand, this drawback can be easily overcome for dopamine and epinine since the β-hydroxylation of both amines is closely related to a specific uptake within the granules.

Finally it must be stressed that after depletion of catecholamines, noradrenaline seems to be preferentially resynthesised in the adrenal medulla (BUTTERWORTH and MANN, 1957). Such a phenomenon is entirely compatible with our model. Indeed, after depletion of catecholamines or throughout an increase of their turnover for some reason, a higher amount of dopamine can be taken up to refill granules with noradrenaline newly formed from dopamine, whereas when this restoration is finished, more dopamine should be set free for the N-methylation process. Afterwards, noradrenaline should be progressively replaced by adrenaline in the granule. In this regard, we put forward this hypothesis according to which a relatively high content of noradrenaline in the adrenal glands of some species, should be the feature of a rapid turnover. Conversely, a slow turnover should indicate a higher content of adrenaline.

N-METHYLATION OF BIOGENIC AMINES IN BRAIN WITH A NEW METHYL DONOR

Although different types of N-methyltransferase in the brain have already been described (see ref. LADURON, 1973), the possibility of N-methylation of dopamine in this tissue has never been considered. Although the origin of schizophrenia still remains unknown, one among numerous hypotheses suggests that the disease should be due to an excessive N-methylation process in the brain. Moreover, since the antipsychotic action of neuroleptics seems to be related to a dopamine-receptor blocking activity, dopamine or other compounds structurally related to it, could either induce schizophrenia or perhaps and more likely, constitute intermediate neurotransmitters for the clinical manifestations of this disease. These reasons prompted us to investigate the N-methylation of dopamine in the brain.

It soon appeared that S-adenosylmethionine might not be the methyl donor of this N-methylation process. Nevertheless the in vitro formation of epinine from a rat brain enzyme became only possible when 5-methyltetrahydrofolic acid replaced S-adenosylmethionine in the incubation mixture.

Our recent work provided the first evidence that a catecholamine can be N-methylated by an enzyme using 5-methyltetrahydrofolic acid (5-MTHF), an unusual methyl-donor for biogenic amines, yet very common in single-carbon transfer reactions (LADURON, 1972b). This new reaction, originally described with dopamine as substrate, may be written as follows:

dopamine + 5-methyltetrahydrofolic acid $\xrightarrow{\text{N-methyltransferase}}$

epinine + tetrahydrofolic acid

Many properties of this reaction have already been reported in detail (LADURON, 1972b; LADURON et al., 1973; LEYSEN et al., 1973). I should only like to stress the fact that the pH optimum for this reaction was recently reported to be equal to 6·4 (LADURON et al., 1973) which differed markedly from that previously found (pH 8·2) but under different experimental conditions, namely in the presence of metabisulphite (LADURON, 1972b). It may be assumed that this reducing agent should interfere with the enzymesubstrate binding unless the enzyme conformation should be modified by it.

The apparent K_m value for 5-MTHF with dopamine as substrate is equal to 6×10^{-5} M. However, attempts to determine K_m values for dopamine or 4-methoxy-3-hydroxyphenylethylamine showed an anomalous type of kinetic behaviour. Instead of a straight line, a concave downward curve with pseudo-linear sections was obtained with dopamine as well as with 4-MDA when the velocity was plotted $1/v$ vs $1/s$. According to TEIPEL and KOSHLAND (1969), such a kinetic behaviour reveals the existence of a negative co-operativity phenomenon. If this regulatory property can be confirmed by more extensive kinetic studies, together with further elucidation of the structural features of this protein, it should be of prime importance in controlling the occurrence of *N*-methylated compounds, namely in schizophrenic disorders.

As reported in Table 1, several substrates have been tested for their ability to be *N*-methylated. Nearly all the catecholamines and indoleamines, (LEYSEN and LADURON, 1973) except the tertiary amines, were found to be active substrates *in vitro* whereas histamine could not be *N*-methylated with 5-MTHF as the methyl donor. For this latter there is a specific histamine *N*-methyltransferase using *S*-adenosyl-methionine as the methyl donor (BROWN *et al.*, 1959). Although the reaction products were not identified for each of these amines, the use of different derivatives substituted for the nitrogen (primary, secondary or tertiary amine) allowed to assess the nature of the *N*-methylation reaction.

TABLE 1. SUBSTRATE FOR *N*-METHYLTRANSFERASE WITH 5-MTHF IN RAT BRAIN

Active	Inactive
Dopamine	*N,N'*-dimethyldopamine
Epinine	3,4-dihydroxyacetic acid
3-hydroxy-4-methoxyphenylethylamine	Bufotenin
3-methoxy-4-hydroxyphenylethylamine	Histamine
3,4-methoxyphenylethylamine	Dimethylamphetamine
Tryptamine	
Serotonin	
N-methylserotonin	
Amphetamine	
N-methylamphetamine	
Noradrenaline	
Adrenaline	
Normetanephrine	
Mescaline	

A simple and rapid method to determine the *N*-methyltransferase using 5-MTHF has been developed with 4-MDA as the substrate (LADURON *et al.*, 1973). Purification of the enzyme was carried out either by sedimentation through a sucrose gradient or by gel filtration through a Sephadex G-200 column. This latter method allowed to yield an *N*-methyltransferase completely free of COMT. In addition to this enzyme preparation, purified COMT, was used to determine the specificity of the methyl donor in the O- and *N*-methylation in the rat brain. As far as we have observed, COMT required *S*-adenosylmethionine as the methyl donor. Most catecholamines and indoleamines however were preferentially *N*-methylated in the rat brain by an *N*-methyltransferase using 5-MTHF. The extremely low and irregular activities measured in the presence of *S*-adenosylmethionine with tryptamine or serotonin were not significant enough to consider this reaction as possible *in vivo* (LEYSEN *et al.*, 1973). Therefore, most of the biogenic amines (except histamine) are *N*-methylated in the brain by means of 5-MTHF whilst *S*-adenosylmethionine seems to be more specifically involved in the O-methylation reactions.

Since catecholamines and indoleamines can be converted *in vitro* into their corresponding *N*-methyl- or *N,N'*-dimethyl derivatives, it is tempting to attribute to folic acid and to its congeners, a prominent role in the metabolism of certain cerebral amines either under normal or pathological conditions. In this regard, a new hypothesis has been formulated to explain the possible origin of schizophrenic disorders

(LADURON, 1972b, 1973). It is based upon the assumption that the enhanced *N*-methylation process observed in schizophrenia should not be related to the methionine but to the folic acid and its main metabolite, 5-methyltetrahydrofolate. An excess of folate coenzymes could give rise to an increased or unusual formation of *N*-methylated amines like epinine, *N*,*N'*-dimethyltryptamine, bufotenine etc. or their methoxylated derivatives. It is worth nothing that 5-MTHF is taken up preferentially in neural tissues (ALLEN and KLIPSTEIN, 1970), which explains why high concentrations of 5-MTHF were found in the spinal fluid (HERBERT and ZALUSKY, 1961; LEVITT *et al.*, 1971). Our hypothesis has already received an indirect support in a preliminary clinical trial where folic acid was given to psychotic patients (VAN LOMMEL; unpublished results quoted by LADURON, 1973).

More recently, an enzyme able to hydrolyse 5-MTHF has been identified in our laboratory (in preparation). This enzyme, most probably a lysosomal one, could play an important role to maintain a constant level of folate congeners in the neural tissue. It has been postulated that this enzyme should be defective or its activity decreased in schizophrenia (LADURON, 1973).

REFERENCES

ALLEN C. D. and KLIPSTEIN F. (1970) *Neurology* 4, 403.
AXELROD J. (1966) *Pharmacol. Rev.* 18, 95–113.
BELPAIRE F. and LADURON P. (1968) *Biochem. Pharmacol.* 17, 411–421.
BRIDGERS W. F. and KAUFMAN S. (1962) *J. Biol. Chem.* 237, 526–528.
BUTTERWORTH K. R. and MANN M. (1957) *Nature, Lond.* 179, 1079–1080.
CONNETT R. J. and KIRSHNER N. (1970) *J. Biol. Chem.* 245, 329–334.
FULLER R. W., WARREN B. J. and MOLLOY B. B. (1970) *Biochem. Biophys. Acta* 222, 210–212.
HERBERT V. and ZALUSKY R. (1961) *Proc. Am. Soc. Exp. Biol.* 20, 453–457.
KIRSHNER N. (1959) *Pharmacol. Rev.* 11, 350–357.
LADURON P. (1969) Biosynthèse, localisation intracellulaire et transport des catécholamines. Thesis, Vander, Louvain.
LADURON P. (1972a) *Arch. Int. Pharmacodyn.* 195, 197–208.
LADURON P. (1972b) *Nature, New Biology* 238, 212–213.
LADURON P. (1973) In: *Advances of Neuropsychopharmacology* (in press).
LADURON P. and BELPAIRE F. (1968) *Biochem. Pharmacol.* 17, 1127–1140.
LADURON P., DE POTTER W. and BELPAIRE F. (1966) *Life Sci.* 5, 2085–2094.
LADURON P. M., GOMMEREN W. R. and LEYSEN J. E. (1973) *Biochem. Pharmacol.* (in press).
LEVITT M., NIXON P. F., PINCUS J. H. and BERTINO J. R. (1971) *J. Clin. Invest.* 50, 1301–1308.
LEYSEN J. and LADURON P. (1973) In: *Abstracts of International Symposium on 5-hydroxytryptamine* (COSTA E. and SANDLER M., Eds.), Raven Press, New-York (in press).
LEYSEN J. E., GOMMEREN W. R. and LADURON P. M. (1973) *Biochem. Pharmacol.* (in press).
MÄRKI F., AXELROD J. and WITKOP B. (1962) *Biochem. Biophys. Acta* 58, 367–369.
NAGATSU T., LEVITT M. and UDENFRIEND S. (1964) *J. Biol. Chem.* 239, 2910–2917.
SMITH A. D. and WINKLER H. (1972) In: *Catecholamines* (BLASCHKO H. and MUSCHOLL E., Eds.), pp. 538–617. Springer-Verlag, Berlin.
TEIPEL J. and KOSHLAND D. E., JR. (1969) *Biochemistry* 8, 4556–4663.
UDENFRIEND S. (1966) *Harvey Lect.* 60, 57–83.

Frontiers in Catecholamine Research 1973, pp. 129 to 131. Pergamon Press. Printed in Great Britain.

RINGDEHYDROXYLATION AND *N*-METHYLATION OF NORADRENALINE AND DOPAMINE IN THE INTACT RAT BRAIN

K. BRANDAU* and J. AXELROD

Laboratory of Clinical Science, National Institute of Mental Health,
Bethesda, Maryland 10014, U.S.A.

INVESTIGATIONS of the metabolism of catecholamines in the rat brain showed that intraventricularly as well as intracisternally injected radioactively labelled noradrenaline, dopamine, and octopamine were metabolised resulting in glycol and ethanol metabolites. Therefore, the main pathway of β-hydroxylated phenylethylamines leads to neutral alcoholic metabolites whereas the acid metabolites appear to be only minor products of the physiological metabolism in the rat brain (BREESE et al., 1969).

In a study concerning the regulation of the octopamine biosynthesis it was observed that after administration of L-Dopa to rats which were pretreated with a monoaminoxidase inhibitor the endogenous concentration of octopamine in the brain increased (BRANDAU and AXELROD, 1972). This result indicated the possibility that octopamine might be formed from L-Dopa or its decarboxylated metabolite dopamine by a ringdehydroxylating reaction.

Furthermore, *N*-methylation of catecholamines is mainly to find in the adrenal glands and in the liver. Presumably it takes place also in the brain (CIARANELLO et al., 1969; MCGEER and MCGEER, 1964).

We wish to report another minor metabolic route for the transformation of the catecholamines dopamine and noradrenaline in the rat brain. This involves ringdehydroxylation to form monophenols and *N*-methylation.

Male Sprague–Dawley rats received dopamine-2-^{14}C or L-noradrenaline-7-^{3}H intracisternally. The animals were pretreated with a monoaminoxidase inhibitor. At either 30 or 45 min after the injection of the radioactive catecholamine the rats were killed, the brains immediately removed, homogenised and centrifuged. After adding non-labelled compounds as carriers to the supernatant the metabolites were separated by column chromatography in a "catechol" fraction (catechols with two free OH-groups) and in a "non-catechol" fraction (*O*-methylated catechols and monophenolic compounds). Further separation and identification was achieved by means of bi-dimensional thin-layer chromatography.

Calculated as a percentage of the intracisternally injected radioactivity the conversion of dopamine as well as of noradrenaline to ringdehydroxylated metabolites is by far less than 1 per cent. Calculated in per cent of the radioactivity of the "non-catechol" fraction about 2–3 per cent of monohydroxyphenol compounds were present. Not only ringdehydroxylation occurred in the rat brain but also *N*-methylation, since the presence of radioactive synephrine, i.e. the *N*-methylated metabolite of

---

* Visiting Scientist from BAYER AG, Wuppertal-Elberfeld Germany
Requests for reprints should be sent to: Dr. K. Brandau, BAYER AG, P.O. Box 13 01 05, D-56 Wuppertal 1, Germany

6

octopamine, could be shown in relatively high amounts. Additionally a very small fraction of *p*-hydroxymandelic acid could be found.

These results clearly demonstrate, that in the rat brain catecholamines can be metabolised not only by deamination and *O*-methylation but on a minor pathway also

TABLE 1. RATE OF METABOLISM OF INTRACISTERNALLY INJECTED DOPAMINE AND NORADRENALINE IN THE RAT BRAIN

		dopamine-^{14}C	noradrenaline-^{3}H
intracisternally injected dose [dpm × 10^{-6}]		17,8	30,2
total radioactivity in supernatant of the brain homogenate after 30 min [dpm × 10^{-6}] [% of injected dose]		4,8 27,0	10,9 36,1
"catechol" fraction [% of injected dose]* [% of supernatant]*		15,7 55,3	13,9 36,6
"non-catechol" fraction [% of injected dose]* [% of supernatant]*		6,9 28,3	11,4 31,8
methoxytyramine		52,3	
normetanephrine	[% of "non-catechol" fraction]	1,3	91
tyramine		2,4	
octopamine		1,5	1,6
synephrine		0,1	0,7

* The radioactivity was determined by liquid scintillation counting. These figures are not quench corrected. This is the reason why the sum of both of these fractions do not give one hundred percent calculated in terms of percentage of the supernatant.

The given figures are average values of three experiments.
The rats were pretreated with a monoaminoxidase inhibitor.

by ringdehydroxylation ("catecholamine ringdehydroxylase") followed by *N*-methylation (phenylethanolamine *N*-methyltransferase). This is shown in Fig. 1.

Independently from our results BOULTON and QUAN (1970) and SANDLER (1971) described the formation of tyramine from dopamine.

It is unlikely that the dehydroxylation reaction reported here is carried out by intestinal bacteria since catecholamines and monophenolic amines cannot cross the blood–brain barrier in either direction. Furthermore we have also demonstrated an increase of endogenous octopamine in rat brain after the administration of L-Dopa to germ free rats (BRANDAU and AXELROD, 1972). Dehydroxylation in mammals represent a unique metabolic pathway and the enzymatic mechanism is presently unknown.

In this presentation it is shown that this conversion is also possible for catecholamines. The opposite way, the formation of catechols from monophenols like octopamine was already demonstrated 10 years ago (AXELROD, 1963). These results lead to an interesting interrelationship between the neurotransmitter (noradrenaline) and the co-neurotransmitter (octopamine) pathways.

However *N*-methylation of octopamine is presumably carried out by phenylethanolamine *N*-methyltransferase (PNMT). Since octopamine is a much better

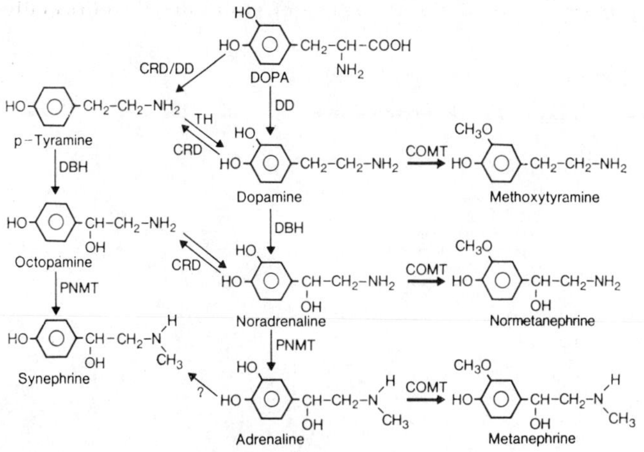

HO

HO—⟨O⟩—CH₂—CH—COOH
              |
             NH₂
DOPA

CRD/DD

DD

HO—⟨O⟩—CH₂—CH₂—NH₂
p–Tyramine

TH

CRD

HO
HO—⟨O⟩—CH₂—CH₂—NH₂   COMT   CH₃O
Dopamine   HO—⟨O⟩—CH₂—CH₂—NH₂
                                  Methoxytyramine

DBH

DBH

HO—⟨O⟩—CH—CH₂—NH₂
              |
             OH
Octopamine

CRD

HO
HO—⟨O⟩—CH—CH₂—NH₂   COMT   CH₃O
              | HO—⟨O⟩—CH—CH₂—NH₂
             OH |
Noradrenaline   OH
                                  Normetanephrine

PNMT

PNMT

HO—⟨O⟩—CH—CH₂—N
              | H
             OH   CH₃
Synephrine

?

HO
HO—⟨O⟩—CH—CH₂—N   H   COMT   CH₃O
              | _   HO—⟨O⟩—CH—CH₂—N   H
             OH   CH₃ | _
Adrenaline   OH   CH₃
                                  Metanephrine

COMT   catecholamine–O–methyltransferase
CRD   "catecholamine ringdehydroxylase"
DBH   dopamine-ß- hydroxylase
DD   dopadecarboxylase
PNMT   phonylethanolamine N–methyltransferase
TH   tyramine hydroxylase

FIG. 1.—Metabolic pathways of dopamine and noradrenaline by ringdehydroxylation in the rat brain.

substrate for PNMT than noradrenaline relatively higher concentrations of synephrine
than of adrenaline can be expected in the brain.

REFERENCES

AXELROD J. (1963) *Science* **140**, 499–500.
BOULTON A. A. and QUAN L. (1970) *Can. J. Biochem.* **48**, 1287–1291.
BRANDAU K. and AXELROD J. (1972) *Naunyn-Schmiedeberg's Arch. Pharmacol.* **273**, 123–133.
BREESE G. R., CHASE T. N. and KOPIN I. J. (1969) *Biochem. Pharmacol.* **18**, 863–869.
CIARANELLO R. D., BARCHAS R. E., BYERS G. S., STEMMLE D. W. and BARCHAS J. D. (1969) *Nature, Lond.* **221**, 368–369.
MCGEER P. L. and MCGEER E. G. (1964) *Biochem. Biophys. Res. Commun.* **17**, 502–507.
SANDLER M., GOODWIN B. L. and RUTHVEN C. R. J. (1971) *Nature Lond.* **229**, 414–415.

Frontiers in Catecholamine Research 1973, pp. 133 to 137. Pergamon Press. Printed in Great Britain,

DEGRADATION OF THE TRANSMITTER AMINES BY SPECIFIC TYPES OF MONOAMINE OXIDASES

N. H. NEFF,[1] H.-Y. T. YANG[1] and C. GORIDIS[2]

[1]Laboratory of Preclinical Pharmacology, National Institute of Mental Health, Saint Elizabeths Hospital, Washington, D. C. 20032, and
[2]Centre National de la Recherche Scientifique Centre de Neurochimie, Strasbourg, France

CHARACTERISATION OF THE MONOAMINE OXIDASES BY SENSIVITY TO DRUGS

MONOAMINE OXIDASE (monoamine: O_2 oxidoreductase (deaminating); (E.C. 1.4.3.4.) exists in multiple forms in mammalian tissue (see COSTA and SANDLER, 1972). JOHNSTON (1968) demonstrated that two forms of enzyme could be revealed in homogenates of rat brain with the drug, clorgyline. For example, with the substrate, tyramine, there was a stepwise inhibition of its deamination in the presence of increasing concentrations of clorgyline (Fig. 1). The enzyme most sensitive to clorgyline was designated as A and the enzyme that was more resistant to clorgyline was designated as B. Other drugs are now known which can, in the proper concentrations, selectively block type A and B enzyme when tested *in vitro* (see COSTA and SANDLER, 1972). Clorgyline, harmine, and Lilly 51641, in low concentrations selectively block type A enzyme (SQUIRES, 1972). In contrast, deprenyl and pargyline in low concentrations selectively block type B enzyme (SQUIRES, 1972). At high concentrations, the specificity of the drugs is lost. SQUIRES (1972) and HALL, LOGAN and PARSONS (1969) have successfully used these drugs to demonstrate the presence of A and B enzyme activities in several species.

Types A and B enzyme have different characteristics. In addition to being more sensitive to clorgyline, type A enzyme has a lower apparent K_m (0.1 mM) for tyramine than type B enzyme (0.8 mM) and it is more heat stable and more readily inactivated by trypsin than type B enzyme (YANG, GORIDIS and NEFF, 1972).

ENDOGENOUS AMINES AND THE SPECIFICITY OF THE MONOAMINE OXIDASES

Figure 1 illustrates the ability of clorgyline to inhibit the deamination of five amines that are normally found in mammalian brain. Using the concentration of clorgyline that blocks the deamination of the amines as an index, it is apparent that serotonin and norepinephrine are specific substrates for type A enzyme whereas β-phenylethylamine is a specific substrate for type B enzyme. Dopamine and tyramine are substrates for both enzymes. Tryptamine, not shown in Fig. 1, is also a substrate for both enzymes.

The concentration of substrate is an important consideration when studying the specificity of the monoamine oxidases. We found that the inhibition of enzyme activity by 0.1 μM clorgyline in a homogenate of pineal gland shifted from about 30 to 50 per cent when the concentration of tyramine was reduced from 2.1 to 0.03 mM. The shift occurs because more of the low K_m enzyme activity (type A enzyme activity) is measured with 0.03 mM tyramine (YANG, GORIDIS and NEFF, 1972).

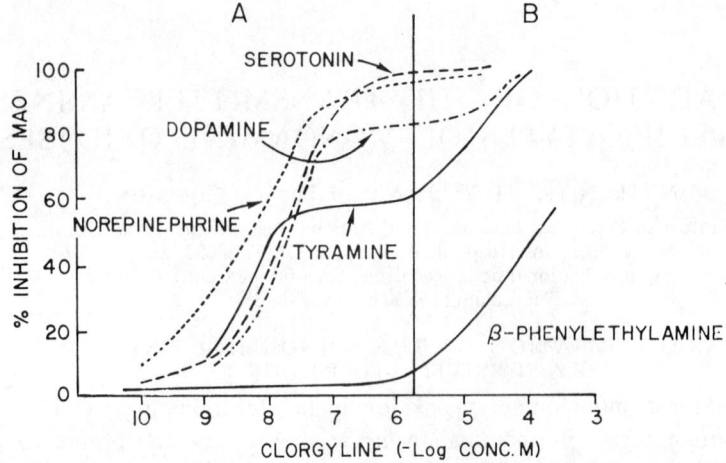

Fig. 1.—The *in vitro* inhibition of the monoamine oxidase activity of rat brain by increasing concentrations of clorgyline using several amine substrates.

THE CONSEQUENCES OF ADMINISTERING DEPRENYL AND CLORGYLINE ON THE METABOLISM OF THE ENDOGENOUS AMINES

The 1000-fold difference in the concentration of clorgyline required to block A and B enzyme activity *in vitro* suggests that it might be possible to block these enzymes selectively *in vivo*. Pargyline, which is similar in structure to clorgyline and deprenyl is an irreversible inhibitor of monoamine oxidase (Erwin and Deitrich, 1971). We found that deprenyl and clorgyline inhibited irreversibly also. When increasing doses of clorgyline or deprenyl were administered intravenously there was a selective blockade of A or B enzyme activity of brain when assayed *in vitro* with the specific substrates for A and B enzymes, serotonin and phenylethylamine, respectively (Fig. 2). Following the administration of clorgyline, which blocks

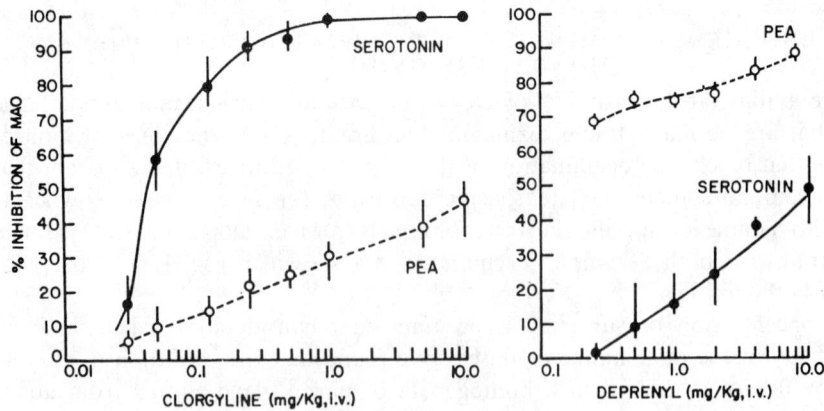

Fig. 2.—Inhibition of monoamine oxidase activity of rat brain versus dose of clorgyline or deprenyl administered intravenously. Animals were killed 2 hr after injection and activity was assayed using serotonin or phenylethylamine as substrate.

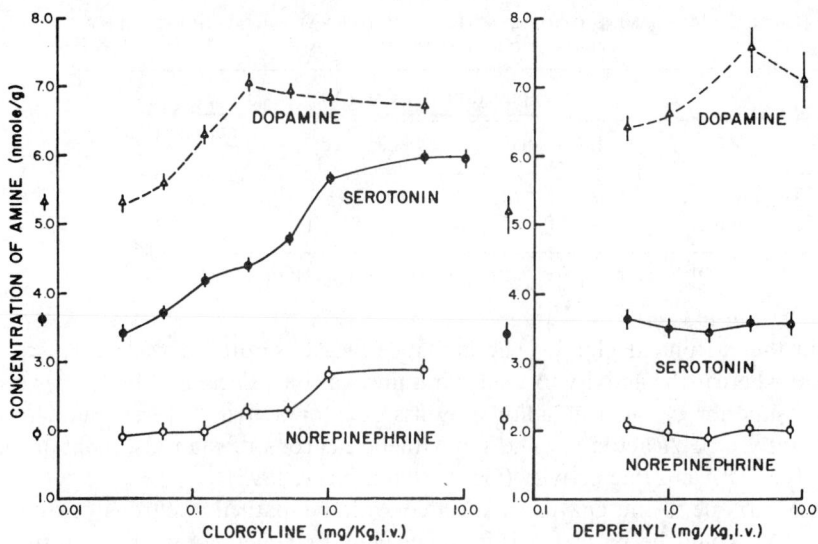

FIG. 3.—Monoamine concentrations in brain 2 hr after administering doses of clorgyline or deprenyl intravenously.

primarily enzyme A activity (Fig. 2), norepinephrine, serotonin, and dopamine concentrations increased in brain (Fig. 3). This was not surprising, as they are all substrates for type A enzyme. In contrast, only dopamine increased in brain following the administration of deprenyl (Fig. 3). Moreover, the metabolism in brain of phenylethylamine, a specific substrate for type B enzyme, was significantly retarded by treatment with deprenyl, but not by clorgyline (Table 1). Thus, the concentrations of the amines can be selectively altered by blocking the enzymes that metabolize the amines in brain.

LOCALIZATION OF THE MONOAMINE OXIDASES

The customary procedure for evaluating the types of monoamine oxidase associated with sympathetic neurons is to assay enzyme activity before and after denervating a tissue and to assay activity in neurons directly using an enzyme-specific substrate such as serotonin (GORIDIS and NEFF, 1971; JARROTT, 1971) (specific for type A enzyme). Table 2 shows the effect of sympathectomy on the enzyme

TABLE 1. RADIOACTIVE β-PHENYLETHYLAMINE IN RAT BRAIN 10 MIN AFTER AN INTRAVENTRICULAR INJECTION TO CLORGYLINE OR DEPRENYL TREATED ANIMALS

Treatment	β-Phenylethylamine (counts/min) $\times$ 10^{-3}/brain sample $\pm$ S.E.M.
None	$2\cdot9 \pm 0\cdot5$
Clorgyline	$4\cdot2 \pm 0\cdot9$
Deprenyl	$9\cdot8 \pm 0\cdot9$ ($P < 0\cdot01$)

Clorgyline or deprenyl in a dose of 1 mg/kg, i.v. was administered 2 hr before injecting 70 nmole of ^{14}C β-phenylethylamine intraventricularly. The animals were killed 10 min later. Each value is the mean for 5–8 samples.

TABLE 2. MONOAMINE OXIDASE ACTIVITY IN THE RAT PINEAL GLAND AFTER
SUPERIOR CERVICAL GANGLIONECTOMY

| Substrate | nmole/pineal/hr $\pm$ s.e.m. (N) | | Loss of activity (%) |
	Innervated	Denervated	
Tyramine	9·4 $\pm$ 0·5 (4)	8·8 $\pm$ 0·8 (5)	6·4
Serotonin	3·4 $\pm$ 0·8 (7)	1·0 $\pm$ 0·05 (7)*	70

* $P < 0.01$ when compared with innervated preparation

activity in the rat pineal gland. The activity towards serotonin was decreased by
70 per cent whereas the activity towards tyramine was only decreased by 6·4 per cent.
When the superior cervical ganglion was assayed for activity it was found to be
predominantly type A activity. The sympathetic neurons of man also contain predominantly type A enzyme activity (Goridis and Neff, 1972).

Type A enzyme is the enzyme associated with sympathetic neurons of animals
and man. As would be predicted it is the enzyme that metabolises the transmitter
amine, norepinephrine (Fig. 1). The presence of a transmitter specific enzyme in
sympathetic nerves is consistent with the hypothesis that oxidation deamination
is the primary metabolic pathway for norepinephrine in neurons (Kopin and Axelrod,
1963). Type A enzyme is not limited to sympathetic neurons, however, it is found
in other tissues as well (Hall, Logan and Parsons, 1969; Squire, 1972).

Monoamine oxidase activity is associated with outer membranes of mitochondria.
Mitochondria are not homogenous; mitochondria isolated from glia and neurons
have different buoyant densities and the ratio of enzyme activities associated with
these mitochondria are different (Hamberger, Blomstrand and Lehninger, 1972).
We have observed that type A and B activity of mitochondria can be partially separated by centrifugation in a continuous sucrose gradient. The ratio of phenylethylamine oxidase activity to serotonin oxidase activity was greater in a 45–50% sucrose
fraction than in a 34–40% sucrose fraction (Table 3). The differences were not due
to fragmentation of mitochondrial membranes or to monoamine oxidase in synaptosomes as freezing and thawing did not change the ratio of activities in the fractions

TABLE 3. EFFECT OF FREEZING AND THAWING ON THE SUBSTRATE SPECIFICITY OF RAT BRAIN MONOAMINE
OXIDASE

| Sample | Control samples (nmole/sample/30 min) | | | Samples after freezing and thawing (nmole/sample/30 min) | | |
	Serotonin	Phenylethyl-amine	Ratio	Serotonin	Phenylethyl-amine	Ratio
Low density mitochondria	3·5	1·8	0·51	3·2	1·7	0·53
High density mitochondria	3·9	6·2	1·6	3·9	6·4	1·6

Data are presented as the mean for duplicate determinations on a single sample. Mitochondria
were layered onto a continuous sucrose gradient (20–50% w/v) and the samples were centrifuged at
100,000 × g for 2 hr. Low density mitochondria (35–40% sucrose fraction) or high density
mitochondria (45–50% sucrose fraction) were frozen three times.

(Table 3). The partial physical separation of the activities in a sucrose gradient supports the hypothesis that there are two forms of the enzyme. They may represent different enzymes or perhaps the presence of various amounts or types of membrane material on a single enzyme modifies the specificity of the enzyme. Moreover, studies *in vivo* indicate that these forms of enzymes are indeed present in the brain of animals.

CONCLUSIONS

It now appears that there are at least two forms of monoamine oxidase in mammalian tissues. These enzymes can be blocked separately with drugs and they have specific substrate requirements. Most exciting for the pharmacologist is the observation that the putative transmitters, norepinephrine and serotonin, are deaminated by type A enzyme and that dopamine is deaminated by type A and type B enzyme. Moreover, these enzymes can be blocked selectively by drugs.

Phenylethylamine is found in tissues of animals (NAKAJIMA, KAKIMOTO and SANO, 1964) and of man (ASATOOR and DALGLEISH, 1959) and large quantities are excreted by phenylketonurics (LEVIN *et al.*, 1964). When large amounts of phenylethylamine are administered alone (NAKAJIMA, KAKIMOTO, and SANO, 1964) or together with a monoamine oxidase inhibitor drug (MANTEGAZZA and RIVA, 1963) animals exhibit an amphetamine-like increase of motor activity. Some investigators have postulated that phenylethylamine may be a physiological stimulating agent in brain (FISCHER *et al.*, 1972). These reports emphasize the importance of type B enzyme as it appears to be the enzyme responsible for the destruction of phenylethylamine in brain. Moreover, there may be other endogenous amines that are specific substrates for type A and B enzyme that have not been identified as yet. The availability of specific inhibitors of type A and B enzyme should make it possible to study the role of these enzymes and, perhaps, allow the neurobiologist to postulate a physiological role for the endogenous amines that these enzymes destroy.

REFERENCES

ASATOOR A. M. and DALGLEISH C. F. (1959) *Biochem. J.* **73**, 26p.
COSTA E. and SANDLER M. (1972) *Adv. Biochem. Psychopharmacol.* **5**,
ERWIN V. G. and DEITRICH R. A. (1971) *Mol. Pharmacol.* **7**, 218–228.
FISCHER E., SPATZ H., HELLER B. and REGGIANI H. (1972) *Experientia* **28**, 307–308.
GORIDIS C. and NEFF N. H. (1972) *Proc. Soc. Exp, Biol. Med.* **140**, 573–574.
GORIDIS C. and NEFF N. H. (1971) *Neuropharmacol.* **10**, 557–564.
HALL D. W. R., LOGAN B. W., and PARSONS G. H. (1969) *Biochem. Pharmacol.* **18**, 1447–1454.
HAMBERGER A., BLOMSTRAND C. and LEHNINGER A. L. (1972) *Cell Biol.* **45**, 221–234.
JARROTT B. (1971) *J. Neurochem.* **18**, 7–16.
JOHNSTON J. P. (1968) *Biochem. Pharmacol.* **17**, 1285–1297.
KOPIN I. J. and AXELROD J. (1963) *Ann. New York Acad. Sci.* **107**, 848–853.
LEVINE R. J., NIRENBERG P. Z., UDENFRIEND S. and SJOERDSMA, A. (1964) *Life Sci.* **3**, 651–656.
MANTEGAZZA P. and RIVA M. (1963) *J. Pharm. Pharmacol.* **15**, 472–478.
NAKAJIMA T., KAKIMOTO Y. and SANO (1964) *J. Pharmacol. Exp. Ther.* **143**, 319–325.
SQUIRES R. F. (1972) *Adv. Biochem. Psychopharmacol.* **5**, 355–370.
YANG H.-Y. T., GORIDIS C. and NEFF N. H. (1972) *J. Neurochem.* **19**, 1241–1250.

Frontiers in Catecholamine Research 1973, pp. 139 to 143. Pergamon Press. Printed in Great Britain.

THE IMPORTANCE OF THE 'HYDROXYLAMINE METABOLIC ROUTE' IN PHARMACOLOGY, TOXICOLOGY AND PHARMACOKINETICS

Arnold H. Beckett

Department of Pharmacy, Chelsea College (University of London),
Manresa Road, London, S.W.3.

HYDROXYLAMINE FORMATION FROM PRIMARY AND SECONDARY ALIPHATIC AMINES—A GENERAL AND IMPORTANT METABOLIC ROUTE

The importance of the metabolism of aromatic amines to hydroxylamines and the implication of these metabolites in ferrihaemoglobin formation and bladder cancer induction is well documented Jenner (1972). However, only recently has the occurrence of hydroxylamines from aliphatic amines been demonstrated.

In our laboratories, we have shown that the *in vitro* metabolism of primary and secondary amines of type I in which R^1 = H or OH, R^2 and R^3 = H or Me, R^4 = H, Me, Et, n-Pr or n-Bu, R^5 = H, OH or CF_3, R^6 = H, OH, OMe or Cl yields hydroxylamines (II). To date, *in vivo* hydroxylamine formation has been obtained with all

(I)   (II)

those compounds investigated. Many of the above compounds are drugs, e.g. amphetamine, methyl- and ethyl-amphetamine, *p*-methoxyamphetamine, fenfluramine, phentermine, ephedrine, norephedrine. The incorporation of hetero atoms, e.g. oxygen between ring and basic centre does not inhibit hydroxylamine formation. The replacement of the phenyl ring of I with two aromatic rings, e.g. as in norpheniramine, norchlorpheniramine, nordiphenhydramine, nororphenadrine, or with condensed ring systems as in chlorpromazine, promethazine, nortriptyline, desipramine still allows hydroxylamine formation. Metabolism to hydroxylamines still occurs when the basic N-atom is part of a ring system e.g. morpholine as in phenmetrazine, piperidine as in anabasine, norpethidine and pipradol, piperazine as in *N*-phenyl-piperazine, or when the ring is part of more complicated structures e.g. as in normorphine and norcodeine. Hydroxylamine formation still occurs when aromatic rings are not present e.g. in propylhexedrine, heptaminol, octamylamine Beckett (1971); Beckett *et al.* (1971); Beckett and Al-Sarraj (1972a,b, and c, and 1973); Beckett and Salami (1972); Beckett *et al.* (1973); Beckett and Essien (1973) and Beckett and others, (unpublished). In many cases the hydroxylamine route is a dominant pathway.

HYDROXYLAMINES ARE LABILE METABOLITES—PROBLEMS OF IDENTIFICATION AND QUANTIFICATION

The above hydroxylamines are converted readily by physical e.g. heat, or chemical, e.g. changes in pH or presence of traces of metal ions, to a variety of other structures during extraction techniques and during methods of analysis. For instance, a secondary hydroxylamine is transferred very quickly in alkaline solution to a nitrone and an oxime; when the solution is then acidified, the nitrone is changed rapidly to an oxime, primary hydroxylamine and ketone (or aldehyde) while the oxime is changed to a ketone (or aldehyde). A primary hydroxylamine in alkaline solution is converted very rapidly into an oxime but in acidic solution is converted slowly into an oxime and a ketone (or aldehyde).

The gas chromatography of many primary hydroxylamines can be accomplished with only slight breakdown to the oxime provided a glass column and the correct support, stationary phase and type of instrument is used. Contact of carrier gas with heated metal has deleterious effects and so has conduction of heat to the metal needle of the syringe.

The stability of the hydroxylamines varies greatly with the substituents e.g. in I, R^5 as CF_3 stabilises, while R^6 as OCH_3 destabilises the primary hydroxylamine relative to R^5 and R^6 being H. In general, secondary hydroxylamine of type I can be gas chromatographed with less decomposition than their primary counterparts. The metabolism of secondary amines is shown in Fig. 1 in which some of the chemical changes which can occur under different conditions are indicated.

The isolation of an oxime or nitrone in metabolic studies is indicative of hydroxylamine formation. A combination of gas chromatography and mass-spectrometry as well as gas chromatography/mass spectrometry is effective in separating and identifying secondary hydroxylamines, oximes and nitrones because these products give mass spectra containing diagnostic fragment ions (BECKETT, COUTTS and OGUNBONA, *J. Pharm. Pharmac.*, in the press). Some primary hydroxylamines can be oxidised quantitatively on certain columns to oximes which can be separated and identified as such but their oxidation can be avoided by direct inlet into the mass spectrometer.

HYDROXYLAMINES UNDERGO CHEMICAL CHANGES BY DIFFERENT ROUTES DEPENDING UPON THE GROUPS ATTACHED TO THE NITROGEN ATOM

As indicated above, the routes of breakdown of the hydroxylamines depend upon the structure. Primary hydroxylamines of structure I are converted rapidly in alkaline solution to oximes provided R^2 or R^3 is H; however, if R^2 and R^3 are alkyl e.g. Me in phentermine, then the hydroxylamine is converted rapidly to the nitroso and nitro compound (BECKETT and BÉLANGER, unpublished).

On the other hand, secondary hydroxylamines are converted to nitrones and oximes in alkaline solution.

Similar differences occur in gas chromatography of the different compounds. Free radical mechanisms are involved in the autoxidation of some hydroxylamines, JOHNSON *et al.* (1956).

THE SIGNIFICANCE OF THE 'HYDROXYLAMINE METABOLIC ROUTE' IN PHARMACOKINETICS AND TOXICOLOGY

It is important in pharmacokinetics to distinguish between a compound which is the product of an enzymatic reaction and one produced by the chemical breakdown of a variety of products. For instance, a pharmacokinetic study of the production of benzylmethylketone by the metabolism of ethylamphetamine would have little relevance concerning metabolic deamination since this ketone is produced by the chemical

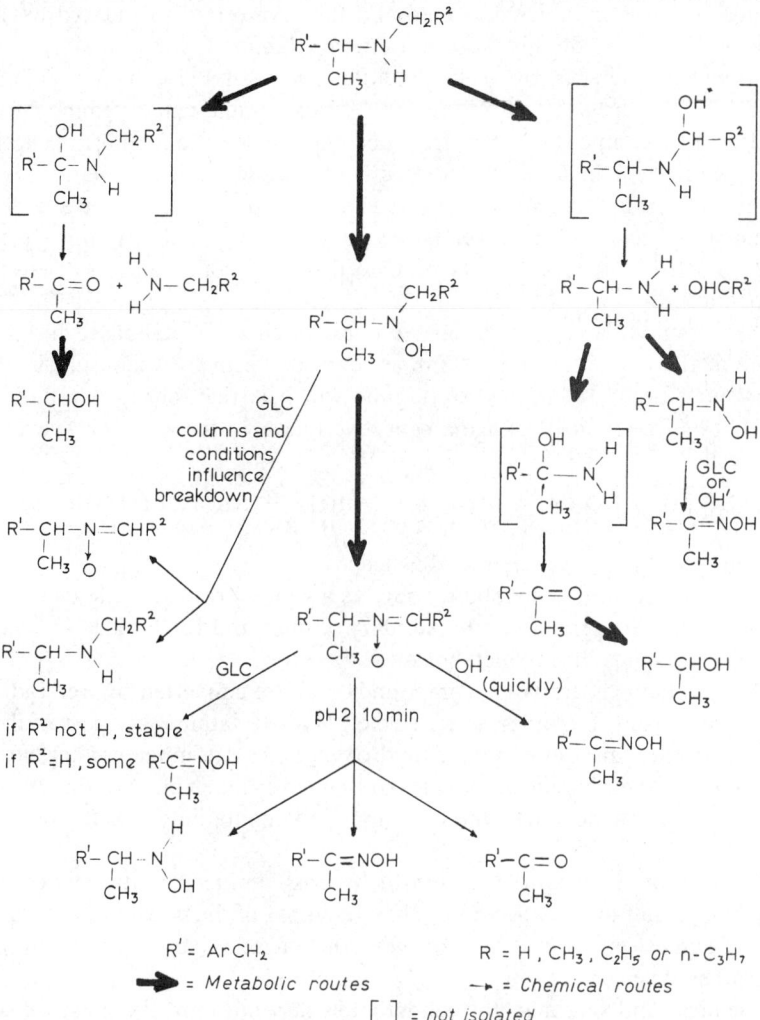

FIG. 1.—The metabolism of aralkyl-secondary amines and the chemical changes of metabolites and metabonates during extraction.

breakdown of the secondary and primary hydroxylamines, the nitrone and the oxime as well as the deamination of the secondary and primary amines, all of which can be present during the extraction of the biological medium and some of which may be produced in the analytical technique employed [See BECKETT and AL-SARRAJ (1973) for quantitative aspects]. Probably very few compounds classed as metabolites and studied as such are as the *direct* result of metabolism (See BECKETT et al. 1971).

Many tertiary amines are metabolised to secondary and primary amines and thus the molecule then exposed to the 'hydroxylamine metabolic route'. This may have important consequences in toxicology; two examples will serve to illustrate the point.

Chlorpromazine in man is demethylated and then converted to the hydroxylamine; a few hours after drug administration the hydroxylamine is found mainly in the red blood cells where it persists for weeks, Beckett and Essien (1973). With continuous treatment with the drug, 20–50 mg of the hydroxylamine can be found in the red blood cells. The compound, when free, decomposes to give 2-chlorophenothiazine which is much more light sensitive and readily oxidised to pigments than is the parent drug. The hydroxylamine can be detected in the blood weeks after a dose of the drug when the latter can no longer be detected. Is the 'hydroxylamine route' responsible for the long term effects of chlorpromazine and also for sensitisation and pigmentation?

In tobacco smoke are some tertiary amines which are metabolised to secondary amines, and also some secondary amines e.g. anabasine. Anabasine is oxidised rapidly by lung tissue to the hydroxylamine which is then changed rapidly to the nitrone. Is this free radical change a causative factor in the abnormal development of certain cells?

IS THE 'HYDROXYLAMINE METABOLIC ROUTE' IMPORTANT IN MECHANISMS OF ACTION OF BIOGENIC AMINES?

The following facts have been established:

(1) The 'hydroxylamine metabolic route' is a general route for the metabolism of aliphatic and alicyclic primary and secondary amines and for tertiary amines which yield these compounds during metabolism.

(2) Hydroxylamines are labile compounds and are converted by free radical and oxidative mechanisms to oximes, nitrones etc. Hydroxylamines are powerful nucleophilic agents in their un-ionised form; hydroxylamines are much weaker bases (2·5–3·0 pK units) than their parent amines and consequently much more of these un-ionised forms exist at physiological pH values. Compounds containing —SH groups add to some nitrones.

(3) The rate of chemical conversion of hydroxylamines to nitrones, oximes etc. under aqueous conditions is affected by changes in pH of the medium, by the presence of traces of some heavy metals, by oxygen concentration, by the characteristics of lipid and other type surfaces etc.

(4) Chemical and pharmacological activities depend upon the presence of labile compounds and activated states rather than upon the presence of stable chemical entities and the ground state of molecules.

(5) The hydroxylamine from amphetamine is a highly active CNS compound in animals. In man, when the urine is acidic and amphetamine is excreted without being metabolised greatly, CNS stimulation is brief; when the urine is alkaline and the drug reabsorbed in kidney tubules to allow extensive metabolism in the body, the subjects exhibit much CNS stimulation and show agressive tendencies especially after about 12 hr. Is the hydroxylamine of amphetamine responsible for amphetamine psychosis?

(6) Preliminary investigations, as expected, indicate that the 'hydroxylamine route' is involved in the metabolism of adrenaline, noradrenaline and dopamine but to date we have not been able to synthesise the resulting hydroxylamines.

It seems logical to pose the question, "Is this 'hydroxylamine route' implicated in CNS transmission involving catecholamines?" Is brain malfuction a result of changes in the involvement of this route and also in changes in the rate of free radical conversion of hydroxylamines to nitrones and oximes?

REFERENCES

BECKETT A. H. (1971) *Xenobiotica.* **1,** 365–384.

BECKETT A. H. and AL-SARRAJ S. (1972a) *J. Pharm. Pharmac.* **24,** 174–176.

BECKETT A. H. and AL-SARRAJ S. (1972b) *Biochem. J.* **130,** No. 1, 14p.

BECKETT A. H. and AL-SARRAJ S. (1972c) *J. Pharm. Pharmac.* **24,** 916–917.

BECKETT A. H. and AL-SARRAJ S. (1973) *J. Pharm. Pharmac.* **25,** 328–334.

BECKETT A. H., COUTTS R. T. and OGUNBONA F. A. (1973) *J. Pharm. Pharmac.* **25,** 190–192.

BECKETT A. H. and ESSIEN E. E. (1973) *J. Pharm. Pharmac.* **25,** 188–189.

BECKETT A. H. and SALAMI M. (1972) *J. Pharm. Pharmac.* **24,** 900–902.

BECKETT A. H., VAN DYK J. M., CHISSICK H. H. and GORROD J. W. (1971) *J. Pharm. Pharmac.* **23,** 809–811.

JENNER P. (1972) In *Biological Oxidation of Nitrogen in Organic Molecules.* (BRIDGES J. W., GORROD J. W. and PARKE D. V. Eds.) pp. 87–106. Taylor & Francis, London. and refs. cited therein.

JOHNSON D. H., ROGERS M. A. T. and TRAPPE G. (1956) *J. Chem. Soc.* 1093–1103.

Frontiers in Catecholamine Research 1973, pp. 145 to 146. Pergamon Press. Printed in Great Britain.

SOME CHARACTERISTICS OF RAT LIVER MITOCHONDRIAL MONOAMINE OXIDASES

A. D'Iorio, J. M. Diaz Borges and C. Kandaswami

Department of Biochemistry, University of Ottawa, Ottawa, Canada K1N 6N5

THE problem of the existence of multiple forms of monoamine oxidase [monoamine oxygen oxidoreductase (deaminating), EC 1.4.3.4] (MAO) of rat liver mitochondria has been extensively reviewed recently (DIAZ BORGES and D'IORIO, 1972; YOUDIM, 1972; SANDLER and YOUDIM, 1972). It appears that while evidence is accumulating for the existence of multiple forms of MAO, many questions concerning methodology still have to be resolved before a precise identification is achieved. Some studies on this problem have reported results using either crude or purified preparations. When separations are obtained with crude preparations the possibility of occurrence of several fractions with different properties due to interaction of MAO with other biological material is increased. Well-controlled experiments with these preparations can however constitute a valuable guide for the establishment of the necessary steps in further purification. In the purified preparations the possibility of interaction with other biological material is diminished but the activity for some substrates could be lost undetected during the course of purification. An additional difficulty arises from the fact that this enzyme is firmly bound to mitochondrial membranes and incomplete liberation or reassociation with other molecules could create the impression that several enzymes exist. Other questions that have been raised are associated with the selective inhibition caused by the solubilization treatment (DIAZ BORGES and D'IORIO, 1972) and the lack of specificity and reliability of detection methods such as tetrazolium staining, a common procedure for the investigation of MAO activity (YOUDIM and LAGNADO, 1972; SANDLER and YOUDIM, 1972; DIAZ BORGES and D'IORIO, 1973).

The data summarized here represent attempts at distinguishing these multiple MAO forms using polyacrylamide gel electrophoresis and some chromatographic separation procedures, detecting the MAO activity with benzylamine, serotonin and tyramine as substrates. For these experiments the mitochondria were treated with Lubrol W in presence of 0·1 M Tris–HCl buffer, pH 8·8, substrates and 30% sucrose w/v (DIAZ BORGES and D'IORIO, 1973). The suspension was then sonicated for 15 min, centrifuged and the precipitate was discarded.

In some experiments the active supernatant was subjected to polyacrylamide gel electrophoresis. The gels containing Lubrol W were prepared by using a modification of the technique described by DAVIS (1964). At the end of each electrophoretic run the acrylamide–agarose (separation) gel was cut into thin discs using a Canalco gel slicer. The acrylamide (spacer) gel was assayed without further division. Each thin disc constituted a fraction in which the MAO activity was detected using radioactive substrates. These procedures are detailed elsewhere (DIAZ BORGES and D'IORIO, 1973). The gel electrophoresis of this crude preparation showed that benzylamine MAO activity migrated as a band to the anode and did not penetrate the acrylamide–agarose gel, remaining in the acrylamide gel. On the contrary, tyramine and serotonin activities were distributed in several bands, some migrated to the anode while others migrated to the cathode. By increasing thé current intensity and the electrophoresis time, it was possible to get the band of benzylamine activity clearly devoid of serotonin activity but not of tyramine activity. Some tyramine and serotonin MAO bands could be isolated absent of the other two activities.

In other experiments, the supernatant obtained after centrifuging the sonicated mitochondrial preparation was treated with DEAE-cellulose. After centrifugation the DEAE-cellulose was discarded and the supernatant was passed through a column of Sephadex G-200. The MAO activity was eluted immediately after the void volume. No significant separation of the MAO activities for the different substrates was obtained in these steps. The fractions containing MAO activity were pooled, concentrated and dialyzed against 0·01 M Tris–HCl buffer, pH 7·4, treated with DEAE-cellulose and centrifuged. The supernatant was filtered (fraction I). The resin was washed with distilled water and the eluate was separated (fraction II). Further elution of the resin with 0·1 M NaCl in 0·01 M Tris–HCl buffer, pH 7·4, yielded another fraction (fraction III). These fractions were found to have some differences in their substrate specificity as measured ·by oxygen consumption (CREASEY, 1956) with an oxygen monitor (Yellow Springs Instrument, Co.). The relation among the MAO activities for benzylamine, serotonin and tyramine were, for example, 15·3: 1·0: 8·2 in fraction I, 1·4: 2·7: 1·0 in fraction II and 1·0: 2·9: 1·2 in fraction III. It is of particular interest to notice the predominance of serotonin activity in fractions II and III.

The data obtained with these two different methods support our earlier findings that MAO can be separated into several fractions with different substrate specificities (SIERENS and D'IORIO, 1970; DIAZ BORGES and D'IORIO, 1972). These experiments are in accordance with the reports of other authors who used different procedures (GORKIN, 1969; RAGLAND, 1968). Our results indicate the occurrence of at least two enzymes with varying substrate specificities. Confirmation would have to await further purification and characterization of the enzymes.

Besides contributing to the comprehension of the role of these enzymes in the metabolism of the biogenic amines, the identification of distinctly different enzymes could lead to the synthesis of specific inhibitors in which undesirable secondary effects could be eliminated.

Acknowledgement—This work was supported by a grant from the Medical Research Council of Canada. We express appreciation to Mrs. B. Betz and Mrs. R. Keren-Paz for their invaluable technical assistance.

REFERENCES

CREASEY N. H. (1956) *Biochem. J.* **64**, 178–183.
DAVIS B. J. (1964) *Ann. N.Y. Acad. Sci.* **121**, 404–427.
DIAZ BORGES J. M. and D'IORIO A. (1972) in *Monoamine Oxidases: New Vistas.* (COSTA E. and SANDLER M., Eds.) *Adv. Biochem. Psychopharmacol.* Vol. **5**, pp. 79–89. Raven Press, New York.
DIAZ BORGES J. M. and D'IORIO A. (1973) *Can. J. Biochem.* In press.
GORKIN V. Z. (1969) *Experientia*, **25**, 1142.
RAGLAND J. B. (1968) *Biochem. Biophys. Res. Commun.* **31**, 203–208.
SANDLER M. and YOUDIM M. B. H. (1972) *Pharmacol. Rev.* **24**, 331–348.
SIERENS L. and D'IORIO A. (1970) *Can. J. Biochem.* **48**, 659–663.
YOUDIM M. B. H. (1972) In *Monoamine Oxidases: New Vistas.* (COSTA E. and SANDLER M. Eds.) *Adv. Biochem. Psychopharmacol.* Vol. **5**, pp. 67–77. Raven Press, New York.
YOUDIM M. B. H. and LAGNADO J. R. (1972) In *Monoamine Oxidases: New Vistas.* (COSTA E. and SANDLER M. Eds.) *Adv. Biochem. Psychopharmacol.* Vol. **5**, 289–292. Raven Press, New York.

Frontiers in Catecholamine Research 1973, pp. 147 to 149. Pergamon Press. Printed in Great Britain.

THE NATURE OF THE ELECTROPHORETICALLY SEPARABLE FORMS OF MONOAMINE OXIDASE

MILES D. HOUSLAY and KEITH F. TIPTON

Department of Biochemistry, University of Cambridge, Tennis Court Road,
Cambridge CB2 1QW, England

A LARGE number of reports have indicated that the enzyme monoamine oxidase (E.C. 1.4.3.4) may exist in more than one form (see e.g. GORKIN, 1966 and YOUDIM, 1972). The activities of monoamine oxidase towards different substrates have been shown to be inhibited to different extents by treatment with the irreversible inhibitors clorgyline (HALL et al., 1969; JOHNSTON, 1968; SQUIRES, 1972) and 2-phenylethyl-hydrazine (TIPTON, 1972) and by heat treatment (OSWALD and STRITTMATTER, 1963). Polyacrylamide gel electrophoresis has been shown to separate monoamine oxidase preparations from several sources into a number of bands of activity, and these separated bands have been reported to differ in their substrate specificities, sensitivities to heat treatment and to certain inhibitors (see e.g. YOUDIM et al., 1970; YOUDIM, 1972). It has been suggested that the multiple forms of monoamine oxidase may represent a single enzyme species to which different amounts of lipid material are attached (VERYOVKINA et al., 1964) and this view is supported by the observation that the electrophoretically separable forms of rat liver and beef adrenal medulla monoamine oxidase have widely different phospholipid contents (TIPTON, 1972; TIPTON et al., 1972). The observations that monoamine oxidase which has been solubilized by treatment with an organic solvent will bind tightly to lipid components of the outer mitochondrial membrane (OLIVECRONA and ORELAND, 1971), and that electrophoresis of monoamine oxidase in the presence of 1·25% Triton causes the abolition of the multiple forms (Tipton, 1972), provide support for the involvement of lipid material in the electrophoretically separable forms of this enzyme.

Chaotropic agents have been shown to disrupt protein-lipid complexes by a weak-ening of hydrophobic bonds (HANSTEIN et al., 1971). We have used the chaotropic agent, sodium perchlorate, to investigate the nature of Triton X-100 solubilised mito-chondrial monoamine oxidase from rat liver (YOUDIM and SANDLER, 1968) and human brain (YOUDIM et al., 1970). Polyacrylamide gel electrophoresis of the preparations (Fig. 1), gave rise to a number of bands of activity, similar to those reported by YOUDIM et al. (1970), although with the rat liver enzyme the band of activity which remained at the origin was found to be an artifact of the loading procedure. After treatment with sodium perchlorate in the manner previously described (HOUSLAY and TIPTON, 1973), only a single band of monoamine oxidase activity could be detected (Fig. 1). In the case of the rat liver enzyme this band had a phospholipid content similar to that of the most mobile cathodically (least phospholipid containing) of the electrophoretically separable forms from the untreated enzyme. Gel filtration of the perchlorate-treated rat liver enzyme on Sepharose 4B indicated that no appreciable change in molecular weight had occurred, however an opalescent fraction could be separated which was shown to be rich in lipids, whilst little such material could be separated from the untreated enzyme.

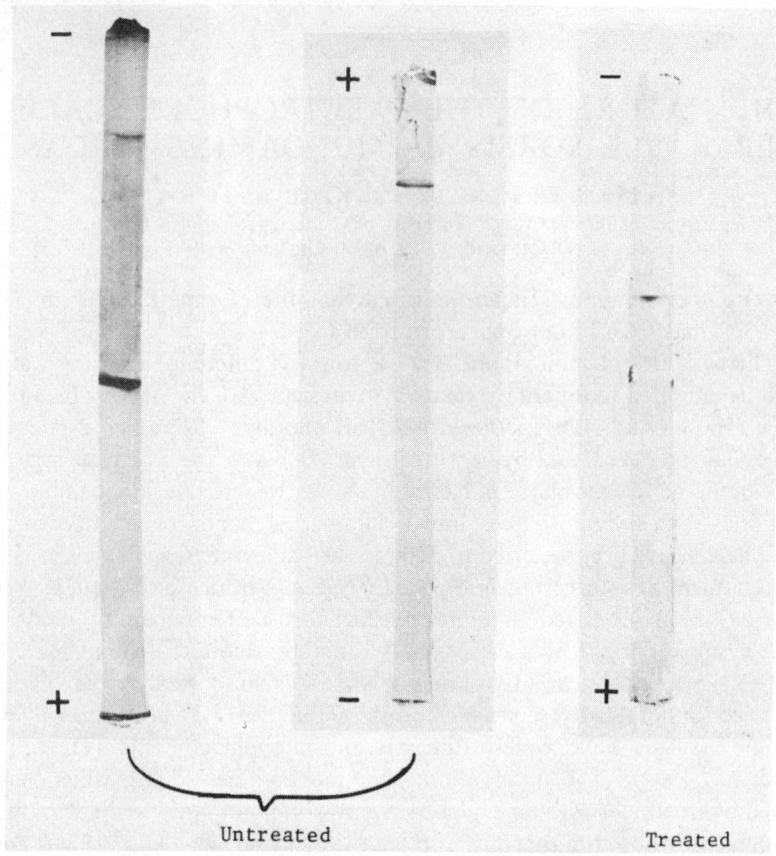

FIG. 1.—Polyacrylamide gel electrophoresis of human brain monoamine oxidase. The methods used for enzyme preparation, electrophoresis and staining for enzyme activity were as described by YOUDIM *et al.* (1970). The treated preparation was incubated for 20 min with sodium perchlorate in the presence of mercaptoethanol and benzylamine in the manner described by HOUSLAY and TIPTON (1973). When the treated preparation was used no band of activity which migrated towards the cathode could be detected.

Perchlorate treated enzymes from both sources were shown to be indistinguishable from the untreated preparations with regard to their K_m and V_{max} values with tyramine, benzylamine and dopamine as substrates. Mixed substrate experiments failed to indicate more than one enzyme to be responsible for the deamination of these three substrates in both the treated and untreated preparations, in agreement with the results of OSWALD and STRITTMATTER (1963). Preincubation with the irreversible inhibitor 2-phenylethylhydrazine has been shown to result in a slower rate of inhibition of the activities of rat liver and human brain monoamine oxidase towards dopamine than towards benzylamine or tyramine (TIPTON, 1972; YOUDIM, 1972).

No such differences could be detected after treatment with perchlorate. After perchlorate treatment the inhibition by clorgyline was also shown to be independent of the substrate used to assay the enzymes from rat liver and human brain, although there was a pronounced difference before this treatment (HOUSLAY and TIPTON, 1973). Preincubation of the enzyme at $40°C$, gave dissimilar inactivation rates when assayed with these three substrates, however after perchlorate treatment, the activity decay curves for all three substrates were superimposable.

Perchlorate treatment results in abolition of the characteristic electrophoretic multiple forms of monoamine oxidase and in the loss of differential effects of a number of inhibitory processes. As this treatment causes the release of lipid material from the preparations and reduces the phospholipid associated with its electrophoretic component, it is tempting to conclude that the multiple forms of monoamine oxidase do indeed represent a single enzyme species with differing amounts of lipid material bound to it.

The significance of these results in terms of the existence of multiple forms of monoamine oxidase *in vivo* is less clear. Since the atttachment of lipid material would appear capable of causing quite significant changes in the heat and inhibitor sensitivities of the enzyme it is tempting to conclude that the existence of the enzyme in different environments, in different regions of an organ such as brain could give rise to forms with different properties *in vivo*. Although the possibility cannot be excluded that these forms are generated by the vigorous purification procedures which are used (YOUDIM and SANDLER, 1968), or the lengthy electrophoresis procedure, which takes no account of local heating effects on the separated forms (YOUDIM, 1972); this could account for the lack of correlation between the inhibition of separated forms and *in vitro* effects on monoamine oxidase reported by COLLINS *et al.* (1972).

Acknowledgements—We are grateful to Mr. N. J. Garrett for his skilled assistance. M. D. H. was supported by an M.R.C. research studentship.

REFERENCES

COLLINS G. G. S., YOUDIM M. B. H. and SANDLER M. (1972). *Biochem. Pharmacol.* **21**, 1995–1998.
GORKIN V. Z. (1966) *Pharmac. Revs.* **18**, 115–120.
HALL D. W. R., LOGAN B. W. and PARSONS G. H. (1969) *Biochem. Pharmacol.* **18**, 1447–1454.
HANSTEIN W. G., DAVIES K. A. and HATEFI Y. (1971) *Arch. Biochem. Biophys.* **147**, 534–544.
HOUSLAY M. D. and TIPTON K. F. (1973) *Biochem. J.* In the press.
JOHNSTON J. P. (1968) *Biochem. Pharmacol.* **17**, 1285–1297.
OSWALD E. O. and STRITTMATTER C. F. (1963) *Proc. Soc. Exptl Biol. N.Y.* **114**, 668–673.
OLIVERCRONA T. and ORELAND L. (1971) *Biochemistry* **10**, 332–340.
SQUIRES R. F. (1972) *Adv. Biochem. Psychopharmacol.* **5**, 355–370.
TIPTON K. F. (1972) *Adv. Biochem. Psychopharmacol.* **5**, 11–24.
TIPTON K. F., YOUDIM M. B. H. and SPIRES I. P. C. (1972) *Biochem. Pharmacol.* **21**, 2197–2204.
VERYOVKINA I. V., GORKIN V. Z., MITYUSHIN V. M. and ELPINER I. E. (1964) *Biophysics (Moscow)* **9**, 503–506.
YOUDIM M. B. H. and SANDLER M. (1968) *Biochimica Applicata* **14**, (*Suppl. 1*), 175–184.
YOUDIM M. B. H., COLLINS G. G. S. and SANDLER M. (1970) *Fedn. Eur. Biochem. Soc. Symp.* **18**, 281–288.
YOUDIM M. B. H. (1972) *Adv. Biochem. Psychopharmacol.* **5**, 67–77.

Frontiers in Catecholamine Research 1973, pp. 151 to 152. Pergamon Press. Printed in Great Britain.

TRANSFORMATION OF MONOAMINE OXIDASES AND DEAMINATION OF FATTY-AROMATIC MONOAMINES

V. Z. GORKIN

Institute of Biological and Medical Chemistry, Academy of Medical Sciences,
Moscow, USSR

TRANSFORMATION of monoamine oxidases [monoamine:O_2 oxidoreductase (deaminating), EC 1.4.3.4] from beef (GORKIN et al., 1971a; AKOPYAN et al., 1971a) or rat (VERYOVKINA et al., 1972) liver mitochondria or from *Sarcina lutea* (TATYANENKO et al., 1971; YAKOVLEV, 1971) is initiated by oxidation of SH groups. This transformation leads to a decrease in deamination of monoamines and appearance of new catalytic properties. An AMP-deaminating activity (VERYOVKINA et al., 1972; TATYANENKO et al., 1971; AKOPYAN et al., 1972) always appeared; ability to deaminate some other nitrogenous compounds were also induced in the modified enzymes. Thus, the bacterial tyramine oxidase, which possesses sharp substrate specificity (YAMADA et al., 1967), after partial oxidation of its SH groups acquired, besides the AMP-deaminating activity, abilities to deaminate lysine, spermine, putrescine, but not histamine. Beef or rat liver mitochondrial monoamine oxidases which have broad substrate specificity after the oxidation of SH groups deaminated histamine (AKOPYAN et al., 1971a; VERYOVKINA et al., 1972). Thus transformation of different monoamine oxidases leads to formation of modified enzymes, which, invariably acquiring an ability to catalyse hydrolytic deamination of AMP, may differ in spectra of nitrogenous compounds oxidatively deaminated by them.

The transformation of monoamine oxidases may take place in mitochondrial fragments (GORKIN et al., 1970; GORKIN and TATYANENKO, 1967). Fe^{2+} (0·01 mM) stimulated lipid peroxidation in mitochondrial membranes (VLADIMIROV and ARCHAKOV, 1972) and caused transformation of monoamine oxidases. In brain the rate of lipid peroxidation is higher then in liver (DI LUZIO and HARTMAN, 1967). Accordingly in both mitochondria and highly purified mitochondrial monoamine oxidases (AKOPYAN et al., 1971b) from beef brain stem the transformation takes place during the isolation and purification procedures.

In pathological states accompanied by increased lipid peroxidation one could expect that the deamination of fatty-aromatic monoamines would be decreased, while the deamination of AMP, histamine, putrescine, cadaverine of lysine would be stimulated (or induced). We did observe these phenomena in irradiation injuries (GORKIN et al., 1968; AKOPVAN et al., 1970a) D_2 — hypervitaminosis (AKOPYAN et al., 1970b) or in liver of tumour-bearing animals (GORKIN et al., 1973; KHUZHAMBERDIEV et al, 1973). These data suggest that the ability of monoamine oxidases to undergo transformation may be realized *in vivo*. The adenylate deaminating activity was inhibited *in vivo* by adenosine-2'(3')-monophosphate (GORKIN, 1971b), which is a competitive inhibitor (GORKIN et al., 1970) of the AMP-deamination induced by transformation of monoamine oxidases.

Adenosine-2'(3')-monophosphate or adenosine-3'-monophosphate (but not adenosine-5'-monophosphate) normalized deamination of fatty-aromatic mono-amines, AMP or diamines in irradiation injuries (GORKIN et al., 1971b) or in tumour-bearing animals (GORKIN et al., 1973; KHUZAMBERDIEV et al., 1973). Increased survival periods in the experimental groups of aminals (as compared with controls similarly treated with AMP) was noted in irradiation injury (GORKIN et al., 1971b) and, especially, in the tumour-bearing animals.

The transformation of monoamine oxidases was prevented by irreversible mono-amine oxidase inhibitors (GORKIN et al., 1971a; AKOPYAN et al., 1971a; VERYOVKINA et al., 1972; TATYANENKO et al., 1971; YAKOVLEV et al., 1971; AKOPYAN et al., 1972; GORKIN et al., 1970). This effect was realized in vivo (AKOPYAN et al., 1972; ABDEL SAMED et al., 1971; GORKIN et al., 1971c); but no prolongation of the survival period in irradiation injury of in tumour-bearing animals was observed. For therapy of these diseases it appears essential not only to normalize the AMP-deamination but also to keep the oxidation of fatty-aromatic nonoamines at the normal levels.

REFERENCES

ABDEL SAMED M. M., AKOPYAN Zh. I., VERYOVKINA I. V., KULYGINA A. A. and GORKIN V. Z. (1971) Biochem. Pharmacol. 20, 2571–2577.

AKOPYAN Z. I., BLAZHEIEVICH N. V., VERYOVKINA I. V., GORKIN V. Z., SYOMINA O. V. and SPIRICHEV V. B. (1970b) Int. J. Vitamin Res., 40, 497–504.

AKOPYAN Z. I., GORKIN V. Z., KUDRYASHOV Yu.B. and SYOMINA O. V. (1970a) Radiobiologiya 10, 826–831.

AKOPYAN Z. I., KULYGINA A. A., TERZEMAN I. I. and GORKIN V. Z. (1972) Biochim. Biophys. Acta 289, 44–56.

AKOPYAN Z. I., STESINA L. N. and GORKIN V. Z. (1971a) J. Biol. Chem. 246, 4610–4618.

AKOPYAN Zh. I., VERYOVKINA I. V., LEVYANT M. I., MOSKVITINA T. A., GORKIN V. Z. and OREK-HOVICH V. N. (1971b) Int. J. Protein Res. 3, 121–130.

DI LUZIO N. R. and HARTMAN A. D. (1967) Fedn. Proc. 26, 1436–1442.

GORKIN V. Z., AKOPYAN Z. I., GONCHARENKO E. N. and KUDRYASHOV Yu. B. (1968) Vop. med. Khim. 14, 538–540.

GORKIN, V. Z. AKOPYAN Z. I., KULYGINA A. A. and ZELNALOV T. A. (1971b) Bull. Exp. Biol. Med. 11, 42–45.

GORKIN V. Z., AKOPYAN Z. I. and STESINA L. N. (1971a) Vop. med. Khim. 17, 444–446.

GORKIN V. Z., AKOPYAN Z. I., VERYOVKINA I. V., GRIDNEVA L. I. and STESINA L. N. (1970) Bio-khimiya 35, 140–151.

GORKIN V. Z., AKOPYAN Zh. I., VERYOVKINA I. V., STESINA L. N. and ABDEL SAMED M. M. (1971c) Biochem. J. 121, 31–32.

GORKIN V. Z., NEYFAKH E. A., ROMANOVA L. A. and KHUZHAMBERDIEV M. (1973) Experientia 29, 22–23.

GORKIN V. Z. and TATYANENKO L. V. (1967) Biochim. Biophys. Res. Commun. 27, 613–617.

KHUZHAMBERDIEV M., ROMANOVA L. A., NEYFAKH E. A. and GORKIN V. Z. (1973) Vop. med. Khim. 19, 345–353.

TATYANENKO L. V., GVOZDEV R. I., LEBEDEVA O. I., VOROBYOV L. V., GORKIN V. Z. and YAKOVLEV V. A. (1971) Biochim. Biophys. Acta 242, 23–35.

VERYOVKINA I. V., ABDEL SAMED M. M. and GORKIN V. Z. (1972) Biochim. Biophys. Acta 258, 56–70.

VLADIMIROV Yu. A. and ARCHAKOV A. I. (1972) Peroxidation of Lipids in Biological Membranes. pp. 69 and 80. Nauka Publishing House, Moscow.

YAKOVLEV V. A., GORKIN V. Z., TATYANENKO L. V., GVOZDEV R. I., LEBEDEVA O. I. and VOROBYOV L. V. (1971) Dokl. Akad. Nauk SSSR 197, 226–228.

YAMADA H., KUMAGAI H., UMAJIMA T. and OGATA, K. (1967) Agr. Biol. Chem. Tokyo 31, 897–901.

Frontiers in Catecholamine Research 1973, pp. 153 to 155. Pergamon Press. Printed in Great Britain.

ON THE MECHANISM OF INHIBITION OF MONOAMINE OXIDASE

E. A. ZELLER and M. HSU

Department of Biochemistry, Northwestern University Medical School,
Chicago, Illinois 60611, U.S.A.

IN THIS laboratory, the first *in vitro* (ZELLER *et al.*, 1952; 1955) and *in vivo* (ZELLER and BARSKY (1952)) inhibition of monoamine:O_2 oxidoreductase [deaminating] (EC 1.4.3.4) and the first pharmacological (GRIESEMER *et al.*, 1953) and toxicological (REBHUN *et al.*, 1954) potentiation of monoamines were carried out with iproniazid. These discoveries were followed by the introduction of a great number of very active inhibitory compounds, e.g. other hydrazine derivatives, phenylcyclopropylamines, and acetylenic amines. In sharp contrast to the widespread use of MAO inhibitors as pharmacological tools stands the lack of understanding of the exact nature of the inhibitory process.

On the basis of isosteric relationships and other principles, the close similarity existing between substrates and inhibitors was stressed (ZELLER (1960)). In fact, CLINESCHMIDT and HORITA (1969), and later TIPTON and SPIRES (1972), demonstrated that one inhibitor, phenylethylhydrazine, was oxidatively degraded by MAO. HELLERMAN and ERWIN (1968) observed spectral changes in purified MAO upon anaerobic addition of pargyline (*N*-methyl-*N*-phenylpropynylamine) to MAO, which indicated a reduction of the flavin moiety by this inhibitor.

In recent years HEMMERICH *et al.* (1970) carried out a great number of oxidation studies with various photochemically activated flavins and, as a result of this extensive work, concluded that the flavin, acting as a nucleophil, accepts an electron pair at the C4a–N5 moiety of the flavin nucleus. When we applied this procedure to the analysis of the photoactivated system, lumiflavin-3-acetic acid and pargyline in phosphate buffer pH 7·2, a dramatic reaction took place: the two peaks representing the quinoid structure of the flavin disappeared and a new band at 395 appeared with a molecular extinction of approximately 24,000! The indane analogue of pargyline, Su 11739 was even more reactive than the former substance. The compound responsible for these spectral changes could be isolated and submitted to IR and NMR spectroscopical analysis (ZELLER *et al.*, 1972) It turned out to be a covalent complex between the flavin and the acetylenic amine. The triple bond was replaced by an olefinic group which was a part of an enamine system. This configuration appeared to be responsible for a deuterium exchange at the β-carbon and for the resonance system causing the 395 nm band. It is feasible that two steps are needed to achieve the formation of the complex: in a first step the deprotonated α-carbon would donate its electron pair to the carbon-4a, while in the second phase, after protonation of the β-carbon, N-5 forms a bond with the γ-carbon of the acetylenic amine. Thus, the double bond between C-4a and N-5 disappears. The discovery of the new chromophor opened a new field in flavin chemistry, which is under intensive investigation by B. Gaertner and P. Hemmerich.

It was of paramount interest to us to find out whether the flavin moiety of MAO is also the target of acetylenic amines. In order to find an answer to this question we isolated an electrophoretically homogeneous MAO from beef liver according to the procedure of Yasunobu et al. (1968). Upon addition of pargyline or Su 11739 to this preparation and in the absence of light, immediately spectroscopic changes at 460 nm took place which indicated reduction of the flavin moiety. In addition a strong band at 410 nm was seen. When a purified preparation which became spontaneously inactive was tested, no spectroscopic changes were seen. Apparently, the interaction between MAO and acetylenic amines is a catalytic event. We found a close parallelship between optical changes at 410 nm and the degree of inhibition. After one mole of Su 11739 per 100,000 g protein was added to the system, a plateau was reached for both parameters (Fig. 1). Furthermore, the number of titratable sulfhydryl groups did not change after inactivation of the enzyme, nor did glutathione protect it. It seems, therefore, unlikely that Belleau and Moran's proposal (1963) that pargyline interacts with sulfhydryl residues, is a major part of the inhibitor mechanism. Finally, purified MAO does not respond by displaying the 410 nm band, when it is treated with substrates or with representatives of other classes of inhibitors, such as phenylcyclopropylamine (tranylcypromine) or iproniazid. The mechanism of MAO inactivation of these inhibitors seems to be quite different from that of acetylenic amines. Thus, the presently known facts can be interpreted by the assumption that the 410 band is related to the new chromophor which is produced photochemically with flavins and acetylenic amines. More work, however, is required to settle this point in a definite manner.

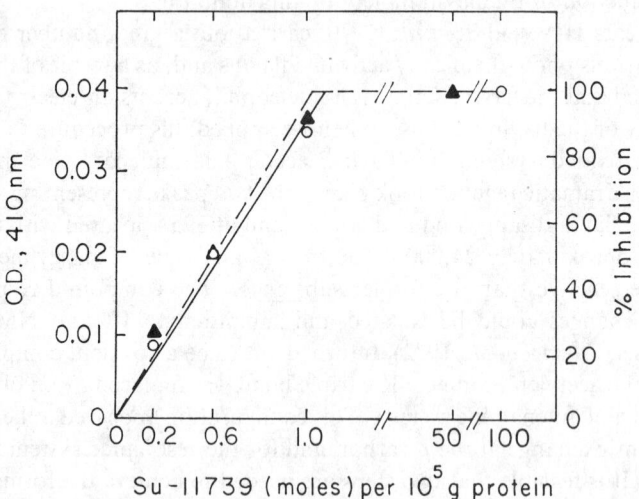

Fig. 1.—The effect of Su-11739 on the absorption spectra and the activity of MAO. Varying amounts of Su-11739 were added to 0·26 mg protein of MAO in 0·1 M phosphate buffer, pH 7·2. The absorption spectra of the treated and untreated enzyme were recorded in Cary Model 14 recording spectrophotometer. The enzyme activity was assayed spectrophotometrically. Triangles and dotted line refer to per cent inhibition; circles and full line to the difference at 410 nm between the treated and untreated enzyme.

With a better insight into the mechanism of inhibitor action, it can be expected that MAO inhibitors, which are used in countless pharmacological studies, may become sharper tools than they have been until now. Our results should also provide some badly needed information about the mechanism of flavin-substrate interaction in MAO.

REFERENCES

BELLEAU B. and MORAN J. (1963) *Ann. N.Y. Acad. Sci.* **107**, 822–839.
CLINESCHMIDT B. V. and HORITA A. (1969) *Biochem. Pharmacol.* **18**, 1011–1020.
GRIESEMER E. C., BARSKY J., DRAGSTEDT C. A., WELLS J. A. and ZELLER E. A. (1953) *Proc. Soc. Exp. Biol. and Med.* **84**, 699–701.
HELLERMAN L. and ERWIN E. V. (1968) *J. Biol. Chem.* **243**, 5234–5243.
HEMMERICH P., NAGELSCHNEIDER G. and VEEGER C. (1970) *FEBS* **8**, 69–83.
REBHUN J., FEINBERG S. M. and ZELLER E. A. (1954) *Proc. Soc. Exp. Biol. and Med.* **87**, 218–220.
TIPTON K. F. and SPIRES I. P. C. (1972) *Biochem. Pharmacol.* **21**, 268–270.
YASUNOBU K. T., IGAUE I. and GOMES B. (1968) *Advan. Pharmacol.* **6A**, 43–59.
ZELLER E. A. (1960) *Experientia* **16**, 399–402.
ZELLER E. A. and BARSKY J. (1952) *Proc. Soc. Exp. Biol. Med.* **81**, 459–461.
ZELLER E. A., BARSKY J. and BERMAN E. R. (1955) *J. Biol. Chem.* **214**, 267–274.
ZELLER E. A., BARSKY J., FOUTS J. R., KIRCHHEIMER W. F. and VAN ORDEN L. S. (1952) *Experientia* **8**, 349.
ZELLER E. A., GAERTNER B. and HEMMERICH P. (1972) *Z. Naturf.* **27b**, 1050–1052.

Frontiers in Catecholamine Research 1973, pp. 157 to 160. Pergamon Press. Printed in Great Britain.

SELECTIVE LOCALISATION OF MONOAMINE OXIDASE FORMS IN RAT MESENTERIC ARTERY

C. GORIDIS[1] and N. H. NEFF[2]

[1]Centre de Neurochimie du CNRS, Faculté de Médecine, 67085 Strasbourg Cedex, France, and [2]Laboratory of Preclinical Pharmacology, National Institute of Mental Health, Saint Elizabeth's Hospital, Washington, D.C. 20032, U.S.A.

THERE is now compelling evidence that different forms of monoamine oxidase (MAO; monoamine: oxygen oxidoreductase (deaminating); EC 1.4.3.4) are associated with specific cell types (NEFF, YANG and GORIDIS, this volume). The MAO associated with the sympathetic nerves (type A MAO) of the pineal gland differs from the extraneuronal form in that it is completely inhibited by 0·1 μM clorgyline (N-methyl-N-propargyl-3-[2,4 dichlorophenoxy] propylamine hydrochloride), is relatively heat-stable and deaminates tyramine, 5-hydroxytryptamine (5-HT) and noradrenaline (NA). By contrast, 0·1 mM clorgyline is required to inhibit the extraneuronal MAO which is heat labile and does not appreciably deaminate 5-HT or NA (GORIDIS and NEFF, 1971; YANG, GORIDIS and NEFF, 1972).

In the rabbit ear artery, the only MAO which can be demonstrated histochemically appears to be located extraneuronally and is distributed throughout the media (DE LA LANDE, HILL, JELLETT and McNEIL, 1970). On the other hand, conflicting results concerning the relative importance of the intra- and extraneuronal MAO in transmitter catabolism have been obtained in physiological experiments on aortic strips (KALSNER and NICKERSON, 1969) or the perfused ear artery (DE LA LANDE and JELLET, 1972).

The present study was undertaken to determine whether selectively-distributed multiple forms of MAO, as found in rat pineal gland, could be also demonstrated in the smaller arteries, and to assess the distribution of these forms between intra- and extraneuronal tissue. The study was carried out on mesenteric arteries because of their dense adrenergic innervation (BERKOWITZ, SPECTOR and TARVER, 1972).

INHIBITION OF MESENTERIC ARTERY MAO BY VARIOUS INHIBITORS

When the inhibition of brain MAO by pargyline using either tyramine or 5-HT as substrate, was plotted semi-logarithmically as a function of time, a straight line was obtained. In contrast, a biphasic curve was obtained when an homogenate of rat mesenteric artery was used as enzyme source with tyramine as substrate. With 5-HT as substrate, the inactivation of arterial MAO could be fitted to a curve virtually identical to that obtained for brain MAO (COQUIL, GORIDIS, MACK and NEFF, 1973). For subsequent investigations we assumed that the biphasic curve was due to the presence of two enzymes acting on tyramine; one rapidly inactivated, and a second either not inhibited by pargyline or with a very low affinity for the inhibitor. By extrapolating the biphasic curve to zero time, 35–40 per cent of the tyramine deaminating activity could be attributed to the pargyline-insensitive MAO. Since a linear relationship was obtained using 5-HT as substrate, apparently only the rapidly inactivated form was able to metabolise 5-HT.

Experiments using the MAO inhibitor clorgyline supported our interpretation of the results with pargyline. When the MAO activity of rat mesenteric artery was determined in the presence of clorgyline using tyramine as substrate, the inhibition of the enzyme could be represented by a sigmoidal shaped curve which levelled off at a maximum of about 60 per cent inhibition. With 5-HT as substrate complete inhibition was attained. Apparently, about 40 per cent of the tyramine metabolising activity present in the arterial wall was insensitive to the inhibitor and this MAO form did not deaminate 5-HT.

When NA was used as substrate and either clorgyline (0·1 mM) or pargyline (0·05 mM) as inhibitors, complete inhibition was attained, as was also the case with 5-HT as substrate. The same concentrations inhibited the tyramine oxidising activity of the arterial wall to 61 and 64 per cent (Table 1). The results are in agreement with our

TABLE 1. INHIBITION OF MONOAMINE OXIDASE ACTIVITIES IN RAT MESENTERIC ARTERY

Mesenteric arteries were pooled and homogenised, and aliquots of the 750 g supernatants assayed for MAO activity as described previously (GORIDIS and NEFF, 1971). The rates of inhibition of MAO were determined by incubation of the reaction mixtures at room temperature for 15 min (clorgyline) or 30 min (pargyline and semicarbazide) before adding substrate. For the rates of inhibition the mean and the range of duplicate determinations on three homogenates is given.

	Substrate		
	Tyramine	5-Hydroxytryptamine	Noradrenaline
Control value nmol/mg protein/hr ± S.E.M. (n)	64·5 ± 5·1 (10)	25·5 ± 1·4 (6)	0·67 ± 0·09 (4)
Clorgyline 10^{-4}M†	61 (58–64)	98 (96–100)	100 (100)
Pargyline 5×10^{-5}M†	64 (61–66)	95 (94–97)	100 (100)
Semicarbazide 10^{-3}M†	33 (29–35)	9 (3–12)	0 (0)
Semicarbazide 10^{-3}M† + pargyline 5×10^{-5}M	100 (100)	98 (97–100)	100 (100)

† Percentage inhibition

previous findings, since in the superior cervical ganglion NA and 5-HT appeared to be metabolised by the same MAO type (GORIDIS and NEFF, 1971b). The carbonyl reagent semicarbazide inhibited only the tyramine metabolising enzyme without affecting appreciably the activity towards 5-HT and NE. Complete inhibition of enzyme activity for all the substrates could be obtained by combining semicarbazide (1·0 mM) and pargyline (0·05 mM) (Table 1). Apparently the MAO type present in the mesenteric artery, which is not inhibited by pargyline or clorgyline, is sensitive to semicarbazide. This form does not appreciably metabolise NA or 5-HT. The other MAO form found in the mesenteric artery has the same characteristics as the type A MAO found in sympathetic nerves (GORIDIS and NEFF, 1971), but which occurs in other tissues as well (JOHNSTON, 1968).

EFFECT OF CHEMICAL SYMPATHECTOMY ON RAT MESENTERIC ARTERY MONOAMINE OXIDASE

Chemical sympathectomy with 6-hydroxydopamine (6OH-DM) was carried out, to assess the distribution of the MAO forms of the arterial wall between intra- and extraneuronal tissue. Chemical sympathectomy by 6OH-DM results in the loss of 70 per cent of the 5-HT oxidising MAO from the arterial wall (Table 2). With tyramine as substrate, the change is less pronounced, but a marked drop of the sensitivity to clorgyline is observed: Whereas in control preparations 0·1 mM clorgyline produced 64·5 per cent inhibition (Table 1), only 31 per cent inhibition was found

TABLE 2. THE EFFECT OF 6-HYDROXYDOPAMINE (6OH-DM) ON MESENTERIC ARTERY MONOAMINE OXIDASE

6OH-DM was injected intravenously into rats twice on day 1 (50 mg/kg) and twice on day 7 (100 mg/kg). The animals were killed on day 9. Four meseneteric arteries were pooled for each determination. The values shown are the mean ± S.E.M. for three determinations.

| | Mesenteric artery 750 g supernatant | | |
	Control	6OH-DM	%Fall
5-HT nmol/hr/artery	12 ± 0·4	3·6 ± 0·7†	70
5-HT nmol/hr/mg protein	26 ± 0·8	9·0 ± 2·0†	65
Tyramine nmol/hr/artery	27 ± 3·3	11·0 ± 2·5	59
Tyramine nmol/hr/mg protein	60 ± 7·0	36·0 ± 4·0	40

† Statistically significant as compared to control animals ($p < 0.01$).

for preparations from sympathectomised animals. These results are consistent with the view that the overwhelming majority of the 5-HT deaminating, clorgyline-sensitive activity (type A MAO) is found within sympathetic nerve endings or in structures dependent upon the nerves. However, we cannot conclude that type A activity of the arterial wall is found exclusively in the adrenergic nerve endings. On the other hand, the clorgyline-resistant activity, which did not change significantly after 6OH-DM (Fig. 1) appears to be localised entirely outside the adrenergic nerves.

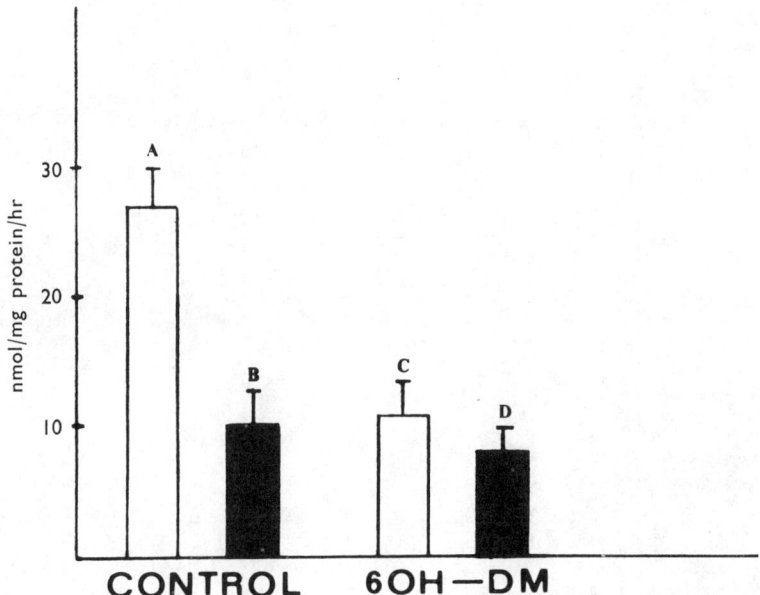

FIG. 1.—Effect of 6OH-DM on mesenteric artery MAO. Aliquots of arterial preparations (750 g supernatants) were incubated with (solid bars) or without (open bars) 10^{-3} M clorgyline for 15 min at room temperature and the remaining activity assayed with tyramine as substrate. A vs C $p < 0.01$; B vs D $p > 0.05$.

We have shown in the present study (Table 1) as well as for other tissues (Neff, Yang and Goridis, this volume) that type A MAO is responsible for the deamination of NA. The selective loss of type A activity after chemical sympathectomy implies that in the arterial wall the extraneuronal activity is of little importance in NA catabolism. De La Lande and Jellett (1972) arrived at essentially the same conclusion based on physiological experiments on the rabbit ear artery.

Our previous work with MAO from normal and denervated pineal gland has shown that almost all of the type A activity, the transmitter specific enzyme, is associated with the sympathetic nerve endings within the gland (Goridis and Neff, 1971) and similar findings have been reported by Jarrott (1971) working on MAO from rat vas deferens. Apparently, the differential distribution of the transmitter specific MAO between the two sites of the neuroeffector junction, we postulated at the example of the pineal gland, is a more general feature of sympathetically innervated tissues.

REFERENCES

Berkovitz B. A., Spector S. and Tarver J. H. (1972) *Brit. J. Pharmacol.* **44**, 10–16.
Coquil J. F., Goridis C., Mack G. and Neff N. H. (1973) *Brit. J. Pharmacol.* in press.
De La Lande I. S., Hill B., Jellet L. B. and McNeil J. M. (1970) *Brit. J. Pharmacol.* **40**, 249–256.
Goridis C. and Neff N. H. (1971) *Brit. J. Pharmacol.* **43**, 814–818.
Jarrott B. (1971) *J. Neurochem.* **18**, 7–16.
Kalsner S. and Nickerson M. (1969) *J. Pharmac. Exp. Theo.* **165**, 152–165.
Yang H. Y. T., Goridis C. and Neff N. H. (1972) *J. Neurochem.* **19**, 1241–1250.

Frontiers in Catecholamine Research 1973, pp. 161 to 166. Pergamon Press. Printed in Great Britain.

OXIDATIVE-REDUCTIVE PATHWAYS FOR METABOLISM OF BIOGENIC ALDEHYDES

V. GENE ERWIN

School of Pharmacy, University of Colorado, Boulder, Colorado 80302, U.S.A

IT IS well known that the catecholamines, norepinephrine, and dopamine, as well as the indoleamine, serotonin, are oxidatively deaminated in brain tissue to the corresponding *aldehyde products* by monoamine oxidase, monoamine:oxygen oxidoreductase (deaminating), (EC 1.4.3.4), (BERNHEIM, 1931; BLASHKO *et al.*, 1937; PUGH and QUASTEL, 1927; CARLSSON and HILLARP, 1962) according to the following reaction:

$$R—CH_2NH_2 + O_2 + H_2O \rightarrow R—CHO + H_2O_2 + NH_3 \qquad (1)$$

In brain and other tissues the aldehydes derived from the catecholamines, i.e., the catecholaldehydes, may be either oxidized to the corresponding acid metabolites or reduced to the alcohol derivative. RUTLEDGE and JONASON (1967) and BREESE *et al.* (1969a, b) observed that formation of predominantly acid or alcohol metabolites of phenylethylamines in brain preparations depended on the specific biogenic amine precursor employed. The β-hydroxy-substituted phenylethylamines are more readily reduced, after the initial deamination to produce the aldehyde, than the phenylethylamines which lack the β-hydroxyl group. These findings, coupled with the observations of KOPIN *et al.* (1961), GLOWINSKI *et al.* (1965) and others (GOLDSTEIN *et al.*, 1961; 1963) have shown that norepinephrine and normetanephrine are metabolised primarily to the corresponding alcohol metabolites, whereas the major product of dopamine metabolism in the brain is the acid derivative (RUTLEDGE and JONASON (1967). These metabolites are formed from the corresponding biogenic aldehyde intermediates either by a reductive of oxidative pathway.

Evidence was presented by ERWIN and DEITRICH (1966) that the oxidation of aldehydes may be catalyzed by aldehyde dehydrogenase. It was observed that brain tissues contain an NAD-dependent aldehyde dehydrogenase, aldehyde:NAD oxidoreductase, (EC 1.2.1.3):

$$R—CHO + H_2O + NAD^+ \rightarrow R—CO_2H + NADH + H^+ \qquad (2)$$

DEITRICH (1966) subsequently found that this enzyme activity was responsible for at least 75 per cent of the total aldehyde oxidising capacity of brain homogenates. The enzyme catalyzed reaction is essentially irreversible.

Recently, TABAKOFF and ERWIN (1970) and ERWIN *et al.* (1972) have reported the isolation, purification and partial characterisation of two aldehyde reductases from bovine brain. One enzyme is NADPH-dependent, while the other is NADH-linked. These enzymes were later also isolated from pig brain (TURNER and TIPTON, 1972). The reversible reaction catalyzed by the enzymes is:

$$R—CHO + NADPH + H^+ \rightleftarrows R—CH_2—OH + NADP^+ \qquad (3)$$
$$(NADH + H^+) \qquad\qquad\qquad (NAD^+)$$

7

Evidence has been obtained (ERWIN *et al.*, 1972, BRONAUGH and ERWIN, 1973; ADLER-GRACHINSKY *et al.*, 1972) which indicates that the NADPH-linked aldehyde reductase, alcohol: NADP oxidoreductase, (EC 1.1.1.2), but not the NADH-dependent enzyme, may be involved in the reduction of biogenic aldehydes including the catecholaldehydes in brain. In Table 1 the cofactor and substrate specificities and inhibitor sensitivities for these aldehyde reductases and for the "classical" liver alcohol dehydrogenase are compared. As shown, the NADH-linked aldehyde

TABLE 1. COMPARISONS OF COFACTOR AND SUBSTRATE SPECIFICITIES AND INHIBITOR SENSITIVITIES OF BRAIN NADPH-LINKED AND NADH-DEPENDENT ALDEHYDE REDUCTASES AND LIVER NAD-LINKED ALCOHOL DEHYDROGENASE

Enzyme	Cofactors	Aldehyde substrates	Inhibitors
NADPH-linked aldehyde reductase*	NADPH	Benzaldehydes Phenylacetaldehydes Catecholaldehydes Indoleacetaldehydes Long and intermediate-chain aliphatic aldehydes	Barbiturates Hydantoins Succinimides Oxazolidinediones Phenothiazines 1,10-orthophenanthroline
NADH-linked aldehyde reductase†	NADH NADPH	Benzaldehydes Phenylacetaldehyde Long and intermediate-chain aliphatic aldehydes	Phenothiazines 1,10-orthophenanthroline
"Classical" NAD-linked alcohol dehydrogenase‡	NADH NADPH	Benzaldehydes Phenylacetaldehydes Catecholaldehydes Indoleacetaldehydes Long, intermediate and short-chain aliphatic aldehydes	Pyrazole Phenothiazines 1,10-orthophenanthroline

* Information compiled from TABAKOFF and ERWIN (1970); ERWIN *et al.*, (1971); and BRONAUGH and ERWIN (1972).
† Information compiled from ERWIN *et al.* (1972).
‡ Information compiled from (BLOMQUIST, 1966; THEORELL and YONETANI, 1963) and from author's own observations, unpublished work.

reductase does not catalyze the reduction of catechol-aldehydes or indoleacetaldehydes. On the other hand, liver NAD-linked alcohol dehydrogenase, alcohol: NAD oxidoreductase, (EC 1.1.1.1), will catalyze the reduction of these aldehydes. In tissues such as liver and kidney, which possess relatively large concentrations of "classical" alcohol dehydrogenase, this enzyme may be partially responsible for reduction of biogenic aldehydes to their corresponding alcohols. RASKIN and SOKOLOFF (1970) have isolated a pyrazole sensitive, NAD-linked alcohol dehydrogenase from brain. It has not been reported whether this enzyme catalyzes the reduction of catecholaldehydes. However, assuming the enzyme does catalyze catecholaldehyde reduction, it posseses such low activity that it would, in all probability, not account for a major portion of biogenic aldehyde reduction in brain.

It has been shown that in various tissues which contain adrenergic innervations, such as brain, heart and lung, the major metabolite of norepinephrine is an alcohol derivative, appearing either as the *O*-methylated derivative, the sulfate conjugate, or the free alcohol (KOPIN *et al.*, 1961; GLOWINSKI *et al.*, 1965; GOLDSTEIN

TABLE 2. COMPARISONS OF ACTIVITIES OF NADPH-LINKED ALDEHYDE REDUCTASE
AND NAD-LINKED ALCOHOL DEHYDROGENASE IN VARIOUS TISSUES.

Tissue	NADPH-linked aldehyde reductase*	NAD-linked alcohol dehydrogenase†
Liver	$6 \cdot 5 \pm 1 \cdot 0$‡	$11 \cdot 7 \pm 2 \cdot 1$
Kidney	$10 \cdot 2 \pm 1 \cdot 8$	$1 \cdot 3 \pm 0 \cdot 3$
Testes	$3 \cdot 2 \pm 0 \cdot 5$	$2 \cdot 1 \pm 0 \cdot 4$
Heart	$2 \cdot 0 \pm 0 \cdot 3$	0§
Brain	$2 \cdot 0 \pm 0 \cdot 3$	0§
Lung	$0 \cdot 7 \pm 0 \cdot 1$	$1 \cdot 2 \pm 0 \cdot 2$

* Activities are expressed as nmoles NADPH oxidized per minute per mg protein with *m*-nitrobenzaldehyde as substrate in the presence of 10^{-2}M pyrazole to inhibit "classical" alcohol dehydrogenase.

† Activities represent differences in rates of NADH oxidation obtained in the absence and presence of 10^{-2}M pyrazole with *m*-nitrobenzaldehyde as substrate.

‡ Represents the one S.D. ($n = 6$).

§ Activity was below measurement by Absorbancy changes, e.g. <$0 \cdot 2$.

et al., 1961, 1963; ADLER-GRACHINSKY, 1972). As shown in Table 2 these tissues and others contain a pyrazole-insensitive, NADPH-linked aldehyde reductase. The enzyme isolated from liver and kidney (ERWIN and DIETRICH, 1972), as well as brain, were inhibited by barbiturates but not by pyrazole. In addition, the data presented in Table 2 indicate that for heart as well as brain NADPH-linked aldehyde reductase is responsible for virtually all of the biogenic aldehyde reduction.

Evidence which further indicates that NADPH-linked aldehyde reductase is responsible for catecholaldehyde reduction is presented in Table 3. Although the

TABLE 3. MICHAELIS CONSTANTS AND MAXIMAL VELOCITIES FOR VARIOUS SUBSTRATES
Values were calculated from Lineweaver–Burk plots. The reaction mixtures contained various concentrations of aldehydes, 0·16 mM NADPH or 1 mM NAD and approximately 1 mg protein from monkey brain (dialyzed ammonium sulfate fraction) in 0·1 M sodium phosphate, pH 7·4, in a final mixture of 3·0 ml.

Substrate	NADPH-linked aldehyde reductase		NAD-linked aldehyde dehydrogenase	
	$K_m \times 10^{-6}$M	V_{max}*	$K_m \times 10^{-6}$M	V_{max}*
3,4-Dihydroxyphenylglycolaldehyde	12·0	0·0043		
	(0·33)†	(0·0011)	5·5	0·0016
3,4-Dihydroxyphenylacetaldehyde	6·1	0·001	10·0	0·005
5-Hydroxyindoleacetaldehyde	27·0	0·005		
	(1·10)	(0·002)	1·0	0·0012
p-Hydroxyphenylglycolaldehyde	7·1	0·011		
	(0·5)	(0·004)	2·7	0·0019
p-Hydroxyphenylacetaldehyde	20·0	0·004		
	(2·2)	(0·0012)	1·5	0·0027
Acetaldehyde	—	0	0·3	0·0030
Propionaldehyde	—	0	0·2	0·0028
NADPH	2·6	—	—	—
NAD	—	—	10·0	—

* Values are expressed as Δ A. at 340 nm per min per mg protein.

† Values in parentheses refer to the K_m and V_{max} values obtained with extremely low concentrations of aldehyde substrates. The Lineweaver–Burk plots are not linear and curve downward abruptly at aldehyde concentrations greater than 5×10^{-6} M. This data is consistent with two enzymes with differing kinetic properties or with an allosteric anzyme. Similar findings have been reported by TURNER and TIPTON (1972).

K_m values for 3,4-dihydroxyphenylglycolaldehyde (DHPGA) and for 3,4-dihydroxy-phenylacetaldehyde (DHPAA) were quite similar, the V_{max} values differed significantly; the value for DHPGA was approximately 3-fold greater than the value for DHPAA. As shown in Table 3 the V_{max} value for the aldehyde derived from dopamine, DHAA, with NAD-linked aldehyde dehydrogenase was 3-fold higher than the value obtained for DHPGA. These kinetic parameters for the reductive and oxidative pathways are consistent with the observations that dopamine is converted in the brain primarily to the acid metabolite and norepinephrine is converted primarily to an alcohol derivative. However, it is unlikely that the distribution of acid and alcohol metabolites of norepinephrine and dopamine can be totally accounted for by these kinetic parameters, particularly since both aldehyde intermediates are good substrates for the oxidative and the reductive pathways. It is probable that subcellular compartmentation or cellular localisation of aldehyde reductase and aldehyde dehydrogenase may be involved in the metabolite formation. The recent observations of DUNCAN et al. (1972) support the concept of a unique cellular localisation of aldehyde dehydrogenase.

In Table 4 the distributions of biogenic aldehyde reducing and oxidising capacities in various areas of bovine brain are listed. As shown, although all areas of the brain

TABLE 4. DISTRIBUTION OF BIOGENIC ALDEHYDE REDUCING AND OXIDIZING CAPACITIES IN VARIOUS AREAS OF BOVINE BRAIN

Area	Aldehyde reducing capacity*	Aldehyde oxidizing capacity*
	(μmoles NADPH oxidized per hr per g tissue)	(μmoles indole-3-acetic acid formed per hr per g tissue)
Midbrain-hypothalamus	9·22 (7·87–11·70)	1·92 (1·86–2·04)
Pons	9·00 (7·65–11·70)	2·28 (1·92–2·64)
Medulla	8·55 (7·87–9·00)	2·28 (1·92–2·64)
Cerebellum	7·42 (6·30–8·10)	1·08 (0·84–1·26)
Caudate†	7·42 (6·97–7·65)	2·94 (2·64–3·30)
Thalamus‡	7·20 (5·62–8·77)	2·04 (1·86–2·28)
Lentiform nucleus§	6·52 (5·85–7·20)	1·74 (1·44–2·04)
Cerebral cortex	6·07 (5·40–6·97)	2·10 (1·92–2·28)

* Values represent the means with the ranges obtained from three bovine brains and were obtained under comparable conditions at pH 7·4. Methods are described (ERWIN and DEITRICH, 1966; TABAKOFF and ERWIN, 1970)
† Head of the caudate nucleus.
‡ Medial nuclei of the thalamus.
§ Includes internal capsule and lateral thalamic nuclei along with globus pallidus and putamen.

possess the enzyme activities, NADPH-linked aldehyde reductase is most concentrated in the midbrain-hypothalamus, pons, and medulla areas, whereas NAD-linked aldehyde dehydrogenase is most concentrated in the caudate nucleus, followed by the pons and medulla. It is of interest that, in keeping with the distribution of aldehyde dehydrogenase, the caudate nucleus contains a higher concentration of dopamine than other brain areas. In addition, those areas highest in norepinephrine concentration contain the highest aldehyde reductase activity. Whether these differences in regional distribution coupled with the above mentioned differences in kinetic parameters for the enzymes can account for the conversion of dopamine or norepinephrine predominantly to their acid or alcohol metabolites is unclear.

Obviously, it would be of interest to know the actual cellular or neuronal localisation of these enzymes. Such studies are presently being conducted in our laboratory.

Recently, we determined the activity of NADH-linked and NADPH-linked aldehyde reductases from brains of various animal species. In these studies, Table 5,

TABLE 5. PHYLOGENETIC STUDY OF BRAIN ALDEHYDE REDUCTASE

Class of vertebrate	Specific activity of aldehyde reductases	
	(NADH-linked nmoles NADH oxidized/min/mg protein)	(NADPH-linked nmoles NADPH oxidized/min/mg protein)
Osteichthyes		
Rainbow trout	1·6	0·61
Amphibia		
Tiger salamander	0	1·91
Reptile		
American chameleon	0	7·10
Avis		
Domestic pigeon	1·4	1·90
Mammals		
Human (fetus)*	1·2	2·80
Rhesus monkey	5·1	4·43
Rat (Sprague–Dawley)	1·4	3·63

* Weighed less than 500 g; approximately 3 months in development.

it was observed that brain tissue from all classes of vertebrates possessed NADPH-linked aldehyde reductase, whereas, the tiger salamander and American chameleon were devoid of the NADH-linked aldehyde reductase activity. Except for the fact that the reptile brain contains primarily epinephrine rather than norepinephrine, a rough correlation exists between the catecholamine levels (BOGDANSKI et al., 1963) and the NADPH-linked aldehyde reductase activities in the brains of the various animals.

Inasmuch as the biogenic aldehydes are highly reactive intermediates which have been shown to alter various metabolic and physiological processes (BOGDANSKI et al., 1963; REHAK and TRUITT, 1955; SABELLI, 1969; ERWIN, 1965), elucidation of various factors involved in regulating the metabolism of these compounds is of importance. Previously (ERWIN et al., 1971; BRONAUGH and ERWIN, 1972) we reported that NADPH-linked aldehyde reductase is markedly inhibited by hypnotic-

TABLE 6. INHIBITORS OF NADPH-LINKED ALDEHYDE REDUCTASE

Inhibitor	K_i values* $\times 10^{-4}$ M
Pentobarbital	0·6
Phenobarbital	1·2
Barbituric acid	40·0
Diphenylhydantoin	1·5
Dimethyloxazolidinedione	4·7
Ethosuximide	5·4
Glutethimide	3·0
Megimide	—
Chlorpromazine	7·0

* Values were determined from double reciprocal plots as described (ERWIN et al., 1971; BRONAUGH and ERWIN, 1972; ERWIN and DEITRICH, 1973).

anticonvulsant barbiturates and to a lesser degree by various phenothiazines. Recently (Erwin and Deitrich, 1973), we have shown that a number of hydantoin, succinimide, and 2,4-oxazolidinedione anticonvulsants are inhibitors of this enzyme, Table 6. The pharmacological significance of these findings remains to be established.

REFERENCES

Adler-Graschinsky E., Langer S. Z. and Rubio M. C. (1972) *J. Pharmac. Exp. Therap.* **180,** 286–301.
Bernheim M. L. C. (1931) *J. Biol. Chem.* **93,** 299–309.
Blashko H., Richter D. and Schlossmann H. (1937) *Biochem. J.* **31,** 2187–2196.
Blomquist C. H. (1966) *Acta Chem. Scand.* **20,** 1747–1757.
Bogdanski D. F., Bonomi L. and Brodie B. B. (1963) *Life Sci.* **1,** 80–84.
Breese G. R., Chase T. N. and Kopin I. J. (1969a) *J. Pharmac. Exp. Ther.* **165,** 9–13.
Ibid.: (1969b) *Biochem. Pharmac.* **18,** 863–869.
Bronaugh R. L. and Erwin V. G. (1973) *J. Neurochem.,* In Press.
Bronaugh R. L. and Erwin V. G. (1972) *Biochem. Pharmac.* **21,** 1457–1464.
Carlsson A. and Hillarp N. A. (1962) *Acta Physiol. Scand.* **55,** 95–100.
Deitrich R. A. (1966) *Biochem. Pharmac.* **15,** 1911–1922.
Duncan R. J. S., Sourkes T. L., Boucher R., Poirier L. J. and Roberge A. (1972) *J. Neurochem.* **19,** 2007–2010.
Erwin V. G. (1965) Thesis, University of Colorado.
Erwin V. G. and Deitrich R. A. (1966) *J. Biol. Chem.* **241,** 3533–3539.
Erwin V. G. and Deitrich R. A. (1972) *Biochem. Pharmac.* **21,** 2915–2924.
Erwin V. G. and Deitrich R. A. (1973) *J. Neurochem.,* In Press.
Erwin V. G., Heston W. D. W. and Tabakoff B. (1972) *J. Neurochem.* **19,** 2269–2278.
Erwin V. G., Tabakoff B. and Bronaugh R. L. (1971) *Molec. Pharmac.* **7,** 169–176.
Glowinski J., Kopin I. J. and Axelrod J. (1965) *J. Neurochem.* **12,** 25–30.
Goldstein M., Freidhoff A. J., Pomerantz S., and Contrera J. F., (1961), *J. Biol. Chem.* **236,** 1816–1821.
Goldstein M. and Gerber H. (1963) *Life Sci.* **2,** 97–100.
Kopin I. J., Axelrod J. and Gordon E. (1961) *J. Biol. Chem.* **236,** 2109–2113.
Pugh C. E. M. and Quastel J. H. (1927) *Biochem. J.* **31,** 2306–2321.
Rehak M. L. and Truitt E. B. (1955) *Quart. J. Stud. Alcohol.* **19,** 399–405.
Rutledge C. O. and Jonason J. (1967) *J. Pharmac. Exp. Ther.* **157,** 493–502.
Raskin N. H. and Sokoloff L. (1970) *J. Neurochem.* **17,** 1677–1687.
Sabelli H. C. (1969) *The Pharmacologist* **11,** 279; *Ibid.:* (1968) **10,** 211.
Tabakoff B. and Erwin V. G. (1970) *J. Biol. Chem.* **245,** 3263–3268.
Theorell H. and Yonetani T. (1963) *Biochem. Z.* **338,** 537–553.
Turner A. J. and Tipton K. F. (1972) *Europ. J. Biochem.* **30,** 361–368.
Turner A. J. and Tipton K. F. (1972) *Biochem. J.* **130,** 765–772.

Frontiers in Catecholamine Research 1973, pp. 167 to 171. Pergamon Press. Printed in Great Britain.

SULFATE CONJUGATES IN THE BRAIN

J. L. MEEK and A. FOLDES
Laboratory of Preclinical Pharmacology, National Institute of Mental Health,
Saint Elizabeths Hospital, Washington, D.C. 20032, U.S.A.

ONE POSSIBLE approach to studying the turnover of norepinephrine (NE) in human brain is to determine the turnover of its metabolites in CSF. The only NE metabolites so far measured in human CSF are 3-methoxy-4-hydroxyphenyl glycol (MOPEG) and its sulfate conjugate (MOPEG-SO$_4$)(SCHANBERG et al., 1968a). Many factors in addition to NE turnover affect the concentration of MOPEG and MOPEG-SO$_4$. Among these are the mechanisms by which the metabolites leave the brain, and the enzyme which converts MOPEG to MOPEG-SO$_4$ (phenolsulfotransferase E.C. 2.8.2.1.). A thorough understanding of these factors is necessary if the turnover of MOPEG or MOPEG-SO$_4$ is to be used as a reliable index of the turnover of NE in man.

EFFLUX OF MOPEG AND MOPEG-SO$_4$ FROM BRAIN

If MOPEG-SO$_4$ is actively transported from the CSF by the choroid plexus, blockade of this transport mechanism should cause MOPEG-SO$_4$ to accumulate in CSF at a rate proportional to its synthesis from NE. Probenecid, a carboxylic acid, competitively inhibits the efflux of other anionic metabolites from the brain. Apparently probenecid can inhibit MOPEG-SO$_4$ transport, since when MOPEG-SO$_4$ is injected into the lateral ventricle of rats (MEEK and NEFF, 1972a) its disappearance is inhibited by probenecid. Administration of probenecid (400 mg/kg) causes the concentration of endogenous MOPEG-SO$_4$ to rise in rat brain (Fig. 1) and in rabbit CSF (EXTEIN et al., 1973). Unfortunately, rather toxic doses of probenecid were required in these experiments to inhibit the efflux of MOPEG-SO$_4$; more than is necessary to cause other anions to accumulate. In a study in humans (GORDON et al., 1973) MOPEG-SO$_4$ concentration in CSF did not rise after probenecid treatment, but the dose they used (100 mg/kg over 18 hr) may have been too low. It is also possible that the mechanisms of elimination of anions from human brain and CSF is different from that of rats and rabbits.

MOPEG-SO$_4$ AS A MEASURE OF NE TURNOVER

Estimation of the accumulation of MOPEG-SO$_4$, and measurement of its steady state concentration have both been used in studies of NE turnover. Measurement of MOPEG-SO$_4$ formation in man may not be possible because of the toxicity of the high doses of probenecid needed to block MOPEG-SO$_4$ efflux. In rats, the steady state concentration of MOPEG-SO$_4$ appears to correlate well with the turnover of NE. For example, both the steady state concentration of MOPEG-SO$_4$ and its rate of formation are increased by phenoxybenzamine (which increases the turnover of rat brain NE)(MEEK and NEFF, 1973a).

Stress also increases in brain the turnover of NE and the concentration of MOPEG-SO$_4$. Stimulation of the locus coeruleus increases the concentration of

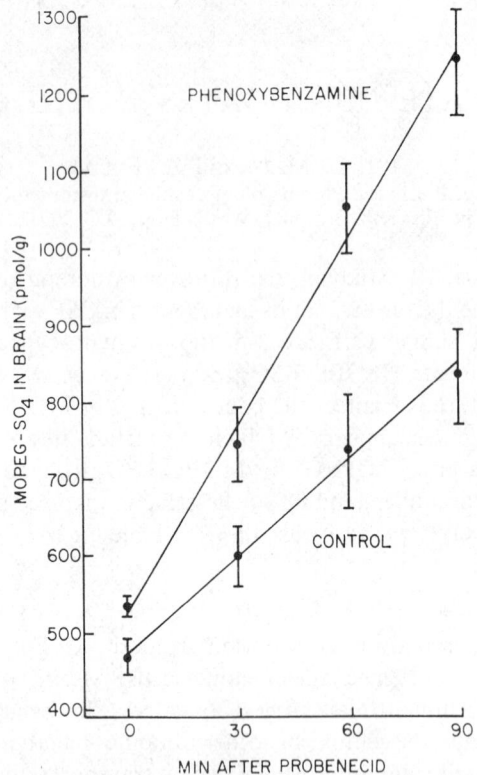

FIG. 1.—Effect of phenoxybenzamine (20 mg/kg i.p.) on the accumulation of MOPEG-SO₄ in brains of rats treated with probenecid (400 mg/kg i.p.). Probenecid was injected 1 hr after phenoxybenzamine. The lines were fitted by the method of least squares ($N = 8$).

MOPEG-SO$_4$ in the ipsilateral cortex, concomitant with an increase of NE turnover (KORF *et al.*, 1973; WALTER and ECCLESTON, 1972). Measurement of the steady state concentration of MOPEG-SO$_4$ is rather simple, and the MOPEG-SO$_4$ concentration in human CSF might correlate well with some disease states.

However, the steady state concentrations of MOPEG-SO$_4$ are probably a less sensitive measure of changes in turnover than estimation of MOPEG-SO$_4$ accumulation. The fraction of NE which is metabolised to MOPEG-SO$_4$ can be estimated since brain MOPEG-SO$_4$ accumulates at a rate of 244 pmol/g/hr after probenecid, while the synthesis rate of NE is 700 pmol/g/hr in the same strain of rats. Assuming that probenecid has blocked completely the efflux of MOPEG-SO$_4$, one can calculate that MOPEG-SO$_4$ is probably not the only metabolite of NE in the rat brain (MEEK and NEFF, 1973a).

SULFATE CONJUGATION IN BRAIN OF THE CATECHOLAMINES AND THEIR METABOLITES

MOPEG-SO$_4$ was the first endogenous sulfurylated biogenic amine metabolite to be identified in the CNS (SCHANBERG *et al.*, 1968a). Earlier, GOLDSTEIN and GERBER (1963) found that conjugated methoxy hydroxy phenylethanol was formed in brain after peripheral injection of labelled dopa or dopamine. After intraventricular injection of labelled tyramine, dopamine, norepinephrine and normetanephrine, conjugation of the amines and their acidic and neutral metabolites occurred (SUGDEN and ECCLESTON, 1971; GOLDSTEIN et al., 1970; TAYLOR and LAVERTY, 1969; SCHANBERG et al., 1968b).

All these studies suggest that the soluble enzyme which conjugates phenols, phenolsulfotransferase, has low specificity in the brain as well as the liver. The similarity of the enzymes from the two tissues has been confirmed in in vitro studies (MEEK and NEFF, 1973b). Brain phenolsulfotransferase has been purified 30-fold and its kinetics studied using phenols related to the biogenic amines (FOLDES and MEEK, 1973) The assay measures the transfer of [^{35}S]-sulfate to an acceptor phenol from the sulfate donor [^{35}S]-3' phosphoadenosine 5' phosphosulfate (PAPS). This preparation of phenolsulfotransferase has roughly similar activity towards the acidic and neutral metabolites of NE (Table 1). The amines themselves are conjugated but at a

TABLE 1. SUBSTRATE SPECIFICITY OF PARTIALLY PURIFIED RAT BRAIN
PHENOLSULFOTRANSFERASE

Substrate	K_m (μM)	Velocity, counts/min/mg/min pH 6·4	pH 9·0
Methoxyhydroxyphenyl glycol	18	17,900	NA
Dihydroxyphenyl glycol	17	18,700	NA
Homovanillic acid	14	14,800	NA
Vanilmandelic acid	100	7700	NA
Dihydroxyphenylacetic acid	10	6700	NA
Dopamine	20	920	4000
Norepinephrine	100	NA	1200

The K_m's for the enzyme with the acidic and neutral phenols as substrates were determined at pH 6·4, while the K_m's for dopamine and norepinephrine were determined at pH 9·0. The enzyme was purified 30-fold by fractionation with ammonium sulfate and alumina gel, and chromatography on DEAE Sephadex.
NA = not assayed.

slower rate, and with a much different pH optimum. Since the K_m's of phenol-sulfotransferase for many of these compounds are low (Table 1), it is probable that conjugates other than MOPEG-SO$_4$ occur in the brain. Their occurrence must be examined if a "balance sheet" is to be constructed comparing the disappearance of the amines, and the formation of metabolites.

ECCLESTON (1973) has introduced a useful technique for studying sulfate conjugation in vivo. If labelled sodium sulfate is injected intraventricularly, together with a phenol, a sulfate ester is formed. The amount formed varies with the amount of phenol injected approximately in accord with Michaelis–Menten kinetics. If rats are injected with phenols and labelled sulfate, and the brains homogenised, protein and unincorporated [^{35}S]-SO$_4$ can be precipitated with barium leaving the arylsulfates in solution. The amines and their acidic and neutral metabolites can be shown by this technique to be conjugated in brain in vivo (MEEK and NEFF, 1973b).

Formation of sulfate conjugates requires phenolsulfotransferase (GREGORY and LIPMANN, 1957) and PAPS (ROBBINS and LIPMANN, 1957). While little is known of the regional or species variation in the occurrence of PAPS (BALASUBRAMANIAN and BACHHAWAT, 1970), we have examined the localisation of phenolsulfotransferase

Table 2. Regional distribution of
phenolsulfotransferase in rat brain

Region	Specific activity (pmol/min/mg protein)
Hypothalamus	$39 \cdot 1 \pm 3 \cdot 3$
Hippocampus	$34 \cdot 0 \pm 1 \cdot 9$
Midbrain	$32 \cdot 4 \pm 2 \cdot 9$
Striatum	$26 \cdot 7 \pm 2 \cdot 8$
Cortex	$17 \cdot 6 \pm 1 \cdot 5$
Medulla	$14 \cdot 0 \pm 1 \cdot 4$
Cerebellum	$3 \cdot 0 \pm 0 \cdot 3$

Phenolsulfotransferase was assayed in the
30,000 g supernatant using phenol as substrate. The results are mean $\pm$ s.e.m. of 6
determinations. The brains were dissected
according to Glowinski and Iversen (1966).

(Table 2). It is unevenly distributed with 13 times more activity in the hypothalamus
than in the cerebellum. This distribution only roughly corresponds to that of the NE
or of MOPEG-SO$_4$ (Glowinski and Iversen, 1966; Meek and Neff, 1972b). We
found that phenolsulfotransferase is not restricted to adrenergic neurons, since its
activity in the spinal cord does not diminish after axonal section. (Table 3). Since

Table 3. Effect of spinal section on activity of
phenolsulfotransferase in spinal cord

	Cranial half	Caudal half
	(counts/min/mg protein/min)	
Control	3394 ± 348	3216 ± 175
Spinal Section	3548 ± 468	4080 ± 580
		(n.s.)

Phenolsulfotransferase activity was determined in the
30,000 g supernatants from the spinal cord of control
rats, and rats whose cord had been transected one week
earlier. $N = 5$

glial cells are probably evenly distributed in the brain but the enzyme is not, we
conclude that it is not primarily a glial enzyme. Phenolsulfotransferase might be
preferentially located post-synaptically to neurons whose transmitters are phenols.

There are species variations in the fraction of MOPEG which occurs as a conjugate (Schanberg et al., 1968a). We have observed that the specific activity of
phenolsulfotransferase also varies among species. The activity is about 3-times
greater in the rabbit than the rat, while mouse and frog brains have only 5–10 per
cent of the activity of rat brains.

In the liver the conjugation of toxic phenols aids in their excretion by converting
them to more polar compounds. Phenolsulfotransferase may have a similar role in
the brain, if it conjugates exogenous as well as endogenous phenols. It is also possible
that phenolsulfotransferase conjugates endogenous non-phenolic compounds.

In summary, the catecholamines and their metabolites can be sulfate conjugated
by brain, in vivo and in vitro. The responsible enzyme does not appear to occur
mainly in glia, or in noradrenergic neurons. One of the conjugated metabolites of
NE, MOPEG-SO$_4$ can probably be a clinically useful index of NE turnover in man.

REFERENCES

BALASUBRAMANIAN A. S. and BACHHAWAT B. K. (1970) *Brain Res.* **20,** 341–360.

ECCLESTON D. (1973) *J. Neurochem.* (In press).

EXTEIN I., KORF J., ROTH R. H. and BOWERS M. B. (1973) *Brain Res.* **54,** 403–407.

FOLDES A. and MEEK J. L. (1973) *Fed Proc.* **32,** 797.

GLOWINSKI J. and IVERSEN L. L. (1966) *J. Neurochem.* **13,** 655–669.

GOLDSTEIN M. and GERBER (1963) *Life Sci.* **2,** 97–100.

GOLDSTEIN M., ANAGNOSTE B., YAMAMOTO A. and FELCH W. C. (1970) *J. Pharmacol. Exp. Ther.* **171,** 196–204.

GORDON E. K., OLIVER J., GOODWIN F. K., CHASE T. N. and POST R. M. (1973) *Neuropharmacol.* **12,** 391–396.

GREGORY J. D. and LIPMANN F. (1957) *J. Biol. Chem.* **229,** 1081–1090.

KORF J., AGHAJANIAN G. K. and ROTH R. H. (1973a) *Eur. J. Pharmacol.* **21,** 305–310.

MEEK J. L. and NEFF N. H. (1972a), *J. Pharmacol. Exp. Ther.* **181,** 457–462.

MEEK J. L. and NEFF N. H. (1972b) *Brit. J. Pharmacol.* **45,** 435–441.

MEEK J. L. and NEFF H. H. (1973a) *J. Pharmacol. Exp. Ther.* **184,** 570–575.

MEEK J. L. and NEFF N. H. (1973b) *J. Neurochem.* **21,** 1–10.

ROBBINS P. and LIPMANN F. (1957) *J. Biol. Chem.* **229,** 837–851.

SCHANBERG S. M., BREESE G. R., SCHILDKRAUT J. J., GORDON E. K. and KOPIN I. J. (1968a) *Biochem. Pharmacol.* **17,** 2006–2008.

SCHANBERG S. M., SCHILDKRAUT J. J., BREESE G. R. and KOPIN I. J. (1968b) *Biochem. Pharmacol.* **17,** 247–254.

SUGDEN R. F. and ECCLESTON D. J. (1971) *J. Neurochem.* **18,** 2461–2468.

TAYLOR K. M. and LAVERTY R. (1969) *J. Neurochem.* **16,** 1367–1376.

WALTER D. S. and ECCLESTON D. J. (1972) *Biochem. J.* **128,** 85P–86P.

Frontiers in Catecholamine Research 1973, pp. 173 to 174. Pergamon Press. Printed in Great Britain.

REPORTER'S NOTES—FIRST SESSION

Dr. O. T. Phillipson

THE TOPIC of substrate inhibition of tyrosine hydroxylase in the presence of tetra-hydropterin *in vitro*, as demonstrated in Dr. Kaufman's paper generated most of the discussion in the first session. Several speakers produced evidence on the physiological significance of these findings. Dr. Weiner found that in isolated intact vas deferens preparations, there was no substrate inhibition at up to three times the plasma concentrations of tyrosine. Dr. Dairman investigated the problem by raising plasma levels of tyrosine by either inhibition of protein synthesis or by feeding tyrosine. He found no change in the levels of NA in the central nervous system of the rats, and no change in the turnover of heart NA. Dr. Karobath found, in synaptosomal preparations, only a 10 per cent inhibition of tyrosine hydroxylation at tyrosine levels up to six times the K_m for tyrosine.

These negative findings tend to cast doubt on the physiological significance of the *in vitro* work and suggest that transport of tyrosine into nerve endings may be controlled to maintain a constant level of intraneuronal amino acid. However, as Dr. Kaufman pointed out, we know very little about the regulation of uptake of tyrosine into nerve cells, or about the intraneuronal concentration of tyrosine. Furthermore, the brain enzyme is less sensitive to inhibition than is the adrenal medullary enzyme, making detection of an effect in intact preparations of this tissue more difficult. In addition, because of the shape of the curve relating tyrosine hydroxylase activity to tyrosine concentration *in vitro*, (Fig. 1 KAUFMAN, this symposim) an unchanged rate of hydroxylation could result from a raised tyrosine concentration, and still be consistent with the inhibitory mechanism proposed. Related to this problem, Dr. Weiner raised the question as to whether phenylalanine could serve as substrate for tyrosine hydroxylase in the presence of tetrahydrobiopterin in intact cells. Using the isolated intact vas deferens preparation, he found that phenylalanine is hydroxylated at only 25 per cent of the rate of tyrosine. When equimolar amounts of tyrosine and phenylalanine are present the conversion of tyrosine to DOPA is not inhibited by the phenylalanine, whereas the conversion of phenylalanine to DOPA is inhibited by the tyrosine. However, in the presence of low tyrosine concentrations, conversion of phenylalanine to tyrosine becomes more efficient. The significance of these findings may be that when tyrosine levels are low, the conversion from phenylalanine may provide a safety mechanism to overcome the deficit. Finally, Dr. Weiner discussed Dr. Nirenberg's finding that neuroblastoma clones in tissue culture showing high tyrosine hydroxylase activity showed good growth in the presence of phenylalanine and no tyrosine, whereas clones with low tyrosine hydroxylase activity showed an absolute growth requirement for tyrosine. The uptake of phenylalanine and tyrosine into cells of different type, and whether or not there is competition between the two, is as yet a largely unstudied problem. More information is clearly needed in this field.

The discussion then turned to the problem of the identity of 5HTP decarboxylase and DOPA decarboxylase. Dr. Thoenen cast doubt on the idea that the two activities represented different substrate utilisation by the same enzyme protein. He found that intracisternal injection of 6-hydroxydopamine caused a fall in activity towards DOPA, but none towards 5HTP. Dr. Goldstein replied by quoting a similar experiment, this time measuring the activities in the striatum (where decarboxylase activity is much higher). In this case, both activities were reduced. The many difficulties involved in the interpretation of work with 6-hydroxydopamine, compared to the immunological studies described in Dr. Dairman's paper, tend to favour the view that there is only one enzyme involved. Dr. Udenfriend re-emphasised the importance of the purity of enzyme preparations in any discussion of differences between the two activities with respect to kinetic measures, and that when purity is ensured the results are consistent with the one-enzyme hypothesis.

Dr. Udenfriend then described the curious phenomenon that in scorbutic animals where tissue ascorbate levels were reduced to as little as 1 per cent of controls, the adrenal medullary cells could manufacture NA at a normal rate, in the face of very low DBH activity. Dr. Kaufman suggested that in these situations the catechol grouping of dopamine may support the hydroxylation reaction. Another explanation is possibly that even in scorbutic animals there is enough ascorbate concentrated, where it is needed, at critical intracellular loci. This argument is, however, difficult to fault or support experimentally.

Commenting on Dr. Hartman's paper, Dr. Osnowman stressed that the finding of an association of adrenergic terminals with intracerebral vessels (a finding for which he too had similar morphological evidence) became interesting only if it could be shown by functional or ultrastructural studies that there was true vasomotor innervation.

Finally, commenting on the immunofluorescent localisation studies for DOPA decarboxylase and DBH reported by Dr. Goldstein, Dr. Geffen strongly argued against accepting this type of technique as a semi-quantitive method for estimating the amount of enzyme present in the cell. Critical was the question of diffusion of antibody to the site of the enzyme in tissue sections, and this involved variables hard to eliminate in the anatomically complex situation seen in the nervous system.

Frontiers in Catecholamine Research 1973, pp. 175 to 176. Pergamon Press. Printed in Great Britain.

REPORTER'S NOTES—SECOND SESSION

Moussa B. H. Youdim

MRC Clinical Pharmacology Unit, University Department of Clinical Pharmacology
Radcliffe Infirmary, Oxford, England.

SINCE the last International Catecholamine Symposium in Milan (1965), almost all the enzymes involved in the metabolism of the catecholamines have been isolated, purified and partially characterized.

The discussion of Dr. Ciarnello's presentation was centered around two major points: (1) whether the genetic control is specific for PNMT or (2) whether the genes regulating all four catecholamine synthesizing enzymes are on the same or different chromosomes. It was pointed out that the chromosomal mapping of the genes is very difficult.

Dr. Oreland, in connection with Dr. Zeller's paper, reported that he has attacked the problem of the mechanism of action of pargyline from another direction. After extensive digestion of purified MAO, previously inhibited by ^{14}C-pargyline, he has purified a pargyline-containing fragment of the enzyme. Analyses of this fragment yielded a pentapeptide, flavin and pargyline in stoichiometric amounts. This result seems to be a confirmation of the biological significance of Dr. Zeller's more basically chemical work.

During open discussion, Dr. Bhagat, Department of Physiology, St. Louis University, presented evidence for the hormonal effect of the activity of monoamine oxidase and other catecholamine enzymes.

In his talk, Dr. Erwin described in detail the properties of purified NAD-dependent aldehyde dehydrogenase and NADH-dependent aldehyde reductase and he presented evidence which suggests that the differences in the metabolism of norepinephrine and dopamine in the brain may be a result of differences in the kinetic parameters of aldehyde reductase and aldehyde dehydrogenase in this tissue. Dr. Sourkes pointed out that, together with Dr. J. Duncan (Transactions of the American Society of Neurochemistry, 1972) he has been studying the kinetics of enzymic dehydrogenation and reduction of aldehydes prepared from many of the currently studied biogenic amines. Which branch of this "aldehyde switch" is followed depends upon many factors, including the presence or absence of a beta-OH on the side chain, relative concentrations of NAD and NADP cofactors, concentration of the aldehyde and physical location of the enzymes. Sourkes and Duncan had previously showed that aldehyde dehydrogenase is largely intraneuronal in the cat striatum; this work has been confirmed by Agid, Javoy and Youdim (Brit. J. Pharmacol, 1973, *in press*) in the rat striatum. In this region, the reductase is unaffected by lesions of the nigrostriatal tract, so that it appears to be extraneuronal. With reference to Dr. Erwin's talk, Dr. Turner pointed out that he has previously shown (Turner, A. J. and Tipton, K. F., Biochem J. (1972) **130**, 765–772) that partially purified preparations of NADPH-linked aldehyde reductase from pig, rat or ox brain exhibit non-linear reciprocal plots, when aldehyde is the variable substrate. This non-linearity is due to the presences of two distinct NADH-linked enzymes which can be separated by DEAE-cellulose chromatography. The separated enzymes appear to give linear reciprocal plots. Both enzymes can reduce the biogenic aldehydes although the kinetic constants are very different. The enzymes also differ in sensitivity to inhibition by phenothiazines and barbiturates. Thus, any results relating to kinetic or inhibition studies of partially purified preparations of brain NADPH-dependent aldehyde reductase should be interpreted with caution.

In relation to Dr. Meek's paper, Dr. Sourkes remarked about the work of Dr. K. P. Wong who studied another conjugative pathway between 1964 and 1966 in his laboratory. Their published work demonstrates that UDPGA transferase in the liver supernatant of many species catalyses the glucuronidation of many catecholamines and their derivatives, but homovanillic acid is not a substrate.

Regulation
(May 22, 1973; 9:00 A.M.–1:00 P.M.)

CHAIRMEN: Leslie L. Iversen and B. R. Belleau
COUNCILLOR: Theodore Sourkes
REPORTER: Norman Weiner

Frontiers in Catecholamine Research 1973, pp. 179 to 185. Pergamon Press. Printed in Great Britain.

TRANS-SYNAPTIC REGULATION OF TYROSINE HYDROXYLASE

Hans Thoenen, Uwe Otten and Franz Oesch

Department of Pharmacology, Biocenter of the University, Basel, Switzerland

NEURONALLY MEDIATED INCREASE OF ENZYME SYNTHESIS AND TRANSPORT OF THE ENZYME FROM THE PERIKARYON TO THE NERVE TERMINALS

THE RESPONSE of an effector neuron to the chemically mediated transmission of a nerve impulse is not confined to immediate effects such as changes in ionic permeability of the cell membrane or changes in the intermediary metabolism. The response also involves more permanent effects, reflected by changes in the synthesis of macromolecular cell constituents.

Investigations over the last few years have shown that the peripheral sympathetic nervous system lends itself favourably to the study of such interneuronal effects and may offer an opportunity to shed more light on the basic problem of the relationship between the functional state of the neuronal membrane and the regulation of the expression of the available genetic information.

After the initial observation of MUELLER et al. (1969a) that the destruction of the sympathetic nerve terminals by 6-hydroxydopamine produces an increase in the *in vitro* activity of tyrosine hydroxylase (L-tyrosine, tetrahydropteridine: oxygen oxidoreductase (3-hydroxylating): EC 1.14.3a) (TH) in the adrenal medulla, it has been demonstrated in many laboratories that a great number of experimental conditions leading to an enhanced activity of the preganglionic cholinergic nerves induce an increased TH activity in the terminal adrenergic neurons and the adrenal medullary cells (KVETNANSKY et al., 1970; MANDELL and MORGAN, 1970; WEINER and MOSIMANN, 1970; BHAGAT and RANA, 1971; BLACK et al., 1971; DAIRMANN and UDENFRIEND, 1971; REIS et al., 1971; CHEAH et al., 1971). This neuronally mediated increase in enzyme activity results from an enhanced synthesis of enzyme protein rather than from the formation of an activator, since the activity of enzyme preparations of controls and treated animals is always additive (MUELLER et al., 1969b) and the increase in the *in vitro* activity can be blocked by pretreatment with inhibitors of protein synthesis (MUELLER et al., 1969c). However, this increased TH synthesis is not a reflection of a generally enhanced formation of protein, since it is not accompanied by a statistically significant increase in the total protein content of sympathetic ganglia (MUELLER et al., 1969b).

Besides arousing interest as a model for investigating the regulation of the synthesis of specific proteins by nerve impulses, the trans-synaptic induction of TH is important for the function of the peripheral sympathetic nervous system. The augmented synthesis of the rate-limiting enzyme of norepinephrine (NE) formation may represent a long-term adaptation to increased transmitter utilization (cf. THOENEN and OESCH, 1973). In the peripheral sympathetic nervous system the adrenergic nerve terminals are not only the main site of NE storage but also of NE synthesis (cf. GEFFEN and LIVETT, 1971). Thus, if trans-synaptic induction of TH represents a relevant factor in the long-term adaptation to increased transmitter utilization, augmented enzyme levels in the nerve terminals rather than in the cell bodies would be of primary importance. Indeed, after administration of reserpine (THOENEN et al., 1970) or after exposure of rats to a swimming stress (Otten, unpublished observations) an increase in TH activity occurs not only in the stellate ganglion but also in the corresponding nerve terminals of the rat heart. However, this increase appears with a delay of 24–36 hr, implying a transport of the induced enzyme protein from the cell body to the periphery. This assumption is supported by the observation that after administration of reserpine the rise in TH activity in the rat lumbar ganglia is followed by a gradual proximo-distal increase in the enzyme activity in the sciatic nerve (THOENEN et al., 1970). The rate of transport estimated by the progress of this "wave" of induced enzyme is slightly faster (2·5 cm/day) than that determined by the rate of TH accumulation (1·8 cm/day) above a ligature of the rat sciatic (OESCH et al., 1973). The rate of transport calculated by the latter procedure represents a net rate of transport including slow and fast moving moieties of the enzyme and is possibly also affected by retrograde transport, whereas the rate of the progress of the "wave of induced enzyme" is mainly determined by the faster moving moieties. Interestingly, in the rat sciatic nerve there is a good correlation between the rate of transport and the subcellular distribution of the enzymes involved in the synthesis of NE. DOPA decarboxylase (3,4-dihydroxy-L-phenylalanine carboxy-lyase: EC 4.1.1.26) which is transported slowest, is exclusively located in the high speed supernatant, whereas dopamine β-hydroxylase (3,4-dihydroxyphenyl-ethylamine, ascorbate: oxygen oxidoreductase (hydroxylating) EC 1.14.2.1) predominantly located in the particulate fraction, is transported at the highest rate (OESCH et al., 1973).

RELATIONSHIP BETWEEN DURATION OF NEURONAL ACTIVITY AND TH INDUCTION: TIME REQUIREMENTS FOR THE SINGLE STEPS OF THE INDUCTION PROCESS

No consistent increase in the *in vitro* activity of TH could be observed earlier than 18 hr after the beginning of the experimental conditions leading to an increased sympathetic activity (MUELLER et al., 1969b,c). To find out whether an increased neuronal activity over this whole time period was necessary or whether a relatively short period of increased firing was sufficient to initiate the induction process we exposed rats to an intense, intermittent swimming stress at a water temperature of 15°C. Exposure of male rats of 100 g body weight to a swimming stress of 1 hr (3 swimming periods of 5–7 min) led to a small but consistent increase in TH activity in both sympathetic ganglia and adrenals 48 hr after the beginning of the stress. Although in the sympathetic ganglia this increase did not regularly reach statistically significant levels, a 2 hr stress produced in both organs a statistically significant ($P < 0·05$) increase as early as 24 hr after stress (16 per cent in the superior cervical ganglia, 38 per cent in the adrenals), and was considerably higher 48 hr after stress (40 per cent and 53 per cent respectively). The increase in TH activity was completely abolished if actinomycin D was given immediately before or after the 2 hr swimming stress and markedly reduced if given 6 hr later (Fig. 1). However, treatment of rats with actinomycin D 12 hr after the beginning of the swimming stress reduced the increase by less than 30 per cent while actinomycin given after 24 hr was without

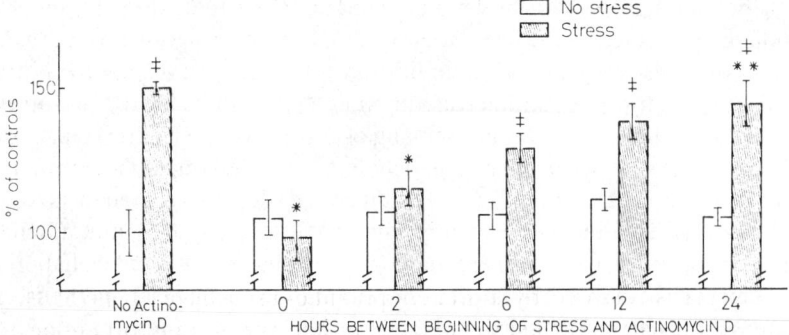

Fig. 1.—Effect of actinomycin D on tyrosine hydroxylase (TH) in stellate ganglia 2 days after stress.

Male Sprague–Dawley rats (100–120 g) were stressed for 2 hr by intermittent swimming to exhaustion. A single dose of actinomycin D (0·8 mg/kg) was injected s.c. at the time indicated. The stellate ganglia were removed for determination of TH (MUELLER *et al.*, 1969b) and protein content (LOWRY *et al.*, 1951). Values represent means ± s.e. of 6 determinations and are expressed as per cent of untreated controls. TH activity in controls was 1·85 ± 0·15 nmol DOPA/hr per mg protein.

‡ $P < 0.01$ compared to non stressed animals which were treated with actinomycin D at the same time intervals respectively.

* $P > 0.1$ compared to non stressed animals which were treated with actinomycin D at the same time intervals respectively.

** $P > 0.1$ compared to activity after stress in the absence of actinomycin D.

statistically significant ($P > 0.1$) effect. Therefore, it seems that the transcription step is terminated after 18–24 hr while the enhanced ribosomal synthesis of TH continues for a further 24–36 hr, indicating that the turnover of messenger RNA, either that of TH itself or that of specific regulatory protein (TOMKINS *et al.*, 1972) is relatively slow.

FIRST AND SECOND MESSENGER OF TRANS-SYNAPTIC INDUCTION OF TH

Trans-synaptic induction of TH can be blocked by ganglionic blocking agents in the rat superior cervical ganglion (MUELLER *et al.*, 1970). This, together with the observations that high doses of acetylcholine (PATRICK and KIRSHNER, 1971) and carbachol (GUIDOTTI and COSTA, 1973) are able to mimic trans-synaptic induction in denervated adrenals, provides good evidence that acetylcholine acts as a first messenger rather than another so far unknown trophic factor liberated from the preganglionic cholinergic nerves.

In view of the well known function of cyclic AMP as a second messenger in many hormonal and neurohumoral systems (cf. PASTAN and PERLMAN, 1971; RALL, 1972) this nucleotide was one of the most obvious candidates to consider to be a second messenger linking the acetylcholine-induced changes in the functional state of the neuronal membrane with the regulation of the synthesis of TH. In fact, dibutyryl cyclic AMP leads to an increase in TH activity in some clones of mouse neuroblastoma (WAYMIR *et al.*, 1972) and to a cycloheximide-sensitive increase in TH activity of mouse sympathetic ganglia in tissue culture (MACKAY and IVERSEN, 1972).

Moreover, the findings of Greengard and collaborators that stimulation of the preganglionic trunk increases the level of cyclic AMP in the superior cervical ganglia of the rabbit could also be taken as an indication for a role of cyclic AMP in trans-synaptic induction. However, this increase in cyclic AMP can be blocked by atropine and phentolamine, whereas ganglionic blocking agents have no effect (cf. Greengard and McAfee, 1972). It seems that acetylcholine liberates dopamine from inter-neurons by a muscarinic mechanism. The liberated dopamine then acts on the adrenergic neurons and increases their cyclic AMP content by an α-adrenergic mechanism. It has very recently been demonstrated by Bloom and collaborators (personal communication) using fluorescence immunocytochemical methods, that the net increase of cyclic AMP in rabbit ganglia results from an increase in neuronal rather than in satellite cells. However, the increase in cyclic AMP levels in the adrenergic neuron seems to be causally related to the slow inhibitory post-synaptic potential (cf. Greengard and McAfee, 1972) and not to the (nicotonic) excitatory potential, which appears to be causally related to the trans-synaptic induction of TH (Mueller et al., 1970) at least in the rat superior cervical ganglion. To further investigate the possible role of cyclic AMP in trans-synaptic induction we studied in the superior cervical ganglion of the rat the correlation between changes in cyclic AMP and the induction of TH. After it was established that the rapidity of the removal of the ganglion was not critical for the level of cyclic AMP—no statistically significant difference between the basic levels of cyclic AMP were observed whether the ganglion was removed from anesthetised animals or from animals killed 1, 2 or 3 min before removal—we studied the effect of cold exposure and swimming stress. Both cold exposure and intermittent swimming stress did not change the level of cyclic AMP in the rat superior cervical ganglion although both types of stress pro-duced an induction of TH which, at least after swimming stress, always reached statistical significance (Fig. 2). Furthermore, decentralisation of this ganglion, which abolished trans-synaptic induction of TH (Thoenen et al., 1969) produced after an initial small decrease a progressive increase in cyclic AMP reaching about 300 per cent 8 days after surgery.

The strongest argument in favor of a role for cyclic AMP as a second messenger in trans-synaptic induction is considerably weakened by the recent finding that dibutyryl cyclic AMP produces in superior cervical ganglia of mice in tissue culture not only an increase in TH but also DOPA decarboxylase and monoamine oxidase (EC 1.4.3.4) (Goodman, unpublished results) while the latter two enzymes are not induced by an increased activity of the preganglionic cholinergic nerves.

All these findings make it rather doubtful whether cyclic AMP acts as a second messenger in the trans-synaptic induction of TH in the sympathetic ganglia. However, it has to be borne in mind that in spite of these findings a small pool of cyclic AMP could be involved, and that its changes could be overshadowed by a much larger pool not reacting in the same way.

Discrepancies between changes in cyclic AMP and subsequent induction of TH are not confined to the rat superior cervical ganglion but are also present in the rat adrenal medulla. Although Guidotti and Costa (1973) reported a generally good correlation between the increase in the level of cyclic AMP in the adrenal medulla and the subsequent increase in TH activity after administration of reserpine or carbachol, we found in recent experiments a series of very marked discrepancies. After cold

a Effect of stress on superior cervical ganglia

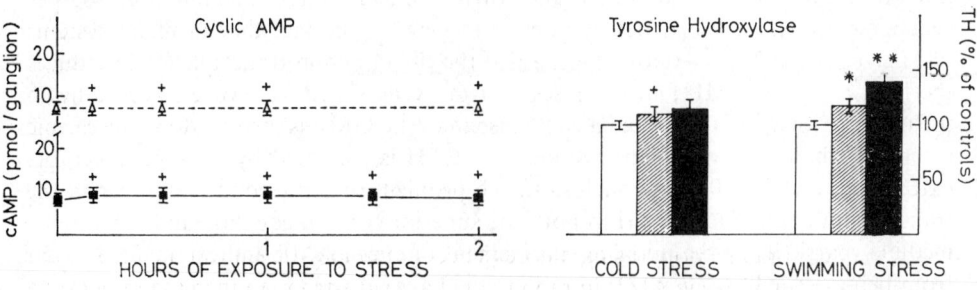

b Effect of stress on adrenal medulla

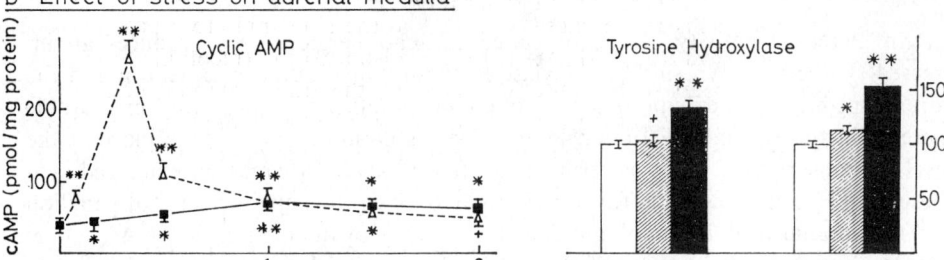

FIG. 2.—Effect of stress on levels of cAMP and tyrosine hydroxylase (TH).
Cyclic AMP was determined according to the method of FISCH et al. (1972) after
exposure of male Sprague–Dawley rats (100–120 g) to cold (0–5°C) (△---△) or
to a swimming stress (■——■). TH was determined according to MUELLER et al.
(1969b) in controls (☐) and in animals exposed to the indicated stress situation
for 1 hr (▨) or for 2 hr (■), 48 hr after the beginning of the stress. Values
represent means ± S.E. of 6–10 determinations. TH activities in controls amounted
to 1·85 ± 0·14 nmol DOPA/hr per mg protein in superior cervical ganglia and 265 ± 14
nmol DOPA/hr per mg protein in adrenal medulla.
+ $P > 0·1$.
* $P < 0·05$.
** $P < 0·001$.

exposure of male Sprague–Dawley rats of 100 g body weight the maximal level of
cyclic AMP in the adrenal medulla was reached after 20 min (Fig. 2), while after
60 min the level approached control values again. Denervation experiments showed
that the changes in cyclic AMP during cold exposure depend on an intact innervation.
If the neuronally mediated changes in the adrenal cyclic AMP levels are causally
related to the induction of TH, an exposure to cold for 60 min should be sufficient to
produce an induction of TH. However, cold exposure for 60 min did not produce a
statistically significant increase in TH within 48 hr whereas an exposure for 2 hr
produced a statistically significant rise to 138 per cent of controls. Thus, the time
period between 60 and 120 min in which the cyclic AMP levels had returned very
nearly to control values seems to be essential for the initiation of changes leading to
a measurable increase in TH activity. Interestingly, swimming stress lead to an
unexpectedly small increase in cyclic AMP (Fig. 2). Thus, the increase in cyclic AMP
produced by swimming to exhaustion is much smaller than that during cold exposure
although the induction of TH is at least as great.

The observation that the decentralisation of the rat superior cervical ganglion leads to a gradual increase in its cyclic AMP level, reaching a maximum of 300 per cent after 8 days, is of particular interest in view of the fact that in many systems (cf. Lee *et al.*, 1972; George *et al.*, 1973) the direct action of acetylcholine reduces the level of cyclic AMP and increases that of cyclic GMP. However, according to the information available so far the increase in cyclic GMP is mediated by muscarinic receptors whereas trans-synaptic induction of TH is mediated by nicotinic receptors (Mueller *et al.*, 1970). In conclusion, the problem of the second messenger in the trans-synaptic induction of TH in both the superior cervical ganglion and the adrenal medulla, particularly the possible involvement of cyclic AMP and cyclic GMP is far from being settled.

SUMMARY

An increased activity of the preganglionic cholinergic nerves produces an increased synthesis of tyrosine hydroxylase in the terminal adrenergic neuron and the adrenal medulla. An enhanced activity of 60 min is sufficient to produce a measurable increase in the enzyme level after 48 hr. The regulation seems to take place at the level of transcription which is terminated after 18–24 hr whereas an enhanced synthesis of TH continues for a further 24–36 hr both in the superior cervical ganglion and the adrenal medulla, implying that the turnover of the messenger RNA (that of TH itself or of a regulator protein) is relatively slow. While there is good evidence that acetylcholine is the first messenger in the trans-synaptic induction of tyrosine hydroxylase the possible involvement of cyclic AMP or cyclic GMP as a second messenger remains to be elucidated.

Acknowledgement—The excellent technical assistance of Mrs. Hilary Wood, Miss Vreni Forster, Mr. Kitaru Suda and Mr. Ueli Jaeggi is gratefully acknowledged. This study was supported by the Swiss National Foundation for Scientific Research (Grant No. 3.653.71).

REFERENCES

Bhagat B. and Rana M. W. (1971) *Brit. J. Pharmacol.* **43**, 250–261.
Black I. B., Hendry I. A. and Iversen L. L. (1971) *Nature, Lond.* **231**, 28–29.
Cheah T. B., Geffen L. B., Jarrott B. and Ostberg A. (1971) *Brit. J. Pharmacol.* **42**, 543–557.
Dairman W. and Udenfriend S. (1971) *Mol. Pharmacol.* **6**, 350–356.
Fisch H. U., Pliška V. and Schwyzer R. (1972) *Europ. J. Biochem.* **30**, 1–6.
Geffen L. B. and Livett B. G. (1971) *Physiol. Rev.* **51**, 98–157.
George W. J., Wilkerson R. D. and Kadowitz Ph. J. (1973) *J. Pharmacol. Exp. Ther.* **184**, 228–235.
Greengard P. and McAfee D. A. (1972) *Biochem. Soc. Symp.* **36**, 87–102.
Guidotti A. and Costa E. (1973) *Science* **179**, 902–904.
Kvetnanski R., Weise V. K. and Kopin I. J. (1970) *Endocrinology* **87**, 744–749.
Lee T. P., Kuo J. F. and Greengard P. (1972) *Proc. Nat. Acad. Sci. U.S.A.* **69**, 3287–3291.
Lowry O. N., Rosebrough N. J., Farr A. L. and Randall R. J. (1951) *J. Biol. Chem.* **193**, 265–275.
MacKay A. V. P. and Iversen L. L. (1972) *Brain. Res.* **48**, 424–426.
Mandell A. J. and Morgan M. (1970) *Nature, Lond.* **227**, 75–76.
Mueller R. A., Thoenen H. and Axelrod J. (1969a) *Science* **163**, 468–469.
Mueller R. A., Thoenen H. and Axelrod J. (1969b) *J. Pharmacol. Exp. Ther.* **169**, 74–79.
Mueller R. A., Thoenen H. and Axelrod J. (1969c) *Mol. Pharmacol.* **5**, 463–469.
Mueller R. A., Thoenen H. and Axelrod J. (1970) *Europ. J. Pharmacol.* **10**, 51–56.
Oesch F., Otten U. and Thoenen H. (1973) *J. Neurochem.* **20**, 1691–1706.
Pastan I. and Perlman R. L. (1971) *Nature, Lond.* **229**, 5–9.
Patrick R. L. and Kirshner N. (1971) *Mol. Pharmacol.* **7**, 87–96.
Rall Th. W. (1972) *Pharmacol. Rev.* **24**, 399–409.
Reis D. J., Moorehead D. T., Rifkin M., Joh T. H. and Goldstein M. (1971) *Nature, Lond.* **229**, 562–563.

Thoenen H., Mueller R. A. and Axelrod J. (1969) *Nature, Lond.* **221**, 1264.
Thoenen H., Mueller R. A. and Axelrod J. (1970) *Proc. Nat. Acad. Sci.* **66**, 58–62.
Thoenen H. and Oesch F. (1973) In: *New Concepts in Neurotransmitter Regulation.* (Mandell A. J. ed.) pp. 33–52, Plenum, New York.
Tomkins G. M., Levinson B. B., Baxter J. D. and Dettilefsen L. (1972) *Nature, Lond.* **239**, 9–14.
Waymir J. C., Weiner N. and Prasad K. N. (1972) *Proc. Nat. Acad. Sci. U.S.A.* **69**, 2241–2245.
Weiner N. and Mosimann W. F. (1970) *Biochem. Pharmacol.* **19**, 1189–1199.

Frontiers in Catecholamine Research 1973, pp. 187 to 189. Pergamon Press. Printed in Great Britain.

MODULATION OF ADRENAL TYROSINE HYDROXY-LASE INDUCTION BY SEROTONIN NERVE FUNCTION

R. A. MUELLER*, R. D. SMITH, G. R. BREESE† and B. R. COOPER‡

Departments of Anesthesiology, Psychiatry, Pharmacology, and the Child Development Institute, University of North Carolina, School of Medicine, Chapel Hill, North Carolina 27514, U.S.A.

THE CONNECTIONS within the brain which control the activity of peripheral sympathetic nerves have not yet been completely elucidated. The hypertensive response to interference with 5-hydroxytryptamine synthesis would imply an inhibitory effect of serotoninergic nerves on peripheral sympatho-adrenal function (ITO et al., 1972). The antihypertensive properties of agents which interfere with central noradrenergic neurons has been interpreted to support the existence of both positive (JAJU et al., 1966) and negative (NAKAMURA et al., 1971) correlations of central and peripheral catecholaminergic activity.

The induction of tyrosine hydroxylase (TH) (EC 1.14.3a) activity in chromaffin and sympathetic ganglion cells, like cardiovascular dynamics, could also be used as an index of peripheral sympathetic function. Adrenal medullary TH can be increased by administration of drugs which produce either a direct (PATRICK and KIRSHNER, 1971) or reflex (MUELLER, 1969a) increase in peripheral sympathetic activity, and by stimulation of certain areas of the brain (REIS et al., 1969). The present investigation explored the effect of agents which interfere with serotonergic nerve function on the increase in adrenal TH produced by insulin and brain stem stimulation. The results suggest that interference with serotonin nerve function can potentiate the induction of adrenal TH produced by insulin or mesencephalic stimulation.

Male Sprague-Dawley rats were housed in a light cycled room for four days after arrival from the supplier before receiving the first dose of p-chlorophenyl-anlanine (pCPA). The group given 75 μg 5,6-dihydroxytryptamine creatinine sulphate monohydrate (DHT) intracisternally and their saline injected controls were given insulin or saline i.v. 1 week later. In a third group bipolar stainless steel electrodes were aimed at the ventral tegmental region of the brain, just medial to the substantia nigra, lying dorsal and lateral to the interpeduncular nucleus. Coordinates for this site were selected from a DeGroot atlas (0·5 mm lateral, 2·25 mm anterior, 1·7 mm below stereotaxic zero). After 1 week to allow healing, the electrodes were attached to swivel connectors and pulse stimulated (1/2 of each second) in order to determine minimal current intensity which produced motor phenomena. The first dose of pCPA was given the following day and the second dose the next day immediately before the first stimulation session. Experimental rats were stimulated for 3 hr at one-half the previously determined current strength. Stimulation in this manner produces a sustained increase in arterial blood pressure. Stimulation was repeated

* Clinical Pharmacology Faculty Development Awardee, Pharmaceutical Manufacturers Assn. Foundation.
† Career Development Awardee (HD-24585).
‡ Postdoctoral fellow of Neurobiology program (MH-11107).

for 3 hr on each of the next two days and the animals were killed 24 hr after the third session. In control rats electrodes were implanted and were connected to the swivel connectors but were not stimulated.

An aliquot of adrenal gland homogenate (0·25 M sucrose) was used to determine catecholamine content after deproteinization with 0·4 N perchloric acid (Anton and Sayer, 1962; Von Euler and Lishajko, 1961). The supernatant obtained after centrifugation (22,000 × g for 10 min) was used to determine TH (Mueller et al., 1969b). An analysis of variance employing Dunnett's procedure to test for significance was employed in all experiments (Steele and Torre, 1960). Brain norepinephrine, dopamine, and serotonin were determined in each animal to confirm the selective effect of pCPA and DHT on serotonin content (Breese and Traylor, 1970).

The administration of insulin produces a profound hypoglycemia which initiates a marked stimulation of the adrenal medulla by a central mechanism (Hokfelt, 1951). The present results (Table 1) confirm the previously reported increase in

Table 1. Effect of pCPA and DHT on insulin-induced adrenal tyrosine hydroxylase activity

Experiment No.	Group	No. of animals	Adrenal TH activity (Mμmoles/ pair/hr)	Adrenal catecholaines (μg/pair)
1	Control	6	11·7 ± 1·1	21·7 ± 1·3
	pCPA	6	13·0 ± 1·1	17·5 ± 1·2*
	Insulin	6	15·8 ± 1·3*	8·8 ± 0·6†
	pCPA + Insulin	6	26·2 ± 1·4†	8·4 ± 0·9†
2	Control	6	19·9 ± 2·3	—
	DHT	6	19·1 ± 1·0	—
	Insulin	6	28·4 ± 4·5*	—
	DHT + Insulin	6	32·2 ± 2·6†	—

Indicated animals received pCPA 100 mg/kg p.o. 48 and 24 hr before receiving saline (pCPA only) or insulin (4 mg/kg i.v.) (pCPA + insulin). Animals given DHT received 75 μg/kg i.c. 7 days before saline (DHT) or insulin (DHT + insulin). Insulin treated groups received 5 ml 20% glucose i.p. 3 hr after insulin to terminate the hypoglycemia. All animals were killed 24 hr after receiving saline or insulin. Values are mean ± s.e.m.

* $p < 0.05$
† $p < 0.01$ } relative to control of that experiment.

adrenal TH which can be observed 24 hr after insulin administration (Viveros et al., 1969). pCPA or DHT alone do not significantly increase the adrenal TH, but when given before insulin, the subsequent increase in in vitro TH activity produced is much larger than that produced by insulin alone. The K_m for pteridine and tyrosine was not altered by pCPA administration but V^*_{max} was increased for both substrate and cofactor.

Insulin was selected to study the possible serotonergic modulation of adrenal TH increase because of its clear central mechanism of action. Because its effects within the brain are comparatively non-specific anatomically, a search was made for sites within the rat brain which produce an increase in adrenal TH after electrical stimulation. The increase in adrenal TH produced by stimulation of the present mesencephalic site (Fig. 1) was even greater if pCPA was administered before stimulation.

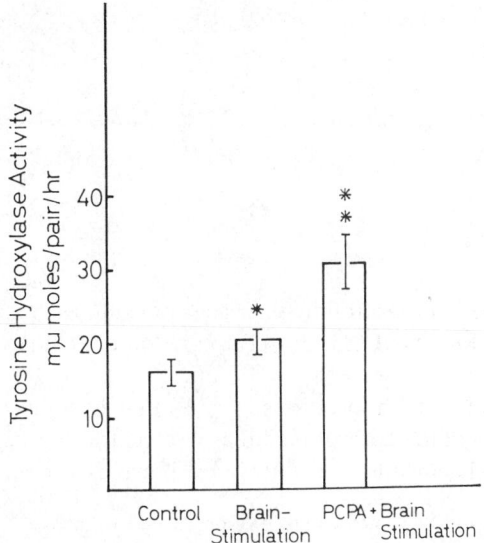

FIG. 1.—Effect of pCPA on increase in adrenal tyrosine hydroxylase after brain stimulation. Each bar represents the mean $\pm$ S.E.M. (brackets) of 6 animals, treated as described in the text.

$\left.\begin{array}{l} * \ P < 0{\cdot}05 \\ ** \ P < 0{\cdot}01 \end{array}\right\}$ compared to control.

The potentiation of the adrenal TH induction produced by agents which interfere with the function of serotonin containing neurones (BAUMGARTEN et al., 1971; KOE and WEISSMAN, 1966) implies that these nerves may normally suppress peripheral sympathetic nervous system activity. Preliminary results indicate 5-hydroxytryptophan can prevent the induction of adrenal TH produced by isoproterenol administration (SMITH et al., unpublished observations). This conclusion is further supported by electrophysiological studies in which intravenous 5-hydroxytryptophan was found to depress sympathetic reflexes in spinal cats (HAVE et al., 1972).

Acknowledgements—This study was supported by USPHS grants (MH–16522), HD–03110, and GM 19088 and N. C. Heart Ass. (1972–73–A–11)

REFERENCES

ANTON A. H. and SAYRE E. D. (1962) *J. Pharmacol. Exp. Ther.* **138**, 360–375.
BAUMGARTEN H. B., BJÖRKLUND A., LACHENMAYER L., NOBIN A. and STENVI U. (1971) *Acta physiol. scand. Suppl.* **373**
BREESE G. R. and TRAYLOR T. D. (1970) *J. Pharmacol. Exp. Ther.* **174**, 413–420.
EULER U. S. VON and LISHAJKO F. (1961) *Acta physiol. scand.* **51**, 348–356.
HAVE B. D., NEUMAYR R. J. and FRANZ D. N. (1972) *Nature* **239**, 336–337.
HÖKFELT B. (1951) *Acta physiol. scand.* **25**, (*suppl. 92*) 1–134.
ITO A. and SCHANBERG S. M. (1972) *J. Pharmacol. Exp. Ther.* **181**, 65–74.
JAJU B. P., TANGI K. K. and BHARGAVA K. P. (1966) *Can. J. Physiol. Pharmacol.* **44**, 687–690.
KOE B. K. and WEISSMAN A. (1966) *J. Pharmacol. Exp. Ther.* **154**, 499–516.
MUELLER R. A., THOENEN H. and AXELROD J. (1969a) *Science* **163**, 468–469.
MUELLER R. A., THOENEN H. and AXELROD J. (1969b) *J. Pharmacol. Exp. Ther.* **169**, 74–79.
NAKAMURA K., GEROLD M. and THOENEN H. (1971) *Naunyn-Schmiedeberg's Arch. Pharmacol.* **268**, 125–139.
PATRICK R. L. and KIRSHNER N. (1971) *Mol. Pharmacol.* **2**, 87–96.
REIS D. A., MOOREHEAD D. T., RIFKIN M., TOH T. H. and GOLDSTEIN M. (1971) *Nature* **229**, 562–569.
STEELE R. G. D. and TORRE J. H. (1960) In: *Principles and Procedures of Statistics*, McGraw-Hill, New York.
VIVEROS O. H., ARQUEROS L., CONNETT R. J. and KIRSHNER N. (1969) *Mol. Pharmacol.* **5**, 69–82.

Frontiers in Catecholamine Research 1973, pp. 191 to 193. Pergamon Press. Printed in Great Britain.

ROLE OF CATECHOLAMINES IN INDUCTION OF ADRENAL TYROSINE HYDROXYLASE

B. Bhagat

Department of Physiology, St. Louis University, School of Medicine,
1402 South Grand Boulevard, St. Louis, Missouri 63104, U.S.A.

Catecholamines (CA) in sympathetic nerves and in the adrenal medulla are stored in dense core vesicles from which they are continuously being released. Under varying physiological conditions, the CA content remains at a steady state level characteristic of each organ. To a considerable extent, this is because of continuous synthesis of CA which keeps pace with the rate of release, and consequently there is only slight or no reduction in CA content of the tissue.

Four enzymes are involved in the biosynthesis of epinephrine from tyrosine: tyrosine hydroxylase (TH), dopa decarboxylase, dopamine β-oxidase and phenyl-ethanolamine-N-methyl transferase (PNMT).

1-Tyrosine, the natural precursor, is present in the circulation in the amount of about 15 mg/l. It enters the adrenergic neuron and is hydroxylated to form 1-dopa by TH; 1-dopa is decarboxylated by dopa-decarboxylase to form dopamine. Dopamine is taken up into the storage particles where it is converted to norepine-phrine (NE) by an enzyme dopamine β-oxidase. For this uptake, ATP and magnesium are essentially required. This requirement for ATP and magnesium disappears when the enzyme is made soluble. In the adrenal medullary cells, NE migrates from the chromaffin granules to the cytoplasm and is N-methylated by PNMT to form epine-phrine. Epinephrine is also stored in the granules.

NE and other catechols have been shown to inhibit the TH, a rate limiting enzyme in the biosynthesis of CA. This inhibition is not competitive with the substrate, but with tetrahydropteridine co-factor (Nagatsu et al., 1964). Since the concentration of co-factor within the cytoplasm of the adrenergic neuron is less than 1 mM (Ikeda et al., 1966), it is possible that a very small amount of CA is able to antagonize its actions and thus may exert considerable influence on the enzymatic activity of TH.

Free intraneuronal NE is considered to be the portion of NE important in the regulation of synthesis of NE in the intact neuron (Alousi and Weiner, 1966). An increase in sympathetic nerve activity due to any stimulus increase the efflux of NE and thereby decreases NE concentration in the axoplasm which in turn diminishes the end-product inhibition of TH. The rate of conversion of tyrosine to dopa is enhanced. Conversely, if the activity of the sympathetic nervous system is reduced, NE accumulates in the neuron inhibiting TH. Thus this mechanism is capable of rapidly adjusting the rate of NE synthesis in response to changes in physiological demand for NE by altering the activity of TH without affecting the amount of enzyme.

When increased sympathetic nerve activity is prolonged over a period of days, TH, as measured in tissue homogenate, is increased (Thoenen et al., 1969). Drugs like reserpine which cause hypotension by interfering with post-ganglionic sympath-etic transmission, cause induction of TH presumably as a result of compensatory reflex increases in sympathetic nerve activity (Bhagat et al., 1971; Mueller et al., 1969a).

The increase in TH activity under these conditions is due to formation of a new enzyme molecule since inhibition of protein synthesis with either cyclohexamide or D-actinomycin can prevent the drug-induced increase in TH in the adrenal gland and ganglia. TH activity in adrenal gland no longer increases after reserpine when splanchnic nerves innervating adrenal gland are sectioned before treatment with

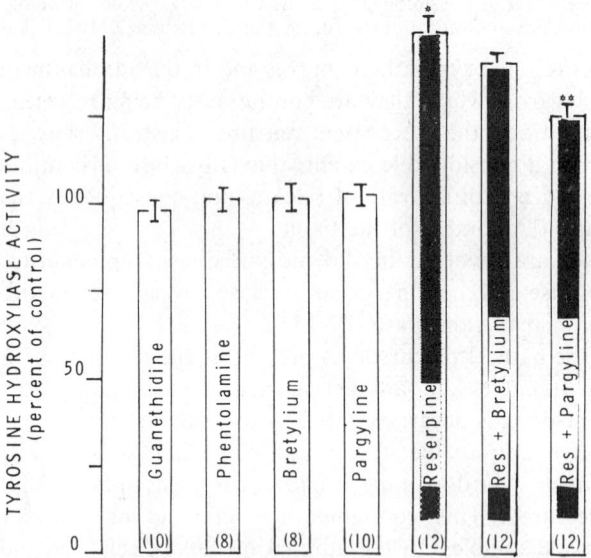

FIG. 1.—Effect of various drugs on TH activity in rat adrenal gland. Mean TH activity is expressed as percentage of control (which averaged 46.9 ± 2.2 nmoles of ^3H_2O formed per hr per pair of adrenal glands); the vertical bar represents standard errors of the means; numbers in parentheses are numbers of animals in group. Guanethidine: 20 mg/kg i.m.; phentolamine: 5 mg/kg i.p. every 4 hr; bretylium: 25 mg/kg i.p. every 8 hr; pargyline: 25 mg/kg i.p. every 12 hr; and reserpine 2·5 mg/kg initially; 1 mg 24 hr later. Control: 0·2 ml 0·9% NaCl. Animals were killed 48 hr after initial injection of reserpine, or 24 hr after initial drug injection. When pargyline or bretylium was given in conjunction with reserpine, the former was injected every 12 hr beginning 1 hr before reserpine, and bretylium every 8 hr after the 2nd injection of reserpine. In a separate series of experiments, treatments of rats with guane-thidine, bretylium or pargyline was continued for 4 days, but adrenal TH activity was not significantly different from control animals injected with 0·9% saline on the same schedule. * $P < 0.01$ compared to control group. ** $P < 0.01$ compared to both the control and reserpine-treated group.

drug, suggesting that induction of TH is neurally mediated (THOENEN et al., 1969). This neurally meditated regulation of TH represents a long term adaptation to chronic increased nerve activity.

It is well known that administration of guanethidine bretylium or pargyline (monoamine oxidase-inhibitor) causes hypotension in normotensive animals and thereby causes reflex increases in sympathoadrenal activity as a result of a decrease in sympathetic postganglionic nerve function. When rats were administered these drugs for 1 day or for 4 days, there was no increase in adrenal TH activity (Fig. 1), suggesting that hypotension (and the subsequent increase in nerve activity) is not sufficient stimulus for induction of adrenal TH.

Reserpine, guanethidine, bretylium and pargyline have a common denominator, namely, the lowering of blood pressure by interfering with adrenergic neuro-transmission. The mechanism of action is, however, different for each drug. Guanethidine, bretylium and pargyline differ from reserpine in their effect on adrenal CA (BHAGAT, 1971). Whereas reserpine depletes the adrenal medulla of CA, these other drugs have no effect on adrenal amines. It is therefore possible that the depletion of adrenal CA may have a role in induction of adrenal TH.

This hypothesis derives support from the following observations: (1) pargyline, which interferes with depletion of adrenal CA by reserpine, reduced reserpine-induced increase in adrenal TH activity (Fig. 1). (2) Bretylium, which has no effect on reserpine-induced depletion of adrenal CA, failed to affect the reserpine-induced induction of adrenal TH (Fig. 1). In contrast, bretylium antagonises the release of NE at postganglionic sympathetic nerve endings and was effective in blocking the inducing effect of reserpine on TH activity in stellate ganglia (MOLINOFF et al., 1972). (3) The decreased receptor response and the consequent hypotension after α-adrenergic blockade results in a stimulation of adrenergic nerve activity to maintain homeostasis. The increased sympathetic nerve activity caused by phenoxybenzamine produces an increase in adrenal TH activity and also decreases the CA content of adrenal gland (THOENEN et al., 1969). In contrast, α-adrenergic blocking drug phentolamine which has no effect on adrenal CA content failed to elicit an increase in adrenal TH activity (Fig. 1).

The hypothesis that, in addition to increased nerve activity, CA may play a role in regulating the rate of synthesis of adrenal TH, leads to a coherent picture of the reported facts. However, the effects of CA appear to be mediated by only a small pool of amine, since 6-hydroxydopamine can cause elevation of adrenal TH without affecting significantly the adrenal CA levels.

REFERENCES

ALOUSI A. and WEINER N. (1966) *Proc. Nat. Acad. Sci. U.S.A.* **56**, 1491–1492.
BHAGAT B. (1971) *Recent Advances in Adrenergic Mechanisms*. Charles C. Thomas, Springfield, Illinois.
BHAGAT B., BURKE W. J. and DAVIS W. J. (1971) *Br. J. Pharmac.* **43**, 819–827.
IKEDA M., LEVITT M. and UDENFRIEND S. (1966) *J. Biol. Chem.* **241**, 4452–4456.
MUELLER R. A., THOENEN H. and AXELROD J. (1969a) *J. Pharmacol. Exp. Ther.* **169**, 74–79.
MOLINOFF P. B., BRIMIJOIN S. and AXELROD J. (1972) *J. Pharmacol. Exp. Ther.* **169**, 74–79.
NAGATSU T., LEVITT B. G. and UDENFRIEND S. (1964) *J. Biol. Chem.* **238**, 2910–2917.
THOENEN H., MUELLER R. A. and AXELROD J. (1969) *J. Pharmacol. Exp. Ther.* **169**, 249–254.

Frontiers in Catecholamine Research 1973, pp. 195 to 200. Pergamon Press. Printed in Great Britain.

DOPAMINE-β-HYDROXYLASE AND THE REGULATION OF CATECHOLAMINE BIOSYNTHESIS*

PERRY B. MOLINOFF† and JAMES C. ORCUTT

Department of Pharmacology, University of Colorado, School of Medicine, Denver,
Colorado 80220, U.S.A.

IT HAS been suggested that tyrosine hydroxylase (EC 1.14.3-, TH) is the rate-limiting step in the biosynthesis of catecholamines (LEVITT, SPECTOR, SJOERDSMA and UDENFRIEND, 1965). This conclusion is based in part on the rates of conversion of various radiolabelled precursors into norepinephrine in the isolated perfused guinea pig heart, and on the fact that the K_m of TH for tyrosine is lower than the concentration of this amino acid in various tissues. It is likely that the *maximum* rate of catecholamine synthesis is limited by the amount of TH in the cell. It is also true that stimulation of sympathetic nerves leads to an increase in the rate of synthesis of catecholamines via an increase in the activity of TH. On the other hand, it is probably an oversimplification to suggest that TH is the only regulated step in this biosynthetic pathway. The idea that one of a series of reactions in a pathway is rate limiting is derived from studies of biosynthetic pathways in bacteria. In order to say that a given step is rate-limiting a number of conditions must be fulfilled. Several of these conditions are not satisfied by the biosynthetic pathway for catecholamines. In the first place, the substrates and enzymes of this pathway do not have free access to one another since dopamine-β-hydroxylase (EC 1.14.2.1, DBH) is sequestered in chromaffin granules and adrenergic vesicles. In the second place, the catecholamine biosynthetic pathway is branched. Some of the dopamine, which is formed from dopa, is taken up into vesicles and converted into norepinephrine and some is deaminated by the intraneuronal enzyme monoamine oxidase (EC 1.4.3.4). Approximately one-third of the catecholamine metabolites found in human urine are derived from dopamine (CAESAR, RUTHVEN and SANDLER, 1969). It is not likely that all of these dopamine metabolites are derived from the CNS. The excretion of dopamine and its metabolites in the urine of patients with Parkinson's disease is decreased only slightly (WEIL-MALHERBE and VAN BUREN, 1969) from normal levels. A competition therefore exists between deamination and uptake/β-hydroxylation. Anything which increases the capacity of the neuron to β-hydroxylate dopamine should increase the rate of norepinephrine biosynthesis. The only circumstance in which changes in uptake/β-hydroxylation would not be reflected by changes in norepinephrine synthesis would be if the capacity of the neuron to deaminate dopamine was limited with deamination normally occurring at a maximal rate. This is clearly not the case since blockade of norepinephrine uptake results in an increase in the excretion of deaminated metabolites (KOPIN and GORDON, 1963; RUTLEDGE and WEINER, 1967).

---

* Supported by a grant from the National Institute of Neurological Disease and Stroke (NS 10206).
† Established Investigator of the American Heart Association.

Additional support for the conclusion that DBH, as well as TH, is subject to regulation comes from the fact that a number of different procedures produce similar changes in the levels of TH and DBH. These procedures include (1) the administration of drugs like reserpine (MUELLER, THOENEN and AXELROD, 1969; MOLINOFF, BRIMI-JOIN, WEINSHILBOUM and AXELROD, 1970) which lead to increases in TH and DBH in sympathetic ganglia, in the adrenal, and in sympathetically innervated organs; (2) the administration of 6-hydroxydopamine (BRIMIJOIN and MOLINOFF, 1972) which leads to similar increases of TH and DBH in the adrenal medulla; (3) the administration of NGF which leads to marked increases in TH and DBH in the superior cervical ganglion of the rat (THOENEN, 1972); and (4) various types of stress including immobilization stress (KVENANSKY, GEWIRTZ, WEISE and KOPIN, 1970a, b) and cold-stress (THOENEN, KETTLER, BURKARD and SANER, 1971) which induce apparently specific changes in DBH and TH.

An enzymatic assay for DBH activity in tissue homogenates has been used to study changes that occurred with pharmacologically induced alterations in sympathetic nervous system activity. This assay is based on the conversion of a β-hydroxylated product into its N-methyl derivative in the presence of radiolabelled S-adenosylmethionine (MOLINOFF, WEINSHILBOUM and AXELROD, 1971). In this assay either phenylethylamine or tyramine is converted into its β-hydroxylated product (phenylethanolamine or octopamine) by DBH in the presence of O_2, ascorbate as an electron donor, and Cu^{2+} which is added to inactivate the endogenous inhibitors of DBH, which are present in nearly all tissues. The pH of the reaction is then changed from 6 to 8·6 and radiolabelled S-adenosylmethionine and a partially purified preparation of bovine adrenal medullary phenylethanolamine N-methyl transferase (EC 2.1.10) are added. The N-methyl derivative thus formed is separated from the radiolabelled S-adenosylmethionine by solvent extraction. The identity of the N-methyl derivative has been verified by thin layer chromatography. At 37°C this assay is linear with respect to time and the amount of product formed is proportional to the amount of homogenate in the incubation. The concentration of tyramine used in this assay is several times greater than the K_m of DBH for tyramine. This accounts, in large part, for the sensitivity of the method. As controls, separate samples are incubated with a known amount of octopamine and with a sample of DBH which has been purified from the bovine adrenal medulla.

The activities of DBH (MOLINOFF et al., 1970) and of TH (MUELLER et al., 1969) were increased in adrenergic organs of the rat by agents, such as reserpine, that are believed to cause a reflex increase in sympathetic tone. After reserpine treatment DBH activity increased by approximately 50% within 24 hr in sympathetic ganglia. In the heart, and as has also been previously reported in the adrenal gland (VIVEROS, ARQUEROS, CONNETT and KRISHNER, 1969), the increase in activity was preceded by a fall that occurred within a few hours of the injection. The activities of several enzymes not involved in catecholamine biosynthesis were not affected by reserpine. Several experimental results suggested that the effect of reserpine was mediated, in at least a permissive sense, by nerve impulse activity. When efferent nerve impulses were blocked by either surgical decentralization or by ganglionic blocking agents the effects of reserpine were abolished. Furthermore, DBH activity was found to be increased within 24 hr when adult sympathetic ganglia were cultured in the presence of increased potassium (SILBERSTEIN, BRIMIJOIN, MOLINOFF and LEMBERGER, 1972). The observed increase in DBH activity appears, from several types of study, to represent the synthesis of new enzyme and not the activation of previously existing enzyme. Thus, the K_m for phenylethylamine, 6×10^{-4} M, was the same for the enzyme from rats which had been treated with reserpine as for that from control animals. Furthermore, the increase in DBH activity was blocked either *in vivo* or in the organ culture system by the protein synthesis inhibitor, cycloheximide. And finally, the incorporation of ^{3}H-leucine into adrenal DBH (measured by immunoadsorption) was increased by approximately three-fold in reserpine pretreated animals (HARTMAN, MOLINOFF and UDENFRIEND, 1970).

It was thus concluded that long-term changes in the level of activity in the sympathetic nervous system result in changes in the amount of DBH in the neuron. These changes, like those observed with TH, appear to be mediated trans-synaptically and to require protein synthesis.

The capacity of the adrenergic neuron to β-hydroxylate dopamine is relatively great compared to its capacity to 3-hydroxylate tyrosine. A number of factors may, however, combine to limit DBH activity *in vivo*. In the first place, since DBH is sequestered in storage granules, dopamine must be transported into these granules before it can be β-hydroxylated. Very little is known at the present time concerning possible restrictions to the transport of dopamine into these granules. It has been shown that inhibition of the uptake of dopamine into granules by reserpine is associated with a marked decrease in the conversion of dopamine to norepinephrine and with an increase in the excretion of the deaminated metabolites of dopamine (RUTLEDGE and WEINER, 1967). Whether or not physiologically significant fluctuations in the uptake of dopamine into granules occur must remain a subject for future investigation. A second potential mechanism for regulating DBH activity *in vivo* involves the endogenous inhibitors of this enzyme. These compounds are found in many different organs and have been partially purified from boiled homogenates of heart (CHUBB, PRESTON and AUSTIN, 1969) and adrenal gland (DUCH and KIRSHNER, 1971). The inhibitor from the heart had a molecular weight between 750 and 1200 and was thought to contain carbohydrate and organic phosphate. The adrenal inhibitor was found to be a polypeptide made up of glutamic acid, cysteine and glycine. Although the composition of the inhibitor resembled that of glutathione it did not appear to be identical with it. It is interesting that the adrenal DBH inhibitor will bind *N*-ethylmaleimide while that which was obtained from the heart will not. It has not thus far been possible to establish whether or not these inhibitors have a physiologic role in regulating DBH activity.

In order to study the endogenous inhibitors of DBH an assay for inhibitory activity was established using dopamine-β-hydroxylase which was purified from the bovine adrenal medulla (FOLDES, JEFFREY, PRESTON and AUSTIN, 1972). Since we intended to use bovine adrenal DBH to study the endogenous inhibitors of DBH, which are present in various rat organs, it was important to establish that the in-inhibitors in the various organs of interest had the same effect on exogenous bovine DBH as they had on the endogenous DBH present in the same tissue. Experiments were therefore carried out in which DBH activity was measured in various tissue homogenates as a function of the concentration of Cu^{2+} (Fig. 1 shows the results obtained with a 1:40 homogenate of brain). It has been previously shown (MOLINOFF, WEINSHILBOUM and AXELROD, 1971), that a very precise concentration of copper is

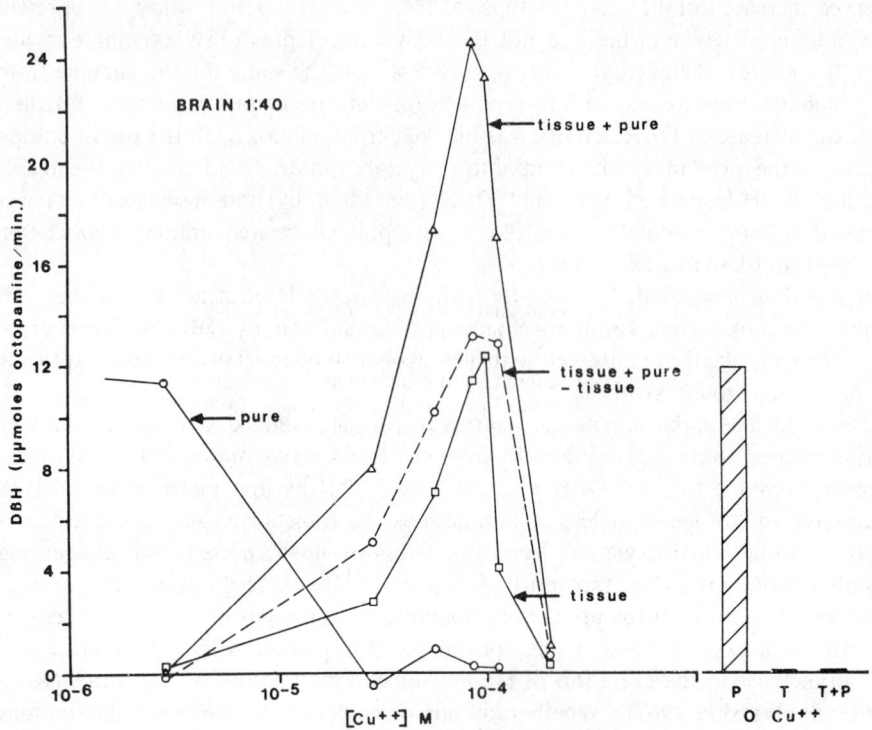

FIG. 1.—The effect of Cu²⁺ on the DBH inhibitor from brain. The curves drawn with
solid lines show the effect of Cu²⁺ on bovine adrenal medullary DBH (pure), on the
endogenous DBH in a 1:40 homogenate of rat brain (tissue) and on the same amount
of adrenal DBH added to the brain homogenate (tissue + pure). The dashed line is
derived by subtracting the activities obtained with tissue from those obtained with
adrenal DBH in the presence of tissue (Pure + tissue − tissue). The bars at the right
show the activity of pure DBH (P), tissue DBH (T) and pure plus tissue DBH (T + P)
in the absence of Cu²⁺.

required to obtain maximal DBH activity in each of several specific organs. If an
insufficient amount of Cu²⁺ is used the endogenous inhibitor will decrease the activity
of the enzyme, while, if too high a concentration of Cu²⁺ is added, the Cu²⁺ will
itself inhibit DBH (Fig. 1). Pure DBH does not require Cu²⁺ for maximal activity
since there is no inhibitor in this preparation. In these experiments we found that
when the activity of pure DBH was measured, in the presence of tissue homogenates,
as a function of the concentration of Cu²⁺, the pure enzyme behaved in exactly the
same way as did the DBH in the homogenate (Fig. 1). A higher concentration of Cu²⁺
is required to inhibit DBH in the presence of a homogenate than in buffer since
Cu²⁺ is non-specifically bound to proteins in such homogenates. These results
suggest that it is valid to use bovine adrenal medullary DBH in an assay of the DBH
inhibitors which are found in rat organs.

A very striking property of the DBH inhibitor or inhibitors is the enormous
amount of inhibitory activity which is present in various organs (Table 1). There is
enough inhibitory activity present in a 1:5000 dilution of homogenates of spleen and
and adrenal to cause a 50 per cent inhibition of the purified bovine DBH.

TABLE 1

Tissue	Dilution (I_{50})	% heat stable (5 min at 95°C)
Heart	1:2950	2·2
Spleen	1:5000	100
Adrenal	1:4800	58
Sciatic nerve	0·3 cm/1·75 ml	50

DBH inhibitor levels in rat tissues. DBH assays were performed with sufficient bovine adrenal medullary DBH to form about 20 pmole of octopamine/min. Tissue homogenates were centrifuged at 12,000 × *g* for 10 min and the ability of various dilutions of the supernatant to inhibit DBH was determined. Dilutions were chosen so that inhibition ranged from 90 to 100% down to less than 10%. A sample of the initial homogenate was heated in a boiling water bath for 5 min and various dilutions of the supernatant, after centrifugation at 16,000 × *g* for 10 min, were tested for their ability to inhibit DBH. I_{50} values were determined from the curves of DBH activity vs dilution. The percentage heat stable was determined by comparing the I_{50} before and after boiling of the homogenates.

In order to establish some of the basic properties of the inhibitory activity in various organs, a number of studies were carried out. It was found that inhibitory activity remains constant if whole organs are stored at −10°C. Approximately half of the inhibitory activity in adrenal, spleen and brain is lost, however, if a homogenate is stored either in the refrigerator or at −10°C. When homogenates of various organs were subjected to temperatures of 95°C, for 5 min or longer, there was no effect on the inhibitory activity in the spleen (Table 1). On the other hand, approximately 98% of the inhibitory activity in the heart was destroyed by this treatment. This type of result suggests that the nature of the inhibitory activity is at least partially organ specific. The inhibitory activity in a number of organs (Table 1) fell by approximately 50% on being subjected to temperatures of 95°C. This suggests that there is more than one inhibitor in each of several adrenergically innervated organs.

In current experiments we are now investigating the possibility that DBH inhibitors are playing a significant physiological role in regulating DBH activity. If the DBH inhibitors are playing such a role then it should be possible to demonstrate that their cellular and subcellular location is such that they have access to the enzyme. The enormous amount of inhibitory activity which is present in various organs suggests that at least some, and probably most, of the inhibitory activity is located in non-neuronal sites. It was thus not surprising that the administration of reserpine or of 6-hydroxydopamine did not result in a consistent fall in the level of inhibitory activity in homogenates of the heart or spleen. On the other hand, there is so much inhibitory activity present that if only a small percentage of it is in the adrenergic nerve terminal, where it has access to the enzyme, then it will be playing a significant physiological role by limiting the activity of DBH and thus modulating norepinephrine biosynthesis. Recent preliminary experiments have suggested that DBH inhibitors do indeed have access to DBH. When homogenates of rat heart are subjected to sucrose density gradient centrifugation, a significant amount of inhibitory activity is found in the same region of the gradient as are noradrenergic storage granules.

REFERENCES

Brimijoin S. and Molinoff P. B. (1971) *J. Pharmacol. Exp. Ther.* **178**, 417–424.
Ceasar P. M., Ruthven C. R. J. and Sandler M. (1969) *Br. J. Pharmacol.* **36**, 70–78.
Chubb I. W., Preston B. N. and Austin L. (1969) *Biochem. J.* **111**, 245–246.
Duch D. S. and Kirshner N. (1971) *Biochem. Biophys. Acta* **236**, 628–638.
Foldes A., Jeffrey P. L. Preston B. N. and Austin L. (1972) *Biochem. J.* **126**, 1209–1217.
Hartman B. K., Molinoff P. B. and Udenfriend S. (1970) *The Pharmacologist* **12**, Ab. 470.
Kopin I. J. and Gordon E. K. (1963) *J. Pharmacol. Exp. Ther.* **140**, 207–216.
Kvetnansky R., Gewirtz G. P., Weise V. K., and Kopin I. J. (1970a) *Endoc.* **87**, 1323–1329.
Kvetnansky R., Gewirtz G. P., Weise V. K. and Kopin I. J. (1970b) *Mol. Pharmacol.* **7**, 81–86.
Levitt M. Spector S., Sjoerdsma A. and Udenfriend S. (1965) *J. Pharmacol. Exp. Ther.* **148**, 1–8.
Molinoff P. B., Brimijoin S., Weinshilboum R. and Axelrod J. (1970) *Proc. Nat. Acad. Sci. U.S.A.* **66**, 453–458.
Molinoff P. B., Weinshilboum R. and Axelrod J. (1971) *J. Pharmacol. Exp. Ther.* **178**, 425–431.
Molinoff P. B., and Axelrod J. (1972) *J. Pharmacol. Exp. Ther.* **182**, 116–129.
Mueller R. A., Thoenen H. and Axelrod J. (1969) *J. Pharmacol. Exp. Ther.* **169**, 74–79.
Ross S. B., Weinshilboum R., Molinoff P. B., Vesell E. S. and Axelrod J. (1972) *Mol. Pharmacol.* **8**, 50–59.
Rutledge C. O. and Weiner N. (1967) *J. Pharmacol. Exp. Ther.* **157**, 290–302.
Silberstein S., Brimijoin S., Molinoff P. B. and Lemberger L. (1972) *J. Neurochem.* **19**, 919–921.
Thoenen H., Kettler R., Burkard W. and Saner A. (1971) *Naunyn-Schmied* **270**, 146–160.
Thoenen H. (1972) *Pharm. Rev.* **24**, 255–267.
Viveros O. H., Arqueros L., Connett R. J. and Kirshner N. (1969). *Mol. Pharmacol.* **5**, 69–82.
Weil-Malherbe H. and Van Buren J. M. (1969) *J. Lab. Clin. Med.* **74**, 305–318.

Frontiers in Catecholamine Research 1973, pp. 201 to 203. Pergamon Press. Printed in Great Britain.

RETROGRADE AXONAL TRANSPORT OF DOPAMINE-β-HYDROXYLASE

STEPHEN BRIMIJOIN

Department of Pharmacology, Mayo Medical School, Rochester, Minnesota 55901, U.S.A.

IT HAS become clear that adrenergic nerves can regulate the synthesis of enzymes involved in formation of the neurotransmitter, norepinephrine. In particular, it seems that the rates of synthesis of tyrosine hydroxylase (L-tyrosine, tetrahydropteridine; oxygen oxidoreductase; EC 1.14.3.a) and dopamine-β-hydroxylase (3,4-dihydroxyphenylethylamine, ascorbate; oxygen oxidoreductase; EC 1,14,2,1) increase in response to a variety of stimuli, especially those producing chronic elevations in nerve activity (MOLINOFF and AXELROD, 1971; THOENEN, 1972).

Another aspect of the regulation of dopamine-β-hydroxylase (DBH) and tyrosine hydroxylase (TH) concerns suppression of the synthesis of these enzymes. BRIMIJOIN and MOLINOFF (1971) examined sympathetic ganglia from rats treated with 6-OH-dopamine, which causes degeneration of adrenergic nerve terminals. These ganglia maintained normal levels of TH activity but lost about half their DBH activity within 3 days. Reserpine, which elevates TH and DBH activity in control ganglia (MOLINOFF et al., 1970), had no effect on ganglia several days after treatment with 6-OH-dopamine or surgical section of the postganglionic nerve trunk. These observations suggest that loss of the terminal portions of the adrenergic neurons might in some way affect the synthesis of enzymes in the cell bodies of these neurons.

An effect of nerve terminals on synthesis of enzymes in cell bodies appears likely on the basis of subsequent experiments in which turnover of DBH was measured by the rate of fall of enzyme activity after inhibition of protein synthesis (BRIMIJOIN, 1972b). For about 40 hr after administration of 6-OH-dopamine, little happened to the apparent rate of synthesis of DBH in ganglia. However, just beyond this time there was a rapid and pronounced fall in synthesis of DBH. The considerable lag between administration of drug and fall of enzyme synthesis suggests that the effect of 6-OH-dopamine on synthesis of DBH is secondary to the destruction of nerve terminals. This lag may also explain why it is possible to induce increases in TH and DBH activity during the first hours after placing sympathetic ganglia in organ culture, a procedure which necessarily separates cell bodies from nerve terminals, (SILBERSTEIN et al., 1972; KEEN and McLEAN, 1972).

For nerve terminals to influence events in cell bodies would require some means of carrying a signal between the two regions of the neuron. One possible mechanism is retrograde axonal transport. DAHLSTRÖM (1965) found that, in addition to the large increase of norepinephrine above a ligature on the sciatic nerve, there was a small increase below the ligature. Likewise, we have observed a large accumulation of DBH activity proximal to a ligature on the sciatic nerve and a small accumulation distal to the ligature (BRIMIJOIN, 1972a). In themselves, these results could be explained either by some retrograde transport in addition to orthograde transport or by some leakage through the ligature. DAHLSTRÖM (1965) ruled out leakage as an explanation for

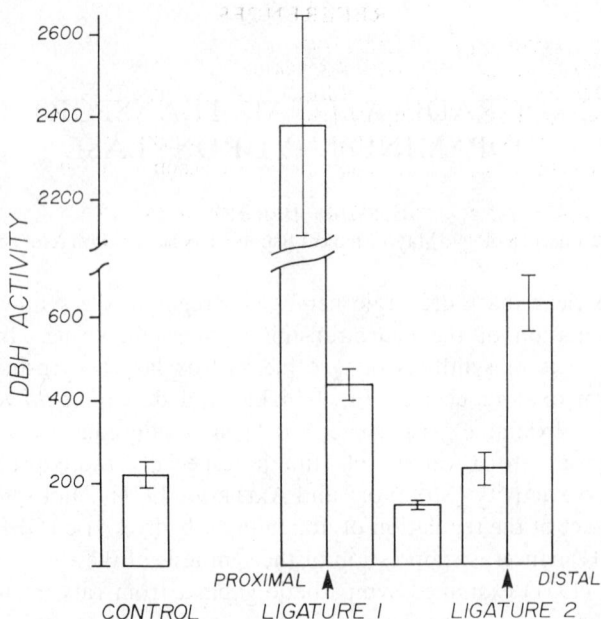

FIG. 1.—Distribution of DBH activity in doubly ligated rat sciatic nerves. DBH activity is given as pmoles octopamine produced per hr incubation per mm of nerve. Means ± S.E.M. (brackets) of 5 determinations of enzyme activity are shown.

distal accumulation of norepinephrine by showing that norepinephrine accumulates primarily *above* the proximal, but *below* the distal of two ligatures. We have performed an analogous experiment with DBH (Fig. 1). Two ligatures were applied, 10–14 mm apart, to the right sciatic nerves of rats under ether anesthesia; the animals were sacrificed 8 hr later. Segments 3 mm long were cut immediately above and below the proximal ligature and the distal ligature; another 3 mm segment was cut from the exact middle of the region between the two ligatures. These tissue samples were homogenised and assayed for DBH activity using tyramine as a substrate as previously described (BRIMIJOIN, 1972a). Compared to control segments from the contralateral nerves, DBH activity increased significantly in the segment above the proximal ligature ($P < 0.001$). DBH activity also increased significantly in the segments *below* both the proximal ($P < 0.005$) and the distal ligature ($P < 0.001$). The segment below the distal ligature had more DBH activity than did either the segment just above it ($P < 0.005$) or the segment just below the proximal ligature ($P < 0.05$). These results are hard to explain by postulating leakage through the ligatures; they strongly suggest a real retrograde transport of DBH. From the relative amounts of enzyme activity accumulating below the distal and above the proximal ligature, it appears that about 20 per cent as much DBH was being transported toward the cell body as toward the nerve terminals.

No evidence presently links retrograde transport of DBH, or any other substance, with the control of enzyme synthesis in the nerve cell body. Nevertheless, the process of retrograde transport deserves further investigation as a possible means for conveying chemical messages of physiological significance from distal to proximal regions of the adrenergic neuron.

REFERENCES

Brimijoin S. (1972a) *J. Neurochem.* **19**, 2183–2193.
Brimijoin S. (1972b) *J. Pharmac. exp. Ther.* **183**, 298–306.
Brimijoin S. and Molinoff P. B. (1971) *J. Pharmac. exp. Ther.* **178**, 417–424.
Dahlström A. (1965) *J. Anat. Lond.* **99**, 677–689.
Keen P. and McLean W. G. (1972) *Arch. Pharmac.* **275**, 465–469.
Molinoff P. B., Brimijoin S., Weinshilboum R. and Axelrod J. (1970) *Proc. Natn. Acad. Sci. U.S.A.* **66**, 453–458.
Molinoff P. B. and Axelrod J. (1971) *Ann. Rev. Biochem.* **40**, 465–500.
Silberstein S., Brimijoin S., Molinoff P. B. and Lemberger L. (1972) *J. Neurochem.* **19**, 919–921.
Thoenen H. (1972) *Pharmac. Rev.* **24**, 255–268.

Frontiers in Catecholamine Research 1973, pp. 205 to 210. Pergamon Press. Printed in Great Britain.

GROWTH AND DEVELOPMENT OF A SYMPATHETIC GANGLION *IN VIVO*

IRA B. BLACK

Laboratory of Neurobiology, Department of Neurology, Cornell University Medical College,
New York, New York, U.S.A.

DEVELOPMENTAL milestones in the mammalian central nervous system have been documented using anatomical (EAYRS and GOODHEAD, 1959; PETERS and FLEXNER, 1950), ultrastructural (AGHAJANIAN and BLOOM, 1967; BUNGE and BUNGE, 1965), electrophysiological (DEZA and EIDELBERG, 1967) and biochemical (LOGNADO and HARDY, 1967; HEBB, 1965) approaches. However, due to the complexity of the central nervous system, such studies have been largely descriptive. Even the simplest brain nuclei contain heterogeneous groupings of cells which differ morphologically, biochemically and probably functionally.

Studies in the periphery (GIACOBINI *et al.*, 1970; GIACOBINI, 1970) provide simpler models of neural ontogeny. Sympathetic ganglia in mouse, rat and cat contain primarily two neural elements in synaptic contact: pre-synaptic cholinergic nerve terminals and post-synaptic adrenergic neurons (GIACOBINI, 1970). Specifically, the well-defined, relatively non-complex superior cervical ganglion (SCG) is ideal for the study of adrenergic growth and development because the SCG is composed of biochemically distinct, well-defined neural elements consisting primarily of the cholinergic-adrenergic neural unit defined above.* Moreover, the SCG is anatomically discrete and easily accessible for manipulation and its bilaterally symmetric nature allows rigorously controlled experiments within a single animal (BLACK *et al.*, 1971a).

In the present report the maturation of mouse SCG *in vivo* is described. Choline acetyltransferase (Acetyl-Co A; choline *O*-acetyltransferase, EC 2.3.1.6.), (ChAc), the enzyme catalysing the conversion of acetyl CoA and choline to the neurotransmitter acetylcholine, is employed as a marker for the development of presynaptic cholinergic fibres. The enzyme is highly localised to these presynaptic terminals (HEBB and WAITES, 1956). Maturation of post-synaptic neurons is followed by measuring the activity of tyrosine hydroxylase (T-OH) (*O*-diphenol: O_2 oxireductase, EC 1.10.3.1), the rate-limiting enzyme in the biosynthesis of noradrenaline (LEVITT *et al.*, 1965), the post-ganglionic neurotransmitter. Visualisation of ganglion synapses with the electron microscope allows estimation of the development of synaptic connections. Correlation of these biochemical and morphological parameters suggests that presynaptic nerve terminals might regulate the development of the post-synaptic neuron. The results of surgical transection of the preganglionic nerve trunk in neonatal animals confirms this hypothesis.

---

* In addition, recent studies have indicated the presence of low numbers of small neurons (WILLIAMS and PALEY, 1969), adrenergic fibres (HAMBERGER *et al.*, 1963) and scattered cholinergic cells (SJOQVIST, 1962) in sympathetic ganglia.

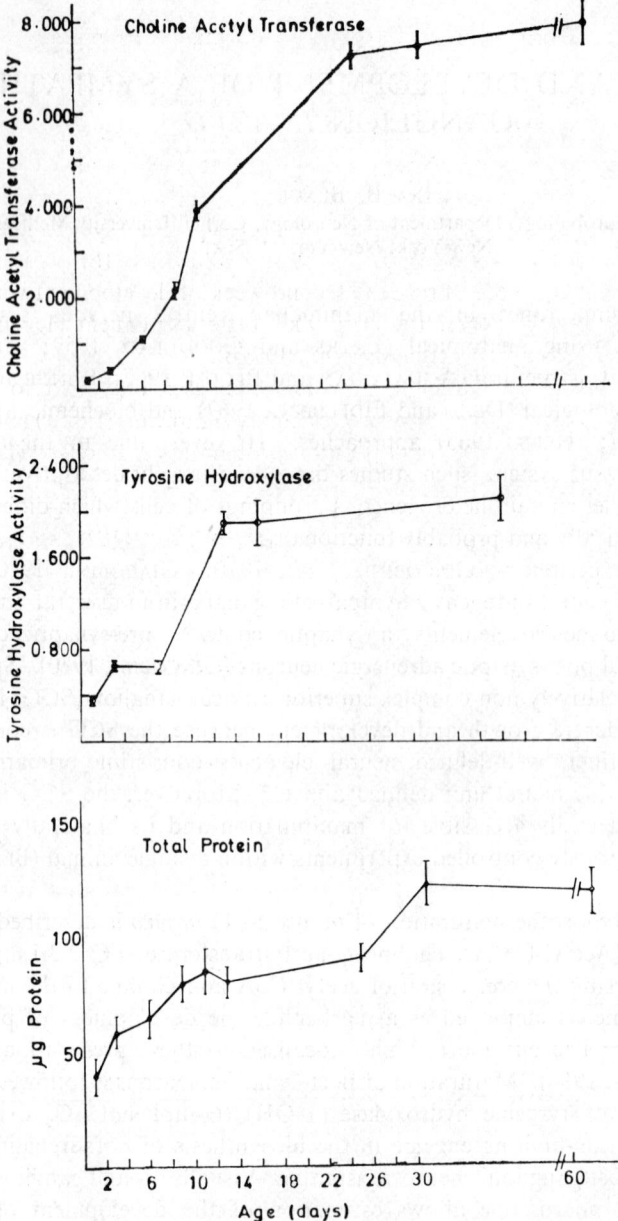

FIG. 1.—Developmental increases of transmitter enzyme activities and total protein in mouse superior cervical ganglia: Groups of six mice were taken from litters of varying ages and ganglion pairs from each animal were assayed for enzyme activites and total protein. Choline acetyltransferase activity is expressed as mean (nmoles product per ganglion pair) per hr $\pm$ s.e.m. (vertical bars). Tyrosine hydroxylase activity is expressed as 10^{-11} moles product per ganglion pair per hr. Total protein is expressed as mean μg per ganglion pair $\pm$ s.e. (BLACK et al., 1971b)

ENZYME DEVELOPMENT IN SCG

ChAc activity increases thirty to forty-fold during the course of development (BLACK *et al.*, 1971b). From low levels on day 1, enzyme activity rises rapidly during the first two weeks of life reaching a hyperbolic plateau by approximately three weeks (Fig. 1) (BLACK *et al.*, 1971b). This increase in enzyme activity may reflect either ongoing invasion of the ganglion by presynaptic nerve endings and/or transport of the enzyme to nerve endings already present in the ganglion.

The developmental curve for T-OH activity differs significantly from that of ChAc. T-OH activity increases six to eight-fold from birth to adulthood (BLACK *et al.*, 1971b). The major increase takes place during the second week of development when enzyme activity undergoes nearly a three-fold rise with little subsequent elevation to the thirty-eighth day of life (Fig. 1) (BLACK *et al.*, 1971b).

During development total ganglion protein increases only two to three-fold, rendering the rises in enzyme specific activities highly significant (Fig. 1) (BLACK *et al.*, 1971a). This relatively modest increase in total protein has been observed previously during the development of ganglia (COHEN, 1960).

MECHANISM OF THE DEVELOPMENTAL INCREASE OF ENZYME ACTIVITY

These increases in enzyme activity could be due either to the activation of pre-existent enzyme molecules, or to the synthesis of new enzyme protein. To distinguish between these alternatives neonatal mice were treated with the protein synthesis inhibitor cycloheximide (TRAKATELLIS *et al.*, 1965). Such treatment prevented the normal developmental increase in T-OH activity, suggesting that the developmental T-OH rise is dependent on the ongoing synthesis of new enzyme molecules (BLACK *et al.*, 1972a).

DEVELOPMENTAL FORMATION OF SYNAPTIC JUNCTIONS

The functional significance of these biochemical correlates of maturation was defined by elucidating their temporal relation to the development of inter-neural connections. Synaptic junctions were identified and counted in ganglia of mice aged one to sixty days. During this period total synapses per ganglion increase from approximately 8000 on day one to 3,000,000 by day sixty (Fig. 2). The number of synapses remains relatively constant during the first two days after birth, but rises dramatically between days five and eleven to an asymptotic plateau (BLACK *et al.*, 1971a). From day eleven to sixty the further increase in the number of synapses just reaches statistical significance.

Comparison of synapse development with the developmental pattern of ChAc activity reveals some similarities. Both functions display roughly hyperbolic curves with relatively modest increases during the first three to four days of life, and rapid rises early in the second week of development. On the basis of the data presented, however, more precise temporal relationships are not evident.

The development of T-OH activity contrasts interestingly with synapse formation. The early rise in T-OH activity precedes the increase in synapses. Immediately following the steep increase in synapse formation T-OH activity increases markedly to adult levels. These observations suggest that development of T-OH activity in post-synaptic neurons may be dependent on contact with presynaptic nerve endings.

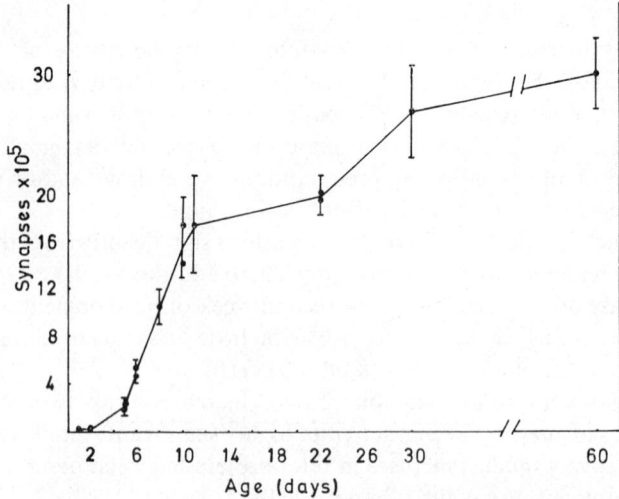

FIG. 2.—Development of ganglion synpases. Total synapses per ganglion were estimated in mice of different ages. Each point represents the estimated total synapses in a superior cervical ganglion from a single mouse. Verticle bars represent s.e. of determinations of synapse numbers in the ten grid squares sampled for each ganglion. (Reproduced by permission from BLACK et al., 1971b)

EFFECT OF DECENTRALISATION OF THE SUPERIOR CERVICAL GANGLION

Ganglia were decentralised in neonatal mice to determine whether the presynaptic cholingeric nerve terminals are necessary for the development of T-OH activity in the post-synaptic neuron. The preganglionic trunk was transected unilaterally in mice aged four days. The contralateral normal ganglion in each mouse served as control. Animals were killed at varying times post-operatively, ipsilateral ptosis and reduced ganglion ChAc activity indicating success of the procedure. As expected, ChAc activity was reduced to approximately 10 per cent of control values. T-OH activity failed to develop normally, remaining at approximately 30–50 per cent of the activity of contralateral unoperated ganglia (Fig. 3) (BLACK et al., 1971a; BLACK et al., 1972a). These findings indicate that the increase in T-OH activity occurring during the second week of development (Fig. 1) is dependent on innervation of the post-synaptic neuron.

THE TRANS-SYNAPTIC MESSAGE

Since treatment of neonatal mice and rats with Nerve Growth Factor (NGF) results in profound hyperplasia and hypertrophy of adrenergic neurons throughout the animal (LEVI-MONTALCINI and ANGELETTI, 1968), this substance was the prime candidate for the trans-synaptic message. However, treatment with NGF does not fully reverse the effects of decentralisation on ganglion T-OH development (BLACK et al., 1972a). Consequently, NGF cannot completely replace pre-synaptic terminals during maturation, and thus cannot be considered alone as the pre-synaptic message.

Another approach to the problem of identifying the transsynaptic factor(s) regulating adrenergic neuron development has involved the use of long-acting ganglionic blocking agents (HENDRY and IVERSEN, 1972; BLACK, 1973). These compounds

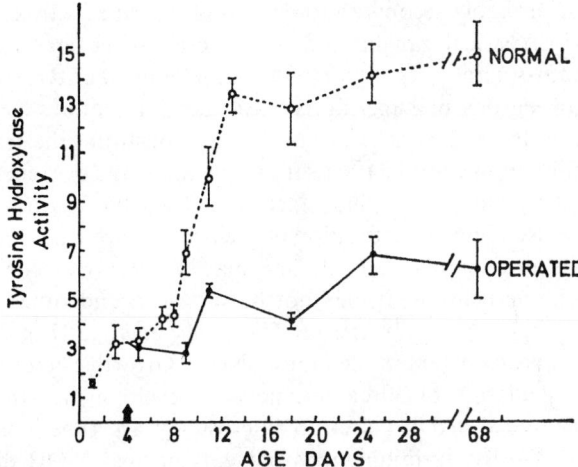

FIG. 3.—Effect of surgical decentralisation on day 4 on development of tyrosine hydroxylase activity in mouse superior cervical ganglion. Groups of six mice were killed at various times post-operatively and tyrosine hydroxylase activity (pmol/ganglion/hr) measured in decentralised and contralateral control ganglia. The value obtained 1 day after surgery does not differ significantly from control, all other values from decentralised ganglia are significantly lower than control values ($P < 0.01$).
(BLACK *et al.*, 1972a)

prevent depolarisation of post-synaptic neurons in sympathetic ganglia by competing with acetylcholine for receptor sites. Indeed, the structurally dissimilar, long-acting ganglionic blocking agents, chlorisondamine and pempidine, prevent the normal development of tyrosine hydroxylase activity in post-synaptic neurons of the superior cervical ganglion (Table 1) (HENDRY and IVERSEN, 1972; BLACK, 1973).

The musarinic blocking agent, atropine, however, does not alter T-OH development (unpublished observation) (Table 1). Such studies suggest that it is the blockade of acetycholine-induced depolarisation of adrenergic neurons which prevents the normal development of tyrosine hydroxylase activity in these cells. Thus, ganglionic blockade reproduces the effects of ganglionic decentralisation in preventing normal adrenergic development. The administration of these agents is not associated with the appearance of inhibitors or disappearance of activators, and chlorisondamine itself does not alter enzyme activity *in vitro* (BLACK, 1973). Consequently, the failure

TABLE 1. EFFECT OF GANGLIONIC BLOCKADE ON THE DEVELOP-
MENT OF TYROSINE HYDROXYLASE ACTIVITY

Group	Tyrosine hydroxylase activity
Control	17.8 ± 2.22
Chlorisondamine	8.5 ± 1.02*
Pempidine	9.1 ± 0.83*
Atropine	20.8 ± 0.93

Groups of 6–8 animals were treated with saline (control), chlorisondamine (5 μg/g), pempidine (50 μg/g) or atropine (5 μg/g) subcutaneously every 12 hr from day 2 to day 7 of life. All animals were killed on day 7. Results are expressed as mean pmoles/ganglion pair hr $\pm$ s.e.m.
  * Differs from control and atropine group at $P < 0.001$. Atropine group does not differ significantly from control ($P > 0.05$).

of maturation is most probably secondary to decreased tyrosine hydroxylase enzyme protein in the superior cervical ganglia and not a result of enzyme inhibition. It would appear that trans-synaptic regulation of the development of tyrosine hydroxylase in the adrenergic neurons of superior cervical ganglia requires depolarisation of these cells. On this basis, acetylcholine itself may constitute the trans-synaptic message. Consequently, it may not be necessary to postulate the existence of some as yet unidentifiable trans-synaptic "trophic" factor(s), since the normal presynaptic neurotransmitter may also regulate maturation of the post-synaptic neuron. It should be stressed that while presynaptic acetylcholine may be *necessary* for the normal development of ganglionic neurons, it may not be *sufficient*, and other unidentified mechanisms may also participate. In the rat (THOENEN *et al.*, 1972), for example, and most probably to a lesser extent in the mouse, Nerve Growth Factor also appears to contribute to the regulation of adrenergic neuron development. In some sense then, the synapses demonstrated by electron microscopy are functional. That is, surgical destruction of the synaptic junctions prevents normal T-OH development. Moreover, prevention of development by ganglionic blockade suggests that synapses demonstrated morphologically may be correlated with the onset of cholinergic synaptic transmission.

POST-SYNAPTIC REGULATION OF PRESYNAPTIC DEVELOPMENT

Selective destruction of ganglionic adrenergic neurons in the neonatal mouse with either 6-hydroxydopamine or Anti-Nerve Growth Factor antiserum prevents the normal development of choline acetyltransferase activity in presynaptic terminals (BLACK *et al.*, 1972b). These observations, in conjunction with those already described, suggest that there is a reciprocal regulatory relationship between cholinergic and adrenergic neurons at the synapse during development.

REFERENCES

AGHAJANIAN G. and BLOOM F. E. (1967 *Brain Res.* **6**, 716–727.
BLACK I. B. (1973) *J. Neurochem.* **20**, 1265–1267.
BLACK I. B., HENDRY I. A. and IVERSEN L. L. (1971a) *Nature, Lond.* **231**, 27–29.
BLACK I. B., HENDRY I. A. and IVERSEN L. L. (1971b) *Brain Res.* **34**, 229–240.
BLACK I. B., HENDRY I. A. and IVERSEN L. L. (1972a) *J. Neurochem.* **19**, 1367–1377.
BLACK I. B., HENDRY I. A. and IVERSEN L. L. (1972b) *J. Physiol.* **221**, 149–159.
BUNGE R. P. and BUNGE M. B. (1965) *Anat. Rec.* **151**, 329.
COHEN S. (1960) *Proc. Natl. Acad. Sci. U.S.A.* **46**, 302–311.
DEZA L. and EIDELBERG E. (1967) *Exp. Neurol.* **17**, 425–438.
EAYRS J. T. and GOODHEAD B. (1959) *J. Anat. Lond.* **93**, 385–402.
GIACOBINI E. (1970) In: *Biochemistry of Simple Neuronal Models, Advances in Biochemical Psycho-pharmacology*, Vol. 2. (COSTA E. and GIACOBINI E., Eds.) Raven Press, New York.
GIACOBINI G., MARCHISIO P. C. GIACOBINI E. and KOSLOW S. H. (1970) *J. Neurochem.* **17**, 1177.
HAMBERGER B., NORBERG K. A. and SJOQVIST F. (1963) *Int. J. Neuropharmacol.* **2**, 279–282.
HEBB C. O. (1956) *J. Physiol.* **133**, 566–570.
HEBB C. O. and WAITES G. M. H. (1956) *J. Physiol.* **132**, 667–671.
HENDRY I. A. IVERSEN L. L. (1972) *Proc. Fifth Int. Cong. Pharmacol. (San Francisco)* **100**.
LEVI-MONTALCINI R. and ANGELETTI P. U. (1968) *Physiol. Rev.* **48**, 534–569.
LEVITT M., SPECTOR S., SJOERDSMA A. and UDENFRIEND S. (1965) *J. Pharmacol. Exp. Ther.* **148**, 1–8.
LOGNADO J. R. and HARDY M. (1967) *Nature, Lond.* **214**, 1207–1210.
PETERS V. B. and FLEXNER L. B. (1950) *Amer. J. Anat.* **86**, 133–157.
SJOQVIST F. (1962) In: *Cholinergic Sympathetic Ganglion Cells* (NORSTEDT P. A. and SONER A., Eds.) Stockholm.
THOENEN H., SANER A., KETTLER and ANGELTTI P. U. (1972) *Brain Res.* **44**, 593–602.
TRAKATELLIS A. C., MONTJAR M. and AXELROD A. E. (1965) *Biochem.* **4**, 2065.
WILLIAMS T. H. and PALAY S. L. (1969) *Brain Res.* **15**, 17–34.

Frontiers in Catecholamine Research 1973, pp. 211 to 221. Pergamon Press. Printed in Great Britain.

STUDIES ON THE MECHANISM OF REGULATION OF TYROSINE HYDROXYLASE ACTIVITY DURING NERVE STIMULATION

N. Weiner, R. Bjur, F.-L. Lee, G. Becker and W. F. Mosimann

University of Colorado School of Medicine, Department of Pharmacology,
4200 East 9th Avenue, Denver, Colorado 80220, U.S.A.

Norepinephrine synthesis occurs in adrenergic neurons both in the central nervous system and in the sympathetic division of the peripheral autonomic nervous system. Although the adrenergic neurons in the brain may synthesise their neurotransmitter at a rate which is quantitatively different from that in the peripheral adrenergic neurons, it would appear that the overall regulatory behaviour of catecholamine biosynthesis in peripheral and central adrenergic neurons is similar. Thus, the enzyme which appears to be most limited in quantity in both central and peripheral adrenergic neurons is tyrosine hydroxylase (Levitt et al., 1965). In both the central nervous system and in the periphery, tyrosine-3-hydroxylase (EC 1.14.3.a) (tyrosine hydroxylase) in adrenergic neurons appears to be a soluble enzyme which requires oxygen, ferrous iron and a reduced pterin cofactor for optimal activity (Nagatsu et al., 1964). The reduced pterin cofactor is probably also similar in central and peripheral adrenergic neurons and may be identical to the cofactor required for liver phenylalanine hydroxylase, tetrahydrobiopterin (Kaufman, 1963). A few years ago, Lloyd demonstrated that tetrahydrobiopterin is present in bovine adrenal medulla tissue at a concentration which was estimated to be approximately 0·01 mM (Lloyd and Weiner, 1971). Lloyd also demonstrated a pterin compound in the caudate region of sheep brain which appeared to be either similar to or identical with the cofactor isolated from bovine adrenal medulla. It is generally assumed that tetrahydrobiopterin is the natural cofactor both in peripheral adrenergic neurons and in central adrenergic neurons, although the cofactor has not been identified in either of these classes of neurons.

In both brain and peripheral adrenergic nervous tissue, soluble tyrosine hydroxylase is inhibited by a variety of catecholamines, including dopamine and norepinephrine. We have observed that dopamine is approximately twice as potent as norepinephrine as an inhibitor of tyrosine hydroxylase. However, since the quantity of norepinephrine in adrenergic neurons is considerably greater than that of dopamine, presumably end-product feedback inhibition of tyrosine hydroxylase activity is mediated primarily by the major adrenergic neuron catecholamine, norepinephrine.

Over the past several years we have been examining a number of the factors which may regulate tyrosine hydroxylase activity, in partially purified enzyme preparations, tissue homogenates and intact tissue, in an effort to elucidate the effects of nerve stimulation and the effects of various drugs which modify the function of the sympathetic nervous system.

We have been assisted considerably in the conduct of these studies, particularly with regard to studies on intact isolated tissues, by the development of a tyrosine hydroxylase assay which is not dependent upon isolation and quantitation of labelled catecholamines formed from tyrosine or phenylalanine. By employing ^{14}C-carboxyl labelled tyrosine we are able to measure tyrosine hydroxylase activity by collection of the labelled CO_2 produced from the aromatic hydroxylation of tyrosine to β-(3,4-dihydroxyphenyl-L-alanine (dopa) and the subsequent decarboxylation of the dopa by aromatic-L-amino acid decarboxylase (dopa decarboxylase) (Waymire et al., 1971). In studies with tissue homogenates or with partially purified enzyme, the addition of excess hog kidney aromatic amino acid decarboxylase plus pyridoxal phosphate ensures the quantitative decarboxylation of dopa so that the reaction which is being measured in the assay is the hydroxylation of tyrosine. When intact tissues are employed, the quantity of dopa decarboxylase within the adrenergic neuron is sufficient to virtually instantaneously decarboxylate all dopa formed from tyrosine and the $^{14}CO_2$ which is produced as a consequence of the hydroxylation of tyrosine may then be quantitatively collected and assayed by liquid scintillation spectrometry as an index of tyrosine hydroxylase activity (Weiner et al., 1972). We have shown that, in intact tissue preparations and with tissue homogenates and partially purified tyrosine hydroxylase, the quantity of ^{3}H-catecholamines produced from 3,5-^{3}H-tyrosine and the quantity of $^{14}CO_2$ produced from ^{14}C-carboxyl labelled tyrosine are equivalent in the presence of a monoamine oxidase inhibitor. If a monoamine oxidase inhibitor is omitted from the incubation system, a significant fraction of the catecholamines are oxidatively deaminated and measurement of catecholamines yields erroneously low values of tyrosine hydroxylase activity. Thus the coupled decarboxylase assay for tyrosine hydroxylase has allowed us to evaluate the regulation of tyrosine hydroxylase activity and norepinephrine synthesis in intact tissue without resorting to the use of monoamine oxidase inhibitors which could themselves affect norepinephrine synthesis (Neff and Costa, 1966; 1968; Spector et al., 1967).

Most of our studies on the regulation of tyrosine hydroxylase activity in adrenergic neurons have been performed with either mouse vas deferens preparations or (for studying the effect of nerve stimulation on tyrosine hydroxylase activity) the hypogastric nerve-vas deferens preparation of the guinea-pig. These tissues contain short adrenergic neurons and the entire cell body, axon and nerve terminals are present intact in the isolated preparations of these tissues. For this reason these tissues may represent a more appropriate model system of in vivo adrenergic neuronal metabolism and, in particular, they may provide a model for central adrenergic neurons, most of which are much shorter than the bulk of the neurons of the peripheral sympathetic nervous system. The properties of the enzymes of the mouse vas deferens preparation and the guinea-pig vas deferens preparation are very similar. Both enzymes have a higher affinity for tetrahydrobiopterin (apparent K_m approximately 50 μM) than for 6,7-dimethyltetrahydropterin (apparent $K_m = 0.22$ mM). The guinea-pig vas deferens tissue is capable of converting both tyrosine and phenylalanine to catecholamines, although tyrosine is much the preferred substrate (Ikeda et al., 1965,1967; Weiner et al., 1973). In the presence of tetrahydrobiopterin, 0.1 mM or higher concentrations of phenylalanine or tyrosine inhibit the tyrosine hydroxylase reaction (Weiner et al., 1973). The chymotrypsin solubilized preparation of the bovine adrenal enzyme appears to be even more sensitive to inhibition by excess tyrosine (Shiman and Kaufman, 1971).

In the intact mouse vas deferens preparation, a variety of drugs known to modify norepinephrine storage or metabolism markedly affect tyrosine hydroxylase activity.

For example amphetamine and tranylcypromine, which both release norepinephrine from storage sites by direct displacement and inhibit monoamine oxidase, inhibit the hydroxylation of tyrosine in intact tissue. At high concentrations, neither of these agents has an effect on dopa decarboxylase activity in the intact tissue and these drugs are without any significant effect on either tyrosine hydroxylase or dopa decarboxylase activities in homogenates prepared from mouse vasa deferentia. Similarly, after acute administration of reserpine, there is a rather marked reduction of tyrosine hydroxylase activity in intact mouse vasa deferentia and in adrenal slices. No effect of this drug on the enzyme is apparent when these preparations are homogenised and assayed with optimal amounts of substrate and cofactor (WEINER et al., 1972). It thus appears that tranylcypromine, amphetamine and reserpine are acting indirectly on the tyrosine hydroxylase of intact tissue to inhibit the activity of the enzyme. Presumably they are exerting their action by increasing the release of norepinephrine from synaptic vesicles (amphetamine and tranylcypromine) or blocking the uptake of norepinephrine into the synaptic vesicles (reserpine). In either event, the effect would be to raise the levels of free intraneuronal norepinephrine which could then inhibit tyrosine hydroxylase. Tyrosine hydroxylase is localised in the cytosol of the nerve terminal (WEINER et al., 1972; WEINER and BJUR, 1972).

NAGATSU et al., (1964) demonstrated that the inhibition of tyrosine hydroxylase by norepinephrine and other catechol compounds was competitively antagonised by reduced pterin cofactors. We have been able to demonstrate that 6,7-dimethyl-tetrahydropterin ($DMPH_4$) can increase the tyrosine hydroxylase activity of intact mouse vas deferens preparations when this synthetic cofactor is added to the incubation medium (WEINER and BJUR, 1972; WEINER et al., 1972). The increase in tyrosine hydroxylase activity is directly proportional to the amount of pterin cofactor added to the medium over a range of concentrations of $DMPH_4$ from 0·05 to 1·0 mM. In view of this effect of added pterin cofactor, it became possible to test the theory that amphetamine, tranylcypromine and reserpine were inhibiting the activity of tyrosine hydroxylase indirectly by increasing the level of free intraneuronal norepinephrine. Thus, if this were the mechanism of action of these agents, the inhibition of tyrosine hydroxylase should be competitively overcome by the addition of reduced pterin cofactor to the medium. With high concentrations of pterin cofactor, the activity of tyrosine hydroxylase in the presence of these agents is restored, an observation which is in support of this hypothesis. However, a linear relationship between the reciprocals of the velocity of reaction and the concentration of added pterin cofactor was not obtained when low concentrations of $DMPH_4$ were examined (WEINER et al., 1972).

The failure of the enzyme reaction to obey Michaelis–Menten kinetics at low cofactor concentrations may be explained by the fact that we neglected in these calculations to include the contribution to tyrosine hydroxylase activity of the endogenous pterin cofactor. If we assume that the endogenous pterin cofactor is equivalent to 0·1 mM $DMPH_4$, and the reciprocals of all cofactor concentrations are recalculated with the addition of this constant factor, then results do fall on a straight line when plotted according to the Lineweaver–Burk relationship. Furthermore, the apparent K_m value obtained for $DMPH_4$ in the intact tissue is 0·22 mM, virtually identical to that obtained with the homogenate prepared from mouse vasa deferentia and assayed under optimal conditions of pH, substrate concentration and in the presence of Fe^{2+} (Table 1). Similarly the apparent K_m for tyrosine, 0·074 mM in the intact tissue, is virtually identical to that obtained for tyrosine when this is derived from studies on homogenates prepared from this tissue. That some correction for endogenous co-factor concentration must be taken into account when attempting to perform these enzyme kinetic analyses on intact tissue is obvious from the observation that a considerable amount of tyrosine hydroxylase activity is demonstrable in isolated, intact tissue preparations when no cofactor is added to the medium.

An analysis of endogenous $DMPH_4$ can best be made using the normalised version of the Lineweaver–Burk plot (Dixon and Webb, 1964). To derive this, the Lineweaver–Burk equation:

$$1/v = \frac{K_m}{V_{max}} \cdot \frac{1}{S} + \frac{1}{V_{max}}$$

is multiplied by V_{max}; where V_{max} is the maximal velocity of the reaction with variable pterin cofactor (S) and constant tyrosine concentration, and K_m is that for the cofactor at a fixed tyrosine concentration (either 0·01 or 0·1 mM).

$$\frac{V_{max}}{v} = \frac{K_m}{S} + 1$$

If V_{max}/v, the reciprocal of the relative velocity of the reaction, is plotted on the ordinate and K_m/S is plotted on the abscissa, a straight line is obtained with slope of one, which intersects the ordinate at $+1$ and the abscissa at -1. Any deviation from this normal line would imply that the enzyme reaction does not obey typical Michaelis–Menten kinetics. By employing the normalised relationship and summing several estimates of the endogenous pterin cofactor concentration and the concentration of $DMPH_4$ present in the medium, a family of lines can be obtained, only one of which coincides with the theoretically correct normal plot. This method of analysis is an exquisitely sensitive means of estimating the endogenous $DMPH_4$ equivalents, assuming that, analogous to the isolated enzyme, Lineweaver–Burk kinetics are obeyed in the intact tissue. The similarity of the kinetic constants obtained from mouse vas deferens homogenates and for intact mouse vasa deferentia indicates that this analysis of intact tissues may be valid (Table 1). A similar coincidence of the apparent K_m for $DMPH_4$ is seen with guinea-pig vas deferens homogenate and intact tissue. In this latter tissue, the modest discrepancy between the apparent K_m of tyrosine for intact tissue and that for the tissue homogenate may be due to the larger size of the organ from this species. Thus, the concentration of tyrosine may be greater in the medium than in the regions of the tissue where the nerve terminals are located, perhaps due either to poor diffusion or to uptake and/or metabolism of the tyrosine prior to reaching the environs of the nerve terminals (Table 1).

The level of endogenous pterin cofactor estimated in the above manner is actually a measure of the *activity* of cofactor in the intact preparation, rather than a measure of its absolute concentration. It is therefore the sum of at least two known variables; (1) the level of reduced pterin cofactor in the cytosol of the nerve ending, and (2) the level of free intraneuronal norepinephrine in that compartment. Thus, if the level of free intraneuronal norepinephrine increases, the apparent level of the endogenous pterin cofactor, as determined in intact tissue by the Michaelis–Menten kinetic analysis, would appear to decrease. In the presence of reduced pterin cofactor with ascorbic acid as reducing agent in the medium (to maintain the cofactor in the reduced

TABLE 1. KINETIC CONSTANTS— TYROSINE HYDROXYLASE
(VAS DEFERENS)

Preparation	Apparent K_m		
	Tyrosine (μM)	DMPH$_4$* (μM)	BH$_4$† (μM)
GUINEA PIG			
Homogenate	16 (BH$_4$) 26 (DMPH$_4$)	220	55
Intact tissue			
Control	61	220	—
Stimulated	65	300	—
MOUSE			
Homogenate	74 (DMPH$_4$)	220	—
Intact tissue	71	220	—

* DMPH$_4$ = 6,7-dimethyltetrahydropterin
† BH$_4$ = tetrahydrobiopterin

state), we have obtained by this analysis the highest estimates of endogenous DMPH$_4$ equivalents, namely 0·15 mM. In the presence of another reducing agent, 2-mercaptoethanol, the apparent level of endogenous pterin cofactor falls to approximately 0·08 mM. The reduced level of pterin cofactor in the presence of 2-mercaptoethanol may be the consequence of the ability of this sulfhydryl compound to release norepinephrine into the axoplasm since, if a similar kinetic analysis is performed on vasa deferentia from mice treated with reserpine, in which the endogenous stores of norepinephrine in the tissue are depleted greater than 95 per cent, the apparent level of endogenous cofactor in the presence of 2-mercaptoethanol equals that found in the presence of ascorbic acid (Table 2).

In view of the stimulatory effect of DMPH$_4$ on tyrosine hydroxylase activity in intact tissue, it would appear that tyrosine hydroxylase is subject to end-product

TABLE 2. ENDOGENOUS DMPH$_4$ EQUIVALENTS IN THE
INTACT MOUSE VAS DEFERENS

Reducing agent added to the assay medium*	DMPH$_4$ equivalents† (mM)
5 mM Ascorbic Acid	0·15
25 mM 2-Mercaptoethanol	0·08
25 mM 2-Mercaptoethanol‡	0·15

* Mouse vasa were assayed intact for tyrosine hydroxylase activity in Krebs–Ringer bicarbonate solution containing 0·1 mM Na$_2$EDTA; 133 mM NaCl; 4·7 mM KCl; 0·1 mM MgCl$_2$; 2·5 mM CaCl$_2$; 16·3 mM NaHCo$_3$; 1·4 mM KH$_2$PO$_4$; 7·8 mM glucose; 0·01 mM 1-^{14}C-L-tyrosine and amounts of DMPH$_4$ ranging from 0·05–1·0 mM. The medium was gassed with 95% O$_2$–5% CO$_2$, the tubes were capped and incubated for 20 min at 37°C. At the end of the reaction, ^{14}CO$_2$ is collected in NCS organic amine solubiliser and counted by liquid scintillation spectrometry (WAYMIRE et al., 1971, WEINER et al., 1972).

† DMPH$_4$ equivalents were derived by best fit analysis of the data to the normalised Lineweaver–Burk relationship.

‡ Mice were treated with 5 mg/kg reserpine, i.p., 24 hr prior to sacrifice.

feedback inhibition in the intact neuron and is probably, under normal circumstances, unable to fully express its enzyme synthesising capacity in the intact tissue either because of insufficient quantities of reduced pterin cofactor in the nerve ending or because of the presence of significant concentrations of free intraneuronal norepinephrine which exerts a persistent end-product feedback inhibition on the tyrosine hydroxylase reaction.

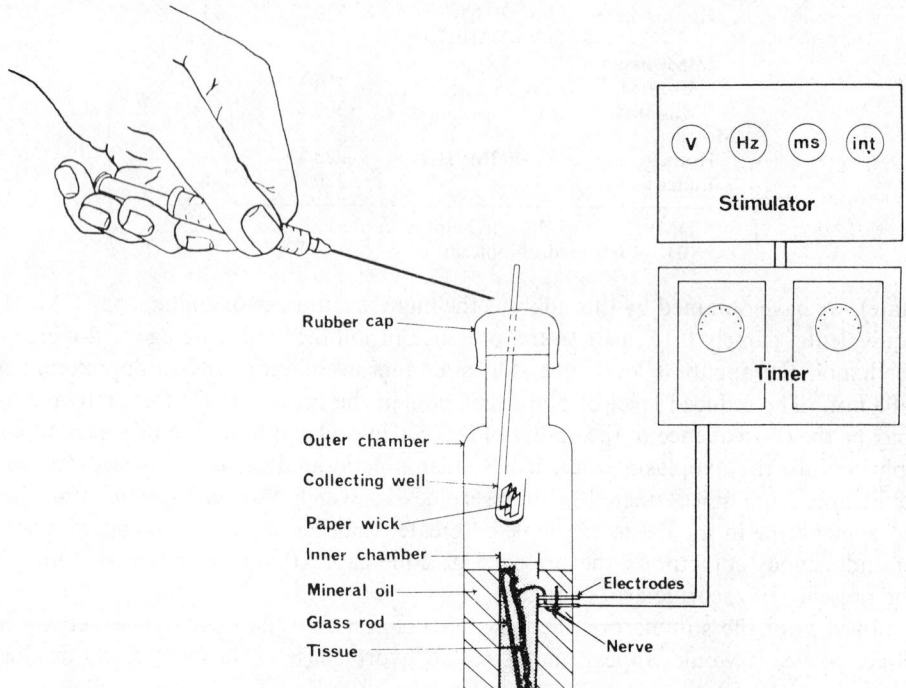

Fig. 1.—The organ bath is located within a larger glass chamber. A pair of platinum electrodes penetrates the wall of the larger chamber and terminates at the wall of the small organ bath at a site adjacent to a small port in the wall of the smaller bath. The vas deferens preparation is inserted in the smaller organ bath and the hypogastric nerve is drawn through the small port and wrapped around the platinum electrode. After the tissue is set up in the organ bath in the Krebs–Ringer bicarbonate medium, the small port is sealed with vaseline and mineral oil is added to the outer glass chamber to a level slightly above the port. The mineral oil prevents the nerve from drying out. The vaseline minimises the possibility of leakage of the organ bath medium from the small vessel into the larger glass chamber. A plastic cup is hung from the rubber septum prior to the incubation. The sham stimulated preparation is set up in an identical fashion in a second organ bath. The outer chamber is sealed off by the rubber septum and the system is gassed with 95% O_2–5% CO_2. Both nerves are stimulated to demonstrate that each tissue responds to the applied voltage. Subsequent to this time the stimulated preparation is randomly selected and stimulated for 10 sec of every min for 40 min. The stimulation is carried out in the presence of carboxyl labelled ^{14}C-tyrosine in the organ bath. At the end of the stimulation period, NCS solubiliser, an organic base compatible with a variety of liquid scintillation cocktails, is injected into the small plastic well protruding from the rubber septum covering the chamber opening. Subsequent to this, 0·2 ml of 50% trichloroacetic acid is injected into the small organ bath to terminate the enzyme reaction and allow the quantitative evolution of the [14]CO_2 which is then trapped in the NCS solubiliser well. The reaction vessels are left in the refrigerator overnight; in the morning the plastic cups are removed from the rubber septum and the contents of the plastic vials are transferred with scintillation fluid into a scintillation vial and counted by liquid scintillation spectrometry.

It has been assumed for many years that nerve stimulation is associated with increased norepinephrine synthesis, since tissue norepinephrine levels are generally quite well maintained in the presence of varying amounts of impulse traffic along sympathetic neurons unless either the storage or the synthesising capacity of the neuron is impaired (WEINER, 1970). Several years ago we were able to demonstrate directly an increased synthesis of catecholamines from labeled 3,5-^{3}H-tyrosine in the isolated hypogastric nerve-vas deferens preparation of the guinea-pig when the hypogastric nerve was stimulated intermittently over a period of 1 hr. Increased synthesis was apparently localised at the tyrosine hydroxylase reaction since no increase in formation of catecholamines from labeled dopa was demonstrable during nerve stimulation. This effect of nerve stimulation on norepinephrine synthesis could be overcome by the addition of exogenous norepinephrine to the medium (ALOUSI and WEINER, 1966; WEINER and RABADJIJA, 1968).

We have recently confirmed that norepinephrine inhibits the nerve stimulated increase in catecholamine synthesis, employing the coupled decarboxylase assay (WAYMIRE et al., 1971) adapted for application to stimulated preparations (Fig. 1). Employing carboxyl-labeled ^{14}C-tyrosine, we have demonstrated that catecholamine synthesis from tyrosine is enhanced by nerve stimulation both in the presence and, to a greater degree, in the absence of pargyline, a monoamine oxidase inhibitor. Norepinephrine and dopamine inhibit tyrosine hydroxylase activity in both stimulated and control preparations. The stimulated preparations appear to be more sensitive to the inhibitory effects of the catecholamines. At the lowest concentration examined (10^{-6} M), norepinephrine appears to be a more effective inhibitor of tyrosine hydroxylase in intact tissue than dopamine (Fig 2). This is most likely because the rate of uptake of norepinephrine into neurons (of rat heart) at low concentrations is greater than that of dopamine (EULER, 1972). Thus, norepinephrine appears to be more potent than dopamine in this system, in spite of the fact that the K_i of dopamine for

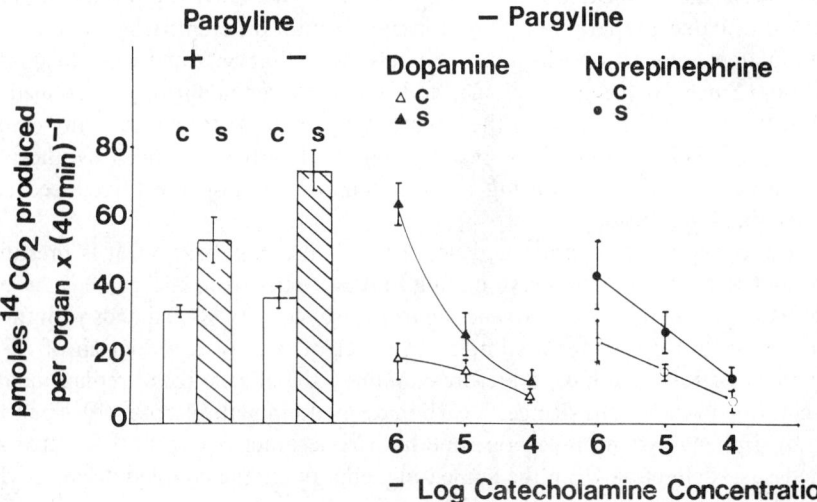

FIG. 2.— The effect of different concentrations of dopamine and norepinephrine and the effect of 0·15 mM pargyline on catecholamine synthesis in control and stimulated preparations of the isolated guinea-pig vas deferens-hypogastric nerve preparation. For details of method, see Fig. 1.

the soluble guinea-pig vas deferens enzyme is approximately one-half that for norepinephrine.

On the basis of these results, we concluded that increased norepinephrine synthesis from tyrosine associated with nerve stimulation was a consequence of reduced end-product feedback inhibition. Presumably this results from a lowering of free intraneuronal norepinephrine as a consequence of nerve terminal depolarisation and the egress of neurotransmitter from the neuron. Our recent observations that reduced pterin cofactor could overcome end-product feedback inhibition in intact adrenergic nervous tissues allowed us to test more directly this earlier proposal (Weiner and Bjur, 1972). Thus, if the increased norepinephrine synthesis during nerve stimulation were indeed the consequence of reduced end-product feedback inhibition, addition of reduced pterin cofactor to the medium would increase the tyrosine hydroxylase activity to a smaller degree in the stimulated preparation than in the sham stimulated preparation. Since the end-product feedback inhibition is competitive with pterin cofactor, one should be able to obtain a similar maximal velocity of norepinephrine synthesis at sufficiently high concentrations of cofactor in both stimulated and sham stimulated preparations. That is, an altered level of end-product enzyme inhibition would be expected to alter the apparent K_m of the reaction for the cofactor, but the V_{max} should not be affected. This hypothesis was tested by comparing, in the presence of different concentrations of DMPH$_4$, ^{3}H-catecholamine synthesis from 3,5-^{3}H-tyrosine in the stimulated hypogastric nerve-vas deferens preparation of the guinea pig with that in sham stimulated controls. The results we obtained failed to support the hypothesis that enhanced norepinephrine synthesis during nerve stimulation is a consequence of reduced end-product feedback inhibition (Weiner et al., 1972; Cloutier and Weiner, 1973). Thus, in the presence of DMPH$_4$, synthesis of catecholamines from tyrosine was increased to an even greater degree during nerve stimulation than in the absence of hypogastric nerve stimulation. A Lineweaver–Burk analysis of these data indicated that the apparent K_m's for DMPH$_4$ in the stimulated and sham stimulated preparations were not significantly different, whereas the V_{max} in the stimulated preparation was approximately twice that obtained with the control preparation. Since, in the intact tissue, addition of norepinephrine to the medium resulted in the expected increase in the apparent K_m of DMPH$_4$ without any change in the V_{max} of the reaction, these results suggest that the enhanced synthesis of norepinephrine from tyrosine during nerve stimulation is not due to reduced end-product feedback inhibition.

Since during nerve stimulation and nerve terminal depolarisation, it is presumed that Na$^+$ is taken up into the nerve ending and K$^+$ exits from the nerve, it is conceivable that the increase in tyrosine hydroxylase activity associated with nerve stimulation results from either a stimulating effect of Na$^+$ or a reduction of an inhibitory effect of K$^+$. We have therefore examined the effects: (a) of replacement of sodium chloride in the Krebs–Ringer bicarbonate medium with sucrose; (b) of ouabain and; (c) of diphenylhydantoin on tyrosine hydroxylase activity in both control and stimulated vasa deferentia from the guinea-pig, employing the coupled decarboxylase assay, modified to allow nerve stimulation of intact tissue (Waymire et al., 1971) (Fig. 1).

Progressive replacement of NaCl in the Krebs–Ringer bicarbonate solution with isotonic sucrose did not significantly influence the basal level of tyrosine hydroxylase activity in the vas deferens preparation. There was also no significant difference in the tyrosine hydroxylase activity in the stimulated preparations when the concentration of NaCl was reduced from 143 mM to 104 mM, 64 mM, or 25 mM. However, when the concentration of Na^+ was reduced to zero, there was a significant increase in the activity of tyrosine hydroxylase in the control preparation, but no significant change in the tyrosine hydroxylase activity of stimulated preparations was observed. The activities of tyrosine hydroxylase in control and stimulated preparations were similar in the absence of Na^+ (Table 3). However, in the absence of Na^+ the tissues

TABLE 3. EFFECT OF TOTAL Na^+ DELETION ON THE TYROSINE
HYDROXYLASE ACTIVITY INCREASE DURING NERVE STIMULATION

Guinea-pig hypogastric nerve vas deferens preparation		
Na^+ concentration pmoles $^{14}CO_2$ produced per organ $\times$ (40 min)$^{-1}$		
	Control	Stimulated
0* (4)	51·3 ± 8·5†	57·3 ± 2·8
118 mM (4)	20·7 ± 2·9	42·9 ± 2·5

Medium contained: 5×10^{-5} M $^{14}CO_2$-L-tyrosine, 10 mCi/mmole; 32 mM imidazole-HCl buffer, pH 7·2; 4·7 mM KCl; 2·5 mM $CaCl_2$; 1·2 mM KH_2PO_4; 1·2 mM $MgSO_4$; 7·8 mM glucose and either 118 mM NaCl or 231 mM sucrose. The medium was equilibrated with 100% O_2. $^{14}CO_2$ produced was collected in plastic cups containing NCS organic base solubiliser and counted by liquid scintillation spectrometry.

* Tissues in absence of Na^+ exhibited very weak spontaneous movements and responses to nerve stimulation were feeble or inconsistent.
† Mean ± S.E.

appeared to contract feebly and spontaneously and the contractions associated with nerve stimulation were weak and irregular. Thus, it is difficult to interpret the abolishment of the differences in tyrosine hydroxylase activity between control and stimulated preparations.

An effect similar to that seen in the absence of Na^+ was observed when these preparations were incubated with 10^{-4} M ouabain. There was a significant increase in tyrosine hydroxylase activity in control preparations and enzyme activities in control and stimulated preparations were similar. The contractile behaviour of the tissues also resembled that seen in the absence of Na^+ (Table 4). The effects of ouabain might suggest that tyrosine hydroxylase *in situ* is either stimulated by Na^+ or, more likely, inhibited by K^+. The absence of extracellular Na^+ or the presence of high concentrations of ouabain would both impair the activity of the Na^+–K^+–ATPase pump and could lead to loss of K^+ from the nerve terminal. Diphenylhydantoin, which presumably enhances the activity of the Na^+–K^+–ATPase pump, did not significantly affect tyrosine hydroxylase activity in either control or stimulated vas deferens preparations (Table 4).

GOLDSTEIN *et al.* (1970) reported that ouabain increased catecholamine synthesis from tyrosine in rat brain cortex slices. These investigators suggested that ouabain may enhance catecholamine synthesis by blocking uptake of released catecholamine, thus reducing end-product feedback inhibition. However, in the vas deferens, other blockers of catecholamine uptake; e.g., cocaine, do not affect catecholamine synthesis in unstimulated preparations (LEE and WEINER, unpublished).

TABLE 4. Effect of ouabain and diphenylhydantoin on
tyrosine hydroxylase activity in stimulated and
control vasa deferentia

| Drug added | pmoles $^{14}CO_2$ produced per organ $\times$ (40 min)$^{-1}$ | |
	Control	Stimulated
None (8)	$37\cdot0 \pm 4\cdot6$	$71\cdot8 \pm 7\cdot4$†
Ouabain 10^{-1} M* (6)	$83\cdot3 \pm 7\cdot6$‡	$90\cdot2 \pm 8\cdot5$§
Diphenylhydantoin		
10^{-4} M (6)	$35\cdot8 \pm 2\cdot7$	$61\cdot2 \pm 3\cdot9$†

* Vasa deferentia exhibited spontaneous activity with ouabain, responses to nerve stimulation were inconsistent.
Tissues were incubated in closed vesssls in Krebs–Ringer bicarbonate buffer previously equilibrated with 95% O_2–5% CO_2. Medium contained 5×10^{-5} M ^{14}C-carboxyl labeled L-tyrosine, 10 mCi/mmole. $^{14}CO_2$ produced was collected in plastic cups containing NCS organic base solubiliser and counted by liquid scintillation spectrometry. Results are expressed as mean $\pm$ s.e.
† Significantly different from control value ($P < 0\cdot01$).
‡ Significantly different from control values of other two groups ($p < 0\cdot01$).
§ Significantly different from value of diphenylhydantoin stimulated group.

In a recent preliminary communication, Gutman and Segal (1972) reported that bovine adrenal medulla soluble tyrosine hydroxylase is stimulated by increasing concentrations of Na^+ in the medium. However, they did not exclude the possibility that the enzyme was inhibited in proportion to the concentration of K^+ since in their studies, Na^+ and K^+ were interchanged. We have reexamined this phenomenon using homogenates from guinea-pig vas deferens preparations. Enzyme activity was assayed under optimal conditions in the presence of 0, 51, 102 and 154 mM NaCl, KCl, or choline chloride. The system was maintained isotonic by replacement of the salt by isotonic amounts of sucrose and the enzyme activity was assayed using either tetrahydrobiopterin or $DMPH_4$ as cofactor. We were unable to demonstrate any significant effect of any of these cations on tyrosine hydroxylase activity.

Gutman and Segal (1972) also presented preliminary evidence indicating that tyrosine hydroxylase activity is stimulated, within a very limited concentration range, by calcium ions. Kuczenski and Mandell (1972a, b), have presented evidence suggesting that high molecular weight sulfated polysaccharides such as heparin may modify the activity of tyrosine hydroxylase by shifting the enzyme from a soluble to a particulate state, with a simultaneous increase in V_{max} and reduction in the K_m of the enzyme for $DMPH_4$. Thus there appears to be a variety of potential mechanisms by which the activity of tyrosine hydroxylase may be regulated within intact adrenergic neurons as a consequence of nervous activity or other metabolic inputs. Whether these or other phenomena are responsible for the enhanced synthesis of norepinephrine from tyrosine which is associated with nerve stimulation is currently under investigation in this laboratory.

Acknowledgements—This work was supported by U.S.P.H.S. Grants NSO7642 and NSO7927. The authors wish to express their gratitude to Dr. R. F. Long, Roche Ltd., Welwyn Gardens, England, for the generous supply of tetrahydrobiopterin used in these studies.

REFERENCES

ALOUSI A. and WEINER N. (1966) *Proc. Nat. Acad. Sci. U.S.* **56**, 1491–1496.
CLOUTIER G. and WEINER N. (1973) *J. Pharmacol. Exp. Ther.* **186**, 75–85.
DIXON M. and WEBB E. C. (1964) In: *Enzymes.* 2nd edn, Academic Press, New York.
EULER U. S. v. (1972) In: *Catecholamines* (BLÁSCHKO H. and MUSCHOLL E,. eds.)
GOLDSTEIN M., OHI Y. and BACKSTROM T. (1970) *J. Pharmacol. Exp. Ther.* **174**, 77–82.
GUTMAN Y. and SEGAL J. (1972) *Biochem. Pharmacol.* **21**, 2664–2666.
IKEDA M., LEVITT M. and UDENFRIEND S. (1965) *Biochem. Biophys. Res. Comm.* **18**, 482–488.
IKEDA M., LEVITT M. and UDENFRIEND S. (1967) *Arch. Biochem. Biophys.* **120**, 420–427.
KAUFMAN S. (1963) *Proc. Nat. Acad. Sci., U.S.A.* **50**, 1085–1093.
KUCZENSKI R. T. and MANDELL A. J. (1972a) *J. Neurochem.* **19**, 131–137.
KUCZENSKI R. T. and MANDELL A. J. (1972b) *J. Biol. Chem.* **247**, 3114–3122.
LEVITT M., SPECTOR S., SJOERDSMA A. and UDENFRIEND S. (1965) *J Pharmacol. Exp. Ther.* **148**, 1–8.
LLOYD T. and WEINER N. (1971) *Mol. Pharmacol.* **7**, 569–580.
NEFF N. H. and COSTA E. (1966) *Life Sci.* **5**, 951–959.
NEFF N. H. and COSTA E. (1968) *J. Pharmacol. Exp. Ther.* **160**, 40–47.
NAGATSU T., LEVITT M. and UDENFRIEND S. (1964) *J. Biol. Chem.* **239**, 2910–2917.
SHIMAN R. and KAUFMAN S. (1971) *J. Biol. Chem.* **246**, 1330–1340.
SPECTOR S., GORDON R., SJOERDSMA A. and UDENFRIEND S. (1967) *Mol. Pharmacol.* **3**, 549–555.
WAYMIRE J. C., BJUR R. and WEINER N. (1971) *Anal. Biochem.* **43**, 588–600.
WEINER N. (1970) *Ann. Rev. Pharmacol.* **10**, 273–290.
WEINER N. and BJUR R. (1972) In: *Monoamine Oxidases: New Vistas.* (COSTA E. and SANDLER M., Eds.) Vol. 5, pp. 409–419. Raven Press, New York.
WEINER N., CLOUTIER G. BJUR R. and PFEFFER R. I. (1972) *Pharmacol. Rev.* **24**, 203–221.
WEINER N., LEE F.-L., WAYMIRE J. C. and POSIVIATA M. (1973) Ciba Foundation Symposium, *Aromatic Amino Acids in the Brain.* In press.
WEINER N. and RABADJIJA M. (1968) *J. Pharmacol. Exp. Ther.* **160**, 61–71.

Frontiers in Catecholamine Research 1973, pp. 223 to 229. Pergamon Press. Printed in Great Britain.

TRANSSYNAPTIC AND HUMORAL REGULATION OF ADRENAL CATECHOLAMINE SYNTHESIS IN STRESS

Richard Kvetňanský

Institute of Experimental Endocrinology, Slovak Academy of Sciences,
Bratislava, Czechoslovakia

ALTHOUGH an activation of sympathetic pathways during emergency situations was emphasized by Cannon, and later by Selye who considered catecholamines to be responsible for the alarm phase of the stress reaction, the role of catecholamines and the extent of their participation in adaptive reactions after repeated stress have not as yet been elucidated. We have therefore attempted to evaluate the reactions of the sympatho-adrenal system in animals during their adaptation to daily repeated stress as well as under acute stress.

In our experiments, immobilization (IMO) was used as the model of stress; rats were restrained daily for 150 min intervals (Kvetňanský et al., 1970a). Such immobilization produces both psychological stress (struggling) and physical stress (muscle work). After a single period of IMO there is a decrease in the content of adrenal epinephrine (EPI) (Kvetňanský and Mikulaj, 1970) (Fig. 1) and an increase in the level of plasma catecholamines (CA) (Kvetňanský et al., 1973 in press) and urinary CA excretion, mainly EPI (Kvetňanský and Mikulaj, 1970). Immobilization, repeated for about 6 weeks, produced a morphologically-apparent increase in the activity of the adrenal medulla (Kvetňanský et al., 1966) associated with a return of the adrenal EPI content to control levels or even higher (Kvetňanský et al., 1966; Kvetňanský and Mikulaj, 1970) and with a manifold increase of urinary EPI excretion. The finding of an increased level of adrenal CA together with increased urinary excretion of CA in repeatedly IMO rats, suggested enhanced CA biosynthesis in the adrenals of such "adapted" animals. The increased formation of adrenal CA was evident in experiments where the synthesis of adrenal CA was evaluated by administration of radioactive precursors; ^{14}C-tyrosine and ^{3}H-dopa (Kvetňanský et al., 1971c). These results clearly showed an increased biosynthesis of adrenal CA after repeated IMO of rats. This technique with double-labelled precursors revealed the rate of CA biosynthesis after IMO to change mainly at steps: tyrosine–dopa and dopamine–norepinephrine, which are catalysed by the enzymes tyrosine hydroxylase (EC 1.14.3.-)(TH) and dopamine-β-hydroxylase (EC 1.14.2.1) (DBH). We measured *in vitro* the activity of these enzymes as well as phenylethanolamine-N-methyltransferase (EC 2.1.1.-)(PNMT) which converts norepinephrine to epinephrine, in the adrenals of immobilized rats. The enzyme activity of TH, DBH and also PNMT proved to be substantially increased in the adrenals of repeatedly immobilized rats (Kvetňanský et al., 1970a, 1971a) (Fig. 1). Adrenal TH was assayed by the method of Nagatsu et al. (1964) and after repeated IMO its activity was found to have risen to over 3 times the control levels (Kvetňanský et al., 1970a). The enhanced synthesis of the enzyme protein is most probably responsible for the elevation of adrenal TH activity after IMO. After cessation of immobilization intervals TH activity returned toward preimmobilization levels with a half-life of

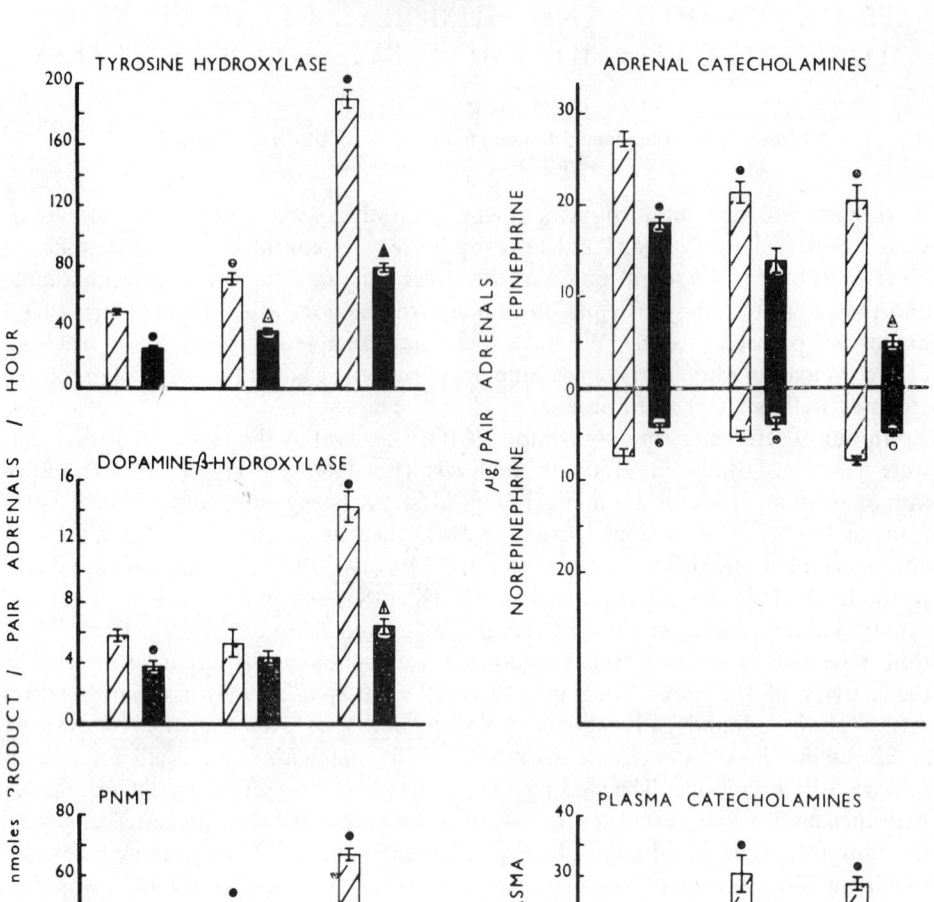

FIG. 1.—Effect of immobilization stress on adrenal and plasma catecholamine levels
and adrenal catecholamine-synthesizing enzymes activity in hypophysectomized and
sham-operated rats. Animals were immobilized for 2·5 hr daily and killed 6 hr after
the last immobilization (14 days after hypophysectomy). Plasma catecholamines were
measured according to Henry (unpublished results). Results are expressed as mean
values of 6–9 animals (± S.E.M.) Statistical significance compared to unstressed sham-
operated group: $\bigcirc = P < 0.05$; $\ominus = P < 0.02$; $\otimes = P < 0.01$; $\bullet = P < 0.001$.
Statistical significance compared to unstressed hypophysectomized group: $\triangle = P <$
0·05; $\triangle = P < 0.02$; $\triangle = P < 0.01$; $\blacktriangle = P < 0.001$.

about 3 days (KVETŇANSKÝ *et al.*, 1970a). Adrenal DBH was assayed by the method of VIVEROS *et al.* (1969), and after repeated IMO rose to over 2 times the control levels (KVETŇANSKÝ *et al.*, 1971a) (Fig. 1). Our results suggest that the turnover rate of DBH is increased during repeated IMO (KVETŇANSKÝ *et al.*, 1971a). Adrenal PNMT activity was assayed by the method of Axelrod (1962) and after repeated IMO was found to have increased by about 50% (KVETŇANSKÝ *et al.*, 1970a) (Fig. 1). These enzyme changes produced by repeated immobilization represent the first evidence of a physiologically-induced elevation of adrenal TH and DBH activity. Evidence is now available of increases in adrenal TH and PNMT activity after exposure to cold (KVETŇANSKÝ *et al.*, 1971b; STANTON *et al.*, 1972; THOENEN, 1970), psychosocial stimulation (AXELROD *et al.*, 1970) or after an attack by another animal (MAENGWYN-DAVIES *et al.*, 1973).

The adrenal medulla was long held to be controlled only by nerve impulses carried to the medulla by the preganglionic sympathetic nerves. In 1966 Wurtman and Axelrod reported that glucocorticoids, which are secreted from the adrenal cortex, enhanced the activity of PNMT in the adrenal medulla by a new synthesis of enzyme protein. This was the first study showing that the biosynthesis of adrenal CA is also regulated by the humoral pathway. It is well known that both the sympathetic nervous and the hypothalamo–pituitary–adrenocortical system are activated during stress. Therefore, it was convenient to use the stress-induced response of the adrenal medulla to examine factors in the regulation of catecholamine-synthesizing enzymes. We have investigated both endocrine and neural regulations by analysing adrenal CA and their synthesizing enzymes in animals after hypophysectomy, adrenal denervation, or lesions to different areas of CNS.

The endocrine status markedly influences stress-induced alterations in the levels of adrenal CA and catecholamine-synthesizing enzymes. In hypophysectomized (hypox.) animals, adrenal EPI levels are below normal values and are further depressed by repeated IMO (KVETŇANSKÝ *et al.*, 1970b) (Fig. 1). In such repeatedly immobilized hypophysectomized rats the total plasma catecholamine levels are also significantly decreased compared to normal animals KVETŇANSKÝ *et al.*, 1973, in press) (Fig. 1). The activities of adrenal TH (KVETŇANSKÝ *et al.*, 1969; 1970b; MUELLER *et al.*, 1970) DBH (GEWIRTZ *et al.*, 1971; KVETŇANSKÝ *et al.*, 1969; WEINSHILBOUM and AXELROD, 1970) and PNMT (WURTMAN and AXELROD, 1966; KVETŇANSKÝ *et al.*, 1970b) decreased following hypophysectomy (Fig. 1). In hypox. rats the TH and DBH activities were significantly increased by repeated IMO, but failed to attain the levels found in the adrenals of sham-operated IMO animals (Fig. 1). PNMT activity, however, was not elevated from the very low level by immobilization of hypox. rats. When ACTH was administered to hypox. rats before each IMO period, adrenal EPI was not depleted and TH (KVETŇANSKÝ *et al.*, 1970b), DBH (GEWIRTZ *et al.*, 1971) and PNMT (KVETŇANSKÝ *et al.*, 1970b) activities approached those found in sham-operated IMO rats. When dexamethasone (9α-fluoro-11β, 17α, 21-trihydroxy-16α-methyl-1,4-pregnandiene-3, 20-dione 21 phosphate) was given prior to IMO, again there was no depletion of adrenal EPI, there was an increase in PNMT activity, but no change in TH activity (KVETŇANSKÝ *et al.*, 1970b).

The neural regulation of adrenal catecholamine synthesis was studied after the left adrenal was denervated by severing the splanchnic nerve; the intact right gland served as control. Denervation did not influence adrenal CA levels in control animals, while in stressed animals no significant decrease of adrenal EPI was seen in the denervated gland (Fig. 2). The immobilization-induced depletion of adrenal EPI was blocked by hexamethonium administration (KVETŇANSKÝ et al., 1971a). Activities of the enzymes TH and DBH, which are increased by repeated IMO in the intact adrenal, were completely blocked in the denervated adrenal, but PNMT activity was elevated in both intact and denervated adrenals (KVETŇANSKÝ et al., 1970a, 1971a) (Fig. 2). In hypophysectomized rats, adrenal denervation also prevented the decrease of adrenal EPI as well as the increase of adrenal TH and DBH usually seen after repeated IMO (GEWIRTZ et al., 1971; KVETŇANSKÝ et al., 1970b) (Fig. 2). PNMT activity in the adrenals of immobilized hypox. animals remained depressed in both intact and denervated adrenals. If ACTH (5 IU, s.c.) was administered to hypox. rats before each IMO, a marked increase in TH activity in the intact adrenal and a small, though significant ($P < 0.01$) increase of TH in the denervated gland was seen (KVETŇANSKÝ et al., 1970b) (Fig. 2). Thus, both humoral and neural factors appear to influence the activity of TH in the adrenal medulla. After ACTH administration DBH activity was increased in the intact adrenal, but not in the denervated adrenal (Fig. 2). Thus, activity of adrenal DBH is also under a dual hormonal-neural regulation, but the neural component appears to be the more important. Treatment of hypox. rats with ACTH restores PNMT in the intact gland to levels produced by repeated IMO in normal rats, but denervation partially interferes with such a restoration of the enzyme (KVETŇANSKÝ et al., 1970b; THOENEN et al., 1970) (Fig. 2). Thus, humoral factors are more important than neural factors in determining the normal levels of PNMT.

Stress-induced adrenaline depletion does not occur in the denervated adrenal, but neither is there any increase in TH and DBH activities. Clearly, elevation of these enzymes depends on neuronal activity, though the question by which mechanism they are regulated remains unknown (PLETSCHER, 1972). It has been shown that the administration of acetylcholine causes an increased TH activity in the denervated gland (PATRICK and KIRSHNER, 1971) and acetylcholine was held possibly to regulate the transsynaptic activation of TH by adrenal medullary adenylate cyclase system (GUIDOTTI and COSTA, 1973).

Our results have demonstrated that nerve impulses associated with stress are required to increase the activity of adrenal enzymes. Presumably the CNS is involved in producing these adrenal medullary changes after stress. It has recently been shown by Reis et al. (1971) that repeated electrical stimulation of the posterior hypothalamus increases the activity of adrenal TH and PNMT but only in animals with behavioural alterations. Hypothalamic stimulation did not increase the enzyme activity in the denervated gland. Animals with lesions to the limbic system also show changes in behaviour; lesions to the septal area produce hyperreactivity, while lesions to the amygdaloid area produce apathy. Such animals would probably have an altered reaction to stress and therefore we undertook to study these areas of the limbic system. In cooperation with Dr. Murgaš and Dr. Dobrakovová from the Institute of Experimental Endocrinology bilateral septal and amygdaloid lesions were made by electrocoagulation with 2 mA for 15 sec. After repeated immobilization of the animals with septal lesions there was an additional decrease in the level of adrenal EPI and an additional increase in the activity of adrenal TH (KVETŇANSKÝ

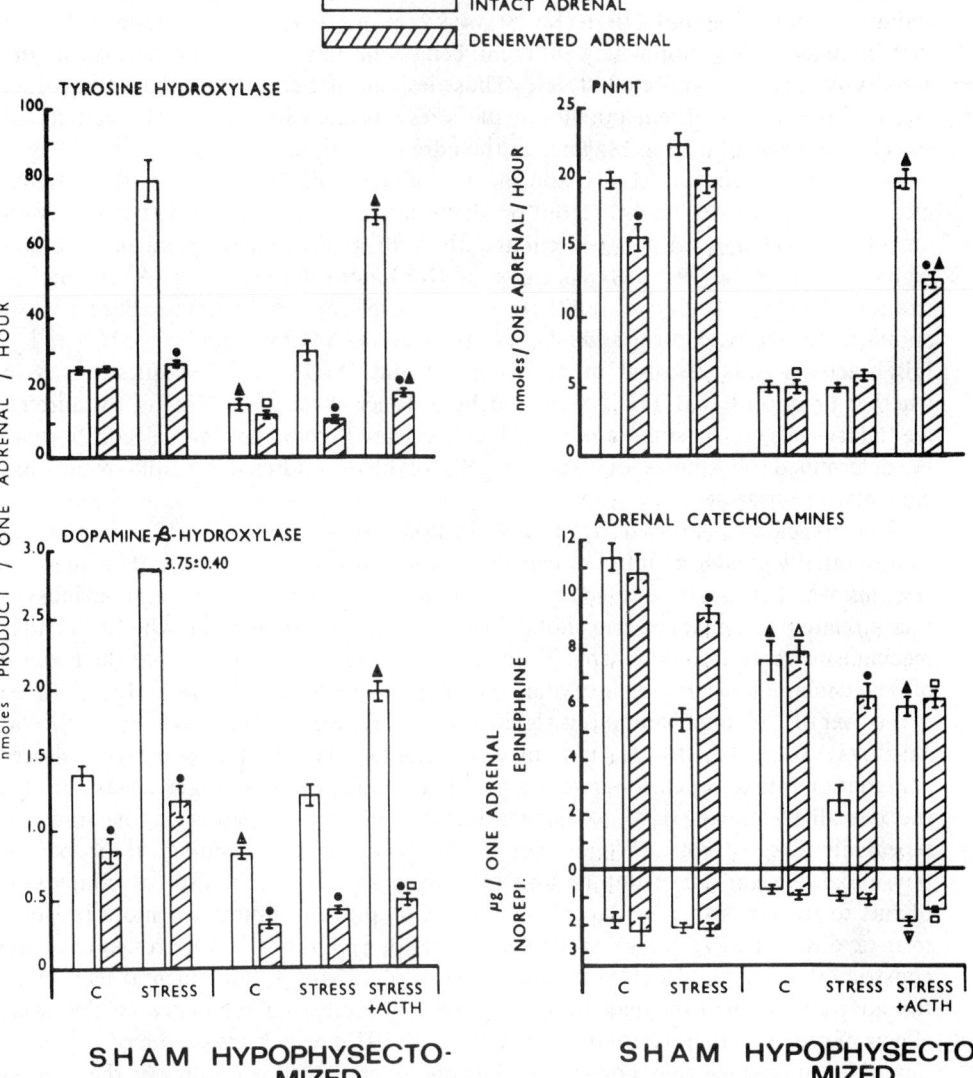

FIG. 2.—Effect of adrenal denervation (severing of nerve fibers from the superior mesenteric ganglia about 0·5 cm above the left adrenal; right adrenal served as intact control) on catecholamine-synthesizing enzymes activity in the adrenals of sham-operated and hypophysectomized rats exposed to stress. Animals were stressed by immobilization for 2·5 hr daily and killed 6 hr after the seventh immobilization (14 days after hypophysectomy and 9 days after denervation). ACTH (Acthar gel, 5 i.u. per rat, s.c.) was administered 60 min before each immobilization. Results are expressed as the mean values of 6–10 animals ($\pm$ S.E.M.). C = control. Statistical significance compared to intact adrenal: $\otimes = P < 0.01$; $\bullet = P < 0.001$. Statistical significance compared to hypophysectomized + immobilized intact adrenal: $\triangle = P < 0.01$; $\blacktriangle = P < 0.001$. Statistical significance compared to hypophysectomized + immobilized denervated adrenal: $\square$ = NS; $\triangle = P < 0.05$; $\triangle = P < 0.001$.

et al., 1972). Repeated IMO of animals with amygdaloid lesions produced increased activity of both TH and DBH (KVETŇANSKÝ *et al.*, 1972). These results showed that animals with a completely different behaviour have a similar increase in the activity of TH after repeated stress. Thus, lesions of both septal and amygdaloid areas of the limbic system potentiate the stress-induced increase in the activity of catecholamine-synthesizing enzymes in the adrenal medulla.

As indicated above, ACTH appears to influence PNMT levels in the adrenal medulla through its effect on steroid production in the adrenal cortex, but its action on adrenal TH appears to be mediated through a different mechanism. Neither high doses of steroids (KVETŇANSKÝ *et al.*, 1970b), nor dibutyryl cyclic-AMP administration (KVETŇANSKÝ *et al.*, 1971d) restored levels of TH in immobilized hypox. animals. In hypox. control animals, dibutyryl cyclic-AMP restored the TH activity. DBH activity was restored in both control and IMO rats. The question as to whether or not adrenal TH is regulated by a direct effect of ACTH on the adrenal medullary cells, or by some unknown factor released from the adrenal cortex, may be determined by studies on isolated cells of bovine adrenal medulla which are currently in progress.

Rats repeatedly stressed by immobilization (so-called "adapted"), dispose of a substantially greater quantity of catecholamines than do the controls. The question remains whether this is due solely to an enhanced adrenal catecholamine biosynthesis related to enhanced catecholamine-synthesizing enzymes, or whether further mechanisms come into play here. From our recent results it appears that the activity of catecholamine-degrading enzyme "catechol-*O*-methyltransferase" (EC 2.1.1.6) in the liver of "adapted" animals is significantly decreased (KVETŇANSKÝ, BABUŠÍKOVÁ, JAHNOVÁ, TORDA, MIKULAJ, unpublished results). This finding suggests a slower degradation rate of the catecholamines which are being produced in increased quantity; the overall effect would be to further increase of the CA levels in the body of repeatedly stressed rats. We do not know as yet what physiological or pathophysiological significance this finding may have for the mechanism of adaptation of rats to stress. Animals adapted to Noble-Collip drum trauma are more resistant to toxic doses of EPI; the splanchnic arterioles in such rats are more resistant to the constrictor action of CA (HRŮZA and ZWEIFACH, 1970). A fundamental feature of the adaptive mechanism may be a progressively increasing resistance to the toxic effects of catecholamines (HRŮZA and ZWEIFACH, 1970), which are produced in greater quantities in adapted rats, possibly as the consequence of this effect. At the present state of knowledge it is not possible to decide whether the increases of catecholamine levels together with resistance to them (which could be different in various organs) is advantageous or injurious for the organism.

CONCLUSIONS

The results presented show that during adaptation of rats to repeated immobilization stress there is an increase in biosynthesis of adrenal medullary catecholamines. This conclusion is supported by the increased levels of adrenal, plasma and urinary catecholamines, by increased synthesis of adrenal catecholamines, measured in vivo after administration of labelled precursors of catecholamines, and by increases in the activity of adrenal catecholamine-synthesizing enzymes: tyrosine hydroxylase, dopamine-β-hydroxylase and phenylethanolamine-N-methyl transferase, measured in vitro. This was the first evidence for a physiologically-induced elevation of these enzymes.

The biosynthesis of adrenal catecholamines in stressed rats is influenced by both neural and endocrine regulators. Increase of adrenal tyrosine hydroxylase activity induced by immobilization stress require both the neural and pituitary-adrenal systems to be intact. Adrenal dopamine-β-hydroxylase is also under this dual control, but the neural regulation appears to be the more important. Adrenal PNMT likewise appears to be controlled by both the systems, but the pituitary-adrenocortical system is the more important.

Electrocoagulation of both septal and amygdaloid areas of limbic system potentiate the stress-induced increases in the activity of adrenal medullary catecholamine-synthesizing enzymes.

The possible physiological meaning of the increased activity of the sympatho-adrenal system in repeatedly stressed-"adapted"-rats is discussed, but remains as yet an open issue.

Acknowledgements—This work was supported in part by Foundation's Fund for Research in Psychiatry, New Haven, Connecticut, U.S.A., Research Grant Number G 71–498.

REFERENCES

AXELROD J. (1962) *J. Biol. Chem.* **237**, 1657–1660.
AXELROD J., MUELLER R. A., HENRY J. P. and STEPHENS P. M. (1970) *Nature Lond.* **225**, 1059–1060.
GEWIRTZ G. P., KVETŇANSKÝ R., WEISE V. K. and KOPIN I. J. (1971) *Molec. Pharmacol.* **7**, 163–168.
GUIDOTTI A. and COSTA E. (1973) *Science* **179**, 902–904.
HRŮZA Z. and ZWEIFACH B. W. (1970) *J. Trauma* **10**, 412–419.
KVETŇANSKÝ R., MITRO A., MIKULAJ L. and HOCMAN G. (1966) *Bratisl. Lek. Listy* **46**, 35–41.
KVETŇANSKÝ R., WEISE V. K. and KOPIN I. J. (1969) *Pharmacologist* **11**, 274.
KVETŇANSKÝ R. and MIKULAJ L. (1970) *Endocrinology* **87**, 738–743.
KVETŇANSKÝ R., WEISE V. K. and KOPIN I. J. (1970a) *Endocrinology* **87**, 744–749.
KVETŇANSKÝ R., GEWIRTZ G. P., WEISE V. K. and KOPIN I. J. (1970b) *Endocrinology* **87**, 1323–1329.
KVETŇANSKÝ R., GEWIRTZ G. P., WEISE V. K. and KOPIN I. J. (1971a) *Mol. Pharmacol.* **7**, 81–86.
KVETŇANSKÝ R., GEWIRTZ G. P., WEISE V. K. and KOPIN I. J. (1971b) *Amer. J. Pysiol.* **220**, 928–931.
KVETŇANSKÝ R., WEISE V. K., GEWIRTZ G. P. and KOPIN I. J. (1971c) *Endocrinology* **89**, 46–49.
KVETŇANSKÝ R., GEWIRTZ G. P., WEISE V. K. and KOPIN I. J. (1971d) *Endocrinology* **89**, 50–55.
KVETŇANSKÝ R., MURGAŠ K., BABUŠÍKOVÁ F., VARGOVÁ E. and DOBRAKOVOVÁ M. (1972) *Abstracts of Intern. Congress of Endocrinology*, No. 552, Washington, D.C.
KVETŇANSKÝ R., HENRY D. P., WEISE V. K., LAMPRECHT F. and KOPIN I. J. (1973) *Physiol. Bohemoslov.* In press.
MAENGWYN-DAVIES G. D., JOHNSON D. G., THOA N. B., WEISE V. K. and KOPIN I. J. (1973) *Psychopharmacologia* **28**, 339–350.
MUELLER R. A., THOENEN H. and AXELROD J. (1970) *Endocrinology* **86**, 751–755.
NAGATSU T., LEVITT M. and UDENFRIEND S. (1964) *Anal. Biochem.* **9**, 122–126.
PATRICK R. L. and KIRSHNER N. (1971) *Molec. Pharmacol.* **7**, 87–96.
PLETSCHER A. (1972) *Pharmacol. Rev.* **24**, 225–232.
REIS D. J., MOORHEAD D. T., RIFKIN M., JOH T. H. and GOLDSTEIN M. (1971) *Nature Lond.* **229**, 562–563.
STANTON H. C., MUELLER R. L. and BAILEY C. L. (1972) *Proc. Soc. Exp. Biol. Med.* **141**, 991–995.
THOENEN H. (1970) *Nature, Lond.* **228**, 861–862.
THOENEN H., MUELLER R. A. and AXELROD J. (1970) *Biochem. Pharmacol.* **19**, 669–673.
VIVEROS O. H., ARQUEROS L., CONNETT R. J. and KIRSHNER N. (1969) *Mol. Pharmacol.* **5**, 60–68.
WEINSHILBOUM R. and AXELROD J. (1970) *Endocrinology* **87**, 894–899.
WURTMAN R. J. and AXELROD J. (1966) *J. Biol. Chem.* **241**, 2301–2305.

Frontiers in Catecholamine Research 1973, pp. 231 to 236. Pergamon Press. Printed in Great Britain.

TRANSSYNAPTIC REGULATION OF TYROSINE HYDROXYLASE IN ADRENAL MEDULLA: POSSIBLE ROLE OF CYCLIC NUCLEOTIDES

A. GUIDOTTI, C. C. MAO and E. COSTA

Laboratory of Preclinical Pharmacology, National Institute of Mental Health, Saint Elizabeths Hospital, Washington, D.C. 20032, U.S.A.

INTRODUCTION

AN INCREASE of impulse flow in afferent nerves to sympathetic ganglia and adrenal medulla enhances the activity of tyrosine hydroxylase (TH) (MULLER et al., 1969; AXELROD, 1971) and dopamine-β-hydroxylase (DBH) (MOLINOFF et al., 1970; AXELROD, 1972). We are now attempting to elucidate the molecular mechanisms whereby an increase rate of transmitter release from afferent nerve terminals promotes an increase synthesis of enzyme molecules in the target cells. Using cold exposure as inducing stimulus and adrenal gland as target tissue we could establish that the induction of TH is composed by two distinct phases each one with an appropriate time constant. The stimulus reaches maximal efficiency in 2 hr and the subsequent elaboration of the stimulus leading to enzyme induction takes approximately 12 hr. This presentation will include a biochemical study of the events associated with stimulus application. By using drugs as tools, we dissected the biochemical events involved and we produced evidence that disruption of certain patterns in the initial events will abolish enzyme induction. Although we are presently working with two neuronal models: adrenal medulla and sympathetic ganglia, most of our data are concerned with the former model. The reason for selecting the adrenal medulla derives from practical consideration and includes the following facts: Adrenal medulla has a uniform cell population (chromaffin cells), a single type of afferent nerves (cholinergic), and a simple type of innervation mostly functioning with an activation of nicotinic receptors.

ACTION OF DRUGS ON cAMP CONCENTRATIONS AND TH ACTIVITY OF ADRENAL MEDULLA

Reserpine (MULLER et al., 1969; THOENEN et al., 1969), aminophylline and carbamylcholine (GUIDOTTI and COSTA, 1973; COSTA and GUIDOTTI, 1973), as well as exposure of rats to 4°C (THOENEN, 1970) are suitable stimuli for inducing TH in adrenal medulla. All these inducing stimuli cause a dramatic increase of adenosine 3',5'-monophosphate (cAMP) concentrations in adrenal medulla (GUIDOTTI and COSTA, 1973; GUIDOTTI et al. 1973). During cold exposure, the time course of this increase parallels the duration of the stimulus for TH induction. When drugs were injected and used as inducing stimulus, we have found that the duration of the increase of cAMP lasts less than two hours. As an example of the temporal sequence of the two events we are showing in Fig. 1 the effects of carbamylcholine and cold exposure on adrenal medulla cAMP concentration and TH activity. The right panel of this figure shows the effects of 2 hr exposure at 4°C. In the intact medulla cyclic

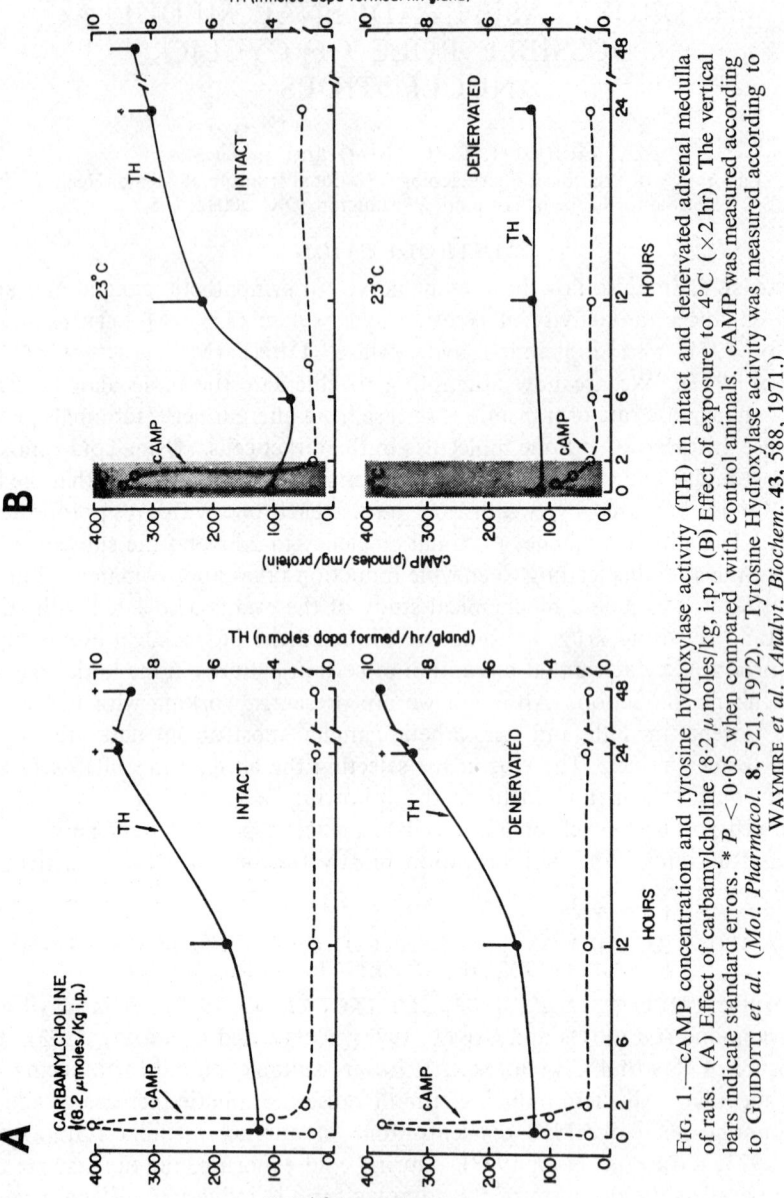

Fig. 1.—cAMP concentration and tyrosine hydroxylase activity (TH) in intact and denervated adrenal medulla of rats. (A) Effect of carbamylcholine (8·2 μmoles/kg, i.p.) (B) Effect of exposure to 4°C ($\times$2 hr) The vertical bars indicate standard errors. *$P < 0.05$ when compared with control animals. cAMP was measured according to Guidotti et al. (Mol. Pharmacol. **8**, 521, 1972). Tyrosine Hydroxylase activity was measured according to Waymire et al. (Analyt. Biochem. **43**, 588, 1971.)

AMP increases dramatically for about 2 hr and TH induction becomes apparently at 12 hrs and reaches maximal level at 24 hr. However in denervated adrenal, the increase of cAMP is insignificant and the TH induction fails to occur. By contrast, the left panel of Fig. 1 shows that carbamylcholine treatment elicits the increase in cAMP concentrations and in TH activity in both denervated and intact adrenal medulla. The contrasting effect of carbamylcholine and cold exposure in denervated gland indicates that a cholinergic mediation is involved in the effects of cold exposure.

LACK OF CORRELATION BETWEEN INCREASE OF cAMP AND RELEASE OR DEPLETION OF ADRENAL CATECHOLAMINES

Reserpine, aminophylline, carbamylcholine, and exposure to 4°C affect the dynamics of catecholamine stores by increasing the release rate either intracellularly or extracellularly. Therefore, it is possible that the increase of cAMP relates to the drug action on the catecholamine stores rather than to their effects on TH activity. The following findings tend to exclude such an association:

(1) Catecholamine turnover rate after tyramine (130 μmoles/kg, i.p.), increase from 0·38 to 0·74 nmoles/pair adrenals/hr but the cAMP concentrations in adrenal are not increased.

(2) Aminophylline (200 μmoles/kg, i.p.) causes a prompt and dramatic (15-fold) increase of cAMP concentrations in medulla, but the turnover rate of catecholamines is not increased.

(3) The increase of blood levels of catecholamines can not be invoked as a cause for the accumulation of cAMP in medulla because epinephrine (130 μmoles/kg, s.c.) and isoproterenol (138 μmoles/kg, i.p.) fail to increase either the adrenal medullary concentration of this cyclic nucleotide or the TH activity at a later time.

These findings indicate that the available information is inconsistent with the view that cAMP increase is associated with changes in the dynamics of catecholamine stores in adrenal medulla or with an increase level of catecholamines in blood. The results of isoproterenol injections are of interest because this drug increases the cAMP content of many other tissues (GUIDOTTI et al., 1972) including sympathetic ganglia and elicit a profound cardiocirculatory response.

REGULATION OF ADRENAL cAMP CONCENTRATION BY EXPOSURE TO COLD AND CARBAMYLCHOLINE INJECTION

The experiments reported in Fig. 1 suggest a transsynaptic control of cAMP in chromaffine cells of adrenal medulla.

The neuronally mediated increase of cAMP presumably involve release of acetylcholine from presynaptic fibers arriving to the adrenal and a subsequent stimulation of a nicotinic receptor in chromaffin cells membrane (DUGLAS et al., 1961).

The data presented in Fig. 2 show that a pretreatment with the nicotinic receptor antagonists, hexamethonium (45 μmoles/kg, i.p.) or mecamylamine (15 μmoles/kg, i.p.) blocks the immediate increase of cAMP concentrations in medulla elicited by either carbamylcholine or by cold exposure. Moreover, while the mecamylamine and hexamethonium fail per se to change the tyrosine hydroxylase activity, they block the increase of TH activity elicited 23 hr after cold exposure or carbamylcholine injection. That this effect of the ganglionic blockers involves a drug action on the stimulus and not on the elaboration of the stimulus is shown by the data reported in

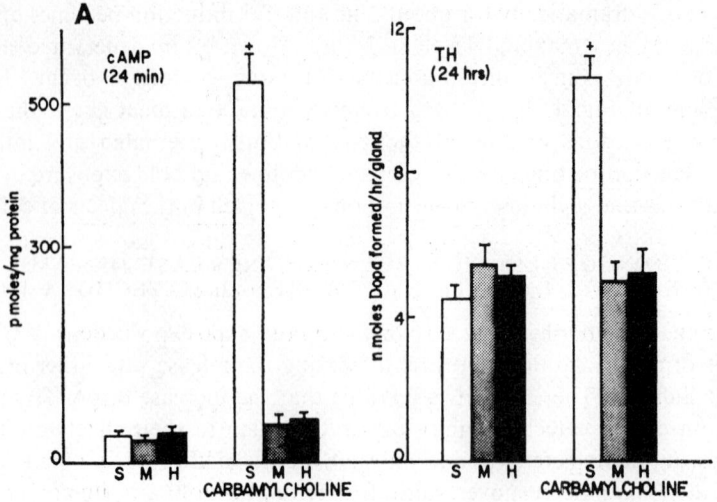

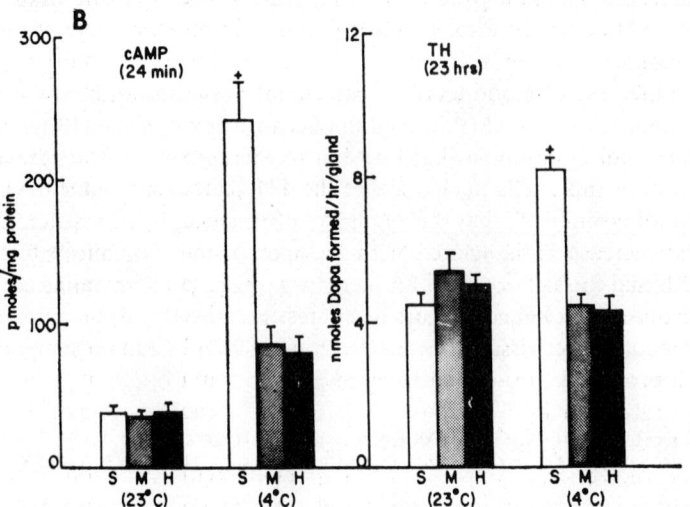

Fig. 2.—Mecamylamine and hexamethonium pretreatment on cAMP concentrations and tyrosine hydroxylase activity (TH) in adrenal medulla of rats.
(A) Receiving carbamylcholine (8·2 μmoles/kg, i.p.)
(B) Exposed to 4°C
cAMP was measured 24 min after either carbamylcholine injection or the beginning of exposure to 4°C. TH was measured 24 hr after carbamylcholine injection or 23 hr after the termination of exposure to 4·C.
(S) Saline; (M) mecamylamine (15 μmoles/kg, i.p.); (H) hexamethonium (45 μmoles/kg, i.p.) were injected 15 min before exposure to 4°C or before carbamylcholine injection. Each column is the mean of 5 experiments.
Vertical bars indicate standard error.
† $P < 0.01$ when compared with the animals pretreated with mecamylamine or hexamethonium.

TABLE 1. TYROSINE HYDROXYLASE ACTIVITY (TH) IN ADRENAL
GLANDS OF RATS TREATED WITH HEXAMETHONIUM (45 μmoles/kg.
i.p.) 2 HOURS AFTER EXPOSURE TO 4°C FOR 1 HR

Treatment	TH (nmoles DOPA formed/hr/gland)
Saline	4·5 ± 0·3
Cold + saline	6·7 ± 0·6*
Cold + hexamethonium	6·8 ± 0·8*

TH was measured 24 hr after the beginning of exposure to 4°C
Each value represents the mean ± S.E. of 5 experiments.
   * $P < 0.05$ when compared with saline treated animals.

Table 1. When the drug is injected 2 hr after cold exposure it does not prevent TH induction.

INVOLVEMENT OF cAMP/cGMP CONCENTRATION RATIOS IN TH INDUCTION

The results described so far strongly implicate cAMP as a promotor of new enzyme protein synthesis in target cells. This implication is in line with a series of observations both in mammalian cells (LANGAM, 1970) and gram negative bacteria (PASTAN and PERLMAN, 1972) providing evidence that intracellular changes of cAMP concentrations are the initial event to promote the synthesis rate of inducible enzyme. However, before concluding that cAMP is the only intracellular second messenger that could be implicated in TH induction, we studied whether changes in medullary guanosine '3,5'-monophosphate (cGMP) concentration could also be implicated in TH induction.

Studies on brain, smooth muscles and heart (FERRENDELLI et al., 1970; GEORGE et al., 1970; LEE et al., 1972). suggest that the occupancy of cholinergic or adrenergic receptors by an agonist changes the cAMP/cGMP concentration ratios. A proposed generalisation is that occupancy of a muscarinic receptor by an agonist lowers the cAMP/cGMP concentration ratios, occupancy of a β-adrenergic receptor by an agonist increases the cAMP/cGMP concentration ratios. Current understanding on the involvement of cyclic nucleotides in adrenergic and cholinergic receptor stimulation involves a change in the cAMP/cGMP ratio as an important event to transduce pertinent extracellular information to a variety of intracellular enzymes, the protein kinases being currently the most widely considered.

We have studied the changes of medullary cGMP concentrations at various times during exposure of rats at 4°C. We found that the concentrations of cGMP progressively decline during cold exposure to a minimal level which is about 30 per cent of that prior to cold exposure.

We report in Table 2 the concentrations in adrenal medulla of cAMP and cGMP at 24 min after various stimuli inducing TH activity. It can be seen that three different inducing stimuli all cause an increase of the cAMP/cGMP concentration ratios greater than 7-fold. However, the effect of the various stimuli on medullary cGMP concentrations is not uniform; exposure to 4°C decreases cGMP concentrations, whereas aminophylline and carbamylcholine increase its concentrations. We have not yet obtained an insight into the mechanisms involved, a regulation through an action on phosphodiesterases can not be excluded for aminophylline and carbamylcholine but could be ruled out for exposure to 4°C. Since the latter represents a physiological event we are inclined to suggest that phosphodiesterase may not be

Table 2. cAMP and cGMP concentration ratios in adrenal medulla of rats after various stimuli inducing TH

| | Response at 24 min | | | |
| | cAMP | cGMP | | |
Stimulus	(pmoles/mg protein)		cAMP/cGMP Ratio	Response at 24 hr (TH induction)
None	39 ± 3.2	3.4 ± 0.41	11	—
4°C	$250 \pm 37^*$	$1.1 \pm 0.35^*$	227	yes
Aminophylline (200 μmoles/kg, i.p.)	$420 \pm 50^*$	$6.1 \pm 0.8^*$	69	yes
Carbamylcholine (5.4 μmoles/kg, i.p.)	$650 \pm 70^*$	$8.6 \pm 1.4^*$	76	yes
Isoproterenol (138 μmoles/kg, i.p.)	32 ± 4.3	2.9 ± 0.4	11	no

* $P < 0.05$ when compared with untreated rats.
cAMP and cGMP were measured in the same tissue sample 24 min after application of the stimulus.
Each point represents the mean value $\pm$ s.e. of 4 determinations.
cAMP was measured according to Guidotti et al. (Mol. Pharmacol. 8, 521, 1972).
cGMP was measured according to Kuo et al. (J. Biol. Chem. 247, 16, 1972).

involved in the regulation of the synaptic transaction. The data reported in Table 2 also suggest that the turnover rate of cAMP may be faster than that of cGMP but this difference is not proved by our experiments. It is interesting that isoproterenol does not change the cAMP/cGMP concentration ratios therefore, excluding a role of β-adrenergic receptors in the second messenger accumulation in adrenal medulla. Since isoproterenol causes hypotension we would be inclined to exclude that this change per se causes an induction of TH. However, we have not tested whether isoproterenol per se inhibits cholinergic receptor response, in adrenal medulla.

In conclusion, our results suggest the involvement of cAMP/cGMP concentration ratios in the cholinergic nicotinic synapses of adrenal medulla. This finding raises a number of questions in the mechanisms of permanent imprinting of synaptic events in target cells which are now being investigated.

REFERENCES

Axelrod J. (1972) Pharmacol. Rev. 24, 233.
Axelrod J. (1971) Science 173, 598.
Costa E. and Guidotti A. (1973) In: New Concepts in Neurotransmitter Regulation (Ed. A. J. Mandell) Plenum Press, New York. p. 135.
Duglas W. and Rubin R. P. (1961) Nature, Lond. 192, 1087.
Guidotti A. and Costa E. (1973) Science 179, 902.
Guidotti A., Zivkovic B., Pfeiffer R. and Costa E. (1973) Naunyn-Schmiedeberg's Arch. Pharmacol. 278, 195.
Guidotti A., Weiss B. and Costa E. (1972) Mol. Pharmacol. 8, 521.
Ferrendelli J., Steiner A. L., Mcdougal D. R. and Kipnis D. M. (1970) Biochem. Biophys. Res. Comm. 41, 1061.
George W. J., Polson J. B., O'Toole A. G. and Goldberg N. D. (1970) Proc. Nat. Acad. Science U.S.A. 66, 398.
Langam T. A. (1970) Role of Cyclic AMP in Cell Function (Greengard P. and Costa E. Eds.) Advances in Biochem. Psychopharmacol. 3, 307.
Molinoff P. B., Brimijoin W. S., Weinshilboum R. M. and Axelrod J. (1970) Proc. Nat. Acad. Sci. U.S.A. 66, 453.
Muller R. A., Thoenen H., and Axelrod J. (1969) Science 163, 468.
Pastan I. and Perlman R. L. (1972) Adv. in Cyclic Nucleotides Research Vol. 1, p 11. Raven Press, New York.
Lee T.-P., Kuo J. K. and Greengard P. (1972) Proc. Nat. Acad. Sci. U.S.A. 69, 3287.
Thoenen H., Mueller R. A. and Axelrod J. (1969) J. Pharmacol. exp. Ther. 169, 249.
Thoenen H. (1970): Nature, Lond. 228, 861.

Frontiers in Catecholamine Research 1973, p. 237. Pergamon Press. Printed in Great Britain.

COMMENT—CHOLINERGIC REGULATION OF TYROSINE HYDROXYLASE

J. S. RICHARDSON

N.I.M.H., Bethesda, Md., U.S.A

IN THE brain, there is a cholinergic modulation of adrenergic activity that appears to be due to a cholinergic action on tyrosine hydroxylase (TH) (EC 1.14.3.a). KAZIC *et al.*, 1973 demonstrated that cholinergic activation by i.v. injection of the acetylcholinesterase inhibitor physostigmine decreases the levels of endogenous norepinephrine and increases the synthesis of [^{14}C]norepinephrine from [^{14}C]tyrosine. We have subsequently demonstrated that shortly after the i.v. injection of physostigmine (200 μg/kg) into rats, TH activity in the hypothalamus is decreased: 10, 15 and 30 min after the injection, TH is reduced by 20–30 per cent. However, this reduction is not yet apparent five minutes after injection and by 60 min TH activity has returned to normal. Dopamine-β-hydroxylase (EC 1.14.2.1) and dopa decarboxylase (EC 4.1.26) activities are not altered at any time point studied. Although physostigmine also reduces TH activity in the pons medulla (30 per cent reduction after 15 min), TH in the neostriatum and in the adrenals is not affected. It is unlikely that physostigmine or a metabolite has a direct action on TH. Furthermore, in the presence of physostigmine (10 μg/100 μl.) hypothalamic TH *in vitro* is not affected. Cholinergic activation by the irreversible acetylcholinesterase inhibitor diisopropylfluorophosphate also produces a 20 per cent decrease in TH activity after intraperitoneal injection (1 mg/kg in salad oil). However, 15 min after cholinergic activation by carbachol, a compound said to have nicotinic receptor-stimulating properties in the central nervous system, TH activity was unaltered. While the mechanism for the decrease in TH activity shortly after administration of physostigmine is unknown, this cholinergic inhibition of adrenergic activity could explain the rapid transient improvement of manic patients treated with i.v. applied physostigmine reported by JANOWSKY *et al.*, 1972).

REFERENCES

JANOWSKY D. S., EL-YOUSEF M. K., DAVIS J. M., HUBBARD B. and SEKERKE H. J. (1972) *Lancet* **i**, 1236–1237.

KAZIC T., RICHARDSON J. S. and KOPIN I. J. (1973) *J. Pharmacol Exp. Ther.*, In press.

Frontiers in Catecholamine Research 1973, pp. 239 to 244. Pergamon Press. Printed in Great Britain.

SYMPATHETIC GANGLIA IN CULTURE

Irwin J. Kopin, Kenneth R. Berv and Jerry G. Webb
Laboratory of Clinical Science, National Institute of Mental Health,
Bethesda, Maryland 20014, U.S.A.

The extraordinary anatomical complexity of the nervous system has compelled investigators interested in unravelling the fundamental processes which govern its development and function to seek simpler systems. The neurones of the sympathetic nervous system are located mainly in the paravertebral ganglia and are innervated by cholinergic neurones from the spinal cord. Sympathetic ganglia are easily accessible, relatively simple nervous structures which contain a limited number of synapses. Adult as well as embryonic ganglia have been shown to survive *in vitro* and have provided a useful model in which to study the development of neurones and to examine some of the factors which control neuronal function.

Larrabee (1970) has reviewed the wide variety of metabolic studies which have been carried out on superior cervical ganglia excised from rats, but, up to the time of his review, relatively few studies on the well-known processes of synthesis, storage, uptake, and release of catecholamines had been reported, although a wealth of information was available on growth of ganglia from embryos, neonates and the effects of nerve growth factor (Levi-Montalcini and Angeletti, 1968) on their survival and growth *in vitro* and *in vivo*. It is the purpose of this review to relate morphological changes in excised sympathetic ganglia maintained in artificial media to alterations in the processes for synthesis, uptake and storage, and release of norepinephrine (NE).

MORPHOLOGICAL CHANGES IN EXCISED GANGLIA MAINTAINED IN ORGAN CULTURE

Freshly excised ganglia contain terminals of preganglionic cholinergic neurones, cell bodies of sympathetic neurones whose axonal processes extend into the post ganglionic nerve trunk or possibly on other ganglion cells and small numbers of other catecholamine-containing cells some of which are intensely fluorescent (the small intensely fluorescent SIF cells first described by Eränkö and Härkönen, 1963). In addition, there are a variety of non-neuronal supporting elements (e.g., Schwann cells, glia). In a recent careful study of neurones in cultured sympathetic ganglia from neonatal rodents and chick embryos, Chamley et al., (1972) distinguished two types of neurones. Most of the neurones, which they called Type II, were relatively large, egg-shaped cells with eccentric round clear nuclei, abundant clear cytoplasm, and one or more processes. The catecholamine fluorescence of these cells was usually low and always evenly distributed. About 5–10 per cent of the cells were small, with scant cytoplasm, and variable intensity of catecholamine fluorescence. These small Type I cells were most outstanding because of their ability to migrate out of the cultured ganglion. Type I neurones are multipolar and differ from the uni- or bi-polar SIF cells which do not migrate. The presence of fluorescent varicosities in the ganglia (Jacobowitz, 1970), indicative of nerve terminals, may provide anatomical evidence in support of functional evidence for dopaminergic interneurones (Libet, 1970).

Nerve fibres begin to grow out of excised ganglia in a few hours and are clearly apparent within a day. The structure of the newly grown fibres from ganglia *in vitro* is similar to sympathetic nerve fibres *in vivo* (PICK, 1970, CHAMLEY *et al.*, 1972). The observation by LEVI-MONTALCINI (1953) that fragments of sarcoma cultured in proximity to sensory or sympathetic ganglia excised from 7- to 9-day old chick embryos resulted in production of a dense halo of nerve fibre outgrowth led to the isolation and characterization of Nerve Growth Factor (NGF). The historical aspects and current status of research on nerve growth factors from several sources has been the subject of a recent monograph (ZAIMIS and KNIGHT, 1972).

Changes which occur in ganglia maintained *in vitro* result from severing of the pre- and post-ganglionic axons and from alterations in the fluid environment (nutritional status, capsule removal, etc.). Severing of the preganglionic nerve axons results in degeneration of their nerve terminals. Most of the neurones in excised ganglion have their axons cut and undergo the well-known morphological and metabolic changes consequent to axonotomy. *In vivo*, severed axons are repaired by a process of regeneration of the nerve fibre leading to reinnervation of the end organs. *In vitro*, fibres which grow out of excised sympathetic ganglia grow into and ramify amongst cells of explanted sympathetically innervated tissues. This phenomenon was first observed in cultures containing cells from embryonic chick sympathetic ganglia and heart (MASUROVSKY and BENITEZ, 1967). Fibres from adult rat superior cervical ganglia transplanted into the anterior chamber of the eye (OLSON and MALMFORS, 1970) or placed in contact with rat iris or pineal glands in organ culture (SILBERSTEIN *et al.*, 1971) also grow into the sympathetically innervated tissue. The rate and extent of such ramifications appears to depend on NGF. The process of ramification does not appear to be species-specific since fibres from mouse ganglia grow into rat or guinea pig iris and fibres from rat ganglia penetrate mouse iris. (SILBERSTEIN *et al.*, 1972a).

Using histochemical methods for a variety of enzymes, (HERMETET *et al.*, 1970) showed that sympathetic neurones in culture maintain carboxylic esterases, dehydrogenases (succinic, glutamic, lactic and alcohol), and monoamine oxidase. NGF was found to promote maintenance of mitochondrial enzymes and acetylcholinesterase. The continued presence of catecholamines is evident from the specific fluorescence found in the neurones and in the fibres growing out of the ganglia (SILBERSTEIN *et al.*, 1971, CHAMLEY *et al.*, 1972). These observations suggest that formation of the enzymes which synthesize catecholamines and of the vesicles which store the neurotransmitter must continue *in vitro*. Excised sympathetic ganglia, therefore, appeared to provide a useful preparation in which to study development, regeneration and function of adrenergic neurones.

UPTAKE OF NOREPINEPHRINE BY SYMPATHETIC NEURONES MAINTAINED *IN VITRO*

The amine uptake mechanism in the membrane of sympathetic nerves is present in embryonic sympathetic ganglionic neurones grown in culture (BURDMAN, 1968; GOLDSTEIN, 1967; ENGLAND and GOLDSTEIN, 1969). Superior cervical ganglia of adult rats incubated *in vitro* in media containing ^{3}H-NE rapidly accumulate the amine (SILBERSTEIN *et al.*, 1972c; HANBAUER *et al.*, 1972). The rate of uptake of NE by the ganglia is increased significantly after only four hours in culture and by two days is increased 4- to 6-fold. The enhanced uptake of NE coincides with the outgrowth of multiple axonal sprouts. Electron-microscopic radioautography of cultured ganglia incubated with ^{3}H-NE proved that the axonal sprouts were the site of uptake of most of the labelled amine (SILBERSTEIN *et al.*, 1972c). Reserpine blocked accumulation of ^{3}H-NE in both fresh and cultured ganglia, but did not block the uptake of ^{3}H-metaraminol. Dense core vesicles are present in the axonal sprouts of the neurones in culture and presumably are the site of reserpine-sensitive amine storage. Culture in media containing either colchicine or vinblastine resulted in diminished uptake of ^{3}H-NE by the ganglia, presumably because interference with neurotubular protein aggregation suppresses the formation of axonal sprouts.

FISCHER and SNYDER (1965) showed that uptake of ^{3}H-NE by sympathetic ganglia *in vivo* was less sensitive to inhibition by cocaine and by tyramine than was uptake of the amine by sympathetic nerve endings. HANBAUER *et al.*, (1972) compared uptake of tritiated norepinephrine by fresh and cultured ganglia with that by sympathetic nerve endings in freshly excised irises. The uptake of ^{3}H-NE by axonal sprouts was assumed to be the difference between the accumulation of tritium by cultured and by fresh ganglia. Uptake of ^{3}H-NE by the axonal sprouts was found to be about as sensitive to inhibition by cocaine, metaraminol, desipramine, and phenoxybenzamine as was uptake of the amine by sympathetic nerve endings in the iris. Uptake of the catecholamine by fresh ganglia was considerably less sensitive to inhibition by these drugs. Thus, the pharmacological characteristics of the amine uptake mechanism in axonal sprouts resembles more closely sympathetic nerve endings than sympathetic cell bodies. Furthermore, the K_m ($1 \cdot 9 \times 10^{-6}$ M) for NE uptake by the axonal sprouts was not significantly different from that of sympathetic nerve endings, but was lower than the K_m ($8 \cdot 0 \times 10^{-6}$) of NE uptake into freshly removed ganglia.

RELEASE OF NOREPINEPHRINE FROM SYMPATHETIC GANGLIA IN CULTURE

The presence of storage vesicles in axonal sprouts and the similarity of the pharmacological and kinetic properties of the amine uptake mechanism in the sprouts to those of the uptake mechanism in sympathetic nerve endings prompted an examination of the possibility that the axonal sprouts might also release NE in response to depolarization. VOGEL *et al.*, (1972) showed that after incubation with ^{3}H-NE, electrical field stimulation of cultured superior cervical ganglia elicited release of the labelled amine, but little release could be obtained from freshly excised ganglia. Release of the transmitter could be produced by depolarisation with elevated levels of potassium (50 mM) as well as by electrical field stimulation. As at sympathetic nerve endings, stimulation-induced release of ^{3}H-NE from cultured ganglia was calcium-dependent. Bretylium blocks release of ^{3}H-NE from sympathetic nerve endings in the cat spleen (HERTTING *et al.*, 1962) and also reduced stimulation-induced release of the amine from the axonal sprouts.

Phenoxybenzamine increases the amount of norepinephrine released in response to sympathetic nerve stimulation and has been hypothesized to block uptake at the receptor, to inhibit reuptake into sympathetic nerve endings, or to enhance release of the neurotransmitter from the nerve ending. Since the drug also increases release of dopamine-β-hydroxylase, a protein constituent of the synaptic vesicle, it appears likely that phenoxybenzamine enhances exocytosis (JOHNSON *et al.*, 1971). Stimulation-induced release of ^{3}H-NE from axonal sprouts of cultured ganglia is also enhanced by phenoxybenzamine. Since there are no post-synaptic sites, the action of the drug must be on the axonal sprouts, presumably prolonging or facilitating exocytosis.

The results of the studies on uptake and release of norepinephrine from cultured ganglia suggest that axonal sprouts, almost as soon as they are formed, have the properties of sympathetic nerve terminals.

CATECHOLAMINE-SYNTHESIZING ENZYMES IN ADRENERGIC TISSUE IN ORGAN CULTURE

It is well established that a variety of procedures which increase nerve impulse activity to the adrenal medulla or to sympathetic ganglia elevate levels of tyrosine hydroxylase (Dairman *et al.*, 1972) and dopamine-β-hydroxylase (Axelrod, 1972). Elevation *in vitro*, of tyrosine hydroxylase by exposure to high, depolarizing concentrations of potassium has been demonstrated in cultures of adrenal medullae (Silberstein *et al.*, 1972d) and cultures of superior cervical ganglia of 6-day or 4-week old mice (Mackay and Iversen, 1972). Dopamine-β-hydroxylase activity is also elevated by maintenance of ganglia in media containing elevated levels of potassium (Silberstein *et al.*, 1972b). The elevation of tyrosine hydroxylase by elevated potassium is blocked by absence of calcium suggesting that release of catecholamines is a prerequisite for induction of the enzyme (Silberstein *et al.*, 1972d).

Kvetnansky *et al.*, (1971) found that administration of dibutyryl cyclic AMP to hypophysectomized rats increased levels of dopamine-β-hydroxylase. McAfee *et al.*, (1971) reported that brief periods of preganglionic nerve stimulation produced a several-fold elevation in the content of cyclic 3',5'-AMP in the postsynaptic cells of rabbit superior cervical ganglia. These observations led several groups of investigators to examine the effects of cyclic AMP and its derivatives on the levels of tyrosine hydroxylase and dopamine-β-hydroxylase in cultured sympathetic ganglia. Mackay and Iversen (1972) found that theophylline significantly enhanced the elevation of tyrosine hydroxylase produced in cultured ganglia by increased levels of potassium and that dibutyryl cyclic AMP was effective in increasing the levels of the enzyme. Similar results have been obtained when dopamine-β-hydroxylase was studied (Keen and McLean, 1972; Silberstein *et al.*, 1972; Webb, Berv and Kopin unpublished observations). The elevations in enzyme activities were prevented by cycloheximide indicating that the increase in enzyme activity represented elevations in the amount of enzyme present.

Kebabian and Greengard (1971) described a dopamine-sensitive adenylcyclase in the bovine sympathetic ganglia and suggested that dopaminergic interneurones, thought to be involved in production of slow inhibitory postsynaptic potentials (see Libet, 1970), might release dopamine at receptor sites which included this adenylcyclase. Atropine, but not hexamethonium, blocks both the preganglionic nerve stimulation-induced increase in cyclic AMP in ganglia (Kebabian and Greengard, 1971) and the inhibitory potential (Libet, 1970). Blockade of nicotine receptors blocks neither the inhibitory potential nor the rise in cyclic AMP, but does prevent the elevation of tyrosine hydroxylase induced by increased neuronal activity (Meuller *et al.*, 1970). Thus, it appears unlikely that the dopaminergic interneurone or the elevation in cyclic AMP produced by preganglionic stimulation is

involved in alterations of catecholamine-synthesizing enzyme activity in the sympathetic neurones.

Levels of dopamine-β-hydroxylase and tyrosine hydroxylase in neurones of
sympathetic ganglia decrease after axonal damage. This may be due to an alteration
in the protein synthetic mechanisms of the cell to produce proteins required for
structure (regeneration and growth) rather than those used to maintain function
(transmitter-synthesizing enzymes). The mechanism for control of growth and
enzyme activities are not known at present, but appropriate use of model systems
in vitro to complement studies *in vivo* may provide some of the answers.

Care must be taken not to equate elevation of the level of an enzyme with induction.
Induction implies a specificity based on alterations of the genetic controls of synthesis
and degradation of the enzyme. Hypertrophy of muscle might be accompanied by an
increase in levels of phosphorylase and reinnervation of a denervated muscle would
result in the appearance of cholineacetylase, but clearly these are not examples of
what is usually meant by enzyme induction.

Ganglia grown with dibutryl cyclic AMP have more tyrosine hydroxylase, more
dopamine-β-hydroxylase, but also take up more ^{3}H-NE (BERV, WEBB and KOPIN,
unpublished observation). The response may be due to non-specific growth rather
than specific enzyme induction.

CONCLUSION

Isolated sympathetic ganglia provide a useful tool for the study of development
and control of neuronal growth and function. The axonal sprouts which grow in
such ganglia resemble sympathetic nerve endings almost as soon as they are formed.
The sprouts synthesize, store, and release NE and the uptake mechanism of the
catecholamine is similar to that found in nerve endings. In studies of enzyme levels,
careful distinction must be made between growth and specific enzyme induction.

REFERENCES

AXELROD J. (1972) *Pharmacol. Rev.* **24**, 233–244.
BURDMAN J. A. (1968) *J. Neurochem.* **15**, 1321–1323.
CHAMLEY J. H., MARK G. E., CAMPBELL G. R. and BURNSTOCK G. (1972) *Z. Zellfursch.* **135**, 287–314.
DAIRMAN W., CHRISTENSON J. G. and UDENFRIEND S. (1972) *Pharmacol. Rev.* **24**, 269–290.
ENGLAND J. M. and GOLDSTEIN M. M. (1969) *J. Cell. Sci.* **4**, 677–691.
ERÄNKÖ O. and HÄRKÖNEN M. (1963) *Acta Physiol. Scand.* **58**, 285–286.
FISCHER J. E. and SNYDER S. (1965) *J. Pharmacol. Exp. Ther.* **150**, 190–195.
GOLDSTEIN M. N. (1967) *Proc. Soc. Exp. Biol. (N.Y.)* **125**, 993–996.
HANBAUER I., JOHNSON D. G., SILBERSTEIN S. D. and KOPIN I. J. (1972) *Neuropharmacol.* **11**, 857–862.
HERMETET J. C., TRESKA J. and MANDEL P. (1970) *Histochemie* **22**, 177–186.
HERTTING B., AXELROD J. and PATRICK R. W. (1962) *Br. J. Pharmacol. Chemother* **18**, 161–166.
JACOBOWITZ D. (1970) *Fed. Proc.* **29**, 1929–1944.
JOHNSON D. G., THOA N. B., WEINSHILBOUM R., AXELROD J. and KOPIN I. J. (1971) *Proc. Nat. Acad. Sci., U.S.A.* **68**, 2227.
KEBABIAN J. W. and GREENGARD P. (1971) *Science* **174**, 1346–1348.
KEEN P. and MCLEAN W. G. (1972) *Naunyn-Schmiedebergs Arch. Pharmacol.* **275**, 465–469.
KVETNANSKY R., GEWIRTZ G. P., WEISE V. K. and KOPIN I. J., (1971) *Endocrinology* **89**, 50–55.
LARRABEE M. G. (1970) *Fed. Proc.* **29**, 1919–1928.
LEVI-MONTALCINI R. and ANGELETTI P. V. (1968) *Physiol. Rev.* **48**, 534–569.
LIBET B. (1970) *Fed. Proc.* **29**, 1945–1956.
MCAFEE D. A., SCHORDERET M. and GREENGARD P. (1971) *Science* **171**, 1156–1158.
MACKAY A. V. P. and IVERSEN L. L. (1972) *Naunyn-Schmiedebergs Arch. Pharmacol.* **272**, 225–229.
MASUROVSKY E. G. and BENITEZ H. H. (1967) *Anat. Rec.* **157**, 285.
MUELLER R. A., THOENEN H., and AXELROD J. (1970) *Europ. J. Pharmacol.* **10**, 51–56.

Olson L. and Malmfors T. (1970) *Acta Physiol. Scand.* **348,** suppl. 1–11.

Pick J. (1970) *The Autonomic Nervous System. Morphological, Comparative, Clinical and Surgical Aspects.* J. B. Lippincott, Philadelphia-Toronto.

Silberstein S. D., Johnson D. G., Jacobowitz D. M. and Kopin I. J. (1971) *Proc. Nat. Acad. Sci. U.S.A.* **68,** 1121–1124.

Silberstein S. D., Berv K. R. and Jacobowitz D. M. (1972a) *Nature, Lond.* **239,** 466–468.

Silberstein S. D., Brimijoin W. S., Molinoff P. B. and Lemberger L. (1972b) *J. Neurochem.* **19,** 919–921.

Silberstein S. D., Johnson D. G., Hanbauer I., Bloom F. E. and Kopin I. J. (1972c) *Proc. Nat. Acad. Sci. U.S.A.* **69,** 1450–1454.

Silberstein S. D., Lemberger L., Klein D. C., Axelrod J. and Kopin I. J. (1972d) *Neuropharmacology* **11,** 721–726.

Vogel S. A., Silberstein S. D., Berv K. R. and Kopin I. J. (1972) *Europ. J. Pharmacol.* **20,** 308–311.

Zaimis E. and Knight J. (1972) *Nerve Growth Factor and its Antiserum.* The Athlone Press, New York.

Frontiers in Catecholamine Research 1973, pp. 245 to 248. Pergamon Press. Printed in Great Britain.

LACK OF RELATIONSHIP BETWEEN NORADRENALINE CONTENT AND TYROSINE HYDROXYLASE ACTIVITY IN CULTURED SYMPATHETIC GANGLIA

A. V. P. Mackay and L. L. Iversen

MRC Neurochemical Pharmacology Unit, Department of Pharmacology, Medical School, Cambridge CB2 2QD, Great Britain

The biosynthetic pathway for the sympathetic transmitter noradrenaline has provided a useful model upon which to investigate the way in which nerve cells may react to environmental stimuli by changes in neurotransmitter synthesis.

Changes in the rate of synthesis of noradrenaline are probably brought about by two distinct mechanisms. Both of these mechanisms are thought to act at the level of the initial step in the biosynthetic sequence, the overall rate-determining enzyme tyrosine hydroxylase (T-OH). The first type of regulation was described by Udenfriend and Dairman (1971) and involves minute to minute variations in tyrosine hydroxylase activity according to the prevailing activity in the sympathetic nervous system. These acute changes in enzyme activity result from intracellular feed-back inhibition of tyrosine hydroxylase by the free cytoplasmic product noradrenaline (NA). The second mechanism is of long-term significance and operates to modify the actual levels of enzyme in response to sustained changes in sympathetic nervous activity. It is well established that trans-synaptic impulse traffic can influence the synthesis of the enzyme T-OH in sympathetic ganglia and adrenal medulla (Molinoff and Axelrod, 1971). However, the precise mechanism by which long-term changes in trans-synaptic traffic regulate synthesis of the enzyme in sympathetic neurones is still obscure.

Silberstein and his colleagues (Silberstein et al., 1972) have recently reported an inverse relationship between T-OH activity and catecholamine content of cultured rat adrenal glands exposed to a prolonged depolarising stimulus in the form of a raised extracellular potassium concentration. We have investigated the effects of long-term environmental manipulations on the T-OH activity and NA content of cultured mouse superior cervical sympathetic ganglia.

Superior cervical ganglia were excised from mice of either 6 or 28 days of age and maintained in organ culture (Mackay and Iversen, 1972a) for 48 hrs. After culture the T-OH activity and NA content of individual ganglia were estimated by sensitive radiochemical techniques (Hendry and Iversen, 1971; Jarrott and Iversen, 1971). We have already established that ganglia from neonatal mice cultured for 48 hr in the presence of a depolarising stimulus in the form of a 10-fold increase in the potassium concentration of the culture medium exhibit a significant rise in their T-OH activity (Mackay and Iversen, 1972a). Figure 1 shows the effects on T-OH activity and NA content of superior cervical ganglia taken from adult (28-day-old) mice and maintained in organ culture for 48 hr in the presence of high potassium and reserpine (25 μM). The potassium stimulus resulted in significant rises in both T-OH activity and NA content. Reserpine had no significant effect on

245

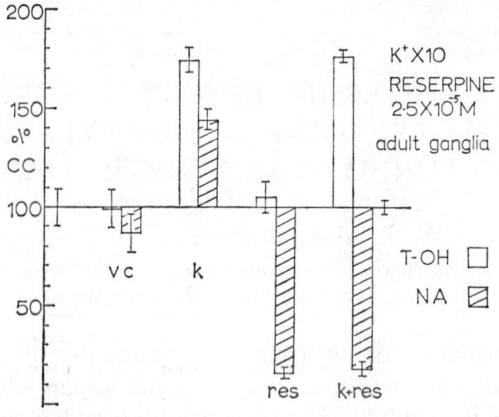

Fig. 1.—Tyrosine hydroxylase activity (T–OH; open bars) and noradrenaline content (NA; hatched bars) of individual superior servical ganglia taken from 28-day-old mice and cultured for 48 hr in the presence of a 10-fold increase in the potassium concentration of the culture medium (K) and reserpine (res; $2.5 \cdot 10^{-5}$ M). Results are means $\pm$ s.e.m. for groups of 6–12 ganglia and are expressed as percentages of the culture control values (100 per cent) obtained from ganglia cultured under normal conditions. VC refers to the T–OH activity (8·5 pmoles dopa formed per hr per ganglion) and NA content (4·2 ng per ganglion) of freshly excised ganglia.

T-OH activity but produced a virtually complete depletion of ganglionic NA Exposure of cultured ganglia to both high potassium and reserpine simultaneously resulted in the usual potassium-induced rise in T-OH activity with severe depletion of NA. So it appeared that responses of cultured superior cervical ganglia to a prolonged depolarising stimulus revealed no constant relationship between the ganglionic content of neurotransmitter and the activity of its biosynthetic enzyme. This lack of relationship was also apparent in ganglia taken from neonatal (6-day-old) mice and exposed to various environmental manipulations over a 48 hr culture period (Table 1). It can be seen from Table 1 that the effects of high potassium and reserpine (25 μM) on the T-OH activity and NA content of neonatal ganglia were very similar to those observed with adult ganglia. In addition, ouabain (1 mM) another form of depolarising stimulus produced a 70 per cent rise in T-OH activity and a 70 per cent fall in NA content. The monoamine oxidase inhibitor clorgyline (1 μM) had no significant effect on T-OH activity but produced a 65 per cent rise in NA content. The T-OH inhibitor α-methyl-p-tyrosine (25 μM) caused a 50 per cent reduction in ganglionic NA, but had no significant effect on T-OH activity, assayed *in vitro* after thorough washing of the cultured ganglia. We have recently reported a rise in the T-OH activity of superior cervical ganglia from neonatal mice cultured in the presence of dibutyryl cyclic AMP (Mackay and Iversen, 1972b), and the results in Table 1 show that this rise in enzyme activity is associated with a doubling of ganglionic NA content.

The increases in ganglionic T-OH activity observed in the presence of high potassium, ouabain and dibutyryl cyclic AMP could all be prevented by the protein synthesis inhibitor cycloheximide at a concentration of 2 μg/ml in the culture medium. At this concentration cycloheximide alone had no significant effect on T-OH activity. Thus these increases in T-OH activity would seem to result from synthesis of new

TABLE 1

	T-OH	NA
CC	100 ± 5	100 ± 7
× 10 K$^+$	180 ± 11	200 ± 4
Reserpine (25 μM)	95 ± 12	5 ± 5
Reserpine (25 μM) + × 10 K$^+$	185 ± 9	7 ± 4
Ouabain (1 mM)	170 ± 4	30 ± 8
Ouabain (1 mM) + × 10 K$^+$	200 ± 8	98 ± 11
Clorgyline (1 μM)	110 ± 7	165 ± 10
α-Me-*para*-tyrosine (20 μM)	110 ± 5	55 ± 10
dibutyryl cyclic AMP (1 mM)	170 ± 6	205 ± 5

Tyrosine hydroxylase (T-OH) activity and noradrenaline (NA) content of individual superior cervical ganglia taken from 6-day-old mice and cultured for 48 hr. Results are means ± s.e.m. for groups of 6–12 ganglia and are expressed as percentages of the culture control (CC) values obtained from ganglia cultured under normal conditions. The control level of T-OH activity was 2·1 pmoles dopa formed per hr per ganglion, and control NA content was 1·7 ng per ganglion. Ganglia were exposed to high potassium (× 10 K$^+$) and drugs for the entire 48-hr culture period.

enzyme protein and thus involve a long-term regulatory mechanism. Our observations that cultured mouse sympathetic ganglia can undergo changes in the rate of synthesis of T-OH quite independently of their ganglionic NA content suggest that NA plays no role in the long term control of T-OH activity in these organs.

A number of workers have reported rises in the cyclic AMP content of various sorts of nervous tissue in response to depolarising stimuli (KAKUICHI et al., 1969; MACAFEE et al., 1971; RALL, 1972). Our own observations (MACKAY and IVERSEN, 1972b) on the effect of this cyclic nucleotide on T-OH activity in cultured sympathetic ganglia and the work of GUIDOTTI and COSTA (1973) who observed correlative rises in the cyclic AMP content and T-OH activity of rat adrenal glands seem to implicate cyclic AMP as a mediator of the long term regulation of T-OH activity in sympathetic ganglia. In recent experiments we have attempted to determine whether the T-OH responses of cultured sympathetic ganglia to both a depolarising stimulus (high potassium) and dibutyryl cyclic AMP operate directly at a ribosomal level to control the rate of transcription of T-OH molecules, or whether the responses are dependent upon nuclear-directed synthesis of messenger RNA. The peptidyl transferase inhibitors anisomycin and sparsomycin inhibited responses to both sorts of stimulus. Responses to both high potassium and dibutyryl cyclic AMP were also abolished in the presence of actinomycin D which specifically inhibits the nuclear synthesis of messenger RNA from DNA templates.

We believe that trans-synaptic events in the form of increased depolarisation can result in a rise in the cyclic AMP content of sympathetic ganglion cells with consequent stimulation of nuclear directed synthesis of tyrosine hydroxylase protein.

REFERENCES

GUIDOTTI A. and COSTA E. (1973) *Science* **179**, 902–904.
HENDRY I. A. and IVERSEN L. L. (1971) *Brain Res.* **29**, 159–162.
JARROTT B. and IVERSEN L. L. (1971) *Biochem. Pharmacol.* **19**, 1841–1843.
KAKUICHI S., RALL T. W. and McIIWAIN J. (1969) *J. Neurochem.* **16**, 485–491.
MACAFEE D. A., SCHORDORET M. and GREENGARD P. (1971) *Science* **171**, 1156–1158.
MACKAY A. V. P. and IVERSEN L. L. (1972a) *N.S. Archiv. Pharmacol.* **272**, 225–229.

MACKAY A. V. P. and IVERSEN L. L. (1972b) *Brain. Res.* **4,** 424–426.
MOLINOFF P. B. and AXELROD J. (1971) *Ann. Rev. Biochem.* **40,** 465–500.
RALL T. W. (1972) *Pharmacol. Rev.* **24,** 399–409.
SILBERSTEIN S. D., LEMBERGER L., KLEIN D. C., AXELROD J. and KOPIN J. (1972) *Neuropharmacol.* **11,** 721–726.
UDENFRIEND S. and DAIRMAN W. (1971) *Adv. Enzyme Regulation* **9,** 145–165.

Frontiers in Catecholamine Research 1973, pp. 249 to 251. Pergamon Press. Printed in Great Britain.

INDUCTION OF TYROSINE HYDROXYLASE IN SUPERIOR CERVICAL GANGLION BY COLD EXPOSURE: EFFECTS OF DECENTRALISATION AND MECAMYLAMINE

INGEBORG HANBAUER

Laboratory of Clinical Science, National Institute of Mental Health,
Bethesda, Maryland 20014, U.S.A.

WE HAVE previously reported that adrenal demedullation facilitates the increase of tyrosine hydroxylase (TH) activity in superior cervical ganglion (SCG) of rats exposed to 4°C (HANBAUER et al., 1973). The aim of this presentation is to demonstrate that during TH induction two phases can be differentiated: (1) the stimulus duration (measured as time of exposure to 4°C) and (2) the time interval between termination of the stimulus and the increase of TH activity. As shown in Table 1, 4 hr at 4°C is the stimulus threshold and the time interval necessary for the increase in TH activity to appear is longer than 20 hr and shorter than 44 hr. It was of interest to examine whether both phases are regulated transynaptically and, if they are, whether either nicotinic or muscarinic receptors are involved. In Table 2, data are reported showing the effects of decentralisation of the TH induction in SCG. Decentralisation inhibits the increase of TH activity when performed at the termination of stimulus or 24 hr later, but it does not reduce TH activity when performed 48 hr after stimulation. The preganglionic nerve fibres are cholinergic, and the function of sympathetic ganglia is regulated by muscarinic and nicotinic receptors. Therefore, from the data in Table 2, it can be inferred that the availability of nicotinic and muscarinic receptors is essential during the second phase of transynaptic induction of TH. THOENEN et al. (1969) have reported that decentralisation before stimulation inhibits the increase of TH activity. The results in Table 3 are consistent with the view that the function of nicotinic receptors is required in phase I but may not be essential for the biochemical processes which are involved in phase II of TH induction. Mecamylamine prevents the increase in TH activity when injected only once before cold exposure, but does not interfere with TH induction when injected several times after termination of stimulus.

In adrenal medulla also, two phases of TH induction can be distinguished. The first phase seems to involve an increase of cyclic-3′,5′-adenosine monophosphate (c-AMP) (GUIDOTTI and COSTA, 1973) and a decrease of cyclic-3′,5′-guanosine monophosphate (c-GMP) (GUIDOTTI et al., 1973). In the adrenal medulla, nicotinic receptors regulate the steady-state concentrations of c-AMP. KEBABIAN and GREENGARD (1971) have reported that sympathetic ganglia of the rabbit contain a dopamine-dependent adenylate cyclase. The activation of this enzyme by nerve stimulation can be prevented by muscarinic-receptor blockers. Hence it can be suggested that decentralisation may impair the function of muscarinic receptors. The activation of muscarinic receptors may increase c-AMP concentrations, and this event may be

INGEBORG HANBAUER

TABLE 1. INDUCTION OF TYROSINE HYDROXYLASE ELICITED IN SUPERIOR CERVICAL GANGLIA AFTER VARIOUS DURATIONS OF COLD EXPOSURE (4°C)

Stimulus duration hours at 4°C	Hours after stimulus at 24°C	Tyrosine hydroxylase activity nmoles/superior cervical ganglion/hour
0	0	35 ± 2 (6)
2	46	42 ± 4 (6)
4	44	45 ± 2* (7)
6	42	43 ± 2* (4)
24	24	56 ± 4* (8)
48	0	58 ± 4* (8)
4	20	31 ± 2 (6)

* $P < 0.01$. Number of assays in parentheses. Tyrosine hydroxylase activity was assayed by the method of SHIMAN et al. (1971).

TABLE 2. EFFECT OF DECENTRALISATION ON TYROSINE HYDROXYLASE INDUCTION ELICITED IN SYMPATHETIC GANGLIA BY COLD EXPOSURE (4°C)

Hours at 4°C	Hours between end of cold exposure and decentralisation	nmoles dopa/SCG/hour	
		Intact superior cervical ganglion	Decentralised superior cervical ganglion
0	—	28 ± 2·4 (8)	25 ± 1·4 (8)
4	0	47 ± 5·0 (10)*	26 ± 1·8 (10)
4	24	37 ± 3·2 (10)*	22 ± 1·9 (10)
4	48	37 ± 4·0 (5)	39 ± 1·2 (5)

* $0.01 > P > 0.001$. Number of experiments in parentheses. Tyrosine hydroxylase activity in superior cervical ganglion was assayed according to SHIMAN et al. (1971).

TABLE 3. EFFECT OF INJECTION OF MECAMYLAMINE ON TYROSINE HYDROXYLASE INDUCTION ELICITED IN SYMPATHETIC GANGLIA BY COLD EXPOSURE (4°C)

Stimulus duration hours	Mecamylamine (15 μmoles/kg, i.p.)		Tyrosine hydroxylase activity (nmoles dopa/SCG/hour
	Hours before stimulus	Hours after stimulus	
6	0·3	—	20·3 ± 1·4 (6)
6	—	—	30·0 ± 1·4 (8)*
6	—	6, 12, 24, 36	27·0 ± 0·7 (10)*
6	—	24, 36	31·6 ± 2·8 (10)*
—	—	—	20·3 ± 1·4 (10)
—	—	6, 12, 24, 36	18·2 ± 3·1 (5)

* $0.01 < P > 0.005$. Number of experiments in parentheses. In all experiments the animals were killed at 48 hr from the beginning of the experiment. The activity was measured according to SHIMAN et al. (1971).

important in the second phase of transynaptic induction. Alternatively, the difference between the effects of mecamylamine and of decentralisation on transynaptic induction of TH may be due to some other factors transported by cholinergic nerves. We are continuing experiments to clarify further the role of both types of receptors in the induction of TH.

REFERENCES

GUIDOTTI A. and COSTA E. (1973) *Science* **179,** 902–904.

GUIDOTTI A., ZIVKOVIC B, PFEIFFER R. and COSTA E. (1973) *Naunyn-Schmiedebergs Arch. Pharmakol.*, In press.

HANBAUER I., COSTA E. and KOPIN I. J. (1973) *Trans. Am. Soc. Neurochem.* 4th Annual Meeting, 143.

KEBABIAN J. W. and GREENGARD P. (1971) *Science* **174,** 1346–1349.

SHIMAN R., AKINO M. and KAUFMAN S. (1971) *J. Biol. Chem.* **246,** 1330–1340.

THOENEN H., MUELLER R. A. and AXELROD J. (1969) *Nature, Lond.* **221,** 1264

Frontiers in Catecholamine Research 1973, pp. 253 to 259. Pergamon Press. Printed in Great Britain.

REGULATION OF CATECHOL SYNTHESIS IN AN ADRENERGIC CLONE OF MOUSE NEUROBLASTOMA

ELLIOTT RICHELSON

Department of Pharmacology and Experimental Therapeutics and Department of Psychiatry and Behavioral Sciences, The Johns Hopkins University School of Medicine, Baltimore, Maryland 21205, U.S.A.

FROM the mouse neuroblastoma cell culture system, we have derived clones which were examined for tyrosine hydroxylase [L-tyrosine, tetrahydropteridine: oxygen oxidoreductase (3-hydroxylating)] (EC 1.14.3.a), choline acetyltransferase (acetyl-CoA:choline O-acetyltransferase) (EC 2.3.1.6), and acetylcholinesterase (acetyl-choline hydrolase) (EC 3.1.1.7) (AMANO, RICHELSON and NIRENBERG, 1972). Three major types of clones were found (1) adrenergic; (2) cholinergic; and (3) clones that do not synthesise acetylcholine or catechols (inactive). All major types of clones had high levels of acetylcholinesterase and electrically excitable membranes. Except for one inactive clone, all clones were able to form neurites. No clone had high levels of both choline acetyltransferase and tyrosine hydroxylase.

Our results showed that mouse neuroblastoma C1300 is a heterogeneous cell population comprising at least three cell types when classified with respect to neuro-transmitter synthesis. In addition acetylcholinesterase is not a marker for cholinergic cells since it is present in high levels in the adrenergic clones as well as the other types of clones. Finally, these results support the hypothesis that a neuron can synthesise only one neurotransmitter (DALE, 1935).

To answer specific questions about the biosynthesis of catecholamines, we have studied the adrenergic neuroblastoma clone (N1E-115) with the highest level of tyrosine hydroxylase. The activity of tyrosine hydroxylase in crude sonicates of N1E-115 is comparable to that found in lysed granules of the adrenal medulla (SHIMAN et al., 1971) which is regarded as a very rich source of this enzyme. This paper reports our work with N1E-115 on transport of neurotransmitter precursors (RICHELSON and THOMPSON, 1973) and regulation of tyrosine hydroxylase (RICHELSON, 1972; RICHELSON, 1973b).

TRANSPORT OF CATECHOL PRECURSORS INTO THE ADRENERGIC CELL: THE FIRST STEP IN THE BIOSYNTHESIS OF CATECHOLAMINES

Evidence in the literature (SHIMAN et al., 1971) supports the precursor role of phenylalanine and tyrosine in the synthesis of catechols. Since these aromatic amino acids are not synthesised by the adrenergic neuron, the first step in the biosynthesis of catecholamines is the transport of the precursor molecule(s) into the cell. The importance of this first step in the biosynthetic pathway is obvious; for here regulation of the entire pathway might occur. However, very little is known about this important first step.

Many studies involving the transport of amino acids into brain tissue have been reported (GUROFF and UDENFRIEND, 1962; ROBERTS, 1968; LOGAN and SNYDER,

1972; NEIDLE *et al.*, 1973). It has not been possible to learn from these reports about transport into the adrenergic neuron, since neurons comprise about 10 per cent of the total number of cells in brain tissue (BLINKOW and GLEZER, 1968) and adrenergic neurons are still another fraction of that number. Furthermore these reports do not give information about specificity of transport of precursor molecules into neurons which do not synthesise catechols.

With cell culture techniques we "dissected" the brain into four of its many component cell types by using adrenergic and cholinergic neuroblastoma clones as well as glial and fibroblastic cell lines. With these four cell types we studied transport of phenylalanine, tyrosine and choline (precursor to acetylcholine). Differences in thermodynamic and kinetic parameters for transport of precursor compounds into the various cell types might reflect differences in utilisation of that compound within the cell.

The transport studies were facilitated by development of a new assay method using microwell tissue culture plates (RICHELSON, 1973a). To deplete endogenous stores and thereby study initial rates, we "starved" the cultured cells for the particular compound before its assay. We also corrected for non-saturable diffusion (non-specific transport) by using the experimental correction described by DIAMOND and KENNEDY (1969) for synaptosomal transport of choline.

The non-specific transport of [^{14}C]-tyrosine is a linear function of the concentration (Fig. 1a) as expected for a process based upon free diffusion. However, the concentration curve for the specific transport of [^{14}C]-tyrosine (Fig. 1a) describes a section of a rectangular hyperbola, that is, this transport is a saturable process. Using the Lineweaver–Burk transformation of the Michaelis equation (Fig. 1b), the K_t (concentration of substrate at half-maximum velocity of transport) and the V_{max} (maximum velocity of transport) from this experiment were $2 \cdot 51 \pm 0 \cdot 38 \times 10^{-5}$ M and $61 \cdot 9 \pm 6 \cdot 8$ pmol min^{-1}, respectively. The V_{max} per cell was $0 \cdot 53 \pm 0 \cdot 06$ fmol min^{-1} cell^{-1}; and per mg protein, 164 ± 18 pmol min^{-1} mg protein^{-1}.

If affinity for transport is approximated by $1/K_t$, it is clear from the data in Table 1 that cells have the highest apparent affinity for transport of choline which is about the same for each cell tested. Our K_t values for choline transport into all four cell types are in close agreement with those found for the higher affinity transport of choline into synaptosomes (WHITTAKER, 1972; YAMAMURA and SNYDER, 1972). Each cell type also transports tyrosine and phenylalanine with about the same affinity, although K_t values for a given compound may vary at most five-fold between cell types (compare K_t values for phenylalanine in the adrenergic vs. cholinergic clones). For all these compounds tested (Table 1) the affinities for transport would be considered very high for transport into brain slices (NEIDLE *et al.*, 1973) or synaptosomes (LOGAN and SNYDER, 1972; WHITTAKER, 1972; YAMAMURA and SNYDER, 1972). These results, therefore, raise the question of whether the demonstration of the high affinity transport of a compound into nervous tissue is supportive evidence for the role of that compound in synaptic transmission.

LEVITT *et al.* (1965) using the isolated perfused guinea pig heart have demonstrated that the rate of synthesis of norepinephrine is dependent upon the concentration of tyrosine in the perfusate; and this process is saturable. From their data it was determined that the half-maximum rate of norepinephrine synthesis occurred at a tyrosine concentration of about 2×10^{-5} M. The K_m value for the overall process of

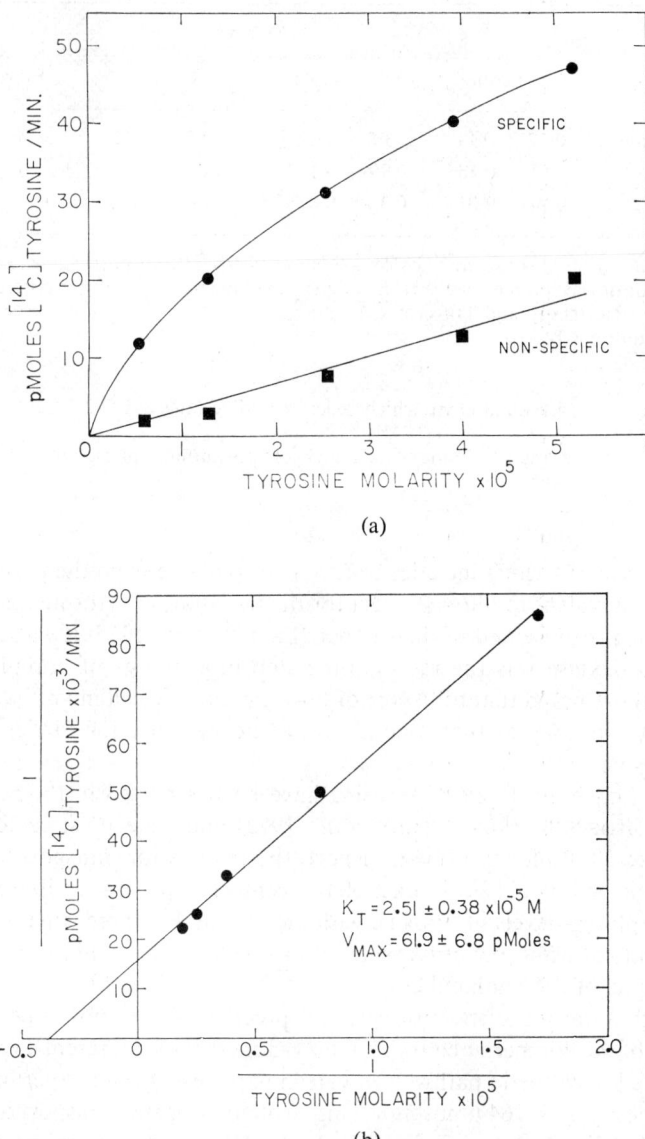

FIG. 1.—(a) Concentration curve for the transport of [¹⁴C]-tyrosine into adrenergic Clone N1E-115. Cells were grown in microwells (Linbro Microwells, Bellco Glass, Inc.) and starved for tyrosine 24 h before assay. The assay technique is described in detail elsewhere (RICHELSON, 1973a). Incubation was at 37° for 4 min. All data represent the averages of duplicate reactions. Specific transport is the calculated difference between total transport of [¹⁴C]-tyrosine into cells minus the transport of [¹⁴C]-tyrosine in the presence of 2·25 mM [¹²C]-tyrosine (non-specific transport). ●, Specific Transport; ■, Non-Specific Transport. (b) Double reciprocal analysis of data presented in (a). (RICHELSON and THOMPSON, 1973.)

TABLE 1. TRANSPORT OF CATECHOLAMINE AND ACETYLCHOLINE PRECURSORS INTO CULTURED CELLS
(K_t VALUES $\pm$ S.E., MOLARITY $\times$ 10^5)*

Precursor	Cell type			
	Adrenergic†	Cholinergic	Glial‡	Fibroblastic
[^{14}C]-phenylalanine	0·77 ± 0·14	3·8 ± 1·6§	0·8 ± 0·22	3·24 ± 0·80‖
[^{14}C]-tyrosine	2·51 ± 0·38	0·79 ± 0·12¶	2·01 ± 0·38	3·16 ± 1·26**
[^{3}H]-choline	0·14 ± 0·01	0·35 ± 0·01§	0·208 ± 0·005	0·33 ± 0·03‖

* Each K_t (concentration of substrate at half-maximum velocity of transport) value represents the weighted computer fit of the specific transport data determined over a range of five different concentrations in duplicate (RICHELSON and THOMPSON, 1973).
† N1E-115, subculture 8-14
‡ C-6, subculture 41-56
§ S-26, subculture 12.
‖ B82, subculture 23-24, a mutant clone which lacks thymidine kinase (EC 2.7.1.21).
¶ S-18, subculture 11.
** A9, subculture 13, a mutant clone which lacks hypoxanthine phosphoribosyltransferase (EC 2.4.2.8).

norepinephrine synthesis which includes transport of tyrosine from the perfusate into the cell is nearly equivalent to our value just for the transport of tyrosine into the cell. From the experiments presented in their paper (LEVITT *et al.*, 1965) it was concluded that tyrosine hydroxylase was the rate-limiting step in synthesis of norepinephrine. Although the authors noted that transport of tyrosine into the cell might be involved, and might be rate-limiting, at that time there was no evidence for carrier-mediated transport of tyrosine.

Since plasma and brain levels of tyrosine have been reported in the range from 4 to 9 $\times$ 10^{-5} M (ROBERTS, 1968; PERRY *et al.*, 1972) under steady-state conditions, with a $K_t = 2\cdot5 \times 10^{-5}$ M for tyrosine transport, the velocity for transport of tyrosine into the adrenergic cell would be at least 60 per cent of V_{max} at 4 $\times$ 10^{-5} M. Thus, with increases in plasma levels of tyrosine with dietary intake, there will not be major increases in the rate of transport of tyrosine into the cell. Such increases might affect the rate of synthesis of catecholamines.

Although high affinity transport of catechol precursors does not appear to be a specific property of the adrenergic cell type, it may, nonetheless, represent a very important aspect of the biosynthetic pathway to catecholamines. In our experiments with clone N1E-115 the $V_{max} = 164$ pmol min^{-1} mg protein^{-1} for the transport of tyrosine into the cell. At confluency, tyrosine hydroxylase activity in crude sonicates of N1E-115 is about 1000 pmol min^{-1} mg protein^{-1} (RICHELSON, 1973b), using the cofactor 2-amino-4-hydroxy-6,7-dimethytetrahydropteridine and a substrate concentration of about three-fold above the K_m (approximately 70% V_{max}) (RICHELSON, unpublished data). In intact N1E-115 cells, the rate of [^{14}C]-catechol formation from [^{14}C]-tyrosine was somewhat less than the V_{max} for the transport of [^{14}C]-tyrosine and therefore significantly less than tyrosine hydroxylase activity in crude sonicates of this clone (RICHELSON, unpublished data). These results suggest that *in this system* the rate-limiting step for catechol biosynthesis is the transport of tyrosine into the cell.

REGULATION OF TYROSINE HYDROXYLASE ACTIVITY IN CLONE
N1E–115

It has been shown that mouse neuroblastoma acetylcholinesterase and choline acetyltransferase are regulated enzymes (AMANO et al., 1972). Therefore, tyrosine hydroxylase activity was examined in clone N1E-115 to determine whether this enzyme is also regulated.

Logarithmically growing cells were incubated for twelve days in petri dishes (150 mm × 25 mm) containing 30 ml of Dulbecco's modified Eagle's medium (DMEM) with 10% foetal calf serum (FCS). From early log phase to late stationary phase, the specific activity of tyrosine hydroxylase increased greater than 30-fold (from 18 to 585 pmoles [^{3}H]OH formed min^{-1} mg protein^{-1}). The increase in tyrosine hydroxylase per cell and per dish over the 12-day period was 12-fold and 2700-fold, respectively. During the log phase of growth, the tyrosine hydroxylase activity per cell remained low and relatively constant. However, after cells had reached the stationary phase of growth, the tyrosine hydroxylase activity per cell rapidly increased. That is, tyrosine hydroxylase activity was higher in cells which had stopped dividing. These results show that tyrosine hydroxylase, a marker for gene expression in adrenergic neurons, is regulated in neuroblastoma clone N1E-115. This regulation may reflect normal maturation processes of an adrenergic neuron (COYLE & AXELROD, 1972).

Adenosine-3′,5′-monophosphate (cyclic AMP) plays a prominent role in sympathetic nervous function (RALL & GILMAN, 1970). $N^6,O^{2'}$-dibutyryl cyclic AMP promotes expression of differentiated properties in mouse neuroblastoma (PRASAD and HSIE, 1971). Therefore dibutyryl cyclic AMP (DBcAMP); sodium butyrate, a breakdown product of DBcAMP; and theophylline, a phosphodiesterase inhibitor were tested for their effects on tyrosine hydroxylase activity in N1E-115 cultured in microwells (RICHELSON, 1973b). Tyrosine hydroxylase activity of the control cultures increased nearly six-fold between day 0 and day 8 for cells which were plated into late log phase (Fig. 2). DBcAMP (1 mM) + theophylline (0·4 mM) markedly enhanced the levels of tyrosine hydroxylase specific activity for all time points after treatment was begun, to levels averaging about twice that of the control. For both conditions, the stationary phase of growth was reached on day 4. For the treated cells, the increase in tyrosine hydroxylase activity over the time period studied is greater than 9-fold, reaching a level of activity nearly twice that reported for lysed granules from bovine adrenal medulla (SHIMAN et al., 1971).

Logarithmically growing cells incubated for eight days in the presence of DBcAMP (1 mM) alone obtained levels of tyrosine hydroxylase activity nearly twice that of the control (Control = 1214 pmoles [^{3}H]OH formed min^{-1} mg protein^{-1}; + DBcAMP = 2042). DBcAMP also caused a slowing of cell division (about 80 per cent of the control value). Sodium butyrate (1 mM) in the presence of FCS also caused an increase in tyrosine hydroxylase activity although the levels of specific activity were not significantly different from that of the control. Sodium butyrate, however, totally arrested cell division in the presence of FCS so that the activity of tyrosine hydroxylase per cell was nearly 2·5 × the control value after 4 days of incubation. When cell division was stopped by the removal of serum, only DBcAMP was able to stimulate tyrosine hydroxylase activity in N1E-115. These results show that this stimulation by DBcAMP of tyrosine hydroxylase activity is independent of its

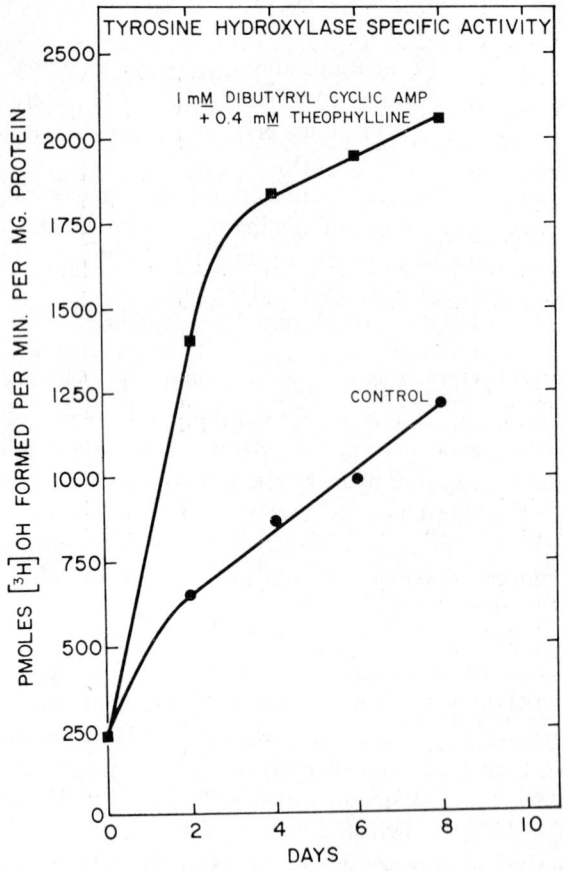

FIG. 2.—Stimulation of tyrosine hydroxylase activity in adrenergic clone N1E-115 by 1 mM dibutyryl cyclic AMP + 0·4 mM theophylline. Cells were plated into microwells at a density of about 9·3 × 10¹ cm⁻² 24 h before Day 0. On Day 0 treatment was begun. Sonicates for enzyme assay were prepared in duplicate for all time points except Day 0 which was prepared in quadruplicate. Tyrosine hydroxylase activity was measured by the release of tritium from L-[3,5-³H]-tyrosine (RICHELSON, 1973b).

ability to slow cell division; whereas the effect of sodium butyrate is related to its ability to arrest cell division.

Pre-ganglionic stimulation, either directly with electrical current or indirectly with reserpine increases tyrosine hydroxylase activity about two-fold (MOLINOFF and AXELROD, 1971). Dopamine which may be located in ganglionic interneurons and released with pre-ganglionic activity stimulates adenyl cyclase derived from sympathetic ganglia (KEBABIAN and GREENGARD, 1971). Preganglionic electrical stimulation causes a post-ganglionic increase in cyclic AMP (McAFEE et al., 1971). Accordingly, it seems possible that cyclic AMP may mediate the post-synaptic increase in tyrosine hydroxylase activity, and that the stimulation in vitro of tyrosine hydroxylase activity in an adrenergic clone of mouse neuroblastoma (N1E-115) by dibutyryl cyclic AMP may reflect events which occur in vivo.

CONCLUSION

Many questions remain unanswered about the regulation of catecholamine biosynthesis. One experimental approach to the problem has involved tissue culture techniques. We have used the cell culture system of mouse neuroblastoma and in particular, have studied the adrenergic clone N1E-115. This clone transports catechol precursors phenylalanine and tyrosine by high affinity mechanisms; and regulates tyrosine hydroxylase activity which is stimulated by dibutyryl cyclic AMP. In addition, this clone forms neurites and has electrically excitable membranes. Since these properties reflect *in vivo* processes, N1E-115 may be a useful model system for the elucidation of some of the mechanisms involved in the regulation of catecholamine biosynthesis.

Acknowledgement—This paper has been supported in part by grant DA 00266 from the United States Public Health Service. Figure 1 and Table 1 were previously published in *Nature, New Biol.* **241**, 201–204 (1973); figure 2 was published in *Nature, New Biol.* **242**, 175–177 (1973).

REFERENCES

AMANO T., RICHELSON E. and NIRENBERG M. (1972) *Proc. Nat. Acad. Sci. U.S.A.*, **69**, 258–263.
BLINKOW S. M. and GLEZER I. I. (1968) *The Human Brain in Figures and Tables*, Plenum Press, New York.
COYLE J. and AXELROD J. (1972) *J. Neurochem.* **19**, 1117–1123.
DALE H. (1935) *Proc. R. Soc. Med.* **28**, 319–332.
DIAMOND I. and KENNEDY E. P. (1969) *J. Biol. Chem.* **244**, 3258–3263.
GUROFF G. and UDENFRIEND S. (1962) *J. Biol. Chem.* **237**, 803–806.
KEBABIAN J. W. and GREENGARD P. (1971) *Science* **174**, 1346–1348.
LEVITT M., SPECTOR S., SJOERDSMA A. and UDENFRIEND S. (1965) *J. Pharmacol. exp. Ther.* **148**, 1–8.
LOGAN W. J. and SNYDER S. H. (1972) *Nature, Lond.* **234**, 297–299.
MCAFEE D. A., SCHORDERET M. and GREENGARD P. (1971) *Science* **171**, 1156–1158.
MOLINOFF P. and AXELROD J. (1971) *Ann. Rev. Biochem.* **40**, 465–500.
NEIDLE A., KANDERA J. and LAJTHA A. (1973) *J. Neurochem.* **20**, 1181–1193.
PERRY T. L., SANDERS H. D., HANSEN S., LESK D., KLOSTER M. and GRAVLIN L. (1972) *J. Neurochem.* **19**, 2651–2656.
PRASAD K. and HSIE A. (1971) *Nature, New Biol.* **233**, 141–142.
RALL T. W. and GILMAN A. G. (1970) *Neurosci. Res. Prog. Bull.* **8**, No. 3.
RICHELSON E. (1972) *Fedn. Proc.* **31**, 898.
RICHELSON E. (1973a) *Anal. Biochem.* **52**, 563–573.
RICHELSON E. (1973b) *Nature, New Biol.* **242**, 175–177.
RICHELSON E. and THOMPSON E. J. (1973) *Nature, New Biol.* **241**, 201–204.
ROBERTS S. (1968) *Progr. Brain Res.* **29**, 235–243.
SHIMAN R., AKINO M. and KAUFMAN S. (1971) *J. Biol. Chem.* **246**, 1130–1340.
WHITTAKER V. P. (1972) *Biochem. J.* **127**, 5P.
YAMAMURA H. I. and SNYDER S. H. (1972) *Science* **178**, 626–628.

Frontiers in Catecholamine Research 1973, pp. 261 to 265. Pergamon Press. Printed in Great Britain.

DEVELOPMENT OF THE CENTRAL CATECHOLAMINERGIC NEURONS OF THE RAT

JOSEPH T. COYLE

Laboratory of Clinical Science, National Institute of Mental Health,
Bethesda, Maryland 20014, U.S.A.

INTRODUCTION

THE ONTOGENESIS of the central nervous system is a complex process involving cell multiplication, cell migration, outgrowth of neuronal processes and synaptogenesis (JACOBSON, 1970). The growth of the brain is not a unitary process; but rather, certain populations of neurons and regions of the brain mature prior to others. Thus, during development, there is a constantly shifting pattern of neural relationships that may even continue throughout adulthood in a more subtle fashion (MOORE et al., 1971). Because of the diverse number of experimental approaches available in catecholamine research, the ontogenesis of the central catecholaminergic neurons of the rat brain is currently the most completely described of all neuronal systems in the developing mammalian brain.

ANATOMY

By means of tritiated thymidine autoradiography in combination with histofluorescent microscopy, it has been possible to determine when the central catecholaminergic neurons undergo cell division in the rat brain (NICHOLSON and BLOOM, 1973). As has been established for other major nuclei in the brainstem, the dopaminergic neurons in the substantia nigra and the noradrenergic neurons in the locus coereleus undergo a brief but intense period of cell division between 12–14 days of gestation. Thus, the total population of catecholaminergic neurons becomes fixed in number quite early in development. At 13 days of gestation, catecholaminergic neurons can first be observed in the rat brain with the histofluorescent technique, and by 15 days of gestation, both noradrenergic and dopaminergic cell bodies can be clearly distinguished (MAEDA and DRESSE, 1969; OLSON and SEIGER, 1972). During this period, the neurons migrate to their ultimate localization for their respective nuclei. This migratory process may be essential for their differentiation; for in the developing peripheral sympathetic nervous system, if the intercellular relationships during the migration of the noradrenergic ganglioblasts are significantly modified, the sympathoblasts fail to develop (COHEN, 1972).

As early as eight days before birth, axonal processes can be seen emanating from the nascent catecholaminergic neurons in the rat brain; this is well before many of the regions of the brain, such as the cerebral cortex and cerebellum, that receive catecholaminergic innervation, have been formed (OLSON and SEIGER, 1972). This observation suggests that early in development the catecholaminergic processes may subserve some organizing or trophic function. By birth, axons and terminals of catecholaminergic neurons are visible in most regions of the brain, although they are sparse (LOIZOU, 1972). These early appearing terminals may, in certain cases, be of a distinctly different functional group than the later appearing terminals. For example, discrete islands of dopaminergic fluorescent terminals are observed in the corpus striatum at birth; in the adult rat, the intense and diffuse fluorescence of the striatum recedes after inhibition of dopamine synthesis revealing discrete islands of fluorescence similar to that of the neonatal striatum (OLSON et al., 1972).

Although catecholaminergic axonal processes are present in the fetal and neonatal rat brain, the major phase of axonal proliferation occurs postnatally. With respect to the noradrenergic neurons, the more caudal regions of the brain develop an adult density of innervation by three weeks of age whereas the rostal regions achieve an adult density by 4–5 weeks of age (LOIZOU, 1972). A distinction must be made, however, between density and total number of terminals in a particular region because of the continued growth of the brain. The cerebellum, for example, exhibits a relatively mature density of innervation soon after birth although its mass increases considerably during subsequent maturation.

BIOCHEMISTRY

The dopaminergic and noradrenergic neurons have in common the first two enzymes in the biosynthetic pathway for their neurotransmitters, tyrosine hydroxylase (L-tyrosine, tetrahydropteridine: O_2 oxidoreductase; EC 1.10.3.1) and dopa decarboxylase (3,4-dihydroxy-L-phenylalanine carboxylase; EC 4.1.1.26). The noradrenergic neurons possess an additional enzyme, dopamine-β-hydroxylase (3, 4-dihydroxyphenylethylamine, ascorbate; O_2 oxidoreductase; EC 1.14.2.1), which catalyzes the conversion of dopamine to norepinephrine (MOLINOFF and AXELROD, 1971). Activity of all three enzymes is demonstrable in vitro in extracts prepared from the fetal rat brain at 15 days of gestation (COYLE and AXELROD, 1972a and b; LAMPRECHT and COYLE, 1972). The total activity of the enzymes in the whole brain increases in a coordinate fashion 500-fold during subsequent development. Since the catecholaminergic neurons have apparently ceased dividing by this point in fetal development, the enormous increase in enzyme activity in the individual neuron during differentiation is evident. Concurrent with the developmental increase in the activity of the biosynthetic enzymes, there is a translocation of their activity from the regions of the brain that contain catecholaminergic cell bodies to the regions that receive their terminals. Also, there is a progressive increase of their activity in the nerve terminal fractions prepared from the developing brain. The regional and subcellular translocation in the activity of tyrosine hydroxylase, dopa decarboxylase and dopamine-β-hydroxylase, correlates with the proliferation of axonal processes as demonstrated by the histofluorescent studies (COYLE, 1973).

The catecholaminergic neurons possess specific high affinity uptake processes for their neurotransmitters, which are localized at the nerve terminal and play an important role in the termination of the effect of the catecholamines released at the synapse (IVERSEN, 1967). In addition, the catecholamines are concentrated within the neuron in storage vesicles, which are involved in the process of release as well as in the protection of the amines from intraneuronal catabolism. A number of studies in the peripheral nervous system indicate that the appearance and subsequent increase in the uptake process for norepinephrine in an organ correlates with the development of sympathetic innervation (SACHS et al., 1970). The specific uptake mechanism for norepinephrine in synaptosomes prepared from rat brain appears at 18 days of gestation (COYLE and AXELROD, 1971). Although there is no subsequent change in the kinetic and pharmacologic characteristics of the uptake processes for norepinephrine, the total number of apparent transport sites in synaptosomal preparations increase over 100-fold in the whole brain from five days before birth through to adulthood. The vesicular storage process in the isolated terminals exhibits a more gradual development. This observation is in agreement with the ultrastructural evidence that immature sympathetic terminals are relatively deficient in storage vesicles (YAMAUCHI and BURNSTOCK, 1969). TENNYSON et al., (1972) have studied the development of the dopaminergic innervation of the corpus striatum. The maturational increase in the uptake process for dopamine correlates with development of adult axonal and terminal profiles as documented by electron microscopy. For both the central noradrenergic and dopaminergic neurons, the uptake process shows a more rapid developmental increase than the concentration of the neurotransmitters. Thus, the specific uptake mechanisms for catecholamines appear to be a sensitive index for the presence and proliferation of the catecholaminergic terminals in the developing brain.

ENDOGENOUS DOPAMINE AND NOREPINEPHRINE

Histofluorescent studies suggest that catecholamines are localized in specific neurons of the brain by 13 days of gestation; and biochemical studies indicate the fetal rat brain possesses all the enzymes in the biosynthetic pathway for the amines by 15 days of gestation. Since a number of factors such as the concentration of cofactors and storage capacity could significantly affect the levels of the catecholamines, the activity of the biosynthetic enzymes in vitro may not accurately reflect the levels of the neurotransmitters. By means of a sensitive enzymatic-radiometric assay, it has been shown that both norepinephrine and dopamine are present in the fetal rat brain at 15 days of gestation, although their concentrations are many fold below that anticipated on the basis of the activity of their biosynthetic enzymes (COYLE and HENRY, 1973). The concentration of the amines increases 15-fold during the subsequent eight days of gestation, attaining one quarter of the adult concentration by birth. During the latter third of the gestational period, the concentrations of the cotransmitters, octopamine and β-phenylethanolamine, are quite high, being over twofold higher than the endogenous norepinephrine; but they decrease to the low levels of the adult brain by birth (SAAVEDRA, COYLE and AXELROD, in preparation). Since octopamine has been implicated as a neurotransmitter in invertebrates, this may represent a situation in which ontogeny recapitulates phylogeny in the developing nervous system (NATHANSON and GREENGARD, 1973). In addition, in the fetal and neonatal brain, the catecholamine levels are less stringently controlled by feedback mechanisms than in the adult brain, since loading with precursors or inhibition of the degradative enzymes result in large increases in the amine levels (COYLE and HENRY, 1973; KELLOGG and LUNDBORG, 1972).

After birth, the concentrations of norepinephrine and dopamine increase in the whole brain in a coordinate fashion with the rise in activity of the biosynthetic enzymes (COYLE and AXELROD, 1972b; BREESE and TRAYLOR, 1972). Adult concentrations of norepinephrine are attained by four weeks after birth whereas dopamine achieves an adult concentration 5–6 weeks post-partum. As implied by the histofluorescent and enzymatic studies, the caudal regions of the brain achieve adult levels of the amines more rapidly than the rostral regions.

PHYSIOLOGY

Of major importance is the communicational relationship between the developing catecholaminergic neurons and the other neurons in the brain. As early as 18 days of gestation in the rat brain, endogenous dopamine and norepinephrine have half-lives of about three hours, which is similar to that observed in the adult brain (Coyle and Henry, 1973). Turnover of the amines has been shown to be proportional to impulse activity in the neurons in the adult brain (Anden, 1967). The earliest electrical recordings from the noradrenergic neurons in the locus coereleus of the rat indicate that they exhibit spontaneous electrical activity at least by birth (Hoffer et al., 1972). Electrophysiologic studies have shown that the Purkinje cells of the neonatal rat cerebellum possess β-adrenergic receptors before any synaptic processes have developed (Woodward et al., 1971). Within a few days after birth, the activity of adenyl cyclase responsive to norepinephrine can be demonstrated in slices prepared from rat brain (Schmidt et al., 1970). Thus, quite early in development, the central catecholaminergic neurons are probably releasing neurotransmitter and, at this stage, certain neurons are capable of responding to the amines. In light of the demonstration of supersensitivity of the β-adrenergic receptor in the denervated pineal gland with a shift in sensitivity down to nanomolar levels of norepinephrine (Deguchi and Axelrod, 1973), it seems feasible that the central catecholaminergic neurons may be able to interact with effector neurons prior to the development of mature synaptic processes.

Behavioural studies have demonstrated functional activity of the central catecholaminergic neurons in the neonatal rat. Both hyperactive and stereotypic behavior, which is mediated by the release of norepinephrine and dopamine, can be induced by amphetamine soon after birth (Fibiger et al., 1970; McGeer et al., 1971). The changing relationship among the neurons in the developing brain is underlined by the fact that the countervailing cholinergic input becomes functional weeks after the amphetamine-induced behavior appears. Although there is little evidence currently available, the maturation of other behavioural as well as certain pituitary functions are probably influenced by the subsequent development of the central catecholaminergic neurons.

CONCLUSION

The histologic, biochemical and physiologic studies indicate that the central catecholaminergic neurons appear extremely early in the development of the mammalian brain. As the brain increases over 60-fold in mass, this fixed population of neurons gradually spins out their axonal processes to interconnect with other regions of the brain. Because of the spatial-temporal changes occurring during the maturation of the brain, there is a constantly shifting pattern of neuronal relationships. The developmental aspects of the central catecholaminergic neurons are of particular relevance to a number of disease states for which the role of catecholamines has been strongly implicated. The hyperkinetic syndrome is primarily a disorder of preadolescence; schizophrenia often becomes apparent in late adolescence, whereas psychotic depression and parkinsonism are disorders of mid and late life, respectively. The pathophysiology of these disorders cannot be properly evaluated with the assumption of a static relationship among neurons but should be examined with a cautious consideration of the subtle, changing relationship among the neurons of the maturing brain.

REFERENCES

ANDEN N.-E. (1967) *Europ. J. Pharmac.* **1**, 1–5.
BREESE G. R. and TRAYLOR T. D. (1972) *Brit. J. Pharmac.* **44**, 210–222.
COHEN A. M. (1972) *J. exp. Zool.* **179**, 167–182.
COYLE J. T. (1973) *The Neurosciences Third Study Program*, in press.
COYLE J. T. and AXELROD J. (1971) *J. Neurochem.* **18**, 2061–2075.
COYLE J. T. and AXELROD J. (1972a) *J. Neurochem.* **19**, 449–459.
COYLE J. T. and AXELROD J. (1972b) *J. Neurochem.* **19**, 1117–1123.
COYLE J. T. and HENRY D. (1973) *J. Neurochem.*, **21**, 61–68.
DEGUCHI T. and AXELROD J. (1973) *Mol. Pharmac.*, in press.
FIBIGER H. C., LYTLE L. D. and CAMPBELL B. A. (1970) *J. comp. physiol. Psych.* **72**, 384–389.
HOFFER B. J., CHU N.-S. and OLIVER A. P. (1972) Abst. Volume Papers, *5th Int. Cong. Pharmac.*, San Francisco, p. 103 (618 abs).
IVERSEN L. L. (1967) *The Uptake and Storage of Noradrenaline in Sympathetic Nerves.* The University Press, Cambridge.
JACOBSON M. (1970) *Developmental Neurobiology.* Holt, Rinehart Winston, New York.
KELLOGG C. and LUNDBORG P. (1972) *Psychopharmacologia (Berl.)* **23**, 187–200.
LAMPRECHT F. and COYLE J. T. (1972) *Brain Res.* **41**, 503–506.
LOIZOU L. A. (1972) *Brain Res.* **40**, 395–418.
MAEDA T. and DRESSE A. (1969) *Acta Neurol. Belg.* **69**, 5–10.
McGEER E. G., FIBIGER H. C. and WICKSON V. (1971) *Brain Res.* **32**, 433–440.
MOLINOFF P. B. and AXELROD J. (1971) *Ann. Rev. Biochem.* **40**, 465–500.
MOORE R. Y., BJORKLUND A. and STEVENI U. (1971) *Brain Res.* **33**, 13–35.
NATHANSON J. A. and GREENGARD P. (1973) *Science* **180**, 308–310.
NICHOLSON J. L. and BLOOM F. E. (1973) *Anat. Rec.* **175**, 398 (abs.).
OLSON L. and SEIGER A. (1972) *Z. Anat. Entwickl.-Gesch.* **137**, 301–316.
OLSON L., SEIGER A. and FUXE K. (1972) *Brain Res.* **44**, 283–288.
SACHS C., deCHAMPLAIN J., MALMFORS T. and OLSON L. (1970) *Europ. J. Pharmac.* **9**, 67–79.
TENNYSON V. M., BARRETT R. E., COHEN G., CÔTÉ L., HEIKKELA R. and MYTILINEOU C. (1972) *Brain Res.* **46**, 251–285.
SCHMIDT M. J., PALMER E. C., DETTBARN W.-D. and ROBINSON G. A. (1970) *Devel. Psychobiol.* **3**, 53–67.
WOODWARD D. J., HOFFER B. J., SIGGINS G. R. and BLOOM F. E. (1971) *Brain Res.* **34**, 73–97.
YAMAUCHI A. and BURNSTOCK G. (1969) *J. Anat.* **104**, 17–34.

Frontiers in Catecholamine Research 1973, pp. 267 to 276. Pergamon Press. Printed in Great Britain.

INTERACTION BETWEEN THE NERVE GROWTH FACTOR (NGF) AND ADRENERGIC BLOCKING AGENTS IN SYMPATHETIC NEURONS

Rita Levi-Montalcini, Luigi Aloe and Eugene M. Johnson Jr.

Department of Biology, Washington University, St. Louis, Missouri
Laboratory of Cell Biology, Via Romagnosi 18A, Rome, Italy

INTRODUCTION

At a symposium of Chemistry and Brain development which took place in Milan in the Fall 1970, the senior author reviewed some aspects of studies performed in our laboratories on growth control mechanisms in the sympathetic nervous system (Levi-Montalcini, 1971). The results to be presented here bring into focus new aspects of the effects elicited on the sympathetic adrenergic neuron by the nerve growth factor and by two drugs, guanethidine and 6-hydroxydopamine (6-OHDA) which, through entirely different mechanisms, interact with the release of the neurotransmitter and cause the destruction of the immature sympathetic nerve cells. While in both instances the simultaneous administration of the NGF and of one of the two drugs results in the operation of the lethal effects caused by the ganglion blocking agents, in the case of 6-OHDA, but not of guanethidine, the combined treatment results in a paradoxical overproduction of axons and collaterals from the adrenergic neurons. The phenomenon as well as its possible cause will be considered in the following pages.

SURVEY OF PREVIOUS WORK

Ever since the discovery, more than two decades ago, that embryonic sensory nerve cells and embryonic and fully differentiated sympathetic nerve cells are selectively stimulated by a protein factor released from various animal tissues (Levi-Montalcini, 1966), a large body of literature has accumulated on the properties of this factor which became known as the Nerve Growth Factor (NGF) and on the structural and biochemical features of the growth response of the target cells. The reader is referred to review articles and to two volumes which appeared in 1972 on the NGF and the antibodies to this factor (Levi-Montalcini and Angeletti, 1968; Levi-Montalcini, Angeletti and Angeletti, 1972; Steiner and Schönbaum, 1972; Zaimis and Knight, 1972). Here we shall only mention that the extensive work directed to the purification and characterisation of the NGF resulted in 1971 in the determination of the primary and secondary structure of the NGF (Angeletti and Bradshaw, 1971). It is instead a prerequisite to the understanding of the results to be considered below to briefly summarize the findings of previous investigations directed to the study of the growth response elicited by the NGF in the sympathetic neuron as well as of the opposite effect, namely the growth inhibition of the same cell by various agents.

The in vitro and in vivo effects of the NGF

Four features characterize the *in vitro* growth response of sensory and sympathetic embryonic ganglia to the NGF. They are: (a) the magnitude of the effect called forth by this agent, which has no parallel in any other known effect and consists in the production of an extremely dense halo of nerve fibers by these ganglia but no other embryonic or adult nerve tissues, (b) the brevity of the time-lag between the stimulus and the growth response, (c) the stereotyped geometrical pattern of nerve fiber outgrowth elicited by the NGF, (d) the potency of the factor which calls forth this effect at a concentration in the order of a few nanograms per ml of culture medium.

While each of these features is unique, all four together represent most reliable and unequivocal evidence for the presence of the NGF in the culture medium. Since, as mentioned, this factor acts at very low doses, it became possible to detect it in tissues and body fluids, where it is usually present in trace amounts. The peculiar change in the features of the growth response evoked by NGF-excess is another unique facet of this stimulus-response effect discussed in previous articles.

The *in vivo* response of the target nerve cells to the NGF also exhibits some unusual features which will be briefly mentioned.

Ever since the first studies, it has become apparent that sympathetic ganglia from newborn or adult birds or mammals undergo marked changes upon subcutaneous administration of NGF. Since almost all the work centered on mammals and in particular on rodents (mice and rats), we shall consider here only the results of observations performed on the latter.

Daily injections of the NGF in newborn mice and rats call forth a striking volume increase of sympathetic ganglia but no effects from sensory ganglia. The former reach a volume up to 10–12 fold larger than controls in a 10-day treatment period. This effect is due to increase in size and number of individual nerve cells. In adult animals nerve cells undergo hypertrophic but no hyperplastic changes. Ultrastructural studies showed that the extraordinary increase in size of individual neurons is due to massive production of neurofibrillar material which fills all available space in the cell perikarya and accounts for the increase in size and number of their axons. This last effect in turn results in the hyperinnervation of all viscera (Levi-Montalcini, Angeletti and Angeletti, 1972). The structural and ultrastructural changes called forth by the NGF are paralleled by an equally impressive increase in all anabolic processes as well as in the synthesis of key enzymes of catecholamine pathway (Angeletti et al., 1965; Thoenen et al. 1971).

Immunosympathectomy

The outstanding role played by the NGF in the life of the sympathetic adrenergic neuron came even more sharply into focus from immunological studies. S. Cohen produced in 1959 a specific antiserum to the NGF isolated from mouse salivary glands (Cohen, 1960). Injection of this antiserum in newborn rodents and other mammals resulted in the selective destruction of sympathetic nerve cells. The process became known as immunosympathectomy. Electron microscopy studies directed to the analysis of the mechanism and site of action of the antiserum to the NGF (AS-NGF) showed that the first detectable lesions caused by antibodies to the NGF take place in the nucleoli and nuclear compartment. Disintegration of the former, clumping and increased electron density of the chromatine material precede in fact any alteration in the cell perikarya. Lesions in the latter become however well apparent and take a precipitous course leading to cell death upon rupture of the nuclear membrane and mixing of nuclear and cytoplasmic content (LEVI-MONTALCINI *et al.*, 1969). Toward the end of the third day, all neurons are in a more or less advanced stage of disintegration while glial and other satellite cells are not affected. It is only in the following days and weeks that a low atrophic process sets in and brings about the gradual and total disappearance of glial and satellite cells. At the end of the first month and in all subsequent life periods para- and prevertebral sympathetic ganglia are reduced to diminutive sclerotic nodules barely detectable at the dissecting microscope.

Chemical sympathectomy

While these studies were in progress another most unforeseeable response of the adrenergic neurons to extrinsic factors was discovered by P.U. Angeletti, who has been associated in this work ever since 1960 and directed the investigations on the chemical nature of the NGF and its biochemical effects.

P.U. Angeletti conceived the idea of exploring the action of a dopamine analog, 6-hydroxydopamine (6-OHDA), on the immature sympathetic nerve cells of newborn rodents. A few years earlier (PORTER *et al.*, 1963) it was discovered that this drug produces a blocking effect on sympathetic function by preventing the release of NE from the adrenergic nerve end terminals. At variance however with other adrenergic blocking agents, 6-OHDA produces a long-lasting effect; restoration of function takes place only four to six weeks after discontinuation of 6-OHDA treatment. Biochemical and E.M. studies by THOENEN and TRANZER (1968) showed that the block of sympathetic function is due to the selective destruction of the synaptic vesicles in adrenergic nerve fibers. This effect may be due to oxidation products of 6-OHDA which become covalently bound to nucleophilic groups of biological macromolecules. As stated by Thoenen, "the high degree of selectivity of damage to adrenergic neurons depends on the efficiency with which 6-OHDA is accumulated by the neuronal membrane pump" (THOENEN, 1972).

Experiments with the antiserum to the NGF had shown that while the antibodies to this factor produce irreversible cytotoxic lesions in the immature sympathetic neuron, temporary and reversible damages are instead caused by the same AS-NGF in the fully differentiated neuron. Thus it seemed possible that the dopamine analog could in a similar way prove to be much more toxic to the immature than to the mature sympathetic neuron. A series of experiments reported in previous articles fully confirmed the expectation. Daily injections of 6-OHDA in newborn mice and rats in the amount of 50–100 μg/g of body weight for a one-week period resulted in the total destruction of sympathetic neurons in the para- and prevertebral chain ganglia (ANGELETTI, 1972; ANGELETTI and LEVI-MONTALCINI, 1970). This effect became known as chemical sympathectomy.

Subsequent studies in our and other laboratories showed that another ganglion blocking agent, guanethidine also causes death of immature nerve cells (ERÄNKÖ and ERÄNKÖ, 1971; ANGELETTI and LEVI-MONTALCINI, 1972). The destructive effects caused by this drug are of the same range as those produced by antibodies to the NFG and by 6-OHDA. It is however of considerable interest to note that the intracellular localisation of the lesions as well as the time course of the process differ markedly from each other. As already mentioned the first detectable lesions called forth by AS-NGF are apparent in the nuclear compartment of the immature sympathetic nerve cells. The lesions produced by 6-OHDA are localised in the synaptic vesicles in the nerve end terminals, and in the endoplasmic reticulum which appears vacuolised and disrupted (Angeletti and Levi-Montalcini, 1970). Guanethidine causes lesions in the endoplasmic reticulum and in the mitochondria. The mitochondria appear enormously dilated and undergo rupture of the cristae and total disintegration. No alterations are apparent in the synaptic vesicles. These findings prompted an investigation of possible interactions between the NGF, guanethidine and the dopamine analog. The results of these studies, to be reported in detail in another article, will be briefly considered below.

INTERACTION BETWEEN NGF AND GUANETHIDINE

Newborn rats from the same litters were divided into 4 groups. The first group was injected daily from the day of birth with the dose of 50 μg of NGF per g of body weight. The second group received daily the same NGF dose but was also injected once a day with 20 μg of guanethidine/g of body weight; the third group received guanethidine alone in the same amount as the second group while the fourth group, serving as control, was injected with saline. Baby rats of the four groups were sacrificed on alternate days between the second and the twelfth day. Superior cervical, stellate and all thoracic chain ganglia were dissected out, mounted *in toto* or used for histological and E.M. studies. Here we will consider two series of superior cervical and stellate ganglia dissected out from baby rats of the four groups sacrificed at the end of the eighth day of treatment.

At the gross inspection, ganglia from the second group (treatment with only NGF) are about three times larger than those of controls at the end of the eight days of treatment; ganglia of the first group (NGF + guanethidine) are only slightly smaller than ganglia of the second group while those of the third group (guanethidine alone) are somewhat smaller than controls (Fig. 1a). Histological studies on stellate and other ganglia of 8-day rats show that the combined NGF and guanethidine treatment results in the perfect preservation of sympathetic nerve cells which compare in size to those of rats treated with NGF alone. The only noticeable difference is that nerve cells are more densely packed in ganglia of guanethidine +NGF treated than NGF treated rats (Fig. 2a,b) and this seems to account for the slight difference in volume of ganglia of the two groups. Histological studies of ganglia of rats treated with guanethidine alone fixed after 8 days of treatment show that the entire cell population consists of glial and satellite cells with, few atrophic neurons scattered among the non-nervous cells (Fig. 2d).

INTERACTION BETWEEN NGF AND 6-HYDROXYDOPAMINE

The only difference between this and the previous experiment is that 6-OHDA rather than guanethidine was used alone or in combination with NGF. 6-OHDA was injected daily in the dose of 100 μg/g of body weight starting from the day of birth. The treatment was continued until the 19th day of life. Specimens of each of the 4 groups were sacrificed starting from the end of the second week on alternate days until the 19th day. Here we shall consider only the results of studies performed on ganglia dissected out from rats sacrificed at 8 and 19 days.

Inspection of these whole mount ganglia shows a striking difference between those injected with NGF alone (II group) or with NGF and 6-OHDA (I group). Figure 1b illustrates the entire series of superior cervical ganglia fixed at 8 days. At

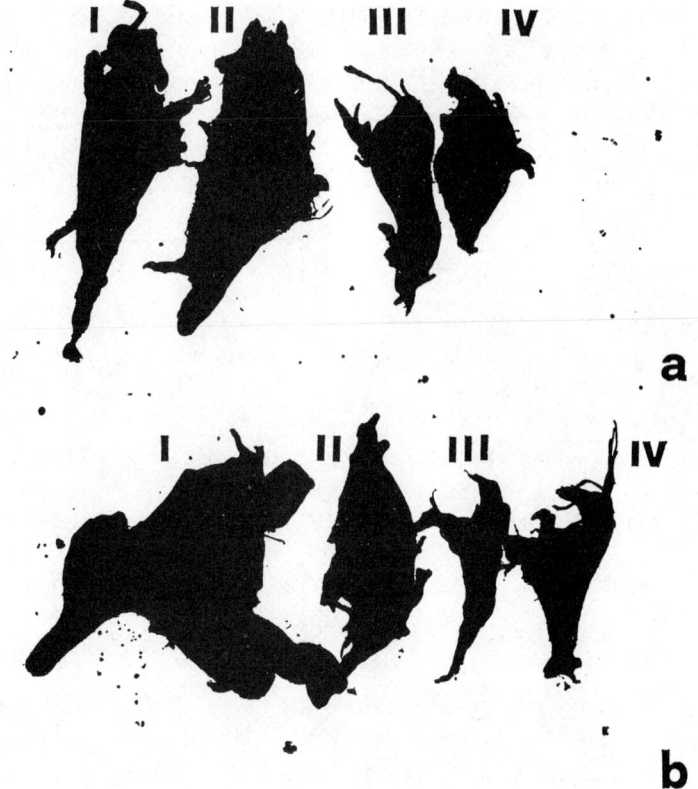

FIG. 1.—(a) whole mounts of superior cervical ganglia from 8-day old littermate
rats injected daily since birth with guanethidine and NGF (I), NGF (II), guanethidine
(III) and saline (IV). × 10 (b) whole mounts of superior cervical ganglia from 8-day
old littermate rats injected daily since birth with 6-OHDA and NGF (I), NGF (II),
6-OHDA (III) and saline (IV). × 9

variance with those of the guanethidine series, ganglia of the first group (combined
NGF and 6-OHDA) are much larger than ganglia of the second group (NGF alone);
the contour of the ganglion is grossly altered and differs markedly from the oval-
shaped profile of ganglia of control or NGF treated rats. Size and shape differences
between ganglia of group I and II become even more pronounced with more extensive
treatment.

Histological studies of sections of superior cervical, stellate and other sympathetic ganglia of rats submitted to the combined 6-OHDA and NGF treatment show
that the tremendous size increase of these ganglia is due to an extraordinary increase
in number and size of postganglionic axons and to a moderate increase in glial and
satellite cells (Figs. 3 and 4).

Studies of ganglia of 8–12 and 19 days give evidence for the distal-proximal
sequences of the process: at first, glial and other sheath cells are seen in numbers far
above the normal, between axons of the postganglionic root and then extend along
the nerve and gain access into the ganglion (Fig. 4, b). The cells are lined along
newly formed axons which emerge from the sympathetic neurons. In spite of this
aberrant situation, nerve cells are found in normal number; they are hypertrophic

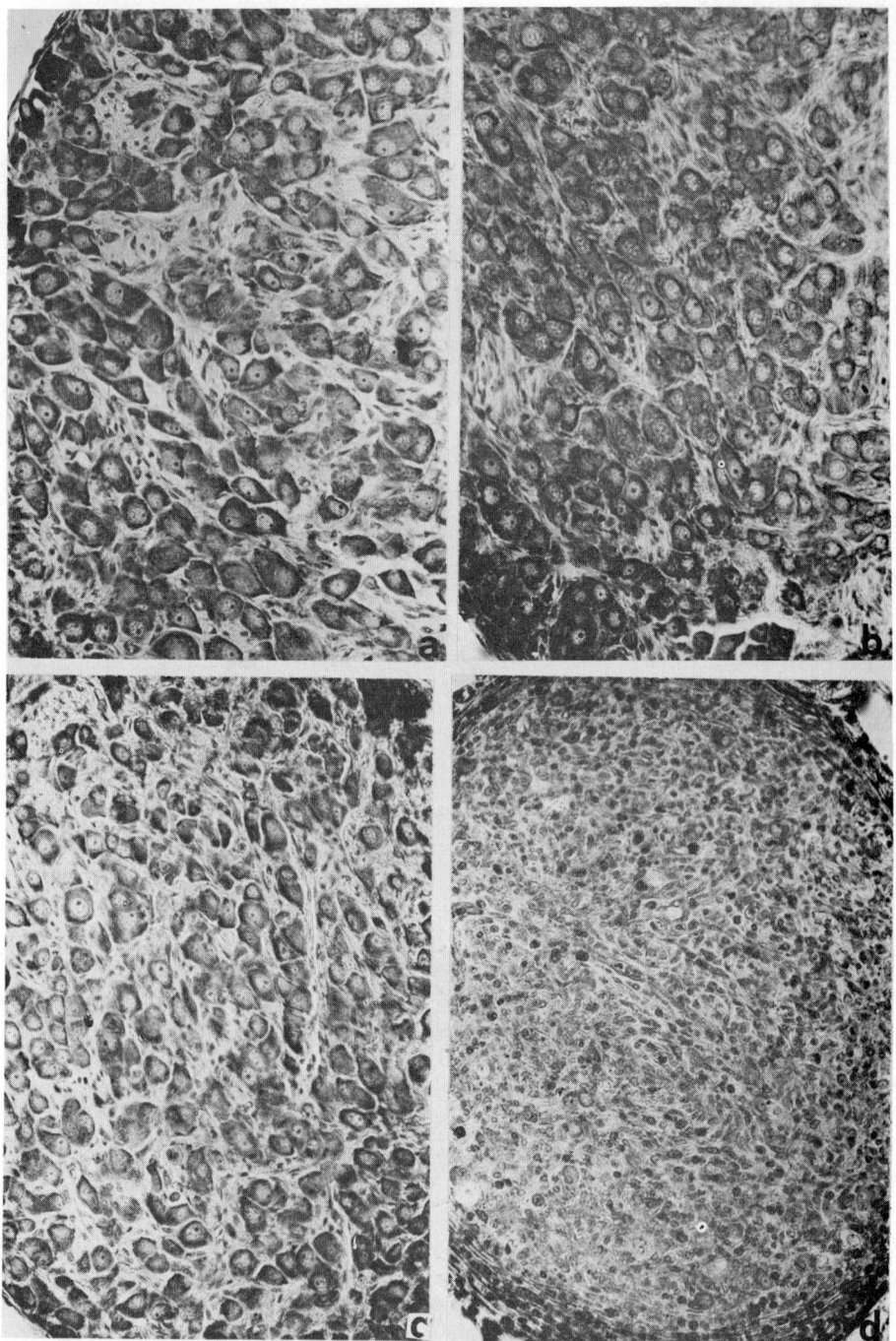

Fig. 2.—Transverse sections through stellate sympathetic ganglia of 8-day old littermate rats injected daily since birth with NGF (a), NGF and guanethidine (b), saline (c), and guanethidine (d). Note that no nerve cells are present in ganglion treated with guanethidine alone. All sections at × 256

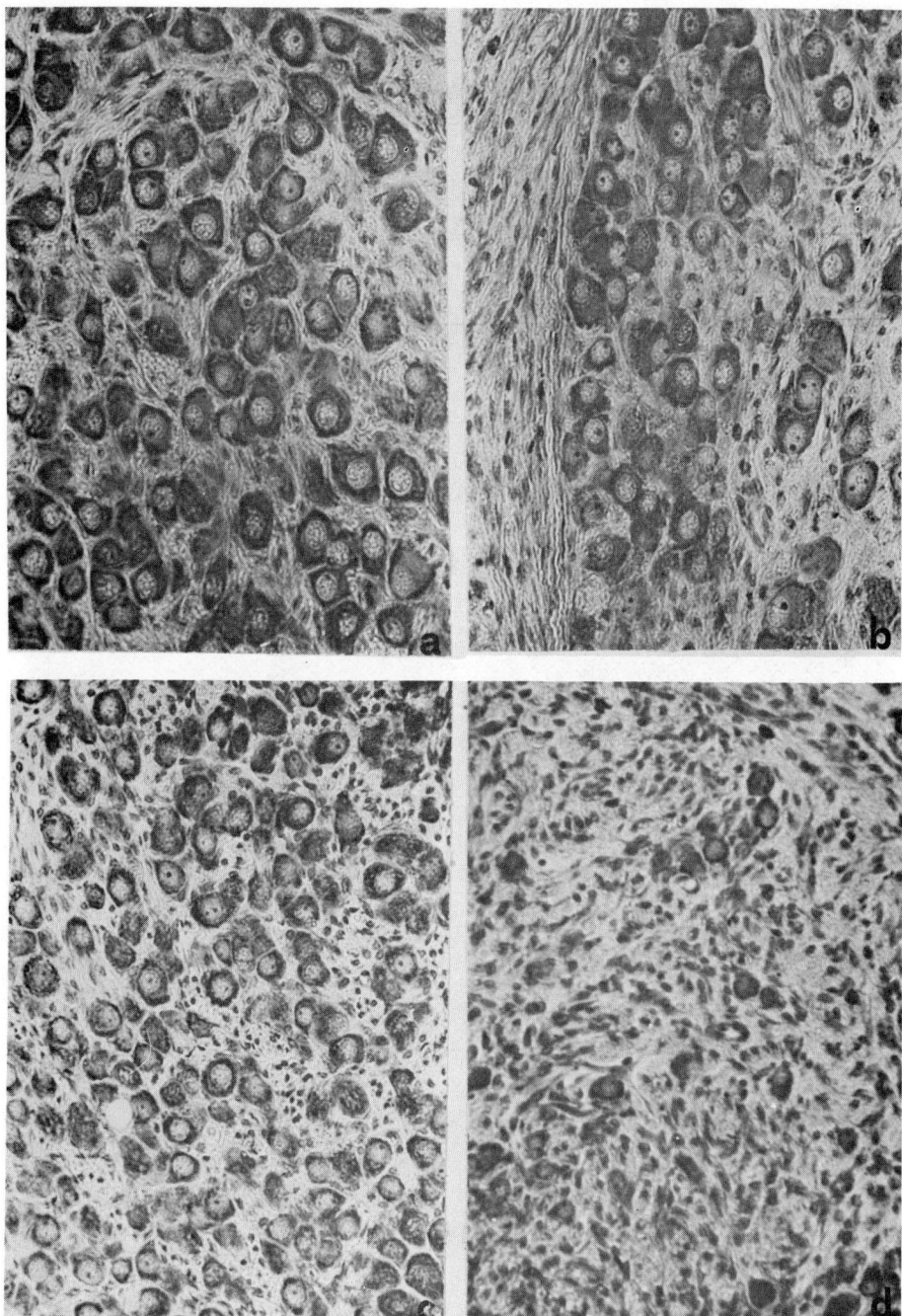

FIG. 3.—Transverse sections through stellate sympathetic ganglia of 19-day old litter-mate rats injected since birth with NGF (a), NGF and 6-OHDA (b), saline (c) and 6-OHDA alone (d). Note massive increase in size and number of postganglionic axons and moderate increase in glial and satellite cells are shown in (b) while only a few atrophic nerve cells are seen in (d) scattered among glial and satellite cells. All sections at × 24

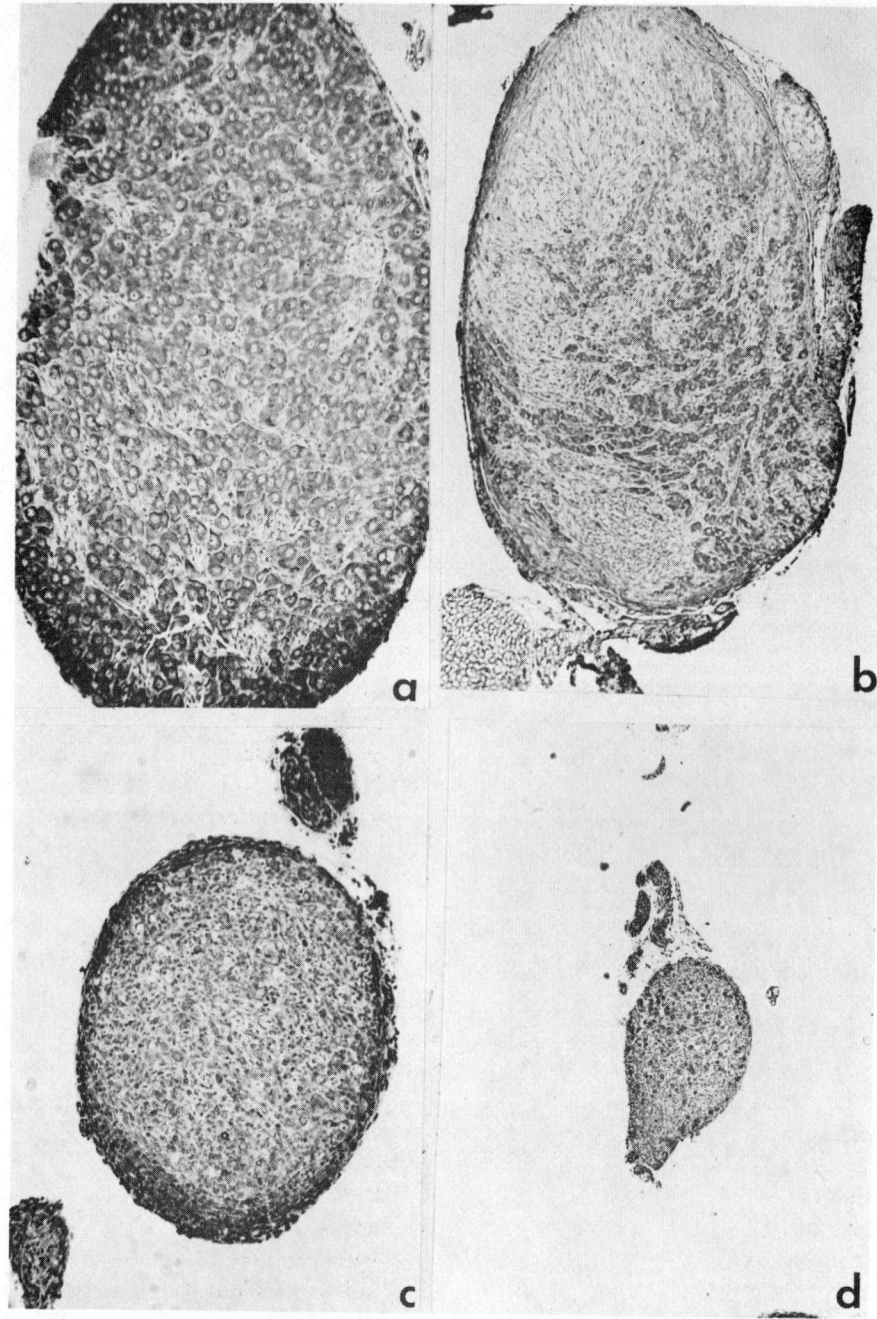

FIG. 4.—Compares the effects of the combined guanethidine and NGF (a), 6-OHDA and NGF (b), guanethidine alone (c) and 6-OHDA alone (d) in littermate rats, injected since birth and sacrificed at 8 days (a, c) and 19 days (b, d). Sections a, c × 128; sections b, d × 48.

like those of littermates treated with NGF alone (Fig. 3b). The typical NGF effect, namely an overproduction of neurofilaments and neurotubules filling the cell perikarya is seen at the electronmicroscope.

These results are at variance with earlier experiments reported by ANGELETTI and LEVI-MONTALCINI (1970), which were performed mainly on mice. Further details will be given in a subsequent paper.

Experiments in progress will answer the question of whether and to what extent this paradoxical situation will be tolerated by the sympathetic neurons and whether or not their functional properties will be impaired. It is of interest to call attention to the fact that neither NGF alone, nor 6-OHDA alone nor guanethidine or the combined NGF and guanethidine treatments, call forth such an increase in size and number of postganglionic axons. Thus it appears that increase in number of axons and, to a much lesser extent, of glial and satellite cells, is responsible for the extraordinary enlargement of the ganglia and also for the changed shape, which are unique features of the combined NGF and 6-OHDA treatment.

A tentative hypothesis, is that this paradoxical effect results from the opposite and conflicting actions of NGF and 6-OHDA. While the former greatly enhances all metabolic processes in the sympathetic neuron, the latter blocks the normal in-and out-traffic at the nerve-end terminals by its selective destruction of the synaptic vesicles which prevent storage and release of the neurotransmitter. Thus the nerve fibers are throttled at their terminals with obvious obstruction of axonal flow and piling up of material synthesised in the cell perikarya. The situation is reminiscent of that which takes place following crushing or ligation of adrenergic and non-adrenergic nerves. Also in these cases, glial and other satellite cells undergo active proliferation and the number of these cells is greatly increased (TORVIK and SKJÖRTEN, 1971; BLINZINGER and KREUTZBERG, 1968). In the present experiments this effect is magnified by the enhanced metabolic processes called forth by the NGF.

CONCLUDING REMARKS

While the most impressive result of these experiments is the tremendous overproduction of axons by sympathetic neurons under the combined treatment of NGF and the dopamine analog, the other effect, namely the preservation of nerve cells and their hypertrophy is no less interesting. This effect, as reported, obtains also in the combined NGF and guanethidine treatment. At present we are not in a position to offer an explanation for these findings. Two alternative hypotheses should be considered: either NGF combines with guanethidine and with 6-OHDA and in this way mitigates their noxious effects by preventing, at least in part, these drugs from gaining access into the adrenergic cells; or the nerve cells, forced to a precocious differentiation by the NGF, are able to withstand the otherwise lethal effects of both drugs. It should be remembered that both guanethidine and 6-OHDA are much more toxic to the immature than to the mature sympathetic neuron.

A third hypothesis which would visualise this effect as a result of a direct competitive action of NGF and one of the two drugs for the same receptor site in the target cell seems most unlikely in view of the entirely different temporal and topographic sequence of events called forth by the NGF and the two agents under consideration.

The NGF elicits in fact its most potent action in the immature neuron, while both guanethidine and the dopamine analog suppress the same neuron by gaining access to the nerve cell through its differentiated end terminals.

Experiments now in progress are aimed to the elucidation of this interaction.

Acknowledgements—This work was supported in part by grants from the National Institute of Neurological Diseases and Stroke (NS-03777), from the National Science Foundation (GB-1633OX) and from the North Atlantic Treaty Organization (449).

REFERENCES

Angeletti P. U. (1972) In: *Immunosympathectomy* (Steiner G. and Schönbaum E., Eds.) pp. 237–250. Elsevier, Amsterdam.

Angeletti P. U., Gandini Attardi D., Toschi G., Salvi M. L. and Levi-Montalcini R. (1965) *Biochim. Biophys. Acta* **95**, 111–120.

Angeletti P. U. and Levi-Montalcini R. (1970) *Proc. Nat. Acad. Sci. USA* **65**, 114–121.

Angeletti P. U. and Levi-Montalcini R. (1972) *Proc. Nat. Acad. Sci. USA* **69**, 86–88.

Angeletti R. H. and Bradshaw R. A. (1971) *Proc. Nat. Acad. Sci. USA* **68**, 2417–2420.

Blinzinger K. and Kreutzberg G. (1968) *Z. Zellforsch.* **85**, 145–157.

Cohen S. (1960) *Proc. Nat. Acad. Sci. USA* **46**, 302–311.

Eränkö O. and Eränkö L. (1971) *Histochem. J.* **3**, 451–456.

Levi-Montalcini R. (1966) *Harvey Lect. ser.* **60**, 217–259.

Levi-Montalcini R. (1971) In: *Chemistry and Brain Development* (Paoletti R. and Davison A. N., Eds.) pp. 185–194. Plenum, New York.

Levi-Montalcini R. and Angeletti P. U. (1968) *Physiol. Rev.* **48**, 534–569.

Levi-Montaclini R., Angeletti R. H. and Angeletti P. U. (1972) In: *Structure and Function of the Nervous Tissue* (Bourne G. H., Ed.) vol. 5, pp. 1–38. Academic Press, New York.

Levi-Montalcini R., Caramia F. and Angeletti P. U. (1969) *Brain Res.* **8**, 347–362.

Porter C. C., Totaro J. A. and Stone C. A. (1963) *J. Pharmacol. Exp. Ther.* **140**, 308–316.

Steiner G. and Schönbaum E. (1972) Eds. *Immunosympathectomy*, Elsevier, Amsterdam (1972).

Thoenen H. (1972) In: *Perspectives in Neuropharmacology* (Snyder S. H., Ed.) pp. 301–338. Oxford University Press, New York.

Thoenen H., Angeletti P. U., Levi-Montalcini R. and Kettler R. (1971) *Proc. Nat. Acad. Sci. USA* **68**, 1598–1602.

Thoenen H. and Tranzer J. P. (1968) *Naunyn-Schmied. Arch. Pharmakol.* **261**, 271–288.

Torvik A. and Skjörten F. (1971) *Acta neuropath. (Berl.)* **17**, 265–282.

Zaimis E. and Knight J., Eds. (1972) *Nerve Growth Factor and its Antiserum*, Athlone Press, London.

Frontiers in Catecholamine Research 1973, pp. 277 to 283. Pergamon Press. Printed in Great Britain.

NEUROBLASTOMA CELLS AS A TOOL FOR NEURONAL MOLECULAR BIOLOGY

P. Mandel, J. Ciesielski-Treska, J. C. Hermetet, J. Zwiller,
G. Mack and C. Goridis

Centre de Neurochimie du CNRS, and Institut de Chime Biologique, Faculté de Médecine,
67085 Strasbourg Cedex, France

INTRODUCTION

CELL CULTURE technology and somatic cell genetics may afford opportunities to investigate molecular aspects of nervous cell differentiation and molecular determinants of neural function in a defined microenvironment. The difficulty of inducing neurons to divide in culture for long periods limited the experiments to the study of primary explants or of surviving neurons. Fortunately AUGUSTI-TOCCO and SATO (1969) showed that from the mouse neuroblastoma C1300 continuous cell lines can be obtained which undergo neuron-like differentiation

NEUROBLASTOMA CELLS IN PROLIFERATING PHASE

General characteristics

The primary cultures or the cloned cells divide rapidly in suspension in a medium supplemented with foetal calf serum and retain their round cell morphology (SCHUBERT *et al.*, 1969). The average cell doubling time is about 18–24 hr. Some of the cells classified as 'active' maintain the ability to produce the enzymes involved in the synthesis and degradation of transmitters as measured by biochemical methods: choline acetyltransferase (Acetyl-CoA:choline *O*-acetyltransferase; EC 2.3.1.6) (ROSENBERG *et al.*, 1971; AMANO *et al.*, 1972), acetylcholinesterase (Acetyl hydrolase; EC 3.1.1.7) (BLUME *et al.*, 1970; SCHUBERT *et al.*, 1971; MINNA *et al.*, 1972; AMANO *et al.*, 1972), tyrosine hydroxylase (L-phenylalanine, tetrahydropteridine:oxygen oxidoreductase; EC 1.14.3.1.) (SCHUBERT *et al.*, 1969; RICHELSON, 1972; WAYMIRE *et al.*, 1972; AMANO *et al.*, 1972; PRASAD *et al.*, 1973), DOPA decarboxylase (3,4-Dihydroxy-L-phenylalanine carboxylase; EC 4.1.1.26) (MACK, ZWILLER and MANDEL, Unpublished data), dopamine-β-hydroxylase (3,4-Dihydroxyphenylethylamine, ascorbate:oxygen oxidoreductase (hydroxylating); EC 1.14.2.1) (ANAGNOSTE *et al.*, 1972), monoamine oxidase (Monoamine:oxygen oxidoreductase (deaminating); EC 1.4.3.4) (GORIDIS *et al.*, 1973). We also found evidence suggesting the presence of enzyme inhibitors; for example inhibitors of DOPA decarboxylase and dopamine-β-hydroxylase.

Ultrastructure

The ultrastructural analysis (SCHUBERT *et al.*, 1969; MANDEL, HERMETET and CIESIELSKI-TRESKA, in preparation) during proliferation in suspension shows round cells with a small number of processes which may contain well characterised microtubules and microfilaments as well as ribosomes. Large dense core vesicles are present in the cytoplasm. Using the staining method of TRANZER and SNIPES (1968) we could show that in some clones during differentiation, these vesicles may perhaps be correlated to the putative catecholamine vesicles.

Unfortunately the neuroblastoma cells contain a virus (SCHUBERT *et al.*, 1969), a factor which must be taken into account when basic cellular molecular events are studied.

Cell heterogeneity

A great heterogeneity exists within cells of the same culture, as can be demonstrated by traditional histochemical methods, by histofluorescence and by karyotype analysis. Thus in different clones, in which during the stationary phase 30 to 40 per cent of the cells have extended processes, only 15 per cent possess acetylcholinesterase activity (MANDEL, CIESIELSKI-TRESKA, GORIDIS, HERMETET and WARTER, in preparation; HERMETET *et al.*, 1973a). Similarly, in several adrenergic clones catecholamine specific histofluorescence can only be detected in 30–40 per cent of cells. The karyotype, as shown by KLEBE and RUDDLE (1969), AMANO *et al.* (1972) and in our laboratory (MANDEL *et al.*, 1972), differs from one clone to another and within the same clone.

Enzymatic activities

The specific activity of acetylcholinesterase did not change appreciably during the period of cell division, there was great variability among cells, as shown by histochemical analysis. By staining for monoamine oxidase we could observe a similar heterogeneity within the same clone. The enzymatic activities involved in glucose metabolism such as aldolase (Fructose-1,6-diphosphate D-glyceraldehyde-3-phosphate lyase; EC 4.1.2.13), glucose-6-phosphate dehydrogenase (D-Glucose-6-phosphate: NADP oxidoreductase; EC 1.1.1.49), lactate dehydrogenase (L-lactate:NAD oxidoreductase; EC 1.1.1.27), glutamate dehydrogenase (L-glutamate:NAD(P) oxidoreductase (deaminating); EC 1.4.1.3), malate dehydrogenase (L-malate:NAD oxidoreductase; EC 1.1.1.37), are twice as high in cultured neuroblastoma cells as in tumoral cells (CIESIELSKI-TRESKA *et al.*, 1972). Striking differences exist between different clones with regard to the enzymes involved in the synthesis and degradation of catecholamines and acetylcholine. We found that in the adrenergic clone, N115s, the monoamine oxidase activity is not detectable by a biochemical method which is sensitive at a level of 0·5 nmole of tyramine metabolised per mg of protein/hr, while in other adrenergic clones the values were about 39 nmole. There is a lot of data concerning tyrosine hydroxylase activity in the literature (SCHUBERT *et al.*, 1969; RICHELSON, 1972; WAYMIRE *et al.*, 1972; AMANO *et al.*, 1972; PRASAD *et al.*, 1973). We have measured dopamine-β-hydroxylase activity in clone M1 and found a value of 9 nmole octopamine produced per mg of protein/hr which is much higher than in rat hypothalamus (2·38 units) while in clone N115s the activity was even lower (1·25 units).

Amines

By a dansylation method which we have developed (ZWILLER *et al.*, 1973) the distribution of amines could be determined. The absloute quantity of amines differed greatly between the clonal lines studied and pronounced differences in the proportion

TABLE 1. BIOGENIC AMINES IN ADRENERGIC NEUROBLASTOMA CLONES

	M1 clonal line proliferative growth	N115s clonal line proliferative growth	N115s clonal line 48 hr after withdrawal of serum
Dopamine	27,100*	230 ± 120	200 ± 46
Norepinephrine	28,600	2740 ± 1020	1902 ± 356
Normetanephrine	12,300	890 ± 360	720 ± 221
Epinephrine	16,500	400 ± 180	402 ± 230
Serotonin		1540 ± 1140	1518 ± 361
Histamine		185 ± 107	184 ± 62

* 10^{-15} M of compounds/mg protein/hr.

dopamine/norepinephrine/normetanephrine/epinephrine could be shown. In addition, serotonin and histamine were present in clone N115s as shown by the dansylation method and using radioactive precursors.

Electrical activity

It has been shown that neuroblastoma cells grown *in vitro* are capable of generating neuron-like action potentials in response to electrical stimulation (NELSON *et al.*, 1969; MINNA *et al.*, 1971, 1972) or iontophoretic application (HARRIS and DENNIS, 1970).

NEUROBLASTOMA CELLS DURING DIFFERENTIATION

Induction of neuron-like differentiation

Under defined conditions the round neuroblastoma cells differentiate in neuron-like structures producing large radiating neurites. This differentiation could be obtained: (1) when the cell suspension in medium containing either 10 or 0·2 per cent serum was transferred to a surface to which the cells could attach, such as glass, collagene, or commercially treated tissue culture dishes (BLUME *et al.*, 1970); (2) by removal of the calf serum from the culture medium (SEEDS *et al.*, 1970); (3) by addition of butyryl cyclic AMP or cAMP phosphodiesterase inhibitors (FURMANSKI *et al.*, 1971; PRASAD and HSIE, 1971; KIRKLAND and BURTON, 1972; RICHELSON, 1972; PRASAD and SHEPPARD, 1972; WAYMIRE *et al.*, 1972; PRASAD *et al.*, 1973), bromodeoxyuridine (an inhibitor of DNA synthesis) (SCHUBERT and JACOB, 1970; KATES *et al.*, 1971), nerve growth factor (NGF) (HERMETET *et al.*, 1972b), prostaglandin E1 (PRASAD, 1972; PRASAD *et al.*, 1973), insulin (HERMETET *et al.*, 1973b), acetylcholine (HARKINS *et al.*, 1972); (4) by X-ray irradiation (PRASAD *et al.*, 1972). Striking differences are observed in the effect of these treatments. By treatment with antibodies against purified acetylcholinesterase of neuroblastoma cells one may select clones with a 100 times lower acetylcholinesterase activity (SIMAN-TOV and SACHS, 1972). The criteria for morphological differentiation were an increase in cellular and nuclear size, and the formation of cytoplasmic processes longer than 50 μm. The induction of the growth of axon-like processes seems to be similar in a serum-free medium and in a medium containing drugs which inhibit cellular DNA synthesis, even in the presence of serum. However dopamine, 6-hydroxydopamine and DL-glyceraldehyde which inhibit cell division do not induce morphological differentiation (PRASAD, 1971a, 1971b).

Ultrastructural investigations were performed by SCHUBERT *et al.* (1969) and by

Mandel, Hermetet and Ciesielski-Treska (in preparation). By electron microscopy we found that cells transferred to monolayer cultures in a medium without serum or in the presence of serum and one of the inducers rapidly adhered to the surface and produced long branched processes in which numerous microtubules, microfilaments and also ribosomes were observed. A large development of Golgi apparatus with accumulation of electron dense material in Golgi vesicles appeared in several clones. The suppression of serum produces cytoplasmic alterations which make the investigation of long term cultures in a serum free medium questionable. However, when differentiation is obtained in a medium containing serum, using the different inducers, a highly developed Golgi apparatus with numerous microfibriles and microtubules may be observed after several days. But individual cells within the same clone exhibit varying susceptibility for spontaneous or induced differentiation.

Gangliosides of neuroblastoma cells

Recently Dawson et al. (1971) and Yogeeswaran et al. (1973) reported marked differences in ganglioside patterns of neuroblastoma cell clones. Similar differences were observed in our laboratory. We expected during differentiation a shift towards a preponderance of polysialic gangliosides which can be observed during ontogenesis of the nervous system. Although an increase in total ganglioside and in the GD_3 content was observed in differentiated neuroblastoma cells, GD_{1B}, GT_1 and GQ_1 gangliosides could not be detected (Rebel et al., 1973). Thus the specific neuronal type distribution was not observed. Since gangliosides are mainly localised in neuronal membranes, and particularly in synaptic membranes, the lack of production of synaptic contacts may be related to the inability to produce the polysialic gangliosides.

Table 2. Gangliosides of neuroblastoma CCL clonal lines

	Undifferentiated cells*	Differentiated cells†	Neurons
Total lipid sialic acid µg/mg dry weight	0·240 (0·239 − 0·241)	0·303 (0·278 − 0·329)	10·7‡
% Total gangliosides			
GM_3	21·3	33·6	Traces
GM_2	44·9	39·9	—
GM_1	Traces	—	42·1
GD_3	—	5·3	—
GD_{1A}	33·8	21·2	35·7
GD_{1B}	—	—	11·5
GT_1	—	—	9·8
GQ_1	—	—	0·7

* Means of 3 experiments; † Means of 5 experiments.
‡ Values from Derry D. M. and Wolfe L. S. (1967) Science 158, 1450–1452, and Hamberger A. and Svennerholm L. (1971) J. Neurochem. 18, 1821–1829.

Oxygen consumption and glucose metabolism

Differentiation of neuroblastoma cells may offer favourable conditions for an investigation of the acquisition of a high energy production characteristic for neurons. In fact, removal of serum from neuroblastoma cultures produced, in parallel with other differentiation phenomena, a striking increase in the oxygen uptake from 0·17 to 0·66 µl per hour/per cell (Nissen et al., 1972). However the change in the oxygen uptake differed greatly from one clone to another. In parallel to the increase in oxygen uptake, we observed an alteration of the activity of some enzymes of the intermediary metabolism: a striking increase of glutamate dehydrogenase and mitochondrial malate dehydrogenase activity whereas lactate dehydrogenase activity decreased (MANDEL et al., 1972). Although the magnitude of change differed from one clone to another, it seems that neuroblastoma cells may be useful for investigating the energetic pathway during neuronal differentiation.

Enzymes involved in transmitter biosynthesis

During the stationary phase of growth or during induced differentiation the specific activity of acetylcholinesterase in some clones, choline acetyltransferase in cholinergic and tyrosine hydroxylase in adrenergic clones increased (for references see page 2).

Our cytochemical findings indicate that some neuroblastoma cells *in vitro* contain both catecholamines and acetylcholinesterase (HERMETET et al., 1972c). Whether the simultaneousness occurrence of catecholamines and acetylcholinesterase is the property of a peculiar cell type or is associated with certain phases of the neuronal differentiation requires further study.

Effect of different inducers

There are numerous reports that butyryl cyclic AMP, analogues of cyclic AMP, phosphodiesterase inhibitors, and bromodeoxyuridine induce differentiation of neuroblastoma cells. Several aspects of this induction were investigated. However neuroblastoma cells vary in their sensitivity to dibutyryl cyclic AMP and bromodeoxyuridine, as has been shown by investigation of enzymatic activities and individual cell morphology by histochemical methods (SCHUBERT and JACOB, 1970; KATES et al., 1971; FURMANSKI et al., 1971; PRASAD and HSIE, 1971; KIRKLAND and BURTON, 1972; RICHELSON, 1972; PRASAD and SHEPPARD, 1972; WAYMIRE et al., 1972; PRASAD et al., 1973).

Using histochemical methods, we found that the relative number of cells possessing processes is similar in different clones either after removal of serum or after addition of bromodeoxyuridine, butyryl cyclic AMP, NGF (HERMETET et al., 1972b) or insulin (HERMETET et al., 1973b) to the serum containing medium. The percentage of processes containing acetylcholinesterase was significantly higher (about twice as much) after addition of the inducer substance than after removal of the serum (MANDEL, CIESIELSKI-TRESKA, GORIDIS, HERMETET and WARTER, in preparation).

After addition of NGF we have also observed an increase of catecholamine fluorescence in a limited number of neuroblasts of a primary culture and in the majority of N115s clonal cells. Acetylcholinesterase activity did not change in NCP-cultures and CCL131 clones but it increased in N9 and N115s clones (HERMETET et al., 1972a; HERMETET, CIESIELSKI-TRESKA, CHAMPY, WARTER and MANDEL, in preparation). The NGF anti-serum did not prevent the appearance of parameters of morphological differentiation, such as the growth of processes. However, cellular alterations developed progressively (vacuolisation, pyknosis) and the specific fluorescence of catecholamines disappeared in all cell lines. Acetylcholinesterase activity was also affected, but unequally.

Neuron-like 'differentiation' was also observed in several clones cultivated in presence of insulin (HERMETET et al.,1973b). The effect was sometimes similar to that of other inducing substances, such as cAMP and bromodeoxyuridine, and higher than that produced by removing calf serum.

Effects of drugs

Incubation of neuroblastoma clones with pargyline was followed by an increased fluorescence of catecholamines. 6-Hydroxydopamine produced a fast degeneration of differentiated adrenergic neuroblasts, leading to vacuolisation, pyknosis, and disappearance of catecholamine fluorescence and of acetylcholinesterase activity. Addition of reserpine to the cultivation medium resulted in a decrease of catecholamine fluorescence in adrenergic clones. Moreover the incubation with reserpine was followed by a net decrease of acetylcholinesterase activity (HERMETET *et al.*, 1972a; HERMETET, CIESIELSKI-TRESKA, CHAMPY, WARTER and MANDEL, in preparation). These results show that neuroblastoma cells can be used to test drugs acting on catecholamine metabolism.

CONCLUSIONS

Neuroblastoma cell cultures afford opportunities to investigate some molecular aspects of nervous cell differentiation and function due to the proliferation and neuron-like differentiation of these cells. However, the different clones present variable-responses to the same effector or inducer, the karyotype is rather complex, the properties of the cells change with the number of passages and a biochemical heterogeneity could be detected within the cells of the same clone by cytochemical methods. Moreover, gangliosides specific for neuronal membranes could not, as yet, be detected and may perhaps explain the lack of synaptic junctions. Although the limitations of biochemical and genetic investigations on the neuroblastoma culture have to be considered, the system can nevertheless be used for investigation of some clearly defined problems like metabolic regulations of putative transmitters, approximate genetic mapping, drug testing.

REFERENCES

AMANO T., RICHELSON E. and NIRENBERG M. (1972) *Proc. Natl. Acad. Sci. U.S.A.* **69**, 258–263.
ANAGNOSTE B., FREEDMAN L. S., GOLDSTEIN M., BROOME J. and FUXE K. (1972) *Proc. Natl. Acad. Sci. U.S.A.* **69**, 1883–1886.
AUGUSTI-TOCCO G. and SATO G. (1969) *Proc. Natl. Acad. Sci. U.S.A.* **64**, 311–315.
BLUME A., GILBERT F., WILSON S., FARBER J., ROSENBERG R. and NIRENBERG M. (1970) *Proc. Natl. Acad. Sci. U.S.A.* **67**, 786–792.
CIESIELSKI-TRESKA J., MANDEL P., THOLEY G. and WURTZ B. (1972) *Nature, New Biology* **239**, 180–181.
DAWSON G., KEMPF S. F., STOOLMILLER A. C. and DORFMAN A. (1971) *Biochem. Biophys. Res. Commun.* **44**, 687–694.
FURMANSKI P., SILVERMAN D. J. and LUBIN M. (1971) *Nature, Lond.* **233**, 413–415.
GORIDIS C., CIESIELSKI-TRESKA J., HERMETET J. C. and MANDEL P. (1973) *C.R. Acad. Sci. Série D*, in press.
HARKINS J., ARSENAULT M., SCHLESINGER K. and KATES J. (1972) *Proc. Natl. Acad. Sci. U.S.A.* **69**, 3161–3164.
HARRIS A. J. and DENNIS M. J. (1970) *Science* **167**, 1253–1255.
HERMETET J. C., CIESIELSKI-TRESKA J., CHAMPY N. and MANDELL P. (1972a) *J. Physiol. (Paris)* **65**, 247–248 A.
HERMETET J. C., CIESIELSKI-TRESKA J. and MANDEL P. (1972b) *C.R. Soc. Biol.* **166**, 1120–1123.
HERMETET J. C., CIESIELSKI-TRESKA J. and MANDEL P. (1972c) *J. Histochem. Cytochem.* **20**, 137–138.
HERMETET J. C., CIESIELSKI-TRESKA J., WARTER S. and MANDEL P. (1973a) *C.R. Soc. Biol.* in press.
HERMETET J. C., CIESIELSKI-TRESKA J., WARTER S. and MANDEL P. (1973b) *J. Physiol. (Paris)* in press.
KATES J. R., WINTERTON R. and SCHLESSINGER K. (1971) *Nature, Lond.* **229**, 345–347.
KIRKLAND W. L. and BURTON P. R. (1972) *Nature, New Biology* **240**, 205–207.

KLEBE R. J. and RUDDLE F. H. (1969) *J. Cell. Biol.* **43,** 69A.

MANDEL P., CIESIELSKI-TRESKA J., HERMETET J. C., HERTZ L., NISSEN C., THOLEY G. and WARTER S. *Intern. Symp.* "*Metabolic Regulation and Functional Activity in the Central Nervous System*", Saint Vincent, Aoste (Italy).

MINNA J., GLAZER D. and NIRENBERG M. (1972) *Nature, New Biology* **235,** 225–231.

MINNA J., NELSON P., PEACOCK J., GLAZER D. and NIRENBERG M. (1971) *Proc. Natl. Acad. Sci. U.S.A.* **68,** 234–239

NELSON P., RUFFNER W. and NIRENBERG M. (1969) *Proc. Natl. Acad. Sci. U.S.A.* **64,** 1004–1010.

NISSEN C., CIESIELSKI-TRESKA J., HERTZ L. and MANDEL P. (1972) *Brain Res.* **39,** 264–267.

PRASAD K. N. (1971a) *Nature, Lond.* **234,** 471–473.

PRASAD K. N. (1971b) *Cancer Res.* **31,** 1457–1460.

PRASAD K. N. (1972) *Nature, New Biology* **236,** 49–52.

PRASAD K. N. and HSIE A. W. (1971) *Nature, New Biology* **233,** 141–142.

PRASAD K. N., MANDEL B., WAYMIRE J. C., LEES G. J., VERNADAKIS A. and WEINER N. (1973) *Nature, New Biology* **241,** 117–119.

PRASAD K. N. and SHEPPARD J. R. (1972) *Exptl. Cell. Res.* **73,** 436–440.

PRASAD K. N., WAYMIRE J. C. and WEINER N. (1972) *Exptl. Cell. Res.* **74,** 110–114.

REBEL G., CIESIELSKI-TRESKA J. and MANDEL P. (1973) *C.R. Acad. Sci. Série D*, in press.

RICHELSON E. (1972) *Nature, New Biology* **242,** 175–177.

ROSENBERG R. N., VANDEVENTER L., DE FRANCESCO L. and FRIEDKIN M. E. (1971) *Proc. Natl. Acad. Sci. U.S.A.* **68,** 1436–1440.

SCHUBERT D., HUMPHREYS S., BARCONI C. and COHN M. (1969) *Proc. Natl. Acad. Sci. U.S.A.* **64,** 316–323.

SCHUBERT D. and JACOB F. (1970) *Proc. Natl. Acad. Sci. U.S.A.* **67,** 247–254.

SCHUBERT D., TARIKAS H., HARRIS A. J. and HEINEMANN S. (1971) *Nature, New Biology* **15,** 79–80.

SEEDS N. W., GILMAN A. G., AMANO T. and NIRENBERG M. (1970) *Proc. Natl. Acad. Sci. U.S.A.* **66,** 160–167.

SIMAN-TOV R. and SACHS L. (1972) *Europ. J. Biochem.* **30,** 123–129.

TRANZER J. P. and SNIPES R. L. (1968) *4th Europ. Regional Conf. on Electron Microscopy, Rome*, Abstr. p. 519.

WAYMIRE J. C., WEINER N. and PRASAD K. N. (1972) *Proc. Natl. Acad. Sci. U.S.A.* **69,** 2241–2245.

YOGEESWARAN G., MURRAY R. K., PEARSON M. L., SANWAL B. D., MCMORRIS F. A. and RUDDLE F. H. (1973) *J. Biol. Chem.* **248,** 1231–1239.

ZWILLER J., CIESIELSKI-TRESKA J., MACK G. and MANDEL P. (1973) *J. Physiol. (Paris)*, in press.

Frontiers in Catecholamine Research 1973, pp. 285 to 288. Pergamon Press. Printed in Great Britain.

REPORTER'S NOTES

Norman Weiner

THE SESSION on regulation of enzymes of biosynthesis was devoted to three main topics: (1) Mechanisms by which increased biosynthetic enzyme levels may be produced; (2) Mechanisms by which the activity of biosynthetic enzymes may be modified; and (3) Ontogenetic development of the enzymes involved in synthesis of neurotransmitters in the sympathetic nervous system.

Dr. Thoenen reviewed the several mechanisms by which increased tyrosine hydroxylase levels may be produced. Each of these is presumably associated with increased transsynaptic impulse traffic in adrenergic neurons, although there is little direct evidence that increased impulse traffic is associated with such experimental situations as cold, swim and immobilisation stress or the administration of drugs such as reserpine, insulin, morphine and α-adrenergic blocking agents. Each of these procedures may be associated with increased levels of tyrosine hydroxylase, generally demonstrated either in the adrenal medulla or in sympathetic ganglia. The increased enzyme levels may be prevented by ganglionic blocking agents or decentralisation.

Dr. Thoenen presented evidence that intermittent swim stress for 1–2 hr to rats was associated with increased tyrosine hydroxylase levels in the superior cervical ganglion, the stellate ganglion and in the adrenal gland which reached maximal levels at about 24 hr. No change in dopa decarboxylase levels was demonstrable. Early administration of actinomycin D blocks the increase. Dr. Thoenen believes the 'first messenger' in this phenomenon is acetylcholine, and the effect is mediated through nicotinic receptors. He challenged the theory proposed by GUIDOTTI and COSTA (*Science* **179**, 902–904, 1973) that the effect is mediated by the 'second messenger' cyclic AMP, since, although cold stress is associated with increased cyclic AMP in the adrenal medulla, swim stress is not. Both are associated with increased tyrosine hydroxylase levels at approximately 24 hr. Guidotti reviewed the evidence in support of a role of cyclic AMP in mediating the transsynaptic increase in tyrosine hydroxylase levels in the adrenal gland. The increases of both cyclic AMP and tyrosine hydroxylase are blocked by denervation, although carbachol, acting directly on nicotinic receptors on the adrenal medulla cell, can bring about the elevations after denervation. Guidotti indicated that after one hour of cold stress, the early rise in cyclic AMP persists a few hours followed by a rise in tyrosine hydroxylase at 24–48 hr. The later rise in tyrosine hydroxylase cannot be blocked by hexamethonium if it is given later than two hours after the cold stress. He presented interesting evidence that cyclic GMP falls after cold stress and, therefore, the ratio of cyclic AMP/cyclic GMP increases markedly. HADDEN, HADDEN, HADDOX and GOLDBERG (*Proc. Natn. Acad. Sci. U.S.A.* **69**, 3024–3027, 1972) have proposed that physiological responses to cyclic nucleotides are determined by the ratio of these two nucleotides in cells, rather than the absolute level of either one. It is thus possible that Dr. Thoenen's results with swim stress, where tyrosine hydroxylase rises, may be explained by a fall in cyclic GMP rather than a rise in cyclic AMP.

Dr. Mueller presented some provocative pharmacological evidence that sympathetic nervous system activity may be modulated by serotoninergic neurons. Thus, although 5,6-dihydroxytryptamine administration did not affect adrenal tyrosine hydroxylase levels in the rat, insulin plus this substance produced a greater effect on the enzyme levels than did insulin alone. Similarly, the rise in adrenal tyrosine hydroxylase produced by electrical stimulation in the brain stem is potentiated by administration of p-chlorophenylalanine. These results, which suggest that 5-HT neurons may antagonise peripheral sympathetic nervous system outflow, are consistent with earlier findings of ANDÉN, CARLSSON and HILLARP (*Acta Pharmacol. Toxicol.* **21,** 183–186, 1964) who observed that 5-hydroxytryptophan inhibited insulin-induced catecholamine depletion of the adrenal gland. Dr. Bhagat briefly reviewed evidence from his laboratory that suggested that only those hypotensive agents which deplete catecholamines from the adrenal gland produce elevations of adrenal tyrosine hydroxylase. Thus, administration of reserpine or phenoxybenzamine was associated with both a depletion of adrenal catecholamines and elevation of tyrosine hydroxylase, whereas guanethidine, pargyline, phentolamine and bretylium produced neither effect. No data on the degree of hypotension produced by these drugs in the rats were presented.

Hanbauer presented evidence suggesting that increased adrenal gland tyrosine hydroxylase levels after cold stress was dependent on two factors: (1) duration of cold stress, which must exceed 4 hr: and (2) duration of time between cold stress and removal of adrenals, which requires about 40 hr. Decentralisation at 24 hr prevents the increase in tyrosine hydroxylase which normally follows the 4 hr of cold stress. She speculated that the first phase may be mediated by muscarinic receptor stimulation whereas the second phase is mediated via nicotinic receptors.

Dr. Kopin reviewed his studies on the behaviour of rat superior cervical ganglia in culture. After 1 week in culture, the cell bodies produce sprouts which may be analogous to axon formation. Correlated with this sprouting is enhanced ^{3}H-norepinephrine uptake which is quite cocaine sensitive, reminiscent of adrenergic nerve ending behaviour. The outgrowth of these processes into irides in culture can be stimulated by nerve growth factor. Field stimulation is associated with enhanced release of ^{3}H norepinephrine from these cultured ganglia. The release is enhanced by phenoxybenzamine which supports the contention of STARKE (NAUNYN–SCHMIEDEBERG's *Arch. Pharmakol.* **274,** 18–45, 1972) that the effect of this drug in enhancing exocytocis is at a presynaptic locus. In this ganglion in culture, the postsynaptic tissue is absent.

Kopin also observed enhanced levels of tyrosine hydroxylase, dopamine-β-hydroxylase and enhanced ^{3}H norepinephrine uptake in cultured ganglia exposed to dibutyryl cyclic AMP. He emphasised that these generalised increases in ganglion activities should be considered more analogous to 'hypertrophy' than 'induction' which has a much more specific connotation. Kopin also pointed out that the increase in cyclic AMP in ganglia demonstrated by GREENGARD and co-workers (*Adv. Cyclic Nucleotide Research* Vol. 1, 337–355, 1972) is different from the increase in tyrosine hydroxylase produced transsynaptically, which also may be mediated by this cyclic nucleotide, since the former is blocked by α-blocking agents and muscarinic blocking agents, but the latter is blocked by nicotinic blocking agents.

McKay has demonstrated that exposure of rat superior cervical ganglia in culture to high K$^+$ or to dibutyryl cyclic AMP is associated with increased tyrosine hydroxylase levels, whereas no such effect is apparent with reserpine. Catecholamine depletion by reserpine does not block the rise in enzyme level produced by K$^+$. He therefore concludes that the level of catecholamines in the ganglion is not a critical factor regulating tyrosine hydroxylase levels.

Richelson summarised his findings on the levels of tyrosine hydroxylase in the 'adrenergic clone' N1E-115. The specific activity of this enzyme in the cells increases about 30-fold during 12 days of incubation. He demonstrated two uptake processes for tyrosine in this clone, a non-saturable process, presumably diffusion, and a saturable process with a K_T of $2 \cdot 5 \times 10^{-5}$ M. He has also confirmed earlier observations (WAYMIRE et al., Proc. Nat. Acad Sci. **69**, 2241–2245, 1972) that dibutyryl cyclic AMP increases levels of tyrosine hydroxylase in adrenergic clones of neuroblastoma. Richelson also showed that sodium butyrate, in the presence of fetal calf serum, produces increased tyrosine hydroxylase levels in this clone and arrests growth, confirming the report of Waymire et al. However when sodium butyrate was added to neuroblastoma cells grown in the absence of fetal calf serum, neither inhibition of cell division nor increased levels of tyrosine hydroxylase were observed. In contrast, dibutyryl cyclic AMP is associated with increased levels of the enzyme without inhibition of cell growth. PRASAD [In: *The Role of Cyclic Nucleotides in Carcinogenesis* (GRATZER and SCHULTZ, Eds.) Academic Press, New York, pp. 207–237, 1973] observed that sodium butyrate can increase endogenous cyclic AMP levels and has suggested that this may be the mechanism by which increased tyrosine hydroxylase levels are mediated. Richelson did not report on cyclic AMP levels in his studies.

P. Mandel reviewed his studies on neuroblastoma cells. He has demonstrated increased levels of many enzymes in these cells during proliferation, including acetylcholinesterase and glucose metabolising enzymes. He rightly cautioned about interpretation of results of changes in the biochemistry of these cells since clones are extremely pleiotropic, with variable chromosome numbers and morphology. Genetic mapping in these cells would be extremely difficult. Furthermore, the distribution of enzymes in these cells is not at all uniform, as demonstrated by histochemical studies of monoamine oxidase and acetylcholinesterase. Thus, increased levels of an enzyme in a cell population could indicate selection of that particular cell type by the drug addition or other procedure employed, rather than enzyme 'induction'. He presented evidence that several amines, including norepinephrine, dopamine, epinephrine, 5-hydroxytryptamine and histamine, are present in the 'adrenergic' clone, NIE-115. He also pointed out that gangliosides do not increase consistently in neuroblastoma cells undergoing differentiation, suggesting that these cells may lack synaptic neurones and, perhaps, other features which are characteristic of nerve cells.

Molinoff proposed that dopamine-β-hydroxylase could be a rate-limiting enzyme reaction in norepinephrine synthesis. He reminded the audience that the concept of rate limiting steps in a biosynthetic pathway apply only when: (1) enzymes and substrates have free access to each other; and (2) there are no branch points in the pathway. Neither of these criteria are met in the norepinephrine biosynthesis path. Molinoff reviewed the evidence that levels of both tyrosine hydroxylase and dopamine-β-hydroxylase are increased after increased sympathetic nervous activity. He presented evidence that large amounts of dopamine-β-hydroxylase inhibitors are present in both heart and spleen, and, presumably, other tissues. Some of these inhibitors may be located in the sites where dopamine β-hydroxylase is located, according to his gradient density centrifugation studies of splenic homogenates. Molinoff showed that the inhibitor in the spleen is heat stable, but most of that in the heart is heat labile, suggesting different types of inhibitors. Kaufman suggested that, if these are sulfhydryl compounds, deficiencies in heat stability may indicate differences in heavy metal ion composition in the two tissues.

Weiner reviewed his earlier observations that tyrosine hydroxylase in intact adrenergic tissue may be regulated by feedback inhibition by catecholamines present in the neuron. This pool of amine would have access to the soluble tyrosine hydroxylase. He presented a means of estimating the concentration of pterin cofactor in the nerve ending, in '6,7-dimethyl-tetrahydropterin equivalents'. The amount in the adrenergic neuron was calculated to be 0·15 mM. Since the putative natural cofactor, tetrahydrobiopterin, is about five times as potent as the synthetic pterin, the concentration of pterin cofactor may be approximately 0·03 mM in the nerve terminals. Weiner presented evidence that the increased tyrosine hydroxylase activity associated with nerve stimulation may not be due to reduced end-product feedback inhibition, but probably involves some other mechanisms, perhaps ion fluxes or metabolic alterations associated with neuron depolarisation and repolarisation. Total Na^+ deletion and ouabain produced similar effects: they abolished the effect of nerve stimulation by increasing the tyrosine hydroxylase activity of control preparations to a level like that of stimulated preparations. Since the responses of the tissue were also abnormal, these results are not easily interpretable. Gutman showed evidence that tyrosine hydroxylase prepared from several tissues exhibited modestly increased activity when Ca^{2+} was increased from 50 to 100 μM. Above this concentration enzyme activity was depressed. Weiner indicated that it would be difficult to evaluate, Ca^{2+} effects on the enzyme with nerve stimulation in intact tissue, since Ca^{2+} is critically required for the neurotransmitter release.

Brimijoin summarised his findings on the accumulation of dopamine-β-hydroxylase distal to either a single or double ligation of the sciatic nerve, an observation suggesting 'retrograde flow' of axoplasm. The physiological relevance of this phenomenon in a nerve which is treated in such a traumatic fashion remains controversial. He also presented evidence suggesting that neuronal damage, e.g. with axotomy or 6-hydroxydopamine administration, may be associated with reduced dopamine β-hydroxylase synthesis, perhaps because the machinery of the neuron is now 'switched over' to making structural, rather than functional, macromolecules.

Black described studies on the ontogenetic development of enzymes in the mouse superior cervical ganglion. He documented a correlation between increased tyrosine hydroxylase and synapse numbers during development. Decentralisation leads to a failure in development of tyrosine hydroxylase, dopamine-β-hydroxylase, dopa decarboxylase and monoamine oxidase. Nerve growth factor could not replace presynaptic innervation in restoring this enzyme development process. Ganglionic blocking agents arrested ganglion development of tyrosine hydroxylase and dopamine-β-hydroxylase without affecting choline acetyltransferase. Coyle summarised his systematic studies on the development of the several enzymes involved in norepinephrine biosynthesis in several regions of the developing rat brain. Lydiard and Sparber reported that reserpine administration to chick embryos resulted in a persistent increase in tyrosine hydroxylase and catecholamine levels in the brain and suggested that some permanent developmental alteration may be produced by affecting neurotransmitter processes during early development.

Levi-Montalcini reviewed her interesting studies on nerve growth factor and nerve growth factor antiserum. She provided evidence that the destructive effects of either guanethidine or 6-hydroxydopamine on rat ganglion cells could be prevented by administration of nerve growth factor.

Synaptic Dynamics of Receptors
(May 22, 1973; 3:00 P.M.–6:30 P.M.)
(May 23, 1973; 9:00 A.M.–12:00 noon)

CHAIRMEN: Robert F. Furchgott and Neil C. Moran
COUNCILLOR: Sydney Spector
REPORTER: G. Alan Robison

Frontiers in Catecholamine Research 1973, pp. 291 to 294. Pergamon Press. Printed in Great Britain.

THE CLASSIFICATION OF ADRENERGIC RECEPTORS*

NEIL C. MORAN

Department of Pharmacology, Division of Basic Health Sciences, Emory University,
Atlanta, Georgia, U.S.A.

TWO ERAS of research into adrenergic receptors overlap in time in this conference. The first era, that of description and classification, began with LANGLEY's (1905, 1906) suggestion of cellular receptive sites for drugs. It has been a period of evolution of concepts of receptors as distinct and specific cellular entities as well as a period of classification and is now coming to a close. The second era, that of isolation, is just beginning. It should bring detailed knowledge of the chemical characteristics of the adrenergic receptors and their molecular interactions with agonists and antagonists.

The concept of adrenergic receptors is based on several established facts. (1) There is a strict relationship between chemical structure of adrenergic agonists and biological activity. This is particularly notable in the marked differences in biological potency of the enantiomorphs of catecholamines such as those of norepinephrine. (2) The number of molecules of potent adrenergic agonists needed to elicit a cellular response is exceedingly small. (3) There is a precise quantitative relationship between the concentration of the drug and the degree of cellular response. (4) The effect of agonists is quickly and completely reversed either by removal of the agonist from the tissue or by addition of a selective antagonist. This suggests simple binding to the receptor molecule with low energy bonds. (5) Reversible, selective antagonism of agonist-induced responses can best be explained by blockade of receptors. (6) Neither agonist nor antagonists are consumed or altered by the reaction with the receptors.

The adrenergic receptors then can be viewed as macromolecules of distinct three dimensional configurations complementary to those of catecholamine molecules with D ($-$) configuration. Interaction between catecholamines and receptors through low energy bonds triggers a cellular response. Both the agonist–receptor interaction and the resulting cellular response can be antagonised reversibly by specific receptor blocking drugs that also bind with low energy bonds.

Attempts to characterise adrenergic receptors have been of three main types. First, attempts to isolate and chemically characterize receptors have as yet not been too fruitful (LEFKOWITZ and associates, 1971, 1972, 1973). Dr. Lefkowitz deals with his attempts to isolate the cardiac *beta* adrenergic receptor in this symposium. Second, workers such as BELLEAU (1967) have speculated on the nature of reactive groups on the receptor on the basis of knowledge of the three dimensional structure of agonists and antagonists. Third, the classical approach, essentially operational, empirical and descriptive, is based on analyses of responses of effector systems to agonists and antagonists and to combinations thereof. The classic method can be represented symbolically by the following relationships

$$Ag + R \rightleftharpoons AgR \rightarrow Effect$$
$$An + R \rightleftharpoons AnR \dashv | No\ Effect$$

---

* Publication number 1142 of the Division of Basic Health Sciences, Emory University.

in which *Ag* is an agonist, *An* an antagonist, and *R* the receptor. This relationship can be treated quantitatively according to the Law of Mass Action by varying the concentrations of *Ag* alone or by using *An* in varying concentrations and measuring the degree of *Effect*. This approach has led to important concepts of drug receptor interaction. However, often it has been oversimplified by failure to recognize fully factors that are antecedent to the drug receptor interaction (i.e., passage of the drug to the receptor) and more importantly the steps consequent to the drug-interaction. The latter can be expressed as follows:

$$Ag + R \rightarrow AgR \rightarrow a \rightarrow b \rightarrow c \rightarrow n \rightarrow Effect.$$

Since few drug effects are fully understood in terms of the steps consequent to drug-receptor interaction, it is difficult to define precisely the site of action, particularly of antagonists (MORAN, 1966, 1973).

Classification and operational analysis of receptors require rigid criteria. As FURCHGOTT (1972) has clearly stated, all secondary processes that influence the concentration of the drug at the receptor must be accounted for. Thus, processes that inactivate a drug (e.g., metabolism and uptake into storage sites) and indirect mimetic effects (e.g., induced release of norepinephrine) must be inhibited. In short, the concentration of the drug in the extracellular fluid should reflect as accurately as possible the concentration at the receptor.

Classification of receptors has two bases, (1) analysis of the orders of potency of a closely related series of agonists on effector systems and (2) selective blockade of responses to agonists.

The first method, analysis of the orders of potency of agonists, is based on the assumption that a given type of receptor, regardless of its distribution in various tissues, will react optimally with one agonist of a given series. That is, there will be a "best fit" between the agonist and the receptor and that the "fit" with other molecules in the same chemical series will be less good. The order of reactivity with a series of agonists, e.g., biological potency of the agonist, will be the same for all systems that have that receptor. However, if two or more types of adrenergic receptors exist in various effector systems, the order of potency of the series of agonists will vary. That is, one receptor type will have an optimum fit with one drug of the homologous series (e.g., norepinephrine), while another receptor type will have an optimum fit with another drug (e.g., isoproterenol). This was the approach that AHLQUIST (1948) employed when he first classified adrenergic receptors. He tested a series of six sympathomimetic amines on various physiological systems and found no uniform single order of potency. Those systems in which isoproterenol was least potent were designated *alpha* and those in which it was the most potent agonist were designated *beta*.

The most powerful tool for the classification of receptors, however, is selective blockade. At the time Ahlquist classified adrenergic receptors only one type of antagonist was known, the type we today label as *alpha* adrenergic blocking drug. It was not until the discovery of dichloroisoproterenol as a selective antagonist of certain adrenergic responses that Ahlquist's classification was validated (MORAN and PERKINS, 1958).

Table 1 shows the schemes for classification of adrenergic receptors into two main classes, *alpha* and *beta*, by comparison of orders of potency of agonists and by

TABLE 1. SCHEME FOR CLASSIFICATION OF ADRENERGIC RECEPTORS

Basis of classification	Receptor types			
	alpha	*beta₁*	*beta₂*	*dopamine*
Potency of agonists	$NE \geq E > D > I$	$I > E \geq NE \gg D$	$I > E \gg NE \gg D$	$D \ggg E, NE, I$
	$S = 0$	$S = +$	$S = + +$	$S = 0$
Selective blockade				
alpha antagonists,				
e.g., POB, PHEN, ERG	Block	No block	No block	No block
beta antagonists				
General, e.g., DCI, PROP	No block	Block	Block	No block
beta₁ type, e.g., PRACT	No block	Block	Weak block	No block
beta₂ type, e.g., BUTOX	No block	Weak block	Block	No block
Dopamine antagonists,				
e.g., HAL	No block	No block	No block	Weak block

NE—norepinephrine, E—epinephrine, I—isoproterenol, D—dopamine, S—salbutamol, POB—phenoxybenzamine, PHEN—phentolamine, ERG—ergot alkaloids, DCI—dichloroisoproterenol, PROP—propanolol, PRACT—practolol, BUTOX—butoxamine, HAL—haloperidol.

selective blockade. It also shows the more recent subdivision of *beta* adrenergic receptors into two subclasses, *beta₁* and *beta₂* (LANDS et al., 1967). The latter sub-classification appears to be valid on the basis of orders of potency and selective blockade. That is, the order of potency of three agonists (isoproterenol, I; epine-phrine, E; norepinephrine, NE) for the *beta₁* receptor is $I > E \geq NE$ (e.g., the heart) and for the *beta₂* receptor $I > E \gg NE$ (e.g., bronchiolar smooth muscle). Further-more, the new agonist, salbutamol, is highly selective for the *beta₂* receptor (FARMER et al., 1970). Selective blockade also allows a distinction between these two subtypes. Thus, practolol is a selective antagonist for *beta₁* receptors in that it is more potent on *beta₁* than on *beta₂* receptors (DUNLOP and SHANKS, 1968). Butoxamine is many times more active as an antagonist of *beta₂* than of *beta₁* receptors (LEVY and WILKENFELD, 1969).

The methods used in the classification of receptors in various tissues should be of use in the identification of isolated receptors. That is, the order of binding constants of a series of sympathomimetic drugs to a putative receptor should be the same as the order of biological potency as agonists in physiological or biochemical systems. The isolated receptor also should show selective binding of the specific antagonist, and the antagonist should prevent binding of the agonists. Finally, although demonstration of stereospecificity is vitally important, this demonstration in itself will not aid the dis-tinction of an *alpha* from a *beta* receptor inasmuch as both receptors are stereospecific for the D (−) form of the catecholamines. MOLINOFF (this volume) has outlined in more detail the criteria needed for isolation and identification of receptors.

The final criterion of the identity of the isolated adrenergic receptor, whether *alpha* or *beta*, will be the demonstration that the receptor, reconstituted with other cellular components, can initiate a response when activated by an agonist. It is important at this time to not assume too adamantly that adenylate cyclase is the universal link with the *beta* adrenergic receptor. Its role as an important link between the receptor and the second messenger in adrenergically induced glycogenolysis and lipolysis seems clear, but it is not established that this enzyme is linked to adrenergically induced augmenta-tion of cardiac contractility (KJEKHUS et al., 1971). Thus, an isolated cellular component meeting the criteria for the cardiac *beta* adrenergic receptor (i.e., appropriate binding constants, stereospecificity, etc.) might not be linked to adenylate cyclase. Another as yet undetected link should be considered.

Careful adherence to these criteria for classification and awareness of the fact that adrenergically-induced effects can be antagonised by drugs that act distal to the receptor can help us avoid indulging in "pharmacological guilt by association". For example, the fact that an adrenergically-induced effect is antagonised by an *"alpha* adrenergic blocking drug"* does not mean that the antagonism is necessarily at the receptor locus. It could be distal to the receptor, i.e.,

$$An$$
$$Ag + R \rightarrow AgR \rightarrow a \rightarrow \| \quad b \quad c \quad n \quad No\ effect$$

It is important to use several drugs of a given class, particularly *alpha* adrenergic blocking drugs, to obtain more conclusive evidence of receptor antagonism. If only one of several different *"alpha"* antagonists blocks a given response it is unlikely that the blockade is at the receptor level. If, however, several different *alpha* blocking drugs all antagonise the effect, but *beta* blocking drugs and other types of antagonists do not, then the attribution of blockade at the receptor level is more certain.

REFERENCES

AHLQUIST R. P. (1948) *Amr. Physiol.* **153**, 586–600.
BELLEAU B. (1967) *Ann. N.Y. Acad. Sci.* **139**, 580–605.
DUNLOP D. and SHANKS R. G. (1968) *Brit. J Pharmacol.* **32**, 201–218.
FARMER J. B., LEVY G. P. and MARSHALL R. J. (1970) *J. Pharm. Pharmacol.* **22**, 945–947.
FURCHGOTT R. F. (1972) In: *Catecholamines* pp. 283–335. (Ed. BLASCHKO H. and MUSCHOLLL E.) Berlin, Springer-Verlag.
KJEKHUS J. K., HENRY P. D. and SOBEL B. E. (1971) *Circulat. Res.* **29**, 486–478.
LANDS A. M., ARNOLD A., MCAULIFF J. P., LUDUENA F, P. and BROWN R. G., Jr. (1967) *Nature, Lond.* **214**, 597–598.
LANGLEY J. N. (1905) *J. Physiol., Lond.* **33**, 347–413.
LANGLEY J. N. (1906) *Proc. Roy. Soc. Lond.* **B78**, 170–194.
LEFKOWITZ R. J. and HABER E. (1971) *Proc. Natl. Acad. Sci. U.S.A.* **68**, 1773–1777.
LEFKOWITZ R. J., HABER E. and O'HARA D. (172) *Proc. Natl. Acad. Sci. U.S.A.* **69**, 2828–2832.
LEFKOWITZ R. J., SHARP G. W. G. and HABER E. (1973) *J. Biol. Chem.* **248**, 342–349.
LEVY B. and WILKENFELD B. E. (1969) *Europ. J. Pharmacol.* **5**, 227–234.
MORAN N. C. and PERKINS M. E. (1958) *J. Pharmacol. Expl. Therap.* **124**, 223–237.
MORAN N. C. (1966) *Pharmacol. Rev.* **18**, 503–512.
MORAN N. C. (1974) In: *Handbook of Physiology, Section 7. Endocrinology, The Adrenal.* In press.

Frontiers in Catecholamine Research 1973, pp. 295 to 299. Pergamon Press. Printed in Great Britain.

ANTAGONISM OF PROPRANOLOL TO ISOPROTERENOL IN GUINEA-PIG TRACHEA: SOME CAUTIONARY FINDINGS

ROBERT F. FURCHGOTT, ARON JURKIEWICZ and N. H. JURKIEWICZ

Department of Pharmacology, State University of New York, Downstate Medical Center, Brooklyn, New York 11203, U.S.A.

IT MAY appear somewhat old-fashioned at a session in which attempts to isolate and to characterise chemically β-adrenergic receptors will be reported, to present an introductory paper on a pharmacologic study of competitive antagonism. However, we believe that findings made in studies such as this are highly relevant to research directed at receptor isolation, since they can provide important criteria for judging whether some isolated material is indeed the receptor.

According to receptor theory for competitive antagonism, the following equation should apply under equilibrium conditions:

$$\log (dr - 1) = \log [B] - \log K_B \qquad (1)$$

where dr (dose ratio) is the ratio of the concentrations of agonist in the presence and absence of the antagonist, respectively, which give an equal response; $[B]$ is the concentration of antagonist, and K_B is the dissociation constant of the antagonist–receptor complex. A plot of $\log (dr - 1)$ against $\log [B]$ should give a straight line with a slope of one. At $\log (dr - 1)$ equal to zero (i.e., for a dr of 2), $\log [B]$ equals $\log K_B$. (Under ideal conditions the pA_2 of SCHILD (1949) would be equal to $-\log K_B$.)

It has been emphasised (FURCHGOTT, 1967, 1970, 1972) that in the experimental determination of the K_B (or pA_a) of a competitive antagonist a number of conditions have to be satisfied in order to ensure reliable values. Assuming that equation (1) does apply under ideal conditions, then one indication that something is wrong with the actual conditions would be the failure of experimental points in a Schild-plot (i.e., $\log (dr - 1)$ vs. $\log [B]$) to fit a straight line with a slope of one. In studies on antagonism to norepinephrine by propranolol in guinea-pig atria (BLINKS, 1967), by pronetholol in the same tissue (FURCHGOTT, 1967), by phentolamine in cat nictitating membrane (LANGER and TRENDELENBURG, 1969), and by propranolol in the guinea-pig trachea (MOORE and O'DONNELL, 1970), the plotted points fell along curves with slopes significantly less than one. The investigators in all of these studies postulated that this discrepancy was due to an active neuronal uptake of norepinephrine within the tissue, and obtained direct evidence for this by showing that data obtained after blockade or elimination of neuronal uptake did fit straight lines with slopes close to one.

In preparing a recent review on the classification of adrenergic receptors, one of us was struck by the fact that in practically all studies on the antagonism between isoproterenol (ISO) and various β-receptor antagonists in isolated guinea-pig trachea, the Schild-plots of the experimental data best fitted curves with slopes significantly less than one (Furchgott, 1972). The low slopes cannot be attributed to a neuronal uptake of ISO. However, since there is evidence for extraneuronal uptake of ISO in trachea (Foster, 1969), it seemed likely that this mechanism might account for the low slopes. The present experimental work was initiated to test this possibility. However, before reporting some of our results, we first would like to introduce briefly a model containing a saturable removal mechanism for the agonist, since this model allows us to derive an equation for the construction of hypothetical log $(dr - 1)$ vs. log [antagonist] curves which reveal how removal of the agonist can influence the slope.

The model is shown in Fig. 1. The equation derived from the model for the case of competitive antagonism under steady-state conditions is (for details see Furchgott, 1972):

$$\frac{[A_a]}{[A_b]_0} = (1 + [B]/K_B)\left\{1 + \frac{V_m/K_{AX} \cdot k}{1 + ([A_b]_0/K_{AX})(1 + [B]/K_B)}\right\} \tag{2}$$

where $[A_b]_0$ is the concentration of free agonist in the region of the receptors required for a given response in the absence of antagonist; $[A_a]$ is the concentration of free agonist in the external solution required for the same response; k is the diffusion constant (see Fig. 1); V_m is the maximal rate and K_{AX} is the Michealis–Menten constant for the removal process; and $[B]/K_B$ is the concentration of free antagonist in the region of the receptors relative to its dissociation constant for the receptor. For any given values of $V_m/K_{AX} \cdot k$ and $[A_b]_0/K_{AX}$, the equation permits a calculation of the dose ratio (dr) for the agonist in the external solution at any value of $[B]/K_B$. Examples of hypothetical curves for log $(dr - 1)$ vs. log $[B]/K_B$ constructed from this equation have been presented elsewhere (Furchgott, 1972). It will suffice to state here that if $V_m/K_{AX} \cdot k$ is sufficiently large and $[A_b]_0/K_{AX}$ is sufficiently small, then the slope of the curve will be considerably less than one over a given range of log $[B]/K_B$.

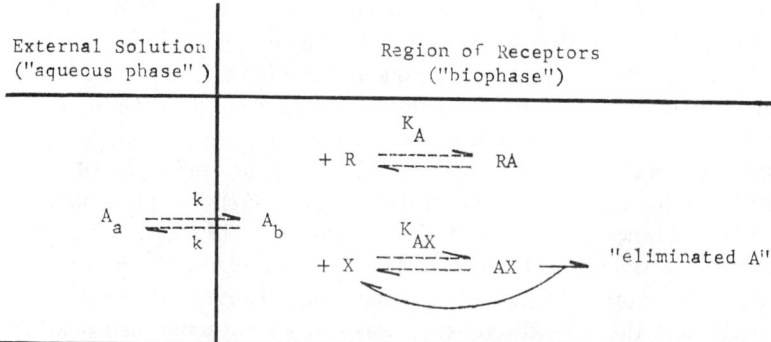

Fig. 1.—Schematic model for an effector system containing a saturable removal mechanism for the agonist. First order kinetics are assumed for diffusion of the agonist, and k is the diffusion constant. A_a and A_b, respectively, represent the free agonist in the external solution and in the region of the receptors. Michealis–Menten kinetics are assumed to apply for elimination of the agonist. R represents the receptor and RA the receptor–agonist complex. Under steady-state conditions, $[A_a]/[A_b] = 1 + (V_m/K_{AX} \cdot k)/(1 + [A_b]/K_{AX})$.

In each of our experiments, four helically-cut strips from the trachea of a reserpine-pretreated guinea pig were mounted in Krebs solution at 37°. Recording was isometric. Each concentration–response curve for relaxation by ISO was obtained with cumulative additions of the drug after the smooth muscle had been brought to a steady level of contraction with carbachol. Each experimental strip was first tested with ISO in the absence, and then successively in the presence, of one or more concentrations of propranolol (PPL). The exposure time to each PPL concentration was close to 75 min. Concentrations of ISO used for calculating dose ratios were those required to produce one-half maximal relaxation. In practically all experiments, one strip received no PPL and served as a control.

In order to block extraneuronal uptake of ISO, some strips were pretreated with 3 μg/ml of dibenamine for 40 min. This pretreatment produced a maximal increase in sensitivity to ISO, which was about 8·5-fold on the average and persisted for many hours. (Blockade of α-adrenergic receptors by phentolamine produced no increase. Pretreatment with phenoxybenzamine could produce essentially the same maximal increase as dibenamine, but was not suitable because it inhibited too greatly the contraction to carbachol on which ISO was tested. The COMT inhibitor, 4-tropolone acetamide, increased sensitivity only about 3-fold, and failed to add to the increase produced by dibenamine pretreatment.)

In the case of both untreated and dibenamine-pretreated strips, PPL produced an essentially parallel shift of the log concentration–response curve of ISO to the right. However, the degree of shift with any given concentration of PPL above 3×10^{-9} M was always less in the case of untreated strips. Schild-plots of the data obtained in a large number of experiments using PPL concentrations from 10^{-9} to 10^{-5} M are shown in Fig. 2. In the case of the untreated strips, the data fit very well a theoretical curve generated with equation (2). Over an intermediate range (3×10^{-9} to 3×10^{-7} M) the average slope of the curve is only 0·7; and thus confirms the findings of others on the low slope of curves in Schild-plots for antagonism between ISO and β-receptor antagonists in guinea-pig trachea. However, as the range of PPL is extended, the slope of the curve steepens and approaches unity both at very low concentrations of PPL (where $[A_b] \ll K_{AX}$) and at very high concentrations (where $[A_b] \gg K_{AX}$). It should be noted that if individual K_B values for PPL are calculated from each of the mean experimental points along the curve, they will progressively increase almost ten-fold in going from the point for the lowest concentration to that for the highest.

In the case of dibenamine-pretreated strips, all the points lie close to a straight line with a slope of one. (The calculated regression line actually has a slope of 0·997.) The regression line gives a pA_2 of 9·15, equivalent to a K_B value of $7·1 \times 10^{-10}$ M. Individual K_B values calculated from each of the plotted points are all very close to this value. Thus, after dibenamine pretreatment, the experimental data are in excellent agreement with the prediction of equation (1); and support the argument that the discrepant data obtained with untreated strips result from the removal of ISO from the region of the receptors by a saturable uptake mechanism.

The results described provide additional evidence for the conclusion that the pharmacologic antagonism between agonists and antagonists acting on adrenergic receptors is satisfactorily explained by the classical receptor theory for competitive antagonism. If and when the isolation and purification of an adrenergic receptor

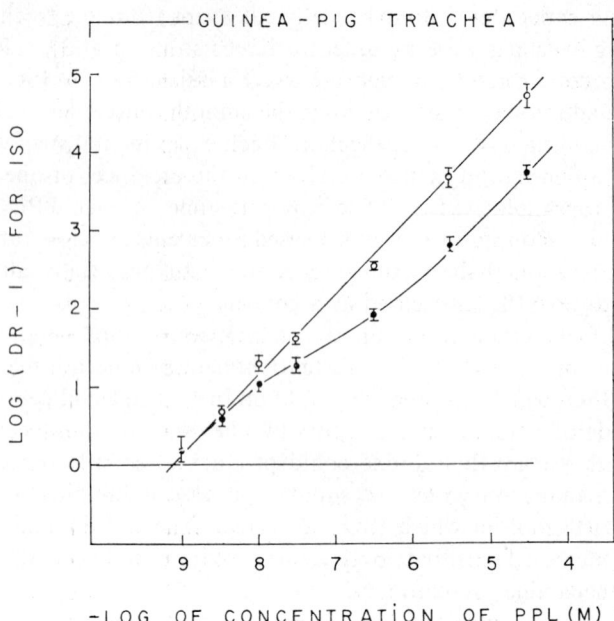

FIG. 2.—Effect of dibenamine pretreatment on competitive antagonism between propranolol (PPL) and isoproterenol (ISO) in the isolated tracheal strip of the guinea pig. Strips were either untreated (●), or pretreated with 3 μg/ml dibenamine for 40 min (○) to block extraneuronal uptake. Each plotted point represents the mean of values obtained with 7 to 47 individual strips. The lower theoretical curve comes from equation (2) with $[A_b]_0 = 0.03$, $V_m/K_{AX} . k = 7.6$, and $K_B = 7.1 \times 10^{-10}$ M. The upper curve (straight line) comes from equation (1), using the same K_B value.

from a given tissue have been achieved, the findings on competition between agonists and antagonists for binding to the receptor should be consistent with the pharmacologic findings on competitive antagonism between these agents in the intact tissue.

We have recently begun to investigate the kinetics of development of blockade by PPL, and of recovery after washout. In six experiments with 3×10^{-7} M PPL, changes in the dose ratio of ISO with time indicate that about 150 min is required to attain essentially complete equilibration of the free concentration of PPL in the region of the receptors with that in the external solution. After 75 min (the exposure time used in the experiments of Fig. 2) equilibration appears to be about 80–90% complete. Recovery from PPL blockade is even slower than development. The decline in the dose ratio of ISO with time after washout of 3×10^{-7} M PPL is consistent with an approximately exponential decline of the free concentration of PPL in the region of the receptors, with a $t_{\frac{1}{2}}$ of the order of 75 min. Our findings indicate that the kinetics of onset and offset of PPL blockade are even slower in the tracheal smooth muscle than in the atria of the guinea pig (Potter, 1967).

For reasons that cannot be considered in this limited space, we do not believe that the very slow kinetics of recovery after washout of PPL reflect a very low rate constant for dissociation of the antagonist–β-receptor complex. Rather, we feel that they can largely be attributed to the presence in the region of the receptors of a large concentration of non-receptor binding sites (acceptors) for PPL. The affinity of these acceptors for this drug would be 2 or 3 orders of magnitude less than that of the β-receptors; but their concentration would be so high that the concentration of PPL reversibly bound to them would greatly exceed the concentration of free PPL in the region of the receptors when the latter concentration was much below the dissociation constant(s) of the acceptor-PPL complex(es). After washout, when free PPL is escaping from the region of the receptors, the dissociable complex(es) would serve as a reservoir or buffer system to supply free PPL in that region, and thus greatly retard the rate of fall of the free concentration there. (This model for explaining the slow kinetics of offset of blockade resembles the "limited biophase" model (see RANG, 1966; THORN and WAUD, 1969; COLQUHOUN, 1972), except that here acceptors rather than receptors are considered responsible for the excess of antagonist in the reversibly bound state.) Since the rate of recovery from PPL block-ade is not accelerated by exposing the tissue to high concentrations of ISO during the washout period, the postulated acceptors, unlike the β-receptors themselves, should have little or no affinity for ISO.

Preliminary results with ^{14}C-PPL are in accord with this acceptor hypothesis. In a 60-min incubation of isolated tracheal smooth muscle with ^{14}C-PPL, the tissue/ solution concentration ratio of the labeled drug attained values well over 15. On washout, the tissue lost less than half of its "bound" ^{14}C-PPL in 60 min. A high concentration of ISO in the washing solution did not alter the rate of loss of ^{14}C-PPL, but a high concentration of unlabeled PPL greatly accelerated it.

In 1967 Potter provided evidence that uptake of PPL by guinea-pig atria and by atrial cell fragments greatly exceeds any uptake that might be attributed to binding of PPL by β-receptors. His findings and the present findings on guinea-pig trachea should serve as a warning to those who are attempting to purify β-receptors that binding of PPL by isolated fractions is very likely to be an unreliable indicator of the presence of these receptors. The findings also raise the possibility of significant non-receptor binding of other β-receptor antagonists and β-receptor agonists by isolated fractions, and emphasise again the need to evaluate the results of studies on competitive binding with criteria derived from careful studies of pharmacologic competitive antagonism.

Acknowledgements—This work was supported by USPHS Grant NS-10195. A. J. and N. H. J. are recipients, respectively, of Brazilian FAPESP and CNPq fellowships; permanent address: Escola Paulista de Medicina, São Paulo, Brazil.

REFERENCES

BLINKS J. R. (1967) *Ann. N.Y. Acad. Sci.* **139**, 673–685.
COLQUHOUN D. and RITCHIE J. M. (1972) *Molec. Pharmac.* **8**, 285–292.
FOSTER R. W. (1969) *Brit. J. Pharmacol.* **35**, 418–427.
FURCHGOTT R. F. (1967) *Ann. N.Y. Acad. Sci.* **139**, 553–570.
FURCHGOTT R. F. (1970) *Fedn. Proc.* **29**, 1352–1361.
FURCHGOTT R. F. (1972) In: *Catecholamines*, pp. 283–335. (BLASCHO H. and MUSCHOLL E. Eds.) Springer-Verlag, Berlin.
LANGER S. Z. and TRENDELENBURG U. (1969) *J. Pharmacol. Exp. Ther.* **167**, 117–142.
MOORE G. E. and O'DONNELL S. R. (1970) *J. Pharm. Pharmacol.* **22**, 180–188.
POTTER L. T. (1967) *J. Pharmacol. Exp. Ther.* **155**, 91–100.
RANG H. P. (1966) *Proc. Roy. Soc. B* **164**, 488–510.
SCHILD H. O. (1949) *Brit. J. Pharmacol.* **4**, 277–280.
THRON C. D. and WAUD D. R. (1968) *J. Pharmacol. Exp. Ther.* **160**, 91–105.

Frontiers in Catecholamine Research 1973, pp. 301 to 306. Pergamon Press. Printed in Great Britain.

REGULATION OF CYCLIC AMP LEVELS IN BRAIN

JOHN W. DALY

National Institute of Arthritis, Metabolism and Digestive Diseases,
National Institutes of Health, Bethesda, Maryland 20014, U.S.A.

INTRODUCTION

THE RATE of formation and degradation of cyclic AMP in the central nervous system depends on a variety of factors: (i) the basal levels of adenylate cyclases (EC 4.6.1.1); (ii) the degree and specificity of activation of this enzyme by biogenic amines such as norepinephrine, dopamine, serotonin and histamine; (iii) the levels and availability of ATP as a substrate for adenylate cyclases; (iv) the degree of stimulation of the system by adenosine, and the extent of synergistic interaction of adenosine with stimulatory biogenic amines; (v) the levels and properties of cyclic nucleotide phosphodiesterases (EC 3.1.4.c); and (vi) the levels of cyclic AMP dependent protein kinases (EC 2.7.1.37) with binding sites that can sequester cyclic AMP.

Two methods have been extensively employed either separately or in consort for the study of the factors involved in the formation of cyclic AMP in brain slices. The first involves the measurement of total endogenous levels of cyclic AMP in slices, while the second measures formation of cyclic AMP from intracellular adenine nucleotides labeled during a prior incubation of slices with radioactive adenine or adenosine (SHIMIZU et al., 1969). The adenine or adenosine is incorporated during this prior incubation into intracellular nucleotides by the action of adenine phosphoribosyltransferase (EC 2.4.2.7) and adenosine kinase (EC 2.7.1.20), respectively. Since the specific activity of the cyclic AMP formed in guinea pig cerebral cortical slices under a variety of stimulatory conditions (norepinephrine, histamine, serotonin, adenosine, veratridine, and combinations of these agents) is three to sixfold higher than that of the total intracellular adenine nucleotides labeled by adenine (SCHULTZ and DALY, 1973a; SHIMIZU and OKAYAMA, 1973), it is apparent that high levels of the enzyme, adenosine-phosphoribosyltransferase, are specifically associated with morphological compartments that contain nucleotides which serve as substrates for adenylate cyclase. Adenosine incorporation into the slice, and levels of adenosine kinase do not appear to be as specifically associated with this compartment, i.e., the specific activities of cyclic AMP is only one to three-fold higher than that of total nucleotides when adenosine instead of adenine has been used to label intracellular nucleotides (cf., SHIMIZU and DALY, 1970). The morphological localization of the labeled nucleotide compartments associated with adenylate cyclase is as yet unknown. However, the ratio of specific activity of the cyclic AMP to that of the total nucleotides indicates that these compartments contain less than 20–30 per cent of the total nucleotides of the slices. The nucleotides of this compartment do not undergo significant equilibration with the remaining nucleotides of the slice during incubations of labeled slices for up to 50 min (SHIMIZU et al., 1970). Depolarization of slices by electrical pulsation results in a selective loss of adenosine of high specific activity from this compartment (PULL and MCILWAIN, 1973). The foregoing result provides tentative evidence that the compartments of nucleotides associated with cyclic AMP generation in cerebral cortex are contained within neuronal structures which are depolarized by electrical pulsation with resultant formation and efflux of adenosine. Indirect evidence that formation of cyclic AMP in brain slices in response to biogenic amines may occur primarily in postsynaptic neuronal structures derives from studies on the amine-mediated increases of cyclic AMP in superior cervical ganglia (MCAFEE and GREENGARD, 1972) and in cerebellar Purkinje cells (SIGGINS et al., 1973), where cyclic AMP is responsible for the alteration in membrane properties that results in membrane hyperpolarization and thereby elevates the threshold for firing of the neuronal cell. The mechanism of this change in membrane properties may involve cyclic AMP-mediated activation of protein kinase-catalyzed phosphorylation of specific membrane proteins (JOHNSON et al., 1972).

BIOGENIC AMINES

The stimulation of cyclic AMP formation by biogenic amines differs among species and brain regions both with respect to magnitude of stimulation and with respect to the properties of the amine regulatory unit of adenyl cyclase, i.e., the receptor (SCHULTZ and DALY, 1973abc and references, therein). In guinea pig cerebral cortical slices, (nor) epinephrine elicits accumulations of cyclic AMP through interaction with a classical α-adrenergic receptor (Table 1). Histamine and serotonin

TABLE 1. RESPONSES OF CYCLIC AMP-GENERATING SYSTEM IN RODENT CEREBRAL CORTICAL SLICES TO VARIOUS AGONISTS AND ANTAGONISTS

Stimulatory agent(s) (0·1 mM)	Accumulation cyclic AMP (pmoles/mg protein ± s.D.)	Degree of inhibition of amine-mediated response* by equimolar concentration of		
		α-Adrenergic antagonist	β-Adrenergic antagonist	Antihistaminic
Guinea pig†				
None	24 ± 11	—	—	—
Epinephrine‡	—	complete	none	partial
Histamine	40 ± 14	none	none	partial
Histamine-norepinephrine	130 ± 32	complete	none	partial
Adenosine	206 ± 30	—	—	—
Adenosine-norepinephrine	568 ± 130	complete	none	none
Adenosine-serotonin§	543	partial	none	none
Adenosine-histamine	682 ± 145	none or partial	none	partial
Rat (Sprague-Dawley)‖				
None	16 ± 3	—	—	—
Norepinephrine	70 ± 13	partial	partial	—
Isoproterenol	30 ± 7	none	complete	—
Adenosine	58 ± 10	—	—	—
Adenosine-norepinephrine	301 ± 80	partial	partial	—
Adenosine-isoproterenol	118 ± 19	none	complete	—

* Antagonists do not reduce basal or adenosine-stimulated levels of cyclic AMP.
† Data from CHASIN et al. (1971); SCHULTZ and DALY (1973b); and HUANG and DALY (1972).
‡ (Nor)epinephrine has been reported to elicit varying degrees of accumulation of cyclic AMP in guinea pig cortical slices, ranging from zero (SCHULTZ and DALY, 1973a) to approximately twice control values (CHASIN et al., 1971).
§ Response partially blocked by methysergide.
‖ Data from SCHULTZ and DALY (1973c).

receptors modulating adenylate cyclase activity are also present in this tissue. In cerebral cortical slices from Sprague-Dawley rats, the responses of the cyclic AMP-generating system to biogenic amines are markedly different (Table 1). Histamine has little effect in this tissue, except in the presence of adenosine, while serotonin has no effect even in the presence of adenosine. The responses to catecholamines either alone or in consort with adenosine appear mediated by *both* α- and β-adrenergic receptors (PERKINS and MOORE, 1973; SCHULTZ and DALY, 1973c). Synergistic interactions between norepinephrine and histamine are not statistically significant in this tissue. Pretreatment of rats with 6-hydroxydopamine causes an enhanced maximal accumulation of cyclic AMP in cerebral cortical slices incubated with norepinephrine (HUANG et al., 1973; PALMER, 1972; KALISKER and PERKINS, 1973; WEISS and STRADA, 1972). Both α- and β- mediated responses to supramaximal concentrations of the stimulatory amine, norepinephrine, are increased twofold during the period investigated, 4–20 days after administration of the 6-hydroxy-dopamine (Table 2). Effects of other stimulatory agents such as histamine, adenosine

TABLE 2. EFFECT OF INTRAVENTRICULAR ADMINISTRATION OF 6-HYDROXYDOPAMINE ON SUBSEQUENT ACCUMULATIONS OF CYCLIC AMP IN SPRAGUE–DAWLEY RAT CEREBRAL CORTICAL SLICES LABELED BY A PRIOR INCUBATION WITH RADIOACTIVE ADENINE (HUANG et al., 1973)

Stimulatory and blocking agents (0·1 mM)	Type of adrenergic receptor stimulated	Radioactive cyclic AMP $\Delta\%$ conversion*	
		Control $\pm$ s.d.	6-Hydroxydopamine $\pm$ s.d.
Norepinephrine	$\alpha + \beta$	2·3 ± 0·2	4·8 ± 0·4†
Norepinephrine-dichloroisoproterenol	α	0·6 ± 0·1	1·1 ± 0·4‡
Isoproterenol	β	0·9 ± 0·2	2·4 ± 0·3†
Adenosine	None	2·8 ± 0·3	3·4 ± 0·4

* $\Delta\%$ Conversion is the per cent conversion of total radioactive adenine nucleotides to radioactive cyclic AMP minus the percentage conversion in control slices.
† $P < 0.01$.
‡ $P < 0.05$.

or veratridine are not enhanced (HUANG et al., 1973). In cerebral cortical slices from NIH general purpose white mice, the effects of norepinephrine are effectively blocked by a β-adrenergic antagonist and only marginally reduced by an α-adrenergic antagonist (SCHULTZ and DALY, 1973c). The foregoing indicates that wide variations in the responses of cyclic AMP-generating systems in cerebral cortical slices from various rodent species will be seen on incubation with biogenic amines and/or adenosine. Indeed, differences have recently been seen in responses of cortical slices from different strains of rats (SKOLNICK, P. and DALY, J., unpublished results) and in the magnitude of responses of cortical slices obtained either from a neurological mutant, the quaking mouse, and from the parent strain of mouse (SKOLNICK and DALY, 1973). Differences in responses to amines have been demonstrated even among functionally distinct regions of squirrel monkey cerebral cortex (SKOLNICK et al., 1973).

The response of the cyclic AMP-generating system in brain slices to biogenic amines is unique in that after one amine-elicited stimulation, followed by washing and return of the cyclic nucleotide to basal levels, the preparation does not respond upon attempted restimulation with the same amine (KAKIUCHI and RALL, 1968; SCHULTZ and DALY, 1973d). However, if adenosine is present with the amine, synergistic responses can be elicited many times. This adenosine effect on restimulation with amines has been demonstrated both in rat cerebral cortical slices (SCHULTZ and DALY, 1973c) and in guinea pig cerebral cortical slices (SCHULTZ and DALY, 1973d), but its molecular basis is unclear. Intermediate incubations with adenosine do not restore the responsiveness of the cyclic AMP-generating system to amines in slices previously stimulated with amines (SCHULTZ and DALY, 1973d).

In rat glioma cells, the response to norepinephrine is mediated by a β-adrenergic receptor (GILMAN and NIRENBERG, 1971; CLARK and PERKINS, 1971), and repetitive sequential stimulations of cyclic AMP formation can be elicited with norepinephrine at least five times (SCHULTZ et al., 1972). The magnitude of the response, however, steadily declines, unless a phosphodiesterase inhibitor is included in the incubations. Thus, in glioma cells, the decline in maximal responses to norepinephrine during restimulations appears mainly due to a cyclic AMP-dependent increase in phosphodiesterase activity. Similar mechanisms may at least in part be responsible for the lack of effect of amines during attempted restimulations in brain slices. Thus, inclusion of a phosphodiesterase inhibitor, isobutylmethylxanthine, with the amines, does result, during a second incubation, in a substantial stimulation of cyclic AMP formation in guinea pig cortical slices (SCHULTZ and DALY, 1973d).

ADENOSINE

The nature and the significance of the interaction of adenosine with the cyclic AMP-generating system of guinea pig cortical slices (SATTIN and RALL, 1970) has been studied extensively. It has been demonstrated that adenosine is "released" from brain slices during electrical pulsation (PULL and McILWAIN, 1972), and during incubation with depolarizing agents and certain metabolic inhibitors (SHIMIZU et al., 1970; SCHULTZ and DALY, 1973a; HUANG, M. and DALY, J., unpublished results), i.e., under conditions where tissue levels of ATP are reduced. The enhanced accumulation of cyclic AMP under such conditions appears due to "released" adenosine (KAKIUCHI et al., 1969; SHIMIZU et al., 1970).

Phosphodiesterase inhibitors, such as N,O-dibutyryl cyclic AMP, papaverine, or isobutylmethylxanthine that potentiate by two to threefold the accumulation of cyclic AMP elicited by histamine or a histamine-norepinephrine combination in brain slices, have no effect on the accumulation of cyclic AMP elicited by adenosine or adenosine-amine combinations (Table 3). A possible role for phosphodiesterase inhibition by adenosine needs, therefore, to be investigated in brain slice preparations. A number of observations (cf., SCHULTZ and DALY, 1973d,e), however, suggest that the stimulatory effects of adenosine are not due to mere inhibition of phosphodiesterases: (i) the efficacy of adenosine in brain slices is not consonant with its known inhibitory activity towards phosphodiesterase, i.e., much potent phosphodiesterase inhibitors have no significant effects on basal levels of cyclic AMP, and potentiate amine-elicited increases to a much lesser extent than adenosine; (ii) theophylline, a known phosphodiesterase inhibitor, blocks the effect of adenosine, instead of further augmenting it (Table 3); (iii) related compounds such as inosine which are equipotent with adenosine as phosphodiesterase inhibitors have no effect on cyclic AMP levels in brain slices; (iv) pulse-labeling studies with radioactive

TABLE 3. EFFECT OF PHOSPHODIESTERASE INHIBITORS ON ACCUMULATION OF RADIOACTIVE CYCLIC AMP IN GUINEA PIG CEREBRAL CORTICAL SLICES LABELED BY INCUBATION WITH RADIOACTIVE ADENINE (SCHULTZ AND DALY, 1973e)

Stimulatory agent (0·1 mM)	Radioactive cyclic AMP (Δ% conversion) in presence of phosphodiesterase inhibitors				
	None	Theophylline (1 mM)	Isobutyl-methylxanthine (1 mM)	Papavarine (0·5 mM)	N,O-dibutryl cyclic AMP (0·1 mM)
Histamine	1·5	1·2	4·2	3·5	3·1
Histamine-norepinephrine	5·6	4·3	8·5	9·2	9·4
Adenosine	6·3	0·1	1·9	5·6	5·7
Adenosine-histamine-norepinephrine	30·1	—	—	26·9	22·6

adenine indicate that the turnover of cyclic AMP under maximal steady-state levels of cyclic AMP elicited by adenosine is quite rapid (SCHULTZ and DALY, 1973d); and (v) compounds which antagonize uptake of adenosine into the cells of the slice do not block the adenosine-elicited rise in cyclic AMP (HUANG, M. and DALY, J., unpublished results). Taken *in toto*, these results suggest that adenosine stimulates cyclic AMP formation, at least in part, by interaction with an adenosine-receptor in the plasma membrane. This receptor would appear to be quite specific for adenosine, since a variety of structurally related nucleosides are without effect or act as adenosine antagonists (SATTIN and RALL, 1970; HUANG et al., 1972).

CONCLUSIONS

At present, the mechanisms involved in the regulation of cyclic AMP levels in brain slices by biogenic amines, adenosine, depolarizing agents, and other parameters have been incompletely and poorly defined. Much further investigation is, therefore, required on both the roles of cyclic AMP in nervous tissue and the physiological regulation of its formation, degradation, and action. The brain slice, in spite of its complexity, has thus far provided a valuable tool for such studies, representing, as it were, a compromise between rather facile biochemical measurements on disrupted enzyme systems present in brain homogenates, and the difficultly accessible biochemical interrelationships that exist within such components of the intact functioning brain.

REFERENCES

CHASIN M., RIVKIN I., MAMRAK F., SAMANIEGO S. and HESS S. M. (1971) *J. Biol. Chem.* **246**, 3037–3041.
CLARK R. B. and PERKINS J. P. (1971) *Proc. Natn. Acad. Sci. U.S.A.* **68**, 2757–2760.
GILMAN A. G. and NIRENBERG M. (1971) *Proc. Natn. Acad. Sci. U.S.A.* **68**, 2165–2168.
HUANG M. and DALY J. W. (1972) *J. Med. Chem.* **15**, 458–462.
HUANG M., HO A. K. S. and DALY J. W. (1973) *Mol. Pharmacol.* In press.
HUANG M., SHIMIZU H. and DALY J. W. (1972) *J. Med. Chem.* **15**, 462–466.
JOHNSON E. M., UEDA T., MAENO H. and GREENGARD P. (1972) *J. Biol. Chem.* **247**, 5650–5652.
KAKIUCHI S. and RALL T. W. (1968) *Mol. Pharmacol.* **4**, 367–378.
KAKIUCHI S., RALL T. W. and McILWAIN H. (1969) *J. Neurochem.* **16**, 485–491.
KALISKER A. and PERKINS J. P. (1973) *Fedn. Proc.* **32**, 680 Abs.
McAFEE D. A. and GREENGARD P. (1972) *Science* **178**, 310–312.
PALMER G. C. (1972) *Neuropharmacol.* **11**, 145–149.

Perkins J. and Moore M. (1973) *J. Pharmacol. Exp. Ther.* **185,** 371–378.
Pull I. and McIlwain H. (1972) *Biochem. J.* **130,** 975–981.
Sattin A. and Rall T. W. (1970) *Mol. Pharmacol.* **6,** 13–23.
Schultz J. and Daly J. W. (1973a) *J. Biol. Chem.* **248,** 843–852.
Schultz J. and Daly J. W. (1973b) *J. Neurochem.* In press.
Schultz J. and Daly J. W. (1973c) *J. Neurochem.* In press.
Schultz J. and Daly J. W. (1973d) *J. Biol. Chem.* **248,** 860–866.
Schultz J. and Daly J. W. (1973e) *J. Biol. Chem.* **248,** 853–859.
Schultz J., Hamprecht B. and Daly J. W. (1972) *Proc. Nat. Acad. Sci. U.S.A.* **69,** 1266–1270.
Shimizu H., Creveling C. R. and Daly J. W. (1970) *Proc. Natn. Acad. Sci. U.S.A.* **65,** 1033–1040.
Shimizu H., Daly J. W. and Creveling C. R. (1969) *J. Neurochem.* **16,** 1609–1619.
Shimizu H. and Daly J. W. (1970) *Biochim. Biophys. Acta* **222,** 465–473.
Shimizu H. and Okayama H. (1973) *J. Neurochem.* **20,** 1279–1284.
Siggins G. R., Battenberg E. F., Hoffer B. J., Bloom F. E. and Steiner A. L. (1973) *Science* **179,** 585–588.
Skolnick P. and Daly J. W. (1973) *Fedn. Proc.* **32,** 680 Abs.
Skolnick P., Huang M., Daly J. W. and Hoffer B. (1973) *J. Neurochem.* In press.
Weiss B. and Strada S. J. (1972) *Advan. Cyclic Nucleotide Res.* **1,** 357–374.

Frontiers in Catecholamine Research 1973, pp. 307 to 310. Pergamon Press. Printed in Great Britain.

REGULATION OF ADENOSINE 3',5'-CYCLIC MONOPHOSPHATE METABOLISM IN HYBRID SOMATIC CELLS

Alfred G. Gilman[1] and John D. Minna[2]

[1]Department of Pharmacology, School of Medicine, University of Virginia, Charlottesville, Virginia 22903, U.S.A. and [2]Laboratory of Biochemical Genetics, National Heart and Lung Institute, National Institutes of Health, Bethesda, Maryland 20014, U.S.A.

Hormones that regulate the rate of synthesis of adenosine 3',5'-cyclic monophosphate (cyclic AMP) are generally presumed to act initially at the plasma membrane by binding to specific macromolecular receptors. A reaction sequence is initiated that must ultimately alter the activity of the membranous adenylate cyclase. Since the nature of the interaction between the receptor and the catalytic site is unknown, it is possible to envision either remote or intimate relationships between the two. Furthermore, the coupling mechanisms that may exist could be rather different and dependent on the type of hormonal stimulus.

Several clonal cultured cell lines display regulation of their intracellular content of cyclic AMP by biogenic amines, peptide hormones, and/or prostaglandins. Genetic analysis of the mechanisms that govern the expression of such function can be undertaken if one fuses clonal cell lines that differ in their phenotype with respect to chemical agents that control their metabolism of cyclic AMP. The inheritance of the differentiated function in question can be studied if one examines the hybrid progeny of appropriate parental clones in the early generations after fusion, when the chromosome composition is relatively stable and complete. In later generations, however, chromosomes may be lost. Unique but genetically related cell lines are then available for genetic linkage analysis, the assignment of phenotypic properties to specific chromosomes, and for biochemical analysis.

This report summarises the first stages of our investigation of the expression of genes controlling the regulation of cyclic AMP accumulation by catecholamines and prostaglandins in parental and hybrid cells.

METHODS

Methodological details have been described previously (Gilman and Minna, 1973; Minna and Gilman, 1973). Briefly, appropriate parental cell lines were fused in the presence of inactivated sendai virus, and hybrid clones were selected by the method of Littlefield (1964). The hybrid clones studied were derived independently and thus arose as the result of individual mating events. Hybrid cell lines were usually studied approximately 25–40 generations after parental cell fusion. The hybrid nature of clones studied were verified by growth in appropriate selective medium, morphology, analysis of total chromosome number and of distinctive parental marker chromosomes, and, in some cases, by analysis of strain specific isozymes by starch gel electrophoresis.

Parent and hybrid cells were analysed for their content of cyclic AMP (Gilman, 1970) after growth to confluency and incubation at 37°C with the chemical agent(s) under study for the indicated period of time.

RESULTS

Seven different parental clones were utilised in six different types of matings. Since approximately 50 hybrid clones have been tested for their responsiveness to catecholamines and prostaglandins, only a small fraction of the actual data can be presented.

Parental cell lines

There were marked differences in the abilities of the parental clones studied to respond to catecholamines with elevated intracellular concentrations of cyclic AMP. Three cell lines C6TG1A (and its subclone C6RC16), BRL30E, and 3T3-4(E) showed highly significant elevations of cyclic AMP concentration following brief exposures to norepinephrine or isoproterenol. All responses were half-maximal at approximately 0·05 μM DL-isoproterenol and were preferentially antagonised by propranolol. Since practolol was relatively ineffective, all responses were hypothesised to be mediated by β_2-adrenergic receptors. The phenotype of the three responsive parental clones was thus designated as β_2^+. Four parental cell lines (RAG, B82, 3T3-4(C2), and N4TG1) failed to show altered concentrations of cyclic AMP in response to catecholamines and were thus designated as β^-.

Responses to prostaglandins were more prevalent. While there were large quantitative differences in the magnitude of response, six of the seven parental clones responded to prostaglandin E_1 (PGE$_1$) with elevated intracellular levels of cyclic AMP. In all cases PGE$_1$ was more effective than PGE$_2$ or PGA$_1$. PGF$_{1\alpha}$ and PGB$_1$ were ineffective at the concentration tested (1 μg/ml). These six lines (C6TG1A, B82, 3T3-4(E), 3T3-4(C2), N4TG1 and BRL30E) were thus classified phenotypically as PGE$^+$. Clone RAG failed to respond to any prostaglandin used. PGE$_1$ was ineffective to a concentration of 10 μg/ml, and this clone is thus designated PGE$^-$.

Hybrid cell lines

Hybrid clones were evaluated that were derived from parents with both qualitative differences in their phenotypes and quantitative differences in their magnitudes of response to catecholamines or prostaglandins. Two general patterns of results were obtained.

When clones derived from $\beta_2^+ \times \beta^-$ matings were studied, all of the hybrids were essentially β^-, with less than 2 per cent of the β_2^+ parental response being apparent. This was particularly clear in the case of the interspecific mating of C6TG1A (derived from rat) and B82 (of mouse origin).

Karyologic analysis of these hybrids (Table 1) showed that the clones contained nearly all the chromosomes expected from the fusion of C6TG1A and B82. In addition, the chromosomes that serve as markers of C6TG1A (the β_2^+ parent) are present.

The catecholamine-stimulated levels of cyclic AMP were determined in 12 hybrid clones and in two pools of hybrids, each containing approximately 100 clones. The pools of hybrids could be studied within 20 generations of fusion. In all hybrid lines the large response to catecholamine characteristic of the C6TG1A parent was

TABLE 1. EFFECT OF NOREPINEPHRINE AND PGE_1 ON C6TGIA × B82 HYBRID CELLS

Cell line	Cyclic AMP* (pmoles/mg protein)			Chromosome number (range)		
	Control	Norepi-nephrine	PGE_1	Mean	C6TGIA Markers†	B82 Markers‡
Parent:						
C6TGIA	10	3000	25	42 (40–43)	20 (18–22)	0
B82	14	13	48	52 (48–55)	0	26 (24–27)
				94§	20§	26§
Hybrid:						
CB4	18	36	121	92 (86–96)	22 (18–26)	28 (26–31)
CB3	15	34	96	92 (90–96)	23 (21–25)	24 (21–27)
CB14	30	32	134	86 (74–94)	21 (19–24)	29 (25–34)
CB1	17	26	90	89 (62–104)	19 (16–23)	22 (19–27)
CB5	17	24	79	90 (74–100)	22 (19–23)	25 (20–30)
CB10	17	24	115	90 (84–98)	21 (19–22)	29 (26–31)
Pool 2	20	24	122	94 (92–98)	21 (19–22)	29 (25–31)
Pool 1	16	23	81	93 (84–96)	22 (20–23)	26 (22–29)
CB2	20	20	120	78 (72–84)	17 (15–22)	20 (18–23)
CB8	10	15	74	93 (84–98)	22 (20–23)	23 (20–25)
CB6	12	14	60	92 (88–98)	24 (22–25)	23 (20–25)
CB12	17	14	84	88 (82–96)	22 (18–26)	32 (27–37)
CB9	9	13	98	92 (90–94)	22 (21–24)	22 (21–24)
CB7	7	12	66	93 (90–96)	22 (20–24)	25 (23–26)

  * Cells incubated with norepinephrine (10^{-4} M) or PGE_1 (2 µg/ml) for 10 min in the presence of 1 mM theophylline.
  † C6TG1A markers are acrocentric and small metacentric chromosomes.
  ‡ B82 markers are large metacentric and large submetacentric chromosomes.
  § Chromosome number expected in hybrid from fusion of two parental cells with mean number of chromosomes.

essentially undetectable. The hybrid clones with the largest responses (CB3 and CB4) showed less than 1 per cent of that apparent with C6TG1A.

When hybrid lines were formed between parental strains with quantitative differences in their magnitudes of response to catecholamine, the hybrid clones showed stimulated levels of cyclic AMP similar to or *less than* that characteristic of the *lower* responding parent.

A completely different pattern was observed when hybrid cell line responses to PGE_1 were studied. Hybrids formed between parents with quantitative differences in their responses to PGE_1 usually showed responses greater than the higher responding parent. This is again illustrated in the cross between C6TG1A and B82 (Table 1), where the concentration of cyclic AMP achieved is roughly consistent with addition of the capabilities of the two parents. Full inheritance of the response of a PGE^+ parent (B82) was also seen in hybrids derived from a $PGE^+ × PGE^-$ (B82 × RAG) mating.

Table 2 presents more complete documentation in summary form of the generalisations illustrated above with the C6TG1A × B82 mating. It can be seen that $\beta_2^+ × \beta^-$ or $\beta_2^+ × \beta_2^+$ matings yielded only hybrids with magnitudes of response similar to or less than the lower responding parent. By contrast, $PGE^+ × PGE^+$ or $PGE^+ × PGE^-$ matings result predominantly in hybrid clones with high levels of PGE_1 responsiveness. While detailed mechanisms are unknown and currently under study, it is clear at this time that the pattern of inheritance of these responses to catecholamines and PGE_1 is strikingly different.

TABLE 2. SUMMARY OF MATING STUDIES PERFORMED TO DETERMINE NATURE OF GENETIC CONTROL OF CYCLIC AMP METABOLISM

Parental phenotypes	Mating	Number of hybrid clones		
		Response similar to or less than lower parent	Intermediate response	Response similar to or greater than higher parent
$\beta_2^+ \times \beta_2^+$	C6RC16 × BRL30E	5	0	0
	C6RC16 × 3T3-4 (E)	7	0	0
$\beta_2^+ \times \beta^-$	C6TG1A × B82	14	0	0
	C6RC16 × 3T3-4 (C2)	8	0	0
PGE⁺ × PGE⁺	N4TG1 × B82*	1	6	2
	C6TG1A × B82	0	0	14
	C6RC16 × BRL30E	1	0	4
	C6RC16 × 3T3-4 (C2)	2	0	6
	C6RC16 × 3T3-4 (E)	3	2	2
PGE⁻ × PGE⁺	RAG × B82	0	0	4

* These hybrids studied after > 80 generations and recloning.

DISCUSSION

While several possible explanations for the loss of the catecholamine response in the $\beta_2^+ \times \beta^-$ hybrids were discussed (GILMAN and MINNA, 1973), it is felt that only two hypotheses require consideration here. One explanation entails the loss of a chromosome from the hybrids coding for a component essential for the β_2-adrenergic response. This hypothesis seems very unlikely in view of the hybrid karyograms described above and reported for the other series of hybrid clones (GILMAN and MINNA, 1973). However, a highly specific loss of a vital chromosome cannot be excluded with certainty. A more appealing hypothesis envisions a negative control mechanism inherited from the β^- parent and operative in the hybrid clones. This type of hypothesis has been advanced on several prior occasions for a variety of differentiated functions whose expression is extinguished in hybrid cells (EPHRUSSI, 1972). If this is the case, it should be possible to obtain β_2^+ hybrid subclones from hybrid cells originally β^- that have lost the chromosome coding for the negative control mechanism. Current experiments are designed to determine the molecular level in the β_2-adrenergic response mechanism at which the hypothesised negative control mechanism is operative.

The difference in inheritance of the responses to catecholamines and PGE_1 suggest differences in genetic mechanisms controlling the synthesis or expression of essential components for these two types of responses. This could imply essential differences in the basic mechanisms by which these agents influence cyclic AMP synthesis. It is felt that the availability of series of genetically related hybrid cell lines will be of material aid in the elucidation of the detailed mechanisms involved in hormonal control of cyclic AMP metabolism.

Acknowledgements—This work was supported by U.S.P.H.S. Grants NS10193 and GRS5S01 RR05431.

REFERENCES

EPHRUSSI B. (1972) *Hybridisation of Somatic Cells.* Princeton University Press, Princeton, N.J.
GILMAN A. G. (1970) *Proc. Natn. Acad. Sci.* 67, 305–312.
GILMAN A. G. and MINNA J. D. (1973) *J. Biol. Chem.*, In Press.
LITTLEFIELD J. W. (1964) *Science* 145, 709–710.
MINNA J. D. and GILMAN A. G. (1973) *J. Biol. Chem.*, In Press.

Frontiers in Catecholamine Research 1973, pp. 311 to 313. Pergamon Press. Printed in Great Britain.

ONTOGENETIC DEVELOPMENT OF THE CATECHOLAMINE-SENSITIVE ADENYL CYCLASE ACTIVITY OF RAT CEREBRAL CORTEX*

JOHN P. PERKINS† and MARILYN M. MOORE

Department of Pharmacology, University of Colorado School of Medicine,
4200 East 9th Avenue, Denver, Colorado 80220, U.S.A.

WE ARE currently investigating factors which influence the responsiveness to catecholamines of the adenyl cyclase activity of rat cerebral cortex. Previously, we reported that norepinephrine (NE) interacts with both α-like and β-like adrenergic receptors to cause a rise in the adenosine 3', 5'-monophosphate (cAMP) content of slices of rat cerebral cortex (PERKINS and MOORE, 1973a). In another report (KALISKER et al., 1973) it was shown that treatment of rats with 6-hydroxydopamine (6-OHDA) leads to two types of alterations in the effect of NE on the cAMP content of slices of rat cortex. First, there is an effect which develops within 24–48 hr which involves an increase in sensitivity to low concentrations of NE. This effect is attributed to destruction of adrenergic nerve endings and the resultant loss of amine uptake sites. Second, there is an effect which is not observed until 72–96 hr after 6-OHDA treatment which involves an increase in responsiveness to high concentrations of NE and which appears to be related to postsynaptic alterations.

This report describes the ontogenetic development of catecholamine sensitive adenyl cyclase activity in rat cerebral cortex. Also, since adenosine is known to interact synergistically with NE to cause marked increases in the cAMP content of brain slices, the development of responsiveness to adenosine and to the combination of adenosine and NE has been examined. The formation of ^{14}C-cAMP in response to NE (30 μM), adenosine (100 μM) or 30 μM NE plus 100 μM adenosine was determined in slices of cerebral cortex from rats from birth to 45 days of age. The procedural details were as described by SHIMIZU, DALY and CREVELING (1969) with minor modifications (PERKINS and MOORE, 1973a; KALISKER et al., 1973) and in essence involved incubation of slices with ^{14}C-adenine to label ATP pools followed by isolation of the ^{14}C-cAMP formed during incubation of slices with the two agonists using ion-exchange chromatography. The validity of this procedure was verified by conducting comparison assays, on the same samples, using the method of GILMAN (1970)

Responsiveness to NE develops abruptly between the 10th and 12th days after birth (Fig. 1). Responsiveness to adenosine is first detected on day 5 and then it gradually increases to a maximum by day 15. Prior to the development of sensitivity to NE, the combination of NE and adenosine resulted in a potentiation of the ability of adenosine to increase ^{14}C-cAMP formation in the slices. In studies not

* This work was supported in part by Grant NS 10233 from the National Institute of Neurological Diseases and Stroke.
† Recipient of a Research Career Development Award from the National Cancer Institute, 6K04 CA 70466.

illustrated here it was shown (PERKINS and MOORE, 1973b) that the catalytic activity of adenyl cyclase in homogenates of rat cerebral cortex is quite high at birth and gradually increases during the succeeding 20 days.

These results are consistent with either of two alternate interpretations. First, if it is assumed that the events shown in Fig. 1 occurred in a homogeneous population

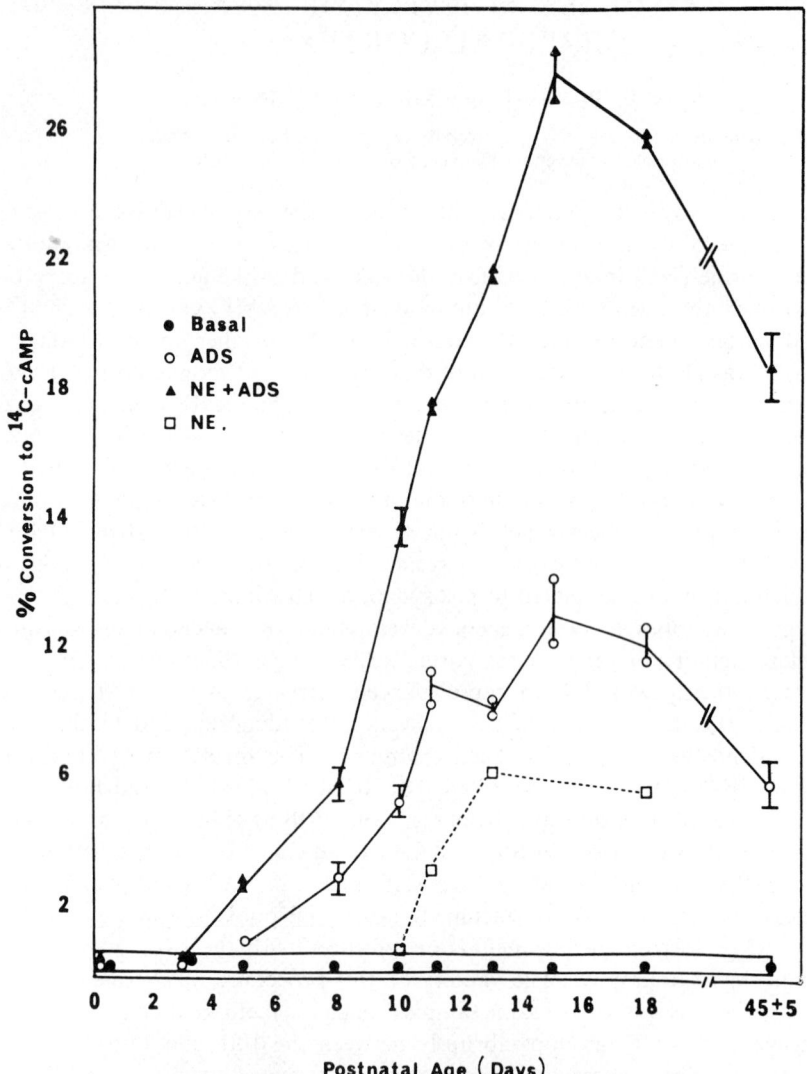

FIG. 1.—The development of catecholamine-sensitive adenyl cyclase activity in rat cerebral cortex. Whether alone or in combination NE was present at 30 μM and adenosine at 100 μM. The bar at the bottom of the graph represents the mean and S.E.M. for 56 determinations of the basal level of ^{3}H-cAMP in 4 separate experiments. Individual data points shown within the bar and data points for the effect of NE alone represent the average of duplicate determinations. Symbols with vertical bars represent the average of 4 determinations ± S.E.M. Symbols connected by a vertical line represent duplicate determinations. Incubation conditions are described elsewere (PERKINS and MOORE, 1973a, 1973b; KALISKER et al., 1973)

of cells, it is clear that the catecholamine-sensitive adenyl cyclase does not develop as an intact functional unit. Certainly the catalytic component of the enzyme system is present prior to day 11 and since NE potentiates the action of adenosine prior to day 11, the catecholamine receptor also must be present. The assumption that NE potentiates the effects of adenosine by interacting with adrenergic receptors is strengthened by the observation that the potentiative effect of NE, but not the effect of adenosine alone, is blocked partially by either propranolol or phentolamine and blocked completely by a combination of both inhibitors (cf. PERKINS and MOORE, 1973a). Thus, if a homogeneous population of cells is involved, it is apparent that the regulatory and catalytic components of the adenyl cyclase system are not appropriately linked prior to day 11.

An alternate interpretation is possible if it is assumed that at least two different cell populations are involved. Thus, an adenyl cyclase system that is not responsive to NE [NE(−)], but responsive to adenosine [adenosine(+)] and which is stimulated in a synergistic manner by a combination of NE and adenosine [synergism(+)] could develop in one type of cell and an enzyme that is NE(+) and possibly adenosine(−) and synergism(−) would develop later, by day 11, in a different population of cells. There are some indications that adenyl cyclases with these two different patterns of regulation may exist in rodent cerebral cortex. First, the cAMP content in slices of adult guinea pig cerebral cortex is not increased significantly by NE alone but it is increased by adenosine, and combination of NE with adenosine potentiates the effect of adenosine. Second, when slices of cerebral cortex from 1-day-old rats are placed in culture in collagen-coated petri dishes, NE alone causes an increase in the cAMP content of the culture after the 4–6th day *in vitro*. The response of such cultures can be characterized as NE(+), adenosine(±) and synergism(−). This result is consistent with the occurrence in normal rat cerebral cortex of an adenyl cyclase that is NE(+) and synergism(−).

There is further evidence that the catecholamine receptors of rat cerebral cortex involved in the synergistic action of NE and adenosine are different from those receptors that are responsive to NE alone. Namely, the potentiative effects of NE are more effectively blocked by phentolamine than by propranolol while the opposite order of effectiveness holds for blockade of the effects of NE alone.

In view of the marked cellular heterogeneity of cerebral cortex and in view of the suggestive evidence mentioned above we currently favor an explanation of the results of Fig. 1 that involves two different cell populations each with adenyl cyclases with different patterns of regulation.

REFERENCES

GILMAN A. G. (1970) *Proc. Natn. Acad. Sci.* **67**, 305–312.
KALISKER A., RUTLEDGE C. O. and PERKINS J. P. (1973) *Mol. Pharmacol.* In Press Sept. Issue.
PERKINS J. P. and MOORE M. M. (1973a) *J. Pharmacol. Exp. Ther.* 185, 371–378.
PERKINS J. P. and MOORE M. M. (1973b) *Mol. Pharmacol.* Submitted for Publication.
SHIMIZU H., DALY J. W. and CREVELING C. R. (1969) *J. Neurochem.* **16**, 1609–1619.

Frontiers in Catecholamine Research 1973, pp. 315 to 319. Pergamon Press. Printed in Great Britain.

ROLE OF β-ADRENERGIC RECEPTOR IN THE CONTROL OF SEROTONIN N-ACETYLTRANSFERASE IN RAT PINEAL

TAKEO DEGUCHI*, M.D. and JULIUS AXELROD, Ph.D.
Laboratory of Clinical Science, National Institute of Mental Health, Bethesda,
Maryland 20014 U.S.A.

CIRCADIAN rhythms of serotonin (SNYDER et al., 1965), melatonin (LYNCH, 1971), hydroxyindole O-methyltransferase (EC 2.1.1.4) activity (AXELROD et al., 1965)) and serotonin N-acetyltransferase (EC 2.3.1.5.) activity (KLEIN et al., 1971) in the rat pineal gland are controlled by sympathetic nerves whose cell bodies are in the superior cervical ganglion. The diurnal rhythms of serotonin and melatonin have been proposed to be driven by N-acetyltransferase activity (KLEIN et al., 1971). In pineal organ culture, the synthesis of melatonin from tryptophan is markedly stimulated by norepinephrine, monoamine oxidase inhibitor and dibutyryl adenosine 3′,5′-cyclic monophosphate (cyclic AMP) (AXELROD et al., 1969; WURTMAN et al., 1971). Norepinephrine or dibutyryl cyclic AMP also induces N-acetyltransferase in organ culture of rat pineal gland (KLEIN et al.,1970).

CONTROL OF CIRCADIAN CHANGE OF N-ACETYLTRANSFERASE IN PINEAL

N-Acetyltransferase activity increases 50-fold after rats were kept in darkness 4 hr at 10 p.m. in confirmation of the previous report (KLEIN et al., 1971). When rats were kept in continuous lighting until 10 p.m. or rats were kept in darkness for 6 hr during daytime (from 10 a.m. to 4 p.m.), N-acetyltransferase activity did not increase (DEGUCHI and AXELROD, 1972b). The result indicates that both the absence of light and proper setting of endogenous clock are necessary for the nighttime elevation of enzyme activity in the pineal.

Reserpine a drug that depletes catecholamines in the nerve endings, propranolol, a β-adrenergic blocking agent, or cycloheximide, a protein synthesis inhibitor injected into rats before the onset of darkness blocked the nighttime increase in enzyme activity, whereas an α-adrenergic blocking agent or actinomycin D had no effect. These observations indicate that the nocturnal increase in N-acetyltransferase activity is induced by neuro-transmitter released from sympathetic nerve via stimulation of postsynaptic β-adrenergic receptor, and that the increase in enzyme activity is due to synthesis of new enzyme protein.

When rats were brought out of darkness into light at night, N-acetyltransferase activity disappeared very rapidly, decreasing to less than 10 per cent of the initial activity in 10 min (DEGUCHI and AXELROD, 1972b; KLEIN and WELLER, 1972). Isoproterenol injected into rats before they were exposed to light completely prevented

* Takeo Deguchi was partially supported by Foundations' Fund for Research in Psychiatry Grant No 70–494 and 72–526.

the light-induced fall of N-acetyltransferase activity, indicating that the light immediately interrupts the release of neurotransmitter from nerve endings in pineals. When propranolol was injected into rats at night in the darkness, N-acetyltransferase activity rapidly decreased to less than 15 per cent within 10 min a similar pattern to that induced by light. Cycloheximide, on the other hand, decreased enzyme activity slowly with a half-life of 60 min. These results indicate that there are two types of inactivation processes of N-acetyltransferase, a rapid fall induced by light or β-adrenergic blocking agent and a slower decrease induced by inhibition of protein synthesis. The slow decrease represents turnover of the enzyme, whereas the rapid fall might be due to the conversion of active form to inactive form of enzyme or to a disaggregation of subunits of the enzyme molecule. During the nighttime, light or propranolol caused an elevation of pineal serotonin (BROWNSTEIN et al., 1973a) and a fall of N-acetylserotonin (BROWNSTEIN et al., 1973b), which corresponds to changes in N-acetyltransferase activity.

INDUCTION OF N-ACETYLTRANSFERASE BY ADRENERGIC DRUGS DURING DAYTIME

To study whether the low level of N-acetyltransferase activity during daytime is due to the lack of responsiveness of β-adrenergic receptor or to the reduced release of neurotransmitter from nerve endings, various adrenergic drugs were injected into rats during daytime. 3'4-Dihydroxy-L-phenylalanine (L-DOPA), the precursor of catecholamine, epinephrine, norepinephrine or isoproterenol increased N-acetyltransferase activity 4 to 60-fold within 3 hr (DEGUCHI and AXELROD, 1972a). Isoproterenol was most effective among these adrenergic drugs. Monoamine oxidase inhibitors, also caused 10- to 20-fold increase in enzyme activity. The increase in enzyme activity by these adrenergic drugs was completely blocked by pretreatment with β-adrenergic blocking agent or protein synthesis inhibitor, indicating that β-adrenergic receptor is involved in the new synthesis of the enzyme molecule.

RELATION BETWEEN CYCLIC AMP AND N-ACETYLTRANSFERASE

Previous observations that dibutyryl cyclic AMP or theophylline increase N-acetyltransferase activity in cultured pineals (KLEIN et al., 1970) and in vivo (DEGUCHI and AXELROD, 1972a) indicate that the induction of enzyme is mediated by adenyl cyclase system. To study the temporal relationship between cyclic AMP and the induction of N-acetyltransferase, both cyclic AMP and N-acetyltransferase activity were measured at various times after injection of β-adrenergic agonist, isoproterenol (DEGUCHI, 1973). After administration of isoproterenol, cyclic AMP levels increased more than 10-fold in 2 min and returned to the baseline level in 30 min. N-acetyltransferase activity started to increase after 1 hr of lag phase and reached the maximal level in 3 hr, and then gradually decreased to the initial level 5–7 hr after injection of isoproterenol. Pretreatment with propranolol, a β-adrenergic blocking agent, blocked the elevation of cyclic AMP and the increase in N-acetyltransferase activity. When propranolol was injected 1 hr after isoproterenol at the time when cyclic AMP level had returned to the baseline level and before N-acetyltransferase activity started to increase, the rise in enzyme activity was blocked. Propranolol injected after enzyme activity had reached its maximal level caused a rapid fall of enzyme activity decreasing to 5 per cent of the initial level in 15 min. Other β-adrenergic blocking agents, practolol or pronethalol, also caused a rapid disappearance of N-acetyltransferase activity, whereas stereoisomer, *d*-propranolol, did not. A comparable rapid decrease in *N*-acetyltransferase activity was also observed in cultured pineals. When pineals were cultured in isoproterenol for 30 min and then transferred into medium without isoproterenol, *N*-acetyltransferase activity increased 20-fold in 6 hr. *N*-Acetyltransferase thus induced decreased to 25–30 per cent of the initial level in 15 min in the presence of 1×10^{-4} M *l*-propranolol or practolol. These observations indicate that the maintenance of the high level of *N*-acetyltransferase activity requires a continuous stimulation of β-adrenergic receptor on the pineal cell, and once β-receptor is blocked, regardless of the level of cyclic AMP, *N*-acetyltransferase activity disappeared very rapidly. Thus cyclic AMP could have two functions; a high level for the initiation of the enzyme induction and the low level for the maintenance of the high level of enzyme activity. The active pool of cyclic AMP might be too small to detect changes after treatment with propranolol, because the β-adrenergic blocking agent does not change the baseline level of cyclic AMP in pineals. An alternate possibility would be that β-adrenergic receptor is linked to other processes than cyclic AMP system that regulates the synthesis and inactivation of *N*-acetyltransferase. This phenomenon would reveal a novel mechanism in the interaction of β-adrenergic agonist or antagonist to their receptors and their effect on the enzyme control.

SUPERINDUCTION OF *N*-ACETYLTRANSFERASE AFTER DEPLETION OF NOREPINEPHRINE

When submaximal doses of either norepinephrine or isoproterenol were injected into rats whose pineal gland had been denervated, there was much higher increase in *N*-acetyltransferase activity compared to intact pineals (DEGUCHI and AXELROD, 1973b). Denervated pineals cultured in the presence of submaximal concentrations of either norepinephrine or isoproterenol also showed much higher increases in *N*-acetyltransferase activity (Table 1). This enchanced induction of *N*-acetyltransferase activity (superinduction) after denervation by submaximal doses of catecholamines is a comparable phenomenon to the supersensitivity observed with contraction of smooth muscles after denervation (TRENDELENBERG, 1963). When pineals were cultured in dibutyryl cyclic AMP, there was no difference in the increase of *N*-acetyltransferase activity between innervated and denervated pineal glands.

TABLE 1. SUPER- AND SUBSENSITIVITY TO ISOPROTERENOL IN CULTURED PINEALS

Concentration of isoproterenol (M)	*N*-Acetyltransferase (pmoles/pineal/10 min)		Isoproterenol treated
	Denervated	Intact	
1×10^{-9}	$329 \pm 90*$	13 ± 2	—
5×10^{-9}	$1330 \pm 208*$	327 ± 41	$26 \pm 7*$
2×10^{-8}	$2190 \pm 333*$	681 ± 155	$69 \pm 24*$
1×10^{-7}	1180 ± 146	943 ± 77	$319 \pm 52*$
1×10^{-8}	1490 ± 173	1720 ± 176	1380 ± 108

Denervated rats were bilaterally ganglionectomized 7 days before they were killed. Isoproterenol-treated rats received L-isoproterenol (2.0 mg/kg) 8, 16 and 24 hr before they were killed. The pineals were cultured for 10 hr in the presence of L-isoproterenol indicated.
* $P < 0.01$ compared to intact rats.

To study whether the absence of neurotransmitter or the absence of nerve endings are responsible for the superinduction of *N*-acetyltransferase, rats were depleted of catecholamines by treatment with reserpine. When the pineals from reserpine treated rats were cultured in submaximal concentrations of isoproterenol, the responsiveness of the pineal with respect to N-acetyltransferase induction was increased, compared to untreated rat pineals. The superinduction by reserpine treatment appeared within 24 hr. The pineals of rats continuously exposed to light also showed superinduction of N-acetyltransferase activity. These observations indicate that depletion of neurotransmitter by denervation or reserpine treatment or reducing the release of neurotransmitter by continuous lighting caused a rapid superinduction of N-acetyltransferase, and the superinduction of the enzyme is due to the changes somewhere between β-adrenergic receptor and the action site of cyclic AMP.

SUPPRESSION OF SUPERINDUCTION OF N-ACETYLTRANSFERASE

Repeated injection of isoproterenol to denervated rats or to reserpine treated rats completely prevented superinduction of N-acetyltransferase (Deguchi and Axelrod, 1973a). When isoproterenol was repeatedly injected into innervated rats, their pineal showed much lower increase in N-acetyltransferase activity than untreated pineals when cultured in the presence of isoproternol. The maximal increase of N-acetyltransferase activity, however, was the same in intact, denervated and isoproterenol treated rat pineals (Table 1).

ENHANCED ELEVATION OF CYCLIC AMP IN DENERVATED OR RESERPINE TREATED PINEALS

Weiss (1969) has reported that sensitivity of adenylate cyclase to norepinephrine and sodium fluoride was increased after chronic denervation and concluded that there was an increase in the amount of adenylate cyclase after chronic denervation. They also showed that continuous lighting induced supersensitivity of adenylate cyclase in rat pineals (Weiss and Strada, 1972). The superinduction of N-acetyltransferase after denervation could be due to the changes of adenylate cyclase. However, superinduction appeared rapidly 1 day after denervation or after reserpine treatment (Deguchi and Axelrod, 1973a), whereas the increase in adenylate cyclase was not manifest until 4 weeks after denervation (Weiss, 1969). To resolve this discrepancy, the response of cyclic AMP level in pineals was assayed *in vivo* and in cultured rat pineal (Deguchi and Axelrod, 1973a,b). When submaximal doses of norepinephrine or isoproterenol was injected into rats, there was much higher increase in cyclic AMP level in denervated pineals or reserpine treated pineals than intact pineals. When pineals from reserpine treated rats were cultured for 10 min in the presence of isoproterenol, the increase in cyclic AMP level was much greater than that in the untreated pineals. These results indicate that superinduction of N-acetyltransferase after denervation or by treatment with reserpine is probably mediated by the enhanced elevation of cyclic AMP. Treatment with cycloheximide, a protein synthesis inhibitor, could not block the enchanced elevation of cyclic AMP in reserpine treated rat pineals, suggesting that the change in the sensitivity of β-adrenergic receptor to its agonist is probably due to the change in the conformation or availability of the receptor rather than the new synthesis of receptor.

Our results demonstrate that when the quantity of neurotransmitter liberated from sympathetic nerve is abolished or reduced, the sensitivity of β-adrenergic receptor on the pineal cell with respect to the response of cyclic AMP or N-acetyltransferase is greatly enhanced. Conversely, if the number of catecholamine molecules reacting with β-adrenergic receptor is increased for a period of time, the pineal cell becomes less sensitive to its agonist. That is, a larger amount of catecholamines is necessary to produce the same amount of increase in *N*-acetyltransferase activity (Table 1). Thus the dose response curve of *N*-acetyltransferase induction and cyclic AMP elevation shifted to the left by depletion of neurotransmitter, and shifted to the right by repeated administration of catecholamines. The maximal response remains essentially the same (Table 1). It thus appears that supersensitivity and tolerance are part but at the opposite poles of the same phenomenon. Supersensitivity in the absence of neurotransmitter and subsensitivity caused by the repeated administration of agonist occurs relatively rapidly within 24 hr. This finding together with the lack of the effect of protein synthesis inhibitor to prevent the development of supersensitivity suggests that the change in sensitivity is probably due to the change in conformation or availability of receptor rather than the new synthesis of receptor sites.

Previously it has been proposed on the basis of enzymatic studies that the tolerance to narcotic drugs is due to unavailability of the receptor sites (AXELROD, 1956). A recent report has also shown that after hypophysectomy adenylate cyclase in adrenal cortex becomes more responsive to ACTH (SAYERS and BEALL, 1973). Thus such a phenomenon might be a general adaptive mechanism for responsive cell to neurotransmitter or hormones.

REFERENCES

AXELROD J. (1956) *Science* **124**, 263–264.
AXELROD J., SHEIN H. M. and WURTMAN R. J. (1969) *Proc. Nat Acad. Sci. U.S.A.* **62**, 544–549.
AXELROD J., WURTMAN R. J. and SNYDER S. H. (1965) *J. Biol. Chem.* **240**, 949–954.
BROWNSTEIN M., HOLZ R. and AXELROD J. (1973a) *J. Pharmacol. exp. Ther.*, in press.
BROWNSTEIN M., SAAVEDRA J. and AXELROD J. (1973b) *Mol. Pharmacol.* in press.
DEGUCHI T. (1973) *Mol. Pharmacol.* **9**, 184–190.
DEGUCHI T. and AXELROD J. (1972a) *Proc. Nat Acad. Sci. U.S.A.* **69**, 2208–2211.
DEGUCHI T. and AXELROD J. (1972b) *Proc. Nat Acad. Sci. U.S.A.* **69**, 2547–2550.
DEGUCHI T. and AXELROD J. (1973a) *Proc. Nat Acad. Sci. U.S.A.*, in press.
DEGUCHI T. and AXELROD J. (1973b) *Mol. Pharmacol.*, in press.
KLEIN D. C., BERG G. R. and WELLER J. (1970) *Science* **168**, 979–980.
KLEIN D. C. and WELLER J. L. (1972) *Science* **177**, 532–533.
KLEIN D. C., WELLER J. L. and MOORE R. Y. (1971) *Proc. Nat Acad. Sci. U.S.A.* **68**, 3107–3110.
LYNCH H. J. (1971) *Life Sci.* **10**, 791–795.
SAYERS G. and BEALL R. J. (1973) *Science* **179**, 1330–1331.
SNYDER S. H., ZWEIG M., AXELROD J. and FISCHER J. E. (1965) *Proc. Nat Acad. Sci. U.S.A.* **53**, 301–305.
TRENDELENBURG U. (1963) *Pharmacol. Rev.* **15**, 225–276.
WEISS B. (1969) *J. Pharmacol. exp. Ther.* **168**, 146–152.
WEISS B. and STRADA S. J. (1972) In: *Advances in Cyclic Nucleotide Research* (P. GREENGARD and G. A. ROBISON, Eds.) Vol. 1, pp. 357–374, Raven Press, New York.
WURTMAN R. J., SHEIN H. M. and LARIN F. (1971) *J. Neurochem.* **18**, 1683–1687.

Frontiers in Catecholamine Research 1973, pp. 321 to 325. Pergamon Press. Printed in Great Britain

BETA-ADRENERGIC REGULATION OF INDOLE METABOLISM IN THE PINEAL GLAND

DAVID C. KLEIN[1] and ARTHUR YUWILER[2]

[1]Section on Physiological Controls, Laboratory of Biomedical Sciences, National Institute of Child Health and Human Development, National Institutes of Health, Bethesda, Maryland 20004, U.S.A. and [2]Neurochemistry Laboratory, Brentwood V. A. Hospital and Department of Psychiatry, University of California at Los Angeles, California, U.S.A.

INTRODUCTION

SHORTLY after a daily rhythm in pineal serotonin was discovered by QUAY (1963), FISKE (1964) made the important observation that this rhythm could be abolished by denervation of the pineal gland. This indicated that the sympathetic nerve fibres in the gland were necessary for the persistence of the rhythm. WOLFE et al. (1962) had previously shown that norepinephrine was localised in the sympathetic nerves in the pineal gland. This observation and the findings of FISKE (1964) were consistent with the hypothesis that norepinephrine was involved in the regulation of the serotonin rhythm. Additional support for this hypothesis was provided by studies of ZWEIG and AXELROD (1969), in which inhibitors of norepinephrine synthesis were used.

In an extension of organ culture studies which indicated that norepinephrine regulated radiolabelled melatonin production from radiolabelled tryptophan (AXELROD, SHEIN and WURTMAN, 1969; KLEIN, 1969), we found that norepinephrine also stimulated the conversion of ^{14}C-serotonin to ^{14}C-N-acetylserotonin (KLEIN and BERG, 1970). This finding led to the discovery that treatment with norepinephrine of pineal organ cultures stimulated the activity of the enzyme responsible for the conversion of serotonin to N-acetylserotonin, serotonin N-acetyltransferase (arylamine: acetyl CoA N-acetyltransferase, E.C. 2.3.1.5) via a mechanism involving protein synthesis (KLEIN, BERG and WELLER, 1970).

There was no information available regarding the activity of this enzyme in vivo at that time. We initiated an investigation of this and found that the activity of this enzyme increased more than 20-fold at night (KLEIN and WELLER, 1970), the time when pineal serotonin was lowest (QUAY, 1963). We also found that the enzyme rhythm persisted in constant darkness and was abolished by constant lighting. These same characteristics had previously been found to be true of the rhythm in pineal serotonin by SNYDER et al. (1965).

On the basis of these similarities, and on the finding that norepinephrine stimulated the conversion of ^{14}C-serotonin to ^{14}C-N-acetylserotonin by cultured pineal glands, we proposed that the nocturnal decrease in serotonin was initiated by the release of norepinephrine from nerve endings. We thought that norepinephrine acted postsynaptically to increase the activity of N-acetyltransferase and thus reduced serotonin content by converting more serotonin to N-acetylserotonin. It appeared that adenosine 3',5'-monophosphate was involved in this mechanism because $N^6,2'O$-dibutyryl adenosine 3',5'-monophosphate mimicked the effects of norepinephrine on both N-acetyltransferase activity and on the conversion of ^{14}C-serotonin to ^{14}C-N-acetylserotonin (KLEIN, BERG and WELLER, 1970). In addition, WEISS and COSTA

321

(1967) had shown that norepinephrine stimulated the activity of adenyl cyclase, the enzyme which forms adenosine 3',5'-monophosphate.

It also appeared quite probable that a beta-adrenergic receptor was involved in this mechanism because WEISS and COSTA (1968) had found that the effects of norepinephrine on adenyl cyclase were blocked by beta-adrenergic antagonists. Later studies made this seem almost certain because it was found that beta-adrenergic antagonists blocked the effects of norepinephrine or isoproterenol on ^{3}H-melatonin production by cultured pineal glands (WURTMAN et al., 1971), cyclic AMP in cultured pineal glands (STRADA et al., 1972) and pineal N-acetyltransferase activity in vivo (DEGUCHI and AXELROD, 1972). However the entire hypothesis of the beta-adrenergic regulation of serotonin had not been tested directly by measuring pineal serotonin. For this reason the studies described below were performed.

STUDIES ON THE REGULATION OF PINEAL SEROTONIN CONTENT

Pineal glands contain about 250 pmol of serotonin after 24 hr of organ culture in medium containing 0.2 mM tryptophan (KLEIN et al., 1973). Treatment with L-norepinephrine (0.1–10 μM) for 6 hr results in a reduction of serotonin to about 40% of control values (Fig. 1). This effect clearly takes place in postsynaptic sites because the effects are also seen in chronically denervated pineal glands (Table 1). That effects of L-norepinephrine are not mimicked by D-norepinephrine (0.1–10 μM) or by dopamine (0.1 μM, Fig. 1) indicates that a highly specific receptor is involved. To examine whether the effects of norepinephrine on serotonin were mediated by a beta-adrenergic receptor, glands were treated with the beta-adrenergic protagonist, isoproterenol (10 μM) or the alpha-adrenergic protagonist, phenylephrine (10 μM). The former produced a large decrease in the serotonin content, but the latter was without effect. Similarly, the effects of L-norepinephrine were blocked by propranolol, a beta-adrenergic antagonist, but neither by phentolamine nor phenoxybenzamine, two beta-adrenergic antagonists. These findings indicate that a beta-adreneric receptor is involved in the regulation of pineal serotonin.

To test whether adenosine 3',5'-monophosphate is involved in regulation of serotonin content, pineal glands were treated with N^6,2'O-dibutyryl adenosine 3',5'-monophosphate (0.1–1.0 mM) or theophylline (10 mM), inhibitors of pineal phosphodiesterase (KLEIN and BERG, 1970). Both compounds lowered the amount of serotonin in the pineal gland in culture (Fig. 1).

TABLE 1. EFFECTS OF L-NE ON 5-HT IN CHRONICALLY DENERVATED GLANDS

| | Treatment Organic culture | | |
Prior to culture	0–24 hr	24–30 hr	5-HT (pmol/gland)
None	None	None	274 ± 17.3
None	None	L-NE (10⁻⁵ M)	114 ± 12.5*
Chronic denervation	None	None	245 ± 13.0
Chronic denervation	None	L-NE (10⁻⁵ M)	108 ± 9.9*

Chronically denervated glands were obtained from animals that had been superior cervical ganglionectomised 2 weeks prior to organ culture. Glands were incubated for 24 hr under control conditions and then were transferred to fresh medium containing no drug. L-norepinephrine (L-NE) was added in a 120 × concentrated solution of 0.01 N HCl 0.24 hr after glands were transferred.

* Significantly less than treated control value, $P < 0.05$ (Taken from KLEIN et al., 1973)

PINEAL SEROTONIN (% OF UNTREATED CONTROL GLANDS)

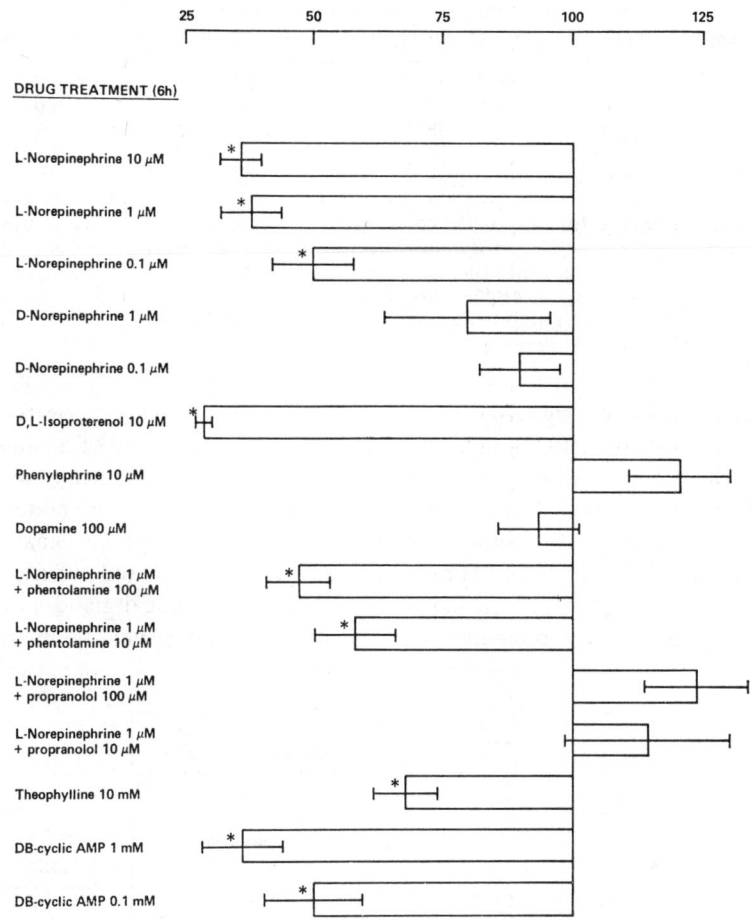

FIG. 1.—Effects of drugs on the serotonin content of pineal glands in organ culture. Each bar represents the mean ($\pm$ s.e.) serotonin content of 4–12 pairs of pineal glands given as the percent of 4–12 untreated control values. The data is a summary of seven experiments in which the average control value was 250 pmoles.
    * Significantly less than untreated controls, $P < 0.05$. DB-cyclic AMP, $N^6,2'0$-dibutyryl adenosine $3',5'$-monophosphate. (Based on data from KLEIN et al., 1973).

The effects of cycloheximide on the regulation of pineal serotonin were examined (Table 2). It was found that cycloheximide alone decreased the amount of serotonin in the pineal gland. This is probably due to inhibition of pineal tryptophan hydroxylation (BENSINGER, KLEIN, WELLER and LOVENBERG, unpublished results). In the presence of cycloheximide, norepinephrine had no effect on serotonin, presumably because stimulation of N-acetyltransferase activity was also blocked (KLEIN, BERG and WELLER, 1970).

CONCLUSION

The results of the above studies are consistent with the model of regulation of serotonin content of the rat pineal gland presented in Fig. 2. Light acting via the

TABLE 2. EFFECT OF CYCLOHEXIMIDE ON PINEAL 5-HT, N-ACETYLTRANSFERASE ACTIVITY, AND PROTEIN SYNTHESIS

Treatment (24–30 hr)	[³H] Radioactivity in TCA precipitate (nCi/gland)	N-Acetyltransferase (pmol/gland/hr)	5-HT (pmol/gland)
None	5·37 ± 0·321	77·5 ± 20·1	297 ± 60·8
Cycloheximide (10⁻⁵ M)	0·36 ± 0·030	50·0 ± 12·2	100 ± 23·3*
L-NE (10⁻⁵ M)	5·01 ± 0·451	2470 ± 321	68 ± 10·8*†
L-NE (10⁻⁵ M) + cycloheximide (10⁻⁵ M)	0·36 ± 0·037	117 ± 9·8	119 ± 14·5*

Glands were incubated for 24 hr under control conditions and then transferred to fresh medium containing L-[³H] 4,5-leucine (4·7 μCi/μmol, 0·38 mM) for a 6 hr treatment period; some media also contained cycloheximide. The L-norepinephrine (L-NE) was added 15 min later as indicated.
* Significantly lower than untreated control $P < 0.05$.
† Significantly lower than group treated with cycloheximide and L-NE, $P < 0.05$. (Taken from KLEIN et al., 1973)

retina sends inhibitory signals to the suprachiasmatic nucleus via the retinohypothalamic tract (MOORE, 1973). This nucleus appears to be the site of an endogenous circadian rhythm generator, a "biological clock," which can periodically send stimulatory signals to the pineal gland, even in a blinded animal. In a sighted animal, light can inhibit this event. Stimulatory signals sent from the suprachiasmatic nucleus course by way of the medial forebrain bundle (MOORE, 1973) and superior cervical ganglia (KLEIN et al., 1971) to reach nerve endings in the pineal gland and there to stimulate the release of norepinephrine. This results in a postsynaptic interaction of

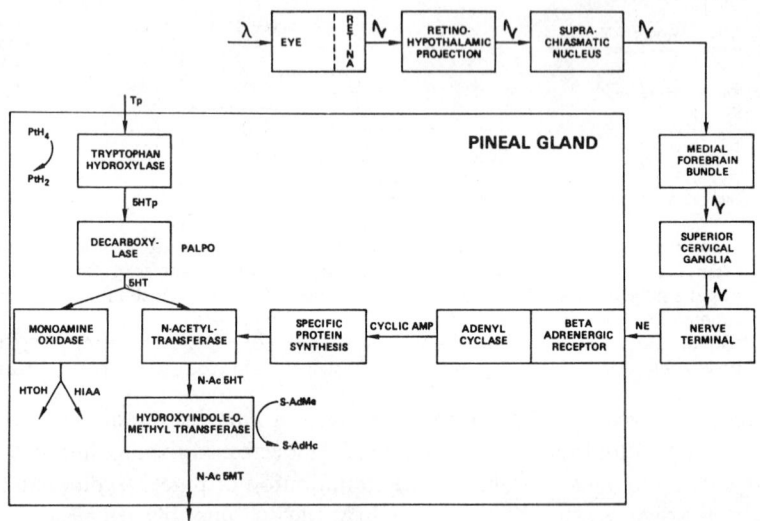

FIG. 2.—A model of the beta-adrenergic regulation of dole metabolism in the pineal gland. Cyclic AMP, adenosine 3′,5′-monophosphate; Tp, tryptophan, PtH₄, tetrohydropteridine; PtH₂, dihydropteridine; 5HTP, 5-hydroxytryptophan; PALPO, pyridoxalphosphate; 5HT, 5-hydroxytryptamine, serotonin; HTOH, hydroxytryptophol; HIAA, hydroxyindole acetic acid; N-Ac 5HT, N-acetyl 5-hydroxytryptamine, N-acetyl serotonin; AcCoA, acetyl coenzyme A; S-Ad Me, S-adenosyl methionine; S-Adhc, S-adenosylhomocysteine; N-Ac 5-MT, N-acetyl 5-methoxytryptamine, melatonin. (After KLEIN, 1973)

norepinephrine with a beta-adrenergic receptor. This event stimulates the activity of adenyl cyclase and causes a net increase in the concentration of adenosine 3′,5′-monophosphate. By a yet undefined mechanism, the synthesis of a specific protein(s) is initiated and thus results in the increased activity of N-acetyltransferase. The increase in enzyme activity causes an increase in the amount of serotonin converted to N-acetylserotonin and this causes the decrease in the serotonin content of the pineal gland because serotonin production does not keep up with the increase in serotonin acetylation.

We have previously (KLEIN and WELLER, 1972) suspected that such a mechanism may exist in other neural structures, i.e., the raphé nuclei, where both serotonergic cell bodies and adrenergic nerve endings occur. However, no data in support of this possibility are yet available.

REFERENCES

AXELROD J., SHEIN H. M. and WURTMAN R. J. (1969) Proc. Natn. Acad. Sci. U.S.A. 62, 544–549.
DEGUCHI T. and AXELROD J. (1972) Proc. Natn. Acad. Sci. U.S.A. 69, 2542–2550.
FISKE V. W. (1964) Science 146, 253–259.
KLEIN D. C. (1969) Fedn. Proc. 28, 734.
KLEIN D. C. (1973) In Serotonin and Behavior (BARCHAS J. and USDIN E., Eds.), Academic Press, New York.
KLEIN D. C. and BERG G. R. (1970) Advan. Biochem. Psychopharmacol. 3, 241–263.
KLEIN D. C., BERG G. R. and WELLER J. (1970) Science 168, 979–980.
KLEIN D. C. and WELLER J. (1970) Science 169, 1093–1095.
KLEIN D. C. and WELLER J. (1972) Science 177, 532–533.
KLEIN D. C., WELLER J. L. and MOORE R. Y. (1971) Proc. Nat. Acad. Sci. U.S.A. 68, 3107–3110.
KLEIN D. C., YUWILER A., WELLER J. L. and PLOTKIN S. (1973) J. Neurochem. 21 (in press).
MOORE R. Y. (1973) In The Neurosciences. Third Study Volume. (PARKER, K. O., Ed.), M.I.T. Press, Cambridge (in press).
QUAY W. B. (1963) Gen. Comp. Endocrinol. 3, 473–479.
SNYDER S. H., ZWEIG M., AXELROD J. and FISCHER E. (1965) Proc. Natn. Acad. Sci. U.S.A. 53, 301–305.
STRADA S. J., KLEIN D. C., WELLER J. and WEISS B. (1972) Endocrinology 6, 1470–1476.
WEISS B. and COSTA E. (1967) Science 156, 1750–1752.
WEISS B. and COSTA E. (1968) J. Pharmacol. Exp. Ther. 161, 310–319.
WOLFE D., POTTER L., RICHARDSON K. and AXELROD J. (1962) Science 138, 440–442.
ZWEIG M. and AXELROD J. (1969) J. Neurobiol. 1, 87–93.

Frontiers in Catecholamine Research 1973, pp. 327 to 333. Pergamon Press. Printed in Great Britain.

SELECTIVE REGULATION OF THE MULTIPLE FORMS OF CYCLIC NUCLEOTIDE PHOSPHODIESTERASE BY NOREPINEPHRINE AND OTHER AGENTS

BENJAMIN WEISS

Department of Pharmacology, Medical College of Pennsylvania,
Philadelphia, Pennsylvania 19129, U.S.A.

THE PHYSIOLOGICAL and biochemical importance of cyclic nucleotides is no longer open to serious question, and there is mounting evidence for their involvement in several clinical diseases (for recent reviews and monographs see RALL et al., 1969; WEISS and KIDMAN, 1969; GREENGARD and COSTA, 1970; ROBISON et al., 1971a, 1971b; GREENGARD et al., 1972; WEISS and STRADA, 1973). However, the fact that these nucleotides are so ubiquitous and have such diverse actions has discouraged many investigators from searching for pharmacologic agents that modify the intracellular concentration of cyclic nucleotides. For how could a drug which acts by modifying the concentrations of adenosine 3', 5'-monophosphate (cyclic 3', 5'-AMP) be specific if this nucleotide has so many actions? The obvious answer to this question is that one must find a drug capable of altering the intracellular concentrations of cyclic 3', 5'-AMP in specific types of cells.

Fortunately, there is great promise for discovering agents that selectively alter the concentration of cyclic nucleotides in specific tissues, for not only are there different nucleotide cyclases for the various nucleotides (HARDMAN and SUTHERLAND, 1969) but there is great specificity with which hormones and neurotransmitters activate a single cyclase such as adenylate cyclase (see SUTHERLAND and RALL, 1960; SUTHERLAND et al., 1968; WEISS and KIDMAN, 1969; RALL and GILMAN, 1970; GREENGARD and COSTA, 1970; ROBISON et al., 1971a, 1971b). Whether this hormonal specificity is due to the presence of several distinct adenylate cyclases or due to the existence of unique regulatory subunits attached to a single catalytic subunit on adenylate cyclase, as originally suggested by ROBISON, BUTCHER and SUTHERLAND (1967), is still not totally resolved. But there is general agreement that the adenylate cyclase systems of various tissues respond differentially to hormones.

MULTIPLE FORMS OF PHOSPHODIESTERASE

A similar situation may exist with the cyclic nucleotide phosphodiesterases. For this enzyme, which catalyzes the hydrolysis of cyclic 3', 5'-AMP and other cyclic nucleotides, exists in several different molecular forms, each of which has its own peculiar properties and substrate specificities (THOMPSON and APPLEMAN, 1971a; KAKIUCHI et al., 1971; BEAVO et al., 1970; UZUNOV and WEISS, 1972; RUSSELL et al., 1973; SCHRODER and RICKENBERG, 1973). If these different isozymes were distributed unequally among the various types of cells and if one could control the activity of each of these isozymes independently, then one would have another pharmacologic tool for manipulating the intracellular concentrations of cyclic nucleotides in each cell type.

Several laboratories have now presented evidence suggesting that the various tissues have different patterns of the multiple forms of phosphodiesterase (Thompson and Appleman, 1971b; Monn and Christiansen, 1971; Campbell and Oliver, 1972). Our own studies, for example, have shown that rat cerebellum has 6 distinct forms of the enzyme (designated as Peaks I to VI according to the order in which they emerge from the electrophoresis column) (Uzunov and Weiss, 1972). In other areas of the brain, the pattern and ratio of the activities of the different forms of phosphodiesterase are markedly different from that of cerebellum. Thus, rat cerebrum has only four forms of phosphodiesterase, Peaks I through IV, and brain stem has only Peaks I, II and III (Uzunov *et al.*, 1972). Other tissues, including lung, pituitary gland and salivary gland have patterns of activity distinct from each other and from that found in brain (Weiss and Fertel, unpublished).

Studies of isolated cell preparations such as fat cells and several cloned cell lines of astrocytoma and neuroblastoma cells indicated that individual cell types have a limited number and characteristic pattern of phosphodiesterase activities. For example, fat cells have only Peaks I and II (Strada and Weiss, unpublished), C–21 astrocytoma cells have only Peaks I and IV, and neuroblastoma (NIE and N18) cells have only only one peak of activity, corresponding to Peak III of rat brain (Uzunov *et al.*, 1972).

Since each type of cell apparently has its own unique ratio of the different cyclic nucleotide phosphodiesterases, by developing effective and specific agents which will selectively alter the activity of each of these isozymes one may be able to regulate the concentration of cyclic 3′, 5′-AMP in discrete cell types.

REGULATION OF PHOSPHODIESTERASES

Recent studies have indicated that two distinct types of control mechanisms are involved in regulating the phosphodiesterases. One such mechanism is characterised by an acute or immediate activation or inhibition of the different forms of phosphodiesterase, an effect which probably does not involve the synthesis of any new protein molecules. The other mechanism is slower in onset and apparently causes a change in the amount of enzyme and therefore requires protein synthesis. An example of each of these mechanisms is presented below.

Acute regulation

The phosphodiesterases in the soluble supernatant fraction of rat cerebral homogenates were separated on a preparative polyacrylamide gel electrophoresis column (Uzunov and Weiss, 1972). The material was eluted from the column and assayed for phosphodiesterase activity (Weiss *et al.*, 1972). Four distinct peaks of enzyme activity were found. The activity of each of these peaks was determined in the presence of several inhibitors of phosphodiesterase and in the presence of a heat-stable, non-dialyzable activator of phosphodiesterase prepared as described by Cheung (1970).

Figure 1 shows the effects of this activator of phosphodiesterase on the four forms of phosphodiesterase of rat cerebrum. As can be seen, the activator stimulated Peak II more than 10-fold but failed to alter the activity of the other peaks. (Uzunov and Weiss, unpublished)

Figure 2 shows the differential inhibition of these multiple forms of phosphodiesterase by three structurally unrelated compounds. The selectivity with which

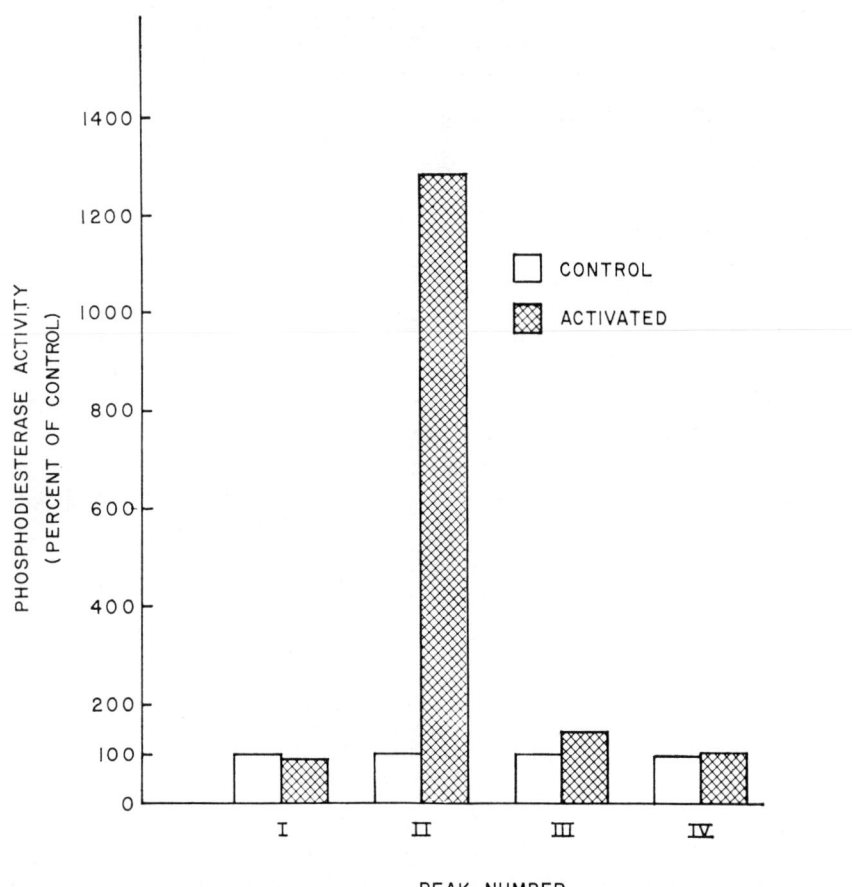

FIG. 1.—Differential activation of the multiple forms of phosphodiesterase of rat cerebrum. The soluble supernatant fraction of rat cerebral homogenates was subjected to electrophoresis on a polyacrylamide gel column. Analysis of the eluted material showed four peaks of phosphodiesterase activity. Each peak was assayed in the absence and presence of an activator of phosphodiesterase. The phosphodiesterase activity of Peaks I to IV in the absence of activator was 0·016; 0·35; 0·90; and 1·02 nmoles cyclic 3′, 5′-AMP hydrolyzed/ml/min, respectively (UZUNOV and WEISS, unpublished).

drugs can inhibit the different forms of the enzyme are demonstrated most clearly by comparing the effects of theophylline and trifluoperazine on Peaks II and III. Trifluoperazine inhibited the phosphodiesterase of Peak II by about 70 per cent in concentrations (0·5 mM) that only produced about a 20 per cent inhibition of Peak III. Theophylline produced just the opposite effects. It inhibited Peak III by about 60 per cent in concentrations (1 mM) that elicited only about a 5 per cent inhibition of Peak II. Of the three compounds shown, papaverine was the most effective inhibitor of Peak IV. No significant differences in the inhibition of Peak I was noted among these three agents (UZUNOV and WEISS, unpublished).

More recent studies have shown that these inhibitors not only acted preferentially on the different enzyme forms but acted by different mechanisms. For example, theophylline inhibits phosphodiesterase by competing with the substrate,

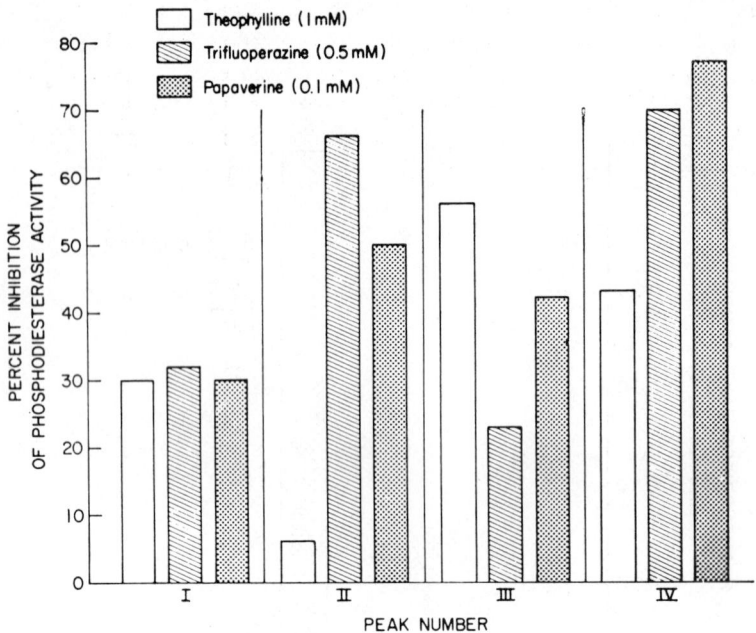

FIG. 2.—Differential inhibition of the multiple forms of phosphodiesterase or rat
cerebrum. The phosphodiesterases of rat cerebrum were prepared and assayed in the
absence and presence of various inhibitors of phosphodiesterase as described in the
legend to Fig. 1. Each bar represents the mean value of 5 experiments (UZUNOV and
WEISS, unpublished).

cyclic 3′, 5′-AMP, whereas trifluoperazine does not. The phenothiazine apparently
inhibits Peak II by blocking the activator of phosphodiesterase (FERTEL et al., 1973).
Since Peak II is the only form of phosphodiesterase in the cerebrum that is stimulated
by the activator, these latter results may explain the relative effectiveness of trifluoperazine against Peak II phosphodiesterase.

That there may also be endogenous inhibitors of phosphodiesterase was shown
by RIEDEL and GERISCH (1971) in their studies of the cellular slime mold.

Chronic regulation

Thus, these studies suggest that the activity of each form of phosphodiesterase
may be selectively increased or decreased by certain endogenous or exogenous agents.
But it is also possible that the organism may have the means of controlling the
synthesis as well as the activity of the phosphodiesterases. The availability of pure
cultures of astrocytoma cells and the finding that these cells have more than one
form of phosphodiesterase (UZUNOV et al., 1972) enabled us to study the long-term
influence of various compounds on the multiple forms of phosphodiesterase in a
single, cloned cell type. In view of our previous studies showing that the activity
of the sympathetic nervous system causes chronic changes in the hormone-sensitive
adenylate cyclase system (see WEISS and STRADA, 1972) we were particularly interested
in determining the possible influence of the adrenergic neurotransmitter on the
multiple forms of phosphodiesterase.

Figure 3 shows that the soluble supernatant fraction of C-21 astrocytoma cells has two peaks of phosphodiesterase activity, corresponding to Peak I and IV of rat brain. Incubation of the cells for 6 hr in the presence of norepinephrine (0·3 mM) caused a 2–3-fold increase in the activity of Peak IV but no significant change in Peak I. This effect of norepinephrine was blocked by incubating the cells with the beta adrenergic blocking drug, propranolol, or with the inhibitor of protein synthesis,

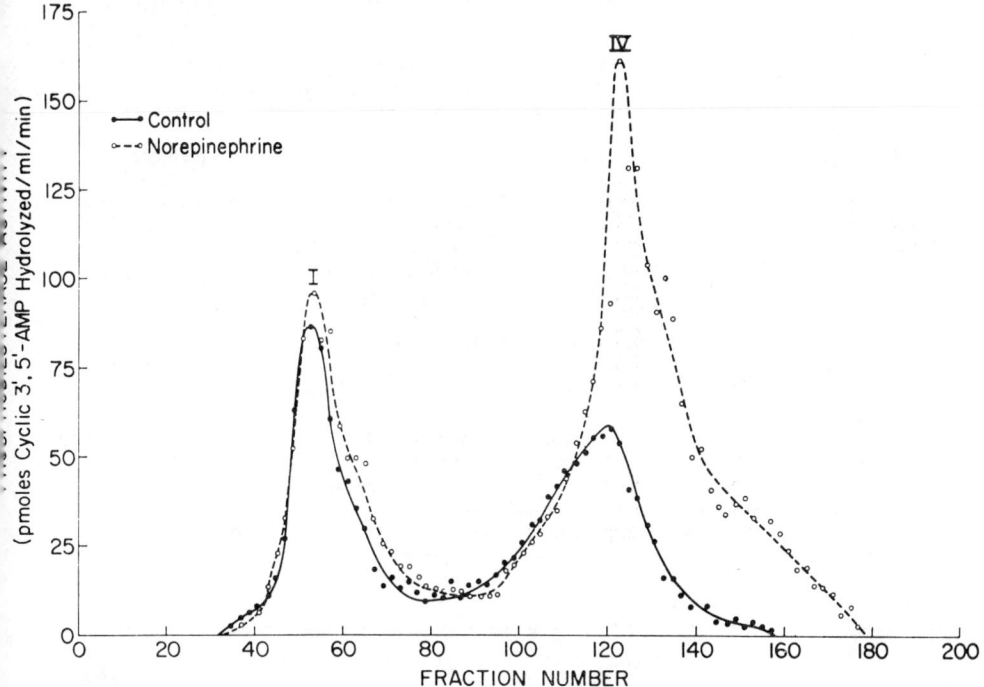

FIG. 3.—Effect of norepinephrine on the two molecular forms of phosphodiesterase of C-21 astrocytoma cells. Cloned C-21 astrocytoma cells were incubated for 6 hr in the absence or presence of 1-norepinephrine (0·3 mM). The cells were then homogenised and the soluble supernatants fluid was subjected to polyacrylamide gel electrophoresis. The eluted fractions were analysed for phosphodiesterase activity. The ratio of the total phosphodiesterase activity of Peak I to that of Peak IV was 1:1·3 for the control sample and 1:2·4 for the cells treated with norepinephrine (UZUNOV et al., 1973).

cycloheximide (UZUNOV et al., 1973). These results suggest that this selective increase in phosphodiesterase activity involves the synthesis of new protein and is a consequence of beta-adrenergic stimulation.

The mechanism by which norepinephrine induced Peak IV phosphodiesterase is as yet unclear. However, the results showing that norepinephrine produces a marked increase in the intracellular concentration of cyclic 3′, 5′-AMP in astrocytoma cells (CLARKE and PERKINS, 1971; GILMAN and NIRENBERG, 1971) and the data showing that dibutyryl cyclic 3′, 5′-AMP itself can induce the formation of phosphodiesterase in fibroblasts (D'ARMIENTO et al., 1972) suggest that in our studies norepinephrine induced a specific phosphodiesterase by raising the concentration of cyclic 3′, 5′-AMP in the C-21 astrocytoma cells.

REGULATION OF ADENYLATE CYCLASE

These acute and chronic effects on the phosphodiesterases are reminiscent of those seen on the adenylate cyclase system. The activity of adenylate cyclase of rat pineal gland homogenates is markedly increased immediately following the acute *in vitro* addition of the adrenergic neurotransmitter, norepinephrine (Weiss and Costa, 1968). Sympathetic activity induces long-term changes in the adenylate cyclase system as well. For example, reducing or abolishing the sympathetic input

TABLE 1. Acute and chronic effects of adrenergic stimulation on the adenylate cyclase–phosphodiesterase systems

Experimental condition	Enzyme system	Effects	Reference
Norepinephrine added acutely (*in vitro*)	Adenylate cyclase	Activation of hormone-sensitive enzyme	Weiss and Costa (1968)
	Phosphodiesterase	Inhibition at high concentrations	Goren and Rosen (1972)
Norepinephrine administered chronically (*in vivo*)	Adenylate cyclase	Reduction of hormone-sensitive adenylate cyclase	Strada and Weiss (Unpublished)
	Phosphodiesterase	Induction of a specific molecular form	Uzunov, Shein and Weiss (1973)
Decreased sympathetic activity	Adenylate cyclase	Increase in nor-epinephrine-sensitive enzyme	Weiss (1969)
	Phosphodiesterase	—	—

to the pineal gland over a several-week period increases the sensitivity of the adenylate cyclase system to norepinephrine (Weiss and Costa, 1967; Weiss, 1969; 1970). This biochemical "denervation supersensitivity" may have been due to a lack of catecholamine at the receptor site since injecting norepinephrine into the animal during the period of reduced sympathetic activity completely abolished the increased responsiveness of the cyclic 3′, 5′-AMP system to the subsequent *in vitro* addition of norepinephrine (Strada and Weiss, 1972).

REGULATION OF CYCLIC 3′, 5′-AMP BY ADRENERGIC ACTIVITY

The various short and long-term effects of adrenergic activity on the enzymes involved in the biosynthesis and hydrolysis of cyclic 3′, 5′-AMP are summarised in Table 1 and indicate that sympathetic nerve activity causes several changes in the adenylate cyclase-phosphodiesterase systems. The immediate effect of adrenergic stimulation is an activation of adenylate cyclase (and perhaps an inhibition of phosphodiesterase) with a concomitant increase in the intracellular concentration of cyclic 3′, 5′-AMP. This acute response may initiate a sequence of events, the ultimate effect of which is to reduce the hormone-sensitive adenylate cyclase system and to increase the activity of a specific cyclic 3′, 5′-AMP phosphodiesterase. (Decreased adrenergic activity would presumably produce the opposite effects.)

These long term changes in adenylate cyclase and phosphodiesterase both apparently involve the synthesis of new protein and may constitute a feedback control mechanism which prevents the biological effects of excessive and prolonged sympathicotonia. Any disruption in this feedback system could lead to several clinical syndromes, the most prominent of which is essential hypertension. The recent publication of AMER (1973) showing that aortas of rats with essential hypertension have a deficiency of the hormone-sensitive adenylate cyclase and an increase in one of the forms of phosphodiesterase suggest that certain clinical abnormalities may be due to long term changes in these enzymes and emphasise the need for studying the multiple forms of adenylate cyclase and phosphodiesterase in a variety of diseased organs.

REFERENCES

AMER M. S. (1973) *Science* **179**, 807–809.
BEAVO J. A., HARDMAN J. G. and SUTHERLAND E. W. (1970) *J. Biol. Chem.* **245**, 5649–5655.
CAMPBELL M. T. and OLIVER I. T. (1972) *Europ. J. Biochem* **28**, 30–37.
CHEUNG W. Y. (1970) *Biochem. Biophys. Res. Comm.* **38**, 533–538.
CLARK R. B. and PERKINS J. P. (1971) *Proc. Natl. Acad. Sci.* **68**, 2757–2760.
D'ARMIENTO M., JOHNSON G. S. and PASTAN I. (1972) *Proc. Natl. Acad. Sci.* U.S.A. **69**, 459–462.
FERTEL R., UZUNOV P. and WEISS B. (1973) *Fedn. Proc.* **32**, 679 abs.
GILMAN A. G. and NIRENBERG M. (1971) *Proc. Natl. Acad. Sci.* **68**, 2165–2168.
GOREN E. N. and ROSEN O. M. (1972) *Molecular Pharmacol.* **8**, 380–384.
GREENGARD P. and COSTA E. (eds.) (1970) *Adv. Biochem. Psychopharmacol.* Vol. 3, pp. 11–381. Raven Press, N.Y.
GREENGARD P., PAOLETTI R. and ROBISON G. A. (Eds) (1972) *Adv. Cyclic Nucleotide Res.* Vol. 1 Raven Press, N.Y.
HARDMAN J. G. and SUTHERLAND E. W. (1969) *J. Biol. Chem.* **244**, 6363–6370.
KAKIUCHI S., YAMAZAKI R. and TESHIMA Y. (1971) *Biochem. Biophys. Res. Comm.* **42**, 968–974.
MONN E. and CHRISTIANSEN R. O. (1971) *Science* **173**, 540–542.
RALL T. W., RODBELL M. and CONDLIFFE P. (eds) (1969) *The Role of Adenyl Cyclase and Cyclic 3', 5'-AMP in Biological Systems*, Fogarty International Center Proceedings No. 4. National Institutes of Health, Bethesda, Md.
RALL T. W. and GILMAN A. G. (1970) *Neurosci. Res. Program. Bull.* **8**, 221–323.
RIEDEL V. and GERISCH G. (1971) *Biochem. Biophys. Res. Comm.* **42**, 119–124.
ROBISON G. A., BUTCHER R. W. and SUTHERLAND E. W. (1967) *Ann. N.Y. Acad. Sci.* **139**, 703–723.
ROBISON G. A., BUTCHER R. W. and SUTHERLAND E. W. (1971a) *Cyclic AMP* Academic Press, N.Y.
ROBISON G. A., NAHAS G. G. and TRINER L. (eds) (1971b) *Cyclic AMP and Cell function.* Ann. N.Y. Acad. Sci. **185**, 1–556.
RUSSELL T. R., TERASAKI W. L. and APPLEMAN M. M. (1973) *J. Biol. Chem.* **248**, 1334–1340.
SCHRODER J. and RICKENBERG H. V. (1973) *Biochem. Biophys. Acta* **302**, 50–63.
STRADA S. J. and WEISS B. (1972) *Soc. for Neuroscience*, p. 229.
SUTHERLAND E. W. and RALL T. W. (1960) *Pharmacol. Revs.* **12**, 265–299.
SUTHERLAND E. W., ROBISON G. A. and BUTCHER R. W. (1968) *Circulation* **37**, 279–306.
THOMPSON W. J. and APPLEMAN M. M. (1971a) *Biochemistry* **10**, 311–316.
THOMPSON W. J. and APPLEMAN M. M. (1971b) *J. Biol. Chem.* **246**, 3145–3150.
UZUNOV P. and WEISS B. (1972) *Biochem. Biophys. Acta* **284**, 220–226.
UZUNOV P., SHEIN H. M. and WEISS B. (1972) *Vth Int. Congr. Pharmacol.* San Francisco, Calif. p. 239.
UZUNOV P., SHEIN H. M. and WEISS B. (1973) *Science* **180**, 304–306.
WEISS B. (1969) *J. Pharmacol. Exptl. Therap.* **168**, 146–152.
WEISS B. (1970) *in: Biogenic Amines as Physiological Regulators.* J. J. BLUM (ed). Prentice-Hall, Englewood Cliffs, N.J. pp. 35–73.
WEISS B. and COSTA E. (1967) *Science* **156**, 1750–1752.
WEISS B. and COSTA E. (1968) *J. Pharmacol. Exptl. Therap.* **161**, 310–319.
WEISS B. and KIDMAN A. D. (1969) *Adv. Biochem. Psychopharmacol.* E. COSTA and P. GREENGARD (Eds). Raven Press, N.Y. **1**, 131–164.
WEISS B. and STRADA S. J. (1973) *Fetal Pharmacology*, Raven Press, N.Y. L. BOREUS (ed) pp. 205–232.
WEISS B. and STRADA S. J. (1972) *Adv. Cyclic Nucleotide Research* (P. GREENGARD, R. PAOLETTI and G. A. ROBISON, (Eds) Raven Press, Vol. 1, pp. 357–374.
WEISS B., LEHNE R. and STRADA S. (1972) *Analyt. Biochem.* **45**, 222–235.

Frontiers in Catecholamine Research 1973, pp. 335 to 337. Pergamon Press. Printed in Great Britain.

STEREOISOMERS OF CATECHOLAMINES

P. N. PATIL*

Division of Pharmacology, College of Pharmacy, The Ohio State University,
Columbus, Ohio 43210, U.S.A.

OPTICAL isomers of substrates for α-chymotrypsin have been very useful in mapping the active site of the enzyme (reviewed by ALWORTH, 1972). By analogy, optical isomers of catecholamines should prove to be a valuable tool to investigate pharmacologic receptors. Stereoselectivity of various drugs for pharmacologic effects has been known for a long time (CUSHNY, 1926). At the adrenergic neuroeffector junction, if not all, many processes are stereoselective for isomers of catecholamines. Under the very best experimental conditions the initial neuronal uptake by rat heart and cortex exhibits no more than four-fold selectivity in the favor of (−)-norepinephrine (IVERSEN, 1971; COYLE and SNYDER, 1969). However, the neuronal membranes of other organs fail to show any great selectivity in the uptake of (−)- and (+)-norepinephrine (DRASKÓCZY and TRENDELENBURG, 1968; KRELL and PATIL, 1972). The retention of isomers presumably by storage vesicles, on the other hand, shows a definite preference for the natural isomer (EULER and LISHAJKO, 1964). After intravenous infusion of equal doses of ^{14}C-labelled (−)- or (+)-norepinephrine, $t_{1/2}$ for the loss of isomers from the mouse heart is 7·6 hr and 2·5 hr, respectively. Reserpine pretreatment drastically reduces the accumulation of isomers and the accumulated isomers disappeared at a similar rate constant with $t_{1/2}$ of approximately 1 hr (GARG et al., 1973).

At the post-junctional sites, either for *alpha* or *beta* adrenergic receptor activation, isomers exhibit marked stereoselectivity. The original study by EASSON and STEDMAN (1933) is particularly useful in explaining the differences in the pharmacologic activity of stereoisomers. They postulated that in an asymmetric molecule like (−)-epinephrine, three of the four groups linked to the asymmetric carbon are involved in the attachment with the receptor (Fig. 1). In the (+)-isomer, only two-point interaction with the receptor is expected. The hypothesis is supported by the fact that desoxyepinephrine, which lacks the alcoholic hydroxyl group, is equiactive with (+)-epinephrine. On several tissues examined for either the *alpha*- or *beta*-adrenergic receptor activity, the pharmacologic activity of a close structural analogue of desoxy-epinephrine, desoxynorepinephrine is identical with (+)-norepinephrine (reviewed by PATIL and LAPIDUS, 1972).

Previously, it has been postulated that from tissue, similarity of the pharmacologic receptors should generate similar isomeric-activity-ratios (PATIL, 1969) (Fig. 1). Optical isomers of norepinephrine were selected for this purpose. On normal tissues the activity difference between the isomers is obscured by various factors that are operative at the nerve terminal. When various routes of drug disposition were blocked, the activity difference between (−)- and (+)-norepinephrine, examined

---

* The author's work is supported by grants from The United States Public Health Service, GM-17859, HE-12215 and NS-08956.

ADRENERGIC NERVE TERMINAL   EFFECT0R CELL   EFFECT

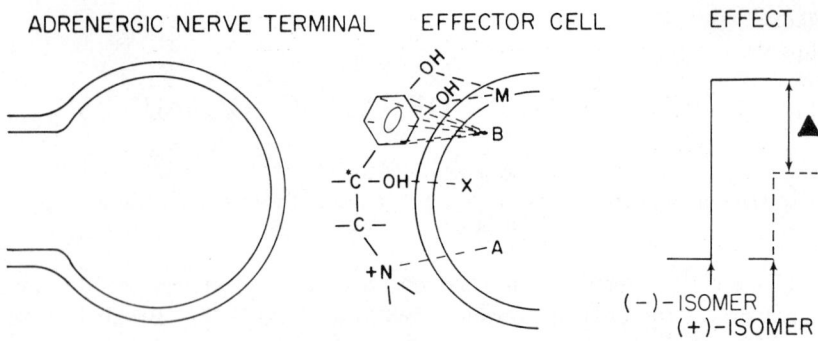

FIG. 1.—Illustrates the possible interactions of (−)-norepinephrine with adrenergic receptor. In (+)-isomer, alcoholic OH-group will be oriented away from the interacting site X. This should result in lesser intensity of effect. Sites A, X and B are unidentified while site M probably represents chelating metal (modified from EASSON-STEDMAN, 1933; BECKETT, 1969 and BELLEAU, 1958). If interacting sites from one tissue to another are similar, it should generate similar ▲ values or similar isomeric activity ratios.

for interaction with α-adrenergic receptors in various tissues were the same. The importance of factors which obscure the pharmacologic activity of the isomers should not be underestimated. If examined for *alpha*-adrenergic effects, the activity differ-ence between (−)- and (+)-norepinephrine on the normal cat spleen is only 0·3 log units (or 2-fold). However, if factors such as uptake, indirect effect, and antago-nistic-acting receptor are blocked, the isomeric-activity-difference is 2·2 log units (unpublished observation). This ratio is the same as that observed for the isomers in other tissues. With the aid of affinity constants of antagonists, FURCHGOTT (1967) also indicated that *alpha*-adrenergic receptors from rabbit fundus, ileum and aorta must be of a single type. *Beta*-adrenergic receptors from different tissues are considered to be of different types. Hence, if follows that molecular configurations of the *beta*-adrenergic sites in various tissues may be dissimilar. The dissimilar sites should generate different isomeric-activity-ratios. As expected, the ratios from six different tissues examined for *beta*-adrenergic receptors were not the same (PATIL *et al.*, 1971).

Some site on the enzyme adenylate cyclase is considered to be synonymous with the *beta*-adrenergic receptors. Recently, BIRNBAUM *et al.* (1973) correlated the bio-chemical and pharmacologic effects of (−)- and (+)-isoproteronol on rat atria. Under proper experimental conditions, the potency differences between the isomers for cyclic AMP formation was 3·50 log units. The potency difference for the chrono-tropic effect was the same. Similar correlation of norepinephrine isomers was examined in rat epididymal fat cells (DESANTIS and PATIL, 1973). Both isomers exhibited the same maxima. In the presence of theophylline, the ED_{50} for lipolysis by the (−)-isomer was $2·5 \times 10^{-8}$ M and for the (+)-isomer it was $8·5 \times 10^{-6}$ M. An isomeric-activity-difference of 2·54 log units (95% C.I.: 2·43 to 2·83) for lipolysis during a 30-min period in the presence of theophylline was obtained. A similar value of 2·52 log units (2·14–2·89) was found for the isomer-induced rise in cAMP during 10 min incubation under the same conditions. Thus, equality of values suggests clearcut evidence for a stereoselective receptor-mediated physiologic effect.

Stereoisomers of catecholamines as well as other substances provide a valuable, perhaps the most important, tool for the study of pharmacologic receptors. It is interesting to note that the proteolipid isolated from bovine spleen, when incorporated in artificial membrane, will show conductance change with (−)-norepinephrine and not with (+)-norepinephrine (OCHOA *et al.*, 1972). PERT and SNYDER (1973) were able to show a high degree of stereoselectivity in binding of narcotic agonist to the subcellular fractions from brain and ileum. This evidence for an opiate receptor is highly attractive. Thus, ideally, if in isolated receptor material the ratio for the affinity constants of isomers is the same as for biochemical, physiological and pharmacological events, proof for the isolated material representing a pharmacologic receptor will be readily accepted.

REFERENCES

ALWORTH W. A. (1972) *Stereochemistry and its application in Biochemistry* p. 163–177, Wiley, New York.

BECKETT A. H. (1959) *Fortschr. Arzneimittel Forsch.* **1**, 455–531.

BELLEAU B. (1958) *Can. J. Biochem.* **36**, 731–753.

BIRNBAUM J., ABLE P. and BUCKNER C. K. (1973) *Fedn. Proc. (Abstract)* **32**, 711.

COYLE J. T. and SNYDER S. M. (1969) *J. Pharmac. Exp. Ther.* **170**, 221–231.

CUSHNY A. R. (1926) *Biological relations of optically isomeric substances*. Williams & Wilkins, Baltimore.

DeSANTIS L. M. and PATIL P. N. (1973) *Fedn. Proc. (Abstract)* **32**, 801.

DRASKÓCZY P. R. and TRENDELENBURG U. (1968) *J. Pharmacol. Exp. Ther.* **159**, 66–73.

EASSON L. H. and STEDMAN E. (1933) *Biochem. J.* **27**, 1257–1266.

EULER U. S. V. and LISHAJKO F. (1964) *Acta Physiol. Scand.* **60**, 217–222.

FURCHGOTT R. F. (1967) *Ann. N. Y. Acad. Sci.* **139**, 553–570.

GARG B. D., KRELL R. D., SOKOLOSKI T. and PATIL P. N. (1973) *J. Pharm. Sci.* in press.

IVERSEN L. L. (1971) *Br. J. Pharmacol.* **41**, 571–591.

KRELL R. D. and PATIL P. N. (1972) *J. Pharmac. Exp. Ther.* **182**, 101–115.

OCHOA E., FISZER DE PLAZAS S. and DE ROBERTIS E. (1972) *J. Pharm. Pharmacol.* **24**, 75–77.

PATIL P. N. (1969) *J. Pharm. Pharmacol.* **21**, 628–629.

PATIL P. N., PATEL D. G. and KRELL R. D. (1971) *J. Pharmacol. Exp. Ther.* **176**, 622–633.

PATIL P. N. and LAPIDUS J. B. (1972) *Ergebn. Physiol.* **66**, 213–260.

PERT C. B. and SNYDER S. H. (1973) *Science* **179**, 1011–1014.

Frontiers in Catecholamine Research 1973, pp. 339 to 343. Pergamon Press. Printed in Great Britain.

EFFECTS OF CATECHOLAMINES AND PROSTAGLANDINS ON CYCLIC AMP LEVELS IN BRAIN *IN VIVO*

Erik Westermann

Institute of Pharmacology, Medical School of Hannover, Hannover, Germany

THERE is no doubt that catecholamines are able to stimulate adenylate cyclase in brain tissue. In experiments with brain homogenates, brain slices and tissue cultures of various brain cells, the addition of norepinephrine and other catecholamines elevated the level of cyclic AMP significantly (for review see Daly *et al.*, 1972; Gilman, 1972).

Surprisingly, this marked effect of catecholamines was not observed in experiments *in vivo*. Neither after intraperitoneal injection of amphetamine nor after injection of noradrenaline directly into the lateral ventrical of rats produced an increase of cyclic AMP in brain (Schmidt *et al.*, 1972). These negative results *in vivo* lead to the suspicion that the observed *in vitro* effects may be to some extent artificial and only a measure of isolated and defective neurons. It seems possible that such large changes as observed *iv vitro* might not occur if the glia were in proper contact with functional neuronal cells. However, after injection of high doses of various catecholamines by intraventricular route (100 μg/rat) and by using a rapid working freeze-stop technique Burkhard (1972) obtained the first evidence that catecholamines are able to increase the cyclic AMP content of the brain.

In our laboratory, Dr. Schwabe and his group improved the microwave radiation technique as described by Schmidt *et al.* (1972) by using a generator with an output of 2·2 kW. This high irradiation power kills small laboratory animals instantaneously and inactivates brain enzymes within a few seconds.

As can be seen in Fig. 1, only 5 sec after exposure of mice in a commercial oven (Philips, model HN 1124) more than 90 per cent of phosphodiesterase was inactivated while in rats it took 15 sec to inactivate the enzyme to the same extent. Phosphodiesterase activity in brain was measured by the method of Thompson and Appleman (1971).

Mice were injected intravenously with catecholamines and other drugs and 1 min later they were killed by the microwave radiation technique. After dissection of the brains the level of cyclic AMP was measured by the protein binding method of Gilman (1970) and expressed as nanomole cyclic AMP/g wet wt. and also as a percentage increase.

Attention should be given to three of the data shown in Table 1: Phenylephrine a stimulant of α-receptors did not change the cyclic AMP level in brain, while Isoproterenol, a stimulant of β-receptors, produced an increase of cyclic AMP by 60 per cent already in a 10-fold lower dose. The central stimulant amphetamine only slightly but significantly ($P < 0.05$) elevated cyclic AMP levels in the brain. The intravenous injection of L-dopa and dopamine (0·1—10 mg/kg i.v.) did not significantly affect the level of cyclic AMP while higher doses of dopamine (20–50 mg/kg i.v.) elevated the level by 50 per cent (Table 1).

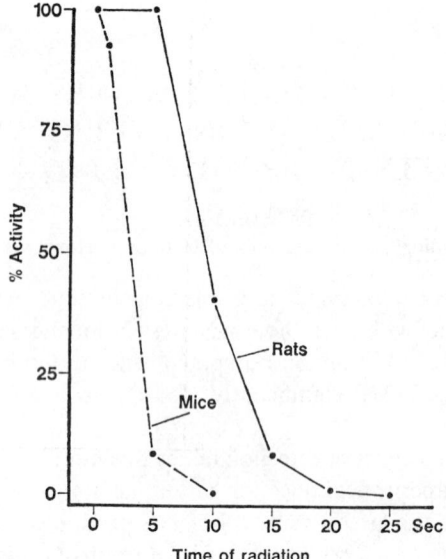

FIG. 1.—Time course of inactivation of cyclic AMP phosphodiesterase in brain by microwave radiation. Tissue fixation in mice was started by radiation with 2·2 kW for 1 sec and continued with 1·1 kW for 9 sec. Rats were irradiated for 15 sec with 2·2 kW and for 10 sec with 1·1 kW (wave length 12·2 cm, frequency 2450 ± 25 MHz). Values are related to phosphodiesterase activity of brains of non-irradiated control animals. Control activity for mice ($N = 4$) was 36 and for rats ($N = 3$) 66 nmoles cyclic AMP/mg protein/min. Data from WELLMANN and SCHWABE (1973).

The time response curve on the left hand side of Fig. 2 shows the very short lasting elevation of cyclic AMP in brain: maximal values were observed 1 min after the injection of isoproterenol and decreased to normal values within 3 min. A possible explanation for the rapid decrease in cyclic AMP levels might be the formation of an inhibitor of adenylate cyclase or of an activator of phosphodiesterase. The dose response curve of isoproterenol (right hand side of Fig. 2) shows a significant

TABLE 1. EFFECT OF VARIOUS SYMPATHOMIMETIC DRUGS ON
THE CYCLIC AMP LEVEL IN MOUSE BRAIN.
Data from STAMM and SCHWABE (unpublished results).

Drugs	N	Cyclic AMP (nmoles/g)	% change
Controls (Saline)	16	0·90 ± 0·02	—
Phenylephrine (1 mg/kg i.v.)	7	0·84 ± 0·05	−7
Isoproterenol (0·1 mg/kg i.v.)	8	1·41 ± 0·03	+57†
Amphetamine (2·5 mg/kg i.v.)	5	1·02 ± 0·05	+13*
Noradrenaline (1 mg/kg i.v.)	7	0·92 ± 0·08	+2
Dopamine (20 mg/kg i.v.)	5	1·30 ± 0·08	+44†

* differs from control ($P < 0·05$)
† differs from control ($P < 0·001$)

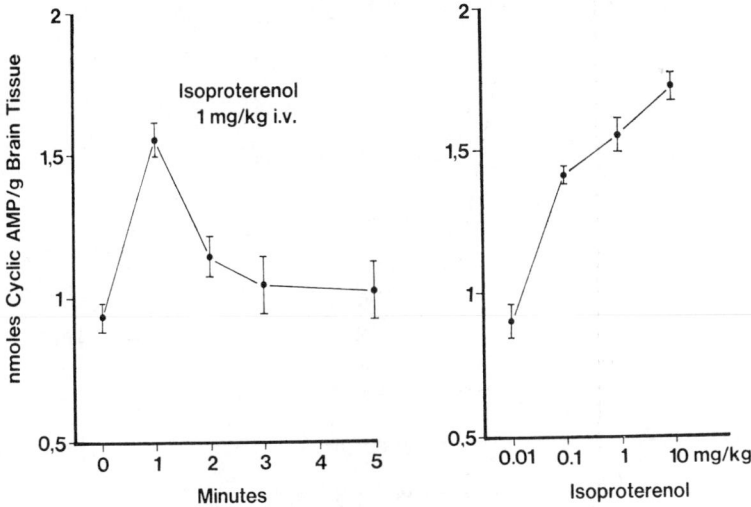

FIG. 2.—Time course and dose response curve of isoproterenol on cyclic AMP levels in mouse brain. The animals were sacrificed by microwave radiation at the time indicated. Each point represents the mean ± S.E. of 5–9 experiments. Data from STAMM and SCHWABE (unpublished results).

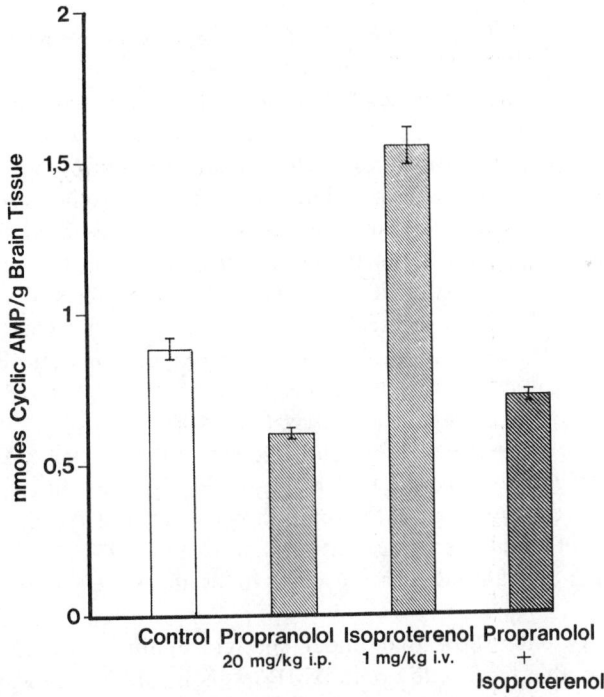

FIG. 3.—Effect of propranolol on the isoproterenol-induced elevation of cyclic AMP level in mouse brain. Propranolol was injected i.p. 30 min before and isoproterenol was injected i.v. 1 min before the animals were sacrificed by microwave radiation. Each column represents the mean ± S.E. of 5–9 experiments. Data from STAMM and SCHWABE (unpublished results).

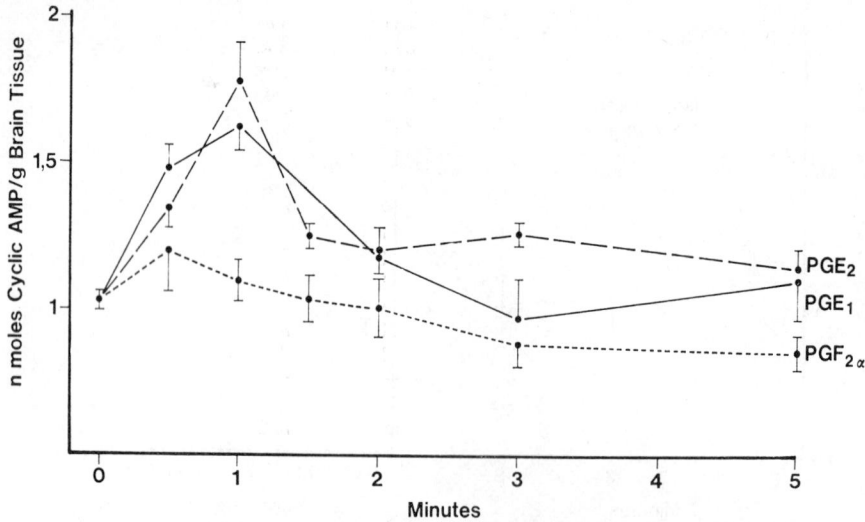

FIG. 4.—Time course of cyclic AMP levels in brains of mice following intravenous injection of 1 mg/kg of PGE_1, PGE_2, and $PGF_{2\alpha}$, respectively. The animals were sacrificed by microwave radiation at the times indicated. The results are the mean ± S.E. of 6–10 experiments. Data from WELLMANN and SCHWABE (1973).

increase of cyclic AMP in brain already in the dose range between 0·01 and 0·1 mg/kg i.v.

As can be seen in Fig. 3 blockade of β-receptors by propranolol significantly reduced the level of cyclic AMP in brain. The increase of cyclic AMP following the injection of isoproterenol was completely blocked by pretreatment of the animals with propranolol.

Prostaglandins are known to stimulate the accumulation of cyclic AMP in many different tissues, while in other tissues inhibitory effects of prostaglandins on hormone-induced responses have been observed. Since cyclic AMP has been shown to play an important role as mediator of hormone and drug effects in the central nervous system (GREENGARD and COSTA, 1970) we have studied the effects of several prostaglandins on cyclic AMP levels in brains of mice and rats. In a similar fashion as catecholamines also the intravenous injection of 1 mg/kg of PGE_1 and PGE_2, respectively, produced a significant but short lasting elevation of cyclic AMP in the brain of mice, while $PGF_{2\alpha}$ had no significant effect (Fig. 4).

Furthermore, we investigated the regional distribution of cyclic AMP levels in rat brain after intravenous injection of 1 mg/kg of PGE_2. The different increase of cyclic AMP in seven discrete areas of the brain shown in Table 2 was closely correlated to the distribution of adenylate cyclase (WEISS and COSTA, 1968). This may indicate that prostaglandins activate adenylate cyclase in brain as well as in several other tissues.

Our preliminary data show that under special conditions catecholamines and prostaglandins are able to elevate cyclic AMP levels in brain significantly and that the adrenergic receptor in mouse brain behaves like a β-receptor. The increase of cyclic AMP in brain, however, was associated with opposite effects on the behaviour of the animals: isoproterenol inducing central excitation and the prostaglandins inducing sedation. Further experiments with various stimulants and blocking agents

TABLE 2. EFFECTS OF PGE₂ (1 mg/kg i.v.) ON CYCLIC AMP LEVELS IN DISCRETE BRAIN AREAS OF RATS THAT WERE SACRIFICED BY MICROWAVE RADIATION 1 min AFTER INJECTION. EACH VALUE REPRESENTS THE MEAN ± S.E. OF 5 EXPERIMENTS. THE INCREASE PROVED TO BE SIGNIFICANT ($P < 0.05$) IN ALL BRAIN AREAS. (Data from WELLMANN and SCHWABE (1973))

Brain area	Control	PGE₂	% increase
Hypothalamus	1·70 ± 0·13	2·39 ± 0·18	41
Cerebellum	1·43 ± 0·07	1·84 ± 0·07	29
Brain stem	1·37 ± 0·10	1·79 ± 0·09	31
Striatum	1·20 ± 0·08	1·61 ± 0·15	34
Thalamus	1·10 ± 0·03	1·62 ± 0·12	47
Hippocampus	1·03 ± 0·07	1·43 ± 0·13	39
Cerebral Cortex	0·94 ± 0·04	1·72 ± 0·12	83

in vivo and the determination of cyclic AMP in better defined parts of the brain will help to get more information on the role of cyclic AMP in brain function.

REFERENCES

BURKHARD W. P. (1972) *J. Neurochem.* **19,** 2615–2619.
DALY J. W., HUANG M. and SHIMIZU H. (1972) *Adv. Cyclic Nucleotide Res.* **1,** 375–387.
GILMAN A. G. (1970) *Proc. Natn. Acad. Sci.* **67,** 305–312.
GILMAN A. G. (1972) *Adv. Cyclic Nucleotide Res.* **1,** 389–410.
GREENGARD P. and COSTA E. (Eds) (1970) *Advances in Biochemical Psychopharmacology* Vol. 3, Raven Press, New York.
SCHMIDT M. J., HOPKINS J. T., SCHMIDT D. E., ROBISON G. A. (1972) *Brain Res.* **42,** 465–477.
THOMPSON W. J. and APPLEMAN M. M. (1971) *Biochemistry* **10,** 311–316.
WEISS B. and COSTA E. (1968) *Biochem. Pharmacol.* **17,** 2107–2116.
WELLMANN W. and SCHWABE U. (1973) *Brain Res.* **59,** 371–378

Frontiers in Catecholamine Research 1973, pp. 345 to 349. Pergamon Press. Printed in Great Britain.

DEVELOPMENT OF ANTIBODIES AGAINST CATECHOLAMINES

SYDNEY SPECTOR, COLIN DALTON and ARTHUR M. FELIX
Roche Institute of Molecular Biology and Hoffmann-La Roche Inc.,
Nutley, New Jersey 07110, U.S.A.

As A consequence of the classical studies of LANDSTEINER (1945) who showed that small molecular weight chemicals (haptens) could induce antibody synthesis when they were conjugated to foreign proteins, it rapidly became evident that a great number of small molecular weight biologically active substances could be made antigenic. One application of the development of antibodies to various biologically active substances has been the quantitative measurement of these agents in biological fluids. The radioimmunoassay technique as developed by BERSON and YALOW (1959) introduced a new dimension not only for the field of endocrinology but every biological discipline. In the course of this symposium, it is also apparent that the immunochemical techniques are being utilised for both the quantitation of various enzymes involved in catecholamine metabolism as well as the localisation of the enzymes.

The immediate objective of our studies was to establish whether dopamine specific antibodies could be elicited in rabbits by immunisation with a dopamine-polyamino acid conjugate. The questions posed in our studies was whether it might not be possible to develop a radioimmunoassay which would offer great practicability because of the exquisite sensitivity of the method. In addition to the applicability of the antibodies for quantitative measurements we considered the question whether the antisera could neutralise the biologic effects caused by either exogenously administered or endogenously released neurotransmitter. The literature is replete with evidence to indicate that the biologic effects of many hormones can be neutralised by an appropriate antisera (SPECTOR, 1973). Since the antigenicity of linear polymers and co-polymers of alpha-amino acids has been demonstrated in many species (MAURER et al., 1963; MAURER and CASHMAN, 1965; GILL and DOTY, 1961; and McDEVITT and SELA, 1965), dopamine was conjugated to a linear co-polymer.

Dopamine was covalently bonded to the gamma-carboxyl group of L-glutamic acid. Co-polymers of epsilon-carbobenzoxy-L-lysine-NCA and gamma-benzyl-L-glutamate-NCA were formed. The co-polymers were then treated with β-(3,4-di-methoxyphenyl) ethylamine followed by BBr_3 (Fig. 1). Sealed tube acid hydrolysis followed by amino acid analysis determined the ratio of lysine: glutamic acid. Infrared analysis and NMR spectra offered functional analysis for the presence of amide linkages, amine groups and the presence of the catechol ring structure. The molecular weight of the co-polymers as determined by ultracentrifugation yielded a figure of about 5000.

The immunogenicity of a peptide is related to a number of factors. One of the critical parameters is its molecular weight and since the immunogen we were using had a molecular weight of about 5000, one would anticipate that it would not be a very good immunogen and that it would give rise to antisera of low titre and avidity. Rabbits were immunised with an emulsion containing the immunogen in complete

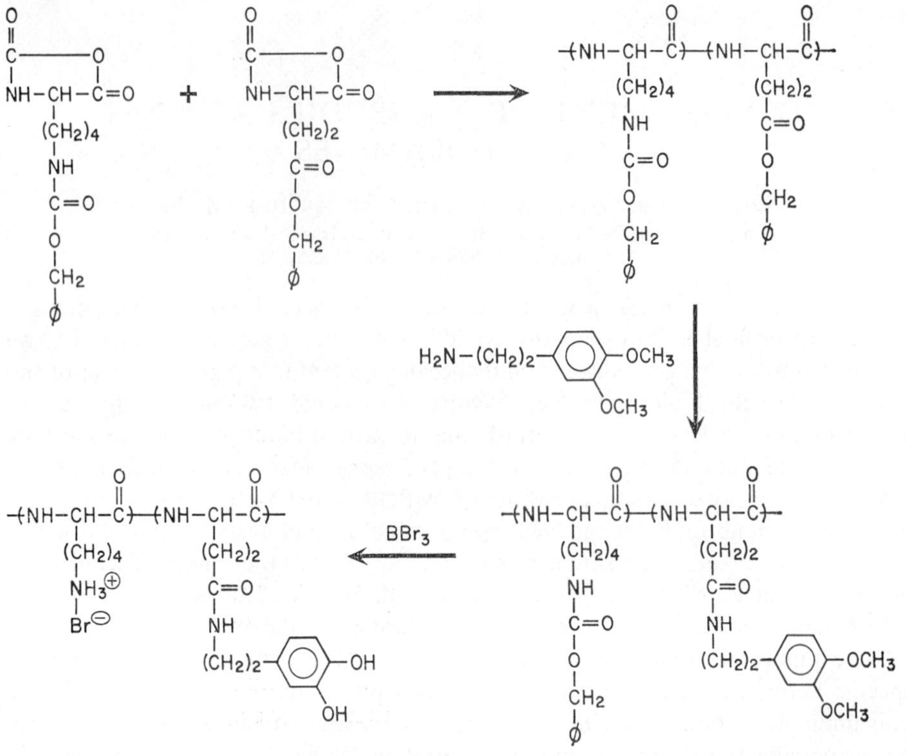

$$H_2N-(CH_2)_2-\langle\ \rangle-OCH_3$$ (with OCH$_3$)

FIG. 1

Freund's adjuvant. The presence of antibodies in immunised animals can be demonstrated by a number of immunologic techniques. The radioimmunoassay approach developed by BERSON and YALOW (1959) involves the incubation of a constant amount of labeled dopamine with varying dilutions of antiserum and then determining the distribution of radioactivity bound to the antibody. The dopamine bound to the antibody can be separated from free dopamine by precipitating the bound form with ammonium sulfate. It was apparent that we had antibodies capable of binding dopamine. A critical consideration in the assessment of the antibody is the specificity of the antibody binding sites. The radioimmunoassay is based on the competition between antigenic determinants of the labeled and unlabeled antigen for a limited number of binding sites on the antibody. Figure 2 shows such a competitive inhibition curve when unlabeled dopamine is added. However, when unlabeled norepinephrine or epinephrine was incubated with the antibody it became apparent that the determinant group which the antibody recognised was the catechol nucleus as it bound the latter as well as dopamine. The antibody failed to bind normetanephrine, phenylethylamine and tyramine.

As mentioned previously, we were interested in the neutralising capacity of the antisera on the biological activity of the catecholamines. It has been shown that several hormones including the catecholamines stimulate adenyl cyclase activity and increase cyclic adenosine 3′, 5′ monophosphate (cAMP) levels in adipose tissue (BUTCHER and SUTHERLAND, 1967, BUTCHER et al., 1968). HUTTUNEN et al., (1970) have shown the relationship of cAMP formation and the activation of lipase in adipose

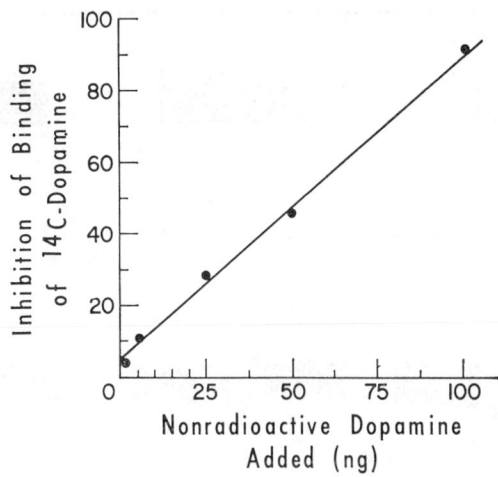

FIG. 2

tissue. Fat cells were isolated by the procedure described by DALTON et al., (1970). Free fatty acid analysis was conducted on a chloroform extract by the procedure of DALTON and KOWALSKI (1967). The data in Table 1 shows that the antibody at a dilution of 1 : 10 inhibited the lipolytic response to norepinephrine by about 85 per cent. If the antibody is diluted 1 : 100 it loses its ability to inhibit the lipolytic action of 10^{-6} M norepinephrine.

KEBABIAN, PETZOLD and GREENGARD (1972) have reported on an adenylate cyclase that is activated specifically by low concentrations of dopamine which can be demonstrated in homogenates of caudate nucleus of rat brain. Following the same procedures of these investigators we have found that 10^{-5} M dopamine stimulated the formation of 37 pmol of cAMP. The increase in enzyme activity caused by 10^{-5} M dopamine was reduced 75 per cent by preincubating dopamine in the presence of the antibody. Studies are currently in progress to ascertain the antagonistic action of the antibody on varying concentrations of dopamine and norepinephrine and to examine the action of the antibody on a compound which mimics dopamine, namely apomorphine.

Since these preliminary in vitro studies indicated that the antibody could neutralise some of the biological activities of catecholamines, studies were inititated to determine its effect on rat blood pressure. At low concentrations of the antibody (less than 1 mg of immunoglobulin) it fails to exert any effect on blood pressure. However, at higher

TABLE 1. EFFECT OF DOPAMINE ANTIBODY ON NOREPINEPHRINE INDUCED
LIPOLYSIS

Conditions	Rate of lipolysis (μeq. FFA/gTG/hr $\pm$ S.E.)
Norepinephrine 10^{-6}M	414 ± 8.7
Norepinephrine (10^{-6}M) + Ab (1:10)	83 ± 2.3 ($p < 0.001$)
Norepineprhine (10^{-6}M) + Ab (1:100)	376 ± 17

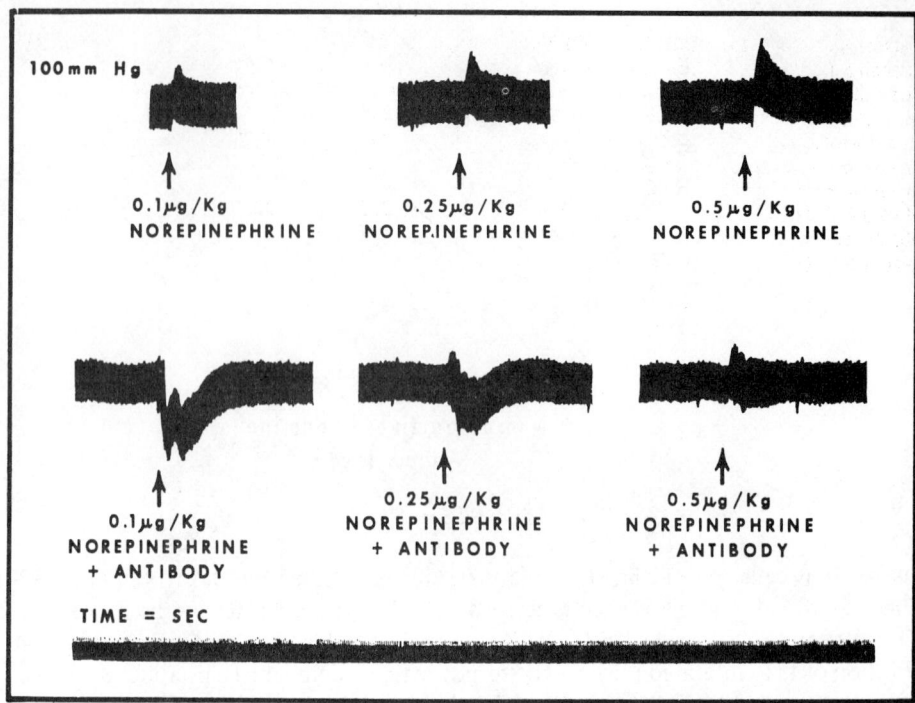

FIG. 3

concentrations (1·5–2·0 mg) it produces a precipitous fall in blood pressure which returns to the previous level within 1 to 2 min. To assess the specificity of this response, comparable quantities of gamma globulin, which were directed against morphine, serotonin or desmethylimipramine, were injected intravenously to rats and none modified the blood pressure. We postulate that the fall in blood pressure is the consequence of the antibody sequesting from circulation catecholamines, and that once the sites on the antibody are saturated the pressure returns to its control level. Figure 3 shows the pressure responses elicited by increasing doses of norepinephrine, and the antagonistic effect the antibody exerts on the norepinephrine pressor action. From these data one might almost be able to calculate the amount of catecholamines the antibody could bind. From the antibody binding studies we knew that the antibody failed to bind tyramine, so the question was then posed as to how effective the antibody could antagonise the elevation in blood pressure produced by the indirect acting sympathomimetic tyramine. As was the case with norepinephrine, the antibody effectively antagonised the hypertensive action of tyramine. Also, if tyramine was initially administered intraveneously, and the pressor action permitted to manifest itself, and if immediately thereafter the antibody was administered, the hypertensive action was immediately reversed.

Although the antibody described lacks the specificity to permit quantitation of dopamine, it might be used to quantitate total catecholamines. It is apparent that the neutralising properties of the biological activities of catecholamines by the antibody merits further exploitation.

REFERENCES

BERSON S. A. and YALOW R. S. (1959) *Nature* **184**, 1648–1649.
BUTCHER R. W., BAIRD C. E. and SUTHERLAND E. W. (1968) *J. Biol. Chem.* **243**, 1705–1712.
BUTCHER R. W. and SUTHERLAND E. W. (1967) *Ann. N.Y., Acad. Sci.* **139**, 849–859.
GILL T. J. and DOTY P. (1961) *J. Biol. Chem.* **236**, 2677–2683.
HUTTUNEN J. K., STEINBERG D. and MAYER S. E. (1970) *Proc. Nat. Acad. Sci.* **67**, 290–295.
LANDSTEINER K. (1945) Cambridge, Harvard.
MAURER P. H. (1964) *Progr. Allerg.* **8**, 1–40.
MAURER P. H., GERULAT B. F. and PINCHUCK P. (1963) *J. Immunol.* **90**, 381–387.
McDEVITT H. D. and SEAL M. (1965) *J. Exp. Med.* **122**, 517–531.
SPECTOR S. (1973) *Pharm. Rev.* **25**, 281–291.

Frontiers in Catecholamine Research 1973, pp. 351 to 356. Pergamon Press. Printed in Great Britain.

METABOLIC RESPONSES OF CARDIAC, SKELETAL MUSCLE, AND FAT CELLS TO CATECHOLAMINES

STEVEN E. MAYER

Division of Pharmacology, Department of Medicine, University of California,
San Diego, La Jolla, Ca. 92037, U.S.A.

CATECHOLAMINES cause increased mobilisation of body energy stores of glycogen and triglycerides. Receptors to these agents have been identified pharmacologically and have been shown to be linked to augmented formation of cAMP* in heart, liver, skeletal muscle and adipose tissue. The only sites of action for cAMP that have been clearly described are the enzymes designated as protein kinases (ATP: protein phosphotransferase, E.C. 2.7.1.36) by KREBS (1972). A large number of proteins have been identified as substrates for cAMP dependent phosphorylation. The most important for the purpose of this discussion are phosphorylase kinase (ATP: phosphorylase phosphotransferase, E.C. 2.7.1.36) (KREBS, 1972), the glucose-6-P dependent form of glycogen synthase (UDP glucose: glycogen α-4-glucosyltransferase, E.C. 2.4.1.11) (LARNER and VILLAR-PALASI, 1971; KREBS, 1972) and hormone-sensitive lipase (glycerol ester hydrolase, E.C. 3.1.1.3) (HUTTUNEN et al., 1970a, 1970b; CORBIN et al., 1970). Glycogen synthase and hormone-sensitive lipase act directly on energy stores. Phosphorylase kinase catalyses the covalent modification by phosphorylation of glycogen phosphorylase (α-1,4-glucan: orthophosphate glucosyltransferase, E.C. 2.4.1.1). The rapid response to catecholamines of enhanced glycogenolysis and triglyceride hydrolysis and the inhibition of glycogen synthesis can thus be explained in terms of well-defined biochemical mechanisms. Reversal to the basal state once the hormone or drug stimulus is removed can be attributed to a series of reactions that include conversion of cAMP to 5'-AMP by 3',5'-cyclic nucleotide phosphodiesterase (E.C. 3.1.4d) and the dephosphorylation of glycogen synthase (which results in its reactivation) and of phosphorylase kinase and phosphorylase (which result in termination of glycogenolysis). The overall scheme of reactions is summarised in Fig. 1. With the identification of troponin as a substrate for phosphorylase kinase (STULL et al., 1972) and for phosphorylase phosphatase (phosphorylase phosphohydrolase, E.C. 3.1.3.17) (ENGLAND et al., 1972) it becomes possible to hypothesize that covalent modification of this protein alters its regulation of actin–myosin interaction and is therefore responsible for the positive inotropic action of catecholamines.

These observations indicate that cAMP is an important mediator of catecholamine action but the question arises: is the elevation of cAMP in response to a catecholamine necessary and sufficient to control these multiple biochemical events in cells? I will address myself to an analysis of biochemical mechanisms that modulate metabolic responses to catecholamines. These mechanisms may exert their actions independently of the role of cAMP or represent supplementary levels of control.

---

* The abbreviation: cAMP is used for adenosine 3',5'-monophosphate.

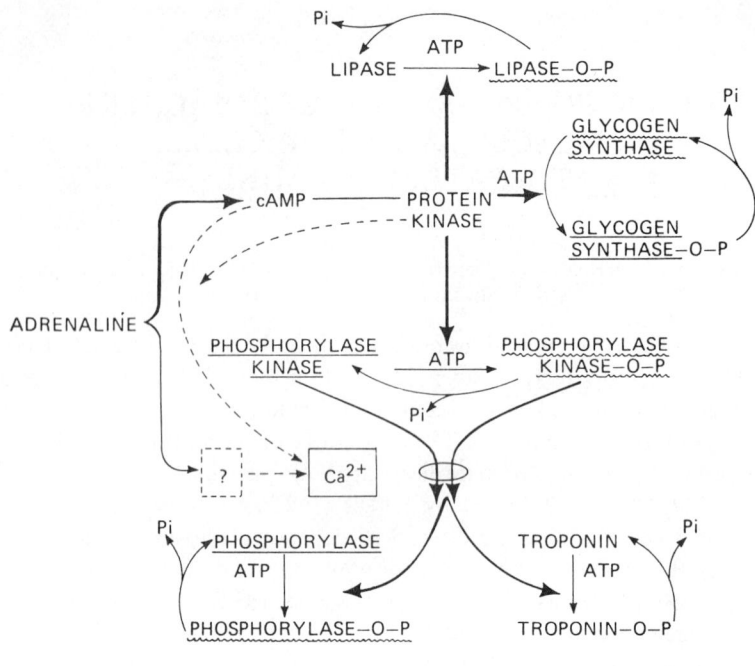

~~~~~ underlining designates activated state of enzyme
———— underlining designates nonactivated state of enzyme

Fig. 1.—Established and hypothetical biochemical mechanisms in the action of
catechol-amines on glycogenolysis and lipolysis.

They are: allosteric mechanisms, the control of phosphorylase kinase activity by
Ca^{2+} and the consequences of the association of rate-limiting enzymes with their
macromolecular substrates.

Allosteric mechanisms controlling the activity of the two forms of muscle or heart
phosphorylase are well known and have been demonstrated to play a significant
role in intact tissue. Formation of phosphorylase a in response to epinephrine was
not found to be sufficient to stimulate glycogenolysis in the intact rat heart (WILLIAMS
and MAYER, 1966) presumably because the energy charge (the relative concentrations
of ATP:ADP:AMP:Pi) favored allosteric inhibition of phosphorylase a. Glycogen
was utilized in the perfused rat heart during ventricular pressure development in the
absence of substrate without conversion of phosphorylase b to the phosphorylated
form of the enzme. Allosteric activation of phosphorylase b by P_i and decreased
glucose-6-P is the most likely mechanism (NEELY et al., 1970).

Allosteric mechanisms thus represent a level of control of metabolic responses to
catecholamines that may be independent of the triggering action of cAMP, and may
in turn reflect oxygen and substrate supply and the demands of performing external
work. Cardiac anoxia or ischemia are powerful stimuli to the rapid utilisation of
glycogen especially if the heart continues to develop ventricular pressure (MAYER
et al., 1967). An adrenergic mechanism, the release of cardiac stores of norepinephrine, has been demonstrated in ischemia with the consequent formation of cAMP,
and transformation of phosphorylase kinase and phosphorylase to their activated

forms (WOLLENBERGER *et al.*, 1969). These responses were prevented by β-adrenergic blockade. The glycogenolytic response to anoxia was however, only slightly attentuated by pronethalol (MAYER *et al.*, 1967). DOBSON and MAYER (in press, 1973) have confirmed that cessation of coronary blood flow probably causes catecholamine release from the heart the effect of which is mediated by cAMP. However, during N_2 breathing a different mechanism appears to be involved. No formation of the activated form of phosphorylase kinase was associated with the rapid and marked transformation of cardiac phosphorylase *b* to *a*. The increase in cyclic AMP that occurred 10 sec after the onset of anoxia could be supressed by propranolol without attentuation of the phosphorylase response (Table 1). This is evidence that the same metabolic response may be obtained through convergent mechanisms, but triggered by different stimuli.

TABLE 1. ACTIVATION OF THE PHOSPHORYLASE PATHWAY IN INTACT HEARTS DURING ANOXIA*

Time (sec)	Cyclic AMP Control	β-block† (μmoles/kg)	Phosphorylase Control	β-block (−AMP: +AMP)	Phosphorylase kinase Control	β-block (pH 6·2:8·2)
0	0·50 ± 0·01	0·44 ± 0·02	0·14 ± 0·01	0·14 ± 0·00	0·08 ± 0·01	0.08 ± 0·00
5	0·48 ± 0·03	—	0.14 ± 0·01	—	0·08 ± 0·02	—
10	0·62 ± 0·04‡	0·52 ± 0·01	0·19 ± 0·04	0·31 ± 0·01‡	0·07 ± 0·01	0·08 ± 0·01
15	0·52 ± 0·03	—	0·43 ± 0·08‡	—	0·08 ± 0·01	—
20	0·53 ± 0·04	0·49 ± 0·01	0·47 ± 0·08‡	0·65 ± 0·01‡	0·08 ± 0·01	0·07 ± 0·01
30	0·44 ± 0·04	0·51 ± 0·03	0·64 ± 0·07‡	0·83 ± 0·02‡	0·08 ± 0·02	0·09 ± 0·01

* Anoxia was produced in open-chest anesthetised rat preparations by switching from artificial respiration with 95% O_2, 5% CO_2 to 100% N_2 at time = 0. Hearts were frozen *in situ* at the times indicated. Analyses were made on ventricular muscle and are expressed on a wet weight basis for cAMP and as the ratio of activities obtained without to with AMP (phosphorylase) or at two different pHs (phosphorylase kinase). These ratios are measures of the activated states of each of the enzymes. Data are presented ± S.E. with $N > 5$. J. G. DOBSON and S. E. MAYER (1973).
† β-adrenergic blockade was produced with practolol, 5 mg/kg.
‡ Significantly different from time = 0 sec.

An important site of control in the phosphorylase pathway in muscle and heart is that exercised by Ca^{2+} on the catalytic activity of phosphorylase kinase. This enzyme has been called the Ca^{2+}-dependent protein kinase by KREBS (1972), because both the phosphorylation of troponin and of phosphorylase *b* require Ca^{2+} in concentrations in the range of 1 μM. This suggests that the availability of free intracellular Ca^{2+} and phosphorylation of the kinase may provide independent mechanisms of regulation, but that Ca^{2+} is both necessary and sufficient and phosphorylation alone is not sufficient to activate the enzyme. Direct and indirect evidence obtained from intact cardiac and skeletal muscle support this concept. The administration of catecholamines to dogs produced an apparent transformation of phosphorylase kinase to the activated (phosphorylated) form that occurred as rapidly as the formation of cAMP and that was demonstrable with any dose of epinephrine that increased cardiac contractile force (MAYER *et al.*, 1971; WASTILA *et al,.* 1972). However, the activation of phosphorylase was delayed both in time (WASTILA *et al.*, 1972) and in terms of the dose of the catecholamine (MAYER *et al.*, 1971). We had previously shown with isolated perfused rat hearts that the omission of Ca^{2+} from the perfusion medium did not diminish the effect of epinephrine on cAMP and formation of the activated form of phosphorylase kinase, but it did prevent the formation of phosphorylase *a*, and of course abolished contractility (NAMM *et al.*, 1968). These results demonstrate the requirement for Ca^{2+} directly. Furthermore, muscle troponin phosphorylation by phosphorylase kinase is Ca^{2+} dependent, but does not require the phosphorylated form of the enzyme (STULL *et al.*, 1972). The observations by DRUMMOND *et al.* (1969) that electrical stimulation of skeletal muscle caused a rapid formation of phosphorylase *a* without cyclic AMP elevation nor phosphorylase kinase transformation is consistent with the concept that the Ca^{2+} released during excitation-contraction coupling is required for both contraction and phosphorylase kinase activity. It is also conceivable that the effects of anoxia reported above are mediated by Ca^{2+}. However, in this case, and perhaps under more physiological conditions as well, one must consider still another mechanism of regulation of this enzyme, altered intracellular pH (for references see KREBS, 1972).

What is the relative importance of cAMP-mediated protein phosphorylation and Ca^{2+}-dependent protein phosphorylation in the metabolic and inotropic actions of catecholamines? In cardiac muscle the evidence cited above suggests that Ca^{2+} is required and that phosphorylase kinase phosphorylation may be necessary for the translation of catecholamine-receptor interaction into phosphorylase *a* formation and perhaps augmented contraction. It is not yet clear whether or not both mechanisms must act in concert to bring about the inotropic response to catecholamines. In skeletal muscle there is evidence that Ca^{2+} plays the dominant role. Small doses of isoproterenol administered into the artery of the blood-perfused rabbit gracilis muscle produced rapid and intensive phosphorylase *b* to *a* transformation without measurable changes in the concentration of cAMP nor of the activation state of phosphorylase kinase (STULL and MAYER, 1971). The lack of change of cAMP cannot be interpreted at its face value, because very small changes of the concentration of the free nucleotide may have occurred, produced an effect but not been detectable. The absence of transformation of phosphorylase kinase does fit a reasonable model. It is consistent with the observations that the total concentration of this enzyme is very high, constituting about 1% of the soluble muscle protein, and that the activity of the non-activated form of the enzyme is sufficient to catalyse phosphorylase phosphorylation provided free Ca^{2+} is made available.

While it is thus clear that Ca^{2+} is of fundamental import in the phosphorylation of substrates of phosphorylase kinase, one must also consider the flux of Ca^{2+} across cell membranes and its uptake by intracellular binding sites. Such processes are presumably important in the action of catecholamines on excitation-contraction coupling. It is not known whether these actions of epinephrine are mediated by cAMP or through some other mechanism coupled to the β-receptor. A lively controversy exists about the effect of cAMP on phosphorylation and Ca^{2+} binding to cardiac sarcoplasmic reticulum. The interrelations between cAMP and Ca^{2+} in secretion, contraction and other cell functions have recently been discussed by RASMUSSEN (1972) and colleagues. (This is a multi author citation).

Metabolic responses by adipose cells to catecholamines do not appear to be dependent on Ca^{2+}. Removal of Ca^{2+} from the incubation medium in which adipocytes were suspended and investigation of the phosphorylase kinase reaction in adipose cell extracts indicated no requirement for this cation (KHOO *et al.*, 1973). In contrast to skeletal muscle and heart the transfromation by epinephrine of phosphorylase and hormone-sensitive lipase to their activated states and of glycogen synthase to the non-activated state appear to be dependent on cAMP as the sole mediator. All these cAMP regulated systems may be simultaneously and automatically triggered by the interaction of catecholamine and receptor. The parallel changes in cAMP concentration and activation states of these enzymes in response to epinephrine suggest that in adipocytes glycogenolysis and lipolysis are coupled with each other and perhaps with other cAMP-dependent processes as well.

Regulation by mechanisms independent of cAMP in fat cells is however likely. The site of insulin's antilipolytic action has been controversial. Our data indicate that not only this action but also the reversal of phosphorylase and glycogen synthase responses to epinephrine cannot be attributed to a direct effect of insulin on the quantity of cAMP (Table 2). It seems likely that insulin acts through a messenger formed as a consequence of the insulin-cell membrane receptor interaction that antagonizes the action of cAMP (for a more extensive discussion see STEINBERG, 1973).

TABLE 2. EPINEPHRINE AND INSULIN INTERACTION ON ISOLATED RAT FAT CELLS*

Addition	Phosphorylase activity (−AMP: +AMP)	Glycogen synthase (−G6P: +G6P)	Glycerol release (μmoles hr^{-1} mg protein^{-1})	cAMP production (pmoles 5 min^{-1} mg protein^{-1})
None	0·19 ± 0·02	0·23 ± 0·03	0·16 ± 0·04	10·0 ± 1·9
Epinephrine 0·5 μM	0·50 ± 0·06†	0·13 ± 0·01†	1·10 ± 0·14†	26·8 ± 5·8† ⎤ ‡
Epinephrine plus insulin	0·24 ± 0·04	0·22 ± 0·03	0·22 ± 0·02	20·1 ± 3·4 ⎦
Insulin 0·4 nM	0·16 ± 0·02	0·35 ± 0·01†	0·10 ± 0·06	8·7 ± 1·7

* Isolated rat epididymal adipocytes were incubated for 5 min at 37 °C in the presence of the indicated concentrations of hormones and 10 mM glucose. Phosphorlyase activity is expressed as that −AMP: +AMP (2 mM); glycogen synthase as that −glucose-6-P: +glucose-6-P (7·5 mM). Each value represents the mean ± S.E. of 4 experiments. Results are from KHOO et al. (1973).

† Differs from "None", $P < 0.05$.

‡ Do not differ significantly as group means but $P < 0.001$ when data were subjected to paired comparisons.

Another site of regulation of catecholamine responses and interaction with other drugs and hormones may be the association of rate-limiting enzymes with macromolecular complexes that contain glycogen or perhaps triglyceride. We have found that starvation depleted adipocyte glycogen and promoted the formation of the activated form of phosphorylase and the non-activated form of glycogen synthase. Insulin was no longer effective in reversing these states (unpublished observations). The binding of most of the glycogen synthase and of part of the fat cell phosphorylase to glycogen observed in cells from fed animals was absent in starvation. We are investigating the hypothesis that the reversible association of these enzymes to glycogen alters their affinity for the enzymes that catalyse their conversion between activated and non-activated forms.

Thus the translation of catecholamine-receptor interaction into physiological responses may be the consequence of cAMP-induced covalent modifications of enzymes. However in a particular tissue other factors may modify or preempt control of the pathway leading to the response. Such factors are the availability of intra-cellular Ca^{2+}, the allosteric regulation of rate-limiting enzymes, and the association of enzymes with macromolecular substrates. These mechanisms may be involved as a consequence of the catecholamine-receptor interaction and also reflect the physiological state of the tissue.

Acknowledgements—The collaboration of Drs. Daniel Steinberg, John C. Khoo and James G. Dobson, Jr. in our current investigations is gratefully recognised. Our research has been supported by a grant, HL 12373, from the National Heart and Lung Institute.

REFERENCES

CORBIN J. D., REIMANN E. M., WALSH D. A. and KREBS E. G. (1970) *J. Biol. Chem.* **245**, 4849–4851.

DOBSON J. G. Jr. and MAYER S. E. (1973) *Circulat. Res.* in press.

DRUMMOND G. I., HARWOOD J. P. and POWELL C. A. (1969) *J. Biol. Chem.* **244**, 4235–4240.

ENGLAND P. J., STULL J. T. and KREBS E. G. (1972) *J. Biol. Chem.* **247**, 5275–5277.

HUTTUNEN J. K., STEINBERG D. and MAYER S. E. (1970a) *Proc. Natl. Acad. Sci. U.S.* **67**, 290–295.

HUTTUNEN J. K., STEINBERG D. and MAYER S. E. (1970b) *Biochem. Biophys. Res. Commun.* **41**, 1350–1356.

KHOO J. C., STEINBERG D., THOMPSON B. and MAYER S. E. (1973) *J. Biol. Chem.* **248**, 3823–3830.

KREBS E. G. (1972) *Curr. Top. Cell. Regul.* **5**, 99–133.

LARNER J. and VILLAR-PALASI C. (1971) *Curr. Top. Cell. Regul.* **4**, 195–236.

MAYER S. E., WILLIAMS B. J. and SMITH J. M. (1967) *Ann. N.Y. Acad. Sci.* **139**, 686–702.

MAYER S. E., NAMM D. H. and HICKENBOTTOM J. C. (1971) *Adv. Enzyme Regul.* **8**, 205–216.

NAMM D. H., MAYER S. E. and MALTBIE M. (1968) *Mol. Pharmol.* **4**, 522–530.

NEELY J. R., WHITFIELD C. F. and MORGAN H. E. (1970) *Am. J. Physiol.* 1083–1088.

RASMUSSEN H., GOODMAN D. B. P. and TENENHOUSE A. (1972) *CRC Crit. Rev. Biochem.* **2**, 95–148.

STEINBERG D. (1973) In: *Protein Phosphorylation in Control Mechanisms* (HUIJING, F., Ed) in press, Academic Press, New York.

STULL J. T. and MAYER S. E. (1971) *J. Biol. Chem.* **246**, 5716–5723.

STULL J. T., BRUSTROM C. O. and KREBS E. G. (1972) *J. Biol. Chem.* **247**, 5272–5274.

WASTILA W. B., SU Y. S., FRIEDMAN W. F. and MAYER S. E. (1972) *J. Pharmacol. Exp. Ther.* **181**, 126–138.

WILLIAMS B. J. and MAYER S. E. (1966) *Mol. Pharmacol.* **2**, 454–464.

WOLLENBERGER A., KRAUSE E.-G. and HEIER G. (1969) *Biochem. Biophys. Res. Commun.* 664–670.

Frontiers in Catecholamine Research 1973, pp. 357 to 360. Pergamon Press. Printed in Great Britain

METHODS OF APPROACH FOR THE ISOLATION OF β-ADRENERGIC RECEPTORS*

P. B. MOLINOFF†

Department of Pharmacology, University of Colorado School of Medicine,
4200 East 9th Avenue, Denver, Colorado 80220, U.S.A.

THE first goal and major problem in beginning to attempt to isolate a receptor for a hormone, for a neurotransmitter or for a drug is finding some means of recognising and assaying for the receptor after the tissue has been disrupted. Receptors are usually defined operationally in terms of the postsynaptic response which is initiated by the interaction of the receptor with an appropriate agonist. Under some circumstances, as in the work with microsacs obtained from homogenates of the electric organ of *Electrophorous* (KASAI and CHANGEUX, 1971), it is possible to obtain a response after homogenization which is reminiscent of that which is observed in the intact tissue. Similarly, hormone sensitive adenylate cyclase has been described in studies of membrane fragments obtained from a number of different tissues. Although β-adrenergic receptors are apparently linked to adenylate cyclase *in vivo*, the receptor site is often easily dissociable from the catalytic activity. Hormone sensitivity may be lost when the tissue is homogenised, or if not then, it is certainly lost when the homogenate is exposed to detergents. This means that in any attempt to isolate β-adrenergic receptors some more or less artificial definition or assay for receptor activity is required.

Probably the easiest way to assay for a receptor *in vitro* is with the use of an irreversible affinity label such as exist for cholinergic receptors in the form of snake venom toxins like α-bungarotoxin (cf. HAMMES, BLOOM and MOLINOFF, 1973). It is somewhat unfortunate that no snake has considered the adrenergic receptor of sufficient importance to have synthesized an irreversible and highly specific blocking agent. On the other hand, the use of irreversible affinity labels has several inherent problems. In the first place, any non-specific sites which bind the ligand, will be purified along with the specific receptor site. In the second place, if the receptor should be degraded or fragmented during the course of the purification then it is likely that purification of a fragment will replace purification of the receptor. A third difficulty concerns the fact that it may be difficult to regenerate intact receptors at the end of the purification.

In any case, these problems are not of particularly great concern with regard to the study of adrenergic receptors since no ligand has been discovered which is irreversible and which possesses the requisite specificity. We are thus forced to the more tedious approach of using reversible ligands to assay for these receptors. Since either agonists or antagonists can be used there are a wide variety of potential ligands. Again, as in the case with irreversible binding agents, specificity is the major difficulty. There are two categories of non-receptor binding that must be considered. The first of these is binding to totally nonspecific sites. Catecholamines are positively charged at neutral pH and they would be expected to interact with any anions which are

* Supported by a Grant-in-Aid from The American Heart Association 72–887.
† The Author is an Established Investigator of The American Heart Association.

present in the tissue preparation. In addition to the totally nonspecific sites, there are a wide variety of specifiable but non-receptor binding sites which are certainly present in any adrenergically innervated organ. Thus, there are four biosynthetic enzymes and four degradative enzymes which would be expected to interact with various catecholamines. There are also two uptake sites, one neuronal (U_1) and one extraneuronal (U_2), and there is a mechanism which is responsible for the uptake of amines into storage vesicles. There may also be one or more storage proteins which will recognize and interact with catecholamines. And, finally, in addition to the specific post-synaptic α- and β-receptors, there is increasing evidence for the existence of pre-synaptic α-receptors (ENERO, LANGER, ROTHLIN and STEFANO, 1972). All of these sites and mechanisms would be expected to recognize and bind to adrenergic amines.

It is fair to say that the use of snake venom toxins, such as α-bungarotoxin (CHANG and LEE, 1963), has provided the greatest impetus to the successful study and isolation of cholinergic receptors (cf. HAMMES et al., 1973). On the other hand, work from several laboratories (CHANGEUX, MEUNIER and HUCHET, 1971; ELDEFRAWI, ELDEFRAWI and O'BRIEN, 1971) has clearly shown that it is possible to isolate cholinergic receptors using reversible labels such as must be used in the case of β-adrenergic receptors. Based largely on the work of Changeux, O'Brien, and their collaborators, it is possible to postulate a number of guidelines which should be followed in the use of reversible ligands to isolate a given β-adrenergic receptor. In the first place, one should take advantage of techniques such as differential and gradient centrifugation to obtain a relatively highly enriched starting material. Approximately a ten-fold enrichment in cholinergic receptor protein was obtained by the use of differential centrifugation of a homogenate of *Torpedo* electric tissue (POTTER and MOLINOFF, 1972). Thus, in a study of β-adrenergic receptors, the first step should probably be to try to obtain membrane fragments which are enriched in epinephrine sensitive adenylate cyclase.

A point which is emphasized by the work of CHANGEUX and of O'BRIEN is that binding studies should be carried out in isotonic media. CHANGEUX et al. (1971) showed, for example, that there was approximately twenty times more binding of decamethonium to membrane fragments from *Electrophorous* when the binding was carried out in hypotonic media than when it was carried out in isotonic media. It is interesting that even in isotonic media there were two sites to which decamethonium was bound. One of these was the acetylcholine receptor, and the other was identified as being the catalytic site of acetylcholinesterase (CHANGEUX et al., 1971).

Since the binding of catecholamines to adrenergic receptors is a rapidly reversible process, then it is important to demonstrate that the binding which is being studied *in vitro* is also a rapidly reversible process. Any binding which does not satisfy this criterion is likely to represent nonspecific binding. It is interesting that there are at least two components in a membrane preparation from rat heart which have the ability to bind epinephrine either irreversibly or at most only very slowly reversibly (MOLINOFF and HECHT, unpublished observations). In addition to being a necessary criterion of binding to a specific β-adrenergic receptor, reversibility of binding is also an important consideration with regard to the exact procedure that is used to assay for the receptor *in vitro*. After incubating membrane fragments with a radioligand, it is often necessary to perform a wash of either a pellet obtained by centrifugation or a filter on which membranes have been collected. This is necessary in order to reduce the blank and to remove nonspecifically occluded ligand. In some cases (PERT and SNYDER, 1973), such a wash procedure can be shown to result in the loss of insignificant amounts of specifically bound ligand. On the other hand, even the very briefest wash can result in the loss of ligand. If, as seems reasonable in the case of the binding of catecholamines to the β-adrenergic receptor, one accepts a value for $K_{diss.}$ of 10^{-7} M and assumes that $k_1 = 10^6$ M^{-1} sec^{-1}, then one can calculate a $t_{1/2}$ for dissociation. Since $K_{diss.} = k_2/k_1$, then $k_2 = 10^{-1}$ sec^{-1} and since $t_{1/2} = 0.693/k_2$, then $t_{1/2} = 6.93$ sec. In this example, then, even a minimal wash, that consumed of the order of 15 sec, would result in the loss of over 3/4 of the specific binding. It is obvious that if the $K_{diss.}$ is less than 10^{-7} M or if k_1 is less than 10^6 M^{-1} sec^{-1}, then the feasibility of using a wash is increased. It is possible to experimentally determine the rate of dissociation at various temperatures. If this rate is sufficiently slow at, for example, 4°C, then it may be possible to carry out a relatively brief wash of the membranes on a filter or in a pellet.

A very useful set of experiments which can be carried out involves testing a number of potential ligands and selecting that one which binds with the greatest affinity to the receptor and to the fewest number of nonspecific binding sites. For example, ELDEFRAWI, ELDEFRAWI and O'BRIEN (1971) examined the binding of a series of cholinergic ligands to 28 nonspecific enzymes and other substances which were available in the laboratory. Muscarone did not bind to any of the nonspecific substances and it bound to only 1 site (a high affinity site, $K = 5.5 \times 10^{-8}$ M) in a particulate preparation of *Electrophorous* electroplax. Other of the potential ligands bound to as many as 10 of the nonspecific substances.

While it is clear that there are a number of problems and difficulties inherent in the use of reversible ligands as tools to study and isolate adrenergic or other receptors, it is also clear that there are a number of potential advantages which can be exploited. For example, kinetic analysis by the method of SCATCHARD (1949) makes it possible to determine the presence of multiple binding sites with different affinities, the equilibrium constant of each site, and the actual number of binding sites of each type. In addition to the intrinsic importance of this information, it thus becomes possible to determine whether anything has occurred during the course of a purification to either alter or destroy one or another of the various binding sites. This is information that is not available if an irreversible affinity ligand is used. A second advantage of the use of reversible ligands concerns the enormous amount of background information which is available regarding the ability of various agonists and antagonists to elicit or block specific receptor mediated responses. The binding that is observed *in vitro* ought to reflect both the relative and absolute magnitude of the affinity of the various ligands as determined with intact preparations. In work with membrane preparations which include β-adrenergic receptors, as well as adenylate cyclase, it should be possible to study binding and activation in the same preparation. If there are no spare receptors, then ED$_{50}$ for the activation of adenylate cyclase should equal the dissociation constant ($K_{diss.}$) for binding. If, however, there are spare receptors then the concentration of amines necessary to achieve half maximal binding ($K_{diss.}$) should be greater than the concentration required to achieve half maximal activation of adenylate cyclase (ED$_{50}$). On the other hand, it is not likely that the $K_{diss.}$ will be less than the ED$_{50}$. What this means in practice is that since we know that the ED$_{50}$

for β-agonists, with regard to the activation of adenylate cyclase, is of the order of 10^{-6} to 10^{-7} M, then we can say that any phenomena which is observed only at 10^{-3} or 10^{-4} M is not likely to be very interesting. On the other hand, if binding is studied at very low concentrations of amine, such as 10^{-9} M, then only a very few percent of a presumably small number of binding sites are going to be occupied. This will, of course, make it even more difficult to detect adrenergic receptors *in vitro*.

An important aspect of many β-receptor mediated responses concerns the fact that the *l*-stereoisomer is considerably more potent than the *d*-stereoisomer (ARIENS, 1964). Where it has been studied the same result has been obtained with regard to the effect of catecholamines on adenylate cyclase (DAVOREN and SUTHERLAND, 1963; ROSEN, ERLICHMAN and ROSEN, 1970). Since the *d*-isomer does not block the effects of the *l*-isomer the binding of amines to β-receptors should also exhibit marked stereospecificity. The ability of *l*-propranolol, but not of *d*-propranolol, to block the binding of amines to a putative β-receptor would be very strong evidence in support of the idea that the binding which is being studied is to the actual physiologically significant β-receptor.

One of the most important advantages which will probably derive from studying various β-adrenergic receptor systems concerns the fact that we already understand in some detail the net result of activation of these receptors. That is to say, the activation of β-receptors *in vivo* results in stimulation of the enzyme adenylate cyclase. Thus, if, or when, we are able to purify not only β-adrenergic receptors, but also adenylate cyclase and the various transducer molecules which may be required to connect the cyclase to the receptor, then we will be in an extremely good position to consider reconstitution experiments in which we will be able to study the interactions of the various components of this receptor system. It is only by carrying out this kind of experiment that we will be able to unravel the precise molecular mechanisms by which receptors modulate cell function.

REFERENCES

ARIENS E. J. (1964) *Mol. Pharmacol.* **3**, 237–238.
BLOOM F., HAMMES A. and MOLINOFF P. B. (1973) *Receptor Biochemistry and Biophysics. A report of an NRP work session*, in press.
CHANG C. C. and LEE C. Y. (1963) *Archs. Int. Pharmacodyn.* **144**, 241–257.
CHANGEUX J.-P., MEUNIER J-C. and HUCHET M. (1971) *Mol. Pharmacol.* **7**, 538–553.
DAVOREN P. R. and SUTHERLAND E. W. (1963) *J. Biol. Chem.* **238**, 3009–3015.
ELDEFRAWI M., ELDEFRAWI A. T. and O'BRIEN R. D. (1971) *Proc. Nat. Acad. Sci., U.S.A.* **68**, 1047–1050.
ENERO M. A., LANGER S. Z., ROTHLIN R. P. and STEFANO F. J. E. (1972) *Br. J. Pharmac.* **44**, 672–688.
HAMMES G. G., MOLINOFF P. B. and BLOOM F. E. (1973) *Receptor Biophysics and Biochemistry, A Report Based on an NRP work session.* NRP Bull. **11**, 155–294.
KASAI M. and CHANGEUX J.-P. (1971) *J. Memb. Biol.* **6**, 1–23.
PERT C. B. and SNYDER S. H. (1973) *Science* **179**, 1011–1014.
POTTER L. T. and MOLINOFF P. B. (1972) In: *Perspectives in Neuropharmacology. A Tribute to Julius Axelrod.* (SNYDER S. H., Ed.) pp. 9–41. Oxford University Press.
ROSEN O. M., ERLICHMAN J. and ROSEN S. M. (1970) *Mol. Pharmacol.* **6**, 524–531.
SCATCHARD G. (1949) *Science* **51**, 660–672.

Frontiers in Catecholamine Research 1973, pp. 361 to 368. Pergamon Press. Printed in Great Britain.

TOWARD ISOLATION OF A β-ADRENERGIC BINDING PROTEIN

ROBERT J. LEFKOWITZ, M.D.

Cardiac Unit, Department of Medicine, Massachusetts General Hospital and Harvard Medical School, Boston, Massachusetts, U.S.A.

IN THE years since the last Catecholamine Symposium significant developments have occurred in the field of receptor isolation LEFKOWITZ, 1973; LEFKOWITZ and HABER, 1973. In particular the emphasis has shifted from a pharmacologic to a biochemical approach. By pharmacologic approach I refer to experiments which have as their goal the elucidation of receptor mechanisms by observation of the effects of a series of drugs on intact tissues, e.g., the potency series of adrenergic amines in increasing cardiac contractility or causing vaso-constriction, etc. Over a period of many years such studies have formed the basis for most of our current definitions of receptors. That this method of approach is intrinsically limited in the information that it can provide is indicated by the following schema taken from the writings of MORAN (1966):

$$A \, \varepsilon \, R \rightarrow a \rightarrow b \rightarrow c \rightarrow d \rightarrow n \rightarrow \text{effect}$$

where A is the drug and R its receptor.

The observed effect is several or many steps downstream from the initial drug–receptor interaction. Thus, conclusions about the receptor or first step are indirect inferences. Actions of "competitive" blocking drugs are also measured under similar circumstances. Although a variety of mathematical formulations can be derived which purportedly "prove" that these blocking drugs are acting by direct competition for the exact sites with which agonist drugs are combining, it is quite possible that additional, distinct receptor sites are involved in their actions or in fact that they might be acting at several points in the sequence of reactions leading to a physiologic response.

The biochemical approach to receptors seeks ultimately to isolate and characterise from responsive tissues the specific macromolecules which function as receptors. The potential difficulties of such an undertaking have been outlined by WAUD (1968), who in fact has raised serious questions about its feasibility. Receptors, when studied as isolated chemical entities, may appear quite different than what might have been expected from studies of their functioning in intact membranes in intact cells within intact tissues.

Conceptually, a good starting point for any attempt to isolate a β-adrenergic receptor is the adenylate cyclase system. As has been reviewed recently, there appears to be a very close parallel between so-called β-adrenergic effects and stimulation of the enzyme adenylate cyclase (ROBISON et al., 1971).

In the heart a number of different drugs and hormones appear to be capable of augmenting cardiac contractility via stimulation of adenylate cyclase and subsequent production of the second messenger (EPSTEIN et al., 1971) cAMP. These include

13*

catecholamines (MURAD *et al.*, 1971), glucagon (LEVEY and EPSTEIN, 1969), histamine (KLEIN and LEVEY, 1971), and possibly the thyroid hormones as well (LEVEY and EPSTEIN, 1968). Each of these appears to act via a distinct set of receptors.

It has been demonstrated repeatedly in the past few years that it is possible to bind a wide variety of radioactively labeled drugs and hormones to subcellular components that have many of the properties which one might expect of receptor binding sites (LEFKOWITZ, 1973).

Recently, this approach has been extended to the catecholamines. We have been engaged in studies of *in vitro* binding of ^{3}H-norepinephrine to (1) microsomal membranes derived from canine ventricular myocardium, and (2) intact chick embryo cardiac myoblasts grown in tissue culture. An advantage of the latter system is that the cells (which respond to catecholamines with an increased rate of spontaneous beating) can be grown in almost pure culture free of any contaminating nerves or blood vessels (WOLLENBERGER, 1964). The characteristics of binding in the two systems, however, appear to be virtually identical. In addition, the binding sites under study have been solubilised, characterised, and purified by the technique of affinity chromatography.

More recently, a number of other laboratories have published findings indicating that very similar, if not identical, sites can be identified and studied in several other tissues which contain catecholamine responsive adenylate cyclase.

The primary data dealing with the characterisation of these β-adrenergic binding sites in intact cells (LEFKOWITZ, O'HARA and WARSHAW, 1973)*, membranes (LEFKOWITZ and HABER, 1971; LEFKOWITZ, SHARP and HABER, 1973), or solubilised preparations (LEFKOWITZ, O'HARA and HABER, 1972; LEFKOWITZ and LEVEY, 1972) have been published previously. Accordingly, the remainder of this presentation will have three major purposes.

(1) To summarise briefly current information about the characteristics of these sites as derived from cardiac tissue.

(2) To tabulate analogous data characterising binding of ^{3}H-catecholamines to membrane fractions from a variety of other adrenergically sensitive tissues.

(3) To evaluate critically the data by (a) indicating how these binding sites differ from all other previously characterised binding sites for catecholamines; (b) examining the ways in which their in vitro properties conform to and differ from what might have been expected of β-adrenergic receptors on the basis of pharmacologic studies; and (c) suggesting a hypothesis for how these sites might function in mediating β-adrenergic effects as part of a β-adrenergic receptor complex.

Properties of isolated β-adrenergic binding sites have been examined by studying the binding of ^{3}H-norepinephrine or ^{3}H-isoproterenol to a membrane fraction (microsomal) derived from canine ventricular myocardium or to intact chick embryo cardiac myoblasts grown in tissue culture, using a simple millipore filter assay (LEFKOWITZ, O'HARA and WARSHAW, 1973; LEFKOWITZ and HABER, 1971; LEFKOWITZ, SHARP and HABER, 1973). Binding is a rapid, reversible process. Bound amine, dissociated by acid, retains full biologic activity and is chromatographically unaltered. The specificity of the binding sites is such that potent β-adrenergic catecholamines effectively compete for the sites with $\frac{1}{2}$ maximum inhibition of ^{3}H-norepinephrine binding at about 10^{-7} M for compounds like isoproterenol, epinephrine, and norepinephrine. The order of potency is isoproterenol

> epinephrine = norepinephrine > dopamine > DOPA. α-adrenergic and indirectly active amines, such as phenylephrine and metaraminol, were less than 1/100th as potent. Phenethylamine, O-methylated metabolites, atropine, glucagon, histamine, and thyroid hormones do not compete. Nonpharmacologically active catechol compounds, such as dihydroxymandelic acid, also compete for binding. Inhibition of binding by β-active catecholamines occurs over the same concentration range as *in vitro* effects on cardiac contractility.

The β-adrenergic antagonist propranolol inhibits binding, whereas phentolamine does not. However, high concentrations are necessary for this effect, $\frac{1}{2}$maximum inhibition occurring at $\sim 3 \times 10^{-4}$ M. The binding reaction is not stereospecific in that d and l forms of catecholamines compete equally well.

Adenylate cyclase, present in the microsomal membranes used for the binding studies, is stimulated by catecholamines, the order of potency being isoproterenol > epinephrine = norepinephrine > dopamine > DOPA (LEFKOWITZ, SHARP and HABER).

SCATCHARD (1969) analysis reveals two distinct orders of binding sites with affinity constants of 10^7 and 10^6 L/M.

The sites appear to be proteins with crucial sulfhydryl groups. Prior digestion of intact cells or membranes with phospholipase A increases the apparent number of available binding sites.

Trace amounts of divalent cations are necessary for the binding reaction, which is inhibited by EDTA and EGTA.

Norepinephrine, covalently linked to agarose beads, competes for binding sites on the myocardial cells, and trypsin bound to agarose beads destroys these sites, suggesting that their predominant location is the cell surface. Studies with cells indicate approximately $2 \cdot 5 \times 10^6$ sites/cell.

The binding sites were solubilised with the detergents Lubrol PX or sodium deoxycholate. The solubilised sites have been purified as much as 10,000 fold by affinity chromatography on columns of norepinephrine linked to agarose via a 30 Å side chain. The characteristics of the solubilised, purified sites are similar to those of the membrane-bound sites, although the affinity constant is an order of magnitude lower.

Gel filtration of soluble sites on agarose columns reveals two molecular weights of 160,000 and 40,000.

It has been pointed out that a wide variety of tissues contain catecholamine responsive cyclase and the β-adrenergic receptor complex (ROBISON et al., 1971). Recently, studies of ^{3}H-catecholamine binding to membrane fractions derived from a number of these tissues have been reported (OCHOA et al., 1972; MARTINETTI et al., 1969; TOMASI et al., 1970; SCHRAMM et al., 1972; BILIZEKIAN et al., 1973; VALLIERES et al., 1973; KELLER et al., 1973; FISZER-de PLAZAS et al., 1972). The results of several of these studies have been summarised in Table 1. The data suggest a striking similarity of the binding sites in each of the tissues studied.

In evaluating the significance of *in vitro* binding studies with catecholamines, it is important to consider a variety of potential biologic binding mechanisms which might conceivably bind the drugs. These include the uptake 1 (neuronal) and uptake 2 (extraneuronal) processes (BURGEN and IVERSEN, 1965), nerve storage vesicle binding (von EULER, 1967), and metabolising enzymes such as catechol-O-methyl

Table 1. Binding of ^{3}H-catecholamines

Tissue from which receptors prepared	Authors	Catecholamine used for binding studies	Association constant $H + R \rightleftarrows HR^*$	Specificity of binding
Canine heart (ventricle) "Microsomal" membranes	Lefkowitz et al.* (13, 14)	^{3}H-norepinephrine	$k_1 = 1 \cdot 04 \times 10^7$ L/M $k_2 = 1 \cdot 33 \times 10^6$ L/M	Isoproterenol > epinephrine = norepinephrine > dopamine > DOPA α-adrenergic drugs much less active Not stereospecific
Cultured chick embryo cardiac myoblasts (whole cells)	Lefkowitz et al.*	^{3}H-norepinephrine	$k_1 = 2 \times 10^6$ L/M $k_2 = 1 \times 10^5$ L/M	β-adrenergic compounds ≫ α-active compounds catechol specific; not stereospecific
Cat heart (ventricle) "Microsomal" membranes	Ochoa et al.	^{3}H-isoproterenol ^{3}H-norepinephrine	NR†	NR
Rat liver Purified plasma membranes	Marinetti et al.	^{3}H-epinephrine	$k_1 = 1 \cdot 85 \times 10^7$ L/M $k_2 = 1 \cdot 36 \times 10^5$ L/M	Catechol group most important Not stereospecific
Turkey erythrocyte "Plasma membrane"	Schramm et al.	^{3}H-epinephrine	3×10^5 L/M	β-adrenergic drugs most potent Importance of catechol group
	Bilizekian and Aurbach	^{3}H-isoproterenol	2×10^5 L/M	β-adrenergic drugs most potent Catechol specific
Skeletal muscle Sarcolemma	Vallieres, Drummond, and Severson	^{3}H-epinephrine	$k_1 = 3 \cdot 7 \times 10^9$ L/M $k_2 = 1 \cdot 8 \times 10^6$ L/M $k_3 = 8 \cdot 2 \times 10^6$ L/M	Isoproterenol > epinephrine > norepinephrine
Rabbit uterus	Keller and Goldfien	^{3}H-norepinephrine	$\sim 10^6$ L/M	Norepinephrine > isoproterenol, epinephrine ≫ metaraminol, methoxamine
Bovine spleen capsule Whole homogenate	Fiszer-dePlazas, DeRobertis	^{3}H-norepinephrine	$k_1 = 3 \cdot 3 \times 10^7$ L/M $k_2 = 1 \cdot 8 \times 10^5$ L/M	NR

*H = hormone; R = receptor
†NR = not reported

TO MEMBRANE FRACTIONS FROM VARIOUS TISSUES

Effects of blocking drugs	Purification and/or solubilisation	Concentration of binding sites	Reactive groups	Catecholamine responsive cyclase in same membrane fraction
Phentolamine inert Propranolol: 50% inhibition of binding at $1-3 \times 10^{-4}$ M	Solubilised with Lubrol PX or sodium deoxycholate Purified by affinity chromatography	Total of all sites: 120 p moles/ mg membranes	-SH—Protein	Yes
Propranolol 10^{-4} M: 40% inhibition of binding Phentolamine: 10^{-4} M inert	—	$2 \cdot 5 \times 10^{6}$ sites/cell ~ 500 sites/μm^2 cell surface ~ 4 pmoles/mg protein	-SH—Protein	Yes
Propranolol: 40% inhibition of binding at 10^{-7} M Phentolamine inert	Soluble "proteolipid" extracted with chloroform-methanol	0·25 pmoles/mg dry tissue	NR	NR
Variable	Solubilised with sodium dodecylsulfate; sodium deoxycholate; DEAE chromatography	71 pmoles/mg protein	-SH—Protein	Yes
Propranolol: 50% inhibition when present in 10 × excess Phentolamine: 50% inhibition at 20 × excess	NR	1000 pmoles/mg protein	NR	Yes
At 10^{-3} M DCl: 85% inhibition Propranolol: 70% inhibition Phentalamine inert	Trypsin solubilises Lubrol PX solubilises	56 pmoles/mg protein	Protein	Yes
Propranolol inhibits binding at concentrations $>5 \times 10^{-6}$ M	NR	NR	NR	Yes
NR	NR	10 pmoles/mg protein	NR	NR
Inhibition of binding: Propranolol– 86%; dibenamine– 83%; phentolamine 56%	"Proteolipid" extracted with chloroform-methanol	0·25 pmoles/mg protein	NR	NR

transferase and monoamine oxidase. It should be noted that each of these differ importantly either in specificity or other major characteristics from the sites being described here as has been discussed fully elsewhere (LEFKOWITZ, 1973). The predominant specificity of the sites is similar to that of catechol-O-methyl-transferase. However, the nonidentity of these sites with catechol-O-methyl transferase is clearly indicated by several considerations.

(1) No structural alteration of catecholamines occurs consequent to the binding reaction.

(2) The sites are primarily membrane bound and appear to be localised pre-dominantly to the cell membrane (LEFKOWITZ, O'HARA and WARSHAW, 1973; OCHOA et al., 1972; MARTINETTI et al., 1969; BILIZEKIAN et al., 1973), whereas catechol-O-methyl transferase is almost exclusively confined to the soluble fractions of tissue (AXELROD and TOMCHICK, 1958). Thus, for example, in the liver, specific activity of binding is 15 × higher in purified plasma membranes than in the cell sap (MARTINETTI, et al., 1969), whereas catechol-O-methyl transferase is found primarily in the cell sap (AXELROD and TOMCHICK, 1958).

(3) Specificities are not identical. Deletion of one OH group from the catechol ring diminishes interaction with the β-adrenergic sites about 100-fold, however displacement curves indicate significant interaction at these higher concentrations. These same compounds do not interact with catechol-O-methyl transferase at all (AXELROD and TOMCHICK, 1958).

(4) K_m of catechol-O-methyl transferase and the binding sites differ by several orders of magnitude (LEFKOWITZ, SHARP and HABER, 1973; AXELROD and TOM-CHICK, 1958).

The question naturally arises as to the biological significance of these adrenergic binding sites. In particular do they play a role as part of the β-adrenergic receptor mechanism, which recognises and binds catecholamines to tissues and triggers subsequent biological processes.

It should be clear from the data presented that in a number of respects the binding characteristics of these sites are strikingly parallel to what might be expected of physiologic β-adrenergic receptors.

(1) The predominant specificity of the binding sites is for the catechol moiety which characterises the most potent β-adrenergic catecholamines isoproterenol, protokylol, epinephrine, and norepinephrine. Interaction with α-agonists which generally lack one or both ring OH's is much weaker or absent.

(2) The sites bind catecholamines in the same concentration ranges over which they normally have their biological effects. Thus, in an isolated beating myocardial cell $\frac{1}{2}$ maximum inhibition of binding and $\frac{1}{2}$ maximum chronotropic effects occur at identical concentrations; in erythrocyte ghosts $\frac{1}{2}$ maximum binding and $\frac{1}{2}$ maximum stimulation of adenylate cyclase occur at identical concentrations (BILIZEKIAN et al., 1973).

(3) Binding is rapid and reversible and does not degrade the amine.

(4) The sites appear to be concentrated in plasma membrane fractions (LEFKOWITZ, O'HARA and WARSHAW, 1973; OCHOA et al., 1972; MARTINETTI et al., 1969; BILIZEKIAN et al., 1973), as is the adenylate cyclase.

(5) The sites interact preferentially with β blockers as opposed to α blockers.

Equally clear is that in two respects the binding characteristics of the sites diverge from what might have been predicted for intact isolated β-adrenergic receptors.

(1) Stereospecificity is not seen. Thus, *d*-norepinephrine is bound as well as *l*, and a variety of other nonphysiologically active catechol compounds are also bound.

(2) Very high concentrations of adrenergic blocking drugs, generally $>10^{-4}$ M, are necessary for inhibition of binding.

What is the significance of these findings? It has been suggested by several authors (BILIZEKIAN *et al.*, 1973; LEFKOWITZ, 1973; BELLEAU, 1963) that the structure activity relationships for β-adrenergic drugs may be best explained by considering that binding of the drug to its "receptor" occurs at several distinct points of contact. One likely point of attachment is at the catechol ring, which might be bound through the two ring hydroxyls. An additional binding site would attach to the side chain and would presumably be relatively specific for the ethanol amine portion of the molecule.

The sites described here would obviously correspond to the catechol specific portion of the binding site. Strong support for such an interpretation has recently been obtained in experiments performed by BILEZEKIAN and AURBACH (1973). They have documented that nonphysiologically active catechol compounds, e.g., dihydroxymandelic acid, dopamine, DOPA, etc., which inhibit binding but which do not activate cyclase in turkey erythrocyte ghosts, are effective competitive inhibitors of isoproterenol-activated cyclase. Thus, binding at a catechol specific site is a necessary, although not a sufficient condition, for cyclase stimulation. Similarly, *d*-norepinephrine at concentrations below those which stimulate the cyclase also inhibit isoproterenol-activated cyclase (BILEZEKIAN and AURBACH, 1973).

Stereospecificity and strong binding of β-adrenergic blockers would be expected to characterise the ethanol amine specific sites or portions of the sites. Compounds like propranolol resemble β-adrenergic agonists at the ethanol amine end of the molecule. They completely lack, however, the catechol function and hence do not stimulate biological processes.

It should be remembered that there is no *a priori* reason to assume and no experimental evidence to prove that β-adrenergic agonists and antagonists bind to receptor structures at completely identical points of contact. Presumably, agonist binding must occur at all necessary sites for biological effects to occur. Disruption of binding at a single point of contact, even though binding at another is intact, would then be sufficient to block biological response. Thus, the competitive inhibition of isoproterenol-stimulated cyclase by propranolol on the one hand and by DOPA on the other (described above) are most likely mediated by blockade of isoproterenol binding at two different and distinct sets of binding points.

Thus, the data suggest that stimulation of biological processes, e.g., adenylate cyclase, by β-active catecholamines requires interaction of the agonists with at least two sets of sites. One set is relatively specific for the catechol portion of the molecule. Although binding here is required, biological activity is seen only if interaction at the "ethanol amine" site also occurs. The term "β-adrenergic receptor" encompasses the entire complex of sites.

FUTURE DIRECTIONS

Future research efforts are likely to proceed in several directions.

(1) Efforts to specifically label sites with a specificity corresponding to the ethanol amine portion of the molecule.

(2) Attempts to isolate and purify the binding sites.

(3) Efforts to reconstitute a functioning receptor-effector system from isolated components, e.g., to show that the addition of one or more adrenergic binding sites to a nonresponsive cyclase renders it catecholamine responsive. This latter demonstration will remain as the ultimate proof of the physiological significance of the isolated binding sites. It is worth underscoring that as of this writing, such a reconstitution has not been achieved in any of the many isolated receptor systems which have been studied in recent years.

REFERENCES

AXELROD J. and TOMCHICK R. (1958) *J. Biol. Chem.* **233,** 702–705.
BELLEAU B. (1963) In: *Modern Concepts in the Relationship Between Structure and Pharmacological Activity.* (BRUNINGS, K. J., ed.) pp. 75–99. Pergamon, Oxford.
BILIZEKIAN J. and AURBACH G. (1973) *J. Biol. Chem.* In press.
BURGEN A. S. V. and IVERSEN L. L. (1965) *Brit. J. Pharm.* **25,** 34–39.
EPSTEIN S. E., LEVEY G. S. and SKELTON C. L. (1971) *Circulation* **43,** 437–450.
ERTEL R. J., CLARKE D. E., CHAO J. C. and FRANKE F. R. (1971) *J. Pharm. Exp. Ther.* **178,** 73–80.
FISZER-dePLAZAS S. and DE ROBERTIS E. (1972) *Biochim. Biophys. Acta* **266,** 246–254.
KELLER D. and GOLDFIEN A. (1973) *Clin. Res.* **21,** 202.
KLEIN I. and LEVEY G. S. (1971) *J. Clin. Invest.* **51,** 1012–1018.
LEFKOWITZ R. J. (1973) *N. Eng. J. Med.* **288,** 1061.
LEFKOWITZ R. J. (1973) *Pharm. Revs.* **25,** 259.
LEFKOWITZ R. J. and HABER E. (1971) *Proc. Nat. Acad. Sci.* **68,** 1773–1777.
LEFKOWITZ R. J. and HABER E. (1973) *Circ. Res.* **23,** Suppl. 1, 46.
LEFKOWITZ R. J. and LEVEY G. S. (1962) *Life Sci.* **11,** (part II), 821–828.
LEFKOWITZ R. J., O'HARA D. and HABER E. (1972) *Proc. Nat. Acad. Sci.* **69,** 2828–2832.
LEFKOWITZ R. J., O'HARA D. and WARSHAW J. (1973) *Nature, New Biology,* **244,** 79.
LEFKOWITZ R. J., SHARP G. and HABER E. (1973) *J. Biol. Chem.* **248,** 342–349.
LEVEY G. S. and EPSTEIN S. E. (1969) *Circ. Res.* **24,** 151–156.
LEVEY G. S. and EPSTEIN S. E. (1968) *Biochem. Biophys. Res. Commun.* **33,** 990–995.
MARINETTI G. V., RAY T. K. and TOMASI V. (1969) *Biochem. Biophys. Res. Commun.* **36,** 185–193.
MORAN N. (1966) *Pharm. Revs.* **18,** 503–512.
MURAD F., CHI Y. M., RALL T. W. and SUTHERLAND E. W. (1962) *J. Biol. Chem.* **237,** 1233–1238.
OCHOA E., DE CARLIN M. C. L. and DE ROBERTIS E. (1972) *Europ. J. Pharm.* **18,** 367–374.
ROBISON A., BUTCHER R. W. and SUTHERLAND E. (1971) *Cyclic AMP.* p. 152. Academic Press, New York.
SCATCHARD G. (1949) *Ann. N.Y. Acad. Sci.* **51,** 660–669.
SCHRAMM M., FEINSTEIN H., NAIM E., LONG M. and LASSER M. (1972) *Proc. Nat. Acad. Sci.* **69,** 523–527.
TOMASI V., KORETZ S., RAY T. K., DUNNICK J. and MARINETTI G. V. (1970) **211,** 31–42.
VALLIERES J., DRUMMOND G. I. and SEVERSON D. L. (1973) *Fedn Proc.* **32,** 773a.
VON EULER U. S. (1967) *Circ. Res.* Suppl. III, **20,** 5–11.
WAUD D. R. (1968) *Pharm. Revs.* **20,** 49–88.
WOLLENBERGER A. (1964) *Circ. Res.* Suppl. II, **14,** 184–201.

Frontiers in Catecholamine Research 1973, pp. 369 to 371. Pergamon Press. Printed in Great Britain.

RECENT ADVANCES IN THE ACTIVE SITE CHEMISTRY OF THE ADRENERGIC ALPHA-RECEPTOR

BRUCE LIPPERT and B. BELLEAU*

Department of Chemistry, McGill University, Montreal, Canada

THE structural basis of the catecholamine α-receptor (CA-R_α) function is of central importance in molecular pharmacology. The active site chemistry and topography of this receptor has remained obscure in spite of the well-documented structure-activity relationships among "affinity blockers" of the Phenoxybenzamine class (NICKERSON, 1957; BELLEAU, 1958; BELLEAU and COOPER, 1963). Several attempts of labelling specifically the CA-R_α with these agents (DIKSTEIN and SULMAN, 1965; LEWIS and MILLER, 1966; MAY et al., 1967; YONG and MARKS, 1969) have not been especially rewarding. Further research with the alkylating blockers is not encouraged by the results of auto-radiographic studies which revealed an absence of specific localization of labelled Phenoxylbenzamine in the smooth muscle cells (GRAHAM et al., 1971). More recently, we have reported on a novel type of irreversible modifying reagent of the CA-R_α possessing powerful CNS depressant activity (BELLEAU et al., 1968; MARTEL et al., 1969). The specific recognition sites for this class of reagents (the prototype being 1-ethoxycarbonyl-2-ethoxy-1, 2-dihydroquinoline or EEDQ†) are carboxyl groups which are rapidly transformed by it into mixed anhydrides (BELLEAU and MALECK, 1968; BELLEAU, et al., 1969; ROBINSON and BELLEAU, 1972). Subsequent attack of the anhydride by proximal nucleophiles (especially amine residues) leads to permanent carbamylation of the attacking group or to cross-linking depending on stereo-chemical factors. The reagent EEDQ† has found wide applications in the field of peptide bone synthesis (BELLEAU and MALECK, 1968). Use of its labelled analog (tritium in the ethoxy carbonyl residue; ROBINSON and BELLEAU, 1972) at 1×10^{-6} M led to the incorporation of irreversibly bound label in rabbit aortic strips. Receptor protection (phentolamine) reduced the incorporation of label but unfortunately, uptake of label by the tissue was linear against EEDQ concentration or with time, thus indicating a lack of receptor specificity. Moreover, radioautography revealed a lack of specific localization of the label in aortic tissue. Accordingly, we have been searching for a new, specific class of affinity blockers for the CA-R_α.

In an extensive search for effective anti-radiation drugs, HERMAN et al., (1971) reported that N-(ω-aminopentyl)-cysteamine and its corresponding disulfide‡ (APC) possessed persistent α-blocking activity in vivo (rat). A subsequent report indicated that at 5×10^{-5} M, APC seemed to act by a competitive mechanism at the CA-R_α level of aortic tissue (G. E. DEMAREE et al., 1971). We have re-investigated this

---

* To whom correspondence should be addressed.
† Available from Aldrich Chemical Company.
‡ N,N′-bis-(ω-aminopentyl)-cystamine (APC); NH_2—$(CH_2)_5$—NH—CH_2CH_2—S—S—CH_2-CH_2NH—$(CH_2)_5$—NH_2 (as the tetrahydrochloride salt).

class of radio-protective agents in depth and have clearly demonstrated that whereas APC at 10^{-4} M is an irreversible α-blocker on aortic tissue, its corresponding thiol (reduced APC) is inactive when testing is performed in the absence of oxygen (Fig. 1). In sharp contrast to the Phenoxybenzamine and EEDQ classes of blockers, APC does not affect responses to 5-hydroxytryptamine at effective α-blocking concentrations. Hence, APC is specific for the CA-R_α which is not normally sensitive to conventional thiol or disulfide reagents (such as mercaptoethanol, dithiothreitol, Ellman's reagent, etc.). We therefore conclude that the four symmetrically disposed cationic charges (at pH 7) of APC‡ interact with complementary negative charges on the Ca-R_α

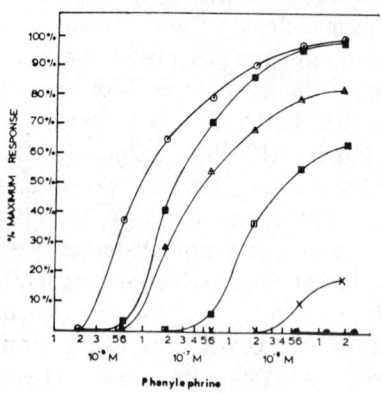

FIG. 1.—Dose–response curves of rabbit aortic strips after exposure to APC at various concentrations and reversal of blockade by cysteamine. —○—: control (phenylephrine at 6×10^{-9} to 6×10^{-5} M); —△—: after APC at 4×10^{-5} M for 20 min; —▢—: after APC at $1·2 \times 10^{-4}$ M for 20 min; —×—: after APC at 4×10^{-4} M for 20 min;
—●—: after APC at 6×10^{-4} M for 20 min;
—■—: after APC at 6×10^{-4} M for 20 min then treatment with cysteamine at 1×10^{-2} M for 30 min.

surface thereby including a conformational change conducive to the unmasking of a buried thiol which can now participate in a disulfide interchange reaction with APC. The resulting S—S bond between drug and receptor also appears to reside below the receptor surface since it is refractory to the reducing action of mercaptoethanol. However, the cationic thiol cysteamine at 10^{-3}–10^{-2} M completely reverses blockade by APC (Fig. 1). Phenylephrine at 10^{-6} M is highly effective in protecting the receptor against APC which in turn affords strong protection against EEDQ inhibition. Significantly, the carbon analog of APC (sulfur replacement by methylene groups) is a competitive inhibitor of the CA-R_α and is effective in protecting it against APC. We conclude that the CA-R_α or one of its key components includes a minimum of four dissymmetrically disposed anionic binding sites on its active surface and a buried thiol at or near the center of symmetry. Finally, using ^{35}S-labelled APC and aortic tissue, non-linear saturation with respect to time or APC concentration (approximating a Michaelis mechanism) was observed (Fig. 2). Curiously, preliminary results indicate that about 1×10^{12} uptake sites per mg of dry tissue may be saturated, a value comparable to those previously observed for alkylating blockers (LEWIS and MILLER, 1966; MAY et al., 1967) and EEDQ (LIPPERT and BELLEAU,

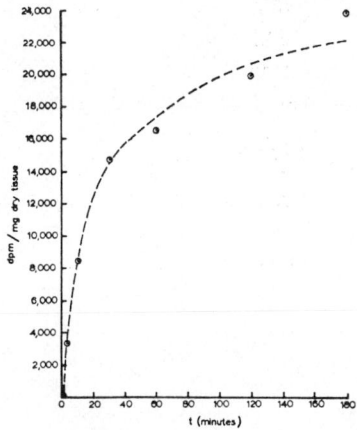

FIG. 2.—Time-course of the uptake of ^{35}S-labelled APC at $1\cdot5 \times 10^{-4}$ M by rabbit aortic strips; receptor blockade reached 100% after 30 min. Dotted line is the calculated curve for a Michaelis saturation mechanism.

unpublished). We believe that the polycationic nature of APC ought to hinder permeation of the plasma membranes. Radioautographic studies are under way and the results will be reported later. It seems quite possible that we may finally have access to a specific class of protein affinity reagents allowing localization and isolation of the CA-R$_\alpha$. In any event, the theory that the CA-R$_\alpha$ surface incorporates multiple control sites (BELLEAU, 1971) as is the case for the nicotinic receptor appears to be confirmed.

REFERENCES

BELLEAU B. (1958) *Can. J. Biochem. Physiol.* **36**, 731–753.
BELLEAU B. (1971) In *Advances in Chemistry Series* (BLOOM B. and ULLYOT G. E., Eds.), Vol. 108, pp. 141–165, American Chemical Society, Washington.
BELLEAU B. and COOPER P. (1963) *J. Med. Chem.* **6**, 579–583.
BELLEAU B., DITULLIO V. and GODIN D. (1969) *Biochem. Pharmacol.* **18**, 1039–1042.
BELLEAU B. and MALECK G. (1968) *J. Am. Chem. Soc.* **90**, 1651.
BELLEAU B., MARTEL R., LACASSE G., MENARD M., WEINBERG N. L. and PEROON Y. G. (1968) *J. Am. Chem. Soc.* **90**, 823–824.
DEMAREE G. E., BROCKENTON R. E., HEIFFER M. H. and ROTHIE W. E. (1971) *J. Pharm. Sci.* **60**, 1743–1745.
DICKSTEIN S. and SULMAN F. G. (1965) *Biochem. Pharmacol.* **14**, 881–885.
GRAHAM J. D. P., IVENS C., LEVER J. D., McQUISTON R. and SPRIGGS T. L. (1971) *Br. J. Pharmacol.* **41**, 278–284.
HERMAN E. H., HEIFFER M. H., DEMAREE G. E. and VICK J. A. (1971) *Archs Int. Pharmacodyn.* **193**, 102–110.
LEWIS J. E. and MILLER J. W. (1966) *J. Pharmacol. Exp. Ther.* **154**, 46–55.
MARTEL R., BERMAN R. and BELLEAU B. (1969) *Can. J. Physiol. Pharmacol.* **47**, 909–911.
MAY M., MORAN J. F., KIMELBERG H. and TRIGGLE D. J. (1967) *Mol. Pharmacol.* **3**, 28–36.
NICKERSON M. (1957) *Pharmacol. Rev.* **9**, 246–259.
ROBINSON W. T. and BELLEAU B. (1972) *J. Am. Chem. Soc.* **94**, 4376–4378.
YONG M. S. and MARKS G. S. (1969) *Biochem. Pharmacol.* **18**, 1619–1626.

Frontiers in Catecholamine Research 1973, pp. 373 to 376. Pergamon Press. Printed in Great Britain.

OPERATION OF TWO EPINEPHRINE RECEPTORS AND AN ACETYLCHOLINE RECEPTOR IN THE EXOCRINE CELL OF THE RAT PAROTID GLAND

MICHAEL SCHRAMM and ZVI SELINGER

Department of Biological Chemistry, The Hebrew University of Jerusalem,
Jerusalem, Israel

RECEPTORS for hormones and neurotransmitters have been widely studied through their response at the physiological level of the whole animal or the perfused intact organ. On the other hand there are recent studies on the binding of hormones to their receptors in plasma membrane preparations (LEFKOWITZ and HABER, 1971; LEFKOWITZ et al., 1970; RODBELL et al., 1971; CHANGEUX et al., 1971; SCHRAMM et al. 1972a). In the wide gap between these two approaches lies the operation of the receptor at the cellular level, about which there is rather limited information. One would especially like to know what happens when two or more different receptors are activated simultaneously in the same cell; whether indeed one cell owns several receptors for neurotransmitters and whether these can be activated independent of each other.

These problems have now been studied using slices of rat salivary parotid gland incubated in a Krebs–Ringer bicarbonate medium. It has been found that epinephrine activates independently and simultaneously two receptors (BATZRI et al., 1971). Activation of the β-receptor results in secretion of amylase and other enzymes into the medium bathing the slices. The response can also be followed at the ultrastructural level since enzyme secretion obligatorily proceeds by sequential fusion of the intracellular secretory granules with the cell membrane at the acinar lumen (AMSTERDAM et al., 1969). The fusion reaction thus serves as a cellular marker for the β-adrenergic response. Activation of the α-receptor causes extensive K^+ release from the cells into the medium. This reaction is associated with the appearance in the Golgi area of the cells, of vacuoles which conveniently serve as cellular markers for the α-adrenergic response (Fig. 1). The release of K^+ ultimately seems to result in secretion of water which serves as the vehicle to drive the concentrated secretory protein away from the cell and into the gland duct (BATZRI et al., 1973).

A third receptor operates in the acinar parotid cell. It is activated by acetylcholine and elicits a response identical with the α-adrenergic response, namely K^+ release and vacuole formation. Using the specific inhibitors atropin and phentolamine it was possible to show that the acetylcholine receptor and the α-adrenergic receptor elicit the response absolutely independent of each other (SCHWALB H., 1972).

The ultrastructural markers, the large amount of K^+ released and the fact that 80–90% of the cells in parotid gland are of the acinar type exclude the possibility that the three receptors reside in different cells.

The β-and the α-adrenergic responses in rat parotid slices were examined in some detail with special reference to their interrelationship within the same cell. Activation of the β-receptor caused formation of cyclic AMP which in turn brought about

373

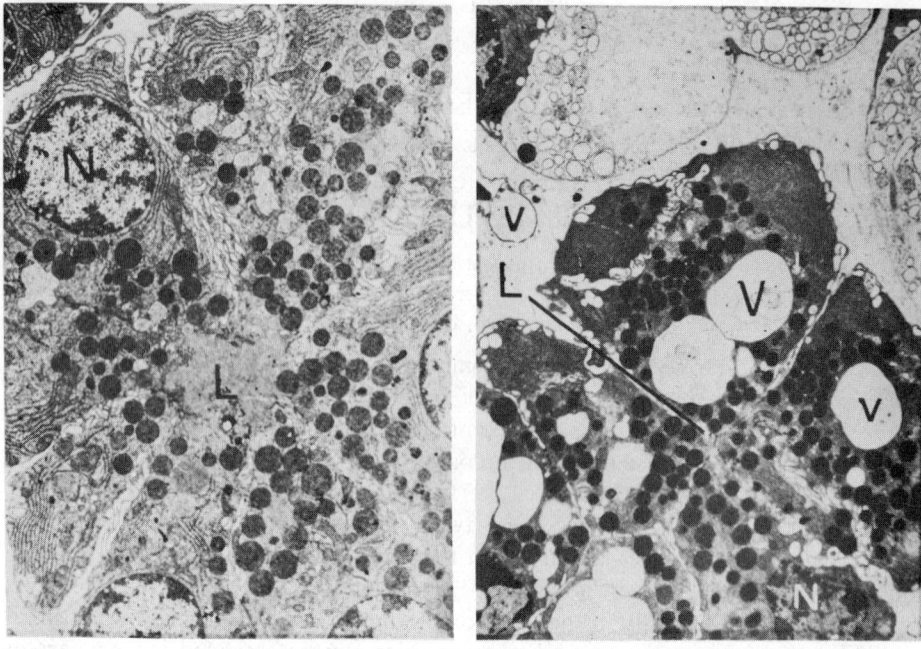

FIG. 1.—α- and β-adrenergic responses at the ultrastructural level in the exocrine cell of rat parotid gland. Slices of the gland were incubated for 10 min with epinephrine (10 μM) in presence of phentolamine (20 μM), left figure, or in presence of propanolol (20 μM), right figure. Left: only the β-receptor is active. The lumen (L) of the acinus is much enlarged through fusion of the secretory granules with the cell membrane and is filled with the secretory product of medium electron density. Right: only the α-receptor is active. Some cells contain a few large vacuoles (V) while others have undergone vacuolation of their entire cellular structure. The lumen is compressed almost beyond recognition.

a transformation of the secretory granule into an amoeba-like structure (SCHRAMM et al., 1972b). The pseudopodia of the activated secretory granule fused with the cell membrane facing the acinar lumen and the content of the granule was thus placed outside the cell. More than 80% of the total amylase in the slices were secreted within 90 min provided that simultaneous activation of the α-receptor was prevented (BATZRI and SELINGER, 1973). This could be achieved by a variety of means: induction of enzyme secretion by isoproterenol which activates only the β-receptor, induction by epinephrine but in the absence of added Ca^{2+} which is required for activation or the α-receptor or, by-passing the receptors altogether by using N^6-monobutyryl 3′, 5′ cyclic AMP.

The α-adrenergic response was studied by measuring the release of intracellular K^+ from the slices into the medium with the aid of a K^+ selective electrode or by atomic absorption. The reaction reached a plateau 3–5 min after epinephrine addition when 20–40% of the total intracellular K^+ had been released (BATZRI et al., 1973). A number of observations indicated that the plateau represents a steady state between K^+ efflux mediated by the α-receptor and K^+ influx mediated by the Na^+, K^+ activated ATPase (SELINGER et al., 1973). Addition of phentolamine which blocked the α-receptor led to reuptake of all the K^+ that had been released. Ouabain which did not by itself induce a rapid K^+ release increased the extent of release caused by epinephrine and prevented reuptake of K^+ upon subsequent addition of phentolamine. The release of K^+ was dependent on the simultaneous and continuous presence of both epinephrine and Ca^{2+}. Addition of a Ca^{2+} chelator during the reaction resulted in reuptake of K^+ into the slices.

Cyclic AMP which is an essential intermediate in the β-adrenergic response of enzyme secretion seemed to have no function in the α-adrenergic response of K^+ release. Isoproterenol and monobutyryl cyclic AMP which readily induced enzyme

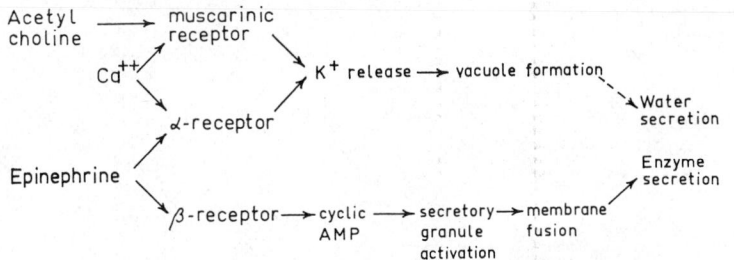

Fig. 2.—Pathways of response of the α- and β-adrenergic receptors and the cholinergic receptor in the exocrine parotid cell of the rat. Catecholamine control is through sympathetic nerve endings and not through the bloodstream. Norepinephrine is about equipotent with epinephrine. Calcium is required for the α-adrenergic as well as the cholinergic response but its exact place in the sequence is not known. While it seems fairly certain that water secretion results from the K^+ release, vacuole formation may be an indirect effect of K^+ loss and Na^+ entry.

secretion failed to cause K^+ release and vacuole formation. Neither did preincubation with these two reagents inhibit K^+ release when epinephrine was subsequently added. Furthermore, the initial rise in intracellular cyclic AMP caused by epinephrine (5 min incubation) was the same whether or not the α-receptor was blocked by phentolamine (Batzri et al., 1973). Thus no support was found for the theory that the α-adrenergic response is mediated through a reduction in the cyclic AMP level in the cell. More prolonged activation of the α-receptor did produce a decline in cyclic AMP and a retardation of enzyme secretion. These changes should probably be ascribed to the indirect effects of ATP depletion through the Na^+, K^+ activated ATPase and to vacuole formation.

The findings suggest that when epinephrine activates simultaneously α- and β-receptors within the same cell it opens two completely independent pathways of response (Fig. 2). The availability, also, of the acetylcholine receptor permits the release of K^+ and probably water flow without concomitant protein secretion. It is apparent that three independent receptors for neurotransmitters within the same cell provide an impressive fine tuning control of the secretory process.

Acknowledgement—This work has been supported by a grant from the National Institutes of Health, U.S.A. (No. AM 10451-07).

REFERENCES

AMSTERDAM A., OHAD I. and SCHRAMM M. (1969) *J. Cell Biol.* **41**, 753–773.
BATZRI S. and SELINGER Z. (1973) *J. biol. Chem.* **248**, 356–360.
BATZRI S., SELINGER Z. and SCHRAMM M. (1971) *Science* **174**, 1029–1031.
BATZRI S., SELINGER Z., SCHRAMM M. and ROBINOVITCH M. R. (1973) *J. biol. Chem.* **248**, 361–368.

376 M. SCHRAMM and Z. SELINGER

CHANGEUX J. P., MEURIER J. C. and HUCHET M. (1971) *J. molec. Pharmacol.* **7,** 538–553.
LEFKOWITZ R. J. and HABER E. (1971) *Proc. Nat. Acad. Sci.* **68,** 1773–1777.
LEFKOWITZ R. J., ROTH J., PRICER W. and PASTAN I. (1970) *Proc. Natn. Acad. Sci.* **65,** 745–752.
RODBELL M. KRANS M., POHL S. L. and BIRNBAUMER L. (1971) *J. biol Chem.* **246,** 1872–1876.
SCHRAMM M., FEINSTEIN H., NAIM E., LANG M. and LASSER M. (1972a) *Proc. Natn. Acad. Sci.* **69,** 523–527.
SCHRAMM M., SELINGER Z., SALOMON Y., EYTAN E. and BATZRI S. (1972b) *Nature, New Biol.* **240,** 203–205.
SCHWALB H. (1972) M.Sc. Thesis, The Hebrew University of Jerusalem.
SELINGER Z., BATZRI S., EIMERL S. and SCHRAMM M. (1973) *J. biol. Chem.* **248,** 369–372.

Frontiers in Catecholamine Research 1973, pp. 377 to 382. Pergamon Press. Printed in Great Britain.

DOPAMINE-, OCTOPAMINE-, AND SEROTONIN-SENSITIVE ADENYLATE CYCLASES: POSSIBLE RECEPTORS IN AMINERGIC NEUROTRANSMISSION

PAUL GREENGARD, JAMES A. NATHANSON and JOHN W. KEBABIAN

Department of Pharmacology, Yale University School of Medicine, New Haven, Connecticut 06510, U.S.A.

THE IMPORTANCE of biogenic amines as potential transmitters in the nervous system makes it desirable to identify the receptors for these compounds and to understand the molecular consequences of the activation of these receptors. In many non-neural tissues the receptors for various hormones are intimately associated with the enzyme adenylate cyclase. In view of the similarities between hormones and neurotransmitters in mediating cell to cell communication, we have studied the effects of various biogenic amines upon the activity of adenylate cyclase in the nervous system. The approach utilized in our laboratory has been to study relatively simple preparations of neural tissue. The results obtained with these preparations have indicated a relationship between various types of aminergic receptors and adenylate cyclase activity.

MAMMALIAN SYMPATHETIC GANGLION

In the mammalian superior cervical sympathetic ganglion, preganglionic cholinergic fibers are known to synapse directly upon postganglionic neurons, and also upon interneurons. These interneurons, in turn, synapse upon the postganglionic neurons. The postganglionic neurons contain norepinephrine, the neurotransmitter of the peripheral sympathetic nervous system; in contrast, the interneurons have been shown to contain dopamine, and are thought to utilize this amine as an inhibitory neurotransmitter within the ganglion. Consistent with this scheme, stimulation of the preganglionic fibers leads both to direct excitation of the postganglionic neurons, as well as to an inhibition of the postganglionic neurons which is most probably mediated indirectly by the dopaminergic internuncial cells.

We have found that preganglionic stimulation of the rabbit superior cervical sympathetic ganglion results in a several-fold increase in the levels of ganglionic cyclic AMP (MCAFEE et al., 1971). Various experiments have indicated that this increase is associated with the process of synaptic transmission within the ganglion, and that the increase can be blocked by alpha adrenergic antagonists as well as by muscarinic cholinergic antagonists (SCHORDERET et al., 1970). These, as well as other results, suggested that the increase in cyclic AMP, following electrical stimulation, might occur as a consequence of dopamine being released from the interneurons and interacting with a receptor on the postganglionic cells. Utilizing slices of bovine superior cervical ganglion (KEBABIAN and GREENGARD, 1971), we investigated more directly the possibility that interaction of dopamine with its receptor causes an activation of ganglionic adenylate cyclase. We indeed found that low concentrations of dopamine stimulated the accumulation of cyclic AMP within this tissue; stimulation

by dopamine occurred at concentrations as low as 1 μM, and half-maximal stimulation at about 7 μM. In contrast, l-norepinephrine was much less effective than
dopamine; half-maximal stimulation by norepinephrine occurred at about 30 μM.
The dopamine-mediated increase in cyclic AMP was antagonized by the alpha adrenergic antagonists, phentolamine and phenoxybenzamine, but was unaffected by the
beta adrenergic antagonists, dichloroisoproterenol, propranolol, or MJ 1999. Using
homogenates, we found that the action of dopamine in causing an increase in ganglionic cyclic AMP was due to a stimulation of adenylate cyclase activity and not to an
inhibition of phosphodiesterase activity. Thus, these results from the bovine ganglion support the contention that dopamine is responsible for the elevation of cyclic
AMP levels within the rabbit superior cervical ganglion which occur following
preganglionic stimulation.

MAMMALIAN CAUDATE NUCLEUS

The highest levels of dopamine within the mammalian central nervous system
occur in the basal ganglia. Moreover, disorders in the biochemistry of dopamine
within these ganglia have been implicated in the aetiology of both naturally-occurring and drug-induced Parkinson's disease. Our studies with the superior cervical
sympathetic ganglion suggested the possibility (KEBABIAN and GREENGARD, 1971)
that the dopamine receptor in the caudate nucleus might, in fact, be the dopamine
binding portion of a dopamine-sensitive adenylate cyclase. Recently (KEBABIAN et al.,
1972), we have demonstrated the presence of a dopamine-sensitive adenylate cyclase
in homogenates of this tissue (Fig. 1). The properties of this enzyme and its sensitivity to various drugs support the possibility that this enzyme is intimately associated with the "dopamine receptor" which has been characterized, indirectly, in a
variety of clinical and laboratory animal studies. Low concentrations of dopamine,
and considerably higher concentrations of l-norepinephrine, caused an approximate
doubling of adenylate cyclase activity, whereas the synthetic beta adrenergic agonist,
l-isoproterenol, was without effect upon this adenylate cyclase (Fig. 1). Low concentrations of apomorphine, a compound which stimulates dopamine receptors in the
caudate nucleus in vivo, also stimulated caudate adenylate cyclase activity (KEBABIAN
et al., 1972). The effects of dopamine, norepinephrine, and apomorphine were non-
additive, which suggests that all three compounds were activating a single class of
receptor. Moreover, dopamine-stimulated activity was blocked by low concentrations
of either haloperidol or chlorpromazine, compounds which also block the caudate
"dopamine receptor". In contrast, promethazine, a compound which does not block
caudate dopamine receptors, was without effect upon the enzyme, except at high
concentrations.

ARTHROPOD GANGLION

Most previous studies concerned with the role of cyclic AMP in neuronal function
have utilized the mammalian nervous system; relatively little is known about the
role of cyclic AMP in invertebrate nervous tissue, despite the fact that many invertebrate neural preparations offer a simplicity not often found in vertebrates. In our
investigations of the possible role of cyclic AMP in the functioning of invertebrate
nervous tissue, we have used, as a model, the thoracic ganglia of the cockroach,
Periplaneta americana. These ganglia are known to contain norepinephrine, dopamine
and serotonin (MURDOCK, 1971). Aminergic nerve endings have been demonstrated

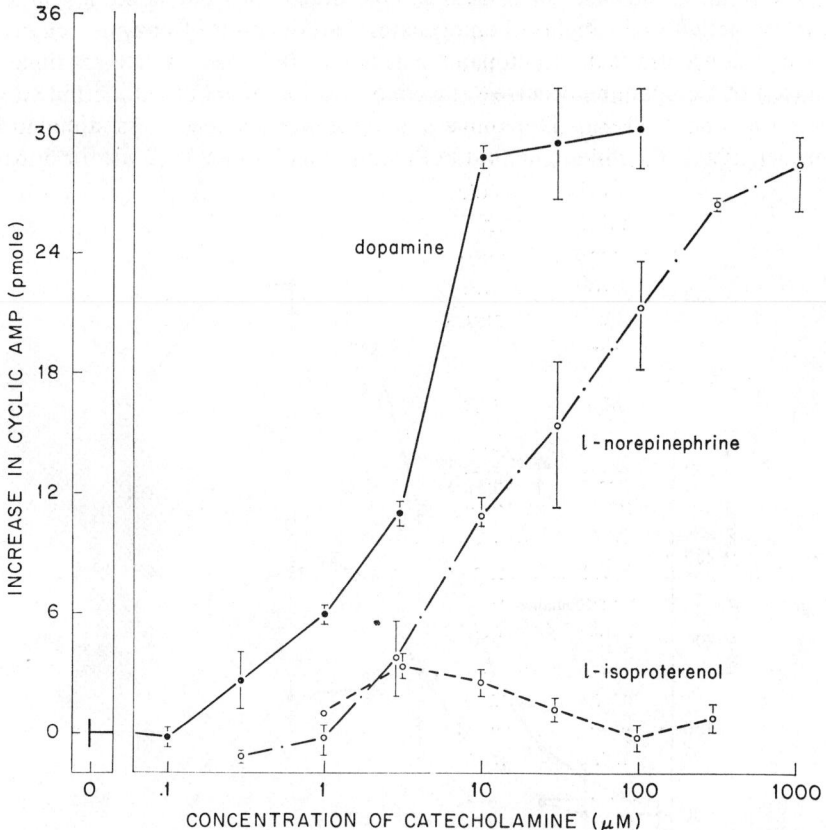

Fig. 1.—Effect of dopamine, *l*-norepinephrine, and *l*-isoproterenol on adenylate cyclase activity in a homogenate of rat caudate nucleus. In the absence of added catecholamine, 27.1 ± 1.0 pmoles (mean $\pm$ s.e.m.) of cyclic AMP was formed. The increase in cyclic AMP above this basal activity is plotted as a function of catecholamine concentration. The data give the mean values and ranges for two to five replicate samples, each assayed in duplicate. (From KEBABIAN, J. W., PETZOLD, G. L. and GREENGARD, P., 1972)

within the ganglia using fluorescence histochemistry (FRONTALI, 1968), and electron microscopic studies have shown the presence of small dense core vesicles within nerve terminals (MANCINI and FRONTALI, 1970). Furthermore, dopamine and norepinephrine are known to facilitate synaptic transmission in the abdominal ganglia, to increase the level of activity in the ventral nerve cord, and, when applied microiontophoretically, to cause a depolarization and firing of identified neurons (HODGSON and WRIGHT, 1963). Injection of the same catecholamines into living cockroaches produces a variety of stereotyped, behavioural effects.

In our studies we found that norepinephrine, dopamine and serotonin each stimulated adenylate cyclase activity in thoracic ganglion homogenates. We found, in addition, that low concentrations of octopamine (an amine known to occur in some invertebrates) caused a marked stimulation of cyclic AMP accumulation in these homogenates, whereas tyramine, phenylethylamine, and phenylethanolamine were relatively ineffective (NATHANSON and GREENGARD, 1973).

Figure 2 shows the effect of *dl*-octopamine, dopamine, and serotonin on adenylate cyclase activity of ganglion homogenates. Stimulation of enzyme activity was observed at concentrations of octopamine as low as 0·03 μM. A half-maximal activation (K_a) by octopamine occurred at a concentration of about 1·5 μM and maximal activation at about 30 μM. Dopamine and serotonin produced less stimulation of enzyme activity than octopamine, but both amines had a low K_a (2 μM for dopamine

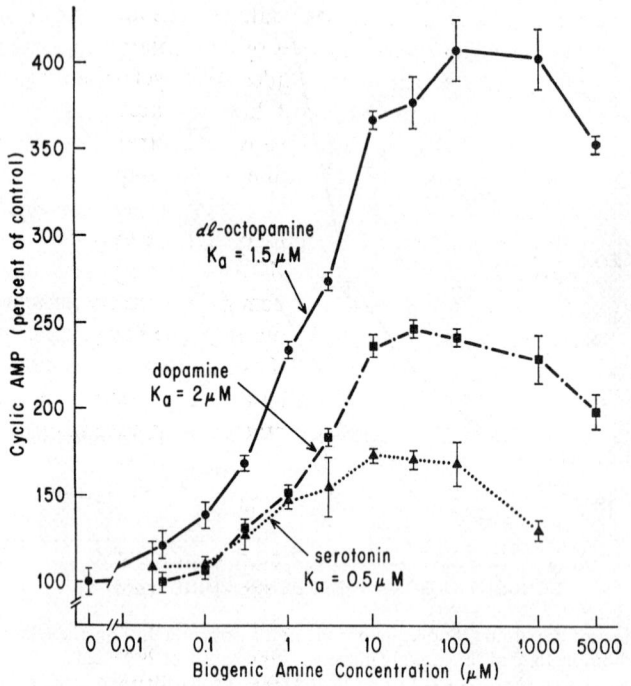

Fig. 2.—Effect of *dl*-octopamine, dopamine, and serotonin on adenylate cyclase activity in a homogenate of cockroach thoracic ganglia. Control activity in the absence of added biogenic amine was 10.0 ± 1.5 pmole/mg protein/min. The data give the mean and ranges for two to three replicate samples, each assayed in duplicate. (From Nathanson, J. A. and Greengard, P., 1973; Copyright 1973 by the American Association for the Advancement of Science.)

and 0·5 μM for serotonin). L-norepinephrine (data not shown) was capable of stimulating enzyme activity of ganglion homogenates to the same degree as octopamine, but much higher concentrations were required to produce this effect: half-maximal activation by norepinephrine occurred at about 30 μM. Octopamine, dopamine, serotonin and norepinephrine did not affect the phosphodiesterase activity of homogenates of the ganglion.

Studies of the additive effects of octopamine, dopamine and serotonin on the adenylate cyclase activity of homogenates of these ganglia showed that various combinations of optimal concentrations of these three amines were fully additive in stimulating enzyme activity, suggesting that each of these three amines activates a separate and distinct class of receptor site. Additivity experiments using combinations of norepinephrine with the various other amines failed to supply evidence for a nor-epinephrine receptor, and suggested instead that the stimulatory effect of norepine-phrine on cockroach ganglion adenylate cyclase may be attributable to a partial activation of the receptor sites for octopamine and dopamine.

The alpha adrenergic antagonist, phentolamine, was highly effective in blocking the stimulatory effect on adenylate cyclase activity of octopamine, norepinephrine, and, to a lesser extent, dopamine. The beta adrenergic antagonist, propranolol, however, had very little effect on the stimulation of enzyme activity by these same three amines. The effect of serotonin on adenylate cyclase activity differed from that of the other amines in that it was affected neither by phentolamine nor by propranolol.

We have found that the hormone-sensitive adenylate cyclase activity of cockroach ganglion homogenates is greater than that of homogenates of the interganglionic nerve connectives. This suggests that, in the insect ventral nerve cord, adenylate cyclase may be localized more in nerve cell bodies and synaptic areas than in axons (of which the connectives are composed).

In addition to our results with ganglion homogenates, we have been able to demon-strate hormone-induced increases in cyclic AMP levels using preparations of intact cockroach hemiganglia. Incubation of intact hemiganglia in the presence of 250 μM octopamine, for example, caused a 4·5- to 7-fold increase in cyclic AMP level com-pared with controls. Increases in cyclic AMP were also seen when ganglia were incubated in the presence of norepinephrine, dopamine, or serotonin, although these increases were somewhat smaller than those obtained with octopamine.

CONCLUDING REMARKS

The results summarized above suggest that adenylate cyclase may be intimately associated with the function of certain types of aminergic receptors in both vertebrate and invertebrate nervous systems. In other work in our laboratory (GREENGARD, 1973; GREENGARD et al., 1972), we have investigated the possibility that cyclic AMP, formed in response to certain neurotransmitters, may actually mediate the electro-physiological response of the postsynaptic membrane to these neurotransmitters. We have also carried out studies on the mechanism by which cyclic AMP may achieve such effects (GREENGARD, 1973; GREENGARD et al., 1972). The data, together with the results summarized above, are compatible with the following hypothetical role and mechanism of action for cyclic AMP in the physiology of synaptic transmission at certain types of synapses: neurotransmitter, released from presynaptic nerve endings, activates an adenylate cyclase in the postsynaptic membrane, and, thereby, causes the accumulation of cyclic AMP in the postsynaptic neuron; this newly formed cyclic AMP activates a protein kinase leading to the phosphorylation of a protein constit-uent of the plasma membrane; this phosphorylation of the plasma membrane of the postsynaptic neuron causes an alteration in the permeability properties of the membrane, resulting in a change in membrane potential of the cell; upon removal of the phosphate from the membrane protein, by a protein phosphatase present in the synaptic membrane, the membrane potential returns to its initial value. This scheme provides a mechanism by which cyclic AMP, acting as the mediator for the actions of some neurotransmitters, may regulate the membrane properties of certain types of neurons and, thereby, modulate their excitability.

A few years ago, we proposed a general mechanism for the action of cyclic AMP in biological tissues (KUO and GREENGARD, 1969). According to that proposal, the diverse effects of cyclic AMP are mediated through regulation of the activity of cyclic AMP-dependent-protein kinases. The mediation of transmission at certain types of neuronal synapses, through regulation of the level of cyclic AMP and the consequent alteration of the activity of cyclic AMP-dependent protein kinase in postsynaptic cells, would represent specific examples of this postulated general mechanism.

REFERENCES

FRONTALI N. (1968) *J. Insect Physiol.* **14,** 881–886.

GREENGARD P. (1973) In: *Protein Phosphorylation and Control Mechanisms,* Miami Winter Symposia, vol. 5, pp 145–174. Academic Press, New York.

GREENGARD P., MCAFEE D. A. and KEBABIAN J. W. (1972) In: *Advances in Cyclic Nucleotide Research* (GREENGARD P., PAOLETTI R. and ROBISON G. A., eds.) Vol. 1, pp. 337–356. Raven Press, New York.

HODGSON E. S. and WRIGHT A. M. (1963) *Gen. Comp. Endocr.* **3,** 519–525.

KEBABIAN J. W. and GREENGARD P. (1971) *Science* **174,** 1346–1349.

KEBABIAN J. W., PETZOLD G. L. and GREENGARD P. (1972) *Proc. Natn. Acad. Sci. U.S.A.* **69,** 2145–2149.

KUO J. F. and GREENGARD P. (1969) *Proc. Natn. Acad. Sci. U.S.A.* **64,** 1349–1355.

MANCINI G. and FRONTALI N. (1970) *Z. Zellforsch. Mikrosk. Anat.* **103,** 341–350.

MCAFEE D. A., SCHORDERET M. and GREENGARD P. (1971) *Science* **171,** 1156–1158.

MURDOCK L. L. (1971) *Comp. Gen. Pharmacol.* **2,** 254–274.

NATHANSON J. A. and GREENGARD P. (1973) *Science* **180,** 308–310.

SCHORDERET M., MCAFEE D. A. and GREENGARD P. (1970) *Experientia* **28,** 743.

Frontiers in Catecholamine Research 1973, pp. 383 to 388. Pergamon Press. Printed in Great Britain.

COMPARISON OF THE EXCITATORY ACTION OF ACETYLCHOLINE AND NORADRENALINE ON THE GUINEA-PIG MYOMETRIUM

JOSEPH H. SZURSZEWSKI

Department of Physiology and Biophysics, Mayo Medical School, Rochester, Minnesota 55901, U.S.A.

INTRODUCTION

THE ACTION of acetylcholine and noradrenaline on smooth muscle are generally antagonistic. For example, in intestinal smooth muscle, acetylcholine produces excitation whereas noradrenaline produces inhibition. The mechanism of the excitatory action of acetylcholine in the guinea-pig ileum has recently been investigated by BOLTON (1972), and the mechanism of the inhibitory action of catecholamines by BÜLBRING and TOMITA (1969a, b). However in certain smooth muscle tissue such as the guinea-pig myometrium under estrogen domination, acetylcholine and noradrenaline both cause excitation (see MARSHALL, 1970). This report describes experiments designed to investigate whether or not the excitatory action of acetylcholine and noradrenaline on the guinea-pig myometrium are due to the same mechanism.

METHODS

The experiments were performed on the myometrium of immature guinea-pigs which had been treated by daily injection of 5 μg estradiol for 8 days. On the ninth day, a strip of myometrium was set up in a double sucrose gap apparatus as described elsewhere (BÜLBRING and SZURSZEWSKI, 1973).

RESULTS

The spontaneous activity recorded with the double sucrose gap consisted of bursts of spikes which were discharged at regular intervals on a plateau depolarisation. In each preparation, the interval between bursts and the frequency of spikes within a burst were remarkably constant. Figure 1 illustrates examples of burst activity from two tissue and alterations of it by isoproterenol (1a) and noradrenaline (1b). In the presence of dihydroergotamine (1×10^{-6} g/ml), isoproterenol inhibited the occurrence of spontaneous activity and produced a hyperpolarisation. Noradrenaline, in the presence of propranalol (1×10^{-6} g/ml), produced a depolarisation, an increase in the frequency of spontaneous spikes and an increase in tension. This excitatory response was followed by a temporary suppression of spontaneous bursts. Thus, in the estrogen dominant guinea-pig myometrium, β-adrenoceptor stimulation inhibits, whereas α-adrenoceptor stimulation increases spontaneous spike activity.

To compare the excitatory effects of noradrenaline and acetylcholine, changes in isometric tension, in membrane potential and membrane resistance were studied. To estimate changes in membrane resistance, constant current pulses of 3 sec duration and of alternating polarity were applied, usually, at 10 sec intervals. Alternating hyperpolarising and depolarising current pulses suppressed the inherent pattern of

383

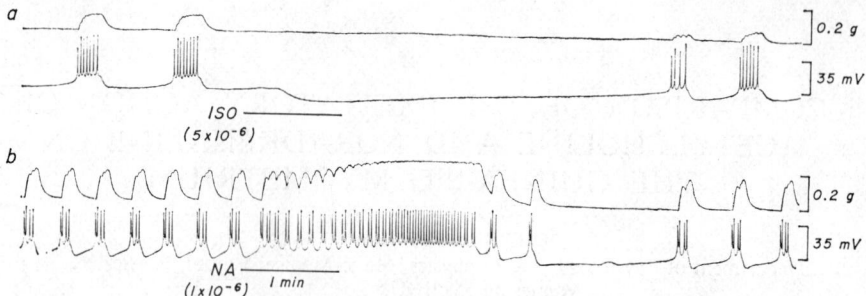

Fig. 1.—Double sucrose gap records of spontaneous activity in two estrogen treated guinea-pig myometria and effect of isoproterenol (ISO) and noradrenaline (NA). (a) effect of isoproterenol in presence of dihydroergotamine (1×10^{-6} g/ml); (b) effect of noradrenaline in presence of propranolol (1×10^{-6} g/ml). Krebs solution 32° C. In this and Fig. 2, black bar indicates period of application of drug. In each panel, top tracing isometric tension, bottom tracing electrical potential.

burst activity in 3–10 min. Spike activity was only observed during the depolarising pulses (Fig. 2).

Noradrenaline caused two types of responses (Fig. 2a and b) while acetylcholine always evoked similar responses (Fig. 2c). In a few preparations, noradrenaline had little effect on the membrane resistance, frequency of spontaneous action potentials and membrane potential. However, it potentiated the phasic contractions (Fig. 2a). In most preparations, noradrenaline decreased the membrane resistance, depolarised the membrane, accelerated spontaneous spike discharge, and initiated

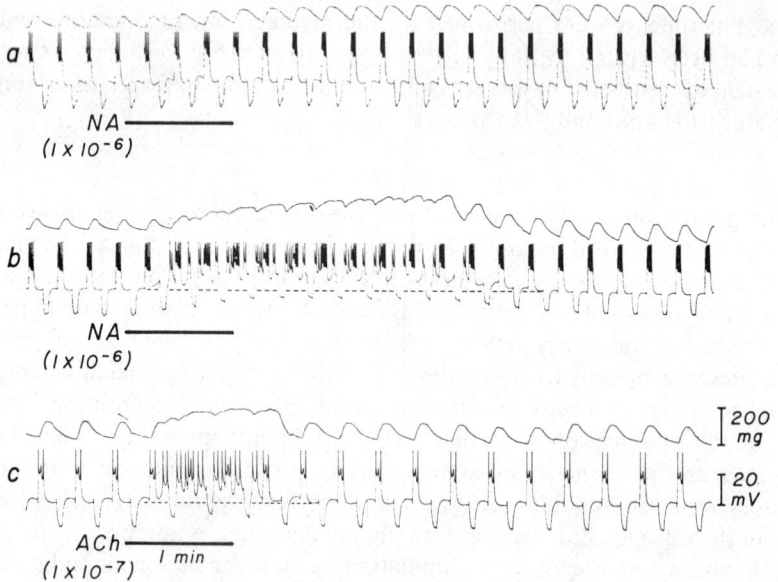

Fig. 2.—Varying responses of strips of myometrium to noradrenaline (a and b) and response to acetylcholine (c). Responses taken from three animals. Propranolol (1×10^{-6} g/ml) present throughout. Electrotonic potentials and evoked spikes evoked by constant current pulses (current strength 4×10^{-7} A, 3 sec duration, every 10 sec, alternately depolarising and hyperpolarising). Krebs solution, 36° C. Some spikes partly off scale. Dashed line indicates original baseline potential.

an increase in resting tension (Fig. 2b). The excitatory effect of noradrenaline resembled that of low doses of acetylcholine (10^{-9} to 4×10^{-7} g/ml) which also caused a decrease in membrane resistance, a depolarisation, an initiation or acceleration of spontaneous spike discharge and a contraction (Fig. 2c). Higher concentrations of acetylcholine (greater than 10^{-6} g/ml) caused a rapid spike discharge which led into a 30 mV sustained depolarisation during which the membrane resistance was reduced nearly to zero; a smooth contracture was maintained throughout the depolarisation.

There were, however, several differences between the effects of the two agonists.
(i) The response of the myometrium to acetylcholine was similar in all preparations and directly related to its dose. By contrast, the pattern of response of the myometrium was not related to the concentration of noradrenaline. When the response to noradrenaline was similar to the one illustrated in Fig. 2a, increasing the concentration of noradrenaline had little additional effect. When the response was similar to the one illustrated in Fig. 2b, increasing the concentration of noradrenaline only slightly augmented the immediate response but prolonged the duration of excitation.
(ii) The response to noradrenaline could be modified by the level of membrane polarisation. When the response to noradrenaline was similar to the one illustrated in Fig. 2a, conditioning depolarising current or additional stretch (50–100 mg) converted it to the response shown in Fig. 2b. Conversely, a conditioning hyperpolarising current converted the response such as in Fig. 2b to one as shown in Fig. 2a. In contrast, responses to acetylcholine were not modified by similar changes in polarisation.
(iii) The duration of action of both agonists differed. The peak excitation produced by noradrenaline usually occurred one minute after stopping its injection and persisted for at least another minute. For acetylcholine, the duration was certainly shorter (Fig. 2c).
(iv) Although its action was short-lasting, acetylcholine was more potent than noradrenaline. When the effect of equal concentrations of acetylcholine and noradrenaline were compared in the range from 1×10^{-6} g/ml to 1×10^{-5} g/ml, it was found that acetylcholine produced a 29 ± 0.7 mV depolarisation (mean $\pm$ S.E.M., $N = 47$) whereas noradrenaline produced a 12 ± 1.2 mV depolarisation (mean $\pm$ S.E.M. $N = 42$).
(v) Unlike the depolarising effect of acetylcholine, the depolarising effect of noradrenaline was inversely related to the temperature of the Krebs solution over the range from 35°C to 6°C. At 6°C, the depolarisation caused by noradrenaline was nearly abolished whereas the depolarisation caused by acetylcholine was not significantly different from that at 35°C.
(vi) The ionic dependence of the depolarisation produced by acetylcholine and noradrenaline were different. The stimulatory action of acetylcholine depended on the external sodium concentration whereas the depolarisation caused by noradrenaline did not. By contrast depolarisation by noradrenaline was highly dependent on the external chloride concentration: reduction of the extracellular chloride concentration from 134 to 7 mM, using either isethionate, glutamate or benzene sulphonate as the substitute, abolished the depolarisation and the increase in conductance caused by noradrenaline but had no significant effect on the stimulatory action by acetylcholine.
(vii) Noradrenaline always potentiated the force of the phasic contractions caused

by depolarising current pulses (Fig. 2a and b), provided the temperature of the bathing solution was not less than 34° C. When noradrenaline accelerated spontaneous spike discharge (Fig. 2b), the potentiating effect was observed after the immediate excitatory response subsided. It was best seen in those tissues that responded to noradrenaline without an increase in spontaneous spike frequency (Fig. 2a). Close inspection of the records in Fig. 2a and 2b reveals that the positive inotropic response occurred without an increase in the number of evoked spikes, and that it outlasted the depolarisation. A positive inotropic influence on the phasic contractions was not seen during or after the administration of acetylcholine (Fig. 2c). Since calcium appears to be a necessary link in the excitation-contraction coupling of contractile tissue, it is tempting to suggest that the long-term positive inotropic response of the myometrium to noradrenaline was due to an increase in the cytoplasmic calcium ion concentration. The following evidence provides some support for this hypothesis. The long-term positive inotropic response is independent of the external concentration of potassium, sodium and chloride since it was observed in solutions with varying potassium concentration (0·1–11·8 mV), in sodium-free solution (tris substitution) and in chloride-deficient solution (7·3 mM chloride). The long-term positive ino-tropic effect of noradrenaline is dependent on the extracellular calcium concentration for it was absent in calcium-free solution and was abolished by lanthanum and verapamil. It has recently been shown in ventricular trabeculae from cats, that verapamil blocks the dynamic calcium current (Kohlhardt et al., 1972). In estrogen dominated guinea-pig myometrium, verapamil blocked the action potential mech-anism, and the mechanical responses and increased spike discharge caused by nora-drenaline. However, verapamil did not effect the depolarisation and decrease in membrane resistance caused by noradrenaline. Similar results were obtained in the presence of lanthanum (see Fig. 8, Szurszewski and Bülbring, 1973). The de-polarisation and increase in conductance caused by noradrenaline however were abolished when the chloride in the verapamil-containing solution was replaced with an impermeant anion. The effects of verapamil and lanthanum thus suggest that while extracellular calcium may not be important for the chloride-dependent de-polarising mechanism, the tension responses to noradrenaline depend either on the movement of calcium across the membrane during the spike mechanism or on the release of calcium from a storage site.

Calcium also participates in the excitatory action of acetylcholine (Szurszewski and Bülbring, 1973; Bülbring and Szurszewski, 1973). In the estrogen domin-ated guinea-pig myometrium, both lanthanum and verapamil in normal Krebs solution significantly reduce the electrical and mechanical excitation caused by acetylcholine.

CONCLUSIONS

Several differences in the way in which the myometrium responds to noradrena-line and acetylcholine were detected. Whereas the response of the myometrium to acetylcholine was dose dependent, the response to noradrenaline was not. In some preparations, noradrenaline failed to initiate or accelerate spontaneous spike frequency whereas in others this response dominated its effect. The experimental evidence suggests that the type of response to noradrenaline depended chiefly on the membrane potential prevailing at the time of noradrenaline administration. When the difference in potential between the threshold for spontaneous spike activity and resting membrane potential was small, as judged by the effect of conditioning hyperpolarisation, the depolarisation caused by noradrenaline easily moved the membrane potential into the "zone of firing" and thereby accelerated spike frequency. When the difference between the spike threshold and resting membrane potention was large, as judged by the effect of stretch and conditioning depolarisation, the depolarisation by noradrenaline did not bring the membrane potential to the firing threshold, and an increase in occurrence of action potential activity did not occur. This suggests that the action of noradrenaline was basically the same for both types of responses, the difference being due to the nearness of the membrane potential to the threshold for spike activity.

There were also differences in the duration of action, potency, temperature dependence and ionic dependence. The effect of cooling on the noradrenaline depolarisation may suggest either that the binding constant for noradrenaline or the processes which link receptor occupancy to changes in the ion permeability of the membrane are temperature-dependent. The temperature dependency of the noradrenaline stimulation might also suggest an effect of noradrenaline on cell metabolism. Regarding the ionic dependence, the primary mechanism for acetylcholine depolarisation was an increase in sodium permeability whereas the depolarisation by noradrenaline was chloride-dependent.

One aspect of the response to noradrenaline which did not conform to the standard model for transmitter action (see GINSBERG, 1967) concerns the ability of noradrenaline to potentiate the force of the phasic contractions caused by depolarising current. There are at least two explanations for this phenomenon. In the first explanation, an intracellular store of calcium may be required and the release of calcium from this store might related to the potential change during the spike. Each time a spike occurs, a certain percentage of the store calcium is released. In the presence of noradrenaline the percent release of calcium by the spike depolarisation is increased, thus making more calcium available for each contraction and thereby potentiating it. The increased amount of calcium released by noradrenaline might be due either to an alteration in the kinetics of calcium release or to an alteration in the kinetics of calcium uptake by the store. Alternately, in the second explanation, an intracellular store of calcium may not be required, the action of noradrenaline being merely to increase the amount of calcium coming into the cell with each action potential. In this sense, the positive inotropic action of noradrenaline due to α-adrenoceptor stimulation in the guinea-pig myometrium would be similar to β-adrenoceptor stimulation in frog atrium (VASSORT et al., 1969).

The blocking action of lanthanum and verapamil on the action of acetylcholine suggests that these agents might interfere either directly with the sodium permeability system in smooth muscle or with a calcium activated sodium system.

Acknowledgement—Part of this study was done with Prof. EDITH BÜLBRING, Department of Pharmacology, University of Oxford, Oxford, England. This investigation was supported in part by NIH Research Grant #RR05530-11 from the Division of Research Resources. Thanks are expressed to Dr. P. M. Vanhoutte for reading and criticising this manuscript.

REFERENCES

BOLTON T. (1972). *J. Physiol., Lond.* **220,** 647–671.
BÜLBRING E. and SZURSZEWSKI J. H. (1973) *Proc. R. Soc., Ser. B.* in press.
BÜLBRING E. and TOMITA T. (1969a) *Proc. R. Soc., Ser. B.* **172,** 89–102.

BÜLBRING E. and TOMITA T. (1969b) *Proc. R. Soc., Ser. B.* **172**, 103–119.
GINSBORG B. L. (1967) *Pharmacol. Rev.* **19**, 289–316.
KOHLHARDT B., BAUER B., KRAUSE A. and FLECKENSTEIN A. (1972) *Pflugers Arch.* **335**, 309–322.
MARSHALL J. M. (1970) *Ergeb. Physiol. Biol. Chem. Exp. Pharmakol.* **62**, 7–67.
SZURSZEWSKI J. H. and BÜLBRING E. (1973) *Proc. R. Soc. Ser. B.* **265**, 149–156.
VASSORT G., ROUGIER O., GARNIER D., SAVVIAT M. P., CORABOEUF E. and GARGOIÜL Y. M. (1969) *Pflugers Arch.* **309**, 70–81.

Frontiers in Catecholamine Research 1973, pp. 389 to 391. Pergamon Press. Printed in Great Britain.

POSSIBLE MECHANISM OF THE ACTION OF CATECHOLAMINES ON SMOOTH MUSCLE

EDITH BÜLBRING

University Laboratory of Physiology, Oxford, England.

SINCE most smooth muscles exhibit spontaneous membrane activity and hence maintain a tone or a regular rhythm of contractions, and since the action potential is due to influx of calcium, the preservation of a low level of ionised calcium in the cytoplasm must be an important function of smooth muscle cells.

Normally, contractions are evoked by action potentials. The amount of calcium entering the cell during the spike may actually be sufficient for E–C–coupling. On the other hand, membrane depolarisation may release calcium bound on the inner surface of the plasma membrane and, in turn, release calcium from the sarcoplasmic reticulum which then activates the contractile mechanism. Some drugs may release calcium from intracellular stores without any change of the membrane potential and thus produce a contraction.

One way of reducing the $[Ca^{2+}]_i$ is the binding of calcium to intracellular membrane structures, i.e., the sarcoplasmic reticulum, or to mitochondria, or to the inner surface of the plasma membrane—an internal sequestration without a reduction of the total calcium content of the cell. Another mechanism may be a passive efflux of calcium coupled to a passive influx of sodium, a Ca–Na–exchange (REUTER, BLAUSTEIN and HAEUSLER, 1973) as in other excitable tissues. A third process may be an active Ca-extrusion (CASTEELS, GOFFIN, RAEYMAEKERS and WUYTACK, 1973).

Recently, it has been suggested that a calcium extrusion pump may be involved in the origin of spontaneous membrane activity (TOMITA and WATANABE, 1973). Thus, the removal of calcium from negative binding sites at the inner surface of the cell membrane would decrease the potassium permeability of the membrane and cause the depolarisation underlying the pacemaker potential which initiates the spontaneous spike.

The high and fairly stable membrane potential of some smooth muscles (including oestrogen and progesterone dominated uterus) is probably due to a large amount of calcium bound at the membrane which results in a high K-permeability (BÜLBRING, CASTEELS and KURIYAMA, 1968). The bursts of activity may be attributed to periodical activity of the Ca-pump. They arise from a slowly developing generator potential, associated with a gradual increase in membrane resistance suggesting a decrease in K-conductance, (BÜLBRING and KURIYAMA, 1973).

The lower membrane potential of intestinal muscle (including the taenia coli) is probably due to a low potassium permeability. Since the pattern of its spontaneous activity is a continuous discharge of action potentials one may assume a high rate of calcium removal from the inner surface of the plasma membrane by a continuously active Ca-pump.

It is well known that the action of catecholamines consists of two components, the α- and β-action (AHLQUIST, 1948). In the uterus the response to the α-action is contraction and that to the β-action is relaxation. In the taenia coli both the α- and β-action cause relaxation. Both components block spontaneous spike activity though by two different mechanisms (BÜLBRING and TOMITA, 1967a,b).

The α-inhibition is the result of membrane hyperpolarisation due to an increase in K-conductance, though there is also a small increase in Cl-conductance. Since firing threshold is not reached, spontaneous activity stops.

The β-inhibition is not due to a change in membrane conductance but to the abolition of the pacemaker potential which normally initiates the spontaneous spikes. Hence spontaneous activity stops though action potentials can be evoked by electrical stimulation.

The β-action reduces the phasic tension response to the spike and frequently also reduces the resting tension. This is generally thought to be due to a greater Ca-uptake into intracellular storage sites. It is the same in the uterus as in the taenia (BÜLBRING, 1973; BÜLBRING and KURIYAMA, 1973) and probably involves cyclic AMP (for references, see ROBISON, BUTCHER and SUTHERLAND, 1970).

The question which remains open is: Why are the responses of the two muscles to the α-action different? Much of the evidence obtained in recent years suggests that the α-action might cause intracellular release of calcium. (HINKE, 1965; VAN BREEMEN and LESSER, 1971; VAN BREEMEN, FARINAS, GERBA and McNAUGHTON, 1972; VAN BREEMEN, FARINAS, CASTEELS, GERBA, WUYTACK and DETH, 1973; PEIPER, GRIEBEL and WENDE, 1971; GOLENHOFEN and HERMSTEIN, 1973).

In contrast to the β-action which decreases the phasic contraction, the α-action produces a long lasting increase of the force of contraction (SZURSZEWSKI and BÜLBRING, 1973). Thus we see an antagonism of the two components on E–C-coupling.

This leads to the possibility that catecholamines act, primarily, on the numerous cellular mechanisms, described above, which control the level of free calcium ions in the cytoplasm (see diagram). While the β-action stimulates Ca-uptake, the α-action stimulates Ca-release and, consequently, it may also stimulate the active extrusion of calcium across the plasma membrane. For this hypothesis it is assumed that the membrane permeability to K depends largely on the amount of Ca bound at the inner surface of the plasma membrane.

If, in spontaneously, very active smooth muscles, (e.g. taenia coli) this membrane bound Ca is kept low by a high rate of Ca extrusion, the sudden intracellular Ca-release by the α-adrenergic action will cause an increase in Ca-uptake at the inner surface of the plasma membrane. This will increase K-permeability and cause a transient hyperpolarisation, cessation of spike discharge and relaxation.

If, on the other hand, Ca bound on the inside of the membrane is high (possibly saturated) and the membrane potential is high (e.g. in the uterus) then a sudden release of Ca into the cytoplasm will cause contraction. It will also stimulate Ca-extrusion and generate spike discharge. Besides, since K-permeability is already high, the increase in Cl-permeability caused by the α-action dominates and produces depolarisation.

The slow time course of the action of catecholamines, the low rate at which the peak of the effect is reached, the long duration of action outlasting the presence of the drug, the marked effects on E-C-coupling and, last not least, the temperature dependence of both the α- and the β-responses to catecholamines indicate a complex

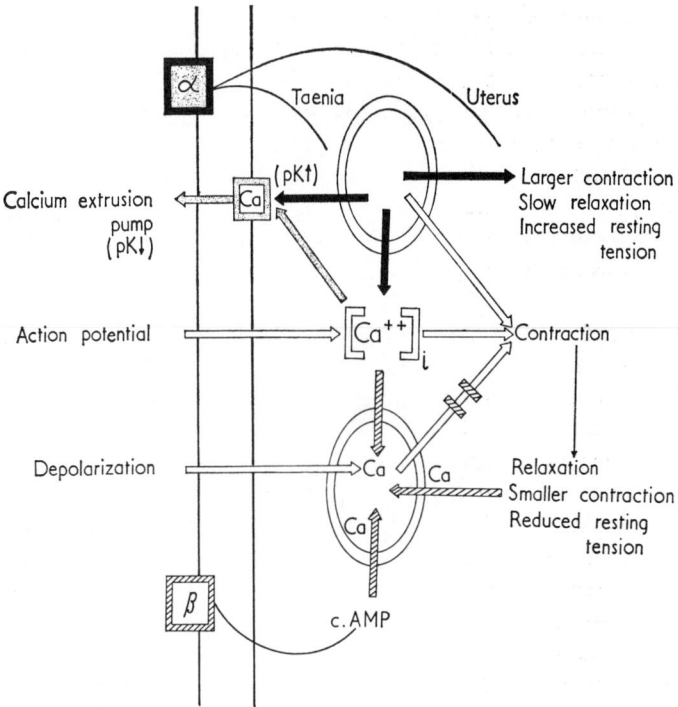

mechanism of action part of which may be energy dependent. The changes in membrane activity which underlie contraction in some, or relaxation in other muscles may be secondary effects produced by, fundamentally, the same action of catecholamines on all smooth muscles, an action on a series of cell functions which regulate the distribution of calcium.

REFERENCES

AHLQUIST R. P. (1948) A study of the adrenotropic receptors. *Am. J. Physiol.* **153**, 586–600.
BREEMEN VAN C., FARINAS B. R., CASTEELS R., GERBA P., WUYTACK F. and DETH R. (1973) *Phil. Trans. Roy. Soc. Lond. B.* **265**, 57–71.
BREEMEN VAN C., FARINAS B. R., GERBA P. and McNAUGHTON E. D. (1972) *Circ. Res.* **30**, 44–54.
BREEMEN VAN C., and LESSER P. (1971) *Microvasc. Res.* **3**, 113–114.
BÜLBRING E. (1973) In: *Drug Receptors.* pp. 1–13. Macmillan, London (in press).
BÜLBRING E., CASTEELS R. and KURIYAMA H. (1968) *Br. J. Pharmac. Chemother.* **34**, 388–407.
BÜLBRING E. and KURIYAMA H. (1973) *Phil. Trans. Roy. Soc. Lond. B.* **265**, 115–121.
BÜLBRING E. and TOMITA T. (1969a) *Proc. Roy. Soc. B.* **172**, 89–102.
BÜLBRING E. and TOMITA T. (1969b) *Proc. Roy. Soc. B.* **172**, 103–119.
CASTEELS R., GOFFIN J., RAEMAEKERS L. and WUYTACK F. (1973) *J. Physiol.* **230**, 85P.
GOLENHOFEN K. and HERMSTEIN N. (1973) *J. Physiol.* **230**, 79P.
HINKE J. A. M. (1965) In: *Muscle* (PAUL, DANIEL, KAY and MONCKTON, Eds.) Pergamon Press, Oxford.
PEIPER U., GRIEBEL L. and WENDE W. (1971) *Pfl. Arch.* **330**, 74–89.
REUTER H., BLAUSTEIN M. P. and HAEUSLER G. (1973) *Phil. Trans. Roy. Soc. Lond. B.* **265**, 87–94.
ROBISON G. A., BUTCHER R. W. and SUTHERLAND E. W. (1971) *Cyclic AMP.* Academic Press, New York.
SZURSZEWSKI J. and BÜLBRING E. (1973) *Phil. Trans. Roy. Soc. Lond. B.* **265**, 149–156.
TOMITA T. and WATANABE H. (1973) *Phil. Trans. Roy. Soc. Lond. B.* **265**, 73–85.

Frontiers in Catecholamine Research 1973, pp. 393 to 395. Pergamon Press. Printed in Great Britain.

REPORTER'S NOTES

G. A. ROBISON

BOTH the presentations and discussions brought together a variety of diverse but complementary approaches to catecholamine receptors, including classical pharmacologic studies, research on cyclic AMP and prostaglandins, and direct binding studies of receptors. Concerning Daly's paper, Perkins (Colorado) presented additional evidence showing that 6-hydroxydopamine (6-OHDA) increased sensitivity and responsivity of brain slices to norepinephrine. The response in these experiments was an increased level of cyclic AMP. Increased sensitivity to low doses of norepinephrine could be seen shortly after 6-OHDA, and could reflect reduced uptake by presynaptic elements. A greater maximal response to high doses of norepinephrine developed more slowly, and appeared to reflect a postsynaptic change.

There was some discussion of whether theophylline might not penetrate brain cells, thereby accounting for its apparent inability to increase cyclic AMP accumulation in brain slices. Regardless of the answer to this question (which appears not to have been experimentally tested), Daly emphasised that he thought theophylline was probably interacting as an antagonist with adenosine receptors on the external surface of the cells. A point which probably *should* have been raised was that theophylline appears to antagonize most if not all of the effects of adenosine in all tissues which are affected by adenosine, regardless of whether cyclic AMP is involved in the response or not. Even in fat cells, where the ability of theophylline to potentiate catecholamines has long seemed explicable in terms of phosphodiesterase inhibition, a different effect may be involved (as suggested by the recent work of Schwabe and Ebert and their colleagues.)

Deguchi was asked whether decapitation leads to a rise in cyclic AMP in the pineal gland as it clearly does in most brain areas. The answer was yes, although Deguchi did not seem to feel that this was an important variable. Weiss pointed out that the speed and magnitude of the postdecapitation rise depends very strongly on the lighting conditions, being much greater in rats previously exposed to constant light. He suggested that this could be understood in terms of the data which he and Strada had previously reported, showing that constant light did lead to pineal supersensitivity.

Westermann was asked whether the administration of L-DOPA had any effect on brain cyclic AMP levels. Apparently it did not, according to the results of preliminary experiments, although the injection of large doses of dopamine apparently did increase the level of cyclic AMP in the mouse brain. Palmer had previously been unable to see an effect of dopamine on rat brain slices, under conditions where norepinephrine produced a large increase. Apparently no one has studied the possible effect of DOPA on pineal cyclic AMP levels.

In connection with attempts to chemically characterize adrenergic receptors, many of the speakers spent a considerable amount of time emphasising the need for caution, especially when considering the possible relation between the binding of catechol-amines to proteins and their interactions with receptors. I should have thought that the desiderata at this stage ought to be imagination and hard work, with caution relatively low on the list, but the point did not seem worth arguing about. Lefkowitz' suggestion that the catecholamine-binding protein which he described might represent part of an adrenergic *beta* receptor was objected to on a number of grounds. The most important objections seemed to be (a) the lack of stereospecificity and (b) the low affinity of the protein for propranolol relative to its affinity for agonists. Lefkowitz had shown a slide in which a binding curve was superimposed on a dose–response curve obtained by studying the chronotropic response of cultured cells to epinephrine, the implication being that the binding constant and the ED_{50} were the same. Patil objected strongly to this kind of analysis, and suggested that a more meaningful parameter would be the ratio of stereoisomers. If the ratio of the binding constants was the same as the ratio of their potencies on intact cells, then this would constitute strong evidence that the binding protein might be related to the *beta* receptor, in Patil's view. Creveling (Bethesda) suggested that Lefkowitz' protein might more likely have been derived from membrane-bound COMT, but Lefkowitz disagreed with this. Snyder mentioned that Pedro Cuatrecasas (Johns Hopkins) had confirmed and ex-tended Lefkowitz' findings, but had come to an entirely different conclusion. Cua-trecasas feels that the protein should be referred to as a catechol-binding protein rather than a catecholamine-binding protein, and that it cannot be related to the adrenergic *beta* receptor in any way. A search for the function of this protein could be very fruitful in Snyder's view.

Lefkowitz gave a spirited defence of his view that the protein could nevertheless be a part of the *beta* receptor, but the consensus appeared to be that the protein should be deemed innocent of this charge until proved guilty.

Stimulated by Spector's paper, Horn (Cambridge) suggested that the problem of the antibodies being specific for catechols rather than catecholamines is most likely due to the fact that the catecholamine had been linked to the polymer via the amino group to produce an amide bond. Unlike an amino group, this function is virtually neutral and will not be charged at physiological pH. Thus one of the most important catechol-amine binding sites is not available for interaction. Horn suggested two possible solutions to the problem: (1) to bind DOPA to the polymer via the carboxyl group (which would introduce an asymmetric carbon atom); or (2) to link the aromatic ring to the polymer via a diazo bond. In response to a question from Lefkowitz, Spector emphasised again that the antibodies were stereospecific, even though the reason for this is not understood.

Following the papers by Mayer and Bulbring, there was some discussion of the possible role of cyclic AMP in *beta* adrenergic responses in which changes in calcium translocation appear to be important. The net conclusion (see also text of Meyer's paper) was that more research will be needed in order to answer this question. The hypothesis that all *beta* responses are mediated by cyclic AMP seems generally well supported, although, as this discussion emphasised, there is still room for doubt. Bulbring's suggestion that *alpha* and *beta* receptors might mediate effects drawing on the same pool of ATP was not strongly supported by the data which Schramm pre-sented, although it was not incompatible with it either. In response to a question from R. H. Levin, Bulbring said she saw no reason to insert the prostaglandins into her general scheme. Van Rossum suggested that the differences between intestinal and uterine smooth muscle, *vis à vis* the nature of their responses to adrenergic stimuli, might be due primarily to differences in the way the two tissues are innervated, but Bulbring thought this was extremely unlikely.

In response to a question from Szurszewski, Greengard emphasised that only muscarinic responses (as opposed to nicotinic responses) could be associated with cyclic GMP, in general, and mentioned several lines of evidence supporting the idea that the slow excitatory post-synaptic potential was mediated by cyclic GMP. An apparent exception to the generalization suggested by Greengard may be the adrenal medullary response to acetylcholine described previously (see elsewhere in this volume) by Costa and his associates.

B. Waldeck (Goteborg) described an experiment which he felt supported Greengard's suggestion of a role for cyclic AMP in regulating synaptic transmission. Motility of reserpinized mice was measured after increasing doses of a constant ratio mixture of ET 495 (to stimulate dopaminergic receptors) and clonidine (to stimulate noradrenergic receptors). Caffeine (25 mg/kg) was ineffective by itself, but greatly reduced the dose of the ET 495-clonidine mixture required to reverse the reserpine-induced depression. Waldeck felt the most probable explanation for this potentiation was phosphodiesterase inhibition by caffeine. However, caffeine and the other methylxanthines produce so many effects that cannot be related to phosphodiesterase inhibition that the experiment still seems hard to interpret.

Synaptic Dynamics
(May 24, 1973; 9:00 A.M.–6:30 P.M.)
(May 25, 1973; 9:00 A.M.–5:30 P.M.)

CHAIRMEN: Hermann K. F. Blaschko, Ulf S. von Euler, Parkhurst A. Shore, Ullrich Trendelenburg

COUNCILLORS: Annica Dahlstrom and George Hertting

REPORTERS: Ross J. Baldessarini and Bertil Hamberger

Frontiers in Catecholamine Research 1973, pp. 399 to 400. Pergamon Press. Printed in Great Britain.

CHAIRMAN'S REMARKS

U. S. VON EULER

Dept. of Physiology, Karolinska Institutet, S-104 01 Stockholm 60, Sweden

TWO RECENT findings have to some extent altered the picture of adrenergic trans-mission processes, viz. the modulating effect of prostaglandins of the E type on the release of transmitter, and the evidence accrued for an inhibitory effect of adrenergic α-agonists on the release process.

The effect of prostaglandins E on the transmitter release was first observed by HEDQVIST (1969) who found that addition of these compounds to the fluid used for perfusion of the isolated cat spleen strongly diminished the amount of labelled noradrenaline released by stimulation of the splenic nerves, and at the same time decreased the response of the vascular bed and the contraction of the splenic capsule. Similar effects were demonstrated on the field-stimulated isolated vas deferens of the guinea pig, in which addition of PGE_1 or PGE_2 to the bath fluid causes an almost complete block of contractile response under suitable stimulation conditions. This effect was particularly evident at low stimulation frequencies (HEDQVIST and EULER, 1972). The same concentrations of PGE_1 and PGE_2 which blocked the response of the vas deferens to transmural postganglionic stimulation, enhanced the contractions elicited by addition of noradrenaline to the bath fluid. The inhibitory effect on transmitter release could not be prevented by the prostaglandin blocker SC-19220, described by SANNER (1969).

The second effect referred to, inhibition of transmitter release by stimulation of adrenergic α-agonists, was first suspected from the increased release of transmitter after phenoxybenzamine (HÄGGENDAL, 1970). Direct evidence was later provided by the finding of diminished release of labelled transmitter from various organs including the isolated superfused vas deferens after addition of α-agonists like methoxamine or noradrenaline itself (ENERO et al., 1972; STARKE, 1972; HEDQVIST, 1973; STJÄRNE, 1973).

As to the mechanisms of these inhibitory actions, exerted by prostaglandins E and by α-agonists, various observations suggest that both are in some way connected with the mobilisation of Ca^{2+} ions, necessary for the release. This will be more fully dealt with in the papers by Hedqvist and Stjärne at this Symposium.

Although the vas deferens apparently contains more noradrenaline per unit weight than any other organ in the body, which has also been corroborated by the rich nerve net seen in fluorescence microscopy, AMBACHE and ZAR (1971) have pointed out some facts speaking against noradrenergic transmission as cause of the muscle twitch, recorded on nerve stimulation of the organ. The main observations leading to this conclusion were the low sensitivity of the organ to added noradren-aline, and the inhibitory effect of noradrenaline on the contractile response to field stimulation. None of these findings preclude by itself adrenergic transmission, however, since added noradrenaline may affect the physiological transmitter response in a way which could be difficult to interpret.

In their thorough analysis of the response of the organ to different kinds of stimuli and to various drugs, AMBACHE and ZAR (1971) report experimental data to the effect that the inhibitory effect of added noradrenaline is counteracted by phentolamine but not by phenoxybenzamine. It was further reported that the β-blocking agent propranolol did not block the inhibitory noradrenaline response.

Recent experiments in our laboratory (EULER and HEDQVIST, 1973) have confirmed earlier findings (HOLMAN and JOWETT, 1964) that noradrenaline tends to inhibit the contractile response of the isolated guinea pig vas deferens to field stimulation in low concentrations, while higher concentrations usually enhance the effect. Tyramine and phenethylamine act similarly. However, in those cases where noradrenaline increases the response to nerve stimulation, phentolamine regularly changes the reaction in such a way that noradrenaline now produces a dose-dependent inhibition of the stimulus response. In no case we have observed that the inhibitory effect of noradrenaline is prevented by phentolamine as reported by AMBACHE and ZAR (1971); on the contrary it is unmasked. On the other hand propranolol, given together with phentolamine, strongly diminishes the inhibition induced by noradrenaline, suggesting that the latter effect is caused by a β-agonistic action (cf. LARGE, 1965). Experiments with practolol and butoxamine indicated that the inhibitory effect of noradrenaline is of the β_2 type. The inhibitory effect of clonidine is, however, blocked by phentolamine (EULER and HEDQVIST, to be published). Phentolamine given alone enhances the twitch response of the vas deferens to field stimulation as also reported by BOYD, CHANG and RAND (1960).

The sometimes strongly enhancing of tyramine or phenethylamine on the contractile response to field stimulation is likewise converted into inhibition by phentolamine, suggesting that the effect is also in this case caused by liberated noradrenaline.

The picture of neurotransmission in the vas deferens apparently differs from that in most other smooth muscle organs. Whereas the inhibitory effect of added noradrenaline on the muscle 'twitch' seems to be one of β-agonistic type, there is still no explanation of the fact that neither α- nor β-blockers prevent the 'twitch' response of the organ to postganglionic nerve stimulation, while prostaglandins E, noradrenaline, indirectly acting amines, isoprenaline and clonidine inhibit the response.

Our results support the conclusion of REIFFENSTEIN and VOHRA (1972) that the guinea-pig vas deferens possesses three types of adrenergic receptors: excitatory α, inhibitory α and inhibitory β by which the motor responses may be influenced.

REFERENCES

AMBACHE N. and ZAR M. A. (1971) J. Physiol. (Lond.) 216, 359–389.
BOYD H., CHANG V. and RAND M. J. (1960) Br. J. Pharmacol. 15, 525–531.
ENERO M. A., LANGER S. Z, ROTHLIN R. P. and STEFANO F. J. E. (1972) Br. J. Pharmacol. 45, 672–688.
EULER U. S. v. and HEDQVIST P. To be published.
FARNEBO L.-O. and MALMFORS T. (1971) Acta physiol. scand. Suppl. 371.
HÄGGENDAL J. (1970) In: New aspects of storage and release mechanisms of catecholamines. Bayer Symposium II. (SCHÜMANN H. J. and KRONEBERG G. Eds.) pp. 100–109. Springer-Verlag, Berlin.
HEDQVIST P. (1969) Acta physiol. scand. 75, 511–512.
HEDQVIST P. (1973) Acta physiol. scand. 1973. In press.
HEDQVIST P. and VON EULER U. S. (1972) Neuropharmacol. 11, 177–187.
HOLMAN M. E. and JOWETT A. (1964) Austral. J. exp. Biol. med. Sci. 42, 40–53.
LARGE B. J. (1965) Br. J. Pharmacol. 24, 194–204.
REIFFENSTEIN R. J. and VOHRA M. M. (1972) Br. J. Pharmacol. 45, 156–157P.
SANNER J. H. (1969) Archs. int. pharmacodyn 180, 46–56.
STARKE K. (1972) Naunyn-Schmiedeberg's Arch. Pharmak. 274, 18–45.
STJÄRNE L. (1973) Nature (Lond.) 241, 190–191.
STJÄRNE L. (1973) Br. J. Pharmacol. In press.

Frontiers in Catecholamine Research 1973, pp. 401 to 402. Pergamon Press. Printed in Great Britain.

CATECHOLAMINE RELEASE AND MOBILISATION OF PROTEIN

H. Blaschko

University Department of Pharmacology, South Parks Road, Oxford OX1 3QT,
England

As so often in the study of events at the synapse, we have profited from findings on protein release from the stimulated adrenal medulla. At Oxford, this began with a study of the soluble protein of chromaffin granules (Blaschko and Helle, 1963), the protein that was purified by Smith and Winkler (1967) and characterised as chromogranin A. Banks and Helle (1965) showed that this protein is released when the adrenal medulla was stimulated and released catecholamine.

Many laboratories have contributed to the study of these phenomena. We now know that upon stimulation of the adrenergic neurones the release of the low-molecular-weight mediator is accompanied also by a mobilisation of the soluble proteins, chromogranin A and dopamine β-hydroxylase. This has been taken as an indication that in the adrenergic neurones as well as in the adrenal medulla liberation of catecholamine occurs by exocytosis (see Blaschko and Smith, editors, 1971). What is as yet not known is if we can attribute to the release of protein a biological significance. This seems to me a question worthy of further study. I may remind you of the observations on the so-called trophic effects of nerves on effector tissues. Some authors believe that these effects are mediated not by the low-molecular-weight transmitter substances but by macromolecules, such as proteins.

The occurrence of 'satellite proteins' (Blaschko, 1971) does not seem to be restricted to adrenergic systems. I should like to remind you of observations on 'vesiculin', a protein present in vesicles that contain acetylcholine (Whittaker, Dowdall and Boyne, 1972). It will be very interesting to find out if this protein is also mobilised upon stimulation. Another type of protein that might be called a satellite protein is neurophysin. This type of protein, that has been studied at Oxford by Dr. D. B. Hope and his colleagues, does appear in the extracellular space together with the biologically active material.

Another interesting type of protein release from the adrenal medulla, related to stimulation, was discovered at Oxford by Dr. F. H. Schneider (1968), who continued these studies at Denver (Schneider, 1970). He found that upon stimulation of the adrenal medulla many lysosomal enzymes are released. At Oxford, Dr. A. David Smith (1969) has found that two lysosomal phospholipases are among the enzymes set free under these conditions.

These challenging observations have not yet been built into our picture of what happens during catecholamine release. Also, we do not as yet know whether lysosomal enzymes are also mobilised when adrenergic neurones release their mediator.

Several interpretations for these observations might be advanced. For instance, lysosomal enzymes may be involved in the events that lead to the fusion of the granule (or vesicle) membrane with the plasma membrane, and to the opening through which the soluble contents are released. Or else, lysosome activation may be related to the removal of the insoluble granule remnants from the plasma membrane or to their subsequent disposal. In other words, lysosome activation may either occur parallel with the release of catecholamines, or it might be an event that happens subsequent to catecholamine release.

These findings clearly show that in the response of the cell to stimulation more than one cell organelle is immediately involved.

REFERENCES

Banks P. and Helle K. (1965) *Biochem. J.* **97**, 40–41C.

Blaschko H. (1971) *Progr. Brain Res.* **34**, 239–242.

Blaschko H. and Helle K. B. (1963) *J. Physiol., Lond.*, **169**, 120–121P.

Blaschko H. K. F. and Smith A. D., editors (1971) *Trans. Roy. Soc., Lond., B,* **261**, 273–440.

Schneider F. H. (1968) *Biochem. Pharmac.* **17**, 848–851.

Schneider F. H. (1970) *Biochem. Pharmac.* **19**, 833–847.

Smith A. D. (1969) *Biochem. J.* **114**, 72P.

Smith A. D. and Winkler H. (1967) *Biochem. J.* **103**, 483–492.

Whittaker V. P., Dowdall M. J. and Boyne A. F. (1972) *Biochem. Soc. Symposia,* **36**, 49–68.

Frontiers in Catecholamine Research 1973, pp. 403 to 408. Pergamon Press. Printed in Great Britain.

NEURONAL AND EXTRANEURONAL CATECHOLAMINE UPTAKE MECHANISMS

L. L. IVERSEN

MRC Neurochemical Pharmacology Unit, Department of Pharmacology,
University of Cambridge, Cambridge, England

THE EXISTENCE and properties of specific uptake processes for catecholamines and their role in terminating the physiological actions of these amines was reviewed by TITUS and DENGLER (1966) at the Second Catecholamines Symposium. Only a few points concerning the progress made in this field since then will be summarised here; more detailed reviews are available elsewhere (IVERSEN, 1967, 1971, 1973; TRENDELENBURG, 1972; HERTTING and SUKO, 1972). The present review will also be restricted to a consideration of the membrane transport processes present in the external membrane of adrenergic neurones and in some post-synaptic cells; other uptake mechanisms also exist within storage vesicles in the axoplasm of adrenergic neurones (for review see VON EULER, 1972; STJARNE, 1972).

NEURONAL UPTAKE MECHANISMS

Noradrenaline

Noradrenergic neurone in the peripheral sympathetic system and in the CNS possess a specific high affinity uptake mechanism for noradrenaline. This system (Uptake$_1$) remains the most thoroughly characterised of the catecholamine uptakes. In brain homogenates ^{3}H-noradrenaline uptake has been demonstrated and studied in synaptosome particles (COYLE and SNYDER, 1969; BOGDANSKI and BRODIE, 1969). Uptake$_1$ appears to have identical properties in the noradrenaline-containing neurones of the peripheral and central nervous systems. It involves a sodium-dependent carrier system dependent on metabolic energy and on the continued functioning of membrane Na$^+$/K$^+$-ATPase. Uptake$_1$ is saturable and has a very high affinity for noradrenaline, the apparent K_m being in the range 0·2 and 1·0 μM in most rat tissues. In the rat, Uptake$_1$ is stereochemically selective for l-noradrenaline (Table 1) although d-noradrenaline is also a substrate; stereochemical specificity may be lacking in other species such as guinea pig and rabbit (IVERSEN et al., 1971; HENDLEY and SNYDER, 1972). In mammalian neurones many other phenolic derivatives of β-phenethylamine can act as substrates for Uptake$_1$; these include adrenaline, metaraminol, the α-methyl analogues of adrenaline and noradrenaline, dopamine, tyramine, octopamine and their α-methyl analogues. The structural requirements for Uptake$_1$ substrates are:—absence of bulky N-substituent groups (isoprenaline is not a substrate), absence of methoxyl groups on phenolic substituents, presence of at least one phenolic hydroxyl group (amphetamine, β-phenethylamine, β-phenylethanolamine and norephedrine are not substrates).

The structure activity relationships for the inhibition of ^{3}H-noradrenaline uptake in rat heart and in synaptosome preparations have also been studied, and are similar in the peripheral and central nervous systems (BURGEN and IVERSEN, 1965; HORN, 1973). Affinity for Uptake₁ is *decreased* by the presence of bulky N-substituents, by methoxylation of phenolic hydroxyls and by the presence of a β-hydroxyl group on the side chain. Affinity for Uptake₁ sites is *increased* by the presence of phenolic hydroxyl groups, especially in the *para* and *meta* positions, and also by α-methylation of the α-carbon of the side chain. In the latter case affinity is highest for the isomer corresponding to *d*-amphetamine. It should be pointed out that the structure–activity relationships for inhibition of Uptake₁ do not correspond exactly to the requirements previously mentioned for compounds in this class to be substrates for the uptake process. Thus, for example, amphetamine and β-phenethylamine are not substrates for Uptake₁ but are nevertheless potent competitive inhibitors. Such compounds, like competitive enzyme inhibitors, may bind strongly to uptake sites but lack the further structural features needed for transport across the membrane to occur.

Uptake₁ is also potently inhibited by many other drugs, notably by the tricyclic antidepressants imipramine and amitriptyline and their analogues. Desipramine is one of the most potent inhibitors so far described, the concentration required for 50 per cent inhibition of Uptake₁ in periphery and CNS being approximately 1×10^{-8} M.

Dopamine

The dopamine-containing neurones in the mammalian CNS possess their own specialised uptake system for dopamine and other catecholamines. This uptake system has been studied in slices or homogenates of dopamine-rich brain areas, such as the corpus striatum (HAMBERGER, 1967; COYLE and SNYDER, 1969; HORN *et al.*, 1971), and also in certain amacrine cells in the retina (EHINGER and FALCK, 1971; HENDLEY and SNYDER, 1972). Although dopamine can be taken up by Uptake₁ and noradrenaline can be taken up by dopaminergic neurones, the uptake process in the latter cells has different kinetic and pharmacological properties from Uptake₁. It has a very high affinity for dopamine and a lower affinity for noradrenaline (Table 1) (COYLE and SNYDER, 1969). Furthermore, the dopamine uptake system, unlike Uptake₁, does not show any stereochemical specificity for the isomers of noradrenaline. The most striking difference in drug sensitivity is the relative ineffectiveness of desipramine and related tricyclics as inhibitors of dopamine uptake; desipramine is about 1000 times less potent on this system than on Uptake₁ (Table 1)(HORN *et al.*, 1971). Dopamine uptake is also potently inhibited by a variety of anti-cholinergic and antihistaminic drugs, including compounds such as benztropine which are used as anti-Parkinsonian agents (HORN *et al.*, 1971).

The selective inhibition of the noradrenaline and dopamine uptake sites in brain by desipramine and benztropine respectively can provide an indication of the presence of these two types of adrenergic neurones in various areas of the CNS. For example CUELLO *et al.* (1973) found that about 40 per cent of the uptake of ³H-dopamine by homogenates of adult male rat median eminence could be inhibited by desipramine; the desipramine-resistant uptake being potently inhibited by benztropine. The existence of a large desipramine-sensitive uptake, together with the finding that median

TABLE 1. KINETIC PROPERTIES AND INHIBITORS OF CATECHOLAMINE UPTAKE PROCESS

	Uptake$_1$ (rat heart)	Dopamine Uptake (rat c. striatum)	Uptake$_2$ (rat heart)
		$K_m \ (-\mu \text{M})$	
dl-isoprenaline	isoprenaline is not a substrate	—	23‖
dl-adrenaline	1·40*	—	52*
l-noradrenaline	0·27*	2·0†	252*
d-noradrenaline	1·39*	1·9‡	252*
dopamine	0·69†	0·31§	590†

(Inhibitors with IC$_{50}$¶, μM)

desipramine (0·01)	d-amphetamine (0·1)	SKF550 (0·08)
l-metaraminol (0·08)		SKF625A (0·25)
d-amphetamine (0·18)	benztropine (0·03)	phenoxybenzamine (2·8)
l-amphetamine (3·7)	chlorpromazine (3·1)	β-oestradiol (2·0)
cocaine (0·38)	desipramine (50·0)	corticosterone (2·7)
phenoxybenzamine (0·78)		dl-metanephrine (2·9)

*Iversen (1967); †Hellmann et al (1971); ‡Iversen et al. (1971); §Coyle and Snyder (1969); ‖Callingham and Burgen (1966). ¶IC$_{50}$ = drug concentration required to cause 50 per cent inhibition of uptake.

eminence tissue contained high concentrations of endogenous noradrenaline, indicated that this brain area may have an unexpectedly dense noradrenergic innervation in addition to the previously described dopaminergic terminals.

EXTRANEURONAL UPTAKE OF CATECHOLAMINES (Uptake$_2$)

The catecholamines are also taken up by a different transport system in various extraneuronal peripheral tissues, such as vascular smooth muscle, cardiac muscle, and certain glandular tissues. In the isolated perfused rat heart catecholamines are taken up by two uptake processes, distinguishable because of their different kinetic and pharmacological properties (IVERSEN, 1965). At low perfusion concentrations catecholamine uptake is mediated largely by the neuronal Uptake$_1$ mechanisms; at high perfusion concentrations, however, a more rapid accumulation of catecholamine occurs and the mechanism responsible for this has been termed "Uptake$_2$". This uptake system is saturable, but has a much lower affinity for noradrenaline and adrenaline than Uptake$_1$ (Table 1). On the other hand, Uptake$_2$ has a very much higher capacity (V_{max}) than Uptake$_1$. Uptake$_2$ is not normally detectable at low perfusion concentrations of catecholamine because the accumulated catecholamine is not firmly retained but is rapidly metabolised by the combined actions of monoamine oxidase and catechol-O-methyl transferase (LIGHTMAN and IVERSEN, 1969). Uptake$_2$ differs from Uptake$_1$ in its substrate specificity (Table 1); the Uptake$_2$ process does not exhibit any stereochemical specificity for d- or l-noradrenaline or adrenaline; it has a higher affinity for adrenaline than noradrenaline; isoprenaline is an even better substrate for Uptake$_2$ although this amine is not taken up by Uptake$_1$ (CALLINGHAM and BURGEN, 1966). Histochemical studies of heart tissue after perfusion with high concentrations of catecholamine have demonstrated that the Uptake$_2$ sites are predominantly localised in the cardiac muscle cells (FARNEBO and MALMFORS, 1969; CLARK et al., 1969). A similar accumulation of exogenous catecholamines in various smooth muscles (blood vessels, splenic capsule, vas deferens, intestinal) has been observed histochemically when these tissues are incubated with high concentrations of catecholamine *in vitro*; this accumulation appears to be mediated by a mechanism similar in properties to Uptake$_2$ (AVAKIAN and GILLESPIE, 1968; GILLESPIE *et al.*, 1970; GILLESPIE and MUIR, 1970; BURNSTOCK, MCLEAN and WRIGHT, 1971; GILLESPIE, 1973). Biochemical studies have also suggested the presence of an extra-neuronal uptake mechanism similar to Uptake$_2$ in the cat nictitating membrane (DRASKOCZY and TRENDELENBURG, 1970), and mouse atrium (SACHS, 1970). In summary, there is considerable evidence for the existence of Uptake$_2$ in various tissues normally innervated by the sympathetic nervous system. Uptake of catechol-amine by this process is rapidly followed by intracellular catabolism; Uptake$_2$ could thus constitute a mechanism for disposing of extracellular catecholamines by an "uptake-followed-by-metabolism" process, as opposed to the "uptake-and-retention" mechanism previously described for Uptake$_1$.

The structure–activity relationships for the inhibition of Uptake$_2$ by sympatho-mimetic amines are almost the converse of these found for Uptake$_1$ (BURGEN and IVERSEN, 1965). Thus, inhibition of Uptake$_2$ was enhanced by N-substitution and by O-methylation, normetanephrine and metanephrine being potent inhibitors of the Uptake$_2$ process. On the other hand, *l*-metaraminol which was one of the most potent inhibitors of Uptake$_1$ had no inhibitory effects on Uptake$_2$. Two other groups of potent Uptake$_2$ inhibitors have since been described. LIGHTMAN and IVERSEN (1969) found that phenoxybenzamine was a very active Uptake$_2$ blocker. Subsequent investigations of other β-haloalkylamines of this type have shown that the com-pounds SKF550 [N-(9-fluorenyl-N-methyl-β-chloroethylamine] and SKF625A [N-(3,4-dimethoxyphenylisopropyl)-N-benzyl-β-chloroethylamine] are even more potent than phenoxybenzamine as Uptake$_2$ inhibitors, although much less potent than phenoxybenzamine as Uptake$_1$ inhibitors (Table 1) (IVERSEN *et al.*, 1972). Various steroids have also been found to be effective Uptake$_2$ blockers in the isolated rat heart (IVERSEN and SALT, 1970). The most potent of these are β-oestradiol and corticosterone, although various other steroids with diverse biological activities, including cholesterol, were also effective (SALT, 1972; SALT and IVERSEN, 1972).

PHYSIOLOGICAL FUNCTIONS OF CATECHOLAMINE UPTAKE

It is now widely accepted that Uptake$_1$ represents the major mechanism for terminating the actions of noradrenaline after its release from adrenergic nerve terminals (IVERSEN, 1967, 1972). LANGER (1970) and HUGHES (1972) found that inhibition of Uptake$_1$ by cocaine led to increases of approximately fourfold in the outflow of labelled noradrenaline and its metabolites from isolated tissues on sym-pathetic nerve stimulation, suggesting that as much as 75–80 per cent of the neurally released noradrenaline may be inactivated by such a recapture mechanism. The enhanced outflow of noradrenaline produced by phenoxybenzamine is now seen to be a complex phenomenon, being a compound of the inhibitory actions of this drug on both Uptake$_1$ and Uptake$_2$, and also in blocking an α-adrenoceptor mediated "auto-inhibition" mechanism by which noradrenaline may normally limit its own release (ENERO *et al.*, 1972; STARKE, 1972).

Uptake$_2$ also appears to play some role in disposing of noradrenaline after its neural release; after inhibition of Uptake$_1$ or in tissues with a low density of sympathetic innervation this may become quantitatively of major importance. Uptake$_2$ also probably plays a major role in the inactivation of circulating adrenaline and noradrenaline. The "Uptake$_2$-and-metabolism" mechanism can be blocked either by compounds that inhibit Uptake$_2$ (steroids, haloalkylamines, metanephrine) or by inhibitors of monoamine oxidase and/or catechol-O-methyl transferase. Thus, the actions of catecholamines on vascular smooth muscle (KALSNER, 1969), nictitating membrane (TRENDELENBURG *et al.*, 1971) and heart muscle (KAUMANN, 1971) are potentiated by steroids and by COMT inhibitors, but these effects are not additive. FOSTER (1969) found that inhibitors of Uptake$_2$ (e.g. metanephrine, phenoxybenzamine) potentiated the actions of isoprenaline on tracheal smooth muscle, and KAUMANN (1971) reported similar potentiating effects of hydrocortisone on isoprenaline actions in heart muscle. We have recently found (SALT and IVERSEN, in preparation) that corticosterone has a similar potentiating effect on the inotropic actions of isoprenaline in the isolated rat heart. Corticosterine also caused an enhanced outflow of ^{3}H-noradrenaline in this preparation in response to potassium-induced depolarisation mainly by preventing the normally extensive extraneuronal metabolism of the released catecholamine.

REFERENCES

AVAKIAN O. V. and GILLESPIE J. S. (1968) *Br. J. Pharmac. Chemother.* **32**, 168–184.
BOGDANSKI D. F. and BRODIE B. B. (1969) *J. Pharmac. exp. Ther.* **165**, 181–189.
BURGEN A. S. V. and IVERSEN L. L. (1965) *Br. J. Pharmac. Chemother.* **25**, 34–49.
BURNSTOCK G., MCLEAN J. R. and WRIGHT M. (1971) *Br. J. Pharmac.* **43**, 180–189.
CALLINGHAM B. A. and BURGEN A. S. V. (1966) *Mol. Pharmac.* **2**, 37–42.
CLARKE D. E., JONES C. J. and LINLEY P. A. (1969) *Br. J. Pharmac.* **37**, 1–9.
COYLE J. T. and SNYDER S. H. (1969) *J. Pharmac. exp. Ther.* **170**, 221–231.
CUELLO A. C., HORN A. S., MACKAY A. V. P. and IVERSEN L. L. (1973) *Nature, Lond.* **243**, 465–466.
DRASKOCZY P. R. and TRENDELENBERG U. (1970) *J. Pharmac. exp. Ther.* **174**, 290–306.
EHINGER B. and FALCK B. (1971) *Brain Res.* **33**, 157–172.
ENERO M. A., LANGER S. Z., ROTHLIN R. P. and STEFANO F. J. E. (1972) *Br. J. Pharac.* **44**, 672–688.
EULER U. S. von (1972) *Handbook of Experimental Pharmacology* XXXIII, pp. 186–230, Springer-Verlag, Berlin.
FARNEBO L. O. and MALMFORS T. (1969) *Europ. J. Pharmac.* **5**, 313–320.
FOSTER R. W. (1969) *Br. J. Pharmac.* **35**, 418–427.
GILLESPIE J. S. (1973) *Br. med. Bull.* **29**, 136–141.
GILLESPIE J. S. and MUIR T. C. (1970) *J. Physiol. Lond.* **206**, 591–604.
GILLESPIE J. S., HAMILTON D. N. H. and HOSIE R. J. A. (1970) *J. Physiol. Lond.* **206**, 563–590.
HAMBERGER B. (1967) *Acta physiol. scand.* **71**, suppl. 295.
HELLMAN G., HERTTING G. and PESKAR B. (1971) *Br. J. Pharmac.* **41**, 256–269.
HENDLEY E. D. and SNYDER S. H. (1972) *Eur. J. Pharmac.* **19**, 56–66.
HERTTING G. and SUKO J. (1972) In: *Perspectives in Neuropharmacology* (SNYDER S. H. Ed.) pp. 267–300, Oxford Univ. Press, N. Y.
HORN A. S. (1972) *Br. J. Pharmac.* **47**, 332–338.
HORN A. S., COYLE J. T. and SNYDER S. H. (1971) *Mol. Pharmac.* **7**, 66–80.
HUGHES J. (1972) *Br. J. Pharmac.* **44**, 472–491.
IVERSEN L. L. (1965) *Br. J. Pharmac.* **25**, 18–33.
IVERSEN L. L. (1967) *The Uptake and Storage of Noradrenaline in Sympathetic Nerves*, Cambridge University Press, London.
IVERSEN L. L. (1971) *Biogenic Amines and Physiological Membranes in Drug Therapy* (J. BIEL, Ed.) Marcel Dekker Inc., New York, Part B, pp. 259–327).
IVERSEN L. L. (1973) *Br. Med. Bull.* **29**, 130–135.
IVERSEN L. L., JARROTT B. and SIMMONDS M. A. (1971) *Br. J. Pharmac.* **43**, 845–855.
IVERSEN L. L. and SALT P. J. (1970) *Br. J. Pharmac. Chemother.* **40**, 528–530.
IVERSEN L. L., SALT P. J. and WILSON H. A. (1972) *Br. J. Pharmac.* **46**, 647–657.
KALSNER S. (1969) *Br. J. Pharmac.* **36**, 582–593.

KAUMANN A. J. (1972) *N.S. Arch. Pharmac.* **273**, 135–153.

LANGER S. Z. (1970) *J. Physiol. Lond.* **208**, 515–546.

LIGHTMAN S. and IVERSEN L. L. (1969) *Br. J. Pharmac.* **37**, 638–649.

SACHS CH. (1970) *Acta physiol. scand.*, **79**, (suppl. 341), 1–67.

SALT P. J. (1972) *Europ. J. Pharmac.* **20**, 329–340.

SALT P. J. and IVERSEN L. L. (1972) *Nature. New Biol.* **238**, 91–92.

STARKE K. (1972) *N.S. Archiv. Pharmac.* **274**, 18–45.

STJARNE L. (1972) *Handbook of Experimental Pharmacology* XXXIII, pp. 231–282, Springer-Verlag, Berlin.

TRENDELENBURG U. (1972) *Handbook of Experimental Pharmacology* XXXIII, pp. 726–761, Springer-Verlag, Berlin.

TRENDELENBURG U., HORN D., GRAEFE K. H. and PLUCHINE S. (1971) *N.S. Archiv. Pharmac.* **271**, 59–62.

Frontiers in Catecholamine Research 1973, pp. 409 to 410. Pergamon Press. Printed in Great Britain.

THE POSSIBLE IMPORTANCE OF YOUNG (LARGE) AMINE STORAGE GRANULES FOR ADRENERGIC NERVE TERMINAL FUNCTION

Annica Dahlström and Jan Häggendal

Institute of Neurobiology and Department of Pharmacology,
University of Göteborg, Sweden

Following one large dose of reserpine (10 mg/kg i.p. to rats) catecholamines and serotonin are depleted centrally and peripherally. The recovery of endogenous noradrenaline (NA) has been studied previously (Dahlström, 1967; Häggendal and Dahlström, 1972). The reserpine effect is considered to be due to an irreversible blockade of the uptake–storage mechanism of the granules (for ref. see Häggendal and Dahlström, 1971; Norn and Shore, 1971). Since functioning amine granules are necessary for nerve terminal functions ^{3}H-NA uptake-retention and transmission are also markedly depressed during the initial period after reserpine. Onset of recovery of the three different parameters (endogenous NA, ^{3}H-NA uptake-retention and transmission) occurs in the rat peripheral tissues 24–36 hr after reserpine (for ref. see Häggendal and Dahlström, 1972). Within the same period of time functioning, NA containing granules start to appear in the nerve terminals of long adrenergic neurons. These granules were probably formed in the cell bodies some hours after the administration of reserpine and transported to the terminals by intra-axonal transport (disc. in Häggendal and Dahlström, 1971). Since these periods coincide it is likely that onset of recovery of the three parameters are due to arrival in the terminals of new functioning amine granules. Recovery to normal, however, proceeds differently; ^{3}H-NA uptake-storage *and* transmission are normalised within 2–3 days, while endogenous NA reach normal levels after 3–5 weeks. This indicated differences in capacities of young and older amine granules.

Due to increased preganglionic impulse activity post-reserpine protein synthesis is stimulated and tyrosine hydroxylase and dopamine-β-hydroxylase (DBH) increase in ganglion cells (cf. Thoenen et al., this volume). The number of new amine storage granules transported down the axons to the terminals is also increased markedly during the 2nd–5th day after reserpine, then declined to subnormal and subsequently normal amounts. (Dahlström and Häggendal, 1969). When studying the recovery of ^{3}H-NA uptake-storage after reserpine, we found a similar, multiphasic event for that parameter, with overshooting during day 3–5, decline to subnormal levels and finally normalisation (Häggendal and Dahlström, 1972).

These results indicate (a) that uptake-retention of ^{3}H-NA (2·5 μg/kg given rapidly i.v. 30 min before death) occurs mainly in "young" amine granules recently arrived to the nerve terminals, and (b) that this capacity of the "young" granules is rather shortlasting. The $T_{\frac{1}{2}}$ for this capacity has been estimated to around 12 hr. (Häggendal and Dahlström, 1972.)

Several investigators have recently shown that axonal storage granules contain much DBH and chromogranin A in relation to NA, are relatively dense in sucrose gradients, and correspond to the large dense core vesicles electronmicroscopically.

The major part of the nerve terminal granules contain probably less protein in relation to NA, are lighter in sucrose gradients, and correspond to small dense core vesicles. (De Potter, this volume; Fillenz, this volume). Part of the nerve terminal particles are, however, of the large, heavy type. The release of NA at nerve activity is closely accompanied by release also of DBH and chromogranin (see e.g. Geffen, this volume). This, together with the findings that newly synthesised or recaptured transmitter is preferentially released suggest to us that the "small, easily releasable" pool of transmitter is contained mainly in the young, large, DBH-rich storage particles, while the 'large stable' pool is stored mainly in old, small, DBH-poor particles. Furthermore, the release of NA by exocytosis occurs mainly from large, young granules not by *total* exocytosis but by *partial* exocytosis (i.e. only a fraction of the DBH content is released each time). The released DBH could hypothetically originate either from the membrane bound form, or from soluble DBH inside the vesicle (Geffen, this volume). With *partial* exocytosis the granule with its DBH can be used for release several times and the DBH used for resynthesis. Repeated release would cause loss of matrix and membrane material and a gradual shrinkage of large vesicles into small ones. The turnover of DBH in nerve terminals, which has been estimated to 40 hr. (De Potter and Chubb, 1971), is in good agreement with the $T_{\frac{1}{2}}$ of 12 hr for the ^{3}H-NA uptake-storage capacity of 'young' granules, and may therefore rather demonstrate turnover of 'young' granules (i.e. their conversion into 'old' small granules) than turnover of all storage granules, as suggested. Since 'light' NA particles of nerve terminals contain very little DBH (De Potter, 1971) we think that NA is a better 'marker' for the 'old' small granules which by suggestion constitute mainly the 'large stable' pool of transmitter. The life span of these small granules seem to be in the order of some weeks (Häggendal and Dahlström, 1971).

(For further details on the presented hypothesis and for references, see Dahlström and Häggendal, 1972; Häggendal and Dahlström, 1972.)

Acknowledgements—Supported by the Swedish Medical Research Council (grants 14x-2204, 04p-4173 and 14x-166)

REFERENCES

Dahlström A. (1967) *Acta physiol. scand.* 69, 167–179.
Dahlström A. and Häggendal J. (1969). *J. Pharm. Pharmacol.* 21, 633–638.
Dahlström A. and Häggendal J. (1972) *Acta physiol. Pol.* XXIII, (3 suppl.) 4, 67–79.
De Potter W. P. (1971) *Phil. Trans. Roy. Soc. Lond. B.* 261, 313–317.
De Potter W. P. and Chubb I. W. (1971) *Biochem. J.* 125, 375–376.
Häggendal J. and Dahlström A. (1971) *J. Pharm. Pharmacol.* 23, 81–89.
Häggendal J. and Dahlström A. (1972) *J. Pharm. Pharmacol.* 24, 565–574.
Norn S. and Shore P. A. (1971) *Biochem. Pharmacol.* 20, 1291–1295.

Frontiers in Catecholamine Research 1973, pp. 411 to 413. Pergamon Press. Printed in Great Britain.

CONFORMATIONAL ASPECTS OF THE INHIBITION OF NEURONAL UPTAKE OF NORADRENALINE BY TRICYCLIC ANTIDEPRESSANTS

A. S. HORN

MRC Neurochemical Pharmacology Unit, Department of Pharmacology, University of Cambridge, Cambridge, England

WHATEVER the exact mode of action of the clinically efficacious tricyclic antidepressants may be, that they are potent inhibitors of noradrenaline uptake *in vitro* and *in vivo* is an established fact (AXELROD *et al.*, 1961; GLOWINSKI *et al.*, 1966). It is known, for example, that a concentration as low as 5.0×10^{-8} M of desipramine will cause *in vitro* a 50 per cent inhibition of the uptake of radiolabelled noradrenaline into homogenates of the rat hypothalamus (HORN *et al.*, 1971) an area containing the highest concentration of noradrenaline in the rat brain (GLOWINSKI and IVERSEN, 1966). It is, therefore, of interest to know what conformational properties of these drugs, which are superfically structurally disimilar to noradrenaline, enable them to be such potent blockers of this uptake process. Information about the possible conformation of the tricyclics at the uptake site may be gained by examining the conformational properties of simpler potent inhibitors of the noradrenaline uptake process. A good example of this is amphetamine, which may be regarded as a noradrenaline analogue. In order to simplify the conformational analysis of the preferred conformation of compounds having relatively mobile side chains, such as amphetamine, structurally more rigid analogues may be studied. Recent work using *cis* and *trans* 2-phenylcyclopropylamine (the *trans* form being tranylcypromine) and 1 and 2-aminoindanes, all of which are more rigid analogues of amphetamine, has shown that the most probable conformation of amphetamine at the noradrenaline uptake sites is with the side chain fully extended and the amino group above the plane of the benzene ring (HORN and SNYDER, 1972). This is known to be the preferred form in the solid state and in solution as shown by X-ray crystallographic (BERGIN and CARLSTROM, 1971) and nuclear magnetic resonance studies (NEVILLE *et al.*, 1971), respectively. The X-ray structure of noradrenaline has also been shown to have the side chain fully extended and the amino group above the plane of the ring (CARLSTROM and BERGIN, 1967) Fig. 1a. Molecular orbital calculations have supported this result (KIER, 1969). Hence it may be reasoned that this might correspond to the actual conformation of noradrenaline at the uptake site. It is known that both amphetamine and the tricyclic antidepressants are competitive inhibitors of the uptake of noradrenaline into the rat hypothalamus (HORN *et al.*, 1971), therefore, it is likely that part of the tricyclic molecule may have a complimentary spatial relationship to the preferred conformation of amphetamine and hence to noradrenaline at the uptake site. Unfortunately the X-ray structure of the imipramine class of antidepressants has not been determined but it is possible that the preferred conformation of imipramine at the uptake site is as shown in Fig. 1b. This is supported by the fact that the imipramine analogue chlorpromazine, which is a potent inhibitor of noradrenaline uptake (HORN *et al.*, 1971) is known to have a similar structure in the solid state

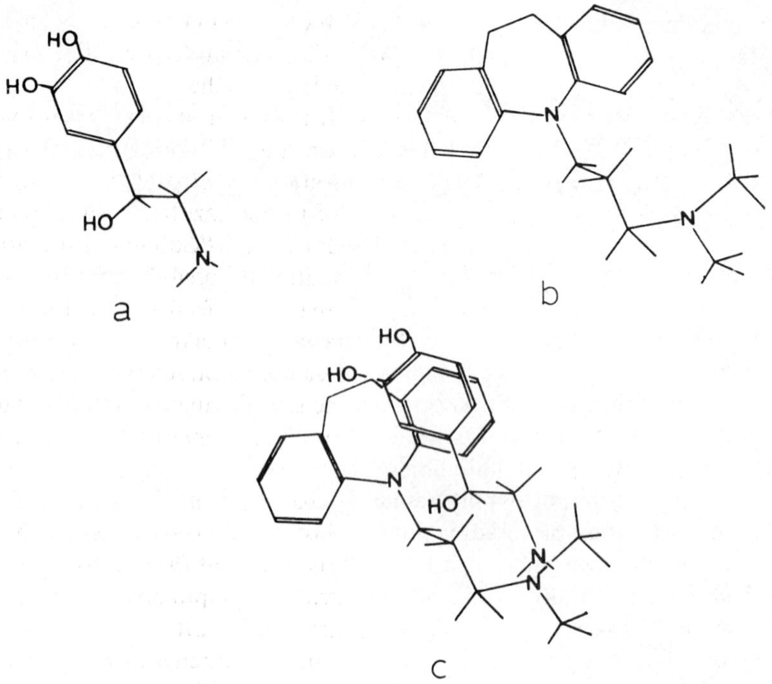

FIG. 1. The conformation of noradrenaline as determined by X-ray crystallography is shown in a. The side chain is fully extended and the amino group is above the plane of the benzene ring i.e. an *anti* conformation. The proposed uptake site preferred conformation of imipramine is shown in b. The ring system has a large dihedral angle between the planes of the outer rings and the side chain will probably be substituted on the nuclear nitrogen in an equatorial position. The terminal amino group is inclined away from the mid-line and points downwards. The superimposition of the two molecules is shown in c.

(McDowell., 1969) as does the analogue *trans* 9,10-dihydro-4-(3-dimethyl-aminopropyliden)-4H-benzo[4, 5] cyclohepta [1, 2-b] thiophene (Bastian and Weber, 1971). It can be seen that it is possible to superimpose the X-ray structure of nor-adrenaline on the proposed uptake site preferred conformation of imipramine such that the two most crucial binding sites, the aromatic ring and the terminal amino group, have the same spatial disposition, Fig. 1c. Thus the tricyclics could bind in this manner and prevent the normal uptake of the natural substrate noradrenaline.

Using this model it is possible to rationalize several of the structure–activity studies that have been reported. There is strong evidence that the structure–activity relations determined for uptake inhibitors in the peripheral nervous system, such as the heart, are also very similar to those found in the CNS (Ferris et al., 1972; Horn, 1973). Thus it is known that in the perfused rat heart (Callingham, 1966) and in the chick brain (Callingham, unpublished observations) increasing or decreasing the length of the side chain by one carbon atom or branching of the chain all lead to a consider-able fall in potency in inhibiting noradrenaline uptake. Molecular models indicate that tricyclics containing 2 or 4 carbon side chains would not fit the proposed con-formation of noradrenaline as well as a tricyclic with a 3 carbon chain. Branching could affect the conformation of the side chain or interfere with its binding to the uptake site. Turning now to the tricyclic ring system it has been known for some time that for inhibitory activity it is necessary for the two outer rings to be at an angle to each other. Several workers have shown that planar tricyclic systems are less potent than those having a large dihedral angle between the planes of the outer rings (MAXWELL et al., 1969; HORN et al., 1971). It is known, however, that several tricyclic ring systems are not rigidly locked in an 'angled' conformation but have a certain amount of conformational freedom (ARONEY et al., 1968; NOGRADI et al., 1970; AIZENSHTAT et al., 1972). In the case of the carbazole and other planar ring systems there is very little conformational freedom. This flexibility of the non-planar ring systems will, of course, affect the conformational possibilities of the side chain possibly leading to a preferred conformation that can effectively block the terminal amino groups binding position at the noradrenaline uptake site more easily. Presumably the planar tricyclics would not have such a pronounced effect on the side chain conformation. It is also possible that the second ring plays a less critical role in binding but must be out of the plane of the first aromatic ring in order not to interfere with the latter's binding.

It is probable that the protonated positively charged form of the tricyclics terminal amino group interacts with the uptake site. This is likely because of the pK_a's of the groups involved (MAXWELL et al., 1969) and the fact that the quaternary compound is active (CALLINGHAM, 1966). It is well known that imipramine is less potent than desipramine in inhibiting noradrenaline uptake (CALLINGHAM, 1966; HORN et al., 1971). Although there will be a small change in basic strength on the addition of a second N-methyl group the fall in potency is most probably due to a steric effect on the correct positioning of the charged nitrogen atom. A slight effect on the conformation of the side chain may also be of some significance.

In order to put the above proposals on a firmer basis further studies using X-ray crystallographic and n.m.r. techniques are required as well as the evaluation of carefully designed rigid analogues of the tricyclics as inhibitors of noradrenaline uptake.

REFERENCES

AIZENSHTAT Z., KLEIN E., WEILER–FEILCHENFELD H. and BERGMANN E. D. (1972) Israel J. of Chem. 10, 753–763.
ARONEY M. J., HOSKINS G. M. and LEFEVRE R. J. W. (1968) J. Chem. Soc. (B) 1206–1208.
AXELROD J., WHITBY L. G. and HERTTING G. (1961) Science 133, 383–384.
BASTIAN J. M. and WEBER H. P. (1971) Helv. Chim. Acta. 54, 293–297.
BERGIN R. and CARLSTROM D. (1971) Acta Cryst., B27, 2146–2152.
CALLINGHAM B. A. (1966) In: Excerpta Med. In. Congr. Series No. 122, Proceedings of the First International Symposium on Antidepresent Drugs, (GARATTINI, S and DUKES, M. N. G., Eds.) pp. 35–43. Excerpta Medica Foundation, Amsterdam 1967.
CARLSTROM D. and BERGIN R. (1967) Acta Cryst. 23, 313–319.
FERRIS R. M., TANG F. L. M. and MAXWELL R. A. (1972) J. Pharmac. exp. Ther. 181, 407–416.
GLOWINSKI J., IVERSEN L. L. and AXELROD J. (1966) J. Pharmac. exp. Ther. 153, 30–41.
HORN A. S., COYLE J. T. and SNYDER S. H. (1971) Mol. Pharmacol. 7, 66–80.
HORN A. S. and SNYDER S. H. (1972) J. Pharmac. exp. Ther., 180, 523–530.
HORN A. S. (1973) Br. J. Pharmac. 47, 332–338.
KIER L. B. (1969) J. Pharm. Pharmac. 21, 93–96.
MAXWELL R. A., KEENAN P. D., CHAPLIN E., ROTH B. and EDKHARDT S. B. (1969) J. Pharmac. exp. Ther. 166, 320–329.
McDOWELL J. J. H. (1969) Acta Cryst. B25, 2175–2181.
NEVILLE, G. A., DESLAURIERS, R., BLACKBURN, B. J. and SMITH, I. C. P. (1971) J. Med. Chem. 14, 717–721.
NOGRADI M., OLLIS W. D. and SUTHERLAND I. O. (1970) Chem. Comm. 158–160.

Frontiers in Catecholamine Research 1973, pp. 415 to 421. Pergamon Press. Printed in Great Britain.

COMPOSITION AND MOLECULAR ORGANISATION OF CHROMAFFIN GRANULES*

H. Winkler and Heide Hörtnagl

Dept. of Pharmacology, University of Innsbruck, Innsbruck, Austria

Twenty years ago (Blaschko and Welch, 1953; Hillarp et al., 1953) the first evidence was obtained that the hormones of the adrenal medulla are stored in subcellular organelles. The subsequent characterisation of these cell particles, the chromaffin granules, depended heavily on the development and use of density gradient centrifugation which made it possible to separate the catecholamine-storing vesicles from mitochondria (Blaschko et al., 1957), from lysosomes (Smith and Winkler, 1966) and from elements of the endoplasmic reticulum (Smith and Winkler, 1968; Winkler, 1969; Kirshner, 1969). Large quantities (10g or more) of highly purified chromaffin granules can be isolated in a few hours by simplified gradient methods (Smith and Winkler, 1967a; Trifaro and Dworkind, 1970; Schneider, 1972).

COMPOSITION OF CHROMAFFIN GRANULES

It is now firmly established that the components listed in Table 1 are genuine constituents of bovine chromaffin granules. This statement is based primarily on biochemical studies in which the density gradient distribution of a particular component was compared with that of known constituents of chromaffin granules. Such studies have been reported for the nucleotides (Falck et al., 1956; Blaschko et al., 1956), for chromogranin A (Sage et al., 1967), for chromomembrin B (Hörtnagl et al., 1973), for dopamine β-hydroxylase (EC 1.14.2.1., Kirshner, 1959; Oka et al., 1967) and for the Mg^{2+}-activated, N-ethyl maleimide-sensitive ATPase (EC 3.6.1.3, Kirshner et al., 1966). These components are in fact specifically localised in the catecholamine-storing vesicles since at least a large portion of each of these constituents is confined to these organelles. This was recently confirmed for chromomembrin B by using immunological methods to localise this protein at the ultrastructural level. Positive immunohistochemical reactions were confined to the membranes of chromaffin granules (Rufener, Nakane, Winkler, Hörtnagl and Schneider, unpublished observation; for an immunohistochemical study at the light microscopic level, see Asamer et al., 1971).

Cytochrome b-559, or b-561 (Flatmark and Terland, 1971) has been reported to be present in both membranes of chromaffin granules and microsomal membranes (Banks, 1965; Ichikawa and Yamono, 1965), however, recent studies (Hörtnagl and Winkler, unpublished observation) indicate that this cytochrome may also be specific for the catecholamine vesicles since it is apparently absent from highly purified elements of the endoplasmic reticulum. Cholesterol and phospholipids are of course constituents of all cellular organelles, whereas the 1-acyl lysolecithin (Winkler and

* The unpublished work quoted in this article was made possible by financial support from the Legerlotz Stiftung and from the Fonds zur Förderung der Wissenschaftlichen Forschung (Austria). This article was written during a stay of H. W. on a Max Kade Fellowship at the Papanicolaou Cancer Research Institute in Miami.

Table 1. Composition of bovine chromaffin granules.
The figures give the amount of each constituent as a percentage of the dry weight of total granules.
The values for the main components (italic) are taken from a paper by Hillarp (1959), those for
the lipids from Blaschko et al. (1967), and those for the nucleotides from Goetz et al. (1972).
Values for the various proteins were calculated from unpublished observations (Winkler and
Hörtnagl) based on immunological data (soluble dopamine β-hydroxylase) and on densitometric
scans of electrophoretic patterns.

Content		Membrane	
Catecholamines	20·5	*Lipids*	22
noradrenaline	5·5	phospholipids	17
adrenaline	15	1-acyl lysolecithin	2
		cholesterol	5
Nucleotides	15	*Insoluble proteins*	8
ATP	10·5	dopamine β-hydroxylase	2
ADP	2	chromomembrin B	1·5
GTP	1·4	cytochrome b-559	?
		ATPase	?
Soluble proteins	27		
Acidic chromogranins	25		
Chromogranin A	11		
dopamine β-hydroxylase	2		

Smith, 1968) found in the adrenal medulla seems mainly confined to chromaffin
granules (Winkler, 1969) where it comprises 16·8% of the total lipid phosphorus
(Blaschko et al., 1967; Winkler et al., 1967; Trifaro, 1969; Da Prada et al.,
1972b). Calcium is a minor component of chromaffin granules (Borowitz et al., 1965;
Philippu and Schümann, 1966). The presence of mucopolysacharides (15 mg/g
protein) has been established recently (Fillion et al., 1971; Da Prada et al., 1972a;
Margolis and Margolis, 1973).

Chromaffin granules of different species have similar properties as far as their
nucleotides (Hillarp and Thieme, 1959), their proteins (Winkler et al., 1966;
Helle, 1966b; Hörtnagl et al., 1971; Hopwood, 1968; Strieder et al., 1968) and
their lipids (Winkler et al., 1967) are concerned. There is also no indication that
chromaffin granules in noradrenaline cells differ in their biochemical composition
from those in adrenaline cells (Hopwood, 1968; Winkler, 1969).

Isolated chromaffin granules can be separated into water soluble components and
an insoluble residue after lysis by hypoosmotic shock. Morphological examination
of this residue shows that it consists of the granule membranes (Winkler et al., 1972a;
Helle, 1971). In Table 1 the various granule components are either assigned to the
content or the membranes of chromaffin granules.

Chromogranins and dopamine β-hydroxylase

The soluble proteins of chromaffin granules can be resolved into several components by starch-or polyacrylamide gel electrophoresis (Smith and Winkler, 1967b;
Hörtnagl et al., 1972, see Fig. 1). These proteins have been collectively called
chromogranins (Blaschko et al., 1967), however, we propose to exclude the slowest
migrating component, the enzyme dopamine β-hydroxylase, since it differs significantly
from the others. The major component of the soluble proteins, chromogranin A, has a
molecular weight of 80,000 (Smith and Winkler, 1967b; Smith and Kirshner, 1967).
Its physicochemical properties indicate that it possesses certain characteristics of a
random coil (Smith and Winkler, 1967b; Kirshner and Kirshner, 1969). The

amino acid composition of chromogranin A is characterised by a preponderance of acidic amino acids (HELLE, 1966a; SMITH and WINKLER, 1967b; SMITH and KIRSHNER, 1967). The other chromogranins apparently possess some of the properties of this main component. All of these proteins are precipitated by the cationic detergent N-cetylpyridinium chloride, which has been attributed (HÖRTNAGL et al., 1972) to the fact that all chromogranins have an acidic amino acid composition (STRIEDER et al., 1968; HELLE and SERCK-HANSSEN, 1969). Furthermore, there is an immunological cross-reaction between chromogranin A and at least several of the other chromo-granins (HÖRTNAGL, WINKLER and LOCHS, unpublished observation). Antibodies used in these immunological studies were produced with chromogranin A which had been eluted from polyacrylamide gels after electrophoresis of the total soluble proteins of chromaffin granules. This simple method gave specific antibodies against the acidic chromogranins which did not cross-react with dopamine β-hydroxylase. It compares, therefore, favourably with previous immunological studies where chromo-granin A was purified by the more tedious procedure of column chromatography (BANKS and HELLE, 1965; SCHNEIDER et al., 1967; SAGE et al., 1967; GEFFEN et al., 1969).

The immunological cross-reaction between the various chromogranins may indicate that these faster moving components are dissociation products of chromo-granin A. However, treatment of the soluble proteins with mercaptoethanol did not change the electrophoretic behaviour (in sodium dodecyl sulfate gels) of the acidic chromogranins, whereas dopamine β-hydroxylase was split into smaller components under these conditions, having a molecular weight of about 80,000, (HÖRTNAGL and WINKLER, unpublished observation; BARTLETT, personal communication). Thus, there is no evidence that chromogranin A consists of subunits linked together by S–S groups. It seems, therefore, that the acidic chromogranins are a closely related family of proteins, but are not members of a dissociating system in which chromogranin A is the largest component.

The slowest moving component (see Fig. 1) of the soluble proteins has been identified as dopamine β-hydroxylase. Enzyme activity in eluates from the gel was confined to this band (DE POTTER et al., 1970; HÖRTNAGL and WINKLER, unpublished observation). Previous studies employing column chromatography (SAGE et al., 1967; GEFFEN et al., 1969) demonstrated that dopamine β-hydroxylase is a minor component which can be separated from the other chromogranins. This enzyme, in contrast to the chromogranins, is not precipitated by N-cetylpyridinium chloride, which has been related to the fact that it does not have an acidic amino acid composition (HÖRTNAGL et al., 1972). Several methods for the isolation of dopamine β-hydroxy-lase have been published (KAUFMAN and FRIEDMAN, 1965; GEFFEN et al., 1969; HARTMAN and UDENFRIEND, 1972; GOLDSTEIN et al., 1972; FOLDES et al., 1972), but for obtaining small amounts for immunological studies the simplest method is to elute the enzyme from a polyacrylamide gel after electrophoresis of the total soluble proteins of chromaffin granules (HÖRTNAGL, WINKLER and LOCHS, unpublished observation).

The membrane proteins of chromaffin granules

The membrane proteins of chromaffin granules differ from the soluble proteins in their amino acid composition (WINKLER, 1969; HELLE and SERCK-HANSSEN, 1969;

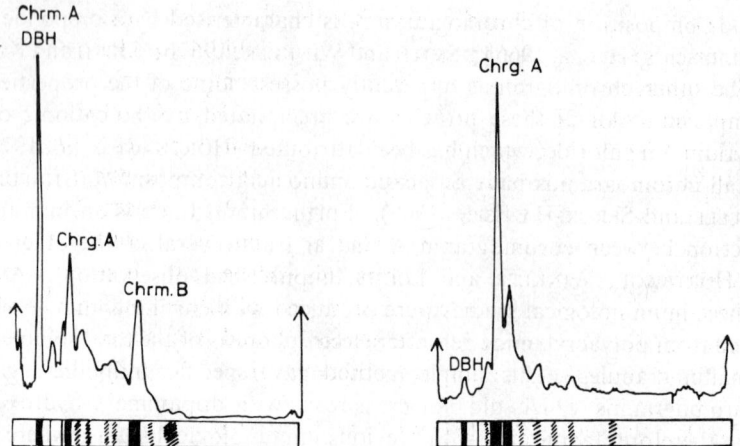

FIG. 1.—Polyacrylamide gel electrophoresis of proteins from chromaffin granules. Membrane proteins (left) and soluble proteins (right) of bovine chromaffin granules (isolated through 1·8 M sucrose, see SCHNEIDER, 1972) were resolved by polyacrylamide gel (5·6%) electrophoresis according to WEBER and OSBORNE (1969) and S. BARTLETT (personal communication). The electrophoretic patterns after staining with Coomassieblue are drawn schematically under the densitometric scannings of the gels. Electrophoretic migration was from left to right. Chrm. A = chromomembrin A; chrg A = chromogranin A; chrm. B = chromomembrin B; DBH = dopamine β-hydroxylase

WINKLER et al., 1970) and their electrophoretic behaviour (WINKLER et al., 1970; see Fig. 1). The component which migrates slowest in polyacrylamide gel electrophoresis (see Fig. 1) has been named chromomembrin A (WINKLER, 1971). Treatment with mercaptoethanol splits chromomembrin A into subunits (A_1: see HÖRTNAGL et al., 1971; A_2 is an artefact, HÖRTNAGL and WINKLER, unpublished observation) which have electrophoretic properties identical to other membrane components. It is, therefore, preferable to run an electrophoresis of the granule membranes without mercaptoethanol present (see Fig. 1).

Chromomembrin A and B have been isolated by Sephadex column chromatography in the presence of sodium dodecyl sulfate (HÖRTNAGL et al., 1971). This detergent inactivates enzymes, however a component having the characteristic spectrum of cytochrome b-559 was eluted from the column together with the lipids (WINKLER et al., 1972a). Since sodium dodecyl sulfate may dissociate the heme from the apoprotein this result does not allow one to identify one of the proteins as the cytochrome. Column chromatography of the membranes solubilised with N-cetylpyridinium chloride made it possible to identify dopamine β-hydroxylase with chromomembrin A (HÖRTNAGL et al., 1972). Thus the major protein of these membranes is an enzyme.

A comparison of dopamine β-hydroxylase purified from the membranes and from the soluble proteins of chromaffin granules indicated that the enzymes from the two different sources were identical (HÖRTNAGL et al., 1972). The same conclusion was reached independently in two other studies (KUZUYA and NAGATSU, 1972; FOLDES et al., 1972).

The ATPase of the granule membranes has recently been solubilised by Lubrol PX, which was found to be the only detergent useful for this purpose (TRIFARO and WARNER, 1972). Preliminary experiments (BAUMGARTNER, HÖRTNAGL and WINKLER, unpublished results) indicate that the solubilised ATPase is eluted from Sephadex columns between chromomembrin A and chromomembrin B.

Based on immunological studies HELLE and SERCK-HANSSEN (1969) claimed that chromogranin A is the major component of the granule membranes, contributing 50% of the total proteins. It is already clear from Fig. 1 that this is highly unlikely. In fact, based on several lines of evidence (WINKLER et al., 1970; HÖRTNAGL et al., 1971; WINKLER et al., 1972a) we concluded that chromogranin A comprises at the most 10% of the membrane proteins. However, even this can be reduced several times if the membranes are washed overnight in a large volume of diluted buffer, indicating that chromogranin A is only adsorbed to the membrane and is not a genuine constituent (HÖRTNAGL and WINKLER, unpublished observation).

THE MOLECULAR ORGANISATION OF CHROMAFFIN GRANULES

At the electron microscopic level the content of chromaffin granules has a fine granular appearance (COUPLAND, 1965) although no regular patterns can be observed and there have been no reports of the presence of crystalline structures similar to those seen in other secretory vesicles. Attempts to crystallise purified chromogranin A *in vitro* have also been unsuccessful (SMITH and WINKLER, unpublished observation). Recent biochemical studies have shown that catecholamines, calcium and ATP can form macromolecular aggregates; however, the question whether the chromogranins can participate in these complexes has not yet been answered (see PLETSCHER, this Symposium).

The membranes of chromaffin granules appear in electron micrographs as triple layered unit membranes (see COUPLAND, 1965). A further characterisation was achieved by the freeze-etching technique (PLATTNER et al., 1969; SMITH et al., 1973). According to present interpretations of results obtained by this method (PINTA DA SILVA and BRANTON, 1972) the basic structure of the granule membrane is a continuous lipid bilayer. No evidence was obtained indicating that the high lysolecithin content of these granules in any way changes this basic arrangement. The inner leaflet of the granule membrane as exposed by the freeze-fracturing technique was covered by particles similar to those which have been observed in membranes from other sources (PINTA DA SILVA and BRANTON, 1972). Such particles are considered to consist of proteins penetrating the two lipid layers. The membranes of chromaffin granules exhibited relatively few such particles, which may be related to the fact that they are relatively poor in protein (WINKLER et al., 1970). What kind of proteins form these particles? Proteins which penetrate the lipid layers might be suitable candidates for transporting molecules through the membranes. There is now good evidence that an ATPase driven pump for catecholamines exists in the membranes (see TAUGNER, 1972). Thus, one might speculate that ATPase or a protein linked with it penetrates the membrane and becomes exposed during freeze-etching. However, the active center of the ATPase containing SH-groups seems to be on the outer face of the granules. SH-groups, as revealed by electron-microscopy after labelling with an electron dense SH-reagent (TAUGNER and HASSELBACH, 1967) are only found on the outside of these vesicles. Chromomembrin B seems also to be present on the outside of these organelles. This was demonstrated with immunological techniques (micro-complement fixation) by comparing the reactivity of intact and lysed granules (WINKLER et al.,

1972b). Chromomembrin B gave a stronger immunological reaction for intact granules suggesting that at least the antigenic groups of this protein are on the outside. Dopamine β-hydroxylase, on the other hand, is localised on the inside of the granule membrane facing the content, a feature indicated by earlier pharmacological studies (KIRSHNER, 1962) and recently confirmed by immunological techniques (HARTMAN and UDENFRIEND, 1972; WINKLER et al., 1972b).

CONCLUSIONS

The composition of chromaffin granules is now well established and we are beginning to understand their molecular organisation. In this article we have considered chromaffin granules as a homogeneous population. However, recent studies (WINKLER et al., 1972c; SLOTKIN and KIRSHNER, 1972) indicate that newly formed chromaffin granules have specific properties. A characterisation of these vesicles should lead to an understanding of how the adrenal medullary cells assemble the various constituents of chromaffin granules until the final molecular organisation of mature granules is achieved.

REFERENCES

ASAMER H., HÖRTNAGL H. and WINKLER H. (1971) Arch. Pharmak. 270, 87–89.
BANKS P. (1965) Biochem. J. 95, 490–496.
BANKS P. and HELLE K. (1965) Biochem. J. 97, 40c–41c.
BLASCHKO H., BORN G. V. R., D'IORIO A. and EADE N. R. (1956) J. Physiol., Lond. 133, 548–557.
BLASCHKO H., COMLINE R. S., SCHNEIDER F. H., SILVER M. and SMITH A. D. (1967) Nature, Lond. 215, 58–59.
BLASCHKO H., FIREMARK H., SMITH A. D. and WINKLER H. (1967) Biochem. J. 104, 545–549.
BLASCHKO H., HAGEN J. M. and HAGEN P. (1957) J. Physiol. Lond. 139, 316–322.
BLASCHKO H. and WELCH A. D. (1953) Arch. exp. Path. Pharmak. 219, 17–22.
BOROWITZ J. L., FUWA K. and WEINER N. (1965) Nature, Lond. 205, 42–43.
COUPLAND R. E. (1965) J. Anat. 99, 231–254.
DA PRADA M., BERLESPSCH K. and PLETSCHER A. (1972a) Arch. Pharmakol. 275, 315–322.
DA PRADA M., PLETSCHER A. and TRANZER J. P. (1972b) Biochem. J. 127, 681–683.
DE POTTER W. P., SMITH A. D. and DE SCHAEPDRYVER A. F. (1970) Tissue and Cell 2, 529–546.
FALCK B., HILLARP N. Å. and HÖGBERG B. (1956) Acta Physiol. Scand. 36, 360–376.
FILLION G., NOSAL R. and UVNÄS B. (1971) Acta Physiol. Scand. 83, 286–288.
FLATMARK T. and TERLAND O. (1971) Biochem. Biophys. Acta. 253, 487–491.
FOLDES A., JEFFREY P. L., PRESTON B. N. and AUSTIN L. (1972) Biochem. J. 126, 1209–1217.
GEFFEN L. B., LIVETT B. G. and RUSH R. A. (1969) J. Physiol. (Lond.) 204, 593–605.
GOETZ U., DA PRADA M. and PLETSCHER A. (1971) J. Pharm. Exp. Ther. 178, 210–215.
GOLDSTEIN M., FUXE K. and HÖKFELT T. (1972) Pharm. Rev. 24, 293–309.
HARTMAN B. K. and UDENFRIEND S. (1972) Pharmac. Rev. 24, 311–330.
HELLE K. (1966a) Molec. Pharmacol. 2, 298–310.
HELLE K. B. (1966b) Biochem. J. 100, 6c.
HELLE K. B. (1971) Biochim. Biophys. Acta. 245, 80–93.
HELLE K. B. and SERCK-HANSSEN G. (1969) Pharmac. Res. Comm. 1, 25–29.
HILLARP N. Å. (1959) Acta Physiol. Scand. 47, 271–279.
HILLARP N. Å., LAGERSTEDT S. and NILSON B. (1953) Acta Physiol. Scand. 29, 251–263.
HILLARP N. Å. and THIEME G. (1959) Acta Physiol. Scand. 45, 328–338.
HOPWOOD D. (1968) Histochemie 13, 323–330.
HÖRTNAGL H., WINKLER H. and LOCHS H. (1972) Biochem. J. 129, 187–195.
HÖRTNAGL H., WINKLER H. and LOCHS H. (1973) J. Neurochem. 20, 977–985.
HÖRTNAGL H., WINKLER H., SCHÖPF J. A. L. and HOHENWALLNER W. (1971) Biochem. J. 122, 299–304.
ICHIKAWA Y. and YAMANO T. (1965) Biochem. Biophys. Res. Commun. 20, 263–268.
KAUFMAN S. and FRIEDMAN S. (1965) Pharmacol. Rev. 17, 71–100.
KIRSHNER A. G. and KIRSHNER N. (1969) Biochim. Biophys. Acta. 181, 219–225.
KIRSHNER N. (1959) Pharmac. Rev. 11, 350–357.
KIRSHNER N. (1969) In Advances in Biochemical Psychopharmacology (COSTA E., GREENGARD P., Eds.) Vol. 1, pp. 71–89. Raven Press, New York.
KIRSHNER N., KIRSHNER A. G. and KAMIN D. L. (1966) Biochim. Biophys. Acta. 113, 332–335.

KUZUYA H. and NAGATSU T. (1972) *Biochem. Pharmacol.* **21,** 740–742.

MARGOLIS R. U. and MARGOLIS K. (1973) *Molec. Pharm.* In press.

OKA M., KAJIKAWA K., OHUCHI T., YOSHIDA H. and IMAIZUMI R. (1967) *Life Sci.* **6,** 461–465.

PHILIPPU A. and SCHÜMANN H. J. (1966) *Arch. Exp. Path. Pharmak.* **252,** 339–358.

PINTA DA SILVA P. and BRANTON D. (1972) *Chem. Phys. Lipids.* **8,** 265–278.

PLATTNER H., WINKLER H., HÖRTNAGL H. and PFALLER W. (1969) *J. Ultrastruct. Res.* **28,** 191–202.

SAGE H. J., SMITH W. J. and KIRSHNER N. (1967) *Molec. Pharmac.* **3,** 81–84.

SCHNEIDER F. H. (1972) *Biochem. Pharmacol.* **21,** 2627–2634.

SCHNEIDER F. H., SMITH A. D. and WINKLER H. (1967) *Br. J. Pharmac. Chemother.* **31,** 94–104.

SLOTKIN T. A. and KIRSHNER N. (1972) *Molec. Pharmac.* **9,** 105–116.

SMITH A. D. and WINKLER H. (1966) *J. Physiol., Lond.* **183,** 179–188.

SMITH A. D. and WINKLER H. (1967a) *Biochem. J.* **103,** 480–482.

SMITH A. D. and WINKLER H. (1967b) *Biochem. J.* **103,** 483–492.

SMITH A. D. and WINKLER H. (1968) *Biochem. J.* **108,** 867–874.

SMITH U., SMITH D. S., WINKLER H. and RYAN J. W. (1973) *Science* **179,** 79–82.

SMITH W. J. and KIRSHNER N. (1967) *Molec. Pharmac.* **3,** 52–62.

STRIEDER N., ZIEGLER E., WINKLER H. and SMITH A. D. (1968) *Biochem. Pharmac.* **17,** 1553–1556.

TAUGNER G. (1972) *Arch. Phamak.* **247,** 299–314.

TAUGNER G. and HASSELBACH W. (1967) *Biochem. J.* **102,** 22P.

TRIFARO J. M. (1969) *Molec. Pharmacol.* **5,** 382–393.

TRIFARO J. M. and DWORKIND J. (1970) *Anal. Biochem.* **34,** 403–412.

TRIFARO J. M. and WARNER M. (1972) *Molec. Pharmac.* **8,** 159–169.

WEBER K. and OSBORN M. (1969) *J. Biol. Chem.* **241,** 4406–4412.

WINKLER H. (1969) *Arch. Exp. Pharmakol. Path.* **263,** 340–357.

WINKLER H. (1971) *Phil. Trans. Roy. Soc. B.* **261,** 293–303.

WINKLER H., HÖRTNAGL H., ASAMER H. and PLATTNER H. (1972a) In *Functional and Structural Proteins of the Nervous System* (DAVISON A. N., MANDEL P. and MORGAN IG., EDS.) pp. 69–81. Plenum Press, New York.

WINKLER H., HÖRTNAGL H., HÖRTNAGL H. and SMITH A. D. (1970) *Biochem. J.* **118,** 340–354.

WINKLER H., HÖRTNAGL H. and LOCHS H. (1972b) *J. Cell Biol.* **55,** 286a.

WINKLER H., SCHÖPF J. A. L., HÖRTNAGL H. and HÖRTNAGL H. (1972c) *Arch. Pharmakol.* **273,** 43–61.

WINKLER H. and SMITH A. D. (1968) *Arch. Exp. Pharmak. Path.* **261,** 379–388.

WINKLER H., STRIEDER N. and ZIEGLER E. (1967) *Arch. Exp. Path. Pharmak.* **256,** 407–415.

WINKLER H., ZIEGLER E. and STRIEDER N. (1966) *Nature, Lond.* **211,** 972–983.

Frontiers in Catecholamine Research 1973, pp. 423 to 425. Pergamon Press. Printed in Great Britain.

A LARGE SECOND POOL OF NOREPINEPHRINE IN THE HIGHLY PURIFIED VESICLE FRACTION FROM BOVINE SPLENIC NERVE

RICHARD L. KLEIN

Department of Pharmacology and Toxicology, University of Mississippi Medical Center,
Jackson, MS. 39216, U.S.A.

LARGE dense-core vesicles (LDV) can now be prepared from bovine splenic nerve at a routine purity of 80–90%. This can be verified by improvements in sedimentable norepinephrine (NE): protein ratios (YEN et al., 1973) and by systematic whole pellet examination at the ultrastructural level (THURESON-KLEIN et al., 1973a, b). The purity compares favorably to that of bovine adrenomedullary vesicles and justifies acceptance of analytical data on the two vesicle types with equal confidence.

An LDV content of 100 nmoles NE/mg protein can be achieved, which depends not only on the purity attained but also on a minimal ∼10 min *post mortem* delay at the slaughterhouse. This compares to 20–30 min delay experienced previously in several laboratories.

The shorter delay reveals a second pool of NE not apparent earlier (YEN et al., 1973). It accounts for 20% of the NE content and it is rapidly lost even in the presence of ATP. This pool is in addition to an ATP-sensitive pool with a much slower NE release rate. ATP increases NE exchange in both pools. However, a specific ATP-enhanced reuptake of NE helps to maintain the slow pool, as first reported by EULER and LISHAJKO (1963, 1967), whereas ATP only hastens net loss of NE from the fast pool.

Highly purified vesicles in fraction $FIII_{M15}$ obtained from homogenates of the whole nerve trunk were used in the recent studies (YEN et al., 1973). It is of interest to learn whether the characteristics of the fraction change during the vesicle maturation and transport process. This is studied by sectioning splenic nerve trunks into proximal, mid-portion and distal segments (Fig. 1, at top), from which purified vesicles are identically prepared. In 12 experiments: the average relative wet wt is ∼2:3:4; the content is 11·9, 14·0 and 19·6 μg NE/g fresh nerve; and the relative number of vesicles in $FIII_{M15}$ is estimated to be 1:2:3·5, respectively, for the proximal, mid-portion and distal segments.

Each of the identically purified vesicle fractions is incubated to determine the time course for net NE loss (Fig. 1). Vesicles from the proximal segment show a small net uptake at 5 min followed by a single exponential loss of NE with half-life of 28 min. Similar initial net uptake in this medium is always indicative of an unfilled potential for NE accumulation, for example: all vesicles obtained after a 20 min *post mortem* delay; vesicles partially depleted in the absence of ATP followed by incubation with NE and ATP; and vesicles supplemented with a soluble "NE pool size enhancing factor", which is present in the high speed supernatant of the nerve homogenate (THURESON-KLEIN et al., 1973c).

Vesicles isolated from the mid-portion clearly show a fast component of NE loss

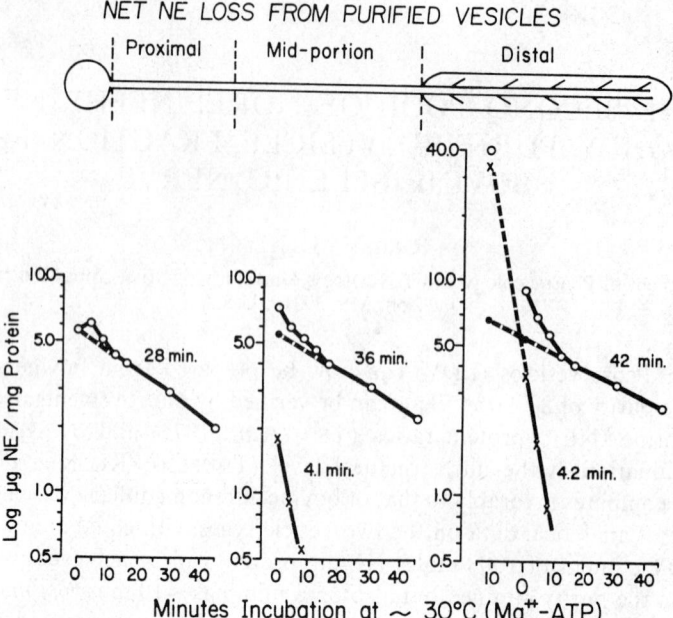

FIG. 1.—The data are averages from 12 experiments including all points, in which highly purified vesicles of fraction FIII$_{M15}$ are compared from proximal and distal nerve trunk segments; 8 experiments include mid-portion vesicles. The incubation medium contains: 5 mM Mg^{2+}-ATP, 1 μg l-NE/ml, 0·1 M each of sucrose and potassium phosphate buffer at pH 7·4 and pargyline at 10 μg/ml.

Curves with '×'s are resolved fast components; dashed-line extensions of curves with circles indicate slow components; the minute term adjacent to each curve is its half-life.

In the distal vesicle fraction, the curves for slow and fast components are extrapolated back 12 min (dashed-lines), which was the mean *post mortem* delay at the slaughterhouse as the tissues cooled from 38 to ~20°C, prior to placing the partially dissected nerves on ice. The delay is arbitrarily taken to average 30°C, which is also the incubation temperature and, thus, justification for use of the same abscissa. Vesicles are considered to be in a steady state during the intervening preparation procedure at 0–6°C. Estimated *in vivo* NE pool sizes are indicated on the ordinate for the slow (dot) and fast (×) components and their sum (circle).

with half-life of 4 min. This pool comprises 25% of the amine content. The half-life of the slow component is increased to 36 min.

Vesicles isolated from the distal (intrasplenic) segment of the nerve trunk possess an even larger fast component of NE loss with an unchanged half-life of 4 min. This pool has increased to 40% of the total NE content. The half-life of the slow component is further increased to 42 min.

It is noteworthy that the NE content of the slower, ATP-sensitive pool remains essentially unchanged, while the sedimentable NE content of the vesicle fraction significantly increases from 5·7 to 7·3 to 9·0 μg NE/mg protein. This increase is due exclusively to the establishment and filling of the fast pool, which contains 0–0·5, 1·8 and 3·6 μg NE/mg protein in the vesicle fraction from the successive segments. The parallel increases in half-life of the slow component can be explained by the two pools being in equilibrium. Thus, as the size of the fast pool increases closer to equilibrium, the rate of filling from the slow pool progressively decreases.

Very recently, it is found that the dopamine β-hydroxylase (DβH)/mg protein content of the FIII$_{M15}$ vesicle fraction from the three segments is unchanged (LAGERCRANTZ et al., unpublished). This is also compatible with a stable pool size for the slower, ATP-sensitive component. The vesicle DβH content is established early in the cell body (CHUBB et al., 1972) and the slow pool appears to be filled to near capacity shortly thereafter.

A number of exciting hypotheses can be made pertaining to the distal vesicle fraction. If one is permitted to extrapolate the fast and slow exponential curves back to zero post-mortem delay (Fig. 1, see legend), the content of the slow pool becomes 6·6 and of the fast pool $\sim 34\cdot0$ μg NE/mg protein. Thus, the fast pool is estimated to comprise 80–90% of the total NE content of the distal vesicles in vivo. We suggested earlier (THURESON-KLEIN et al., 1973b) that the role of ATP to maintain vesicle NE, as evidenced by enhanced reuptake, may actually reflect the maintenance and uptake of a soluble component of the binding complex, rather than of NE per se. This could be lipoprotein or DβH macromolecules (THURESON-KLEIN et al., 1973c). Based on present data, we can further hypothesize that the ATP-sensitive NE complex in the slow release pool functions primarily to synthesize NE in order to maintain the much larger fast pool, rather than to complex all the vesicle NE.

The LDV fraction in the distal segment can be calculated to contain 200–250 nmoles NE/mg protein and the absolute content to be $1-2 \times 10^{-6}$ pg NE/vesicle in vivo. The latter estimate approaches $2-4 \times 10^{-6}$ pg that can be calculated from the data of DAHLSTRÖM et al., (1966) for the smaller terminal vesicle of iris and vas deferens. Thus, the generally held concept that the smaller terminal vesicle contains 10–100 times more NE than the LDV can be seriously challenged.

The most simple and supported interpretation of the present data is that both NE pools are associated with LDV, this being the principal vesicle type present in fraction FIII$_{M15}$, after separation on the sucrose-D$_2$O density gradient. The vesicles occur in a single band (FIII$_M$) at a density of $\sim 1\cdot2$ g/cm^3. However, it is possible that a very minor unidentified population of smaller vesicles, with similar density and much higher NE content than the LDV, could be present to account for the fast pool. The least likely alternative, in our opinion, is that the fast pool represents a relatively labile extravesicular NE complex.

Acknowledgement—This research was supported by a grant from the National Institute of General Medical Sciences, U.S.P.H.S., 2-RO1 GM-15490-06.

REFERENCES

CHUBB J. W., DEPOTTER W. P. and DESCHAEPDRYVER A. F. (1972) Life Sci. 11, 323–333.
DAHLSTRÖM A., HÄGGENDAHL J. and HÖKFELT T. (1966) Acta physiol. scand. 67, 289–294.
EULER U. S. VON and LISHAJKO F. (1963) Acta physiol. scand. 59, 454–461.
EULER U. S. VON and LISHAJKO F. (1967) Acta physiol. scand. 71, 151–162.
KLEIN R. L., YEN S. -S., CHEN-YEN S. -H. and THURESON-KLEIN Å. (1973) Proc. EMSA, 532–533.
THURESON-KLEIN Å., KLEIN R. L. and YEN S. -S. (1973a) J. Ultrastruct. Res. 43, 18–35
THURESON-KLEIN Å., KLEIN R. L. and LAGERCRANTZ H. (1973b) J. Neurocytol. 2, in press.
THURESON-KLEIN Å., KLEIN R. L., YEN S. -S. and LAGERCRANTZ H. (1973c) Proc. EMSA, 530–531.
YEN S. -S., KLEIN R. L. and CHEN-YEN S. -H. (1973) J. Neurocytol. 2, in press.

Frontiers in Catecholamine Research 1973, pp. 427 to 429. Pergamon Press. Printed in Great Britain.

EFFECT OF ELECTROLYTES ON AMINE TRANSPORT AND RELEASE MECHANISMS IN NERVE ENDINGS

DONALD F. BOGDANSKI

Laboratory of Chemical Pharmacology, National Heart and Lung Institute, National Institutes of Health, Bethesda, Maryland, 20014, U.S.A.

THE presence or absence of extracellular electrolytes has a profound effect on the retention of bound amine. The rate of depletion of bound NE from rat heart slices or synaptosomes is greatly stimulated by a medium deficient only in Na^+. In both preparations, a K^+-free medium containing Na^+ has a much smaller effect (BOGDANSKI, 1965; BOGDANSKI and BRODIE, 1969; BOGDANSKI et al., 1970). The profound effect of a Na^+-deficient medium might be of importance in terms of the reversed extracellular/intracellular Na^+-gradient which might energise outward transport of intraneuronal amine. To test this hypothesis, the electrolyte composition of the cytosol of nerve endings was modified by prolonged immersion of the tissue (BOYLE and CONWAY, 1941), in a modified Krebs–bicarbonate solution containing 1% of the usual $[Ca^{2+}]$ and $[Mg^{2+}]$ and 10% of the dextrose, at $0°$. This treatment has been called autodialysis (BLASZKOWSKI and BOGDANSKI, 1972). Usually the predominant electrolyte was Na^+, but Na^+ was sometimes replaced by $choline^+$, Li^+ or K^+.

The rate of efflux of NE from slices incubated in a Na^+-free ($choline^+$) medium was most rapid if the slices were previously autodialysed with a high $[Na^+]$. The rate of efflux was less rapid if the slices were previously autodialysed with $choline^+$, Li^+ or K^+, which cannot substitute for Na^+ as substrates for cotransport. Desipramine (3×10^{-6} M), a competitive inhibitor of the uptake of NE (MAXWELL et al., 1969), inhibited the efflux of amine from slices autodialysed with Na^+, but not the efflux of amine from the slices previously autodialysed with Li^+ or $choline^+$. The evidence thus suggests that the membrane transport mechanism can mediate the translocation of amine in an outward direction.

How does outward transport relate to Na^+, Ca^{2+} and synaptic transmission? Although Ca^{2+} is required for the release of stored NE, it has been thought that the antagonism between Na^+ and Ca^{2+} occurs at a transport mechanism for Ca^{2+} at the outer membrane surface of cholinergic nerve endings (GAGE and QUASTEL, 1966). It has also been suggested that the action of Ca^{2+} is terminated by intracellular binding (DIAMOND et al., 1971) or by being released from the cell by a Na^+–Ca^{2+} exchange mechanism (BLAUSTEIN and WIESMANN, 1970). BIRKS and COHEN (1968) postulated that intracellular Na^+ releases Ca^{2+} from a carrier, permitting free Ca^{2+} to trigger the release of acetylcholine. A relationship between Ca^{2+}, Na^{2+} and release of hormone from a variety of organs has been reviewed by RUBIN, 1970.

The results of our experiments employing the autodialysis technique indicate that intracellular Na^+ inhibits the release of ^3H-NE that occurs after Ca^{2+} enters the cell. Thus, the Ca^{2+} in Krebs medium releases NE from slices autodialysed with the high $choline^+$ solution. By contrast, the Ca^{2+} in Krebs medium helps to retain NE (relative to the Ca^{2+}-free Krebs medium) in slices autodialysed with the high

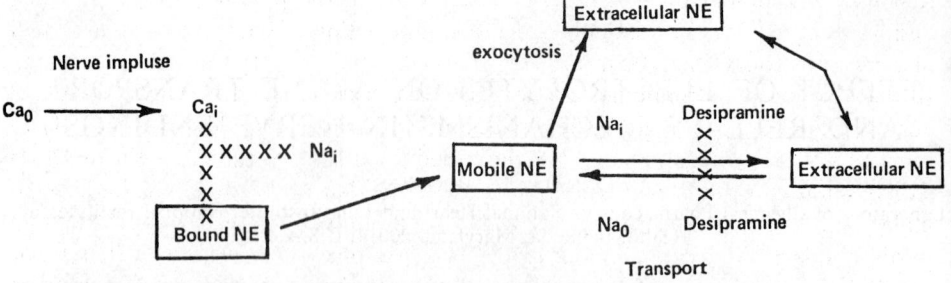

FIG. 1.—Role of membrane transport in release and re-uptake of NE. In response to the nerve impulse, Ca^{2+} enters the cell and mobilises amine in vesicles attached to the cell membrane. Intracellular Na prevents a more generalised mobilisation and helps to remove Ca^{2+} from its site of action. Outward transport (probably equilibration) of mobilised amine can occur as a result of the concentration gradient for NE, and, perhaps a local accumulation of intracellular Na^+ during the nerve impulse. Transport might occur directly from vesicles attached to the cell membrane. Between impulses, inward (active) transport would be favoured by the reestablishment of an inward Na^+ gradient and dissipation of the outward NE gradient. Transport in either direction is inhibited by desipramine. The diagram shows the predominant direction of reactions during synaptic transmission, but transport in both directions is illustrated. $\times \times \times \times \times$ = inhibitory reactions. Subscripts o and i refer to outside and inside the plasma membrane, respectively.

Na^+-solution. The antagonism between Ca^{2+} and Na^+ must, therefore, occur intracellularly, or at the inner surface of the plasma membrane, since Ca^{2+} enhanced the release of NE from those slices that were deficient in intracellular Na^+. Moreover, the Na^+-free (choline$^+$) medium enhanced Ca^{2+}-dependent release of NE from Na^+-deficient slices. These findings suggest that Na^+ antagonised the effects of Ca^{2+} in two ways: (1) upon entering the nerve ending it inhibits the effect of Ca^{2+} (BLASZKOWSKI and BOGDANSKI, 1972) and (2) extracellular Na^+ may interfere with the movement of Ca^{2+} through the plasma membrane into the nerve ending. Moreover, the above experiments suggest that exocytosis is not essential for the egress of amine from the nerve ending, a fact which may be of some interest for theories of synaptic transmission.

There are parallels between the physiological release of transmitters from nerve endings and the efflux of NE from heart slices incubated in Na^+-free (choline$^+$ or K^+) media. Desipramine can inhibit the physiological release of NE (THEONEN et al., 1964 BOULIN, 1966) as well as the release from heart slices. Release requires extracellular Ca^{2+}, which apparently must enter the cell in order to act (GAGE and QUASTEL, 1966). Calcium appears to act on a saturable process and Na^+ inhibits the Ca^{2+}-dependent release of NE by a noncompetitive action (BLASZKOWSKI and BOGDANSKI, 1971). Apparently, the action of Ca^{2+} is not limited to binding vesicles to the plasma membrane. Bretyllium-like drugs first potentiate, then inhibit the release of NE from heart slices, although the latter action is only moderate (unpublished).

HAGGENDAL and MALMFORS (1969) have indicated that there may be temporary inhibition of the amine transport mechanism during the nerve impulse, but this seems unlikely because the accumulation of ^{3}H-NE in the presence of 149 mM Na^+ was inhibited only 26% by K^+ at 100 mM (BOGDANSKI and BRODIE, 1969), which should depolarise adrenergic nerve endings. Not only will depolarised nerve endings transport exogenous NE, but the transport mechanism helps to retain intracellular NE that has been mobilised by partial depolarisation in the presence of Ca_2^+. Thus, desipramine at $3\cdot3 \times 10^{-6}$ M doubles the rate of depletion of NE from slices incubated in a medium containing 100 mM Na^+, 75 mM K^+ and $2\cdot5$ mM Ca^{2+} (unpublished). In the absence of desipramine, amine must have been retained by active re-uptake against a concentration gradient. In accord with this view, desipramine fails to increase the rate of depletion of NE mobilised from slices incubated with the Na^+-free (choline$^+$) medium in which re-uptake by the Na^+-dependent system is negligible. Since desipramine potentiates rather than inhibits the depletion of NE from partially depolarised nerve ending in the presence of extracellular Na^+, it must be concluded either that the transport mechanism at the inner and outer membrane surfaces is not symmetrical with respect to desipramine or that NE can utilise alternate pathways of egress.

Thus, available data do not allow the conclusion that outward transport of NE is a major aspect of synaptic transmission. However, transport of amine in both directions apparently can occur in depolarised nerve endings. The model incorporates some of the ideas discussed in the text.

REFERENCES

BIRKS R. I. and COHEN M. W. (1968) *Proc. Rl. Soc. B*, **170**, 401–421.
BLASZKOWSKI T. P. and BOGDANSKI D. F. (1971) *Biochem. Pharmacol.* **20**, 3281–3294.
BLASZKOWSKI T. P. and BOGDANSKI D. F. (1972) *Life Sci.* Part 1, 867–876.
BLAUSTEIN M. P. and WIESMANN W. P. (1970) *Proc. Nat. Acad. Sci.*, **66**, 664–671.
BOGDANSKI D. F. (1965) *Pharmacologist*, **7**, 168.
BOGDANSKI D. F. and BRODIE B. B. (1966) *Life Sci.* **5**, 1563–1569.
BOGDANSKI D. F. and BRODIE B. B. (1969) *J. Pharmacol. Exp. Ther.* **165**, 181–189.
BOGDANSKI D. F., BLASZOWSKI T. P. and TISSARI A. H. (1970) *Biochem. Biophys. Acta.* **211**, 521–532.
BOULLIN D. J. quote by COSTA *et al.*, (1968) *Pharm. Rev.* **18**, 591–
BOYLE P. J. and CONWAY E. J. (1941) *J. Physiol.* **100**, 1–63.
DIAMOND I. and GOLDBERG A. L. (1971) *J. Neurochem.* **18**, 1419–1431.
GAGE P. W. and QUASTEL D. M. J. (1966) *J. Physiol. Lond.* **185**, 95–123.
HAGGENDAL J. and MALMFORS T. (1969) *Acta Physiol. Scand.* **75**, 28.
MAXWELL R. A., KEENAN P. D., CHAPLIN E., ROTH B. and ECKHARDT S. B. (1969) *J. Pharmacol. Exp. Ther.* **166**, 320–329.
RUBIN R. P. (1970) *Pharm. Rev.* **22**, 384–428.
THEONEN H., HURLIMANN A. and HAEFFELY (1964) *J. Pharmacol. Exp. Ther.* **144**, 405.

Frontiers in Catecholamine Research 1973, pp. 431 to 437. Pergamon Press. Printed in Great Britain.

SMALL INTENSELY FLUORESCENT (SIF) CELLS *IN VIVO* AND *IN VITRO**

OLAVI ERÄNKÖ and LIISA ERÄNKÖ

Department of Anatomy, University of Helsinki, Siltavuorenpenger,
Helsinki 17, Finland 00170

CHROMAFFIN CELLS AND SYMPATHETIC GANGLIA

KOHN (1899) pointed out that chromaffin cells can be found in sympathetic ganglia and nerves; he considered chromaffin cells as the second specific cell type of the sympathetic system. The common origin of sympathetic neurons and chromaffin cells is now generally accepted (WATZKA, 1943; LEMPINEN, 1964 and COUPLAND, 1965). In certain instances chromaffin cells have been shown to participate in the function of the sympathetic ganglion. Thus, MUSCHOLL and VOGT (1964) correlated the high catecholamine content in the venous effluent of the hypogastric ganglion of the dog with the presence of chromaffin cells in it. Neurophysiological observations led ECCLES and LIBET (1961) to postulate participation of preganglionically innervated chromaffin cells in the production of a positive potential change by sympathetic nerve cells.

SMALL INTENSELY FLUORESCENT CELLS

Fluorescence induced by formaldehyde was shown to demonstrate noradrenaline-containing chromaffin cells in the adrenal medulla (ERÄNKÖ, 1955). With an improved modification of this method, catecholamines were demonstrated not only in the cytoplasm of sympathetic nerve cells but also in small cells in sympathetic ganglia which exhibited an extremely bright fluorescence (ERÄNKÖ and HÄRKÖNEN, 1963). These cells are now commonly called 'small intensely fluorescent' cells or SIF cells. Against expectations, these cells were observed to be non-chromaffin (ERÄNKÖ and HÄRKÖNEN, 1965). The presence of SIF cells in various sympathetic ganglia of several species has been firmly established by fluorescence microscopy (NORBERG and HAMBERGER, 1964; ERÄNKÖ and HÄRKÖNEN, 1965; OWMAN and SJÖSTRAND, 1965; NORBERG et al., 1966; NORBERG and SJÖQVIST, 1966; CSILLIK et al., 1967; JACOBOWITZ, 1967; OLSON, 1967; VAN ORDEN et al., 1970; ERÄNKÖ and ERÄNKÖ, 1971b; KANERVA, 1971). Clusters of SIF cells are especially common near blood vessels, a position suggesting either endocrine function (ERÄNKÖ and HÄRKÖNEN, 1965) or regulation of the SIF cells by hormonal factors via the blood stream (ERÄNKÖ and ERÄNKÖ, 1971b). On the other hand, SIF cells and their processes have been found on sympathetic nerve cell bodies and fibres (ERÄNKÖ and ERÄNKÖ, 1971b) in a manner which suggests that the SIF cells may influence the nervous transmission by catecholamine liberation, as was proposed by ECCLES and LIBET (1961).

---

* Work supported by a grant from the Sigrid Jusélius Foundation to O. E. and by a grant from the Emil Aaltonen Foundation to L. E.

431

CATECHOLAMINES IN THE SIF CELLS

While ERÄNKÖ and HÄRKÖNEN (1965) tentatively suggested that the SIF cells contain 5-hydroxytryptamine, microspectrophotometric studies by NORBERG *et al.* (1966) showed that catecholamines are responsible for the intense fluorescence. BJÖRKLUND *et al.* (1970) using hydrochloric acid for spectrofluorometric differentiation between dopamine and noradrenaline (BJÖRKLUND *et al.*, 1968) concluded that SIF cells of the pig, the cat and the rat contain dopamine but ERÄNKÖ and ERÄNKÖ (1971b) found only noradrenaline in the SIF cells of the rat with an essentially similar method (ERÄNKÖ and ERÄNKÖ, 1971c). The possibility cannot be excluded that some SIF cells may contain dopamine, while others contain noradrenaline or indeed another amine, and further studies are obviously required.

FINE STRUCTURE OF THE SIF CELLS

ERÄNKÖ and HÄRKÖNEN (1965) reported that cells of the same size and shape as the SIF cells, though non-chromaffin, electron microscopically closely resembled chromaffin cells, containing numerous intensely osmiophilic granules. The presence of granular vesicles in cells apparently corresponding to SIF cells was soon confirmed by several workers in careful electron microscopical studies, in which the size of the granular vesicles was also reported. The diameter of the vesicles in the superior cervical ganglion of the rat was reported to be approximately 100 nm (HÖKFELT, 1969; MATTHEWS and RAISMAN, 1969; TAXI, 1969; WILLIAMS and PALAY, 1969; L. ERÄNKÖ, 1972) that is larger than that of most granular vesicles in the sympathetic nerve cells which is about 50 nm (HÖKFELT, 1969; MATTHEWS and RAISMAN, 1969; O. ERÄNKÖ, 1972), but smaller than those of the adrenal medulla (whose diameter is commonly 150 nm or larger) (COUPLAND, 1965; D'ANZI, 1969). However, granular vesicles in many peripheral ganglia are larger, the following diameters having been reported: 150–250 nm in the inferior mesenteric ganglion of the rat (VAN ORDEN *et al.*, 1970), 80–300 nm in the paracervical ganglion of the rat (KANERVA and TERÄVÄINEN, 1972), 90–170 nm in the inferior mesenteric ganglion of the rabbit (ELFVIN, 1968) and 100–250 nm in the hypogastric ganglion of the guinea pig (WATANABE, 1971). Thus, these larger vesicles are of the same size as the granular vesicles in the adrenal medulla.

Most likely the size of granular vesicles is a crucial factor for the outcome of the chromaffin reaction, which is positive only in cells with sufficiently large granular vesicles, since the aqueous solvent displaces the amines before precipitation reaction in cells with smaller granular vesicles. Accordingly, it is understandable why small cells in some ganglia show a positive chromaffin reaction (KOHN, 1903; IWANOW, 1932; MUSCHOLL and VOGT, 1964), while other SIF cells are non-chromaffin, in spite of their essentially similar fine morphology. SIEGRIST *et al.* (1968) and JACO-BOWITZ (1970) feel that all cells with granular vesicles resembling the adrenal medul-lary cells should be called 'chromaffin cells'. WILLIAMS and PALAY (1969), MAT-THEWS and RAISMAN (1969) as well as ERÄNKÖ and ERÄNKÖ (1971b) feel that non-chromaffin cells should not be called chromaffin and use the names 'SIF cells' or 'small, granule-containing cells'.

SYNAPTIC CONNEXIONS OF SIF CELLS

There is a general agreement about the presence of afferent synapses on the SIF cells (SIEGRIST *et al.*, 1968; MATTHEWS and RAISMAN, 1969; WILLIAMS and PALAY,

1969). Since these synaptic boutons contain 'empty' vesicles, they are likely to be cholinergic, presumably originating from the preganglionic fibres.

The apposition of slender SIF cell processes on nerve cell bodies observed by fluorescence microscopy (ERÄNKÖ and ERÄNKÖ, 1971b) raises the question whether sympathetic nerve cells receive efferent synapses from the SIF cells. While synapses from SIF cell processes to nerve cell perikaryon have not been reported, evidence is available for the presence of efferent synapses from the SIF cells to nerve fibres (SIEGRIST *et al.*, 1968; WILLIAMS and PALAY, 1969; MATTHEWS and RAISMAN, 1969). Thus, SIF cells may serve as interneurons in the ganglion. This function does not necessarily exclude secretory activity of these cells by which means they can reach more nerve cells (TAXI *et al.*, 1969). There is some uncertainty of the nature of the postsynaptic structures, and ELFVIN (1971) has emphasised the need of tracing them to their origin by serial sections. Analysing in an extensive series of ultra-thin sections the relations of a small granule-containing cell in the inferior mesenteric ganglion of the cat he failed to find any close contact between this cell and processes from the surrounding nerve cells.

CHEMICAL SYMPATHECTOMY AND SIF CELLS

Sympathetic terminals (THOENEN and TRANZER, 1968) and, in newborn rats, almost all sympathetic nerve cells (ANGELETTI and LEVI-MONTALCINI, 1970) can be destroyed by injections of 6-hydroxydopamine. However, in the small residual ganglion the total number of SIF cells is the same as that in the ten times as large control ganglion (ERÄNKÖ and ERÄNKÖ, 1972b).

Guanethidine also damages sympathetic neurons (JENSEN-HOLM and JUUL, 1970), causing in newborn rats an equally complete chemical sympathectomy as 6-hydroxy-dopamine (ERÄNKÖ and ERÄNKÖ, 1971a). However, the effect of guanethidine on the SIF cells is different: there are 3–5 times as many SIF cells as those in the corresponding ganglion of an untreated rat (ERÄNKÖ and ERÄNKÖ, 1971a). The presence of unusually large clusters with finger-like protrusions of SIF cell chains suggests proliferation of the SIF cells by division under the influence of guanethidine.

When administered to newborn rats, the adrenergic blocking agent bretylium tosylate has been shown to cause irreversible damage to sympathetic neurons much in the same way as 6-hydroxydopamine and guanethidine (CARAMIA *et al.*, 1972). It would be interesting to know how bretylium tosylate affects the SIF cells, which were not examined by CARAMIA *et al.* (1972).

EFFECT OF GLUCOCORTICOIDS ON SIF CELLS

LEMPINEN (1964) observed in our laboratory that administration of cortisone or hydrocortisone to newborn rats not only prevents the normal postnatal degeneration of extra-adrenal chromaffin tissue but also causes an increase in the size of the Organ of Zuckerkandl and in the intensity of its chromaffin reaction; moreover, chromaffin cells appeared in sympathetic chain ganglia, which normally contained no or few chromaffin cells. ERÄNKÖ *et al.* (1966) demonstrated that the increase in the intensity of the chromaffin reaction of the Organ of Zuckerkandl due to hydrocortisone was associated with the appearance of adrenaline in this Organ, which normally contained only noradrenaline. At the same time, it was shown by AXELROD (1966) that the level of phenylethanolamine N-methyltransferase (EC 2.1.1) in the adrenal medulla is regulated by glucocorticoids, providing an explanation for the appearance of adrenaline in the Organ of Zuckerkandl.

Recently, ERÄNKÖ and ERÄNKÖ (1972a) re-examined the effect of hydrocortsone on newborn rats. They confirmed the increase in the size and fluorescence intensity of the Organ of Zuckerkandl and observed an about ten-fold increase in the number of SIF cells in the superior cervical, thóracic, lumbar and coeliac ganglia. While SIF cells and their electron microscopic equivalents, small granule-containing cells, are readily found in sympathetic ganglia of newborn rats (L. ERÄNKÖ, 1972), it appeared that they were not the only source of new SIF cells observed after hydrocortisone administration, because individual SIF cells and small clusters were scattered everywhere in the ganglia of hydrocortisone-treated rats, including areas not previously showing SIF cells. It was therefore concluded that the new SIF cells were probably formed by differentiation from non-fluorescent cells under the influence of glucocorticoids (ERÄNKÖ and ERÄNKÖ, 1972a). HERVONEN et al. (1972) confirmed the increase in the number of SIF cells in the paracervical ganglion of the rat uterus using prednisolone, and CIARANELLO et al. (1973) in the superior cervical ganglion and the Organ of Zuckerkandl of newborn rats using dexamethasone. The latter authors also made the important observation that dexamethasone treatment caused an about 10-fold increase in the phenylethanolamine N-methyltransferase activity of these structures. They proposed that the SIF cells may be formed not only from previously undifferentiated stem cells but from pre-existing ganglion cell bodies as well.

Electron microscopic studies on sympathetic ganglia of newborn rats treated with hydrocortisone have shown that not only is the number of small granule-containing cells increased but also the number of small granular vesicles in each cell (ERÄNKÖ et al., 1973), correcting preliminary observations (ERÄNKÖ and ERÄNKÖ, 1972a) which failed to provide evidence for the presence of granular vesicles in the newly formed SIF cells. However, it is still possible that a part of the newly formed catecholamines is situated in the endoplasmic reticulum of the SIF cells, in addition to the granular vesicles, as has been shown to be the case in sympathetic nerve cells (O. ERÄNKÖ, 1972) and terminals (TRANZER, 1972).

SIF CELLS IN CULTURED GANGLIA

SIF cells have also been demonstrated in cultures of sympathetic ganglia (LEVER and PRESLEY, 1971; ERÄNKÖ et al., 1972a,b; CHAMLEY et al., 1972). Addition of hydrocortisone into the culture medium has been shown to result in a dramatic increase in the number of SIF cells, an increase in the intensity and shift in the colour of the formaldehyde-induced fluorescence these cells, and formation of large clusters of them (ERÄNKÖ et al., 1972a). Electron microscopically, the SIF cells in control cultures resembled those in ganglia directly taken from animals, except that in vitro the granule inside the vesicle was often oblong or hour-glass-shaped (Fig. 1.). After addition of hydrocortisone into the culture medium, numerous small cells with granular vesicles were found, and the number of granular vesicles in each cell was also increased (ERÄNKÖ et al., 1972b). The granules of the SIF cells in cultures with hydrocortisone showed large size variations from one cell to another and within many cells. Furthermore, there were very large granular vesicles, and the electron density of the granules after glutaraldehyde and osmium tetroxide fixation varied from

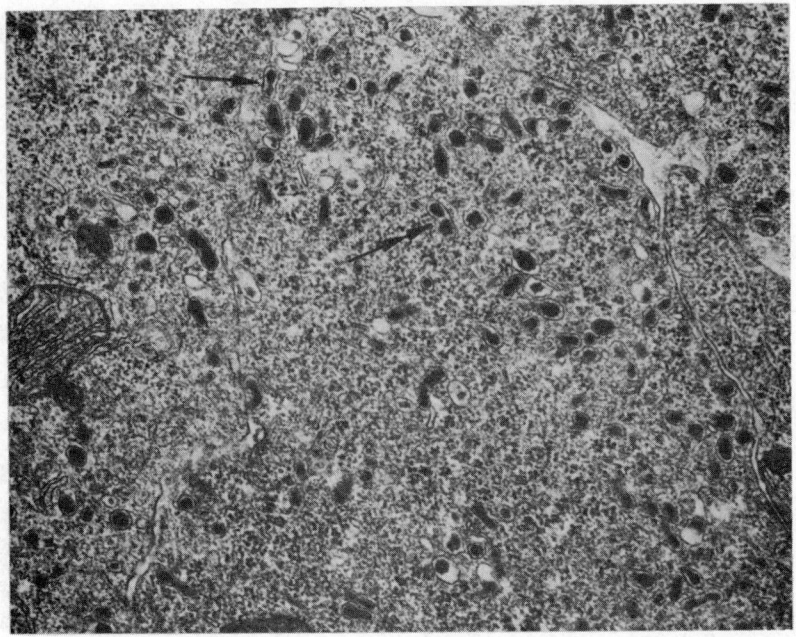

FIG. 1.—Electron micrograph of a SIF cell cultured in a control medium for 7 days. Fixation in glutaraldehyde and osmium tetroxide. Magnification ×27,000. Single arrow indicates an hour-glass-shaped granule, double arrow two separate granules inside a vesicle. From ERÄNKÖ *et al.* (1972b), courtesy *Z. Zellforsch.*, Springer-Verlag.

low to high (Fig. 2), suggesting that some granules were noradrenaline-, the others adrenaline-containing.

SIF CELLS, CAROTID BODY CELLS AND PARAGANGLION CELLS

ERÄNKÖ and HÄRKÖNEN (1965) reported a well vascularized organ composed of SIF cells near the superior cervical ganglion of the rat, and OLSON (1967) found a large group of SIF cells near the nerve trunk leaving the superior cervical ganglion of the mouse. Moreover, ELFVIN (1971) cites personal communication by Ross according to which 'it is known that in several rodents, e.g. the rat, the superior cervical ganglion and the carotid body are often fused'. These observations raise the question of the relations between the SIF cells and the glomus cells of the carotid body. Glomus cells have closely similar fluorescence microscopic (KORKALA *et al.*, 1973) and electron microscopic characteristics (ZAPATA *et al.*, 1969; MASCORRO and YATES, 1970) as SIF cells and so do also the paraganglion cells (COUPLAND, 1965; DUNCAN and YATES, 1967; HERVONEN, 1971; ERÄNKÖ and ERÄNKÖ, 1972a). Glomus cells also respond to glucocorticoids with an increased fluorescence (KORKALA *et al.*, 1973). It has long been known that some glomus cells are chromaffin, while others are not and the same applies to paraganglion cells (WATZKA, 1943) and to SIF cells in sympathetic ganglia, most likely depending on the size of the granular vesicles (see above). Since evidence is available that all these cells are derivatives of the neural crest (LEMPINEN, 1964; PEARSE *et al.*, 1973), there are good reasons to think that they are

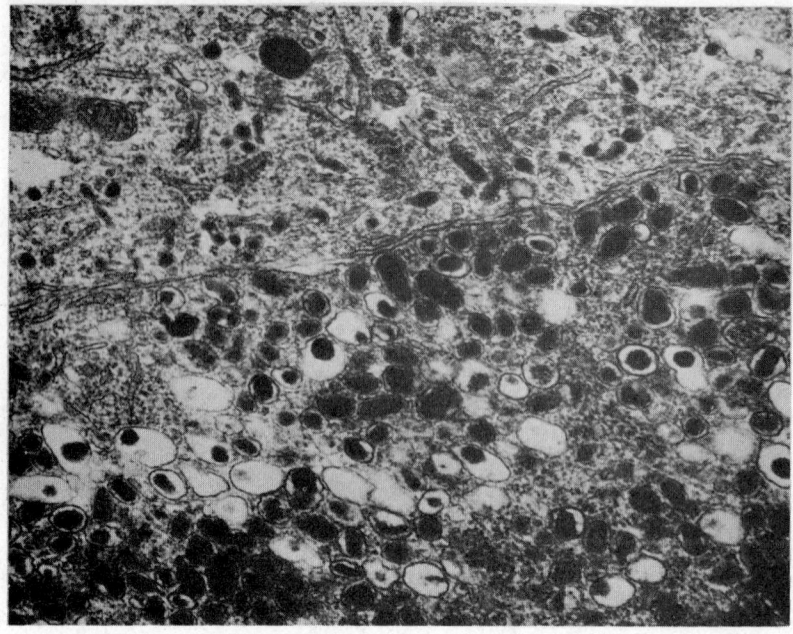

FIG. 2.—Border of two SIF cells from a hydrocortisone-containing culture. In the
upper cell the granules are essentially similar to those in the control cultures. In the
lower cell most granules are much larger and their electron density is variable. Technique
and magnification as in Fig. 1. From ERÄNKÖ et al. (1972b), courtesy Z. Zellforsch.,
Springer-Verlag.

closely related and may perform related functions. While release of catecholamines
is one likely task, liberation of chromogranins (BANKS and HELLE, 1971) and polypeptide hormones (PEARSE et al., 1973) stored in the granular vesicles may prove
to be of functional significance in sympathetic ganglia.

REFERENCES

ANGELETTI P. U. and LEVI-MONTALCINI R. (1970) Proc. Nat. Acad. Sci. 65, 114–121.
AXELROD J. (1966) Pharmacol Rev. 18, 95–113.
BANKS P. and HELLE K. B. (1971) Phil. Trans. R. Soc. Lond. B. 261, 305–310.
BJÖRKLUND A., CEGRELL L., FALCK B., RITZEN M. and ROSENGREN E. (1970) Acta physiol. scand. 78,
   334–338.
BJÖRKLUND A., EHINGER B. and FALCK B. (1968) J. Histochem. Cytochem. 16, 263–270.
CARAMIA F., ANGELETTI P. U., LEVI-MONTALCINI R. and CARRATELLI L. (1972) Brain Res. 40, 237–246.
CHAMLEY J. H., MARK G. and BURNSTOCK G. (1972) Z. Zellforsch. 135, 315–327.
CIARANELLO R. D., JACOBOWITZ D. and AXELROD J. (1973) J. Neurochem. 20, 799–805.
COUPLAND R. E. (1965) The Natural History of the Chromaffin Cell. Longmans, London.
CSILLIK B., KALMAN G. and KNYIHAR E. (1967) Experientia (Basel) 23, 477–478.
D'ANZI F. A. (1969) Am. J. Anat. 125, 381–398.
DUNCAN D. and YATES R. D. (1967) Anat. Rec. 157, 667–681.
ECCLES R. M. and LIBET B. (1961) J. Physiol. Lond. 157, 484–503.
ELFVIN L. G. (1968) J. Ultrastructure 22, 37–44.
ELFVIN L. G. (1971) J. Ultrastructure Res. 37, 432–448.
ERÄNKÖ L. (1972) Brain Res. 46, 159–175.
ERÄNKÖ L. and ERÄNKÖ O. (1971a) Acta pharmacol. et toxicol. 30, 403–416.
ERÄNKÖ L. and ERÄNKÖ O. (1972a) Acta physiol. scand. 84, 125–133.

ERÄNKÖ L. and ERÄNKÖ O. (1972b) *Acta physiol. scand.* **84**, 115–124.
ERÄNKÖ O. (1955) *Acta endocrinol.* **18**, 174–179.
ERÄNKÖ O. (1972) *Histochem. J.* **4**, 213–224.
ERÄNKÖ O. and ERÄNKÖ L. (1971b) in *Histochemistry of Nervous Transmission* (ERÄNKÖ O., ed.) *Progr. Brain Res.*, Vol. 34, pp. 39–51. Elsevier, Amsterdam.
ERÄNKÖ and ERÄNKÖ L. (1971c). *J. Histochem. Cytochem.* **19**, 131–132.
ERÄNKÖ O. and HÄRKÖNEN M. (1963) *Acta physiol. scand.* **58**, 285–286.
ERÄNKÖ O. and HÄRKÖNEN M. (1965) *Acta physiol. scand.* **63**, 511–512.
ERÄNKÖ O., ERÄNKÖ L., HILL C. E. and BURNSTOCK G. (1972a) *Histochem. J.* **4**, 49–58.
ERÄNKÖ O., HEATH J. and ERÄNKÖ L. (1972b) *Z. Zellforsch.* **134**, 297–310.
ERÄNKÖ O., HEATH J. and ERÄNKÖ L. (1973) *Experientia* **29**, 457–459.
ERÄNKÖ O., LEMPINEN M. and RÄISÄNEN L. (1966) *Acta physiol. scand.* **66**, 253–254.
HERVONEN A. (1971) *Acta physiol. scand. Suppl.* 368.
HERVONEN A., KANERVA L., LIETZEN R. and PARTANEN S. (1972) *Z. Zellforsch.* **134**, 519–527.
HÖKFELT T. (1969) *Acta physiol. scand.* **76**, 427–440.
IWANOW G. F. (1932) *Z. ges. Anat. 3. Ergebn. Anat. Entwickl. Gesch.* **29**, 87–280.
JACOBOWITZ D. (1967) *J. Pharmacol exp. Ther.* **158**, 227–240.
JACOBOWITZ D. (1970) *Fedn. Proc.* **29**, 1929–1944.
JENSEN-HOLM J. and JUUL P. (1970) *Acta pharmacol. et toxicol.* **28**, 283–298.
KANERVA L. (1971) In *Histochemistry of Nervous Transmission* (ERÄNKÖ, O., ed.) *Progr. Brain Res,.* Vol. 34, pp. 433–444. Elsevier, Amsterdam.
KANERVA L. and TERÄVÄINEN H. (1972) *Z. Zellforsch.* **129**, 161–177.
KOHN A. (1899) *Anat. Anz.* **15**, 394–400.
KOHN A. (1903) *Arch. mikr. Anat.* **62**, 263–365.
KORKALA O., ERÄNKÖ O., PARTANEN S., ERÄNKÖ L. and HERVONEN A. (1973) *Histochem. J.* in press.
LEMPINEN M, (1964) *Acta physiol. scand.* **62**, Suppl. 231, pp. 1–91.
LEVER J. D. and PRESLEY R. (1971) In *Histochemistry of Nervous Transmission* (ERÄNKÖ O., ed.) *Progr. Brain Res.*, Vol. 34, pp. 499–512. Elsevier, Amsterdam.
MASCORRO J. A. and YATES R. D. (1970) *Tex. Rep. Biol. Med.* **28**, 59–68.
MATTHEWS M. R. and RAISMAN G. (1969) *J. Anat., Lond.* **105**, 255–282.
MUSCHOLL E. and VOGT M. (1964) *Br. J. Pharmac. Chemother.* **22**, 193–203.
NORBERG K. A. and HAMBERGER B. (1964) *Acta physiol. scand.* **63**, Suppl. 238.
NORBERG K. A., RITZEN M. and UNGERSTEDT U. (1966) *Acta physiol. scand.* **67**, 260–270.
NORBERG K. A. and SJÖQVIST F. (1966) *Pharmacol. Rev.* **18**, 743–751.
OLSON L. (1967) *Z. Zellforsch.* **81**, 155–173.
OWMAN O. and SJÖSTRAND N. O. (1965) *Z. Zellforsch. mikrosk. Anat.* **66**, 300–320.
PEARSE A. G. E., POLAK J. M., ROST F. W. D., FONTAINE J., LE LIEVRE C. and LE DOUARIN N. (1973) *Histochemie* **34**, 191–203.
SIEGRIST G., DOLIVO M., DUNANT Y., FOROGLOV-KERAMEUS C. and RIBAUPIERRE F. (1968) *J. Ultrastruct. Res.* **25**, 381–407.
TAXI J., GAUTRON J. and L'HERMITE P. (1969) *C.r. hebd. Séanc. Acad. Sci., Paris* **269**, 1281–1284.
THOENEN H. and TRANZER J. P. (1968) *Naynyn-Schmiedebergs Arch. exp. Path. Pharmak.* **261**, 271–288.
TRANZER J. P. (1972) *Nature, New Biology* **237**, 57–58.
ZAPATA P., HESS A., BLISS E. L. and EYZAGUIRRE C. (1969) *Brain Res.* **14**, 473–496.
VAN ORDEN L. S., BURKE J. P., GEYER M. and LODOEN F. V. (1970) *J. Pharmacol. exp. Ther.* **174**, 56–71.
WATANABE H. (1971) *Am. J. Anat.* **130**, 305–330.
WATZKA M. (1943) in *Handbuch der mikroskopischen Anatomie* (MÖLLENDORF, M.v., ed.) Band VI, pp. 262–308. Springer-Verlag, Berlin.
WILLIAMS T. H. and PALAY S. L. (1969) *Brain Res.* **15**, 17–34.

Frontiers in Catecholamine Research 1973, pp. 439 to 446. Pergamon Press. Printed in Great Britain.

NEURONAL CATECHOLAMINE STORAGE VESICLES

Tomas Hökfelt

Department of Histology, Karolinska Institute, S-104 01 Stockholm 60, Sweden

In the very early electron microscopic studies on nervous tissues small electron lucent vesicles with a diameter of about 500 Å were recognised as the outstanding intracellular component of presynaptic nerve endings in general (De Robertis and Bennet, 1954; Palade and Palay, 1954; Sjöstrand, 1954). They were termed synaptic vesicles and ascribed a functional role as storage sites for the neurotransmitters and as the possible morphological basis for quantal release of these substances at the synapses (see Katz, 1969). Direct proof for this hypothesis in the form of a histochemical demonstration of the transmitter in the vesicles is, however, till now lacking for all types of neurotransmitter suspects, except for certain monoamines. Increasing evidence indicate that under appropriate experimental and fixation conditions peripheral noradrenaline (NA), central dopamine (DA), NA and 5-hydroxytryptamine (5-HT) neurons are characterised by their content of synaptic vesicles containing an electron dense core—*dense core* or *granular vesicles* (De Robertis and Pellegrino de Iraldi, 1961a, b,; Lever and Esterhuizen, 1961; Richardson, 1962; Wolfe et al., 1962, Hökfelt, 1968; Tranzer et al., 1969). A large number of studies suggest that the dense core in fact is correlated to the amine levels. This fortunate situation has offered possibilities to identify and study the ultrastructural characteristics of peripheral and central monoamine neurons. In fact, these neurons are so far the only ones which could be identified in the microscope on the basis of their transmitter substance although recent attempts to map "GABA" and "glycine neurons" with autoradiography seem promising (for ref., see Hökfelt and Ljungdahl, 1972; Iversen and Schon, 1973). Since the last catecholamine (CA) symposium and the review article by Grillo (1966) on electron microscopy of sympathetic tissues the combined efforts of morphological and biochemical techniques have considerably increased our knowledge of transmitter storage vesicles and transmitter mechanisms. Due to the limitations in the allotted space, however, the intention of this paper can only be to discuss mainly some ultrastructural and histochemical aspects of intraneuronal CA storage vesicles. We refer to other papers presented at this meeting and recent reviews for more complete information (Gray, 1969; Tranzer et al., 1969; Bloom, 1970; Gray and Willis, 1970; Jaim-Etcheverry and Zieher, 1971; Burnstock and Iwayama 1971; Geffen and Livett, 1971; Hökfelt, 1971; Akert et al., 1971, 1972; Hökfelt and Ljungdahl, 1972; Pappas and Waxman, 1972; Smith, 1971, 1972; Berl et al., 1973; Livett, 1973; Pfenninger, 1973).

THE TERM "DENSE CORE OR GRANULAR VESICLE"

The term dense core or granular vesicle can be used for any round, membrane bound structure containig a core of a more or less strong electron density and has been widely used in electron microscopy especially of the nervous system and particularly of adrenergic neurons. Several authors have introduced a number of subdivisions (small, very small, intermediate, large, very large, types 1–3, etc.) to differentiate between various types of granular vesicles. It seems therefore necessary to specify exactly the experimental conditions under which the granular vesicles were produced and of course their size, appearance and localisation. It may be added that future ultrastructural work hopefully will introduce even more types of granular visicles in attempts to trace e.g. neurotransmitters, since this is the principle by which substances are visualised in the electron microscope, namely by an electron dense precipitate.

DIFFERENT CATECHOLAMINE STORAGE COMPARTMENTS

GRILLO and PALAY (1962) recognised at least two different types of granular vesicles often described as *small* (diameter about 500 Å) and *large* (diameter about 1000 Å) *granular vesicles* (SGV and LGV, respectively) in addition to classical agranular synaptic vesicles. Subsequent experiments indicate that either by adding exogenous amines (TRANZER and THOENEN, 1967; HÖKFELT, 1968) or by applying a sensitive fixation procedure like potassium permanganate ($KMnO_4$) (RICHARDSON, 1966) practically all vesicles in adrenergic nerves contain a dense core (HÖKFELT, 1969; WOODS, 1970; FILLENZ, 1971). Both SGV and LGV can according to histochemical techniques store and take up the amine (HÖKFELT, 1968; TRANZER and THOENEN, 1968; FILLENZ, 1971). Subcellular fractionation studies have during the years indicated the presence of an *extravesicular compartment* containing "free" CA with a size depending on tissue and amine studied (IVERSEN, 1967; VON EULER, 1969). At least part of these soluble amines may be derived from a new amine storing compartment recently described by TRANZER (1972) consisting of "tubular reticulum-like structures", possibly corresponding to the (*axonic*) *smooth endoplasmic reticulum* and containing a reserpine-sensitive, electron dense precipitate. From autoradiographic studies on uptake of ^{3}H-NA it has furthermore been suggested that a *macromolecular complex* of probable protein nature, synthesised in the perikaryon, could represent a binding site for CA (DESCARRIES and DROZ, 1970; SOTELO, 1971; GEFFEN *et al.*, 1971). It is possible that this macromolecular complex could correspond to the compartment described by TRANZER (1972).

INTRANEURONAL DISTRIBUTION OF GRANULAR VESICLES

SGV and LGV were first and most consistently demonstrated in the *nerve endings* of adrenergic neurons where they are found in large amounts. Calculations on

---

FIG. 1.—Electron micrograph of the rat iris (dilator muscle). $KMnO_4$ fixation. Two nerve endings (g) containing granular vesicles are seen. The majority of the vesicles are of the small type and only a few large granular vesicles (arrow) are seen. Note that almost all vesicles in these two nerve endings contain an electron dense core. Two other nerve endings (a) contain only agranular vesicles. Magnification 48,000×.

FIG. 2.—Electron micrograph of the rat iris (dilator muscle) after chronic treatment with reserpine. Rat sacrificed 3 days after the last dose (see HÖKFELT, 1973). $KMnO_4$ fixation. In one nerve ending several tubular structures (short arrow) are seen together with small round granular (long arrow) or agranular vesicles. Two other nerve endings (a) contain only agranular vesicles. Magnification 58,000×.

FIG. 3.—Electron micrograph of an adrenal medullary transplant to the anterior eye chamber. $KMnO_4$ fixation. A process of a transformed cell is seen. Many tubular structures sometimes containing an electron dense precipitate (short arrow) and granular vesicles (long arrows) are seen (from HÖKFELT and OLSON, unpublished data). Magnification 62,000×.

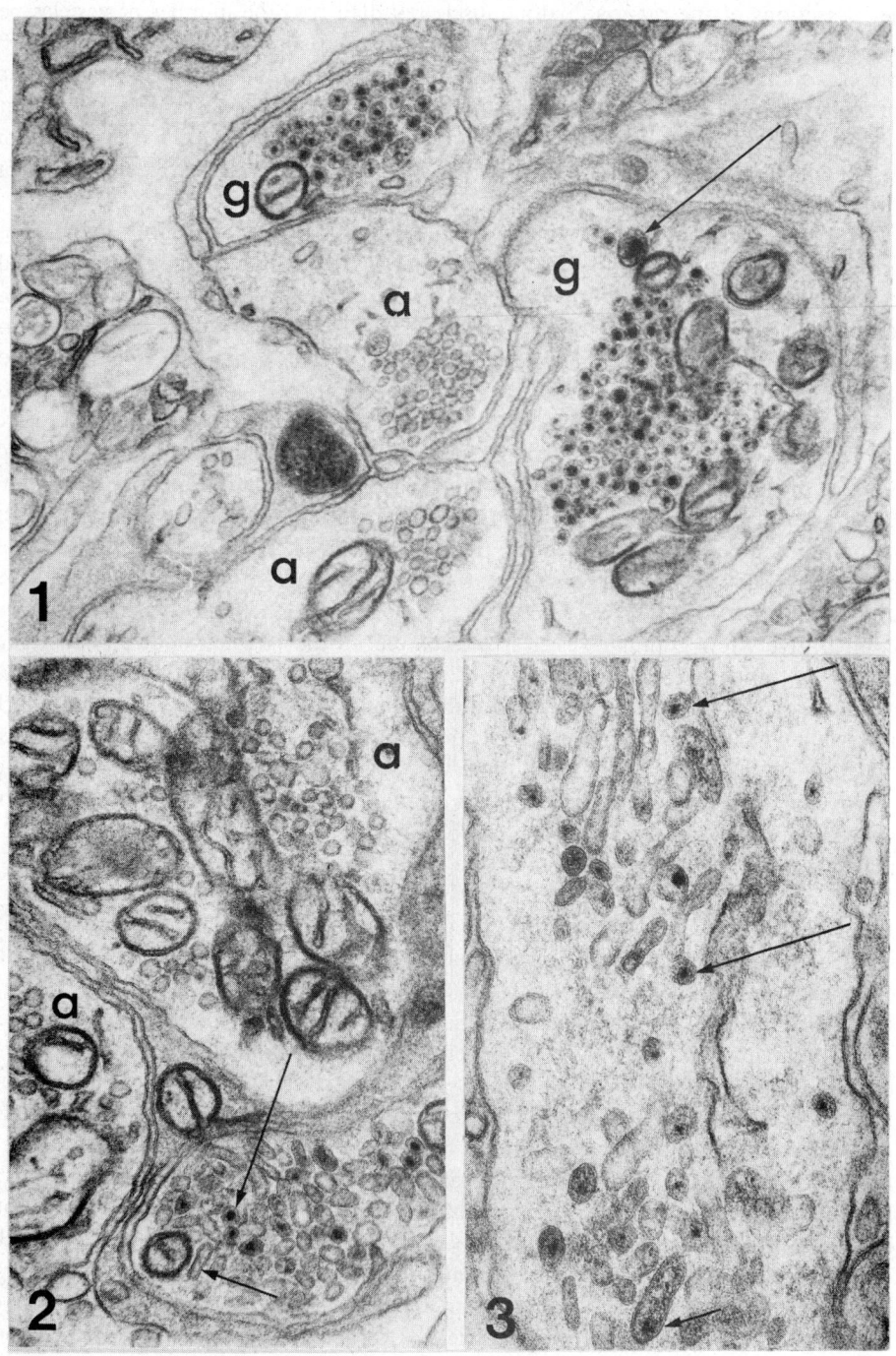

FIGS. 1–3.—See captions opposite.

serial sections of the rat iris dilator muscle revealed that the number of vesicles per varicosity (as well as the size of the varicosity) varied considerably (from 50 to 800) but that a content of 300–500 vesicles may represent a reasonable mean in this special organ (HÖKFELT, 1969). About two per cent of the vesicles were of the large type, the rest were SGV. In the vas deferens the corresponding figure for LGV is four per cent (FARRELL, 1968) whereas a considerably higher proportion (20 per cent) of LGV is found in the spleen (BISBY and FILLENZ, 1971).

SGV and LGV are also found in the axons and the cell bodies of CA neurons. In the *axons* the proportion of LGV seems to be higher than in the terminals (GEFFEN and OSTBERG, 1969; HÖKFELT, 1969) although clusters of SGV, almost resembling a varicosity (HÖKFELT, 1969; FILLENZ, 1971), occasionally can be seen in the axons. In the *cell bodies* mainly LGV have been found after OsO_4 or glutaraldehyde-OsO_4 fixation technique. However, both TAXI (1965) and GRILLO (1966) showed clusters of vesicles, a few of which contained an electron dense core, and after $KMnO_4$ fixation of the rat superior cervical ganglion large amounts of vesicles present in cell bodies contain an electron dense core (HÖKFELT, 1969; VAN ORDEN et al., 1970). These SGV are preferably localised in the peripheral parts of the cytoplasm and are often found where the cell processes leave the soma. Also fluorescence microscopy when properly performed show this distribution of NA (ERÄNKÖ and HÄRKÖNEN, 1963; VAN ORDEN et al., 1970). In the central nervous system comparatively few granular vesicles are seen e.g. in the NA cell bodies of the locus coeruleus (HÖKFELT, 1967). However, after intraventricular injections of colchicine an increased number of SGV are seen in these cell bodies (unpublished data) similarly to results with local application of colchicine or vinblastine on the superior cervical ganglia demonstrating increased numbers of SGV and LGV in the NA cell bodies (HÖKFELT and DAHLSTRÖM, 1971). Interestingly, increased numbers of vesicles are seen in the perinuclear cytoplasm possibly indicating arrest of transport from this region to more peripheral parts of the cell body.

The existence of SGV in cell bodies has been discussed, and the lack of these vesicles described by many authors have led to the opinion that LGV are the NA storage sites in the cell bodies and/or that the amines are bound to macromolecular complexes not directly demonstrable in the electron microscope (DESCARRIES and DROZ, 1970). In glutaraldehyde fixed tissue SGV could not be demonstrated even after administration of 5-OH-DA in superior cervical ganglia (TRANZER et al., 1971) or in chick embryo cultures (TEICHBERG and HOLTZMAN, 1973). Two explanations may be advanced: (1) 5-OH-DA is not taken up in the small agranular vesicles present in the cell body to the same extent as in the nerve endings. This could be the case since the surface/volume ratio is small, leading to much lower intraneuronal concentrations of 5-OH-DA than in nerve endings. (2) The SGV demonstrated in the cell bodies mainly after $KMnO_4$ fixation represent an artefact in the sense that e.g. tubules and sacs of the smooth endoplasmic reticulum (see paragraph on "formation of granular vesicles") under the influence of the vigorous action of the fixative are transformed into vesicle-like structures. It is difficult to exclude any of the two possibilities. The fact that after zinc–iodide–OsO_4 (ZIO) impregnation of *glutaraldehyde prefixed tissue* (AKERT et al., 1971), ZIO positive vesicles with the same distribution pattern as the SGV after $KMnO_4$ fixation are found in the superior cervical ganglion (unpublished data; cf. MATUS, 1971) may speak against the possibility of a $KMnO_4$ induced transformation of smooth endoplasmic reticulum into SGV.

COMPOSITION OF THE CATECHOLAMINE STORAGE VESICLES

Our knowledge in this field stems from histochemical and biochemical data. Unfortunately, the histochemical methods mostly only give hints as to the chemical composition whereas the specific biochemical techniques available sometimes may be difficult to interpret due to problems to isolate the various types of vesicles in a pure form. Both SGV and LGV are clearly surrounded by a triple layered membrane. Earlier biochemical investigations have indicated that the chemical composition of this membrane differs from that of the cell membrane (WHITTAKER, 1966; LAPETINA *et al.*, 1968), which has been considered as a serious argument *against* the view that synaptic vesicles are formed by pinocytosis from the cell membrane. Recent studies, however, seem to indicate less obvious differences (MORGAN *et al.*, 1973). The content of the granular vesicles has been the subject of many studies (see SMITH, 1971, 1972). Considerably more is known about the LGV than about the SGV due to the fact that it has been possible to isolate the former vesicles in increasingly purer fractions (DE POTTER *et al.*, 1970; THURESON-KLEIN *et al.*, 1970; 1973). From histochemical studies it seems likely that both SGV and LGV in monoamine neurons (LGV are present also in non-monoamine neurons) contain endogenous amines (and have the ability to take up exogenous amines) (TRANZER and THOENEN, 1967; 1968; HÖKFELT, 1968; 1969; THURESON-KLEIN *et al.*, 1970; 1973; WOODS, 1970; FILLENZ, 1971). The NA content of SGV seems to be very high approaching one molar concentrations as judged from rough estimations in the rat iris (HÖKFELT and LJUNGDAHL, 1972; see also FOLKOW *et al.*, 1967). Histochemical studies indicate clear differences in the composition of the non-amine content of SGV and LGV. Thus, the interior of the LGV but not of the SGV is stained by bismuth-iodide (PFENNINGER *et al.*, 1969) whereas impregnation with zinc-iodide results in a precipitate in the SGV but not in the LGV (AKERT and SANDRI, 1968). These findings have been interpreted to indicate that the interior of the LGV is mainly built up of proteins, which actually is also indicated by the pronounced core density of the LGV after glutaraldehyde fixation. Biochemical studies have demonstrated the presence of dopamine-β-hydroxylase, chromogranin A and ATP in LGV isolated from nerve trunks (HÖRTNAGL *et al.*, 1969; DE POTTER *et al.*, 1970; HELLE *et al.*, 1971).

Improvements in the isolation procedures have made it possible to isolate two populations of NA storing vesicles (CHUBB *et al.*, 1970; BISBY and FILLENZ, 1971) which in the electron microscope contained SGV and LGV respectively (BISBY and FILLENZ, 1971). It could be shown with these fractions that both the LGV and the SGV contain dopamine-β-hydroxylase (BISBY *et al.*, 1973).

THE ORIGIN OF THE GRANULAR VESICLES

The origin of synaptic vesicles in general has been discussed in several papers, the main questions being, (1) from which cellular component are they formed and (2) where in the neuron does the formation take place. Ultrastructural and cytochemical studies have pointed to microtubules (DE ROBERTIS, 1964; PELLEGRINO DE IRALDI and DE ROBERTIS, 1968), mitochondria (SPRIGGS *et al.*, 1967), nerve cell membrane and complex vesicles (ANDRES, 1964; WESTRUM, 1965; BIRKS, 1966; KANASEKI and

Kadota, 1969; Gray and Willis, 1970; Douglas *et al.*, 1971; Holtzman *et al.* 1971), LGV (Geffen and Ostberg, 1969; Smith, 1972), the Golgi apparatus (Aker *et al.*, 1971; Stelzner, 1971; Iijima and Awazi, 1973) and the smooth endoplasmi reticulum (Rodriguez Echandia *et al.*, 1971; Machado, 1971; Stelzner, 1971 Korneliussen, 1972; Tranzer, 1972; Hökfelt, 1973; Teichberg and Holtzman 1973) as possible precursors.

The view on the process of vesicle formation in CA neurons has been influence by fluorescence and electron microscopic studies. Pharmacological experiment demonstrating that the recovery of fluorescence after depletion by reserpine starts i the perinuclear zone (Dahlström *et al.*, 1965) and experiments on ligated nerve demonstrating accumulation of fluorescence and granular vesicles above the ligatur (Dahlström, 1965; 1971; Kapeller and Mayor, 1967) have been the basis for th hypothesis that the storage vesicles are formed in the perikaryon (Golgi apparatus? and transported with the axonal flow to the nerve terminals (see also Smith *et al.*, 1970) Recent studies have been interpreted to indicate that only LGV are formed in the ce body, transported to the axon terminals and after release of the amines (and vesicula proteins) by exocytosis transformed into SGV (Geffen and Ostberg, 1969; Smith 1971). Evidence for such a process has also been obtained by analysis of homogenate of the spleen after electrical stimulation of the splenic nerve (De Potter *et al.*, 1973) According to this hypothesis LGV are "young" granules whereas the SGV represent a more mature state.

Recent studies with an improved fixation technique (Tranzer, 1972) and studie on adrenergic neurons under conditions where an increased vesicle formation coul be expected such as developing axons in the pineal gland (Machado, 1970), culture embryonic chick sympathetic neurons (Teichberg and Holtzman, 1973), transplant of the adrenal gland to the anterior eye chamber (Olson, 1970; Hökfelt and Olson unpublished data) and adrenergic axons after chronic reserpine treatment (Hökfelt 1973) indicate, however, that the smooth endoplasmic reticulum may be at least on type of precursor to amine storage vesicles. This could be compatible with man experimental data on various intraneuronal mechanisms such as vesicle formation Since the endoplasmic reticulum is present in all parts of the neuron and is transporte with the axoplasmic, flow, SGV may at least under certain conditions be forme both in cell bodies, axons and nerve endings, and maybe above a ligature.

CONCLUSIONS

Under appropriate experimental and fixation conditions peripheral and centra monamine neurons are chacterised by their content of small granular vesicles. Afte permanganate fixation the distribution of the SGV closely parallels the distribution o amines as revealed with fluorescence histochemistry. Other amine storage compart ments such as large granular vesicles and possibly the smooth and axonic endoplasmi reticulum have also been described. Biochemical studies have shown that the LGV contain dopamine-β-hydroxylase, chromogranin and ATP. Various hypotheses fo the site of formation of the storage vesicles have been forwarded. For adrenergi neurons formation of granular vesicles in the cell body (Golgi apparatus) an subsequent axonal transport to the nerve endings and formation from the smoot endoplasmic reticulum in various parts of the neuron has been discussed.

Acknowledgements—Supported by grants from the Swedish Medical Research Council (04X-2887, 4P-3262) and by grants from Karolinska Institutets Forskningsfonder and Magn. Bergvalls Stiftelse.
The skilful technical assistance of Mrs. W. Hiort and Miss. A.-C. Swensson is gratefully acknowledged.

REFERENCES

AKERT K., MOOR H. and PFENNINGER K. (1971) In: *Int. Symp. Cell Biol. Cytopharmacol. 1st, Venice 1969* (CLEMENTI F. and CECCARELLI B., Eds.) Raven Press, New York. Also *Adv. Cytopharmacol.* **1**, 273–290.
AKERT K., PFENNINGER K., SANDRI C. and MOOR H. (1972) In: *Structure and Function of Synapses* (PAPPAS G. D. and PURPURA D. P., Eds.) pp. 67–86. Raven Press, New York.
AKERT K. and SANDRI C. (1968) *Brain Res.* **7**, 286–295.
ANDRES K. H. (1964) *Z. Zellforsch.* **64**, 63–73.
BERL S., PUSZKIN S. and NICKLAS W. J. (1973) *Science* **179**, 441–446.
BIRKS R. I. (1966) *Ann. N.Y. Acad. Sci.* **135**, 8–19.
BISBY M. A. and FILLENZ M. (1971) *J. Physiol.* **215**, 163–179.
BISBY M. A., FILLENZ M. and SMITH A. D. (1973) *J. Neurochem.* **20**, 245–248.
BLOOM F. E. (1970) *Int. Rev. Neurobiol.* pp. 27–66.
BURNSTOCK G. and IWAYAMA T. (1971) In: *Histochemistry of Nervous Transmission. Progr. Brain Res.* (ERÄNKÖ O., Ed.) Vol. 34, pp. 389–404. Elsevier, Amsterdam.
CHUBB I. W., DE POTTER W. P. and DE SCHAEPDRYVER A. F. (1970) *Nature, Lond.* **128**, 1203–1204.
DAHLSTRÖM A. (1965) *J. Anat. (Lond.)* **99**, 677–689.
DAHLSTRÖM A. (1971) *Phil. Trans. Roy. Soc. Lond. B.* **261**, 325–358.
DAHLSTRÖM A., FUXE K. and HILLARP N.-Å. (1965) *Acta pharmac. toxic.* **22**, 277.
DE POTTER W. P., SMITH A. D. and DE SCHAEPDRYVER A. F. (1970) *Tissue & Cell* **2**, 529–546.
DE ROBERTIS E. (1964) *Histophysiology of Synapses and Neurosecretion.* Pergamon, Oxford.
DE ROBERTIS E. and BENNETT H. S. (1954) *Fedn. Proc.* **13**, 35.
DE ROBERTIS E. and PELLEGRINO DE IRALDI A. (1961a) *Experientia* **17**, 122–124.
DE ROBERTIS E. and PELLEGRINO DE IRALDI A. (1961b) *J. biophys. biochem. Cytol.* **10**, 361–372.
DESCARRIES L. and DROZ B. (1970) *J. Cell Biol.* **44**, 385–399.
DOUGLAS W. W., NAGASAWA J. and SCHULTZ R. A. (1971) *Nature, Lond.* **232**, 340.
ERÄNKÖ O. and HÄRKÖNEN M. (1963) *Acta physiol. scand.* **58**, 285–286.
EULER U. S. VON (1969) *Science* **173**, 202–206.
FARRELL K. E. (1968) *Nature (Lond.)* **217**, 279–281.
FILLENZ M. (1971) *Phil. Trans. Roy. Soc. Lond. B.* **261**, 319–323.
FOLKOW B., HÄGGENDAL J. and LISANDER B. (1967) *Acta physiol. scand. Suppl. 307.*
GEFFEN L. B., DESCARRIES L. and DROZ B. (1971) *Brain Res.* **35**, 315–318.
GEFFEN L. B. and LIVETT B. G. (1971) *Physiol. Rev.* **51**, 98–157.
GEFFEN L. B. and OSTBERG A. (1969) *J. Physiol.* **204**, 583–592.
GRAY E. G. (1969) In: *Mechanisms of Synaptic Transmission. Progr. Brain Res.* (AKERT K. and WASER P. G., eds.) Vol. 31, pp. 141–155. Elsevier, Amsterdam.
GRAY E. G. and WILLIS R. A. (1970) *Brain Res.* **24**, 149–168.
GRILLO M. A. (1966) *Pharmacol. Rev.* **18**, 387–399.
HELLE K. B., LAGERCRANTZ H. and STJÄRNE L. (1970) *Acta physiol. scand.* **80**, 7A–8A.
HÖKFELT T. (1967) *Z. Zellforsch.* **79**, 110–117.
HÖKFELT T. (1968) *Z. Zellforsch.* **91**, 1–74.
HÖKFELT T. (1969) *Acta physiol. scand.* **76**, 427–440.
HÖKFELT T. (1971) In: *Histochemistry of Nervous Transmission. Progr. Brain Res.* (ERÄNKÖ O., Eds.) Vol. 34, pp. 213–222. Elsevier, Amsterdam.
HÖKFELT T. (1973) *Experientia (Basel)* **29**, 580–582.
HÖKFELT T. and DAHLSTRÖM A. (1971) *Z. Zellforsch.* **119**, 460–482.
HÖKFELT T. and LJUNGDAHL Å .(1972) In: *Advances in Biochemical Psychopharmacology* (COSTA E. and IVERSEN L. L., Eds.) Vol. 6, pp. 1–36. Raven Press, New York.
HOLTZMAN E., FREEMAN A. R. and KASHNER L. A. (1971) *Science* **173**, 733–736.
HÖRTNAGAL H., HÖRTNAGL H. and WINKLER H. (1969) *J. Physiol. (Lond.)* **205**, 103–114.
IJIMA K. and AWAZI N. (1973) *Z. Zellforsch.* **136**, 329–348.
IVERSEN L. L. (1967) *The Uptake and Storage of Noradrenaline in Sympathetic Nerves.* Cambridge Univ. Press, London, p. 253.
IVERSEN L. L. and SCHON F. (1973) In: *New Concepts in Transmitter Regulation* (MANDELL A. and SEGAL D., Eds.), in press.
JAIM-ETCHEVERRY G. and ZIEHER L. M. (1971) In: *Advances in Cytopharmacology* (CLEMENTI F. and CECCARELLI B., Eds.) Vol. 1, pp. 343–361. Raven Press, New York.

Kanaseki T. and Kadota K. (1969) *J. Cell Biol.* **42**, 202–220.

Kapeller K. and Mayor D. (1967) *Proc. Roy. Soc. (Lond.) Ser. B.* **167**, 282–292.

Katz B. (1969) *The Sherrington Lectures X*, University Press, Liverpool.

Korneliussen H. (1972) *Z. Zellforsch.* **130**, 28–57.

Lapetina E. G., Soto E. F. and De Robertis E. (1968) *J. Neurochem.* **15**, 437–445.

Lever J. D. and Esterhuizen A. C. (1961) *Nature (Lond.)* **192**, 566–567.

Livett B. G. (1973) *Brit. Med. Bull.* **29**, 93–99.

Machado A. B. M. (1971) In: *Histochemistry of Nervous Transmission, Progr. Brain Res.* (Eränkö O., Ed.) Vol. 34, pp. 171–185, Elsevier, Amsterdam.

Matus A. I. (1970) *Brain Res.* **17**, 195–203.

Morgan I. G., Zanetta J.-P., Vincendon G. and Gombos G. (1973) *Brain Res.*, in press.

Palade G. E. and Palay S. L. (1954) *Anat. Rec.* **118**, 335–336.

Pappas G. D. and Waxman S. G. (1972) In: *Structure and Function of Synapses* (Pappas G. D. and Purpura D. P., Eds.) pp. 1–43. Raven Press, New York.

Pellegrino De Iraldi A. and De Robertis E. (1968) *Z. Zellforsch.* **87**, 330–344.

Pfenninger K. H. (1973) *Progr. Histochem. Cytochem.* **5**, (1), 1–83.

Pfenninger K., Sandri C., Akert K. and Eugster C. H. (1969) *Brain Res.* **12**, 10–18.

Richardson K. C. (1962) *J. Anat. (Lond.)* **96**, 427–442.

Richardson K. C. (1966) *Nature, Lond.* **210**, 756.

Rodriguez Echandia E. L., Zamora A. and Prezzi R. S. (1970) *Z. Zellforsch.* **104**, 419–428.

Sjöstrand F. (1954) *Z. wiss. Mikr. mikr. Tech.* **62**, 65.

Smith A. D. (1971) *Phil. Trans. Roy. Soc. Lond. B.* **261**, 363–370.

Smith A. D. (1972) *Pharmacol. Rev.* **24**, 435–457.

Smith D. S., Järlfors U. and Beranek R. (1970) *J. Cell Biol.* **46**, 199–219.

Sotelo C. (1971) *J. Ultrastruct. Res.* **36**, 824–841.

Spriggs T. L. B., Lever J. D. and Graham J. D. P. (1967) *J. Microscopie* **6**, 425–430.

Stelzner D. J. (1971) *Z. Zellforsch.* **120**, 332–345.

Taxi J. (1965) *Ann. Sci. Nat. Zool. (Paris)* (**12**) 7, 413–674.

Teichberg S. and Holtzman E. (1973) *J. Cell Biol.* **57**, 88–108.

Thureson–Klein A., Klein R. L., Lagercrantz H. and Stjärne L. (1970) *Experientia (Basel)* **26**, 994–995.

Thureson-Klein Å., Klein R. L. and Yen S.-S. (1973) *J. Ultrastruct. Res.*, **43**, 18–35.

Tranzer J. P. (1972) *Nature, New Biol.* **237**, 57–58.

Tranzer J. P. and Thoenen H. (1967) *Experientia (Basel)* **23**, 123–124.

Tranzer J. P. and Thoenen H. (1968) *Experientia (Basel)* **24**, 484–486.

Tranzer J. P., Thoenen H., Snipes R. L. and Richards J. G. (1969) In: *Mechanisms of Synaptic Transmission. Progr. Brain Res.* (Akert K. and Waser P. G., Eds.) Vol. 31, pp. 33–46. Elsevier, Amsterdam.

Van Orden III L. S., Burke J. P., Geyer M. and Lodoen F. V. (1970) *J. Pharmacol. exp. Ther.* **174**, 56–71.

Westrum L. E. (1965) *J. Physiol. (Lond.)* **179**, 4P–6P.

Whittaker V. P. (1966) *Ann. N.Y. Acad. Sci.* **137**, 982–998.

Wolfe D. E., Potter L. T., Richardson K. C. and Axelrod J. (1962) *Science* **138**, 440–442.

Woods R. I. (1970) *Proc. Roy. Soc. Lond. B.* **176**, 63–68.

Frontiers in Catecholamine Research 1973, pp. 447 to 452. Pergamon Press. Printed in Great Britain.

SECRETION AND RECOVERY OF CATECHOLAMINES BY THE ADRENAL MEDULLA*

N. KIRSHNER and T. A. SLOTKIN†
Department of Biochemistry and Department of Physiology and Pharmacology,
Duke University Medical Centre, Durham, N.C., U.S.A.

IT IS generally accepted that secretion from the adrenal medulla occurs by exocytosis, a process which involves temporary fusion of the catecholamine storage vesicles with the plasma membrane and extrusion of the entire soluble contents of the vesicles directly to the exterior of the cell. Only a few of the pertinent experiments which contributed to the elucidation of this phenomenon will be presented here. This discussion will be concerned largely with the quantal aspects of secretion and the events which occur during recovery of the catecholamine stores. For detailed reviews the reader is referred to DOUGLAS (1968), SMITH and WINKLER (1972), and VIVEROS (1973).

Evidence that secretion occurs by exocytosis was obtained by demonstrating that concurrent with the release of catecholamines other soluble components of the storage vesicles are also released into perfusates of stimulated glands in the same relative amounts as are present in the intact vesicles. DOUGLAS et al., (1965) and DOUGLAS and POISNER (1966a, b) showed that ATP or its metabolites and catecholamines are simultaneously released upon stimulation of perfused cat adrenal glands. Shortly thereafter BANKS and HELLE (1965), KIRSHNER et al., (1966, 1967) and SCHNEIDER et al., (1967) demonstrated the simultaneous release of the soluble protein of the storage vesicle and catecholamines while cytoplasmic proteins such as lactic dehydrogenase and phenylethanolamine-N-methyltransferase were not released. Subsequently DUCH et al., (1968) found that dopamine-β-hydroxylase was present in the storage vesicle both in a readily soluble form and in a form tightly bound to the membrane, and VIVEROS et al., (1968) showed that the soluble form of the enzyme was also released upon stimulation. The release of chromogranins and catecholamines was also shown to occur in vivo (BLASCHKO, 1967).

Taking an alternate experimental approach and measuring what remained in the gland after secretion, VIVEROS et al., (1969a, b, c; 1971a,b) conclusively demonstrated that the total soluble content of the vesicles were released in an "all-or-none" fashion while the membranes were retained. Those demonstrations were made possible by the fortunate distribution of dopamine-β-hydroxylase as both a membrane-bound and soluble enzyme. The membrane-bound enzyme serves as a marker for the membrane while release of the soluble enzyme is an indicator of neurogenically stimulated release and enables one to distinguish this process from mere catecholamine depletion such as that caused by reserpine in the presence of ganglionic blocking agents (VIVEROS et al., 1971a). Evidence for the quantal nature of secretion was

* The work reported here was supported by grants from the United States Public Health Service AM05427 and MH23278 and from the Council for Tobacco Research—U.S.A.
† Recipient of Faculty Development Award in Pharmacology from the Pharmaceutical Manufacturers Association Foundation.

448   N. KIRSHNER and T. A. SLOTKIN

TABLE 1. TIME DEPENDENT CHANGES IN DOPAMINE-β-HYDROXYLASE
AND CATECHOLAMINES IN RABBIT ADRENAL STORAGE VESICLES
AFTER INSULIN TREATMENT

Time (hr)	DBO	CA	DBO/CA
0	853 ± 65	58 ± 5	15 ± 2
3	116 ± 13	10 ± 1	12 ± 2
24	168 ± 11	10 ± 1	17 ± 5
48	745 ± 26	18 ± 3	41 ± 15
96–144	896 ± 104	49 ± 6	18 ± 5

For experimental details see VIVEROS et al., (1972b). DBO is
expressed as nmoles × 100/gland pair/hr; CA in µg/gland pair.

obtained by comparing the dopamine-β-hydroxylase contents of purified vesicles
obtained after stimulation to those obtained from control animals based on the
following rationale: If the vesicles released only a part of their content upon stimulation then one would expect an increase in the dopamine-β-hydroxylase:catechol-
amine ratio since all of the membrane-bound dopamine-hydroxylase is retained by
the gland. Alternatively if release were all-or-none, then the ratio dopamine-β-
hydroxylase:catecholamines in the remaining vesicles would be unchanged. Neural
stimulation of the adrenal medulla was evoked in rabbits by the administration of
insulin; 3 hr later (Table 1) there was a massive depletion of both the catecholamine
stores and the intact vesicles, and those vesicles which remained had the same
DBH:CA ratios as vesicles from the control animals, indicating that all-or-none
secretion had taken place. On the other hand, non-neurogenic depletion of catecholamine stores by reserpine in the presence of ganglionic blocking agents caused no
loss of DBH and resulted in a marked increase in the DBH:CA ratio of the isolated
vesicles (VIVEROS et al., 1971a, b).

Studies of rat secretory vesicles 1, 2 and 3 hr after hypoglycemic doses of insulin
provided additional evidence for quantal secretion (SLOTKIN and KIRSHNER, 1973a).
Figure 1 shows the correlation between the DBH and CA contents of intact vesicles.
Despite the high correlation ($r = 0.89$) the regression line indicates that a significant

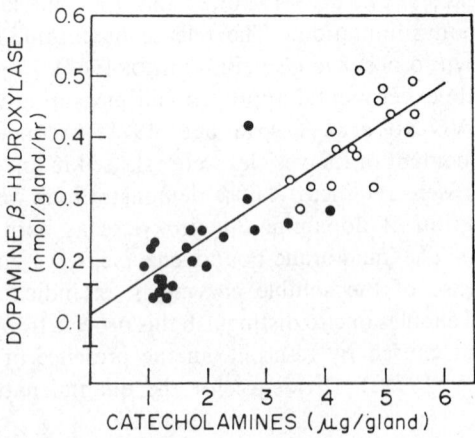

FIG. 1.—Dopamine-β-hydroxylase activity and catecholamine content of purified
storage vesicles from adrenal glands of control (○) and insulin treated (●) rats (SLOTKIN
and KIRSHNER, 1973a).

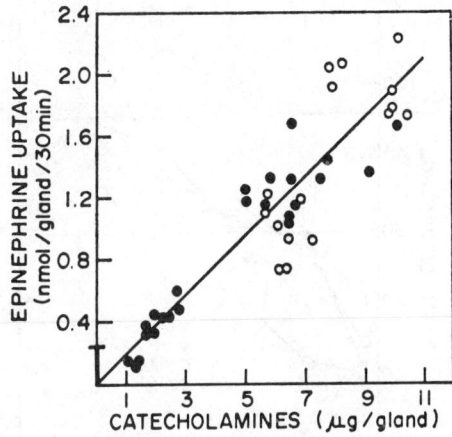

FIG. 2.—Epinephrine uptake and catecholamine content of purified storage vesicles from adrenal of control (○) and insulin treated rats (●) (SLOTKIN and KIRSHNER, 1973a).

amount of DBH would still remain in the intact vesicle fraction after complete catecholamine depletion. The larger-than-expected amounts of DBH at the 3 hr period may be due to several factors including greater contamination with vesicle membrane fragments and inclusion of newly formed vesicles with lower than normal catecholamine contents (SLOTKIN and KIRSHNER, 1973b). However, most significantly, at 1 and 2 hr after insulin when the glands contained 65 and 50 per cent respectively of the normal catecholamine content, the ratio of DBH:CA of the vesicles was the same as that of the controls.

Further evidence for disruption of the storage vesicles and loss of their physiological function upon secretion is the correlation shown in Fig. 2 between the loss of catecholamines and the loss of ability to take up exogenous amines *in vitro*. The high correlation coefficient ($r = 0.91$) and the extension of the least square line through the origin indicate a one to one relationship between catecholamine content and uptake. When the uptake was expressed in nanomoles per 100 μg of CA in the vesicles there was no difference between insulin-treated and control animals indicating that uptake per intact vesicle remained unchanged. A similar correlation between loss of catecholamines and ATP of the vesicles was also obtained. The totality of evidence thus indicates that immediately after secretion one has a reduced population of vesicles which are indistinguishable from vesicles obtained from unstimulated glands and a second population of membranes or "empty" vesicles derived from those vesicles which have responded to the secretory stimulus and which are non-functional.

RECOVERY OF CATECHOLAMINE STORES

Recovery of the catecholamine stores requires not only the synthesis of the amines but also synthesis of ATP and the proteins released and the reformation or *de novo* synthesis of new secretory vesicles. Table 1 and Fig. 3 summarise the changes which occur in purified storage vesicles of rabbit and rat during recovery. Changes in catecholamine stores of the isolated vesicle accurately reflect changes in the entire gland. In rabbit storage vesicles there is complete recovery of DBH at 48 hr but only about a 30 per cent recovery of catecholamine stores. Recovery of rat adrenals

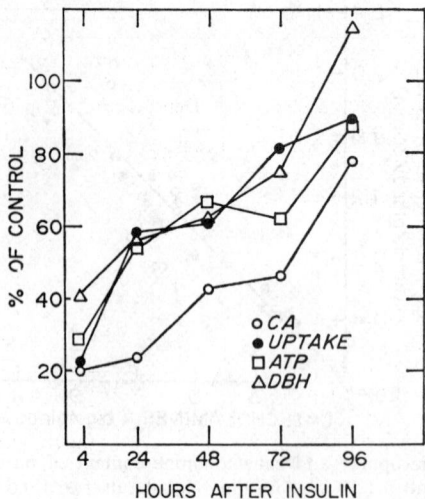

FIG. 3.—Recovery of catecholamines, ATP epinephrine uptake, and DBH activity
in the purified storage vesicle fraction of rat adrenal glands following insulin-induced
secretion (Slotkin and Kirshner, 1973b).

showed both similarities to and differences from rabbit adrenal (Viveros *et al.*,
1969b, Slotkin and Kirshner, 1973b). In the rat, as in the rabbit the recovery of
DBH preceded the recovery of catecholamines, but in the rabbit the recovery of
uptake closely paralleled the recovery of catecholamine stores while in the rat the
recovery of both uptake and ATP content more closely paralleled the recovery of
DBH. Despite these differences studies of both species indicate that vesicle formation
with its complement of proteins is not the rate limiting step in recovery, and suggest
that catecholamine synthesis itself may be rate limiting. In this regard, it is important
to note that, as first demonstrated by Viveros *et al.*, (1969b) and subsequently
confirmed by many others, there is a marked increase in the activity of tyrosine
hydroxylase between 24 and 48 hr after insulin administration.

An intact neural supply also plays a significant role in the recovery of catecholamine stores. Hökfelt (1951) and Kroneberg and Schumann (1959) showed that
denervation of the adrenal gland following insulin or reserpine treatment delayed
the recovery of the catecholamine stores. Denervation of the adrenal gland prior
to reserpine (Mueller *et al.*, 1969) or insulin administration (Weiner and Mosimann, 1970; Patrick and Kirshner, 1971a) prevents the increase in tyrosine
hydroxylase observed in intact glands. However, increases in tyrosine hydroxylase
activity can be produced in denervated glands by administering acetylcholine to
eserine and atropine-treated rats (Patrick and Kirshner, 1971a). As shown in
Table 2 treatment of reserpinised rats with acetylcholine increases both the tyrosine
hydroxylase activity and rate of recovery of the catecholamine stores. Four days
after repeated reserpine treatment the denervated glands of rats not treated with
acetylcholine had only partially recovered their catecholamine stores and showed
no increase in tyrosine hydroxylase activity while in the intact gland there was a
threefold increase in tyrosine hydroxylase activity and full recovery of its stores.
Treatment with acetylcholine resulted in a doubling of the tyrosine hydroxylase

TABLE 2. EFFECT OF RESERPINE AND ACETYLCHOLINE ON INTACT AND DENERVATED RAT ADRENAL GLANDS

Group	Treatment	Tyrosine hydroxylase		Catecholamines	
		Intact	Denervated	Intact	Denervated
		(mμmoles/hr/gland)		(μg/gland)	
	None	9·1 ± 0·5 (17)	8·8 ± 0·5	12·2 ± 0·8 (21)	9·6 ± 0·6
		(%controls)		(%controls)	
A	Reserpine	332 ± 33 (9)	103 ± 6	43 ± 5 (13)	49 ± 5
	Reserpine + acetylcholine	549 ± 26 (15)	180 ± 10	39 ± 2 (14)	43 ± 3
B	Reserpine	360 ± 27 (13)	99 ± 4	91 ± 8 (13)	66 ± 3
	Reserpine + acetylcholine	400 ± 24 (14)	198 ± 16	102 ± 6 (14)	98 ± 6

Animals were treated with reserpine for 3 consecutive days. Twenty-four hours after the last reserpine injection the animals received acetylcholine (PATRICK and KIRSHNER, 1972b). The animals in groups A and B were killed 2 and 4 days respectively after the last reserpine treatment.

activity and full recovery of the catecholamine stores of the denervated glands at four days (PATRICK and KIRSHNER, 1971b). These data suggest that the nerve-dependent increase in the rate of recovery of catecholamine stores is mediated through increases in tyrosine hydroxylase activity and subsequently increased rates of catecholamine formation.

From these studies and others (not described here) we can reconstruct the events which most probably occur during physiological secretion from the adrenal medulla. Stimulation of the splanchnic nerve causes release of acetylcholine, which, in the presence of Ca^{2+} results in the quantal, exocytotic release of the contents of the secretory vesicles. Immediately after secretion the vesicle membranes are detached from the plasma membrane and retained by the cell. During recovery new vesicles are formed which contain their normal complements of soluble proteins but which are deficient in ATP and catecholamines. Subsequently the vesicles regain their normal content of ATP and catecholamines. As a further consequence of neural stimulation tyrosine hydroxylase activity increases which most likely accelerates the rate of recovery of the catecholamine stores.

REFERENCES

BANKS P. and HELLE K. (1965) Biochem. J. 97, 40c–41c.
BLASCHKO H., COMLINE R. S., SCHNEIDER F. H., SILVER M. and SMITH A. D. (1967a) Nature, Lond. 215, 58–59.
DOUGLAS W. W. (1968) Brit. J. Pharmacol. 34, 451–474.
DOUGLAS W. W. and POISNER A. M. (1966a) J. Physiol. 183, 236–248.
DOUGLAS W. W. and POISNER A. M. (1966b) J. Physiol. 183, 249–256.
DOUGLAS W. W., POISNER A. M. and RUBIN R. P. (1965) J. Physiol. 179, 130–137.
DUCH D. S., VIVEROS O. H. and KIRSHNER N. (1968) Biochem. Pharmacol. 17, 255–264.
HÖKFELT B. (1951) Acta Physiol. Scand. 25, Suppl. 92.
KIRSHNER N., SAGE H. J., SMITH W. J. and KIRSHNER A. G. (1966d) Science 154, 529–531.
KRONEBERG G. and SCHUMANN H. J. (1959) Experientia 15, 234–235.
MUELLER R. A., THOENEN H. and AXELROD J. (1969a) J. Pharmacol. Exp. Ther. 169, 74–79.
PATRICK R. L. and KIRSHNER N. (1971a) Mol. Pharmacol. 7, 87–96.
PATRICK R. L. and KIRSHNER N. (1971b) Mol. Pharmacol. 7, 389–396.
SCHNEIDER F. H., SMITH A. D. and WINKLER H. (1967) Brit. J. Pharmacol. 31, 94–104.
SLOTKIN T. A. and KIRSHNER N. (1973a) Biochem. Pharmacol. 22, 205–219.
SLOTKIN T. A. and KIRSHNER N. (1973b) Mol. Pharmacol. 9, 105–116.

Smith A. D. and Winkler H. (1972) in *Handbook of Experimental Pharmacology, New Series*. (Blaschko, H. and Muscholl, E., Eds.) pp. 537–617. Springer-Verlag, Berlin.
Viveros O. H. In: *Handbook of Physiology*, Williams and Wilkins, Balt. (In press)
Viveros O. H., Arqueros L. and Kirshner N. (1968) *Life Sci.* **7,** 609–618.
Viveros O. H., Arqueros L., Connett R. J. and Kirshner N. (1969a) *Mol. Pharmacol.* **5,** 60–68.
Viveros O. H., Arqueros L., Connett R. C. and Kirshner N. (1969b) *Mol. Pharmacol.* **5,** 69–82.
Viveros O. H., Arqueros L. and Kirshner N. (1969c) *Mol. Pharmacol.* **5,** 342–349.
Viveros O. H., Arqueros L. and Kirshner N. (1971a) *Mol. Pharmacol.* **7,** 434–443.
Viveros O. H., Arqueros L. and Kirshner N. (1971b) *Mol. Pharmacol.* **7,** 444–454.
Weiner N. and Mosimann W. F. (1970) *Biochem. Pharmacol.* **19,** 1189–1199.

Frontiers in Catecholamine Research 1973, pp. 453 to 458. Pergamon Press. Printed in Great Britain.

NEW ASPECTS OF THE LOCALISATION OF CATECHOLAMINES IN ADRENERGIC NEURONS

J. P. TRANZER

Department of Experimental Medicine, F. Hoffmann-La Roche & Co. Ltd.,
Basle, Switzerland

ELECTRON microscopic studies have revealed that in adrenergic nerve terminals norepinephrine (NE) is localised in two types of vesicles, the large dense core (l.d.c.) vesicles and the small dense core (s.d.c.) vesicles whereas in the axon of the sympathetic nerve trunk only l.d.c. vesicles seem to exist. Biochemical subcellular fractionation studies have led to similar conclusions. Thus, in sympathetically innervated organs two types of NE storing particles, the heavy and the light were found, whereas in the nerve trunk only heavy particles could be detected. Moreover, it is now considered likely that the heavy particles are the equivalent of the l.d.c. vesicles and the light ones that of the s.d.c. vesicles. For current reviews see GEFFEN and LIVETT (1971) and SMITH (1972).

Recent re-investigations of the fine structure of adrenergic axons indicate that there may exist additional particulate amine storing compartments.

1. TUBULAR RETICULUM STORING AMINES

When tissues were prepared for electron microscopy with an improved technique for the localisation of biogenic amines (TRANZER and SNIPES, 1968; TRANZER and RICHARDS (in prep.), it appeared that in addition to l.d.c. and s.d.c. vesicles there existed in the adrenergic nerve terminals of various tissues of rats, a peculiar tubular structure which contained an electron dense, reserpine sensitive material (TRANZER, 1972). In addition, the content of this tubular reticulum showed a positive chromaffin reaction in the absence of OsO_4 treatment, similar to the content of the l.d.c. and of the s.d.c. vesicles. Both of these unrelated findings strongly suggest that in adrenergic nerve terminals, this electron dense material stored in the tubular reticulum represents catecholamines, since these techniques are commonly considered to be specific for the demonstration of biogenic amines at a fine structural level (BLOOM, 1970). No such electron dense fine structures could be observed in cholinergic axons or in any other cells. A very similar tubular reticulum was found in adrenergic axons of animals pretreated with 3,4,5-trihydroxyphenyl-ethylamine (5-hydroxydopamine), an amine which optimally traces amine storing sites (TRANZER and THOENEN, 1967b). In addition it appeared that strong variations exist from one organ to another. Thus a tubular reticulum filled with endogenous amines is very apparent in adrenergic nerve terminals of the rat mesenteric artery and iris (RICHARDS, unpublished results) but is quite rare in those of the rat vas deferens (Fig. la,b).

At the present time the functional meaning of the tubular reticulum storing amines is not clear. However, it seems reasonable to think that this fine structure, which seems to be rather labile, would liberate its amine content during homogenisation of the tissue and thereby may account for at least a part of the "pool of free amines" found in all subcellular fractionation studies.

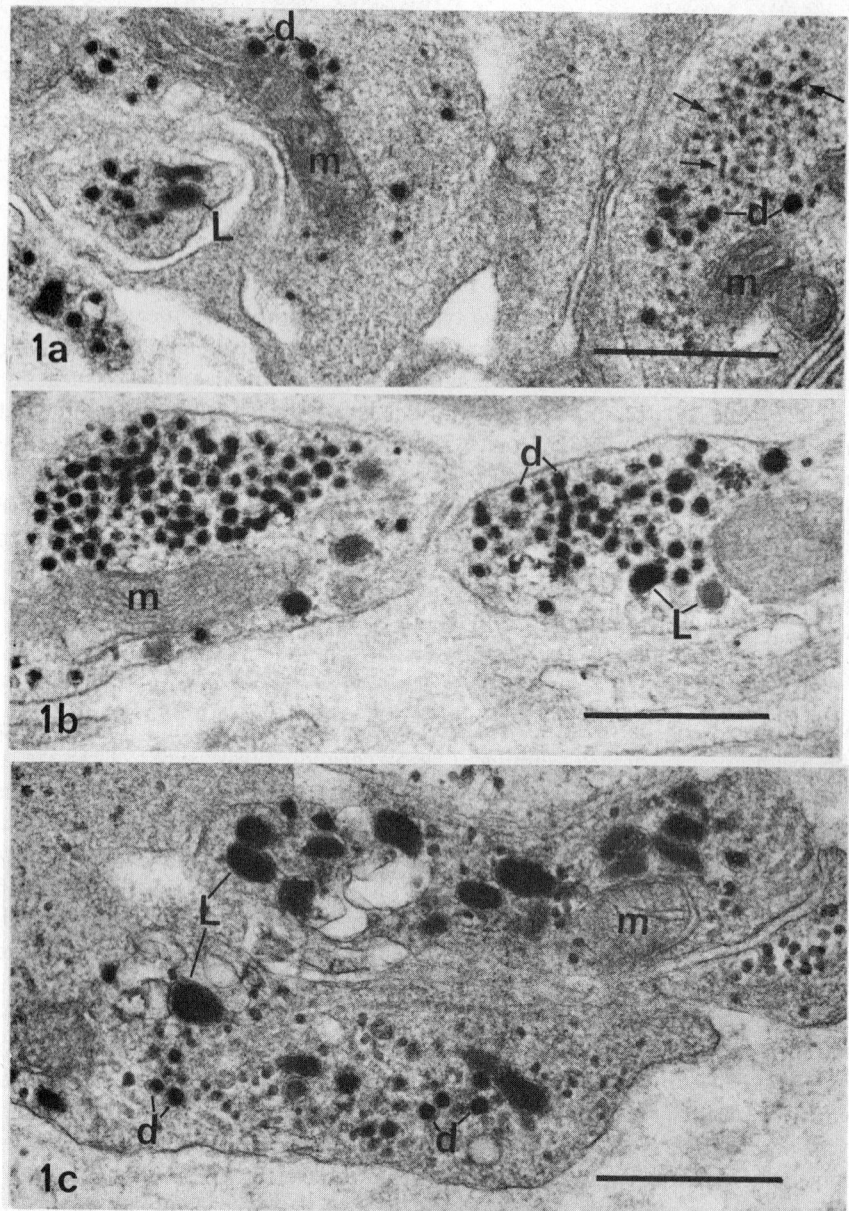

Fig. 1.—Adrenergic nerve terminals of various organs: rat mesenteric artery (a),
rat vas deferens (b) and bovine spleen capsule (c). Note the frequent occurrence of:—
a tubular reticulum storing amines (→) in (a);— small dense core vesicles in (b);—
large dense core vesicles in (c). Mitochondria (m); small dense core vesicles (d); large
dense core vesicles (L). ×49,400 (calibration bar = 0·5 μm).

Thus, fine structural evidence, both cytochemical and pharmacological, is brought for a third compartment storing amines in adrenergic axons. This compartment was predominantly apparent in varicosities, although it also occurred in non-varicose regions of the axon. It is described here as a tubular reticulum storing amines which is morphologically quite distinct from the microtubules as well as from the hitherto classical amine storing sites, the small and the large dense core vesicles.

2. AMINE STORING VESICLES IN SYMPATHETIC NERVE FIBRES

Parallel fine structural investigations of the axons of the postganglionic nerve fibres of the vas deferens (Fig. 2a) and the mesentery of rats revealed that these axons contained a noticeable number of vesicles which, from their size and shape, were very similar to those found in axon terminals of the same organ. These vesicles were indeed much less frequent in the former than in the latter (about 140 times for the vas deferens and 70 times for the mesentery), whether one equates their number per surface area of axoplasm or per axon profile. However, the proportion of the two main types of vesicles in the non terminal axon was comparable to that in the terminals. Thus, in the axons of the nerve fibres of both organs approximately 10 per cent l.d.c. vesicles (900 Å in diameter) and 90 per cent small vesicles (500 Å in diameter) were found, whereby about half of the small vesicles contained a dense core and the other half appeared empty. By comparison the axon terminals contained about 5 per cent l.d.c. vesicles (900 Å in diameter), 88 per cent l.d.c. vesicles (500 Å in diameter) and 7 per cent small empty vesicles (500 Å in diameter). Furthermore, it could be shown that the electron dense cores of the various vesicles of the nerve fibres were reserpine sensitive similarly to the dense cores of the vesicles of the axon terminals, which strongly suggests that this reserpine sensitive electron dense material represents catecholamines, most probably NE. That only approximately half of the small vesicles of the nerve fibres contained a dense core is not surprising if one takes the commonly accepted view that most of the NE is synthesised in the nerve terminals rather than in the nerve trunk. On the other hand, it has previously been shown that the s.d.c. vesicles and the small empty vesicles of the axon terminals quite probably represent the same type of vesicle, differing only in their degree of filling with amines (TRANZER and THOENEN, 1967a and b).

Since up to now almost all subcellular fractionation studies were done on bovine splenic nerve (SMITH, 1972), this nerve too was re-investigated by electron microscopy. It was found that the axon terminals (Fig. 1c) located in the spleen capsule between the smooth muscle cells contained an unusually high proportion of l.d.c. vesicles, about 40 per cent, 55 per cent s.d.c. vesicles and 5 per cent small empty vesicles. Although the axons of the nerve fibres (Fig. 2b) contained about 50 times fewer vesicles per surface area of axoplasm, the proportion of large and small vesicles was comparable to that found in the axon terminals i.e. 38 per cent l.d.c. vesicles and 62 per cent small vesicles, whereby again about half of the small vesicles contained a dense core and the other half appeared empty. That, like in the rat, the small empty vesicles represent functionally the same type of vesicles as the s.d.c. vesicles is most probable but not yet demonstrated. In any case these results indicate that the s.d.c. vesicles present in the bovine splenic nerve fibres represent at least 30 per cent of the total number of amine storing vesicles.

In addition it appeared in this study that the post mortem delay preceding the

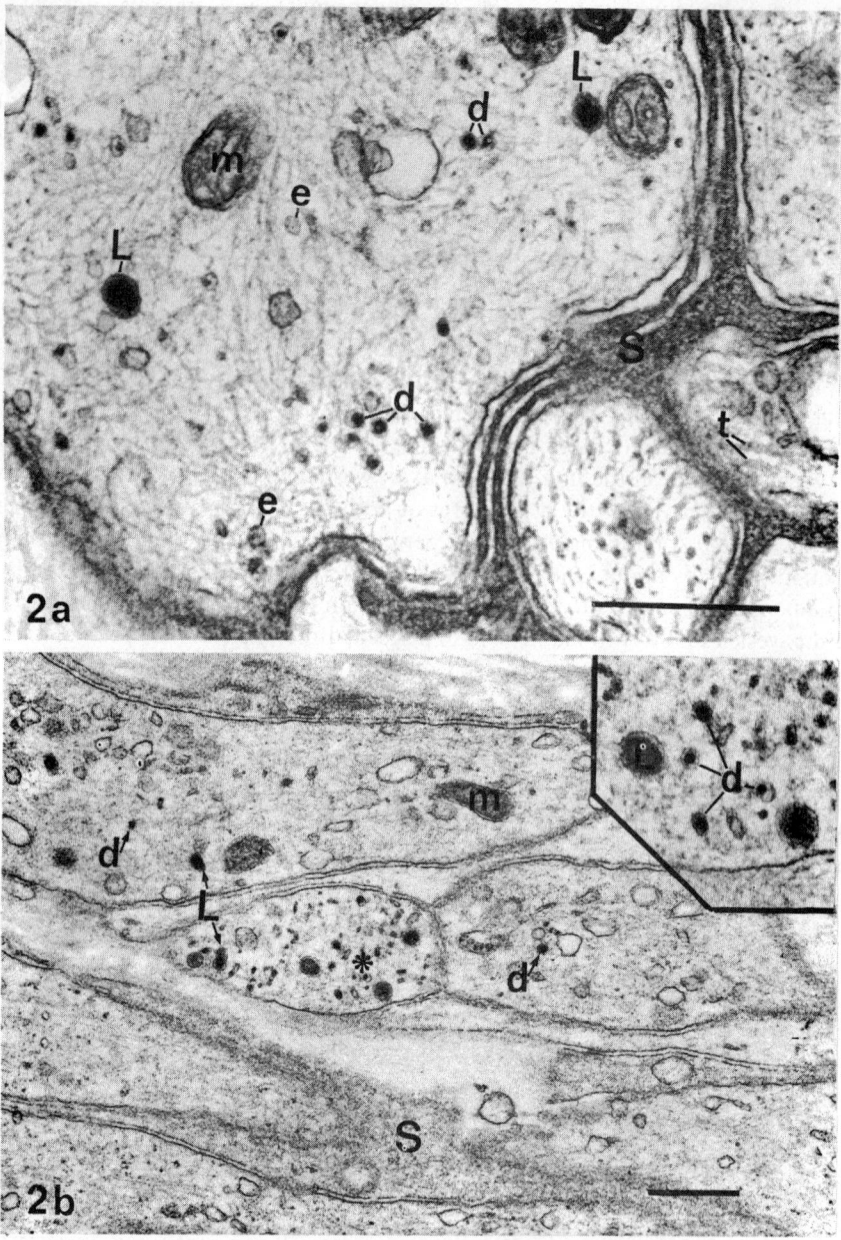

Fig. 2.—Axon profiles of adrenergic nerve fibres: rat vas deferens (a) and bovine
splenic nerve (b). The inset represents at higher magnification, a part of the axon
labelled(*). Note the presence of small dense core vesicles (d), small empty vesicles
(e) and of large dense core vesicles (L) in some of the axons. Mitochondria (m);
microtubules (t); Schwann's cell (S). Fig. 2a ×49,400; Fig. 2b ×25,200; inset ×49,400
(calibration bar = 0·5 μm).

fixation may be of great importance. The presently reported observations were made on bovine splenic nerves taken from calves as fresh as possible, i.e. less than five minutes post mortem. A similar fine structural examination of the bovine splenic nerve taken from an animal about 45 min post mortem revealed that almost no small dense core vesicles could be found in the nerve fibres, only some large dense core vesicles could be detected.

In conclusion the major finding of this investigation claims that in the nerve fibres, as in the terminals, there exist a comparatively high proportion of small vesicles capable of storing NE. This may represent as much as 80–90 per cent of the total number of vesicles of rat mesentery and vas deferens sympathetic nerves fibres and about 60 per cent of the vesicles of bovine splenic nerve fibres.

Although these results confirm and extend earlier findings (ELFVIN, 1958; HÖKFELT, 1969) they are in obvious contradiction to most of the recent fine structural investigations (ROTH et al., 1968; GEFFEN and OSTBERG, 1969; FILLENZ, 1970; KLEIN and THURESON-KLEIN, 1971) and biochemical subcellular fractionation studies of sympathetic nerves (ROTH et al., 1968; BURGER et al., 1969; DE POTTER et al., 1970; LAGERCRANTZ, 1971; KLEIN and THURESON-KLEIN, 1971). As stated above until now only heavy NE storing particles i.e. large dense core vesicles have been found in the sympathetic nerve. However, all these subcellular fractionation studies were done on bovine splenic nerve. The post mortem changes reported above may give at least a partial explanation for the apparent discrepancy in the results. On the other hand the use of fixation techniques with too low a sensitivity for the detection of biogenic amines, probably explains why in most previous fine structural investigations the s.d.c. vesicles were not detected in sympathetic nerve fibres.

Whether or not small vesicles able to store NE exist in the non terminal axon may be of some importance not only for further work attempting to isolate, purify and analyse them but also for a better comprehension of the various functional aspects of the adrenergic neuron.

Preliminary results (RICHARDS and TRANZER, in preparation) seem to indicate that the various vesicles as well as the tubular reticulum storing amine also exist in the cell body and in the dendrites of the adrenergic neuron, which confirms and extends earlier findings of HÖKFELT (1969), and speaks strongly in favour of their origin at the level of the cell body. On the other hand the present results do not favour the view that the small vesicles are formed locally in the axon terminals, directly or indirectly from the large vesicles as proposed recently by various authors but that they may migrate by neuroplasmic flow from the cell body down to the axon terminals, as is known to occur with various other subcellular organelles.

SUMMARY

The amine storing sites of adrenergic axons have been re-investigated with an improved technique for the detection and localisation of biogenic amines with the electron microscope. In addition to the well documented large and small dense core vesicles, the axons, especially their terminals, were revealed to contain a distinct third compartment, a tubular reticulum storing amines.

On the other hand, the nerve fibres while containing much less amine storing vesicles per surface area compared to their terminals, show a similar proportion of large and small vesicles able to store NE. Thus small vesicles storing amines do not only exist in the nerve terminals but also in the whole axon, and presumably in the whole neuron. It is suggested that the three different compartments storing amines are formed in the cell body and probably migrate to the axonal terminals by means of neuroplasmic flow.

REFERENCES

Burger A., Phillipu A. and Schümann H. J. (1969) *Naunyn-Schmiedebergs Arch. Pharmakol. Exp. Pathol.* **262**, 208–220.

De Potter W. P., Smith A. D. and De Schaepdryver A. F. (1970) *Tissue and Cell.* **2**, 529–546.

Bloom F. E. (1970) *Int. Rev. Neurobiol.* **13**, 27–66.

Elfvin L. G. (1958) *J. Ultrastruct. Res.* **1**, 428–454.

Fillenz M. (1970) *Proc. Roy. Soc. Ser. B Biol. Sci.* **174**, 459–468.

Geffen L. B. and Livett B. G. (1971) *Physiol. Rev.* **51**, 98–157.

Geffen L. B. and Ostberg A. (1969) *J. Physiol. (London)* **204**, 583–592.

Hökfelt T. (1969) *Acta physiol. Scand.* **76**, 427–440.

Klein R. L. and Thureson-Klein A. (1971) *J. Ultrastruct. Res.* **34**, 473–491.

Lagercrantz H. (1971) *Acta Physiol. Scand.* **82** (suppl. 366), 1–44.

Roth R. H., Stjärne L., Bloom F. E. and Giarman N. J. (1968) *J. Pharmacol. Exp. Ther.* **162**, 203–212.

Smith A. D. (1972) *Pharmacol. Rev.* **24**, 435–457.

Tranzer J. P. (1972) *Nature, New Biol.* **237**, 57–58.

Tranzer J. P. and Snipes R. (1968) *Proc. 4th Europ. Reg. Conf. Elect. Microscop., Rome*, **2**, 519–520.

Tranzer J. P. and Thoenen H. (1967a) *Experientia (Basel)* **23**, 123–124.

Tranzer J. P. and Thoenen H. (1967b) *Experientia (Basel)* **23**, 743–745.

Frontiers in Catecholamine Research 1973, pp. 459 to 461. Pergamon Press. Printed in Great Britain.

PLASTIC CHANGES OF CENTRAL NORADRENALINE NEURONS AFTER 6-HYDROXYDOPAMINE

Gösta Jonsson, Chris Pycock* and Charlotte Sachs
Department of Histology, Karolinska Institutet,
S-104 01 Stockholm, Sweden

It is generally believed that systemic injection of 6-hydroxydopamine (6-OH-DA) does not affect central catecholamine (CA) neurons which has been related to 6-OH-DA not easily passing the blood–brain barrier (BBB). However, systemic administration of 6-OH-DA (3 × 100 mg/kg, s.c.) to newborn rats leads to a permanent and selective reduction of endogenous NA and ^{3}H-NA uptake in the forebrain, e.g. cerebral cortex, when analysed in the adult stage (see Fig. 1, Sachs and Jonsson, 1972; see also Clark et al., 1972; Singh and de Champlain, 1972). 6-OH-DA thus seems to be able to pass the BBB in newborn rats and elicit its degenerative effects on NA nerve terminals. This is also supported by fluorescence histochemical results showing marked axonal accumulations of NA (a degenerative sign) in the dorsal NA bundle in animals treated with 6-OH-DA at birth. Analysing ^{3}H-NA uptake in cerebral cortex from animals of different ages (5-70 days) treated with 6-OH-DA at birth, revealed that the relative reduction in ^{3}H-NA uptake was approximately constant at all ages, being about 70 per cent. These data thus indicate that 6-OH-DA has no apparent effect on the outgrowth of those NA nerves that did not undergo degeneration in cerebral cortex (Sachs, 1973).

After administration of 6-OH-DA (3 × 100 mg/kg, s.c.) at various periods of time after birth (1–18 days) and evaluating the neurotoxic action by measuring the *in vitro* uptake of ^{3}H-NA in hypothalamus and cerebral cortex in the adult stage, it was found that the ^{3}H-NA uptake had completely recovered in animals receiving 6-OH-DA later than one week after birth (Sachs, 1973). This is certainly related to the postnatal development of BBB. A puzzling observation was, however, that 6-OH-DA (100 mg/kg, i.v.) given to adult rats led to an about 25 per cent decrease in endogenous whole brain NA and ^{3}H-NA uptake in cerebral cortex and the spinal cord which was long-lasting (Sachs and Jonsson, unpubl.). These data thus favour the view that BBB is not completely protective to the neurotoxic action of *systemically injected 6-OH-DA to adult rats* and that 6-OH-DA can *cause a damage of a certain number of NA nerve terminals in the cerebral cortex and the spinal cord*.

The obvious question arises, why 6-OH-DA does not seem to damage NA nerve terminals when injected on the 9th to 18th day, postnatally. In order to elucidate this problem, 11 days old rats were treated with 6-OH-DA (3 × 100 mg/kg, s.c.), sacrificed 3, 14 and 42 days after the first 6-OH-DA injection and ^{3}H-NA uptake in cerebral cortex slices was measured. Three days after the 6-OH-DA injection, ^{3}H-NA uptake was diminished about 25 per cent while no change was seen 14 and 28 days after the 6-OH-DA injection. Thus, there was a *transient reduction* in ^{3}H-NA uptake which may be related to degeneration followed by regeneration of NA nerve terminals.

---

* Wellcome Trust Fellow.

Further analysis showed that this recovery in ³H-NA uptake is lost between the 28th and 42nd day, postnatally. It thus seems reasonable to assume that NA nerves to the cerebral cortex have a certain regenerative capacity up to about 28 days, post-natally, which later disappears. Whether this is related to maturation of the NA neurons themselves or of extraneuronal tissue structures is at present unknown. The reason why there is a permanent damage of a large number of NA nerve terminals when the rats are treated at birth, might possibly be related to a more pronounced damage of the NA axons.

A very remarkable observation in rats treated with 6-OH-DA at birth and analysed as adults was the increase (approx. 75 per cent) in endogenous NA and ³H-NA uptake in the pons-medulla (Fig 1). Analysis of the ³H-NA uptake kinetics in the pons-

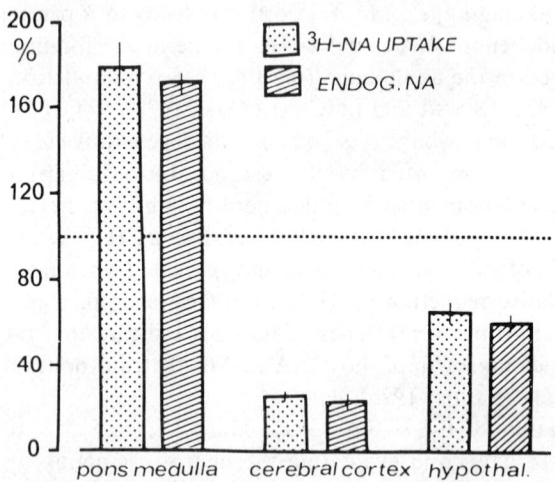

Fig. 1.—Effect of 6-OH-DA (3 × 100 mg/kg, s.c., 24 hr intervals) on the endogenous NA concentration and on the *in vitro* uptake of ³H-NA in homogenates of *pons-medulla* and slices of *cerebral cortex* and *hypothalamus*. The rats were injected at birth and sacrificed when 70–100 days old. The pons-medulla homogenates were incubated in 0·05 μM³H-NA for 5 min, while the slices from cerebral cortex and hypothalamus were incubated in 0·1 μM³H-NA for 10 min. Each column represents the mean ± S.E.M. (*n* = 6–8) and the values expressed as per cent of untreated litter mate control.

medulla showed similar K_m-values for 6-OH-DA treated and control rats. Investigating the subcellular distribution of endogenous NA, revealed that the most pronounced NA increase was seen in the microsomal fraction, which is considered to contain the amine storage granules. Fluorescence histochemistry showed a markedly reduced number of NA nerve terminals in many brain regions, e.g. cerebral cortex and hypothalamus, while the NA cell bodies located in the pons-medulla, appeared unaffected. The fluorescence histochemical data support the view that NA nerve terminals originating from the locus coeruleus are preferentially attacked. However, an increased number of NA nerve terminals, many of which displayed an increased NA fluorescence intensity, was observed in regions of the pons-medulla normally innervated by NA nerve terminals. 6-OH-DA treatment at birth thus leads to a marked NA nerve terminal denervation in the forebrain, most pronounced in the cerebral cortex, whereas in the pons-medulla this treatment induces an increase in the transmitter concentration of NA nerve terminals and probably also of a growth of NA terminals. This latter statement is supported by the observed increase in ^{3}H-NA uptake and also by the fluorescence histochemical data. The most attractive explanation to the increase in NA concentration in pons-medulla terminals is a "collateral accumulation" of NA granules (see ANDÉN et al., 1966). Assuming that the synthesis and transport of NA storage granules in the locus coeruleus cell bodies is unaltered after the 6-OH-DA treatment, which has markedly reduced their number of nerve terminals in the forebrain, this would lead to redistribution of granules and increase in NA concentration in the remaining NA nerve terminals, e.g. in the pons-medulla.

The present results show that central NA neurons may undergo plastic changes following 6-OH-DA treatment. The functional significance of this is at present unclear, but it is obvious that here mentioned processes as growth and "collateral accumulation" of NA should be considered when damaging NA neurons in the brain.

Acknowledgements—Supported by the Swedish Medical Research Council (04X-2295 and 04X-3881), Karolinska Institutet, C-B. Nathorsts Stiftelse and Svenska Livförsäkringsbolaget.

REFERENCES

ANDÉN N.-E., FUXE K. and LARSSON K. (1966) *Experientia* **22,** 842–847.
CLARK D. W., LAVERTY R. and PHELAN E. L. (1972) *Brit. J. Pharmacol.* **44,** 233–243.
SACHS CH. (1973) *J. Neurochem.* **20,** 1753–1760.
SACHS CH. and JONSSON G. (1972) *Res. Comm. Chem. Path. Pharm.* **4,** 203–220.
SINGH B. and DE CHAMPLAIN J. (1973) *Brain Res.* **48,** 432–437.

Frontiers in Catecholamine Research 1973, pp. 463 to 465. Pergamon Press. Printed in Great Britain.

CHARACTERISATION OF CORTICAL NORADRENERGIC AXON TERMINALS WITH HIGH RESOLUTION RADIOAUTOGRAPHY

LAURENT DESCARRIES and YVES LAPIERRE

Centre de Recherche en Sciences Neurologiques, Université de Montréal,
Montréal 101, Canada

NORADRENERGIC (NA) axons, demonstrated throughout mammalian cerebral iso- and allo- cortex by means of the fluorescence technique of FALCK and HILLARP (FUXE et al., 1968), presumably originate from "central norepinephrine neurons" situated in the lower brain stem. In the rat, UNGERSTEDT (1971) has shown that most cortical NA fibres reach their destination via an ascending "dorsal NA bundle" arising mainly from the locus coeruleus, that is, from a homogeneous group of some 1400 norepinephrine-containing nerve cell bodies closely packed on each side of the fourth ventricle (DESCARRIES and SAUCIER, 1972). These findings actually constitute the first anatomical description of an entire cortical afferent system identified in terms of its neurotransmitter content.

High-resolution radioautography may be used to identify and study cortical axon terminals of rat frontal cortex labeled after topical application of tritiated DL-nore-pinephrine (NA-^{3}H) (DESCARRIES and LAPIERRE, 1973; LAPIERRE et al., 1973). Three hours after administration of 300–450 μCi of NA-^{3}H (4·18–12·5 Ci/mM) in adult rats pretreated with monoamine oxidase inhibitor, the tracer is highly concentrated within minute portions of the cortical neuropil beneath areas of application. In light microscope radioautographs, these reactive sites appear as small, dense aggregates of silver grains, prominent against a ubiquitous diffuse reaction. Electron microscope radioautography demonstrates that all identifiable structures where the radioactivity is accumulated correspond to unmyelinated axons exhibiting enlargements containing synaptic vesicles. When viewed in consecutive thin sections, labeling of this type is invariably confined to the same axonal processes.

The intense axonal reactions can be attributed to specific uptake and retention of NA-^{3}H within presynaptic terminals normally containing endogenous norepine-phrine. Indeed, the configuration and repartition of the reactive sites bear a close resemblance with the distribution pattern of fluorescent NA fibres previously described in the frontal cortex (FUXE et al., 1968). Furthermore, earlier administration of 6-hydroxydopamine, in a dosage known to produce selective destruction of all cortical NA terminals (UNGERSTEDT, 1971) and eventual disappearance of most NA nerve cell bodies in locus coeruleus (DESCARRIES and SAUCIER, 1972), results in the complete absence of all axonal reactions in the frontal cortex.

In electron microscope radioautographs of material fixed by perfusion with glutaraldehyde and post-fixed in OsO$_4$, the mean diameter of cortical reactive ter-minals, calculated from planimetric measurements, is 1·15 μm. Following conven-tional development with Microdol-X, the ultrastructure of the labeled nerve endings is sometimes difficult to discern beneath dense accumulations of silver grains. However preliminary examination of radioautographs processed with the physical developer

464 L. DESCARRIES and Y. LAPIERRE

paraphenylenediamine clearly reveals typical junctional zones of synaptic contact often present between labeled cortical nerve endings and small dendritic profiles. No reactive terminals are observed in close apposition to neuronal perikarya or making synaptic contact with trunks of apical dendrites. Most labeled presynaptic enlargements contain clear synaptic vesicles in association with several large dense-core vesicles, but the latter are also seen in unlabeled synaptic terminals.

Following optimal labeling, the distribution of reactive cortical nerve endings shows a relative predominance in molecular layer I (Fig. 1). This has already been

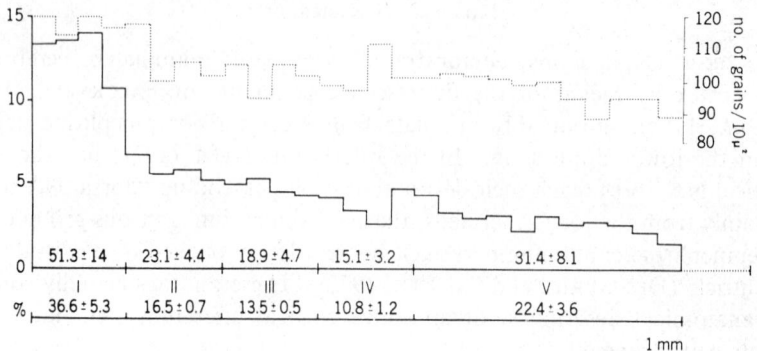

FIG. 1.—Distribution of presynaptic axons labeled with topical NA-^{3}H in the frontal cortex of rat. Data collected from 21 0·5 μm-thick sections processed by light microscope radioautography (mean numbers $\pm$ s.e. for 3 animals corrected for 1 mm²). The dotted line represents the intensity of diffuse reaction at increasing distance from the cortical surface. The greater proportion of labeled NA terminals occupies molecular layer I. In layers II to IV, reactive nerve endings are less numerous and appear somewhat equally distributed. Some are present in the upper portion of layer V, but few are found in the lower part of this layer or within layer VI. For most practical purposes, reactive terminals are confined within 1 mm of the cortical surface. In contrast with the number of labeled presynaptic axons, the intensity of diffuse reaction shows slight and gradual diminution with increasing distance from the cortical surface.

found using fluorescence histochemistry (FUXE et al., 1968) and interpreted to indicate that NA fibres could be mainly in contact with arborisations of apical dendrites issuing from pyramidal nerve cell bodies. It also suggests that scattered NA afferents whose synapses are situated far from pyramidal nerve cell bodies or axon hillocks might exert a modulatory rather than direct influence on cortical activity.

From combined data obtained by light and electron microscope radioautography, the total number of NA nerve endings in the frontal cortex is approximated at 96,600 per mm³, assuming that most NA terminals are detected. Based on current evaluations of the total number of synapses in the neocortex of rodents, this figure corresponds to a mean incidence of 1 NA terminal per $8·8–14·5 \times 10^3$ cortical synapses. In molecular layer I, the proportion increases to 1 NA nerve ending per $3·6–5·9 \times 10^3$ synapses. Such values represent a mean content of $2·33 \times 10^{-3}$ pg of endogenous norepinephrine per cortical NA terminal or an average concentration of 2900 μg (0·3%) of norepinephrine per g wet wt. of NA terminals.

Although these estimates must be interpreted with caution, they do provide an order of magnitude potentially useful in the understanding of cortical NA mechanisms operative at the synaptic level. The relative scarcity of NA terminals in cortex does not preclude an important function for widely dispersed afferents; it could in fact imply that these have strategic synaptic relationships and/or a particular mode of action. Generalised to the entire cerebral cortex, the present values remain compatible with the proposition that most cortical NA terminals have a common origin in the nerve cell bodies of the locus coeruleus: each of these neurons would then supply an average of 19,300 NA terminals in rat cerebral cortex ($5 \cdot 6$ cm^2).

Further characterisation of cortical NA axon terminals, defined in terms of their origin, morphological constituents, fine structural relationships and neurotransmitter content, should provide an excellent model for functional studies of a synaptic organisation in mammalian cerebral cortex.

Acknowledgements—The authors are indebted to Miss L. Farley, Messrs. K. C. Watkins, E. Rupnik and P. Dumouchel for skilful technical assistance, and to the Medical Research Council of Canada (L.D.) and the Conseil de la recherche médicale du Québec (Y.L.) for financial support.

REFERENCES

DESCARRIES L. and SAUCIER G. (1972) *Brain Res.* **37**, 310–316.
DESCARRIES L. and LAPIERRE Y. (1973) *Brain Res.* **51**, 141–160.
FUXE K., HAMBERGER B. and HÖKFELT T. (1968) *Brain Res.* **8**, 125–131.
LAPIERRE Y., BEAUDET A., DEMIANCZUK N. and DESCARRIES L. (1973) *Brain Res.* in press.
UNGERSTEDT U. (1971) In: *6-Hydroxydopamine and Catecholamine Neurons.* (MALMFORS T. and THOENEN H., Eds.) pp. 101–127. North-Holland, Amsterdam.
UNGERSTEDT U. (1971) *Acta physiol. scand. (Suppl.)* **367**, 1–48.

Frontiers in Catecholamine Research 1973, pp. 467 to 469. Pergamon Press. Printed in Great Britain.

MATRIX PHASE ORGANISATION IN THE ADRENAL CHROMAFFIN GRANULE

K. B. Helle and G. Serck-Hanssen

Institute of Physiology, University of Bergen, and Institute of Pharmacology,
University of Oslo, Norway

The chromaffin granule of the bovine adrenal medulla has served as a useful model for studies of storage and release mechanisms in adrenergic tissues (Stjärne, 1972). Although it has proved difficult to reveal the molecular organisation of the granule constituents, the hypothetical storage complex of catecholamines and ATP has been assumed to be stabilised by the matrix chromogranins, possibly in the form of lipoproteins (Helle, 1968; Helle and Serck-Hanssen, 1969; Mylroie and Koenig, 1971). The localisation of the granule phospholipids to the matrix phase of the lysed granules has however been obscured by their tendency to form multilayered liposomal structures which co-sediment with the membrane phase (Agostini and Taugner, 1973; Helle, 1973a). These artificial structures which readily formed in frozen and thawed preparations (Helle, 1973a), were on the other hand indications of the labile associations between phospholipids and the granule protein. Furthermore, as the liposomal structures accounted for a major source of latent chromogranin A and dopamine-β-hydroxylase (EC 1.14.2.1) activity in the lysis-resistant membrane phase, it became apparent that incorporation of these granules constituents in multilayered liposomes might give a clue to their possible organisation in the intragranular matrix of the intact amine storage granule.

In order to elucidate this possibility, the formation of chromogranin-containing phospholipid particles was carried out (Helle, 1973b), as given in Fig. 1, by application of a method primarily designed for the formation of unilayered phospholipid vesicles, i.e. liposomes (Bangham, 1968; Calissano and Bangham, 1971). The sonicated suspensions as well as the isolated particles were highly potentiated with respect to dopamine-β-hydroxylase by the presence of phospholipids in the preparations. Furthermore, the stimulating effect was proportional to increases in the lipid P/protein ratio in a manner which pointed to cooperative effects of the phospholipids on the enzymic protein. Thus the phospholipids appeared to interact with the enzyme and thereby regulate the rate of enzyme activity, in keeping with the higher specific activity of dopamine-β-hydroxylase in the liposomal fraction of the lysis-resistant membrane phase (Helle, 1973a).

The artificial phospholipid–chromogranin particles exhibited a certain capacity for uptake of noradrenaline (NA) in presence of ATP (2 nmoles NA/min/mg particle protein at 2×10^{-4} M NA and 1×10^{-3} M ATP at 37°C) and the uptake was 10 times higher than that observed for Ca^{2+}, suggestive of a preference for the amine. A maximal value of 300 nmoles NA/mg protein could be observed in a preparation containing 2 μmoles ATP/mg protein (Helle, 1973b), and the storage capacity of the artificial particles seemed therefore less than that of the intact chromaffin granules by a factor of 10. Release studies indicated that the amount of noradrenaline which was firmly bound to the particle protein was of the order of 30 nmoles/mg protein, a level

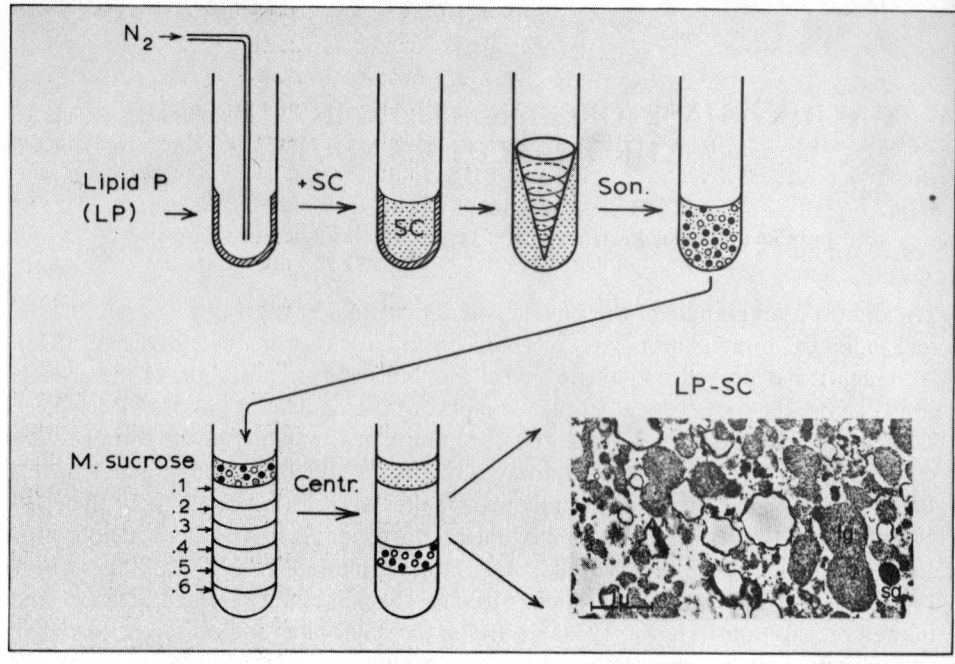

FIG. 1.—Particle formation between phospholipids and chromogranins. The inner surface of a glass tube was coated with a film of 4μmoles phospholipids under a stream of N_2. The aqueous phase contained 2mg/ml chromogranins (SC) in 60 mM KCL, 5 mM Na-succinate pH 6·0. The two phases were mixed by vortex, sonicated and fractionated on a sucrose gradient at 72 × $10^3 g_{av}$min. The ultrastructure of the major particle fraction (0·4 M sucrose layer) showed "empty" vesicles (liposomes, L) and granular aggregates (large, lg, and small, sg).

which paralleled that described for the uptake of noradrenaline in the lysis-resistant membrane phase of the granules (AGOSTINI and TANGNER, 1973). However, the presence of noradrenaline or adrenaline had a pronounced inhibitory effect on the dopamine-β-hydroxylase activity of the sonicated suspensions as well as of the isolated particles; at 3 × 10^{-6} M the inhibition was 50 per cent while at 2 × 10^{-4} M it was 75-90 per cent. Such an inhibition of the product and its N-methylated derivative on the enzymic activity suggested a binding of these components to the enzyme itself, seemingly excreting a negative, feed-back regulation of the enzyme activity. Such a mechanism has however not been described for the purified enzyme. Yet such a mechanism might serve a useful purpose if the enzyme protein also were to function as the storage matrix for the product of the enzyme reaction.

Ultrastructurely the artificial particles (Fig. 1) consisted of numerous granular aggregates between vesicles of concentric bilayers. The granular aggregates, assumed to account for the protein in the fraction, were however devoid of enveloping bilayers and the phospholipids bound to the protein would therefore have to be part of the granular organisation pattern. On the whole, the artificial granular aggregates bore close resemblance to the matrix phase of intact chromaffin granules (HELLE et al.,1971; COUPLAND, 1971). The morphological appearance of the latter changed upon influx of aqueous phase during lysis and the earlier studies have indicated that bilayers may also occur in the matrix phase. Granule-bound Ca^{2+} may possibly play a role in the stabilisation of phospholipid bilayers in the matrix phase, as the lamellar and microvesicular organisation patterns, indicative of phospholipid/water phase interactions, could be clearly seen in sections of bovine adrenal medulla fixed in presence of oxalate ions (SERCK-HANSSEN and HELLE, Unpublished observations). The observations of lamellar organisation patterns in the matrix phase of the intact chromaffin granules thus point to the presence of phospholipids, not only in the outer membrane but also intragranularly. By analogy to the artificial particles of phospholipids and chromogranins, the phospholipids of the intact matrix phase might serve unspecifically as aggregating agents and specifically as an allosteric regulator of the dopamine-β-hydroxylase at low amine concentrations.

The present comparison between the properties of the artificial phospholipid-chromogranin particles and the intact chromaffin granules has led us to propose a model for the amine storage granule founded on the principles of phase organisation between phospholipids and aqueous solutes (BANGHAM, 1968). Such a model offers explanations for a number of the characteristic features of the adrenergic storage organelles (STJÄRNE, 1972), e.g. the rapid sequestering of the hydrophobic catecholamines in contrast to the impermeability towards the hydrophilic nucleotides, the high Q_{10} of the storage mechanism and the lytic effects of hypotonic solvents. Furthermore, the stability of the intragranular store can be envisaged if one assumes that diffusion has to proceed across closed, concentric phospholipid bilayers separated by the aqueous compartments containing the soluble chromogranins, ATP and catecholamines in a multilamellar liposomal structure.

Acknowledgements—The authors are indebted to the Norwegian Research Council for Science and the Humanities for financial support.

REFERENCES

AGOSTINI B. and TAUGNER G. (1973) *Histochemie* **33**, 255–272.
BANGHAM A. D. (1968) In: *Progress in Biophysics and Molecular Biology* (BUTLER J. A. V. and NOBLE D., Eds) Vol. 18, 31–95.
CALISSANO P. and BANGHAM A. D. (1971) *Biochem. Biophys. Res. Commun.* **43** 504–509.
COUPLAND R. E. (1971) *Mem. Soc. Endocrinol.* **19**, 611–633.
HELLE K. B. (1968) *Biochem. J.* **109**, 43P–44P.
HELLE K. B. (1973a) *Biochim. Biophys. Acta* **318**, 167–180.
HELLE K. B. (1973b) *Biochim. Biophys. Acta* **318**, 181–196.
HELLE K. B. and SERCK-HANSSEN G. (1969) *Pharmacol. Res. Commun.* **1**, 25–29.
HELLE K. B., FLATMARK T., SERCK-HANSSEN G. and LÖNNING S. (1971) *Biochim. Biophys. Acta* **226**, 1–9.
MYLROIE R. and KOENIG H. (1971) *FEBS Letters* **12**, 121–124.
STJÄRNE L. (1972) In: *Handbook of Experimental Pharmacology* (BLASCHKO H. and MUSCHOLL E., Eds.) Vol. 33, pp. 231–269, Springer Verlag, Heidelberg.

Frontiers in Catecholamine Research 1973, pp. 471 to 475. Pergamon Press. Printed in Great Britain.

RELEASE OF AMINES FROM SYMPATHETIC NERVES

W. P. De Potter

Heymans Institute of Pharmacology, University of Ghent Medical School, Belgium

CELL FRACTIONATION and electron microscopic studies on adrenergic nerves and adrenal medullary cells have shown that catecholamines are present in particles or granules and indicated that some may also be free in the cell sap. Because of this dual localisation and according to which of the two pools were believed to be immediately involved in the release process, several hypotheses for the release have been put forward; including:

(i) noradrenaline may be released from the granules into the cytosol and the amines may then diffuse from the cytosol in the extracellular space;

(ii) extrusion of whole vesicles;

(iii) exocytosis, a process in which the entire soluble content of the storage granules is released directly to the exterior of the cell leaving the membrane of the granules within the cell.

Biochemical studies from several laboratories have now firmly established that, at least for the adrenal gland, the release of catecholamines occurs by exocytosis. Evidence for this secretion mechanism comes from studies which showed that: (i) granule specific proteins (chromogranins)—which occur in a soluble state within the particle—are released stoichiometrically with ATP and catecholamines, (ii) enzymes which are located in the cell sap are not released and (iii) specific components of the storage particle membrane were retained within the cell.

That this mechanism holds for adrenergic nerves has been more difficult to establish. Considerable evidence has, however, accumulated which strongly indicates that exocytosis does occur in adrenergic nerve terminals. The evidence has been extensively reviewed by SMITH and WINKLER (1972), GEFFEN and LIVETT (1971), DE POTTER et al. (1972), SMITH (1973) and therefore it is not necessary to go into great detail here. Instead, after a brief survey of our present knowledge of both storage and release characteristics, additional evidence for an exocytotic mechanism will be presented.

STORAGE CHARACTERISTICS

(i) Morphological

In the peripheral sympathetic nerve terminal two types of vesicles are found : small (40–60 nm) and large (80–120 nm), both of which are dense cored and contain catecholamines (FILLENZ, 1971). While terminals contain both types the other parts of the neuron contain almost exclusively large granular vesicles (SMITH, 1972).

(ii) Biochemical

Centrifugation studies on homogenates of sympathetic nerves and noradrenergically innervated organs have given the following information on noradrenaline storage characteristics:

(a) Axons and ganglia show a unimodal distribution of the noradrenaline on density gradients. In the ganglia the noradrenaline peak is partially paralleled by dopamine-β-hydroxylase (EC. 1.14.2.1), the rest of the enzyme activity being attributable to broken or immature vesicles. The distribution of axonal particles is paralleled quite closely by dopamine-β-hydroxylase and chromogranin A.

(b) After homogenisation and centrifugation of tissues containing nerve terminals a bimodal distribution of noradrenaline is obtained. In the rat heart (Roth et al., (1968) these particles have been termed 'light' and 'heavy' according to their position in a density gradient. The spleen shows a similar distribution of noradrenaline but when dopamine-β-hydroxylase was estimated in these gradients an interesting fact emerged. Only the 'heavy' particle was found to contain considerable amounts of dopamine-β-hydroxylase whereas the other had only little if any enzyme activity. From these experiments it was suggested that the 'heavy' particle in the terminals was the same as the one found in the axons while the other was specific to the nerve ending. These experiments on the spleen were not supported by morphological evidence but since the splenic nerve axon contains only large dense cored vesicles it seemed acceptable to equate 'heavy' with large granular vesicles.

(c) The apparent correlation between 'light' and small and 'heavy' and large granular vesicles has been further substantiated by electron microscopic analysis of gradient fractions [from splenic nerve axons and vas deferens (Lagercrantz, 1971; Bisby and Fillenz, 1971)] which were enriched on noradrenaline.

RELEASE

As has been pointed out there were several biochemical and pharmacological observations which suggested that (from adrenergic nerves) the vesicles must be immediately involved in the release process; e.g. only those amines which are stored in granules can be released by electrical stimulation of the nerve; the release of amines by electrical stimulation requires the presence of calcium (whereas tyramine induced release does not); there are several secretory systems, but in those for which biochemical or morphological evidence exists for secretion by exocytosis, calcium has always been an absolute prerequisite. Such a correlation, without of course being a proof, is at least strong indicative that noradrenaline may be secreted by exocytosis; in electrophysiological studies on the vas deferens Burnstock and Holman (1966) were able to measure spontaneous excitatory function potentials of smooth muscle cells. They interpreted their observations as being indicative of quantal release. One obvious candidate for spontaneous release of a preset number of molecules is the storage vesicle.

Despite these arguments, which perhaps look more convincing a posteriori, most of the workers believed the free or axoplasmic pool to be the pool from which the neurotransmitter was released; probably one of the reasons being that such a mechanism allowed an explanation for the feedback inhibition of tyrosine hydroxylase. A series of biochemical studies, analogous to those which have established exocytosis as the release mechanism for the adrenal medulla, have also been carried out on adrenergically innervated tissues. De Potter et al. (1969) have shown that stimulation of the splenic nerves of isolated calf and dog spleens results in a simultaneous appearance of noradrenaline and two vesicle specific proteins (dopamine-β-hydroxylase and chromogranin A). Although the amounts of the two proteins (relative to noradrena-line) were much less than those values found for lysed vesicles of splenic axons, it was shown that the proteins did come from the adrenergic neuron and not from any extra-neuronal source. It was also found that there was a striking correlation between the amounts of the two proteins and the catecholamines found in the perfusate. The release of the two proteins was calcium dependent.

Studies by others have since shown that dopamine-β-hydroxylase is also released upon stimulation of splenic nerves of the cat (GEWIRTZ and KOPIN, 1970) and of the hypogastric nerves to the vas deferens of the guinea pig (JOHNSON et al., 1971).

Again, at least for the cat spleen, there was an inflated ratio of dopamine-β-hydroxylase to noradrenaline in the perfusate as compared to the ratio in axons and probably this was also the case for release of dopamine-β-hydroxylase and chromo-granin A from the sheep spleen, as shown by GEFFEN et al. (1969), but in view of the qualitative nature of their immunochemical analyses, this information is lacking.

How do these proteins get into the extracellular space? That dopamine-β-hydroxy-lase and chromogranin A could diffuse out of the vesicle into the axoplasma and then through the plasmamembrane is very unlikely and is made even more remote in view of the fact that tyrosine hydroxylase (EC. 1.14.3.1) and dopa decarboxylase (EC. 4.1.1.26), two cytoplasmic enzymes and the latter a smaller molecule than dopamine-β-hydroxylase, were not detectable in the perfusate after stimulation of the splenic nerves. It was therefore concluded that the most likely mode of release of the vesicle proteins is exocytosis.

That the proteins are released by exocytosis is not in doubt, the question is, what proportion of the noradrenaline content released is also secreted in this way.

In view of the discrepancy between the dopamine-β-hydroxylase-noradrenaline ratio in a soluble lysate of vesicles and the ratio found in the perfusate it could be argued that exocytosis is only a minor mode of release of noradrenaline. However, it must be emphasised that this ratio was measured on axonal vesicles and not on the vesicles from which the release would occur i.e. those in the terminal. When allowance was made for the filling of the vesicle in the terminal (SMITH and WINKLER, 1972) it was found that the two ratios were closer together. It should also be pointed out however that they could never be the same. There is a terminal particle containing noradrenaline but no dopamine-β-hydroxylase, there is a re-uptake of some released noradrenaline and there is no way of knowing what rate small and large molecules will cross into the perfusate. Keeping these points in mind it seems unnecessary to attribute a minor role to exocytosis solely on grounds of differing ratios of nor-adrenaline to protein. The fact that dopamine-β-hydroxylase is released from nerves is only evidence of exocytosis from the large vesicles. Until specific soluble proteins are identified within the small vesicles, it will not be possible to use this approach to look for exocytosis from these vesicles. However, much circumstantial evidence exists which is consistent with the release of noradrenaline by exocytosis from small vesicles, e.g. electron microscopic studies, the calcium dependence of release, the parallelism between release of dopamine-β-hydroxylase and noradrenaline under many experi-mental conditions.

However, it is striking that in one experimental situation the release of nor-adrenaline evoked by tyramine, dopamine-β-hydroxylase is not released (CHUBB et al., 1972a). This finding indicates that tyramine does not evoke release by exocytosis and is consistent with other studies which have shown that tyramine acts in a different way to the nerve impulse (SMITH, 1973).

So far we have described observations where use has been made of the soluble property of dopamine-β-hydroxylase. However, since some 80 per cent of the enzyme is bound to the large vesicle membrane it should be possible to use that property to look for any changes that might occur after exocytosis. A simple diffusion of noradrenaline out of the vesicle would only result in slight changes in density; in other words the vesicle would still be intact but empty (CHUBB et al., 1972b). Exocytosis on the other hand must in one or other way directly affect the particle. Preliminary evidence suggests that the distribution of dopamine-β-hydroxylase on sucrose density gradient was altered by electrical stimulation of the splenic nerve prior to homogenisation (CHUBB et al., 1972b). In these experiments only a small depletion of noradrenaline was found. It was therefore thought better to use conditions where a massive depletion would occur and since α-adrenergic blocking agents are known to increase the release of both transmitter and dopamine-β-hydroxylase, density gradients experiments on spleens stimulated in the presence of phentolamine or phenoxybenzamine were performed. It was found that in the presence of these drugs the spleens could be depleted of their noradrenaline content by 50 and 80 per cent respectively.

Sucrose density gradients of spleens treated with both drugs during nerve stimulation show a marked shift of dopamine-β-hydroxylase to the low density region of the gradient. The amount of noradrenaline was too low to be measured accurately after phenoxybenzamine but phentolamine treated spleens showed a major peak of noradrenaline associated with the low density peak of dopamine-β-hydroxylase. Omission of Ca^{2+} from the perfusion medium completely prevented these changes.

In experiments where phenoxybenzamine was present during nerve stimulation, subsequent infusion of L-dopa led to a partial restoration of the noradrenaline content of the spleen. It was striking that this newly-formed noradrenaline was found only in the less dense type of vesicle. This is in contrast to the situation in untreated spleens where newly-synthesised noradrenaline can be found in both large and small vesicles. Furthermore, infusion of L-dopa did not result in a return of the low density dopamine-β-hydroxylase to the position in the gradient of unstimulated spleens.

These results show that after prolonged stimulation the storage function of the large vesicles is lost, presumably because they have disappeared. Which vesicles are present in the spleen after stimulation? Under the experimental conditions described above, it seems clear that a distinct population of vesicles has been formed from the normal large vesicles found in the spleen. These vesicles have at least the capacity to synthesise noradrenaline since they contain dopamine-β-hydroxylase and it is obvious that they can store the transmitter (in view of their noradrenaline content, after density gradient centrifugation.)

These experiments show that a second population of vesicles (small vesicles?) can be formed from the large particle but probably only under exceptional circumstances and by an all or none process. There is no evidence from these experiments for a gradual transformation from large to small vesicles which, when coupled with size distribution studies of GEFFEN and OSTBERG (1969) argue against such a gradual change.

Acknowledgements—It is a pleasure to thank Drs. A. D. Smith and I. Chubb for the help in the preparation of this manuscript.

REFERENCES

BISBY M. A. and FILLENZ M. (1971) *J. Physiol. (Lond.)* **215,** 163–179.

BURNSTOCK G. and HOLMAN M. E. (1966) *Pharmacol. Rev.* **18,** 481–493.

CHUBB I. W., DE POTTER W. P. and DE SCHAEPDRYVER A. F. (1972a) *Naunyn-Schmiedeberg's Arch. Pharmacol.* **274,** 281–286.

CHUBB I. W., DE POTTER W. P. and DE SCHAEPDRYVER A. F. (1972b) *Experientia* **28,** 293–294.

DE POTTER W. P., DE SCHAEPDRYVER A. F., MOERMAN E. J. and SMITH A. D. (1969) *J. Physiol. (Lond.)* **204,** 102P–104P.

DE POTTER W. P., CHUBB I. W. and DE SCHAEPDRYVER A. F. (1972) *Arch. int. Pharmacodyn. Ther.* Suppl. **196,** 258–287.

FILLENZ M. (1971) *Phil. Trans. Roy. Soc. Lond. (Ser. B)* **261,** 319–323.

GEWIRTZ G. P. and KOPIN I. J. (1970) *Nature (Lond.)* **227,** 406–407.

GEFFEN L. B. and LIVETT B. G. (1971) *Physiol. Rev.* **51,** 98–157.

GEFFEN L. B. and OSTBERG A. (1969) *J. Physiol. (Lond.)* **204,** 583–592.

GEFFEN L. B., LIVETT B. G. and RUSH R. A. (1969) *J. Physiol. (Lond.)* **204,** 58P–59P.

JOHNSON D. G., THOA N. B., WEINSHILBAUM R., AXELROD J. and KOPIN I. (1971) *Proc. Natn. Acad. Sci., U.S.* **68,** 2227–2230.

LAGERCRANTZ (1971) *Acta Physiol. Scand. Suppl.* **366,** 1–44.

ROTH R. A., STJÄRNE L., BLOOM F. E. and GIARMAN N. J. (1968) *J. Pharmac. exp. Ther.* **162,** 203–212.

SMITH A. D. (1972) *Pharmacol. Rev.* **24,** 435–457.

SMITH A. D. (1973) In: *Catecholamines* (IVERSEN L. L., Ed.) *Br. Med. Bull.* **29,** 123–129.

Frontiers in Catecholamine Research 1973, pp. 477 to 482. Pergamon Press. Printed in Great Britain.

MECHANISMS OF EXOCYTOSIS

ALAN M. POISNER

Dept. of Pharmacology, University of Kansas, Medical Center,
Kansas City, Kansas 66103, U.S.A.

EXOCYTOSIS is a method of discharge of secretory products in which there is selective release of granule-bound material. Other forms of secretion may cause non-selective release of cellular constituents. In the adrenal medulla, for instance, elevated pH causes the release not only of catecholamines but also of a cytoplasmic marker, lactic dehydrogenase (LDH) (SCHNEIDER, 1969). This is probably indicative of cell damage. Reserpine can release catecholamines without discharge of other soluble granule constituents (VIVEROS et al., 1969). This represents another mechanism of catecholamine release and is also different from exocytosis. Evidence that exocytosis occurs in the adrenal medulla in response to the physiological transmitter, acetylcholine, and other drugs has been reviewed elsewhere (cf. DOUGLAS, 1968; RUBIN, 1970; KIRSHNER and VIVEROS, 1972). The present review will discuss some of the factors believed to be involved in exocytisis in the adrenal medulla.

CALCIUM

Extracellular versus intracellular calcium: Extracellular calcium (or a rapidly equilibrating pool) is required for the release of catecholamines induced by cholinergic agents and potassium chloride (DOUGLAS and RUBIN, 1961, 1963) and by biogenic amines and polypeptides (POISNER and DOUGLAS, 1966). Agents which can initiate catecholamine release in the absence of extracellular calcium include caffeine (POISNER, 1971, 1973a; RAHWAN et al., 1973), aminophylline (POISNER, 1971, 1973b), thymol (POISNER, 1973c), and cyclic AMP (PEACH, 1972). The methyl xanthines probably stimulate exocytosis since catecholamine release in calcium-free media is accompanied by the release of high concentrations of ATP without release of LDH. The ratio of catecholamines to ATP approaches 5·0 in some experiments (POISNER, 1973d). Vinblastine can also release catecholamines and ATP in calcium-free media without discharging LDH (POISNER, unpublished). Caffeine is known to mobilise calcium in muscle and this is thought to be the basis for its action in facilitating muscle contraction. Studies on the adrenal medulla indicate that caffeine can mobilise calcium during catecholamine release (RAHWAN et al., 1973). Further similarities between catecholamine release and muscle contraction will be discussed later. The role of intracellular calcium in catecholamine release evoked by various agents is discussed elsewhere (POISNER, 1970a, 1973a, b, c, d, e; POISNER and HAVA, 1970; RAHWAN et al., 1973).

Intracellular sites of action of calcium: There are a number of potential sites of action of calcium in promoting exocytosis. Calcium is taken up into the adrenal medulla during secretory activity (DOUGLAS and POISNER, 1962; DOUGLAS et al., 1967) and is distributed among various cell organelles (BOROWITZ, 1969). Mitochondria and microsomes prepared from the adrenal medulla show an energy-dependent uptake of calcium (POISNER and HAVA, 1970). These organelles may operate in the intact cell to keep the cytoplasmic calcium at a low level as has been suggested for other tissues. It has been proposed that the influx of sodium and calcium may trigger the release of calcium from membrane stores (POISNER, 1973c, e). One site of action of calcium could be the chromaffin granule. Calcium binds to chromaffin granules and neutralises their fixed charges (BANKS, 1966). It has been suggested that calcium facilitates the binding of chromaffin granules to the plasma membrane preparatory to the discharge process (BANKS, 1966; POISNER and TRIFARO, 1967). Other actions of calcium may relate to effects on contractile proteins, microtubule protein, and the fusion reaction. These will be discussed later.

ATP

Inhibitors of ATP synthesis increase the spontaneous release of catecholamines from the adrenal gland while inhibiting the evoked release (RUBIN, 1969). These findings have been interpreted along with the evidence that ATP activates calcium uptake by cell organelles (POISNER and HAVA, 1970) and that ATP can release catecholamines from isolated chromaffin granules (POISNER and TRIFARO, 1967) (see below). Thus low ATP could lead to elevated cytoplasmic calcium as the energy-dependent accumulation (and/or pumping) of calcium is reduced. This elevated level of calcium could then trigger the secretory process. A subsequent requirement for ATP in drug-evoked catecholamine release would then be inhibited by the low level of this nucleotide. ATP may be a key factor in other facets of catecholamine release. First, it may provide the energy utilised in intracellular transport in concert with microtubular elements. Thus for the ultimate movement of granules from their site of synthesis to the cell surface an energy-dependent mechanism may be involved as has been suggested for fast axoplasmic transport (OCHS, 1971). Secondly, ATP could be an important factor in the fusion reaction occurring at the plasma membrane (POISNER and TRIFARO, 1967; see also below). Finally, ATP serves as the precursor of cyclic AMP which may support some secretory activity.

CYCLIC AMP

Three of the criteria for second messenger status have been met for cyclic AMP in adrenal medullary secretion: (1) cyclic AMP is present in the adrenal medulla (in high concentrations) and its level increases when secretion is stimulated *in vivo* (GUIDOTTI and COSTA, 1972a, b), (2) cyclic AMP itself can initiate catecholamine release from the adrenal medulla (PEACH, 1972), and (3) aminophylline, a phosphodiesterase inhibitor which raises the level of cyclic AMP in the adrenal medulla, stimulates catecholamine release *in vivo* (GUIDOTTI and COSTA, 1972a, b) and *in vitro* (POISNER, 1973b). An important missing criterion is activation of adenyl cyclase in adrenal medullary membrane fragments by secretogogues. In fact, SERCK-HANNSEN et al. (1972) have reported lack of activation of adrenal medullary adenyl cyclase by acetylcholine. Although this negative result may have been related to activation of adenyl cyclase during preparation, this important piece of evidence implicating cyclic AMP as a second messenger in cholinergic stimulation of the adrenal medulla is still lacking. A likely possibility is that muscarinic stimulation of the adrenal medulla may activate guanyl cyclase with production of cyclic GMP. Stimulation of cyclic GMP accumlation by muscarinic agonists has been reported for brain, heart and intestinal smooth muscle (LEE et al., 1972). In preliminary experiments we have found that cyclic GMP stimulates catecholamine release from the perfused bovine adrenal.

How could cyclic nucleotides promote exocytosis? One way is by causing a redistribution of intracellular calcium. This has been discussed previously (POISNER, 1970a, 1973c). Evidence that cyclic AMP can mobilise calcium from other cells has been presented (BORLE, 1971; CHI and FRANCIS, 1971; SOMLYO et al., 1971; PRINCE et al., 1972). Another way in which cyclic AMP could promote exocytosis is through activation of protein kinases. Endogenous substrates for cyclic AMP-activated protein kinases include microtubule protein (GOODMAN et al., 1970), secretory granules (LABRIE et al., 1971), and synaptic vesicles and plasma membrane (JOHNSON et al., 1971; LABRIE et al., 1971). Chromaffin granules seem to possess protein kinase activity and also serve as substrate for phosphorylation by ATP (TRIFARO and DWORKIND, 1971; TRIFARO, 1972). Whether this kinase is activated by cyclic AMP remains to be seen. Once these various sites are phosphorylated, they may bind calcium and/or lead to interaction between cell components: granule-granule; granule-plasma membrane; granule-microtubule. For futher discussion of cyclic AMP and catecholamine release, see POISNER (1970a, 1973c).

MICROTUBULES

Morphological, biochemical, and functional evidence suggest that microtubules play a role in catecholamine release (POISNER, 1973f; REDBURN et al., 1972; TOWMEY and POISNER, 1972; POISNER and BERNSTEIN, 1971), possibly in facilitating transport of chromaffin granules to the cell surface. Antimitotic agents which inhibit catecholamine release include colchicine, vinblastine, vincristine (POISNER and BERNSTEIN, 1971), hexylene glycol, griseofulvin and podophyllotoxin (POISNER, 1973f). Deuterium oxide acts in the opposite direction, potentiating catecholamine release (POISNER and BERNSTEIN, 1971). Similar results have been obtained on histamine release from mast cells (GILLESPIE et al., 1968). KCl and aminophylline are relatively resistant to colchicine and vinblastine (POISNER, unpublished), which may reflect a site of action of the antimitotic agents other than microtubule protein. Alternatively there may be different mechanisms of action by various secretogogues (such as mobilisation of different pools of calcium). Vinblastine does not act like hexamethonium in blocking nicotine; it may be acting at a calcium binding site (POISNER, 1973f; WILSON et al., 1970). Calcium itself can control microtubule assembly (WEISENBERG, 1972). High concentrations of vinblastine (more than 5×10^{-5}M) cause release of catecholamines from the bovine adrenal gland and from isolated chromaffin granules. This could be related to its interaction with microtubule protein or actin-like protein in the granule and plasma membrane. Membrane-bound proteins with properties similar to tubulin and actin have been reported in a wide variety of tissues, including the adrenal medulla (REDBURN et al., 1972; TWOMEY and POISNER, 1972; POISNER, 1973c; POISNER, unpublished). There may be an interaction between tubulin and contractile protein in the adrenal medulla (TWOMEY and POISNER, 1972) as has been suggested for other tissues (PUSZKIN and BERL, 1970; PUSZKIN et al., 1971). The possibility that microfilaments are involved in catecholamine release has been suggested since nicotine-induced catecholamine release is inhibited by cytochalasin B (POISNER, 1973f). This agent (which disrupts microfilaments) also inhibits KCl-evoked catecholamine release (POISNER, unpublished) with 50 per cent inhibition at 2×10^{-5}M. Since cytochalasin B acts at more than one site, the precise role of microfilaments in catecholamine release is far from clear. It is of interest, however, that cytochalasin B and the antimitotic agents also block the release of norepinephrine from sympathetic nerves induced by electrical stimulation (THOA et al., 1972), nicotine, and potassium chloride (SORIMACHI et al., 1973).

CONTRACTILE PROTEINS

Similarities between secretion and contraction of muscle have led to the proposal that a true contractile event is involved in catecholamine release (POISNER and TRIFARO, 1967; POISNER, 1970a, 1973c). This view is supported by the fact that contractile-like protein can be found in the adrenal medulla and in chromaffin granules (POISNER, 1970a, b, 1973c). ATP-evoked release of catecholamines from isolated chromaffin granules, which is inhibited by hypertonicity, high pH, high concentrations of ATP, and inhibitors of granule ATPase (POISNER and TRIFARO, 1967, 1968, 1969; OKA et al., 1967), has been compared to ATP-induced contraction of isolated actomyosin (POISNER, 1970a). Both processes are ordinarily independent of calcium in vitro in contrast to the in vivo requirements. However, native tropomyosin sensitises actomyosin to calcium (EBASHI and ENDO, 1968) and a protein extracted from the adrenal medulla in a similar way also sensitises isolated chromaffin granules to ATP-induced catecholamine release (POISNER, unpublished). The possible role of tropomyosin-like factors has been discussed (POISNER, 1970a, 1973c). OKA et al., (1972) recently reported that a fraction of adrenal medullary supernatant also sensitises chromaffin granules to calcium in ATP-induced release. N-ethylmaleimide (NEM), an inhibitor of granule ATPase, inhibits release of catecholamines from granules and acetylcholine-induced release from the intact gland (POISNER and TRIFARO, 1967; FERRIS et al., 1970). Thus the parallels on all levels between secretion and contraction continue to accumulate. In addition to the contractile event considered to occur in chromaffin granules, other contractile activity may take place in conjunction with intracellular transport (with microtubules) and in the plasma membrane.

FUSION PROCESS

The process by which the granule membrane becomes transiently fused with the surface membrane is not well understood. Lysolecithin can cause membrane fusion (POOLE et al., 1970; LUCY, 1970) and it has been suggested that it acts in this way during catecholamine release (WINKLER, 1971). Calcium under some conditions can promote membrane fusion (TOISTER and LOYTER, 1971). It has previously been proposed that calcium along with ATP and ATPase participates in the fusion process in the chromaffin cell (POISNER and TRIFARO, 1967), A recent review on theories of membrane fusion suggests that calcium, ATP, and membrane ATPase are important regulators (POSTE and ALLISON, 1971). It is of interest that ATP and ATPase have been implicated in endocytosis (BEN-BASSAT et al., 1972; PENNISTON, 1972).

CONCLUSION

Since calcium, ATP, ATPase and cyclic AMP are known to regulate a large number of biological processes, it is to be expected that many of the individual steps in the secretory process are modulated by these factors. It is also likely that various secretogogues may have differential effects on these cell components and thus initiate exocytosis by different routes. This possibility allows for alternative mechanisms for stimulation or inhibition of catecholamine release.

REFERENCES

BANKS R. (1966) *Biochem. J.* **101**, 18c–20c.
BEN-BASSAT I., BENSCH K. G. and SCHRIER S. L. (1972) *J. Clin. Invest.* **51**, 1833–1844.
BORLE A. B (1971) *Fedn. Proc.* **30**, 41.
BOROWITZ J. L. (1969) *Biochem. Pharmacol.* **18**, 715–723.
CHI Y. Y. and FRANCIS D. (1971) *J. Cell. Physiol.* **77**, 169–173.
DOUGLAS W. W. (1968) *Brit. J. Pharmacol.* **34**, 451–474.
DOUGLAS W. W., KANNO T. and SAMPSON S. R. (1967) *J. Physiol.* **191**, 107–121.
DOUGLAS W. W. and POISNER A. M. (1962) *J. Physiol.* **162**, 385–392.
DOUGLAS W. W. and RUBIN R. P. (1961) *J. Physiol.* **159**, 40–57.
DOUGLAS W. W. and RUBIN R. P. (1963) *J. Physiol.* **167**, 288–310.
EBASHI S. and ENDO M. (1968) *Progr. Biophys. Mol. Biol.* **18**, 123–183.
FERRIS R. M., VIVEROS O. H. and KIRSHNER N. (1970) *Biochem. Pharmacol.* **19**, 505–514.
GILLESPIE E., LEVINE R. J. and MALAWISTA S. E. (1968) *J. Pharmacol. Exp. Ther.* **164**, 158–165.
GOODMAN D. B. P., RASMUSSEN H., DIBELLA F. and GUTHROW Jr., C. E. (1970) *Proc. Natn. Acad. Sci. USA.* **67**, 652–659.
GUIDOTTI A. and COSTA E. (1972a) *Fedn. Proc.* **31**, 555.
GUIDOTTI A. and COSTA E. (1972b) *Proc. 5th Int. Congr. Pharmacol.* 9 Abs.
JOHNSON E. M., MAENO H. and GREENGARD P. (1971) *J. Biol. Chem.* **246**, 7731–7739.
KIRSHNER N. and VIVEROS O. H. (1972) *Pharmacol. Rev.* **24**, 385–398.
LABRIE F., LEMAIRE S., POIRIER G., PELLETIER G. and BOUCHER R. (1971) *J. Biol. Chem.* **246**, 7311–7317.
LEE T. P., KUO J. F. and GREENGARD P. (1972) *Proc. Natn. Acad. Sci. USA.* **69**, 3287–3291.
LUCY J. A. (1970) *Nature Lond.* **227**, 815–817.
OCHS S. (1971) *Proc. Natn. Acad. Sci. USA* **68**, 1279–1282.
OKA M., OHUCHI T., YOSHIDA H. and IMAIZUMI R. (1967) *Jap. J. Pharmacol.* **17**, 199–207.
OKA M., IZUMI F. and KASHIMOTO T. (1972) *Jap. J. Pharmacol.* **22**, 207–214.
PEACH M. J. (1972) *Proc. Natn. Acad. Sci. USA* **69**, 834–836.
PENNISTON J. T. (1972) *Arch. Biochem. Biophys.* **153**, 410–412.
POISNER A. M. (1970a) *Adv. Biochem. Psychopharmacol.* **2**, 95–108.
POISNER A. M. (1970b) *Fedn. Proc.* **29**, 545.
POISNER A. M. (1971) *Fedn. Proc.* **30**, 445.
POISNER A. M. (1973a) *Proc. Soc. Exp. Biol. Med.* **142**, 103–105.
POISNER A. M. (1973b) *Biochem. Pharmacol.* **22**, 469–476.
POISNER A. M. (1973c) In: *Frontiers in Neuroendocrinology 1973.* (GANGONG, W. F. and MARTINI L., Eds.) pp. 33–59. Oxford University Press, New York.
POISNER A. M. (1973d) *Proc. 8th Midwest Conf. Endocrinol. Metab.* (in press).
POISNER A. M. (1973e) *Proc. 5th Internat. Congr. Pharmacol.* (in press).
POISNER A. M. (1973f) *Proc. 4th Internat. Congr. Endocrinol.* (in press).
POISNER A. M. and BERNSTEIN J. (1971) *J. Pharmacol. Exp. Ther.* **177**, 102–108.
POISNER A. M. and DOUGLAS W. W. (1966) *Proc. Soc. Exp. Biol. Med.* **123**, 62–64.
POISNER A. M. and HAVA M. (1970) *Mol. Pharmacol.* **6**, 407–415.
POISNER A. M. and TRIFARO J. M. (1967) *Mol. Pharmacol.* **3**, 561–571.
POISNER A. M. and TRIFARO J. M. (1968) *Mol. Pharmacol.* **4**, 196–199.
POISNER A. M. and TRIFARO J. M. (1969) *Mol. Pharmacol.* **5**, 294–299.
POOLE A. R., HOWELL J. I. and LUCY J. A. (1970) *Nature, Lond.* **227**, 810–813.
POSTE G. and ALLISON A. C. (1971) *J. Theor. Biol.* **32**, 165–184.
PRINCE W. T., BERRIDGE M. J. and RASMUSSEN H. (1972) *Proc. Natn. Acad. Sci. USA* **69**, 553–557.
PUSZKIN E. and BERL S. (1970) *Nature, Lond.* **225**, 558–559.
PUSZKIN E., PUSZKIN S. and ALEDORT L. M. (1971) *J. Biol. Chem.* **246**, 271–276.
RAHWAN R. G., BOROWITZ J. L. and MIYA T. S. (1973) *J. Pharmacol. Exp. Ther.* **184**, 106–118.
REDBURN D. A., POISNER A. M. and SAMSON Jr., F. E. (1972) *Brain Res.* **44**, 615–624.
RUBIN R. P. (1969) *J. Physiol.* **202**, 197–209.
RUBIN R. P. (1970) *Pharmacol. Rev.* **22**, 389–428.
SCHNEIDER F. H. (1969) *Biochem. Pharmacol.* **18**, 101–107.
SERCK-HANSSEN G., CHRISTOFFERSEN T., MØRLAND J. and OSNES J. B. (1972) *Europ. J. Pharmacol.* **19**, 297–300.
SOMLYO A. P., SOMLYO A. V. and FRIEDMAN N. (1971) *Ann. N.Y. Acad. Sci.* **185**, 108–114.
SORIMACHI M., OESCH F. and THOENEN H. (1973) *Naunyn-Schmiedeberg Arch. Pharmacol.* **276**, 1–12.

THOA N. B., WOOTEN G. R., AXELROD J. and KOPIN I. J. (1972) *Proc. Nat. Acad. Sci. USA* **69**, 520–522.
TOISTER J. and LOYTER A. (1971) *Biochem. Biophys. Acta.* **241**, 719–724.
TRIFARO J. M. (1972) *FEBS Letters* **23**, 237–240.
TRIFARO J. M. and DWORKIND J. (1971) *Mol. Pharmacol.* **7**, 52–65.
TWOMEY S. L. and POISNER A. M. (1972) *Brain Res.* **46**, 341–347.
VIVEROS O. H., CONNET R. J. and KIRSHNER N. (1969) *Mol. Pharmacol.* **5**, 69–82.
WEISENBERG R. C. (1972) *Science* **177**, 1104–1105.
WILSON L., BRYAN J., RUBY A. and MAZIA D. (1970) *Proc. Nat. Acad. Sci. USA* **66**, 807–814.
WINKLER H. (1971) *Phil. Trans. Roy. Soc. Lond. B.* **261**, 293–303.

Frontiers in Catecholamine Research 1973, pp. 483 to 489. Pergamon Press. Printed in Great Britain.

IMMUNOLOGICAL STUDIES OF DOPAMINE β-HYDROXYLASE AS A MARKER FOR ADRENERGIC SYNAPTIC VESICLES

L. B. GEFFEN and R. A. RUSH

School of Medicine, Flinders University of South Australia, Bedford Park, S. A. 5042, Australia and Roche Institute of Molecular Biology, Nutley, N.J. 07110, U.S.A.

SEVERAL alternative life cycles for adrenergic synaptic vesicles have been proposed that differ regarding their origin, transport and fate within the neurone (GEFFEN and LIVETT, 1971; DAHLSTRÖM, 1971). Among the alternatives that have been proposed are that synaptic vesicles are formed either in the neurone cell body or locally in axon terminals; that vesicles release transmitter either through the axon membrane by exocytosis or within the axon terminal from which it then diffuses extracellularly; and that each vesicle either repetitively releases transmitter or is annihilated in the transmission of a single nerve impulse.

Study of these questions has been facilitated in adrenergic axons by the availability of a relatively homogenous and accessible source of homologous organelles, the adrenal medullary chromaffin vesicles. The subcellular dynamics of their life cycle have been established in some detail (KIRSHNER and KIRSHNER, 1971). The enzyme dopamine β-hydroxylase (DBH) (EC 1.14.2.1) is the principal protein constituent of the membrane of chromaffin vesicles and, on the basis that a variable proportion of DBH can be solubilized by the lysis of these vesicles, a distinction has been made between 'soluble' and 'particulate' DBH (KIRSHNER and KIRSHNER, 1971) or 'membrane bound' and 'membrane enclosed' forms of the enzyme (FOLDES, JEFFERY, PRESTON and AUSTIN, 1972). However, no biochemical differences have been detected between the two forms (FOLDES et al., 1972) and it is conceivable that in a heterogenous population of vesicles, the enzyme may be incorporated into vesicle membranes to different degrees, depending on the state of maturity of the vesicle.

DBH has been assayed both enzymatically and immunologically as a marker for adrenergic synaptic vesicles. Enzymatic assays of DBH are complicated by the lability of the enzyme activity under various extraction conditions and do not permit either histochemical localization of the enzyme nor detection of its precursor and degradation subunits. Accordingly, we have employed immunological techniques that are independent of the activity of the enzyme by using antibodies directed against the apoenzyme. This article reviews past and present work in our laboratory on the life cycle of adrenergic synaptic vesicles in sympathetic neurones and is not intended as a comprehensive review of the field.

ORIGIN OF ADRENERGIC SYNAPTIC VESICLES

Immunofluorescence histochemistry showed that DBH was distributed throughout sympathetic neurones and identified DBH as one of the proteins whose rapid axoplasmic transport had been detected earlier by radioactive precursor labelling (LIVETT, GEFFEN and AUSTIN, 1968; GEFFEN, LIVETT and RUSH, 1969a and 1970). Subsequent biochemical studies showed that DBH and noradrenaline were transported more rapidly than tyrosine hydroxylase (EC 1.14.3a) and L-aromatic aminoacid decarboxylase (EC 4.1.1.26), that in turn were transported faster than the cytoplasmic enzyme phosphoglucoisomerase (EC 5.3.19). Thus the first two enzymes in the biosynthetic pathway appear to be transported on separate particle(s) from DBH (Jarrott and Geffen, 1972; Dairman, Marchelle and Geffen, 1973). In preterminal axons DBH is localized in large granular vesicles and it is also present in the small granular vesicles that are the predominant form of vesicle in axon terminals and may contain up to 50 per cent of the DBH present (Bisby, Fillenz and Smith, 1973).

The cellular distribution and transport of DBH in relation to that of noradrenaline and of electron-dense vesicles provides strong support for the view that protein constituents of adrenergic synaptic vesicles are synthesized in the soma of the neurone, incorporated into large granular vesicles and transported rapidly to the axon terminals where they are transformed into small granular vesicles (Geffen and Ostberg, 1969). It is assumed that small granular vesicles represent mature synaptic vesicles by virtue of their distribution and high noradrenaline content, but there is no agreement on whether large granular vesicles have additional functions such as a role in trophic regulation.

MECHANISM OF TRANSMITTER RELEASE

The most likely mechanism for the release of transmitter from synaptic vesicles is considered to be exocytosis (Geffen and Livett, 1971; Smith and Winkler, 1972). The evidence that chromaffin vesicles undergo exocytosis during stimulation of the adrenal medulla is based on biochemical studies showing proportional release of catecholamines, adenine nucleotides and chromogranins (Kirshner and Kirshner, 1971), immunohistochemical evidence for depletion of DBH protein (Frydman and Geffen, 1973), and electron microscopic observations of exocytotic figures (Grynszpan-Winograd, 1971).

While comparable quantitative data are not yet available for the release of vesicle constituents from sympathetic nerves, the essential evidence that vesicle specific proteins are released together with transmitter has been established. The release of proteins from stimulated nerves was first detected in spleen perfusates after labelling *in vivo* with radioactive precursors (Geffen and Livett, 1968; Geffen *et al.*, 1970). These proteins were subsequently identified as chromogranins by the same immunological and enzymatic methods as had been used in adrenal perfusates (De Potter, De Schaepdryver and Smith, 1969; Geffen, Livett and Rush, 1969b). The ratio of DBH to noradrenaline in spleen perfusates was much lower than in splenic nerve vesicles (Smith, De Potter, Moerman and De Schaepdryver, 1970) and an attempt has been made to account for the discrepancy by applying corrections for (i) the relative proportions of membrane bound to membrane enclosed DBH; (ii) the decrease in the ratios of DBH to noradrenaline in large granular vesicles as they are transported towards the nerve terminals; (iii) the relative clearances of released DBH and noradrenaline into the perfusate modified by differences in diffusion coefficients, capillary membrane permeability and active uptake processes (Smith and Winkler, 1972).

Alternative hypotheses involving exocytosis

It is doubtful whether meaningful calculations can be made from these results that will distinguish between the three main alternative hypotheses for the release of transmitter from adrenergic neurones, all of which are based on some form of exocytosis and involve transformation of large granular vesicles to small granular vesicles (Fig. 1). The ratio of DBH to noradrenaline in spleen perfusates has been interpreted as favouring hypothesis B on the assumptions that DBH is released only from the soluble pool in large granular vesicles, and that noradrenaline is released predominantly from small granular vesicles.

In splenic nerve vesicles, 49 per cent of whose volume is occupied by membrane thickness, up to 90 per cent of the DBH is firmly bound in the membrane (SMITH and WINKLER, 1972). The proportion that is firmly bound in the membrane of the small granular vesicles, that occupies 68 per cent of their volume, is not known but may be even higher. It is therefore possible, as proposed in hypothesis A, that DBH is released from these vesicle membranes when a portion fuses with the axon membrane during exocytosis. Other evidence that favours the view that large granular vesicles may be gradually transformed into small granular vesicles is that no qualitative biochemical differences in composition between the two vesicle forms have yet been detected; and that analysis of vesicle sizes in adrenergic terminals does not confirm the visual impression of two distinct vesicle populations as required by hypotheses B and C (GEFFEN and OSTBERG, 1969). On the other hand, the frequency distribution of vesicle diameters in axon terminals does not provide positive support for hypothesis A unless assumptions are made about the relative rates of transformation of large to small vesicles and the rates of arrival and degradation of vesicles at either end of the spectrum. One test of whether DBH is released from vesicle membranes during exocytosis is whether small amounts of other components of vesicle membranes such as phospholipids are also released.

Re-use of synaptic vesicles

There is also no agreement on whether synaptic vesicles are re-used during their life cycle. It is implicit in hypothesis A that the vesicles are replenished with transmitter by synthesis or re-uptake of noradrenaline (BLAKELEY, BROWN and GEFFEN, 1967) and that the vesicles finally become effete by virtue of the loss of DBH and other vesicle proteins involved in the uptake, synthesis and storage of transmitter. In hypotheses B and C, the synaptic vesicles may be annihilated by a single exocytosis (DE POTTER and CHUBB, 1971) or re-used (SMITH and WINKLER, 1972). If each varicosity released the contents of one vesicle per nerve impulse, then during tonic firing at 1 per sec, a neurone with 25,000 varicosities would release 9×10^7 vesicles per hr. However, it has been calculated that only 4×10^3 large granular vesicles are exported from a typical sympathetic neurone cell body per hour (BANKS and MAYOR, 1972). Even if one large vesicle gave rise to three small vesicles, and the probability was less than one that every varicosity releases transmitter from a vesicle each nerve impulse, the exigencies of biological economy appear to be such as to necessitate the re-use, not only of transmitter, but also of vesicles. Axoplasmic transport contributes daily less than 1 per cent of the peripheral stores of noradrenaline (GEFFEN and RUSH, 1968) whereas transmitter stores can turnover in as little as 3–4 hr by local synthesis. The most rapid turnover of DBH in nerve terminals, estimated from the rate of replenishment by axoplasmic transport, is about 40 hr (DE POTTER and CHUBB, 1971) suggesting that each vesicle exchanges its transmitter

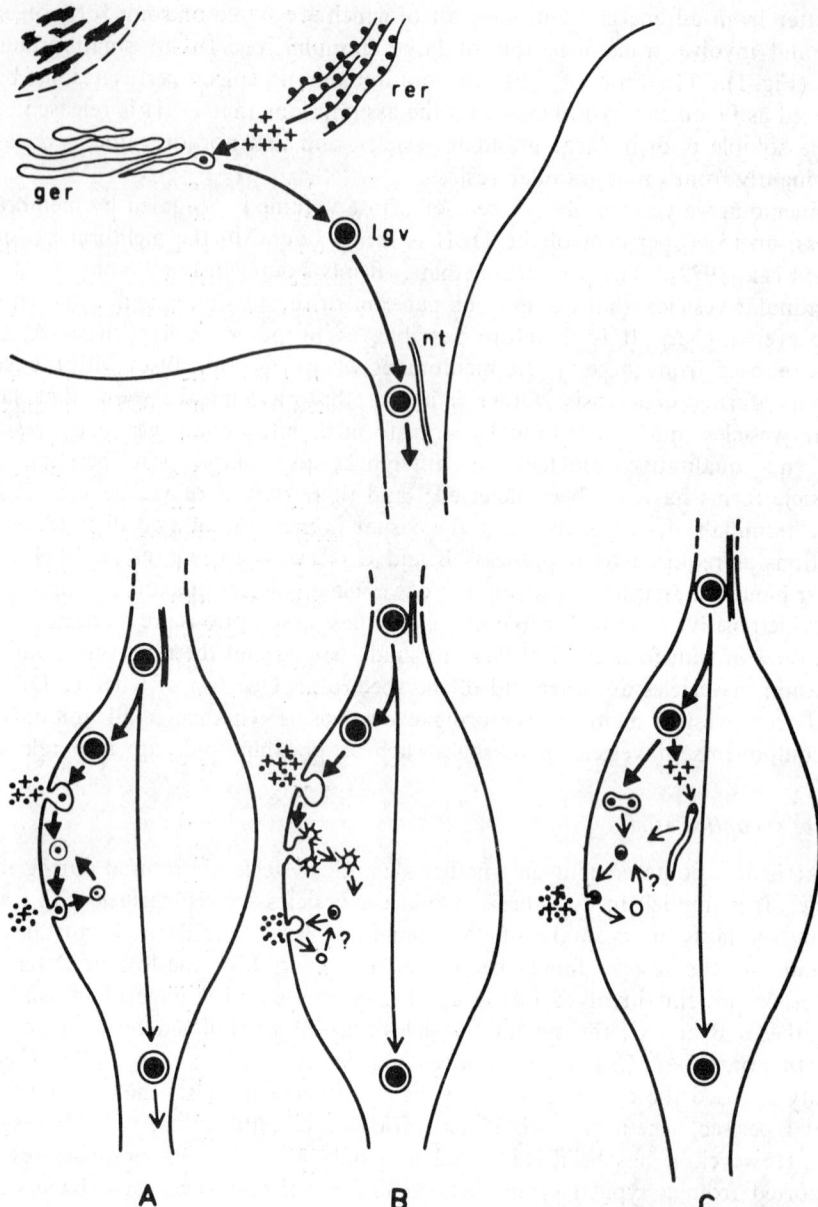

Fig. 1.—Life cycle of adrenergic synaptic vesicles. Their macromolecular components
(+) are synthesised by ribosomes in the endoplasmic reticulum (rer) of the cell body
and the vesicles are assembled in the Golgi endoplasmic reticulum (ger). The large
granular vesicles (lgv) formed that can synthesize and store noradrenaline (●) are
exported to the axon terminals by an axoplasmic transport system involving neurotubules (nt). In the axon varicosities the large granular vesicles are transformed into
small granular vesicles in one of three possible ways: hypothesis A postulates that
large granular vesicles are transformed into small granular vesicles as a result of repetitive
participation in transmission, with gradual loss of vesicle membrane proteins such
as DBH; hypothesis B proposes that large granular vesicles undergo a single exocytosis,
releasing 'soluble' DBH from their cores and that small granular vesicles are formed
from the retrieved membrane; in hypothesis C, small granular vesicles are the only
vesicles involved in exocytosis and are formed locally from large granular vesicles by
budding, or some other form of transfer of contents.

at least 10 times in its life span. If only a fraction of the content of a vesicle is released each time, it may participate in transmission even more frequently.

CIRCULATING DBH AS AN INTEGRATIVE INDEX OF SYMPATHETIC CARDIOVASCULAR FUNCTION

Irrespective of the precise mechanism of release of DBH from synaptic vesicles, the detection of DBH enzymatic activity in plasma (WEINSHILBOUM and AXELROD, 1971; GOLDSTEIN, FREEDMAN and BONNAY, 1971; NAGATSU and UDENFRIEND, 1972) raised the possibility that serum DBH concentration may be a useful integrative index of sympathetic function and dysfunction. In sheep, DBH has a much longer half-life than catecholamines and does not appear to be taken up locally after release as are catecholamines (RUSH and GEFFEN, 1972).

Results obtained thus far in human studies using enzymatic assays have given puzzling results. Mean values for normal human sera are up to 1000 times higher than in other species examined, including primates. There is a 200-fold range of variation between normal individuals but within subjects values are consistent over long periods (NAGATSU and UDENFRIEND, 1972). Individual differences in serum DBH activity cannot be due to the presence of endogenous inhibitors in some sera since identical recovery of activity was achieved when a preparation of purified enzyme was added to the sera with high and low intrinsic DBH activities. Since variations in the proportion of active and inactive DBH rather than in the total amount of enzyme protein released could account for some of these differences, we developed a radioimmunoassay for serum DBH that is independent of the activity of the enzyme (RUSH and GEFFEN, 1972).

Separation of free from bound ^{125}I labelled tracer DBH is achieved by a solid phase method (CATT and TREGEAR. 1967) since the molecular size of DBH makes liquid separation difficult. Antibodies to DBH are coated on the inner surface of polystyrene tubes in which the incubation with tracer and standard or test solutions is then carried out. After washing, the dry tube is counted and the proportion of bound to added counts determined. The entire assay is thus performed in a single disposable tube. The assay is linear in the range 1–100 ng DBH protein/ml, sufficient to assay DBH in 10 μl of human plasma with a coefficient of variation of 5 per cent. The specificity of the assay was tested in the following ways: (i) tubes coated with serum from unimmunized rabbits bound less than 1 per cent of the tracer whereas immune serum bound 70–80 per cent, (ii) increasing dilutions of antiserum produced corresponding decreases in binding, (iii) addition of up to 100 μg of chromogranin-A/ml did not affect the binding of DBH in ng/ml concentrations, (iv) serially diluted samples of sheep and human plasma showed parallel reductions in immunoreactivity.

Radioimmunoassay of circulating DBH in humans

The following preliminary results have been obtained by combined enzymatic and radioimmunoassay of human sera (RUSH, NAGATSU, GEFFEN and UDENFRIEND, 1973). In 34 normal human sera, DBH enzymatic activities ranged from 0·6 to 85·0 units/ml, whereas the amount of cross-reacting protein present is more constant, varying from 68 to 200 μg/ml. DBH rapidly loses enzymatic activity during purification and the amount of purified enzyme required to inhibit the binding of labelled enzyme to the antibody had only 20 per cent of the enzymatic activity of a crude adrenal homogenate producing an equivalent amount of inhibition of binding.

This indicates that immunoreactivity was not dependent upon enzymatic activity as was further demonstrated by the crude adrenal extract retaining its full immuno-reactivity after being heated sufficiently to inactivate all its enzymatic activity. It is possible that much of the DBH protein present in serum has been inactivated in the process of release and that measurement of enzymatic activity does not necessarily reflect the amount of enzyme released from nerve endings. These findings with DBH can be extrapolated as a general principle that when tissue enzymes with no known function in blood are measured in the circulation as an index of the activity or in-tegrity of their parent tissues, the relevant parameter may not be enzymatic activity.

In collaboration with W. J. Louis and A. E. Doyle (Austin Hospital, Melbourne) we have measured catecholamines and DBH plasma levels in normotensive and hypertensive patients, including some with phaeochromocytoma. In patients with essential hypertension, both plasma catecholamines and DBH concentrations were significantly elevated, whereas in phaeochromocytoma patients catecholamine levels were raised disproportionately (Table 1) (Geffen, Rush, Louis and Doyle, 1973a

TABLE 1. Plasma catecholamines (CA) and dopamine β-hydroxylase (DBH) after 48 hr bed rest

Patients	N	CA (ng/ml)	DBH (ng/ml)
Normotensive	8	0·16 ± 0·04	133 ± 17
Hypertensive	20	0·40 ± 0·05*	265 ± 33*
Phaeochromocytoma	8	5·75 ± 1·39*†	329 ± 81*

Mean ± s.d. Significance of difference on Student's t test ($P < 0.01$), * from normotensive, † from hypertensive.

and b). In patients without tumors there were positive correlations between resting basal 'diastolic' blood pressure and both catecholamine and DBH plasma levels. In eight of these hypertensive patients the effect of acute sympathetic ganglion blockade on blood pressure, plasma noradrenaline and DBH has also been studied. There were significant falls in all three parameters within 30 min that persisted for at least 2 hr (Geffen, Rush, Louis and Doyle, 1973c); for procedures and full catecholamine results, see Louis, Doyle and Anavekar, 1973).

Preliminary conclusions on the basis of these experiments are as follows. Plasma DBH, as measured by radioimmunoassay, appears to be a useful integrative index of sympathetic activity in essential hypertension. It correlates closely with diastolic blood pressure and plasma catecholamines both at rest and after ganglion blockade. The parallel elevation in plasma DBH and catecholamine levels in the patients with hypertension suggests that increased sympathetic activity contributed to their elevated blood pressure. The dissociation of catecholamine and DBH secretion in the phaeo-chromocytoma patients, on the other hand, supports the hypothesis that release of catecholamines from the tumor does not occur by exocytosis of storage vesicles, but from an excess pool of newly synthesized catecholamines that bypass the normal storage and secretory mechanisms and diffuse into the circulation (Winkler and Smith, 1968).

Acknowledgement—This work was supported by grants from the National Heart Foundation and National Health and Medical Research Council of Australia.

REFERENCES

BANKS P. and MAYOR D. (1872) *Biochem. Soc. Symp.* **36**, 133–149.

BISBY M. A., FILLENZ M. and SMITH A. D. (1973) *J. Neurochem.* **20**, 245–248.

BLAKELEY A. G. H., BROWN G. L. and GEFFEN L. B. (1969) *Proc. R. Soc. B.* **174**, 51–68.

CATT K. J. and TREGEAR G. W. (1967) *Science* **158**, 1570–1572.

DAHLSTRÖM A. (1971) *Phil. Trans. R. Soc. Lond. B.* **261**, 325–358.

DAIRMAN W., MARCHELLE M. and GEFFEN L. B. (1973) *J. Neurochem.* (in press)

DE POTTER W. P. and CHUBB I. W. (1971) *Biochem. J.* **125**, 375–376.

DE POTTER W. P., DE SCHAEPDRYVER A. F. and SMITH A. D. (1969) *J. Physiol.* 102–104 p.

FOLDES A., JEFFREY P. L., PRESTON B. N. and AUSTIN L. (1972) *Biochem. J.* **126**, 1209–1217.

FRYDMAN R. and GEFFEN L. B. (1973) *J. Histochem. Cyctochem.* **21**, 164–172.

GEFFEN L. B. and LIVETT B. G. (1968) *Proc. Int. Union. Physiol. Sci.* **7**, 152.

GEFFEN L. B. and LIVETT B. G. (1971) *Physiol. Rev.* **51**, 98–157.

GEFFEN L. B., LIVETT B. G. and RUSH R. A. (1969a) *J. Physiol.* **204**, 593–604.

GEFFEN L. B. LIVETT B. G. and RUSH R. A. (1969b) *J. Physiol.* **204**, 58–59 p.

GEFFEN L. B., LIVETT B. G. and RUSH R. A. (1970) In *New Aspects of Storage and Release Mechanisms of Catecholamines* (SCHÜMANN H. S. and KRONEBERG G. Eds.) pp. 58–72, Springer-Verlag, Berlin.

GEFFEN L. B. and OSTBERG A. (1969) *J. Physiol.* **204**, 583–592.

GEFFEN L. B. and RUSH R. A. (1968) *J. Neurochem.* **15**, 925–930.

GEFFEN L. B., RUSH R. A., LOUIS W. J. and DOYLE A. E. (1973a) *Clin. Sci.* **44**, 421–423.

GEFFEN L. B., RUSH R. A., LOUIS W. J. and DOYLE A. E. (1973b) *Clin. Sci.* (in press).

GEFFEN L. B., RUSH R. A., LOUIS W. J. and DOYLE A. E. (1973c) *Biochem. J.* (in press).

GOLDSTEIN M., FREEDMAN L. S. and BONNAY M. (1971) *Experientia* **27**, 632–633.

GRYSZPAN-WINOGRAD O. (1971) *Phil. Trans. R. Soc. B.* **261**, 291–292.

JARROTT B. and GEFFEN L. B. (1972) *Proc. Nat. Acad. Sci. U.S.A.* **69**, 3440–3442.

KIRSHNER N. and KIRSHNER A. G. (1971) *Phil. Trans. R. Soc. B.* **261**, 279–290.

LIVETT B. G., GEFFEN L. B. and AUSTIN L. (1968) *J. Neurochem.* **15**, 931–939.

LOUIS W. J., DOYLE A. E. and ANAVEKAR S. (1973) *New Eng. J. Med.* (in press).

NAGATSU T. and UDENFRIEND S. (1972) *Clin. Chem.* **18**, 980.

RUSH R. A. and GEFFEN L. B. (1972) *Circulat. Res.* **31**, 444–452.

RUSH R. A., NAGATSU T., GEFFEN L. B., and UDENFRIEND S. (1973) *Fedn Proc.* (in press).

SMITH A. D. (1971) *Phil. Trans. R. Soc. Lond. B.* **261**, 423–437.

SMITH A. D., DE POTTER W. P., MOERMAN E. H. and DE SCHAEPDRYVER A. F. (1970) *Tissue and Cell* **2**, 547–568.

SMITH A. D. and WINKLER H. (1972) In *Handbook of Experimental Pharmacology* (Edited by BLASCHKO H. and MUSCHOLL E.) **33**, 539–617. Springer-Verlag, Berlin.

WEINSHILBOUM R. and AXELROD J. (1971) *Circulat. Res.* **28**, 307–315.

WINKLER H. and SMITH A. D. (1968) *Lancet* **i**, 793–795.

Frontiers in Catecholamine Research 1973, pp. 491 to 496. Pergamon Press. Printed in Great Britain.

MECHANISMS OF CATECHOLAMINE SECRETION

DUAL FEEDBACK CONTROL OF SYMPATHETIC NEUROTRANSMITTER SECRETION; RÔLE OF CALCIUM

L. STJÄRNE

Department of Physiology, Karolinska Institutet, Stockholm, Sweden

IN a general sense the function of motor nerves is obviously to convey signals, ultimately from the central nervous system, to adjust the activity state of effector cells. This control is exerted by regulation of the local concentration of neurotransmitter at the level of specific receptors of the effector cells. Regulation can be achieved by centrally evoked variation in the nerve impulse frequency of individual motor neurons, and in the degree of recruitment of different motor neurons converging on the same receptor function. However, in the autonomic nervous system considerable recent evidence indicates that there are in addition mechanisms whereby the effects of any given centrally elicited nerve impulse activity can be modified by local factors modulating the amount of neurotransmitter secreted from individual "varicosities" on arrival of each propagated nerve impulse. Thus activation of the parasympathetic neurons of the autonomic ground plexus of the rabbit heart depresses the simultaneous nerve stimulation induced secretion of noradrenaline (NA) from adjacent adrenergic nerve terminals (LÖFFELHOLZ and MUSCHOLL, 1970). Moreover, experiments with isolated organs show that circulating angiotensin may enhance the secretion of NA from sympathetic nerves (HUGHES and ROTH, 1971). Clearly specific environmental changes may be of importance for the secretory efficiency of sympathetic nerve terminals, and thus for the central control of the concentration of NA at the receptor level. However, there is in addition recent evidence that the very process of NA secretion from sympathetic nerves may in itself trigger two separate local feedback control systems, operating to modulate further secretion of NA.

One such control mechanism appears to be triggered *via* an alpha-adrenoceptor function on the nerve terminals themselves. It seems that this type of control was originally proposed by HÄGGENDAL (1970); on the basis of recognition of the superior enhancing effect of alpha-adrenoceptor blocking agents on the secretion of NA, as compared to that of drugs merely blocking reuptake of NA, he proposed that the alpha-adrenoceptor mediated changes in the contractile state of innervated tissues normally evoked by nerve stimulation, are somehow fed back to the nerves to modify further secretion of transmitter. Later work is clearly in support of the existence of alpha-adrenoceptor mediated feedback control of NA secretion from sympathetic nerves. However, it appears that this is not a mechano-chemical feedback; the specific receptors triggering the effect are probably prejunctional (FARNEBO and HAMBERGER, 1970; LANGER, 1970; WENNMALM, 1971a; ENERO *et al.*, 1972). Particularly the recent work of STARKE (1972) indicates that the sympathetic neuron is indeed capable of adjusting its secretory activity according to changes in the local perineuronal NA concentration.

A different research line led HEDQVIST (1969a) to propose that prostaglandins of the E type (PGE) released as a result of sympathetic nerve stimulation (DAVIES,

HORTON and WITHRINGTON, 1967) may have the physiological function of modulating further nerve stimulation induced secretion of NA. The original basis for this hypothesis was the finding that exogenous PGE, at concentrations well within the physiological range, distinctly, dose-dependently and reversibly depressed nerve stimulation induced secretion of NA from the perfused cat spleen (HEDQVIST, 1969a). This has been found to apply to several sympathetically innervated tissues from various species (cf. HEDQVIST, 1970; WENNMALM, 1971b). In his thesis work WENNMALM made the vital additional observation that specific inhibition of the synthesis of PGE with 5,8,11,14-eicosatetraynoic acid (ETA) enhanced the nerve stimulation induced secretion of NA in the isolated rabbit heart (SAMUELSSON and WENNMALM, 1971), thus strongly supporting HEDQVIST's hypothesis that the secretion of NA from sympathetic nerves is normally restricted by a PGE-mediated feedback control. This important effect of inhibition of PGE synthesis has later been observed in several isolated organs, such as cat spleen (HEDQVIST, STJÄRNE and WENNMALM, 1971) and guinea-pig vas deferens (SWEDIN, 1971; HEDQVIST and EULER, 1972; STJÄRNE, 1972a, 1973a). The effects obtained with ETA, in various isolated tissues, could be reproduced qualitatively as well as quantitatively with a different inhibitor of PGE synthesis, indomethacin (JUNSTAD, CHANH and WENNMALM, 1973; STJÄRNE, 1973a). Even *in vivo* indomethacin treatment of rats resulted in enhanced secretion of NA from sympathetic nerves, as reflected in the urinary excretion of NA, at least during exposure to cold (STJÄRNE, 1971, 1972b) or other stress conditions (JUNSTAD and WENNMALM, 1973).

There are thus two different experimental lines of evidence for local feedback control of NA secretion from sympathetic nerves. One obvious question is whether they represent two different aspects of the same control function, i.e. whether locally formed PGE might be the chemical mediator of feedback control triggered *via* prejunctional alpha-adrenoceptors. However, this does not seem to be the case, since alpha-adrenoceptor antagonists enhance, and alpha-adrenoceptor agonists depress, nerve stimulation induced secretion of ^{3}H-NA even after inhibition of PGE synthesis, in isolated guinea-pig vas deferens (STJÄRNE, 1973b).

Thus it appears that there are indeed two separate mechanisms, one dependent and one independent of PGE, for local feedback control of the probability for secretion of NA on arrival of each propagated nerve impulse. This in harmony with previous electrophysiological (cf. BURNSTOCK and HOLMAN, 1966) as well as biochemical evidence indicating that on short trains of nerve stimulation the average amount of NA secreted per nerve impulse is not constant, but increases with stimulation frequency, in the sympathetic nerves of cat spleen (STJÄRNE, 1970) as well as in vas deferens from rabbit (HUGHES, 1972), mouse (BENNETT, 1973) and guinea-pig (STJÄRNE, 1973c). This in itself shows that the efficiency of excitation-secretion coupling in sympathetic nerves is not constant, but subject to facilitation/summation processes, and thus implies that regulation of the probability for secretion is feasible, particularly at low frequency, where the slope of the curve is steep. As expected, both the enhancing effect on nerve stimulation induced secretion of ^{3}H-NA of inhibition of PGE synthesis, and of subsequent additional alpha-adrenoceptor blockade, and the depressing effect of exogenous PGE_2, were inversely related to nerve stimulation frequency (STJÄRNE, 1973c). Interestingly, alpha-adrenoceptor blockade enhanced, and alpha-adrenoceptor agonists or exogenous PGE_2 depressed, the secretion of ^{3}H-NA evoked by direct depolarisation with high potassium (STJÄRNE, 1973c,d). This shows that the mechanisms involved in alteration of nerve stimulation induced secretion of NA are not, or not to any major extent, concerned with the invasion of the nerve terminals by propagated nerve impulses, but with regulation of excitation-secretion coupling in individual secretory "varicosities".

All the evidence thus indicates that there are two different feedback controls of sympathetic neurotransmitter secretion. This raises a number of questions: Which is the relative physiological rôle of each control system? What factor is the physiological trigger for each? By what mechanism does each control system modify excitation-secretion coupling?

In guinea-pig vas deferens, under the experimental conditions, the capacity of the PGE-independent control was found to be much greater than that of the PGE-dependent control (STJÄRNE, 1973 a, b, c). Concerning the trigger mechanism for the alpha-adrenoceptor mediated control there does not seem to be much evidence in favour of the original assumption that it represents a mechano-chemical feedback, evoked by alpha-adrenoceptor mediated alteration in the contractile state of effector organs (cf. HÄGGENDAL, 1970). On the contrary, all the evidence indicates that this mechanism, which is probably independent of PGE, is highly sensitive to changes in the NA concentration around the neuron. Thus STARKE (1972) reported that NA concentrations as low as 10 ng/ml depressed the nerve stimulation induced secretion of ^{3}H-NA from rabbit heart by 50 per cent.

Concerning the trigger mechanism and the source of PGE in the PGE-mediated control the possibility has been considered that it is triggered by alpha-adrenoceptor mediated muscular contraction (HEDQVIST, 1970), and that the PGE is derived from extra-neuronal sources such as muscle (GILMORE, VANE and WYLLIE, 1968). However, the ^{3}H-NA secretion induced by stimulation of the sympathetic nerves of guinea-pig vas deferens is enhanced by blocking PGE synthesis even in the complete absence of muscular contraction (STJÄRNE, 1972a); thus this does not seem to be a mechano-chemical feedback either. Alpha-adrenoceptor blockade dose-dependently depresses the enhancing effect of ETA on ^{3}H-NA secretion; this indicates that the PGE-mediated control, as well as the PGE-independent one, is triggered by alpha-adrenoceptors (STJÄRNE, 1973e). At present it is not known with certainty whether the two systems are triggered by the same alpha-adrenoceptors (cf. below), or whether the receptors triggering PGE release are pre- or postjunctional. If they are prejunctional, it seems that the particular fraction of the total PGE release evoked on sympathetic nerve stimulation, which is specifically involved in the feedback control of NA secretion, may be neural in origin (STJÄRNE, 1972a).

Experiments to study the effect of addition of the physiologically occurring alpha-adrenoceptor agonist, 1-NA, on the nerve stimulation induced secretion of ^{3}H-NA in guinea-pig vas deferens confirmed the observations in rabbit heart reported by STARKE (1972); in the presence of drugs blocking reuptake of NA exogenous 1-NA even at 2×10^{-7} M (33·8 ng/ml) significantly depressed nerve stimulation induced secretion, in vas deferens. At NA 10^{-6} M ^{3}H-NA secretion was depressed by 50% (STJÄRNE, 1973f). Surprisingly, blocking PGE synthesis with ETA did not appreciably affect the relative inhibitory effect of exogenous NA on ^{3}H-NA secretion (Fig. 1), in spite of the fact that addition of ETA had significantly enhanced ^{3}H-NA secretion during stimulation in the absence of exogenous NA in the medium (by $41 \pm 3\%$, $n = 23$).

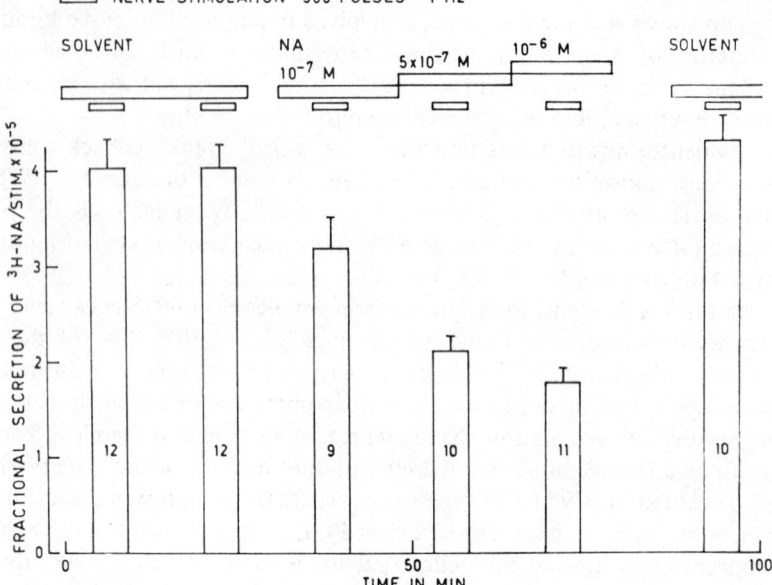

Fig. 1.—Inhibitory effect of exogenous 1-NA on nerve stimulation induced secretion of ^{3}H-NA from isolated superfused field stimulated guinea-pig vas deferens. Desipramine 6×10^{-7} M and normetanephrine 10^{-5} M were present in the medium to block uptake of NA. Preparation preincubated with ETA 6×10^{-5} M for 10 min to block PGE synthesis. Vertical bars: s.e. Figures in columns: Number of observations.

This result as such may be due to masking; as mentioned above the PGE-dependent control appears to be quantitatively less impressive than the PGE-independent control. However, there is also the possibility that triggering of the alpha-adrenoceptors involved in the PGE-mediated control requires NA concentrations much higher than $2–20 \times 10^{-7}$ M; this level is certainly far lower than that intermittently occurring in the narrow synaptic cleft of guinea-pig vas deferens, during nerve stimulation (cf. JOHANSSON et al., 1972). Finally there is of course the possibility that the interpretation of experimental results leading to the conclusion that the PGE-dependent control is triggered via alpha-adrenoceptors, and is thus sensitive to changes in the perineuronal concentration of NA (STJÄRNE, 1973e), is wrong. It is not possible as yet to state with certainty which alternative is correct. However, an improved method for quantitative determination of nerve stimulation induced secretion of ^{3}H-NA in guinea-pig vas deferens indicates that extremely high concentrations of phentolamine, at least $7·5 \times 10^{-5}$ M, are required to completely block the enhancing effect of inhibition of PGE synthesis with ETA on ^{3}H-NA secretion (STJÄRNE, 1973f). This may be taken as a possible indication for the existence of two different functional groups of, possibly prejunctional, alpha-adrenoceptors specifically involved in the control of NA secretion, "set" to discriminate two different concentration levels of NA. Such a concept is in line with similar considerations of separate "junctional" and "extra-junctional" alpha-adrenoceptors on the muscle membrane (HOTTA, 1969). If this is correct it is of course conceivable that each alpha-adrenoceptor mediated feedback system preferentially triggers one or the other group of receptors (Fig. 2).

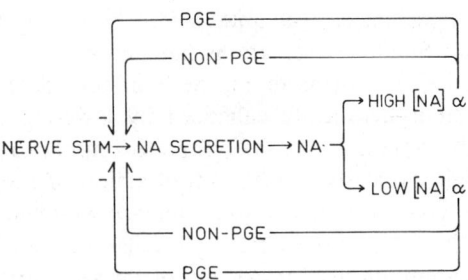

FIG. 2.—Diagram to summarise the observations concerning trigger mechanisms for local PGE-dependent and -independent feedback control of the secretion of ^{3}H-NA from the sympathetic nerves of guinea-pig vas deferens. For further explanation see Text.

In the search for mechanisms whereby the different control systems alter the probability for secretion of NA, on arrival of propagated nerve impulses, it appeared logical to study the rôle of calcium ions, since the secretion of neurotransmitters, including NA (HUKOVIĆ and MUSCHOLL, 1962; KIRPEKAR and MISU, 1967), has an absolute requirement for calcium (cf. RUBIN, 1970).

The curve describing the calcium dependence of nerve stimulation induced secretion of ^{3}H-NA from isolated guinea-pig vas deferens turned out to be sigmoid in shape, and closely similar to that describing the calcium dependence of the secretion of acetylcholine, as reflected in the amplitude of recorded endplate potentials, e.g. in frog neuromuscular junction (DODGE and RAHAMIMOFF, 1967). Progressive release of ^{3}H-NA secretion from alpha-adrenoceptor mediated feedback control by graded alpha-adrenoceptor blockade moved the curve to the left and shifted it into a rectangular hyperbola (STJÄRNE, 1973g), suggesting that in the disinhibited state ^{3}H-NA secretion is a simple function of the calcium concentration in the medium. This led to tentative application of Michaelis–Menten kinetic analysis to characterise the calcium dependence of the secretory mechanism, largely in analogy with similar previous trials for cholinergic secretion (cf. DODGE and RAHAMIMOFF, 1967).

The basis for this approach is the proposal of DEL CASTILLO and KATZ (1954) that the probability for secretion of transmitter on arrival of propagated nerve impulses is ultimately determined by entry of calcium ions from the medium to reversibly form a complex with a hypothetical carrier substance X in the membrane. Replacing, in the classical Michaelis–Menten equation, velocities with fractional ^{3}H-NA secretion per nerve impulse, and substrate concentration with calcium in the medium, it turned out that the calcium dependence of ^{3}H-NA secretion, after disinhibition from negative feedback control, obeyed Michaelis–Menten kinetics. The Lineweaver–Burk double reciprocal plot of ^{3}H-NA secretion against calcium in the medium yielded a straight line (STJÄRNE, 1973g). From the intercept with the abscissa a K_m value of 2·96 mM calcium was obtained; the closeness of this value to the physiologically occurring concentration of free calcium ions in tissue fluid seems to speak in favour of the validity of the method. Moreover, the results indicate that the PGE-independent feedback control may be exerted *via* interference with calcium availability for the secretory mechanism.

Interestingly, similar possibilities have been considered for the PGE-dependent control. HEDQVIST (1969b) observed that the inhibitory effect of exogenous PGE_2 on the secretion of NA from cat spleen was antagonised by raising the calcium concentration in the medium. Similar observations have been made in guinea-pig vas deferens (JOHNSON et al., 1971). Furthermore, the inhibitory potency of exogenous PGE_2 on ^3H-NA secretion is drastically enhanced by lowering the calcium concentration in the medium (STJÄRNE, 1973h). Tentative application of the kinetic method to test the possibility that the mechanism of action of exogenous PGE_2 on ^3H-NA secretion in this preparation involves interference with calcium showed that PGE_2 not only depresses the V_{max} of the secretory mechanism, but in addition raises K_m, i.e. depresses the apparent affinity for calcium of the secretory mechanism, and progressively more with falling calcium concentration in the medium (STJÄRNE, 1973i).

REFERENCES

BENNETT M. R. (1973) J. Physiol. (Lond.) 229, 515–531.
BURNSTOCK G. and HOLMAN M. E. (1966) Pharmacol. Rev. 18, 481–493.
DAVIES B. N., HORTON E. W. and WITHRINGTON P. G. (1967) J. Physiol. (Lond.) 188, 38P–39P.
DEL CASTILLO J. and KATZ B. (1954) J. Physiol. (Lond.) 124, 560–573.
DODGE F. A., JR. and RAHAMIMOFF R. (1967) J. Physiol. (Lond.) 193, 419–432.
ENERO M. A., LANGER S. Z., ROTHLIN R. P. and STEFANO F. J. E. (1972) Br. J. Pharmac. 44, 672–688.
FARNEBO L.-O. and HAMBERGER B. (1970) J. Pharm. Pharmacol. 22, 855–857.
GILMORE N., VANE J. R. and WYLLIE J. H. (1968) Nature (Lond.) 218, 1135–1140.
HEDQVIST P. (1969a) Acta physiol. scand. 75, 511–512.
HEDQVIST P. (1969b) Acta physiol. scand. 80, 269–270.
HEDQVIST P. (1970) Acta physiol. scand. Suppl. 345.
HEDQVIST P., STJÄRNE L. and WENNMALM Å. (1971) Acta physiol. scand. 83, 430–432.
HEDQVIST P. and EULER U. S. von (1972) Nature, New Biol. 236, 113.
HOTTA Y. (1969) Agents and Actions 1, 13–21.
HUGHES J. (1972) Br. J. Pharmac. 44, 472–491.
HUGHES J. and ROTH R. H. (1971) Br. J. Pharmac. 41, 239–255.
HUKOVIĆ S. and MUSCHOLL E. (1962) Naunyn-Schmiedebergs Arch. exp. Path. Pharmak. 244, 81–96.
HÄGGENDAL J. (1970) In Bayer Symposium II (SCHÜMANN, H. J. and KRONEBERG, G., eds.) pp. 100–109. Springer-Verlag, Berlin.
JOHANSSON B., JOHANSSON S. R., LJUNG B. and STAGE L. (1972) J. Pharmacol. exp. Ther. 180, 636–646.
JOHNSON D. J., THOA N. B., WEINSHILBOUM R., AXELROD J. and KOPIN I. J. (1971) Proc. Natn. Acad. Sci. (Wash.) 68, 2227–2230.
JUNSTAD M. and WENNMALM Å. (1972) Acta physiol. scand. 85, 573–576.
JUNSTAD M., CHANH P. H. and WENNMALM Å. (1972) Acta physiol. scand. 86, 563.
KIRPEKAR S. M. and MISU Y. (1967) J. Physiol. (Lond.) 188, 219–234.
LANGER S. Z. (1970) J. Physiol. (Lond.) 208, 515.
LÖFFELHOLZ K. and MUSCHOLL E. (1970) Naunyn-Schmiedebergs Arch. Pharmak. 267, 181–184.
RUBIN R. P. (1970) Pharmacol. Rev. 22, 389–428.
SAMUELSSON B. and WENNMALM Å. (1971) Acta physiol. scand. 83, 163–168.
STARKE K. (1972) Naunyn-Schmiedebergs Arch. Pharmak. 275, 11–23.
STJÄRNE L. (1970) Bayer Symposium II (SCHÜMANN H. J. and KRONEGERG G., eds.) pp. 112–127. Springer-Verlag, Berlin.
STJÄRNE L. (1971) Acta physiol. scand. 83, 574–576.
STJÄRNE L. (1972a) Acta physiol. scand. 86, 574–576.
STJÄRNE L. (1972b) Acta physiol. scand. 86, 388–397.
STJÄRNE L. (1973a) Europ. J. Pharmacol. 22, 233–238.
STJÄRNE L. (1973b) Nature, New Biol. 243, 190.
STJÄRNE L. (1973c) Br. J. Pharmac. In press.
STJÄRNE L. (1973d) Prostaglandins, 3, 421–426.
STJÄRNE L. (1973e) Prostaglandins 3, 111.
STJÄRNE L. (1973f) Prostaglandins In press.
STJÄRNE L. (1973g) Naunyn-Schmiedebergs Arch. Pharmacol. 278, 323–327.
STJÄRNE L. (1973h) Prostaglandins 3, 105–110.
STJÄRNE L. (1973i) Acta physiol. scand. 87, 428–430.
SWEDIN G. (1971) Acta physiol. scand. 83, 473–485.
WENNMALM Å. (1971a) Acta physiol. scand. 82, 532.
WENNMALM Å. (1971b) Acta physiol. scand. Suppl. 365.

Frontiers in Catecholamine Research 1973, pp. 497 to 499. Pergamon Press. Printed in Great Britain.

THE EFFECT OF TRANSMITTER RELEASE ON THE DOPAMINE β HYDROXYLASE CONTENT OF SYMPATHETIC NERVE TERMINALS

Marianne Fillenz and D. P. West

University Laboratory of Physiology, Oxford

The enzyme dopamine β hydroxylase (EC 1.14.2.1) (DBH) appears to be uniquely associated with noradrenaline storage particles. In axonal vesicles 18 per cent of the enzyme is soluble and is released by hypo-osmotic shock (De Potter et al., 1970). In the terminals of the dog spleen DBH is reported to be associated with the high density but not the low density noradrenaline storage particles (Chubb et al., 1970); however in the rat vas deferens there is evidence that both the large and the small vesicles contain DBH (Bisby et al., 1973). It is not known what percentage of the DBH in the small vesicles is soluble, but it is likely to be less than in the large vesicles, since the volume to surface ratio decreases steeply with vesicle diameter (Smith, 1972).

The demonstration that sympathetic nerve stimulation leads to the release of soluble DBH and noradrenaline in the same proportions as they are found in the tissue (Weinshilboum et al., 1971) is the most important evidence in support of the hypothesis that noradrenaline release occurs by exocytosis. However, the subsequent fate of the noradrenaline storage vesicle is not known at present. Estimates of vesicle life span range from 4 weeks (Dahlström, 1971) to 40 hr (De Potter and Chubb, 1971). The former figure implies repeated refilling and reuse of the vesicles; the latter is compatible with a single cycle of filling and emptying and with non-reuse of vesicles.

The hypothesis of repeated reuse of vesicles is based on evidence of the rate of accumulation of noradrenaline in tied nerves (Dahlström, 1971), on the turnover rate of noradrenaline in sympathetically innervated tissues (Burack and Draskozsy, 1964) and the demonstration that noradrenaline synthesis occurs in nerve terminals. The very much lower figure for vesicle life span is based on measurements of DBH accumulation in tied nerves (De Potter and Chubb, 1971), on turnover rates of DBH in sympathetic ganglion cells (Thoenen et al., 1971; Brimijoin, 1972) and on the finding that the noradrenaline uptake capacity of vesicles is reduced after release (Fillenz and Howe, 1971b).

In order to investigate this problem further, we have measured changes in DBH content of sympathetic nerve terminals resulting from transmitter release (Fillenz and West, in preparation). Rats were exposed to cold for periods ranging from 5 to 30 min. The resulting changes in DBH and noradrenaline concentration are shown in Fig. 1: the DBH content falls, whereas the noradrenaline content rises briefly and then returns to control levels. The mean decrease in DBH over the 30 min period is 28 per cent (Table 1); if this were to be accounted for by release of soluble DBH by exocytosis, it would imply the release of more than the total vesicular store. Since this is improbable it suggests that some of the vesicles are lost or destroyed after release. The early rise in vesicular noradrenaline suggests a very rapid increase in

498 M. FILLENZ and D. P. WEST

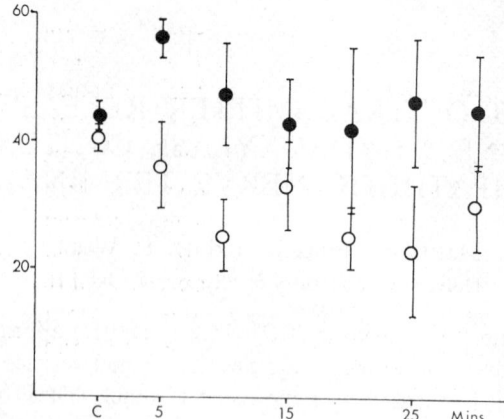

FIG. 1.—Changes in DBH and noradrenaline content of the vesicular fraction of rat heart. ● Noradrenaline content ○ DBH content. C = control value. Abscissa: time in minutes during which rats were exposed to 3°C. Ordinate: noradrenaline in ng/mg protein: DBH in ng phenylethanolamine formed/30 min incubation/mg protein. Each figure is the mean ± S.E.M.

synthesis rate which, after an initial overswing, stabilises at a rate equal to the rate of release.

A rise in vesicular noradrenaline may seem difficult to reconcile with a simultaneous loss or destruction of vesicles. It could be explained if the remaining vesicles could increase their noradrenaline content. There is evidence that the vesicular noradrenaline storage capacity is normally not saturated: the administration of exogenous noradrenaline or the inhibition of monoamine oxidase leads to a 50 per cent increase in vesicular noradrenaline (FILLENZ and HOWE, 1971a).

We have used the ratio of noradrenaline to DBH (NA:DBH ratio) as a measure of saturation of the vesicular noradrenaline storage capacity (FILLENZ and WEST, in preparation). Table 1 shows the changes produced by cold treatment in the mean

TABLE 1. CHANGES IN MEAN DBH CONTENT, MEAN NORADRENALINE CONTENT AND MEAN NA:DBH RATIO OF RAT HEART VESICULAR FRACTION, RELATIVE TO SIMULTANEOUS CONTROL VALUES, RESULTING FROM COLD EXPOSURE OVER THE 30 MIN PERIOD OF COLD TREATMENT (see Fig. 1). DBH EXPRESSED IN ng OF PHENYLETHANOLAMINE/30 MINUTE INCUBATION/ mg PROTEIN   NORADRENALINE EXPRESSED IN ng NA/mg PROTEIN; NA:DBH RATIO IS THE MEAN OF INDIVIDUAL VALUES COMPARED TO SIMULTANEOUS CONTROL VALUES. *$P = <0.01$, †$P = <0.001$.

	DBH	NA	NA/DBH
Cold/Control	0·72 ± 0·067† n = 18	1·12 ± 0·07 n = 22	1·56 ± 0·16* n = 17

DBH content, mean noradrenaline content and the mean NA:DBH ratio. It will be seen that the increased release rate associated with cold stress causes a rise in the NA: DBH ratio. This is not as great as can be achieved with the administration of exogenous noradrenaline (FILLENZ and WEST, unpublished observations).

These results suggest that the supply of vesicles as well as the rate of noradrenaline synthesis may be factors which limit and control the rate of transmitter release (FILLENZ, 1972). The finding (THOENEN et al., 1971) that prolonged cold exposure leads to induction of tyrosine hydroxylase, the enzyme representing the rate limiting step in noradrenaline synthesis, and induction of DBH, the marker enzyme for vesicles, would seem to lend support to this hypothesis. In these experiments the increase in DBH activity was considerably less than that of tyrosine hydroxylase. However, DBH activity gives no indication of the turnover rate of the enzyme or of the subcellular particles with which it is associated. Turnover rates of DBH in cold treated animals have so far been measured in the cell bodies but not in the terminals of the neurons (THOENEN *et al.*, 1971) In order to assess the role of vesicle supply in transmitter release one would also need to know the rate of conversion of the large axonal vesicles to the small terminals vesicles, since the latter are the main storage form of transmitter (SMITH, 1972) and probably the main source of released transmitter (FILLENZ and HOWE, 1970; BISBY *et al.*, 1971).

REFERENCES

BISBY M. A., CRIPPS H. and DEARNALEY D. P. (1971) *J. Physiol.* **214**, 13–14P.
BISBY M. A., FILLENZ M. and SMITH A. D. (1973) *J. Neurochem.* **20**, 245–248.
BRIMIJOIN S. (1972) *J. Neurochem.* **19**, 2183–2193.
BURACK W. R. and DRASKOZSY P. R. (1964) *J. Pharmac. exp. Ther.* **144**, 66–75.
CHUBB I. W., DE POTTER W. P. and DE SCHAEPDRYVER A. F. (1970) *Nature, Lond.* **228**, 1203–1204.
DAHLSTRÖM A. (1971) *Phil. Trans. Roy. Soc. B Lond.* **261**, 325–358.
DE POTTER W. P. and CHUBB I. W. (1971) *Biochem. J.* **125**, 375–376.
DE POTTER W. P., SMITH A. D. and DE SCHAEPDRYVER A. F. (1970) *Tissue & Cell* **2**, 529–546.
FILLENZ M. (1972) *Nature, Lond.* **238**, 41–43.
FILLENZ M. and HOWE P. R. C. (1970) *J. Physiol.* **212**, 42–43P.
FILLENZ M. and HOWE P. R. C. (1971a) *J. Physiol.* **217**, 27–28P.
FILLENZ M. and HOWE P. R. C. (1971b) *J. Physiol.* **218**, 67–68.
SMITH A. D. (1972) *Pharmacol. Rev.* **24**, 435–457.
THOENEN H., KETTLER R., BURKARD W. and SANER A. (1971) *Naunyn-Schmiedebergs Arch. Pharmakol.* **270**, 146–160.
WEINSHILBOUM R. M., THOA N. B., JOHNSON D. C., KOPIN I. J. and AXELROD J. (1971) *Science* **174**, 1349–1351.

Frontiers in Catecholamine Research 1973, pp. 501 to 503. Pergamon Press. Printed in Great Britain.

PHOSPHOLIPIDS OF THE CHROMAFFIN GRANULE MEMBRANE AND CATE-CHOLAMINE RELEASE*

J. M. TRIFARÓ

Department of Pharmacology and Therapeutics, McGill University,
Montreal, Canada

DURING stimulation, the adrenal medulla releases the soluble content of the chromaffin granules to the cell exterior. The soluble substances (catecholamines, ATP and soluble protein) can be quantitatively recovered in the effluent from perfused glands (SMITH and WINKLER, 1972).

When bovine adrenal glands are perfused *in vitro* or *in situ* with Locke's solution and are stimulated with acetylcholine (ACh), the catecholamine efflux rises sharply, but there is no change in the efflux of phospholipids or cholesterol, substances that are known to be present in chromaffin granule membranes. The amount of these lipids recovered in the perfusates during stimulation is only 0·5–1·0 per cent of that expected if whole chromaffin granules were extruded from the cell (TRIFARÓ et al., 1967). "Crude granule preparations" from medullae that have been stimulated to secrete amines contain less catecholamines (59 ± 4%), less protein (71 ± 5%), but as much phospholipids (105 ± 5%) and cholesterol (104 ± 9%) as do granule preparations from non-stimulated medullae (POISNER et al., 1967). Further resolution of "crude granule fractions" by density gradient centrifugation reveals a redistribution of the phospholipids and cholesterol after stimulation from denser to lighter layers (POISNER et al., 1967). Similar results are obtained when the distribution of dopamine β-hydroxylase is determined in density gradients prepared from adrenal medullae of rabbits after insulin administration (VIVEROS et al., 1971). Membranes prepared from hypo-osmotic shocked granules are also recovered in the upper layers of the gradient (POISNER et al., 1967). These results are interpreted to mean that the membranes of the chromaffin granules remain as discrete particles within the chromaffin cells after catecholamine secretion and that they are either less dense (reduced specific gravity) or are smaller than chromaffin granules and may not have reached their equilibrium density (POISNER et al., 1967). Experiments in which granule fractions from control and ACh-stimulated medullae are compared strengthen this latter possibility. Stimulated medullae show a large number of electron-translucent vesicles of smaller diameter than normal granules (MALAMED et al., 1968). These observations also seem to indicate that the attachment of the granule membranes to the plasma membrane does not involve longterm incorporation. However, the fate of granule membranes after secretion is unknown. It is possible that the emptied vesicles can be recharged with catecholamines and other soluble components, or alternatively, that emptied granules may be digested, perhaps by lysosomes (TRIFARÓ et al., 1967). Preliminary experiments show that if immediately after stimulation

* This research was supported in part by Grant MT-3214 from the Medical Research Council of Canada.

a pulse of ^{32}P is given, the amounts of ^{32}P-nucleotides and ^{32}P-phospholipids present 4 hr later in granule fractions from stimulated glands are larger than those from control glands (unpublished observations).

Since release by exocytosis involves the interaction of both the granule and plasma membranes, it is possible that some kind of rearrangement of the lipids of the two membranes takes place during the interaction. When the distribution of individual phospholipids of the granule membrane is measured, a large amount (12·9 ± 1·4%) of lysolecithin is found (DOUGLAS et al., 1966; BLASCHKO et al., 1967). This lipid, which shows membrane-lytic properties, is not likely formed during the stimulation of the glands, because no differences are found between control and ACh-stimulated medullae in either the content or the specific activity of lysolecithin when this phospholipid was previously labelled with ^{32}P or ^{14}C-glycerol (TRIFARÓ 1969a). In addition, no enzyme has been found in chromaffin granules that can be responsible for the synthesis or breakdown of lysolecithin (SMITH and WINKLER, 1972). These results indicate that lysolecithin is not formed during secretion, but they do not

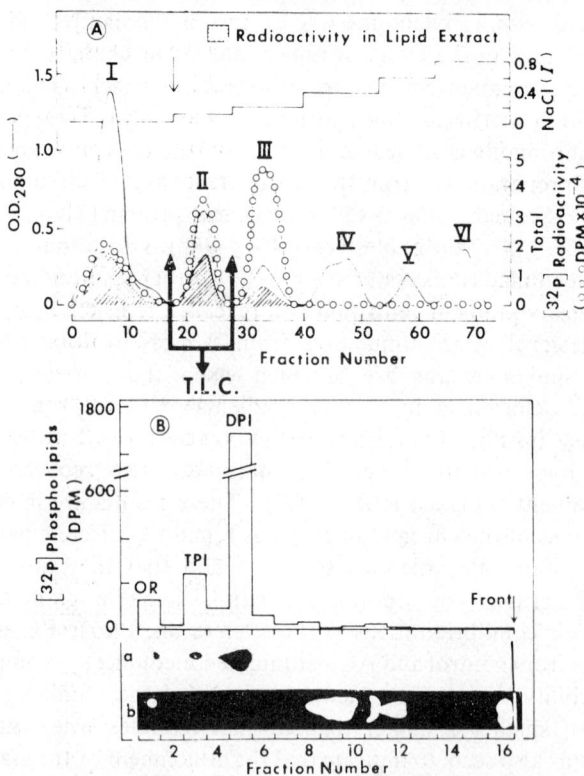

FIG. 1.—Chromaffin granules were incubated in the presence of 2 mM Mg-[γ-^{32}P]ATP (TRIFARÓ, 1972). Granule membrane components were solubilized by 2% Lubrol PX, and the Lubrol-extracted material was applied to a DEAE-Sephadex A-25 column. The column (A) was eluted as previously described (TRIFARÓ and WARNER, 1972). Radioactivity was determined in the protein and lipid extracts of the fractions. The lipid extract from peak II was subjected to T.L.C. (B). Radioactivity was mainly recovered with DPI as indicated in (a), which is the autoradiography of the chromatogram shown in (b) after iodine vapour identification of the main phospholipids. For further experimental details see TRIFARÓ and WARNER (1972) and TRIFARÓ (1972).

rule out the hypothesis that this lipid is involved in membrane fusion (LUCY, 1970). The incorporation of lysolecithin into lipid membranes should produce transition from bilayer to micellar configuration, and fusion should occur when a relatively high proportion of the lipids of the membranes are arranged in micelles (LUCY, 1970). During the ACh-induced release of catecholamines from the adrenal medulla, there is also an increase in ^{32}P incorporation into phosphatidic acid ($340 \pm 12\%$) and phosphatidyl inositol ($270 \pm 10\%$) (TRIFARÓ, 1969a). However, this effect of ACh is slower than its action in inducing secretion (TRIFARÓ, 1969a). This, and the fact that ^{32}P incorporation is not affected by Ca^{2+} omission from the extracellular environment (TRIFARÓ, 1969b) argue against a direct relation of phospholipid turnover to stimulus-secretion coupling. This effect of ACh is probably related to the synthesis of new membranes, since there is a large ^{32}P incorporation into the microsomal fraction (unpublished observations).

Finally, it is known that when isolated chromaffin granules are incubated under certain conditions, Mg-ATP induces the release of the soluble granule components (TRIFARÓ and POISNER, 1967). When ATP exerts this action, it is hydrolyzed by enzyme(s) present in granule membranes. Part of the Pi liberated from ATP is transferred to the granule membranes (TRIFARÓ and DWORKIND, 1971). The phosphorylated granule components have been recently solubilised and isolated by ion exchange chromatography (TRIFARÓ and WARNER, 1972; TRIFARÓ, 1972). There are at least two phosphorylated components (Fig. 1A), one of them is a phosphoprotein (peak III, Fig. 1A), the other (peak II, Fig. 1A) can be extracted with chloroform:methanol + [HCl] ($2:1+0.25\%$) and upon thin layer chromatography, the radioactivity can be recovered mainly with diphosphatidyl inositol (DPI) (Fig. 1B). This indicates the presence of phosphatidyl inositol kinase (4.0 ± 0.2 nmoles DPI formed/mg protein/hr) in chromaffin granule membranes. Synthesis of DPI has also been detected in microsomal and granule fractions prepared from glands previously stimulated by ACh (unpublished observations). It has been shown recently that membrane calcium binding is directly related to the levels of membrane DPI (BUCKLEY and HAWTHORNE, 1972). However, it is too premature to assign a role to DPI in secretion, and more research is needed to determine whether this increase in the DPI synthesis is related to the role of Ca^{2+} in the secretory process.

REFERENCES

BLASCHKO H., FIREMARK H., SMITH A. D., WINKLER H. (1967) *Biochem. J.* **104**, 545–549.
BUCKLEY J. T. and HAWTHORNE J. N. (1972) *J. Biol. Chem.* **247**, 7218–7223.
DOUGLAS W. W., POISNER A. M. and TRIFARÓ J. M. (1966) *Life Sci.* **5**, 809–815.
LUCY J. A. (1970) *Nature, Lond.* **227**, 815–817.
MALAMED S., POISNER A. M., TRIFARÓ J. M. and DOUGLAS W. W. (1968) *Biochem. Pharmacol.* **17**, 241–246.
POISNER A. M., TRIFARÓ J. M. and DOUGLAS W. W. (1967) *Biochem. Pharmacol.* **16**, 2101–2108.
SMITH A. D. and WINKLER H. (1972) In: *Handbook of Experimental Pharmacology* (BLASCHKO H. and E. MUSCHOLL, Eds.) Vol. 33, pp. 538–617, Springer-Verlag, N.Y.
TRIFARÓ J. M. (1969a) *Mol. Pharmacol.* **5**, 382–393.
TRIFARÓ J. M. (1969b) *Mol. Pharmacol.* **5**, 420–427.
TRIFARÓ J. M. (1972), *F.E.B.S. Lett.* **23**, 237–240
TRIFARÓ J. M. and DWORKIND J. (1971) *Mol. Pharmacol.* **7**, 52–65.
TRIFARÓ J. M. and POISNER A. M. 1967) *Mol. Pharmacol.* **3**, 572–580.
TRIFARÓ J. M., POISNER A. M. and DOUGLAS W. W. (1967) *Biochem. Pharmacol.* **16**, 2095–2100.
TRIFARÓ J. M. and WARNER M.(1972) *Mol. Pharmacol.* **8**, 159–169.
VIVEROS O. H., ARQUEROS L. and KIRSHNER N. (1971) *Mol. Pharmacol.* **7**, 444–454.

Frontiers in Catecholamine Research 1973, pp. 505 to 507. Pergamon Press. Printed in Great Britain.

EXOCYTOSIS AT SYMPATHETIC NERVE TERMINALS

G. Frederick Wooten

Laboratory of Clinical Science, National Institute of Mental Health,
Bethesda, Maryland 20014, U.S.A.

Norepinephrine and dopamine-β-hydroxylase (3,4-dihydroxyphenylethylamine, ascorbate: O_2 oxidoreductase; EC 1.14.2.1), the enzyme which catalyses the conversion of dopamine to norepinephrine, are highly localised in vesicular structures at noradrenergic nerve terminals (Potter and Axelrod, 1963) and in chromaffin granules of the adrenal medulla (Kirshner, 1957). Dopamine-β-hydroxylase is released with catecholamines from the isolated perfused adrenal gland by stimulation with acetylcholine (Viveros, Arqueros and Kirshner, 1968), and from the spleen by electrical stimulation of the splenic nerve (Geffen, Livett and Rush, 1969; Smith, De Potter, Moerman and De Schaepdryver, 1969). Recently, utilising a sensitive new radiometric assay for dopamine-β-hydroxylase devised by Molinoff et al. (1971), Weinshilboum et al. (1971) demonstrated in vitro a proportional release of norepinephrine and dopamine-β-hydroxylase from the isolated guinea pig vas deferens following electrical stimulation of the hypogastric nerve. Further, the ratio of norepinephrine to dopamine-β-hydroxylase which was released into the bathing medium was similar to their ratio in the soluble fraction of vas deferens homogenates. These results suggested that exocytosis was the mechanism whereby norepinephrine is released from sympathetic nerves. These workers also demonstrated that stimulation of the hypogastric nerve to the guinea pig vas deferens in the presence of phenoxybenzamine or supraphysiologic concentrations of calcium caused a marked increase in the amounts of enzyme and norepinephrine released into the bath (Johnson, Thoa, Weinshilboum, Axelrod and Kopin, 1971). The increased release occurring in the presence of high extracellular calcium concentrations or phenoxybenzamine was partially reversed by $1\cdot 8\ \mu M$ prostaglandin E_2. Thus, prostaglandins might exert some control over peripheral norepinephrine release by either modulating the intracellular accumulation of calcium or its subsequent function in initiating the process of exocytosis.

Studies of the possible involvement of microtubules and microfilaments in various processes of cell secretion have been facilitated by the use of the alkaloids colchicine and vinblastine which bind to and disaggregate microtubules, and cytochalasin-B, a fungal metabolite that causes disaggregation of microfilaments. Microtubules have been implicated in the release of insulin from β-cells of the pancreas, ^{131}I from isolated thyroid glands, histamine from mast cells and catecholamines from the adrenal medulla, while cytochalasin-B also inhibited release of ^{131}I from the thyroid gland (for reference see Thoa et al., 1972). We have studied the effects of colchicine, vinblastine and cytochalasin-B on release of norepinephrine and dopamine-β-hydroxylase from the isolated guinea pig vas deferens (Thoa, Wooten, Axelrod and Kopin, 1972). Colchicine (1 mM), vinblastine (1 mM) and cytochalasin-B (3 μg/ml) markedly inhibited the stimulation-induced exocytosis. Thus microtubules and microfilaments may play a role in exocytosis at sympathetic nerve terminals and exocytosis may involve a contractile mechanism similar to that occurring in muscle.

Cyclic 3'5'-AMP and calcium appear to play a physiologic role in excitation-secretion coupling in several neural and endocrine tissues. Cyclic 3'5'-AMP accumulated in tissues following specific stimuli during release of thyrocalcitonin, insulin, salivary secretions and gastric HCl. The dibutyryl derivative of cyclic 3'5'-AMP as well as the methyl xanthines which inhibit inactivation of cyclic-AMP enhanced acetylcholine release at the neuromuscular junction and caused catecholamine release from heart–lung preparations, brain, and adrenal medulla. Not only was release of neurotransmitter from both cholinergic nerve endings and the adrenal medulla associated with an influx of extracellular calcium, but stimulus-induced release of norepinephrine from sympathetic nerves has also been found to vary directly with extracellular calcium concentration (for reference see Wooten et al., 1973). In order to study the relationship between calcium and cyclic-AMP in stimulus-secretion coupling in noradrenergic neurons, we have examined the effects of dibutyryl cyclic-AMP (DC-AMP) and theophylline on exocytosis in the guinea pig vas deferens (Wooten, Thoa, Kopin and Axelrod, 1973).

In the presence of a normal (2·2 mM) extracellular calcium concentration both DC-AMP (0·1 mM) and theophylline (1 mM) enhanced stimulation-induced release of dopamine-β-hydroxylase and norepinephrine with DC-AMP also increasing spontaneous release. However, in a bath medium in the absence of calcium ion, the stimulation-induced release of dopamine-β-hydroxylase and norepinephrine was completely blocked. Stimulation-induced release was partially restored when DC-AMP or theophylline was added to the medium. DC-AMP enhanced spontaneous release as in the normal calcium medium. These results suggest that cyclic 3'5'-AMP may have a role in release, acting either together with calcium or indirectly by mobilizing intracellular, bound calcium.

A general mechanism may be proposed for the events that couple excitation with exocytosis at sympathetic nerve terminals. Depolarization of the neuronal membrane by an electrical impulse causes an influx of extracellular calcium and activation of membrane-bound adenyl cyclase resulting in increased intracellular cyclic 3'5'-AMP levels. Besides mobilising intracellular calcium and thus acting in series with calcium, cyclic-AMP may act directly in parallel with calcium to activate the presumed neurotubular–and/or neurofilament–vesicle complex resulting in fusion of the vesicular membrane with the cell membrane and the rapid extrusion of the soluble vesicular contents by a contractile process (Poisner, 1970; Thoa, Wooten, Axelrod and Kopin, 1972).

REFERENCES

Geffen L. B., Livett B. G. and Rush R. A. (1969) *J. Physiol.* **204,** 58–59P.
Johnson D. G., Thoa N. B., Weinshilboum R., Axelrod J. and Kopin I. J. (1971) *Proc. Natn Acad. Sci.* **68,** 2227–2230.
Kaufman S. and Friedman S. (1965) *Pharmacol. Rev.* **17,** 71–100.
Kirshner N. (1957) *J. Biol. Chem.* **226,** 821–825.
Molinoff P. B., Weinshilboum R. and Axelrod J. (1971) *J. Pharmacol. Exp. Ther.* **178,** 425–431.
Poisner A. M. (1970) In: *Advances in Biomedical Psychopharmacology.* (Costa E. and Giacobini E., Eds.) Vol. 2, pp. 95–108. Raven Press, New York.
Potter L. T. and Axelrod J. (1963) *J. Pharmacol. Exp. Ther.* **142,** 299–305.
Smith A. D., De Potter W. P., Moerman E. J. and De Schaepdryver A. F. (1970) *Tissue Cell* **2,** 547–568.

THOA N. B., WOOTEN G. F., AXELROD J. and KOPIN I. J. (1972) *Proc. Natn Acad. Sci.* **69,** 520–522.
VIVEROS O. H., ARQUEROS L. and KIRSHNER N. (1968) *Life Sci.* **7,** 600–608.
WEINSHILBOUM R. M., THOA N. B., JOHNSON D. G., KOPIN I. J. and AXELROD J. (1971) *Science*
   174, 1349–1351.
WOOTEN G. F., THOA N. B., KOPIN I. J. and AXELROD J. (1973) *Mol. Pharmacol.* **9,** 178–183.

Frontiers in Catecholamine Research 1973, pp. 509 to 511. Pergamon Press. Printed in Great Britain.

ON THE DYNAMICS OF CATECHOLAMINES STORAGE STRUCTURES IN SYMPATHETIC NEURONS OF THE RAT AS STUDIED BY RADIOAUTOGRAPHY

JACQUES TAXI and CONSTANTINO SOTELO

Laboratoire de Biologie animale Université, Paris VI, 12 Rue Cuvier,
75005 Paris, France

THE ultrastructural studies on the ligated splenic nerve of the cat have led to the conception according to which noradrenaline (NA) containing vesicles are formed in the cell body of the sympathetic neurons and then transported along the axon towards the sites of NA release, i.e. the distal part of the axon (DAHLSTROM and HAGGENDAL, 1970; GEFFEN and OSTBERG, 1969; KAPELLER and MAYOR, 1967, 1969). Our observations on the ligated sciatic nerve of the rat, in which an accumulation of NA proximal to a ligature was first described (DAHLSTROM, 1965), were performed in order to determine to what extent this scheme is applicable to the sympathetic neurons of the rat. In our experiments, radioautography at the ultrastructural level after injection of NA-^{3}H or a precursor-^{3}H (for technical details, see SOTELO and TAXI, 1973) enables visualisation of the storage properties of various structures encountered in ligated nerves.

It was demonstrated that only a very small part of NA accumulated proximal to a ligature may come by migration from the perikaryon, a result in agreement with that of GEFFEN et al. (1972). But, since the axons are devoid of ribosomes, the enzymes intervening in NA synthesis and the storage proteins must be synthesized in the perikaryon and carried along the axon. Then the question arises: what structures are involved in this transport? Are the storage structures of the distal part of axons manufactured in the cell body and then transported to the periphery, or are they locally assembled?

The response must take in consideration that the labeling of perikarya by exogenous NA-^{3}H and that of the distal part of sympathetic axons require quite different conditions (TAXI and DROZ, 1969). In the distal part of axons, the labeling is largely independent of the action of an inhibitor of monoamine-oxydases (IMAO) and generally related to 'small granulated vesicles', 45 nm in diameter (SGV). On the contrary, good labeling of perikarya of sympathetic neurons is only obtained using an IMAO; in this case, the silver grains of the radioautographic reaction are apparently located at random over the cell organelles, suggesting a binding of NA to some hyaloplasmic component. As NA is not firmly maintained as in closed units like vesicles, it is sensitive to MAO and diffusion artefacts may occur. Without IMAO treatment, neuronal perikarya exhibit only a poor labeling, limited mainly to clusters of SGV. It seems a priori unlikely that two ways of storage having so distinctive features involve the same organelles.

However, it is well established that neuronal perikarya contain the same two morphological types of vesicles as the distal part of axons, SGV and LGV (large granulated vesicles), 80 nm in diameter. But SGV are mainly located in clusters close to the plasmalemma, either in the cell body or, more frequently, in the dendrites. Opposite to these clusters, the plasmalemma often exhibits on its internal face dense patches mimicking a synaptic 'active zone', suggesting a possible release of NA at this place (Fig. 1). Thus SGV seem more involved in local release than in migration phenomena. LGV are not numerous in the perikarya, where they are located in the vicinity of the Golgi apparatus. These zones are not preferentially labeled after NA-³H injection in IMAO treated animals; moreover, LGV are not modified under the action of reserpine, even after two injections (5 mg/kg i.p.) made at an interval of some hours, a treatment which completely abolish the fluorescence reaction of Falck and Hillarp. Thus it seems highly doubtful that LGV seen in the perikaryon in our technical conditions (SOTELO and TAXI, 1973) contain catecholamines.

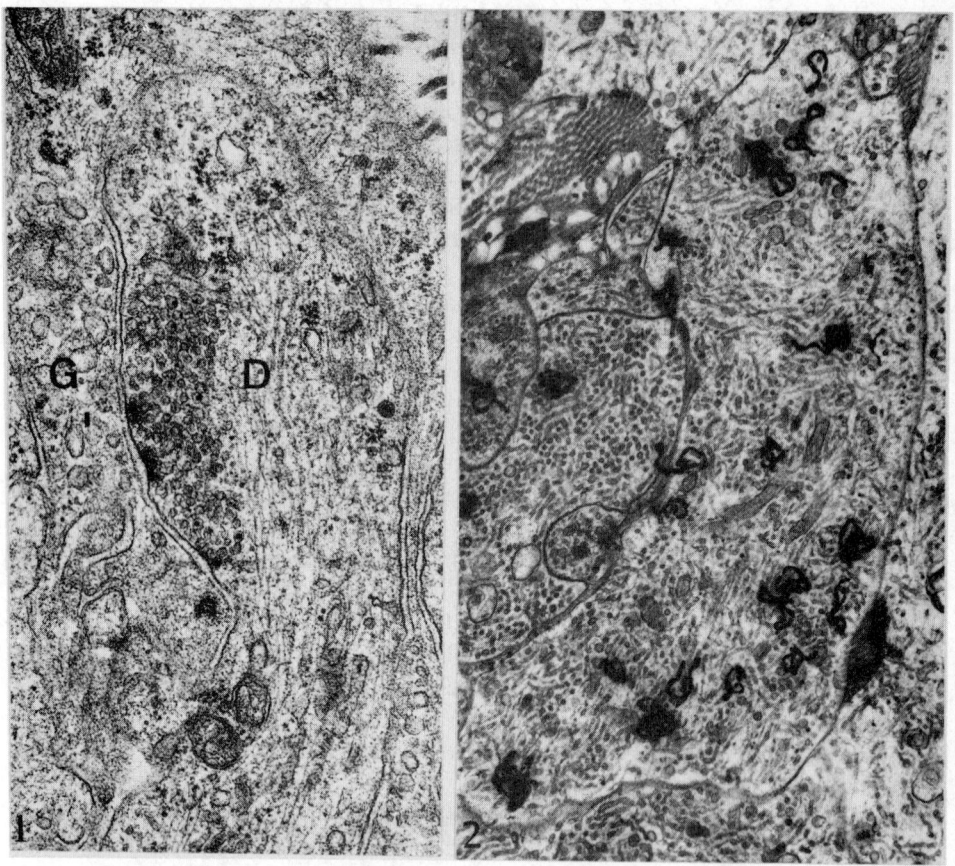

FIG. 1.—Rat. Superior cervical ganglion. D, dendrite containing a cluster of SGV, opposite dense patches attached to the plasmalemma.

FIG. 2.—Rat. Sciatic nerve 22 hr after a ligation; DOPA-³H was injected 30 min before fixation. Labeled fibres contain empty vesicular and tubular profiles.

Let us consider now the situation of SGV and LGV in normal sympathetic fibres and in the segment proximal to a ligature. In normal postganglionic axons fixed by 3% MnO$_4$K, a fixative especially suitable for preservation of granulated vesicles (HOKFELDT, 1968; RICHARDSON, 1966), LGV are by far more numerous than SGV. In the ligated fibres, radioautography shows that the storage properties of sympathetic fibres are restricted to a short, highly modified proximal segment, the morphology of which is largely variable. This segment may contain a moderate number of LGV, some SGV and intermediary forms of granulated vesicles, but empty vesicular or tubular profiles belonging to endoplasmic reticulum often are predominant (Fig. 2). The present limitation of the resolution of the radioautographic technique does not allow us to determine if the labeled molecules are located inside or outside closed profiles, but experiments on animals untreated by IMAO show that a large part of NA is in a protected pool. Experiments with reserpine showed that LGV constitute a heterogene population, composed by both reserpine-sensitive and reserpine-unsensitive vesicles. The second type probably corresponds to LGV encountered around the dictyosomes in the perikaryon. It seems that the abnormal conditions realised proximal to a ligature, which combine accumulation of material brought by axonal flow and remodeling processes, rapidly lead to an explosive outgrowth of endoplasmic reticulum, giving rise locally to numerous vesicular profiles (SOTELO and TAXI, 1973). During this process, enzymes of NA synthesis and storage proteins can be packed in various types of vesicles, in which NA protected from MAO is accumulated.

In fact nerve segment proximal to a ligature appears as a rather specific system of local morphogenesis, and could not be taken as a model of the peripheral part of normal axons. About the storage structures they are either assembled in the periphery or, if they migrate, they are in an immature form until they reach the terminals. This is the way to explain the difference of storage properties in perikarya and terminals.

REFERENCES

DAHLSTROM A. (1965) *J. Anat.* (*Lond.*) **99**, 677–689.
DAHLSTROM A. and HAGGENDAL J. (1970) In: *Biochemistry of simple neuronal models* (E. COSTA and E. GIACOBINI eds.) pp. 65–93, Raven Press, New York.
GEFFEN L. B., DESCARRIES L. and DROZ B. (1972) *Brain Res.*, **35**, 315–318.
GEFFEN L. B. and OSTBERG A. (1969) *J. Physiol.* (*Lond.*) **204**, 583–592.
HOKFELT T. (1968) *Z. Zellforsch.* **91**, 1–74.
KAPELLER K. and MAYOR D. (1967) *Proc. Roy. Soc.* B **167**, 282–292.
KAPELLER K. and MAYOR D. (1969) *Proc. Roy. Soc.* B **172**, 39–51.
RICHARDSON K. C. (1966) *Nature* (*Lond.*) **210**, 756.
SOTELO C. and TAXI J. (1973) *Z. Zellforsch.* **138**, 345–370.
TAXI J. and DROZ B. (1969) In: *Cellular dynamics of the neuron;* (H. BARONDES ed.), *Symp. Int. Soc. Cell Biol.*, **8**, 175–190, Academic Press, New York.

Frontiers in Catecholamine Research 1973, pp. 513 to 521. Pergamon Press. Printed in Great Britain.

NEW INVESTIGATIONS OF THE CARDIOVASCULAR ACTIONS OF DOPAMINE

Leon I. Goldberg[1], T. Budya Tjandramaga[1]
Aaron H. Anton[2] and Noboru Toda[3]

[1]Clinical Pharmacology Program, Emory University School of Medicine, Atlanta, Georgia 30322, the [2]Department of Anaesthesiology, Case Western Reserve University School of Medicine, Cleveland, Ohio 44106, and the [3]Department of Pharmacology, Faculty of Medicine, Kyoto University, Kyoto, Japan.

INTRODUCTION

In the cardiovascular system, as in the brain, the effects of dopamine are qualitatively different from those produced by the other endogenous catecholamines. Furthermore, these unusual actions of dopamine appear to be due to action on a specific dopamine vascular receptor. The cardiovascular and renal actions of dopamine have recently been comprehensively reviewed (Goldberg, 1972). Older references included in that review will not be cited unless necessary for emphasis.

Dopamine exerts three distinctive actions in the cardiovascular system. First, *alpha*-adrenergic vasoconstriction can be produced by dopamine in all blood vessels if a sufficiently large dose is used. This action can be completely blocked by *alpha*-adrenergic blocking agents, and these agents in fact must be employed in certain animal species and in certain vascular beds to demonstrate dopamine-induced vasodilation. Second, dopamine stimulates the heart by an action on *beta*-adrenergic receptors. This effect is due in part to a direct action on the receptors, and in part to release of norepinephrine. Dopamine has extremely weak *beta*-adrenergic vasodilating actions. Third, of more interest dopamine causes vasodilation in certain vascular beds by a mechanism which appears to be due to action on specific dopamine receptors

Evidence supporting the existence of a specific dopamine vascular receptor may be summarized briefly as follows: (a) The vasodilation is selective, being found only in the renal, mesenteric, coronary and more recently the intracerebral vascular beds (Essen von, 1972). (b) The vasodilation is not blocked by usual antagonists, but is attenuated by haloperidol, phenothiazines, apomorphine and bulbocapnine. Preliminary data (Goldberg and Musgrave) indicate that 3-methoxy-4-hydroxy phenylethylamine is also a weak antagonist. (c) The vasodilating action is extremely structure-specific. Only the *N*-methyl analogue of dopamine, epinine, displays unequivocal dopamine-like activity. Apomorphine also appears to have weak dopamine-like activity suggesting that it is both an agonist and an antagonist. More recently, a large number of aminoacyl derivatives of dopamine have been screened in a search for an orally effective dopamine analogue. Initial results indicated that *L*-alanyl-dopamine and *N-L*-isoleucyldopamine were active in increasing the renal blood flow (Goldberg, Musgrave and Biel, unpublished data). These results were subsequently confirmed in both dogs and monkeys (Biel *et al.*, in press). Administration of the *L*-alanine amide of dopamine to patients with essential hypertension was found to result in increments of the renal clearance of para

513

aminohippurate (C_{PAH}) and the urinary excretion of sodium (Tjandramaga, Velasco and McNay, unpublished data).

EFFECTS OF DOPAMINE ON ISOLATED BLOOD VESSELS

Most of the previous studies were conducted in the intact dog. Such a preparation, however, cannot be used to prove the existence of a dopamine receptor because the vasodilating action could have been due to the release of another endogenous vasodilating substance such as a kinin or a prostaglandin, or to production of an unknown metabolite of dopamine. Although there was no evidence for either of these arguments, it was essential to find an isolated blood vessel which would specifically relax to dopamine in a dose-related manner. Despite many attempts, such an isolated system has not been previously described.

We initially investigated canine renal and mesenteric arteries since these vessels are responsive to dopamine in the intact animal (TODA and GOLDBERG, in press). Helical strips of arteries ranging in size from 0·3 to 3 mm (outside diameter) were suspended in an isolated organ bath containing nutrient solution through which O_2 (95%) and CO_2 (5%) were constantly bubbled. The method used has been previously described (TODA et al., 1972). After a period of equilibration, dopamine was added to the bath in increasing concentrations. In most preparations, contraction was the predominant effect, but in some vessels relaxation was also observed. In order to eliminate variability in the spontaneous tone and *alpha*-adrenergic activity, subsequent preparations were treated with phenoxybenzamine in concentrations varying from 10^{-6} to 10^{-5} M, for periods ranging from 20 to 60 min, and then contracted with potassium (10–30 mM). With the lower concentrations of phenoxybenzamine both relaxation and contraction were observed with dopamine. With the 10^{-5} M concentration and treatment for 60 min, most renal and mesenteric arteries exhibited relaxation. In many of these preparations, however, contractions were also observed with some doses. The mechanism of this contraction is not known, since the dose of phenoxybenzamine used should have been sufficient to block both the specific and non-specific contracting actions. The resistance to *alpha*-adrenergic block may have been related to the potentiation of contractions by potassium (FURCHGOTT and BHADRAKOM, 1953). Figure 1 illustrates the contractions produced by dopamine in an untreated mesenteric arterial strip, and the marked potentiation of the contractions after the strip was treated with potassium. More important, Fig. 1 illustrates the dose-related relaxation caused by dopamine when the contracting actions are eliminated by phenoxybenzamine treatment, 10^{-5} M, for 1 hr. The relaxation produced by dopamine is characteristically gradual in onset and in reaching a plateau. In most preparations the effect of doses greater than 2×10^{-5} M persisted for 30–60 mins. This is in marked contrast to isoproterenol which rarely persists more than 10 mins. Also the relaxing action of dopamine is not blocked by the *beta*-adrenergic blocking agents, propranolol (10^{-6} M) and sotalol (10^{-4} M), which markedly attenuate the relaxing actions of isoproterenol.

As in the intact animal, there were marked differences in the sensitivity of different arteries. Despite repeated applications of phenoxybenzamine, only contractions were observed in ten experiments with large (greater than 1 mm outside diameter) femoral arteries. Small femoral arteries were much less responsive than renal and

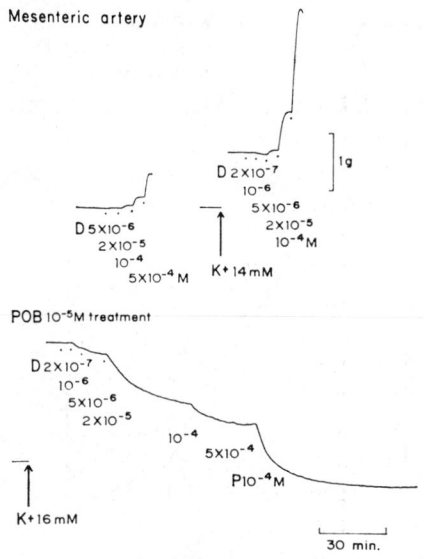

FIG. 1.—Effects of dopamine (D) on an isolated canine mesenteric arterial strip. Upper tracing (left)—contractions produced by dopamine in the untreated strip. Upper tracing (right)—contractions produced by dopamine after the strip was partially contracted with potassium. Lower tracing dose-related relaxation produced by dopamine in the same arterial strip after treatment with phenoxybenzamine (POB), 10^{-5} M, for 1 hr. Final relaxation produced by papaverine (P), 10^{-4} M Horizontal lines indicate base lines prior to contraction of the strip with potassium. Vertical scale denotes 1 g tension. Horizontal scale signifies 30 min.

mesenteric arteries (Fig. 2). No difference was found in the responses of large or small renal and mesenteric arteries.

In most experiments papaverine (10^{-4} M) caused further relaxation after the maximum effect of dopamine (Fig. 1). Since this dose of papaverine usually returned the strip to at least the resting tension before application of potassium, the papaverine effect was accepted as 100% in constructing the dose–response relationship shown in Fig. 2. Unlike with dopamine, there was no detectable difference in the responses of the renal, mesenteric and femoral arteries to papaverine.

Preliminary data indicate that canine coronary and basilar arteries respond to dopamine in the same manner as do renal and mesenteric vessels. On the other hand, the response of rabbit aortic strip appears to resemble that of the large femoral artery. Initial structure-activity studies have shown that epinine appears to have similar relaxing actions as dopamine, but the dopamine metabolite, 3-methoxy-4-hydroxy phenylethylamine, is effective only with very large doses. Studies of potential dopamine antagonists have indicated that haloperidol, chlorpromazine, apomorphine and bulbocapine in estimated effective concentrations cause marked relaxation of the isolated arteries. Thus, further studies are required. Finally, because of the potentiation of the contracting actions of dopamine by potassium, other agents are being studied. Dopamine-induced relaxation has been observed in strips contracted by barium, vasopressin and prostaglandin $F\alpha_2$. The latter agent appears to be superior to potassium.

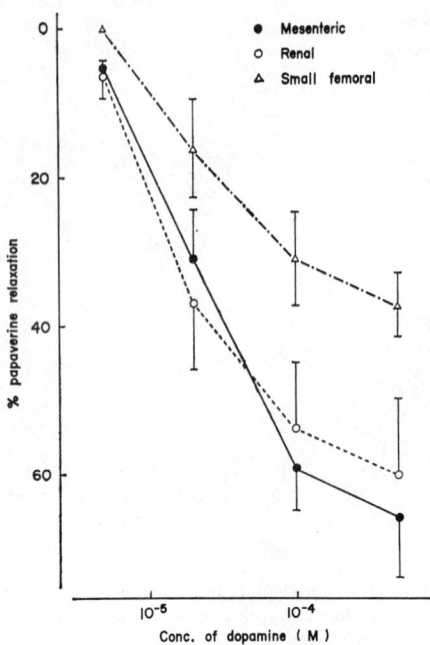

Fig. 2.—Dose–response curves of dopamine-induced relaxation observed in 23 mesenteric, 12 renal and 8 small femoral artery strips, after treatment with phenoxybenzamine, 10^{-5} M, 1 hr, and contraction with potassium. Relaxation caused by papaverine, 10^{-4} M, is considered as 100%. Values are mean ± s.e.

In summary, a method has been described by which the relaxing actions of dopamine can be investigated in isolated blood vessels. Relaxation has been observed in canine renal, mesenteric, coronary and basilar arteries. Dopamine is less effective in small femoral arteries and apparently ineffective in large femoral arteries. The relaxing actions of dopamine are not blocked by *beta*-adrenergic blocking agents. These results support the concept of a specific dopamine vascular receptor.

CORRELATION STUDIES OF THE CARDIOVASCULAR ACTIONS AND PLASMA LEVELS OF DOPAMINE

Because of the reproducible dose–response characteristics of cardiovascular and renal actions of dopamine in the intact animal and man, we have recently attempted to correlate these pharmacological actions with plasma levels of dopamine (Tjandramaga *et al.*, 1973a; Tjandramaga *et al.*, 1973b). Our interest in carrying out these studies was prompted by observations of tolerance to the cardiac actions of L-dopa in patients treated for 3 months or longer with the amino acid (Whitsett and Goldberg, 1972). By simultaneously obtaining measurements of plasma dopa and dopamine, we hoped to demonstrate that the tolerance was due to inhibition of dopa decarboxylase in peripheral tissues. In a few patients, there appeared to be a relative reduction of dopamine to dopa ratio after continuous administration (Fig. 3), but in other patients, this relationship was not observed. Furthermore, we were suprised to find that some patients had dopamine plasma levels approaching 100 μg% without exhibiting any signs of excessive cardiovascular activity such as hypertension

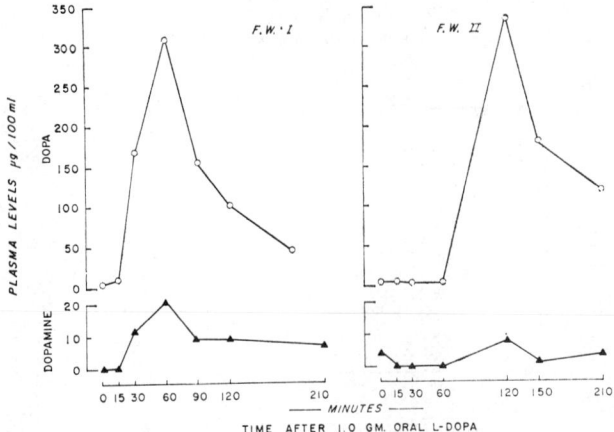

Fig. 3.—Dopa and dopamine plasma levels in Patient F. W. following 1 g morning test-doses of oral L-dopa. The first and the second studies were performed four months apart during continuous L-dopa treatment.

or arrhythmias. WHITSETT and GOLDBERG (1972) had previously estimated that in order to produce the same positive inotropic actions as after 1 g of L-dopa, a 5 min dopamine infusion of 4 μg/kg/min was required. This infusion rate should certainly not produce plasma levels in the range of 100 μg%. Therefore, systematic studies of this apparent lack of correlation were instituted.

Initial studies were carried out using the method previously described (WHITSETT and GOLDBERG, 1972). Patients with Parkinson's disease and hypertension who were on continuous oral L-dopa were given 1 g test doses of L-dopa after an overnight fast. Each patient was maintained in the supine position for at least 15 min before initiation of recording. Serial blood specimens were collected through an indwelling polyethylene catheter placed in the antecubital vein for the determination of plasma dopa and dopamine (ANTON and SAYRE, 1964). Systolic time intervals for left ventricular systole were measured by simultaneously recording the electrocardiogram, phonocardiogram and the carotid arterial pulse tracing. From these tracings, the systolic pre-ejection period (PEP) was derived. HARRIS et al. (1967) had previously demonstrated that shortening of the PEP measurement accurately reflected the positive inotropic action of drugs including isoproterenol and cardiac glycosides. WHITSETT and GOLDBERG (1972) demonstrated that L-dopa produced consistent shortening of this interval in patients during the initial period of L-dopa treatment and that this effect was blocked by the *beta*-adrenergic blocking agent, propranolol (Fig. 4).

Our recent studies, however, demonstrated that there was no correlation between the shortening of the pre-ejection period and measured dopamine plasma levels (TJANDRAMAGA et al., 1973a). In nine studies from eight patients, pre-ejection period was significantly shortened (10 msec or more) and the average dopamine peak plasma level was $12 \cdot 4 \pm 3 \cdot 2$ μg% (mean ± S.E.M.). In 11 studies from 10 patients in whom the pre-ejection period was not shortened, a higher average peak plasma level of dopamine ($20 \cdot 7 \pm 6 \cdot 1$ μg%) was recorded. In one patient, dopamine was infused intravenously at increasing rates of 1, 2 and 4 μg/kg/min during 15 min each, and pre-ejection period shortened in a dose-related manner. Comparison of the

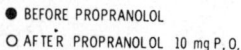

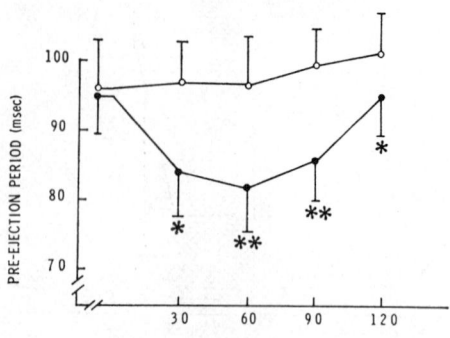

FIG. 4.—PEP-changes in six patients following L-dopa (1·5 g) before and after pre-treatment with propranolol. Values are mean ± standard error.   * — $P < 0.05$; ** — $P < 0.01$.   (T. L. Whitsett and L. I. Goldberg, *Circulation* **35**:97, 1972. Reprinted by permission from the American Heart Association.)

pre-ejection period changes induced by dopamine with those produced by oral L-dopa suggested that the equivalent of roughly 1–2 µg/kg/min of dopamine was being generated 30–60 min after oral L-dopa administration in the group of patients showing shortening of the pre-ejection period. Measurements of plasma dopamine levels in a patient receiving dopamine infusion of 1–2 µg/kg/min indicated that the dopamine plasma levels were less than 2 µg% rather than 20 µg% as occurred with L-dopa.

This lack of correlation was then examined in the anesthetised dog in whom more precise measurements of cardiac contractility and renal blood flow could be made (Tjandramaga *et al.*, 1973b). Dogs were anesthetised with pentobarbital, and after appropriate surgical procedures, cardiac contractile force was measured by, a Walton–Brodie strain gauge arch, renal blood flow by an electromagnetic flow meter, and arterial blood pressure by Statham transducer. Again, a pronounced lack of correlation was demonstrated (Fig. 5). As may be seen, intravenous infusions of dopamine at rates of 1·5, 3 and 6 µg/kg/min produced peak dopamine levels of less than 5 µg%. Intravenous administration of L-dopa at single injections of 1·25, 2·5 and 5 mg/kg produced average dopamine plasma levels ranging from approximately 8 to 25 µg%. The increase in renal blood flow and cardiac contractile force produced by dopamine, however, was markedly greater than that produced by L-dopa. These effects were not due to non-specific depression of the myocardium, or block of *beta*-adrenergic receptors, because the intravenous infusion of dopanime in the animals at a time when L-dopa was exerting its maximal effects, consistently produced a further increase in the cardiac contractile force (Fig. 6). These marked further increments in cardiac action during dopamine infusion occurred with only a very small additional increment in dopamine plasma levels.

Several mechanisms could be responsible for this apparent lack of correlation. First, it is possible that the measurements of dopamine in the plasma *in vitro* does not accurately reflect the level of free dopamine in the plasma *in vivo*. Previous studies have demonstrated that known metabolites of dopamine do not interfere with the specific measurement of this catecholamine (Anton and Sayre, 1964; Tjandramaga

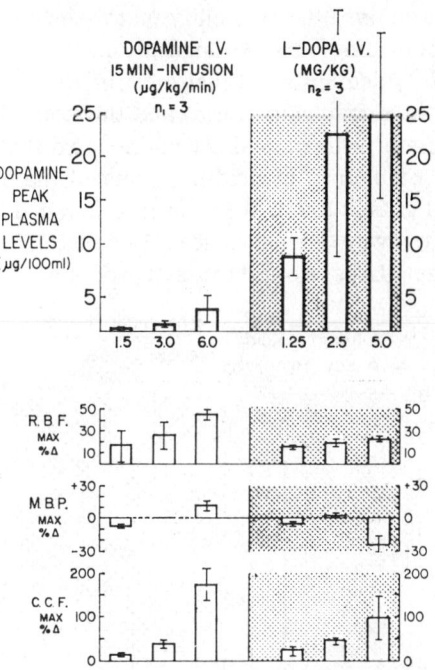

FIG. 5.—Dopamine peak plasma levels and associated maximum responses of renal blood flow (RBF), mean blood pressure (MBP) and cardiac contractile force (CCF) after serial intravenous dopamine infusions and L-dopa injections. Values are mean ± S.E. n_1—3 dogs receiving dopamine infusions; n_2—3 dogs receiving L-dopa injections. (T. B. TJANDRAMAGA, L. I. GOLDBERG and A. H. ANTON, *Proc. Soc. Exp. Biol. Med.* **142**:424, 1973. Reprinted by permission from the Society for Experimental Biology and Medicine.)

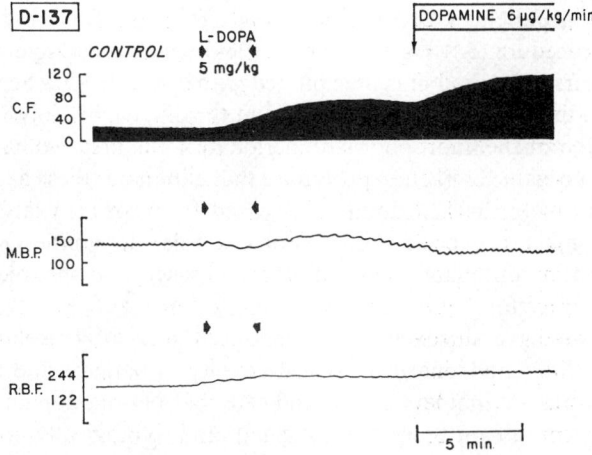

FIG. 6.—Marked further increase in cardiac contractile force from an anaesthetised open-chest dog, induced by super-imposed intravenous dopamine infusion (6 μg/kg/min) following L-dopa injection (5 mg/kg). CF—cardiac contractile force; MBP—mean blood pressure; RBF—renal blood flow.

et al., 1973b). Thus, we examined the possibility that the method used in some way caused the release of free dopamine from a biologically inactive conjugate. IMAI (1970) and his associates previously demonstrated that most of the dopamine circulating in normal human plasma is in a conjugated form, and this could be released by harsh treatment with acid. These investigators also demonstrated that similar increases in plasma dopamine could be obtained by acid treatment of the plasma from patients treated with L-dopa (IMAI *et al.*, 1971). We also have found that acid hydrolysis as well as enzymatic hydrolysis with β-glucuronidase (Ketodase, Warner-Chilcott), and sulfatase (Sulfatase-II, Sigma Laboratories) causes a pronounced increase in measured

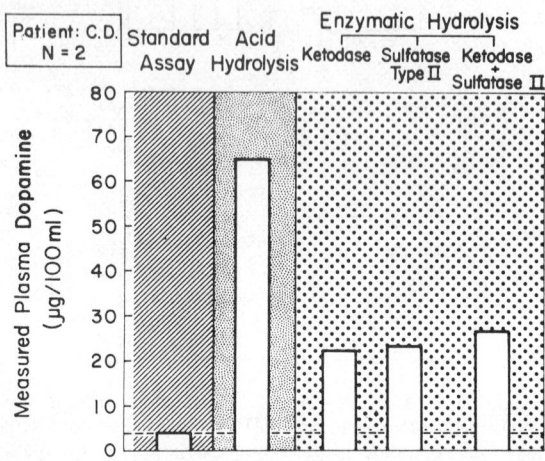

FIG. 7.—Comparison of the yields in measured plasma dopamine using the regular dopamine assay procedure alone, with preliminary acid hydrolysis, and with enzymatic hydrolysis. The bars represent the average values of two measurements from a pair of pooled plasma samples collected during the first and during the second plus third hour following administrations of 1·0 g L-dopa test doses.

plasma dopamine in patients treated with L-dopa (Fig. 7). There are several steps in our analytical procedure (ANTON and SAYRE, 1964) which might hydrolyse an acid-labile conjugate. First, the protein is precipitated with 0·4 N $HClO_4$. Secondly, dopamine is eluted from the Al_2O_3 with 0·05 N HCl, and thirdly, the hydrolysis might occur during the formation of the fluorophor with periodate. The first two have been ruled out by the use of an organic extraction procedure that eliminates these acid treatments. Although it is still likely that dopamine is released from an acid labile conjugate, other mechanisms must also be explored. For example, dopamine *in vivo* may be partially bound to albumin, platelets or red blood cells, whereas the free dopamine is somehow released *in vitro*.

In summary, we have uncovered a pronounced lack of correlation between dopamine plasma levels and pharmacological activity in patients and anaesthetised dogs receiving L-dopa. Preliminary studies indicate that plasma dopamine generated from L-dopa is predominantly in a conjugated and biologically inactive form. Although it is possible that an acid-labile conjugate could explain the apparent discrepancy, alternate explanations such as binding of dopamine to some plasma constituent is not excluded. Thus, further investigation is required to elucidate the basis for this discrepancy.

Acknowledgements—This research is supported in part by Grant GM-14270 from the National Institutes of General Medical Sciences.

REFERENCES

ANTON A. H. and SAYRE D. F. (1964) *J. Pharmacol. Exp. Ther.* **145,** 326–336.
BIEL J., SOMANI P., JONES P. H. and MINARD F. N. (1973) *Frontiers in Catecholamine Research,* Pergamon Press.
FURCHGOTT R. F. and BHADRAKOM S. (1953) *J. Pharmacol. Exp. Ther.* **108,** 129–143.
GOLDBERG L. I. (1972) *Pharmacol. Rev.* **24,** 1–29.
HARRIS W. S., SCHOENFELD C. D. and WEISSLER A. M. (1967) *J. Clin. Invest.* **46,** 1704–1714.
IMAI K., SUGIURA M. and TAMURA Z. (1970) *Chem. Pharm. Bull.* **18,** 2134.
IMAI K., SUGIURA M., TAMURA Z., HIRAYAMA K. and NARABAYASHI H. (1971) *Chem. Pharm. Bull.* **19,** 439–440.
TJANDRAMAGA T. B., GOLDBERG L. I. and ANTON A. H. (1973) *Fedn. Proc.* **32,** 798.
TJANDRAMAGA T. B., GOLDBERG L. I. and ANTON A. H. (1973) *Proc. Soc. Exp. Biol. Med.* **142,** 424–428.
TODA N., USUI H., NISHINO N. and FUJIWARA M. (1972) *J. Pharmacol. Exp. Ther.* **181,** 512–521.
TODA N. and GOLDBERG L. I. (1973) *J. Pharm. Pharmacol.* In press.
ESSEN VON C. (1972) *J. Pharm. Pharmacol.* **24,** 668.
WHITSETT T. L. and GOLDBERG L. I. (1972) *Circulation* **35,** 97–106.

Frontiers in Catecholamine Research 1973, pp. 523 to 526. Pergamon Press. Printed in Great Britain.

DRUG ACTIONS ON α-METHYL-m-TYRAMINE, A FALSE DOPAMINERGIC TRANSMITTER IN THE RAT STRIATUM*

P. A. SHORE and R. L. DORRIS

Department of Pharmacology, University of Texas Southwestern Medical School, Dallas, Texas, U.S.A.

WE HAVE investigated the actions of several drugs on the rat striatum by observing their effects on striatal α-methyl-m-tyramine (MMTA), a false dopaminergic transmitter (DORRIS and SHORE, 1971, 1972). As shown in Table 1, neuroleptics enhance the disappearance of striatal MMTA, an action reversed by apomorphine. This illustrates the opposing actions of these agents on the DA receptor to reflexly alter nigro-striatal impulse flow and consequent amine release. Amphetamine and amfonelic acid, a powerful CNS stimulant, (ACETO et al., 1970) on the other hand appear to cause release largely by a direct action since release still occurs in the presence of apomorphine.

We found that the tyrosine hydroxylase inhibitor, α-methyltyrosine (αMT) inhibits normal release, but even more strikingly, release caused by a neuroleptic; while amphetamine and, especially, amfonelic acid releasing actions are still present despite αMT administration (Table 2). The effects of αMT on normal and neuroleptic-induced release might mean that MMTA is displaced from the storage pool by newly formed DA in exact proportion to the extent of DA synthesis, but the results also suggest that αMT, in addition to enzyme blockade, also inhibits transfer of amine from the storage pool to the impulse-releasable site. This latter effect would be especially important during a high rate of nerve firing (such as after haloperidol), when pool as well as newly synthesised amine is released. An action of αMT on inter-pool amine movement has previously been suggested (ENNA et al., 1973). Consistent with this possibility is evidence that the half-life of striatal DA is considerably prolonged by the presence of αMT (IVERSEN and GLOWINSKI, 1966). Such an effect of αMT would also be consistent with the marked potentiation by αMT of the antipsychotic effect of phenothiazines (CARLSSON et al., 1972).

In marked contrast to αMT action, amantadine (ATD), in low dosage, markedly inhibited the MMTA-releasing actions of the direct releasers amphetamine and amfonelic acid, and showed a tendency to inhibit normal release, but had no effect on haloperidol-induced MMTA release (Table 3). ATD also greatly inhibited MMTA lowering by the reserpine-like drug RO 4-1284. Similar effects of ATD were seen on amfonelic acid-induced release of striatal m-tyramine after m-tyrosine treatment in the presence of MAO blockade. ATD had no effect on the release of heart metaraminol by RO 4-1284. Thus ATD may stabilise striatal amine storage granules in such a way as to inhibit amine efflux. This possibility is further strengthened by the observation that while MMTA accumulation in the striatum is greatly reduced by prior reserpine treatment, injection of amantadine largely overcame this effect of

* Supported by USPHS Grant MH-05831.

TABLE 1. INTERACTIONS OF APOMORPHINE, NEUROLEPTICS, AND STIMULANTS ON RAT STRIATAL MMTA

	0 time control	3 hr control	6 hr control	Presence or absence of drugs						
Apomorphine	−	−	−	+	−	+	−	+	+	+
Haloperidol	−	−	−	−	+	+	+	−	−	−
Trifluoperazine	−	−	−	−	−	−	+	+	−	−
Amphetamine	−	−	−	−	−	−	−	−	+	−
Amfonelic acid	−	−	−	−	−	−	−	−	−	+
MMTA conc. (μg/g ± s.e.)	0·75 ± 0·03	0·63 ± 0·02	0·55 ± 0·01	0·73 ± 0·04	0·38 ± 0·00	0·70 ± 0·03	0·36 ± 0·05	0·66 ± 0·03	0·58 ± 0·03	0·44 ± 0·01

Zero time in the above chart denotes start of the experiment and corresponds to a time 17 hr after injection of the MMTA precursor, $dl\alpha$MMT (100 mg/kg s.c.). Apomorphine dosage was 2 mg/kg s.c. hourly. Rats were killed 6 hr after haloperidol (0·1 mg/kg i.p.) or trifluoperazine (1 mg/kg ip) or 3 hr after d-amphetamine sulphate (3 mg/kg i.p.) or amfonelic acid (2 mg/kg i.p.). Figures denote mean striatal MMTA (μg/g) ± s.e. in 4–15 experiments. Note that experiments involving haloperidol are referable to 6 hr control; those with stimulants to apomorphine alone.

TABLE 2. INTERACTIONS OF α-METHYLTYROSINE, HALOPERIDOL AND STIMULANTS ON RAT STRIATAL MMTA

	0 time control	2 hr control	6 hr control	Presence or absence of drugs								
Haloperidol	—	—	—	+	+	+	+	+	+	—	—	—
Amphetamine	—	—	—	—	—	—	—	—	—	—	+	+
Amfonelic acid	—	—	—	+	—	—	—	—	+	+	—	—
DL-αMT	—	—	—	—	+*	—	—	—	—	—	—	—
L-αMT	—	—	—	—	—	+	+	+	—	—	—	—
D-αMT	—	—	—	—	—	—	+	—	+	—	—	—
MMTA conc. μg/g ± S.E.	0·78±0·03	0·75±0·03	0·67±0·02	0·37+0·01	0·83±0·04	0·64±0·01	0·61±0·04	0·40±0·01	0·45±0·04	0·23±0·02	0·64±0·04	0·32±0·02

Details same as Table 1 except in some cases striatal MMTA introduced directly by injection of d-MMTA (10 mg/kg s.c.) in which case zero time is 6 hr after MMTA injection. Rats killed 2 hr after d-amphetamine sulfate (5 mg/kg i.p.) or amfonelic acid (5 mg/kg i.p.). DL αMT was the methyl ester hydrochloride (250 mg/kg i.p.) followed where appropriate by 125 mg/kg 3 hr later. D or L αMT was the free amino acid (100 mg/kg + 50 mg/kg) 3 hr later. Note that experiments with haloperidol are referable to 6 hr control, those with stimulants to 2 hr control.

* 6 hr

TABLE 3. INTERACTION OF AMANTADINE WITH OTHER AGENTS ACTING ON RAT STRIATAL MMTA

	2 hr control	6 hr control	Presence or absence of drugs									
Apomorphine	—	—	—	+	+	+	+	+	+	+	+	+
Haloperidol	—	—	+	—	—	—	—	—	—	—	—	—
Amantadine	—	+	—	+	+	+	+	+	+	—	+	+
Amphetamine	—	—	—	—	—	—	—	—	—	—	—	—
Amfonelic acid	—	—	—	—	—	—	+	+	—	—	+	—
Ro 4-1284	—	—	—	—	—	—	—	—	—	+	—	+
MMTA conc. μg/g ± S.E.	0·74±0·03	0·66±0·02	0·76±0·04	0·28±0·01	0·35±0·02	0·78±0·04	0·55±0·03	0·72±0·00	0·29±0·02	0·67±0·07	0·35±0·00	0·78±0·08

Amantadine (10 mg/kg i.p.) and/or Ro 4-1284 (5 mg/kg i.p.) given at zero time. Rats killed 2 hr after amphetamine or amfonelic acid. Other details same as Tables 1 and 2.

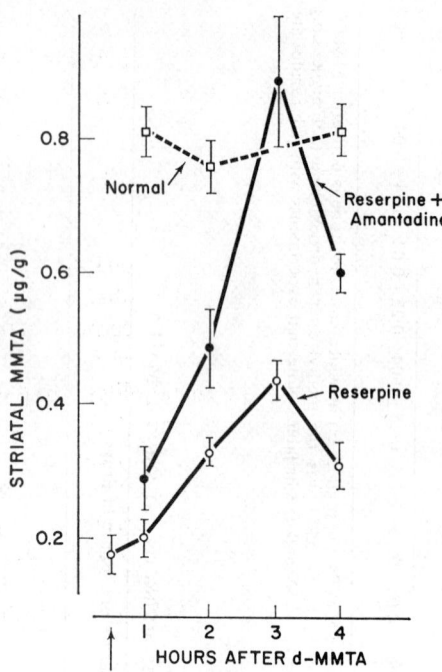

FIG. 1.—Concentrations of MMTA in rat striatum at various times after *d*-MMTA injection (10 mg/kg s.c.). Some rats were untreated, others had received reserpine (2 mg/kg s.c.) 18 hr before MMTA. Amantadine (10 mg/kg i.p.) was given 30 min (arrow) after MMTA injection into reserpine-treated rats. Points denote mean ± S.E. of 3–8 experiments.

reserpine (Fig. 1). The value of ATD in Parkinsonism may be to help correct a leaky amine granule, an effect consistent with the reported stabilisation of DA after intracisternal administration of DA (SYMCHOWICZ *et al.*, 1973). This action of ATD is also consistent with the drug's special usefulness when used in combination with L-Dopa.

REFERENCES

ACETO M. D., BOTTON I., LEVITT M., MARTIN R., BENTLEY H. C. and SPEIGHT P. T. (1970) *Europ. J. Pharmacol.* **10**, 344–354.
CARLSSON A., PERSSON T., ROOS B.-E. and WALINDER J. (1972) *J. Neural Transmiss.* **33**, 83–90.
DORRIS R. L. and SHORE P. A. (1971) *J. Pharmacol.* **179**, 10–14.
DORRIS R. L. and SHORE P. A. (1972) *J. Pharm. Pharmacol.* **24**, 581–583.
ENNA S. J., DORRIS R. L. and SHORE P. A. (1973) *J. Pharmacol.* **184**, 576–582.
IVERSEN L. L. and GLOWINSKI J. (1966) *J. Neurochem.* **13**, 671–682.
SYMCHOWICZ S., KORDUBA C. A. and VEALS J. (1973) *Europ. J. Pharmacol.* **21**, 155–160.

Frontiers in Catecholamine Research 1973, pp. 527 to 530. Pergamon Press. Printed in Great Britain.

EXTRANEURONAL UPTAKE OF CATECHOLAMINES*

U. Trendelenburg

Department of Pharmacology and Toxicology, University of Würzburg, Germany

PERFUSED HEARTS

Most of the studies of the neuronal uptake of noradrenaline were designed to measure the accumulation of the amine in the tissue. Such studies may well provide a distorted picture of the role of the metabolising enzymes, since noradrenaline is well retained after its uptake (first across the neuronal membrane and then across the vesicular membrane) into storage vesicles, while the metabolites formed intraneuronally are not retained, leave the nerve endings and accumulate in the perfusion fluid instead of in the tissue. This disadvantage of studies of accumulation is avoided by the use of the method of Lindmar and Muscholl (1964) in which the removal of the amine from the perfusion fluid is measured. With this method, it is possible to study the time course of events during a perfusion with a constant concentration of a sympathomimetic amine (see, for instance, Graefe et al., 1971). If one uses radio-actively labelled amine, it is, in addition, possible to determine the appearance of the metabolite(s) in the venous effluent.

For studies of the extraneuronal uptake the same problem arises. Most former studies dealt with the accumulation of the amine (or its metabolites) in the tissue, irrespective of whether measurements were made fluorimetrically, by scintillation counting or by fluorescence microscopy. In order to assess the relative importance of extraneuronal storage (of the unchanged amine) and of extraneuronal metabolism, extraneuronal removal, accumulation and O-methylation were measured during the perfusion of isolated rat and guinea-pig hearts with ^{3}H-isoprenaline. This amine was chosen, since it is not taken up by adrenergic neurones (Hertting, 1964).

When isolated hearts were perfused with 0·95 μM ^{3}H-isoprenaline for 30 min, the rate of removal of the amine fell quickly during the first 1–2 min of the perfusion, in order to fall slowly thereafter. While the first phase of removal was due to the filling of extracellular space and right ventricle, the second phase represented an approach to a steady-state rate of removal (with a half time of 7·2 and 5·6 min for rat and guinea-pig hearts, respectively). The metabolite, ^{3}H-O-methyl-isoprenaline (^{3}H-OMI), on the other hand, appeared in the venous effluent at rapidly increasing rates, in order to approach a steady-state level of efflux with a half time of about 1 min. The steady-state levels for the removal of ^{3}H-isoprenaline and for the efflux of ^{3}H-OMI are identical. Or in other words, while removal initially is due to both accumulation and O-methylation, only O-methylation is observed after attainment of steady-state conditions. It is of interest that a pronounced species difference was observed: the steady-state rate for removal and O-methylation was 10 times higher in rat than in guinea-pig hearts. Since the COMT activity of homogenates of rat and guinea-pig hearts is roughly the same (Jarrott, 1970), the observed species difference suggests that the rate limiting step is found in an uptake process rather than in the enzyme.

The accumulation of ^{3}H-OMI in the heart approached steady-state levels as quickly as the efflux of the metabolite form the heart did: a steady-state content of ^{3}H-OMI was reached within less than 10 min. The accumulation of ^{3}H-isoprenaline, on the other hand, approached a steady-state level much more slowly, just as the removal approached steady-state levels more slowly than the efflux of ^{3}H-OMI did. Again, a species difference was observed: rat hearts accumulated eight times more ^{3}H-OMI and about three times more ^{3}H-isoprenaline than guinea-pig hearts did.

Thus, we observed a constant rate of O-methylation virtually throughout the experiment, while accumulation continued to increase. This non-parallelism between O-methylation and accumulation of isoprenaline is incompatible with a one-compartment system. On the contrary, the two processes seem to occur in different compartments: while the COMT-containing compartment equilibrates very quickly with the perfusion fluid, the isoprenaline-storing compartment equilibrates much more slowly.

When COMT was inhibited by the presence of 100 μM U-0521, no ^{3}H-OMI was detected in either venous effluent or heart. In that case removal of isoprenaline fell towards zero during the second phase of removal, and the half time for approach to equilibrium was increased (13·5 and 8·0 min for rat and guinea-pig hearts, respectively). Similarly, the accumulation of ^{3}H-isoprenaline in the rat heart reached higher levels than in the absence of the enzyme inhibitor and had an increased half time.

Within the framework of the two-compartment system proposed above, the effect of block of COMT suggests that the two compartments are arranged in series. If the rate of O-methylation is normally high (rat heart), block of the enzyme increases accumulation; if, on the other hand, the rate of O-methylation is normally low (guinea-pig heart), block of the enzyme fails to cause a significant increase in accumulation of the unchanged amine. Further evidence for differences between the accumulation and the O-methylation of ^{3}H-isoprenaline were obtained by kinetic analysis. Accumulation (K_m 20 μM, V_{max} 38 nmoles g^{-1} min^{-1}) differed from O-methylation (apparent K_m 2·7 μM; V_{max} 1·7 nmoles g^{-1} min^{-1}). The values for accumulation are in agreement with those obtained by CALLINGHAM and BURGEN (1966).

Corticosterone is known to inhibit extraneuronal uptake (IVERSEN and SALT 1970). In the presence of 0·5–10 μM corticosterone both accumulation and O-methylation of ^{3}H-isoprenaline in the rat heart were reduced.

O-methylated amines represent another group of agents able to impair extraneuronal uptake (IVERSEN, 1967). In the presence of 1–10 μM O-methyl-isoprenaline the accumulation and O-methylation of ^{3}H-isoprenaline were reduced in a way that was qualitatively similar to the effect of corticosterone.

In a second series of experiments rat hearts were perfused with 0·95 μM ^{3}H-isoprenaline for 30 min. During subsequent perfusion with an amine-free solution, the efflux of the amine and its metabolite was studied.

When COMT was intact, the rate of efflux of ^{3}H-OMI fell from the level observed during steady-state conditions. This fall was initially much slower than that of the rate of efflux of ^{3}H-isoprenaline. In fact, a convex shaped curve was obtained for the efflux of ^{3}H-OMI. This unexpected shape of the efflux curve seems to be due to the fact that the rate of O-methylation is close to (but still below) the V_{max} of the O-methylating system. Thus, in spite of a rapidly declining efflux rate of ^{3}H-isoprenaline (as an index of the quick loss of the amine from the stores), the rate of efflux of ^{3}H-OMI declines slowly as the enzyme (or access to the enzyme) is desaturated. This view is supported by the observation that the convexity of the efflux curve is much more pronounced during efflux subsequent to a 30 min perfusion with a 25 times higher concentration of ^{3}H-isoprenaline (23·8 μM): the rate of efflux of ^{3}H-OMI falls very slowly during the first 50 min of the wash out period, in order to show a rapid decline thereafter. The interpretation presented here is consistent with the K_m and V_{max} values given above.

Because of the unexpected shape of the efflux curves, an analysis of efflux had to be restricted to experiments in which COMT had been blocked. Efflux from the extra-neuronal stores was then biphasic (with half times of 11 and 24 min, respectively). When corticosterone (20 μM) was present during the filling of the heart, the compart-ment with the short half time was filled poorly. When corticosterone was added to the wash out solution only, efflux became monophasic (half time 29 min). Apparently, corticosterone delayed part of the efflux, namely that from the compartment with the short half time.

In additional experiments, O-methyl-isoprenaline (36 μM) was added to the wash out solution only. As with corticosterone, efflux became monophasic (half time: 27 min). The delaying effect of O-methyl-isoprenaline is consistent with the obser-vation of IVERSEN (1965) that the presence of another O-methylated compound, metanephrine, in the wash out solution prevents the efflux of noradrenaline from the extraneuronal stores.

The results obtained with corticosterone and O-methyl-isoprenaline indicate that both agents act on one of two extraneuronal compartments, and that they reduce the rate of both influx and efflux.

RABBIT AORTIC STRIPS

Strips were made nerve-free by removal of the adventitia. Cocaine was present throughout to prevent neuronal uptake into nerve endings that had not been removed. Strips were incubated with 1·18 mM ($\pm$)^{3}H-noradrenaline for 30 min and washed thereafter with amine-free solution. Efflux of the amine and its metabolites was followed for up to 240 min of wash out.

The main metabolite resulting from extraneuronal uptake was normetanephrine. Extraneuronal accumulation was increased by block of COMT (100 μM U-0521) and not by block of MAO (pretreatment of the strip with 513 μM pargyline). In the presence of corticosterone (87 μM) both the accumulation of ^{3}H-noradrenaline and the formation of ^{3}H-normetanephrine (measured in tissue *and* incubation medium) were reduced.

Efflux curves indicated that ^{3}H-noradrenaline efflux came from at least two extraneuronal compartments, only one of which was sensitive to corticosterone. Again, it was the compartment with the shorter half time of efflux which was cort-icosterone-sensitive. When corticosterone was present during the incubation with ^{3}H-noradrenaline, the compartment filled poorly. When corticosterone was present during wash out only, the efflux from the corticosterone-sensitive compartment was greatly delayed. This second effect of corticosterone was observed only when the compartment discussed here was filled well, i.e., when COMT was inhibited. In the absence of U-0521 the delay of the efflux from this compartment was not demonstrable because of the poor filling.

Hence, the results suggest that the corticosterone-sensitive compartment has a very high COMT activity, because of which there is very poor filling of this compartment with the unchanged amine as long as the enzyme is intact. Corticosterone prevents access to this compartment and, hence, O-methylation. However, if the corticosterone-sensitive compartment is filled well e.g., after block of COMT), corticosterone clearly delays the efflux of the unchanged amine from the compartment.

ISOLATED NICTITATING MEMBRANE

Denervated nictitating membranes treated with pargyline and U-0521 (to block the metabolising enzymes) take up ^{3}H-noradrenaline on incubation with this amine. This extraneuronal accumulation of the amine is reduced by hydrocortisone when incubations are of short duration (2·5 min); however, no inhibitory effect of the steroid is observed for longer incubation periods (10 min). Apparently, the nictitating membrane resembles the other tissues discussed here in having a steroid-sensitive and a steroid-resistant extraneuronal uptake mechanism; the former seems to be of the quickly equilibrating type. Preliminary results from experiments with preparations whose COMT was not inhibited suggest that the nictitating membrane differs from the other two tissues in having COMT activity in *both* compartments.

CONCLUSIONS

The experiments described here provide evidence for the view that extraneuronal storage is multicompartmental. In rat and guinea-pig heart and in rabbit aorta most, if not all, COMT activity is associated with the steroid-sensitive compartment; however, in the cat nictitating membrane COMT may be localised in the steroid-resistant compartment as well. In heart and aorta, the COMT activity of the steroid-sensitive compartment seems to be so high that filling of the amine with unchanged amine is poor as long as the enzyme is intact. In both tissues the effect of corticosterone consisted of a decrease in O-methylation, a decrease in the rate of influx of the amine and a decrease in the rate of efflux of the amine from this compartment. In all three tissues evidence was also obtained for a steroid-resistant storage compartment.

Preliminary results of these studies have been published (Bönisch and Uhlig, 1973; Eckert and Henseling. 1973). The experiments with isolated nictitating membranes were carried out by K. H. Graefe.

Acknowledgements—This study was supported by the Deutsche Forschungsgemeinschaft

REFERENCES

Bönisch H. and Uhlig W. (1973) *Naunyn-Schmiedeberg's Arch. Pharmacol.* **277**, R6.
Callingham B. A and Burgen A. S. V. (1966) *Mol. Pharmacol.* **2**, 37–42.
Eckert E. and Henseling M. (1973) *Naunyn-Schmiedeberg's Arch. Pharmacol.*, **277**, R14.
Graefe, K. H., Bönisch H. and Trendelenburg U. (1971) *Naunyn-Schmiedebergs Arch. Pharmakol.* **271**, 1–28.
Hertting G. (1964) *Biochem. Pharmacol.* **13**, 1119–1128.
Iversen L. L. (1965) *Brit. J. Pharmacol.* **25**, 18–33.
Iversen L. L. (1967) *The uptake and storage of noradrenaline in sympathetic nerves.* Cambridge University Press, Cambridge. 253 pages.
Iversen L. L. and Salt P. J. (1970) *Brit. J. Pharmacol.* **40**, 528–530.
Jarrott B. (1970) *Brit. J. Pharmacol.* **38**, 810–821.
Lindmar R. Muscholl E. (1964) *Naunyn-Schmiedebergs Arch. exp. Path. Pharmak.*, **247**, 469–492.

Frontiers in Catecholamine Research 1973, pp. 531 to 535. Pergamon Press. Printed in Great Britain.

REGULATION OF CATECHOLAMINE RELEASE

JAN HÄGGENDAL

Department of Pharmacology, University of Göteborg, Göteborg, Sweden

KNOWLEDGE about the amount of transmitter released per nerve impulse from the nerve ending, is of great importance for understanding the transmitter economy of the neuron. This question has been studied particularly in the peripheral noradrenaline-(NA) containing neurons, where the overflow of released NA to the blood stream or perfusate has been measured after stimulation with a known number of nerve impulses. The problem with the presence of the blood–brain barrier in the central nervous system (CNS), preventing the transmitter released from central mono-aminergic neurons from entering into the blood stream, has been overcome with field stimulation of brain slices as used e.g. by FARNEBO and HAMBERGER (1971a).

The transmitter that is released from the nerve terminals can be affected by several inactivating mechanisms at the nerve terminal–effector cell level. The quantitative role of these mechanisms has to be considered when estimating the 'true release' of transmitter from the nerve terminals. The inactivating mechanisms are: (1) binding to receptors, (2) enzymatic destruction by catechol-O-methyl-transferase (COMT, E.C. 2.1.1.6.) and monoamine oxidase (MAO, E.C. 1 4.3.4.), (3) re-uptake into the nerve terminals by the 'membrane pump', (4) extra-neuronal uptake, and (5) diffusion into the perfusate (e.g. blood).

Variations in the amount of NA overflow after a certain number of sympathetic stimuli were earlier often discussed to be due to quantitative variations in the role of these inactivating mechanisms. A typical example is the effect on the overflow of NA obtained after α-receptor blocking drugs, such as phenoxybenzamine (PBZ). The increase is often found to be 4–10 times the normal NA overflow. This effect has been explained as due to blockade of one or several of the local inactivating mechanisms, thus permitting the non-inactivated NA to reach the perfusate (for review see e.g. BOULLIN, COSTA and BRODIE, 1967). Hence, PBZ was considered to be a tool for revealing the 'true release' of transmitter.

However, when the quantitative role of the local inactivating mechanisms is investigated (see below) it seems that a blockade of one, or several of them, could not explain the very high increase in NA overflow found after α-receptor blockers. It has therefore been proposed that the amount of transmitter released from the nerve terminals per nerve impulse is increased after an α-receptor blocker, such as PBZ (HÄGGENDAL, 1969, 1970; FARNEBO and HAMBERGER, 1970, 1971b; LANGER, 1970; KIRPEKAR and PUIG, 1971, STARKE, MONTEL and WAGNER, 1971; ENERO, LANGER, ROTHLIN and STEFANO, 1972; For history see STARKE, this Symposium.)

What may be called a more 'direct' way of showing an increased release from the nerve terminals after PBZ has been presented: It now seems generally accepted that during the normal release of NA at nerve activity some dopamine-β-hydroxylase (DBH, E.C. 1.14.2.1.), a granular component, is also released. No inactivating mechanisms for released DBH are known to be present at the nerve terminal-effector cell level. An increased release not only of NA but also of DBH following PBZ has been demonstrated by De Potter, Chubb, Put and De Schaepdryver (1971) and Johnson, Thoa, Weinhilsboum, Axelrod and Kopin (1971). This increased release is thus likely to be due to an increased release from the amine granules in the nerve terminals.

The different types of studies indicate that the amount of transmitter released from the nerve terminals is not constant but may vary. With quite another approach, using prostaglandin, Hedqvist (1969a,b,) discussed a variable release from the nerve terminals at nerve activity. The effect of different types of prostaglandins on the release has now been investigated by many authors (e.g. Hedqvist, 1970; Wennmalm, 1971; Hedqvist and V. Euler, 1972; Bergström, Farnebo and Fuxe, 1973). For further information, see Farnebo, Hedqvist, Langer, Starke and Stjärne, (this Symposium).

The studies on PBZ-induced NA overflow have provided an increased knowledge on the role of the different inactivation mechanisms:

(1) *Binding to the receptors.* After formation of a NA-α-receptor complex, the complex seems to be broken very soon. The NA is likely either to be set free to the synaptic gap again, (where it will be dealt with by the other inactivating mechanisms) or metabolised at the receptor level. However, if this is the case, the NA metabolites are probably not trapped in the tissue, but will instead contribute to the sum of metabolites found in the perfusate.

(2) *Metabolism.* The sum of NA metabolites in most tissues seems to be rather low during stimulation, about 30 per cent or less of the total release of NA plus its metabolites (e.g. Rosell, Kopin and Axelrod, 1963; Häggendal, Johnsson, Jonason and Ljung, 1970). Therefore, a total block of the NA metabolism will increase the free NA levels only moderately. Furthermore, inhibitors of MAO or COMT do not appear to increase the NA to any marked extent (e.g. Brown, 1965; Folkow, Häggendal and Lisander, 1967; Farnebo and Hamberger, 1971b).

In tissues preincubated with ^{3}H-NA, the total amount of ^{3}H-NA plus its metabolites in the overflow were increased after PBZ (e.g. Farnebo and Hamberger, 1970; Häggendal et al., 1972a). This indicates that the increased overflow of free NA (after PBZ) can not be explained as due to a block of the NA metabolism. If the increase of free NA had been due mainly to a blockade of the NA metabolism the sum of free NA plus metabolites should have been about constant.

(3) *Uptake via the membrane pump.* A number of investigators have found the increase in NA overflow to be only small after administration of membrane pump blockers (see e.g. Farnebo and Hamberger, 1971b, for results and references). This is in agreement with our results (Häggendal, 1970; Häggendal et al., 1970, 1972).

Furthermore, after an apparently optimal treatment with blockers of the membrane pump, the addition of PBZ markedly increased the NA overflow (e.g. Geffen, 1965; Häggendal, 1970; Farnebo and Hamberger, 1971b). Therefore, it is not possible to explain the effect of PBZ on the NA overflow in terms of membrane pump inhibition. Furthermore, an increase in NA release takes place already at low doses of PBZ, where its membrane blocking effect is not observed, e.g. Enero et al. (1972).

(4) *Extra-neuronal uptake.* Under physiological conditions, NA is probably only taken up extra-neuronally in small amounts. This NA will probably be metabolised (cf. Iversen, 1971). The NA metabolites formed are not likely to be trapped in the

tissue, but to contribute to the sum of metabolites in the overflow. However, as indicated above, the sum of metabolites is rather small under normal conditions. Also, after normetanephrine, a blocker of the extra-neuronal uptake, the NA over-flow was increased only very slightly (FARNEBO and HAMBERGER, 1971b). The increased NA overflow after PBZ can thus not be explained as due to inhibition of extra-neuronal uptake.

(5) *Overflow to the circulation*. The vaso-constriction normally observed at sympathetic stimulation is not present after α-receptor blockade. The maintained high blood flow has therefore been thought to facilitate the NA overflow. This may be correct within some limits (cf. ROSELL, KOPIN and AXELROD, 1963; CARLSSON, FOLKOW and HÄGGENDAL, 1964). However, it does not explain the increase in NA overflow found after PBZ when the blood flow was kept at the same level as in normal (e.g. BOULLIN *et al.*, 1967; HÄGGENDAL, 1970). Also, in thin preparations, such as isolated rat portal vein and rat iris, where the possibilities for diffusion of the released NA ought to be good and not markedly changed if PBZ is present or not, the overflow of NA is highly increased after PBZ.

In summary: it seems not adequate to explain the PBZ-induced large increase in NA overflow on the basis only of blockade of inactivating mechanisms at the nerve terminal-effector cell level. The results rather indicate that after PBZ more trans-mitter than normally is released from the nerve terminals per nerve impulse. An increased transmitter release after receptor blocking has also been reported for cholinergic neurons in CNS (POLAK, 1971).

The increased NA release after α-receptor blockers indicates that α-receptors or α-receptor-like structures are involved in a local feed-back regulation of transmitter release during nerve activity. These α-receptors may be localised either pre- or post-synaptically; or on both places.

(1) A regulation via presynaptic receptors has been discussed by FARNEBO and HAMBERGER (1971b), ENERO *et al.* (1972), STARKE (1972). Experiments on the heart showed that β-receptor blockers did not markedly change the transmitter over-flow while α-receptor blockers increased the release. Since the proportion of α-receptors in the heart has been considered as small or negligible, the results may support the view of a presynaptic localisation of α-receptor-like structures. Recent results seem to demonstrate α-receptors in different structures of the heart (GOVIER, 1968; HOLMBERG, SVEDMYR and ÅBERG, 1972; EKSTRÖM-JODAL, HÄGGENDAL, MALMBERG and SVEDMYR, 1972; KUNOS, 1973; BENFEY, 1973), but this does not exclude the possibility of presynaptic α-receptor like structures being involved in the regulation of the release.

Some support for the presynaptic localisation of α-receptors is obtained from studies indicating that other receptors of importance for the transmitter release are presynaptically localised. Thus this seems to be the case for the muscarinic receptor that inhibits the NA release (LINDMAR, LÖFFELHOLZ and MUSCHOLL, 1968; LÖFFEL-HOLZ and MUSCHOLL, 1970).

A presynaptic regulation of the transmitter release means that the feed-back mechanism is not involving the effector cell. Such a mechanism may appear to be of only limited interest in relation to the main function of the neuron, which is to give a signal to the effector cell. It therefore was tempting to look for a feed-back mechanism that is triggered from the effector cell.

(2) Regulation via postsynaptic mechanisms have been discussed by Häggendal (1970); Farnebo and Hamberger (1971b); Farnebo and Malmfors (1971). In this case it is not necessary to postulate any pre-ganglionic α-receptor like structure. Both in the periphery and in CNS it has been shown that receptor activators decrease the transmitter release in monoaminergic neurons (Farnebo and Hamberger, 1971). After methoxyamine the contraction of mouse vas deferens was potentiated and the release of NA decreased (Farnebo and Malmfors, 1971). In other tissues, such as the heart, it seems to be difficult to see such a connection (Starke, 1972). In studies with receptor activators, pre- as well as postsynaptic receptors may be activated.

A weak point in the construction of a feed-back mechanism operating from the effector cell is that the message has to over-bridge a distance that sometimes can be large, e.g. up to 1000 Å or more in blood vessels.

More information seems to be necessary until the question can be solved about pre- or postsynaptic localisation of the receptor-like structures that are involved in the regulation of transmitter release. Most data seem; however, to indicate a pre-synaptic localisation, (see this Symposium).

Also the nature of this mechanism needs to be clarified. It may involve such difficult problems as the mechanisms behind α-receptor activation as well as the detailed machinery for transmitter release. Prostaglandins may be of particular importance. However, recently evidence has been given for a separation of the mechanisms by which prostaglandins and α-receptors are involved. These problems have been discussed, as well as the role of Ca^{2+} ions, at this Symposium by Farnebo, Hedqvist, Starke and Stjärne.

Even if the detailed mechanisms behind a feed-back regulation at the nerve terminals-effector cell level still are not very well known, the presence of such mechanisms dealing with α-receptors seems to be rather well established. Due to the above-discussed feed-back mechanism, the following points must be considered:

(1) The amount of transmitter released from the nerve terminals per time unit is dependent not only on the nerve impulse rate, but also on the amount of transmitter released per nerve impulse.

(2) Following α-receptor blockade, transmitter may be released in such large amounts that other monoaminergic receptors can be affected either via the blood stream or directly.

The blood NA levels in patients on treatment with high doses of chlorpromazine can be more than 8–10 μg/l. under physical exercise, while untreated controls only show a level of about 2 μg/l. (Carlsson, Dencker, Grimby and Häggendal, 1964). These levels are certainly high enough to directly stimulate the β-receptors in the heart. Also before reaching the circulation the released NA might activate β-receptors in the vicinity of the releasing nerve terminals. Thus a stimulation-induced vasodilation due to β-receptor activation was observed in cats' skeletal muscle after PBZ treatment (Folkow and Häggendal, unpublished).

In CNS similar mechanisms may also be of importance after receptor blockade. The transmitter e.g. NA released in increased amounts after blockade of α-receptor like structures may have a direct effect on other neurons or receptors. Therefore the possibility exists that following the administration of a 'pure' receptor blocker the picture may be obscured by a simultaneous activation of other types of receptors.

(3) The recognition of a variable transmitter release implies that α-receptor

blockers, e.g. PBZ, are not suitable for studying 'true' transmitter release, since artificially high values will be obtained. In fact the disclosed feed-back mechanism demonstrates that 'true release' (i.e. from the nerve terminals) is not constant but variable. Consequently the figures of about 2×10^{-5} obtained for fractional release (released NA amount per stimulus in relation to the total tissue stores of NA) by FOLKOW et al. (1967), who did not use PBZ but maintained the high blood flow by concomitant motor nerve stimulation, may be within the physiological range, but not represent the lowest figure for fractional release that can occur under normal conditions.

Acknowledgements—Part of the studies were supported by grants from the Swedish Medical Research Council (14X-166).

REFERENCES

ANDERSSON R., HOLMBERG S., SVEDMYR N. and ÅBERG G. (1972) *Acta med. scand.* **191**, 241–244.
BENFEY B. G. (1973) *Fedn. Proc.* **32**, No. 3, 2694.
BERGSTRÖM S., FARNEBO L.-O. and FUXE K. (1973) *Europ. J. Pharmacol.* **21**, 362–368.
BOULLIN J., COSTA E. and BRODIE B. B. (1967) *J. Pharmacol. exp. Ther.* **157**, 125–134.
BROWN G. L. (1965) *Proc. Roy. Soc. B.* **162**, 1–19.
CARLSSON A., FOLKOW B. and HÄGGENDAL J. (1964) *Life Sci.* **3**, 1335–1341.
CARLSSON C., DENCKER S. J., GRIMBY G. and HÄGGENDAL J. (1967) *Acta Pharmacol. Toxicol.* **25**, 97–106.
DE POTTER W. P., CHUBB I. W., PUT A. and DE SCHAEPDRYVER A. F. (1971) *Archs. int. Pharmacodyn* **193**, 191–197.
EKSTRÖM-JODAL B., HÄGGENDAL E., MALMBERG R. and SVEDMYR N. (1972) *Acta med. scand.* **191**, 245–248.
ENERO M., LANGER S. Z., ROTHLIN R. P. and STEFANO F. J. E. (1972) *Br. J. Pharmac.* **44**, 672–688.
FARNEBO L.-O. and HAMBERGER B. (1970) *J. Pharm. Pharmac.* **22**, 855–857.
FARNEBO L.-O. and HAMBERGER B. (1971a) *Acta physiol. scand. Suppl.* **371**, 35–44.
FARNEBO L.-O. and HAMBERGER B. (1971b) *Br. J. Pharmac.* **43**, 97–106.
FARNEBO L.-O. and MALMFORS T. (1971) *Acta physiol. scand. Suppl.* **371**, 1–18.
FOLKOW B. and HÄGGENDAL J. (1970) In: *Bayer-Symposium II.* pp. 91–97. Springer-Verlag, Heidelberg.
FOLKOW B., HÄGGENDAL J. and LISANDER B. (1967) *Acta physiol. scand. Suppl.* **307**, 1–38.
GEFFEN L. B. (1965) *J. Physiol. Lond.* **181**, 69–70P.
GOVIER W. (1968) *J. Pharmacol. Exp. Ther.* **159**, 82–90.
HÄGGENDAL J. (1969) *Acta physiol. scand. Suppl.* **330**, 29.
HÄGGENDAL J. (1970) In: *Bayer-Symposium II.* pp. 100–109. Springer-Verlag Heidelberg.
HÄGGENDAL J., JOHANSSON B., JONASON J. and LJUNG B. (1970) *Acta physiol. scand. Suppl.* **347**, 17–32.
HÄGGENDAL J., JOHANSSON B., JONASON J. and LJUNG B. (1972) *J. Pharm. Pharmac.* **24**, 161–164.
HÄGGENDAL J., JOHANSSON B., JONASON J. and LJUNG B. (1972) *J. Pharm. Pharmacol.* **24**, 557–564.
HEDQVIST P. (1969a) *Acta physiol. scand.* **75**, 511–512.
HEDQVIST P. (1969b) *Acta physiol. scand.* **76**, 383–384.
HEDQVIST P. (1970) *Acta physiol. scand. Suppl.* **345**, 1–40.
IVERSEN L. L. (1971) *Br. J. pharmac.* **41**, 571–591.
JOHNSON D. G., THOA N. B., WEINSHILSBOUM R., AXELROD J. and KOPIN I. (1971) *Proc. Natn. Acad. Sci. U.S.A.* **63**, No. 9, 2227–2230.
KIRPEKAR S. M. and PUIG M. (1971) *Br. J. Pharmac.* **43**, 359–369.
KUNOS G., VERMES-KUNOS I., BOYD G. and NICKERSON M. (1973) *Fedn. Proc.* **32**, No. 3, 2692.
LANGER S. Z. (1970) *J. Physiol. Lond.* **208**, 515–546.
LINDMAR R., LÖFFELHOLZ K. and MUSCHOLL E. (1966) *Br. J. Pharmac.* **32**, 280–294.
LÖFFELHOLZ K. and MUSCHOLL E. (1970) *Naunyn-Schmiedebergs Arch. Pharmak.* **267**, 181–184.
POLAK R. L. (1971) *Br. J. Pharmac.* **41**, 600–606.
ROSELL S., KOPIN I. J. and AXELROD J. (1963) *Am. J. Physiol.* **205**, 317–321.
STARKE K. (1972) *Naunyn-Schmiedebergs Arch. Pharmak.* **274**, 18–45.
STARKE K., MONTEL H. and WAGNER J. (1971) *Naunyn-Schmiedebergs Arch. Pharmak.* **271**, 181–192.
WENNMALM Å (1971) *Acta physiol. scand. Suppl.* **365**, 1–36.

Frontiers in Catecholamine Research 1973, pp. 537 to 542. Pergamon Press. Printed in Great Britain.

REGULATION OF CATECHOLAMINE RELEASE. THE MUSCARINIC INHIBITORY MECHANISM

E. MUSCHOLL

Department of Pharmacology, University of Mainz D-65 Mainz, F. R. Germany

THE ELUCIDATION of a muscarinic mechanism which inhibits the release of noradrenaline from peripheral adrenergic nerve fibres started from the observation, on the perfused rabbit heart, that atropine greatly enhanced the noradrenaline output produced by infusion of acetylcholine (LÖFFELHOLZ, LINDMAR and MUSCHOLL, 1967). It was suggested that the peripheral adrenergic nerve fibre contained muscarinic inhibitory receptors in addition to the well-known nicotinic receptors which mediate noradrenaline release. In a more detailed paper (LINDMAR, LÖFFELHOLZ and MUSCHOLL, 1968) it was shown that a high concentration of acetylcholine ($2 \cdot 1 \times 10^{-4}$ M) increased the resting output of noradrenaline by only a small amount; this output was enhanced up to tenfold if atropine ($2 \cdot 9 \times 10^{-9}$–$2 \cdot 9 \times 10^{-6}$ M) was added to the perfusion medium of rabbit or guinea-pig hearts. Conversely, these concentrations of atropine did not alter the noradrenaline output produced by dimethylphenylpiperazinium (DMPP, $3 \cdot 1 \times 10^{-5}$ M) which releases the transmitter from the rabbit heart by a nicotinic action. These findings suggested the possibility that the noradrenaline release by acetylcholine which is mediated through activation of nicotinic receptors, is at the same time depressed by the muscarinic activity of acetylcholine. Correspondingly, the action of acetylcholine was mimicked by that of a combination of DMPP and methacholine; that is, atropine increased the noradrenaline output after this drug combination, just as it increased the output after acetylcholine. It was concluded that the peripheral adrenergic nerve fibre contains inhibitory muscarinic receptors in addition to the nicotinic receptors mediating the release of noradrenaline. Since the muscarinic receptors are stimulated by much lower concentrations of acetylcholine than the nicotinic receptors, a substantial release can be produced by acetylcholine only if its muscarinic action is blocked.

Most of the evidence for the muscarinic inhibitory mechanism, as this phenomenon was called, has been published in papers quoted below. For experimental details the reader is referred to the original papers. There are several review articles dealing with the interrelations between adrenergic and cholinergic mechanisms (KOSTERLITZ and LEES, 1972) and with the muscarinic inhibitory mechanism (MUSCHOLL, 1970; 1973). Recent work from the author's laboratory has shown that the muscarinic drug, methacholine, inhibits the release of a false transmitter, α-methyladrenaline, as well as that of the endogenous noradrenaline. These experiments were carried out in collaboration with H. FUDER and R. WEGWART and the results are summarised in Table 1.

MUSCARINIC INHIBITION OF CATECHOLAMINE RELEASE EVOKED BY ELECTRICAL STIMULATION OF SYMPATHETIC NERVES

On the perfused rabbit heart acetylcholine ($5 \cdot 5 \times 10^{-8}$–$5 \cdot 5 \times 10^{-6}$ M) dose-dependently inhibited the noradrenaline release evoked by electrical stimulation of

TABLE 1. OUTPUTS OF NORADRENALINE (NA) AND α-METHYLADRENALINE (αMA) FROM HEARTS OF RABBITS PREVIOUSLY INFUSED WITH αMA. EFFECT OF METHACHOLINE AND OF LOWERING OF THE CALCIUM CONCENTRATIONS IN THE PERFUSION MEDIUM.

Release evoked by	First period		Treatment	Second period¶ (% of output during first period)		Number of hearts
	(ngNA)	(ngαMA)		NA	αMA	
Symp. nerve stim. (10 Hz, 1 msec, max. voltage for 1 min)	40 ± 7·0	155 ± 25	Control	58 ± 6·0	63 ± 9·7	6
	72 ± 20	222 ± 55	MCh†	11 ± 2·5*	15 ± 5·1*	6
	82 ± 46	179 ± 69	low Ca‡	1·7 ± 1·2*	5·3 ± 2·2*	6
p-Aminophenethyl-trimethylammonium iodide (PAPETA) (3·16 × 10⁻⁵ M for 2 min)	460 ± 100	1280 ± 239	Control	79 ± 8·2	73 ± 6·5	8
	420 ± 92	1000 ± 155	MCh†	17 ± 3·8*	16 ± 2·5*	6
	250 ± 39	803 ± 98	Atr., MCh§	72 ± 4·0	67 ± 7·2	6
	320 ± 46	890 ± 102	low Ca‡	9·0 ± 1·5*	9·4 ± 1·8*	10
High potassium (54 mequiv./l. K⁺ 17 mequiv./l. Na⁺ for 4 min)	171 ± 41	497 ± 57	Control	77 ± 12	55 ± 2·4	6
	137 ± 35	449 ± 43	MCh†	19 ± 10*	18 ± 3·5*	6
	305 ± 37	657 ± 26	Atr., MCh§	48 ± 9·1	50 ± 4·3	3
	124 ± 44	404 ± 48	low Ca‡	14 ± 13*	27 ± 1·8*	6
Tyramine (3·6 × 10⁻⁵ M for 5 min)	(ng)	(ng)		(ng)	(ng)	
	59 ± 35	1040 ± 122	Control	34 ± 20	620 ± 60	6
	27 ± 16	1065 ± 120	MCh†	7·5 ± 15	630 ± 48	6
	12 ± 63	1340 ± 126	Atr., MCh§	37 ± 16	750 ± 51	3
	53 ± 27	932 ± 122	low Ca‡	0 ± 9·0	645 ± 73	6

Given are means ± s.e.m.; * indicates a significant difference ($P < 0.005$) from the relevant control value (t-test).
¶ Second period of nerve stimulation carried out 10 min after beginning of first period; with PAPETA, high potassium and tyramine second period 15 min after first period.
† Methacholine 4×10^{-5} M infused 3 min before and during second period.
‡ Ca^{2+} of Tyrode solution (3·6 mequiv./l.) lowered to 0·2 mequiv./l. 7 min before and during second period.
§ Atropine $1·44 \times 10^{-6}$ M added 8 min before first period and perfused throughout; MCh 4×10^{-5} M infused 3 min before and during second period.

postganglionic sympathetic nerves with pulses of 10 Hz at maximal strength (LÖFFELHOLZ and MUSCHOLL, 1969). Likewise, dose-dependent depression of the transmitter release elicited by nerve stimulation was observed with a series of seven muscarinic compounds including acetylcholine (FOZARD and MUSCHOLL, 1972). The inhibitory effects of the muscarinic agents were short in onset (1 min pre-perfusion before nerve stimulation) and readily reversed by atropine or by perfusion with drug-free solution. In the experiments with acetylcholine the inhibitory action can be assigned to the muscarinic rather than to the nicotinic effect since atropine but not hexamethonium fully antagonised the inhibition of noradrenaline release.

The concentration of hexamethonium administered prevented the noradrenaline release in response to DMPP (LINDMAR and MUSCHOLL, 1961; FOZARD and MUSCHOLL, 1972) but did not alter the output of the transmitter after sympathetic nerve stimulation (LÖFFELHOLZ and MUSCHOLL, 1969; FOZARD and MUSCHOLL, 1972). The rate of noradrenaline release caused by electrical nerve stimulation was not depressed when tested 4–8 min after starting an infusion of nicotine (4×10^{-5} M) or acetylcholine ($2 \cdot 1 \times 10^{-4}$ M, in the presence of atropine $1 \cdot 44 \times 10^{-6}$ M) (LÖFFELHOLZ, 1970). It has been reported that the vasoconstrictor action of periarterial nerve stimulation is enhanced, though not consistently, by small doses of acetylcholine or muscarine (MALIK and LING, 1969a; RAND and VARMA, 1970). On the other hand, HUME, DE LA LANDE and WATERSON (1972) failed to observe facilitation of the effect of periarterial nerve stimulation by small doses of acetylcholine. On the rabbit heart no enhancement of noradrenaline release was observed when the concentration of acetylcholine was lowered from $5 \cdot 5 \times 10^{-9}$ to $5 \cdot 5 \times 10^{-11}$ M (MUSCHOLL, 1973). There is, however, a distinct facilitation of noradrenaline release 1–2 min after acetylcholine in the presence of atropine, i.e. by the nicotinic action of the drug (LÖFFELHOLZ, 1970). This is in harmony with the observation that nicotine facilitated the contractile response of the rabbit pulmonary artery to nerve stimulation (NEDERGAARD and BEVAN, 1969). The experiments in which noradrenaline output was determined demonstrate that muscarinic and nicotinic effects on the release of the adrenergic transmitter can be clearly separated (for a contrary view see MALIK and LING, 1969b).

Recently several authors obtained evidence for a muscarinic inhibition of noradrenaline release on organs other than heart. On the cat spleen the venous outflow of noradrenaline evoked by sympathetic nerve stimulation was markedly reduced by carbachol (KIRPEKAR, PRAT, PUIG and WAKADE, 1972) and on the rabbit ear artery, acetylcholine decreased the efflux of radioactivity evoked by stimulation of periarterial nerves after labeling with ^{3}H-noradrenaline (ALLEN, GLOVER, RAND and STORY, 1972; STEINSLAND, FURCHGOTT and KIRPEKAR, 1973). All these effects were antagonised by atropine.

Excitation of muscarinic receptors also inhibits the release of a false transmitter by sympathetic nerve stimulation. α-Methyladrenaline was incorporated into the amines stores of the heart by intravenous infusion into rabbits according to the procedure described by LINDMAR, MUSCHOLL and SPRENGER (1967b). The release of both noradrenaline and α-methyladrenaline was significantly decreased by methacholine added to the perfusion fluid prior to the nerve stimulation (Table 1).

MUSCARINIC INHIBITION OF CATECHOLAMINE RELEASE EVOKED BY NICOTINIC DRUGS AND POTASSIUM IONS

The nicotinic drug used in most of the following experiments was DMPP. It was established that the various non-nicotinic actions described for DMPP did not interfere with the assessment of muscarinic inhibition on the rabbit heart (for a detailed discussion see MUSCHOLL, 1970; 1973). In agreement with the proposition of a muscarinic inhibitory mechanism, the noradrenaline output of the heart evoked by DMPP was gradually diminished and finally abolished when increasing concentrations of muscarinic drugs (methacholine, acetylcholine, pilocarpine: LINDMAR et al., 1968; oxotremorine, acetylcholine, methacholine, carbachol, furtrethonium, pilocarpine: FOZARD and MUSCHOLL, 1972) were added. The inhibitory responses to furtrethonium and pilocarpine were only partially antagonised, but those to all other muscarinic agents were fully antagonised by atropine. None of the drugs in the maximally inhibitory concentrations caused a significant rise of the noradrenaline output above the resting value.

Recently p-aminophenethyltrimethylammonium (PAPETA) was introduced as a specific nicotinic drug (BARLOW and FRANKS, 1971). The noradrenaline output of the rabbit heart elicited by infusion of PAPETA was decreased by methacholine in a dose-dependent manner and the inhibition was fully antagonised by atropine (MUSCHOLL, 1973). Similar observations were made when PAPETA was used to evoke release of the false transmitter, α-methyladrenaline (Table 1).

Muscarinic inhibition probably affects a process which in the sequence of events leading to release of transmitter occurs after excitation of nicotinic receptors. Blockade of nicotinic receptors with hexamethonium does not prevent the acetylcholine-induced inhibition of noradrenaline output evoked by sympathetic nerve stimulation (LÖFFELHOLZ and MUSCHOLL, 1969). The question arises as to whether muscarinic inhibition of transmitter release is due to blockade of propagation of impulses travelling along the terminal axons. Propagation of orthodromic impulses is a prerequisite for noradrenaline release by electrical nerve stimulation. Antidromic impulses have been recorded from postganglionic sympathetic nerves after injection of nicotinic drugs into the blood supply of cat spleen and heart (FERRY, 1963; CABRERA, TORRANCE and VIVEROS, 1966) though they may not be causally involved in transmitter release (HAEUSLER, THOENEN, HAEFELY and HUERLIMANN, 1968). However, the noradrenaline release evoked by high concentrations of potassium is independent of propagation of impulses (KIRPEKAR and WAKADE, 1968; HAEUSLER et al., 1968). On the rabbit heart methacholine caused a dose-dependent and atropine-sensitive depression of the noradrenaline output after infusion of 135 mequiv./l. potassium (DUBEY and MUSCHOLL, 1973). Likewise, release of α-methyladrenaline by high potassium was inhibited by methacholine and this effect was reversed by atropine (Table 1). In the latter experiments the potassium concentration of Tyrode solution was elevated to 54 mequiv./l., the sodium concentration was lowered to 17 mequiv./l. and isotonicity was maintained by adding sucrose 43·8 g/l. Thus, the muscarinic inhibition of catecholamine release probably occurs at the site of the physiological stimulus-secretion coupling process.

THE RELATION BETWEEN MUSCARINIC INHIBITION AND DEPENDENCE ON CALCIUM FOR CATECHOLAMINE RELEASE

The catecholamine releasing procedures susceptible to muscarinic inhibition discussed above are invariably dependent on the presence of extracellular calcium ions.

This has been demonstrated for noradrenaline release by sympathetic nerve stimulation (HUKOVIC and MUSCHOLL, 1962; BOULLIN, 1967), nicotinic drugs (LINDMAR, LÖFFELHOLZ and MUSCHOLL, 1967a; LÖFFELHOLZ, 1967) and high potassium (KIRPEKAR and WAKADE, 1968). On the other hand, tyramine-induced noradrenaline release which does not require calcium (LINDMAR et al., 1967a; THOENEN, HUERLIMANN and HAEFELY, 1969) is not decreased by methacholine (LÖFFELHOLZ and MUSCHOLL, 1969).

The correlation between calcium-dependency of transmitter release and presence of muscarinic inhibition was confirmed by studies utilising a false transmitter amine. The methacholine-induced decrease in output of α-methyladrenaline in response to nerve stimulation, PAPETA and high potassium has been mentioned above. The release of both α-methyladrenaline and noradrenaline evoked by nerve stimulation, PAPETA and potassium was also significantly inhibited when the calcium concentration of the Tyrode solution was lowered to 0·2 mequiv./l. (Table 1). However, tyramine-induced release of α-methyladrenaline was unaltered after methacholine or low calcium. The experimental evidence is thus in favour of the idea (LÖFFELHOLZ and MUSCHOLL, 1969) that mechanisms of transmitter release that are closely linked to the entry of calcium ions into the nerve terminal are sensitive to muscarinic inhibition.

Confirming the facilitation by atropine of the noradrenaline release evoked by acetylcholine on the cat heart HAEUSLER et al. (1968) suggested that muscarinic receptors of the adrenergic nerve terminals mediate hyperpolarisation of the membrane and that the depolarisation by nicotinic drugs and the consecutive influx of calcium may thus be decreased. This is in line with the observations of FOZARD and MUSCHOLL (1972) who studied the effects of 9 muscarinic drugs and found that the neuronal receptors mediating inhibition of noradrenaline release after nerve stimulation or DMPP correspond to those mediating hyperpolarisation rather than depolarisation of the superior cervical ganglion.

REFERENCES

ALLEN G. S., GLOVER A. B., RAND M. J. and STORY D. F. (1972) Br. J. Pharmacol. 46, 527–528.
BARLOW R. B. and FRANKS F. (1971) Br. J. Pharmacol. 42, 137–142.
BOULLIN D. J. (1967) J. Physiol. 189, 85–99.
CABRERA R., TORRANCE R. W. and VIVEROS H. (1966) Br. J. Pharmacol. 27, 51–63.
DUBEY M. P. and MUSCHOLL E. (1973) Naunyn-Schmiedebergs Arch. Pharmacol. 277, (Suppl.) 12.
FERRY C. (1963) J. Physiol. 167, 487–504.
FOZARD J. R. and MUSCHOLL E. (1972) Br. J. Pharmacol. 45, 616–629.
HAEUSLER G., THOENEN H., HAEFELY W. and HUERLIMANN A. (1968) Naunyn-Schmiedebergs Arch. Pharmak. exp. Path. 261, 389–411.
HUKOVIC S. and MUSCHOLL E. (1962) Naunyn-Schmiedebergs Arch. exp. Path. u. Pharmak. 244, 81–96
HUME W. R., DE LA LANDE I. S. and WATERSON J. G. (1972) Europ. J. Pharm. 17, 227–233.
KIRPEKAR S. M., PRAT J. C., PUIG M. and WAKADE A. R. (1972) J. Physiol. 221, 601–615.
KIRPEKAR S. M. and WAKADE A. R. (1968) J. Physiol. 194, 595–608.
KOSTERLITZ H. W. and LEES G. M. (1972) In: Handbook Exp. Pharmacol. (BLASCHKO H. and MUSCHOLL E., Eds.) Vol. 33, pp. 762–812. Springer, Berlin.
LINDMAR R., LÖFFELHOLZ K. and MUSCHOLL E. (1967a) Experientia 23, 933–934.
LINDMAR R., LÖFFELHOLZ K. and MUSCHOLL E. (1968) Br. J. Pharmacol. 32, 280–294.
LINDMAR R. and MUSCHOLL E. (1961) Naunyn-Schmiedebergs Arch. exp. Path. u. Pharmak. 242, 214–227.
LINDMAR R., MUSCHOLL E. and SPRENGER E. (1967b) Naunyn-Schmiedebergs Arch. Pharmak. exp. Path. 256, 1–25.
LÖFFELHOLZ K. (1967) Naunyn-Schmiedebergs Arch. Pharmak. exp. Path. 258, 108–122.
LÖFFELHOLZ K. (1970) Naunyn-Schmiedebergs Arch. Pharmak. 267, 64–73.
LÖFFELHOLZ K., LINDMAR R. and MUSCHOLL E. (1967) Naunyn-Schmiedebergs Arch. Pharmak. exp. Path. 257, 308 (abstr.).

Löffelholz K. and Muscholl E. (1969) *Naunyn-Schmiedebergs Arch. Pharmak.* **265**, 1–15.

Malik K. U. and Ling G. M. (1969a) *Circulat. Res.* **25**, 1–9.

Malik K. E. and Ling G. M. (1969b) *J. Pharm. Pharmacol.* **21**, 514–519.

Muscholl E. (1970) In: *New Aspects of Storage and Release Mechanisms of Catecholamines.* (Schümann H. J. and Kroneberg G., Eds.) pp. 168–186. Springer, Berlin.

Muscholl E. (1973) *Proc. 5th. Int. Congr. Pharmacology* **4**, 440–457, Karger, Basel.

Nedergaard O. A. and Bevan J. A. (1969) *Abstr. 4th. Int. Congr. Pharmacology*, 144.

Rand M. J. and Varma B. (1970) *Br. J. Pharmacol.* **38**, 758–770.

Steinsland O. S., Furchgott R. F. and Kirpekar S. M. (1973) *J. Pharmacol. exp. Ther.* **184**, 346–356.

Thoenen H., Huerlimann A. and Haefely W. (1969) *Europ. J. Pharm.* **6**, 29–37.

Frontiers in Catecholamine Research 1973, pp. 543 to 549. Pergamon Press. Printed in Great Britain.

THE REGULATION OF TRANSMITTER RELEASE ELICITED BY NERVE STIMULATION THROUGH A PRESYNAPTIC FEED-BACK MECHANISM

S. Z. LANGER

Instituto de Investigaciones Farmacológicas CONICET, Buenos Aires, Argentina

THE PRESENCE of presynaptic alpha receptors in adrenergic nerve endings has been recently postulated (LANGER et al., 1971; FARNEBO and HAMBERGER, 1971; STARKE, 1971; ENERO et al., 1972; STARKE, 1972a). While stimulation of the presynaptic alpha receptors inhibits the release of norepinephrine (FARNEBO and HAMBERGER, 1971; STARKE, 1972b; LANGER et al., 1972) blockade of these receptors increases the release of the neurotransmitter regardless of the nature of the adrenergic receptor responsible for the response of the effector organ (LANGER, 1970; LANGER et al., 1971; STARKE et al., 1971; ENERO et al., 1972).

These results have led to the hypothesis that the presynaptic alpha receptors form part of a negative feed-back control mechanism through which the transmitter may inhibit its own release once a threshold concentration is achieved in the synaptic gap.

The overflow of ^{3}H-norepinephrine elicited by nerve stimulation was studied in the cat's nictitating membrane, in isolated guinea-pig atria and in the cat's perfused spleen. The methodological details are described by ADLER-GRASCHINSKY et al. (1972) for guinea-pig atria, by ENERO et al. (1972) and by ENERO and LANGER (1973) for the nictitating membrane and by DUBOCOVICH and LANGER (1973) for the perfused spleen.

In experiments in which ^{3}H-norepinephrine was separated from the ^{3}H-metabolites the method described by GRAEFE et al. (1973) was employed.

Dose–response curves to norepinephrine and to dopamine were obtained in the nictitating membrane with the cumulative method as described by LANGER and RUBIO (1973). In all experiments in which the addition of cocaine elicited the development of tone, the response to nerve stimulation or exogenous catecholamines was added to the underlying tone as described by LANGER (1966).

INFLUENCE OF RESERPINE-INDUCED DEPLETION OF NOREPINEPHRINE ON THE NEGATIVE FEED-BACK MECHANISM FOR TRANSMITTER RELEASE

According to the hypothesis of the negative feed-back mechanism for the regulation of norepinephrine release during nerve stimulation it might be expected that this mechanism will operate most effectively when the quantity of transmitter released by each impulse is high. Therefore, after depletion of the endogenous transmitter by the administration of reserpine, the release of norepinephrine might be reduced near or below the threshold necessary to stimulate the presynaptic receptors. This possibility was tested by studying the overflow of ^{3}H-norepinephrine elicited by nerve stimulation in the cat's nictitating membrane after reserpine-pretreatment

(0·3 mg/kg, 4 days prior to the experiment). Under these experimental conditions there is a pronounced reduction in the endogenous norepinephrine levels while at this time granular retention of ^{3}H-norepinephrine had already recovered to approximately 40 per cent of the controls (Table 1).

In the reserpine-pretreated tissue the fraction release per shock of ^{3}H-norepinephrine, induced by nerve stimulation, was significantly higher than in the controls (Table 1). It is likely that in the reserpine-pretreated group the total release of the transmitter was reduced near the threshold concentration required for the activation of the negative feed-back mechanism mediated through the presynaptic alpha receptors. The latter increased the fractional release per shock of ^{3}H-norepinephrine because of the reduced presynaptic inhibition. The analysis of the individual values revealed that the increase in the fractional release of ^{3}H-norepinephrine was bigger in the tissues in which the depletion of endogenous norepinephrine was more pronounced.

If the interpretation of these results were correct, the blockade of the alpha receptors by phenoxybenzamine should be less effective in increasing release of ^{3}H-norepinephrine in reserpine-pretreated tissues when compared with the controls.

Table 1 shows that in the presence of $2·9 \times 10^{-6}$ M of phenoxybenzamine the decrease in responses to nerve stimulation was of the same magnitude in the controls and in reserpine-pretreated tissues, indicating a similar degree of blockade of the alpha receptors in the effector organ. Yet, while phenoxybenzamine increased transmitter overflow about 13-fold in the controls, the increase was only 3-fold in the reserpine-pretreated group (Table 1).

The decrease in the effectiveness of phenoxybenzamine to enhance the overflow of the labelled transmitter after reserpine-pretreatment appears to be related to the reduction in the total release of norepinephrine in the tissues depleted of the neurotransmitter. The effectiveness of phenoxybenzamine in increasing transmitter overflow could be restored when reserpine-pretreated nictitating membranes were incubated with exogenous norepinephrine and pargyline to increase the tissue norepinephrine levels to the range of the controls.

These results support the view that in tissues depleted of norepinephrine decreased transmitter output fails to fully activate the presynaptic alpha-receptors which mediate the negative feed-back mechanism that regulates transmitter release by nerve stimulation.

INHIBITION OF ^{3}H-NOREPINEPHRINE RELEASE INDUCED BY NERVE STIMULATION IN THE PRESENCE OF (−) NOREPINEPHRINE OR DOPAMINE

Both in the cat's spleen and in the cat's nictitating membrane the response of the effector organ is mediated through the activation of alpha receptors. In both tissues exogenous (−) norepinephrine, in the presence of cocaine, $2·6 \times 10^{-5}$ M, inhibited in a concentration dependent manner the release of ^{3}H-norepinephrine elicited by nerve stimulation.

In the perfused spleen and in the presence of the same concentration of cocaine, dopamine inhibited in a concentration dependent manner the release of the labelled transmitter elicited by nerve stimulation. Figure 1a shows that dopamine was practically equipotent with (−) norepinephrine in reducing transmitter release in the perfused spleen. However, in the presence of the same concentration of

TABLE 1. EFFECTS OF RESERPINE-PRETREATMENT ON NOREPINEPHRINE LEVELS AND ON ^{3}H-NOREPINEPHRINE RELEASE INDUCED BY NERVE STIMULATION IN THE CAT'S NICTITATING MEMBRANE

Experimental groups	n	Endogenous norepinephrine levels (μg NE/g) (a)	Neuronal retention (ng ^{3}H-NE/g) (b)	Fractional release per shock ($\times 10^{-5}$) (c)	n	Responses $\frac{\text{PBA}}{\text{CONT}}$ (d)	Fractional release per shock $\frac{\text{PBA}}{\text{CONT}}$ (e)
Controls	8	7·28 ± 0·96	784 ± 96	1·16 ± 0·24	4	0·48 ± 0·06	12·9 ± 2·5
Reserpine-pretreated (0·3 mg/kg 4 days prior)	12	0·94 ± 0·17***	287 ± 21***	2·51 ± 0·39*	5	0·58 ± 0·03	3·1 ± 0·4**

(a) Endogenous levels of norepinephrine in the medial muscle of the nictitating membrane determined fluorimetrically.
(b) ^{3}H-norepinephrine retained in the medial muscle of the nictitating membrane 180 min after the end of a 30 min incubation with 242 ng/ml of ^{3}H-norepinephrine.
(c) Fractional release of total radioactivity elicited by nerve stimulation (10 Hz, during 2 min, 0·5 msec duration and supramaximal voltage).
(d) Ratio of responses to nerve stimulation obtained in the presence of $2\cdot9 \times 10^{-6}$ M phenoxybenzamine and in the corresponding controls.
(e) Fractional release of total radioactivity elicited by nerve stimulation: ratio between $2\cdot9 \times 10^{-6}$ M phenoxybenzamine and the corresponding controls.
NE: norepinephrine   PBA: phenoxybenzamine   CONT: controls n = number of experiments   Shown are mean values ± S.E.M.
* $p < 0.05$
** $p < 0.01$
*** $p < 0.001$

19

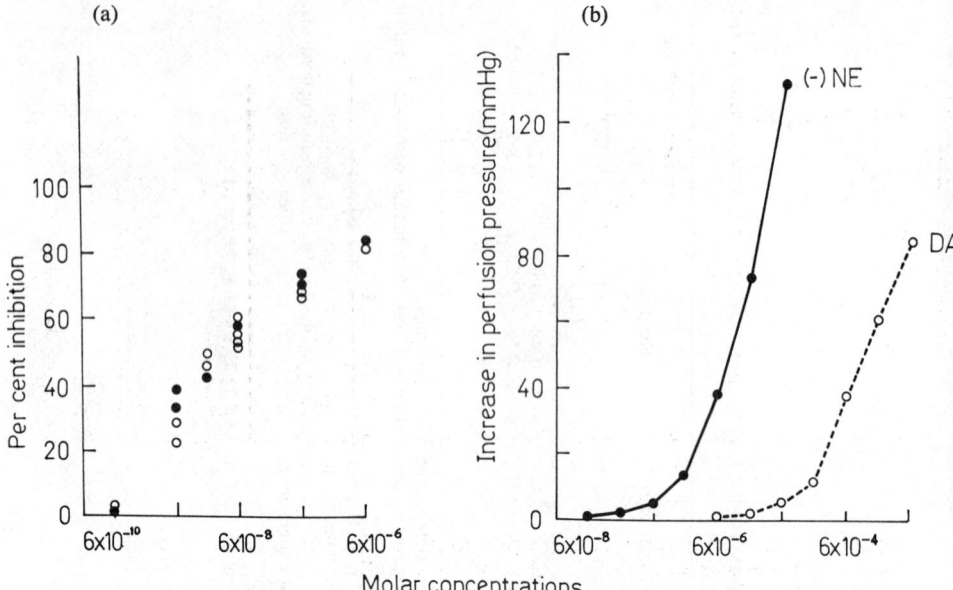

FIG. 1.—Effects of (−) norepinephrine and of dopamine on transmitter release and on changes in perfusion pressure in the perfused cat's spleen. (a). Inhibition of ^{3}H-norepinephrine overflow elicited by nerve stimulation (10 Hz, 30 sec, 0·1 msec duration, supramaximal voltage). Cocaine, 2·6 × 10^{-5} M, was present throughout.
●: (−) norepinephrine; ○: dopamine
Shown are individual data from different experiments.
(b) Increase in perfusion pressure in response to (−) norepinephrine or dopamine added to the perfusion fluid. Cocaine, 2·6 × 10^{-5} M, was present throughout.
●——●: (−) norepinephrine (NE); ○----○: dopamine (DA).
Shown are the mean values of 3 experiments in each dose–response curve.

cocaine (2·6 × 10^{-5} M), (−) norepinephrine was 100 times more potent than dopamine in activating the alpha receptors of the effector organ (Fig. 1b).

In the isolated nerve-muscle preparation of the nictitating membrane dopamine decreased the responses to postganglionic nerve stimulation obtained in the presence of 2·6 × 10^{-5} M cocaine (Fig. 2a). The effect was concentration-dependent and readily reversible when dopamine was removed from the organ bath. With a similar experimental design exogenous (−) norepinephrine failed to reduce responses to nerve stimulation in concentrations up to 6·0 × 10^{-7} M.

In the presence of cocaine (2·6 × 10^{-5} M) or in the presence of cocaine (2.6 × 10^{-5} M) plus phentolamine (1·0 × 10^{-6} M) dopamine reduced ^{3}H-norepinephrine overflow induced by nerve stimulation. A typical experiment obtained in the presence of cocaine and phentolamine is shown in Fig. 2b. Dopamine, 6·5 × 10^{-7} M, reduced transmitter overflow to approximately one fourth of the controls; the effect was readily reversed when dopamine was washed out. While exogenous (−) norepinephrine was nearly equipotent with dopamine in reducing ^{3}H-norepinephrine overflow elicited by nerve stimulation, the relative potencies of these two catecholamines differed by about 50-fold when activation of the alpha receptors of the effector organ was determined in the presence of the same concentration of cocaine (Table 2).

Both in the spleen and in the nictitating membrane, dopamine was equipotent with

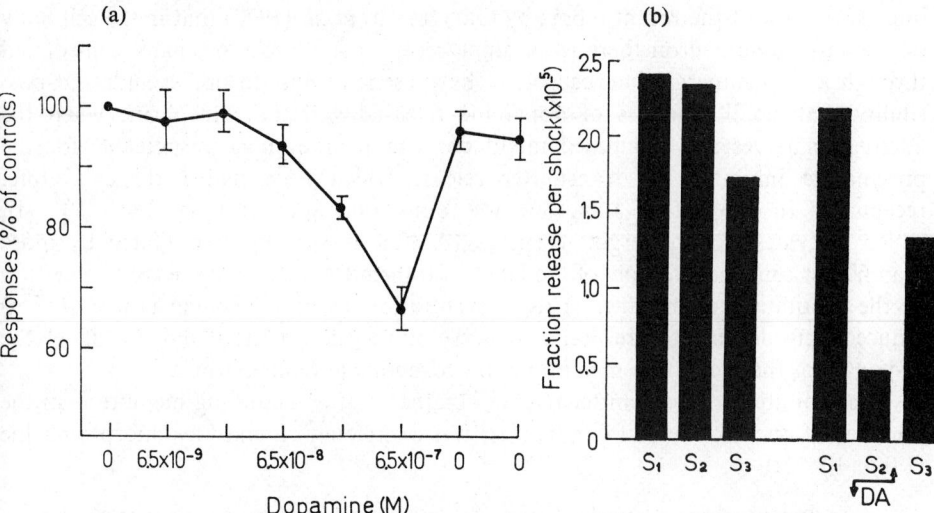

Fig. 2.—Effects of dopamine on responses to nerve stimulation and on transmitter overflow in the isolated cat's nictitating membrane. (a) Responses to nerve stimulation (1 Hz, 60 sec, 0·5 msec duration, supramaximal voltage) expressed as percent of the first period of stimulation. Cocaine, $2·6 \times 10^{-5}$ M was present throughout. Shown are mean values ± S.E.M.
(b) Inhibition by dopamine of the overflow of labelled transmitter elicited by nerve stimulation (10 Hz, 120 sec, 0·5 msec duration, supramaximal voltage). Shown are results of a typical paired experiment obtained in the presence of $2·6 \times 10^{-5}$ M cocaine plus $1·0 \times 10^{-6}$ M phentolamine. S_1, S_2 and S_3 indicate the three consecutive periods of stimulation. In the contralateral membrane, dopamine (DA) $6·5 \times 10^{-7}$ M was added during the second period of stimulation.

(−) norepinephrine in reducing transmitter overflow induced by nerve stimulation. Yet, dopamine was considerably less potent than (−) norepinephrine in activating the alpha receptors of the effector organ in these two tissues. Consequently, these results are compatible with the view that the presynaptic and the postsynaptic alpha receptors differ with regard to their relative affinities for sympathomimetic amines. In support of this view, low concentrations of phenoxybenzamine ($2·9 \times 10^{-9}$ M) which blocked by 65 per cent the responses to exogenous (−) norepinephrine and to nerve stimulation, failed to increase significantly the overflow of endogenous norepinephrine when the nerves were stimulated at either 5 Hz or 30 Hz in the perfused spleen.

The high potency of dopamine on the presynaptic alpha receptor is of great

TABLE 2. RELATIVE POTENCIES OF (−) NOREPINEPHRINE AND OF DOPAMINE IN THE ISOLATED CAT'S NICTITATING MEMBRANE IN THE PRESENCE OF COCAINE

Experimental groups	n	pD_2 (a)	Tension (g) (b)
(−) Norepinephrine	5	6·91 ± 0·05	18·3 ± 0·5
Dopamine	6	5·25 ± 0·14*	17·9 ± 1·6

(a) pD_2: mean negative log molar ED_{50} ± S.E.M. obtained in the presence of $2·6 \times 10^{-5}$ M cocaine.
(b) Tension: maximum development of tension in grams (mean ± S.E.M.).
* $p < 0·001$.

interest in view of the recent report by GREENGARD *et al.* (1972) that in the cell body of the adrenergic neuron there is an alpha-receptor inhibitory mechanism mediated through a dopaminergic interneuron. The present results do not exclude the possibility that small amounts of dopamine released with norepinephrine when the adrenergic nerves are stimulated might contribute or even be responsible for the presynaptic inhibition of transmitter release which is mediated through alpha receptors. In support of this view we found that apomorphine ($1 \cdot 0 \times 10^{-7}$ M), which activates dopaminergic receptors (ANDÉN *et al.*, 1967) inhibited by more than 60 per cent the overflow of the labelled transmitter elicited by nerve stimulation in the nictitating membrane. This concentration of apomorphine ($1 \cdot 0 \times 10^{-7}$ M) reduced considerably the response to nerve stimulation while it did not affect the responses of the nictitating membrane to exogenous norepinephrine.

The inhibition of transmitter release obtained in the nictitating membrane in the presence of apomorphine ($1 \cdot 0 \times 10^{-7}$ M) was completely reversed by chlorpromazine ($1 \cdot 0 \times 10^{-6}$ M).

EFFECTS OF PAPAVERINE ON TRANSMITTER OVERFLOW ELICITED BY NERVE STIMULATION

FARNEBO and MALMFORS (1971) have recently postulated that the response of the effector organ regulates the release of norepinephrine by nerve stimulation through a transynaptic mechanism. In other words, a decrease in the response of the effector organ would lead to an increase in transmitter release by nerve stimulation. Papaverine ($8 \cdot 0 \times 10^{-5}$ M) decreases the responses of the nictitating membrane to nerve stimulation without blocking the adrenergic alpha receptors. In the presence of this concentration of papaverine transmitter overflow was increased 3 to 4-fold in the nictitating membrane. Yet, the same concentration of papaverine increased by nearly 5-fold the overflow of transmitter elicited by nerve stimulation in isolated guinea-pig atria without affecting the positive chronotropic effect induced by nerve stimulation. Since in guinea-pig atria papaverine increased transmitter overflow without affecting the response of the effector organ, these results are not compatible with the hypothesis of the transynaptic regulation of transmitter release.

In all these experiments, exposure to papaverine increased selectively the spontaneous outflow of the ^{3}H-deaminated catechol glycol (3-4,dihydroxyphenylglycol: DOPEG). This effect indicates that papaverine has a granular site of action because the ^{3}H-deaminated glycol, DOPEG, results from the presynaptic deamination of ^{3}H-norepinephrine (ADLER-GRASCHINSKY *et al.*, 1972; GRAEFE *et al.*, 1973; DUBOCOVICH and LANGER, 1973). The granular effect of papaverine appears to be related to the increase of transmitter overflow obtained in the presence of this drug.

Since papaverine is a potent inhibitor of phosphodiesterase (WEINRYB *et al.*, 1972), the effects of other inhibitors of this enzyme on transmitter release by nerve stimulation were determined. Dipyridamole ($1 \cdot 0 \times 10^{-5}$ M) increased the spontaneous outflow of ^{3}H-DOPEG and the overflow of the labelled transmitter by nerve stimulation in the cat's nictitating membrane.

SUMMARY

The results obtained support the hypothesis of a presynaptic regulation for the release of norepinephrine by nerve stimulation. This regulatory mechanism seems to be mediated through alpha adrenergic receptors located in the outer surface of the adrenergic nerve endings. Thus, a negative feed-back mechanism results in which norepinephrine inhibits its own release once a threshold concentration is achieved in the synaptic gap.

Our results suggest that the presynaptic alpha receptor may differ from the alpha receptor that mediates the response of the effector organ. The high potency of dopamine in reducing norepinephrine overflow and the fact that low concentrations of apomorphine elicit the same effect support this view. In addition these results suggest that the presynaptic receptor that regulates norepinephrine release by nerve stimulation may be of the dopaminergic type.

A negative feed-back mechanism for acetylcholine released by nerve stimulation similar to the one described for norepinephrine has been suggested, and it appears to be mediated through muscarinic receptors (SZERB and SOMOGYI, 1973).

It is possible that presynaptic inhibition of the release of the neurotransmitter through a negative feed-back mechanism mediated through the neurotransmitter itself represents a more general type of phenomenon and might occur not only for norepinephrine and acetylcholine but for other neurotransmitters as well.

Acknowledgements—Some of the studies reported in this paper were carried out in collaboration with Drs. Ma. A. Enero, E. Adler-Graschinsky and M. Dubocovich.

REFERENCES

ADLER-GRASCHINSKY E., LANGER S. Z. and RUBIO M. C. (1972) *J. Pharmac. Exp. Ther.* **180,** 286–301.
ANDÉN N. E., RUBENSON A., FUXE K. and HÖKFELT T. (1967) *J. Pharmac. Pharmacol.* **19,** 627.
DUBOCOVICH M. and LANGER S. Z. (1973) *Naunyn-Schmiedeberg's Arch. Pharmacol.* **278,** 179–194.
ENERO MA. A., LANGER S. Z., ROTHLIN R. P. and STEFANO F. J. E. (1972) *Br. J. Pharmac.* **44,** 672–688.
ENERO MA. A. and LANGER S. Z. (1973) *Br. J. Pharmac.* (In press).
FARNEBO L. O. and HAMBERGER B. (1971) *Br. J. Pharmac.* **43,** 97–106.
FARNEBO L. O. and MALMFORS T. (1971) *Acta physiol. scand. Suppl.* **371,** 1–18.
GRAEFE K. H., STEFANO F. J. E. and LANGER S. Z. (1973) *Biochem. Pharmac.* **22,** 1147–1161.
GREENGARD P., MCAFEE D. A. and KEBABIAN J. W. (1972) *Advances in Nucleotide Research* Vol **1,** p. 337–355. Raven Press, N.Y.
LANGER S. Z. (1966) *J. Pharmac. Exp. Ther.* **154,** 14–34.
LANGER S. Z. (1970) *J. Physiol. (Lond)* **208,** 515–546.
LANGER S. Z., ADLER E., ENERO MA. A. and STEFANO F. J. E. (1971) *XXVth. Int. Congr. Physiol. Sciences,* p. 335.
LANGER S. Z., ENERO MA. A., ADLER-GRASCHINSKY E. and STEFANO F. J. E. (1972) *Vth. Int. Congr. Pharmacology,* p. 134.
LANGER S. Z. and RUBIO M. C. (1973) *Naunyn-Schmiedeberg's Arch. Pharmac.* **276,** 71–88.
STARKE K. (1971) *Naturwiss.* **58,** 420.
STARKE K., MONTEL H. and SCHÜMANN H. (1971) *Naunyn-Schmiedeberg's Arch. Pharm.* **270,** 210–214.
STARKE K. (1972a) *Naunyn-Schmiedeberg's Arch. Pharmac.* **274,** 18–45.
STARKE K. (1972b) *Naunyn-Schmiedeberg's Arch. Pharmac.* **275,** 11–23.
SZERB J. C. and SOMOGYI G. T. (1973) *Nature, Lond.* **241,** 121–122.
WEINRYB I., CHASIN M., FREE C. A., HARRIS D. N., GOLDENBERG H., MICHEL I. M., PAIK V. S., PHILLIPS M., SAMANIEGO S. and HESS S. M. (1972) *J. Pharmac. Sciences* **61,** 1556–1567.

Frontiers in Catecholamine Research 1973, pp. 551 to 556. Pergamon Press. Printed in Great Britain.

THE NEUROGENIC SHORT-TERM CONTROL OF CATECHOLAMINE SYNTHESIS AND RELEASE IN THE SYMPATHO-ADRENAL SYSTEM, AS REFLECTED IN THE LEVELS OF ENDOGENOUS DOPAMINE AND β-HYDROXYLATED CATECHOLAMINES

A. Carlsson, S. R. Snider, O. Almgren and M. Lindqvist

Department of Pharmacology, University of Göteborg, Sweden

This paper will focus on dopamine (DA) in the peripheral sympatho-adrenomedullary system. Relatively little work has been done in this area, largely due to insufficient sensitivity of the methods generally used for the detection and quantitative measurement of this compound. Recently several modifications of the isolation and fluorimetric assay of this amine have been introduced, resulting in a considerably increase in sensitivity (Atack and Magnusson, 1970; Atack, 1973). New possibilities have thus been opened up for investigating the role of DA in tissues where it occurs in low concentration.

Data will be reported on the occurrence and behaviour of DA in the adrenals and in peripheral tissues innervated by adrenergic fibres. It will be shown that under various experimental conditions DA very often shows changes opposite to those of the β-hydroxylated catecholamines. The explanation appears to be that whereas the β-hydroxylated catecholamines mainly reflect changes in release, DA appears to be mainly influenced by variations in the rate of catecholamine synthesis. Information on the coupling between release and synthesis can thus be obtained by measuring DA and β-hydroxylated catecholamines simultaneously.

ADRENAL DOPAMINE: PRECURSOR OF THE ADRENOMEDULLARY HORMONES

Evidence for precursor role

DA constitutes about one percent of the total catecholamines of the rat adrenals. It is a substrate for the ATP-dependent and reserpine-sensitive uptake mechanism of amine storage granules (Carlsson et al., 1963). Thus, a major fraction of adrenal DA is particle-bound (Eade, 1958, unpublished data of this laboratory) and is brought to disappear by reserpine (Snider and Carlsson, 1972). The bulk of adrenal DA appears to occur in one homogenous pool, as indicated by its monoexponential decline (half-life 70–80 min) after inhibition of tyrosine hydroxylase (EC 1.14.3.2, Snider and Carlsson, 1972, Fig. 1). Previously a half-life of 9 days has been reported for the adrenaline of the rat adrenals (Udenfriend et al., 1953). The main pathway for adrenal DA appears to be β-hydroxylation; an inhibitor of DA-β-hydroxylase (EC 1.14.2.1), i.e. FLA-63, (= bis-(4-methyl-l-homopiperazinyl-thiocarbonyl) disulfide, Astra, Södertälje) markedly retarded the disappearance of adrenal DA induced by an inhibitor of tyrosine hydroxylase (Fig. 1), whereas inhibitors of other

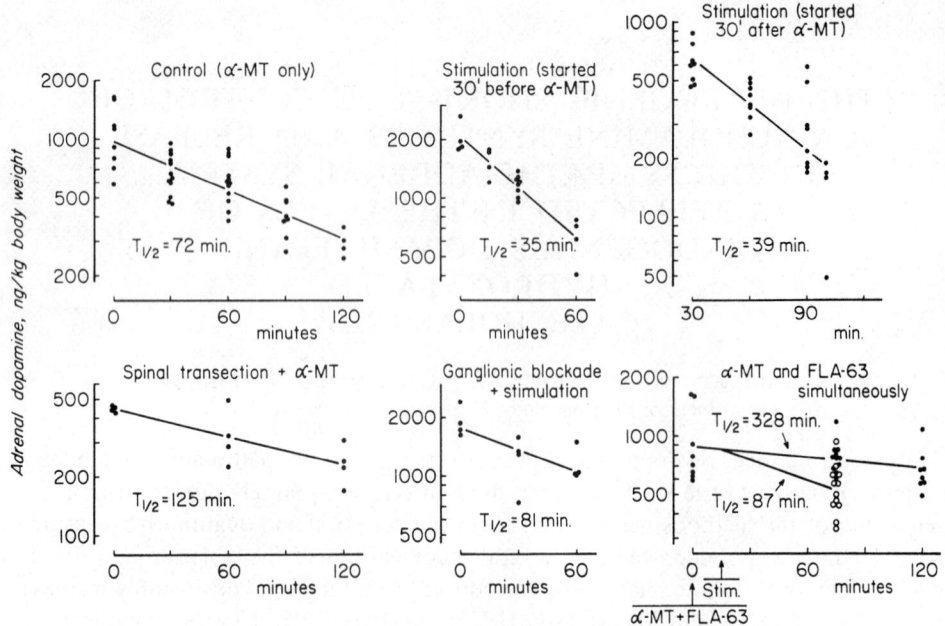

FIG. 1.—Semilogarithmic plots of adrenal dopamine at various times after α-methyltyrosine (α-MT, 400 mg/kg of the methylester HCl given i.p. at zero time). The different groups were treated as indicated in the respective graphs. Spinal transection was performed at C_7 24 hr before α-MT. Stimulation was in all cases performed by physostigmine 1 mg/kg + phentolamine 10 mg/kg, given i.p. at the times indicated. FLA-63 was given i.p. in a dose of 40 mg/kg. Ganglionic blockade was induced by chlorisondamine, 5 mg/kg, given i.p. $2\frac{1}{2}$ hr before α-MT; in this group stimulation was started 30 min before α-MT. Regression lines calculated by method of least squares; difference from control: spinal transection $P < 0.05$, stimulation before or after α-MT $P < 0.005$. Ganglion blockade + stimulation vs. stimulation (before α-MT) $P < 0.005$. α-MT + FLA-63, unstimulated vs. stimulated $P < 0.005$.

enzymes involved in catecholamine metabolism, i.e. monoamine oxidase (EC 1.4.3.4, inhibited by pargyline, Abbott, Chicago) and catechol-O-methyltransferase (EC 2.1.1.6, inhibited by α-propyldopacetamide, H 22/54, Hässle, Mölndal) had no significant influence.

These observations support the view that DA mainly, if not entirely, serves as a precursor of the β-hydroxylated catecholamines in the adrenals of the rat.

Effect of variations in the nerve impulse flow to the adrenal medulla

Insulin (SNIDER and CARLSSON, 1972) or physostigmine (unpublished data) causes a marked elevation of adrenal DA and a decrease of β-hydroxylated catecholamines (A + NA); the effect of physostigmine is markedly potentiated by phentolamine, an α-adrenergic receptor antagonist, presumably by interfering with a feedback mechanism (Fig. 2). Disconnection of supraspinal control by a high spinal transection causes the opposite effects, i.e. a decrease in DA and an increase in β-hydroxylated catecholamines. Moreover, spinal transection blocks the above-mentioned actions of insulin (SNIDER and CARLSSON 1972) and physostigmine + phentolamine, suggesting that these effects are mediated via nervous pathways from supraspinal centres (Table 1).

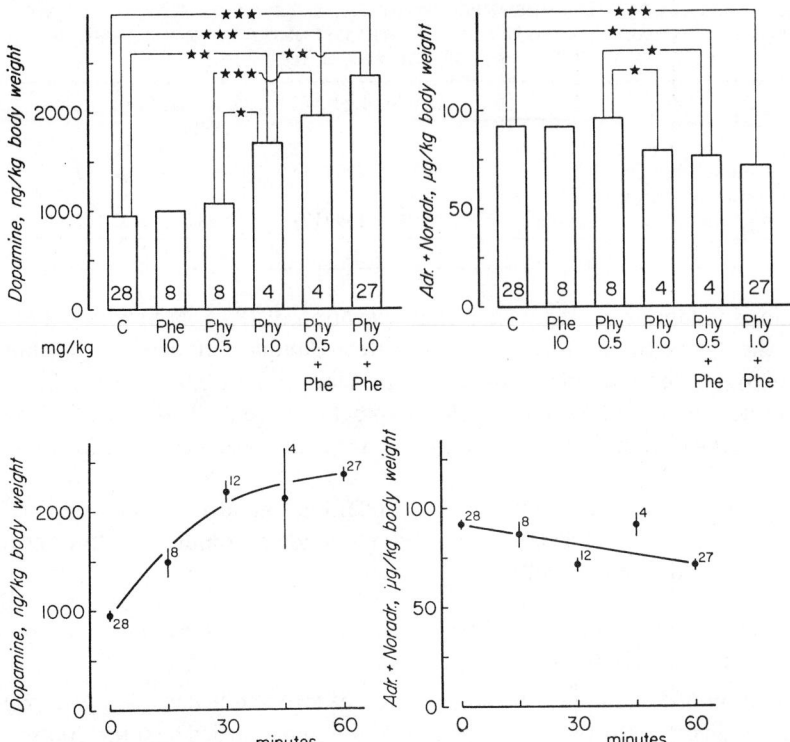

FIG. 2.—Adrenal catecholamines, 60 min after different doses of physostigmine (Phy), phentolamine (Phe), or both drugs combined (*upper graphs*), *or* various intervals after Phy 1 mg/kg and Phe 10 mg/kg i.p. (lower graphs). Left graphs: dopamine ng/pair per kg body weight. Right graphs: adrenaline + noradrenaline μg/pair per kg body weight). Significance levels were calculated by analysis of variance followed by *t*-test (all data of upper graph). Variance within groups: dopamine 173859; adrenaline + noradrenaline 175.8. No. of experiments shown in bars (upper graph) or at the means $\pm$ s.e.m. (lower graph). $*P < 0.05$, $**P < 0.01$, $***P < 0.001$.

TABLE 1. EFFECTS OF SPINAL TRANSECTION ON ADRENAL CATECHOLAMINES: DECREASE IN DOPAMINE (DA), INCREASE IN β-HYDROXYLATED CATECHOLAMINES (A + NA) AND BLOCKADE OF CHANGES INDUCED BY PHYSOSTIGMINE + PHENTOLAMINE.

	DA (ng/kg b.w.)	A + NA (μg/kg b.w.)
Control	948 ± 54 (28)*	91.9 ± 2.7 (28)
Transected, 18–24 hr†	$544 \pm 26\ddagger$ (18)	$119.7 \pm 4.1\ddagger$ (18)
Transected, 18–24 hr + Physostigmine 1 mg/kg, 1 hr + Phentolamine 10 mg/kg, 1 hr	467 ± 21 (10)	109.2 ± 7.8 (10)

 * mean $\pm$ s.e.m. (*n*); values refer to amount of amine in one pair of adrenals, divided by the body weight.
 † Time interval between treatment and death.
 ‡ Differs from control ($P < 0.001$)

Effect of a ganglionic blocking agent

 Chlorisondamine (5 mg/kg) caused a transient increase in adrenal DA, lasting about 2 hr. This increase was also blocked by spinal transection (SNIDER and CARLSSON,

TABLE 2. FAILURE OF CHLORISONDAMINE, A GANGLIONIC BLOCKING AGENT, TO PREVENT THE RISE IN ADRENAL DOPAMINE INDUCED BY PHYSOSTIGMINE + PHENTOLAMINE. FOR EXPLANATION AND CONTROL VALUES, SEE TABLE 1.

	DA (ng/kg b.w.)	A + NA (μg/kg b.w.)
Physostigmine 1 mg/kg, 1 hr + Phentolamine 10 mg/kg, 1 hr	2288 $\pm$ 218 (11)	81·8 $\pm$ 4·00 (12)
Chlorisondamine 5 mg/kg, 3 hr + Physostigmine + Phentolamine	2150 $\pm$ 139 (12) N.S.	98·7 $\pm$ 2·43 (12) $P < 0·005$

1972). Chlorisondamine caused a significant elevation of (A + NA) within 2 hr and prevented the decrease in β-hydroxylated catecholamines induced by insulin or physostigmine + phentolamine. However, the elevation of DA, caused by these agents, was not prevented by the ganglionic blocking agent (Table 2). Analogous observations using insulin for neurogenic stimulation have been published (SNIDER and CARLSSON, 1972).

Similarly combined treatment with chlorisondamine and atropine at least partially prevented the elevation of DA induced by neurogenic stimulation (SNIDER and CARLSSON, 1972 and unpublished data).

Effect of variations in the nerve impulse flow on the turnover of adrenal dopamine, studied by means of synthesis inhibition

The depletion of DA induced by an inhibitor of tyrosine hydroxylase was significantly accelerated by physostigmine + phentolamine and retarded by spinal transection (Fig. 1).

The elevation of DA occurring after neurogenic stimulation in spite of an increased turnover rate, and the decrease caused by spinal transection are thus presumably due to an increased and decreased synthesis respectively, induced by changes in the activity of the rate-limiting tyrosine hydroxylase. Table 3 shows the rates of adrenal DA synthesis under normal conditions, after neurogenic stimulation and spinal transection, calculated under the assumption of a single homogenous DA pool. The existence of one or several small DA pools with rapid turnover cannot be excluded. If they exist, the calculations probably underestimate the true synthesis rates. Nevertheless, they may give useful approximations of relative synthesis rates. It can be seen that the calculated synthesis rate is higher after neurogenic stimulation and lowest after spinal transection, the ratio between these two values being about 20.

TABLE 3. TURNOVER RATE CONSTANTS (k), STEADY STATE LEVELS (DA$_0$) AND CALCULATED SYNTHESIS RATES $k \times$ (DA$_0$) OF ADRENAL DOPAMINE.

	k (hr^{-1})	(DA$_0$) (ng/kg b.w.)	$k \times$ (DA$_0$) (ng/kg b.w. hr^{-1})
Untreated rats	0·58*	970*	560
Spinal transection, 18–24 h	0·33*	445*	149
Physostigmine 1 mg/kg + Phentolamine 10 mg/kg	1·20*	2362†	2840

* Data obtained from Fig. 1.

† Data obtained from Fig. 3, assuming that the dopamine level at the 1 hr interval comes closest to a "steady state".

In animals pretreated with chlorisondamine, followed by physostigmine + phentolamine the disappearance of DA after α-methyltyrosine appeared to occur at an
approximately normal rate. The increase in DA induced by physostigmine + phentolamine and persisting after chlorisondamine treatment must therefore be assumed
to reflect an increased rate of synthesis.

Availability of adrenal dopamine for release by neurogenic stimulation

Small amounts of DA were found to be released into the perfusate after exposing
the perfused sheep adrenal gland to acetylcholine (LISHAJKO, 1969). The effect of
neurogenic stimulation on the adrenal DA level was therefore investigated after the
simultaneous inhibition of tyrosine hydroxylase and DA-β-hydroxylase. As shown in
Fig. 1, physostigmine + phentolamine treatment under these conditions accelerated
the efflux of DA. Percentagewise the effect on DA was comparable to that on the
β-hydroxylated catecholamines, which were reduced by about one-third. These
observations suggest that in the adrenals of the intact rat DA, like the β-hydroxylated
catecholamines, is available for release by nerve impulses.

DOPAMINE IN PERIPHERAL ADRENERGIC NERVES

In normal rat submaxillary glands and hearts we found DA levels of about 15 ng/g,
which is 1–2 per cent of the NA in these tissues. The DA is probably located mainly
within the adrenergic nerves, as indicated i.a. by a marked decrease after treatment
with 6-hydroxydopamine (SNIDER et al., 1973).

Electrical stimulation of the cervical sympathetic tract caused a marked elevation
of DA and a decrease in NA in the submaxillary glands. These effects could be
prevented by chlorisondamine. Both in the resting and stimulated glands the DA-β-
hydroxylase inhibitor FLA-63 caused an elevation of DA, whereas inhibition of
monoamine oxidase by pargyline had no significant effect (SNIDER et al., 1973).
Apparently newly formed DA is efficiently protected against this enzyme, presumably
by being rapidly incorporated into the storage granules. In these respects there is a
close similarity between the adrenal medulla and the adrenergic nerves, except for the
response to chlorisondamine. As in the adrenal glands physostigmine + phentolamine
caused a marked elevation of DA and a decrease in NA.

CATECHOLAMINES IN RAT BLOOD PLASMA

In some of the experiments mentioned above measurements of adrenaline and
noradrenaline in blood plasma were performed, the blood being collected after decapitation. In untreated controls about 8 ng/ml, predominantly adrenaline, was found.
In spinal rats insignificant levels were found, even after treatment with physostigmine
+ phentolamine. Also after chlorisondamine the catecholamine levels were markedly
reduced and remained so after treatment with physostigmine + phentolamine.
Without any pretreatment this drug combination markedly increased plasma catecholamine levels.

COMMENT

The rapid elevation of DA demonstrated here after neurogenic stimulation of the
adrenal medulla is probably due to increased synthesis, because the turnover of DA,
studied by means of synthesis inhibition was found to be accelerated by neurogenic
stimulation and retarded by interruption of the nerve impulse flow from supraspinal

centres. In all probability the changes in DA levels are due to some kind of modulation of the rate-limiting tyrosine hydroxylase; appreciable changes in the total number of enzyme molecules are unlikely in these short-term experiments.

It has been proposed that tyrosine hydroxylase is normally controlled by end-product inhibition. The present observations cannot be entirely explained by this mechanism. Chlorisondamine proved unable to prevent the neurogenic DA elevation while efficiently blocking the release of β-hydroxylated catecholamines into the blood stream. After chlorisondamine the A + NA levels generally tended to be higher than normal, and in some experiments this increases reached statistical significance. Moreover, after treatment with chlorisondamine, followed by physostigmine + phentolamine subnormal plasma levels of β-hydroxylated catecholamines persisted. It would appear that tyrosine hydroxylase can be controlled by neurogenic influences via a hitherto unknown mechanism not involving end-product inhibition. This receives some further support by unpublished data of this laboratory, indicating that chlorisondamine has no significant influence on the subcellular distribution of adrenal catecholamines.

Probably the neurogenic control of adrenal tyrosine hydroxylase is not entirely mediated via ordinary nicotinic receptors, because the activation of the enzyme appeared to be at least partly resistant to chlorisondamine in doses which efficiently prevented the release of β-hydroxylated catecholamines. Neither do ordinary muscarinic receptors seem to be involved as indicated by the experiments with atropine. Further work is needed to clarify this problem.

There is a close similarity between the present observations and those made in long-term experiments demonstrating induction of tyrosine hydroxylase by neurogenic stimulation of the adrenal medulla. For example, also this induction proved relatively resistant to treatment with ganglionic blocking agents (MUELLER et al., 1970).

In the peripheral adrenergic system the behaviour of DA was in many respects similar to that observed in the adrenal medulla. Thus, the ratio of DA to β-hydroxylated catecholamines was similar in the two types of tissue, and so was the response to nerve stimulation. Moreover, in both cases DA-β-hydroxylase seems to be considerably more important than monoamine oxidase for the metabolism of DA. However, chlorisondamine more or less completely blocked the rise in the DA of adrenergic nerves, in contrast to the findings in the adrenals. Interestingly, an analogous discrepancy appears to exist with respect to long-term experiments on the tyrosine hydroxylase induction (MUELLER et al., 1970). Possibly the receptors mediating neurogenic short- or long-term stimulation of catecholamine synthesis are structurally different in the two tissues.

REFERENCES

ATACK C. V. (1973) Br. J. Pharmac. (in press).
ATACK C. V. and MAGNUSSON T. (1970) J. Pharm. Pharmac. 22, 625–627.
CARLSSON A., HILLARP N. -Å. and WALDECK B. (1963) Acta physiol. scand. 59, (Suppl. 215), 1–38.
EADE N. R. (1958) J. Physiol., (Lond.) 141, 183–192.
LISHAJKO F. (1970) Acta physiol scand. 79, 405–410.
MUELLER R. A., THOENEN H. and AXELROD J. (1970) Europ. J. Pharmac. 10, 51–56.
SNIDER S. R., ALMGREN O. and CARLSSON A. (1973) Naunyn-Schmiedeberg's Arch. Pharmac. 278, 1–12.
SNIDER S. R. and CARLSSON A. (1972) Naunyn-Schmiedeberg's Arch. Pharmac. 275, 347–357.
UDENFRIEND S., COOPER J. R., CLARK C. T. and BAER J. E. (1953) Med. Sci. 117, 663–665.

Frontiers in Catecholamine Research 1973, pp. 557 to 559. Pergamon Press. Printed in Great Britain.

IN VIVO RELEASE OF ³H-DA FROM THE CAT CAUDATE NUCLEUS

M. J. Besson, A. Cheramy, J. Glowinski and C. Gauchy

Groupe NB (INSERM U. 114), Collège de France, Laboratoire de Biologie Moléculaire, 11, place Marcelin Berthelot, Paris 5e, France

Among the criteria necessary to demonstrate that dopamine (DA) found in the corpus striatum can be considered as a neurotransmitter in nigro striatal DA containing neurons, evidence of its release into the extracellular space is of great importance. To demonstrate an *in vivo* release of DA from the cat caudate nucleus different techniques have been employed such as the push pull cannula (Mc Lennan, 1964; Riddel and Szerb, 1971) and the cerebroventricular perfusion (Portig and Vogt, 1969; Von Voigtlander and Moore, 1971). In our studies (Besson *et al.*, 1971; Chéramy *et al.*, 1973), an area of the caudate nucleus was exposed by a local decortication and superfused by a modification of the cup technique generally used at the cortical level. The area delimited by the cup was superfused throughout the experiment with a physiological medium containing L-3·5-³H-tyrosine. Consequently, the dopaminergic terminals were continuously labelled with ³H-DA synthetised endogenously. The estimation of the ³H-DA content in superfusates allowed the study of the characteristics of the releasing processes of this amine.

INHIBITION OF ³H-DA RELEASE

Among the different factors able to reduce the spontaneous release of ³H-DA we examined the inhibition both of DA synthesis and of neuronal activity.

(1) *Effect of DA synthesis inhibition*

³H-DA synthesis was inhibited by α-methyl-paratyrosine (α-MpT) administered intravenously (100 mg/kg) or added to the superfusing medium (10^{-4} M) (Fig. 1) 2 or 3 hr after the onset of superfusion with ³H-tyrosine. Under these conditions, the spontaneous ³H-DA release markedly decreased immediately after the α-MpT administration, despite the relatively large amount of ³H-DA present in the tissue. The DA being located in different storage compartments (Javoy and Glowinski, 1971), it appears from these results that the newly synthetised amine is immediately available for release.

(2) *Effect of blockade of neural activity*

The addition of tetrodotoxin (10^{-5} g/ml) to the medium rapidly inhibited (60 per cent) the spontaneous release of ³H-DA (Fig 1).

Interruption of the nerve impulses was also obtained by a section of the nigro striatal neurons along the frontal plane A7, L1 to L7, H-5. Immediately after the section, the spontaneous release of ³H-DA decreased markedly and was completely blocked 30 min after the transection despite the persistence of the ³H-DA synthesis (Besson *et al.*, 1973). (Fig. 1).

All these results demonstrate that the spontaneous release of ³H-DA is dependant on nerve impulses and does not represent a simple diffusion process.

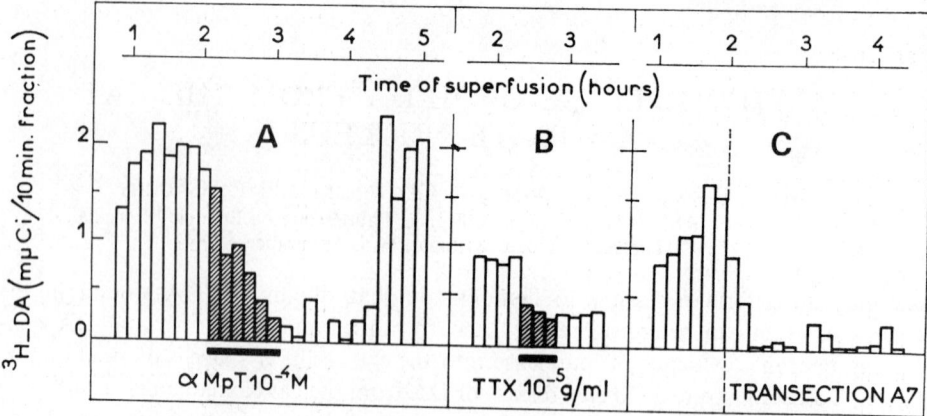

Fig. 1.—In awake cat locally anaesthetised, a part of the ventricular surface of the left caudate nucleus was delimited by a cup and continuously superfused with a physiological medium containing ^{3}H-tyrosine (40–50 μCi/ml). ^{3}H-DA release was estimated in serial collected fractions. Part A of the figure illustrates the effect of α-MpT (10^{-4} M) added to the medium during the period indicated by the black bar. Part B illustrates the action of adding tetrodotoxin (TTX 10^{-5}g/ml) to the superfusing fluid (period of application is indicated by the black bar). Part C shows the effect of a frontal transection made with a metallic spatula (coordinates A7, L1 to 7, H-5).

ACTIVATION OF ^{3}H-DA RELEASE

The release of ^{3}H-DA can be increased by activating the nigro striatal system or by interfering with the amine metabolism.

(1) *Activation of the nigro striatal neurons*

The stimulation of the substantia nigra (20 Hz, 0·2 msec, 5 V) induced an increased release of ^{3}H-DA (50 per cent) in the fraction following the 10 min stimulation period. An increase release of ^{3}H-DA (100 per cent) was observed during the stimulation of the red nucleus (A 4; L 1·5; H-4). However it is possible that in this case a direct stimulation of the DA fibres near this structure enhanced the release.

Direct depolarisation of the caudate nucleus terminals effected by increasing the K$^+$ concentration of the superfusing medium (30 mM) induced a marked output of ^{3}H-DA (9–10 times the spontaneous release). On the other hand, blockade of the DA receptors probably induced an interneuronal activation of the nigro striatal DA neurons. The action of thioproperazine (TZ) a neuroleptic of the phenothiazine type was tested. We found that the ^{3}H-DA release increased progressively after an intravenous injection of TZ and reached about 300 per cent of the spontaneous level 3 hr after the injection (Chéramy et al., 1973). However, this effect was not observed after a transection of the nigro striatal pathway. Therefore adding the drug to the superfusing fluid (10^{-6} M) even during a long application period (2 hr) does not appear to activate ^{3}H-DA release. This observation provides further support in favour of an interneuronal activation of DA neurons.

(2) *Effects of drugs which interfere with the DA metabolism*

D amphetamine locally applied 10^{-6} M induced an immediate increase in the release of ^{3}H-DA (400–450 per cent) which disappeared rapidly after omission of the drug. D amphetamine may exert its effect by a direct releasing action.

The blockade of the main catabolic pathway of DA by inhibition of MAO also affected the amount of ^{3}H-DA present in the extracellular space. In fact, the effects induced by the MAOI depend on the chemical structures of these drugs. Pargyline added to the medium (10^{-5} M) induced a progressive and persistent increase (50 per cent) of the quantities of ^{3}H-DA collected even after omission of the drug.

Tranylcypromine (10^{-5} M) enhanced markedly and immediately the release of ^{3}H-DA (400 per cent). After elimination of the drug, this effect is partially reduced (50 per cent of maximal effect). It appears that tranylcypromine may have a releasing effect in addition to its MAOI properties.

The inhibition of the DA reuptake mechanisms in DA terminals by benzotropine (10^{-6} M) (COYLE and SNYDER, 1969) progressively enhanced the extraneuronal quantities of ^{3}H-DA (400–500 per cent 1 hr after the onset of drug administration). An increase in ^{3}H-DA synthesis (40 per cent) was simultaneously observed CHÉRAMY *et al.*, 1973). These results reveal the importance of the reuptake process in the inactivation of DA and the influence of the newly taken up amine in the intraneuronal regulation of DA synthesis.

CONCLUSION

The amount of ^{3}H-DA collected in the superfusates reflects to some extent the quantities of the amine present in the extracellular space, and probably in the synaptic clefts as well. The DA available at the receptor sites is directly dependant on nerve impulses as suggested by the experiments with tetrodotoxin, transection and electrical stimulation.

Furthermore, the extraneuronal concentration of the amine released under spontaneous nerve impulses can be modified by drugs that interfere with DA metabolism at different levels.

REFERENCES

BESSON M. J., CHERAMY A., FELTZ P. and GLOWINSKI J. (1971) *Brain Res.* **32**, 407–424.
BESSON M. J., CHERAMY A., GAUCHY C. and GLOWINSKI J. (1973) *Naunyn-Schmiedeberg's Arch. Pharmacol.* **278**, 101–105.
CHERAMY A., JAVOY F., BESSON M. J., GAUCHY C. and GLOWINSKI J. (1972) CINP Meeting: 'Striatum and Neuroleptics' Copenhagen; in press.
CHERAMY A., GAUCHY C., GLOWINSKI J. and BESSON M. J. (1973) *Europ. J. Pharmacol.* **21**, 246–248.
COYLE J. T. and SNYDER S. H. (1969) *Science* **166**, 899–902.
JAVOY F. and GLOWINSKI J. (1971) *J. Neurochem.* **18**, 303–307.
MC LENNAN H. (1964) *J. Physiol.* **174**, 152–161.
PORTIG P. J. and VOGT M. (1969) *J. Physiol.* **204**, 687–715.
RIDDELL D. and SZERB J. C. (1971) *J. Neurochem.* **18**, 989–1006.
VON VOIGTLANDER P. F. and MOORE K. E. (1971) *Neuropharmacol.* **10**, 733–741.

Frontiers in Catecholamine Research 1973, pp. 561 to 565. Pergamon Press. Printed in Great Britain.

REGULATION OF CATECHOLAMINE RELEASE: α-RECEPTOR MEDIATED FEED-BACK CONTROL IN PERIPHERAL AND CENTRAL NEURONES

KLAUS STARKE

Institute of Pharmacology, University of Essen, Germany

THE OUTFLOW of noradrenaline from organs in response to stimulation of their sympathetic nerves is greatly increased by α-adrenolytic drugs (BROWN and GILLESPIE, 1957; BROWN et al., 1958). FURCHGOTT (1959) suggested that these agents not only block the inactivation of liberated noradrenaline within the tissue, but also "in some manner cause a much greater release of transmitter from the sympathetic nerve endings". Considerable evidence now supports this view. The research concerning the mode of action of the adrenolytic drugs has led to the detection of a local, α-receptor mediated, feed-back control of the release of noradrenaline (HÄGGENDAL, 1970; FARNEBO and HAMBERGER, 1971a, b; FARNEBO and MALMFORS, 1971; KIRPEKAR et al., 1972; ENERO et al., 1972). The present report summarizes our evidence for the functioning of this mechanism in the rabbit heart and the rat brain cortex.

When the postganglionic sympathetic nerves of isolated perfused rabbit hearts are stimulated electrically, part of the released noradrenaline escapes into the venous effluent (HUKOVIĆ and MUSCHOLL, 1962). α-adrenolytic drugs greatly enhance the stimulation-induced overflow. How do they achieve this effect?

Phenoxybenzamine, phentolamine and dihydroergotamine increase the stimulation-induced overflow at low concentrations, which do not inhibit the uptake of extracellular noradrenaline across the adrenergic nerve membrane (STARKE et al., 1971 a, b; STARKE, 1972a). At these low concentrations, phentolamine enhances the stimulation-evoked overflow of both noradrenaline and its metabolites (Fig. 1); phentolamine does not favour the overflow of the transmitter by reducing its biotransformation. The method described in Fig. 1 was used to test the influence of phentolamine, after both the neuronal and the extraneuronal binding of noradrenaline had been blocked by high concentrations of cocaine (10^{-4} M) and ($\pm$)-metanephrine (3×10^{-4} M), which were added immediately after pre-perfusion with ^{3}H-noradrenaline. Phentolamine (10^{-6} M) enhanced the stimulation-induced overflow of tritium (i.e. ^{3}H-noradrenaline + ^{3}H-metabolites) even in the presence of these drugs and, thus, by a mechanism other than an inhibition of the neuronal or extraneuronal binding of noradrenaline.

In summary, α-adrenolytic drugs do not enhance the stimulation-induced overflow of noradrenaline by blocking any of the known pathways of its inactivation within the tissue. Per exclusionem we may infer that these drugs increase the amount of transmitter secreted per impulse from the nerve terminals. The hypothesis offered itself that the secretion process is normally inhibited through an *activation* of α-receptors by liberated noradrenaline. Accordingly, the influence of α-receptor stimulant drugs on the release of noradrenaline was investigated.

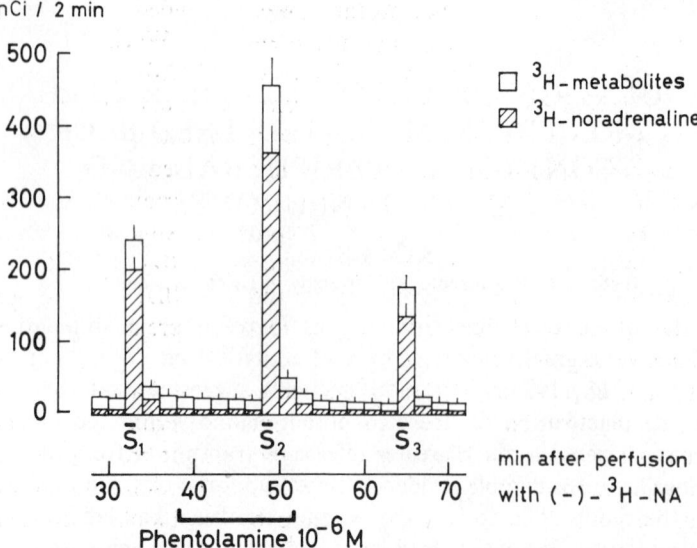

Fig. 1. Effect of phentolamine on the stimulation-induced overflow of noradrenaline and its metabolites from rabbit hearts. The hearts were pre-perfused for 15 min with 10^{-7} M (−)-^{3}H-noradrenaline, specific activity 1·6 Ci/mmole. From 28 min after the labelling perfusion onwards, the venous effluent was collected in 2 min samples for the determination of ^{3}H-noradrenaline and ^{3}H-metabolites (Starke, 1972b). The accelerans nerves were stimulated three times (S_1–S_3) for 1 min by rectangular pulses of 3 msec duration, 8 mA (supramaximal) and 5 Hz. Means ± s.e.m. of four experiments. In the absence of phentolamine, the overflow caused by S_2 was lower than that caused by S_1 (not shown). Phentolamine increased the overflow of both ^{3}H-noradrenaline and ^{3}H-metabolites.

In the rabbit heart, the following sympathomimetic agents cause a concentration-dependent decrease of the of the stimulation-induced overflow of noradrenaline: clonidine, naphazoline, oxymetazoline, phenylephrine, noradrenaline and (only at high concentrations) orciprenaline (Werner et al., 1970; Starke, 1971, 1972a, b, 1973; Starke et al., 1972). Many of these drugs are known to inhibit rather than to enhance the neuronal and extraneuronal binding of noradrenaline (Burgen and Iversen, 1965; Salt, 1972; Starke, 1972 a; Starke et al., 1972).

In hearts pre-perfused with ^{14}C-noradrenaline, the sympathomimetic agents depress the stimulation-induced overflow of both ^{14}C-noradrenaline and total ^{14}C; the outflow of noradrenaline is not reduced because of a greater metabolic transformation (Starke, 1972b, 1973, and unpublished observations). Thus, these drugs do not decrease the overflow of noradrenaline by favouring any of the known pathways of its inactivation within the tissue. We conclude that they diminish the amount of transmitter secreted per impulse from the nerve endings.

Preinfusion of phenoxybenzamine prevents the inhibition of the secretory response to stimulation caused by sympathomimetic agents. On the other hand, oxymetazoline and clonidine at high concentrations prevent the facilitatory effect of phentolamine and phenoxybenzamine; in contrast, the increase of the stimulation-induced overflow caused by cocaine is unchanged. The specific antagonism between lytic and mimetic drugs favours a common, viz. α-receptive, site of action on the release of noradrenaline (Starke, 1972a, b, 1973; Starke and Altmann, 1973).

From the experiments described so far it was concluded that α-receptors exist in the vicinity to the cardiac adrenergic nerve endings. When these receptors are activated, the release of noradrenaline declines. A local feed-back inhibition curtails the secretory response to stimulation. α-Adrenolytic agents deny the transmitter access to the receptors and set its secretion free from restriction. Where are these receptors located? α-Receptors are probably not involved in the response of the myocardium to sympathetic nerve impulses. At concentrations which greatly influence the release of noradrenaline, α-receptor blocking and stimulant drugs have no effect of their own on the force and rate of contraction. This dissociation of myocardial and neurotropic effects permits the view that the α-receptors which modulate the secretion of noradrenaline are localised prejunctionally. However, direct evidence is lacking.

FARNEBO and HAMBERGER (1971b) presented evidence that a similar mechanism operates in the brain. The following experiments concerning the effect of drugs with affinity to α-receptors on the efflux of tritium from rat brain cortex slices loaded with ^{3}H-noradrenaline confirm this view. The stimulation-induced overflow of tritium can be considered as a good measure of the biophase concentration of ^{3}H-noradrenaline built up during stimulation (BALDESSARINI and KOPIN, 1967).

Clonidine and oxymetazoline cause a concentration-dependent decrease of the overflow of tritium evoked by 2 min stimulation at 5 Hz, 10^{-7} M of either drugs being significantly effective. The interaction between oxymetazoline and drugs which enhance the stimulation-evoked overflow is demonstrated in Fig. 2. In the absence of oxymetazoline, phentolamine, phenoxybenzamine and cocaine increase the response to stimulation. At the high concentration of 10^{-5} M, oxymetazoline prevents the effect of the adrenolytic agents, but not that of cocaine. Similar results are obtained, if clonidine is used instead of oxymetazoline (STARKE and MONTEL, 1973b). Thus, the actions and interactions of mimetic and lytic drugs on central noradrenaline neurones closely resemble their effects on peripheral ones.

If liberated noradrenaline and exogenous sympathomimetic drugs depress the release of noradrenaline by one mechanism, then the inhibitory effect of exogenous sympathomimetics should disappear, when the biophase concentration of liberated noradrenaline is very high, and the endogenous brake mechanism maximally activated. To test this prediction, we varied the biophase concentration of noradrenaline by different stimulation frequencies, or by the addition of cocaine. Both in the rabbit heart and the rat brain cortex an inverse relation was found between the overflow of noradrenaline (or tritium) under control conditions and the inhibitory effect of sympathomimetic drugs. The negative relation confirms the assumption that exogenous agonists mimic the role of endogenous noradrenaline in a feedback restriction of transmitter release (STARKE, 1972a; STARKE and ALTMANN, 1973; STARKE and MONTEL, 1973b).

During sympathetic nerve stimulation, prostaglandins are released from unknown tissue components and put a brake on the release of noradrenaline (HEDQVIST, 1970). Are the prostaglandin mediated and the α-receptor mediated control of noradrenaline release linked? To test this possibility, we blocked the formation of prostaglandins by indomethacin.

Rats were given a daily s.c. injection of 5 mg/kg of indomethacin for the 3 days before the experiment. Slices of their cerebral cortex were pre-incubated with

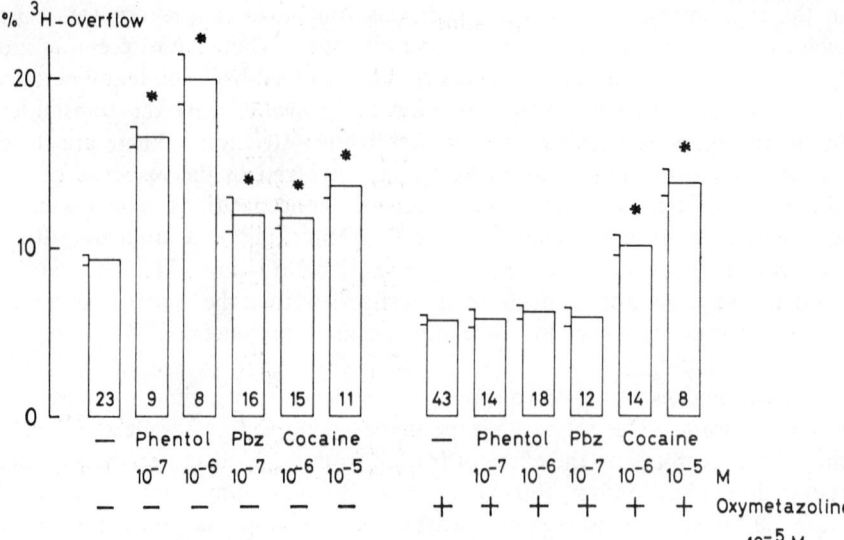

FIG. 2. Interaction between oxymetazoline, phentolamine, phenoxybenzamine and cocaine on the stimulation-induced overflow of tritium from brain slices preincubated with ^{3}H-noradrenaline. Slices of the rat cerebral cortex, diameter 3 mm, were incubated for 30 min with 10^{-7} M of $(-)$-^{3}H-noradrenaline, specific activity 6·4 Ci/mmole. They were then superfused with fresh Krebs–Ringer solution containing the drugs to be tested, and after 30 min stimulated by an electrical field for 2 min at 10 Hz (2 msec, 12 mA). The stimulation-induced tritium overflow is expressed as per cent of the tritium content of the slices at the onset of stimulation (FARNEBO and HAMBERGER, 1971b). Means ± S.E.M. of the number of experiments indicated at the bottom of the columns. *, significant differences from corresponding experiments in the absence of phentolamine, phenoxybenzamine and cocaine ($P < 0·01$).

$(-)$-^{3}H-noradrenaline, and 3×10^{-5} M indomethacin was added to the superfusion fluid after preincubation (for the doses used, cf. FLOWER and VANE, 1972; JUNSTAD and WENNMALM, 1972). The overflow of tritium from these slices caused by 2 min stimulation at 10 Hz was slightly higher than in control experiments. 10^{-5} M oxymetazoline decreased, and 10^{-6} M phentolamine increased, the stimulation-induced overflow of tritium to a similar extent, whether the synthesis of prostaglandins was blocked or not. The absence of an influence of indomethacin on the effects of oxymetazoline and phentolamine indicates that prostaglandins are not involved in the α-receptor mediated modulation of noradrenaline release. Analogous results were obtained in the rabbit heart (STARKE and MONTEL, 1973a).

In conclusion, our experiments demonstrate that the release of noradrenaline from both peripheral and central neurones can be modulated via α-receptors. In the heart, there is no evidence for postjunctional effects triggered via these receptors; they may be located prejunctionally both in peripheral organs and the brain. Liberated noradrenaline activates the receptors and thereby depresses the secretory response to forthcoming nerve impulses. A local feed-back mechanism controls the amount of noradrenaline secreted per nerve impulse. Exogenous sympathomimetic drugs mimic the effect of noradrenaline, while α-adrenolytic drugs block the receptors and interrupt the feed-back restriction. In this mechanism, prostaglandins are not involved as a chemical link. A dual local control of transmitter secretion seems to exist: one, mediated by prostaglandins, the other, independent of prostaglandins, and mediated by the effect of noradrenaline on, possibly prejunctional, α-receptors.

REFERENCES

BALDESSARINI R. J. and KOPIN I. J. (1967) *J. Pharmacol. Exp. Ther.* **156**, 31–38.

BROWN G. L. and GILLESPIE J. S. (1957) *J. Physiol. (Lond.)* **138**, 81–102.

BROWN G. L., DAVIES B. N. and GILLESPIE J. S. (1958) *J. Physiol. (Lond.)* **143**, 41–54.

BURGEN A. S. V. and IVERSEN L. L. (1965) *Br. J. Pharmacol. Chemother.* **25**, 34–49.

ENERO M. A., LANGER S. Z., ROTHLIN R. P. and STEFANO F. J. E. (1972) *Br. J. Pharmacol.* **44**, 672–688.

FARNEBO L. O. and HAMBERGER B. (1971a) *Br. J. Pharmacol.* **43**, 97–106.

FARNEBO L. O. and HAMBERGER B. (1971b) *Acta Physiol. Scand.* Suppl. **371**, 35–44.

FARNEBO L. O. and MALMFORS T. (1971) *Acta Physiol. Scand.* Suppl. **371**, 1–18.

FLOWER R. J. and VANE J. R. (1972) *Nature (Lond.)* **240**, 410–411.

FURCHGOTT R. F. (1959) *Pharmacol. Rev.* **11**, 429–441.

HÄGGENDAL J. (1970) In *New Aspects of Storage and Release Mechanisms of Catecholamines* (SCHÜMANN H. J. and KRONEBERG G., eds.), pp. 100–109. Springer-Verlag, Berlin.

HEDQVIST P. (1970) *Acta Physiol. Scand.* Suppl. **345**.

HUKOVIĆ S. and MUSCHOLL E. (1962) *Naunyn-Schmiedebergs Arch. exp. Path. Pharmakol.* **241**, 81–96.

JUNSTAD M. and WENNMALM A. (1972) *Acta Physiol. Scand.* **85**, 573–576.

KIRPEKAR S. M., WAKADE A. R., STEINSLAND O. S., PRAT J. C. and FURCHGOTT R. F. (1972) *Fedn Proc. Fedn Am. Soc. Exp. Biol.* **31**, 566 Abs.

SALT P. J. (1972) *Europ. J. Pharmacol.* **20**, 329–340.

STARKE K. (1971) *Naturwissenschaften* **58**, 420.

STARKE K. (1972a) *Naunyn-Schmiedebergs Arch. Pharmacol.* **274**, 18–45.

STARKE K. (1972b) *Naunyn-Schmiedebergs Arch. Pharmacol.* **275**, 11–23.

STARKE K. (1973) *Experientia* **29**, 579–580.

STARKE K. and ALTMANN, K. P. (1973) *Neuropharmacology* **12**, 339–347.

STARKE K. and MONTEL, H. (1973a) *Naunyn-Schmiedebergs Arch. Pharmacol.* **278**, 111–116.

STARKE K. and MONTEL H. (1973b) *Neuropharmacology* (in press)

STARKE K., MONTEL H. and SCHÜMANN H. J. (1971a) *Naunyn-Schmiedebergs Arch. Pharmakol.* **270**, 210–214.

STARKE K., MONTEL H. and WAGNER J. (1971b) *Naunyn-Schmiedebergs Arch. Pharmakol.* **271**, 181–192.

STARKE K., WAGNER J. and SCHÜMANN, H. J. (1972) *Archs Int. Pharmacodyn. Ther.* **195**, 291–308.

WERNER U., STARKE K. and SCHÜMANN H. J. (1970) *Naunyn-Schmiedebergs Arch. Pharmakol.* **266**, 474–475.

Frontiers in Catecholamine Research 1973, pp. 567 to 574. Pergamon Press. Printed in Great Britain.

EFFECT OF IMPULSE FLOW ON THE RELEASE AND SYNTHESIS OF DOPAMINE IN THE RAT STRIATUM

Robert H. Roth, Judith R. Walters,
and George K. Aghajanian

Departments of Pharmacology and Psychiatry, Yale University School of Medicine, and the
Connecticut Mental Health Center, New Haven, Connecticut 06510, U.S.A.

IT IS generally accepted that an alteration of impulse flow in a peripheral sympathetic noradrenergic neuron leads to a predictable change in the turnover and synthesis of the norepinephrine (NE) associated with that neuron (COSTA, 1970). An increase in sympathetic activity results in an increase in NE turnover and a decrease or block in impulse flow leads to a reduction in NE turnover. Recently, similar observations have also been made in central noradrenergic and serotonergic neurons. Stimulation of noradrenergic neurons originating in the locus coeruleus and projecting to the cerebral cortex and hippocampus leads to an increase in NE turnover (ARBUTHNOTT et al., 1970; KORF et al., 1973a), as well as a frequency dependent increase in the accumulation of 3-methoxy-4-hydroxyphenylglycol sulfate (KORF et al., 1973a). Cessation of impulse flow produced by acute destruction of the locus coeruleus results in a reduction in NE turnover without producing a significant alteration in the steady state levels of NE in the cerebral cortex (KORF et al., 1973b). Likewise, stimulation of serotonergic neurons in the midbrain raphe results in a frequency-dependent increase in serotonin synthesis (SHIELDS and ECCLESTON, 1972) and accumulation of 5-hydroxyindole acetic acid in the forebrain (SHEARD and AGHAJANIAN, 1968b). If impulse flow in serotonergic neurons is abolished by placement of a lesion in the midbrain raphe, the turnover of serotonin in the forebrain is retarded (HERR and ROTH, 1972).

Dopaminergic neurons, on the other hand, do not respond in a completely analogous fashion. Stimulation of the dopamine (DA) neurons originating in the zona compacta of the substantia nigra does result, similar to the NE and serotonin systems, in a release of transmitter (VONVOIGTLANDER and MOORE, 1971) and in a frequency-dependent increase in DA synthesis and accumulation of a DA metabolite, dihydroxyphenylacetic acid (DOPAC) in the striatum (MURRIN and ROTH, unpublished data). However, differences between the DA system and other monoamine systems in the CNS become apparent when one begins to investigate the effects of a reduction or abolition of impulse flow in the nigro-neostriatal pathway.

For a number of years our laboratory has been interested in the biochemical and pharmacological actions of a compound called γ-hydroxybutyrate (GHB), which occurs as a natural metabolite of mammalian brain (ROTH and GIARMAN, 1970). When administered in anaesthetic doses GHB or its lactone precursor γ-butyrolactone (GBL) has been shown to cause a rapid increase in neostriatal DA levels without having a very significant effect on other brain monoamines such as NE or serotonin (GESSA et al., 1966; ROTH and SURH, 1970). The time course of the accumulation of DA is correlated with the behavioural effects and the brain levels of the drug

(Walters and Roth, 1972). GHB does not have a direct effect on the enzymes synthesising or metabolising DA (Gessa et al., 1968), but it does cause a marked decrease in the utilisation of DA (Roth and Surh, 1970).

In order to try to gain further insight into the mechanism responsible for the increase in DA produced by GHB, the effect of GHB on the firing of dopaminergic neurons in the zona compacta of the substantia nigra was investigated by means of single unit recording techniques. The recent identification of DA neurons based on both histochemical and neurophysiological characteristics made it possible to make extracellular recordings from these neurons (Bunney et al., 1973a,b). These unit recording experiments revealed that GHB or GBL administered intravenously or intraperitoneally in doses of 100–200 mg/kg routinely decrease the firing of units localised to the DA containing cells present in the zona compacta of the substantia nigra (Fig. 1). Serotonin and NE containing neurons in the midbrain raphe and locus coeruleus respectively were much less responsive to the inhibitory properties of GHB. This action of GHB seems to be unique since other classes of hypnotics and anaesthetics do not inhibit the firing of these DA units.

If impulse flow is interrupted in the DA system by placement of a high frequency electrothermic lesion in the nigro-neostriatal pathway or by direct injection of local anaesthetics into this pathway an increase in neostriatal DA similar to that produced by GHB treatment is also observed (Andén et al., 1971; Walters et al., 1973; Roth, unpublished data). Histochemical observations indicate that the increase in fluorescence seen with both treatments is localised to the nerve terminals of the DA containing neurons (Walters et al., 1973). Thus the DA system when compared to other monoamine systems appears to respond in a rather unique fashion to an inhibition of impulse flow by rapidly increasing the transmitter content of its terminals. A similar increase in the steady state levels of NE or serotonin is not observed when impulse flow in these respective neuronal systems is interrupted pharmacologically or by placement of lesions in the locus coeruleus or the midbrain raphe (Korf et al., 1973; Herr and Roth, 1972; Carlsson et al., 1972). However, cholinergic neurons in the hippocampus do respond to a lack of impulse flow in a fashion similar to dopaminergic neurons by increasing the content of acetylcholine in their terminals (Sethy et al., 1973). This rapid increase in the accumulation of DA seen after pharmacological or mechanical interruption of impulse flow in the nigro-neostriatal pathway appears to be explainable, in part, due to a marked increase in DA synthesis. The incorporation of labelled tyrosine into DA (Walters et al., 1973) and the accumulation of dihydroxyphenylalanine (DOPA) after inhibition of DOPA decarboxylase (Fig. 2) are both increased by treatment with GHB or by placement of a lesion in the nigro-neostriatal pathway. An increase in DOPA accumulation in the striatum after transverse cerebral hemisection has also been reported by Carlsson et al. (1972).

The above findings suggest that great caution should be used in interpreting data obtained from techniques employing administration of labeled precursors such as tyrosine to assess the functional activity of dopaminergic neurons in the CNS. It is quite apparent that the rate of DA synthesis increases in this system both in response to a lack as well as to an increase in impulse flow. Recent extracellular recording studies in our laboratories (Bunney et al., 1973a) have indicated that antipsychotic drugs of the phenothiazine and butyrophenone class cause an increase

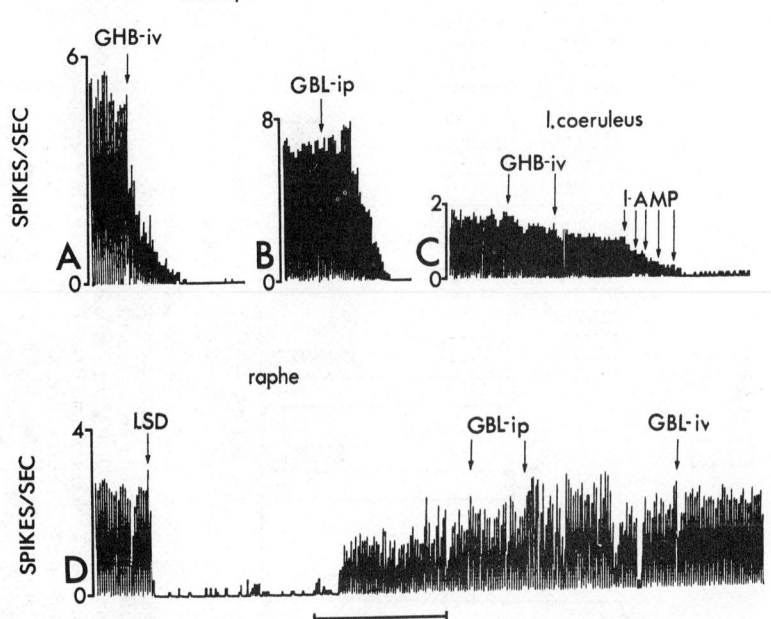

FIG. 1(a).—Effect of GHB (100 mg/kg i.v.) on the firing rate of a DA cell in the pars compacta of the substantia nigra of the rat. These cells typically respond to an i.v. injection of GHB by an abrupt decrease in firing rate followed by a further, gradual decline. The cells are generally inhibited completely by doses of 100–350 mg/kg GHB i.v. Recovery is dose-dependent and very gradual; 35 min being required for 25 per cent recovery after 150 mg/kg GHB i.v. All animals were anesthetised with chloral hydrate. Methods as described in BUNNEY et al. (1973).

(b) Inhibition of firing of a DA cell in the pars compacta of the substantia nigra by i.p. injection of GBL (200 mg/kg), a precursor of GHB. Fifty per cent of the DA cells recorded are inhibited with this dose of GBL administered i.p. Other classes of "anaesthetics" do not inhibit the firing of these cells; in fact many cause an increase in the rate of firing. The inhibitory effect of GBL was not specific for DA cells, however. Other units recorded in the reticular formation dorsal to the substantia nigra and in the pars reticulata of the substantia nigra were also inhibited by doses of 200–400 mg/kg GBL i.p.

(c) Effect of GHB (200 + 300 mg/kg i.v.) on firing of a NE cell in the locus coeruleus. In contrast with the DA cells, no abrupt decline in firing rate occurred and only a slight decrease in the rate was observed. Typically locus cells showed either no effect or only a slight decrease in response to either i.p. injections of GBL or i.v. injections of GHB in doses up to 1200 mg/kg. A total of 1·6 mg/kg of l-amphetamine (AMP) (in doses of 0·1, 0·1, 0·2, 0·4 and 0·8 mg/kg i.v.) inhibited the firing of this cell providing additional indication of the noradrenergic nature of the cell (WALTERS et al., unpublished data). Coordinates as described in GRAHAM and AGHAJANIAN (1971).

(d) Administration of 600 mg/kg GBL (in doses of 200 + 200 mg/kg i.p. and 200 mg/kg i.v.) to a serotonergic cell in the raphe nucleus. GBL produced no significant change in the firing of the cell. LSD (20 μg/kg i.v.) caused a rapid inhibition of the firing of this unit. This sensitivity to LSD has been shown to be an identifying characteristic of serotonin-containing raphe cells (AGHAJANIAN et al., 1970). Coordinates as described by AGHAJANIAN et al. (1970). Raphe cells are perhaps slightly more responsive to GBL than those of the locus coeruleus, but much less responsive than the DA cells. Of the raphe cells recorded from, 25 per cent were partially inhibited by doses of 200–600 mg/kg GBL, administered either i.p. or i.v., the rest were unaffected.

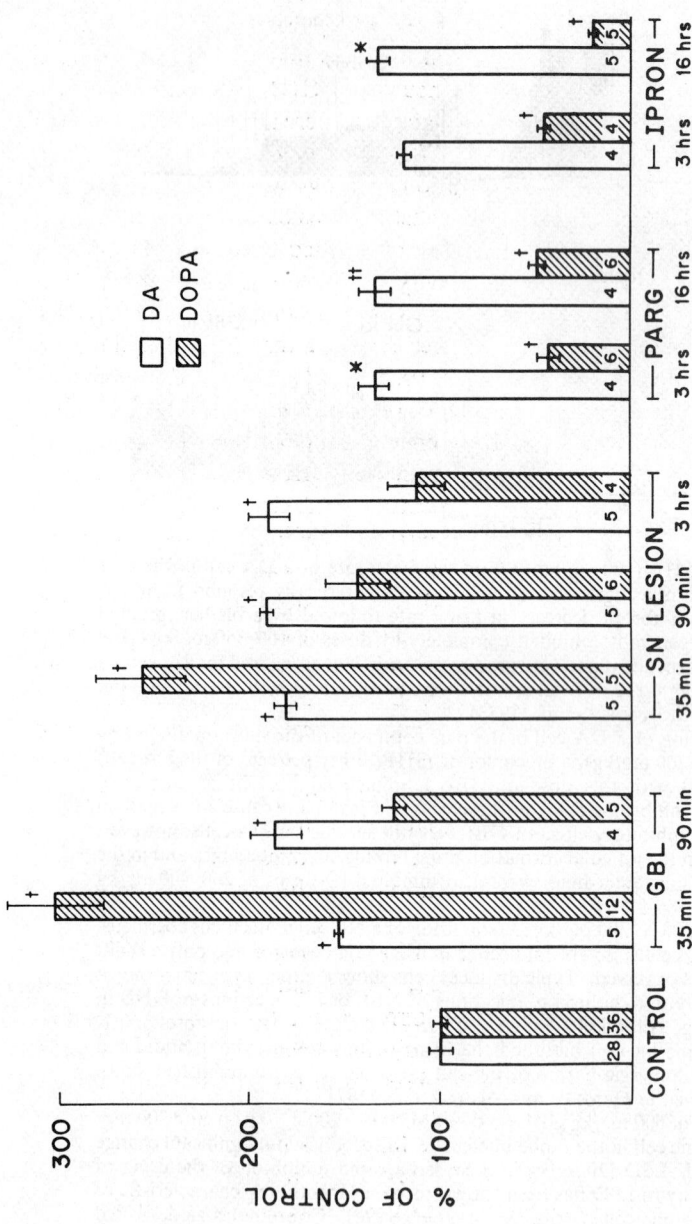

Fig. 2.—GBL (750 mg/kg i.p.) was administered 35 or 90 min prior to sacrifice. Electrothermic lesions were placed in the nigro-striatal pathway under halothane anaesthesia 35, 90 or 180 min before sacrifice, as described by WALTERS et al., 1973. Pargyline (75 mg/kg) or iproniazid (100 mg/kg) were administered 3 hr or 16 hr prior to sacrifice. For all DOPA determinations, R04-4602 (800 mg/kg) was administered 30 min before sacrifice. DOPA was isolated and measured fluorometrically, essentially by the technique of KEHR et al. (1972), as modified by WALTERS and ROTH (manuscript in preparation). DA was isolated and measured fluorometrically as described previously (WALTERS and ROTH, 1972). Results are expressed as a percentage of individual controls. The bars represent S.E.M. and the numbers in the columns indicate the number of individual experiments. The pooled control value for striatal DA was 9.42 ± 0.37 μg/g, $n = 14$; the pooled value for striatal DOPA was 864 ± 35 ng/g, $n = 36$ (mean $\pm$ S.E.M.).

* Significantly different from individual control $P < 0.01$
† Significantly different from individual control $P < 0.001$
†† Significantly different from individual control $P < 0.05$

in firing of dopaminergic neurons. It is already a well known fact that these drugs also cause an increase in DA turnover measured by a variety of independent techniques (CARLSSON and LINDQVIST, 1963; NYBÄCK and SEDVALL, 1968; CORRODI et al., 1967). Biochemical measurements of DA turnover thus appear to provide a valid index for determining if a psychotropic drug increases impulse flow in dopaminergic neurons. However, what happens when a given drug treatment leads to a decrease or abolition of impulse flow in dopaminergic neurons? GHB blocks impulse flow in the neostriatal pathway (Fig. 1) yet this drug causes an increase in the accumulation of labelled DA after tyrosine administration (WALTERS et al., 1973), in a fashion very similar to that observed after administration of drugs which increase impulse flow in this system. Thus it is clear that turnover estimations based on the calculation of the rate of DA synthesis from the conversion of labelled tyrosine to DA do not provide a useful index of the functional activity of dopaminergic neurons. For example, recent biochemical studies employing ^{14}C-DA accumulation after labelled tyrosine administration as an index of turnover revealed that d-amphetamine, in doses which inhibit firing of DA neurons (BUNNEY et al., 1973a, b) produces an *increase* rather than a decrease in DA "turnover" (COSTA et al., 1972).

It has been our experience that a useful biochemical index as to whether a given drug increases or decreases functional activity in the nigro-striatal neurons is to follow the decrease in DA after inhibition of synthesis with AMPT or the short term accumulation of DOPAC in the striatum. An interruption of impulse flow by GHB or electrothermic lesions placed in the nigro-striatal pathway leads to a dramatic decrease in the rate of DA disappearance after AMPT (ROTH and SURH, 1970; WALTERS et al., 1973; ANDÉN et al., 1971) as well as to a rapid decrease in DOPAC (Table 1). Stimulation of the pathway on the other hand results in a frequency-dependent increase in the accumulation of DOPAC (Table 1). Drugs which increase impulse flow in DA systems such as the antipsychotic drugs, chlorpromazine and haloperidol and the anaesthetics, chloral hydrate and halothane, all lead to a significant increase in the accumulation of DOPAC. A phenothiazine lacking antipsychotic potency, promethazine, has no effect on the activity of dopaminergic neurons and produces no alteration in DOPAC levels in the striatum even when administered in high doses. Drugs such as amphetamine which blocks impulse flow in the nigro-striatal pathway cause a significant decrease in DOPAC.

It is noteworthy that despite the fact that both pharmacological and mechanical lesions of the DA system lead to a large increase in the endogenous levels of DA, the rate of DA synthesis determined *in vitro* (BUSTOS et al., 1972) and *in vivo* (Fig. 2) after the establishment of this increased steady-state level of DA is not significantly different from the rate of DA synthesis observed in the striatum of untreated control animals. These observations are suggestive that either striatal tyrosine hydroxylase is relatively insensitive to feedback inhibition by accumulated DA or perhaps, more likely, that the newly formed DA does not have free access to tyrosine hydroxylase. If endogenous levels of DA are increased by treatment with MAO inhibitors such as pargyline or iproniazid, however, the synthesis of DA is markedly retarded even though the endogenous levels of DA do not rise nearly as high as after GHB treatment or placement of electrothermic lesions in the nigro-neostriatal pathway (Fig. 2). This may in part be explained by the intraneuronal distribution of the increased DA. Preliminary experiments have in fact indicated that the increase in

Table 1. Effect of alteration of impulse flow in the nigro-striatal pathway on the accumulation of DOPAC in the rat striatum

Treatment	Effect on unit activity of dopaminergic neurons	n	DOPAC levels (% of control)
None	no change	19	100 ± 3
Sham lesion	no change	4	102 ± 3
30 min SN lesion	decreased	4	75 ± 4^b
GBL (750 mg/kg, i.p.)	decreased	4	76 ± 11^c
d-Amphetamine* (1·25 mg/kg, i.p.)	decreased	4	47 ± 2^a
Chloral Hydrate (400 mg/kg, i.p.)	increased	8	140 ± 5^a
Halothane	increased	4	128 ± 4^a
Haloperidol* (0·1 mg/kg)	increased	4	206 ± 14^a
Chlorpromazine* (1·25 mg/kg, i.v.)	increased	4	168 ± 11^a
Promethazine* (10 mg/kg, i.p.)	no change	4	102 ± 5
*Electrical stimulation**			
Sham—(Chloral hydrate Anaesthesia)	no change	9	105 ± 5
5 Hz—(Chloral hydrate Anaesthesia)	increased	3	169 ± 11^a
15 Hz—(Chloral hydrate Anaesthesia)	increased	3	308 ± 7^a
30 Hz—(Chloral hydrate Anaesthesia)	increased	3	256 ± 45^a

Lesions of the nigro-striatal pathway (SN lesion) were performed under halothane anaesthesia as described by Walters et al., 1973. In the stimulation experiments rats were anaesthetised with chloral hydrate and the nigro-striatal pathway stimulated for periods of 20 min at the indicated frequency. Results are expressed in terms of percentage of the unstimulated side. Rats were killed 30 min after drug treatment and DOPAC isolated and analysed fluorometrically as described previously (Walters and Roth, 1972). *—Rats were anaesthetised with chloral hydrate and results are expressed as % of chloral hydrate anaesthetised controls. All results are expressed as the mean ± s.e.m. The level of DOPAC in striatum of untreated rats was 935 ± 27 ng/g, $n = 19$. Significantly different from appropriate control: a—$P < 0.001$; b—$P < 0.002$; c—$P < 0.01$. Data taken in part from Bunney et al., 1973b and Murrin and Roth, unpublished data.

DA produced by GHB is largely particulate while the increase in DA observed after pargyline pretreatment is largely soluble.

The above data are consistent with the concept that under some circumstances (i.e. after MAO inhibition) end product inhibition of tyrosine hydroxylase by DA occurs in striatal DA neurons. However, the data also indicate that another mechanism is perhaps more important in some cases for the control of synthesis. The actual mechanism responsible for this increase in DA synthesis observed in response to a lack of impulse flow in the nigro-striatal pathway is unknown but it appears likely that the mechanism may be ultimately linked to the transmitter release process or to other events which occur during depolarisation. It appears possible that tyrosine hydroxylase activity may be controlled by changes in the degree of DA receptor activation which is ultimately dependent upon the amount of DA reaching the synaptic cleft (Kehr et al., 1972). In an effort to further investigate this possibility we tested to see if drugs which would stimulate DA receptors either directly or indirectly would alter the increase in DA synthesis observed when impulse flow was blocked in the nigro-striatal pathway. Pretreatment with apomorphine or amphetamine was found to markedly antagonise the increase in DOPA accumulation observed after inhibition of impulse flow pharmacologically with GHB or by placement of a lesion in the nigro-striatal pathway (Table 2). It is possible that treatment with amphetamine leads to some change in the interneuronal distribution of DA making more free DA available to act as an endogenous inhibitor of tyrosine hydroxylase. However, an alternate explanation, consistent with the theory of receptor mediated changes in tyrosine

TABLE 2. EFFECT OF PRETREATMENT WITH AMPHETAMINE AND APOMORPHINE ON THE ACCUMULATION OF DA AND DOPA IN THE RAT NEOSTRIATUM AFTER GBL OR SN LESION

Treatment	n	Dopamine (% of control)	n	DOPA (% of control)
None	15	100 ± 5	26	100 ± 4
GBL (750 mg/kg)	4	198 ± 11^a	12	302 ± 26^a
Amphetamine (5 mg/kg)	4	129 ± 16	4	157 ± 11^b
GBL + amphetamine	4	141 ± 5^a	4	151 ± 11^c
Apomorphine (2 mg/kg)		—	5	62 ± 9^c
GBL + apomorphine		—	4	125 ± 24
SN lesion	5	180 ± 6^a	5	256 ± 24^b
SN lesion + amphetamine	4	135 ± 7^c	4	110 ± 13

GBL was administered 35 min before sacrifice and where indicated amphetamine and apomorphine were administered 50 and 42 min respectively prior to sacrifice. Electrothermic lesions of the nigro-neostriatal pathway were made under halothane anaesthesia 35 min before sacrifice. DOPA and DA determinations were as described in Fig. 2. Results are expressed as a percentage of individual controls (mean ± S.E.M.). The levels of DA and DOPA in the striatum of untreated rats were $8 \cdot 74 \pm 0 \cdot 54 \, \mu g/g$, $n = 15$ and $844 \pm 36 \, ng/g$, $n = 26$, respectively.
Significantly different from control a—$P < 0 \cdot 001$; b—$P < 0 \cdot 005$; c—$P < 0 \cdot 02$; d—$P < 0 \cdot 05$.

hydroxylase activity, is that amphetamine causes an increase in the amount of DA present in the synaptic cleft which then interacts with pre- or postsynaptic DA receptors. Apomorphine presumably has a direct stimulatory action on these receptors. This increase in the activation of DA receptors might then set in motion the changes which ultimately result in a decrease in tyrosine hydroxylase activity.

In conclusion, cessation of impulse flow in DA neurons results in a decrease in release and catabolism of DA and an increase in synthesis. These changes are reflected by a dramatic increase in the steady state levels of DA in the neuronal terminals. Once this new steady state level of DA is established the rate of synthesis returns to normal but is not inhibited as it is when DA levels are increased by inhibition of MAO. Drugs which are capable of stimulating DA receptors block this increase in DA observed after inhibition of impulse flow, suggesting that DA synthesis may in part be controlled by alterations in receptor activity.

Acknowledgements—This research was supported by NIMH Grant MH-14092 and the State of Connecticut.

REFERENCES

AGHAJANIAN G. K., FOOTE W. E. and SHEARD M. H. (1970) *J. Pharmacol. Exp. Ther.* **171**, 178–187.
ANDÉN N.-E., CORRODI H., FUXE K. and UNGERSTEDT U. (1971) *Europ. J. Pharmacol.* **15**, 193–199.
ARBUTHNOTT G. W., CROW T. J., FUXE K., OLSON L. and UNGERSTEDT U. (1970) *Brain Res.* **24**, 471–483.
BUNNEY B. S., AGHAJANIAN G. K. and ROTH R. H. (1973a) *Fedn. Proc.* **32**, 200.
BUNNEY B. S., WALTERS J. R., ROTH R. H. and AGHAJANIAN G. K. (1973b) *J. Pharmacol. Exp. Ther.*, **185**, 560–571.
BUSTOS G., KUHAR M. J. and ROTH R. H. (1972) *Biochem. Pharmacol.* **21**, 2649–2652.
CARLSSON A., BEDARD P., LINDQVIST M. and MAGNUSSON T. (1972) *Biochem. Soc. Symp.* **36**, 17–32.
CARLSSON A. and LINDQVIST M. (1963) *Acta Pharmacol. Toxicol.* **20**, 140–144.
CORRODI H., FUXE K. and HÖKFELT T. (1967) *Life Sci.* **6**, 767–772.
COSTA E. (1970) *Adv. Biochem. Psychopharmacology* **2**, 169–204.
COSTA E., GROPPETTI A. and NAIMZADA M. K. (1972) *Brit. J. Pharmacol.* **44**, 742–751.
GESSA G. K., SPANO P. F., VARGIU L., CRABAI F. and MAMELI L. (1968) *Life Sci.* **7**, 289–298.
GESSA G. L., VARGIU L., CRABAI F., BOERO G. C., CABONI F. and CAMBA R. (1966) *Life Sci.* **5**, 1921–1930.

GRAHAM A. W. and AGHAJANIAN G. K. (1971) *Nature, Lond.* **234**, 100–102.
HERR B. and ROTH R. H. (1972) *Proc. Fifth Int. Cong. Pharmacol.* p. 100, No. 600.
KEHR W., CARLSSON A. and LINDQVIST M. A. (1972) *Naunyn Schmied. Arch. Pharmacol.* **274**, 273–280.
KEHR W., CARLSSON A., LINDQVIST M., MAGNUSSON T. and ATACK C. (1972) *J. Pharm. Pharmacol.* **24**, 744–747.
KORF J., AGHAJANIAN G. K. and ROTH R. H. (1973a) *Europ. J. Pharmacol.* **21**, 305–310.
KORF J., ROTH R. H. and AGHAJANIAN G. K. (1973b) *Europ. J. Pharmacol.*, in press.
NYBÄCK H. and SEDVALL G. (1968) *J. Pharmacol. Exp. Ther.* **162**, 294–301.
ROTH R. H. and GIARMAN N. J. (1970) *Biochem. Pharmacol.* **19**, 1087–1093.
ROTH R. H. and SURH Y. (1970) *Biochem. Pharmacol.* **19**, 3001–3012.
SETHY V. H., KUHAR M. J., ROTH R. H., VANWOERT M. H. and AGHAJANIAN G. K. (1973) *Brain Res.* **55**, 481–484.
SHEARD M. H. and AGHAJANIAN G. K. (1968) *J. Pharmacol.* **163**, 425–430.
SHIELDS P. J. and ECCLESTON D. (1972) *J. Neurochem.* **19**, 265–272.
VONVOIGTLANDER P. F. and MOORE K. E. (1971) *Brain Res.* **35**, 580–583.
WALTERS J. R., AGHAJANIAN G. K. and ROTH R. H. (1972) *Proc. Fifth Int. Cong. Pharmacol.* p. 245, No. 1472.
WALTERS J. R. and ROTH R. H. (1972) *Biochem. Pharmacol.* **21**, 2111–2121.
WALTERS J. R., ROTH R. H. and AGHAJANIAN G. K. (1973) *J. Pharmacol. Exp. Ther.*, in press.

Frontiers in Catecholamine Research 1973, pp. 575 to 577. Pergamon Press. Printed in Great Britain.

EFFECTIVENESS OF ADRENERGIC TRANSMITTER CONCENTRATIONS IN SYNAPTIC CLEFTS OF DIFFERING DIMENSIONS

JOHN A. BEVAN and CHE SU

Department of Pharmacology UCLA School of Medicine Los Angeles,
California 90024, U.S.A.

THE DIMENSIONS of the adrenergic neuromuscular synapse show considerable variation in different tissues. In the arterial tree, the synaptic cleft size appears to diminish with vessel diameter. The shortest neuromuscular distance ranges from approximately 20,000 Å in the elastic arteries to 1000 Å in the smaller arteries and arterioles. Intermediate size vessels show intermediate cleft dimensions. In the cat nictitating membrane and the vas deferens of various species the closest neuromuscular cleft is approximately 300 Å. A difference in cleft width of almost two orders of magnitude would be expected to lead to considerable differences in synaptic concentrations of the transmitter, l-norepinephrine.

Two parameters of adrenergic transmitter concentration have been measured experimentally in isolated tissues in which quantitative information of neuromuscular relationships is available (BEVAN and SU, 1973). Measurements were made during transmural electrical stimulation of the neural elements at 10 Hz. Using a method similar to that of LJUNG (1969), peak transmitter concentrations at the α-adrenergic receptor were determined. Since it can be assumed that peak transmitter concentrations must occur at those α-receptors closest to the nerve endings, these values may be taken to indicate the order of intrasynaptic transmitter concentration. The second transmitter concentration measurement is presumed to be that within the plexus of nerves at the adventitio-medial junction, but outside the immediate confines of the synapse itself, i.e., perisynaptic. It has been estimated as the concentration 'head' that causes the transmitter to move through the wall of the vessel at the rate measured experimentally during continuous nerve stimulation.

The results of these studies made on vessels with mean neuromuscular distances of 10,000, 5000, 2000 and 1000 Å, respectively, are shown in Fig. 1. When the neuromuscular cleft is wide, the intra- and perisynaptic transmitter concentrations are similar and relatively low. As the neuromuscular cleft narrows, an increasing concentration differential occurs between the inside and outside of the synapse. The concentration within the narrow synapse rises to an extremely high level, whereas that outside the synapse becomes very low. Experimental data are not available from a tissue in which the cleft is less than 1000 Å. However it would be expected that in such tissues this trend would be continued.

Some of the functional implications of these transmitter concentrations can be understood when related to l-norepinephrine dose-response curves for different tissues. Also included in Fig. 1 are dose–response curves for ten tissues or organs from a variety of animal species determined under circumstances in which neuronal uptake is minimised. The l-NE ED$_{50}$ for all tissues lies within the range of 10^{-7} and 10^{-6} M. When the determined transmitter concentration values are projected upon

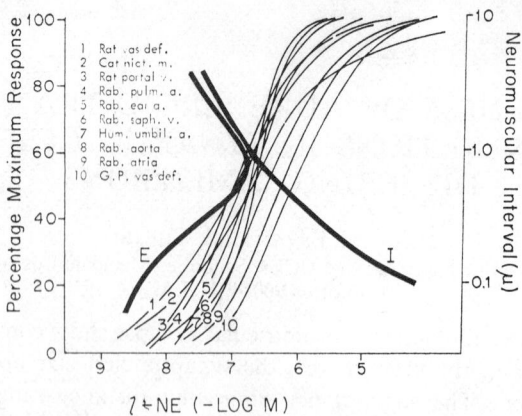

FIG. 1.—Plot of relationship between *l*-NE concentration and percentage maximum contractile response of various tissues numbered (1–10), and between shortest neuro-muscular interval and 1–NE concentration equivalent to intra- (1) and extra- (E) or perisynaptic transmitter concentration during neural activity at 10 Hz.

[1] KRELL and PATIL (1972).	[7] ALTURA *et al.*, (1972).
[2] TSAI *et al.*, (1968).	[9] BLINKS (1966).
[3] JOHANSSON *et al.*, (1970).	[10] WESTFALL *et al.*, (1972).

[4,5,6,8] BEVAN, PEGRAM, SU (unpublished data).

these dose–response curves, the following relationships can be seen: When the neuromuscular cleft is wide the transmitter concentrations achieved within and without the synapse are at the lower end of the dose–response curve. As the cleft becomes narrower, the intrasynaptic concentration rises to maximal or supramaximal levels. At the same time the perisynaptic transmitter concentrations fall to almost subthreshold values. This suggests that as the cleft becomes narrower, the action of the transmitter becomes confined more and more to the synapse itself.

These differing patterns of transmitter distribution and concentration have functional correlates in the contractile responses to nervous activity. In elastic arteries for example (cleft 20,000 Å), there is no evidence of any special contractile effect or response resulting from transmitter action on cells closest to the nerve elements. Effector response is the sum total of local transmitter effects on the individual smooth muscle cells in the vessel wall. In muscular arteries, such as the ear artery of the rabbit (cleft 5000 Å), there are two phases of response associated with two modes of transmitter excitation. One phase is associated with the action of the transmitter on the muscle cells closest to the neural plexus; the other with the transmitter that overflows into the medial coat of the vessel and its action on successive layers of smooth muscle cells (BEVAN *et al.*, 1970). The intra- and perisynaptic transmitter concentration parameters can account for these two effects. In vessels with a narrow cleft, there is increasing evidence associating most of the neural response to a transmitter action within the synapse. In the rat portal vein (cleft 1000 Å), the α-adrenergic receptors may be confined to intrasynaptic sites. There may be no functional α-adrenergic receptors associated with muscle cells at a distance from the nerve endings (JOHANSSON *et al.*, 1970).

Evidence is emerging that variation in the size of the neuromuscular cleft is part of an important functional pattern of neuromuscular excitation, regulation and integration.

REFERENCES

ALTURA B., MALAVIYA D., REICH C. F. and ORKIN L. R. (1972) *Am. J. Physiol.* **222,** 345–355.
BEVAN J. A., GARSTKA W., SU C. and SU M. (1973) *Europ. J. Pharmacol.* **22,** 47–53.
BLINKS J. R. (1966) *J. Pharmacol. Exp. Ther.* **151,** 221–235.
JOHANSSON B., LJUNG B., MALMFORS T. and OLSON L. (1970) *Acta Physiol. Scand. (Suppl.)* **349,** 5–16.
KRELL R. D. and PATIL P. N. (1972) *J. Pharmacol. Exp. Ther.* **182,** 101–115.
LJUNG B. (1969) *Acta Physiol. Scand.* **77,** 212–223.
TSAI T. H., DENHAM S. and MCGRATH W. R. (1968) *J. Pharmacol. Exp. Ther.* **164,** 146–157.
WESTFALL D. P., MCCLURE D. C. and FLEMING W. W. (1972) *J. Pharmacol. Exp. Ther.* **181,** 328–338.

Frontiers in Catecholamine Research 1973, pp. 579 to 581. Pergamon Press. Printed in Great Britain.

PULSE-TO-PULSE MODULATION OF NORADRENALINE RELEASE THROUGH A PREJUNCTIONAL α-RECEPTOR AUTO-INHIBITORY MECHANISM

M. J. Rand, D. F. Story, G. S. Allen, A. B. Glover and
M. W. McCulloch

Department of Pharmacology, University of Melbourne, Parkville, Victoria
3052, Australia

CONSIDERABLE evidence has been amassed in support of the concept of prejunctional α-adrenoreceptors through which noradrenaline (NA) release is inhibited. It has been postulated that the effect of α-receptor antagonists in increasing the amount of NA released by adrenergic nerve stimulation is due to blockade of the prejunctional receptors (HOTTA, 1969; HÄGGENDAL, 1970; KIRPEKAR and PUIG, 1971; ENERO, LANGER, ROTHLIN and STEFANO, 1972; McCULLOCH, RAND and STORY, 1972). More direct support for the concept is provided by the finding that exogenous (−)-NA decreases the release of transmitter NA (McCULLOCH et al., 1972).

These observations suggest that the NA released in response to nerve impulses would inhibit the release in response to subsequent impulses when there was a sufficient concentration in the neuro-effector gap to exert the inhibitory action through the presynaptic α-receptor; that is, transmitter NA produces pulse-to-pulse modulation of NA release.

The existence of the phenomenon has been investigated in two isolated tissue preparations: guinea-pig atria and rabbit ear artery. The artery preparation has been described by ALLEN, RAND and STORY (1973). The tissues were incubated in high specific activity (^3H)-$(−)$-NA and then washed with fresh Krebs solution until the resting rate of efflux of radioactivity had fallen to approach a steady resting state. Then, stimulation of the sympathetic nerves within the tissues evoked an increased efflux of radioactivity. With successive periods of stimulation, the stimulation-induced (S-I) efflux gradually declined. The effect of drugs on S-I efflux may be tested by comparison with that evoked in corresponding periods of stimulation in the absence of drugs. Observations were made on efflux from the two preparations with respect to: (i) the enhancement with an α-receptor antagonist (phenoxybenzamine, PBZ); (ii) the reduction with α-receptor agonists ((−)-NA; (+)-NA; dopamine, DA) in the presence of cocaine (1×10^{-4} M for the artery and 5×10^{-5} M for the atria) to prevent displacement of (^3H)-NA from the transmitter pool by exogenous amine; (iii) the effect of a pulse of stimulation on succeeding pulses, in which the S-I efflux with a single pulse was compared to S-I efflux/pulse with a train of pulses.

GUINEA-PIG ATRIA

The S-I efflux evoked by stimulation at 5 Hz for 30 sec was increased by PBZ (1×10^{-7} to 1×10^{-5} M) in a concentration-dependent manner: with 1×10^{-5} M PBZ, the mean increase was $792 \pm 140\%$ of that in control periods. Exogenous

* This research was supported by grants awarded by the National Health and Medical Research Council and by the National Heart Foundation of Australia.

579

(−)-NA (5×10^{-7} to 5×10^{-5} M) reduced the S-I efflux in a concentration-dependent manner: with 5×10^{-7} M NA the mean decrease was $36 \pm 9\%$. The weaker agonists, (+)-NA and DA, had no effect at 5×10^{-7} M.

The mean S-I efflux with a single pulse was 1984 ± 416 d/min, being 73% above the resting efflux of 2714 ± 60 into the same volume of bathing solution in the same period of time. The S-I efflux/pulse fell with increasing number of pulses (2, 4, 8 and 16 at 1 Hz), that with 16 pulses being $14 \pm 2\cdot5\%$ of that with a single pulse. In the presence of PBZ (1×10^{-5} M) the S-I efflux/pulse did not fall with increasing numbers of pulses to the same extent, and with 16 pulses the S-I efflux/pulse was $51 \pm 7\%$ of that with a single pulse. PBZ has no effect on the S-I efflux with one pulse. Cocaine (1×10^{-4} M) had little effect on S-I efflux/pulse with one or multiple pulses, so the effect of PBZ was not due to blockade of neuronal re-uptake of (^{3}H)-NA. These observations support the concept that transmitter release from the adrenergic neurones in the guinea-pig atria is subject to pulse-to-pulse modulation by an inhibitory feedback mechanism exerted through α-receptors.

RABBIT EAR ARTERIES

The S-I efflux evoked by stimulation at 5 Hz for 30 sec was increased by 1×10^{-5} M PBZ to a lesser extent than in atria: with 1×10^{-5} M PBZ the increase in S-I efflux was $274 \pm 13\%$. The decrease with 5×10^{-7} M exogenous (−)-NA was similar to that with atria and (+)-NA had no effect. It has recently been found that DA (5×10^{-7} M) decreases S-I efflux by about 40 per cent in the artery but not the atrial preparation (McCulloch, Rand and Story, 1973).

In striking contrast to the findings with atria, the S-I efflux/pulse did not fall with increasing numbers of pulses (2, 4, 8, 16, 64 and 128 at 1 Hz) below that with a single pulse: in fact, it was increased in the range of 2–16 pulses. The finding suggests that transmitter NA was not accumulating in sufficient concentration to exert an inhibitory effect on subsequent release with the particular regime of stimulation. Further experiments were carried out, therefore, with stimulation for 200 sec at 0·5, 1, 2, 5, 10, 20 and 50 Hz. The S-I efflux/pulse rose to a peak of 77 ± 10 d/min at 5 Hz and decreased with increasing frequencies to $30\cdot5 \pm 4\cdot6$ d/min at 20 Hz and $6\cdot9 \pm 0\cdot98$ d/min at 50 Hz. A possible explanation is that transmitter NA accumulates in sufficient concentration to exert its inhibitory presynaptic effect on further release only when the stimulation frequency is above 5 Hz; however, other mechanisms (e.g., failure of impulse conduction) may also account for decreased transmitter release with high frequencies of stimulation.

CONCLUSIONS

Pulse-to-pulse modulation of transmitter release through the prejunctional α-receptor inhibitory mechanism is manifested at adrenergic neuro-effector junctions in a guinea-pig atrial preparation with transmitter noradrenaline released by a single pulse; however, in the rabbit ear artery preparation, it does not operate except possibly with trains of pulses at a frequency of more than 5 Hz.

REFERENCES

Allen G. S., Rand M. J. and Story D. F. (1973) *Cardiovascular Res.* **7**, 423–428.
Enero M. A., Langer S. Z., Rothlin R. P. and Stefano F. J. E. (1972) *Br. J. Pharmac.* **44**, 672–688.
Hotta Y. (1969) *Agents and Actions* **1**, 13–21.

HÄGGENDAL J. (1970) In: *New Aspects of Storage and Release of Catecholamines.* (SCHÜMANN
   H. J. and KRONEBERG G., Eds.) pp. 100–109. Springer-Verlag, Berlin.
KIRPEKAR S. M. and PUIG M. (1971) *Br. J. Pharmac.* **43**, 359–369.
MCCULLOCH M. W., RAND M. J. and STORY D. F. (1972) *Br. J. Pharmac.* **46**, 523–524P.
MCCULLOCH M. W., RAND M. J. and STORY D. F. (1973) *Br. J. Pharmac.* In press.

Frontiers in Catecholamine Research 1973, pp. 583 to 587. Pergamon Press. Printed in Great Britain.

ASPECTS ON PROSTAGLANDIN AND α-RECEPTOR MEDIATED CONTROL OF TRANSMITTER RELEASE FROM ADRENERGIC NERVES

PER HEDQVIST

Department of Physiology I, Karolinska Institute, Stockholm, Sweden

INTRODUCTION

DURING recent years much interest has focused on mechanisms controlling the release of transmitter from adrenergic nerves. Although ultimately regulated by the central nervous system it has become increasingly evident that the terminal network of peripheral neurons is susceptible to control by local factors operated to modulate the release of noradrenaline (NA). Based on the almost general presence of prostaglandins E (PGE) in mammalian tissues, their availability for increased formation and release by sympathetic nerve activity, and their capacity, even in oligodynamic concentrations, to inhibit the release of transmitter from adrenergic nerve endings, they have been attributed a control function at peripheral neuroeffector junctions (HEDQVIST, 1970). α-adrenoceptors, possibly located at both the nerve endings and at the effector cells, have been postulated to exert a similar control function (HÄGGENDAL, 1970).

This paper describes some actions of PGs and α-receptors on adrenergic neuroeffector junctions, suggesting that they at least in part represent 2 independent means for control of transmitter release.

PROSTAGLANDIN ACTION ON TRANSMITTER RELEASE FROM ADRENERGIC NERVES

The NA stores of the superfused guinea pig vas deferens can be labelled with ^{3}H-NA, and the release of radioactivity can be used as an appropriate indicator of the release of endogenous NA. (HEDQVIST, 1973a). Addition of PGE_2 to the medium superfusing the vas deferens inhibits the release of ^{3}H-NA in response to postganglionic nerve stimulation. (Fig. 1). The inhibition increases with the dose of PGE_2, and is quickly reversed when the compound is omitted. When the number of pulses given are kept constant the inhibitory effect of PGE_2 decreases with increasing stimulation frequency. These observations indicate that PGE_2 in very low concentrations inhibits transmitter release from the vas deferens, particularly at a low stimulation frequency. The results are also in harmony with closely similar effects of PGE_1 and PGE_2 in the cat spleen and rabbit heart (HEDQVIST, 1970; WENNMALM, 1971).

Concerning the mechanism by which PGEs act prejunctionally on sympathetic neuroeffector transmission there is good reason to believe that they do not alter physical or chemical inactivation of the transmitter. Nor do they seem to affect NA biosynthesis or to act directly on NA storage particles. On the other hand, an interaction between PGE_2 and Ca^{2+} has been demonstrated in the cat spleen (HEDQVIST, 1970). In this tissue, increasing the calcium concentration in the perfusion medium counteracts the inhibitory effect of PGE_2 on transmitter release. On the other hand, the NA releasing effect of tyramine, which is a calcium independent process (BURN

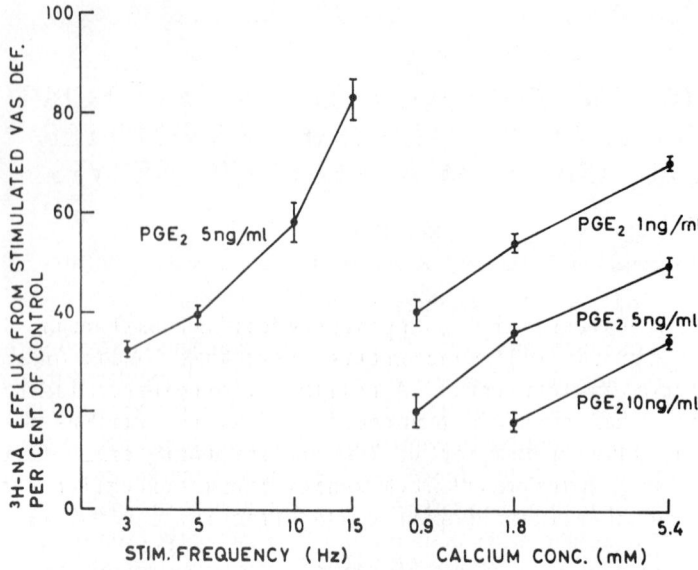

FIG. 1.—Inhibitory effect of PGE_2 on ^3H-NA efflux from superfused guinea pig vas deferens, transmurally stimulated (450 pulses, 1 msec pulse duration, supramaximal voltage). To the left: calcium concentration in superfusion medium 1·8 mM, stimulation frequency 3–15 Hz. To the right: calcium concentration in superfusion medium 0·9–5·4 mM, stimulation frequency 5 Hz. Values presented in per cent of corresponding control stimulation without PGE_2 and given as means ± S.E.M., $n = 4$–6.

and GIBBONS, 1965), is not affected by PGE_2. Also in the guinea pig vas deferens, an increased calcium concentration counteracts the inhibitory effect of PGE_1 and PGE_2 on transmitter release (HEDQVIST, 1973b) (Fig. 1). These observations are particularly interesting in view of the NA release process being critically dependent upon Ca^{2+} (cf. HUBBARD, 1970), and implies that PGEs restrict stimulated release of NA by inhibiting influx of Ca^{2+} into the neuron.

In this context it is worth noticing that the PGEs have a depolarising effect on cell membranes and that therefore the presumed decreased influx of Ca^{2+} in the presence of PGEs may be secondary to a reduced amplitude of the nerve action potential. Support for this view is given also by the inability of PGEs to affect the Ca^{2+}-dependent hormone secretion from the adrenal medulla (MIELE, 1969). In this tissue the resting membrane potential is rather low and stimulation does not lead to reversal and overshoot but only to a minimal local depolarisation (DOUGLAS et al., 1967).

In the cat spleen, the rabbit heart and the guinea pig vas deferens, the doses of PGEs needed to significantly depress NA release seem to be of the same order of magnitude as those that can be released from respective organ in response to nerve activity (cf. HEDQVIST, 1973a). Moreover, when local PG formation in these tissues is blocked by eicosatetraynoic acid (ETA) or indomethacin, the release of transmitter in response to nerve stimulation is consistently increased (HEDQVIST et al., 1971; WENNMALM, 1971; CHANH et al., 1972; FREDHOLM and HEDQVIST, 1973; HEDQVIST, 1973a). Indomethacin has also been shown to increase urinary excretion of NA in rats (STJÄRNE, 1972; JUNSTAD and WENNMALM, 1972). Therefore, although by no means conclusively demonstrated, ample experimental evidence suggests that locally formed PGEs represent a significant mechanism for control of transmitter release from adrenergic nerves.

In sympathetic neuromuscular junctions the effector cells probably represent an important source for the PG overflowing from stimulated tissues. Thus in the spleen, nerve stimulation and injection of adrenaline cause contraction of the organ and release of PGs, while the efflux of PGs is blocked after denervation or α-receptor blockade, when no contraction is visible (DAVIES *et al.*, 1968; GILMORE *et al.*, 1968). However, these observations do not exclude a contribution from the nerve terminals. In particular, splenic nerves contain a factor tentatively identified as PGE_2 since it behaves identically with PGE_2 when subjected to silicic acid column and thin layer chromatography and it cannot be distinguished from PGE_2 by quantitative parallel bioassay. It is therefore reasonable to assume that both the nerve endings and the effector cells can manufacture PGEs operated to affect local processes in the adrenergic neuroeffector junction.

α-RECEPTOR ACTIVATION AS A MEAN FOR CONTROL OF TRANSMITTER RELEASE

The reactivity of the effector cell has been suggested to control transmitter release from sympathetic nerves (HÄGGENDAL, 1970). Considerable support for this view is given by observations on the nerve stimulated mouse vas deferens and rabbit heart (FARNEBO and MALMFORS, 1971; STARKE, 1972). In these tissues NA overflow is increased after administration of α-receptor blockers while decreased in the presence of α-receptor stimulating agents. α-receptor blockers increase transmitter overflow also in tissues in which the proportion of α-receptors in the effector organ is very small or absent, and in doses which do not affect the neuronal uptake mechanism (WENNMALM, 1971, ENERO *et al.*, 1972). Therefore α-receptors of the effector cells and the neuronal uptake mechanism need not be primarily involved in the promoting action on transmitter release by α-receptor blockers. As a consequence a mechanism for control of transmitter release has been envisaged by postulating the presence of α-receptors on the nerve endings (FARNEBO and MALMFORS, 1971; ENERO *et al.*, 1972). Part of the reduced NA release after activation of these receptors could be explained as the result of increased local formation of PGs since PG formation and efflux is increased by α-receptor activation and abolished after α-receptor blockade (DAVIES *et al.*, 1968, GILMORE *et al.*, 1968).

Do α-receptors in part represent a discrete and not PG-mediated mean for control of transmitter release from adrenergic nerves? This aspect has been considered in recent experiments on the superfused guinea pig vas deferens preloaded with ^{3}H-NA (HEDQVIST, 1973c, d). Uptake 1 and 2 were inhibited by desmethylimipramine and normetanephrine ($0\cdot1$–$0\cdot4$ μg/ml) and local PG formation by ETA or indomethacin (10 μg/ml). With these precautions phentolamine, hydergine, and phenoxybenzamine (100 ng/ml) were still effective and significantly increased ^{3}H-NA overflow in response to nerve stimulation. With the same technique, increasing doses of NA and methoxamine progressively diminished the overflow of ^{3}H-NA. These observations seem to indicate that α-receptors, apart from acting as trigger mechanism for PG formation, presumably represent a discrete mechanism for control of transmitter release from the neurons.

HYPOTHETICAL MODEL FOR PG AND α-RECEPTOR MEDIATED CONTROL OF TRANSMITTER RELEASE

The following model may be advanced (Fig. 2). The nerve action potential causes an inward movement of membrane bound Ca^{2+}, which in turn promotes focal and probably quantal release of NA. Subsequent binding of the transmitter to the effector cell membrane triggers increased synthesis of PG of which at least a fraction escapes metabolic degradation and leaves the effector cell in intact form. This PG should have at least the same opportunity as exogenously administered PG to reach the nerve terminal membrane and inhibit further release of transmitter. It is assumed

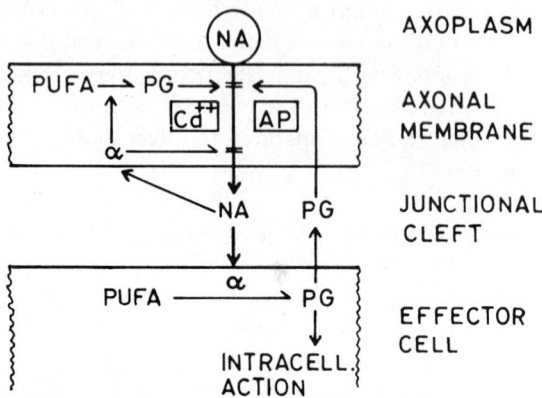

FIG. 2.—Hypothetic model for PG and α-receptor mediated control of NA release from adrenergic nerves. For explanation, see text.

that PG inhibits transmitter release by restricting Ca^{2+} influx into the neuron, and that this may be achieved by a direct action on membrane bound calcium or that calcium influx is reduced secondary to a diminished amplitude of the nerve action potential.

Also NA attachment to presumed α-receptors at the nerve terminal membrane may be of importance as trigger mechanism for PG formation. In addition, when activated, these receptors *per se* seem capable to inhibit further release of transmitter. Presently it is not known how presynaptic α-receptors attack the process of excitation secretion coupling, and one can only speculate if the target is the same as for PG.

Acknowledgements—Supported by grants from the Swedish Medical Research Council, Project No. 04X-4342, and from the Medical Faculty, Karolinska Institute.

REFERENCES

BURN J. H. and GIBBONS W. R. (1965) *J. Physiol. (Lond.)* **181**, 214–223.
CHANH P. H., JUNSTAD M. and WENNMALM Å. (1972) *Acta physiol. scand.* **86**, 563–567.
DAVIES B. N., HORTON E. W. and WITHRINGTON P. G. (1968) *Brit. J. Pharmacol.* **32**, 127–135.
DOUGLAS W. W., KANNO T. and SAMPSON S. R. (1967) *J. Physiol. (Lond.)* **188**, 107–120.
ENERO M. A., LANGER S. Z., ROTHLIN R. P. and STEFANO F. J. E. (1972) *Brit. J. Pharmacol.* **44**, 672–688.
FARNEBO L-O. and MALMFORS T. (1971) *Acta physiol. scand.* (Suppl.) **371**, 1–18.
FREDHOLM B. and HEDQVIST P. (1973) *Acta physiol. scand.* **87**, 570–572.
GILMORE N., VANE J. R. and WYLLIE J. H. (1968) *Nature (Lond.)* **218**, 1135–1140.
HÄGGENDAL J. (1970) *Bayer-Symposium II*, pp. 100–109, Springer-Verlag, Berlin.
HEDQVIST P. (1970) *Acta physiol. scand.* **79**, (Suppl. 345), 1–40.
HEDQVIST P. (1973a) *Advances in the Biosciences* **9**, 461–473.
HEDQVIST P. (1973b) *Acta physiol. scand.* In press.

HEDQVIST P. (1973c) *Acta physiol. scand.* **87,** 42A–43A.
HEDQVIST P. (1973d) *Acta physiol. scand.* In press.
HEDQVIST P., STJÄRNE L. and WENNMALM Å. (1971) *Acta physiol. scand.* **83,** 430–432.
HUBBARD J. I. (1970) *Progr. Biophys. and molecular biology* **21,** 33–124.
JUNSTAD M. and WENNMALM Å. (1972) *Acta physiol. scand.* **85,** 573–578.
MIELE E. (1969) In: *Prostaglandins, peptides and amines.* (MANTEGAZZA P. and HORTON E. W., Eds.) pp. 85–93. Academic Press, New York.
STARKE K. (1972) *Naunyn-Schmiedeberg's Arch. Pharmacol.* **274,** 18–45.
STJÄRNE L. (1972) *Acta physiol. scand.* **86,** 388–397.
WENNMALM Å. (1971) *Acta physiol. scand.* (Suppl. 365), 1–36.

Frontiers in Catecholamine Research 1973, pp. 589 to 593. Pergamon Press. Printed in Great Britain.

CATECHOLAMINE RELEASE AND RECEPTORS IN BRAIN SLICES

LARS-OVE FARNEBO and BERTIL HAMBERGER

Department of Histology, Karolinska Institutet, S-104 01 Stockholm, Sweden

MECHANISMS for release and release regulation of catecholamines are still subject for considerable speculation. In peripheral adrenergic nerves the occurrence of noradrenaline (NA) release has been known for several years EULER, 1946). During the last 5 years evidence has accumulated that the release of NA per nerve impulse is not constant (HÄGGENDAL, 1970) and can be influenced pharmacologically in numerous ways. It has been shown that α-receptor blocking drugs increase NA release while α-receptor stimulant drugs inhibit NA release (FARNEBO and HAMBERGER, 1970a, 1971a; HÄGGENDAL, 1970; STARKE, 1971; 1972; ENERO et al., 1972 and this symposium).

In the brain, demonstration and quantitation of catecholamine release from nerve terminals is technically more difficult. Field stimulation of isolated tissue proved to be a good model for studies on monoamine release in the brain (BALDESSARINI and KOPIN, 1967, FARNEBO and HAMBERGER 1971b). In these studies it is of great importance to choose brain regions where only one type of monoamine nerve terminal is studied at a time. The most widely used areas are cerebral cortex for NA and 5-hydroxytryptamine nerve terminals and the neostriatal region for dopamine (DA) nerve terminals. The tissue slices are with the present technique incubated with radioactively labelled monoamine in vitro to allow the amine to be taken up and stored in the storage granules in the nerve terminals. Unspecific labelling is kept low by using low concentrations of the radioactive monoamine in the incubation medium. The tissue is after incubation superfused with a Krebs–Ringer bicarbonate buffer and stimulated by an electrical field which is generated between two platinum electrodes. The efflux of radioactivity caused by stimulation can be determined (for further details, see FARNEBO and HAMBERGER, 1971b). Field stimulation performed in this way is known to induce transmitter release which is due to depolarisation of the nerve cell membrane. This release is calcium dependent and is very similar to direct nerve stimulation (BALDESSARINI and KOPIN, 1967, FARNEBO and HAMBERGER, 1970b, FARNEBO, 1971).

The pharmacological results obtained with stimulation of brain slices are summarised in Table 1. This technique to stimulate brain slices is identical to that used on peripheral tissues (FARNEBO and HAMBERGER, 1971a). Thus, comparisons between adrenergic nerve terminals and the central monoamine nerve terminals can be performed.

A general characteristic for all three central monoamine nerve terminals is the effect of receptor blocking and stimulating drugs on transmitter release. Receptor blocking drugs like phentolamine, phenoxybenzamine and chlorpromazine increase ^{3}H-NA release while pimozide and chlorpromazine increase ^{3}H-DA release. The receptor-stimulant drugs clonidine and FLA137 decrease ^{3}H-NA release while apomorphine and LSD-25 decrease the release of ^{3}H-DA and ^{3}H-5-hydroxytryptamine

TABLE 1. EFFECT OF DRUGS ON FIELD STIMULATION-INDUCED OVERFLOW OF
RADIOACTIVITY FROM BRAIN SLICES PREINCUBATED WITH ^{3}H-MONOAMINE

Monoamine	Drug	M	Stimulation-induced overflow	n	
Noradrenaline	α-*Receptor stimulating*				
	Clonidine	10^{-8}	83 ± 5·3	6 x	b
		10^{-7}	77 ± 7·0	8 x	a
		10^{-6}	97 ± 9·2	9 NS	a
	FLA 137	10^{-8}	76 ± 7	9 xx	c
		10^{-7}	75 ± 5	12 xxx	c
	α-*Receptor blocking*				
	Phentolamine	10^{-7}	126 ± 8·0	4 NS	a
		10^{-6}	171 ± 14·1	11 xxx	a
		10^{-5}	158 ± 12·5	8 xxx	a
	Phenoxybenzamine	10^{-7}	131 ± 6·9	6 x	a
		10^{-6}	189 ± 11·7	8 xxx	a
		10^{-5}	180 ± 16·2	6 xxx	a
	Chlorpromazine	10^{-8}	104 ± 17·3	4 NS	a
		10^{-7}	117 ± 4·2	4 NS	a
		10^{-6}	142 ± 10·4	12 xxx	a
		10^{-5}	142 ± 9·2	8 xxx	a
	β-*Receptor stimulating*				
	Isoprenaline	10^{-7}	106 ± 6·0	4 NS	b
		10^{-6}	100 ± 3·0	4 NS	b
		10^{-5}	120 ± 5·1	4 x	b
	β-*Receptor blocking*				
	Propranolol	10^{-7}	96 ± 1·0	4 NS	b
		10^{-6}	100 ± 6·3	5 NS	b
		10^{-5}	86 ± 6·0	4 NS	b
	Inhibitor of NA uptake				
	Desipramine	10^{-8}	128 ± 8·8	8 x	a
		10^{-7}	121 ± 7·7	7 NS	a
		10^{-6}	144 ± 13·0	7 xxx	a
		10^{-5}	149 ± 7·7	4 xxx	a
	Barbiturates				
	Pentobarbitone	10^{-6}	101 ± 5·4	6 NS	d
		10^{-5}	106 ± 6·1	6 NS	d
		10^{-4}	94 ± 6·4	6 NS	d
	Phenobarbitone	10^{-7}	96 ± 4·1	4 NS	d
		10^{-6}	89 ± 12·5	6 NS	d
		10^{-5}	102 ± 2·9	6 NS	d
		10^{-4}	110 ± 6·7	4 NS	d
	Minor tranquilliser				
	Chlordiazepoxide	10^{-7}	96 ± 6·2	10 NS	d
		10^{-6}	97 ± 4·7	16 NS	d
		10^{-5}	97 ± 4·7	22 NS	d
		10^{-4}	91 ± 5·9	10 NS	d
	Prostaglandin				
	Prostaglandin E₂	3×10^{-6}	81 ± 4·8	10 x	e
Dopamine	*DA Receptor stimulating*				
	Apomorphine	10^{-8}	87 ± 8·0	8 NS	a
		10^{-7}	77 ± 10·3	12 x	a
		10^{-6}	74 ± 14·9	4 NS	a
		10^{-5}	86 ± 11·5	8 NS	a
	DA Receptor blocking				
	Pimozide	10^{-7}	83 ± 11·5	4 NS	a
		10^{-6}	140 ± 14·9	8 xx	a
		10^{-5}	132 ± 11·5	9 xx	a

TABLE 1. (*continued*)

Monoamine	Drug	M	Stimulation-induced overflow	n		
Dopamine	Chlorpromazine	10^{-7}	$146 \pm 13\cdot8$	12	xxx	a
		10^{-6}	$152 \pm 14\cdot9$	12	xxx	a
	α-Receptor blocking					
	Phentolamine	10^{-6}	$94 \pm 17\cdot2$	4	NS	a
		10^{-5}	$82 \pm 24\cdot1$	4	NS	a
	Inhibitor of DA uptake					
	Cocaine	10^{-6}	$84 \pm 18\cdot4$	4	NS	a
		10^{-5}	$161 \pm 18\cdot4$	10	xxx	a
		10^{-4}	$143 \pm 20\cdot7$	8	xx	a
	Prostaglandin					
	Prostaglandin E_2	3×10^{-6}	$77 \pm 7\cdot1$	16	x	e
5-Hydroxy-tryptamine	*5-HT Receptor stimulating* LSD-25	10^{-8}	$91 \pm 14\cdot0$	4	NS	a
		10^{-7}	$91 \pm 11\cdot6$	8	NS	a
		10^{-6}	$53 \pm 14\cdot0$	4	xx	a
		10^{-5}	$28 \pm 11\cdot6$	4	xxx	a
	Ergocornine	10^{-7}	$58 \pm 5\cdot6$	12	xxx	c
		10^{-6}	$56 \pm 6\cdot4$	12	xxx	c
	α-Receptor stimulating Clonidine	10^{-7}	$106 \pm 28\cdot1$	6	NS	b
	α-Receptor blocking Phentolamine	10^{-6}	$84 \pm 17\cdot4$	4	NS	a
		10^{-5}	$116 \pm 17\cdot2$	4	NS	a
	Chlorpromazine	10^{-7}	$98 \pm 15\cdot3$	4	NS	a
		10^{-6}	$91 \pm 4\cdot7$	8	NS	a
	Inhibitor of 5-HT uptake Chlorimipramine	10^{-7}	$112 \pm 7\cdot9$	9	NS	a
		10^{-6}	$165 \pm 12\cdot8$	8	xxx	a

Slices of rat cerebral cortex were preincubated with ³H-NA or ³H-5-hydroxytryptamine and slices of rat neostriatum were preincubated with ³H-DA. The slices were then superfused with buffer containing the drug to be tested and stimulated by an electrical field (10 Hz) for 2 min. (see Fig. 1). The stimulation-induced overflow of radioactivity was determined. The results are expressed as per cent of the appropriate drug-free control.

x $P < 0\cdot05$, xx $P < 0\cdot01$, xxx $P < 0\cdot001$ compared with the control.
a Results from FARNEBO and HAMBERGER, 1971
b FARNEBO and HAMBERGER, unpublished
c FARNEBO, FUXE and HAMBERGER, unpublished
d LIDBRINK and FARNEBO, 1973
e BERGSTRÖM, FARNEBO and FUXE, 1973.

respectively. These results are in full agreement with what has been found in adrenergic nerves and speak in favour of a negative feed-back mechanism for release-regulation in monoamine nerve terminals also in the brain.

Another general characteristic is the increase of transmitter overflow following blockade of the uptake mechanism obtained with desipramine, cocaine and chlorimipramine. However, the increase is not so pronounced as could have been expected, considering that the drugs used are very potent blockers of transmitter uptake. This is probably due to the fact that the released transmitter causes an increased receptor activation, leading to a decreased release (HÄGGENDAL, 1970, FARNEBO and HAMBERGER, 1971a, b; FARNEBO and MALMFORS, 1971).

The β-receptor blocking drug propranolol and the stimulant drug isoprenaline were without marked effects on NA release although β-receptors could be expected

CEREBRAL CORTEX (³H-NA)

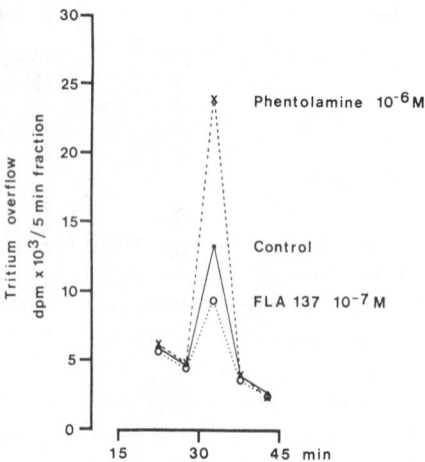

FIG. 1.—Tritium overflow during superfusion and stimulation for 2 min of cerebral cortex slices preincubated with ³H-NA. Effect of the α-receptor blocking drug phentolamine and the α-receptor stimulant drug FLA 137 is demonstrated.

to exist in the cerebral cortex. These findings are, however, in good agreement with results on isolated atria, where β-receptor blockade and stimulation did not seem to affect NA release (WERNER et al., 1971, unpublished). On the other hand NA release from adrenergic nerves in isolated atria was regulated by α-receptor-active drugs just as in other NA nerve terminals. This would indicate that release regulation is a property of NA nerve terminals and not dependent on the receptor tissue.

To increase the reliability of the present results control experiments were also made. Thus, phentolamine did not influence DA release while chlorpromazine, clonidine and phentolamine were without any marked effects on 5-hydroxytryptamine release. These data give support to the view that the results obtained with these drugs are not unspecific.

Drugs known to influence monoamine receptors are also known to influence synthesis and turnover of monoamines. Thus, receptor stimulant agents are claimed to decrease NA, DA and 5-hydroxytryptamine turnover while receptor blocking drugs induce an increase of catecholamine turnover (CARLSSON and LINDQVIST, 1963; ANDÉN et al., 1970; NYBÄCK et al., 1970; BESSON et al., 1971). This was believed to be related to changes in nerve impulse flow but the present results indicate that the effects obtained with these drugs may at least partly be due to a local feed-back in the synaptic area (FARNEBO and HAMBERGER, 1971a). This is further supported by experiments demonstrating that synthesis changes of neostriatal DA occur also after acute denervation (KEHR et al., 1972).

Phenobarbitone, pentobarbital and chlordiazepoxide do not influence NA release in isolated brain slices (LIDBRINK and FARNEBO, 1973) although these drugs can prevent stress-induced increase of NA turnover in the cerebral cortex (CORRODI et al., 1971). These data indicate that there are drugs affecting turnover which do not affect release of monoamines and thus ought to influence nerve impulse flow.

Experiments on release regulation in several laboratories with various techniques seem to indicate that there in fact exists a local regulation of monoamine release which can be demonstrated with receptor blocking and stimulating agents. This mechanism can be found in peripheral adrenergic nerves as well as in central NA, DA and 5-hydroxytryptamine nerve terminals, and for each type of monoamine neuron it seems to be mediated via receptors sensitive for the transmitter of the neuron.

At present the localisation of these release regulating receptors is not clear. Stimulation of the postsynaptic receptor of the effector cell may induce a transsynaptic regulation of transmitter release (HÄGGENDAL, 1970). Prostaglandins cause an inhibition of NA release and may well be involved in release regulation (HEDQVIST, 1970). However, there is no evidence that prostaglandins are solely responsible for release regulation. Calcium which is well known to be crucial for monamine release has been suggested to be the mediator of release regulation but there is very little experimental evidence to support this view.

Presynaptic α-receptors localised on the adrenergic nerve terminals have also been suggested to regulate NA release (FARNEBO and HAMBERGER, 1971a; KIRPEKAR and PUIG, 1971; STARKE, 1971, 1972; ENERO et al., 1972). The demonstration of receptor-mediated release regulation also in central NA, DA and 5-hydroxytryptamine nerve terminals indicates that ability for release regulation is a characteristic of the mono-amine nerve terminal. It thus seems probable that monoamine nerve terminals have presynaptic receptors for release regulation which are sensitive to the transmitter of the nerve terminal.

Acknowledgements—The present study was supported by grants from the Swedish Medical Research Council (14X-2330). M. Bergvalls Foundation and Karolinska Institutet.

REFERENCES

ANDÉN N.-E., BUTCHER S. G., CORRODI H., FUXE K. and UNGERSTEDT U. (1970) *Europ. J. Pharmacol.* **11**, 303–314.
BALDESSARINI R. J. and KOPIN I. J. (1967) *J. Pharmacol. exp. Ther.* **156**, 31–38.
BERGSTRÖM S., FARNEBO L.-O. and FUXE K. (1973) *Europ. J. Pharmacol.* **21**, 362–368.
BESSON M. J., CHERAMY A. and GLOWINSKI J. (1971) *J. Pharmacol. exp. Ther.* **177**, 196–205.
CARLSSON A. and LINDQVIST M. (1963) *Acta pharmacol. toxicol.* **20**, 140–144.
CORRODI H., FUXE K., LIDBRINK P. and OLSON L. (1971) *Brain Res.* **29**, 1–16.
ENERO M. A., LANGER S. Z., ROTHLIN R. P. and STEFANO F. J. E. (1972) *Brit. J. Pharmacol.* **44**, 672–688.
EULER U. S. v. (1946) *Acta physiol. scand.* **12**, 73–97.
FARNEBO L.-O. and HAMBERGER B. (1970a) *J. Pharm. Pharmacol.* **22**, 855–857.
FARNEBO L.-O. and HAMBERGER B. (1970b) *J. Pharmacol. exp. Ther.* **172**, 332–341.
FARNEBO L.-O. (1971) *Acta physiol. scand.* (Suppl. 371), 19–27.
FARNEBO L.-O. and HAMBERGER B. (1971a) *Brit. J. Pharmacol.* **43**, 97–106.
FARNEBO L.-O. and HAMBERGER B. (1971b) *Acta physiol. scand.* (Suppl. 371), 35–44.
FARNEBO L.-O. and MALMFORS T. (1971) *Acta physiol. scand.* (Suppl. 371), 1–18.
HEDQVIST P. (1970) *Acta physiol. scand.* (Suppl. 345), 1–40.
HÄGGENDAL J. (1970) In: *New Aspects of Storage and Release Mechanisms of Catecholamines.* (SCHÜMANN H. J. and KRONEBERG G., Eds.) Bayer Symposium II, pp. 100–109. Springer-Verlag, Berlin.
KEHR W., CARLSSON A., LINDQVIST M., MAGNUSSON T. and ATACK C. (1972) *J. Pharm. Pharmacol.* **24**, 744–747.
KIRPEKAR S. M. and PUIG M. (1971) *Brit. J. Pharmacol.* **43**, 359–369.
LIDBRINK P. and FARNEBO L.-O. (1973) *Neuropharmacology*, in press.
NYBÄCK H., SCHUBERT J. and SEDVALL G. (1970) *J. Pharm. Pharmacol.* **22**, 622–624.
STARKE K. (1971) *Naturwissenschaften* **58**, 420.
STARKE K. (1972) *Naunyn-Schmiedebergs Arch. Pharmakol.* **275**, 11–23.
WERNER U., WAGNER J. and SCHÜMANN H. J. (1971) *Arch. Pharmakol.* **268**, 102–113.

Frontiers in Catecholamine Research 1973, pp. 595 to 600. Pergamon Press. Printed in Great Britain.

RECEPTOR REGULATION STUDIES: SIGNIFICANCE OF CENTRAL CATECHOLAMINE RECEPTORS IN CARDIOVASCULAR CONTROL

MATTS HENNING

Department of Pharmacology, University of Göteborg, S-400 33 Göteborg 33, Sweden

IN THE central as well as in the peripheral nervous system catecholamine (CA) receptors regulate the function in postsynaptic structures. In recent years considerable evidence has accumulated indicating that receptor mechanisms are also involved in the regulation of several presynaptic processes in CA neurons; e.g. synthesis and release of the transmitter.

No attempt will be made to cover the broad field suggested by the caption; this presentation will be restricted to (1) some aspects of postsynaptic CA receptor mechanisms with special regard to the significance of central CA receptors in cardiovascular control; (2) a brief discussion on the relationship between CA receptor mechanisms and CA turnover with regard to the mode of action of the antihypertensive drug, 2-(2,6-dichlorophenylamino)-2-imidazoline (clonidine).

In his comprehensive review in the First Symposium of Catecholamines, ROTHBALLER (1959) summarises a number of early results of blood pressure recordings after local applications of CA in the central nervous system which, retrospectively may be taken as indicative of central CA receptor regulation of cardiovascular function. The interpretation of such studies was greatly facilitated when the central synaptic transmitter function of CA became evident. Of considerable significance was the observation that the blood pressure lowering drug, reserpine also affected CA metabolism in the brain. In the course of studies related to the central nervous actions of reserpine, McCUBBIN, KANEKO and PAGE (1960) observed that injection of small amounts of noradrenaline (NA) into the cerebral ventricles reduced mean arterial blood pressure in dogs; significantly, the CA precursor, L-3,4-dihydroxyphenylalanine (L-dopa) was also active and its action was augmented following inhibition of monoamine oxidase (MAO). Subsequent work has indicated some species variation but largely substantiates the results of McCUBBIN et al. (cf. VAN ZWIETEN, 1973). It should be emphasised that the distribution of intracerebroventricularly administered CA may differ from that of endogenous CA (FUXE and UNGERSTEDT, 1966) and local concentrations may attain highly unphysiological levels. Moreover, there is some evidence that CA may have qualitatively different actions on blood pressure at different levels in the brain (PHILIPPU, PRZUNTEK, HEYD and BURGER, 1971).

Our studies on central CA regulation of blood pressure were based on the well-known fact that the CA precursor, L-dopa, in contrast to the CA themselves readily penetrates the blood–brain barrier for CA. Of great significance was the previous observation that selective inhibition of dopa decarboxylase in peripheral tissues allows a dissociation between central and peripheral actions of the L-dopa analogue, L-α-methyl-3,4-dihydroxyphenylalanine (methyldopa; HENNING 1968, 1969a). By

596 M. HENNING

analogy, it was thought possible to differentiate actions of L-dopa in the peripheral sympathetic system from actions in the central nervous system.

When given alone to conscious, normotensive rats, L-dopa markedly elevates mean arterial blood pressure as recorded from an in-dwelling catheter (HENNING and RUBENSON, 1970a, b). There is also an accumulation of dopamine (DA) in sympathetically innervated tissues, a marked increase in the brain DA concentration and a reduction of the NA content of the heart. In animals pretreated with the peripheral decarboxylase inhibitor, α-hydrazino-α-methyl-β-(3,4-dihydroxyphenyl) propionic acid (hydrazinomethyldopa; PORTER, WATSON, TITUS, TOTARO and BYER, 1962), the formation of DA in peripheral tissues is almost completely prevented. In addition, the lowering of heart NA is blocked, indicating that this effect is due to a displacement of NA by DA formed from L-dopa.

The increase in blood pressure evoked by moderate doses of L-dopa is completely blocked after pretreatment with hydrazinomethyldopa. Further, there is a significant fall in blood pressure when the dose of L-dopa is increased after the same pretreatment. The concentration of brain DA is now markedly increased but there is no effect on the brain NA level. This may be related to a displacement of endogenous NA by the large amounts of DA formed. After pretreatment with large doses of the decarboxylase inhibitor, seryl-2,3,4-trihydroxybenzylhydrazine (Ro 4-4602; PLETSCHER and GEY, 1963; BARTHOLINI and PLETSCHER, 1969) the same high dose of L-dopa has no effect on blood pressure. In the doses used, Ro 4-4602 inhibits both central and peripheral decarboxylase activity (see ref. cited above). This pretreatment does not entirely prevent the increase of brain DA induced by L-dopa but the increase is significantly less than after treatment with hydrazinomethyldopa. The hypotensive action of L-dopa after peripheral decarboxylase inhibition is greatly potentiated following pretreatment with nialamide (HENNING, unpublished observations).

The modifications by the two decarboxylase inhibitors of the blood pressure response to L-dopa may be interpreted as follows. The pressor effect of L-dopa is probably mediated through peripheral adrenergic mechanisms. The hypotensive response to L-dopa which is revealed after inhibition of peripheral decarboxylase is most likely due to a central action of DA or NA or both. It appears unlikely that the small amounts of DA still formed after the inhibition of peripheral decarboxylase can account for the hypotensive effect of L-dopa since the intravenous infusion of DA over a wide dose range consistently results in elevation of blood pressure (RUBENSON, 1971a).

Following a severe depletion of central CA by pretreatment with the tyrosine hydroxylase inhibitor, α-methyl-p-tyrosine methyl ester (H 44/68) in combination with large doses of α-methyl-m-tyrosine (MMT), L-dopa fails to produce a hypotensive response in combination with hydrazinomethyldopa (RUBENSON, 1971b). This finding suggests that the metabolites of L-dopa mediate the hypotensive effect indirectly by a displacement of endogenous CA. Since the formation of DA in the brain in these experiments was about the same as in those without H 44/68 and MMT it is unlikely that MMT exerted any significant degree of decarboxylase inhibition in the brain. Further support for an indirect effect of the L-dopa metabolites was provided by the observation that the hypotensive action cannot be reproduced after repeated injections of L-dopa ('tachyphylaxis'; RUBENSON, 1971b). Taken together, these findings indicate that the hypotensive effect of L-dopa after peripheral decarboxylase inhibition is mainly mediated via displacement of central NA by the DA formed from L-dopa.

Pretreatment with the central DA receptor blocking agent, spiroperidol (ANDÉN, BUTCHER, CORRODI, FUXE and UNGERSTEDT, 1970b) is without effect on the hypotensive reaction induced by L-dopa (HENNING and RUBENSON, 1970), suggesting that this is not mediated via central DA receptor mechanisms. The importance of noradrenergic mechanisms is indicated by the finding that L-dopa lacked a hypotensive effect after pretreatment with the DA β-hydroxylase inhibitor, bis-(4-methyl-1-homopiperazinyl-thiocarbonyl)disulphide (FLA 63; FLORVALL and CORRODI, 1970; CORRODI, FUXE, HAMBERGER and LJUNGDAHL, 1970a); as before, L-dopa was given in combination with hydrazinomethyldopa (HENNING and RUBENSON, 1970a, b). This may be related to an impairment by FLA 63 of the synthesis of NA from DA originating from L-dopa. However, since FLA 63 may cause depletion of endogenous NA (CORRODI et al., 1970a) the lack of effect of L-dopa after FLA 63 may alternatively be due to a decreased availability of endogenous brain NA for displacement by the L-dopa metabolites.

As briefly mentioned above, the approach used to study central effects of L-dopa was originally applied in the case of methyldopa. In animals pretreated with an inhibitor of both central and peripheral decarboxylase the hypotensive action of methyldopa is inhibited but it persists after pretreatment with an inhibitor of peripheral decarboxylase only (HENNING 1968, 1969a, b). Recently, DAY, ROACH and WHITING (1972) came to the same conclusion using essentially the same approach. These observations indicate that the hypotensive action of methyldopa is elicited by its decarboxylated metabolites acting in the central nervous system. The relative importance of direct and indirect components in the effect of methyldopa has been assessed in analogy with L-dopa as described above, i.e. after CA depletion by MMT in combination with tyrosine hydroxylase inhibition. In spite of a pronounced depletion of endogenous brain CA, the hypotensive effect of methyldopa is not altered (HENNING and RUBENSON, 1971). Moreover, no 'tachyphylaxis' to repeated injections of methyldopa is observed (RUBENSON, 1971b). This is in contrast to the results obtained for L-dopa and suggests a direct action on the CA receptors of methyldopa metabolites. In order to determine whether methyl-DA or methyl-NA is the active amine metabolite in mediating the hypotensive effect of methyldopa, we have examined this following pretreatment with the DA β-hydroxylase inhibitor, FLA 63. This pretreatment prevents the hypotensive effect of methyldopa as well as the synthesis of methyl-NA while there is a significant increase in the amounts of methyl-DA formed (HENNING and RUBENSON, 1971). This was later confirmed by DAY et al. (1972) with regard to blood pressure effects using a different β-hydroxylase inhibitor. Assuming a predominantly direct action of the amines formed from methyldopa, these findings ascribe to methyl-NA the most important role in mediating the hypotensive action of methyldopa.

In this connection it is pertinent to note that the amine metabolites of L-dopa as well as of methyldopa are effective agonists at spinal NA receptors while metabolites of MMT are ineffective in this respect (ANDÉN, BUTCHER and ENGEL, 1970a; cf. ANDÉN, this Symposium). Interestingly, MMT does not lower blood pressure (HENNING, 1967). The stimulation of spinal NA receptors by methyldopa metabolites is prevented following inhibition of DA β-hydroxylase (Andén et al., 1970a) and the action of L-dopa is indirect, mediated presumably through displacement of NA by DA formed from L-dopa (Andén, Engel and Rubenson, 1972).

The indications of central NA receptors regulating blood pressure have interesting relations to the possible mechanisms underlying the antihypertensive action of clonidine. The administration of this drug evokes a transient elevation of blood pressure followed by a prolonged hypotension which is accompanied by bradycardia and a decreased cardiac output. The initial increase in blood pressure results from a vasoconstrictor action via stimulation of peripheral α-adrenergic receptors (Kobinger and Walland, 1967; Constantine and McShane, 1968). The second, hypotensive phase is most probably due to a central nervous action of clonidine, leading to a decrease in central sympathetic outflow and an increase in parasympathetic activity on the heart (Hoefke and Kobinger, 1966; Kobinger, 1967; Sattler and van Zwieten, 1968; Constantine and McShane, 1968; Schmitt and Schmitt, 1969). In this context it may be mentioned that also L-dopa seems to decrease the directly recorded impulse flow in sympathetic nerves (Schmitt, Schmitt and Fenard, 1972).

A significant contribution to the understanding of the action of clonidine was the observation that this drug increases the flexor reflex in spinal rats in spite of a pronounced NA depletion induced by pretreatment with reserpine and a tyrosine hydroxylase inhibitor; this strongly suggests a direct stimulation of NA receptors in the spinal cord (Andén, Corrodi, Fuxe, Hökfelt, Hökfelt, Rydin and Svensson, 1970c; cf. Andén, this Symposium). Further, clonidine has been found to decelerate the depletion of NA in the central nervous system after tyrosine hydroxylase inhibition (Andén et al., 1970c). This is also true of the NA depletion after inhibition of DA β-hydroxylase (Corrodi, Fuxe, Ljungdahl and Ögren, 1970b). Since there seems to exist feedback mechanisms from CA receptors regulating the turnover of CA (Andén, Corrodi and Fuxe, 1969; Andén, Carlsson and Häggendal, 1969; cf. Andén, this Symposium), these results may be taken as further evidence that clonidine is an agonist at central nervous NA receptors. With regard to the cardiovascular actions of clonidine it is interesting to note that these are blocked by agents known to act as central NA receptor antagonists (Kobinger and Walland, 1967; Schmitt, Schmitt and Fénard, 1971; Haeusler and Finch, 1972). Preliminary reports indicate that this may be the case for the hypotensive effect of L-dopa (Schmitt, Schmitt and Fenard, 1972) as well as of methyldopa (Haeusler and Finch, 1972).

There is an obvious parallellism between the actions of clonidine and those of L-dopa and methyldopa, both with regard to their blood pressure effects and their actions on the spinal cord NA receptors. This invites the speculation that these three compounds may owe their hypotensive action to a common mechanism, i.e. activation of central NA receptors involved in the control of blood pressure. This would lead to a decrease in central sympathetic outflow and hence a decrease in blood pressure. It is tempting to speculate further that such mechanisms could be involved in the antihypertensive action of MAO inhibitors but so far no experimental corroboration of this seems to be available. Concerning the localisation of such NA receptors, decerebration experiments suggest structures in the brain stem or the spinal cord in the case of methyldopa and L-dopa (Henning, Rubenson and Trolin, 1972; cf. Henning and van Zwieten, 1968) as well as clonidine (Shaw, Hunyor and Korner, 1971). Histochemical studies (Dahlström and Fuxe, 1965; Fuxe, 1965) indicate a very dense NA terminal network innervating the brain stem nuclei, n. tractus solitarii and n. dorsalis nervi vagi which are known to be involved in medullary vaso-motor control (SCHERRER, 1966; CRILL and REIS, 1968). The cell bodies of the preganglionic sympathetic nerves in the lateral column of the spinal cord also receive noradrenergic innervation but preliminary results from our laboratory using local applications of 6-hydroxydopamine indicate that spinal NA nerves are probably not involved in the actions of L-dopa and methyldopa.

In conclusion, there is now considerable evidence that CA receptors are involved in central nervous regulation of cardiovascular function. In this review, the interest has mainly been focussed on noradrenergic mechanisms and medullary structures but, as mentioned in the introduction, other levels may also be involved and there are reasons to believe that CA receptors may subserve different functions at different levels of central cardiovascular control.

Acknowledgements—Supported by the Swedish State Medical Research Council (grant no. 14X-2863).

REFERENCES

ANDÉN N. E., BUTCHER S. G. and ENGEL J. (1970a) *J. Pharm. Pharmac.* **22**, 548–550.
ANDÉN N. E., CARLSSON A. and HÄGGENDAL J. (1969) *Ann. Rev. Pharmacol.* **9**, 119–134.
ANDÉN N. E., CORRODI H. and FUXE K. (1969) In: *Metabolism of amines in the brain* (HOOPER G., ed.) pp. 38–47. Macmillan, London.
ANDÉN N. E., ENGEL J. and RUBENSON A. (1972) *Naunyn-Schmiedeberg's Arch. Pharmacol.* **273**, 1–10.
ANDÉN N. E., BUTCHER S.G., CORRODI H., FUXE K. and UNGERSTEDT U. (1970b) *Europ. J. Pharmacol.* **9**, 513–523.
ANDÉN N. E., CORRODI H., FUXE K., HÖKFELT B., HÖKFELT T., RYDIN C. and SVENSSON T. (1970c) *Life Sci.* **9**, 513–523.
BARTHOLINI G. and PLETSCHER A. (1969) *J. Pharm. Pharmacol.* **21**, 323–324.
CONSTANTINE J. W. and MCSHANE W. K. (1968) *Europ. J. Pharmacol.* **4**, 109–123.
CORRODI H., FUXE K., HAMBERGER B. and LJUNGDAHL A. (1970a) *Europ. J. Pharmacol.* **12**, 145–155.
CORRODI H., FUXE K., LJUNGDAHL A. and ÖGREN S. O. (1970b) *Brain Res.* **24**, 451–470.
CRILL W. E. and REIS D. J. (1968) *Am. J. Physiol.* **214**, 269–276.
DAHLSTRÖM A. and FUXE K. (1965) *Acta physiol. scand.* **64**, suppl. 247.
DAY M. D., ROACH A. G. and WHITING R. L. (1972) *Br. J. Pharmacol.* **45**, 168P–169P.
FLORVALL L. and CORRODI H. (1970) *Acta Pharmaceut. suec.* **7**, 7–22.
FUXE K. (1965) *Z. Zellforsch.* **65**, 573–596.
FUXE K. and UNGERSTEDT U. (1966) *Life Sci.* **5**, 1817–1824.
HAEUSLER G. and FINCH L. (1972) *Journ. de Pharmacol.* **3**, 544–545.
HENNING M. (1967) *J. Pharm. Pharmacol.* **19**, 775–779.
HENNING M. (1968) *Br. J. Pharmacol.* **34**, 233P–234P.
HENNING M. (1969a) *Acta pharmacol. et toxicol.* **27**, 135–148.
HENNING M. (1969b) *Acta physiol. scand. suppl.* 322.
HENNING M. and RUBENSON A. (1970a) *J. Pharm. Pharmacol.* **22**, 241–243.
HENNING M. and RUBENSON A. (1970b) *J. Pharm. Pharmacol.* **22**, 553–560.
HENNING M. and RUBENSON A. (1971) *J. Pharm. Pharmacol.* **23**, 407–411.
HENNING M. and VAN ZWIETEN P. A. (1968) *J. Pharm. Pharmacol.* **20**, 407–419.
HENNING M., RUBENSON A. and TROLIN G. (1972) *J. Pharm. Pharmacol.* **24**, 447–451.
HOEFKE W. and KOBINGER W. (1966) *Arzneimittelforsch.* **16**, 1038–1050.
KOBINGER W. (1967) *Arch. exp. Path. Pharmakol.* **258**, 48–58.
KOBINGER W. and WALLAND A. (1967) *Europ. J. Pharmacol.* **2**, 155–162.
MCCUBBIN J. W., KANEKO Y. and PAGE I. H. (1960) *Circulat. Res.* **8**, 849–858.
PHILIPPU A., PRZUNTEK H., HEYD G. and BURGER A. (1971) *Europ. J. Pharmacol.* **15**, 200–208.
PORTER C. C., WATSON L. S., TITUS D. C., TOTARO J. A. and BYER S. S. (1962) *Biochem. Pharmacol.* **11**, 1067–1077.
PLETSCHER A. and GEY K. F. (1963) *Biochem. Pharmacol.* **12**, 223–228.
ROTHBALLER A. B. (1959) *Pharmacol. Rev.* **11**, 494–547.
RUBENSON A. (1971a) *Acta pharmacol. et toxicol.* **29**, 135–144.
RUBENSON A. (1971b) *J. Pharm. Pharmacol.* **23**, 228–230.

SATTLER R. W. and VAN ZWIETEN P. A. (1967) *Europ. J. Pharmacol.* **2**, 9–13.
SCHMITT H. and SCHMITT H. (1969) *Europ. J. Pharmacol.* **6**, 8–12.
SCHMITT H., SCHMITT H. and FENARD S. (1971) *Europ. J. Pharmacol.* **14**, 98–100.
SCHMITT H., SCHMITT H. and FENARD S. (1972) *Europ. J. Pharmacol.* **17**, 293–296.
SHAW J., HUNYOR S. N. and KORNER P. (1971) *Europ. J. Pharmacol.* **14**, 101–111.
SCHERRER H. (1966) *Acta neuroveg.* **29**, 56–74.
ZWIETEN P. A. VAN (1973) *J. Pharm. Pharmacol.* **25**, 89–95.

Frontiers in Catecholamine Research 1973, pp. 601 to 604. Pergamon Press. Printed in Great Britain.

RECEPTOR REGULATION OF DOPAMINE TURNOVER

Henrik Nybäck, Frits-Axel Wiesel and Göran Sedvall

Department of Pharmacology, Karolinska Institutet, Stockholm, Sweden

It is now 10 years since Carlsson and Lindqvist (1963) formulated the hypothesis that antipsychotic drugs, through their receptor blocking property, activate brain catecholamine (CA) neurons by a compensatory feed-back mechanism. Since then several investigators using different techniques have obtained evidence for a regulation of brain CA turnover from central CA receptors. Strong support for the receptor regulation theory has recently been provided by direct measurements of single CA neuron activity following receptor blocking and stimulating drugs (Graham and Aghajanian, 1971). In the present article some experiments undertaken to investigate the receptor regulation of brain dopamine (DA) turnover will be reviewed.

METHODS

The turnover of brain DA was studied by measuring the accumulation and disappearance of DA formed from ^{14}C-labelled tyrosine in rat and mouse brain *in vivo* (Nybäck and Sedvall, 1970). The radioactive precursor was administered by i.v. injection or constant rate i.v. infusion for 20 min. Drugs were given 20–40 min before the start of the ^{14}C-tyrosine infusion or 2 hr after the ^{14}C-tyrosine injection. Animals were killed immediately after the precursor infusion or 2 and 6 hr after the precursor injection. Endogenous and labelled tyrosine and DA were isolated in brain extracts by ion exchange chromatography and determined by liquid scintillation counting.

RESULTS AND DISCUSSION

When ^{14}C-tyrosine was infused i.v. to rats or mice, labelled CA accumulated in the brain. Since some CA was concomitantly released and catabolized during the infusion the accumulation reflects a minimal synthesis rate. When the precursor was administered as a single i.v. injection, labelled CA reached a maximum after about 1 hr. Between 2 and 10 hr after the precursor administration ^{14}C-CA disappeared exponentially. Since there may be some persisting synthesis of DA from remaining ^{14}C-tyrosine during the disappearance, the "true" turnover rate may be somewhat more rapid than indicated by the slope of the disappearance.

In mice pretreated with chlorpromazine (CPZ) in increasing doses, successively more ^{14}C-DA was formed and retained in the brain. If the drug was given during the disappearance of ^{14}C-DA, the rate of disappearance was accelerated. These results indicate that CPZ, in a dose-dependent manner, accelerates the rate of turnover of brain DA (Nybäck and Sedvall, 1970).

In another experiment we studied the effect of a new antipsychotic drug, pimozide, and apomorphine, which is known to stimulate DA receptors in the brain (Ernst and Smelik, 1966, Ernst, 1967). Pimozide accelerated the accumulation and disappearance of ^{14}C-DA (Fig. 1), whereas apomorphine decelerated the turnover of labelled DA, which points to the existence of a sensitive feed-back control of DA turnover regulated from the receptor site.

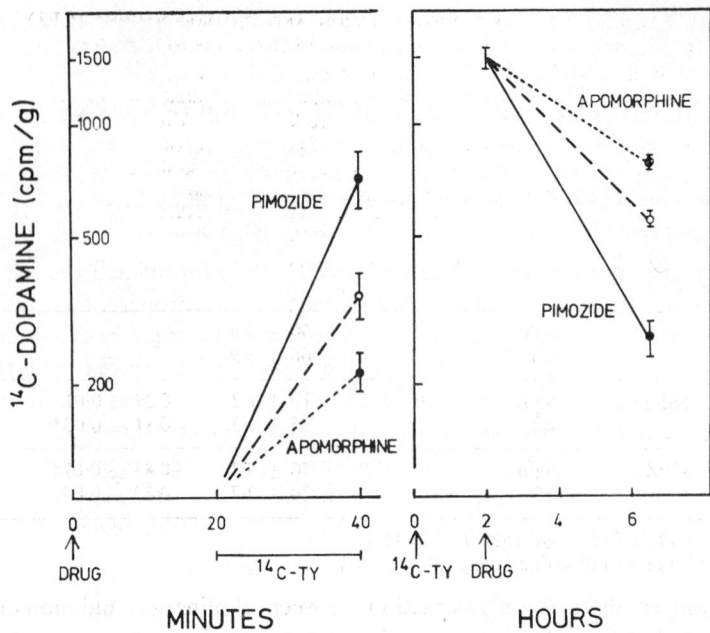

FIG. 1.—Effect of pimozide (————), saline (– – – –) and apomorphine (· · · · ·) on accumulation and disappearance of DA formed from ^{14}C-tyrosine in mouse brain.

In a study on different rat brain regions it was found (NYBÄCK and SEDVALL, 1969) that the stimulatory effect of CPZ on DA turnover was significant only in the striatum. Therefore we assumed that the nerve impulse flow in the nigro-striatal DA pathway was increased by the drug. To investigate this possibility a stereotaxic lesion was made in the left nigro-striatal DA pathway to inhibit the nerve impulse flow on this side. About 18 hr after the lesion, before the DA neurons had started to degenerate and loose their DA content, the influence of CPZ on the disappearance of ^{14}C-DA was studied (Table 1). It was found (NYBÄCK, 1972) that the stimulatory effect of CPZ on DA turnover was antagonized on the lesion side. This indicates that the regulatory process from DA receptors needs intact nerve fibres through the lateral hypothalamus where the lesion was placed.

When studying other psychotropic drugs it was found (NYBÄCK et al., 1968) that all antipsychotic drugs with different molecular structures, phenothiazines, thioxanthenes and butyrophenones accelerated the turnover of brain DA. To further investigate the molecular requirements for the DA receptor blockade we studied the effect of various metabolites of CPZ on accumulation and disappearance of ^{14}C-DA (NYBÄCK and SEDVALL, 1972). We found (Fig. 2) that several of the metabolites, as desmethyl-CPZ, didesmethyl-CPZ, 7-hydroxy-CPZ and CPZ-N-oxide, like the parent compound, accelerated the turnover of brain DA. However, CPZ-sulphoxides and desmethyl-7-hydroxy-CPZ were without effect. Thus sulphoxidation and molecular changes in more than one position seems to diminish the access of the drug to the DA receptors and/or inhibit the interaction of the drug with DA receptors. This finding is of interest in view of the recent finding of SAKALIS et al., (1972) that schizophrenic patients who did not improve on CPZ treatment had higher plasma levels of CPZ-sulphoxide than those who improved on the treatment.

TABLE 1. EFFECT OF A LEFT NIGRAL LESION AND CHLORPROMAZINE (CPZ) ON THE DISAPPEARANCE OF DA FORMED FROM ^{14}C-TYROSINE IN RAT STRIATUM.
(About 18 hr after the lesion ^{14}C-tyrosine was injected i.v., 2 hr later one group of animals was killed and the remainder were injected with saline or CPZ (10 mg/kg). The latter two groups were killed 6 hr after ^{14}C-tyrosine administration. Rate constants (k) for the disappearance of ^{14}C-DA were calculated according to the method of least squares of amines (ln cpm/g) at 2 and 6 hr after precursor administration.)

Treatment		Time after ^{14}C-tyrosine (hr)	^{14}C-tyrosine spec. act. (cpm/μg)	k^{14}C-DA
—	right	2	50 ± 6·5	—
	left		66 ± 6·9	—
Saline	right	6	19 ± 1·2	0·29 ± 0·01
	left		25 ± 4·3	0·11 ± 0·03*
CPZ	right	6	26 ± 5·2	0·43 ± 0·02†
	left		24 ± 2·7	0·10 ± 0·04

* Differs from right side ($P < 0.001$).
† Differs from saline group ($P < 0.001$).

In the striatum there is an interaction between cholinergic and dopaminergic neurons. Several authors have recently found that anticholinergics can counteract the stimulatory effect of antipsychotics on brain DA turnover (O'KEEFFE et al., 1970; ANDÉN and BÉDARD, 1971; BOWERS and ROTH, 1972; CORRODI et al., 1972a). These findings indicate that the feed-back regulation of nigro-striatal DA neurons involves a cholinergic tract. With the ^{14}C-tyrosine method we were unable to confirm this finding (Table 2), even with high doses of atropine. However, atropine counteracted the decrease in DA turnover induced by apomorphine which may indicate that a cholinergic mechanism is involved in the reduction of DA turnover. The new DA

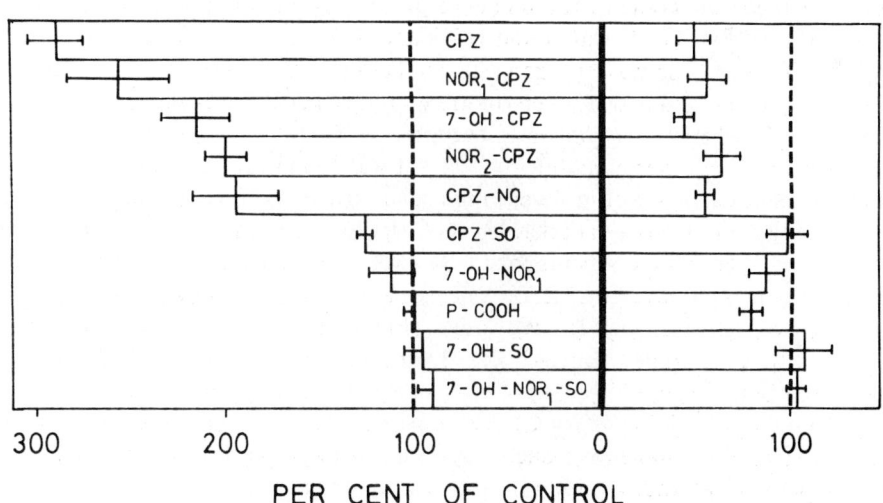

FIG. 2.—Effect of chlorpromazine (CPZ) and some of its metabolites on accumulation (left) and disappearance (right) of DA formed from ^{14}C-tyrosine in mouse brain. Experimental design as in Fig. 1.

TABLE 2. EFFECT OF DRUGS ON ACCUMULATION OF DA FORMED FROM [14] C-TYROSINE IN RAT BRAIN.

(Animals were treated with drugs 40 min before an i.v. infusion of ^{14}C-tyrosine for 20 min. Immediately after the infusion the animals were killed. and the contents of endogenous and labelled tyrosine and DA were determined. Figures represent mean values $\pm$ s.e. of 4–6 animals.)

Drug	Dose (mg/kg)	^{14}C-tyrosine (cpm/μg)	^{14}C-dopamine (cpm/g)
Saline	—	876 ± 86	45 ± 5·6
Chlorpromazine	10	1180 ± 129	75 ± 3·3‡
Atropine	100	1300 ± 220	42 ± 4·8
Atr + CPZ		1250 ± 136	87 ± 8·2†
Apomorphine	10	959 ± 75	29 ± 3·7*
Atr + Apom		1370 ± 137*	45 ± 6·3
ET 495	100	1360 ± 52†	24 ± 2·9*
Atr + ET 495		1280 ± 116*	29 ± 2·6*

Differs from saline group; * $P < 0.05$,   † $P < 0.01$,   ‡ $P < 0.001$.

receptor stimulating agent ET 495 (CORRODI *et al.*, 1972b) reduced the accumulation of ^{14}C-DA. However, this reduction was not significantly antagonized by atropine. Thus there seems to be differences in the sensitivity to atropine antagonism between different receptor blocking and stimulating drugs.

Acknowledgements—The present study was supported by the Swedish Medical Research Council (40X-3560) and the Swedish Delegation for Applied Medical Defence Research (U 59/1972).

REFERENCES

ANDÉN N. -E. and BÉDARD P. (1971) *J. Pharm. Pharmac.* **23**, 460–462.
BOWERS M. B. and ROTH R. H. (1972) *Br. J. Pharmac.* **44**, 301–306.
CARLSSON A. and LINDQVIST M. (1963) *Acta pharmacol. toxicol.* **20**, 140–144.
CORRODI H., FUXE K. and LIDBRINK P. (1972a) *Brain Res.* **43**, 397–416.
CORRODI H., FARNEBO L.-O., FUXE K., HAMBERGER B. and UNGERSTEDT U. (1972b) *Europ. J. Pharmacol.* **20**, 195–204.
ERNST A. M. (1967) *Psychopharmacologia* **10**, 316–323.
ERNST A. M. and SMELIK P. G. (1966) *Experientia* **22**, 837–838.
GRAHAM A. W. and AGHAJANIAN G. K. (1971) *Nature, Lond.* **234**, 100–102.
NYBÄCK H. (1972) *Acta physiol. scand.* **84**, 54–64.
NYBÄCK H., BORZECKI Z. and SEDVALL G. (1968) *Europ. J. Pharmacol.* **4**, 395–403.
NYBÄCK H. and SEDVALL G. (1969) *Europ. J. Pharmacol.* **5**, 245–252.
NYBÄCK H. and SEDVALL G. (1970) *Europ. J. Pharmacol.* **10**, 193–205.
NYBÄCK H. and SEDVALL G. (1972) *Psychopharmacologia (Berl.)* **26**, 155–160.
O'KEEFFE R., SHARMAN D. F. and VOGT M. (1970) *Br. J. Pharmac.* **38**, 287–304.
SAKALIS G., CURRY S. H., MOULD G. P. and LADER M. H. (1972) *Clin. Pharmac. Therap.* **13**, 931–946.

Frontiers in Catecholamine Research 1973, pp. 605 to 611. Pergamon Press. Printed in Great Britain.

REGULATION OF DA SYNTHESIS IN THE NIGRO NEOSTRIATAL SYSTEM

J. Glowinski, Y. Agid, M. J. Besson, A. Chèramy
C. Gauchy and F. Javoy

Groupe NB (INSERM U. 114), Laboratoire de Biologie Moléculaire, Collège de France,
11, place Marcelin Berthelot, Paris 5e, France

INTRODUCTION

Various methods have been used to appreciate changes in catecholamines (CA) synthesis occurring in the CNS in different pharmacological or physiological states. A method generally chosen is to estimate the initial accumulation of labelled dopamine (DA) or norepinephrine (NE) formed from radioactive tyrosine, a conversion index of ^{3}H–CA over the specific activity of tyrosine in tissues can then be calculated. In most cases this technic provides valuable qualitative information but presents some limitations. The specific activity of tyrosine in tissues and in CA neurons are assumed to be similar. As suggested by experiments made on peripheral NE neurons, this may not be the case (Costa et al., 1972). Furthermore, ^{3}H–CA accumulation in tissues underestimates the total formation of amines since part of the newly synthesised CA are rapidly utilised. On the other hand, changes in ^{3}H-transmitter release may also be involved in modifications of ^{3}H–CA initial accumulation in tissues (30–60 min) and may lead in some cases to erroneous conclusions about synthesis.

The initial rate of accumulation of NE or DA in tissues seen after the injection of various MAO inhibitors can also be used to appreciate the synthesis rate of CA in NE and DA terminals. We found with this method that the synthesis rate of NE and DA was respectively 0·6 μg/g/hr and 20–25 μg/g/hr in the cortex (Thierry et al., 1973a) and in the striatum of the rat (Javoy et al., 1972). These high rates of CA synthesis agree with estimations of turnover rates based on the analysis of dynamic properties of the two main compartments in which amines are stored in their respective terminals (Thierry et al., 1973a; Javoy and Glowinski, 1971). The rate of the initial accumulation of dopa after the injection of dopa-decarboxylase inhibitors has recently been used as an index of the first limiting step of CA synthesis (Carlsson et al., 1972). This latter approach as the former presents some inconvenients: 5-hydroxytryptamine metabolism is also affected; moreover, changes in amines or dopa levels induced by MAO or dopa-decarboxylase inhibitors may rapidly (15–30 min respectively) reduced the rate of the first step of CA synthesis (Javoy, Agid, Bouvet, unpublished observations).

In the present report, various regulatory processes involved in the control of DA synthesis in DA terminals of the nigro neostriatal system will be discussed. They have been investigated with a new methodology allowing the measurement of events occurring at the first and rate limiting step of the transmitter synthesis.

THE ^{3}H–H$_2$O *IN VITRO* AND *IN VIVO* METHODS

During the last few years, concomitantly with the three methods described above, we have used another approach to investigate the regulatory processes involved in

CA synthesis. It consists in measuring the accumulation of tritiated water formed during the conversion of L–3·5–^{3}H-tyrosine into ^{3}H-dopa. The ratio of the total quantity of ^{3}H–H$_2$O synthesised over tyrosine specific activity in tissues provides a good index of the first step of CA synthesis since identical quantities of ^{3}H–H$_2$O and ^{3}H-dopa are formed. The estimation of ^{3}H–H$_2$O avoids the use of drugs which could by themselves interfere with CA synthesis. This very sensitive approach presents as other technics some limitations. (1) As already discussed for another method, the specific activity of tyrosine in tissues may not be identical to that found in CA terminals (2) The estimation of ^{3}H–H$_2$O, as well as that of dopa after inhibition of dopa-decarboxylase, may reflect changes in tyrosine hydroxylation in tissues occurring simultaneously in NE and DA terminals. Indeed, structures which were considered to be mainly NE seem to contain numerous DA terminals. This was shown in the rat cortex in our laboratory (Thierry et al., 1973b, see Thierry et al., this symposium). but may also occur in other NE structures such as the spinal cord and the brain stem. Therefore most of our investigations on the regulatory processes of CA synthesis have been made upon the rat striatum and upon the cat caudate nucleus, structures particularly rich in DA terminals and containing relatively few NE terminals.

The determination of ^{3}H–H$_2$O was used in three different situations. (1) *in vitro*, striatal slices are incubated with L–3·5–^{3}H-tyrosine for 15–30 min and the total formation of ^{3}H–H$_2$O formed is estimated (Besson et al., 1971). (2) *in vivo*, the ^{3}H-amino-acid is introduced by a microinjection stereotaxic procedure locally into the caudate nucleus of the rat, ^{3}H–H$_2$O levels in tissues are estimated immediately at the end of the continuous injection which lasts for three minutes (Javoy et al., 1973; Javoy, Agid, Bouvet and Glowinski, in prep.). (3) *in vivo*, in release studies performed on the ventricular surface of the cat caudate nucleus with the help of a cup technique (Besson et al., 1971). During the continuous superfusion of the tissues with L–3·5–^{3}H-tyrosine, ^{3}H–H$_2$O formed diffused in superfusates and is estimated parallely to released ^{3}H–DA in serial collected fractions (Besson et al., 1973a, b). Undoubtedly, absolute amounts of ^{3}H–H$_2$O formed are measured in *in vitro* studies; as ^{3}H–H$_2$O diffused rapidly from its site of formation, estimations made *in vivo* in tissues or superfusates may be slightly lower than the absolute rate of conversion of L-3·5–^{3}H-tyrosine into ^{3}H–dopa.

The specificity of these "^{3}H–H$_2$O methods" has been examined in two ways. Inhibition of CA synthesis induced by an injection of α-methylparatyrosine (α-Mp T) blocked the formation of ^{3}H–H$_2$O parallely to that of ^{3}H–DA both in striatal slices (Besson et al., 1971) and in the two *in vivo* models (Javoy et al., 1973; Besson et al., 1973b). Degeneration of the nigro striatal pathway induced by local injection of 6-OHDA into the substantia nigra was associated with a complete disappearance of ^{3}H–H$_2$O as well as ^{3}H–DA formation from L-3·5–^{3}H-tyrosine in the rat striatum both *in vitro* and *in vivo* (Javoy et al., 1973; Agid, Javoy, Bouvet, in prep.). Moreover, the ^{3}H–H$_2$O methods gave similar information than other methods available in pharmacological experiments classically known to affect DA synthesis.

RAPID INHIBITING PROCESSES OF DA SYNTHESIS

Changes in DA synthesis do not necessarily parallel those of DA release as assumed in many cases. Rapid inhibition of synthesis can be seen with drugs enhancing DA release such as MAO inhibitors or amphetamine, or with drugs decreasing synaptic quantities of the transmitter such as reserpine. Blockade of MAO with tranylcypromine (10 mg/kg, intravenous) rapidly increased the release of newly synthesised ^{3}H–DA in the cat caudate nucleus (BESSON et al., 1972; BESSON et al., 1973a), this effect was associated with a progressive reduction in ^{3}H–H_2O formation suggesting an inhibition of the transmitter synthesis (GAUCHY, BESSON, CHÉRAMY and GLOWINSKI, in preparation). Intraperitoneal injections of pargyline (75 mg/kg) or pheniprazine (5 mg/kg) in the rat resulted in a rapid rise in DA levels in the striatum; the initial accumulation of striatal ^{3}H–DA measured 15 min after the local injection of L-3·5–^{3}H-tyrosine was markedly reduced in animals treated 1 or 2 hr earlier with these drugs when compared to those injected with the ^{3}H-amino-acid 15 min after MAO inhibition (JAVOY et al., 1972). As confirmed by the estimation of ^{3}H–H_2O formation in slices of rats pretreated one or two hours earlier with MAO inhibitors, this effect was related to an inhibition of DA synthesis occurring at the first limiting step (JAVOY et al., 1972). The increase in intraneuronal concentrations of DA induced by incubating striatal slices with exogenous DA also markedly inhibited ^{3}H–H_2O formation; inhibition of DA synthesis could be partially prevented by benzotropine, an inhibitor of DA uptake (JAVOY et al., 1972). Inhibition of DA synthesis as indicated by the reduced ^{3}H–H_2O/tyrosine specific activity conversion index, was also seen in striatal slices of rats pretreated with amphetamine (5 mg/kg, 90 min i.p.) or reserpine (2 mg/kg, 24 hr i.p.) (BESSON et al., 1971). These various treatments may enhance DA cytoplasmic levels at the tyrosine hydroxylation site and reduced synthesis by end-product inhibition.

In some experiments, inhibition of DA synthesis could be seen in striatal slices of rats injected only 10 min earlier with MAO inhibitors (SCATTON, BESSON, unpublished observations). Such rapid effects were also seen in vivo, in the rat striatum very shortly (5–10 min) after the local or intraperitoneal injection of pargyline (AGID, BOUVET and JAVOY, in prep.), or in the cat caudate nucleus during its direct superfusion with tranylcypromine (10^{-5} M) (GAUCHY, BESSON, CHÈRAMY and GLOWINSKI, in prep.). A short extraneuronal feed-back process linked to the rapid elevation of transmitter concentrations at post or presynaptic DA receptor sites may be involved in these rapid fluctuations of DA synthesis. Such a mechanism, not induced initially by modifications of intraneuronal DA levels, but by the synaptic concentrations of the transmitter could explain the brutal interruption in the raise of DA content observed in striatal terminals as soon as 10–15 min after MAO inhibition (JAVOY et al., 1972).

RAPID ACTIVATING PROCESSES OF DA SYNTHESIS

Two situations can be distinguished in rapid regulatory processes implicated in the activation of DA synthesis. This is comparable to that already mentioned for inhibitory processes. Acceleration of DA synthesis may be associated to an activation of the transmitter release, such a compensatory intraneuronal process has been observed in many cases. However, increased DA synthesis may also be induced by interruption of the transmitter release. This effect was demonstrated more recently in a few experiments.

The introduction of benzotropine (10^{-6} M) into the cup during the continuous superfusion of the cat caudate nucleus with L-3·5–^{3}H-tyrosine rapidly induced a marked enhancement of quantities of newly synthesised ^{3}H–DA in superfusates (CHÈRAMY et al., 1973). The increased release of DA, probably related to the inhibitory effect of the drug on the amine uptake process, was rapidly associated with a stimulation of $^3H-H_2O$ formation, suggesting an acceleration of the labelled transmitter synthesis from L-3·5-^3H-tyrosine (Chèramy et al., 1973). Various drugs given in vivo may accelerate DA turnover in DA terminals. Compounds such as morphine (60 mg/kg, 30 min i.p.) (Gauchy et al., 1973) amantadine (40 mg/kg, 120 min i.p.) (Scatton et al., 1970) and thioproperazine (5 mg/kg 90 min i.p.) (Javoy et al., 1970), which exhibit different pharmacological effects, increased the activity of DA neurons. These effects induced in vivo likely by different processes, can still be detected in vitro in striatal slices. The rate of formation of $^3H-H_2O$ and ^3H-DA were parallely accelerated and the conversion index of tyrosine into dopa were increased. As indicated by the ^3H-DA levels stored in tissues and released into the incubating medium, in all cases changes in DA synthesis were associated with an increase in transmitter release (Chèramy et al., 1970; Scatton et al., 1970; Gauchy et al., 1973). In all situations, the increased rate of the first step of DA synthesis may be mediated by a positive intraneuronal feed back process triggered by the reduction of cytoplasmic DA levels at the site of tyrosine hydroxylation. This effect is the reverse of the end product inhibition seen after MAO inhibitors or amphetamine injections. It is of interest to recall that the competitive inhibitory action of amphetamine on DA synthesis (Besson et al., 1971) can be reversed by thioproperazine pretreatment (Chèramy et al. 1970). On the other hand, the benzotropine activating action and the exogenous DA inhibitory effect on DA synthesis emphasised the importance of the newly taken up transmitter in the intraneuronal control of its synthesis. If the mechanisms by which morphine and amantadine activate DA neurons are still not well elucidated, it seems that the effect of thioproperazine is mainly related to the neuroleptic action on DA postsynaptic receptors. An interneuronal feed-back is probably involved in thioproperazine action since DA synthesis was stimulated both in DA cell bodies and terminals (Javoy et al., 1970). Nerve activity is required since thioproperazine added in vitro did not activate synthesis in slices (Chèramy et al., 1970).

As Faul and Laverty (1969), we recently observed that striatal DA levels in the rat increased rapidly shortly after an electrolytic lesion. A microinjection of 6-OHDA (8 μg) into the substantia nigra (Javoy et al., 1973), as well as a peripheral administration of the anaesthetic, γ-hydroxybutyrate (750 mg/kg) (Walters and Roth, 1972; Javoy et al., 1973) produced the same effect. When ^3H-tyrosine was injected intravenously or locally shortly after these treatments, a marked increase in striatal ^3H-DA accumulation was observed 10 min after the ^3H-amino-acid injection (Javoy et al., 1973). These effects may be related both to the inhibition of the newly synthesised transmitter release and to the acceleration of ^3H-DA synthesis. Indeed, transection of the nigro neostriatal system in the cat immediately blocked the release of newly synthesised ^3H-DA (Besson et al., 1973b). Electrolytic lesion of the substantia nigra (Javoy et al., 1973) as well as 6-OHDA local injection (Ungerstedt, 1971) rapidly reduced the rate of DA utilisation . A blockade of DA release has also been seen after the injection of γ-hydroxybutyrate (Bustos and Roth, 1972). Furthermore, as revealed by the estimation of $^3H-H_2O$ in tissues at the end of a 3-min micro injection of ^3H-tyrosine into the striatum, the three treatments described above increased the conversion of tyrosine into dopa (Javoy et al., 1973). Since the latter effect was associated in all cases with a blockade of the transmitter release, the temporary increase in DA synthesis may be indirectly related to the amine disappearance at DA receptor sites hypothetically localised on DA terminals or on postsynaptic neurons. These data are in agreement and support those of KEHR *et al.* (1972). These authors have demonstrated, with the help of dopa-decarboxylase inhibitors, an acceleration of the first step of CA synthesis in the rat fore-brain after transection of the nigro neostriatal pathway. Such a short extraneuronal feed-back process for DA synthesis was originally suggested by Farnebo and Hamberger studies on the effect of neuroleptics on CA release in brain slices submitted to field stimulation (FARNEBO and HAMBERGER, 1971). The activation of DA synthesis seen shortly after the interruption of the amine release may be finally induced directly by rapid subcellular changes in the distribution of DA.

LONG TERM CHANGES IN THE REGULATION OF DA SYNTHESIS

Rapid events occurring in the regulation of DA synthesis after acute treatment with reserpine, amphetamine and thioproperazine, which can still be detected in slices, were not associated with changes in the activity of purified tyrosine hydroxylase (BESSON *et al.*, 1973c). It was thus interesting to test if the long term changes occurring in the activity of DA neurons after repeated injections of drugs were able to affect the activity of the purified enzyme. Surprisingly, repeated daily injections to rats of pargyline (40 mg/kg, s.c.) amphetamine (1 mg/kg, s.c.), reserpine (0·5 mg/kg, s.c.) and thioproperazine (1·6 mg/kg, s.c.) over 8–13 days failed to change the activity of semi-purified tyrosine hydroxylase prepared from striatal tissues. However these treatments still induced effects on DA synthesis in striatal slices (BESSON *et al.*, 1973c). The synthesis was reduced after the chronic treatment with pargyline, amphetamine and reserpine. Curiously, in opposite to the activation of synthesis seen after acute treatment, repeated injections of thioproperazine inhibited DA synthesis (BESSON *et al.*, 1973c; SCATTON, GARRET and JULOU, in prep.). Long term alterations of interneuronal processes involved in the control of the nigro neostriatal system may be responsible for this inhibition of DA synthesis. The unchanged activity of striatal purified tyrosine hydroxylase still found after repeated drug treatments must be underlined since the enzyme activity in the rat brain stem was significantly increased after chronic reserpine treatment on one hand, and decreased after repeated thioproperazine injections on the other hand. Similar effects were also seen in the cortex (BESSON *et al.*, 1973c). This may reflect differences in the long term regulatory processes of CA synthesis in NE and DA neurons. It cannot be excluded that changes in enzyme activity detected in the brain stem and the cortex revealed differences in regulatory processes among various types of DA neuronal systems, since DA terminals have already been identified in one of these structures.

CONCLUSION

The ^{3}H–H_2O method, used *in vitro* or adapted to *in vivo* studies seems to be a precious tool to elucidate regulatory processes involved in the amine synthesis in DA terminals. In all cases examined, similar and complementary information were obtained with this technique in *in vitro* and *in vivo* experiments. As already mentioned, rapid changes in synthesis do not always parallel those of the amine release. Measurements of DA synthesis made by classical methods should be completed by *in vivo* and *in vitro* estimations of the rate of conversion of tyrosine into dopa to clearly differentiate effects on release and synthesis.

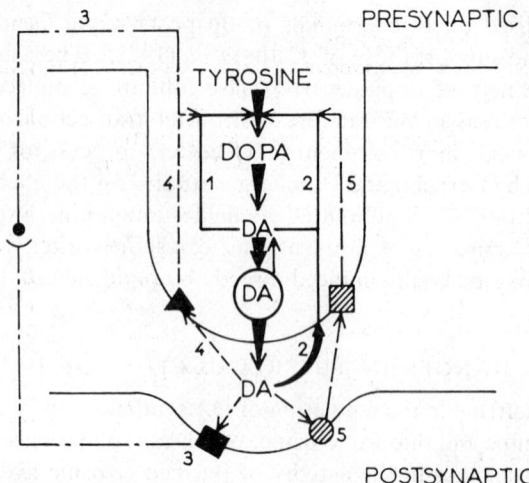

FIG. 1.—Possible regulatory processes involved in drugs action on the control of the first limiting step of DA synthesis in DA terminals. (1) Direct modification of cytoplasmic DA levels. (2) Fluctuations of the DA reuptake process. (3) Interneuronal mono or polysynaptic feed-back processes linked to action on postsynaptic receptor sites. (4) Short feed-back process linked to action on presynaptic receptor sites. (5) Transynaptic effect by release of hypothetical chemical factors from postsynaptic cells.

In all pharmacological situations examined, the changes in the rate of conversion of tyrosine into dopa can finally be attributed to modifications of cytoplasmic DA levels at tyrosine hydroxylation sites which may result in fluctuations in the availability of the enzyme cofactors. However, the mechanism by which these changes occur may be quite different. Drugs may directly affect the intraneuronal distribution of the amine or act indirectly on this subcellular distribution, through their effect on neuronal membrane permeability (Fig. 1). In most cases, receptors sensitive to fluctuations in extraneuronal transmitter levels may be implicated in drugs action on DA synthesis. Drugs may act (1) on postsynaptic receptors and change the overall activity of DA neurons by monosynaptic or polysynaptic interneuronal processes; (2) on presynaptic receptors and induce local changes in DA metabolism (Fig. 1). New experiments are required to further explore these phenomenons and to appreciate their relative importance in rapid and long term changes in synthesis regulation. Furthermore, it can also be postulated that chemical factors released from postsynaptic cells, under changes in synaptic DA levels, act on specific receptors localised on DA terminals (Fig. 1) Such transynaptic processes already proposed by Hedqvist (1973) to explain the effects of prostaglandins on NE release could also be involved in the local control of DA metabolism.

REFERENCES

Besson M. J., Chéramy A., Feltz P. and Glowinski J. (1971) *Brain Res.* **32,** 407–424.
Besson M. J., Chéramy A., Gauchy C. and Glowinski J. (1972) (*5th Congress on Pharmacol.,* San Francisco)
Besson M. J., Chéramy A., Gauchy C. and Glowinski J. (1973a) *2nd L-DOPA Conference,* Princeton, 11-13 *April* 73, Raven Press.
Besson M. J., Chéramy A., Gauchy C. and Glowinski J. (1973b) *Naunyn-Schmiedebergs' Arch Pharmacol.* **278,** 101–105.
Besson M. J., Chéramy A., Gauchy C. and Mussachio J. M. (1973c) *Europ. J. Pharmacol.,* **22,** 181–186.

BESSON M. J., CHÉRAMY A. and GLOWINSKI J. (1971) *J. Pharmacol. exp. Ther.* **177,** 196–205.
BUSTOS G. and ROTH R. H. (1972) *Brit. J. Pharmacol.* **46,** 101–115.
CARLSSON A., DAVIS N. J., KEHR W., LINDQVIST M. and ATACK C. V. (1972) *Naunyn-Schmiedebergs, Arch Pharmacol.* **275,** 153–168.
CHÉRAMY A., BESSON M. J. and GLOWINSKI J. (1970) *J. Pharmacol.* **1,** 151–152.
CHÉRAMY A., GAUCHY C., GLOWINSKI J. and BESSON M. J. (1973) *Europ. J. Pharmacol.* **21,** 246–248.
COSTA E., GREEN A. R., KOSLOW S. H., LEVEFRE H. F., REVUELTA A. V. and WANG C. (1972) *Pharmacol. rev.* **24,** (2), 161–167.
FAUL R. L. M. and LAVERTY R. (1969) *Exp. Neurol.* **23,** 332–340.
FARNEBO L. O. and HAMBERGER B. (1971) *Acta Physiol. Scand.* (suppl) **371,** 35–44.
GAUCHY C., AGID Y., GLOWINSKI J. and CHÉRAMY A. (1973) *Europ. J. Pharmacol.,* **22,** 311–319.
HEDQVIST (1973) In: *Dynamic aspects of the Synapse in press Brain Res.* (abs p. 34).
JAVOY F., AGID Y., BOUVET D. and GLOWINSKI J. (1972) *J. Pharmacol. exp. Ther.* **182,** (3) 454–463.
JAVOY F., AGID Y., SOTELO C. and GLOWINSKI J. (1973) Symposium on "*Dynamics of degeneration and growth in neurons*", May 1973, Stockholm.
JAVOY F. and GLOWINSKI J. (1971) *J. Neurochem.* **18,** 1305–1311.
JAVOY F., HAMON M. and GLOWINSKI J. (1970) *Europ. J. Pharmacol.* **10,** 178–188.
KEHR W., CARLSSON A., LINDQVIST M., MAGNUSSON T. and ATACK C. V. (1972) *J. Pharm. Pharmacol.* **24,** 744–747.
SCATTON B., CHÉRAMY A., BESSON M. J. and GLOWINSKI J. (1970) *Europ. J. Pharmacol.* **13,** 131–133.
THIERRY A. M., BLANC G. and GLOWINSKI J. (1973a) *Archs. Pharmacol.* (in press).
THIERRY A. M., STINUS L., BLANC G. and GLOWINSKI J. (1973b) *Brain Res.* **50,** 230–234.
UNGERSTEDT U. (1971) In: *6-hydroxydopamine on monamines neurons in the rat brain* MALMFORS T THONEN H. Eds.) North Holland Amsterdam, 101–127.
WALTERS J. R. and ROTH R. H. (1972) *Biochem. Pharm.* **21,** 2111–2121.

Frontiers in Catecholamine Research 1973, pp. 613 to 614. Pergamon Press. Printed in Great Britain.

SYNTHESIS AND PREFERENTIAL RELEASE OF NEWLY FORMED NOREPINEPHRINE DURING INCREASED SYMPATHETIC ACTIVITY

NGUYEN B. THOA

Laboratory of Clinical Science, National Institute of Mental Health, Bethesda, Maryland 20014, U.S.A.

SYMPATHETIC nerve stimulation accelerates synthesis of norepinephrine in cat spleen and adrenal gland (EULER and HELLNER-BJORKMAN, 1955; HOLLAND and SCHUMANN, 1958), mouse heart (OLIVERIO and STJARNE, 1965) and rat salivary gland (SEDVALL and KOPIN, 1967). In the adrenergic terminals of the isolated guinea pig vas deferens, stimulation of the hypogastric nerve was associated with an increase in the conversion of labeled norepinephrine from [14C]L-tyrosine (ROTH et al., 1966; ALOUSI and WEINER, 1966; THOA et al., 1971a). Although the level of tyrosine hydroxylase activity of vasa deferentia was unaltered following 60 min of stimulation (THOA et al., 1971a), the rate-limiting step appeared to be the conversion of tyrosine to dopa presumably controlled via feedback regulation of tyrosine hydroxylase activity by norepinephrine. Thus, exogenous norepinephrine inhibited the synthesis of [3H]-norepinephrine from [3H]tyrosine (WEINER and RABADJIJA, 1968). The conversion of tyrosine to norepinephrine, but not of dopa to norepinephrine, was increased during nerve stimulation (WEINER and RABADJIJA, 1968). When dopa decarboxylation was inhibited by drugs, a sharp increase in [14C]dopa formation from [14C]-tyrosine was still observed (THOA et al., 1972a).

With increased adrenergic activity, normal levels of norepinephrine in sympathetic nerve endings are maintained through a balance between the release of the transmitter and its *de novo* synthesis and reuptake. In the vas deferens, stimulation of the hypogastric nerve was not accompanied by any alteration of the endogenous level of norepinephrine (WEINER and RABADJIJA, 1968; THOA et al., 1971a). However, when the reuptake of norepinephrine was inhibited with phenoxybenzamine, a significant decrease in endogenous levels of norepinephrine was observed and there was a corresponding marked increase in norepinephrine concentration in the incubation medium (THOA et al., 1971b). Inhibition of tyrosine hydroxylase with α-methyl-para-tyrosine (α-MPT) prevented the formation of [14C]norepinephrine from [14C]tyrosine in the organs as well as in the incubation media. Levels of endogenous norepinephrine, however, were not altered significantly. When both α-MPT and phenoxybenzamine were added to the incubation medium, levels of norepinephrine in the tissue were only decreased as much as with phenoxybenzamine alone. Also, the amount of amine released was similar to that released by phenoxybenzamine alone (THOA et al., 1971b). Thus, in periods of acute increase in adrenergic activity, reuptake of norepinephrine appears to play a major role in the maintenance of a normal norepinephrine storage level.

Newly formed [14C]norepinephrine appears to be released preferentially. When vasa deferentia were incubated with phenoxybenzamine and [14C]tyrosine, stimulation resulted in release of [14C]norepinephrine into the incubation medium. The specific activity (counts/min per ng) of the released catecholamine was three times as high as that found in the tissue (Thoa *et al.*, 1971b). The mechanisms responsible for preferential release of newly formed transmitters are unclear.

During the past 2 years, it was shown that stimulation of the hypogastric nerve resulted in a release of norepinephrine from sympathetic nerve endings by exocytosis. Following depolarization, the norepinephrine vesicles appear to fuse with the nerve cell membrane and discharge their soluble contents (Weinshilboum *et al.*, 1971; Johnson *et al.*, 1971; Thoa *et al.*, 1972b; Wooten *et al.*, 1973). In preparations treated with phenoxybenzamine, the ratio of norepinephrine to dopamine-beta-hydroxylase (EC 1.14.2.1) in the incubation medium (49·2 ± 3·4) was slightly higher than the ratio in the soluble fraction of tissue homogenates (28·4 ± 3·4) (Weinshilboum *et al.*, 1971). In preparations treated with both phenoxybenzamine and β-MPT, stimulation induced a release of norepinephrine and dopamine-β-hydroxylase in a ratio identical with that found in the soluble fraction of vasa deferentia homogenates (29·3 ± 2·3). These results suggest the following mechanism for preferential release of newly synthesized norepinephrine: Following exocytosis, the norepinephrine vesicles are again reformed by pinching off from the nerve cell membrane (Holtzman *et al.*, 1973). These post-exocytotic vesicles have a normal amount of membrane-bound dopamine-β-hydroxylase activity but their soluble contents are diminished or absent. These emptied vesicles, however, can still take up dopamine from the cytoplasm and synthesize norepinephrine. However, this newly formed norepinephrine might be weakly bound. When depolarization occurs, this labile norepinephrine would be released first, as a result of weak binding and/or the proximity of post-exocytotic vesicles to the nerve cell membrane. During stimulation of the hypogastric nerve, when an increase in the rate of $[^{14}C]$norepinephrine formation occurs, norepinephrine and dopamine-β-hydroxylase are released from normal vesicles, but post-exocytotic vesicles release only norepinephrine. In α-MPT treated preparations, release must result from mobilisation of intact vesicles so that the ratio of released norepinephrine to dopamine-β-hydroxylase in incubation media closely approximates that found in the soluble fraction of the vesicles.

REFERENCES

Alousi A. and Weiner N. (1966) *Proc. Nat. Acad. Sci. USA* **56**, 1491–1496.
Euler U. S. v. and Hellner-Bjorkman S. (1955) *Acta Physiol. Scand.* **33**, (Suppl. 118) 17–20.
Holland W. C. and Schumann H. (1956) *Brit. J. Pharmacol.* **11**, 449–453.
Holtzman E., Teichberg S., Abraham S. J., Citkowitz E., Crain S. M., Kawai N. and Peterson E. R. (1973) *J. Histochem. Cytochem.* **21**, 349–385.
Johnson D. G., Thoa N. B., Weinshilboum R. M., Axelrod J. and Kopin I. J. (1971) *Proc. Nat. Acad. Sci. USA* **68**, 2227–2230.
Oliverio A. and Stjarne L. (1965) *Life Sci.* **4**, 2339–2343.
Roth R. H., Stjarne L. and Euler U. S. v. (1966) *Life Sci.* **5**, 1071–1075.
Sedvall G. C. and Kopin I. J. (1967) *Life Sci.* **6**, 45–51.
Thoa N. B., Johnson D. G., Kopin I. J. and Weiner N. (1971a) *J. Pharmacol. Exp. Ther.* **178**, 442–449.
Thoa N. B., Johnson D. G. and Kopin I. J. (1971b) *Eur. J. Pharmacol.* **15**, 29–35.
Thoa N. B., Johnson D. G. and Kopin I. J. (1972a) *J. Pharmacol. Exp. Ther.* **180**, 71–77.
Thoa N. B., Wooten G. F., Axelrod J. and Kopin I. J. (1972b) *Proc. Nat. Acad. Sci. USA* **69**, 520–522.
Weiner N. and Rabadjija M. (1968) *J. Pharmacol. Exp. Ther.* **160**, 61–71.
Weinshilboum R. M., Thoa N. B., Johnson D. G., Kopin I. J. and Axelrod J. (1971) *Science* **174**, 1349–1351.
Wooten G. F., Thoa N. B., Kopin I. J. and Axelrod J. (1973) *Molec. Pharmacol.* **9**, 178–183.

Frontiers in Catecholamine Research 1973, pp. 615 to 616. Pergamon Press. Printed in Great Britain.

DECENTRALISATION INCREASES NORADRENALINE RELEASE FROM FIELD-STIMULATED RAT IRIDES

Bertil Hamberger and Lars-Ove Farnebo

Department of Histology, Karolinska Institutet, S-104 01 Stockholm, Sweden

There are now several reports indicating that noradrenaline (NA) release from adrenergic nerves can be varied by α-receptor blocking and stimulating drugs (Häggendal, 1970; Farnebo and Hamberger, 1970, 1971; Kirpekar and Puig, 1971; Starke, 1971; Enero *et al.*, 1972 and this symposium). These data indicate that adrenergic nerve terminals may have a possibility to regulate its own release of NA via feedback either through the effector cell or directly on the nerve terminal from the concentration of the transmitter in the synapse. The localisation of this feedback mechanism, is however, not known and to try to find out more about this, unilateral decentralisation of the superior cervical ganglion of the rat was performed to find out how NA release is influenced by supersensitive effector cells. The technique used for field stimulation is described elsewhere (Farnebo and Hamberger 1971, 1973). It is obvious from Fig. 1 that stimulation of decentralised irides results in an increased ^{3}H-NA overflow. This increase becomes obvious one week after decentralisation and is persistent at least after 8 weeks. One likely explanation to this increase is that decentralisation has decreased the endogenous feed-back inhibition of NA release. This is further supported by pharmacological results demonstrating that the α-receptor stimulating drug clonidine cannot induce a decrease of ^{3}H-NA overflow in decentralised irides as it does in control irides (Farnebo and Hamberger, 1973). The α-receptor blocking drug phentolamine induces in decentralised irides an increase of ^{3}H-NA release which is less pronounced than in control irides. If decentralisation

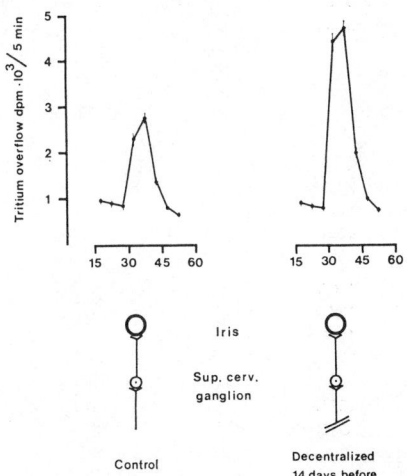

Fig. 1.—Field stimulation of isolated irides preincubated with ^{3}H-NA. After preincubation with ^{3}H-NA (10^{-7} M) the irides were superfused for 30 min and then stimulated at 10 Hz for 10 min. Mean $\pm$ s.e.m. of 10 experiments.

(TRENDELENBURG, 1966) only induced postsynaptic events, a trans-synaptic release regulating mechanism ought to cause a decreased release. Furthermore, decentralisation supersensitivity does not seem to be due to supersensitive postsynaptic receptors (FLEMING, *et al.*, 1973). The present results could thus be taken as evidence against trans-synaptic release regulation. While most tissues are known to become supersensitive after decentralisation, GREEN (1969) reported data indicating that the superior cervical ganglion becomes subsensitive. Subsensitive α-receptors on the adrenergic nerve terminal could provide a reasonable explanation for the present findings of an increased NA release after decentralisation as well as a decreased sensitivity to the α-receptor stimulating drug clonidine.

Acknowledgements—Supported by the Swedish Medical Research Council (14X-2330).

REFERENCES

ENERO M. A., LANGER S. Z., ROTHLIN R. P. and STEFANO F. J. E. (1972) *Brit. J. Pharmacol.* **44,** 672–688.
FARNEBO L.-O. and HAMBERGER B. (1970) *J. Pharm. Pharmacol.* **22,** 855–857.
FARNEBO L.-O. and HAMBERGER B. (1971) *Brit. J. Pharmacol.* **43,** 97–106.
FARNEBO L.-O. and HAMBERGER B. (1973) *Brain Research.* In press.
FLEMING W. W., MCPHILLIPS J. J. and WESTFALL D. P. (1973) *Ergebnisse der Physiology* (in press).
HÄGGENDAL J. (1970) In *New Aspects of Storage and Release Mechanisms of Catecholamines* (SCHUMANN H. J. and KRONEBERG G., eds.) Bayer Symposium, pp. 100–109, Springer-Verlag. Berlin.
KIRPEKAR S. M. and PUIG M. (1971) *Brit. J. Pharmacol.* **43,** 359–369.
STARKE K. (1972). *Naturwissenschaften* **58,** 420.
TRENDELENBURG U. (1966) *Pharmacol. Revs.* **18,** 629–640.

Frontiers in Catecholamine Research 1973, pp. 617 to 618. Pergamon Press. Printed in Great Britain.

EFFECT OF ACETYLCHOLINE ON THE RELEASE OF ^{3}H-NOREPINEPHRINE BY NICOTINE AND POTASSIUM CHLORIDE FROM RAT BRAIN SLICES

THOMAS C. WESTFALL

Department of Pharmacology, University of Virginia School of Medicine, Charlottesville-Virginia 22093, U.S.A.

IT HAS recently been shown that muscarinic agonists produce a decrease in the release of norepinephrine (NA) following nerve stimulation or nicotine administration in the rabbit heart (LÖFFELHOLZ and MUSHOLL, 1967; LINDMAR et al, 1968 and LÖFFELHOLZ et al., 1969), rabbit ear artery (STEINSLAND et al., 1973) and perfused guinea pig heart (WESTFALL and BRASTED, 1972). This has given rise to the suggestion that low doses of acetylcholine modulates the release of NE by decreasing its release to physiological or pharmacological stimulation while high concentrations increase release. This represents an additional control site located at the level of the presynaptic adrenergic nerve terminal. The purpose of this present study was to investigate the effect of muscarinic agonists on the release of NE from central adrenergic neurons.

METHODS

Studies were carried out in Sprague–Dawley rats weighing 150–200 g. Animals were killed by decapitation and the brain dissected into various regions. Tissue of various brain regions were then chopped into 0·3 mm square sections by means of a MacIlwain tissue chopper and incubated with 10^{-7} M l^3H-norepinephrine (S.A. 6·2 Ci/mmol) for 30 min. The slices were washed and placed in jacketed chambers (2 ml volume) containing a mesh wire which held the slices in place. The slices were superfused with Besson's solution at constant rate of 1 ml/min by a pump. The

TABLE 1. EFFECT OF NICOTINE IN THE PRESENCE AND ABSENCE OF Ca^{2+} OR ACETYLCHOLINE ON THE EFFLUX OF ^{3}H-NE FROM HYPOTHALAMUS

Drugs	Collection period (min)									
	10	20	30	40	50	60	70	80	90	100
None	43·5	12	9·2	8·4	8·3	8·5	7·8	7·3	7·2	7·1
Nicotine (+Ca)	45·6	13	11	9·5	8·7	12·3	18·5	14·1	8·5	7·9
Nicotine (−Ca)	43·3	12	11	10	8·5	8·4	8·2	7·9	7·5	7·5
Acetylcholine (10^{-5} M) plus Nicotine	47	13	10	9·5	8·8	8·6	8·2	7·3	7·4	7·2

Data is expressed as dis/min/pre mm of ^{3}H-NE ($\times 10^{-3}$)

Nicotine was administered for 20 min in a concentration of 1 mM from 60–80 min of collection

Acetylcholine was administered in a concentration of 10^{-5} M starting 20 min before the addition of nicotine.

618

T. C. WESTFALL

TABLE 2. THE EFFECT OF POTASSIUM CHLORIDE IN THE ABSENCE OR PRESENCE OF
ACETYLCHOLINE ON THE RELEASE OF ³H-NE FROM RAT CEREBELLAR CORTEX

Drugs	Collection period*											
	10	20	30	40	50	60	62	64	68	70	80	90
None	41·5	13	10	9·6	8·5	8·4	8·4	8·2	7·9	7·5	7·4	7·0
KCl†	39·8	14	12	9·8	9·3	8·9	16·4	18·5	13	9·5	8·5	8·0
ACh‡ + KCl	44·4	12	9·9	8·5	8·4	8·0	7·9	7·9	7·5	7·5	8·1	7·4

* Data is expressed as dis/min per 5 mm × 10⁻³.
† KCl was administered in a concentration of 50 mM at 60–64 min.
‡ Acetylcholine was administered in a concentration of 10⁺⁵ M 20 min before KCl.

efflux from the slices was continuously collected and the ³H-NE determined by liquid
scintillation spectrometry following purification on alumina columns. Muscarinic
agonists were added in various concentrations to the perfusion solution 10–20 min
before inducing release by 1 mM nicotine or 50 mM potassium chloride.

RESULTS AND DISCUSSION

Table 1 shows the effect of nicotine in a concentration of 1 mM on the release of
³H-NE from the superfused slice of rat hypothalamus alone; in the absence of extracellular Ca²⁺ and presence of acetylcholine. It can be seen that nicotine produces
a marked increase in the efflux of ³H-NE which is blocked by the removal of extracellular Ca²⁺ or in the presence of acetylcholine. A similar effect is seen in slices
obtained from the cerebellar cortex when potassium chloride is the releasing agent
(Table 2). The present data is quite consistent with observations made on peripheral
adrenergic neurons and suggests that acetylcholine can decrease the release of NE to
depolarizing and pharmacological influences from neurons in the central nervous
system. This extends the hypothesis that acetylcholine may modulate the release of
NE by decreasing its output to various forms of stimulation.

Acknowledgements—This work was supported by PHS grant No. NS 10260.

REFERENCES

LINDMAR R., LÖFFELHOLZ K. and MUSCHOLL E. (1968) *Br. J. Pharmacol.* **32**, 280.
LÖFFELHOLZ K., LINDMAR R. and MUSCHOLL E. (1967) *Naunyn-Schmiedebergs Arch. Pharmak.* **257**, 308.
LÖFFELHOLZ K. and MUSCHOLL E. *Naunyn-Schmiedelbergs Archs. Pharmak.* (1969) **265**, 1.
STEINSLAND O. S., FURCHGOTT R. F. and KIRPEKAR (1973) *J. Pharmacol. Exp. Ther.* **184**, 346.
WESTFALL T. C. and BRASTED M. J. (1972) *Pharmacol. Exp. Ther.* **182**, 409.

Frontiers in Catecholamine Research 1973, pp. 619 to 621. Pergamon Press. Printed in Great Britain.

REPORTER'S NOTES—FIRST SESSION

Ross J. Baldessarini

Harvard University, Boston, Mass. 02114

SEVERAL important issues were discussed in the presentations and discussions of this session. They include the morphology and life-cycle of neuronal and adrenal amine storage vesicles or granules the details of the 'release' process and its regulation.

Dr. Hökfelt suggests that the presence of electron-dense cores in neuronal vesicles may be a useful morphologic index by which to evaluate the distribution of amine-containing nerve terminals in a complex tissue such as the CNS by electron microscopy. The main alternatives for electron microscopy at present are autoradiographic methods which are becoming much better as demonstrated by Dr. Descarries. Possibly specific histo-immunologic methods, which are as yet in their infancy, can also be adapted to electron microscopy. Methodological modifications of tissue preparation for electron microscopy by Dr. Tranzer have led to a very marked increase incidence of visualisation of small electron-dense-core vesicular and perhaps pre-vesicular endoplasmic reticular structures in adrenergic neurons where amines may be stored. The generally accepted notion that large dense-core vesicles occur in greater frequency closer to the perikaryon and less toward the nerve-terminals may not be correct as large ($0 \cdot 1$ μm) dense-core vesicles may not represent more than 10 per cent of total vesicles in many adrenergic neurons and most are small ($0 \cdot 05$ μm), although the commonly studied bovine splenic nerve preparation may have an unusually high incidence of larger vesicles.

The structure of adrenal granules was also discussed by many participants. There seems to be general agreement now that they and probably adrenergic vesicles as well contain a 'soluble' intra-vesicular phase made up of catecholamines, ATP, Mg^{2+} and a variety of acidic proteins (the chromogranins) as well as DBH (dopamine-β-hydroxy-lase). Immunologic studies of Dr. Winkler and his collaborators have led to the conclusion that another fraction of DBH is bound to the inner layer of the granule membrane (identical to 'chromomembrin A'), while another protein chromomem-brin B) is on the outer surface. Mg^{2+}-ATP'ase is also present in these membranes. The exact molar ratio of catecholamines : ATP is still controversial, but methodologic artifacts are now known which when considered, strongly suggest that commonly accepted values of 4:1 or so are much too small (probably really greater than 10:1). In the soluble phase, DBH represents only a small proportion ($<$ 10 per cent) of the total proteins.

Of the several possible mechanisms for physiologic release of catecholamines from adrenal granules or adrenergic neurons (escape via cytoplasm, expulsion of storage particles or exocytotic emptying of particle contents), excoytosis now seems to be the most popular hypothesis. The main evidence for this was reviewed by Drs. Kirshner, DePotter, Poisner and Geffin and it includes the observations that the soluble components of granules or vesicles (amines, chromogranins and some DBH) are expelled into perfusing media during stimulation mediated by neuronal activity. Presumed cytoplasmic 'markers' (LDH, PNMT and tyrosine hydroxylase) are not released. The similarity of molar ratios of the stored and released catecholamines: proteins is a fundamental argument in favour of the theory of exocytosis. Dr. Geffin pointed out that the biochemical assay of DBH activity is subject to many variables and for such studies is perhaps better done by sensitive and relatively specific immuno-logic methods such as his new radioimmunoassay for DBH. An important point repeatedly emphasised is that 'release' in the physiologic sense is a Ca^{2+}-dependent process following nerve-depolarisation which leads to the expulsion of catecholamines along with soluble components of the storage vesicles, presumably by exocytosis. In contrast, expulsion of amines by drugs such as reserpine or the indirect sympatho-mimes does not require Ca^{2+}, apparently does not involve exocytosis, and probably involves leakage from storage sites into the cytoplasm and thus out of the cell.

A rough summary of the state of understanding of the specific molecular mechan-isms involved in the release process is that very little is known. Speculations include a role of ATP or even cyclic phosphonucleotides as some pharamacologic evidence suggests that ATP or cAMP can 'release' amines and that putative phosphodiesterase inhibitors such as the xanthines may also facilitate or provoke 'release'. Since tubular and filamentous proteins, some resembling actin, occur in sympathetic tissues and since antimitotic alkaloids such as cholchicine and vinblastine can prevent 'release', these proteins may be involved in the fusion of vesicles to the cell membrane or the expulsion of vesicular contents. It is clear that the rationalisation of these and other pharmacologic observations will require increasingly complex, multi-step models of the molecular physiology of release of amines. It is also clear that this process and the post-synaptic events at synapses or other amine-receptive cells remain among the most poorly understood aspects of catecholamine physiology.

The modulation of the release is also becoming increasingly complex according to pharmacologic observations reviewed by Drs. von Euler and Stjärne. An hypoth-esis with growing support is that pre-synaptic 'receptors' of amines may exist. Thus α-blockers can enhance the release of norepinephrine. Furthermore, the release of prostaglandins (PGE), probably from post-synaptic cells, may feed back to the throttle the release of norepinephrine in some tissues.

The physiologic importance of inactivation of released amines by uptake processes was emphasised by Dr. Iversen and other speakers. These include a high-affinity stereospecific neuronal uptake process believed to be important at synapses ('uptake$_1$') and a low affinity, high-capacity process into muscle, vascular and other tissues ('uptake$_2$') which is inhibited by 3-O-methylated-catecholamines and steroids. Dr. Gillis pointed out that the lung is likely to be an important organ offering a huge surface area that may provide a means of clearing amines from the 'venous' blood by uptake mechanisms more or less similar to 'uptake$_2$'.

The fate of the contents of the presynaptic vesicles other than the amines is poorly understood. There is no evidence that the proteins and other components are reaccumulated by a conserving mechanism comparable to amine reuptake. There is more or less consensus that most of the outer membrane of the vesicles is retained and may be 'refilled' by unknown mechanisms for later reuse in storage or release processes. Whether the axonal transport of fresh young vesicles from the region of the perikaryon proposed by Dr. Dahlström is the only means of obtaining vesicles at sympathetic nerve terminals is still unclear. Dr. Blaschko suggested that sympathetic neurotransmission seems to be associated with increased availability of lysozymes and that these may be involved in the scavenging of 'left-over' protein.

In summary, while better methods of histology, electron microscopy, radio-autography supplemented by modern immunologic techniques and a more firmly biochemically-based approach to classical agonist–antagonist–receptor analysis have contributed many new approaches and insights, the precise life-cycle of storage particles and the precise molecular basis of the release process remain to be explored further.

Frontiers in Catecholamine Research 1973, p. 623. Pergamon Press. Printed in Great Britain.

REPORTER'S NOTES

BERTIL HAMBERGER

Dept. of Histology, Karolinska Institutet, S–104 01 Stolkholm, Sweden

IT IS evident from several papers in this session that acetylcholine (ACh) can both inhibit and, in high concentrations, increase NA release from adrenergic nerves. This got further support by Dr. Furchgott reporting that ACh also inhibits NA release from the spleen. He suggested that vasodilatation could include an inhibition of NA release. Furthermore, Dr. Westfall also reported that NA release in the brain could be inhibited by ACh.

Dr. Furchgott further suggested that phentolamine might have an effect on ^{3}H-5HT release in Dr. Farnebo's model of field stimulation.

Dr. Trendelenburg agreed with the suggestion by Dr. Jarrott that ^{3}H-isoproterenol uptake in the slowly equilibrating pool could be localised in collagen.

Dr. Muscholl wanted to know whether the anti-hypertensive drug clonidine had its major effects on post- or presynaptic receptors. Dr. Henning thought that both were possible while Dr. Anden had evidence to support that the effect of clonidine was postsynaptic. In connection with Dr. Roth's paper, Dr. Stock and Anden presented almost identical results, also demonstrating that 6HB inhibits DA cell bodies in the substantia nigra by mimicking the effect of GABA. Finally Dr. Carlsson closed the discussion by expressing his satisfaction with Dr. Thoa's work.

Most authors in this section thus seemed to agree on most points and there is strong emphasis on local regulation of monoamine release both in peripheral and central nervous system, in addition to earlier described mechanisms. Among all other questions which still remain after the meeting two will be elaborated:

1. Is the effect of neuroleptic drugs on monoamine synthesis and turnover due entirely to local effects on the monoamine neuron or are other mechanisms also involved?

2. What is the physiological role of release regulation and what mechanism (s) is (are) of importance under physiological conditions?

Catecholamines in Central Nervous System
(May 21, 1973; 10:30 A.M.–6:30 P.M.)
(May 22, 1973; 9:00 A.M.–6:30 P.M.)

CHAIRMEN: Joel Elkes, Marthe Vogt, Albert Sjoerdsma and Hans Weil-Malherbe
COUNCILLORS: Donald Reis and Christer Owman
REPORTERS: Barry Hoffer and Merton Sandler

Frontiers in Catecholamine Research 1973, pp. 627 to 632. Pergamon Press. Printed in Great Britain.

INTRODUCTORY REMARKS: REGIONAL ECONOMY AND FAMILIES OF NEUROREGULATORY SUBSTANCES

JOEL ELKES

The Johns Hopkins University and Hospital, Baltimore, Md., U.S.A.

THE PROGRAM Committee have indeed been generous to its Chairmen. In an inordinately hard pressed and concentrated program they have given them the most precious commodity of time: An unusual concession, for which I am grateful.

Looking at this program, I suspect that we are passing through a historic phase. The information explosion alone suggests that something important is afoot: Never has the growth curve been so steep. Yet, size and complexity do not change the givens and basic purposes of a field. Let me explain:

Nearly twenty years ago, in April 1954, the First International Neurochemical Symposium was convened at Magdalen College, Oxford, England. Drs. Seymour Kety, Jordi Folch-Pi, the late Heinrich Waelsch (whose death was such a loss to Neurochemistry), and Louis Flexner were the American partners of the Committee. In Britain, there were Drs. Derek Richter, Geoffrey Harris, and myself (who acted as organising secretary). The field was sparse in those days; yet we located sixty-nine colleagues from nine countries to attend what proved a most useful exchange.

Writing in the introduction to the Proceedings, we put our purpose this way:

'We agreed also that from the start it would be well to consider the brain as a biological entity in all its complexity of morphology and function, rather than as a homogenate, or an engineering problem. For that reason, we felt that the most useful contribution of a Symposium of this kind would be an attempt to reintegrate biochemical process with structure and function, particularly with respect to the chemical topography of the brain, which, to us, seemed of greatest moment in an understanding of function. The program thus not only represents the framework of a conference, but also expresses an attitude: and of necessity includes discussion of structural, genetic, and pathological aspects, as well as subject matter that, in the more limited sense, may be termed 'neurochemical'. We feel that this approach may be helpful in slowly building the foundations for a rational therapy of disorders of the nervous system'. (WAELSCH and ELKES, 1963)

We seem to be still at it; but it is significant that a little point on the horizon in 1954—what we called the CNS 'Sympathins' has now enlarged to a whole continent. The forces which brought us to this point are, in part, accidental and empirical, and in part, deliberate. It required the boot of empiricism (the discovery of some of the psychoactive drugs) to give us new facts. The dialectic began when new facts began to hunt for explanations; and here it is the evolution of new *methods* which brought us to the position we are in today.

These methods were, in part, anatomical and physiological, in part chemical, in part behavioural and in part clinical. In this respect, our field brooks no compromise, and compels conversation between various approaches like no other. It was a combination of Anatomy, Physiology and Behaviour which slowly bared the mosaic of the subsystems governing Sleep, Wakefulness, Arousal and focussed Attention. It gave us an anatomical scaffolding for an understanding of the control of Hunger, Thirst, Sexuality, Social play and Predatory Attack. It was the same combination, coupled with microphysiological approaches, which led to an understanding of the role of convergence, filtering and funnelling of signals in peripheral sense organs (such as Retina and Olfactory Bulb), integration at high central level (such as the cerebellum) or the assignment of significance and value in what we call Affective terms in pathways and structures of the Limbic System. It is here, too, that the organising power of the Reward and Punishment systems emerged; with still another field (the sequential organisation of motor performance, including speech) being related to the striatum and the dominant hemispheres.

If all this sounds like the language of the nineteenth century cerebral 'localisers' it both is, and isn't. The brain, to be sure, is an assembly of sub organs, a system of systems, a mosaic of biassed homeostats. It has its nodal areas, its cables, and connections; and presumably in certain respects is 'wired' for a purpose. But the wiring,' and control centers, should not mislead. For equally clear, and emerging in the light of recent experimental evidence, are diffuse and generalised effects in large areas of the brain, or in the brain as a whole; phenomena of modulation, gating, and slow potential drifts. It is these which may be related to the now established fact of differential cell populations in even small areas of the brain; and the influence, by way of widespread projections of some nodal midline areas (such as the locus coeruleus) over large areas.

While these developments were proceeding, a chemical map was being quietly superimposed upon functional anatomical findings. It started with the painstaking dissection of gross areas of the brain and bioassay. I still remember the excitement of reading the great paper of FELDBERG and VOGT (1948). It proceeded to the sampling of ever smaller areas, by the application of the LINDERSTROM-LANG techniques to the analysis of single cells (POPE, 1955; HYDEN and LANGEE, 1961). The use of radio isotopes added to this dimension, and further elucidated the regional chemical organisation within the C.N.S. But it did more. For by taking down the regional process to the organisation of sub-cellular components, by showing up the astonishing precision and economy of the processes governing release, reuptake, and storage, it related *spatial* sub-cellular organisation to turnover in *time*. It added the concept of quantal discharge of transmitters by differential synaptosomal populations (IVERSEN and SNYDER, 1968; KUHAR et al., 1970; SHASKAN and SNYDER, 1970) thus introducing a much needed statistical, probabilistic element into our thinking. It also made much more real the concept of heterogeneous cell populations surmised by some of us early on. This proposition was literally illuminated by the advent of the path-breaking fluorescent and immunofluorescent techniques (FALCK, 1964; HOKFELT 1973) a proposition carried further by the iontophoretic application of drugs to single cells by multibarrel micro-pipette in several laboratories, including my own stamping ground of the NIMH, St. Elizabeths Hospital (SALMOIRAGHI and BLOOM, 1964).

It is, however, the combination of methods which again yielded special benefits: The combination for example of micro-pipette, radioisotope techniques in electro-microscopy (BLOOM and HOFFER, 1973) or the immunochemical approach and electron microscopy (HOKFELDT, 1973). Regional Neurochemistry, the subject of a special symposium at Varenna (KETY and ELKES, 1961) thus seemed to be coming of age.

If events studied by the above methods in the C.N.S. are very fast, the integer which we know as behaviour, including speech, is a very slow readout of such events. Here a refinement of methods, all the way from an automated approach to Skinnerian operant work, to the use of the Skinner Box as a metabolic cage, to self-stimulation, self-injection and self-infusion techniques can give one fine-grain analysis of behaviour over time. Ethological study of behaviour of animals in their natural surroundings, or in a controlled open environment, form the counterpart of such observations. They are particularly valuable in the study of the internalisation of early social cues in neurochemical terms. If, indeed, environment can produce chemical change in the brain of the developing animal, the first outlines of an environmental neurobiology may be at hand. Ability to cope, and inability to cope may have its early chemical counterparts, and thus open a direct link to the study of clinical states. Two days of the symposium are devoted to advances in this latter field alone. Here the combination of measurement of behaviour (and of subjective phenomena), with the refinement in the chemical estimation of hormones, neurohormones and metabolites in tissues, body fluids (SHARMAN, 1973) and post mortem material opens truly breathtaking possibilities. The use of stable isopopes in man (SEDVALL et al., 1973) should add a further dimension. I believe that already we have clearer reference points for an understanding (and rational treatment) of affective disorders than we have, for ex-ample, for the genesis of coronary disease. Grant agencies, investing in scientific 'Growth Stock' might well take note of these facts.

You will have noted that I have gone on so far without mentioning Dopamine, which is to occupy us a good deal this morning. The reason for this is because I am, once again, reminded of the way mental habit operate in our field.

Once again, a personal example may serve by way of illustration: Between 1949 and 1953, Bradley and I had studied the effects of cholinergic and noncholinergic drugs, including the amphetamines and LSD-25 on the electrical activity of the brain, and on behaviour in the conscious animal, and in acute preparations. We had surmised that amphetamine acted on amine receptors placed high up in the so-called Arousal system which was being developed at that time by MORUZZI and MAGOUN (1949) and that LSD-25 was acting elements peculiarly related to a medial afferent system whose function was primarily inhibitory (BRADLEY and ELKES, 1953, 1957). Imagine our excitement when at a meeting of the British Physiological Society we saw for the first time Dr. MARTE VOGT's (1954) diagram on the distribution of sympathins in the CNS; or when, in a personal communication, the late J. H. GADDUM (1953) told me of the antiserotonin action of LSD-25; an observation he published the following April. Between 1954 and 1957, I proposed (ELKES, 1953; 1958; BRADLEY and ELKES, 1957) a view of the existence of three families of neuroregulatory substances within the C.N.S. As we said at the time:

'Perhaps rather than thinking in unitary terms, it may at this stage, be ad-visable to think in terms of the possible selection by chemical evolution of small families of closely related compounds, which by mutual interplay would govern the phenomena of excitation and inhibition in the central nervous system. Acetylcholine, nor-adrenaline and 5-hydroxytryptamine may be parent molecules of this kind; but one has only to compare the effects of acetylcholine with succinylocholine, or nor-adrenaline with its methylated congener to realise how profound the effects of even slight changes of molecular configuration can be. The astonishing use which chemical evolution has made of the steroids is but another example of the same economy. It is likely that neurons possessing slight but definite differences in enzyme constitution may be differentially susceptible to neurohumoral agents. Such neurons may be unevenly distributed in topographically close, or widely separated areas in the central nervous system; these differences probably extending to the finest level of histological organisation. Phylogenetically older parts, and perhaps, more particularly, the mid-line regions and the periventricular nuclei may, in terms of cell population and chemical constitution be significantly different from parts characteristic of late development.

As yet, little information is available of the chemistry of the mosaic of cells, and cell groups making up the so-called reticular activating system. The neurons of this system, which is really a system of systems, bear a somewhat special and reciprocal relation to the afferent pathways which impinge upon them by way of collaterals. They are activated by these, but equally, through their activity, determine the ultimate perception of the signal arriving at the cortex by way of the sense-linked pathways. The translation of afferent signals into perception may well depend on the interaction of cortical and reticular elements, and may have its neural counterpart in the three dimensional apposition, and patterning of excitory and inhibitory states in a very large cell population. The reticular formation are distinctive for the diffuseness of their connections and of their effects. Equally, in this dense reticular field self-excitory phenomena may predominate, and the powerful operation of vectorial and spatial influences is likely. Slight variation in local titre of a neurohumoral agent in these key mid-line areas may thus profoundly affect the excitability of large neurone pools at a distance. It would perhaps be permissible to speak of the operation of chemical fields in these regions, which would depend on the rate of liberation, diffusion and destruction of locally-produced neurohumoral agents. The agents in question may be either identical with or, more likely, derived from neuro-effector substances familiar to us at the periphery. Their number is probably small, but their influence upon integrative action of higher nervous activity may be profound. The basic states of consciousness may well be determined by variations in the local concentration of these agents'. (BRADLEY and ELKES, 1957.)

The suggestion of the existence of three families of neuroregulating substances, one, related to cholinesters, of diffuse distribution; a second to catecholamines, related to the brain stem mid-line structures regulating attention and affect, and a third, an indole related to the afferent systems, exerting a regulating inhibitory, filtering role on afferent signals, was regarded a first crude approximation in a study in which cytochemistry, pharmacology, and electrophysiology are mutually complementary and interdependent. Since them many facts have appeared. Behind Serotonin there is Melatonin; behind Norepinephrine there is Dopamine; Histamine and GABA, even glycine and other membrane active amino acids may play a part in neuroregulatory substances.

All substances mentioned have one thing in common, namely, the extraordinary care and precision of their intra and pericellular economy. New members continue to be added. They may perhaps make up a sort of alphabet which the nervous system uses to construct its membrane-located specific sensors and recognisers; possibly by conferring conformational changes onto glycolypids or glyco proteins in bimolecular leaflets. Inhibition is at the heart of information transfer in the C.N.S.; it is the nervous system's stupendous capacity to selectively store and to *ignore* which makes precise performance (the so-called 'readout' of programs) possible. It is the informed silence which carries the message, and tells the tale.

I do not think, therefore, that we should be unduly swayed by our own current interests (such as Dopamine), fully justified though it may be. The epochs of Acetylcholine, Norepinephrine Serotonin, GABA and Dopamine, which we have lived through confirm the value of each. The field has been tilled well, and the findings lie there, to be used wisely and conjointly. That we should pass through these special preoccupations attests both to our curiosity and to our need—our very human need—for the scaffolding and support of an explanation (ELKES, 1970). Explanations provide psychological safety—for a time. But they are also hazardous and transient; for I suspect, while using the '*Either/Or*', the language of our brain really is the language of the '*Also*' and the '*And*'.

This brings me to my last point: I have a feeling that our field may be passing through a phase not unlike that undergone by physics at the turn of the century; and it is the evolution of new and precise mathematical languages which made modern physics what it is. As yet, the languages we are using are clumsy and qualitative; but when phenomena accumulate beyond a certain point (and we may perhaps be reaching it at this very Symposium), a new look is needed at relationships. The linear or binary view no longer suffices; and a topology of multiple phenomena in time will require linguistic models not met by our present old verbal hardware. The mathematicians may be waiting in the antechambers, and may even be looking over our shoulder. One does not know how long it will be before mathematics interacts with neuro- and psychopharmacology. When it does, however, new and precise configurations will become apparent. Psychobiology will then have come of age; and, quite possibly, enlarge the realm of science itself in the process.

REFERENCES

BLOOM F. E. and HOFFER B. J. (1973) Norepinephrine as a Central Synaptic Transmitter. This volume.
BRADLEY P. B. and ELKES J. (1953) *J. Physiol.* (*London*) **120**, 13.
BRADLEY P. B. and ELKES J. (1957) *Brain* **80**, 113–114.
ELKES J. (1958) Ciba Foundation Symposium on the Neurological Basis of Behavior, p. 303.
ELKES J. (1970) In: *The Psychopathology of Adolescence.* Grune & Stratton, New York.
ELKES J. (1953) In: *Prospects in Psychiatric Research* (TANNER J. M., Ed.), p. 126, Blackwell, Oxford.
FALCK B. (1964) In: *Biogenic Amines* (HIMWICH H. E. and HIMWICH N. A. Eds.) *Progressive Brain Research*, Vol. **8**, p. 28.
FELDBERG W. and VOGT M. (1948) *J. Physiol.* (*London*) **107**, 372.
GADDUM J. H. (1953) *J. Physiol.* (*London*) **121**, 15P.
HÖKFELT T. (1973) In: *Neuropathology of Schizophrenia* (S. S. KETY Ed.) 1973 (In Press).
HÖFELT T. (1973) Localisation of Catecholamines with special reference to synaptic vesicles. This volume.

Hyden H. and Lange P. (1961) In: *Regional Neurochemistry*, (Kety S. S. and Elkes J. Eds.) Pergamon Press, Oxford, p. 190.

Iversen L. L. and Snyder S. H. (1968) *Nature*, (*Lond.*) **220**, 796.

Kety S. S. and Elkes J. (1961) *The Regional Chemistry, Physiology and Pharmacology of the Nervous System*, Pergamon Press, Oxford.

Kuhar M. J., Green Alan I., Snyder S. H. and Gfeller E. (1970) *Brain Res.* **21**, 405.

Moruzzi G. and Magoun H. W. (1949) *Electroencephalogr. Clin. Neurophysiol.* **1**, 455.

Pope A. (1955) In: *Biochemistry of the Developing Nervous System*, (H. Waelsch Ed.) Newport Academic Press, p. 350.

Salmoiraghi G. C. and Bloom F. E. (1964) *Science* **144**, 493.

Sedvall G. C., Mayersky A., Samuel D., and Fri C. G. (1973) O^{18} Measurement of Dopamine Turnover in Rat Brain. This volume.

Sharman D. F. (1973) Catecholamine Metabolites in C.S.F. This volume.

Shaskan E. G. and Snyder S. H. (1970) *J. Phar. Exp. Ther.* **175**, 404.

Vogt M. (1954) *J. Physiology.* **123**, 451.

Waelsch H. and Elkes J. (1963) In: *Biochemistry of the Developing Nervous System.* (H. Waelsch Ed.), New York, Academic Press p. 5.

Frontiers in Catecholamine Research 1973, pp. 633 to 634. Pergamon Press. Printed in Great Britain.

CHAIRMAN'S INTRODUCTORY REMARKS

MARTHE VOGT

Institute of Animal Physiology, Babraham, Cambridge, England

ON READING the abstracts of to-day's session, and, even more so, on surveying the current literature, one is struck by the contradictions in the interpretation of results relating catecholamine activity to behaviour. It may be helpful to try and analyse some of these discrepancies. In order not to complicate matters, I will take it for granted that there is universal agreement about the need to keep the distinction between dopaminergic and adrenergic neurones clearly in mind.

(1) Many statements about function are based on the results of damage to neuronal activity. The degree of damage done by what appears to be the same procedure can vary with fine distinctions in technique. Owing to the remarkable capacity of the brain to maintain crucial activities when an enzyme, an amine, or a group of neurones responsible for a particular function have been severely reduced in quantity, results will differ according to whether the experimenter has just reached or just failed to reach the threshold beyond which cerebral compensatory mechanisms fail. A good example of this is found in the work by Breese, Cooper and their co-workers (COOPER, BREESE, HOWARD and GRANT, 1972; BREESE and COOPER, this session). They found it necessary to reinforce the damaging effect of intracisternal 6-hydroxy-dopamine (6-OHDA) with either pargyline or α-methyltyrosine in order to reduce aminergic activity sufficiently to obtain an impairment of the acquisition of avoidance responses in rats.

(2) Hardly ever do different experimenters use the same procedures for testing behaviour. This in itself should be a useful thing, permitting one to pin down the particular feature of a test which either can, or cannot, be compensated for after what is presumed to be the infliction of identical damage. This requires much consultation between groups of workers, and one would hope that a symposium like the present one will offer opportunities to resolve some of the apparent contradictions.

(3) The site of the damage is, needless to say, all-important. This, however, is the most difficult feature to assess, particularly when the comparison concerns the results of electrolytic lesions, or of local, or even of intraventricular injections of damaging substances like 6-OHDA. Since intraventricular injection of 6-OHDA is both useful and frequently used, I wish to report on some observations with the injection, into the lateral ventricle of rats, of a somewhat related compound, 5,6-dihydroxytryptamine (5,6-DHT). This substance does not seriously interfere with catecholamine-containing neurones, but damages selected 5-HT containing neurones. For reasons not yet known, the spinal cord appears to be the most permanently affected region (BAUMGARTEN et al., 1972). 5,6-DHT leaves a brown pigment on the surface of those ventricular regions which are reached by high concentrations of the drug. It was surprising to see that, in spite of keeping injection time, and volume and concentration of the injected solution, constant, the distribution of the brown stain was variable. Sometimes it only covered the caudate nucleus and septum on the injected side, but occasionally it also stained the aqueduct and the third ventricle.

It is unlikely that such differences did not produce variations in the groups of neurones affected. Another point is of importance in the interpretation of early behavioural changes after the intraventricular injection of 6-OHDA. During the first days which follow the injection of 5,6-DHT, the rats are excitable, tend to fight if not caged singly, and are aroused by the slightest noise. The same signs, however, are familiar consequences of the injection, by the same route, of 6-OHDA (e.g. JACKS, De CHAMPLAIN and CORDEAU, 1972). To try and relate these emotional changes to either release or lack of NA or DA is obviously not going to be compatible with the production of the same syndrome by 5,6-DHT.

We know a great deal about correlations between effects of drugs on behaviour and on brain biochemistry, but the transformation of mere correlations into causal relationships is in its mere infancy. The same holds for the subject which is sometimes called "manipulation of cerebral catecholamines" and its effects on behaviour; causal relationships are extremely difficult to establish.

REFERENCES

BAUMGARTEN H. G., EVETTS K. D., HOLMAN R. B., IVERSEN L. L., VOGT M. and WILSON G. (1972) *J. Neurochem.* **19,** 1587–1597.

COOPER B. R., BREESE G. R., HOWARD G. L. and GRANT L. D. (1972). *Physiol. Behav.* **9,** 727–731.

JACKS B. R., DE CHAMPLAIN J. and CORDEAU J.-P. *Eur. J. Pharmacol.* (1972) **18,** 353–360.

Frontiers in Catecholamine Research 1973, pp. 635 to 636. Pergamon Press. Printed in Great Britain.

CENTRAL CATECHOLAMINES AND HYPERTENSION

ALBERT SJOERDSMA

Centre de Recherche Merrell International, 16, rue d'Ankara, 67000 Strasbourg,
France

THERE appear to be two main themes in the session to ensue: (1) the central adrenergic actions of antihypertensive drugs such as α-methyl-dopa and clonidine; and (2) involvement of central and peripheral catecholamine mechanisms in blood pressure control and the pathophysiology of hypertension. I wish to begin the introduction of these subjects with a philosophic overview based on a personal long range interest in the ultimate target of our remarks, namely, the human patient with essential hypertension. I recall that early-on, at a time when no effective therapy was available and hypertension was a clinically-obvious killer, that visitors to our laboratory would frequently suggest that we should concentrate on finding the cause of hypertension, after which the cure would certainly follow. I had a number of angry responses to this naive suggestion, to whit, that people were dying because their blood pressure was too high and there was not time to wonder why, and besides we did not need to know the cause(s) in order to develop effective therapy, further that if we succeeded in controlling the blood pressure we would know more about etiology as a consequence, and finally that species variation in blood pressure responses were such that we intended to ignore animal pharmacology and follow directly into man leads provided us by catecholamine biochemists. This latter approach paid off in 1959 with our discovery in hypertensive patients of blood pressure lowering and sedative effects following administration of α-methyl-dopa (OATES et al., 1960). It is of interest that, several years later, the same pharmacologic effects of clonidine were also first observed in man (GRAUBNER and WOLF, 1966). By 1966 we became sufficiently satisfied with the clinical therapeutic status to become interested in studying catecholamine metabolism in the spontaneously hypertensive rat (SHR) of Okamoto and Aoki and established a parent colony in the United States. Thereby, and with trepidation, we joined others on the primrose path of investigating the role of catecholamines in experimental hypertension.

In 1966 I expressed the opinion, based on extensive research findings in laboratory animals and man that "the effects of methyldopa are due to a combination of effects—including tranquilising and possibly as yet unknown actions in the central nervous system, peripheral transmitter depletion, possibly a substitute transmitter component, and perhaps even a slight inhibitory effect on catecholamine biosynthesis" (SJOERDSMA, 1967). It was of course clear by this time that α-methyl-dopa lowers blood pressure, not because it is a decarbozylase inhibitor, but because it is itself decarboxylated to α-methyl-dopamine which in turn is β-hydroxylated to the active product, α-methyl-norepinephrine. Subsequently, a mass of evidence has accumulated permitting the conclusion that acute hypotensive effects of methyldopa in animals are due to the formation of α-methyl-noradrenaline which acts at an α-adrenergic site in the brain, probably in lower brainstem. Findings leading to this conclusion were achieved by a variety of techniques including pretreatment with centrally-active decarboxylase and β-hydroxylase inhibitors and α-receptor antagonists, and central routes of compound administration. Breakthrough evidence was provided by HENNING and RUBENSON (1971) and associates in Sweden, with independent parallel and confirmatory studies provided by HEISE and KRONEBERG (1972) in Germany, FINCH and HAEUSLER (1973) in Switzerland and others. Supportive evidence for central noradrenergic stimulation as a basis for blood pressure reduction comes from studies on clonidine, which will be discussed by Prof. Schwartz and others shortly.

I would now like to hark back to my earlier statement that improved treatment of hypertension with drugs might have etiologic implications. Certainly, the importance of surgical and later medical "sympathectomy" in therapy has been a potent stimulus to studies on catecholamine metabolism in the peripheral sympathetic system, both in human and experimental hypertension. Some of our speakers will address themselves to this problem and we can then judge the etiologic implications. But now that two of our best drugs, methyldopa and clonidine, have been shown to act on central catecholamine mechanisms, a huge effort focused on the role of catecholamines in the brain in blood pressure control and the etiology of hypertension is clearly warranted. What we hear to-day will only be the beginning. We began to focus on this latter area a few years ago and proposed in 1970 (YAMORI et al., 1970) that catecholamines, while being pressor peripherally, may participate in a central depressor system. This hypothesis was based in part on suggestive evidence of decreased norepinephrine synthesis rates and concentrations in brainstem of the spontaneously hypertensive rat and was supported by the demonstration (YAMORI et al., 1972) of a highly significant inverse correlation between levels of norepinephrine in brainstem, and blood pressure, in genetically hypertensive animals treated with various combinations of three drugs. These included a monoamine oxidase inhibitor, a peripheral decarboxylase inhibitor, and L-dopa.

Recalling the historical aspects again, I would like to conclude by re-emphasising the importance of studies in patients with hypertension. The burden is now on the clinical pharmacologist to ascertain whether the central actions demonstrated in laboratory animals account for blood pressure responses to clonidine and methyldopa in patients. It should be rather simple to ascertain whether the hypotensive effects of these drugs in man are antagonised by central receptor antagonists such as chlorpromazine, haloperidol and phenoxybenzamine. While I cannot recommend use of the vertebral artery or intracranial route of drug administration which has been so revealing in animal experiments, some of the same biochemical approaches used in patients with Parkinsonism could certainly be applied. In a similar vein, thought should be given to the possibility of a localised central catecholamine deficiency as an etiologic factor in patients with hypertension and an attempt made to study this in post-mortem material.

REFERENCES

FINCH L. and HAEUSLER G. (1973) *Br. J. Pharmac.* **47**, 217–228.
GRAUBNER W. and WOLF M. (1966) *Arzneim-Forsch. (Drug Res.)* **16**, 1055–1058.
HEISE A. and KRONEBERG G. (1972) *Europ. J. Pharmac.* **17**, 315–317.
HENNING M. and RUBENSON A. (1971) *J. Pharm. Pharmacol.* **23**, 407–411.
OATES J. A., GILLESPIE L., UDENFRIEND S. and SJOERDSMA A. (1960) *Science* **131**, 1890–1891.
SJOERDSMA A. (1967) In: *Supplement III to Circulat. Research*, **XX** and **XXI**, 119–125.
YAMORI Y., DE JONG W., YAMABE H., LOVENBERG W. and SJOERDSMA A. (1972) *J. Pharm. Pharmac.* **24**, 690–695.
YAMORI Y., LOVENBERG W. and SJOERDSMA A. (1970) *Science* **170**, 544–546.

Frontiers in Catecholamine Research 1973, pp. 637 to 642. Pergamon Press. Printed in Great Britain.

NOREPINEPHRINE AS A CENTRAL SYNAPTIC TRANSMITTER

FLOYD E. BLOOM and BARRY J. HOFFER

Laboratory of Neuropharmacology, Division of Special Mental Health Research,
National Institute of Mental Health, St. Elizabeths Hospital,
Washington, D.C. 20032, U.S.A.

THE ULTIMATE test of the function of central noradrenergic synapses must result from the selective activation of these terminals and from the detection of their activity patterns in the unrestrained experimental animal. That such higher order testing can even be conceived indicates the conceptual and experimental advances which have occurred since the last major catecholamine review (SALMOIRAGHI, 1966; BLOOM, 1968; BLOOM et al., in press; HOFFER and BLOOM, 1972). At that time emphasis was rightly placed upon data which demonstrated that iontophoretically administered norepinephrine (NE) could produce actions with sufficient potency and regularity to merit consideration as a synaptic transmitter. At present, many differences in interpretation are still associated with data obtained by applying the iontophoretic method to CNS areas in which the precise anatomy of the NE-containing terminals has yet to be related to the cells undergoing testing. Beyond these interpretive problems however, it has been possible to determine the actions of NE, fibre systems arising from the pontine nucleus, locus coeruleus (LC), and projecting to the cerebellar Purkinje cells (HOFFER et al., 1971a) and to the hippocampal pyramidal cells (SEGAL and BLOOM, in press) and to record from the LC neurons themselves during unrestrained observations of sleeping and waking behaviour (CHU and BLOOM, 1973). All of these actions which reflect upon the function of central noradrenergic synapses have been recently reviewed, (BLOOM et al., in press; HOFFER and BLOOM, 1973; HOFFER et al., 1971a) and in the space allotted for the present survey we will concentrate upon two questions: what are the principal effects of iontophoretically applied NE; what is the nature of the physiological central NE receptor and the molecular mechanisms by which NE synapses can exert these effects.

Principle actions of iontophoretically applied NE. The extensive work of the past 5 years has amply supported the view that NE is able to effect the discharge of neurons in all portions of the neuraxis (see Table 1 of HOFFER and BLOOM, 1973). Present work is more realistically directed at determining the functional significance of the presence or absence of responses to NE and the qualitative nature of the response, i.e., excitatory or inhibitory. Retrospective analysis of experimental inconsistencies suggests that major variables were uncontrolled. Thus, excitatory responses to NE in the cerebral cortex were found to be more frequent when the animal was unanesthetised or under the influence of halothane anesthesia while inhibitory cortical responses were more frequent with barbiturate anesthesia (JOHNSON et al., 1969). Similarly, the pH of the NE solution to be iontophoretically applied also takes on critical importance, as pH values less than 4 are associated with major increases in the frequency of excitatory NE responses (FREDERICKSON et al., 1972). An extreme view of the anesthesia-pH controversy is that all excitatory responses to NE are the result of vasoconstriction artefacts (STONE, 1971), a suggestion which was quickly rejected (BOAKES et al., 1972).

The eclectic view of these controversies must await additional observations on several other factors such as the cytological or functional heterogeneity of the cell population being tested: it is known that when defined poulations of neurons are tested, the responses are far more reproducible than when all randomly encountered neurons in a given CNS region are lumped together (SALMOIRAGHI and STEFANIS, 1967). An additional cytological index which must be considered is whether or not the population of cells to be tested receives a demonstrable synaptic input of NE-containing terminals. Thus, the highly variable responsiveness of hypothalamic neurons to NE (BLOOM et al., 1963) reduces to clear cut patterns of primarily excitatory responses when the data are restricted to those border cells of the ventro-medial nucleus which receives NE-containing synapses (KREBS and BENDRA, 1971). The same degree of variable results can also be obtained on random populations of medullary neurons, and the data here reduce to primarily inhibitory actions when analysis is restricted to cells identified as cranial motoneurons (OLIVER et al., 1972) all of which are the recipient of relatively dense NE-containing terminals. The latter two observations indicate that even when anesthesia and solution pH are held constant, variable results can still be obtained when no significance is attached to the neuron to be tested or the input it receives.

When we restrict our condensation of reported results to those populations of defined post-synaptic cells for which NE-containing nerve terminals have been demonstrated by light or electron microscopic techniques, the somewhat shorter list obtained shows almost universally inhibitory results (BLOOM, 1968; BLOOM et al., in press; HOFFER and BLOOM, 1972; SALMOIRAGNI and STEFANIS, 1967). Thus, cerebellar Purkinje cells, (BLOOM et al., in press; HOFFER and BLOOM, 1972; HOFFER et al., 1971a) olfactory bulb mitral cells, (BLOOM et al., 1964) supraoptic and para-ventricular hypothalamic neurosecretory neurons (BARKER et al., 1971), spinal motoneurons (WRIGHT and SALMOIRAGNI, 1967), primary neurons of the medial geniculate (TEBECIS, 1970) and polysensory neurons of the primate frontal cortex (NELSON et al., in press) show reproducible inhibitory effects, in addition to the populations of defined neurons mentioned above in the hypothalamus (BLOOM et al., 1963) and medulla (KREBS and BENDRA, 1971). The most serious obstacle to interpretation of even these results has been that the NE-containing pathway has not been amenable to selective electrical activation for comparisons of qualitative and pharmacological results on the specified post-synaptic population of test neurons. Conversely, synaptic effects which were subjectible to selective activation were neither exclusively NE-containing nor sensitive to NE antagonists. Thus, in neither the olfactory mitral cells (SALMOIRAGNI et al., 1964) nor the hypothalamic neurosecretory cells (NICOLL and BARKER, 1971) could the recurrent antidromic synaptic inhibition be removed by the results of acute or chronic NE depletion, or NE receptor blockade.

Over the past several years we have pursued the mechanism by which NE slows the discharge of cerebellar Purkinje cells (HOFFER et al., 1971b). We have used light and electron microscopy to establish that these cells receive NE-containing synapses onto their dendrites (BLOOM et al., 1971), and that these NE fibres arise from the LC (OLSON and FUXE, 1971; BLOOM et al., 1972a). By electrophysiological methods, we have analysed the pharmacological receptors of the Purkinje cells (SIGGINS et al.,

1971a) and the effects of electrical activation of the pathway (HOFFER et al., 1971a; SIGGINS et al., 1971b). Briefly, these experiments indicate that NE slows Purkinje cells by interaction with a beta receptor, by prolonging the pauses between bursts of single spikes without effect on climbing fibre responses. By intracellular recordings, NE hyperpolarises the membrane of Purkinje cells and this hyperpolarisation is generally accompanied by increased membrane resistance (but never by increased membrane conductance). The actions of norepinephrine on the Purkinje cell are blocked by iontophoretic application of a beta antagonist (MJ 1999), of prostaglandins of the E series and by nicotinate; the latter both inhibit adenylate cyclase in some autonomic tissues. The effects of norepinephrine on discharge rate and membrane parameters are precisely emulated by iontophoretic application of cyclic AMP, and the effects of both the applied cyclic AMP and of NE are potentiated by any of several phosphodiesterase inhibitors. On the basis of these data, we proposed (HOFFER et al.,1971b; SIGGINS et al., 1971a) that the synaptic action of NE was mediated by an interaction via the adenyl cyclase of the cerebellar cortex, known to be highly responsive to NE (see BLOOM et al., in press).

With the anatomical information that the cerebellar NE fibres arose from the LC (OLSON and FUXE, 1971; BLOOM et al., 1972a), it was possible to test this proposal by activating and analysing the effects of the pathway on Purkinje cell properties. These experiments disclosed that stimulation of the pathway inhibited Purkinje cell discharge, especially single spike bursts, that the inhibitory effects of stimulating LC required active synthesis of NE, and that no effects on cerebellar neuronal discharge were observed when the area of the locus was stimulated in animals pretreated with 6-hydroxydopamine to eradicate the adrenergic projection to the cerebellum (HOFFER et al., 1971a). By intracellular recording during the activation of the LC, Purkinje cells were found to be hyperpolarised and this hyperpolarisation was usually accompanied by a definitive increase in the resistance of the membrane (SIGGINS et al., 1971b). Similar effects of NE have been observed on motoneurons (ENGBORG and MARSHALL, 1971). Pharmacologically, activation of the LC led to an inhibition of spontaneous discharge which could be potentiated by local iontophoresis of phosphodiesterase inhibitors onto the Purkinje cell and could be blocked by local iontophoretic administration of prostaglandins of the E series. All these results supported the concept that this adrenergic projection could be operating by the trans-synaptic elevation of cyclic AMP in Purkinje cells. The latter observation has now been documented by application to tissue sections of an immunocytochemical method for cyclic AMP (BLOOM et al., 1972b). Using this method we have observed that topical application of NE or electrical activation of the LC will elevate the number of Purkinje cells showing positive immunocytological staining for cyclic AMP from resting frequencies of 5–15 per cent to levels greater than 75 per cent (SIGGINS et al., 1973). Neither topical application of GABA, glycine, histamine, or acetylcholine, or electrical activation of other cerebellar pathways has this effect on Purkinje cell cyclic AMP (SIGGINS et al., 1973). A generally similar set of observations, although less extensively analysed have been obtained by Greengard and his co-workers for, the dopaminergic intraganglionic synapses in rabbit sympathetic ganglia (GREENGARD, et al., 1973).

Finally, it is worthwhile considering whether the paradigm of NE effects at the synaptic receptor examined on cerebellar Purkinje cells is to be considered as a general example of the central synaptic actions of NE or rather as a specific and somewhat unique type of receptor. Based on biochemical experiments, it is known that the actions of transmitters on adenylate cyclase activation are specific, within a species, for certain regions of the brain, and within regions, for certain transmitters (Rall, 1971). In the cerebellum of the rat, NE is the most active transmitter substance in activating cyclase activity (Rall, 1971) but in the cerebral cortex, NE-induced adenylate cyclase activation is highly species dependent, being relatively poor in rat and quite potent in primate (Skolnick et al., in press) and human cortex (Fumagilli et al., 1971; Shimazu et al., 1971).

Viewed from a different perspective, we may ask two testable questions: (1) need all actions of NE be explicable in terms of activation of cyclic AMP synthesis and the subsequent actions of the cyclic nucleotide; (2) are all actions of cyclic AMP at central synapses due to activation of the adenylate cyclase by catecholamines. The first action might be tested by analysing whether the ability of NE to affect neuronal discharge correlates with the ability of NE to activate adenylate cyclase of neurons in a given region of a given species. Such a test has been made by Phillis and collaborators (Jordan et al., 1972) who found that the actions of NE were uniformly inhibitory across species lines regardless of the neurochemically determinable stimulatory potency of NE on the adenylate cyclase. On the other hand, the squirrel monkey shows marked receptivity to NE (Nelson et al., in press) and a very striking elevation of cyclic AMP synthesis in vitro to NE (Skolwick et al., in press). Part of the solution to these apparently paradoxical data may depend upon the same considerations with which we began this survey, namely that receptivity to NE may not be assumed to infer the presence of NE-mediated synapses. In fact, a large number of NE terminals in an area might even result in apparent poor receptivity due to rapid accumulation of iontophoretically released NE enroute to post-synaptic receptors. If that assumption be granted, then clearly receptivity to NE also need not imply an NE-sensitive adenylate cyclase underlying the action. Furthermore, the ability to demonstrate the capacity of a putative transmitter to activate adenylate cyclase activity is also in a stage of continual technical development, in which cofactors and ionic conditions are extremely important for the result observed. Therefore, when faced with a negligible apparent rise in the cyclic AMP synthesis rate in vitro for a brain slice, one must consider that the system being used to test for the rise in cyclase activity may have been inadequate or that the cells within the slice which exhibited a stimulatory effect by the exogenous transmitter represented such a small proportion of the total tissue mass of the slice that the rise went undetected by the assay. In the case of the cerebellar cortex, the use of the immunocytochemical staining method offers a direct demonstration of increased cyclic AMP content of Purkinje cells, a system where the synapses and the effects of the synaptic pathway can be equated with the action of applied NE. For other regions of the brain, particularly where the cytology is less amenable to resolution of precise synaptology to cells identifiable during extracellular recording, these same tests may be difficult to complete.

It has also been observed that the actions of NE in the cerebral cortex (Jordan et al., 1972) and in the brainstem (Anderson et al., 1973) does not show clearcut potentiation when phosphodiesterase inhibitors are administered either iontophoretically or parenterally. Lake et al. (1972) did in fact find that aminophylline and papaverine could potentiate the inhibitory actions of NE in approximately 70% of the unidentified neurons they tested in the feline cerebral cortex. However, these investigators interpreted this response to be a non-specific summation of inhibitory actions since the phosphodiesterase inhibitors were directly inhibitory on the same cells. While these results may be interpreted as non-specific, it perhaps should be expected that phosphodiesterase inhibition—which would be expected to elevate cyclic AMP levels generally—might produce more widespread actions than NE, particularly since the action of phosphodiesterase is independent of the hormone which activated cyclic AMP synthesis. Both of these considerations should also be weighed in evaluation of the observations on prostaglandin antagonisms of NE responsiveness in the cat cortex (JORDAN et al., 1972) and the cat brainstem (ANDERSON et al., 1973).

Perhaps it would be most fair to ask just what evidence would be needed to determine that the actions of NE in a given region were unrelated to an effect on adenylate cyclase. Given the limitations imposed by present techniques, one would have to demonstrate the following: (1) NE synapses, arising from a defined NE cell group, are present on cells electrophysiologically identifiable during electrical recording; (2) the action of NE and of the NE-pathway is not emulated by cyclic AMP, either on the rate or pattern of discharge or on membrane properties; (3) the actions of phosphodiesterase inhibitors do not potentiate the NE-related actions, but do selectively potentiate the actions of exogenous cyclic AMP; (4) the biochemical or immunocytochemical estimation of cyclase activity is independent of the actions of applied NE or activation of the NE synapses.

In conclusion, despite the great progress which has been made in the electrophysiological documentation of NE as a central synaptic transmitter in selected regions of the CNS, many technical and interpretive pitfalls lie in the path of experiments designed to extend these observations and approaches to other CNS regions. A fruitful approach may be difficult to obtain, but still it must be sought.

Acknowledgements—We thank Dr. G. R. Siggins for critical advice on the manuscript, and Mrs. O. T. Colvin for meticulous, rapid typing.

REFERENCES

ANDERSON E. G., HAAS H. L. and HOSLI L. (1973) *Brain Res.* **49**, 467.
BARKER J. L., CRAYTON J. C. and NICOLL R. A. (1971) *J. Physiol.* **218**, 19.
BLOOM F. E. (1968) In: *Psychopharmacology* (EFFRON D. Ed.) p. 355. Government Printing Office, Washington.
BLOOM F. E., OLIVER A. P. and SALMOIRAGHI G. C. (1963) *Int. J. Neuropharmac.* **2**, 181.
BLOOM F. E., COSTA E. and SALMOIRAGHI G. C. (1964) *J. Pharmac. exp. Ther.* **146**, 16.
BLOOM F. E., HOFFER B. J. and SIGGINS G. R. (1971) *Brain Res.* **25**, 501.
BLOOM F. E., HOFFER B. J. and SIGGINS G. R. (1972a) *Biol. Psychiat.* **4**, 157.
BLOOM F. E., HOFFER B. J., BATTENBERG E. F., SIGGINS G. R., STEINER A. L., PARKER C. W. and WEDNER H. J. (1972b) *Science* **177**, 436.
BLOOM F. E., CHU N-s., HOFFER B. J., NELSON C. N. and SIGGINS G. R. (1973) *Neurosci. Res.* **5**, 53.
BOAKES R., BRADLEY P., CANDY J. and DRAY A. (1972) *Nature, New Biol.* **239**, 151.
CHU N-s. and BLOOM F. E. (1973) *Science* **179**, 908.
ENGBORG I. and MARSHALL K. C. (1971) *Acta. physiol. scand.* **83**, 142.
FREDERICKSON R., JORDAN L. and PHILLIS J. (1972) *Brain Res.* **35**, 556.
FUMAGILLI R., BERNAREGGI V., BERTI F. and TRABUCCHI M. (1971) *Life Sci.* **10**, 1111.
GREENGARD P., KEBABIAN J. W. and McAFEE D. A. (1973) In: *Proc. Vth Int. Congr. Pharmac.* (ACHESON G Ed.), Karger, Basel.

HOFFER B. J. and BLOOM F. E. (1973) In: *The Influence of the Limbic System on Autonomic Function* (HOCHMAN C. Ed.) p. 91. Charles Thomas, Springfield. (in press).

HOFFER B. J., SIGGINS G. R., OLIVER A. P. and BLOOM F. E. (1971a) *J. Pharmac. exp. Ther.* **184**, 553.

HOFFER B. J., SIGGINS G. R., OLIVER A. P. and BLOOM F. E. (1971b) *Ann. N.Y. Acad. Sci.* **185**, 531.

JOHNSON E., ROBERTS M. and STRAUGHAN D. (1969) *J. Physiol.* **203**, 261.

JORDAN L. M., LAKE N. and PHILLIS J. (1972) *Europ. J. Pharmac.* **20**, 381.

KREBS H. and BENDRA B. D. (1971) *Nature, Lond.* **229**, 178.

LAKE N., JORDAN L. M. and PHILLIS J. (1972) *Nature, New Biol.* **240**, 249.

NELSON C., HOFFER B. J., CHU N-s. and BLOOM F. E. *Brain Res.* (in press).

NICOLL R. A. and BARKER J. L. (1971) *Brain Res.* **35**, 501.

OLIVER A. P., SIMS K. L. and BLOOM F. E. (1972) *Abst. Vth Int. Congr. Pharmac.* 171.

OLSON L. and FUXE K. (1971) *Brain Res.* **28**, 165.

RALL T. W. (1971) *Ann. N.Y. Acad. Sci.* **185**, 520.

SALMOIRAGHI G. C. (1966) *Pharmac. Rev.* **18**, 717.

SALMOIRAGHI G. C. and STEFANIS C. (1967) *Int. Rev. Neurobiol.* **10**, 1.

SALMOIRAGHI G. C., COSTA E. and BLOOM F. E. (1964) *Am. J. Physiol.* **207**, 1417.

SEGAL M. and BLOOM F. E. *Abst. 3rd Ann. Meet. Soc. Neurosci.* (in press).

SHIMIZU H., TANAKA S., SUZUKI T. and MATSUKADO Y. (1971) *J. Neurochem.* **18**, 1157.

SKOLNICK P., HUANG M., DALY J. and HOFFER B. J. (1973) *J. Neurochem.* (in press).

SIGGINS G. R., HOFFER B. J. and BLOOM F. E. (1971a) *Brain Res.* **25**, 535.

SIGGINS G. R., HOFFER B. J., OLIVER A. P. and BLOOM F. E. (1971b) *Nature, Lond.* **233**, 481.

SIGGINS G. R., BATTENBERG E. F., HOFFER B. J., BLOOM F. E. and STEINER A. L. (1973) *Science* **179**, 585.

STONE T. (1971) *Nature, Lond.* **234**, 145.

TEBECIS A. (1970) *Neuropharmac.* **9**, 381.

WEIGHT F. F and SALMOIRAGHI G. C. (1967) *Nature, Lond.* **213**, 1229.

Frontiers in Catecholamine Research 1973, pp. 643 to 648. Pergamon Press. Printed in Great Britain.

CENTRAL DOPAMINERGIC NEURONS: NEUROPHYSIOLOGICAL IDENTIFICATION AND RESPONSES TO DRUGS*

GEORGE K. AGHAJANIAN and BENJAMIN S. BUNNEY

Departments of Psychiatry and Pharmacology, Yale University School of Medicine
and the Connecticut Mental Health Center, New Haven, Conn. 06508, U.S.A.

INTRODUCTION

FLUORESCENCE histochemical studies have shown that the soma of central dopamine (DA)-containing (dopaminergic) neurons are mainly located in the zona compacta (ZC) of the substantia nigra and adjacent ventral tegmental (VT) area (DAHLSTRÖM and FUXE, 1965; UNGERSTEDT, 1971a). Neurons in the zona reticulata (ZR) of the substantia nigra, on the other hand, do not contain DA and are thus distinguishable from ZC and VT cells by fluorescence histochemical methods. Despite these histochemical differences, in neurophysiological studies involving either stimulation of or recording from the substantia nigra it has not been customary to distinguish between dopaminergic and non-dopaminergic neurons. In view of the histochemical studies, however, it was of interest to determine if ZC and VT neurons could be distinguished from ZR and other cells according to their pattern of single unit activity and whether they would show selective responses to drugs and other substances known to affect dopaminergic mechanisms.

IDENTIFICATION OF DOPAMINERGIC NEURONS

To establish the dopaminergic identity of neurons observed in single unit recordings conducted in rat brain, three independent procedures were employed (BUNNEY et al., 1973b). As a first step, on the basis of standard histological examination, cells were presumptively identified as dopaminergic if a dye spot ejected from a stereotaxically placed recording electrode was located within the ZC or VT. This criterion, however, is by itself insufficient, since some non-dopaminergic neurons (e.g., in the ZR) are found on the immediate border of the ZC and VT. Therefore, two other criteria, both involving combined neurophysiological and fluorescence histochemical methods were employed. It was found that when an electrode tip was located in the ZC or VT, microiontophoresis of L-dopa led to a markedly enhanced histochemical fluorescence of neurons surrounding the electrode tip; in other areas, only the neuropil or capillaries became fluorescent under these conditions. In a counterpart of the preceding experiment, small amounts of 6-hydroxydopamine were injected in the vicinity of the substantia nigra, a treatment which results in a selective destruction of dopaminergic neurons (cf. UNGERSTEDT, 1971b). 1–4 days following this procedure, there was a selective loss of single units which had the firing pattern and responses to drugs which, by the above histological and histochemical techniques, were determined to be characteristic of the neurons in the ZC and VT (see below).

---

* This research was supported by NIMH Grant MH-17871, USPHS Research Scientist Development Award MH-14459 (to G. K. A.), and the State of Connecticut.

NEUROPHYSIOLOGICAL PROPERTIES OF DOPAMINERGIC NEURONS

Neurons in the rat brain identified as "dopaminergic" by the above criteria were found to be remarkably homogeneous with respect to their firing pattern and responses to drugs and other substances administered systemically (Bunney et al., 1973a,b) or applied iontophoretically. In these same respects, dopaminergic neurons were clearly distinguishable from neurons in nearby areas such as the ZR of the substantia nigra, the red nucleus, or the reticular formation. Dopaminergic cells typically show extracellular action potentials with a positive-negative wave form of unusually long duration ($\sim$2msec). They tend to have a regular rhythm and a slow rate of firing (2–6 spikes/sec) in unanaesthetised, gallamine-paralysed preparations, but curiously exhibit an increase in firing rate and a bursting pattern when the animal is anaesthetised with chloral hydrate or halothane. In contrast, ZR cells tend to have a relatively rapid rate of firing and are either unaffected or depressed by anaesthesia.

Microiontophoretic studies show still further differences between the dopaminergic neurons of the ZC and VT and the non-dopaminergic neurons of the ZR. Dopaminergic cells showed little or no response to acetylcholine whereas neurons in the ZR were consistently excited by acetylcholine even at very low microiontophoretic currents (Fig. 1). On the other hand, with one exception, ZR neurons tested ($N = 24$) showed no response to DA applied microiontophoretically (Fig. 1, bottom trace), whereas all dopaminergic neurons tested showed at least an initial depressant response to DA at low ejection currents (Fig. 1, top trace). Interestingly, after repeated applications there tended to be an attenuation of the response to DA. As can be seen from Fig. 1 (top trace) the initial application of DA at an ejection current of 10 nA temporarily produced a total inhibition of firing but at later times as much as 60 nA was required to cause approximately this same degree of inhibition. At this later time, 10 nA of DA had no appreciable effect In postsynaptic areas (i.e., those receiving dopaminergic input, such as the caudate nucleus, accumbens nucleus and the olfactory tubercles (Fuxe, 1965; Anden et al., 1966; Ungerstedt, 1971a) no attenuation of response to DA was seen after repeated applications (Aghajanian and Bunney, unpublished data).

ZC and VT cells as well as ZR cells were inhibited by γ-aminobutyric acid and excited by glutamate applied microiontophoretically; thus these substances did not serve to distinguish dopaminergic from non-dopaminergic neurons.

DOPAMINERGIC NEURONS: EFFECT OF DA-RECEPTOR BLOCKADE

Drugs believed to have actions upon the dopaminergic system in the CNS have been found to have dramatic effects upon the rate of firing of ZC and VT neurons (Bunney et al., 1973a,b). Such drugs fall into two general classes: those which appear to block DA receptors and those which either increase DA availability or mimic the action of DA at receptors. The antipsychotic drugs of the phenothiazine and butyrophenone type have been shown to have a structural relationship to DA and thus might competitively block DA receptors (Horn and Snyder, 1971). These drugs have been shown to markedly increase DA turnover in the neostriatum (Carlsson and Lindqvist, 1963; Nybäck et al., 1968; Corrodi et al., 1967; Gey and Pletscher, 1968), an effect requiring an anatomically intact nigrostriatal pathway (Cheramy et al., 1970; Andén et al., 1971; Nybäck and Sedvall, 1971). It has been hypothesised that a blockade of DA receptors produced by these drugs might

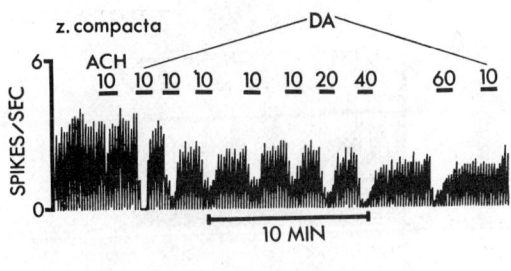

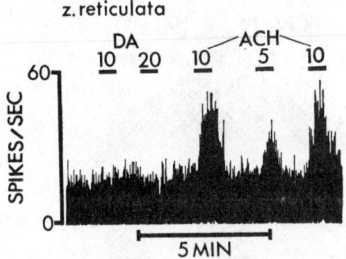

FIG. 1.—Comparison of responses of neurons in zona compacta (ZC) and zona reticulata (ZR) of the substantia nigra to the microintophoretic application of dopamine (DA) and acetylcholine (ACh).

In the *top trace*, ACh ejected at a current of 10 nA upon a ZC neuron is seen to produce no effect. There was no appreciable response in any of the ZC and VT cells tested with ACh ($N = 8$), even with ejection currents up to 40 nA. On the other hand, all ZC and VT cells tested with DA ($N = 16$) responded, at least initially, with a depression in firing rate. Typically as is shown in the *top trace*, there was an attenuation of discrete responses to DA with repeated applications, associated with a downward drift of baseline rate.

The *bottom trace* illustrates that ejection of DA usually causes no change in firing rate of ZR neurons. In only 1 of 24 ZR cells tested was there a response (depression) to microiontophoretic DA. In contrast, ACh was consistently excitatory on all ZR cells tested ($N = 24$) and was usually effective at low ejection currents (5–10 nA, as shown in *bottom trace*).

Methods for recording and microiontophoretic drug applications through 5–barreled micropipettes were as previously described (AGHAJANIAN *et al.*, 1972; HAIGLER and AGHAJANIAN, 1973). The duration of ejecting current is indicated by length of bar and intensity of ejection current by number (in nA) above bar. Rate of firing is displayed by on-line integrated rate record calibrated on ordinate in spikes/sec. Concentration of substances in micropipettes: ACh, 0·2 M (pH 6) ; DA. 0·2 M (pH 4).

lead to a compensatory increase in the activity of dopaminergic neurons via a neuronal-feedback mechanism (CARLSSON and LINDQVIST, 1963; ANDÉN *et al.*, 1964). In direct support for this hypothesis it has been found that the systemic administration of antipsychotic compounds including chlorpromazine (CPZ) and haloperidol (HAL) causes an increase in the rate of firing of ZC and VT neurons (BUNNEY *et al.*, 1973a,b). However, when CPZ is applied directly to dopaminergic neurons it has little or no effect of its own and does not block the inhibitory action of DA (Fig. 2A). In contrast small intravenous doses (0·5–1·0 mg/kg) of CPZ produced a doubling in the rate of firing of ZC and VT neurons (Fig. 2B). HAL has been found to accelerate the firing of dopaminergic cells to the same extent as does CPZ (BUNNEY *et al.*, 1973b). When HCL is given after CPZ it produces no further effect (Fig. 2B), suggesting that these drugs act upon a common set of receptors within the dopaminergic system.

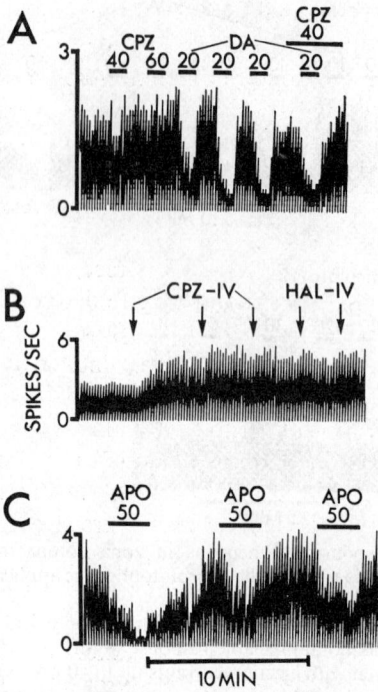

Fig. 2.—Effects of chlorpromazine (CPZ), dopamine (DA) and apomorphine (APO) on the rate of firing of dopaminergic neurons of the zona compacta (ZC) and ventral tegmental (VT) areas.

In "A", CPZ is seen to have no effect on firing of a dopaminergic cell nor does it block DA inhibition when applied concurrently by microiontophoresis. This was the case with all ZC and VT cells tested with CPZ ($N = 8$) despite the fact that the drug was ejected up to levels sufficient to produce some degree of local anaesthetic effect.

In "B", CPZ given intravenously (IV) in a sequence of low doses (0·5, 0·5 and 1·0 mg/kg) is seen to maximally accelerate the firing of a dopaminergic neuron. This result is in accord with previous studies (Bunney et al., 1973b). Once a maximal effect is achieved with CPZ no further effect is seen with haloperidol (HAL: 0·1 and 0·5 mg/kg).

In "C", APO given iontophoretically is seen to inhibit the firing of a dopaminergic neuron. This effect was found with all dopaminergic neurons tested with APO ($N = 10$). As can be seen in this example, there was some attenuation of response with repeated applications.

Methods as described in Fig. 1. Concentrations of drugs used in the 5-barreled pipettes: CPZ, 0·2 M (pH4); APO, 0·06 M (pH 3·5).

DOPAMINERGIC NEURONS: EFFECTS OF DA RECEPTOR STIMULATION

Drugs that are believed to directly or indirectly stimulate DA receptors have been found to cause a depression of the firing rate of dopaminergic neurons (Bunney et al., 1973a,b). Amphetamine has been shown to increase the release and/or block the reuptake of central catecholamines (CA) (McLean and McCartney, 1964; Moore and Larivière, 1963; Stein, 1964; Farnebo, 1971; Glowinski and Axelrod, 1965; McKenzie and Szerb, 1968; Coyle and Snyder, 1969; Besson et al., 1969, 1971a,b; Tilson and Sparber, 1972; Voigtlander and Moore, 1973). It has been suggested that amphetamine, by increasing the concentration of CA post-synaptically might initiate a neuronal-feedback inhibition of CA-containing neurons (Corrodi et al., 1967). Consistent with this hypothesis, systematically administered *d*-amphetamine has been found to inhibit the firing of dopaminergic neurons (BUNNEY *et al.*, 1973b); this effect can be reversed by CPZ or HAL. A direct inhibitory mechanism seems unlikely since *d*-amphetamine has little effect on ZC and VT cells when applied by microiontophoresis and the effect of systemic *d*-amphetamine is lost when connections between the substantia nigra and neostriatum are interrupted (BUNNEY and AGHAJANIAN, this volume). Taken together, these results point to a neuronal-feedback circuit as the mechanism by which amphetamine inhibits dopaminergic neurons.

Apomorphine, a drug thought to stimulate DA receptors directly (ERNST, 1967; ANDÉN *et al.*, 1967; PERSSON, 1970) was also found to depress the firing of dopaminergic neurons when it is administered systematically (BUNNEY *et al.*, 1973a). However, unlike amphetamine, apomorphine has a powerful inhibitory effect when applied by microiontophoresis directly to dopaminergic neurons (Fig. 2C). As with DA, there appears to be an attenuation of this effect with repeated ejections. Thus, in addition to its postulated effects upon "postsynaptic" DA receptors, apomorphine appears to mimic the action of DA on "presynaptic" DA receptors (i.e., receptors upon the dopaminergic neuron itself). These results are consistent with the findings of KEHR *et al.*, (1972) which show that apomorphine can inhibit DA synthesis in the neostriatum even following interruption of the nigrostriatal pathway. It was suggested that there may be a receptor-mediated feedback inhibition of tyrosine hydroxylase by apomorphine. An alternative explanation for these findings would be that apomorphine can directly inhibit the enzyme tyrosine hydroxylase (GOLDSTEIN *et al.*, 1970). In any case, our electrophysiological results give support for the notion that DA and DA agonists can exercise a local feedback control upon synthesis or release of this putative transmitter by acting upon pre-synaptic DA receptors.

SUMMARY AND CONCLUSIONS

(1) By combined neurophysiological and histochemical methods, dopaminergic neurons of the ZC and VT areas were found to be clearly distinguishable from adjacent non-dopaminergic neurons (e.g., in ZR). Of particular interest was the finding that ZC and VT cells are inhibited by microiontophoretic DA but are insensitive to acetylcholine. On the other hand, ZR neurons are relatively insensitive to DA but are excited by acetylcholine at very low ejection currents. This suggests that if there are cholinergic afferents to the substantia nigra they would need to impinge upon ZR rather than upon ZC cells to produce a physiological effect.

(2) Drugs which either increase DA availability at DA receptors (e.g., amphetamine) or have a direct DA-agonist action (e.g., apomorphine) inhibit the firing of dopaminergic neurons; these results are consistent with the operation of a compensatory negative feedback system. Conversely, presumed DA-receptor blockers (e.g., CPZ and HAL) increase the rate of firing of dopaminergic neurons; these results are consistent with the operation of a compensatory positive feedback system.

(3) The soma of dopaminergic neurons appear to have DA receptors since they are responsive to the direct, microiontophoretic application of either DA or the DA-agonist, apomorphine. If the terminals of dopaminergic neurons also have such DA-receptors (i.e., "presynaptic DA-receptors"), then this might explain the postulated receptor-mediated feedback control of striatal tyrosine hydroxylase activity at dopaminergic synapses (cf. KEHR *et al.*, 1972).

REFERENCES

AGHAJANIAN G. K., HAIGLER H. J. and BLOOM F. E. (1972) *Life Sci.* **11**, Pt. 1, 615–622.
ANDÉN N.-E., DAHLSTRÖM A., FUXE K., LARSSON K., OLSON L. and UNGERSTEDT U. (1966) *Acta Physiol. Scand.* **67**, 313–326.
ANDÉN N.-E., ROOS B.-E. and WERDINIUS D. (1964) *Life Sci.* **3**, 149–158.
ANDÉN N.-E., RUBENSON A., FUXE K. and HÖKFELT T. (1967) *J. Pharm. Pharmacol.* **19**, 627–629.
BESSON M. J., CHERAMY A., FELTZ P. and GLOWINSKI J. (1969) *Proc. Natn. Acad. Sci.* **62**, 741–748.
BESSON M. J., CHERAMY A., FELTZ P. and GLOWINSKI J. (1971a) *Brain Res.* **32**, 407–424.
BESSON M. J., CHERAMY A. and GLOWINSKI L. (1971b) *J. Pharmacol. Exp. Ther.* **177**, 196–205.
BUNNEY B. S., AGHAJANIAN G. K. and ROTH R. H. (1973a) *Nature*, in press.
BUNNEY B. S., WALTERS J. R., ROTH R. H. and AGHAJANIAN G. K. (1973b) *J. Pharmacol. Exp. Ther.*, **185**, 560–571.
CARLSSON A., FUXE K., HAMBERGER B. and LINDQVIST M. (1966) *Acta Physiol. Scand.* **67**, 481–497.
CARLSSON A. and LINDQVIST M. (1963) *Acta Pharmacol. Toxicol.* **20**, 140–144.
CHERMAY A., BESSON M. J. and GLOWINSKI J. (1970) *Eur. J. Pharmacol.* **10**, 206–214.
CORRODI H., FUXE K. and HÖKFELT T. (1967) *Europ. J. Pharmacol.* **1**, 363–368.
COYLE J. T. and SNYDER S. H. (1969) *J. Pharmacol. Exp. Ther.* **170**, 221–251.
DAHLSTRÖM A. and FUXE K. (1965) *Acta Physiol. Scand.* **62**, (Suppl. 232), 1–55.
ERNST A. M. (1967) *Psychopharmacologia* **10**, 316–323.
FUXE K. (1965) *Acta. Physiol. Scand.* **64**, (Suppl. 247), 41–85.
GEY K. F. and PLETSCHER A. (1968) *Experientia* **24**, 335–336.
GLOWINSKI J. and AXELROD J. (1965) *J. Pharmacol. Exp. Ther.* **149**, 43–49.
GOLDSTEIN M., FREEDMAN L. S. and BACKSTRÖM T. (1970) *J. Pharm. Pharmacol.* **22**, 715–717.
HAIGLER H. J. and AGHAJANIAN G. K. (1973) *Europ. J. Pharmacol.* **21**, 53–60.
HORN A. S. and SNYDER S. H. (1971) *Proc. Natn. Acad. Sci.* **68**, 2325–2328.
KEHR W., CARLSSON A., LINDQVIST M., MAGNUSSON T. and ATACK C. (1972) *J. Pharm. Pharmacol.* **24**, 744–746.
MCKENZIE G. M. and SZERB J. C. (1968) *J. Pharmacol. Exp. Ther.* **162**, 302–308.
MCLEAN J. R. and MCCARTNEY M. (1961) *Proc. Soc. Exp. Biol. Med.* **107**, 77–79.
MOORE K. E. and LARIVIERE E. W. (1963) *Biochem. Pharmacol.* **12**, 1283–1288.
NYBÄCK H., BORZECKI Z. and SEDVALL G. (1968) *Europ. J. Pharmacol.* **4**, 395–403.
NYBÄCK H. and SEDVALL G. (1971) *J. Pharm. Pharmacol.* **23**, 322–325.
PERSSON T. (1970) *Acta. Pharmacol. Toxicol.* **28**, 378–390.
STEIN L. (1964) *Fedn. Proc.* **23**, 836–850.
TILSON H. A. and SPARBER S. B. (1972) *J. Pharmacol. Exp. Ther.* **181**, 387–398.
VON VOIGTLANDER P. F. and MOORE K. E. (1973) *J. Pharmacol. Exp. Ther.* **184**, 542–552.
UNGERSTEDT U. (1971a) *Acta Physiol. Scand.* (Suppl. 267), 1–48.
UNGERSTEDT U. (1971b) In *6-Hydroxydopamine and Catecholamine Neurons*. (MALMFORS T. and THEONEN H., Eds.) pp. 101–127. North-Holland Publishing Company, Amsterdam.

Frontiers in Catecholamine Research 1973, pp. 649 to 651. Pergamon Press. Printed in Great Britain.

EXISTENCE OF DOPAMINERGIC NERVE TERMINALS IN THE RAT CORTEX

A. M. Thierry and J. Glowinski

Groupe NB (Inserm U.114), Laboratoire de Biologie Moléculaire, Collège de France,
11, place Marcelin Berthelot, Paris 5e, France

CATECHOLAMINERGIC nerve terminals in the cerebral cortex of mammals have been assumed to be mainly represented by arborisations of the dorsal noradrenergic (NE) pathway (FUXE, 1965; OLSON and FUXE, 1972; UNGERSTEDT, 1971). However, the relative amounts of dopamine (DA) and norepinephrine (NE) found in the cortex of different strains of rats (VALZELLI and GARATTINI, 1968; THIERRY et al.,1973) or in various cortical area of the cat (MCGEER et al.,1963) are much higher than those estimated in the cerebellum (KOSLOW et al., 1972). This latter structure is, as the cortex, innervated by NE neurons originating from the locus coeruleus. The high DA levels found in the cerebral cortex cannot be only explained by the role of precursor played by DA in NE terminals. We thus postulated the occurrence of DA neurons or nerve endings in the cerebral cortex. To test this hypothesis, the effect of various lesions of ascending catecholaminergic pathways on the endogenous cortical DA and NE levels were estimated in rats of the Charles River strain.

Two types of lesions were made. Electrolytic bilateral lesions were performed with high frequency current (100 kHz, 2 mA, 10 sec) in the locus coeruleus (group A_6) or in the area ventralis tegmenti (ATV; group A_{10}). Such lesions destroy respectively the cell bodies of the dorsal NE pathway and of the DA mesolimbic system. Chemical lesions of the ventral and dorsal ascending NE pathways were performed by local microinjections of 6-OH-DA (2 μg in 1 μl injected in 5 min) made laterally to the pedunculus cerebellaris superior (L-PCS) or in the medial forebrain bundle (MFB) in the lateral hypothalamus. In all cases catecholamine levels were estimated three to five weeks later with the help of biochemical techniques (THIERRY et al., 1971).

Cortical NE levels were markedly reduced (75–98 per cent) after lesions of the locus coeruleus (5 weeks), the L-PCS (4 weeks) and the MFB (3,5 weeks) (Fig. 1). These effects were expected since the dorsal NE pathway originates or passes through all the lesion sites selected. The selective degeneration of the ventral NE pathway induced by electrolytic lesion of the ATV did not affect NE content in the cortex. On the other hand, surprisingly, DA cortical levels were not significantly reduced either after lesion of the locus coeruleus or after degeneration of the ascending NE pathway induced by an injection of 6-OH-DA into the L-PCS (Fig. 1). Cortical DA levels were also slightly but not significantly reduced after chemical lesion of the MFB although this lesion affected not only NE systems but also the nigrostriatal pathway. Indeed DA levels in the neostriatum were markedly reduced ($12 \cdot 8 \pm 1 \cdot 9$ μg/g in sham operated; $1 \cdot 3 \pm 0 \cdot 04$ μg/g in lesioned rats). Finally as revealed by the electrolytical lesion of the ATV the DA mesolimbic system does not innervate the cerebral cortex since DA cortical levels were not affected.

In vivo and *in vitro* formation of DA from tyrosine should persist in the cortex, after degeneration of the dorsal NE pathway if DA is synthesised in sites independent

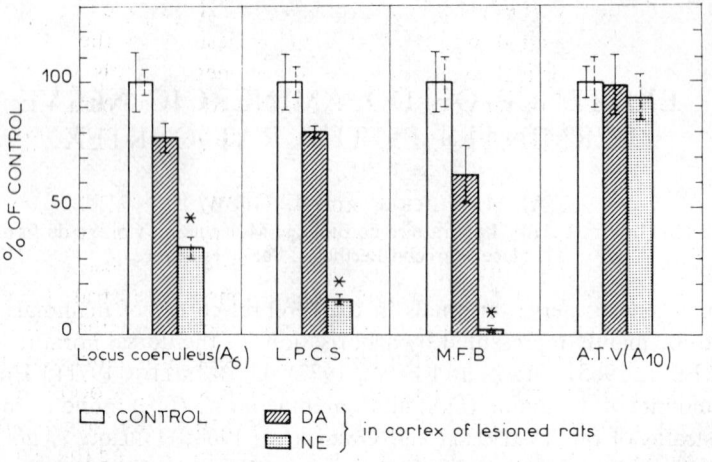

Fig. 1.—Effect of bilateral lesions of the NE ascending pathways on DA and NE content of the rat cerebral cortex. Lesions were made as described in the text. Degeneration of the dorsal NE bundle was induced by electrolytic lesion of the locus coeruleus; degeneration of both ventral and dorsal NE bundle was induced by microinjection of 6-OH-DA either laterally to the PCS (L-PCS) or into the MFB; degeneration of the ventral NE bundle was induced by electrolytic lesion of the ATV. In all cases controls were sham operated animals. Control values for each group of operated rats were respectively for DA: 0.200 ± 0.027; 0.133 ± 0.015; 0.145 ± 0.099; $0.164 \pm 0.010 \,\mu g/g$ and for NE: 0.244 ± 0.15; 0.154 ± 0.008; 0.204 ± 0.018 and $0.157 \pm 0.017 \,\mu g/g$. Results are the means of groups of eight rats and are expressed as percentage of respective control values $\pm$ SEM; $\times P < 0.001$ when compared with control values.

of NE terminals. Therefore, sham operated rats and animals injected with 6-OH-DA (2 μg in 1 μl) into the L-PCS received 4 weeks later an intracisternal injection of L-3·5-³H-tyrosine (42 Ci/mM, 80 μCi) and were sacrificed 20 min after the ³H-amino-acid administration. ³H-NE formation was markedly reduced (57 per cent) in 6-OH-DA lesioned rats (control: 4.05 ± 0.25 nCi/g; lesioned: 1.77 ± 0.38 nCi/g) whereas ³H-DA levels were slightly but not significantly affected (control: 15.9 ± 1.5 nCi/g; lesioned 13.0 ± 1.4 nCi/g).

In a second group of experiments, purified synaptosomes were prepared from cortex of sham operated and 6-OH-DA pretreated rats according to the technique of Gray and Whittaker (1963). As previously 6-OH-DA was injected bilaterally in the L-PCS 4 weeks before the experiment. Synaptosomes were incubated for 15 min with L-3·5-³H-tyrosine (40 Ci/mM; 53 μCi) and total ³H-DA and ³H-NE content of the tissues and incubating medium were analysed. As observed *in vivo* the synthesis of ³H-DA (24.8 ± 0.8 nCi) from ³H-tyrosine exceeded that of ³H-NE (14.7 ± 0.5 nCi) in sham operated animals. ³H-NE formation was completely abolished in cortical synaptosomes prepared from lesioned rats since it represented only 11 per cent (1.58 ± 0.01 nCi) of control values (14.7 ± 0.5 nCi). DA synthesis was only slightly affected and ³H-DA formation (16.8 ± 0.7 nCi) from ³H-tyrosine in lesioned animals still represented 67 per cent of the control levels (24.8 ± 0.8 nCi). In complementary experiments the persistence of ³H-DA formation could also be demonstrated in cortical synaptosomes 14 days after bilateral injection of 6-OH-DA into the L-PCS. Finally, the ^{3}H-catecholamines from ^{3}H-tyrosine in cortical synapto-somes was also examined 4 weeks after bilateral lesions of the locus coeruleus. ^{3}H-NE synthesis in cortical synaptosomes of lesioned animals (1.64 ± 0.19 nCi) was markedly reduced (74 per cent) when compared to sham operated rats (6.25 ± 0.13 nCi), but ^{3}H-DA synthesis was not modified (control: 12.3 ± 0.2 nCi; lesioned: 11.4 ± 0.3 nCi).

In conclusion degeneration of the dorsal NE pathway was associated with a disap-pearance of cortical NE synthesis both *in vivo* and *in vitro*. The selective persistence, in lesioned animals, of cortical DA content and of ^{3}H-DA synthesis from ^{3}H-tyrosine *in vivo* as well as *in vitro* in synaptosomal preparations, demonstrate the existence of dopaminergic nerve endings independent of noradrenergic terminals in the cerebral cortex. Innervation of the cortex by collateral sprouts of dopaminergic neurons from neighbouring structures can be excluded since ^{3}H-DA synthesis occurred in cortical synaptosomes at short (14 days) or at long (4 weeks) time after degeneration of the dorsal NE pathway. Moreover the endogeneous levels of DA and the synthesis of the transmitter are comparable in sham operated and lesioned animals. The DA meso-limbic system does not appear to be involved in the DA cortical innervation. The localisation of the cell bodies of the dopaminergic neurons which innervate the cerebral cortex must still be determined. The occurrence of DA interneurons in the cortex can be postulated since some cortical cell bodies have been shown to take up exogenous catecholamines (ARBUTHNOTT, 1969; DESCARRIES and LAPEIRRE, 1973).

REFERENCES

ARBUTHNOTT G. W. (1969) *J. Neurochem.* **16,** 1599–1604.
DESCARRIES L. and LAPIERRE Y. (1973) *Brain Res.* **51,** 141–160.
FUXE K. (1965) *Acta physiol. scand.,* **64** (suppl 247), 38–120.
McGEER P. L., WADA J. A. and McGEER E. G., (1963) In *Recent advances in Biological Psychiatry,* Vol 5, pp. 228–236. Plenum Press, New York.
GRAY E. G. and WHITTAKER V. P. (1963) *J. Anat.* **96,** 79–84.
KOSLOW S. H., CATTABENI F. and COSTA E. (1972) *Science* **176,** 177–179.
OLSON L. and FUXE K. (1972) *Brain Res.,* **42,** 289–295.
THIERRY A. M., BLANC G. and GLOWINSKI J. (1971) *Europ. J. Pharmacol.,* **14,** 303–307.
THIERRY A. M., STINUS L., BLANG G. and GLOWINSKI J. (1973) *Brain Res.* **50,** 230.
UNGERSTEDT U. (1971) On the anatomy, pharmacology and function of the nigro striatal dopamine system. Thesis, Stockholm.
VALZELLI L. and GARATTINI S. (1968) *J. Neurochem.* **15,** 259–261.

Frontiers in Catecholamine Research 1973, pp. 653 to 656. Pergamon Press. Printed in Great Britain.

EXCITATORY EFFECTS OF CATECHOLAMINES IN THE C.N.S.

(DISCUSSION OF PAPERS BY G. K. AGHAJANIAN AND B. J. HOFFER)

P. B. BRADLEY

Department of Pharmacology (Preclinical), Medical School, Birmingham, England

I SHOULD like to discuss some excitatory actions of catecholamines in the central nervous system, particularly since the two previous presentations may give the impression that these substances produce only depression of neuronal activity in the brain. In fact, Dr. Hoffer goes as far as to suggest that excitatory effects of microiontophoretically applied noradrenaline 'may be at least partly artefactual'. Whilst this may be true for some effects of noradrenaline, we think that it is unlikely to be true for all; furthermore we consider that some excitatory effects may be of physiological significance.

In 1958, Mollica and I (BRADLEY and MOLLICA, 1958) showed that adrenaline and noradrenaline could modify the activity of single neurones in the brain stem reticular formation in decerebrate cats and that both excitation and inhibition of neuronal activity could be observed. However, these effects were obtained with systemic injections of the amines and could have been indirect. Subsequently (BRADLEY and WOLSTENCROFT, 1962) it was found that similar effects could be obtained with microiontophoretic applications of noradrenaline to spontaneously active single neurones in the brain stem of unanaesthetised decerebrate animals. Thus, the firing rate of some neurones was depressed, whilst others were excited, and there was a third group of neurones which were unaffected by iontophoretically applied noradrenaline.

In a further study in which the effects of the (+)- and (−)-isomers of noradrenaline, together with various agonists and antagonists were examined, it was found that there were four distinct types of response of brain stem neurones to iontophoretically applied noradrenaline (BOAKES, BRADLEY, BROOKES, CANDY and WOLSTENCROFT, 1971). These consisted of an excitatory response, as had been observed previously (Fig. 1D), two types of inhibitory response, one of which was short-lasting and with a rapid onset (Fig. 1A), whilst the other was long-lasting and often delayed in onset (Fig. 1C), and a biphasic response consisting of excitation preceded by inhibition (Fig. 1B). The excitatory response showed a tendency to desensitisation with repeated applications and it also demonstrated stereochemical specificity in that the (−)-isomer was more potent than the (+)-isomer (Fig. 2). This was less marked with the inhibitory responses. Other sympathomimetic amines tested showed agonistic effects but with varying potency (Fig. 2), and it was found that dopamine, for example, had no independent actions, i.e. it did not affect the activity of neurones which did not respond to noradrenaline. Many of the classical noradrenaline antagonists of the α- and β-type were found to be ineffective but α-methyl-noradrenaline (BOAKES, CANDY and WOLSTENCROFT, 1968) and chlorpromazine (BRADLEY, WOLSTENCROFT, HÖSLI and AVANZINO, 1966), consistently antagonised noradrenaline excitation. We therefore concluded (BOAKES et al., 1971) that the excitatory actions of (−)-noradrenaline in the

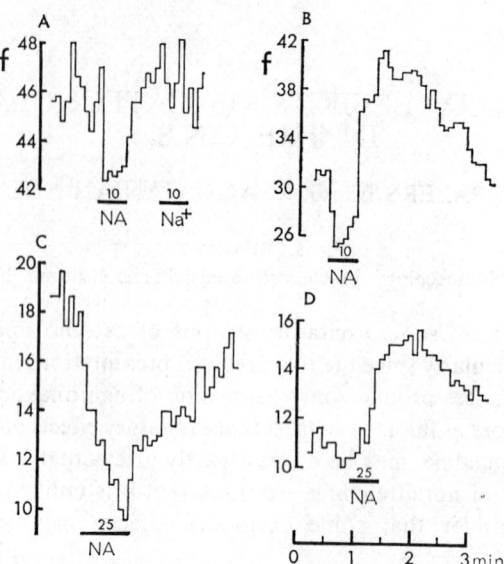

FIG. 1.—Effects of (−)-noradrenaline, applied iontophoretically, on the spontaneous activity of four neurones in the brain stem reticular formation of an unanaesthetised decerebrate cat. The mean firing rate of each neurone, in impulses/sec (f) in successive 5 sec epochs, is plotted against the time in minutes. Iontophoretic applications of (−)-noradrenaline (NA), or of a current control (Na⁺) are shown by horizontal bars. Short-lasting and long-lasting inhibitory responses are shown in A and C. An excitatory response is shown in D and a biphasic response in B. The iontophoretic currents used were 10–25nA. (From: BOAKES *et al.*, 1971.)

brain stem are probably mediated by a specific mechanism and also that the receptors for noradrenaline in this region of the brain do not correspond to peripheral receptors of the classical α- and β-types. Further studies are in progress in order to characterise, both physiologically and morphologically, the neurones which show these different responses to noradrenaline.

Excitatory effects of noradrenaline have also been observed in other regions of the brain. For example, JOHNSON, ROBERTS, SOBIESCEK and STRAUGHAN (1969) reported excitation of neurones in the cerebral cortex with iontophoretic applications of noradrenaline and found that this response was especially sensitive to anaesthetic agents. Excitatory effects by noradrenaline on cortical neurones have also been reported by other workers (GIARDINA, PEDEMONTE and SABELLI, 1973). It is necessary to use acid solutions in order to expel noradrenaline by microiontophoresis and pH has been implicated as a possible factor in the effects observed since it has been found that hydrogen ions alone can cause excitation of single neurones in the cerebral cortex (KRNJEVIĆ and PHILLIS, 1963). Whilst pH may be a factor affecting the proportions of neurones in the cerebral cortex showing excitatory responses to iontophoretically applied noradrenaline (FREDERICKSON, JORDAN and PHILLIS, 1971), it is unlikely to account for all excitations observed, since KRNJEVIĆ and PHILLIS (1963) found that 'excitation by hydrogen ions is very unlikely if the internal pH exceeds 2·5', and most workers have used pH's of 3·5–4·0. Our own experiments were carried out with noradrenaline solutions at pH 4·0 (BRADLEY and WOLSTENCROFT, 1962) or pH 5·0–6·0 (BOAKES *et al.*, 1971). Since excitation of brain stem neurones was consistently

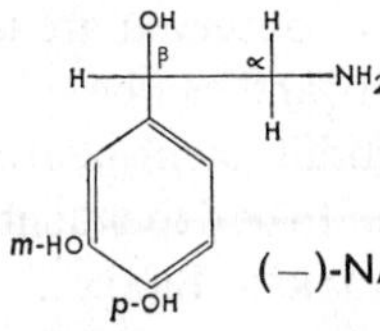

OH H
H— | β | α —NH₂
        | H

m-HO

p-OH   (−)-NA

AGONIST	MOLECULAR CHANGE		EFFECT	
	RING	SIDE-CHAIN	EXCITATION	INHIBITION
(−)Noradrenaline			+++++	-----
(−)Adrenaline		N-CH₃	++++	-----
(−) Isoprenaline		N-C₃H₇	o	−
(−)α-MethylNA		α-CH₃	++	-----
(+)Noradrenaline		Stereoisomer	++	-----
Dopamine		no β-OH	+++	---
(−)Phenylephrine	no p-OH	N-CH₃	++	---
(−)Metaraminol	no p-OH	α-CH₃	+	---
(±)p-Sympatol	no m-OH	N-CH₃	o	---
Tyramine	no m-OH	no β-OH	++	---
(+)Amphetamine	no p-, m-OH	no β-OH, α-CH₃	++++	-----
(−)Ephedrine	no p-, m-OH	N-CH₃; α-CH₃	++++	---
2-Aminoheptane			+++	--

FIG. 2.—The potencies of various sympathomimetic amines as agonists of (−)-noradrenaline. The number of symbols (+ for excitation and − for inhibition) allotted to each compound is based on a comparison of the effects of the compound with those of (−)-noradrenaline on the same neurone. N.B. Inhibitory actions includes both short- and long-lasting effects. (From: BOAKES et al., 1971.)

observed with these solutions, we do not consider pH effects to be an important factor in our experiments.

Another possible 'artefactual' explanation for excitatory responses by central neurones to iontophoretically applied noradrenaline is related to the suggestion that these effects may be due to an indirect action by noradrenaline on blood vessels (STONE, 1971). The main support for this is the parallel time course of constriction of arterioles as a result of noradrenaline application and the time course of excitatory effects on neurones in the central nervous system. However, interesting as this suggestion is, the comparison has so far only been made with arterioles in intestine and mesentary and there is no evidence that small cerebral vessels respond in the same way, or that if they do, their indirect influence on neighbouring neurones is likely to be excitatory.

The interpretation of the effects of microiontophoretically applied putative neurotransmitters and centrally active drugs, to single neurones in the central nervous system is certainly very difficult and we must not assume that the effects of a substance, released iontophoretically in the vicinity of a neurone, are necessarily the same as those of the endogenous transmitters. Some effects, which have been observed and reported in the literature, may well be due to artefacts. On the other hand, it must not be assumed that, because certain effects can be mimicked by hydrogen ions, or because there are similarities between the responses of peripheral arterioles and central neurones, that all noradrenaline excitations are necessarily artefacts. As far as neurones in the brain stem reticular formation are concerned it is our view that the excitatory responses to noradrenaline we have observed are unlikely to be due to artefacts since they show stereochemical specificity and can be specifically antagonised. Furthermore, the actions of noradrenaline on brain stem neurones are closely mimicked by (+)-amphetamine whose actions appear to be dependent upon the presence of noradrenaline in presynaptic nerve terminals (Boakes, Bradley and Candy, 1972).

REFERENCES

Aghajanian G. K. and Bunney B. S. (1973) This volume.

Boakes R. J., Bradley P. B., Brookes N., Candy J. M. and Wolstencroft J. H. (1971) *Br. J. Pharmac.* **41,** 462–479.

Boakes R. J., Bradley P. B. and Candy J. M. (1972) *Br. J. Pharmac.* **45,** 391–403.

Boakes R. J., Candy J. M. and Wolstencroft J. H. (1968) *Brain Res.* **11,** 450–452.

Bloom F. E. and Hoffer B. J. (1973) This volume.

Bradley P. B. and Mollica A. (1958) *Arch. ital. Biol.* **96,** 168–186.

Bradley P. B. and Wolstencroft J. H. (1962) *Nature, Lond.* **196,** 840 & 873.

Bradley P. B., Wolstencroft J. H., Hösli L. and Avanzino G. L. (1966) *Nature, Lond.* **212,** 1425–1427.

Frederickson R. C. A., Jordan L. M. and Phillis J. W. (1971) *Brain Res.* **35,** 556–560.

Giardina W. J., Pedemonte W. A. and Sabelli H. C. (1973) *Life Sci.* **12** pt. I, 153–161.

Johnson E. S., Roberts M. H. T., Sobieszek A. and Straughan D. W. (1969) *Int. J. Neuropharmac.* **8,** 549–566.

Krnjević K. and Phillis J. W. (1963) *J. Physiol.* **165,** 274–304.

Stone T. E. (1971) *Nature, Lond.* **234,** 145–146.

Frontiers in Catecholamine Research 1973, pp. 657 to 659. Pergamon Press. Printed in Great Britain.

GENETIC FACTORS IN THE CONTROL OF BRAIN LEVELS OF BIOGENIC AMINES AND BRAIN EXCITABILITY

G. M. EVERETT

Dept. of Pharmacology, University of California Medical Centre, San Francisco, Calif., U.S.A.

SEVEN crosses among three strains of mice, ICR (low amines), C57B16J (high amines) and BALB (high amines), have been studied in regard to general behaviour and brain excitability as measured by electroshock seizure latency, and maximal electroshock seizure threshold. Seizure latency is measured as the time in seconds from the initiation of a supramaximal electroshock (via corneal electrodes) until the beginning of the tonic extensor thrust. Seizure latency and seizure threshold are closely correlated parameters and are constant for each strain and F_1 cross studies. Levels of brain dopamine (DA), norepinephrin (NE) and serotonin (5HT) were also determined. As shown in Table 1, the amine levels for each strain give unique profiles and are remarkably constant for each strain studied. The behavioural differences are also shown. The BALB strain has a high level of dopamine and the males fight each other. This rarely occurs with the males of the other strains unless drug induced.

The seizure latencies and thresholds of the two high amine strains (C57 and BALB) are approximately double the threshold of ICR (low amine strain) mice (Table 2). All crosses of the ICR (low amine) strain with either of the high amine strains showed brain levels of DA, NE and 5HT approximately the mean of the two strains involved. Seizure latencies and thresholds of these crosses also were approximately the mean of the strains involved. Crosses of the two high amine strains resulted in high brain amines, long seizure latencies and high convulsive thresholds. The data are summarised in Table 3. Males and females gave similar data on amine levels and brain excitability tests. The genetic factors controlling levels of brain biogenic amines in the three strains of mice studied are isogenetic in the crosses studied. The seizure latency and electroshock threshold studies confirm the close correlation of brain amine levels with brain excitability and suggest a powerful inhibitory role for these neuromodulators in the control of brain excitability.

TABLE 1. BEHAVIOURAL RESPONSES AND BIOGENIC AMINES IN THREE VARIETIES OF MICE

Mouse type	DA	NE	5HT	Behaviour
ICR	0.94 ± 0.04	0.39 ± 0.01	0.61 ± 0.02	Benign, low muscle tone. DOPA-marked irritation, fighting
657B16J	1.04 ± 0.08	0.59 ± 0.02	0.71 ± 0.03	Benign, high muscle tone. DOPA-moderate irritation, no fighting
BALB	1.81 ± 0.12	0.43 ± 0.02	0.68 ± 0.02	Low muscle tone, fighting.

Table 2. Seizure latency, seizure threshold and brain biogenic amines in three strains of mice

Strain	Latency (sec)	T_x Est$_{50}$ (Volts) (CL)*	Brain conc. of amines μg/g $\pm$ S.E.		
			DA	NE	5HT
ICR	1·69 ± 0·02	45(44–47)	0·94 ± 0·02	0·45 ± 0·01	0·53 ± 0·01
BALB	2·49 ± 0·08	63(60–65)	1·59 ± 0·06	0·54 ± 0·01	0·68 ± 0·01
C57BL6/J	3·44 ± 0·11	74(71–77)	1·36 ± 0·03	0·65 ± 0·01	0·69 ± 0·01

* Tonic extensor electroshock threshold$_{50}$ and confidence limits.

Table 3. Seizure latency and brain biogenic amine levels in various strains and crosses of mice

Strain or cross	No. detn.	E-shock latency	No. detn.	Brain amines μg/g		
				DA	NE	5HT
ICR	206	1·56 ± 0·04	59	0·98 ± 0·02	0·48 ± 0·01	0·61 ± 0·01
C57BL6J	79	3·35 ± 0·16	22	1·39 ± 0·02	0·65 ± 0·01	0·75 ± 0·01
BALB	70	2·55 ± 0·10	22	1·44 ± 0·04	0·52 ± 0·01	0·66 ± 0·01
BALB + C57BL6J	27	2·61 ± 0·11	10	1·47 ± 0·04	0·57 ± 0·01	0·74 ± 0·01
		(2·80 ± 0·13)*		(1·42 ± 0·03)	(0·59 ± 0·01)	(0·71 ± 0·01)
ICR + C57BL6J	10	2·10 ± 0·14	19	1·14 ± 0·03	0·53 ± 0·02	0·64 ± 0·02
		(2·46 ± 0·10)		(1·19 ± 0·02)	(0·57 ± 0·01)	(0·68 ± 0·01)
ICR + BALB	9	1·92 ± 0·07	2	1·29	0·47	0·66
		(2·06 ± 0·07)		(1·2 ± 0·03)	(0·50 ± 0·01)	(0·69 ± 0·01)

* Calculated isogenetic cross values = ( ).

SUMMARY

Brain concentrations of DA, NE and 5HT are highly constant for a given mouse strain. However there are large differences in amine levels in different strains and amine profiles of the three amines are unique to a given strain. All three amines are high in high amine strains and low in low amine strains. F_1 crosses of ICR, C57B16 and BALB mice gave amine levels that are the mean of the levels in the parent strains suggesting an isogenetic inheritance of brain amine level control.

Electroshock seizure latency and threshold are markedly higher in high brain amine strains suggesting a role of the biogenic amines as an inhibitory system in the control of brain excitability.

Acknowledgements—These studies were carried out at Abbott Laboratories with the assistance of Mrs. P. Morse and Mr. J. Borcherding.

Frontiers in Catecholamine Research 1973, pp. 661 to 665. Pergamon Press. Printed in Great Britain.

CATECHOLAMINE RECEPTOR MECHANISMS IN VERTEBRATES

NILS-ERIK ANDÉN

Department of Pharmacology, University of Göteborg, Göteborg, Sweden

TWO CENTRAL catecholamine (CA) neuron systems can conveniently be completely lesioned without damaging other CA neurons, i.e., the dopamine (DA) neurons to the corpus striatum and the noradrenaline (NA) neurons to the spinal cord. The functions of these neuron systems can also be studied easily and selectively. They are, therefore, suitable for investigations of central DA and NA receptor mechanisms.

THE STRIATAL DOPAMINE SYSTEM

The DA in the corpus striatum belongs to neurons with the cell bodies in the substantia nigra and the terminals in the neostriatum, i.e., the caudate nucleus plus putamen (ANDÉN, CARLSSON, DAHLSTRÖM, FUXE, HILLARP and LARSSON, 1964). The function of the uncrossed nigro-neostriatal DA neurons can be revealed as asymmetries in body position after one of the following unilateral manipulations: (1) chronic lesion of the ascending axons, causing loss of the DA terminals on the operated side (ANDÉN, 1966; ANDÉN, DAHLSTRÖM, FUXE and LARSSON, 1966a; UNGERSTEDT, 1968); (2) acute removal of the corpus striatum, causing loss of both the DA terminals and the effector cells with the DA receptors on the operated side (ANDÉN et al., 1966a); (3) administration of drugs into the neostriatum on one side (UNGERSTEDT, BUTCHER, BUTCHER, FUXE and ANDÉN, 1969). Strangely enough, unilateral removal of the corpus striatum or unilateral lesion of the ascending DA pathway chronically (more than 7 days) does not induce any clear-cut asymmetry of the body position and movements, perhaps due to compensatory mechanisms or to damage of antagonistic or efferent pathways (ANDÉN et al., 1966a). In general, the results are easier to interpret after acute removal of the corpus striatum on one side.

Injection of L-dopa after inhibition of the monoamine: oxygen oxidoreductase (EC 1.4.3.4, monoamine oxidase) or the peripheral 3,4-dihydroxy-L-phenylalanine carboxy-lyase (EC 4.1.1.26, dopa decarboxylase) causes a rat with a unilateral striatectomy to turn its head and tail to the operated side (ANDÉN et al., 1966a). Simultaneously, it becomes very excited and rotates vigorously to the operated side. Since striatal DA terminals and receptors are present only on the unoperated side, the effect is evoked via stimulation of DA receptors in the intact neostriatum. This view is strengthened by the finding that administration of DA (after inhibition of the monoamine oxidase) into the neostriatum on one side induces turning of the head and tail to the contralateral side (UNGERSTEDT et al., 1969). Release of endogenous DA, by treatment with reserpine after inhibition of the monoamine oxidase, also induces turning to the operated side, both after acute removal of the corpus striatum and after chronic axotomy (ANDÉN et al., 1966a). This effect must be evoked from the intact neostriatum since there is no striatal DA to release on the lesioned side. However, treatment with L-dopa after chronic lesion of the ascending DA pathway produces turning to the unoperated side, i.e., the opposite effect to that seen after unilateral

striatectomy. The formation of DA from L-dopa in the neostriatum is markedly
reduced but not completely inhibited on the axotomized side, indicating that the
smaller amount of DA on the operated side has a more pronounced functional effect
than the greater amount in the intact neostriatum, perhaps due to denervation supersensitivity (ANDÉN, 1966; ANDÉN et al., 1966a; UNGERSTEDT, 1971).

An effect similar to that produced by L-dopa in unilaterally acutely striatectomized rats is also observed after treatment with amphetamine and with apomorphine
(ANDÉN, RUBENSON, FUXE and HÖKFELT, 1967a). The amphetamine-induced asymmetry is inhibited by pretreatment with α-methyltyrosine (plus reserpine), indicating
that amphetamine acts via release of DA from the nerve terminals. On the other hand,
apomorphine acts also in the absence of DA and it probably directly stimulates the
DA receptors on the effector cells (ERNST, 1967; ANDÉN et al., 1967a). After chronic
axotomy, apomorphine induces the opposite effect, i.e., turning to the unoperated side,
whereas treatment with amphetamine causes turning to the operated side (UNGER-
STEDT, 1971). Again, the effect of apomorphine can be explained by a denervation
supersensitivity of the DA receptors on the operated side and the effect of amphetamine is probably due to release of DA in the intact neostriatum. Piribedil (Trivastal®,
ET 495) appears to directly stimulate the DA receptors in the above-mentioned
models, although it also has some releasing action (CORRODI, FARNEBO, FUXE,
HAMBERGER and UNGERSTEDT, 1972). The neostriatal DA receptors are stimulated
also by metatyramine (formed from metatyrosine) and by α-methyl-DA (formed from
α-methyl-dopa) (ANDÉN, BUTCHER and ENGEL, 1970a), but these drugs are less
potent than DA. Treatment with 5-hydroxytryptophan does not produce any turning
of unilaterally operated rats, despite a considerable formation of 5-hydroxytryptamine
(5-HT) and marked behavioural changes such as tremor and extension of the hindlegs,
showing that 5-HT and DA act on different receptors (ANDÉN et al., 1966a). Likewise,
unilateral injections of 5-HT into the neostriatum of rats pretreated with reserpine,
α-methyltyrosine and nialamide do not cause any asymmetry. On the other hand,
intrastriatal administration of NA induces in such a preparation a moderate turning
to the contralateral side, indicating that NA can stimulate the DA receptors though
more weakly than DA (UNGERSTEDT et al., 1969).

When reserpine is given to a rat with a striatectomy or with a chronic axotomy of
the nigro-neostriatal DA neurons on one side, the head and tail are turned to the
unoperated side (ANDÉN et al., 1966a). Injection of L-dopa or apomorphine to a
unilaterally striatectomized, reserpine-treated rat causes turning from the unoperated
to the operated side, showing that the reserpine-induced asymmetry is due to a lack
of DA on the receptors of the effector cells in the neostriatum on the unoperated side.
The asymmetry produced by reserpine is the result of a muscle rigidity on the unoperated side caused by an increased activity of the α-motoneurons and a decreased
activity of the γ-motoneurons (ANDÉN, LARSSON and STEG, 1971).

The same asymmetry of unilaterally operated rats as produced by reserpine is also
seen after treatment with neuroleptic drugs such as chlorpromazine, haloperidol and
pimozide (ANDÉN et al., 1966a; ANDÉN, BUTCHER, CORRODI, FUXE and UNGERSTEDT,
1970b). The turning seen after injection of these drugs is, however, unchanged by the
subsequent administration of L-dopa, apomorphine or amphetamine, indicating that
the neuroleptic drugs block the postsynaptic DA receptors whereas reserpine prevents
the release of DA from the presynaptic nerve terminals. Adrenergic α-receptor

blocking agents such as phenoxybenzamine, and adrenergic β-receptor blocking agents such as propranolol, neither induce any asymmetry of operated rats nor reduce the L-dopa-induced turning, showing that the DA receptors are different from the NA receptors in the periphery. Nor do sedative drugs such as barbiturates and promethazine produce any asymmetry.

The neuroleptic-induced asymmetry of unilaterally operated rats is reduced by anticholinergic drugs. Thus, an acetylcholine mechanism must antagonize the effects of DA beyond the DA receptors on the effector cells in the neostriatum (ANDÉN and BÉDARD, 1971). Also, the unilateral intrastriatal administration of potassium ions can influence the striatal function after treatment with neuroleptic drugs, causing the rats to turn to the contralateral side (STOCK, MAGNUSSON and ANDÉN, 1973). This effect may be the result of a persistent depolarization of the effector cells in the neostriatum. Systemic treatment with apomorphine antagonizes this potassium-induced turning (KELLER, BARTHOLINI, PIERI and PLETSCHER, 1972; STOCK et al., 1973). This effect can be explained as a hyperpolarization of the effector cells due to DA receptor stimulation, in agreement with electrophysiological data.

THE SPINAL NORADRENALINE SYSTEM

The NA in the spinal cord is localized in nerve terminals of neurons whose cell bodies are present in the lower brain stem (ANDÉN, HÄGGENDAL, MAGNUSSON and ROSENGREN, 1964; DAHLSTRÖM and FUXE, 1965). Hence, a mid-thoracic spinal cord transection separates the NA nerve terminals in the caudal part of the spinal cord from the cell bodies of the same neurons.

Injection of L-dopa induces a marked increase in flexor reflex activity of acutely spinalized animals (CARLSSON, MAGNUSSON and ROSENGREN, 1963; ANDÉN, JUKES, LUNDBERG and VYKLICKÝ, 1966b). L-dopa has been shown to act via the formation of CAs and the release of NA which diffuses to and stimulates NA receptors on the effector cells (ANDÉN, JUKES and LUNDBERG, 1966c). The NA may inhibit inhibitory interneurons, thereby releasing a massive and longlasting flexor reflex which is normally concealed in the spinal animal (ANDÉN et al., 1966b).

Treatment either with amphetamine or the antihypertensive agent clonidine (Catapresan®) induces an increase in flexor reflex activity, similar to that observed after injection of L-dopa (ANDÉN, CORRODI, FUXE, HÖKFELT, HÖKFELT, RYDIN and SVENSSON, 1970). Pretreatment with reserpine plus α-methyltyrosine blocks this effect of amphetamine but not that of clonidine, indicating an indirect action of amphetamine but a direct action of clonidine on the NA receptors. Incidently, apomorphine has no effect on the spinal NA receptors whereas clonidine has no effect on the neostriatal DA receptors. The flexor reflex activity is markedly enhanced after treatment with α-methyl-dopa due to the formation of α-methyl-NA which is a strong stimulating agent of these NA receptors (ANDÉN et al., 1970a). After inhibition of the DA-β-hydroxylase, the L-dopa-induced increase in flexor reflex activity is clearly but not completely inhibited, suggesting that DA has only a weak effect on the central NA receptors (ANDÉN, ENGEL and RUBENSON, 1972). Injection of 5-hydroxytryptophan induces no increase in flexor reflex activity, but induces other reflex changes, suggesting that 5-HT and NA act on different receptors (ANDÉN, 1968).

The increase in flexor reflex activity seen after treatment with L-dopa or clonidine is eliminated by many neuroleptic drugs, e.g., chlorpromazine and haloperidol, thus indicating a blockade of the NA receptors (ANDÉN *et al.*, 1966c; ANDÉN, CORRODI, FUXE and HÖKFELT, 1967b). There are, however, neuroleptic drugs without any apparent NA receptor blocking ability, e.g., the diphenylbutylpiperidine derivative pimozide (ANDÉN *et al.*, 1970b). On the other hand, the α-adrenergic blocking agent phenoxybenzamine efficiently inhibits the L-dopa- or clonidine-induced reflex changes, whereas β-adrenergic blocking agents such as propranolol appear to lack this effect (ANDÉN *et al.*, 1966c; 1967b). On the whole, the NA receptors in the spinal cord are influenced by drugs in about the same manner as the peripheral α-receptors, provided that the compounds can enter the central nervous system. Also in the brain, NA receptors of the α-type may be of importance, judging from the inhibitory effect of phenoxybenzamine on the L-dopa- or clonidine-induced behavioural changes (ANDÉN, 1967). These findings do not exclude the existence of other types of NA receptors, e.g., of the β-type, in the brain and in the spinal cord.

REGULATION OF CATECHOLAMINE TURNOVER FROM CATECHOLAMINE RECEPTORS

The drugs acting on the CA receptors of the effector cells normally change the activity of the presynaptic CA neuron: when the CA receptors are blocked, the turnover of the CA is increased, and the reverse chemical effect is observed when the receptors are stimulated (HENNING, this symposium). If the receptor activity is changed for only one of the two CAs, the turnover of the corresponding CA can be influenced selectively. Thus, the DA receptor stimulating agent apomorphine decelerates the turnover of DA but not that of NA, and the NA receptor blocking agent phenoxybenzamine accelerates the turnover of NA but not that of DA. There are also exceptions, e.g., a slight deceleration of the DA turnover after treatment with the NA receptor stimulating agent clonidine and an acceleration of the NA turnover after treatment with the DA receptor blocking agent pimozide.

Generally, the changes in turnover may be considered as compensatory to the alterations of the receptor activity. They may be evoked via a neuronal or humoral feedback mechanism from the effector cells. There is, however, also the possibility that the CA neuron possesses CA receptors, either on its terminals or on its cell body or both, which regulate the turnover of the transmitter. These hypothetical presynaptic CA receptors should then have properties similar to those on its innervated post-synaptic cell, to judge from the ability of some drugs to act selectively, both functionally and chemically, on only one CA.

REFERENCES

ANDÉN N.-E. (1966) In: *Mechanisms of Release of Biogenic Amines*. (EULER U. S. von, ROSELL S. and UVNÄS B., Eds.) pp. 357–359. Pergamon Press, Oxford.

ANDÉN N.-E. (1967) In: *Progress in Neuro-Genetics*. (BARBEAU A. and BRUNETT J. R., Eds.) pp. 265–271. Excerpta Medica Foundation, Amsterdam.

ANDÉN N.-E. (1968) *Adv. Pharmacol.* **6A,** 347–349.

ANDÉN N.-E. and BÉDARD P. (1971) *J. Pharm. Pharmacol.* **23,** 460–462.

ANDÉN N.-E., BUTCHER S. G., CORRODI H., FUXE K. and UNGERSTEDT U. (1970b) *Europ. J. Pharmacol.* **11,** 303–314.

ANDÉN N.-E., BUTCHER S. G. and ENGEL J. (1970a) *J. Pharm. Pharmacol.* **22,** 548–550.

ANDÉN N.-E., CARLSSON A., DAHLSTRÖM A., FUXE K., HILLARP N.-Å. and LARSSON K. (1964) *Life Sci.* **3,** 523–530.

ANDÉN N.-E., CORRODI H., FUXE K. and HÖKFELT T. (1967b) *Europ. J. Pharmacol.* **2,** 59–64.

ANDÉN N.-E., CORRODI H., FUXE K., HÖKFELT B., HÖKFELT T., RYDIN C. and SVENSSON T. (1970) *Life Sci.* **9,** 513–523.

ANDÉN N.-E., DAHLSTRÖM A., FUXE K. and LARSSON K. (1966a) *Acta pharmacol. toxicol.* **24**, 263–274.
ANDÉN N.-E., ENGEL J. and RUBENSON A. (1972) *Naunyn-Schmiedeberg's Arch. Pharmacol.* **273**, 1–10.
ANDÉN N.-E., HÄGGENDAL J., MAGNUSSON T. and ROSENGREN E. (1964) *Acta physiol. scand.* **62**, 115–118.
ANDÉN N.-E., JUKES M. G. M. and LUNDBERG A. (1966c) *Acta physiol. scand.* **67**, 387–397.
ANDÉN N.-E., JUKES M. G. M., LUNDBERG A. and VYKLICKÝ L. (1966b) *Acta physiol. scand.* **67**, 373–386.
ANDÉN N.-E., LARSSON K. and STEG G. (1971) *Acta physiol. scand.* **82**, 268–271.
ANDÉN N.-E., RUBENSON A., FUXE K. and HÖKFELT T. (1967a) *J. Pharm. Pharmacol.* **19**, 627–629.
CARLSSON A., MAGNUSSON T. and ROSENGREN E. (1963) *Experientia* **19**, 359–360.
CORRODI H., FARNEBO L.-O., FUXE K., HAMBERGER B. and UNGERSTEDT U. (1972) *Europ. J. Pharmacol.* **20**, 195–204.
DAHLSTRÖM A. and FUXE K. (1965) *Acta physiol. scand.* **(64)**, Suppl. 247, 39–85.
ERNST A. M. (1967) *Psychopharmacologia* **10**, 316–323.
KELLER H. H., BARTHOLINI G., PIERI L. and PLETSCHER A. (1972) *Europ. J. Pharmacol.* **20**, 287–290.
STOCK G., MAGNUSSON T. and ANDÉN N.-E. (1973) *Naunyn-Schmiedeberg's Arch. Pharmacol.* **278**, 347–361.
UNGERSTEDT U. (1968) *Europ. J. Pharmacol.* **5**, 107–110.
UNGERSTEDT U. (1971) *Acta physiol. scand.* Suppl. 367, 69–93.
UNGERSTEDT U., BUTCHER L. L., BUTCHER S. G., ANDÉN N.-E. and FUXE K. (1969) *Brain Res.* **14**, 461–471.

Frontiers in Catecholamine Research 1973, pp. 667 to 671. Pergamon Press. Printed in Great Britain.

EXCITATORY EFFECTS OF DOPAMINE ON MOLLUSCAN NEURONES

P. ASCHER

Laboratoire de Neurobiologie, Ecole Normale Supérieure, 46, Rue d'Ulm
Paris 5e.

DOPAMINE (DA) is the only catecholamine which has been detected in appreciable amounts in the nervous system of Molluscs (SWEENEY, 1963; see GERSCHENFELD, 1973, for a review). It is particularly abundant in certain ganglia, and can sometimes be localised in identifiable neurones (MARSDEN and KERKUT, 1970). The demonstration that DA is a transmitter in the molluscan nervous system is, however, still incomplete. Liberation of DA by stimulation of the ganglia has been observed, but in experimental conditions where one could fear serious—and still unexplained—artefacts (ASCHER, GLOWINSKI, TAUC and TAXI, 1968). Application of DA on molluscan neurones has shown that it could trigger a variety of effects (see GERSCHENFELD, 1973) but, primarily because of an incomplete knowledge of the pharmacological properties and of the ionic mechanisms of DA responses, no identifiable synaptic potential could be securely claimed to be due to DA, until very recently (BERRY and COTTRELL, 1973).

In the experiments to be described, an attempt was made to characterise the responses of identified *Aplysia* neurones to electrophoretic applications of DA. Special emphasis will be placed on the excitatory effects of DA, since they have been only rarely observed elsewhere. The methods used in these experiments have been described in a previous article (ASCHER, 1972).

DA EFFECTS CAN BE EXCITATORY, INHIBITORY, OR EXCITATORY-INHIBITORY

The effects of DA differ according to the cell studied. For example, in the pleural ganglia of *Aplysia*, the "anterior" cells are excited by DA while the "medial" cells are inhibited, and most "posterior" cells show a biphasic, excitatory-inhibitory response to locally applied DA.

Most previous studies of DA effects, even in Molluscs, reported mainly or exclusively inhibitory effects. The frequent observation of excitatory effects of DA in these studies on *Aplysia* can probably be ascribed to differences in experimental conditions.

First, it appears that DA excitatory effects are by far more prone to desensitisation than inhibitory responses. For example, in cells presenting a biphasic response to a short (200 msec) electrophoretic pulse of DA, repetition of the injection at frequencies as low as 1/min often produces a progressive decrease of the excitatory component, while the inhibitory one is hardly affected. On these same cells, the response to a longer pulse will appear as a very prolonged inhibition following a short excitation, and DA applied in perfusion will show only the inhibitory component. It thus appears that the mode of application of DA is a critical requirement for the observation of DA excitatory effects.

But even when DA is applied with short electrophoretic pulses appropriately spaced, the occurrence of an excitatory component will depend critically on the positioning of the pipette. Considering again the posterior pleural cells of *Aplysia*, it was observed that if the DA pipette was progressively withdrawn from a position where the injection elicited a biphasic response, the excitatory component was the first to disappear. This may explain that if the pipette is positioned "blindly", a variety of positions will be found from which the response will appear purely inhibitory.

PHARMACOLOGICAL PROPERTIES OF DA EXCITATORY RESPONSES

Various antagonists were applied to the preparation in an attempt to demonstrate, in a purely qualitative way, that DA receptors mediating excitatory responses were (1) different from the DA receptors mediating inhibitory responses and (2) different from the acetyl-choline (ACh) receptors mediating excitatory responses.

For the first purpose, the most convenient cells were again the posterior cells of the pleural ganglia, on which DA produces a biphasic response. It was observed that ergometrine (WALKER, WOODRUFF, GLAIZNER, SEDDEN and KERKUT, 1968) and methyl-ergometrine produced a selective elimination of the inhibitory component, whereas curare and strychnine selectively eliminated the excitatory component. These pharmacological differences were confirmed in other cells, and in particular in cells of the visceral ganglion where DA usually elicits only inhibitory responses, but where an excitatory component could be unmasked after application of methyl-ergometrine.

Although these data show that different receptors mediate the excitatory and inhibitory DA responses, they did not discriminate between the receptors mediating DA and ACh excitatory effects. These however, could be also differentiated. On the anterior pleural cells, on which both DA and ACh elicit predominantly excitatory effects, hexamethonium blocked selectively the ACh response, whereas DA in perfusion abolished selectively the DA response.

The possible use of the above data is exemplified by the recent observations of BERRY and COTTRELL (1973) in the pleural ganglion of *Planorbis*. Stimulation of giant neuron presumed to contain DA triggered, in various "post-synaptic" cells, synaptic potentials which were excitatory or excitatory-inhibitory. The pharmacological properties of these synaptic potentials resemble those of the DA responses, thus strengthening the case for the presence of dopaminergic synapses in molluscan ganglia.

COMPARISON OF THE IONIC MECHANISMS INVOLVED IN DA AND ACh EXCITATORY EFFECTS

The electrophysiological characterisation of the ionic permeability changes underlying a transmitter action usually requires the measurement of an "inversion potential" in various ionic environments. In the case of DA effects on *Aplysia* neurones, this could be done for many *inhibitory* responses, which were shown to be due to an increased K^+ permeability (ASCHER, 1972). But attempts to measure an inversion potential for *excitatory* DA responses were not as successful. Delayed rectification, which causes a dramatic fall in the cell's input resistance, precluded the observation of responses on strongly depolarised neurones, and thus impeded the direct measurement of an inversion potential. Anomalous rectification (KANDEL and TAUC, 1966), on the other hand, interfered with the estimation of an inversion potential by extrapolation from the responses studied on hyperpolarised neurons.

This last difficulty could sometimes be overcome by manipulation of the external

K^+ concentration, or by addition of Cs^+ to the sea water. The first series of records in Fig. 1 show that in normal sea water, bringing the membrane potential from -60 to -90 mV caused a decrease in membrane resistance, no change in the amplitude of the ACh response, and a reduction in the amplitude of the DA response. After reducing the external K^+ concentration, however, the reduction of input resistance was no longer observed when going from -60 to -90 mV, and in these conditions both the DA and ACh responses increased with hyperpolarisation of the membrane.

These findings indicate that both the DA and the ACh responses result from an increase in membrane conductance. In addition, the figure shows that the effect of polarisation is more marked on the ACh response, which indicates a more negative value of the (extrapolated) inversion potential. Indeed for ACh responses an inversion potential could sometimes be directly measured at membrane potentials between 0 and -30 mV. Further studies showed, however, that these cases corresponded to ACh responses resulting from the superposition of two effects : the "true" excitatory effect of ACh—defined by its sensitivity to hexamethonium (TAUC and GERSCHENFELD, 1962)—and an hexamethonium-resistant increase in chloride conductance. This latter conductance change pulls the "global" inversion potential towards the chloride equilibrium potential, and accounts for the sensitivity of the ACh response to changes in chloride concentrations (ASCHER, GERSCHENFELD and KEHOE,1972).

A similar problem was encountered when an extrapolated value of the inversion potential was calculated for DA responses. The results showed an important scatter (between $+30$ and -10 mV), most of which is probably due to the superposition of "true" excitatory DA effects (defined by their curare sensitivity) and of inhibitory DA effects—the relative contribution of the two depending on the position of the pipette, the duration of the injection. . . (see above).

These complications prevented a full analysis of the ionic mechanisms of DA and

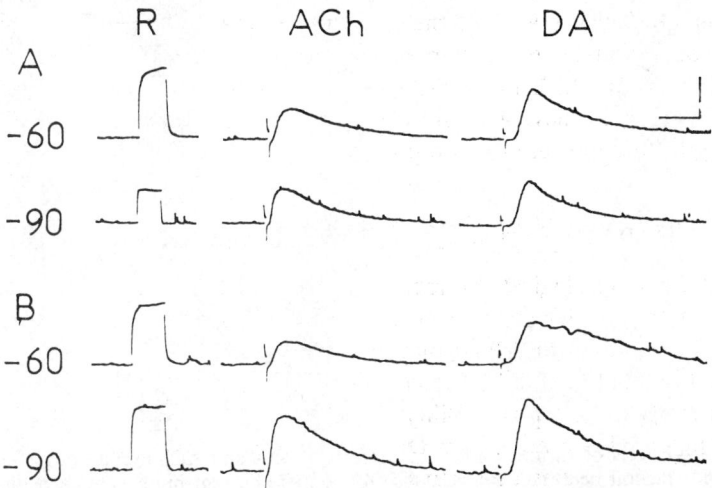

FIG. 1.—Effects of membrane polarisation on the amplitude of DA and ACh responses. Anterior pleural neurone. R: response to a square current pulse. ACh, DA: responses to electrophoretic applications of DA and ACh. The membrane potential was changed from -60 to -90 mV in normal sea-water (A) and in low K^+ (1 mM) sea-water (B). The suppression of anomalous rectification in (B) "normalised" the behavior of DA and ACh responses as a function of membrane polarisation. Calibration 4 sec, 5 mV.

ACh excitatory effects. However, these two types of effects were observed to react differently to various ionic substitutions in the sea water perfusing the ganglion, as shown by the examples illustrated in Fig. 2.

Replacement of Na+ by Li+ markedly reduced the DA response, but only slightly affected that of ACh. After returning to Na+ sea-water, the ACh respose was temporarily increased, while the DA response remained depressed for many minutes.

The effects of Li+ are difficult to interpret. In squid axon, Li+ sea-water may lead

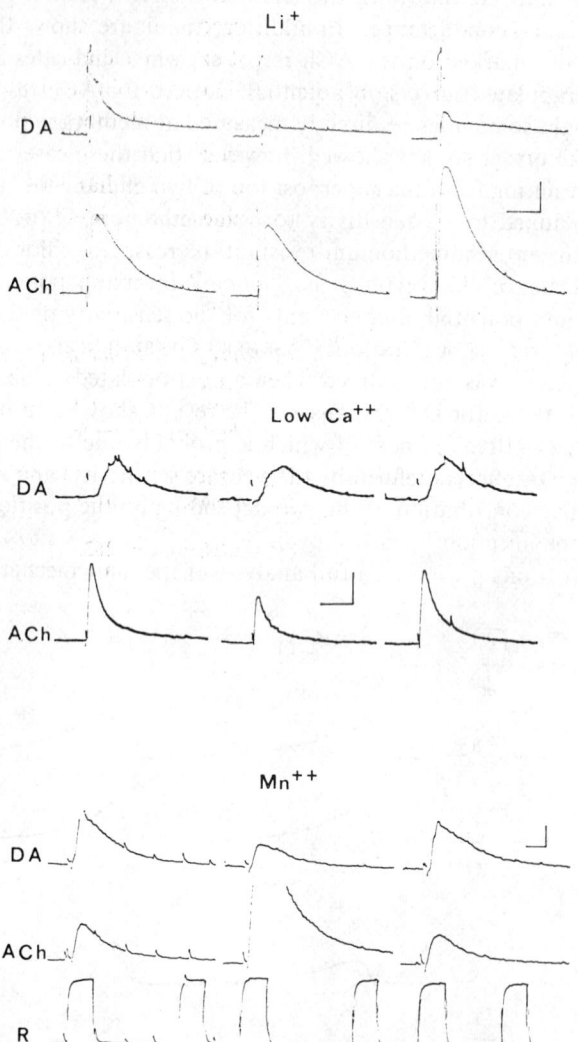

FIG. 2.—Effects of various ionic substitutions on DA and ACh excitatory responses. Anterior pleural neurone. R: response to a square current pulse. On each line the left record represents the initial response (in normal sea-water); the center one the response in modified sea-water (Li+: after replacement of half external Na+ by Li+; low Ca2+: after reduction of the external Ca2+ concentration from 10 to 0·3 mM; Mn2+: after replacement of 50 mM/1 MgCl₂ by 50 mM/1 MnCl₂); the right record represents the response 3 min after returning to normal sea-water. Calibration 5mV; (Li+): 10 sec; (Low Ca2+): 4 sec; (Mn2+): 1 sec.

to a reduction of the Ca^{2+} electrochemical gradient (BAKER, BLAUSTEIN, HODGKIN and STEINHARDT, 1969). However, in the case studied, reduction of the Ca^{2+} concentration did not mimick the effects of Li^+ : on the contrary, it caused a reduction of the ACh response, but had little effect of the DA response (Fig. 2, Center).

It could tentatively be suggested that the DA excitatory action is due to a very selective increase of Na^+ permeability (in which neither Li^+ nor Ca^{2+} could replace Na^+), whereas the ACh excitatory action would correspond to the opening of a less specific channel. Such an interpretation is probably premature however. It does not account for the slow (and often only partial) recovery of the DA response after returning from Li^+ to Na^+ sea-water. It does not explain the complex effects observed when Mg^{2+} ions were substituted by various divalent ions—in particular Co^{2+} or Mn^{2+} : for example, as can be seen in Fig. 2, Mn^{2+} caused simultaneously a reduction of the DA response and a marked increase of the ACh response.

These results do not completely reveal the ionic mechanisms of DA and ACh responses. They do, however, offer additional clues for the identification of dopaminergic excitatory synaptic potentials.

REFERENCES

ASCHER P. (1972) *J. Physiol. Lond.* **225**, 173–209.
ASCHER P., GERSCHENFELD H. M. and KEHOE J. S. (1972) *J. de Physiol.* **65**, 92A.
ASCHER P., GLOWINSKI J., TAUC L. and TAXI J. (1968) *Adv. Pharmacol.* **6A**; 365–368.
BAKER P. F., BLAUSTEIN M. P., HODGKIN A. L. and STEINHARDT R. A. (1969) *J. Physiol.* **200**, 431–458
BERRY M. S. and COTTRELL G. A. (1973) *Nature New Biol.* **242**, 250–253.
GERSCHENFELD H. M. (1973) *Physiol. Rev.* **53**, 1–119.
KANDEL E. R. and TAUC L. (1966) *J. Physiol.* **183**, 287–304.
MARSDEN C. and KERKUT G. A. (1970) *Comp. Gen. Pharmacol.* **1**, 101–116.
SWEENEY D. (1963) *Science N.Y.* **139**, 1051.
TAUC L. and GERSCHENFELD H. M. (1962) *J. Neurophysiol.* **25**, 236–262.
WALKER R. J., WOODRUFF G. N., GLAIZNER B., SEDDEN C. B. and KERKUT G. A. (1968) *Comp. Biochem. Physiol.* **24**, 455–470.

Frontiers in Catecholamine Research 1973, pp. 673 to 674. Pergamon Press. Printed in Great Britain.

ANALYSIS OF DOPAMINE SPECIFIC, EXCITATORY AND INHIBITORY ACTIONS ON NEURONS OF THE SNAIL

H. A. J. Struyker Boudier, W. Gielen and J. M. Van Rossum

Department of Pharmacology, University of Nijmegen, Nijmegen, The Netherlands

Dopamine fulfills a neurotransmitter function in several areas of the mammalian brain as well as in nervous system of several invertebrates.

A number of psychopharmacological agents induces behavioural changes in animals and man by interfering with dopaminergic transmission processes. Amphetamine and apomorphine produce stereotyped behaviour by indirect and direct activation of dopamine receptors in the neostriatum while neuroleptic agents block dopamine receptors and thereby abolish the behavioural effects of apomorphine, amphetamine and dopamine (Van Rossum, 1970).

Dopamine when applied directly onto neurons of the neostriatum causes inhibition of spontaneous activity of many sensitive neurons but excitation of others (Bloom et al., 1965; York, 1970).

Most dopamine sensitive neurons of the brain of the snail *Helix aspersa* show inhibition of activity after application of dopamine, although there are neurons that are excited by dopamine. The neurons inhibited by dopamine have been analysed in detail (Woodruff, 1971). It appeared, however, that the notorious dopamine-mimetic apomorphine does not cause dopamine like inhibition but rather antagonised dopamine induced inhibition (see Fig. 1C). Furthermore neuroleptics, like haloperidol, did not antagonise dopamine in inhibitory dopamine sensitive neurons (see Fig. 1A). On the other hand ergometrine appeared to act as a selective antagonist in these neurons (Woodruff, 1971).

From behavioural studies evidence is accumulating that stereotyped behaviour is mediated by activation of dopamine receptors on neurons that are excited by dopamine rather than inhibited (Cools, 1973).

We therefore analysed in the brain of the snail those dopamine sensitive neurons that are excited by dopamine. (See Figs. 1D and E). These neurons which are excited by dopamine also show excitation following application of apomorphine. Further the neuroleptics as for instance haloperidol are blocking agents of dopamine induced excitation. In most of such "excitatory" dopamine sensitive neurons, haloperidol by itself induced inhibition but in others it was inactive or even might produce excitation.

It is concluded that the excitatory dopamine neurons of the snail serve as a model for dopamine sensitive neurons of the mammalian brain which are involved in behavioural effects elicited by apomorphine and blocked by neuroleptics.

REFERENCES

Bloom F. E., Costa E. and Salmoiraghi G. C. (1965) *J. Pharmacol. exp. Ther.* **150,** 244–252.
Cools A. R. (1973) *Brain Res.* In press.
Van Rossum J. M. (1970) *Int. Rev. Neurobiol.* **12,** 307–383.
Woodruff G. N. (1971) *Comp. gen. Pharmac.* **2,** 439–455.
York D. H. (1970) *Brain Res.* **20,** 233–249.

23

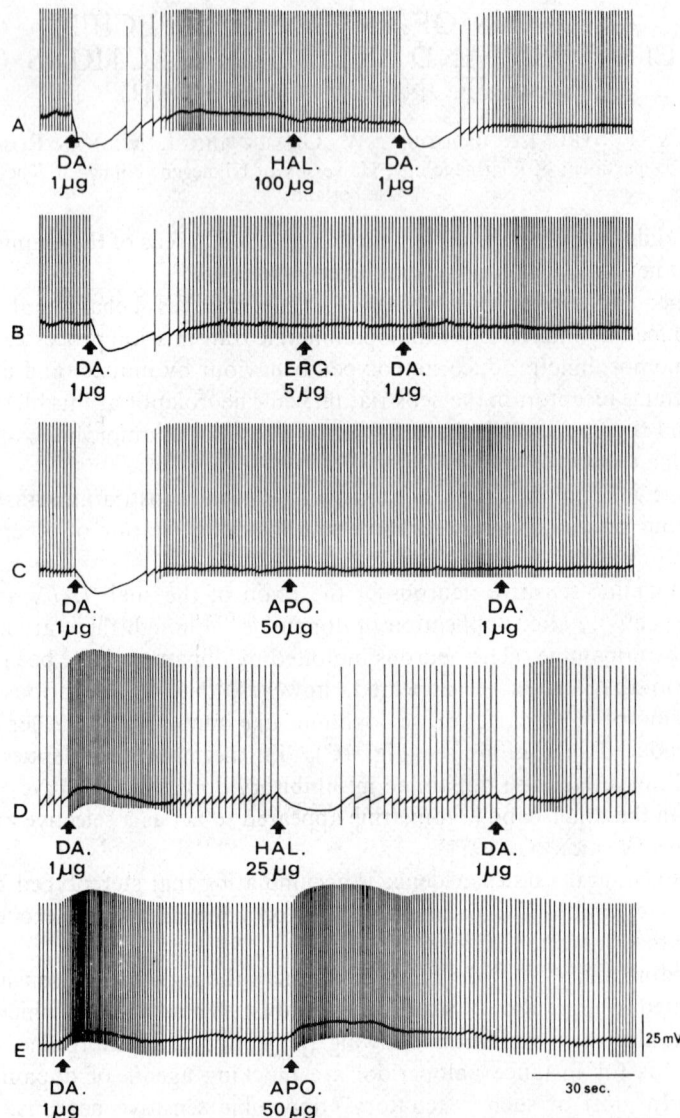

FIG. 1.—Registration of electrical activity of "inhibitory" (A–C) and "excitatory" (D, E) dopamine sensitive neurons. (A) Dopamine is inhibitory in sensitive neurons of the visceral and pleural ganglion cells as shown by hyperpolarisation and a diminution of action potential frequency. The neuroleptic haloperidol is ineffective in antagonizing this inhibition. (B) Ergometrine blocks the inhibitory action of dopamine. (C) Apomorphine does not mimic dopamine on inhibitory cells but rather acts as an antagonist. (D) Dopamine is excitatory on some neurons in the visceral ganglion and haloperidol antagonises this excitation. (E) Apomorphine mimics the dopamine response on neurons excited by dopamine.

Frontiers in Catecholamine Research 1973, pp. 675 to 676. Pergamon Press. Printed in Great Britain.

DOPAMINERGIC TRANSMISSION IN RELATION TO MECHANISMS UNDERLYING STEREOTYPED BEHAVIOUR

C. L. E. Broekkamp, A. J. J. Pijnenburg and J. M. Van Rossum

Department of Pharmacology, University of Nijmegen, Nijmegen, The Netherlands

Stereotyped behaviour as continuously sniffing, gnawing, bizarre social behaviour in rats, pecking in birds, head-shaking in monkeys, "pudning" in man, etc. occurs in all animal species as a result of a disease (rabies, psychosis) or the administration of certain drugs such as amphetamine and apomorphine (Randrup and Munkvad, 1970).

Neuroleptic drugs selectively inhibit stereotyped behaviour in all animal species because of their dopamine antagonizing properties presumably on neurons that are excited by dopamine (Cools, 1973; Struyker Boudier et al., 1973). On the other hand ergometrine, a potent dopamine antagonist on neurons where dopamine acts inhibitory does not abolish stereotyped behaviour but rather induces a strong increase of a kind of stereotyped locomotor activity when injected in the N. accumbens of the rat (Pijnenburg et al., 1973).

Stereotyped behaviour depends on the behavioural repertoire of the individuals and species involved, the environmental setting and the past experience of the individual. The common characteristic of stereotyped behaviour is that it is built up of behavioural elements which recur in an abnormal high frequency. In behaviouristic terminology it could be stated that a drug which induces stereotyped behaviour, reinforces particular behavioural elements to an abnormal high frequency of occurrence. The dopaminergic system may be involved in the reinforcement of behaviour as it can support self-stimulation (Phillips and Fibiger, 1973; Crow, 1972). We therefore investigated the reinforcing action of apomorphine with regard to the reinforcing action of electrical stimulation in rats with electrodes in the dopaminergic cell groups of the midbrain (A9-A10). Apomorphine (0·2 mg/kg s.c.) increased the rate of self-stimulation in 7 out of 13 rats up to 270 per cent of the base line rate,

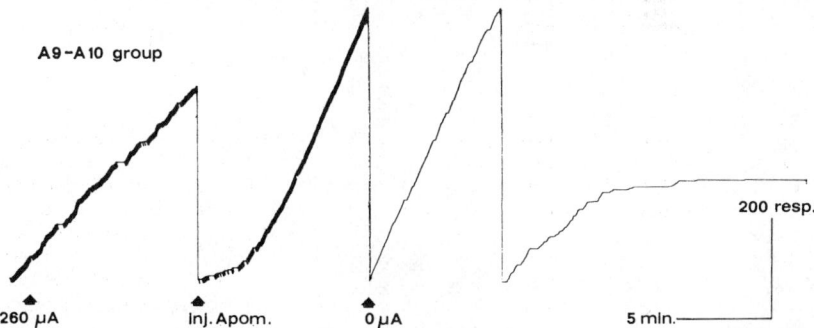

A9–A10 group

200 resp.

260 μA   Inj. Apom.   0 μA   5 min.

Fig. 1.—Apomorphine (0·2 mg/kg s.c.) on lever pressing for electrical stimulation in the A9–A10 cell group. The rat continued to press the lever when the current is switched off.

but in 5 rats the stimulation rate decreased substantially. Such individual differences were found to be highly reproducible.

In contrast to untreated rats in which reduction of the current to zero resulted in a rapid extinction of lever pressing for the stimulation, the apomorphine treated rats continued to press the lever when the current is switched off.

In conclusion the hypothesis is proposed that the apomorphine induced stereotyped behaviour is based in part on the reinforcing action of apomorphine by activation of dopamine receptors. The behavioural elements that are in operation accidentally or by other effects of the drug, during the onset of the apomorphine action will be reinforced and due to the continuing reinforcement repeated in a stereotyped manner.

REFERENCES

COOLS, A. R. (1973) *Brain Res.* in press.
CROW T. J. (1972) *Brain Res.* **36**, 265–273.
PHILLIPS A. G. and FIBIGER H. C. (1973) *Science* **179**, 575–577.
PIJNENBURG A. J. J., WOODRUFF G. N. and VAN ROSSUM J. M. (1973) *Brain Res.* In press.
RANDRUP A. and MUNKVAD I. (1970) In: *Amphetamines and related Compounds.* (COSTA E. and GARATTINI S., Eds.) pp. 695–713.
STRUYKER BOUDIER H. A. J., GIELEN W. and VAN ROSSUM, J. M. (1973) this symposium.

Frontiers in Catecholamine Research 1973, pp. 677 to 681. Pergamon Press. Printed in Great Britain.

HISTOCHEMICAL DEMONSTRATION OF MONOAMINE NEURON SYSTEMS IN THE HUMAN FETAL BRAIN

ANDERS NOBIN and ANDERS BJÖRKLUND

Department of Histology, University of Lund, Lund, Sweden

CONSIDERING the great interest in the functional significance of the monoamine neurons in the CNS of man it is remarkable how so little is known about the cellular localisation of biogenic amines in the human brain. The major reason for this lack of knowledge is that this material offers special technical problems. Although the brain levels of catecholamines and serotonin decline rather slowly during the first hours after death (BERTLER and ROSENGREN, 1959; JOYCE, 1962; McGEER and McGEER, 1962) their optimal fluorescence histochemical visualisation in central nervous tissues is possible only when the tissue is processed within the first 30–60 min after death (BJÖRKLUND, FALCK and ROSENGREN, 1967; DE LA TORRE, 1972, BJÖRKLUND and NOBIN, unpublished observations). We have in our studies on human brain circumvented this obstacle by investigating fresh brains from 3–4 month-old human fetuses (NOBIN and BJÖRKLUND, 1973) and this paper is a brief account of these studies.

In chemical investigations on human fetal brain significant but rather low concentrations of dopamine, noradrenaline and serotonin have been reported (BERTLER, 1961; HYYPPÄ, 1972; NOBIN and BJÖRKLUND, 1973). Nevertheless abundant systems of fluorescent catecholamine-containing and indolamine-containing neurons have been demonstrated (NOBIN and BJÖRKLUND, 1973; OLSON, BOREUS and SEIGER, 1973). Thus, green-fluorescent (CA-containing) and yellow-fluorescent (IA-containing) cell bodies occurred in abundant systems in medulla oblongata, pons and mesencephalon and some green-fluorescent cell bodies occurred also in the hypothalamus (NOBIN and BJÖRKLUND, 1973). The cell body fluorescence was either rather diffuse, covering also the nucleus, or more distinct and confined to the cytoplasm. The fluorescent cell bodies could be referred to seven larger cell formations four CA-containing and three IA-containing: (1) A ventral and ventro-lateral system of CA-containing cells in the medulla oblongata; (2) CA-containing cells in the dorsal and dorsomedial regions of the medulla oblongata, partly within the area postrema; (3). A system of pontine CA-containing cells in the dorsal and lateral regions of the pons, mainly within the locus coeruleus and the subcoeruleus area (see Fig. 1); (4) An extensive system of CA cells in the ventral and medial regions of the mesencephalon (extending caudally into the rostral pons) mainly within the substantia nigra and the ventromedial tegmentum; (5 and 6). Two formations of IA-containing cells in the midline raphe regions, one in the rostral part of the medulla oblongata, and another more extensive system, extending in the raphe region of the pons and the caudal mesencephalon; (7) A less abundant group of IA-containing cells in the lateral reticular formation, in the caudal mesencephalon and the rostral pons. In our material the CA-containing cells were confined to the infundibular nucleus of the hypothalamus, whereas OLSON, BOREUS and SEIGER (1973) were able to demonstrate

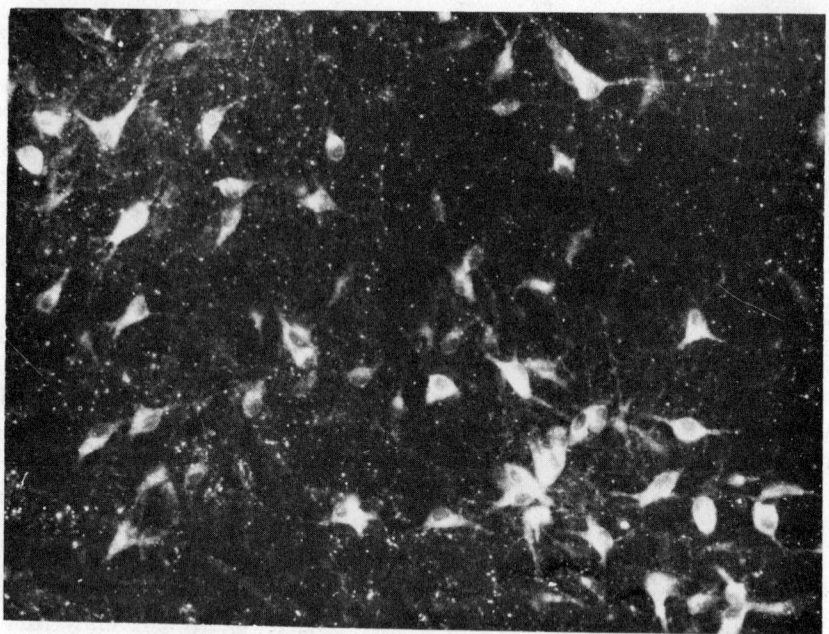

Fig. 1.—Frontal section through the subcoeruleus area from a human fetus (CRL about 15 cm) showing green fluorescent cell bodies. Fluorescent processes and non-fluorescent dark nuclei are clearly seen. Fluorescent varicose fibres are distributed between the cell bodies (×210).

the existence of more extensive diencephalic CA neuron systems in the human fetal brain corresponding to groups A11, A12 and A13 in the rat brain (cf. Fuxe et al., 1969; Björklund and Nobin, 1973).

The monoamine axon bundles had a notably high fluorescence in the fetal brain (Nobin and Björklund, 1973; Olson et al., 1973) and could thus be traced for long distances through the lower brain stem and the hypothalamus up to the basal ganglia and the septal region. One major system of CA fibres could be followed from the region of the lower medulla oblongata through pons and mesencephalon; at the level of the red nucleus, it joined the medial forebrain bundle (MFB). Within the MFB, one system of CA fibres could be further traced up to the septal region. Along the course of this long, probably ascending fibre system it received fibres from the CA cell systems in medulla, pons and mesencephalon.

A second large system of ascending CA fibres was observed to originate in the fluorescent cell bodies in the substantia nigra and via the tegmental fields of Forel, the lateral hypothalamus, and the zona incerta, it ran into the internal capsule, towards the basal ganglia. This ascending pathway is probably identical with the so-called nigrostriatal dopaminergic pathway. Although there is much indirect evidence for such a system also in man, e.g., from observations in parkinsonian patients (see Hornykiewicz, 1966), its actual demonstration has been lacking. The course of the nigrostriatal fibres revealed in the human fetal brain has clear similarities to that described by Carpenter and Peter (1972) in the monkey, and by Moore et al.,

(1971) in the cat. Fluorescent varicose, probably terminal, axon parts were, in the fetal material studied, detected to a more limited extent and only in certain regions. The most abundant CA terminal systems had developed in the basal ganglia (Fig. 2) and the olfactory and septal regions. In the basal ganglia not only the caudate nucleus and putamen (Fig. 2) but also the developing globus pallidus exhibited a dense CA innervation. This conforms well to the biochemical findings of moderate to high concentrations of dopamine in these regions in adults (BERTLER, 1961; BERTLER and ROSENGREN, 1959, EHRINGER and HORNYKIEWICZ, 1960; FAHN et al., 1971; HORNYKIEWICZ, 1966; SANO et al., 1959) and in a 8-months-old human fetus published by BERTLER (1961). As the globus pallidus appears to be devoid of CA-containing terminals in the rat (FUXE, 1965, FUXE et al., 1969) it has been questioned whether the dopamine present in this structure in man was confined to non-terminal fibres passing through the nucleus (HORNYKIEWICZ, 1966). The present findings provide evidence for a direct CA (probably mainly dopamine-containing) fibre supply to the globus pallidus, suggesting that in man the mesencephalic dopamine containing neurons might project also to globus pallidus. Nigropallidal fibre connexions have been presumed to exist in the human brain (HORNYKIEWICZ, 1966) but neither in cat (MOORE et al., 1971) nor in monkey (CARPENTER and PETER, 1972) a substantial termination of nigral fibres has been possible to demonstrate in the globus pallidus with silver staining of degenerating terminals. Thus, until direct evidence for a nigro-pallidal projection has been obtained in man, the possibility has to be considered that the CA-fibres in the globus pallidus originate in other areas, e.g. in mesencephalic CA cell bodies outside the substantia nigra.

Green-fluorescent varicose, probably terminal, fibres occurred also in several hypothalamic and hypophyseal areas whereas in the thalamus, CA fluorescence was observed only in some scattered varicose fibres in ventricle-near regions. A variable density of fluorescent varicose fibres was detected in the subthalamic nucleus. The pineal was supplied with CA fibres organised into thick bundles at the surface or within interlobular septa. OLSON et al. (1973) reported in fetuses with a CRL about 15 cm scattered parenchymal cells exhibiting a yellow fluorescence.

Significant CA fibre supplies were found in the posterior hypothalamic area, the dorsomedial nucleus, the perifornical area, the mamillary body, the tuberal nucleus, the periventricular nucleus, the infundibular nucleus, the paraventricular nucleus, the suprachiasmatic nucleus, and in the median eminence-pituitary region. Some scattered fibres were also observed in the dorsal, lateral and preoptic hypothalmic areas, and in the ventromedial nucleus, where they appeared as parallely arranged fibres traversing the nucleus in ventral and ventrolateral directions. In the rostral infundibulum two systems of fibres extended between the infundibulum and the region of the MFB in a manner suggestive of a commisural arrangement.

In the median eminence-pituitary region varicose CA fibres were demonstrated both in the internal and external layers of the developing median eminence, in the stalk, and in the neural lobe. In the external layer of the median eminence, some of the fibres were oriented perpendicular to the surface (for details see NOBIN and BJÖRKLUND, 1973).

The demonstration of CA-containing cell bodies and axons in the tuberohypophyseal region of the human brain (NOBIN and BJÖRKLUND, 1973) is of particular interest in view of the important role these neurons play in the release of

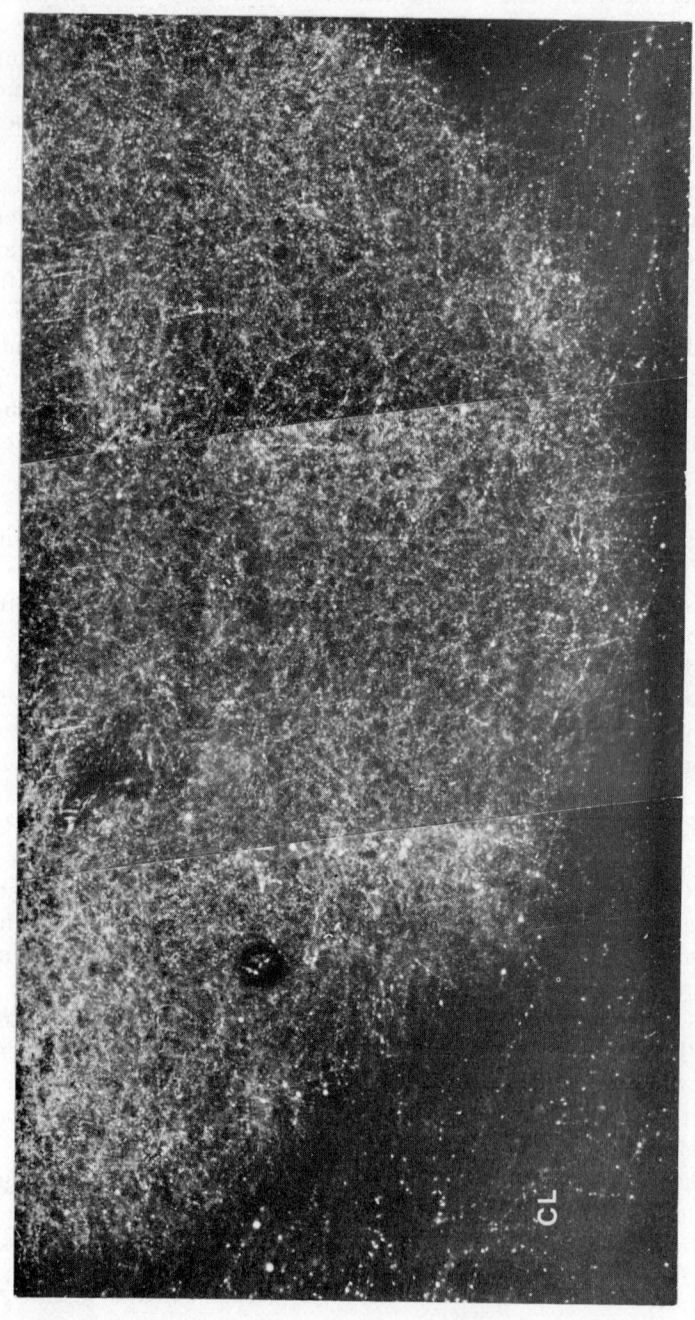

Fig. 2.—Frontal section at the level of caudal neostriatum of a human fetal brain (CRL about 12 cm). Photomontage of the ventral part of the putamen showing the very dense CA-innervation of this nucleus. Some CA fibres are also present in the claustrum (CL) (×130).

hypophysiotrophic hormones from the median eminence in the rat (for reviews see HÖKFELT and FUXE, 1972; McCANN *et al.*, 1972). The morphological basis for such a function thus seems to exist also in man.

It can be concluded that although there are many differences with respect to details the principles of organisation of monoamine neuron systems are notably similar in rat and man.

Acknowledgements—This study was supported by grants from Svenska Livförsäkringsbolags Nämnd för Medicinsk Forskning, Harald och Greta Jeanssons Stiftelse, the Ford Foundation and from the National Institutes of Health, U.S. Public Health Service (grant No. NS 06701-07). The skilful assistance of Annika Borgelin and Christina Wedin is gratefully acknowledged.

REFERENCES

BERTLER Å. (1961) *Acta physiol. scand.* **51**, 97–107.
BERTLER Å. and ROSENGREN E. (1959) *Acta physiol. scand.* **47**, 350–361.
BJÖRKLUND A., FALCK B. and ROSENGREN E. (1967) *Life Sci.* **6**, 2103–2110.
BJÖRKLUND A. and NOBIN A. (1973) *Brain Res.* **51**, 193–205.
CARPENTER M. B. and PETER P. (1972) *J. Comp. Neurol.* **144**, 93–116.
DE LA TORRE J. C. (1972) *Acta neuropath. (Berl.)* **21**, 165–168.
EHRINGER H. and HORNYKIEWICZ O. (1960) *Wien. Klin. Wschr.* **38**, 1236–1239.
FAHN S., LIBSCH L. R. and CUTLER R. W. (1971) *J. neurol. Sci.* **14**, 427–455.
FUXE K. (1965) *Acta physiol. scand.* **64**, (suppl. 247), 37–85.
FUXE K., HÖKFELT T. and UNGERSTEDT U. (1969) In: *Metabolism of Amines in the Brain* (HOOPER E. Ed.). Macmillan, London. pp. 10–22.
HORNYKIEWICZ O. (1966) *Pharmacol. Rev.* **18**, 925–964.
HYYPPÄ M. (1972) *Neuroendocr.* **9**, 257–266.
HÖKFELT T. and FUXE K. (1972) In: *Brain-Endocrine Interaction. Median Eminence: Structure and Function* (KNIGGE K. M., SCOTT D. E. and WEINDL A. eds.). Karger, Basel. pp. 181–223.
JOYCE D. (1962) *Brit. J. Pharmacol.* **18**, 370–380.
McCANN S. M., KALRA P. S., DONOSO A. O., BISHOP W., SCHNEIDER H. P. G., FAWCETT C. P. and KRULICH L. (1972) In: *Brain-Endocrine Interaction. Median Eminence: Structure and Function* (KNIGGE K. M., SCOTT D. E. and WEINDL A., Eds.). Karger, Basel. pp. 224–235.
McGEER E. G. and McGEER P. L. (1962) *Canad. J. Biochem. Physiol.* **40**, 1141–1151.
MOORE R. Y., BHATNAGAR R. K. and HELLER A. (1971) *Brain Res.* **30**, 119–135.
NOBIN A. and BJÖRKLUND A. (1973) *Acta physiol. scand.* (Suppl. 388), 1–40.
OLSON L., BOREUS L. O. and SEIGER Å. (1973) *Z. Anat. Entwickl.-Gesch.* **139**, 259–282.
SANO I., GAMO T., KAKIMOTO Y., TANIGUCHI K., TAKESADA M. and NISHINUMA K. (1959) *Biochim. Biophys. Acta (Amst.)* **32**, 586–587.

Frontiers in Catecholamine Research 1973, pp. 683 to 687. Pergamon Press. Printed in Great Britain.

GLYOXYLIC ACID CONDENSATION: A NEW FLUORESCENCE HISTOCHEMICAL METHOD FOR SENSITIVE AND DETAILED TRACING OF CENTRAL CATECHOLAMINE NEURONS

Olle Lindvall, Anders Björklund and Bengt Falck

Department of Histology, University of Lund, Lund, Sweden

The histochemical formaldehyde fluorescence method (Falck, 1962; Falck et al., 1962; Corrodi and Hillarp, 1963, 1964) demonstrates central adrenergic neurons with high sensitivity and specificity. However, whereas the visualization of the cell bodies and axon terminals can be accomplished with the Falck–Hillarp method, demonstration of the non-terminal parts of the axons is, due to their very low amine concentration, usually not possible in intact and untreated adult animals. This requires localised mechanical or chemical lesions of the neuron in order to increase the intraaxonal amine concentration (Andén et al., 1966; Ungerstedt, 1971; Jacobowitz and Kostrzewa, 1971; Sachs and Jonsson, 1972).

Glyoxylic acid (GA) was recently introduced as a fluorescence histochemical reagent for biogenic monoamines and related compounds (Axelsson et al., 1972; Björklund et al., 1972; Axelsson et al., 1973), and in model experiments (Axelsson et al., 1973) it was demonstrated that this reagent has a considerably higher capacity than formaldehyde to form fluorophores with biogenic indolamines and catecholamines. When the GA method is applied to Vibratome sections of GA perfused brain tissue (Lindvall et al., 1973; Lindvall and Björklund, 1973; Björklund et al., 1973b) the catecholamine (CA) neurons are demonstrated with a sensitivity and richness in details that is superior to the standard Falck-Hillarp method. Thus, the GA method has proved extremely useful for sensitive and detailed neuroanatomical studies on CA neurons in the CNS. The following is a brief presentation of the principal features of the new method.

CHEMISTRY

The fluorescence induced from primary and secondary phenylethylamines and indolylethylamines in the GA method has been shown to be due to the efficient formation of highly fluorescent isoquinoline and β-carboline fluorophores, similar to the Falck–Hillarp formaldehyde method (Björklund et al., 1972, 1973a). The GA reaction proceeds in two steps (Björklund et al., 1972). In the first step, the phenylethylamine or indolylethylamine reacts with GA in an acid catalyzed Pictet-Spengler condensation yielding the 1,2,3,4-tetrahydroisoquinoline-1-carboxylic acid or 1,2,3,4-tetrahydro-β-carboline-1-carboxylic acid, respectively, via a Schiff's base. These very weakly fluorescent compounds can be transformed into strongly fluorescent molecules in two alternative ways: via autoxidative decarboxylation to the 3,4-dihydroisoquinoline or 3,4-dihydro-β-carboline, or through a second, intra-molecularly acid catalyzed reaction with another GA molecule to the 2-carboxymethyl-3,4-dihydroisoquinolinium or 2-carboxymethyl-3,4-dihydro-β-carbolinium compound.

The high fluorescence yields in the GA reaction can probably be referred mainly to the efficient formation of the strongly fluorescent 2-carboxymethyl-3,4-dihydro compounds through intramolecular acid catalysis exerted by the carboxyl group on the 1-carbon of the tetrahydroisoquinoline or tetrahydro-β-carboline molecules. The interested reader is referred to the paper by BJÖRKLUND et al. (1972) for further details and discussions on the reaction mechanisms.

METHODOLOGY

For a detailed methodological description of the GA fluorescence method applied to Vibratome sections the reader should consult the paper by LINDVALL and BJÖRKLUND (1973). The principal steps of the GA procedure are as follows: Adult rats are perfused via the ascending aorta with ice-cold 2% GA in a Krebs–Ringer bicarbonate buffer. The pH of the perfusion solution is adjusted to 7·0 with NaOH. Fifty ml is perfused during about 3 min; thereafter the brains are rapidly dissected out and cooled in the buffer. Pieces of brain tissue are then cut on a Vibratome® instrument (Oxford Instruments, San Mateo, Calif., USA) in 30 μm thick sections, according to the principles introduced by HÖKFELT and LJUNGDAHL (1972). During the sectioning procedure, the tissue piece is immersed in Krebs–Ringer bicarbonate buffer (pH 7·0) kept at a temperature between 0 and +5°C by metal blocks cooled to a very low temperature by a dry ice–ethanol mixture. These metal blocks are changed at suitable intervals. The sections are immersed for 5 min in the ice-cold GA perfusion medium and then transferred to glass microscope slides. Excess buffer is removed with a filter paper, and the sections are then put under the warm air-stream from a hair-dryer for 15 min. The sections are kept overnight in vacuo in a desiccator containing fresh phosphorous pentoxide. All sections are then treated with GA for 2 min according to the procedure described by LINDVALL and BJÖRKLUND (1973). Briefly, this GA treatment is performed as follows: 2g GA monohydrate (dried over phosphorous pentoxide for about 24 hr) are heated at +100°C for at least 1 hr in a closed vessel connected to the reaction vessel, which is placed together with the GA vessel in the oven. After evacuation of the reaction vessel with a vacuum pump, hot GA saturated air (from the GA vessel) is introduced into the hot reaction vessel—containing the tissue specimens—to a partial pressure of 300 torr (mm Hg). Hot air is then let into the reaction vessel to atmospheric pressure. Thus, the temperature is approximately +100°C throughout the reaction. As controls, non-GA treated sections from non-perfused brains are used. The sections are mounted in liquid paraffin and examined in a fluorescence microscope equipped with Schott BG 12 as primary lamp filter and Zeiss 50+47 as secondary barrier filters.

COMMENTS

When the GA method is applied according to the description above dopamine-containing and noradrenaline-containing neurons become strongly fluorescent, whereas the fluorescence induced in indolamine-containing structures is lower and more variable, and the present procedure does not appear useful for the visualisation of intraneuronal indolamines. The CA-containing neurons emit a green fluorescence, which differs from the brownish-yellow fluorescence induced in the indolamine-containing cells. In the control non-GA treated sections from non-perfused tissue, no fluorescent monoamine structures are observed.

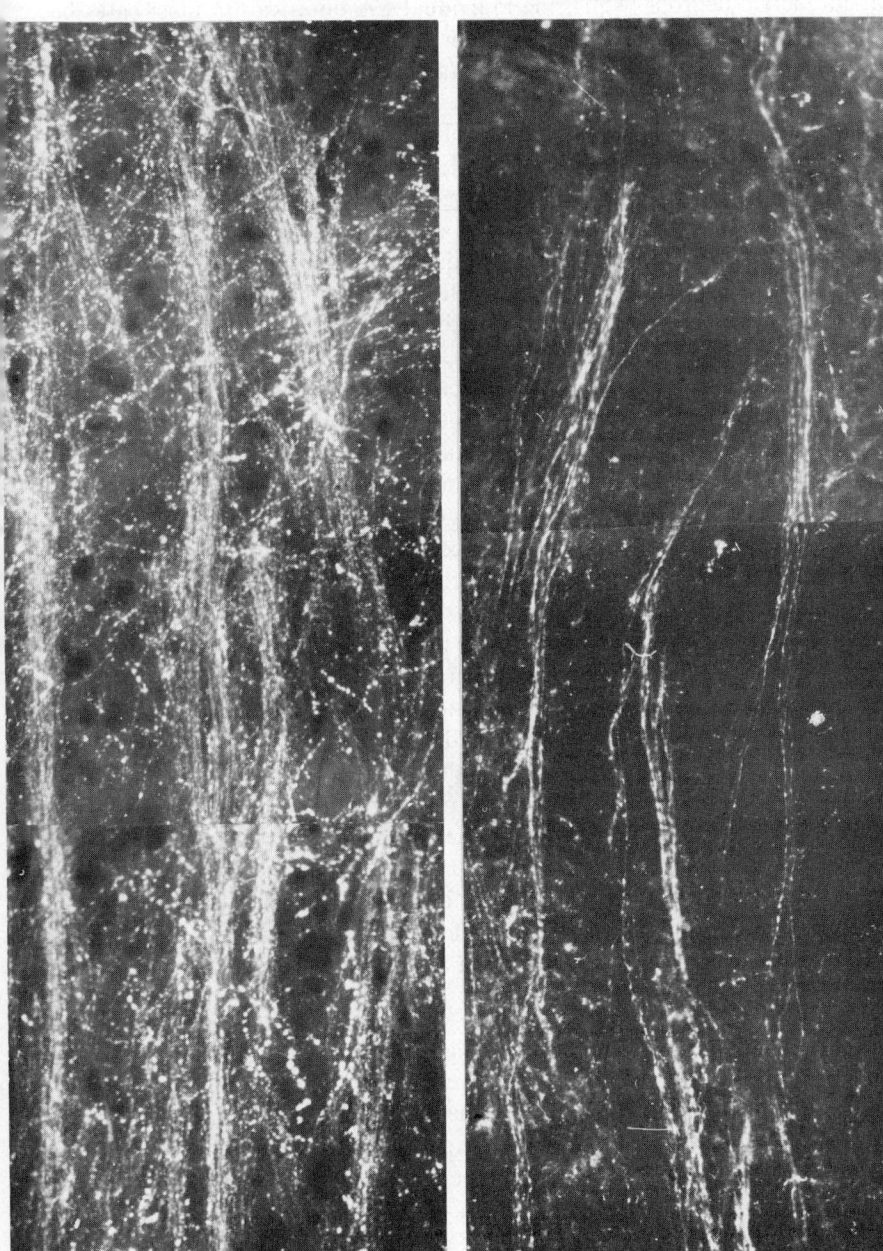

FIG. 1.—Bundles of non-terminal catecholamine axons in the rat brain. (top) Sagittal section through the primarily dopamine-containing nigro-striatal pathway in the lateral hypothalamus. The axons which are likely to be dopamine-containing have an extremely delicate varicose appearance. ×210. (bottom) Sagittal section through the noradrenaline-containing dorsal catecholamine bundle at the level of the interpenduncular nucleus. ×210.

The fluorescence microscopical picture obtained in the GA-treated sections is characterized by an extreme sensitivity and precision in the visualization of the central CA neurons. The high sensitivity is valid above all for the axonal part of the central adrenergic neuron. Thus, the individual CA fibres can in most cases be followed from their cell bodies of origin all the way to the terminal area. This is e.g. the case for the CA axons within the nigro-striatal pathway (Fig. 1), the medial fore-brain bundle and the dorsal CA bundle (Fig. 1). Moreover, systems of non-terminal axons not previously described have been demonstrated in many areas of the brain.

Systems of CA-containing axon terminals demonstrable with the Falck-Hillarp technique (Fuxe, 1965) are observed also in the GA treated sections. In most terminal fibres the smooth intervaricose segments are clearly visible despite their low fluorescence intensity. With the present technique many areas show a more prominent adrenergic innervation than observed earlier, and furthermore in many areas considered to be practically devoid of CA-containing structures the new method has revealed abundant CA terminal systems. Thus, in the thalamus, whose CA innervation previously has been considered insignificant in comparison with that of the hypothalamus (Fuxe, 1965; Fuxe et al., 1969), prominent supplies of delicate varicose axon terminals are demonstrated (Björklund et al., 1973b).

The usefulness of the GA method for the demonstration of adrenergic cell bodies is limited because of their variable morphology in the Vibratome sections. The fluorescence often covers the nucleus and the cell bodies sometimes have a diffuse outline. The cellular fluorescence is, however, in many cases stronger than in formaldehyde-treated, freeze-dried specimens.

The high precision and richness in details in the visualization of the CA neurons are most probably due not only to the increased sensitivity (see above), but also to the absence of diffusion in the GA treated sections. Thus, the fluorescence in the caudate nucleus is distinct and confined to very delicate, smooth and varicose fibres. In addition, scattered fibres with coarser varicosities are observed, possibly identical with noradrenaline-containing terminals. In the median eminence, where the fluorescence also often appears diffuse after formaldehyde treatment, each of the very delicate, beaded palisade fibres can be traced through the external layer up to the surface. The richness in details of the CA-axons in the GA treated sections is further demonstrated in e.g. the septal nuclei, where a fine structural picture is obtained of the terminals located around neuronal perikarya and dendrites.

CONCLUSIONS

In the GA method, the intraneuronal CA:s are very efficiently transformed into strongly fluorescent molecules in a well-defined reaction with GA. When the method is applied to sections of GA perfused brain tissue, the CA neurons are demonstrated with a sensitivity and richness in details that is superior to the standard formaldehyde method. Thus, the individual CA fibres can be followed from their cell bodies to the terminal area and the detailed morphology of the axon can be studied. Consequently, the GA method seems to be *ideal for studies on the details of organization of the CA neuron systems in the CNS.*

REFERENCES

Andén N.-E., Dahlström A., Fuxe K., Larsson K., Olson L. and Ungerstedt U. (1966) *Acta physiol. scand.* **67**, 313–326.

AXELSSON S., BJÖRKLUND A., FALCK B., LINDVALL O. and SVENSSON L. Å. (1973) *Acta physiol. scand.* **87**, 57–62.

AXELSSON S., BJÖRKLUND A. and LINDVALL O. (1972) *J. Histochem. Cytochem.* **20**, 435–444.

BJÖRKLUND A., FALCK B., LINDVALL O. and SVENSSON L. Å. (1973a) *J. Histochem. Cytochem.* **21**, 17–25.

BJÖRKLUND A. HÅKANSON R., LINDVALL O. and SUNDLER F. (1973b) *J. Histochem. Cytochem.* **21**, 253–265.

BJÖRKLUND A., LINDVALL O., NOBIN A. and STENEVI U. (1973c) *Comm. Dept. Anat. Univ. Lund.* no. 4, 1–21.

BJÖRKLUND A., LINDVALL O. and SVENSSON L. Å. (1972) *Histochemie* **32**, 113–131.

CORRODI H. and HILLARP N.-Å. (1963) *Helv. Chim. Acta* **46**, 2425–2430.

CORRODI H. and HILLARP N.-Å (1964) *Helv. Chim. Acta* **47**, 911–918.

FALCK B. (1962) *Acta physiol. scand.* **56**, (suppl. 197), 1–25.

FALCK B., HILLARP N.-Å., THIEME G. and TORP A. (1962) *J. Histochem. Cytochem.* **10**, 348–354.

FUXE K. (1965) *Acta physiol. scand.* **64**, (suppl. 247), 39–85.

FUXE K., HÖKFELT T. and UNGERSTEDT U. (1969) In: *Metabolism of Amines in the Brain* (HOOPER E. Ed.). Macmillan, London. pp. 10–22.

HÖKFELT T. and LJUNGDAHL Å. (1972) *Histochemie* **29**, 325–339.

JACOBOWITZ D. and KOSTRZEWA R. (1971) *Life Sci.* **10**, 1329–1342.

LINDVALL O. and BJÖRKLUND A. (1973) *Acta physiol. scand.* In press.

LINDVALL O., BJÖRKLUND A., HÖKFELT T. and LJUNGDAHL Å. (1973) *Histochemie* **35**, 31–38.

SACHS CH. and JONSSON G. (1972) *J. Neurochem.* **19**, 1561–1575.

UNGERSTEDT U. (1971) In: *6-Hydroxydopamine and Catecholamine Neurons* (MALMFORS T. and THOENEN H. Eds.). North-Holland, Amsterdam. pp. 101–127.

Frontiers in Catecholamine Research 1973, pp. 689 to 693. Pergamon Press. Printed in Great Britain.

BEHAVIOURAL–ANATOMICAL CORRELATES OF CENTRAL CATECHOLAMINE NEURONS

U. UNGERSTEDT and T. LJUNGBERG

Department of Histology, Karolinska Institutet, 104 01 Stockholm, Sweden

THE ASCENDING catecholamine (CA) pathways may be subdivided into a few major neuron systems (FUXE, 1965; UNGERSTEDT, 1971a; MAEDA and SHIMIZU 1972). The ventral noradrenaline (NA) bundle (including the intermediate bundle) arising from cell groups in the medulla and pons and terminating in the mesencephalon, the hypothalamus and the preoptic area. The dorsal NA bundle arising from the locus coeruleus proper and terminating in the cerebellar and cerebral cortices, the hippocampus and probably also in several subcortical areas, e.g. the reticular formation, the thalamus and the hypothalamus. The nigro-striatal dopamine (DA) system originating in the zona compacta of the substantia nigra and terminating in the striatum (nucl. caudatus and putamen) and the meso-limbic DA system arising from cell bodies dorsal and lateral to the interpeduncular nucleus and terminating in the nucleus accumbens and the olfactory tubercle.

The anatomy of the CA cell groups and axon bundles permits the localisation of lesions interrupting certain pathways while leaving others intact. The ventral and dorsal NA bundles may be lesioned separate from each others and caudal to the DA cell groups. The striatal and the limbic DA systems may also be separated by lesions but the lesions will always cause some damage to the ascending NA axons. However, the extent of the DA lesion may be kept constant by positioning lesions along the DA pathway, while the involvement of the NA lesion is varied in this way.

The specificity of the lesion technique may be considerably increased by the use of intracerebral, stereotaxic injections of 6-hydroxydopamine (6-OH-DA) (UNGERSTEDT, 1968; 1971a, 1973). The NA and DA neurons are lesioned with great selectivity while the unspecific lesion is limited to a narrow zone around the tip of the cannula (UNGERSTEDT, 1971b; HÖKFELT and UNGERSTEDT, 1973). We have used stereotaxic injections of 6-OH-DA into the rat brain with drugs affecting CA transmission and quantitative analysis of behaviour in order to gain an understanding for the involvement of CA neurons in the control of behaviour. The combination of drugs and lesions has permitted us to study the effects of hyper- as well as hypofunction in a certain CA neuron system.

DOPAMINE NEURONS

A unilateral 6-OH-DA induced lesion of the nigro-striatal DA pathway induces profound changes in movements and posture (UNGERSTEDT, 1968). The rat deviates towards the side of the lesion and amphetamine induced release of DA from the non-lesioned DA system changes the deviation into continuous circling movements toward the lesioned side. We have constructed a "rotometer" consisting of a bowl, shaped as a hemisphere, where the animal moves connected by a thin wire to the recording device (UNGERSTEDT and ARBUTHNOTT, 1970). Pharmacologically induced DA release induces a dose dependent rotational behaviour toward the lesioned side (UNGERSTEDT, 1971c) while DA receptor stimulating drugs, like apomorphine, (ERNST, 1967, ANDÉN *et al.*, 1967) induce a rotation in the opposite direction (UNGERSTEDT, 1971d). The direction of the rotational behaviour seem to reveal on which side the DA receptors are most activated. DA releasing drugs as well as DA receptor stimulating drugs cause rotation towards the lesion after unilateral damage of the striatum (ANDÉN *et al.*, 1965) and unilateral injection of DA into the striatum induces rotation towards the side opposite the injection, i.e. away from the most activated side (UNGERSTEDT *et al.*, 1969). In the unilaterally 6-OH-DA denervated animal apomorphine induces rotation towards the intact side indicating that the denervated side is "supersensitive" to apomorphine. In fact, iontophoretic application of apomorphine (HOFFER *et al.*, 1973) or DA (FELTZ and CHAMPLAIN, 1973) shows that the denervated striatal cells are more sensitive to apomorphine and DA than the innervated.

Bilateral 6-OH-DA induced degeneration of the nigro-striatal DA system induces a state of akinesia, catalepsia, adipsia and aphagia. The animals die if not supported by tube feeding. The syndrome is closely similar to the "lateral hypothalamus lesion" syndrome which has been attributed to the degeneration of an eating and drinking centre in the lateral hypothalamus (ANAND and BROBECK, 1951). However, recent studies with histochemical technique show that lateral hypothalamic lesions interrupt the nigro-striatal DA system and thus denervates the striatum (UNGERSTEDT, 1970, 1971e). It seems highly probable that the lateral hypothalamic adipsia and aphagia is due to interference with striatal rather than hypothalamic function. The severity of the syndrome seems related to the completeness of the DA denervation. The less DA that remains in the striatum the less likely is the animal to resume eating and drinking. Animals with no DA nerve terminals detectable in the striatum and no DA cell bodies in the substantia nigra show no sign of recovery even after 6 months of tube feeding (AVEMO *et al.*, 1973). Such animals show no response to amphetamine (2 mg/kg) but a supersensitive response to apomorphine which is in accordance with the findings in the unilaterally denervated animal.

In order to further link the syndrome of akinesia, adipsia and aphagia to the degeneration of DA neurons, we have substituted the lost DA with a low dose of apomorphine: normal animals were trained to run a maze for water reward. When fully trained the DA neurons were bilaterally degenerated with 6-OH-DA. The lesion completely abolished all responses in the maze. However, after a low dose of apomorphine (0·1 mg/kg) the animals were able to run the maze and consume the water (LJUNGBERG and UNGERSTEDT 1973). Too high a dose of apomorphine introduced stereotyped behaviour and decreased or abolished the maze performance. The fact that bilateral removal of the DA system abolished a previously learned response may in this case be related to the akinesia and not to any interference with learning processes. To test this, we developed an under-water Y-maze (RANJE and UNGERSTEDT, 1973). Rats were forced under the water surface in the start box, swam to the choice point and had to select the lit up arm in order to reach the surface. The position of the arm was altered. Fully trained animals were bilaterally DA denervated. They still performed in the maze being slightly slower than controls but not significantly different as regards errors. It is interesting that the animals were, in fact, able to overcome their akinesia in this situation of "high motivation".

There is an obvious parallel in the "paradoxical kinesia" observed in similar situations in Parkinson degeneration of DA neurons.

In search for the functional deficit common to the rotating, unilaterally denervated animal, as well as the bilaterally denervated akinetic, adipsic and aphagic animals, we have examined their ability to react, i.e. orient towards sensory stimuli. A unilaterally DA denervated animal does not orient towards any visual, olfactory, auditory or tactile stimuli applied to the side of the body contralateral to the degeneration of the DA neurons (LJUNGBERG and UNGERSTEDT, 1973). He gradually recovers his ability to orient towards stimuli, however, the inability to orient towards tactile stimuli seems permanent. The fact that the animals are able to orient towards certain sensory stimuli but not towards others indicates that the deficit is not purely motor but an inability to integrate certain sensory stimuli, with motor output. A similar syndrome of sensory neglect has recently been described in lateral hypothalamic lesioned animals (MARSCHALL et al., 1972). It is obvious that a severe deficit in sensory-motor integration may account for the state of adipsia and aphagia.

The state of akinesia is associated with a state of rigidity which is, e.g. well known in Parkinson's disease. STEG (1964) explained this rigidity as an increased α tonus on the basis of studies after reserpine. We have now studied discharge from ventral roots during spontaneous activity as well as after electrical stimulation of the dorsal roots (STEG and UNGERSTEDT, 1973). Bilaterally DA denervated animals showed increased spontaneous α discharge as well as α reflex discharge. Treatment with apomorphine or DOPA in the same doses that abolished akinesia in the behavioural models normalised the α discharge.

Electrical self stimulation is elicited from regions in the brain that anatomically overlap with the ascending DA fibers. In order to test the involvement of DA pathways in the behaviour fifteen unilaterally DA denervated animals were implanted with electrodes bilaterally in the hypothalamus (CHRISTIE et al., 1973). All animals self-stimulated from the electrode on the intact side. Four animals did not self-stimulate on the lesioned side. These four animals showed the highest rotation score on apomorphine and thus the most pronounced supersensitivity which is an indication of the extent of the denervation. The 6-OH-DA injection also damages the NA fibres, however, there was no correlation between failure to self-stimulate and extent of NA degeneration.

A lesion of the nigro-striatal DA pathway may to a varying extent involve the meso-limbic DA pathway. It is difficult to attribute a behavioural effect only to the striatal system. In terms of the rotational behaviour histochemical analysis reveals that the behaviour is elicited even when there is no degeneration of the limbic DA system. Limbic lesions may cause a short period of hypodipsia and hypophagia, while striatal DA degeneration causes the serious syndrome of adipsia and aphagia. Finally the stereotyped behaviour after apomorphine is greatly potentiated by 6-OH-DA induced degeneration of the DA nerve terminals in the nucl. accumbens. The denervation probably induces postsynaptic supersensitivity and the increased stereotyped behaviour may, thus, be due to increased stimulation of DA receptors in this area.

Several authors have argued that DA is involved in stereotyped behaviour while NA is involved in locomotion or exploratory behaviour. In order to quantify different aspects of the drug induced behaviour we designed an automatic "hole-board" where movements over the bottom of the box as well as the looking into the holes in the bottom were recorded. A normal animal explores the open field as well as the holes. The apomorphine induced behaviour varies considerably with the dose. A low dose of apomorphine (0·1–0·5 mg/kg) causes an interruption of all ongoing behaviour while increased doses induce intense locomotion over the open field with no looking into holes. At still higher doses, sniffing, licking and gnawing develops, while loco-motion may decrease. Amphetamine increases locomotion but causes intense looking into holes as well. High doses induce stereotyped behaviour and decreased loco-motion similarly to apomorphine. It is tempting to suggest that the apomorphine induced behaviour is an expression of DA receptor stimulation alone, while the amphetamine response is due to a combined increase in DA and NA release. When the effect is confined to DA receptors a series of non-directed patterns of behaviour are triggered. When both NA and DA transmission is increased the motor patterns remain, but they are more directed towards real objects, e.g. the holes.

NORADRENALINE NEURONS

While degeneration of the DA neurons caused profound and evident changes in behaviour comparable changes after degeneration of the ascending NA axon bundles do not occur. After lesions of the dorsal and ventral NA bundles there are no obvious changes in spontaneous behaviour apart from a slight increase in sponta-neous activity in the hole-board. However, when the animals are tested in an operant situation (lever pressing for water reward) several changes may be detected (Ljung-berg and Ungerstedt, 1973). The NA lesioned animals are slower to learn, continue pressing for water longer during a session, and show less extinction if water is removed. However, if the operant testing is discontinued for a longer period and then resumed, the NA lesioned animals show a decreased retention (Ljungberg and Ungerstedt, 1973).

The behavioural changes are correlated with lesions of the ventral NA bundle innervating the hypothalamus while lesions of the dorsal NA bundle alone do not induce the changes in operant responses. The ventral NA bundle lesion also induces an increased weight gain and moderate overeating (Ljungberg and Ungerstedt, unpublished). However, the weight increase is not as pronounced as after ventro-medial hypothalamic lesions.

Dorsal NA bundle lesions, denervating the cortex and the hippocampus, do not produce evident changes in behaviour *per se*. In a maze task the animals are performing as well as controls. However, after amphetamine (2 mg/kg) the dorsal bundle lesioned animal is unable to run the maze, while the control animal is only slightly impaired. The deficit after dorsal NA bundle lesions seems to be related to an increased stereotyped response to amphetamine (Ljungberg and Ungerstedt, 1973).

REFERENCES

Anand B. K. and Brobeck J. R. (1951) *Yale J. Biol. Med.* **24,** 123–140.
Andén N.-E., Dahlström A., Fuxe K. and Larsson K. (1966) *Acta pharmacol. (Kbh.)* **24,** 263–274.
Andén N.-E., Rubenson A., Fuxe K. and Hökfelt T. (1967) *J. Pharm. Pharmacol.* **19,** 627–629.
Avemo E., Ljungberg T. and Ungerstedt U. (1973) unpublished work.
Christie J., Ljungberg T. and Ungerstedt U. (1973) unpublished work.
Ernst A. M. (1967) *Psychopharmacologia (Berl.)* **10,** 316–323.
Fuxe K. (1965) *Acta physiol. scand.* **64,** (Suppl. 247), 36–85.

HOFFER B., SIGGINS G., UNGERSTEDT U. and BLOOM F. (1973) unpublished work.
HÖKFELT T. and UNGERSTEDT U. (1973) *Brain Res.* Vol. 60, 269–297.
LJUNGBERG T. and UNGERSTEDT U. (1973) *Brain Res.* (in press).
MAEDA T. and SHIMIZU N. (1972) *Brain Res.* **36**, 19–36.
MARSHALL J. F., TURNER B. H. and TEITELBAUM P. (1971) *Science* **174**, 523–525
STEG G. (1964) *Acta physiol. scand.* **61**, (Suppl. 22), 1–53.
UNGERSTEDT U. (1968) *Europ. J. Pharmacol.* **5**, 107–110.
UNGERSTEDT U. (1970) *Acta physiol. scand.* **80**, 35A–36A.
UNGERSTEDT U. (1971a) *Acta physiol. scand.* **82**, (Suppl. 367), 1–48.
UNGERSTEDT U. (1971b) in *6-Hydroxydopamine and catecholamine neurons* (MALMFORS T. and
   THOENEN H., eds.) pp. 101–127. North-Holland Amsterdam.
UNGERSTEDT U. (1971c) *Acta physiol. scand.* **82**, (Suppl. 367), 49–68.
UNGERSTEDT U. (1971d) *Acta physiol. scand.* **82**, (Suppl. 367), 69–93.
UNGERSTEDT U. (1971e) *Acta physiol. scand.* **82**, (Suppl. 367), 95–122.
UKGERSTEDT U. (1973) In: *Neurosciences Research.* (EHRENPREIS S. and KOPIN I., Eds.). In press.
UNGERSTEDT U. and ARBUTHNOTT G. (1970) *Brain Res.* **24**, 485–493.
UNGERSTEDT U., BUTCHER L. L., BUTCHER S. G., ANDÉN N.-E. and FUXE K. (1969) *Brain Res.*
   14, 461–471.

Frontiers in Catecholamine Research 1973, pp. 695 to 700. Pergamon Press. Printed in Great Britain.

BEHAVIOURAL PHARMACOLOGY OF 6-HYDROXYDOPAMINE

R. I. Schoenfeld and M. J. Zigmond

Dept.'s of Psychiatry and Pharmacology,
Children's Hospital Medical Center and Harvard Medical School,
Boston, Mass., U.S.A.
and
Dept.'s of Biology and Psychology,
University of Pittsburgh, Pittsburgh, U.S.A.

The ability of 6-hydroxydopamine (6-HDA) to produce a specific degeneration of central catecholamine-containing neurons offers a unique opportunity to study the consequences of loss of these neurons on the performance of a conditioned response. At the same time we are provided with a useful experimental tool for testing present theories of the mechanism of action of drugs whose behavioural effects are thought to be mediated via an effect on these neurons. The present report will discuss several experiments utilizing 6-HDA treated rats, which were designed with this dual significance in mind.

6-HDA AND OPERANT BEHAVIOUR

Studies of the effects of drugs on operant behaviour have shown that the schedule of reinforcement maintaining behaviour is a prime determinant of drug action (Kelleher and Morse, 1968). The principle of schedule-dependency extends to the behavioural effects of amine-depleting drugs. Reserpine (Smith, 1964), tetrabenazine (McKearny, 1968) and α-methyltyrosine (Schoenfeld and Seiden, 1967, 1969) have all been shown to decrease responding maintained by some schedules at a dose which had less or no effect on responding maintained by other schedules. In light of the accumulating evidence that the turnover of brain dopamine (DA) and norepinephrine (NE) is increased during performance of a conditioned response (Fuxe and Hanson, 1967; Schoenfeld and Seiden, 1969; Arbuthnott et al., 1971; Lewy and Seiden, 1972), it has been suggested that behaviour maintained by schedules generating higher turnover rates would be most sensitive to these drugs (Harvey, 1971). Although any proposed mechanism is speculative, it remains essential that we recognise the existence of schedule-dependency. It is for this reason that we have studied the effects of 6-HDA on behaviour maintained by several different schedules of water reinforcement.

(1) *Fixed ratio* (FR)

This schedule generates a high response rate that can be maintained throughout the experimental session. After performance had stabilised on an FR-20 schedule, two doses of 250 μg 6-HDA or vehicle were administered intraventricularly to groups of six rats. Twenty-four hours after the second injection responding was reduced to $51 \cdot 8 \pm$ of the pre-injection control rate in the 6-HDA group, while the vehicle group responded at control rate. Responding during the next session (48 hr post-injection) was decreased to $67 \cdot 1 \pm 19\%$ of control in the 6-HDA group, but by the third day (72 hr post-injection) responding had returned to control level and remained at this level for the duration of the experiment (Fig. 1a). Acquisition of performance on this schedule was unimpaired by administration of 6-HDA prior to training.

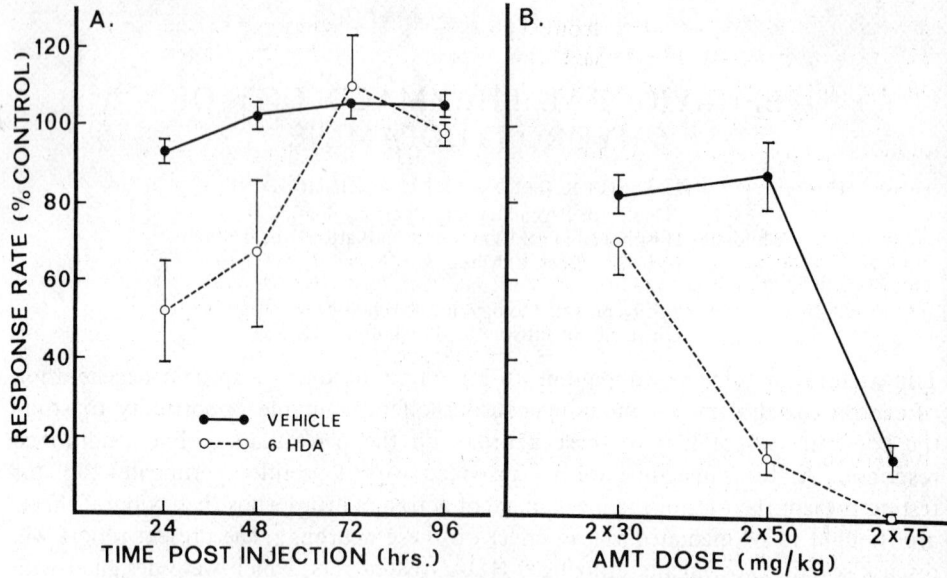

Fig. 1a.—Effect of 6-HDA on FR-20 performance. 250 μg of 6-HDA or vehicle
was injected into alternate lateral ventricles (48 hr apart) under ether anesthesia.
Twenty-four hr after the second injection testing was continued. Each point represents
the mean (±S.E.M.) response output as a percentage of each group's mean response
output for three days prior to injection ($N = 6$).
b. Effect of α-methyltyrosine (AMT) on FR-20 performance of 6-HDA- and
vehicle-treated rats. D, L- α-methyl-tyrosine or saline was administered 6 and 3 hr prior
to the session. Each point represents the mean (±S.E.M). response output as a percentage
of the previous day's response output ($N = 6$).

Next we examined the effect of repeated injection (intracisternal) of 200 μg
6-HDA, given at weekly intervals, on FR 20 performance. When tested 1 hr after
the first injection, responding was completely abolished in the 6-HDA-treated group.
As after the 2×250 μg intraventricular treatment, this was followed by recovery
to control levels of responding within 3 days. One week after the first injection (when
responding was at control levels), a second 200 μg was administered followed by a
third injection one week later. Again, after each injection, responding was abolished
1 hr post-injection, followed by recovery to control levels within a few days (Schoenfeld and Zigmond, 1970).

(2) Fixed interval (FI)

On this schedule, responding is very low immediately following reinforcement
and then rapidly accelerates to a final high rate that is sustained until the next reinforcement. This pattern of low followed by high response rates is repeated throughout
the session. Rats treated with 6-HDA prior to training on an FI 3 min schedule
developed a pattern of responding identical to control rats.

(3) Variable interval (VI)

Rats trained on this schedule have a moderate rate of responding that is sustained
throughout the session. When 6-HDA was administered prior to training, the

response rate stabilised after 6 weeks (30 sessions) at a level approximately four times higher than control rats. If 6-HDA was administered to rats already trained on a VI 1·5 min schedule, responding increased over a period of 6 weeks after an initial decrease, until it again stabilised at a level approximately four times higher than control rats. Superimposing a shock contingency on the VI schedule (every 30th response results in administration of an electric shock) was effective in suppressing responding of the 6-HDA-treated rats. Introduction of a time-out (TO) period into the VI schedule (during the middle 9 min of the session responding had no schedule consequences) also diminished responding, although it took several days for the 6-HDA-treated group to reach the same low level of responding observed in the control group (SCHOENFELD and URETSKY, 1972a).

TABLE 1. EFFECT OF DIFFERENT 6-HDA TREATMENTS ON VARIABLE INTERVAL PERFORMANCE AND BRAIN AMINE LEVELS

Treatment	Responses/Session*	Dopamine†	Norepinephrine
Vehicle	557 ± 107	0·46 ± 0·027	0·30 ± 0·009
3 × 50μg 6-HDA	525 ± 93	0·49 ± 0·021 (106·5%)	0·15 ± 0·007 (51·4%)
2(25mg/kg DMI + 250μg 6-HDA)	1459 ± 106	0·09 ± 0·008 (14·2%)	0·27 ± 0·019 (87·4%)

*Values represent mean (±S.E.M.) number of responses during the *fortieth* daily session following treatment ($N = 5$).
†Values represent mean (±S.E.M.) brain amine level (uncorrected for recovery) of groups of rats ($N = 5$ or more) treated identically to those used in behaviour experiment.
Values in parenthesis indicate percentage of control group.

The 6-HDA treatment used in the VI experiment, two injections of 250 μg 6-HDA, produced an 80–90 per cent decrease in brain NE and a 70–80 per cent decrease in brain DA. We have since attempted to reproduce the effect of 6-HDA on VI performance in rats selectively depleted of either DA or NE. Groups of five rats were treated with either three injections of 50 μg 6-HDA, to destroy NE-containing neurons, or two injections of 250 μg 6-HDA 1 hr after 25 mg/kg of desmethylimipramine (DMI), to affect only DA-containing neurons. The group treated with 3 × 50 μg 6-HDA, which lowered whole brain NE by 50 per cent, responded at control rates while the group treated with 2 × DMI + 250 μg 6-HDA, which decreased brain DA to 14 per cent of control, increased their response rate so that after 40 sessions they were responding at a level three times higher than the control group (Table 1).

In summary, 6-HDA produced an initial decrease in responding when administered to rats trained on FR and VI schedules. A few days after administration, responding on the FR schedule returned to normal. Normal response patterns also developed on the FI schedule as well as when as time-out or shock contingency was added to the VI schedule. These results demonstrate that except for the initial deficits observed, the 6-HDA treatment had little effect on the control of behaviour by environmental stimuli. In these animals reinforcing stimuli still maintain responding, discriminative stimuli still control responding (TO schedule) and aversive stimuli still suppress responding (shock schedule). Consequently, it might be concluded that catecholamine-containing neurons are not involved in these behavioural processes, which would be at variance with much experimental evidence to the contrary (see above). Another possibility is that compensatory mechanisms exist to maintain normal response patterns in spite of the marked destruction of DA- and NE-containing neurons produced by 6-HDA.

Only when rats were responding on a VI schedule, where the relationship between external events and responding provides few cues to the organism, was a prolonged effect observed. Under these conditions 6-HDA-treated rats did not maintain the low response rate of control rats but gradually reached a new level of responding several times higher than controls. This occurred when DA neurons were selectively destroyed but not when only NE neurons were affected, suggesting that dopamine-containing neurons are involved in controlling the level of performance.

6-HDA AND DRUG EFFECTS ON OPERANT BEHAVIOUR

α- Methyltrosine

This drug has been shown to produce a specific depletion of DA and NE as a consequence of inhibition of catecholamine (CA) synthesis (SPECTOR et al., 1965). It has been shown to decrease responding maintained by several different schedules of reinforcement (POSCHEL and NINTEMAN, 1966; FUXE and HANSON, 1967; SCHOENFELD and SEIDEN, 1967). That the effect of α-methyltyrosine (AMT) on operant behaviour is a consequence of CA depletion is supported by the finding that L-dopa can restore lever-pressing in AMT-treated rats (SCHOENFELD and SEIDEN, 1969).

We have found that AMT is effective in decreasing responding of 6-HDA-treated rats. In fact, 6-HDA-treated rats (2×250 μg) responding on an FR-20 schedule were affected by AMT at a dosage level that had no effect on control rats (Fig. 1b). This increased sensitivity to AMT has also been demonstrated in 6-HDA-treated rats responding on a VI schedule (SCHOENFELD and URETSKY, 1972a), on an FI schedule (SCHOENFELD, unpublished results) and on a CRF schedule (COOPER et al., 1972). Considering the selectivity of the depletion produced by AMT and the fact that dopa can reverse its effects in normal rats, we take the effect of AMT on 6-HDA-treated rats as evidence that CA-containing neurons surviving 6-HDA treatment are involved in maintaining responding in these rats.

d-Amphetamine

d-Amphetamine has both rate-increasing and rate-decreasing effects on lever-pressing. These effects depend on the schedule of reinforcement maintaining responding (see KELLEHER and MORSE, 1968). Pre-treatment with AMT blocks the rate-increasing effect of d-amphetamine (RECH, 1970; MAICKEL et al., 1970) but not the rate-decreasing effect (HEFFNER and ZIGMOND, unpublished observations), suggesting that the former effect is mediated via release of catecholamines while the latter effect is due to a direct action on CA receptors or involves some other neuronal population. We have studied both the rate-increasing and rate-decreasing effects of d-amphetamine in 6-HDA-treated rats trained on different schedules of reinforcement.

(1) FR: d-Amphetamine decreased responding on this schedule. There was no difference in the dose–response curves obtained from vehicle- and 6-HDA-treated rats (Fig. 2a).

(2) VI: A dose of 0·56 μg/kg of d-amphetamine increased responding of vehicle-treated rats on this schedule to approximately 160 per cent of the control rate. No

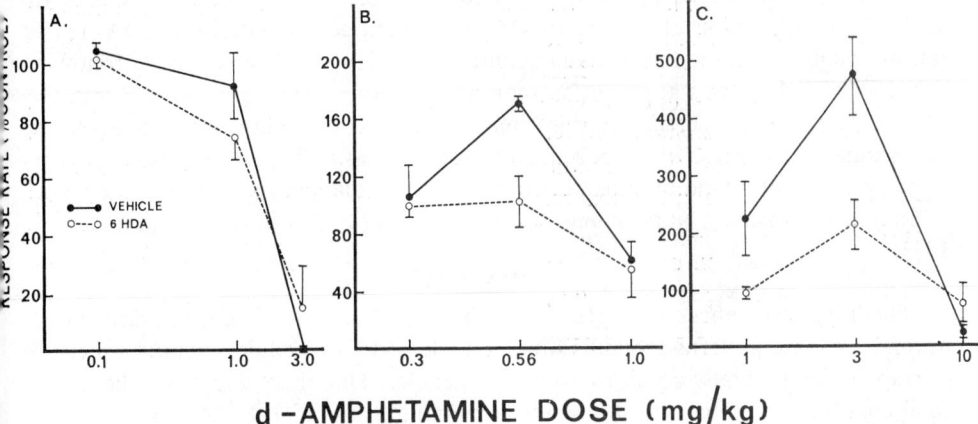

d -AMPHETAMINE DOSE (mg/kg)

FIG. 2.—Effect of 6-HDA pre-treatment on the rate-increasing and rate-decreasing effect of d-amphetamine. d-Amphetamine sulphate was administered immediately before the test session to 6-HDA- and vehicle-treated rats ($N = 4$–6) performing on FR-20 (a), VI-1·5 min. (b), or FI-3 min. (c) schedules of reinforcement. Each point represents the mean ($\pm$ s.e.m.) response output as a percentage of control for the entire session, except FI where the effect of d-amphetamine on responding during the initial 1·5 min of each 3 min interval is reported as a percentage of control responding during this time.

increase was observed in 6-HDA-treated rats at this dose or lower doses. Both groups showed a similar decrease in responding after 1 mg/kg d-amphetamine (Fig. 2b).

(3) FI: In vehicle-treated rats, d-amphetamine produced a large increase in the low rates of responding which occur during the initial part of each interval. Although an increase was observed in 6-HDA-treated rats, it was much less than that obtained in vehicle-treated rats (Fig. 2c).

The reduced effectiveness of d-amphetamine to increase responding on the FI schedule, in spite of the low response rate of 6-HDA-treated rats suggests that the loss of CA-containing neurons prevents the expression of its rate-increasing effect. This is supported by the inability of d-amphetamine to increase responding on the VI schedule. In this case, however, the higher response rate of 6-HDA-treated rats may make this schedule more like the FR schedule, in which case only rate-decreases would be expected. The unaltered rate-decreasing effect obtained on all schedules tested supports the suggestion that some action other than the release of catecholamines mediates this effect of d-amphetamine.

6-HDA AND CENTRAL DENERVATION SUPERSENSITIVITY

Several mechanisms may participate in the recovery of performance following 6-HDA treatment. One possibility is that changes in the synaptic region produce an increased sensitivity to released catecholamines. Based on the effects of apomorphine and L-dopa in 6-HDA-treated rats, it has been suggested that 6-HDA produces a post-synaptic type of supersensitivity in the CNS (UNGERSTEDT, 1971, SCHOENFELD and URETSKY, 1972b). However, determination of the dose–response curve for the effects of dopa on motor activity indicated that 6-HDA shifts the curve by a factor of ten toward lower doses. The time course of the increase in activity and brain catecholamines produced by 100 mg/kg L-dopa, suggested that the effect was mediated by the conversion of L-dopa to DA, which did accumulate in 6-HDA-treated rats although to a lesser extent than in control rats. The increase in activity following this dose of L-dopa was apparent as soon as 24 hr after the second dose of 6-HDA $(2 \times 250 \mu g$, 48 hr apart). At this time the uptake of ^{3}HDA was decreased by approximately 50 per cent. Consequently, on the basis of these results, it appears that a pre-synaptic type of supersensitivity to catecholamines is also a consequence of 6-HDA treatment (URETSKY and SCHOENFELD, 1971; SCHOENFELD and URETSKY, 1973).

SUMMARY

The long-term effect of 6-HDA on operant behaviour is schedule-dependent; normal response patterns are maintained on FR and FI schedules, while an increase in responding gradually develops on a VI schedule. This effect appears to be related to altered function of central dopamine-containing neurons. 6-HDA-treated rats are more sensitive to AMT, suggesting that CA-containing neurons surviving 6-HDA treatment participate in maintaining the normal patterns of responding observed on the FR and FI schedules. Maintenance of normal function may also involve the development of central denervation supersensitivity which, as indicated by experiments with L-dopa, occurs soon after 6-HDA treatment. The rate-decreasing effect of d-amphetamine is unaltered in 6-HDA-treated rats while the rate-increasing effect is attenuated, consistent with proposals that the rate-increasing effect is mediated by release of catecholamines.

Acknowledgements—This work was supported in part by the following grants: HD-033773 (NICHD), MH-18349 and MH-20620 (NIMH). The authors would like to acknowledge Dr. N. J. Uretsky for his contribution to many of these experiments and specifically for the amine data presented in Table 1.

REFERENCES

ARBUTHNOTT G., FUXE K. and UNGERSTEDT U. (1971) *Brain Res.* **27**, 406–413.
COOPER B. R., BREESE G. R., HOWARD J. L. and GRANT L. D. (1972) *Psychopharmacologia* **27**, 99–110.
FUXE K. and HANSON L. C. F. (1967) *Psychopharmacologia* **11**, 439–447.
HARVEY J. (1971) *Behavioral Analysis of Drug Action.* p. 172. Scott Foresman, Glenview, Illinois.
KELLEHER R. T. and MORSE W. H. (1968) *Ergeb. Physiol.* **60**, 1–56.
LEWY A. J. and SEIDEN L. S. (1972) *Science* **175**, 454–456.
MAICKEL R. P., COX R. H., KSIR C. J., SNODGRASS W. R. and MILLER F. P. (1970) In: *Amphetamines and Related Compounds.* (COSTA E. and GARATTINI S., Eds.) pp. 747–759. Raven Press, New York.
MCKEARNEY J. W. (1968) *J. Pharmacol. Exp. Therap.* **159**, 429–439.
POSCHEL B. P. H. and NINTEMAN F. W. (1966) *Life Sci.* **5**, 11–16.
RECH R. H. (1970) In: *Amphetamines and Related Compounds.* (COSTA E. and GARATTINI S., Eds.) pp. 385–413. Raven Press, New York.
SCHOENFELD R. I. and SEIDEN L. S. (1967) *J. Pharm. Pharmacol.* **19**, 771–772.
SCHOENFELD R. I. and SEIDEN L. S. (1969) *J. Pharmacol. Exp. Ther.* **167**, 319–327.
SCHOENFELD R. I. and URETSKY N. J. (1972a) *Europ. J. Pharmacol.* **20**, 357–362.
SCHOENFELD R. I. and URETSKY N. J. (1972b) *Europ. J. Pharmacol.* **19**, 115–118.
SCHOENFELD R. I. and URETSKY N. J. (1973) *J. Pharmacol. Exp. Ther.* In press.
SCHOENFELD R. I. and ZIGMOND M. J. (1970) *Pharmacologist* **12**, 227.
SMITH C. B. (1964) *J. Pharmacol. Exp. Ther.* **146**, 167–174.
SPECTOR S., SJOERDSMA A., and UDENFRIEND S. (1965) *J. Pharmacol. Exp. Ther.* **147**, 86–95.
UNGERSTEDT U. (1971) *Acta Physiol. Scand. (Suppl.)* **367**, 69–93.
URETSKY N. J. and SCHOENFELD R. I. (1971) *Nature (New Biol.)* **234**, 157–159.

Frontiers in Catecholamine Research 1973, pp. 701 to 706. Pergamon Press. Printed in Great Britain.

BIOCHEMICAL AND BEHAVIOURAL ALTERATIONS FOLLOWING 6-HYDROXYDOPAMINE ADMINISTRATION INTO BRAIN*

GEORGE R. BREESE, BARRETT R. COOPER and RONALD D. SMITH

Departments of Pharmacology and Psychiatry, Biological Sciences Research Center,
Child Development Institute, University of North Carolina School of Medicine,
Chapel Hill, North Carolina 27514, U.S.A.

SINCE the discovery that 6-hydroxydopamine (6-OHDA) destroys peripheral adrenergic fibers (THOENEN and TRANZER, 1968), several laboratories have investigated the possibility that 6-OHDA might produce selective destruction of central catecholamine neurons when introduced into brain (UNGERSTEDT, 1968; BREESE and TRAYLOR, 1970; URETSKY and IVERSEN, 1970). Following such treatment with 6-OHDA, BLOOM et al., (1969) described degenerating fibers in areas rich in adrenergic neurons. The reduced brain tyrosine hydroxylase activity and reduced synthesis of catecholamines from tyrosine after 6-OHDA administration provided further evidence for the view that 6-OHDA causes a "central sympathectomy" (BREESE and TRAYLOR, 1970; URETSKY and IVERSEN, 1970).

In initial studies, intracisternal injection of 6-OHDA was found to cause a greater depletion of brain norepinephrine (NE) than dopamine (DA) (BREESE and TRAYLOR, 1970). Subsequent work indicated that administration of multiple small doses of 6-OHDA provided a method to reduce brain NE which had little effect on DA. In contrast, treatment with desipramine prior to administration of 6-OHDA protected noradrenergic fibers while permitting relatively selective destruction of dopaminergic neurons (BREESE and TRAYLOR, 1971). Equivalent depletions of both NE and DA could be produced by 6-OHDA in animals pretreated with pargyline (BREESE and TRAYLOR, 1970). These procedures to produce chronic reduction of brain NE, DA, or both catecholamines have been used by our laboratory to investigate physiological, behavioural and pharmacological responses postulated to be dependent upon central sympathetic neural systems (BREESE et al., 1971; BREESE et al., 1972; BREESE and TRAYLOR, 1972; COOPER et al., 1972; COOPER et al., 1973a,b; SMITH et al., 1973a,b; BREESE et al., 1973; HOLLISTER et al., 1973; Table 1).

EFFECTS OF 6-OHDA TREATMENT ON OPERANT BEHAVIOUR

Early studies with 6-OHDA indicated that depletions of brain catecholamines were not accompanied by a chronic behavioural effect (TAYLOR and LAVERTY, 1972; COOPER et al., 1972). In general, following an initial period of behavioural depression, performance of 6-OHDA treated animals returned to control levels. This apparent dissociation of behavioural and biochemical effects may be similar to that described after treatment with reserpine. In 1967, PIRCH et al. found that reserpine produced depression of avoidance responding followed by recovery within a few days, although at the time of recovery biogenic amine concentrations remained severely reduced.

* Supported by USPHS grants MH-16522 and HD-03110. G.R.B. is a Career Development Awardee (HD-24585). B.R.C. and R.D.S. are post-doctoral fellows (MH-11107).

Subsequently, RECH et al. (1968) demonstrated that such animals showed an enhanced sensitivity to a small dose of α-methyltyrosine. Recently, 6-OHDA treated rats, which showed no chronic behavioural deficits, were also found to display enhanced sensitivity to doses of α-methyltyrosine and reserpine which did not affect active avoidance or bar-press responding in control animals (COOPER et al., 1972).

Several possibilities may account for the absence of permanent behavioural deficits following the destruction of brain catecholamine fibres by 6-OHDA treatment (COOPER et al., 1972). These include considerations of increased activity of surviving fibres and denervation supersensitivity. It is also possible that anatomical location of certain catecholamine-containing fibres make them more resistant to destruction by 6-OHDA. Such proposals suggest that greater depletion of catecholamines with 6-OHDA would produce chronic behavioural deficits, as remaining adrenergic fibres became so sparse that catecholamines could no longer be released in sufficient quantity to maintain function. In support of this view, 6-OHDA in combination with pargyline was found to cause drastic depletion of brain catecholamines which chronically reduced acquisition and performance of an active avoidance task (COOPER et al., 1973a).

In an effort to examine the relative role of each of the brain catecholamines in shuttle-box avoidance responding, animals were employed in which either brain NE or DA was preferentially reduced. In initial experiments, no behavioural deficits were observed following these treatments. In fact, animals depleted of NE were more active than controls and displayed a significant increase of avoidance responding during acquisition of this task (COOPER et al., 1973a), such as obtained following treatment with p-chlorophenylalanine (TENEN, 1967). Therefore, any interpretation of these findings must consider the significant reduction of serotonin following this treatment (Table 1).

Since 6-OHDA treated rats were found to be more sensitive to the behavioural depressant effects of α-methyltyrosine, this treatment was applied to animals preferentially depleted of NE or DA. In addition, the dopamine-β-hydroxylase inhibitor, 1-phenyl-3-[2-thiazolyl]-2-thiourea [U-14,642], was also administered to reduce only brain NE. It was reasoned that the additional catecholamine depletion produced by these drugs might reveal deficits related to prior reduction of NE or DA in the groups depleted of either brain amine. Doses of α-methyltyrosine which did not alter performance of control animals resulted in decreased responding from rats depleted of brain DA (Fig. 1a). Rats depleted of NE also showed reduced avoidance responding following α-methyltyrosine, but their performance remained above the group in which brain DA was depleted. Administration of U-14,642, which further reduced brain NE by approximately 75 per cent in all animals, produced no behavioural depression (Fig. 1a). These findings support the view that DA plays a critical role in the maintenance of avoidance behaviour (SEIDEN and CARLSSON, 1963).

Since multiple injections of 6-OHDA have been shown to produce greater depletion of brain catecholamines, two doses of 6-OHDA (240 μg) were given to desipramine-treated rats to increase the depletion of brain DA. This procedure antagonised acquisition of the shuttle-box avoidance response as well as acquisition of a one-way avoidance task (Table 1). In other experiments, 6-OHDA was injected into various brain sites to destroy specific NE and DA fibre tracts. In contrast to results after depletion of NE with intracisternal injections, destruction of dorsal and ventral NE

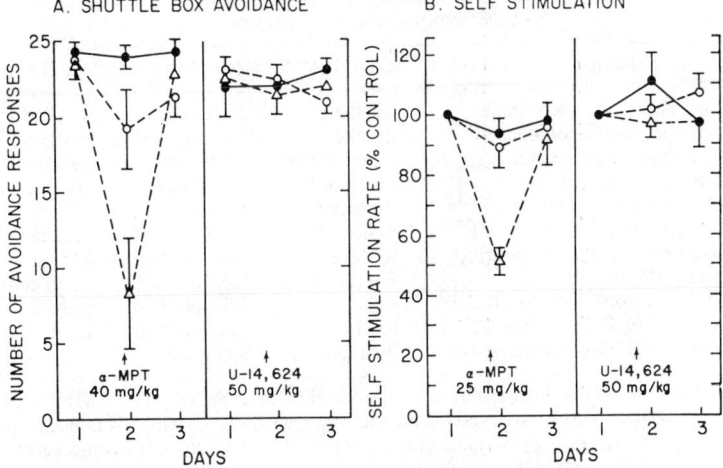

A. SHUTTLE BOX AVOIDANCE   B. SELF STIMULATION

FIG. 1.—Shuttle box avoidance (a) and self-stimulation (b) responding from control animals (●——●) or from rats depleted of DA (Δ————Δ) or of NE (○————○) after α-methyltyrosine (α-MPT) or U-14,624. Doses of α-MPT and U-14,624 were administered 4 hr and 6 hr, respectively, before behavioural testing. Methods to deplete NE or DA are described by BREESE et al. (1972).

Brain catecholamines 4 hr after α-MPT (40 mg/kg).

Control rats:	NE = 51 ± 2 and DA =	40 ± 2% control
Rats depleted of NE:	NE = 23 ± 2 and DA =	39 ± 7% control
Rats depleted of DA:	NE = 38 ± 4 and DA =	11 ± 3% control

Brain catecholamines 4 hr after α-MPT (25 mg/kg).

Control rats:	NE = 62 ± 7 and DA =	55 ± 7% control
Rats depleted of NE:	NE = 21 ± 1 and DA =	53 ± 6% control
Rats depleted of DA:	NE = 52 ± 4 and DA =	18 ± 4% control

Brain catecholamines 6 hr after U-14,624 (50 mg/kg).

Control rats:	NE = 27 ± 6 and DA =	117 ± 19% control
Rats depleted of NE:	NE = 8 ± 2 and DA =	91 ± 11% control
Rats depleted of DA:	NE = 27 ± 3 and DA =	24 ± 3% control

pathways with 6-OHDA did not facilitate acquisition of the avoidance task, even though NE was reduced by 85 per cent in the forebrain. Administration of 6-OHDA into the nigro-striatal pathway as well as into the caudate was found to block acquisition of the shuttle-box avoidance response providing further evidence for the proposal that dopaminergic neurons are essential for maintaining avoidance responding.

Evidence suggests that central catecholamine-containing fibres also play a crucial role in self-stimulation of brain (POSCHEL and NINTEMAN, 1963). In accord with this view, 6-OHDA in combination with pargyline was found to cause a chronic depression of responding in this task (BREESE et al., 1971). Whereas preferential depletion of NE or DA alone did not alter self-stimulation, administration of α-methyltyrosine to these groups caused drastic reduction of self-stimulation rates in rats depleted of brain DA, but did not significantly alter responding of rats depleted of brain NE (Fig. 1b). Furthermore, depression of self-stimulation was not observed in any group following inhibition of dopamine-β-hydroxylase to reduce brain NE (Fig. 1b). Thus, brain DA may play a role in maintaining electrical self-stimulation of brain.

TABLE 1. BEHAVIOURAL RESPONSES IN RATS FOLLOWING VARIOUS
6-HYDROXYDOPAMINE TREATMENTS

Behaviour	"6-OHDA"*	"NE Down"†	"DA Down"‡
Acquisition: shuttle-box avoidance	Blocked	Enhanced	Blocked
Acquisition: one-way avoidance	Blocked	—	Blocked
Acquisition: passive avoidance	Like control	Reduced	Like control
Habituation of motor activity	Like control	Enhanced	Like control
Self-stimulation of brain	Reduced	Like control	—
Sucrose consumption	Reduced	Like control	Reduced
Saline consumption (DOCA-induced)	Reduced	Like control	Reduced
Motor activity (Dopa-induced)	Enhanced	Like control	Enhanced
Motor activity (amphetamine-induced)	Reduced	Like control	Reduced
Stereotypy (amphetamine-induced)	Reduced	Like control	Reduced
Hypothermia (6-OHDA-induced)	Blocked	Reduced	—

* "6-OHDA" refers to intracisternal administration of $2 \times 200 \mu g$ of 6-OHDA, one with pargyline (50 mg/kg) pretreatment and the other without. Brain NE concentration has been shown to be $15 \cdot 2 \pm 3 \cdot 7 \%$ of control, DA $7 \cdot 1 \pm 1 \%$ of control and serotonin $87 \cdot 2 \pm 3 \cdot 4 \%$ of control after this treatment.

† "NE Down" refers to intracisternal injection of $3 \times 25 \mu g$ of 6-OHDA. Brain NE has been found to be $41 \cdot 9 \pm 3 \cdot 7 \%$ of control, DA $94 \cdot 2 \pm 6 \cdot 7 \%$ of control and serotonin $81 \cdot 3 \pm 6 \cdot 1 \%$ of control after this treatment.

‡ "DA Down" refers to intracisternal administration of $2 \times 240 \mu g$ of 6-OHDA to desipramine (30 mg/kg) pretreated rats. Brain NE has been shown to be $80 \pm 3 \cdot 8 \%$ of control, DA $12 \pm 1 \cdot 4 \%$ of control and serotonin $79 \pm 5 \%$ of control after this treatment.

EFFECT OF 6-OHDA ON CONSUMMATORY BEHAVIOUR

Following administration of 6-OHDA into brains of rats pretreated with pargyline, an acute period of aphagia and adipsia occurs which lasts several days (ZIGMOND and STRICKER, 1972). Furthermore, UNGERSTEDT (1971) has reported acute depression of food and water consumption after injection of 6-OHDA into the substantia nigra, implicating disruption of dopaminergic systems in this response. Intracisternal treatment with 6-OHDA to reduce brain DA also causes an acute decrease in eating and drinking further supporting this view (BREESE et al., 1973).

Recovery of consummatory function after 6-OHDA treatment has been likened to the syndrome occurring after lesioning of the lateral hypothalamus (UNGERSTEDT, 1971). Similar to animals "recovered" from lateral hypothalamic lesions(TEITELBAUM and EPSTEIN, 1962), rats surviving treatment with 6-OHDA in combination with pargyline failed to increase food intake in response to 2-deoxyglucose (ZIGMOND and STRICKER, 1972) and insulin (BREESE et al., 1973). Enhanced saline preference found in control rats following desoxycorticosterone (DOCA) treatment is also markedly reduced in rats treated with 6-OHDA. In addition, it has been observed that control rats drink large volumes of sucrose solution substituted for water, while 6-OHDA treated rats show little increase in consumption of this solution. Investigation of sucrose consumption in rats preferentially depleted of NE or DA suggest that the failure of 6-OHDA treated rats to increase fluid intake in response to a sucrose solution is related to depletion of brain DA (BREESE et al., 1973).

TREATMENT OF DEVELOPING RATS WITH 6-OHDA

Administration of 6-OHDA to immature rats has been found to produce marked reductions of brain catecholamine levels and tyrosine hydroxylase activity. Accompanying the destruction of central catecholamine-containing fibres in developing rats is a marked deficiency in growth (BREESE and TRAYLOR, 1972; LYTLE *et al.*, 1972). Furthermore, immature rats treated with 6-OHDA not only fail to increase fluid consumption when a sucrose solution is substituted for water, but also fail to increase preference for saline when treated with desoxycorticosterone (SMITH *et al.*, 1973b). Food and water intake of 6-OHDA treated rats is also reduced when compared with intake of controls. Such evidence would suggest that the growth deficiency may be related to a permanent change in consummatory behaviour. In addition, rats treated with 6-OHDA when immature show a significant deficit in acquisition of the shuttle-box avoidance response consistent with previous findings from adult rats that received 6-OHDA in combination with pargyline.

Treatments developed in adult animals to deplete NE or DA were also applied to the neonate. Rats depleted of brain DA displayed deficits in growth, consummatory behaviour and acquisition of avoidance responding (SMITH *et al.*, 1973a). These deficits were not observed in animals in which brain NE was preferentially reduced. However, depletion of brain NE in immature rats facilitated performance early in acquisition of the shuttle-box avoidance task and produced hyperactivity during habituation to circular activity cages. Depletion of NE in neonates did not alter brain serotonin, suggesting that these behavioural changes may indeed be due to altered noradrenergic function.

AMPHETAMINE-INDUCED BEHAVIOUR AFTER 6-OHDA TREATMENT

In accord with the view that catecholamines are important for the behavioural actions of amphetamine (HANSON, 1967), HOLLISTER *et al.* (1973) recently found that amphetamine induced motor activity was reduced following depletion of both catecholamines in brain with 6-OHDA (Table 1). Stereotypic behaviour produced by amphetamine was likewise reduced. While depletion of NE did not reduce amphetamine-stimulated motor activity and stereotypies, depletion of DA did antagonise these activities. In contrast to antagonism of the pharmacological actions of amphetamine, preferential depletion of DA with 6-OHDA markedly potentiated DOPA stimulated motor activity (Table 1).

SUMMARY

Adult and neonatal rats treated with 6-OHDA to reduce NE, DA or both catecholamines have been used to examine the role of brain catecholamines in several behaviours (Table 1). The data implicated DA-containing fibres in the maintenance of several diverse functions including consummatory behaviour, active avoidance responding and self-stimulation of brain. Motor activity and stereotypies induced by amphetamine also appear to be dependent upon brain DA. Table 1 shows that brain NE at this time has been associated only with temperature control. Present findings are consistent with proposals (EVERETT and WIEGAND, 1962; SEIDEN and CARLSSON, 1963) suggesting alternative roles for DA not clearly related to its usual association with extrapyramidal function.

REFERENCES

BLOOM F. E., ALGERI S., GROPPETTI A., REVUELTA A. and COSTA E. (1969) *Science* **166**, 1284–1286.
BREESE G. R. and TRAYLOR T. D. (1970) *J. Pharmacol. Exp. Ther.* **174**, 413–420.
BREESE G. R. and TRAYLOR T. D. (1971) *Br. J. Pharmacol.* **42**, 88–99.
BREESE G. R. and TRAYLOR T. D. (1972) *Br. J. Pharmacol.* **44**, 210–222.

BREESE G. R., HOWARD J. L. and LEAHY J. P. (1971) *Br. J. Pharmacol.* **43**, 255–257.
BREESE G. R., MOORE R. A. and HOWARD J. L. (1972) *J. Pharmacol. Exp. Ther.* **180**, 591–602.
BREESE G. R., SMITH R. D., COOPER B. R. and GRANT L. D. (1973) *Pharmacol. Biochem. Behav.* In press.
COOPER B. R., BREESE G. R., HOWARD J. L. and GRANT L. D. (1972) *Psychopharmacol.* **27**, 99–110.
COOPER B. R., BREESE G. R., GRANT L. D. and HOWARD J. L. (1973a) *J. Pharmacol. Exp. Ther.* **185**, 358–370.
COOPER B. R., GRANT L. D. and BREESE G. R. (1973b) *Psychopharmacol.* In press.
EVERETT G. M. and WIEGAND R. G. (1962) *Proc. 1st Int. Pharmacol. Meeting* **8**, 85–92.
HANSON L. C. F. (1967) *Psychopharmacol.* **10**, 289–297.
HOLLISTER A., COOPER B. R. and BREESE G. R. (1973) *Pharmacologist* In press.
LYTLE L. D., SHOEMAKER W. J., COTTMAN K. E. and WURTMAN R. J. (1972) *J. Pharmacol. Exp. Ther.* **183**, 56–64.
PIRCH J. H., RECH R. H. and MOORE K. E. (1967) *Int. J. Neuropharmacol.* **6**, 375–385.
POSCHEL B. P. H. and NINTEMAN F. W. (1963) *Life Sci.* **1**, 782–788.
RECH R. H., CARR L. A. and MOORE K. E. (1968) *J. Pharmacol. Exp. Ther.* **160**, 326–335.
SEIDEN L. S. and CARLSSON A. (1963) *Psychopharmacol.* **4**, 418–423.
SMITH R. D., COOPER B. R. and BREESE G. R. (1973a) *J. Pharmacol. Exp. Ther.* **185**, 609–619.
SMITH R. D., COOPER B. R. and BREESE G. R. (1973b) *Pharmacologist* In press.
TAYLOR K. M. and LAVERTY R. (1972) *Europ. J. Pharmacol.* **17**, 16–24.
TEITELBAUM P. and EPSTEIN A. N. (1962) *Psychol. Rev.* **69**, 74–90.
TENEN S. (1967) *Psychopharmacol.* **10**, 204–219.
THOENEN H. and TRANZER J. P. (1968) *Arch. Pharmakol. Exp. Pathol.* **261**, 271–288.
UNGERSTEDT U. (1968) *Europ. J. Pharmacol.* **5**, 107–110.
UNGERSTEDT U. (1971) *Acta Physiol. Scand. (suppl.)* **367**, 95–122.
URETSKY N. J. and IVERSEN L. L. (1970) *J. Neurochem.* **14**, 269–278.
ZIGMOND M. J. and STRICKER E. M. (1972) *Science* **177**, 1211–1214.

Frontiers in Catecholamine Research 1973, pp. 707 to 708. Pergamon Press. Printed in Great Britain.

INHIBITION OF THE BLOOD PRESSURE RESPONSE TO α-METHYLDOPA IN UNANESTHETIZED RENAL HYPERTENSIVE RATS BY NEONATAL 6-HYDROXYDOPAMINE (6-OHDA) TREATMENT

WYBREN DE JONG, ABRAHAM P. PROVOOST and FRANS P. NIJKAMP
Rudolf Magnus Institute for Pharmacology, Medical Faculty,
University of Utrecht, Utrecht, The Netherlands

NEONATAL treatment with 6-hydroxydopamine (6-OHDA), given subcutaneously in a dose of 100 $\mu g/g$ on day 1 and 2 and of 250 $\mu g/g$ on day 8 and 15, caused a permanent depletion (88–99 per cent) of noradrenaline (NA) in heart, spleen and kidneys, but not in adrenals. At 26 weeks of age no evidence of regeneration of NA levels was observed (PROVOOST, DE KEMP and DE JONG, 1973). Effectiveness of this treatment was also indicated by increased sensitivity to the NA pressor response. In 10–12-week-old rats anesthetized with pentobarbital there was a five-fold increased sensitivity to various doses of intravenously (i.v.) administered NA. In contrast, the pressor response to i.v. administered tyramine was greatly diminished. In the pithed and adrenalectomized 6-OHDA-treated rat only minimal pressor responses were obtained by electrical stimulation (supramaximal voltage) of the sympathetic outflow of the spinal cord; i.e. less than 5 per cent of the value obtained in controls. These data probably indicate that a high degree of functional sympathetic denervation was achieved by the administration of 6-OHDA to the new-born rat.

This treatment failed to prevent the development of renal hypertension induced in 6-week-old rats by application of a solid silver clip (LEENEN and DE JONG, 1971)

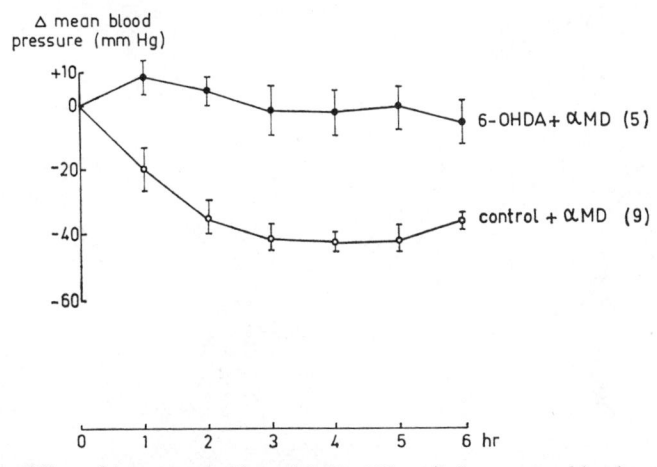

FIG. 1.—Effect of i.p. α-methyldopa (αMD, 200 mg/kg) on mean blood pressure of unanesthetized renal hypertensive rats. Animals had been treated after birth with 6-hydroxydopamine (6-OHDA, ●——●) or vehicle (control, ○——○). The number of rats used is indicated in parentheses, means ± s.e.m. are given.

708   W. DE JONG, A. P. PROVOOST and F. P. NIJKAMP

(internal diameter 0·20 mm) on one renal artery. Blood pressure measurements in unanesthetized rats were done by means of an indwelling chronic iliac cannula. Four weeks after surgery the blood pressure level reached was as high in 6-OHDA-treated renal hypertensive rats as in those treated with vehicle only after birth. Presumably the catecholamines released from the adrenal medulla have an important role in maintaining hypertension in the 6-OHDA-treated rats. The effect of α-methyldopa (α-MD; 200 mg/kg intraperitoneal (i.p.)) was examined in both groups of hypertensive rats, since previous studies (HENNING, 1969; DAY, ROACH and WHITING, 1973 and NIJKAMP and DE JONG, unpublished data) indicated that in the unanesthetized renal hypertensive rat the decrease in blood pressure induced by α-MD totally depends on the central action of this drug. As shown in Fig. 1 the decrease in blood pressure after α-MD was virtually absent in the rats treated with 6-OHDA neonatally, while a decrease of 43 ± 3 mm Hg was found in controls.

These preliminary data are interpreted as indicating a role of inhibition of sympathetic nervous function in the blood pressure decrease induced by α-MD in the awake renal hypertensive rat.

REFERENCES

DAY M. D., ROACH A. G. and WHITING R. L. (1973) *Europ. J. Pharmacol.* **21,** 271–280.
HENNING M. (1969) *Acta Pharmacol. Toxicol.* **27,** 135–148.
LEENEN F. H. H. and DE JONG W. (1971) *J. appl. Physiol.* **31,** 142–145.
PROVOOST A. P., DE KEMP H. and DE JONG W. (1973) *Europ. J. Pharmacol.* in press.

Frontiers in Catecholamine Research 1973, pp. 709 to 710. Pergamon Press. Printed in Great Britain.

BIOGENIC AMINES AND THEIR METABOLISM IN THE STUDY OF AGGRESSIVE BEHAVIOUR IN RATS

BURR EICHELMAN

Department of Psychiatry, Stanford University, California 94305, U.S.A.

VARIOUS models of animal aggression may involve different central mechanisms and separate neurotransmitters. Shock-induced aggression, spontaneous aggression, and predatory aggression are facilitated or inhibited differentially by altering central brain amine systems. Shock-induced aggression is an experimental model which pairs two rats in a small enclosure and subjects them to footshock (in these studies: 50 footshocks of 2 mA intensity and 0·4 sec duration given every 7·5 sec). In this paradigm the numbers of attacks made are tabulated (EICHELMAN, 1971). Predrug baseline levels of aggression are established and compared with levels following drug treatment. Spontaneous aggression is defined as a category of aggression where rats when placed together in a small enclosed space will attack each other without further stimuli being necessary to elicit aggressive behaviour. Predatory aggression was studied by placing mice in cages with rats whose mouse-killing behaviour was previously determined and observing whether this behaviour changed over a 24 hr period (i.e. whether non-killers commenced killing, or whether killers no longer were muricidal).

Alteration of serotonin metabolism with the tryptophan hydroxylase inhibitor para-chlorophenylalanine induced muricide in 9 of 34 rats when used in doses ranging from 100 mg/kg to 600 mg/kg. There were no conversions in the vehicle-treated group. Even though there were increases in footshock sensitivity in the pCPA-treated rats, their shock-induced fighting level changed insignificantly from 19 to 26·8% (EICHELMAN and THOA, 1973). In these studies no marked changes in spontaneous aggression occurred.

Stimulation of dopamine receptors with the drug apomorphine (1 mg/kg, i.v.) elicited spontaneous fighting in rats, which was markedly exaggerated in rats which had previously had brain norepinephrine and dopamine lowered by intracisternal injections of 200 μg of 6-hydroxydopamine (THOA et al., 1972a). However, in spite of facilitated spontaneous aggression, levels of shock-induced fighting remained insignificantly altered (pre-drug baseline: 11 per cent; post-drug mean: 8·4 per cent) (THOA et al., 1972b).

In contrast, alteration predominantly of norepinephrine metabolism appears to greatly affect shock-induced fighting without altering the other models of aggression. Depletion of brain norepinephrine and dopamine and damage to central catecholamine nerve terminals with intracisternal 6-hydroxydopamine (200 μg) elevates shock-induced fighting from a baseline level of 9·3 per cent to 37·9 per cent (EICHELMAN et al., 1972a). This same effect can be achieved by treating rats with intraventricular injections of 6-hydroxydopa (90 μg) which depletes only brain norepinephrine. This also raises fighting levels, from 6·4 up to 23·3 per cent (THOA et al., 1972c). The effect of these two drugs appears to relate to more than amine depletion, possibly involving the development of supersensitivity to endogenously released neuro-transmitter, since the increase occurs over several days and is not observed when other central amine lowering drugs are used (alpha-methylparatyrosine or FLA 63) (THOA et al., 1972b). This would imply that activation of central noradrenergic systems facilitates shock-induced fighting.

Further evidence that shock-induced fighting is facilitated by activation of a central noradrenergic system is furnished by ancillary studies. Treatment of rats with rubidium chloride (1·5 mequiv, bid, i.p.) for 15 days raises shock-induced fighting from 17·4 to 40·4 per cent while norepinephrine turnover is increased by 300 per cent (EICHELMAN, 1972b). Conversely lithium chloride decreases functional norepinephrine by increasing re-uptake and deamination (SCHANBERG, et al., 1967) and decreases shock-induced aggression (EICHELMAN et al., 1973). Chronic immobilization stress to rats for over one month (2 hr/day) markedly increases shock-induced fighting. These stressed rats have elevated levels of hypothalamic tyrosine hydroxylase suggesting in-creased central adrenergic metabolism (LAMPRECHT et al., 1972). REM deprivation for five days raises shock-induced fighting from a baseline of 41·8 to 67·4 per cent. This stress is also associated with increased catecholamine and serotonin turnover rates (STERN et al., 1971). Administration of monoamine oxidase inhibitors in an attempt to facilitate central amine activity also facilitates shock-induced fighting. Pargyline (20 mg/kg/4 da), nialamide (100 mg/kg/4 da) and iproniazid (150 mg/kg/4 da) all raise shock-induced fighting significantly from baseline levels of 17–24 per cent to postdrug levels of 40–50 per cent. Lastly, imipramine, a tricyclic antidepressant, when given over repeated doses increases norepinephrine turnover (SCHILDKRAUT et al., 1970). Given similarly (10 mg/kg/bid/5 da) imipramine increases shock-induced fighting from baseline levels of 18·8 to postdrug levels of 33·9 per cent.

In contrast, rats more aggressive in terms of shock-induced fighting do not show alterations in predatory behavior. Rats treated with 6-hydroxydopamine, rubidium, or immobilization stress do not become muricidal.

Thus, it appears that different models of aggression are more closely related to specific putative neurotransmitters. Alteration of serotonin metabolism can induce muricidal behavior, stimulation of dopaminergic terminals can induce spontaneous aggression, and alteration of adrenergic, probably noradrenergic, systems facilitates irritable or shock-induced aggression.

REFERENCES

EICHELMAN B. (1971) J. Comp. Physiol. Psychol. 74, 331–339.
EICHELMAN B. and THOA N. B. (1973) Biological. Psychiat. 6, 143–164.
EICHELMAN B., THOA N. B. and NG K. Y. (1972a) Physiol. and Behav. 8, 1–3.
EICHELMAN B., THOA N. B. and PEREZ-CRUET J. (1972b) Fedn. Proc. 31, 289 Abs.
EICHELMAN B., THOA N. B. and PEREZ-CRUET J. (1973) Alkali metal cations: effects on aggression and adrenal enzymes. Pharmacol. Biochem. Behav. 1, 121–123.
LAMPRECHT F., EICHELMAN B., THOA N. B., WILLIAMS R. B. and KOPIN, I. J. (1972) Science 177, 1214–1215.
SCHANBERG S. M. SCHILDKRAUT J. J. and KOPIN I. J. (1967) Biochem. Pharma. 16, 393–399.
SCHILDKRAUT J. J., WINOKUR A. and APPLEGATE C. W. (1970) Science 168, 867–869.
STERN W. C., MILLER F. P., COX R. H. and MAICKEL R. P. (1971) Psychopharmacologia 22, 50–55.
THOA N. B., EICHELMAN B. and NG K. Y. (1972a) J. Pharmacy Pharmacol. 24, 337–338
THOA N. B., EICHELMAN B. and NG K. Y. (1972b) Brain Res. 43, 467–475.
THOA N. B., EICHELMAN B., RICHARDSON J. S. and JACOBOWITZ D. (1972c) Science 178, 75–77.

Frontiers in Catecholamine Research 1973, pp. 711 to 713. Pergamon Press. Printed in Great Britain.

BRAIN NOREPINEPHRINE AND INGESTIVE BEHAVIOUR

SARAH FRYER LEIBOWITZ

The Rockefeller University, New York, N.Y. 10021, U.S.A.

OVER the past decade, catecholamines have received considerable attention as possible mediators in brain-behavioural mechanisms. More specifically, there is evidence to suggest that brain norepinephrine (GROSSMAN, 1962) and dopamine (UNGERSTEDT, 1971) may have a functional role in the regulation of ingestive behaviour. Most studies concerned with this problem have administered drugs directly into the brain via chronically implanted cannulas and have observed their effects on behaviour. The studies to be described in this paper (LEIBOWITZ, 1972, 1973, and unpublished) examined, in the rat, the effects on feeding and drinking of centrally administered norepinephrine (NE), epinephrine (EPI), and dopamine (DA). Numerous areas of the brain were tested, in an attempt to determine a possible linkage between the catecholamine-induced behavioural changes and specific regions in the brain.

This work may be summarised as follows:

(1) Adrenergic stimulation of the 'perifornical' hypothalamus, at the level of the anterior hypothalamus, had profound effects on rat ingestive behaviour. Two of these effects, stimulation of feeding and suppression of drinking, were found to be mediated by alpha receptors. A third effect, the suppression of feeding, was found to be mediated by beta receptors.

(2) Examination of other ventral structures of the brain, extending from the pontine tegmentum to the caudate nucleus, revealed a consistent pattern of results with respect to ingestive behaviour. From the rostral midbrain to the preoptic area (and possibly the septum), the medial portion of the brain exhibited sensitivity only to alpha-receptor stimulation, whereas the lateral portion appeared to be sensitive only to the effect of beta-receptor stimulation.

(3) In my studies, dopaminergic stimulation of various diencephalic structures failed to have reliable effects on feeding or drinking.

For these investigations, albino rats were each stereotaxically implanted with a unilateral brain cannula under Nembutal anesthesia. After a week of post-operative recovery, the rats were tested while either food- and water-satiated, food-deprived, or water-deprived. With the use of a microsyringe, the drug or its vehicle, in a volume of 0·2 to 0·5 μl, was administered directly into the brain through the implanted cannula. After injection, the rats were given measured food and/or water, and their consumption was recorded at frequent intervals during the next 2hr.

Injection of NE or EPI into the 'perifornical' hypothalamus, immediately dorsal to the anterior hypothalamus, induced eating in fully satiated rats and potentiated feeding in already hungry rats. EPI, which proved somewhat more potent than NE, was effective in eliciting this response at a dose as low as 200 pmoles (66 ng). The magnitude of the response increased monotonically as the dose of EPI or NE increased to approximately 50 nmoles. This stimulation of feeding effect appears to be mediated by alpha-adrenergic receptors, since it could be blocked by alpha-receptor antagonists but not by beta-receptor antagonists.

In addition to stimulating feeding behaviour, adrenergic stimulation of the perifornical hypothalamus had a strong suppressive effect on the water consumption of rats. This suppression of drinking phenomenon, which also increased monotonically with increase in dose, was reliably observed at a dose of EPI as low as 5 pmoles (1·7 ng). This dose, which is by far the lowest ever found to reliably alter behaviour, is much closer to probable physiological levels. This finding gives support to the hypothesis that adrenergic receptor mechanisms in the brain are physiologically active in the regulation of ingestive behaviour. Like the stimulation of feeding effect induced by NE or EPI, this suppression of drinking effect was found to be mediated by alpha receptors.

Under certain conditions, a third effect could also be induced by perifornical hypothalamic injections of EPI or NE. This effect was a suppression of feeding which these agonists were able to produce in hungry rats. This suppression was most readily seen at the higher doses of EPI or NE (20–100 nmoles) and was never reliably seen at doses below 5 nmoles (1·7 μg). Furthermore, in contrast to the alpha feeding stimulation effect, this suppression of feeding effect appears to be mediated by beta receptors, since it was blocked only by beta-receptor antagonists. The finding that alpha-receptor blockers could enhance the beta suppression effect, as well as lower its threshold dose, suggests that the relatively high doses required to observe the suppression may in part be due to the antagonism caused by the simultaneously occurring, and apparently more predominant, alpha stimulation of feeding effect.

Further work at a variety of central sites has indicated that different regions of the brain may be differentially sensitive to the effects of adrenergic stimulation. In experiments similar to those described above, NE and EPI, at a wide range of doses, were tested at seven different levels of the brain: (1) pontine tegmentum, (2) rostral midbrain tegmentum, (3) middle hypothalamus, (4) anterior hypothalamus, (5) preoptic area, (6) septum, (7) nucleus accumbens septi and caudate nucleus. At each level, two groups of rats were tested; one with a medial cannula (0·0–0·5 mm lateral, as determined histologically) and one with a lateral cannula (1·3–2·0 mm lateral, as determined histologically).

The results obtained in this study were remarkably consistent in differentiating the medial and lateral parts of the brain. It was found that at each of the levels extending from the rostral midbrain to the preoptic area, the two alpha-receptor phenomena, facilitation of feeding and suppression of drinking, could be elicited only with *medial* adrenergic stimulation; whereas the beta suppression of feeding phenomenon could be elicited only with *lateral* adrenergic stimulation. While the septum showed a tendency towards this pattern of sensitivity, the levels rostral to the septum and caudal to the rostral midbrain were found to be generally insensitive to adrenergic stimulation.

Although at several brain levels medial adrenergic stimulation was found to facilitate feeding in hungry rats, only at the level of the hypothalamus could such stimulation reliably initiate feeding in satiated rats. The most effective placement for eliciting feeding in satiated rats was found to be the paraventricular nucleus, which lies medial to the fornix at the level of the anterior hypothalamus. From these results on feeding behaviour, it becomes apparent that an on-going response can be modulated by adrenergic stimulation at a wide range of medial sites, but that the initiation of a new response arises more specifically from adrenergic stimulation of the medial hypothalamus.

CONCLUSIONS

(1) It appears that the brain's sensitivity to alpha-adrenergic stimulation, which has reciprocal effects on ingestive behaviour (feeding facilitation and drinking suppression), follows a medial course from the rostral midbrain through the preoptic area (and possibly into the septum). The periventricular zone, which runs medially through each of these regions and which is very heavily innervated by adrenergic terminals, may possibly assume an important role in the mediation of these alpha-receptor effects on ingestive behaviour.

(2) In response to medial adrenergic stimulation, both the hypothalamic and the extrahypothalamic structures effectively *facilitated* the feeding of hungry rats. In satiated rats, however, only the hypothalamus was found to be reliably effective in *initiating* a feeding response. Within the hypothalamus, the paraventricular nucleus proved to be the most sensitive site for feeding initiation. This structure is part of the periventricular zone, an area generally sensitive for feeding facilitation. One can speculate that the paraventricular nucleus is a focal point for the initiation of a new feeding response.

(3) In addition to stimulating feeding behaviour, central adrenergic stimulation, under certain conditions, was found to suppress feeding behaviour. Areas of the brain sensitive to this suppressive effect, a possible beta-receptor phenomenon, are different from those sensitive to the alpha-receptor stimulation of feeding effect. The most effective regions were found to follow a lateral course from the rostral midbrain through the preoptic area and possibly into the septum.

Acknowledgements—This research was supported by U.S. Public Health Service research grant MH 13189 and by grants from Hoffmann-La Roche and Smith Kline & French.

REFERENCES

GROSSMAN S. P. (1962) *Am. J. Physiol.* **202,** 872–882.
LEIBOWITZ S. F. (1972) In: *Neurotransmitters.* Res. Publ. A.R.N.M.D., Vol. 50, pp. 327–358.
LEIBOWITZ S. F. (1973) In: *The Neurosciences: Third Study Program* (SCHMITT, F. O., Ed.) M.I.T. Press, Cambridge, Mass. (in press).
UNGERSTEDT U. (1971) *On the Anatomy, Pharmacology and Function of the Nigro-striatal Dopamine System.* Kungl, Boktryckeriet P. A. Norstodt and Söner, Stockholm.

Frontiers in Catecholamine Research 1973, pp. 715 to 717. Pergamon Press. Printed in Great Britain.

EFFECTS ON FEEDING BEHAVIOUR AFTER INTRAHYPOTHALAMIC INJECTIONS OF 6-HYDROXY-DOPAMINE IN RATS

Joseph L. Slangen

Rudolf Magnus Institute for Pharmacology, Medical Faculty, University of Utrecht, Vondellaan 6, Utrecht, The Netherlands

Direct application of norepinephrine, epinephrine and clonidine to the perifornical region of the rat hypothalamus induces eating in food satiated rats (Booth, 1967; Slangen and Miller, 1969; Broekkamp and Van Rossum, 1972). This effect is prevented by pretreatment with alpha adrenergic receptor blocking agents and is not antagonised by the beta adrenergic blocker propranolol (Booth, 1968; Slangen and Miller, 1969). The norepinephrine induced eating is potentiated by prior intrahypothalamic administration of desmethylimipramine (Booth, 1968; Slangen and Miller, 1969). Eating can also be elicited by intrahypothalamic application of a monoamine oxidase inhibitor followed by a catecholamine depleting agent (Slangen and Miller, 1969). The feeding response can not be induced however by beta-adrenergic agonists, dopamine or serotonin (Booth, 1968; Slangen and Miller, 1969). The site where noradrenaline is most effective in eliciting eating is the mediolateral hypothalamus near the fornix approximately 1·3 mm from the midline (Booth, 1967). Ungerstedt has reported that bilateral destruction of the nigrostriatal dopamine system is associated with a syndrome of adipsia and afagia. In particular the bilateral injections of 6–8 μg of 6-hydroxy-dopamine (6-OHDA) in the dopamine axons assembled in the lateral hypothalamus resulted in prolonged afagia and adipsia. Afagia and adipsia also occur after interruption of the nigrostriatal system outside the hypothalamus (Ungerstedt, 1971a). Obviously this effect cannot be attributed to the destruction of the postulated "feeding system" of noradrenergic fibres in the hypothalamus. In the lateral hypothalamus the dopaminergic nigro-striatal system passes very closely to the perifornical area in which noradrenergic neurons are assumed to mediate the eating response (Ungerstedt, 1971a). It is not unlikely that the results obtained with chemical stimulation of the perifornical area are dependent upon interaction with the nigro-striatal system and therefore the question to be resolved is whether an intrahypothalamic adrenergic feeding system can be distinguished from the dopaminergic nigro-striatal system. In all experiments male albino rats (200–350 g) of an inbred Wistar strain were used. Under anaesthesia the animals were placed in a stereotaxic instrument and received bilaterally an injection of 8 μg 6-OHDA and 0·8 μg ascorbic acid dissolved in 1 μl of distilled water. Control rats received 0·8 μg ascorbic acid in 1 μl of distilled water only. The injections were aimed at an A–P level of 4620 according to the atlas of König and Klippel and at different lateral positions ranging from 0·4 to 3·0 mm from the midline. Up to 1·4 mm lateral the depth of the injection site was kept constant at 3 mm below horizontal zero. From 1·4 mm lateral onwards the depth was decreased in order to keep the injection sites parallel to the optic tract. Body weight and food and water intake were recorded

715

for two weeks. Animals were assigned to 10 different groups depending on the site of injection. The only animals included in this report are those in which the location of the injection site on the left and right sides of the brain corresponded exactly.

Afagia occurred only when 6-OHDA was injected in an area 1·6 to 2·1 mm from the midline. Ungerstedt has localised the axons of the nigro-striatal system to this area (UNGERSTEDT, 1971a). Injecting 6-OHDA into the more medially located peri-fornical area does not result in any impairment of feeding behaviour. If a noradren-ergic feeding system had existed in this area an impairment would have been expected after 6-OHDA. The lack of any effect may be attributed to an uncomplete destruction of noradrenergic terminals. We therefore repeated the experiment in a series of points between 0·4 and 1·6 mm lateral to the midline. On each side of the brain two injections were given into the hypothalamus but at a vertical distance of 1 mm from each other. After histological examination animals were assigned to 6 different groups depending on the sites of injections. Afagia and adipsia were observed only in the group which received injections 1·6 mm lateral from the midline. Injections made in the area extending from 0·4 to 1·4 mm laterally caused a transient loss of body weight lasting for a few days only.

In order to ascertain that 6-OHDA reached the exact site from which noradrener-gic eating could be elicited cannulas were implanted bilaterally in the perifornical region of 30 rats. Per cannula rats were tested three times. Injections were given every other day. In 13 rats the food intake after injection of 30 nmol norepinephrine (NE) was $2·0 \pm 1·1$ g (mean $\pm$ S.D.). Seven rats were weak eaters. Their mean eating response to 30 nmol NE was $1·1 \pm 0·7$ g. After the NE-test 8 μg 6-OHDA was injected bilaterally. It was found that 6-OHDA still was without an effect on eating in these 20 rats. When 30 nmol NE was injected 7 days after the 6-OHDA treatment an enhanced feeding response was obtained in two tests per cannula. The mean food intake of the strong eaters was $3·1 \pm 1·3$ g and of the weak eaters $3·2 \pm 1·4$ g. The difference between the mean response to NE before and after the 6-OHDA injection is statistically significant in both groups at the 0·1 % level (two tailed Student's t-test). This enhanced eating response to NE suggests that some change in the noradrenergic terminals had occurred as a result of the 6-OHDA treatment. In the same animals a second intrahypothalamic injection with 16 μg 6-OHDA was given. Although this treatment must have caused an even greater presynaptic degeneration no impairment of feeding was observed in 14 days.

In another series of experiments cannulas were implanted bilaterally in the nigro-striatal system of the lateral hypothalamus. 24 animals were tested three times per cannula with saline, 30 nmol NE and 30 nmol dopamine (DA) in a simple crossover design. Results are reported for 15 rats in which the placements of the cannula tips were found at the point where lateral hypothalamus and capsula interna meet. The response to NE ($0·8 \pm 0·6$ g) and to DA ($0·7 \pm 0·6$ g) was not different from the control response to saline ($0·6 \pm 0·4$ g). The bilateral administration of 6-OHDA (8 μg) in these animals did not result however in any feeding deficits. In a new attempt to stimulate the nigro-striatal system cannulas were implanted bilaterally in 15 other rats. In 10 rats histological verification showed that both cannulas had penetrated at a reasonably correct depth the lateral hypothalamic region extending from 1·8 to 2·2 mm lateral. Each implant was tested three times with 30 nmol NE and saline. No eating responses were seen. After the bilateral administration of 8 μg 6-OHDA no changes in consummatory behaviour were observed. The 6-OHDA treatment was repeated after 7 days and again no effects on feeding were obtained. Finally these animals were tested with 30 nmol NE again but no eating response was found that differed from the response to NE before the 6-OHDA treatment. The fact that 6-OHDA caused no afagia when given via cannulas that were aimed at the nigro-striatal system may be caused by technical difficulties. It is our feeling that with the cannula technique the site of the nigro-striatal system is more difficult to localise than with the single injection technique. The reported results support entirely the concept of an anatomical and functional distinction between the hypothalamic nigro-striatal DA system and the perifornical NE system (UNGERSTEDT, 1971b). NE can elicit eating only in the perifornical region and has no such effect in the directly adjacent nigro-striatal DA system. A single injection of 8 μg 6-OHDA causes afagia and adipsia when administered in an area of about half a millimeter width in which the nigro-striatal DA system has been localised and has no such effect when administered in the more medial lateral hypothalamus in which the NE "eating" system has been localised. After administration of 6-OHDA in the NE system a supersensitivity to NE stimulation has been observed whereas no such effect has been seen in the DA system. Since we have not analysed in sufficient detail the food intake of our rats we may have overlooked small effects of 6-OHDA injected into the perifornical region. But the lack of any great effect on the regulation of food intake after degeneration of the perifornical NE system suggests that this system is not a major control system for food intake. It may be however that the antagonistic relationship as proposed by LEIBOWITZ (1970) and GOLDMAN et al. (1971) between an alpha adrenergic "feeding" system and a beta adrenergic "satiety" system, both localised in the perifornical area, is such a completely complementary one that the observed ineffectiveness of 6-OHDA may have been due to an equivalent catecholamine depletion in the two opposing systems. This hypothesis predicts the dominance of one system over the other when these systems are not destructed to the same extent.

Acknowledgement—This work was supported by the Foundation for Medical Research FUNGO. I am indebted to Mrs. Marya Hein-Jongepier and Mr. Kees Broekman for their skillfull technical assistance.

REFERENCES

BOOTH D. A. (1967) *Science* **158**, 515–517.
BOOTH D. A. (1968) *J. Pharmacol. Exp. Ther.* **160**, 336–348.
BROEKKAMP C. and VAN ROSSUM J. M. (1972) *Psychopharmacologia (Berl.)* **25**, 162–168.
GOLDMAN W., LEHR D. and FRIEDMAN E. (1971) *Nature, Lond.* **231**, 453–455.
LEIBOWITZ S. F. (1970) *Nature, Lond.* **226**, 963–964.
SLANGEN J. L. and MILLER N. E. (1969) *Physiol. Behavior* **4**, 543–552
UNGERSTEDT U. (1971a) *Acta Physiol. scand. Suppl.* **367**, 1–48.
UNGERSTEDT U. (1971b) *Acta physiol. scand. Suppl.* **367**, 95–122.

Frontiers in Catecholamine Research 1973, pp. 719 to 722. Pergamon Press. Printed in Great Britain.

BRAIN STEM NOREPINEPHRINE: BEHAVIOURAL AND BIOCHEMICAL DIFFERENTIATION OF RESPONSES TO FOOTSHOCK IN RATS

JON M. STOLK[1] and JACK D. BARCHAS[2]

[1]Departments of Pharmacology and Psychiatry, Dartmouth Medical School, Hanover, New Hampshire 03755, U.S.A. and [2]Department of Psychiatry, Stanford University School of Medicine, Stanford, California 94305, U.S.A.

VARIOUS procedures falling under the general category of "stress" have demonstrated effects on brain norepinephrine content and metabolism (see BARCHAS et al., 1972). Dynamic measures of cerebral catecholamine function generally reveal that norepinephrine utilisation is facilitated in stressed subjects; conversely, drugs that reduce the levels of stored catecholamines in brain generally result in a behavioural depression. These responses have been interpreted as an attempt by the animal to maintain neurochemical homeostasis, and have been applied to various aspects of behavioural dysfunction in humans (KETY et al., 1967; SCHILDKRAUT and KETY, 1967). Bi-directional mutability is at least implied by the term homeostasis. However, only two isolated instances of depressed norepinephrine utilisation after behavioural manipulation have been described (WELCH and WELCH, 1969; STONE, 1970). Additionally, there have been few instances where norepinephrine responses have been intimately related to any given behaviour (STEIN and WISE, 1969). The present report will describe experiments suggesting (a) that brain stem norepinephrine metabolism undergoes rapid, bidirectional changes, and (b) that these biochemical changes may be related to a specific constellation of behavioural responses.

We have studied norepinephrine turnover and metabolism in the brain stem of male Long–Evans rats subjected to electric footshock either with or without another rat. Shock in the presence of another rat results in the reliable appearance of aggressive responses, whereas shock alone causes prominent escape attempts. Details of the shock regimen, and of the shock-elicited fighting paradigm may be found in STOLK et al. (1971). To study norepinephrine turnover and metabolism, we have employed intracisternal injections of ^{3}H-dopamine, a procedure which circumvents many of the methodological difficulties associated with the cerebroventricular administration of radioactive norepinephrine itself (details of the labelled dopamine procedure are contained in STOLK, 1973). All animals in the study received the dopamine intracisternally 4 hr before commencing with behavioural manipulation; since the rats were behaviourally naïve at the time of the injection, it was assumed that all subjects were identical up to the start of behavioural manipulation. Control rats remained in their home cages. Rats receiving footshock without a partner (referred to as Shock rats) were placed into the shock chamber 4 hr after the dopamine injection. Rats shocked with another rat (referred to as Fighting rats) were placed into the chamber at the same time after the dopamine injection, and were subjected to the same sequence of shock. In all cases, the duration of footshock was 5 min. Animals were sacrificed at various times after receiving shock, and brain stem tissue was analysed at various times after receiving shock, and brain stem tissue was analysed for levels of radioactive norepinephrine (formed from the ^{3}H-dopamine) and normetanephrine.

Immediately after rats were removed from the shock chamber, significant alterations in ^{3}H-norepinephrine levels were found only in Shock subjects (18 per cent reduction in labelled norepinephrine content compared to either Fighting or Control groups: $P < 0.05$). Parenthetically, other brain regions failed to show any significant alterations at this time, regardless of which group was studied. Thus, Shock, but not Fighting, groups reveal a marked increase in the rate of norepinephrine turnover in brain stem during the period of shock presentation. This change was accompanied by a profound decrease in the levels of ^{3}H-normetanephrine (Table 1); again, despite exposure to identical shock parameters, Shock rats biochemically were markedly different from Fighting rats.

TABLE 1. ^{3}H-NORMETANEPHRINE LEVELS IN BRAIN STEM FOLLOWING EXPOSURE OF SHOCK RATS TO 5 MINUTES OF ELECTRIC FOOTSHOCK. [Absolute radioactivity and per cent of Control group normetanephrine levels in two individual experiments are compared at the time rats were removed from the shock chamber. Changes in normetanephrine radioactivity are contrasted with altered ^{3}H-norepinephrine levels obtained in the same rats. Values represent the means ($\pm$ S.E.M.) obtained in groups of from 6 to 12 rats each. An asterisk (*) denotes a significant difference from the indicated control value. A double asterisk (**) denotes a significant difference from the ^{3}H-norepinephrine content measured in the same group.

Exp. No.	Group	^{3}H-Normetanephrine content		^{3}H-Norepinephrine % of control
		(dis/min per g)	% of control	
1†	Control	43,900 ± 6800	100 ± 14	100 ± 5
	Shock	27,200 ± 1500 (*)	62 ± 4 (*) (**)	86 ± 4 (*)
2	Control	8350 ± 700	100 ± 9	100 ± 4
	Shock	5100 ± 500 (*)	61 ± 8 (*) (**)	79 ± 9 (*)

† 3·1 μCi ^{3}H-dopamine injected in Experiment 1; 0·8 nCi were injected in Experiment 2. The dopamine specific activity in both experiments was 8·8 Ci/mmole.

Turnover of brain stem norepinephrine was estimated over the one hour period after Shock and Fighting rats were removed from the shock chamber. In contrast to the lack of change in Fighting rats observed immediately following shock presentation, norepinephrine turnover over the subsequent hour was increased substantially (Table 2). Conversely, ^{3}H-norepinephrine turnover in brain stem of Shock rats, which increased during the shock period itself (see previous paragraph), was significantly slower than both Control and Fighting group norepinephrine turnover (Table 2). These results clearly differentiate the two groups exposed to electric footshock, whether on a biochemical or a behavioural basis, and suggest that brain stem norepinephrine containing neurons participate actively in the behavioural responses evoked by electric footshock.

TABLE 2. FRACTIONAL RATE CONSTANTS (k, IN RECIPROCAL HOURS) FOR THE RATE OF DECLINE IN BRAIN STEM OF CONTROL, SHOCK AND FIGHTING GROUPS. [Rate constants were determined by least square regression analysis following logarhythmic transformation of the norepinephrine radioactivity levels obtained at selected times during the one hour period following footshock. Values represent k ± standard deviation. An asterisk (*) denotes a significant difference from all other groups.]

	Control group	Shock group	Fighting group
k:	0·373 ± 0·065 ($N = 32$)	0·246 ± 0·105 ($N = 31$)*	0·616 ± 0·104 ($N = 32$)*

Further evidence both for the increased norepinephrine turnover rate in Fighting rats and for the behavioural–biochemical relationship between shock-elicited fighting behaviour and norepinephrine metabolism is present in Fig. 1. These data demonstrate that the number of fighting episodes measured for each subject correlates extremely well with the levels of ³H-norepinephrine isolated in brain stem one hour after the fighting pairs were removed from the shock chamber.

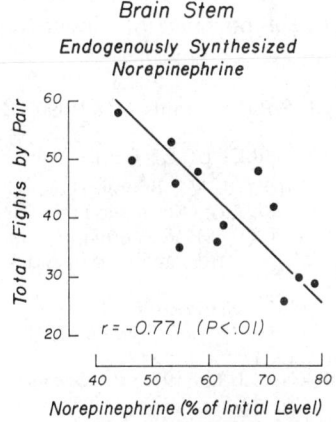

Brain Stem

*Endogenously Synthesized
Norepinephrine*

FIG.1.—Relationship between brain stem ³H-norepinephrine content and magnitude of fighting behaviour in rats. Catecholamine determinations were made 1 hr after the fighting period; radioactive norepinaphrine content is expressed as per cent of level obtained prior to beginning of behavioural manipulation.

Apart from the behavioural relationships to brain stem norepinephrine metabolism documented in the studies presented above, we have also observed a significant differentiation of normetanephrine production and norepinephrine turnover. Thus, ³H-normetanephrine levels are markedly reduced at a time when ³H-norepinephrine turnover is increased (immediately following shock; Table 1). There are several ways to interpret this finding. One interpretation would place the results in the context of generally accepted current models of central nervous system noradrenergic function, which predict that increased release of norepinephrine into the synaptic cleft manifests itself in increased rates of normetanephrine formation (since catechol-0-methyl-transferase is localised predominantly extraneuronally). By such models, these data could be taken as an indication that norepinephrine release decreases both during and after shock, regardless of whether catecholamine turnover increases or falls.

On the other hand, several other investigators also were unable to document normetanephrine accumulation during periods of increased norepinephrine utilisation following shock (STEIN and WISE, 1969; TAYLOR and LAVERTY, 1969; BLISS et al., 1968). Thus, an alternate interpretation of the findings in the latter references as well as in the present study is that changes in normetanephrine alone may not be an indicator of transynaptic utilisation. This possibility is further strengthened by the fact that the model, from which the usual interpretation of transsynaptic utilisation of norepinephrine involving normetanephrine is based, is derived primarily from pharmacological studies, and information thus obtained may not be representative of

natural noradrenergic function. Hypothetically, increased utilisation may be manifest by increases in intracellular norepinephrine metabolism involving a neuroregulatory role of norepinephrine, as well as by the more popular neurotransmitter
role. (We do not mean to imply that the standard model does not apply to natural
function in transsynaptic transmission, since normetanephrine accumulation during
accelerated turnover of brain stem norepinephrine following shock in the Fighting
groups is increased, as also indicated by the data of THIERRY et al., (1968); however,
we raise the question of whether this is the only valid interpretation of norepinephrine's
role in the brain.) The relative importance of these two alternatives in behavioural
states remains to be elucidated.

Acknowledgements—Supported by U.S.P.H.S. grants MH 21090, MH 16632 and ONR 102–715.

REFERENCES

BARCHAS J. D., CIARANELLO R. D., STOLK J. M., BRODIE H. K. H. and HAMBURG D. A. (1972)
   In: *Hormones and Behavior*, (S. LEVINE, Ed) (Academic Press, New York. p. 325.
BLISS E. L., AILION J. and ZWANZIGER J. (1968). *J. Pharmacol. Exp. Ther.* **164**, 122.
KETY S. S., JAVOY F., THIERRY A. M., JULOU L. and GLOWINSKI J. (1967) *Proc. Natn. Acad. Sci.
   U.S.A.* **58**, 1249.
SCHILDKRAUT J. J. and KETY S. S. (1967). *Science* **156**, 21.
STEIN L. and WISE C. D. (1969). *J. Comp. Physiol. Psychol.* **67**, 189.
STOLK J. M. (1973). *J. Pharmacol. Exp. Ther.*, in press.
STOLK J. M., CONNER R. L. and BARCHAS J. D. (1971) *Psychopharmacologia* **22**, 250.
STONE E. A. (1970). *Psychosomat. Med.* **32** 51.
TAYLOR K. M. and LAVERTY F. (1969). *J. Neurochem.* **16**, 1361.
THIERRY A. M., JAVOY F., GLOWINSKI J. and KETTY S. S. (1968). *J. Pharmacol. Exp. Ther.* **163**, 163.
WELCH B. L. and WELCH A. S. (1969). In: *Aggressive Behavior*, (S. GARATTINI and E. B. SIGG,
   Eds.) Excerpta Medica Fdn. Basle, p. 179.

Frontiers in Catecholamine Research 1973, pp. 723 to 726. Pergamon Press. Printed in Great Britain.

THE COERULO-CORTICAL NOREPINEPHRINE SYSTEM AND LEARNING

T. J. CROW

Department of Psychiatry, University Hospital of South Manchester,
West Didsbury, Manchester, M20 8LR England

FROM observations that electrical self-stimulation responding is reduced or abolished by drugs which deplete monoamine stores or inhibit catecholamine synthesis, and is enhanced by the amphetamines and cocaine, several workers have suggested that this behaviour results from activation of central noradrenergic mechanisms. However, responding can continue after administration of the dopamine-β-oxidase inhibitor, disulfiram (ROLL, 1970), and self-stimulation can be obtained not only with electrode tips located in relation to the norepinephrine-containing cells of the locus coeruleus (CROW, SPEAR and ARBUTHNOTT, 1972), but also in relation to the dopamine-containing cells in the ventral mesencephalon (CROW, 1972a). A modified version of the original hypothesis is therefore that electrical self-stimulation results from activation of either of two catecholamine-containing systems, the dopamine neurones arising from the A9 and A10 cell body groups, or the noradrenaline neurones arising from the A6 group, the locus coeruleus (CROW, 1972b).

Recent observations on the effects of (+) and (−) amphetamine on self-stimulation response rates (PHILLIPS and FIBIGER, 1973) are consistent with the possibility that both dopaminergic and noradrenergic mechanisms are involved. An association between contraversive turning and self-stimulation through lateral hypothalamic (GRASTYAN et al., 1969) and ventral mesencephalic (ARBUTHNOTT et al., 1970; ANLEZARK et al., 1971) electrodes, can be explained if activation of dopaminergic neurones underlies both stimulation-induced turning (ARBUTHNOTT and CROW, 1971) and some cases of electrical self-stimulation. On the other hand, the involvement of the coerulo-cortical norepinephrine system in self-stimulation with electrodes in the locus coeruleus is consistent with the observation that this behaviour is accompanied by increased norepinephrine turnover in the ipsilateral cortex (ANLEZARK et al., 1973).

The motor behaviours associated with activation of these two catecholamine systems are, however, quite distinct. Marked increases in forward locomotor activity together with elements of the sniffing, licking, and gnawing syndrome accompany self-stimulation with electrodes in the A9 and A10 cell-body areas (CROW, 1972a), but are not seen with electrodes in the region of the locus coeruleus (CROW et al., 1972). The functions of the two reward systems are thus clearly different, and the behavioural concomitants are compatible with a hypothesis (CROW and ARBUTHNOTT, 1972; CROW, 1973) that the dopamine neurones constitute a "motor activating" system which mediates the effects of rewarding environmental stimuli on the organisms immediate behaviour, while the noradrenaline neurones of the locus coeruleus register the success of preceding motor behaviours. According to this hypothesis the dopamine system facilitates appetitive behaviours by transmitting the "incentive motivational" effects of rewarding stimuli, and the noradrenaline system conveys the "reinforcing" effects of such stimuli (for a historical review of the concept of "reinforcement" see WILCOXON, 1969).

IS CORTICAL NOREPINEPHRINE NECESSARY FOR LEARNING?

Recent experiments (ANLEZARK, CROW and GREENWAY, 1973a,b) have been designed to test whether learning is possible in the absence of the coerulocortical norepinephrine-containing system. Rats with bilateral lesions in the region of the locus coeruleus were compared with a group of rats with bilateral lesions in the cerebellum, a group with bilateral lesions in the brainstem ventral to the locus coeruleus, and a group with burr holes alone. Three weeks after operation all rats were food-deprived to 90 per cent of body weight, and were given five trials on each of sixteen consecutive days in an L-shaped runway. A food reward was available in the short-arm goal box, and the time taken to run the long initial arm was assessed by two photocells. Most rats showed a rapid decrease in running time with increasing runway experience.

After behavioural testing the rats were killed and the brains removed for brainstem histology and cortical norepinephrine assay. In the experimental group the

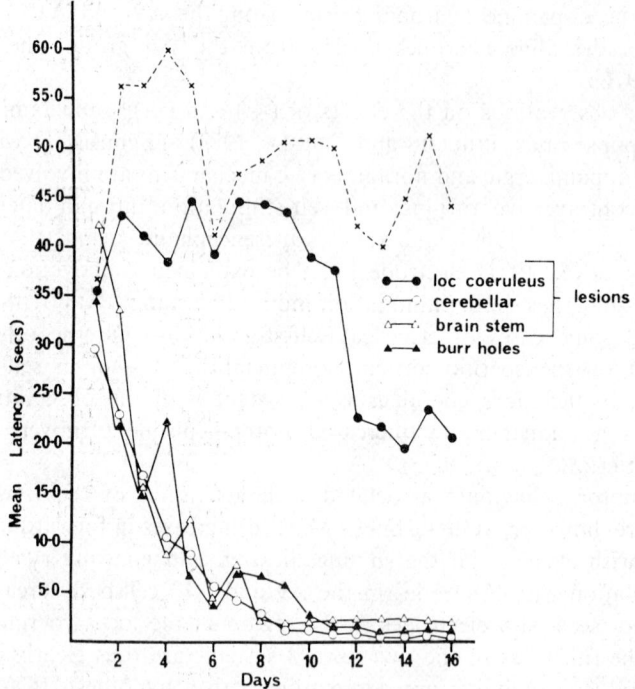

FIG. 1.—Mean running time in the initial arm of an L-shaped runway over the 16-day test period. The dotted line shows the performance of the three rats with apparently complete bilateral ablations of the locus coeruleus as assessed on histological examination. [From ANLEZARK G. M., CROW T. J. and GREENWAY A. P. (1973b) copyright, American Association for the Advancement of Science]

locus coeruleus was only partially ablated on one or both sides in many cases and a group of six rats with the most precise bilateral lesions was selected (from the total group of 28) for comparison with the six rats in each of the three control groups. In these rats cortical norepinephrine was substantially diminished by comparison with controls (mean 102ng/g vs. 325ng/g; $P < 0.001$).

Behavioural testing (Fig. 1) showed a rapid decline in running time with increasing experience in each of the three control groups. The rate of decrease was much slower in the six rats with locus coeruleus lesions (analysis of variance; $F = 19.629$, d.f. 3/20, $P < 0.001$), and in three rats with apparently complete lesions of the locus coeruleus (dotted line) no decrease in running time was observed over the 16-day test period. In other experiments we have found that such rats do not differ from controls in exploratory activity, food intake or weight gain, nor in ability to discriminate sucrose solution from water. The deficit associated with loss of the cortical noradrenergic innervation therefore appears to be an impairment of the capacity to learn. In this respect the effects of ablations of the locus coeruleus may be contrasted with those of even large lesions of the cerebral cortex itself, which have little effect on the animal's ability to learn simple tasks (LASHLEY, 1929).

THE ROLE OF NOREPINEPHRINE IN LEARNING

KETY (1970) and I (CROW, 1968) independently proposed hypotheses concerning the possible role of the cortical adrenergic innervation in learning. The common element in these hypotheses is that the noradrenergic system is envisaged as acting upon particular configurations of recently-active cells to bring about some long-term change in their synaptic interconnexions.

Both hypotheses require the existence within the cortex of a "short-term trace" mechanism with a decay time of perhaps a few seconds. If norepinephrine release occurs within this time the synaptic interconnexions between the cells so identified will be permanently enhanced. Thus neural pathways which have led to "biologically satisfactory" motor outputs will be successively facilitated, and the role of the noradrenergic reinforcement system is to convert short-term trace changes into the structural synaptic alterations postulated to underly long-term memory. This conception of the action of a reinforcement mechanism is closely similar to THORNDIKE'S (1933) theory of "the action of the after-effects of a connection upon it", and to the theory developed by J. Z. YOUNG (1964) in relation to his studies on the octopus, of a neural mechanism for delivering a "results of action signal." Such a concept would account for the striking effects on the organism's learning capacity which we have observed following removal of the coerulo-cortical norepinephrine system.

REFERENCES

ANLEZARK G. M., ARBUTHNOTT G. W., CHRISTIE J. E. and CROW T. J. (1971) *Br. J. Pharmac.* **41**, 406P.
ANLEZARK G. M., ARBUTHNOTT G. W., CROW T. J., ECCLESTON D. and WALTER D. S. (1973) *Br. J. Pharmac.* **47**, 645P.
ANLEZARK G. M., CROW T. J. and GREENWAY A. P. (1973a) *J. Physiol. Lond.* **231**, 119–120P.
ANLEZARK G. M., CROW T. J. and GREENWAY A. P. (1973b) *Science* **181**, 682–684.
ARBUTHNOTT G. W. and CROW T. J. (1971) *Expl. Neurol.* **30**, 484–491.
ARBUTHNOTT G. W., CROW T. J., FUXE K. and UNGERSTEDT U. (1970) *J. Physiol., Lond.* **210**, 61–62P.
CROW T. J. (1968) *Nature Lond.* **219**, 736–737.
CROW T. J. (1972a) *Brain Res.* **36**, 265–273.
CROW T. J. (1972b) *Psychol. Med.* **2**, 414–421.

CROW T. J. (1973) *Psychol. Med.* **3**, 66–73.
CROW T. J. and ARBUTHNOTT G. W. (1972) *Nature New Biol.* **238**, 245–246.
CROW T. J., SPEAR P. J. and ARBUTHNOTT G. W. (1972) *Brain Res.*, **36**, 275–287.
GRASTYAN E., SZABO I., MOLNAR P. and KOLTA P. (1969) *Comm. Behav. Biol.* **2**, 235–266.
KETY S. S. (1970) *Assn. Res. Nerv. Ment. Dis. Publn.* **50**, 376–388.
LASHLEY K. S. (1929) *Brain Mechanisms and Intelligence.* University of Chicago Press, Chicago.
PHILLIPS A. G. and FIBIGER H. C. (1973) *Science* **179**, 575–577.
ROLL S. K. (1970) *Science* **168**, 1370–1372.
THORNDIKE E. L. (1933) *Psychol. Rev.* **40**, 434–439.
YOUNG J. Z. (1964) *A model of the Brain.* Oxford University Press, Oxford.
WILCOXON H. C. (1969) *Reinforcement and Behaviour* (TAPP J. T., Ed.) pp. 1–46 Academic Press, New York.

Frontiers in Catecholamine Research 1973, pp. 727 to 728. Pergamon Press. Printed in Great Britain.

ELECTRICAL SELF-STIMULATION IN THE LOCUS COERULEUS AND NOREPINEPHRINE TURNOVER IN CEREBRAL CORTEX

T. J. CROW

University Hospital of South Manchester, Manchester, England

IN RECENT experiments (ANLEZARK *et al.*, 1973) we have demonstrated that electrical self-stimulation with electrodes in the locus coeruleus is associated with increased turnover of norepinephrine in the ipsilateral cerebral cortex. We measured the cortical content of the norepinephrine metabolite H.M.P.G. in rats with locus coeruleus electrodes supporting electrical self-stimulation and in control rats with electrodes nearby. Increased H.M.P.G. accumulation by comparison with controls could be demonstrated both in rats stimulated under anaesthesia, and in rats self-stimulating for the 1-hr period before being killed (Fig. 1). In the latter case there there was a lesser increase in H.M.P.G. in the contralateral cortex.

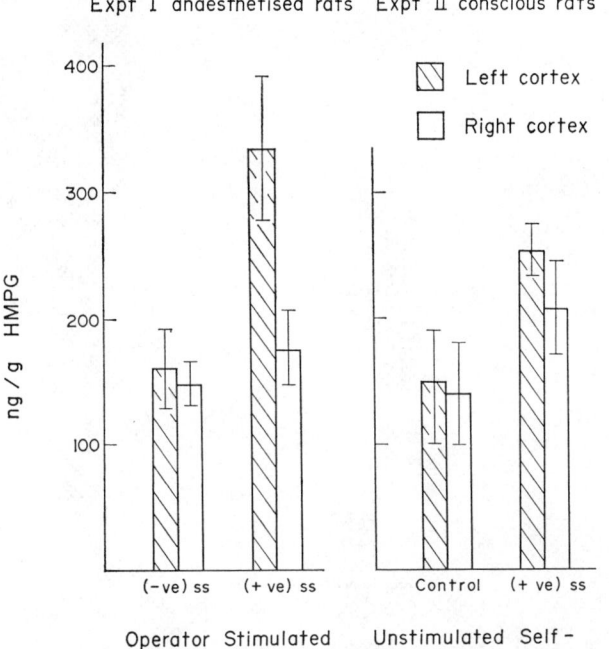

FIG. 1.—All rats had electrodes implanted in the region of the left locus coeruleus and were tested for self-stimulation responding. When stimulated under anaesthesia for 1 hr before being killed (experiment I) those animals whose electrodes had previously supported self-stimulation ((+ve)ss) showed increased H.M.P.G. in the ipsilateral cortex. Rats self-stimulating for 1 hr before being killed (experiment II) were found to have an increase in H.M.P.G. in the ipsilateral cortex, and also had a lesser increase in the contralateral (right) cortex.

This data therefore supports the involvement of the coerulo-cortical norepineph-rine system in self-stimulation.

REFERENCE

ANLEZARK G. M., ARBUTHNOTT G. W., CROW T. J., ECCLESTON D. and WALTER D. S. (1973) *Br. J. Pharmac.* **47**, 645 P.

Frontiers in Catecholamine Research 1973, pp. 729 to 739. Pergamon Press. Printed in Great Britain.

EFFECTS OF 6-HYDROXYDOPA

DAVID M. JACOBOWITZ

Laboratory of Clinical Science, National Institute of Mental Health
Bethesda, Maryland 20014, U.S.A.

6-HYDROXYDOPA (2, 4, 5-trihydroxyphenylalanine)(6-OHDOPA), the carboxyl analog of 6-hydroxydopamine, is a drug whose usefulness lies in its ability to enter the brain after peripheral injection and cause a selective destruction of noradrenergic terminals (JACOBOWITZ and KOSTRZEWA, 1971; SACHS and JONSSON, 1972a). Studies on the pharmacologic effects of 6-OHDOPA show that this agent is capable of reducing the myocardial and the brain norepinephrine (NE) content of the mouse for several days (JACOBOWITZ and KOSTRZEWA, 1971; SACHS and JONSSON, 1972a; SACHS and JONSSON, 1972b; KOSTRZEWA and JACOBOWITZ, 1973; ONG et al., 1969; BERKOWITZ et al., 1970; KOSTRZEWA and JACOBOWITZ, 1972; CORRODI et al., 1971).

PHARMACOLOGICAL STUDIES

A comparison of various routes of administration (KOSTRZEWA and JACOBOWITZ, 1973; BERKOWITZ et al., 1970) clarifies the differences observed in NE brain levels after a single injection of 6-OHDOPA (SACHS and JONSSON, 1972a; BERKOWITZ et al., 1970; CORRODI et al., 1971). An intravenous (i.v.) injection was clearly far superior in depleting the brain NE content in a 24-hr period than the intraperitoneal (i.p.) or the subcutaneous route of administration (KOSTRZEWA and JACOBOWITZ, 1973). An i.v. dose of 100 or 150 mg/kg 6-OHDOPA caused a 30 or 50 per cent reduction, respectively, of NE in the brain after 24 hr (Fig. 1) (JACOBOWITZ and KOSTRZEWA, 1971; KOSTRZEWA and JACOBOWITZ, 1973). The brain dopamine content is not significantly changed for this time period after administration of 100 mg/kg and is reduced by 20 per cent 24 hr after 150 mg/kg. Therefore, the 100 mg/kg dose is the effective dose which is selective for noradrenergic neurons.

In peripheral organs, 6-OHDOPA, like 6-hydroxydopamine, produces variations in tissue susceptibility (KOSTRZEWA and JACOBOWITZ, 1972). After 100 mg/kg (24 hr i.v.), there was a 60 per cent and a 20 per cent reduction of the NE concentration in the ventricle and the submaxillary gland, respectively, whereas no significant changes were observed in the spleen or vas deferens for this time period. After three hours, reductions in the amine content are observed in the spleen, vas deferens and iris, which indicate a transient reversible depletion. It appears that a critical concentration of 6-OHDOPA is necessary for prolonged depletion of NE. Differences in blood flow could account for unequal effects on the NE content in various tissues.

The time course of depletion and recovery of brain amines after a single i.v. injection of 6-OHDOPA is indicated in Fig. 1 (JACOBOWITZ and KOSTRZEWA, 1971). A dose of 100 mg/kg reduced the NE content by 32 per cent after 1 day. After 66 days, the brain NE was still significantly decreased by 21 per cent. A larger dose (150 mg/kg) resulted in a 32 per cent decrease after 66 days (JACOBOWITZ and KOSTRZEWA, 1971). With the 100 mg/kg dose, the brain dopamine content was

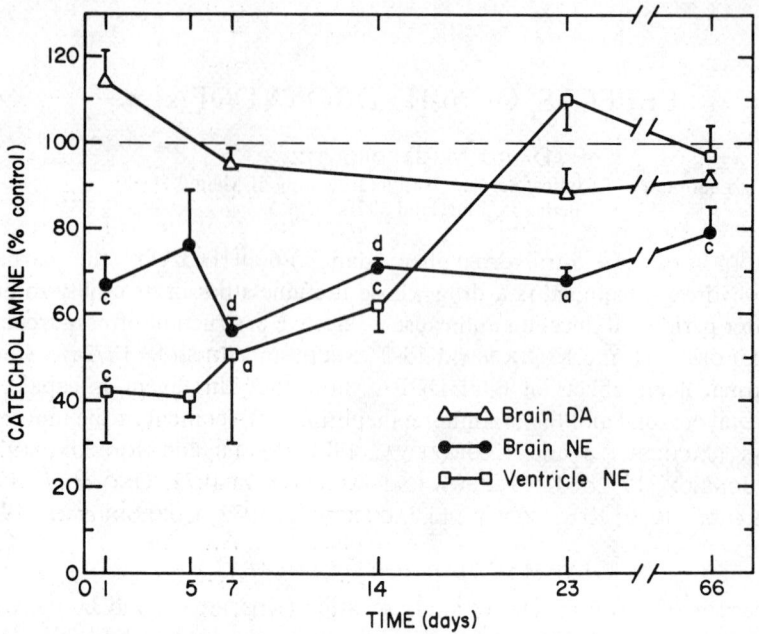

FIG. 1.—Time-course curves after a single intravenous injection of 100 mg/kg 6-hydroxydopa (6-OHDOPA). Each point represents the mean norepinephrine (NE) or dopamine content (μg/g $\pm$ S.E.M.) of an average of five treated and six control mice. Results are expressed as a percentage of the mean of control values for each time period. The mean control values of NE and dopamine in the brain are 0·38 and 0·61 μg/g, respectively, and 0·76 μg/g in the ventricle. Letters indicate significance (p values: a $< 0·05$, b $< 0·01$, c $< 0·005$, d $< 0·001$).

not significantly altered. In the ventricle, the NE content was reduced by 38 per cent by 14 days after 6-OHDOPA; at this dose level, it was normal by 23 days (Fig.1).

The effects of various drugs on the NE-depleting action of 6-OHDOPA were studied (KOSTRZEWA and JACOBOWITZ, 1973; ONG et al., 1969; KOSTRZEWA and JACOBOWITZ, 1972). The action of 6-OHDOPA (100 mg/kg) in depleting the NE content of the brain and of the heart could be prevented by D- and L-amphetamine, desipramine and chlorpromazine. These agents have been shown to be inhibitors of the membrane pump for uptake of catecholamines (AXELROD et al., 1961). Since amino acids such as tyrosine and dopa are not inhibited by blockers of the amine uptake mechanism, it would seem that 6-OHDOPA, an amino acid, is converted extraneuronally to 6-hydroxydopamine, which is then inhibited from uptake into adrenergic nerves by the above blocking agents.

Tranylcypromine, a monoamine oxidase inhibitor, failed to alter the effect of 6-OHDOPA on the NE content in the ventricle and in the brain, as compared with saline-injected controls. However, this could be misleading in terms of potential destructive effects of 6-OHDOPA on adrenergic neurons. An increase in the amine content of the remaining neurons after administration of a monoamine oxidase inhibitor might suggest a failure of this drug to influence the organ content of NE (KOSTRZEWA and JACOBOWITZ, 1972; KOSTRZEWA and JACOBOWITZ, 1973). However, SACHS and JONSSON (1972a, 1972b) showed that pretreatment of mice with nialamide

(100 mg/kg, i.p.) 2 hr prior to an i.p. injection of 6-OHDOPA resulted in a significant reduction of [³H]NE uptake in the atrium, cortex and iris after 1 and 7 days. Although an i.p. injection of 6-OHDOPA was not effective in altering the [³H] amine uptake, the combination of nialamide (100 mg/kg, i.p.) with 6-OHDOPA (100 mg/kg, i.p., three injections at 24-hr intervals) 2 hr before the latter drug appears to result in a long-lasting reduction of endogenous brain NE and of [³H] amine uptake in the cortex and atrium. It would therefore appear that monoamine oxidase is involved in the inactivation of 6-OHDOPA. Administration of tranylcypromine (5 mg/kg, i.p.) 1 hr prior to 6-OHDOPA (100 mg/kg, i.v.) was useful for histochemical studies in that an enhancement of catecholamine fluorescent intensity of preterminal axons was observed (JACOBOWITZ and KOSTRZEWA, 1971).

The peripherally acting dopa decarboxylase inhibitor, MK-486, protected the mouse ventricle against the NE-depleting effect of 6-OHDOPA. The NE level of the brain was significantly reduced by an additional 14 per cent after the combination of MK-486 and 6-OHDOPA. It would thus appear that inhibition of decarboxylation of 6-OHDOPA to form 6-hydroxydopamine would make more 6-OHDOPA available in the brain and would thereby result in a greater reduction of NE. MK-485, another peripheral decarboxylase inhibitor, did not potentiate the NE depletion in the rat brain after an i.p. injection of 6-OHDOPA (250–400 mg/kg) (CORRODI et al., 1971) or decrease the uptake of [³H]NE after nialamide and 6-OHDOPA in slices of mouse cortex (SACHS and JONSSON, 1972a). The inability of MK-485 to alter the 6-OHDOPA effects in the latter studies may be due to the appearance of enlarged, distorted nonterminal adrenergic axons (JACOBOWITZ and KOSTRZEWA, 1971; SACHS and JONSSON, 1972a), which could conceivably serve to store or take up increased amounts of NE at a time when there are fewer terminals. Thus, a compartmental relocation of adrenergic nerve uptake and storage sites may serve to mislead in the interpretation of the effects of both dopa decarboxylase and monoamine oxidase inhibitors.

6-OHDOPA had no effect on the monoamine oxidase and choline acetyltransferase activities in the mouse telencephalon (cortex and striatum regions), brain stem and cerebellum after two and 14 days. Catechol-O-methyltransferase activity is reduced slightly in the brain stem after 14 days. Acetylcholinesterase activity is reduced by 20 and 30 per cent in the telencephalon 2 and 14 days, respectively, after 6-OHDOPA. Since choline acetyltransferase activity is not altered, it would appear that cholinergic nerves are not affected by 6-OHDOPA. It is suggested that the acetylcholinesterase, which is reduced by 6-OHDOPA, is present within the adrenergic nerves of the cortex and/or the striatum.

The serotonin content of the brain has been reported to be unchanged (BERKOWITZ et al., 1970), slightly increased 4 hr after a massive dose (CLARKE et al., 1972) or slightly reduced in the cortex after nialamide and 6-OHDOPA treatment (SACHS and JONSSON, 1972a). After the latter treatment, however, the uptake of [¹⁴C] serotonin was 35 per cent above the nialamide control value. It would therefore appear that serotonergic nerves are not impaired after 6-OHDOPA treatment.

It is of interest that 6-OHDOPA failed to decrease the tyrosine hydroxylase activity in rat brain parts one day or one week after injection although reductions in NE content ranging from 54 to 24 per cent of the control value were observed (THOENEN, 1972). This suggested that 6-OHDOPA administration does not result in destruction of adrenergic neurons. However, as appeared above in those studies which showed a lack of the NE-depleting effects of 6-OHDOPA caused by dopa decarboxylase and monoamine oxidase inhibitors, a misleading interpretation could arise if a buildup of tyrosine hydroxylase occurs in nonterminal adrenergic axons. Such an accumulation of tyrosine hydroxylase results after a ligation of adrenergic nerve trunks (WOOTEN and COYLE, 1973).

In an effort to demonstrate an actual destruction or impairment of neuronal function rather than a simple depletion of NE stores, hearts of control rats and of those treated with 6-OHDOPA (100 mg/kg, i.v., followed 1 hr later by 100 mg/kg, i.p.) and killed 24 hr after the first injection were perfused with 1 μg/ml NE (KOSTRZEWA and JACOBOWITZ, 1972). The hearts of 6-OHDOPA-treated rats with normally undetectable amounts of NE took up 46 per cent less NE than the control rat hearts. In addition, perfusion of the 6-OHDOPA-treated rats with NE did not increase the number of histochemically observable adrenergic nerve terminals. A buildup of catecholamine fluorescence was observed in swollen preterminal axons of the iris of these rats. This buildup is very similar to that observed with 6-hydroxydopamine (MALMFORS and SACHS, 1968; GOLDMAN and JACOBOWITZ, 1971) and suggests that actual destruction of adrenergic nerves can occur after 6-OHDOPA administration.

A more recent interesting development is the ability of 6-OHDOPA, injected into either pregnant rats or their newborn offspring, to produce long-lasting depletion of whole brain NE. No change was found in the brain dopamine content (ZIEHER and JAIM-ETCHEVERRY, 1973). NE in the tele-diencephalon and in the hypothalamus was significantly reduced three to seven months after birth. The NE content of the brain stem, i.e. the region containing the noradrenergic cell bodies, was significantly increased during this time period. There was no reduction in the NE content of the heart and of the salivary glands of rats whose mothers received 6-OHDOPA during the period of gestation (18–20 days). This observation gives promise of a method whereby separation of central from peripheral effects on noradrenergic neurons can be accomplished with 6-OHDOPA.

BEHAVIORAL STUDIES

To facilitate the study of the various parameters of behavior, rats were injected with 6-OHDOPA in the lateral cerebral ventricles in order to avoid the peripheral effects on adrenergic neurons (RICHARDSON and JACOBOWITZ, 1973). Intraventricular injection of 6-OHDOPA produced a graded dose-dependent reduction of NE in various brain parts. Dopamine levels were not altered by the doses of 6-OHDOPA used (45–180 μg). A working dose of 90 μg was chosen, and a time course of depletion and recovery was studied in various regions of the brain and of the spinal cord (Fig. 2). Telencephalic NE remained at 60 per cent of the control levels for 14 days and had returned to normal levels by 70 days after a single injection. The NE content continued to fall in the diencephalon for 14 days and was still significantly reduced by 33 per cent after 70 days. NE levels in the cerebellum continued to fall for 14 days and remained reduced by 57 per cent after 70 days. Hindbrain NE was at its lowest level two days after injection and had returned very gradually to control values after 70 days. The spinal cord NE levels had progressively declined to 33 per cent of the control values by the end of 14 days. Brain dopamine levels were not affected for up to 70 days after the intraventricular injection.

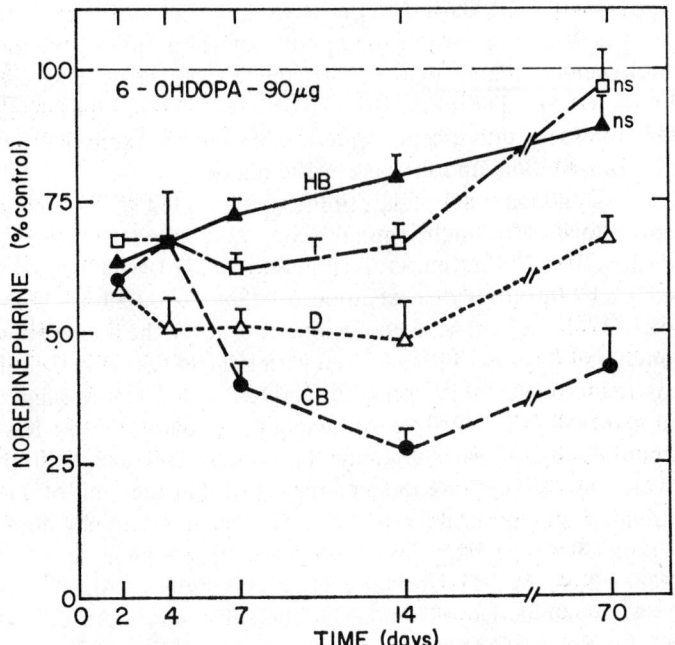

Fig. 2.—Time-course curves after a single intravenous injection of 90 μg 6-hydroxy-dopa (6-OHDOPA) of the norepinephrine (NE) content in the hindbrain (HB), telencephalon (T), diencephalon (D) and cerebellum (CB) of the rat brain. Each point represents the mean ± s.e.m. of 6–16 drug-treated rats as compared with 6–11 diluent-injected control rats; "n.s." refers to those points that are not statistically different from controls.

Injection of 90 μg 6-OHDOPA reduces food and water consumption for three to five days, but returns to normal at a time when NE levels in the diencephalon are still greatly reduced.

The level of emotionality was quantified by a modification of the rating scale devised by Brady and Nauta (1955) to measure the septal rage syndrome. Emotionality scores were increased in rats after an intraventricular injection of 6-OHDOPA and returned to normal after two months, when NE levels in the telencephalon and hindbrain had also returned to normal (Fig. 2). Further studies are needed to determine whether the telencephalon, which includes the limbic system, and/or hindbrain, which includes the reticular formation, may be involved with the inhibition of emotional reactivity.

A significant increase in shock-induced aggression occurred in rats four days after an intraventricular injection of 90 μg 6-OHDOPA (Thoa et al., 1973). It may be suggested that shock-induced aggression is modulated through a central noradrenergic system.

HISTOCHEMICAL STUDIES

The catecholamine histofluorescence method of Falck and Hillarp (Falck, 1962; Falck et al., 1962; Falck and Owman, 1965) was used to examine the effects of 6-OHDOPA on the brain. Mice were treated with 6-OHDOPA (100 or 150 mg/kg, i.v.) and observed after 1–23 days (JACOBOWITZ and KOSTRZEWA, 1971). After 24 hr, there was a variable decrease in the number of varicosities in regions known to contain noradrenergic fibers (DAHLSTRÖM and FUXE, 1964; DAHLSTRÖM and FUXE, 1965; FUXE, 1965). The most striking observation was the appearance of an abundance of intensely fluorescent nonterminal smooth axons in the reticular formation of the pons-medulla and mesencephalic regions.

Pretreatment with monoamine oxidase inhibitors resulted in an increase in the intensity of fluorescence of the nonterminal axons (JACOBOWITZ and KOSTRZEWA, 1971; SACHS and JONSSON, 1972a; SACHS and JONSSON, 1972b), which was reminiscent of that observed after central regional ablation or stereotaxic lesions (ANDEN et al., 1965; ANDEN et al., 1966). There were no obvious changes in the fluorescence content of dopamine-containing regions. This, coupled with the fact that only the NE content of the brain was reduced after injection of 6-OHDOPA, led to the suggestion that the nonterminal axons were primarily noradrenergic (JACOBOWITZ and KOSTRZEWA, 1971). The accumulation of catecholamine in nonterminal axons in the brain, which normally are not visible or are in some regions just at the limit of fluorescence observation, provided an opportunity to map out the nonterminal noradrenergic axons of the brain. Such pathways were obtained by mapping onto projections of cresylviolet-stained slides. The noradrenergic cell bodies (Al, A2, A7, locus coeruleus) and the dopaminergic cell bodies in the substantia nigra did not appear to be different from those of the controls. A major trunkline of large axons was observed bilaterally through the reticular formation that descends towards the median forebrain bundle (MFB), which carries axons to the hypothalamus, septal region and cerebral and olfactory cortices. This major reticular formation tract in the mouse is probably equivalent to the dorsal noradrenergic bundle described by UNGERSTEDT (1971) in the rat. Nonterminal processes were no longer seen two weeks after injection of 6-OHDOPA. Noradrenergic terminals appeared normal in most regions except the cortex, hippocampus and cerebellum. The reappearance of varicose terminals suggests that regeneration of noradrenergic nerves occurs.

Histochemical observations of rat brains were made after intraventricular injections of 90 μg 6-OHDOPA (RICHARDSON and JACOBOWITZ, 1973). After 24 hr, a reduction in the number and intensity of fluorescence of varicose nerve fibers was observed in regions containing noradrenergic terminals. There was a greater decrease in varicose terminals in the regions more proximal to the ventricular system (e.g. periventricular, paraventricular and dorsomedialis nuclei of the hypothalamus). Regions containing dopamine nerve terminals and all monoaminergic cell bodies (NE, serotonin, dopamine) in the hindbrain appeared normal. As seen in the mouse brain one day after an i.v. injection of 6-OHDOPA, the most prominent observation was the appearance of many nonterminal axons with an intense fluorescence (RICHARDSON and JACOBOWITZ, 1973; THOA et al., 1973). The axonal trunks contained swollen and enlarged segments. The biochemical and histochemical evidence indicates that these nonterminal axons are noradrenergic processes. After 4 and 7 days, small axonal sprouts of varicose nerve fibers were observed to be budding off the main trunks. After 14 days, regions of terminal regeneration were observed in the forebrain. It therefore appears that the regrowth of noradrenergic nerves results from axonal and/or perhaps terminal sprouting and probably accounts for the return to normal levels of telencephalic and hindbrain NE 70 days after injection of 6-OHDOPA. It also seems that reinnervation to the cerebellum is very inefficient after an intraventricular injection of 6-OHDOPA in the rat. Possibly, the cerebellar penduncles are somehow not conducive to regeneration of adrenergic neuronal processes. The suggestion that individual NE cell bodies of the locus coeruleus can innervate both the cerebral cortex and the cerebellum (ANDEN *et al.*, 1967; OLSON and FUXE, 1971; UNGERSTEDT, 1971) may explain the increased rate of regeneration of noradrenergic nerves to the telencephalon. An increased transport of NE in the remaining axonal trunks could serve to facilitate the regeneration of nerves to the telencephalon.

Because of the accumulation of fluorescent catecholamine in the nonterminal axons, noradrenergic neuronal pathways of the brain were capable of being mapped (JACOBOWITZ and RICHARDSON, in preparation). Rat brains were studied two to seven days after a single intraventricular injection of 6-OHDOPA (90 μg) with or without pretreatment with tranylcypromine two hours prior to sacrifice. The dorsal-ascending noradrenergic bundle (UNGERSTEDT, 1971) was revealed in the mesen-cephalon and followed to its point of union with the ventral and dopaminergic tracts in the MFB (Fig. 3). This bundle has previously been cited as giving off branches to the cerebral cortex, hippocampus, colliculus, thalamus and geniculate bodies (UNGERSTEDT, 1971; OLSON and FUXE, 1971).

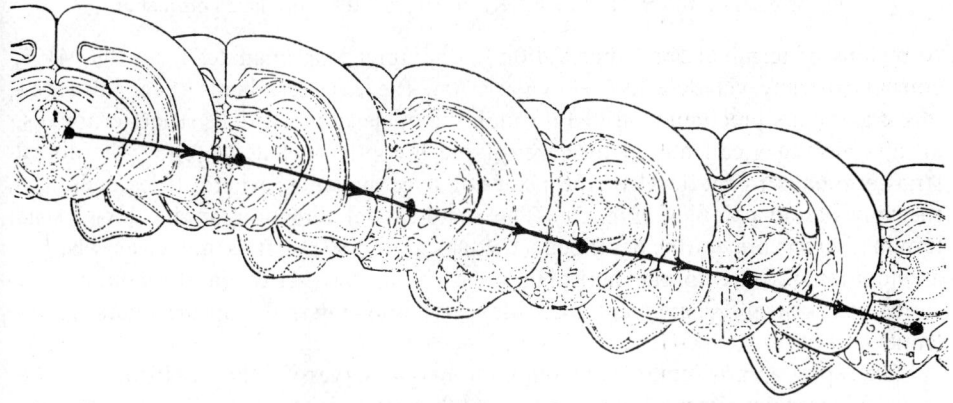

FIG. 3.—Projection of the dorsal adrenergic projection. Black circles indicate exact localization of the dorsal bundle. See KÖNIG and KLIPPEL (1963) brain atlas for co-ordinate information.

The fasciculus longitudinalis dorsalis (Schütz), pars tegmentalis, was found to contain a noradrenergic tract that appeared at the level of the mesencephalic-metencephalic junction (Fig. 4). This tract was followed rostrally through the substantia grisea as the fasciculus longitudinalis dorsalis (Schütz) in the mesence-phalon. The latter bundle descends as the fibrae periventriculares hypothalami at the level of the posterior mamillary body and continues rostrally to the nucleus dorsomedialis (hypothalami), where it can no longer be followed. It would seem that this tract contributes noradrenergic fibers to the nucleus dorsomedialis and possibly to the periventricular nucleus.

Preterminal axonal accumulation of amine can be observed along the length of the MFB. At various levels, a small number of distorted axons can be followed

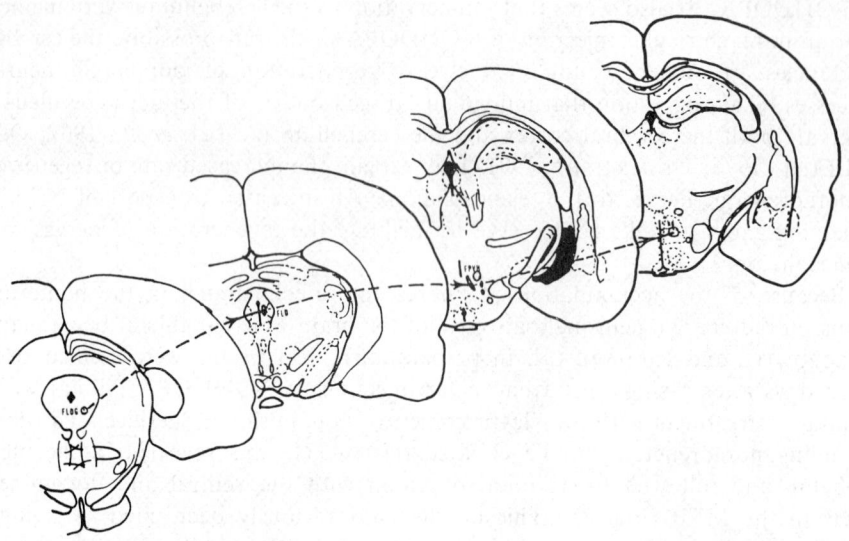

FIG. 4.—Projection of the fasciculus longitudinalis dorsalis (Schütz), pars tegmentalis (FLDG). Dashed-arrow line projects rostrally to the levels of the fasciculus longitudinalis dorsalis (Schütz), fibrae periventriculares hypothalami (FPVH) and nucleus dorsomedialis (hd). See König and Klippel (1963) for coordinate information.

to regions of terminations. Innervation of the amygdala emanates from the MFB, courses rostrally via the ansa lenticularis and the stria terminalis (pars infracommissuralis, pars precommissuralis) through the nucleus interstitialis striae terminalis, all of which turn caudally at the level of the preoptic nucleus to follow the dorsal stria terminalis to the level of the posterior hypothalamus, where it descends into the amygdaloid complex (Fig. 5). The rosto-caudal direction of the dorsal stria terminalis was previously described (Ungerstedt, 1971). It is not clear whether an additional route to the amygdala exists from the MFB via the commissura supraoptic dorsalis, pars ventralis (Meynert) and internal capsule (pars retrolenticularis).

The septal area is innervated by noradrenergic nerves via the MFB through the tractus diagonalis (Broca) and tractus septohypothalamicus (Fig. 5). The nucleus accumbens appears to contain noradrenergic processes from the MFB by way of the tractus striohypothalamicus (Fig. 5). It would be of interest to know whether the latter processes provide terminal ramifications of noradrenergic terminals or are merely axons on passage through the nucleus accumbens, a region currently regarded as primarily dopaminergic (Dahlström and Fuxe, 1965; Ungerstedt, 1971).

The hippocampus receives innervation via the MFB by way of the tractus diagonalis through the septum along the fornix superior in a caudal direction beneath the corpus callosum (Fig. 6). Another possible path is along the tractus corticohabenularis medialis from the stria medullaris into the fornix column and caudally with the fornix superior to the hippocampus (Fig. 6). A tract appears to traverse the stria medullaris in a caudal direction. The destination of these processes is not clear, although it is suggested here that these axons innervate the thalamus.

The cortex receives innervation by several routes: (1) Via the MFB and tractus

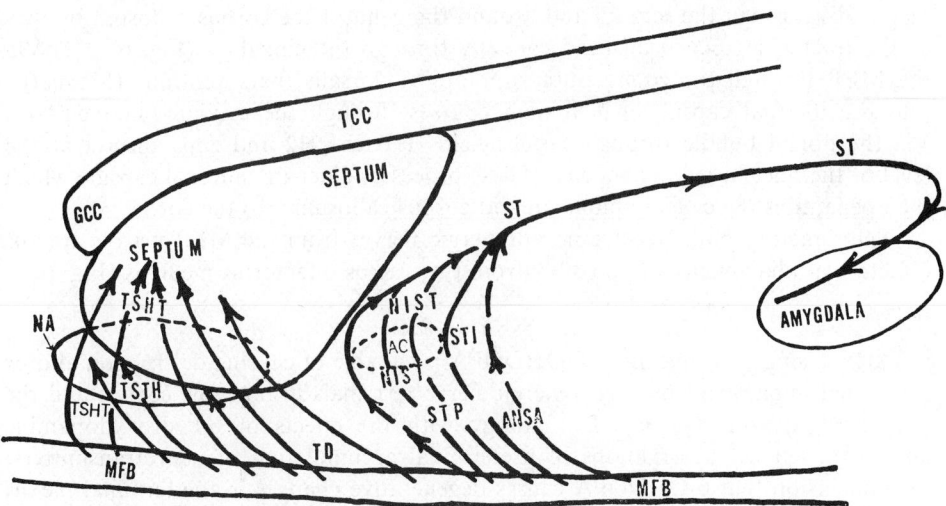

FIG. 5.—Schematic parasagittal projection of noradrenergic fiber tracts to the amygdala, septum and nucleus accumbens (NA). Abbreviations: anterior commissure (AC); ansa lenticularis (ANSA); genu corporis callosi (GCC); median forebrain bundle (MFB); nucleus interstitialis striae terminalis (NIST); stria terminalis (ST); stria terminalis, pars infracommissuralis (STI); stria terminalis, pars precommissuralis (STP); truncus corporis callosi (TCC); tractus diagonalis Broca (TD); tractus septohypothalamicus (TSHT); tractus striohypothalamicus (TSTH).

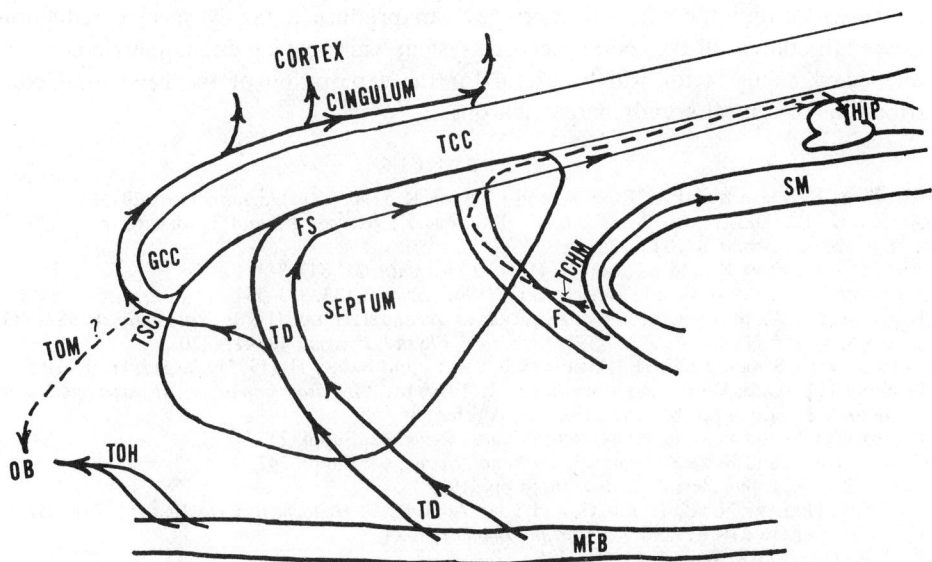

FIG. 6.—Parasagittal projection of noradrenergic projection fiber tracts to the cortex, hippocampus and olfactory bulbs. Abbreviations: columna fornicis (F); fornix superior (FS); genu corporis callosi (GCC); hippocampus (HIP); median forebrain bundle (MFB); olfactory bulbs (OB); stria medullaris thalami (SM); truncus corporis callosi (TCC); tractus corticohabenularis medialis (TCHM) ; tractus diagonalis (Broca) (TD); tractus olfactohypothalamicus (TOH); tractus olfactorius medialis (TOM); tractus septocorticalis (TSC).

diagonalis through the septum and around the genu of the corpus callosum by way of the tractus septocorticalis and caudally through the cingulum (Fig. 6). (2) Via the MFB through the commissura supraoptica dorsalis, pars ventralis (Meynert), into the internal capsule, which then courses through the caudate-putamen. (3) Via the dorsal bundle through Forel's fields H1 and H2 and zona incerta at the level of the nucleus subthalamicus. These processes enter the internal capsule which then penetrates the caudate-putamen and corpus callosum into the cortex.

The olfactory bulbs receive noradrenergic nerves from the MFB via the tractus olfacto-hypothalamicus and possibly from the tractus olfactorius medialis (Fig. 6).

SUMMARY

After a single i.v. injection, 6-OHDOPA is capable of causing destruction and/or functional impairment of noradrenergic nerve terminals in both the central and the peripheral nervous system. By analogy with the effects of 6-hydroxydopamine and histochemical observations of preterminal accumulation of neurotransmitters, the conclusion that 6-OHDOPA causes degenerative changes in the terminal plexus is derived. There is no question that the final determination of neuronal degeneration should come from electron microscopic studies. The accumulation of histochemically observable catecholamine fluorescence in nonterminal axons, the time course of depletion and recovery of NE within various peripheral organs and impairment of uptake of exogenously administered NE support the conclusion that 6-OHDOPA has the potential to cause selective destruction of noradrenergic terminals.

6-OHDOPA is a valuable tool for the investigation of central pathways and functional organization of the noradrenergic neurons in the brain. Injection of 6-OHDOPA into the cerebral ventricles can produce a highly specific reduction in the NE content of the central nervous system while leaving dopaminergic neurons unaffected. This factor will be useful for the exploration of the behavioral contributions of central noradrenergic neurons.

REFERENCES

Anden N. E., Dahlström A., Fuxe K. and Larsson K. (1965) Am. J. Anat. 116, 329–333.
Anden N. E., Dahlström A., Fuxe K., Larsson K., Olson L. and Ungerstedt U. (1966) Acta Physiol. Scand. 67, 313–326.
Anden N. E., Fuxe K. and Larsson K. (1966) Experientia 22, 842–843.
Axelrod J., Whitby L. G. and Hertting G. (1961) Science 133, 383–384.
Berkowitz B. A., Spector S., Brossi A., Focella A. and Teitel S. (1970) Experientia 26, 982–983.
Brady J. V. and Nauta W. J. H. (1955) J. Comp. Physiol. Psychol. 48, 412–420.
Clarke D. E., Smookler H. H., Hadinata J., Chi C. and Barry H. (1972) Life Sci. 11, 97–102.
Corrodi H., Clark W. G. and Masuoka D. I. (1970) In: 6-Hydroxydopamine. (Malmfors T. and Thoenen H., Eds.) pp. 187–192. Elsevier, Amsterdam.
Dahlström A. and Fuxe K. (1964) Acta Physiol. Scand. 62, Suppl. 232.
Dahlström A. and Fuxe K. (1965) Acta Physiol. Scand. 64, Suppl. 247.
Falck B. (1962) Acta Physiol. Scand. 56, Suppl. 197.
Falck B., Hillarp N. A., Thieme G. and Torp A. (1962) J. Histochem. Cytochem. 10, 348–354.
Falck B. and Owman C. (1965) Acta Univ. Lund. 7, 1–23.
Fuxe K. (1965) Z. Zellforsch. 65, 573–596.
Goldman H. and Jacobowitz D. (1971) J. Pharmacol. Exp. Ther. 176, 119–133.
Jacobowitz D. and Kostrzewa R. (1971) Life Sci. 10, 1329–1342.
König J. F. R. and Klippel R. A. (1963) The Rat Brain: A Stereotaxic Atlas of the Forebrain and Lower Parts of the Brain Stem. Williams & Wilkins Co., Baltimore.
Kostrzewa R. and Jacobowitz D. (1972) J. Pharmacol. Exp. Ther. 183, 284–297.
Kostrzewa R. and Jacobowitz D. (1973) Europ. J. Pharmacol. 21, 70–80.
Malmfors T. and Sachs, C. (1968) Europ. J. Pharmacol. 3, 89–92.

Olson L. and Fuxe K. (1971) *Brain Res.* **28,** 165–171.
Ong H. H., Creveling C. R. and Daly J. W. (1969) *J. Med. Chem.* **12,** 458–461.
Richardson J. S. and Jacobowitz D. M. (1973) *Brain Res.*, In press.
Sachs C. and Jonsson G. (1972a) *J. Neurochem.* **19,** 1561–1575.
Sachs C. and Jonsson G. (1972b) *Brain Res.* **40,** 563–568.
Thoa N. B., Eichelman B., Richardson J. S. and Jacobowitz D. (1972) *Science* **178,** 75–77.
Thoenen H. (1970) In: *Perspectives in Neuropharmacology.* (Snyder S. H., ed.) pp. 302–338. Oxford University Press, New York.
Ungerstedt U. (1971) *Acta Physiol. Scand.*, Suppl. 367.
Wooten G. F. and Coyle J. T. (1973) *J. Neurochem.*, **20,** 1361–1371.
Zieher L. M. and Jaim-Etcheverry G. (1973) *Brain Res.*, In press.

Frontiers in Catecholamine Research 1973, pp. 741 to 745. Pergamon Press. Printed in Great Britain.

CHOLINERGIC–DOPAMINERGIC RELATION IN DIFFERENT BRAIN STRUCTURES

G. Bartholini, H. Stadler and K. G. Lloyd

Department of Experimental Medicine, F. Hoffmann-La Roche & Co. Ltd.,
Basle, Switzerland

Neostriatal neurons respond in an opposite manner to the iontophoretic application of dopaminergic and cholinergic compounds and to their respective antagonists. Thus, dopamine (DA) depresses, whereas acetylcholine (ACh) or physostigmine enhance the firing rate of caudate cells (Bloom et al., 1965). Facilitation by physostigmine is not only inhibited by atropine but also by L-DOPA (Steg, 1969). Depletion of striatal DA (e.g. by reserpine) also increases the discharge frequency of striatal units (Steg, 1969). It seems, therefore, that at least some striatal cells receive antagonistic cholinergic (activating) and dopaminergic (inhibitory) inputs.

This antagonism probably results in a functional balance determining the efferent motor control of the striatum (Barbeau, 1962). Unbalance towards a cholinergic preponderance—as for example due to impaired dopaminergic transmission—leads to parkinsonian symptoms. Several observations support this concept of a functional balance between antagonistic dopaminergic and cholinergic activities. Thus, cholinergic drugs (e.g. physostigmine) exacerbate (Duvoisin, 1967), anticholinergic compounds, however, ameliorate parkinsonism (Sigwald, 1971). In addition, anticholinergic drugs counteract the parkinsonism induced by neuroleptic blockade of dopaminergic transmission in both animals (Zetler et al., 1960) and man (Klawans, 1968). Finally, L-DOPA, which restores the dopaminergic activity, has an outstanding therapeutic effect in parkinsonian patients whose striatal DA is greatly reduced (Lit. see Hornykiewicz, 1972). The question which arises is whether or not such a functional balance implies an interconnection between striatal dopaminergic and cholinergic neurons resulting in a mutual regulation.

It is likely that the striatal dopaminergic neurons are regulated by a cholinergic system. Thus, physostigmine enhances (Perez-Crouet et al., 1971), whereas anticholinergic compounds diminish the turnover of cerebral DA (Bartholini and Pletscher, 1971). A possible interpretation of these findings is that an increase or decrease in cholinergic activity leads to a compensatory activation or inhibition of DA neurons respectively, suggesting a link between the two systems. However, no direct evidence exists for this view; there is even less indication for the inverse regulation of cholinergic neurons by the DA system.

A direct approach to the problem of the interconnection and the mutual regulation of striatal DA and ACh systems is provided by the measurement of drug-induced changes in the amounts of transmitters continuously released within the striatum. This has been made possible by means of a push–pull cannula implanted into the head of the caudate nucleus of the gallamine immobilized cat. ACh released into the perfusate was measured radioenzymatically in 10 min samples collected throughout several hours (Stadler et al., 1973). The output of ACh was markedly increased by neuroleptic drugs such as chlorpromazine (CPZ) (Fig. 1, Table 1) or haloperidol

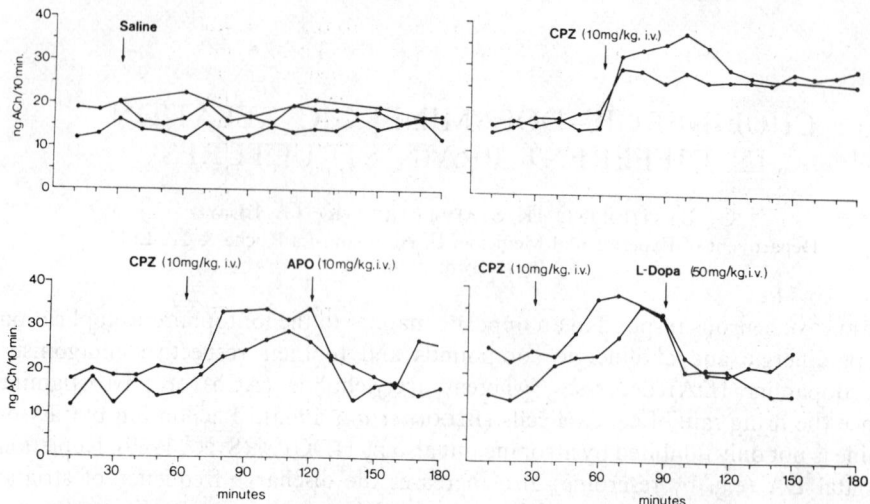

Fig. 1.—Output of acetylcholine (ACh) into 10 min samples of the perfusate of cat caudate nucleus during the control (preinjection) period and after various treatments. Each curve indicates one experiment. The points represent averages of duplicate determinations. Time 0 indicates the beginning of collection.
CPZ = chlorpromazine; APO = apomorphine.

(Table 2). This effect of CPZ is not due to a non specific interaction with gallamine since similar data were obtained in chronically implanted, unrestrained cats (STADLER et al., 1973). The increase in ACh output seems to result from enhanced turnover since in preliminary experiments the concentration of ACh was not changed by CPZ in the perfused caudate nucleus of two cats (controls: 2·46; CPZ: 2·63 ng/mg tissue). Inhibition of ACh-esterase (acetylcholine acetyl-hydrolase, EC3.1.1.7) by CPZ can be excluded as the enzyme activity was unaffected in striatal tissue (controls: 40·5; CPZ: 41·9 mU/mg tissue). Neither is the enhanced output of ACh by CPZ specific for the phenothiazine configuration since promethazine, a non-neuroleptic phenothiazine with antihistaminic properties, did not modify the ACh release during 2 hr perfusions (Table 1) (STADLER et al., 1973). In preliminary experiments atropine (25 mg/kg i.v.) increased the striatal ACh output in agreement with similar findings in the cortex (MITCHELL, 1963; PEPEU, 1971). However, for CPZ its anticholinergic properties can be at most partially responsible for the increased ACh release since also haloperidol, a neuroleptic butyrophenone devoid of both anticholinergic and antihistaminic properties, markedly enhanced the output of the transmitter.

It is therefore likely that the enhancement of striatal ACh release by neuroleptics is connected with the blockade of DA receptors by these drugs. This hypothesis is supported by the fact that apomorphine—a DA receptor stimulating compound (ANDÉN et al., 1966)—or L-DOPA injected 1 hr after CPZ reversed the neuroleptic-induced increase in ACh output (Fig. 1). In addition, apomorphine alone diminished the release of ACh and, injected before CPZ, prevented the enhancement of the transmitter due to the latter drug (Table 1). Finally, the enhancement of ACh release by CPZ or haloperidol in the perfused cat striatum occurred concomitantly to the acceleration of DA turnover as evidenced by the rise of homovanillic acid (Table 2). This effect of neuroleptics on DA turnover (ANDÉN et al., 1964), which

TABLE 1. RELEASE OF ACETYLCHOLINE (ACh) INTO THE PERFUSATE OF THE CAT CAUDATE NUCLEUS. The ACh contents of the 10-min samples (ng/10 min) collected throughout the time periods indicated were averaged and statistically evaluated by the Student's t test. Saline, chlorpromazine and promethazine when given alone, were injected at 60 min. In the combined treatment apomorphine was administered at 60 min followed by chlorpromazine at 120 min. The samples collected from 60 to 70 and 120 to 130 min were omitted. In parentheses: at left number of determinations, at right number of cats. *$P < 0.001$ vs control period.

Treatment (mg/kg i.v.)	Time after beginning of collection (min)		
	control	post injection	
	0–60	70–120	130–180
Saline	14·5 ± 0·6 (12) (2)	15·2 ± 0·7 (10) (2)	15·8 ± 0·6 (10) (2)
Chlorpromazine (10)	16·2 ± 0·9 (30) (5)	25·5 ± 0·8* (25) (5)	26·3 ± 1·2* (20) (4)
Promethazine (10)	15·1 ± 1·1 (12) (2)	14·6 ± 0·6 (10) (2)	16·3 ± 1·1 (10) (2)
Apomorphine (10) followed by chlorpromazine (10)	14·1 ± 0·42 (12) (2)	10·6 ± 0·74* (12) (2)	13·72 ± 1·4 (12) (2)

probably results from the blockade of DA receptors, did not occur after promethazine (unpublished results) which also failed to enhance ACh output (see above).

All of the above-mentioned results indicate that a striatal cholinergic system may be under a regulatory dopaminergic influence (BARTHOLINI et al., 1973). It is possible that the nigro-striatal DA pathway modulates the activity and function of some cholinergic neurons by a tonic inhibitory input. Thus, impairment of dopaminergic transmission would result in an increased activity of cholinergic neurons, ACh release and in the

TABLE 2. CONCENTRATIONS OF DOPAMINE (DA), HOMOVANILLIC ACID (HVA) IN THE CAT CAUDATE NUCLEUS AND OF ACETYLCHOLINE (ACh) IN THE PERFUSATE. DA and HVA have been measured in the perfused caudate nucleus of the cat 2 hr after injection of saline or of the drugs. The values for ACh in perfusate represent the averages with SEM of the concentrations in 10 min samples collected during the second hour after injection. In parentheses at left number of determinations, at right number of cats. *$P < 0.01$ vs saline treated animals.

Treatment (mg/kg i.v.)	Caudate nucleus		Perfusate
	DA ($\mu g/g$)	HVA ($\mu g/g$)	ACh (ng/10 min)
Saline	11·97 ± 0·51 (15) (15)	3·94 ± 20 (15) (15)	15·8 ± 0·6 (12) (2)
Chlorpromazine (10)	12·03 ± 1·11 (6) (6)	5·87 ± 0·62* (7) (7)	26·3 ± 1·2* (24) (4)
Haloperidol (2)	9·04 ± 0·30 (12) (12)	6·13 ± 0·51* (12) (12)	35·2 ± 4·7* (12) (2)

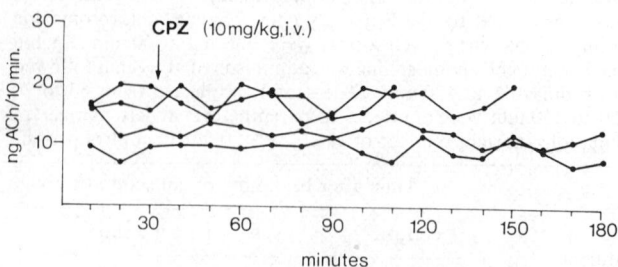

Fig. 2.—Output of acetylcholine (ACh) into 10 min samples of the perfusate of cat pre-motor cortex during the control period (0–30) and after chlorpromazine (CPZ). For details see Fig. 1.

appearance of parkinsonian symptoms. This is supported by the observations that anticholinergic drugs ameliorate rigidity and tremor in Parkinson's disease and extrapyramidal dysfunction caused by neuroleptic compounds (see above).

Based on these findings it must be assumed that the regulation of a striatal cholinergic activity by dopaminergic neurons occurs via by a link between DA and ACh systems. The possibility that dopaminergic neurons act on the cholinergic system by a presynaptic input should be considered.

In conclusion, it is likely that in the neostriatum dopaminergic and cholinergic neurons are interconnected and mutually regulating in such a way that cholinergic activity is under an inhibitory dopaminergic influence whereas the DA system is activated by a cholinergic input:

$$DA \underset{+}{\overset{-}{\rightleftarrows}} ACh$$

This interregulation might explain the mechanism by which, in the striatum, DA turnover is enhanced as a consequence of neuroleptic blockade of DA receptors. Thus, impairment of dopaminergic transmission activates a cholinergic system which in turn would cause a compensatory increase in the activity of DA neurons.

The question arises whether or not such a link between DA and ACh neurons exists in areas of the brain other than the striatum. In our cat preparation neuroleptics did not change the ACh release from the perfused pre-motor cortex (Fig. 2). Other areas were also investigated, for instance limbic structures, which may be of particular interest for the therapeutic effect of neuroleptic drugs. In fact, it has been hypothesized that blockade of DA receptors in the limbic system could be involved in the antipsychotic action of neuroleptics. In our experiments, haloperidol—although it markedly enhanced the DA turnover—did not modify the release of ACh from the nucleus accumbens septi—is the limbic region with the most dense DA network. This suggests that, as for cortical areas, a cholinergic-dopaminergic connection similar to that in the striatum does not exist in the nucleus accumbens septi of the cat and that the antipsychotic action of neuroleptic drugs may not be mediated by cholinergic neurons in this structure. For details on the effect of neuroleptic drugs on limbic DA and ACh neurons see: Lloyd et al. (this book).

SUMMARY

The influence of the dopamine (DA) neurons on a cholinergic system in the neostriatum has been investigated in experiments in which acetylcholine (ACh) liberated from the cat caudate nucleus was collected by means of a push-pull cannula and measured radioenzymatically. Neuroleptic drugs (e.g. chlorpromazine or haloperidol, i.v.) enhanced the output of striatal ACh and this effect was prevented or reversed by dopaminergic agents such as apomorphine or L-DOPA. Neuroleptics also increased DA turnover. Promethazine, a non-neuroleptic phenothiazine, did not affect either ACh or DA. From these and other findings it is concluded that in the neostriatum a cholinergic mechanism, which might be involved in parkinsonism, is under the influence of a tonic dopaminergic inhibitory input whereas the DA system is activated by cholinergic neurons. This mechanism is possibly involved in the feed-back regulation of DA turnover. As neuroleptics failed to affect ACh output in cortex and limbic structures—although in the latter these drugs markedly increased DA turnover—such a link and mutual regulation between DA and ACh systems seems to be specific for the neostriatum.

REFERENCES

ANDÉN N. E., ROOS B. E. and WERDINIUS B. (1964) *Life Sci.* **3**, 149–158.
ANDÉN N. E., DAHLSTRÖM A., FUXE K. and LARSSON K. (1966) *Acta Pharmacol. Toxicol.* **24**, 263–274.
BARBEAU A. (1962) *Can. Med. Ass. J.* **87**, 802–807.
BARTHOLINI G. and PLETSCHER A. (1971) *Experientia* **27**, 1302–1303.
BARTHOLINI G., STADLER H. and LLOYD K. G. (1973) In: *Symposium on the treatment of Parkinsonism* (Calne, D. B., Ed.), *Advances in Neurology*, Vol. 3, Raven Press, New York.
BLOOM F. E., COSTA E. and SALMOIRAGHI G. C. (1965) *J. Pharmacol. Exp. Ther.* **150**, 244–252.
DUVOISIN R. G. (1967) *Arch. Neurol. Psychiat.* (*Chicago*) **17**, 124–136.
HORNYKIEWICZ O. (1972) In: *Handbook of Neurochemistry* (LAJTHA A., Ed.) Vol **6**, pp. 465–501, Plenum Press, New York.
KLAWANS H. L. (1968) *Dis. Nerv. System*, **29**, 805–816.
MITCHELL J. F. (1963) *J. Physiol.* (*Lond.*) **165**, 98–118.
PEPEU G. (1971) In: *Chemistry and Brain Development* (PAOLETTI R. and DAVIDSON A. N., Eds.) pp. 195–205. Plenum Press, New York.
PEREZ-CROUET J., GESSA G. L., TAGLIAMONTE A. and TAGLIAMONTE P. (1971) *Fedn. Proc.* **30**, 216.
SIGWALD J. (1971) In: *Monamines et noyaux gris centraux* (AJURIAGUERRA J. de and GAUTHIER G. Eds.) pp. 369–378, Georg & Masson, Geneva and Paris.
STADLER H., LLOYD K. G., GADEA-CIRIA M. and BARTHOLINI G. (1973) *Brain Res.* **55**, 476–480.
STEG G. (1969) In: *Third Symposium on Parkinson's disease* (GILLINGHAM F. J. and DONALDSON I. M. L., Eds.) pp. 26–29, Livingstone, Edinburgh.
ZETLER G., MAHLER K. and DANIEL F. (1960). *Arch. Pharmak.* **238**, 486–501.

Frontiers in Catecholamine Research 1973, pp. 747 to 749. Pergamon Press. Printed in Great Britain.

CHANGES IN STRIATAL DA METABOLISM INDUCED BY CHOLINERGIC AND ANTI-CHOLINERGIC NIGRAL STIMULATION IN THE RAT

F. Javoy, Y. Agid and J. Glowinski

Groupe NB (INSERM U.114), Laboratoire de Biologie Moléculaire, Collège de France,
11, place Marcelin Berthelot, Paris 5e, France

INTRODUCTION

Numerous reports have demonstrated the existence of an ascending dopamine (DA) nigro-neostriatal pathway (see review of Hedreen and Chalmers, 1972). On the other hand, high levels of choline acetyl transferase were found in the neostriatum and could also be detected in the substantia nigra (SN) (Fahn and Coté, 1968). Furthermore, confirming the anatomical existence of a descending caudato-nigral pathway (Szabo, 1962) with the help of electro-physiological procedures, Frigyezie and Purpura (1967) reported that caudate cells influence nigral cells activity. Since a histochemical study revealed the presence of cholinesterasic striato-nigral efferents (Olivier et al., 1970), a cholinergic system was presumed to act on the DA nigral cells and to regulate the activity of the DA neurons (Corrodi et al., 1972). The finding of Smelik and Ernst (1966) support such a mechanism. These authors reported that cholinergic stimulation of the SN activated the DA nigro-neostriatal neurons and concluded to the existence of cholinergic fibers ending on the DA nigral cells.

Therefore, besides the striatal acetylcholine (ACh) and DA balance involved in extrapyramidal motor control (Hornykiewicz, 1971) there seem to exist interactions between DA and ACh neurons at the level of the SN. In view of these data, it was interesting to investigate if chemical stimulation of the compacta layer of the SN by muscarinic (carbachol) or antimuscarinic (atropine) drugs rapidly affected DA metabolism in DA nerve endings of the NCP (caudate nucleus + putamen).

RESULTS

Experiments were performed on male Charles Rivers rats (250 g) positioned in a stereotaxic apparatus, and anaesthetised with a mixture of Halothane, oxygen and nitrousoxide. The pars compacta of the SN was infused bilaterally during 10 min with a physiological solution (2 μl of NaCl 9%). The solution administered in the right SN (treated side) contained either carbachol or atropine at $5 . 10^{-4}$ M. The left side (sham operated side) was used as a control nigro-neostriatal system. Five-min after the onset of the nigral infusions a pulse injection 164 μCi of L-3·5-^{3}H-tyrosine (49 Ci/mM, Amersham) was performed intravenously. The animals were killed 10 min later (15 min after the beginning of the drug infusion). The NCP of the treated and control systems were analysed separately (Javoy et al., 1972). The values are the mean of groups of eight animals.

The administration of carbachol induced a small but significant increase in DA levels in the ipsilateral NCP when compared to the control side whereas atropine

747

injection was ineffective (DA μg/g: control: 7·0 ± 0·2; carbachol: 7·9 ± 0·3, $P < 0·05$; atropine: 7·5 ± 0·2). In addition, both drugs increased the initial ac-cumulation of newly synthesised ^{3}H-DA in the NCP corresponding to the treated nigro-neostriatal system (^{3}H-DA mμCi/g: control: 35 ± 1; carbachol: 57 ± 3 $P < 0·001$; atropine: 48 ± 5 $P < 0·05$). The effect was more pronounced after carbachol (60%) than after atropine (35%). Specific activity of tyrosine was not changed. Therefore, both drug treatments affected striatal DA metabolism although the mechanisms involved seemed to differ.

After either pharmacological treatment, DA synthesis was estimated with ·a method adapted from the experimental approach originally described by Carlsson *et al.* (1972). The initial rate of accumulation of ^{3}H-DOPA(L-3,4-dihydroxy-phenylalanine) synthesised from L-3·5-^{3}H-tyrosine was estimated after inhibition of central DOPA-decarboxylase (aromatic amino acid decarboxylase) with RO 4-4602 [N-(DL-seryl-N-2,3,4-trihydroxybenzyl)hydrazine] (800 mg/kg) intraperi-toneally. Rats were injected with RO 4-4602, 15 min before the beginning of the nigral infusion with carbachol or atropine. The injection of L-3·5-^{3}H-tyrosine was performed as in previous experiments, 5 min after the onset of the 10 min nigral infusions. Rats were sacrificed 10 min after the injection of the labelled amino-acid. RO 4-4602 pretreatment inhibited the formation of ^{3}H-DA. The accumulation of ^{3}H-DOPA in the NCP was unaffected by carbachol application on the contrary atropine treatment increased the levels of ^{3}H-DOPA in the corre-sponding NCP when compared to normal (^{3}H-DOPA nCi/g: control: 26·5 ± 2·4; carbachol: 30 ± 3; atropine: 33·2 ± 1·7 $P < 0·05$). Thus the anticholinergic drug, through its action at the SN level, stimulates the first step of DA synthesis in the ipsilateral NCP. The cholinomimetic drug is uneffective on this process.

Simultaneously in these experiments, the rate of DA disappearance after in-hibition of the amine synthesis with RO 4-4602 was used as an index of the rate of striatal utilization. Such a treatment, induces a (25%) decrease in striatal DA levels (5·9 ± 0·3 μg/g $P < 0·01$) within 30 min when compared to untreated rats (7·8 ± 0·4 μg/g). When carbachol was applied in the SN 15 min after the injec-tion of RO 4-4602 striatal DA levels in the treated side were 15% higher (6·8 ± 0·2 μg/g $P < 0·05$) than those of the sham operated side. In the contrary after atropine treatment DA disappearance was enhanced, since DA levels in the ipsilateral side (5·3 ± 0·08 μg/g $P < 0·02$) were significantly lower than those of the contra-lateral NCP. These effects observed in a short-time interval (15 min) suggest that DA utilisation in the NCP was counteracted by carbachol and stimulated by the atropine nigral infusion. Therefore, the increased DA and ^{3}H-DA levels seen in the NCP after carbachol nigral infusion in animals untreated with the synthesis inhibitor, most likely reveal a reduced rate of DA utilisation. On the other hand, the unchanged DA levels associated with an increased accumulation of ^{3}H-DA observed in the NCP of normal rats treated with atropine may be explained by the simultaneous activation of DA utilisation and synthesis induced by the anticholinergic agent.

CONCLUSION

Nigral applications of cholinergic or anticholinergic drugs, induced immediate changes in the metabolism of striatal DA. However, they seem to have antagonistic effects: (1) carbachol did not affect DA synthesis but inhibited DA release. These results disagree with the conclusions of SMELIK and ERNST (1966): on the basis of behavioural observations these authors provided evidence for the activation of DA nigro-neostriatal neurons after physostigmine implantation in the SN; (2) atropine on the contrary, stimulated both synthesis and utilisation of striatal DA.

Our data strongly suggest the presence of cholinergic receptors in the SN, and consequently an input of cholinergic fibers. These cholinergic neurons innervating the SN seem to be involved in the regulation of DA transmission. The origin of these cholinergic neurons is not known. It has been postulated that striato-nigral cholinergic fibers may contribute to the maintenance of normal DA levels in the NCP (OLIVIER et al., 1970; CORRODI et al., 1972). However, McGEER et al. (1973) failed to demonstrate a cholinergic striato-nigral pathway by lesions studies made in the cat. Therefore, the occurrence of cholinergic interneurons in the SN or the presence of cholinergic terminals originating from other brain structures cannot be excluded. On the other hand McGEER et al. (1973) described a descending striato-nigral gabaminergic pathway in agreement with previous reports of PRECHT and YOSCHIDA (1971) and KIM et al. (1971). Studies are in progress to examine the respective contribution of gabaminergic and cholinergic pathways in the regulation of nigro-neostriatal DA neurons in normal and pharmacological states.

REFERENCES

CARLSSON A., DAVIS J. N., KEHR W., LINDQVIST M. and ATACK C. V. (1972) Naunyn-Schmiedeberg's Arch. Pharmacol. 275, 153–168.
CORRODI H., FUXE K. and LIDBRINK P. (1972) Brain Res. 43, 397–416.
FAHN S. and COTE L. J. (1968). Brain Research, 7, 323–325.
FRIGYESI T. L. and PURPURA D. P. (1967) Brain Res. 6, 440–456.
HEDREEN J. C. and CHALMERS J. P. (1972) Brain Res. 47, 1–36.
HORNYKIEWICZ O. (1971) Recent advances in Parkinson's disease. Ed. by McDOWELL F. H. and MARKHAM C. H. Eds. F. A. Davis, Philadelphia, 33–65.
JAVOY F., AGID Y., BOUVET D. and GLOWINSKI J. (1972) J. Pharmacol. exp. Ther. 182, 454–463.
KIM J. S., BAK I. J., HASSLER R. and OKADA Y. (1971) Exp. Brain Res. 14, 95–104.
McGEER E. G., FIBIGER H. C., McGEER P. L. and BROOKE S. (1973) Brain Res. 52, 289–300.
OLIVIER A., PARENT A., SIMARD H. and POIRIER L. J. (1970) Brain Res. 18, 273–282.
PRECHT W. and YOSHIDA M. (1971) Brain Res. 32, 229–233.
SMELICK P. G. and ERNST A. M. (1966) Life Sci. 5, 1485–1488.
SZABO J. (1962) Exp. Neurol. 5, 21–36.

Frontiers in Catecholamine Research 1973, pp. 751 to 757. Pergamon Press. Printed in Great Britain.

ROLE OF CATECHOLAMINES IN THE CONTROL OF THE SLEEP-WAKING CYCLE

Michel Jouvet

Department of Experimental Medicine, Claude-Bernard University, 8, avenue Rockerfeller,
Lyon, France

The study of the role of catecholamine (CA) containing neurons of the central nervous system of the cat is difficult: the pharmacological approach (inhibition of synthesis of CA or injection of precursor) is easy, but it lacks the topographical dimension which is of paramount importance since histochemical methods have shown that the CA containing neurons belong to different anatomical systems (Dahlström and Fuxe, 1964). Since the topography of CA containing neurons is now relatively well known in the cat (Maeda et al., 1973), a direct approach upon these systems is possible. This can be done by coagulation. But this method is not selective enough since other neurons may be destroyed by the lesion. Recently, 6-hydroxydopamine (6-OHDA) has appeared to be a very selective tool for destroying CA containing neurons. However, our results suggest that 6-OHDA is not as selective for CA neurons in the cat as it is in the rat.

We shall restrict this review to the neuropharmacological, neurophysiological and biochemical approaches for studying some groups of CA containing neurons in the regulation of the sleep-waking cycle in the cat. All the experiments were made in chronically implanted cats, polygraphically recorded 23 hr a day.

ANATOMICAL ORGANISATION OF THE CA NEURONS IN THE CAT BRAIN STEM

Two main ascending pathways originating from the pontine group of CA neurons (nucleus locus coeruleus, subcoeruleus, parabrachialis) which correspond to groups $A_6 A_7$ of the rat have been mapped out by Maeda et al. (1973): the ascending dorsal CA bundle originates from the perikarya of the dorsal group of CA neurons of the pons, ascends in the mesencephalon and contributes to the innervation of the ipsilateral cortex with thin terminals. The intermediate CA bundle originates from perikarya located in the ventral group of CA pontile neurons (n. subcoeruleus), ascends in the mesencephalic reticular formation, enters the subthalamus and the hypothalamus. Some fibres cross in the supra-optic decussation. This system contributes to the innervation of the hypothalamus, septal region and ipsi and control lateral cortices with thick terminals. The ventral bundle, originating from the medulla, ascends ventrally to the perikarya of the locus coeruleus where it sends some collaterals. It contributes more rostrally to a 'nebula' like formation composed of terminal varicosities in the mesencephalic reticular formation. This formation receives also numerous collaterals from the dorsal and the intermediary bundles. The most rostral part of the ventral bundle has not yet been mapped out. Finally, the DA containing neurons have been mapped out also in the substantia nigra and the nigro striatal system is similar to that described in the rat.

Most of the work which has been carried out in our laboratory has been directed to the effect of destruction of either DA or NA containing perikarya or of the dorsal and intermediary CA ascending pathways.

NEUROPHARMACOLOGICAL APPROACH TO THE CA NEURONS

The following experiments have been selected in order to demonstrate the complexity of the effects induced by a theoretically simple experiment, i.e. the selective inhibition of CA synthesis.

In normal cats, the inhibition of the biosynthesis of CA, at the level of tyrosine hydroxylase, with alpha-methyl-p-tyrosine (AMPT) induces a temporary hypersomnia with increase of both slow wave sleep (SWS) and paradoxical sleep (PS) (KING and JEWETT, 1971). The interpretation of this effect is not easy since it has been shown that the decrease of endogenous cerebral CA which follows the administration of AMPT is accompanied by an acceleration of serotonin (5 HT) turnover (as shown by the increase of 5 HIAA and of 5 HT synthesis) (STEIN et al., 1973). Was the increase of sleep after AMPT due to the inactivation of some CA neurons or to the activation of 5 HT neurons? In order to answer this question, AMPT was given to insomniac cats which raphe system has been previously destroyed (JOUVET, 1969, 1972). In these conditions, the inhibition of CA synthesis was followed by a dramatic temporary decrease of waking (but not by hypersomnia since PS did not appear). Thus, there is some indirect evidence that, even when the 5 HT system is inactivated by lesion, AMPT, by decreasing the turnover of CA neurons (see PUJOL et al., 1973) may suppress temporarily waking. Thus it appears that the hypnogenic effect of AMPT may be mediated by (at least) two synergetic mechanisms: the inhibition of some CA mechanism related to waking and the increase of the turnover of the 5 HT neurons related to SWS and the priming of PS.

EFFECTS OF 6-HYDROXYDOPAMINE

Effect of micro-injection of 6-hydroxydopamine in the dorso-lateral part of the pontine tegmentum

The micro-injection of 6-OHDA in the locus coeruleus and subcoeruleus, with the technique and the dose described by UNGERSTEDT (1969), has given the following results: during the first 4 days there is an increase of the PGO activity (reserpinic syndrome) during waking and SWS, whereas the frequency of PGO during PS decreased together with the daily percentage of PS. After the 6th or 7th day, PS and PGO are totally suppressed. In the cats sacrificed on the 8–10th day, there is a 30–40 per cent decrease of NA in the mesencephalon and telediencephalon but also a similar decrease of 5 HT which indicates that some 5 HT neurons of the raphe system might have taken up 6-OHDA. The long delay (3–4 days) which is necessary for suppressing PS suggests that some neurons of the locus coeruleus or subcoeruleus might be relatively resistant to the effect of 6-OHDA. In control cats in which microinjections of Ringer solution at the same pH are given, PS decreases only by 30 per cent as compared with preinjection controls (BUGUET et al., 1970).

These experiments do not present the crucial proof that only CA-containing neurons of the pontine tegmentum are responsible for PS since the mechanical lesions which follows any microinjection into the brain might have affected other neurons.

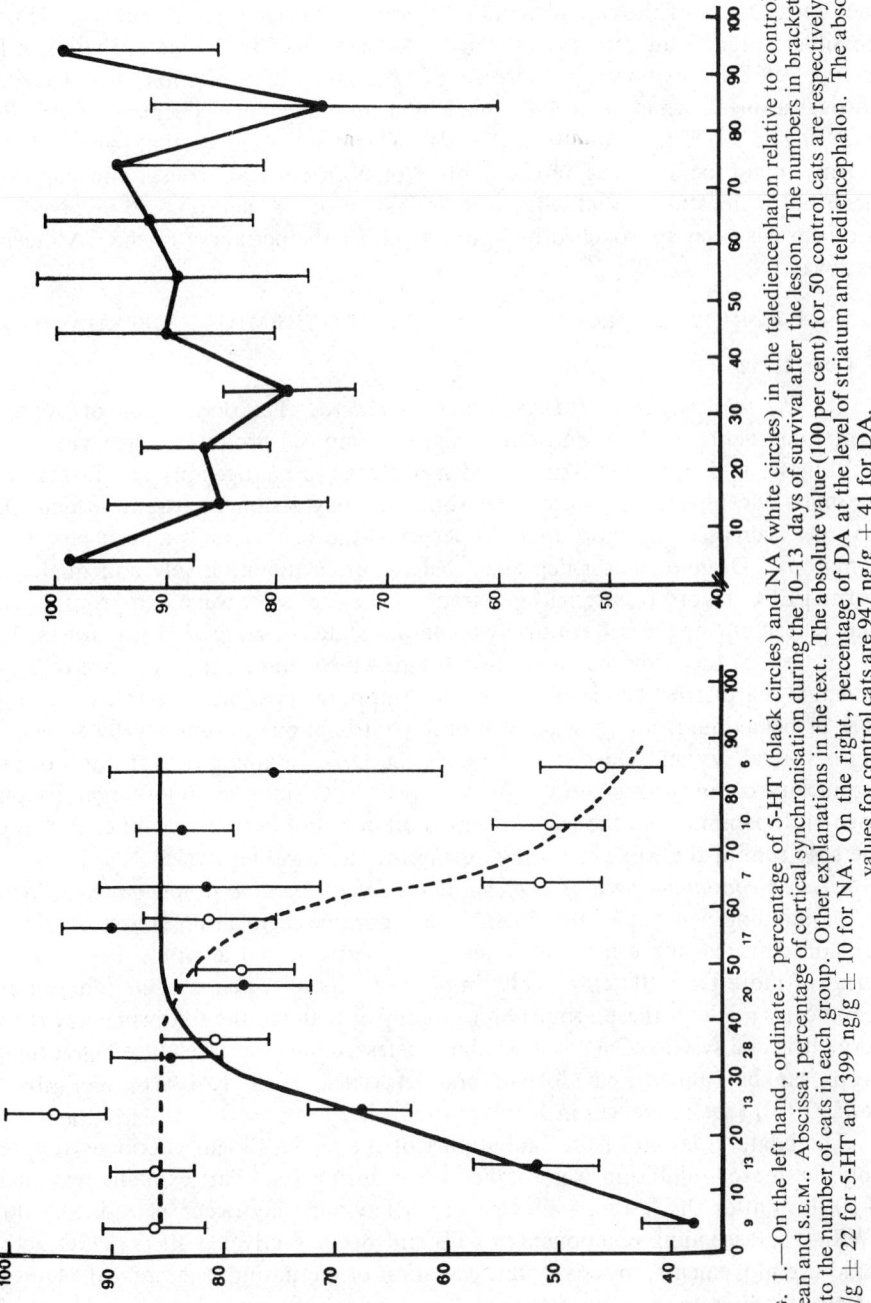

FIG. 1.—On the left hand, ordinate: percentage of 5-HT (black circles) and NA (white circles) in the telediencephalon relative to control cat (mean and s.e.m.. Abscissa: percentage of cortical synchronisation during the 10–13 days of survival after the lesion. The numbers in brackets refer to the number of cats in each group. Other explanations in the text. The absolute value (100 per cent) for 50 control cats are respectively 622 ng/g ± 22 for 5-HT and 399 ng/g ± 10 for NA. On the right, percentage of DA at the level of striatum and telediencephalon. The absolute values for control cats are 947 ng/g ± 41 for DA.

Effect of intraventricular injection of 6-OHDA

Contrary to the rat, 6-OHDA when injected intra-ventricularly in the cat, at a dose of 0·3–2·5 mg induces a decrease of both CA, 5 HT and 5 HIAA. However, the pretreatment of the cats with chlorimipramine protects 5 HT neurons. Thus it is possible to inactivate 'selectively' only CA terminals. In this case, there is a long-lasting (at least two weeks) increase of cortical synchronisation, i.e. decrease of cortical arousal, which is accompanied by a dose-dependent decrease of PS (PETIT-JEAN *et al.*, 1972, LAGUZZI *et al.*, 1972). This result suggests that some CA mechanisms might be involved in the control of both cortical arousal and paradoxical sleep but it does not reveal what specific system of CA neurons are involved. This problem has been approached by lesioning either the perikarya or the CA ascending pathway.

NEUROPHYSIOLOGICAL AND NEUROCHEMICAL APPROACH

CA perikarya

(i) *The mesencephalic DA-containing perikarya.* The destruction of dopamine-containing neurons of the substantia nigra (group A_9) produces a behavioural state of akinesia and unresponsiveness and a decrease in endogenous DA in the rostral brain (telencephalon, striatum). In some severely lesioned cats, in which DA is severely decreased (by more than 90 per cent) the behaviour is almost permanently comatous. Despite such a depressed behaviour, a quantitatively and qualitatively normal EEG record of alternating paradoxical sleep, slow wave and arousal activity may persist during the behaviourally comatous state (JONES *et al.*, 1968, JONES, 1969).

In view of the numerous anatomical data which suggest the existence of a dopaminergic nigro-striatal ascending system, it appears possible that DA could play a role in behavioural alterness, and motor coordination presumably by way of this nigro-striatal system. On the other hand, the DA-containing neurons do not play a significant role in the regulation of the cortical EEG since an almost total disappearance of dopamine from the tele-diencephalon does not induce a significant change in the spectrum of the EEG recording during the sleep-waking cycle.

(ii) *CA-containing perikarya of the dorso-lateral pontine tegmentum and PS.* The CA-containing neurons of the dorso lateral pontine tegmentum (group A_5, A_6, A_7) are concentrated in the locus coeruleus, subcoeruleus and adjacent nuclei. From a series of more than 100 chronically implanted cats in which coagulations were systematically made in these groups and in control regions, the following pictures of a very intricated system of neurons, leading to descending and ascending CA-containing axons can be summarised (JOUVET and DELORME, 1965, ROUSSEL, 1967, BUGUET *et al.*, 1970) (See references in JOUVET, 1972).

The bilateral lesion of the caudal part of the nucleus locus coeruleus suppresses only the motor inhibition which takes place during PS. This explains why around 8–10 days after the lesion, hallucinatory behaviour may occur periodically during SWS. The ascending components of PS still occur (activated EEG, PGO activity, rapid eye movements, myosis, total relaxation of nictitating membranes). However, the cat standing up looks awake since it may attack unknown enemies, play with an absent mouse, or display a flight behaviour. There exist orienting movements of the head, whereas the animal does not respond to visual or auditory stimuli. These 'pseudo-hallucinatory' episodes (which are good arguments that 'dreaming' may occur during PS in the cat) last for 3–5 min and result in sudden awakening and return to SWS.

After lesion of the caudal part of the locus coeruleus there are no significant alterations of CA endogenous content in the mesencephalon or telediencephalon, but a 30–40 per cent decrease of NA occurs in the cervical spinal cord. Thus, it is possible that some descending CA neurons might be implicated in the control of total decrease of muscle tone which occurs during PS.

More extensive bilateral lesions, involving the caudal half of the locus coeruleus and the N. subcoeruleus suppress definitively the occurrence of PS and 'pseudo-hallucinatory behaviour'. However, some PGO activity may still occur during slow wave sleep. This lesion induces a decrease of about 30 per cent of NA in the tele-diencephalon without any alteration of the 5 HT or DA levels.

Control lesions located either ventrally to these groups of neurons, or medially, or laterally or caudally (the entire group of vestibular nuclei) do not result in significant alterations of SWS, PS or of the amine content of the rostral brain.

From these data it may be concluded that most of the neurons located in the dorso lateral part of the pons play a role in the execution of PS and that there is a very complicated organisation which is not yet understood. The most caudal part of the pontine CA neurons is related to descending mechanisms, whereas phasic PGO activity and the fast EEG activity appears to depend upon most of the neurons located in the ventral and caudal part of the locus coeruleus and most of the subcoeruleus. Only the group of CA neurons located in the anterior third of the locus coeruleus does not participate in the PS mechanisms, but this group which sends neurons in the dorsal NA bundle is apparently concerned with the control of waking.

CA ascending pathway

(i) *Dorsal NA pathway.* The bilateral destruction of the dorsal NA pathway *at the level of the isthmus* induces a temporary dramatic hypersomnia with an increase of SWS and PS (up to 400 per cent). It is likely that this effect could be due to the suppression of some CA fibres 'controlling' directly or indirectly the anterior part of the raphe system, since there is an increase of 5 HT turnover after this lesion (PETITJEAN and JOUVET, 1970) (see PUJOL *et al.*, 1973)·

The lesion of the dorsal NA pathway more rostrally than the isthmus (in the mesencephalon) decreases the duration of cortical arousal without interfering with the phasic short-lasting arousal which usually follows sensory stimulation. This decrease of cortical arousal (or increase of cortical synchronisation) is not accompanied by any increase of PS. There is a significant correlation between the decrease of telencephalic NA which follows the lesion of the dorsal NA pathway and the increase of cortical synchronisation (JONES, 1969). Thus the CA-containing neurons ascending in the dorsal NA bundle play a very subtle role in the regulation of the sleep-waking cycle: some of them are involved in the process of the tonic cortical arousal during waking (hence their inactivation leads to 'unwaking' state), others may control tonically the 5 HT neurons of the rostral raphe system (hence their selective inactivation leads to hypersomnia with increase of both SWS and PS).

(ii) *Lesion of the ascending intermediary CA pathway.* The destruction of the region of the intermediary pathway by electrocoagulation, at the level of the isthmus suppresses the PGO activity in the lateral geniculate and occipital cortex (Laurent et al., 1972). This finding is in accordance with the role played by the N. subcoeruleus in the generation of PGO waves. However, the direct proof that only the NA fibres ascending in the region of the intermediary pathway would be responsible for initiating PGO activity has not yet been obtained.

SUMMARY

Three different systems of CA-containing neurons appear to be involved in the regulation of the sleep-waking cycle.

(1) The system originating from the central and caudal part of locus coeruleus and subcoeruleus is concerned with the executive mechanisms of PS, whereas the 'priming mechanisms' of PS (probably serotoninergic) are apparently located in the medial and caudal raphe system (see references in Jouvet, 1972).

(2) The anterior NA pontile waking system originates from the anterior part of the locus coeruleus. This system sends axons to the dorsal NA bundle which ascends near the grey matter. The destruction of the dorsal NA bundle at the level of the isthmus is followed by a significant increase of both states of sleep, whereas the decrease of NA is accompanied by an increase of 5 HT turnover. It is thus possible that some collaterals issued from the anterior nucleus locus coeruleus might exert a control upon the rostral raphe system during waking. The suppression of this control by a lesion would cause an increase in activity of the raphe system leading to a temporary increase of both SWS and PS.

(3) Other neurons ascending in the dorsal NA system in the mesencephalon contribute to the control of cortical arousal. It is not yet known if this effect is directly mediated at the cortical level, or through some connections with the terminal varicosities located in the mesencephalic reticular formation. The destruction of this system decreases cortical arousal but there is no increase of PS. Thus the destruction of the NA neurons in the mesencephalon decreases EEG waking but does not induce a true hypersomnia.

(4) The substantia nigra and the nigro striatal system are apparently concerned with the control of waking behaviour through the extra-pyramidal system, but are not concerned with the regulation of cortical activity during the sleep-waking cycle.

Thus the NA anterior pontile and mesencephalic systems are antagonistic to the 5 HT-containing neurons of the raphe system which is strongly implicated in sleep mechanisms. This is illustrated in the figure which summarises the biochemical data obtained from 133 cats subjected to lesions either of the midline raphe, or of the bulbar, pontine mesencephalic tegmentum or of the substantia nigra. Control lesions were also made outside CA-containing neurons, while 12 sham operated cats served as control (groups of animals are classified according to the percentage of EEG synchronisation during the 10–13 days of the postoperative survival and the mean value of the percentage of 5 HT NA and DA in the telencephalon and diencephalon for each group is given. It is clear that 5 HT and NA fit two opposite curves, while DA does not change significantly. Cats with insomnia have a decreased telediencephalic 5 HT and normal telencephalic NA content. Cats with normal or subnormal levels of synchronisation have normal or subnormal values for both 5 HT and NA, whereas cats with increased synchronisation have a decreased NA with normal or subnormal 5 HT level in the telediencephalon.

These results demonstrate the complexity of the study of the role of CA containing neurons in the brain stem of the cat. More and more converging data from neurophysiology, neuropharmacology and from biochemistry strongly suggest however that the CA containing neurons located in the pons or in the mesencephalon are implicated in the regulation of the sleep-waking cycle in the cat, either directly or indirectly by acting upon serotoninergic mechanisms.

REFERENCES

BUGUET A., PETITJEAN F. and JOUVET M. (1970) *C.R. Soc. Biol.* **164,** 2293–2300.
DAHLSTROM A. and FUXE K. (1964) *Acta Physiol Scand.* **62,** (suppl. 232).
JONES B. (1969) Catecholamine-containing neurons in the brain stem of the cat and their role in waking. M.A. thesis, Lyon, pp. 87.
JONES B., BOBILLIER C., PIN C. and JOUVET M. (1973) *Brain Res.* **58,** 157–177.
JOUVET M. (1969) *Science* **163,** 32–41.
JOUVET M. (1972) *Ergebn. Physiol.* **64,** 166–307.
JOUVET M. and DELORME J. F. (1965) *C.R. Soc. Biol.* **159,** 895–899.
KING D. and JEWETT R. E. (1971) *J. Pharmacol. Exp. Ther.* **177,** 188–194.
LAGUZZI R., PETITJEAN F., PUJOL J. F. and JOUVET M. (1972) *Brain Res.* **48,** 295–310.
LAURENT J. P., CESPUGLIO R. and JOUVET M. (1972) *Experientia, Busel* **28,** 1174–1176.
MAEDA T., PIN C., SALVERT D., LIGIER M. and JOUVET M. (1973) *Brain Res.* **57,** 119–152.
PETITJEAN F. and JOUVET M. (1970) *C.R. Soc. Biol.* **164,** 2288–2293.
PETITJEAN F., LAGUZZI R., SORDET F., JOUVET M. and PUJOL J. F. (1972) *Brain Res.* **48,** 281–193.
PUJOL J. F., STEIN D., BLONDAUX C., PETITJEAN F., FROMENT J. L. and JOUVET M. This volume.
ROUSSEL B. (1967) Monamines et sommeils: suppression du sommeil paradoxal et diminution de la noradrénaline cérébrale par les lésions des noyaux locus coeruleus. Thèse Médecine, Lyon, Tixier ed. 141 pp.
STEIN D., JOUVET M. and PUJOL J. F. (1973) *Brain Res.* (in press).
UNGERSTEDT U. (1969) *Europ. J. Pharmacol.* **5,** 107–111.

Frontiers in Catecholamine Research 1973, pp. 759 to 762. Pergamon Press. Printed in Great Britain.

CATECHOLAMINE MECHANISMS: THEIR PRESUMPTIVE ROLE IN THE GENERATION OF REM SLEEP PGO WAVES

STEVEN HENRIKSEN, BARRY JACOBS*, WILLIAM DEMENT and JACK BARCHAS

Department of Psychiatry, Stanford University School of Medicine,
Stanford, Calif. 94305, U.S.A.

IT CAN no longer be doubted that the neurochemical processes underlying mammalian sleep and waking behaviour at least partially utilise brainstem monoamine systems. This level of understanding of the role of monoamines in state processes has ultimately been due to the convergent findings of a variety of neurobiological disciplines: neurohistology (DAHLSTROM and FUXE, 1964, 1965; PIN et al., 1968 and JONES, 1969); neuropharmacology (KING, 1972 and JOUVET, 1972); and neurochemistry (see JOUVET, 1972). The undisputed leader in the experimental realm as well as in the theoretical conceptualisation of monamine systems and how they relate to state, is Dr. Michel Jouvet. For over a decade Jouvet and his colleagues in Lyon have steadily marshalled incontrovertible evidence implicating these systems in the state processes. Largely through their efforts the relationships between 5HT and sleep, NE and cortical arousal, and DA and behavioural arousal have given rise to a general construct that might be termed the "monamine theory of state" (JOUVET, 1972). As with other bonafide paradigms that are relatively new and as broad as this, the monoamine theory has been hotly contested and debated throughout its ascendancy. Yet, presently, as a measure of its viability and value, it remains essentially intact despite its rather defiant scope. Whether or not the monoamine construct continues to remain intact as a basic descriptor of state processes in the future is irrelevant to its value as a starting point and impetus for research in the present. Clearly, monoaminergic mechanisms are now among the most vigorously investigated processes in the mammalian brain. Research into these processes has offered the unique opportunity to unify lesion, neuropharmacological, and neurochemical studies in an attempt to describe and analyse the underlying mechanisms of behaviour. Indeed, the interdisciplinary study of state (perhaps the most basic mammalian behaviour) has fostered the concept that monoamines may mediate *specific* waking as well as sleeping sub-processes within the very general concept of states.

With reference to the present catecholamine symposium, Dr. Jouvet has expertly outlined the currently viewed role of these specific monoamines (i.e. NE and DA) in waking phenomenon and REM sleep processes. With reference to the latter, Dr. Jouvet suggests that a noradrenergic mechanism located in the nucleus subcoeruleus is at the basis of the *sine qua non* of feline REM phasic phenomena, the ponto-geniculo-occipital (PGO) wave. The evidence for this suggestion he derived from

---

* Present address Department of Psychology, Princeton University, Princeton, New Jersey 08540, U.S.A.

the observation that in cats, both neuropharmacological (6-OH DA) as well as electrolytic lesions in this norepinephrine containing locus result in the long lasting suppression of REM sleep, including the ubiquitous PGO waves. In addition, results from other studies utilising the administration of a variety of other neuropharmacological agents have indirectly implicated noradrenergic mechanisms in the generation of PGO waves (Jones, 1972).

Work in our laboratory at Stanford bears on this specific problem raised by Jouvet and his colleagues: the neurochemical basis of the PGO wave. Our approach to this problem with respect to catecholaminergic (CA) mechanisms has utilised two experimental manipulations in cats: (1) The direct and selective neuropharmacological impairment of CA mechanisms through administration of alphamethyl-p-tyrosine (AMPT); and (2) the administration of CA antagonists secondary to chronic administration of p-chlorophenylalanine (PCPA).

STUDIES USING ALPHA-METHYL-*p*-TYROSINE (AMPT)

The CA anti-synthesis agent AMPT induces a severe depletion of brain CA when administered to cats (King and Jewett, 1971; Henriksen and Dement, 1972). In our hands, AMPT infused by way of an intravenous drip to avoid nephrotoxic side effects, depleted regional brain NE over 70 per cent. This depletion was accompanied by an initial increase (during the first 24 hr following initiation of the drip) in both the per cent REM sleep of total sleep as well as the absolute amount of REM sleep. As administration was continued, slow-wave sleep (SWS) time increased dramatically such that at the height of the sleep effects (24—48 hr following initial administration) SWS could itself account for over 60 per cent of the 24 hr recording time. During this peak effect on sleep no change in either the frequency, amplitudes or temporal relationships of the REM PGO waves could be seen (See Fig. 1). On the

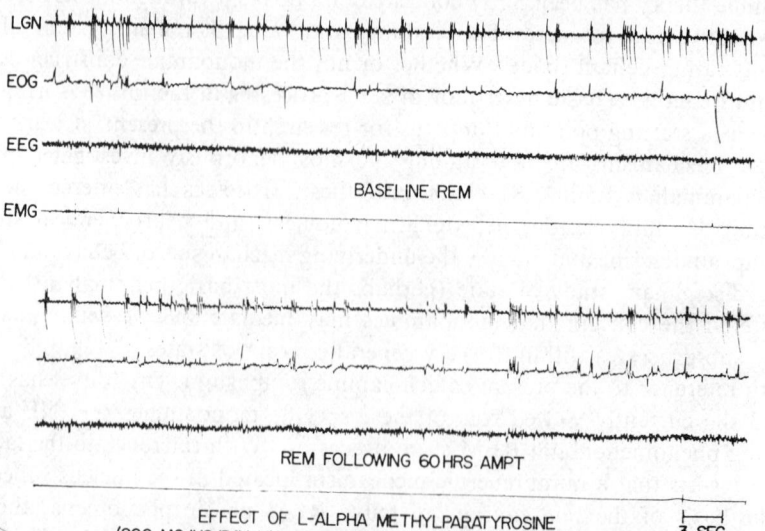

Fig. 1.—The top 4 and bottom 4 polygraphic tracings show approximately 1 min of REM sleep in each case. Note the essentially identical nature of the LGN record. Abbreviations: LGN, lateral geniculate nucleus; EOG electroculogram; EEG, cortical electroencephologram; EMG, neck electromyogram.

other hand, a significant increase in the absolute number of REM PGO waves was seen due to the initial increase in REM sleep time. Similar results, using slightly different experimental procedures have been reported by others (KING, C. and JEWETT, R., 1971).

STUDIES USING ANTI-CA AGENTS SECONDARY TO PCPA TREATMENT

Daily injections of 150 mg/kg PCPA to cats result in the emergence of PGO waves into the waking state by the fourth or fifth treatment day. These waves are by all criteria identical to the PGO waves observed during REM sleep in the same cats (see JACOBS et al., 1973). The emergence of PGO waves into the waking state makes these waves more amenable to study as their occurrence is no longer dependent upon the concomitant appearance of other events (i.e. tonic muscle inhibition) definitive of REM sleep. To this PCPA "preparation" we have administered the following anti-CA agents: the DA receptor blocker—pimozide (2·5–4·0 mg/kg); alpha-CA blocking agents—phentolamine (4·0 mg/kg) and phenoxybenzamine (7·5–10·0 mg/kg); the beta-CA blocking agent—propranolol (4·0 mg/kg) and the CA anti-synthesis agent—AMPT(150 mg/kg i.p., split doses). Figure 2 compares sample baseline waking PGO records with those following the anti-CA agents above. In no case did any of the CA antagonists alter the discharge frequency of waking PGO waves when compared to the baseline PGO rate. When AMPT was administered at the height of the PCPA insomnia (at the maximum point of waking PGO's [PCPA days 4–5]), an *increase* in PGO wave frequency was observed in spite of the partial return of behavioural and electrographic SWS.

These results, although far from definitive, argue against the generation of the PGO waves of REM sleep through the participation of CA mechanisms. On the

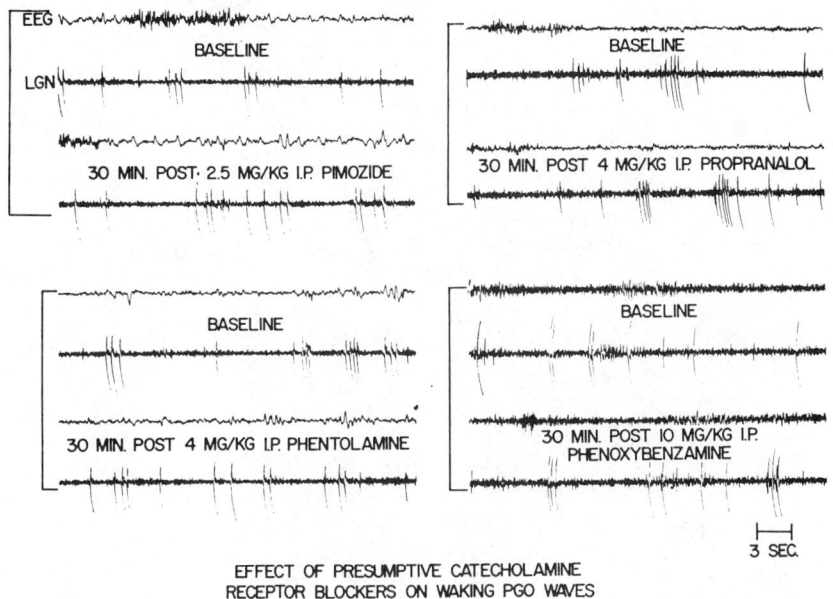

EFFECT OF PRESUMPTIVE CATECHOLAMINE RECEPTOR BLOCKERS ON WAKING PGO WAVES

FIG. 2.

contrary these results suggest as other reports have (JALFRE *et al.*, 1972), that CA mechanisms may in fact act to inhibit PGO mechanisms.

Although highly speculative, it is possible that the CA neuronal mechanisms proposed by Dr. Jouvet to be related to electrocortical waking (supported by our results) may themselves directly or indirectly act to regulate pontine areas involved with PGO generation and/or projection.

REFERENCES

DAHLSTROM A. and FUXE K. (1964) *Acta Physiol. Scand.*, **62**, (Suppl.) 232.
DAHLSTROM A. and FUXE K. (1965) *Acta Physiol. Scand.*, **64**, (Suppl.) 247.
HENRIKSEN S. and DEMENT W. (1972) *Sleep Res.*, **1**, 55.
JACOBS B., HENRIKSEN S. and DEMENT W. (1972) *Brain Res.*, **48**, 406–411.
JALFRE M., MONACHON M-A and HAEFELY W. (1972) In: *Proceedings of the First Canadian International Symposium on Sleep* (MCCLURE, D. J., Ed.), pp. 155–185. Roche Scientific Service, Vaudrevil, Quebec.
JONES B. MA Thesis, Lyon: Tixier (1969), 87 pp.
JONES B. (1972) *Brain Res.*, **39**, 121–136.
JOUVET M. (1972) In: *Ergeb. der Physiol.*, volume 64, pp. 166–307, Springer-Verlag, New York.
KING D. and JEWETT R. (1971) *J. Pharmacol. Exp. Ther.*, **177**, 188–194.
PIN C., JONES B. and JOUVET M. (1968) *C.R. Soc. Biol. (Paris)* **162**, 2136–2141.

Frontiers in Catecholamine Research 1973, pp. 763 to 766. Pergamon Press. Printed in Great Britain.

EFFECTS ON PRECISE MOTOR RESPONDING OF BILATERAL 6-HYDROXYDOPAMINE INFUSION INTO THE SUBSTANTIA NIGRA

LARRY L. BUTCHER, GORDON K. HODGE, GUY K. BRYAN
and SHEILA M. EASTGATE

Department of Psychology, University of California, Los Angeles, California, 90024, U.S.A.

INTRODUCTION

UNILATERAL injection of 6-hydroxydopamine (6-OHDA) into the substantia nigra (SN) produces turning ipsilateral to the lesioned side (UNGERSTEDT, 1971). Since the zona compacta of the SN has a high proportion of neuron somata containing dopamine, this "rotation" or "turning" model has been extensively used to assess the effects of pharmacological agents thought to affect predominately dopaminergic neurotransmission mechanisms (e.g. see UNGERSTEDT, 1971). Although this approach has yielded a considerable volume of interesting data, the question can legitimately be raised as to what the "rotation" model tells us about the fundamental functional significance for motor behaviour of the SN and neuroanatomically interconnected structures. Indeed, this model may only reflect the consequences of producing a functional imbalance in the extrapyramidal motor system. For example, ipsilateral turning is also observed after hemicerebellectomy (MANNI and DOW, 1963). In this paper we describe a behavioural technique, lever-positioning, which may permit a more precise evaluation of possible functional relationships between motor behaviour and decrements in nigral function produced by direct injection of 6-OHDA into the zona compacta of the SN.

METHODS

Behavioural procedures

Male Sprague–Dawley rats were reduced to 80 per cent of their *ad libitum* weights by means of restricted feeding. They were then trained to hold a constant force (2 g $\times$ gravity) lever within a displacement band 6° wide beginning at 17° and ending at 23° from the horizontal for 2·0 sec in order to obtain one 45-mg food pellet. The total possible displacement of the lever was 40°. Responses were recorded on a polygraph, on running time meters, and on digital counters. The rats were tested in daily consecutive sessions. A session was terminated (1) after 100 45-mg food pellets had been received and consumed or (2) after 1 hr, whichever came first. Quite stable motor performance baselines were obtained with this procedure, both within and between sessions. Following stabilization of the lever-positioning behaviour, each rat was injected bilaterally in the SN with 6-OHDA or its vehicle. Testing in the lever-positioning task commenced 24 hr after surgery and continued until the behavioural baseline again stabilized. Food intake and locomotor activity were also quantified for each experimental animal both before and after surgery.

Neurotoxic specificity of 6-OHDA and intranigral infusion procedures

Although initial experimental evidence suggested that 6-OHDA was a neuro-toxic drug selective for catecholamines at low doses (e.g., 8 μg 6-OHDA in 4 μl vehicle; Ungerstedt, 1971), more recent studies (Poirier *et al.*, 1972) have cast considerable doubt on this hypothesis. In our laboratory we have confirmed the basic conclusion of Poirier *et al.* (1972) that intracerebrally administered 6-OHDA, using a vertical approach to the structure studied (cf. Ungerstedt, 1971), possesses considerable non-specific activity: (1) unilateral infusion into nucleus ruber of 8 μg 6-OHDA in 4 μl Ringer's–ascorbic acid solution produced within 48 hr a complete loss of neuron somata on the injected side. The zone of destruction was approximately 2 mm in diameter. Although this dose has been reported to produce selective degener-ation of catecholamine neurons (Ungerstedt, 1971), the cell bodies of nucleus ruber contain neither dopamine nor noradrenaline (Dahlström and Fuxe, 1964). (2) Bilateral administration of 8 μg/4 μl 6-OHDA into the SN produced almost com-plete loss of neuron somata in the zona compacta correlated with the production of catalepsy and rigidity. Both the histopathology and consequent motor symptomat-ology could be duplicated by bilateral intranigral injection of 8 μg/4 μl copper sulfate, a relatively non-specific cytotoxic agent.

These data strongly suggest that "specificity" of neuron destruction can be obtained with 6-OHDA only to the extent that the drug is injected into brain regions which are neurochemically homogeneous. For this reason and because we were particularly interested in dopaminergic mechanisms of motor behaviour, we have attempted to produce 6-OHDA lesions confined as much as possible to the zona compacta of the SN. We have been able to achieve almost complete and relatively selective destruction of zona compacta neuron somata by injecting 6-OHDA through a cannula inserted at an angle of 53° from the horizontal. The cannula shaft lies between the row of zona compacta neurons and the medial lemniscus. This lateral entry obviates damage to the nucleus ruber and ventral noradrenaline bundle, both of which may be destroyed, either mechanically or by 6-OHDA, if the traditional vertical approach to the SN is employed. In the current experiments solutions of 6-OHDA were stereotaxically delivered to the SN through a stainless steel cannula (external diameter = 0.32 mm) at a rate of 1 μl/min and in amounts of either 4 μg/2 μl or 1 μg/2 μl. Solutions of 6-OHDA were prepared immediately before injection and were kept at 0°C until infusion. The Ringer's vehicle contained 0.2 mg ascorbic acid per ml to retard 6-OHDA auto-oxidation. The cannula was kept in place for 1 min after termination of the infusion period to allow for pressure equalization. The lateral approach was used for both the histological and behavioural studies described in the following section.

RESULTS AND DISCUSSION

Histology

Animals ($N = 10$) injected unilaterally with 4 μl/2 μl 6-OHDA and sacrificed 48 hr later showed almost complete loss of zona compacta neuron somata with the exception of a few cell bodies in the lateral portion of the compacta (Luxol fast blue-basic fuchsin and also thionin stained sections). Infusion of 1 μg/2 μl 6-OHDA produced approximately 30 per cent loss of cell somata in the zona compacta on the injected side ($N = 10$ rats). Most of the degenerated cell bodies lay in the medial portion of the compacta. With both doses of 6-OHDA there was no apparent loss of neuron somata in pars reticulata and only slight, if any, damage to pars lateralis. Furthermore, there did not appear to be any loss of myelinated fibres in the region of the SN.

Behaviour

Rats ($N = 6$) receiving 4 μg/2 μl 6-OHDA in each SN showed lever-positioning behaviour characterized by long holding times (Fig. 1). This motor pattern reflected the relatively cataleptic posture that the rats typically assumed. Locomotor activity was markedly reduced over a 6-day period such that statistically significant motility decrements to 51 per cent ($P < 0.05$) and 26 per cent ($P < 0.01$) of control were observed on days 1 and 6 after surgery, respectively. Similarly, food intake was decreased to 8 per cent ($P < 0.01$) of control on the first post-operative day and to 5 per cent ($P < 0.01$) on the sixth. Passive movement of the limbs revealed that they were markedly rigid. The hindlimbs were extended and appeared more rigid than the forelimbs.

The lever-positioning records of animals ($N = 6$) bilaterally injected with 1 μg/2 μl 6-OHDA showed oscillatory, tremor-like movements in addition to long holding times (Fig. 1). The tremor had both a regular amplitude and periodicity ($\simeq$ 5 Hz).

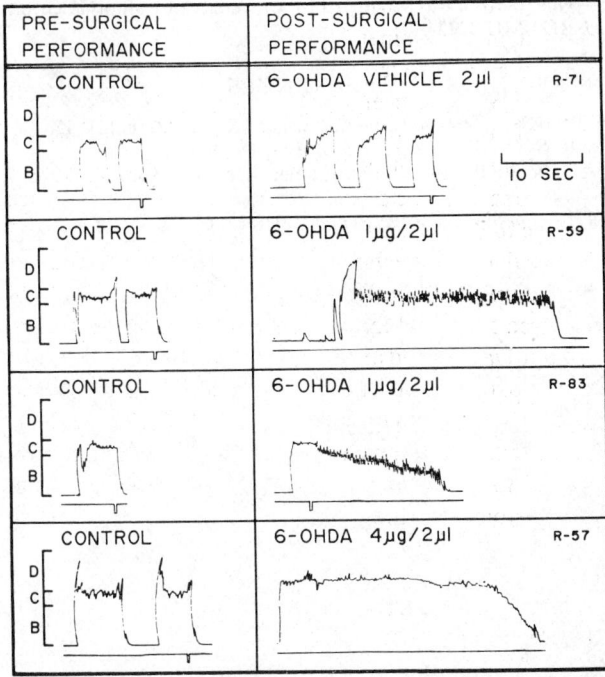

FIG. 1.—Effects on lever-positioning behaviour of bilateral 6-hydroxydopamine infusion into the substantia nigra. The post-operative performance is shown for rat R-71 on the second day after surgery, for rat R-59 on day 6, for rat R-83 on day 3, and for rat R-57 on day 6. The downward deflections on the lower trace in each pair of records represents the delivery of a food pellet. B = the lever displacement zone from 0 to 17°, C = the reinforced zone from 17 to 23°, and D = the zone from 23 to 40°.

These rats also displayed rigidity and hypokinesia but less severe than animals given 4 μg/2 μl 6-OHDA. Food intake was 7 per cent ($P < 0.01$) of control on the first post-surgical day but improved to 41 per cent ($P < 0.05$) of control by the sixth daily testing session. The initial decrease in motility to 49 per cent ($P < 0.05$) of control 24 hr after surgery was replaced by an increase to 71 per cent ($P =$ N.S.) of control on the sixth post-operative day. Nonetheless, marked effects on lever-positioning were still observed 6 days after the operation (Fig. 1) indicating that the technique of lever positioning was capable of detecting motor deficits which were not readily apparent using more traditional behavioural observation methods. No motor abnormalities (Fig. 1) or deficiencies in food intake and locomotor activity were observed in animals ($N = 4$) injected with 2 μl of the vehicle alone.

Conclusions

Our data demonstrate that, depending on the dose of 6-OHDA bilaterally injected into the SN and also on the behavioural measuring technique, a preparation can be obtained which displays motor symptomatology similar to the syndrome observed in human Parkinsonism, taking into account obvious differences in motor expression due to species differences. Furthermore, our results suggest that the motor pathologies of rigidity, tremor, and hypokinesia may have a common neurochemical basis, possibly a deficiency of dopamine in the nigro-striatal system.

Acknowledgements—This research was supported by University of California Grant No. 2637 and by NIMH Grant No. 1 RO3 MH 22284-01.

REFERENCES

Dahlström A. and Fuxe K. (1964) *Acta physiol. scand., Suppl.* **232(62),** 1–55.
Manni E. and Dow R. S. (1963) *J. comp. Neurol.* **121,** 189–194.
Poirier L. J., Langelier P., Roberge A., Boucher R. and Kitsikis A. (1972) *J. neurol. Sci.* **16,** 401–416.
Ungerstedt U. (1971) *Acta physiol. scand., Suppl.* **367(82),** 69–93.

Frontiers in Catecholamine Research 1973, pp. 767 to 769. Pergamon Press. Printed in Great Britain.

TELENCEPHALIC DISTRIBUTION OF TERMINALS OF BRAINSTEM NOREPINEPHRINE NEURONS*

ROBERT Y. MOORE

Departments of Pediatrics, Medicine (Neurology) and Anatomy and the
Joseph P. Kennedy, Jr. Mental Retardation Research Center, The University
of Chicago, Chicago, Illinois 60637, U.S.A.

DURING the past 10 years there have been remarkable advances in the understanding of the organization and function of central monoamine-containing neuron systems. The first evidence associating monoamines with a specific group of central neurons was a demonstration that destruction of the medial forebrain bundle, or areas contributing axons to that pathway in the lateral hypothalamus, produced significant decreases in brain serotonin and norepinephrine in the rat (cf. HELLER et al., 1968; HELLER and MOORE, 1965; MOORE, 1970). Subsequent studies showed that the loss of amines and their synthetic enzymes following medial forebrain bundle lesions occur uniformly throughout the telencephalon but that degenerating axons and terminals cannot be demonstrated by conventional neuroanatomical methods in all affected telencephalic areas (MOORE and HELLER, 1967; HELLER and MOORE, 1968). This discrepancy between the loss of amines and enzymes and the apparent distribution of anterograde axonal degeneration following the lesion led to introduction of the concept that the amine and enzyme changes occur, at least in part, as transsynaptic neurochemical effects within otherwise intact neurons. In contrast to this, studies employing the Falck–Hillarp histochemical method indicate that all monoamine-producing neuron cell bodies are located in the diencephalon and brainstem with axons and terminals distributed throughout the neuraxis (ANDEN et al., 1966; FUXE et al., 1970; UNGERSTEDT, 1971). Ungerstedt's (1971) work, in particular, has suggested that the nucleus locus coeruleus of the pontine tegmentum gives rise to most, if not all, of the norepinephrine terminals innervating telencephalic structures. The studies to be presented here were undertaken, first, to determine if evidence could be obtained for the existence of norepinephrine-producing neurons in denervated cerebral cortex and, second, to attempt providing evidence for a direct projection from the norepinephrine neurons of locus coeruleus throughout the telencephalon.

In the first study the auditory cortex of the cat was isolated from subcortical and other cortical areas with its blood supply intact and subsequently analyzed for norepinephrine content and the capacity to produce norepinephrine from intravenously administered, labelled tyrosine. If norepinephrine-producing neurons are present, both cell bodies and axons, in cortex, they should be preserved after isolation of the cortex and should continue to have the capacity to produce norepinephrine (cf. MOORE, 1970). In isolated cat auditory cortex, however, there is no assayable norepinephrine present and none is produced from labelled tyrosine even though the tyrosine enters cortex (Table 1). The greater tyrosine content of denervated cortex probably reflects the loss of white matter in these samples. Thus, no evidence is obtained from this study for the presence of norepinephrine-producing cells in cat neocortex.

---

* Supported by grants NS-05002 and HD-04583 from the National Institutes of Health, USPHS.

Table 1. Norepinephrine biosynthesis from labelled tyrosine in the
isolated cat cerebral cortex

Sample analysed	Tyrosine		Norepinephrine	
	(μg/g tissue)	(dis/min per n mole)	(μg/g tissue)	(dis/min per n mole)
Control Cortex	2.70	825	0.21	69
Isolated	4.65	665	—	—

Two cats were subjected to bilateral superior cervical ganglionectomy followed by complete unilateral isolation of auditory cortex (all cortex from the superasylvian gyrus to the temporal pole). Thirty days later each received an intraveneous injection of tyrosine-^{14}C and after one hour the auditory cortex from each side was removed and analysed for tyrosine and norepinephrine content and specific activity (Zigmond and Wurtman, 1970). No fluorimetrically identifiable norepinephrine was present in isolated cortex nor were any counts above background present in the eluate off alumina.

In the second study the effect of restricted unilateral lesions in the locus coeruleus on telencephalic and diencephalic amines was analyzed. As can be seen in Table 2, localized locus coeruleus destruction produces a nearly total unilateral depletion of telencephalic norepinephrine without significantly affecting thalamic or hypothalamic norepinephrine levels. Similar lesions also were found to markedly diminish the formation of labelled norepinephrine from tyrosine in the telencephalon without affecting diencephalic synthesis (Moore, unpublished observations). These observations, in accord with those of Ungerstedt (1971), support the concept of a direct axonal projection from the locus coeruleus to the entire telencephalon. Further corroboration for this is obtained from the third study in which tritiated leucine was injected into the locus coeruleus. The available evidence indicates that the amino acid will be incorporated into protein in the cell bodies of the nucleus and transported to terminals of the axons arising from the cells (Cowan et al., 1972; Moore and Lenn, 1972). After injection of the locus coeruleus, labelled protein is transported to all areas of the telencephalon (Table 3). Because of the temporal parameters of the experiment, the labelled material would be transported within the rapid phase of axonal transport predominantly to axon terminals and without significant transneuronal transport (Cowan et al., 1972; Grafstein, 1971). As with the lesion effects on norepinephrine levels, the transport of labelled material takes place largely unilaterally and throughout

Table 2. Locus coeruleus lesion—effect on regional norepinephrine levels
in rat brain

Region analysed	Norepinephrine content (μg/g $\pm$ s.e.)		Percentage difference	P
	Control side	Lesion side		
Telencephalon	0·200 $\pm$ 0·017	0·007 $\pm$ 0·001	−97	< 0·001
Thalamus	0·071 $\pm$ 0·063	0·562 $\pm$ 0·072	−2	N.S.
Hypothalamus	2·415 $\pm$ 0·382	2·140 $\pm$ 0·396	−11	N.S.

Unilateral locus coeruleus lesions were produced using a radiofrequency lesion maker. Animals were sacrificed 60 days later and brain samples were analysed for norepinephrine content by a modification of the method of Anton and Sayre (1962). P values were obtained using a two-tailed t-test for differences; N.S. refers to differences with P > 0·05.

TABLE 3. RAPID AXONAL TRANSPORT OF LABELLED PROTEIN TO TELENCEPHALON FOLLOWING
LOCUS COERULEUS INJECTION OF LABELLED AMINO ACID

Sample counted	Telencephalic region (dis/min per/mg tissue ± s.e.)							
	Olfactory bulb	Septum	Amygdala	Hippo-campus	Caudate-putamen	Frontal cortex	Temporal cortex	Occipital cortex
Injected side	91 ± 14	411 ± 43	212 ± 43	121 ± 24	170 ± 22	134 ± 28	82 ± 5	102 ± 5
Control side	48 ± 4	185 ± 23	82 ± 20	61 ± 8	66 ± 11	50 ± 5	45 ± 4	62 ± 12

Unilateral injections of tritated leucine (25 μCi in 1 μl saline were placed stereotaxically in the locus coeruleus. Animals were sacrificed 48 hr later, brains dissected and telencephalic regions counted. Site of injection was verified autoradiographically. Unilateral injections of an identical amount of tritiated leucine into the restiform body gave a homogeneous background bilaterally in all telencephalic structures with a mean of approximately 50 dis/min per/mg tissue.

the telencephalon. These data, then, are all in support of the view that norepinephrine in telencephalic structures is exclusively located in axons and terminals arising from neuronal cell bodies of the ipsilateral locus coeruleus.

REFERENCES

ANDEN N.-E., DAHLSTRÖM A., FUXE K., LARSSON K., OLSEN L. and UNGERSTEDT U. (1966) *Acta physiol. scand.* **67**, 313–326.
COWAN W. M., GOTTLIEB D. I., HENDRICKSON A. E., PRICE J. L. and WOOLSEY T. A. (1972) *Brain Res.* **37**, 21–51.
FUXE K., HÖKFELT T. and UNGERSTEDT U. (1970) *Int. Rev. Neurobiol.* **13**, 93–126.
GRAFTSTEIN B. (1071) *Science* **172**, 177–179.
HELLER A., HARVEY J. A. and MOORE R. Y. (1962) *Biochem. Pharmacol.* **11**, 859–866.
HELLER A. and MOORE R. Y. (1965) *J. Pharm. Exp. Therap.* **150**, 1–9.
HELLER A and MOORE R. Y. (1968) *Adv. Pharm.* **6A**, 191–206.
MOORE R. Y. (1970) *Int. Rev. Neurobiol.* **13**, 67–92
MOORE R. Y. and HELLER A. (1967) *J. Pharm. Exp. Therap.* **156**, 12–22.
MOORE R. Y. and LENN N. J. (1972) *J. Comp. Neurol.* **146**, 1–14
UNGERSTEDT U. (1971) *Acta. physiol. scand. Suppl.* **367**, 1–122.
ZIGMOND M. J. and WURTMAN R. J. (1966) *J. Pharm. Exp. Therap.* **172**, 416–422.

Frontiers in Catecholamine Research 1973, pp. 771 to 772. Pergamon Press. Printed in Great Britain.

BIOCHEMICAL EVIDENCES FOR INTERACTION PHENOMENA BETWEEN NORADRENERGIC AND SEROTONINERGIC SYSTEMS IN THE CAT BRAIN

J. F. PUJOL, D. STEIN, CH. BLONDAUX, F. PETITJEAN,
J. L. FROMENT and M. JOUVET

Departement de Medicine Experimentale, Université Claude Bernard, Lyon, France

RECENTLY, functional interactions between monoaminergic systems were suggested to explain some aspects of the sleep-waking cycle regulation (JOUVET, 1971; JOUVET and PUJOL, 1972). Indirect evidence for a noradrenergic control of serotonin (5-HT) synthesis was previously obtained in the cat PETITJEAN and JOUVET, 1970 and in the rat (BLONDAUX et al., 1973; JOHNSON et al.,1972; JUGE et al., 1972). In this paper we describe two experimental models of reciprocal biochemical mechanisms of regulation between some noradrenergic and serotoninergic neurons.

In a first experimental situation an inactivation of dorsal noradrenergic pathways was obtained by bilateral coagulation of the initial part of the bundle at the level of the ponto-mesencephalic junction. 48-Hours later, a global study of the rate of synthesis of 5-HT was performed in different brain areas by following the initial accumulation of ^{3}H-5HT and ^{3}H-5-hydroxy indole acetic acid (5-HIAA) endogenously synthetised from exogenous ^{3}H-Tryptophan (Trp) (1 Ci/mM The Radiochemical Centre, Amersham) injected intravenously (1·5 mCi) 20 min before sacrifice. An important and significant increase of the *in vivo* conversion index of Trp was observed in the raphe region as well as in the cortex (Fig. 1) and in other areas of projection of the serotoninergic system (e.g. in the thalamus, mesencephalon or medulla oblongata).

A reciprocal interaction between the two aminergic systems was found by studying the effect of inactivation of the anterior part of the raphe system upon the rate of disappearance of Norepinephrine (NE) after blockade of its synthesis by alpha-methyl-*p*-Tyrosine (AMPT 2000 mg/kg i.p. 8 hr before the sacrifice). 48 Hours after coagulation of the raphe dorsalis and the raphe centralis nuclei the following observations were made:

(1) The destruction of the anterior part of the raphe system did not change the NE concentrations measured in the different brain structures examined 48 hr after the lesion.

(2) At this time, 8 hr after administration of AMPT, the NE concentrations measured in the cortex and cerebellum (Fig. 1) were significantly lower as compared to control animals. The rate of disappearance of NE was also higher in the mesencephalon but not in the other areas of the brain.

These results show that the surgical inactivation of the anterior raphe system induces a significant increase of NE turnover at the level of dorsal noradrenergic bundle terminals. Our observation of activation of both cortical and cerebellar terminals agrees with the increase in green fluorescence observed by Olson and

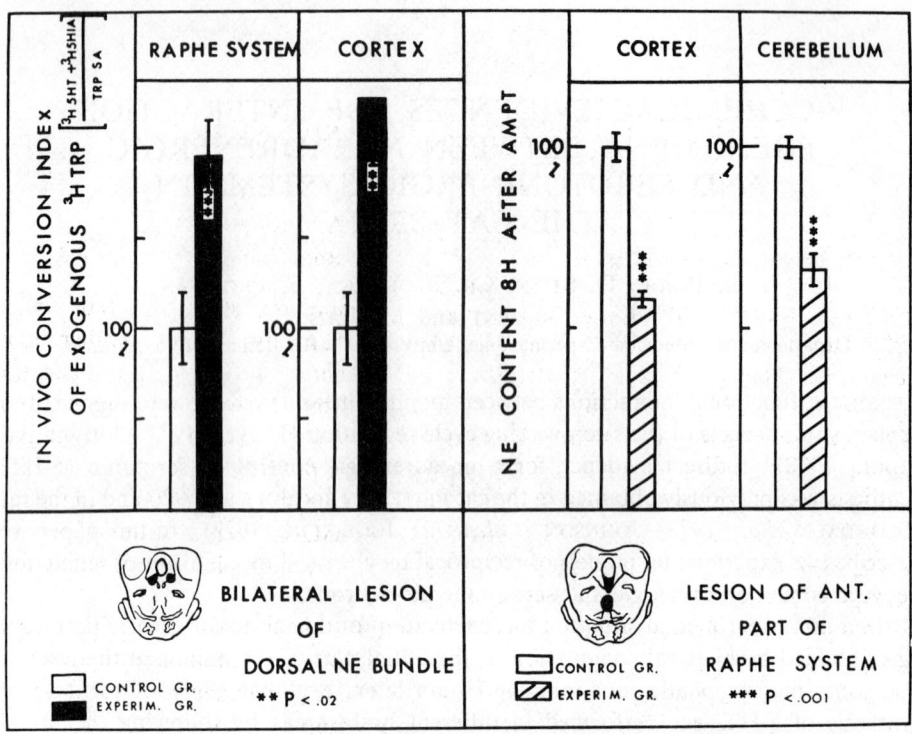

Fig. 1.—Biochemical interaction between serotonergic and dorsal ascending noradrenergic systems. In the left part of the figure results obtained 48 hr after bilateral coagulation of the dorsal noradrenergic bundles are represented The rate of synthesis of 5-HT was estimated by following the initial accumulation of ^{3}H-5HT and ^{3}H-5HIAA endogenously synthetised from ^{3}H-Trp and measuring the conversion index of Trp by the ratio:

$$\frac{^3\text{H-5HT} + \,^3\text{H-5HIAA}}{\text{Trp S. A.}}$$

Results for the experimental group (black bars) are expressed as a percentage of the mean value for the control group (white bars) $\pm$ s.e.m. (5 determinations). P-values were calculated by a "t" test. The right part of the figure shows NE concentrations in the cortex and cerebellum of control cats (white bars) and raphe destroyed cats (hatched bars) 8 hr after blockade of NE synthesis by AMPT (200 mg/kg i.p.). Results are expressed as described above.

Fuxe in the cerebellum after lesion of dorsal noradrenergic pathway in the rat (Olson and Fuxe, 1971).

These facts strongly suggest the existence of interaction phenomena between the 5HT-containing neurons of the raphe system and the dorsal noradrenergic neurons. Anatomical demonstrations of these interconnections as well as their biochemical mechanisms remain to be elucidated.

REFERENCES

Blondaux C., Juge A., Sordet F., Chouvet G., Jouvet M. and Pujol J. F. (1973), Brain Res. 50, 101–114.
Johnson G. A., Kim E. G. and Boukma S. S. (1972) J. Pharmocal. Exp. Ther. 180, 539–546.
Jouvet M. (1971) Ergebn. Physiol. 64, 166–307.
Jouvet M. and Pujol J. F. (1972) Rev. Neurol. 127, 115–138.
Juge A., Sordet F., Jouvet M., and Pujol J. F. (1972) C.R. Acad. Sci. 274, 3266–3268.
Olson L. and Fuxe K. Brain Res. 28, 165–171.
Petitjean F. and Jouvet M. (1970) C.R. Soc. Biol. 164, 2288–2293.

Frontiers in Catecholamine Research 1973, pp. 773 to 775. Pergamon Press. Printed in Great Britain.

NE-NEURONS AND PHASIC SLEEP PHENOMENA

W. Haefely, M. Jalfre and M.-A. Monachon

Department of Experimental Medicine, F. Hoffman-La Roche Ltd., Basle, Switzerland

Ponto-geniculo-occipital (PGO) waves are monophasic slow potentials which originate in the dorsal pontine reticular formation and also occur in other brain structures. They are a characteristic phasic phenomenon of REM sleep but can also be induced by drugs. The involvement of cerebral monoamines in the generation of PGO waves has been postulated by Delorme et al. (1965). While it is now widely accepted that a serotoninergic raphe system exerts a physiological inhibitory effect on the elaboration of PGO-waves (Jouvet, 1972; Jalfre et al., 1973), the possible participation of catecholamines in the regulation of PGO-waves is a matter of dispute. We present results providing strong evidence for a modulating inhibitory action of a NE-neuron system in the cat.

The experimental set-up used (Jalfre et al., 1970, 1973; Monachon et al., 1973) is briefly as follows: PGO-waves are recorded and counted in both lateral geniculate bodies in unanaesthetised curarised cats. The animals are either given the benzo-quinolizine Ro 4-1284 (20 mg/kg i.p.) 1 hr after completion of the surgery or are pretreated with p-chlorophenylalanine (PCPA) in doses of 300 mg/kg i.p. 3 and 2 days before the acute experiment. The aim is to partially deplete either preferentially brain serotonin (Keller, 1972) or both serotonin and NE. Partial depletion of one or both monoamines releases PGO-generator cells from tonic inhibition. The density of PGO-waves thus produced can be further augmented or diminished by drugs or procedures decreasing or increasing, respectively the activity of the monoamines at the postsynaptic membrane.

The most relevant findings pointing to a modulating influence of NE on PGO-waves are the following: *Stimulation of adrenergic receptors* by small doses of cloni-dine i.v. decreased the PGO-waves in a dose-dependent way; they were virtually abolished after 0·3 mg/kg. High doses of NE injected into a lateral ventricle had the same effect. DA by the same route had less than 1/10 the potency of NE. Dex-amphetamine (10 mg/kg i.v.) almost abolished PGO-waves. *Inhibitors of the re-uptake of NE*, e.g. desipramine, reduced the density of PGO-waves probably by in-creasing the concentration of NE in synapses. These drugs were, however, less potent than the tertiary tricyclic amines known to inhibit also the uptake of serotonin. *Blockade of adrenergic receptors* by phenoxybenzamine markedly increased the PCPA-induced PGO-waves: a maximum effect was obtained after a cumulative dose of 4 mg/kg i.v. Thioridazine (3 mg/kg i.v.) which has been shown to increase the turn-over of cerebral NE probably by blocking NE receptors (Keller et al., 1973), significantly increased the number of PGO-waves. The β-adrenergic blocker pro-nethalol was inactive. *Inhibitors of NE synthesis* such as α-methyltyrosine or disul-firam, when given in addition to PCPA produced a higher density of PGO-waves than PCPA alone. Acute bilateral electrolytic *lesion of the locus coeruleus* increased PGO-waves in PCPA-treated cats. *Benzodiazepines* (chlordiazepoxide, flurazepam, flunitrazepam) in small doses increased the PGO-waves in PCPA-treated animals but had only a border-line effect on Ro 4-1284-induced PGO-waves. The effect of the benzodiazepines was prevented by prior bilateral lesions either in the amygdala, the septum or the medial forebrain bundle and by atropine (0·4 mg/kg i.v. in cumulative doses).

Most agents studied have multiple effects on the CNS and alter the degree of alertness. Furthermore, it seems unlikely that the PGO-waves depend exclusively upon the activity of monoamine neuron systems. Hence, no single isolated observation reported above can be accepted as evidence in favour of the role of NE in PGO-wave regulation. However, when all findings are taken together, the conclusion appears inevitable that a NE neuron system exerts a tonic inhibitory influence on PGO-waves. Probably this system originates in the locus coeruleus. Figure 1 provides a hypothetical neuronal model for the generation and modulation of PGO-waves.

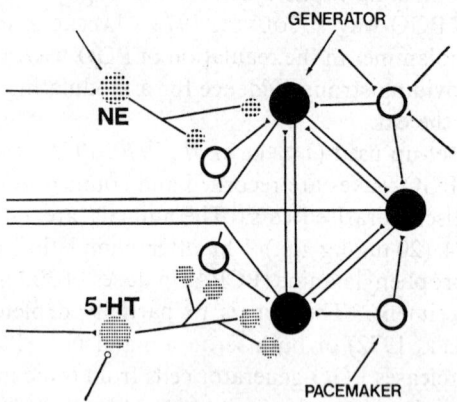

Fig. 1.—Hypothetical neuronal scheme for the generation and regulation of PGO-waves.

(a) Diffusely scattered pacemaker cells with spontaneous asynchronous activity and hypothetical excitatory inputs from different structures impinge on generator cells. (b) Generator cells for PGO-waves, located in the dorsal pontine reticular formation are not spontaneously active; when the input from the pacemaker cells is strong enough, they start discharging synchronously because of reciprocal connections. (c) Serotoninergic raphe neurons tonically damp the generator cells; the control of the former cells is unknown. (d) The postulated NE neurons (from the locus coeruleus) exert an additional inhibitory effect on generator and/or pacemaker cells. A descending inhibitory pathway projecting onto the locus coeruleus neurons is suggested to be activated by benzodiazepines and to contain a muscarinic link.

REFERENCES

Delorme F., Froment J.L. and Jouvet M. (1966) *C.r. Soc. Biol.* **160**, 2347–2351.

Jalfre M., Monachon M.-A. and Haefely W. (1970) *Pharmacological modifications of benzo-quinolizin-induced geniculate spikes. Experientia* **26**, 691.

Jalfre M., Monachon M.-A. and Haefely W. (1973) *Drugs and PGO-waves in the cat.* McClure D. J., (Ed.): *Proc. First Canadian Int. Symposium on Sleep, Montreal*, April 1972, Roche Scientific Service, Montreal, pp. 155–184.

Jouvet M. (1972) The role of monoamines and acetylcholine-containing neurons in the regulation of the sleep-waking cycle. *Rev. Physiol.* **64**, 166–307, (Springer, Berlin).

KELLER H. H. (1972) *Experientia* **28**, 177.
KELLER H. H., BARTHOLINI G. and PLETSCHER A. (1973) *Experientia* **29**, 26.
MONACHON M-A., JALFRE M. and HAEFELY W. (1973) A modulating effect of chlordiazepoxide on drug-induced PGO-spikes in the cat. In: GARATTINI S., MUSSINI E., RANDALL L. O. (eds.): *The Benzodiazepines*. Raven Press, New York, p. 513–529.

Frontiers in Catecholamine Research 1973, pp. 777 to 779. Pergamon Press. Printed in Great Britain.

DOPAMINE AND ACETYLCHOLINE NEURONES IN STRIATAL AND LIMBIC STRUCTURES: EFFECT OF NEUROLEPTIC DRUGS

K.G. LLOYD, H. STADLER and G. BARTHOLINI

Department of Experimental Medicine, F. Hoffmann-La Roche & Co. Ltd.,
Basle, Switzerland

INVOLVEMENT of dopamine (DA) in schizophrenic states is suggested by several observations (METTLER and CRANDELL, 1959; ROSSUM et al., 1970 GOODWIN, 1971; SNYDER, 1973) but mainly by the fact that neuroleptic drugs block DA receptors whereas structurally related compounds devoid of such an effect do not show therapeutic activity (NYBÄCK, 1968). DA terminals occur mainly in the neostriatum, thus the mechanism of action of neuroleptics may be connected with this structure. Neuroleptic blockade of striatal DA receptors leads to parkinsonian symptoms (HORNYKIEWICZ, 1972), not all of which, however, are directly due to impaired DA transmission, but rather to a preponderance of a striatal cholinergic system (BARTHOLINI et al., 1973). Thus, it has been shown that neuroleptic drugs enhance the acetylcholine (ACh) liberation from the cat caudate nucleus (STADLER et al., 1973), an effect which is reversed or prevented by dopaminergic agents (BARTHOLINI et al., this volume). This is the first direct evidence that a striatal cholinergic mechanism— possibly involved in parkinsonism—is regulated by an inhibitory dopaminergic input. Amelioration of some parkinsonian symptoms by anticholinergic drugs and their exacerbation by physostigmine (DUVOISIN, 1967) supports this view. Opposite regulation of the DA system by ACh neurons is likely since anticholinergics counteract the neuroleptic-induced enhancement of DA turnover (ANDÉN, 1972) indicating an interregulation of the two systems. The question arising is whether or not the activation of striatal cholinergic neurons mediates also the antipsychotic action of neuroleptics. However, anticholinergic compounds ameliorate the parkinsonism caused by neuroleptics without affecting their therapeutic activity. In this respect, therefore, striatal cholinergic activation may not be involved.

A possible extrastriatal site of action of neuroleptic compounds is the limbic system, which is also rich in DA and ACh neurons (FUXE, 1965; HERNÁNDEZ-PEÓN et al., 1963). As for the effect of neuroleptics, however, a fundamental difference exists between striatal and limbic structures. Thus, chlorpromazine or haloperidol (Fig. 1) failed to enhance the liberation of ACh collected by means of a push–pull cannula from the septum or nucleus accumbens of the gallamine-immobilized cat. This is in contrast to their marked effect on the caudate nucleus (BARTHOLINI et al., this volume; STADLER et al., 1973) although the DA turnover in both regions was markedly accelerated, as estimated by the rise of homovanillic acid (Table 1). These findings indicate that in septum and nucleus accumbens of the cat the activation of cholinergic neurons is not regulated by a DA input. Conversely, the lack of a cholinergic regulation on limbic DA neurons is known since anticholinergic agents do not diminish the neuroleptic-induced enhancement of limbic DA turnover (ANDÉN, 1972).

In conclusion, some extrapyramidal effects of neuroleptic drugs depend on the

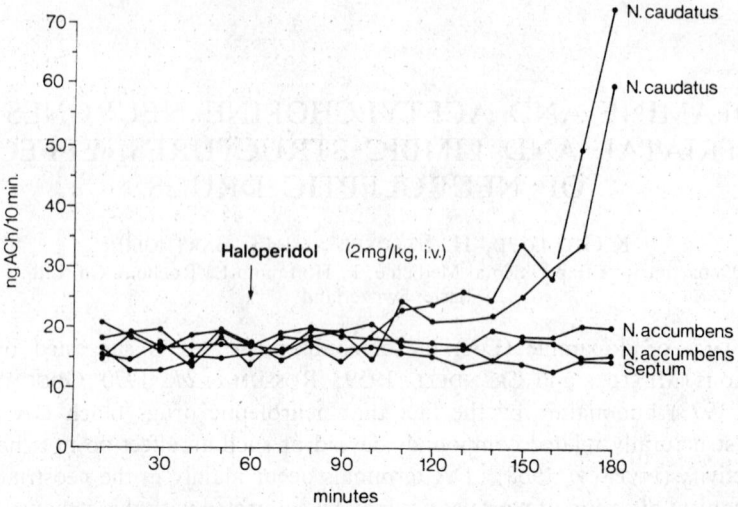

FIG. 1.—Output of acetylcholine (ACh) into 10-min samples of perfusate of various brain regions of the cat during the control period (0–60) and after haloperidol. Each curve indicates one experiment. The points represent averages of duplicate determinations. Time 0 indicates the beginning of collection.

TABLE 1. CONCENTRATION OF DOPAMINE (DA) AND HOMOVANILLIC ACID (HVA) IN DIFFERENT BRAIN REGIONS OF THE CAT. Caudate nucleus and nucleus accumbens were perfused by means of push-pull cannulae as in the experiments for acetylcholine (see Fig. 1). Saline (control animals) or haloperidol (2 mg/kg) was injected intravenously 2 hr before sacrifice. Number of determinations in parentheses. $P < 0.01$ vs controls.

Brain region		DA		HVA	
		$(\mu g/g)$	(% control)	$(\mu g/g)$	(% control)
Caudate n.	controls	11·41 ± 0·58 (9)		4·08 ± 0·26 (9)	
	haloperidol	9·04 ± 0·30 (12)	79·2 ± 2·6*	6·13 ± 0·51 (12)	150·2 ± 12·5*
N. accumbens Septum Preoptic area	controls	3·42 ± 0·28 (9)		2·15 ± 0·17 (9)	
	haloperidol	3·41 ± 0·22 (12)	99·7 ± 6·4	3·07 ± 0·24 (12)	142·8 ± 11·1*

activation of a striatal cholinergic mechanism due to impairment of DA transmission. The antipsychotic effect of these drugs may not be connected with cholinergic neurons in either striatum or limbic system; however, a limbic DA involvement is conceivable.

REFERENCES

ANDÉN N. E. (1972) J. Pharm. Pharmac. 24, 905–906.
BARTHOLINI G., STADLER H. and LLOYD K. G. (1973) In: The Treatment of Parkinsonism (CALNE D. B., Ed.) Raven Press, New York.
DUVOISIN R. G. (1967) Archs Neurol. 17, 124–126.
FUXE K. (1965) Acta Physiol. Scand. 64, (Suppl.) 247.

Goodwin F. K. (1971) *J. Am. Med. Assc.* **218**, 1915–1920.

Hernández–Peón R., O'Flaherty J. J. and Mazzuchelli–O'Flaherty A. L. (1967) *Brain Res.* **4**, 243–267.

Hornykiewicz O. (1972) In: *Handbook of Neurochemistry* (Lajtha A., Ed.) Vol. **6**, pp. 465–501, Plenum Press, New York.

Mettler F. A. and Crandell A. (1959) *J. Nerv. Ment. Dis.* **129**, 551–563.

Nybäck H. (1968) *Europ. J. Pharmacol.* **14**, 395–403.

Rossum J. M. Van, Boissier J. R., Janssen P. A. S., Julou L., Loew D. M., Nielsen I. M., Munkvad I., Randrup A., Stille G. and Tedeschi D. H. (1970) In: *Modern Problems in Pharmacopsychiatry* (*Neuroleptics*) (Bobon J. and Janssen P., Eds), pp. 23–33, Karger, Basel.

Snyder S. H. (1973) *Am. J. Psychiat.* **130**, 61–67.

Stadler H., Lloyd K. G., Gadea–Ciria M. and Bartholini G. (1973) *Brain Res.*, **55**, 476–480.

Frontiers in Catecholamine Research 1973, pp. 781 to 785. Pergamon Press. Printed in Great Britain.

ROLE OF CATECHOLAMINES IN NEUROENDOCRINE FUNCTION

R. J. WURTMAN

Laboratory of Neuroendocrine Regulation, Massachusetts Institute of Technology,
Cambridge, Massachusetts 02139, U.S.A.

THE BRAIN AND THE REGULATION OF THE EXTRACELLULAR FLUID*

THE CELLS of mammals are bathed in a medium, the extracellular fluid, whose temperature and chemical composition are kept within a constant range by regulatory systems. Each such system is composed of multiple organs, usually including the brain. The extracellular fluid is divided into two portions which are in equilibrium: a smaller intravascular portion known as serum, or plasma, and a larger portion that surrounds the cells. When the level of any regulated function in the intravascular portion falls too low (or, in the case of temperature, osmolarity, or calcium concentration, rises too high), this is sensed by specialised cells within or outside the brain, and mechanisms are activated that either add more of the regulated compound to the circulation or decrease the rate at which it is being removed from the blood. The only regulated compound whose plasma concentration does not involve the brain is calcium. Three aspects of the extracellular fluid are regulated: (a) the volume, pressure and temperature of its intravascular compartment: (b) its concentrations of certain essential compounds, such as glucose, water, various ions; and, (c) its concentrations of three hormones, i.e., thyroxine, hydrocortisone (or corticosterone, in the rat) and estradiol or testosterone. These compounds differ in several important ways from the other large class of hormones, the spurt hormones, whose plasma levels are not independently regulated. The regulated hormones are lipid-soluble, and travel in plasma bound to specific globulin proteins; they act to coordinate changes in most cells, including the brain itself, and not in only one or two target organs; their secretion is controlled by hormones released by the anterior pituitary gland.

One result of the brain's involvement in regulation of the extracellular fluid is that the designated concentration (or set-point) of the regulated compound (or the intravascular volume, pressure, or temperature) can be increased or decreased as a function of time or age, or in response to stress or a particular external environment. For example, the apparent set-points for plasma hydrocortisone and body temperature exhibit characteristic daily rhythms; the concentration for plasma hydrocortisone that the system maintains is increased by certain kinds of stress; the set-point for plasma testosterone is very low prior to puberty and rises thereafter (indeed, this change in the sensitivity of the brain may well be the mechanism that *causes* puberty).

---

* Some of the concepts presented in the following discussions have been described in greater detail elsewhere (Wurtman, 1970a, 1970b, 1971, 1973; ANTON-TAY and WURTMAN, 1971).

NEUROENDOCRINE TRANSDUCERS

The brain has at its disposal three types of output channels for increasing or decreasing the levels of regulated components of the extracellular fluid: The purely neuronal channels of the sympathetic and parasympathetic nervous systems, behavioural channels (for example, hunger in response to hyperglycemia; thirst in response to hyperosmolarity), and neuroendocrine channels. The latter all utilize a special kind of cell, termed the neuroendocrine transducer. These cells differ from both neurons and true endocrine cells in that they convert a neuronal-type input (that is, a neurotransmitter, received at a synapse) to a humoral or endocrine output (a hormone released into the circulation). At least five sets of neuroendocrine transducers have been identified. Two reside within the brain; they are: (1) the cells of the supraoptic and paraventricular nuclei, which respond to such neuronal inputs as pain and suckling by causing the release of vasopressin and oxytocin; and (2) the releasing-factor cells, presumably located within the median eminence of the hypothalamus, which secrete hormones into the hypothalamo-hypophyseal portal vascular apparatus which either cause or inhibit the secretion of anterior pituitary hormones. Neuroendocrine transducers residing outside the brain include: (a) the pineal organ, which synthesises melatonin in response to norepinephrine released from sympathetic nerve terminals; (b) the adrenal medulla, which secretes epinephrine in response to a preganglionic cholinergic input; and (c) the juxtaglomerular cells of the kidney, which, like the pineal, are stimulated by norepinephrine released from sympathetic nerve endings, and respond by releasing renin into the circulation. (Recent evidence suggests that the beta cells of the pancreas receive a sympathetic innervation that physiologically suppresses their release of insulin; this suggests that these cells may constitute a sixth neuroendocrine transducer.) The functional activity of all of the neuroendocrine transducers seems to depend on humoral, as well as neural inputs; thus, the rate at which vasopressin is secreted varies with plasma osmolarity, probably sensed within the hypothalamus: corticotropin-releasing-factor secretion depends on the concentration of plasma hydrocortisone, sensed within the median eminence; epinephrine synthesis depends upon the induction of an adreno-medullary enzyme, phenyl-ethanolamine-N-methyltransferase (PNMT), by the very high concentrations of glucocorticoids delivered to the medulla by the intra-adrenal portal circulation. The uniqueness of these cells, however, resides in their capacity to convert synaptic inputs to humoral outputs. Whenever the brain can be shown to influence directly the concentration of a compound within the extracellular fluid, there is good reason to postulate the existence of an intermediary neuroendocrine transducer cell.

All of the neuroendocrine transducers identified thus far utilise a catecholamine as one of their input or output signals. Thus, norepinephrine causes the pineal to synthesise melatonin, the juxtaglomerular cells to secrete renin, and suppresses insulin secretion from the pancreas; the adrenal medulla secretes the catecholamine epinephrine as its hormone. The precise roles of catecholamines in the functional activities of the neuroendocrine transducers within the hypothalamus (the supraoptic and paraventricular nuclei, and the median eminence hypophysiotropic cells) have been very difficult to characterise, inasmuch as the presynaptic inputs to these cells cannot easily be isolated. However, considerable evidence, described below, suggests that catecholamines participate in controlling the secretion of releasing factors and, thereby, the hormones of the anterior pituitary, thyroid, adrenal cortex and gonads.

INVOLVEMENT OF BRAIN CATECHOLAMINES IN CONTROL OF ANTERIOR PITUITARY FUNCTION

The roles of catecholamine-containing brain neurons in the control of pituitary function have been studied experimentally in several ways: (1) Animals or human subjects have received drugs that are presumed to act selectively at catecholaminergic synapses, and changes in endocrine function have been measured. For example, L-dopa administration has been shown to stimulate the secretion of growth hormone, and suppress that of prolactin. Even assuming that a particular drug did act selectively on brain synapses utilising dopamine or norepinephrine (an assumption that clearly is unwarranted in the case of L-dopa), its specific site of action in producing its neuroendocrine effect would still be difficult to identify, inasmuch as it could act on monoaminergic neurons at some distance from the releasing-factor cells (i.e., within the medial forebrain bundles or on peripheral sympathetic synapses) as well as on the releasing-factor cells themselves. Studies on the endocrine effects of drugs presumed to act on brain catecholamines are discussed in other papers in this Symposium. (2) Animals have been given hormones, or glands have been removed surgically, and changes in the synthesis, turnover, or levels of brain mono-amines have been measured. For example, as described below, ovariectomy has been shown to accelerate the synthesis and turnover of catecholamines in rat brain. (3) Changes in endocrine function and in the synthesis or turnover of brain catecholamines have been studied in parallel in unoperated animals (e.g., during the various phases of the vaginal estrous cycle). This "natural history" approach can be used to determine whether, for example, major changes in brain catecholamine synthesis normally *precede* puberty or coincide with the critical period that occurs on the day of proestrus.

BRAIN CATECHOLAMINES AND GONADAL FUNCTION IN RATS

The rates at which the brain of the female rat synthesises and turns over cate-cholamines are influenced by the secretion of pituitary gonadotropins and gonadal steroid hormones. Ovariectomy has been shown to elevate the norepinephrine content (DONOSO et al., 1967) and tyrosine hydroxylase activity (BEATTIE et al., 1972) of the anterior hypothalamus, and to accelerate the accumulation of brain ³H-catechols following systemic administration of ³H-tyrosine (ANTON-TAY et al., 1970); all of these effects can be suppressed by treating the ovariectomised animals with estradiol, or with estradiol plus progesterone (BEATTIE et al., 1972; DONOSO and STEFANO, 1967; BAPNA et al., 1971). Ovariectomy also accelerates the turnover of ³H-norepinephrine taken up into the brain from the cerebrospinal fluid (ANTON-TAY and WURTMAN, 1968; ANTON-TAY et al., 1969). Since this acceleration is not observed in hypophysectomised animals, but can be reproduced in intact or hy-pophysectomised-ovariectomised rats by treating them with follicle-stimulating hormone (FSH), it has been suggested that FSH may affect the catecholamine-containing neurons directly, as well as by controlling the rates of secretion of gonadal steroids. (See Fig. 1.)

The rate of brain ³H-catechol synthesis during the afternoon of proestrus (i.e., during the critical period), measured in untreated female rats killed 10 min after

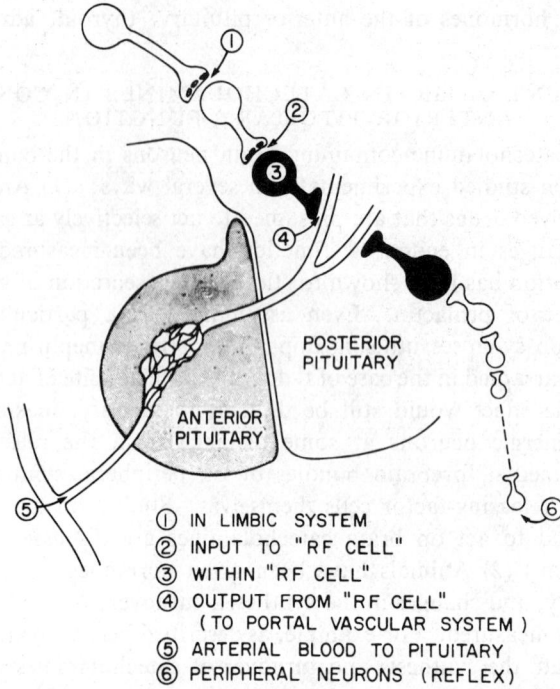

① IN LIMBIC SYSTEM
② INPUT TO "RF CELL"
③ WITHIN "RF CELL"
④ OUTPUT FROM "RF CELL"
   (TO PORTAL VASCULAR SYSTEM)
⑤ ARTERIAL BLOOD TO PITUITARY
⑥ PERIPHERAL NEURONS (REFLEX)

FIG. 1.—Some possible loci at which drugs might act via catecholaminergic neurons to affect the secretion of pituitary hormones. Releasing-factor (RF) cells, shown in dark, presumably secrete hormones into the hypothalamo-hypophyseal portal circulation, which preferentially delivers them to the anterior pituitary. The anterior pituitary also receives arterial blood from the systemic circulation. (1) The drug could act at a catecholaminergic synapse within the limbic system (e.g., in the medial forebrain bundle). (2) It could modify the release, reuptake, or receptor interactions of norepinephrine or dopamine liberated by neurons synapsing with RF cells. (3) It could act within the RF cells (i.e., it is possible that these cells contain both monoamines and nonmonoaminergic releasing factors, and that the former compounds affect the secretion of the latter. (4) It could cause or suppress the release of a catecholamine from the RF cells (i.e., the identity of one or more of the releasing factors might be norepinephrine or dopamine). (5) It could increase the amounts of norepinephrine or epinephrine released into the general circulation from sympathetic neurons or the adrenal medulla, and thereby increase the concentration reaching the pituitary via its arterial blood supply. (6) It could act peripherally to alter the level of one of the regulated aspects of the extracellular fluid (e.g., blood pressure; blood glucose), leading to reflex changes in central catecholaminergic function and pituitary secretion. (Figure is redrawn from Fig. 12.1 in "Brain catecholamines and the control of secretion from the anterior pituitary gland" by R. J. WURTMAN, in *Hypophysiotropic Hormones of the Hypothalamus*, J. MEITES, ed., Williams & Wilkins, Baltimore, 1970, p. 186.

receiving a tracer dose of ^{3}H-tyrosine, is almost four times as great as during diestrus and twice that during estrus (ZSCHAECK *et al.*, 1973). The location of the specific groups of catecholaminergic Neurons affected by the estrous cycle awaits identification, as does their neurotransmitter (dopamine or norepinephrine, both of which are found in the hypothalamus). Presumably, these cyclic changes in brain catecholamine synthesis could be part of the mechanism that causes the estrous cycle, or could be a consequence of the changes in concentrations of circulating hormones. They might underlie some of the behavioral and autonomic correlates of cyclic ovarian function.

REFERENCES

ANTON-TAY F., ANTON S. and WURTMAN R. J. (1970) *Neuroendocrinology* **6**, 265–273.

ANTON-TAY F., PELHAM R. W. and WURTMAN R. J. (1969) *Endocrinology* **84**, 1489–1492.

ANTON-TAY F. and WURTMAN R. J. (1968) *Science* **159**, 1245.

ANTON-TAY F. and WURTMAN R. J. (1971) Brain monoamines and endocrine function. In *Frontiers in Neuroendocrinology*. (MARTINI L. and GANONG W. F., eds.) pp. 45–66. Oxford University Press, New York.

BAPNA J., NEFF N. H. and COSTA E. A. (1971) *Endocrinology* **89**, 1345–1349.

BEATTIE C. W., RODGERS C. H. and SOYKA L. F. (1972) *Fedn. Proc.* **31**, 211.

DONOSO A. O. and STEFANO F. J. E. (1967) *Experientia* **23**, 655–666.

DONOSO A. O., STEFANO F. J. E., BISCARDI A. M. and CUKIER J. (1967) *Am. J. Physiol.* **212**, 737–739.

WURTMAN R. J. (1970a) Brain catecholamines and the control of secretion from the anterior pituitary gland. In *Hypophysiotropic Hormones of the Hypothalamus: Assay and Chemistry*. (MEITES J., Ed.) pp. 184–194. Williams & Wilkins, Baltimore.

WURTMAN R. J. (1970b) Neuroendocrine transducer cells in mammals. In *The Neurosciences: Second Study Program*. (SCHMITT F. O., Ed.-in-Chief) pp. 530–538. Rockefeller University Press, New York.

WURTMAN R. J. (1971) Brain monoamines in endocrine function. *Neurosciences Research Program Bulletin* **9** (2), 172–297.

WURTMAN R. J. (1973) Biogenic Amines and Endocrine Function. Introduction: Neuroendocrine Transducers and Monoamines. *Fed. Proc.* In press.

ZSCHAECK L. L. and WURTMAN R. J. (1973) *Neuroendocrinology* **11**, 144–149.

Frontiers in Catecholamine Research 1973, pp. 787 to 794. Pergamon Press. Printed in Great Britain.

RECENT MORPHOLOGICAL AND FUNCTIONAL STUDIES ON HYPOTHALAMIC DOPAMINERGIC AND NORADRENERGIC MECHANISMS

KJELL FUXE, TOMAS HÖKFELT, GÖSTA JONSSON and ANDERS LÖFSTRÖM

Department of Histology, Karolinska Institutet, S-104 01 Stockholm 60, Sweden

DURING recent years there has been an ever increasing interest in the role that hypothalamic catecholamines (CA) play in regulating anterior pituitary function. We have been particularly concerned with the functional role of the tubero-infundibular dopamine (TIDC) neurons (FUXE et al., 1967; FUXE and HÖKFELT, 1969; FUXE et al., 1970; HÖKFELT and FUXE, 1972). The evidence so far obtained suggest that the TIDA neurons mainly are involved in the control of gonadotrophin secretion (FUXE et al., 1967; FUXE and HÖKFELT, 1969; HÖKFELT and FUXE, 1972; SCHNEIDER and MCCANN, 1970; KAMBERI et al., 1970). The groups have agreed that the dopamine (DA) in the median eminence may enhance the secretion of prolactin inhibitory factor (PIF) (see also MEITES et al., 1972). Our own work based on turnover studies of the TIDA neurons in various endocrine states and on pharmacological work suggest that the TIDA neurons also could inhibit luteinising hormone releasing factor (LRF) and possibly follicle stimulating hormone releasing factor (FRF) secretion (FUXE et al., 1969a, b; FUXE et al., 1970; HÖKFELT and FUXE, 1972). Studies using intraventricular injections of CA, on the other hand, have suggested that DA could increase the secretion of LRF and FRF (SCHNEIDER and MCCANN 1970; KAMBERI et al., 1970). The present paper will mainly review recent new work from our group that relate to the functional role of the CA neurons.

MORPHOLOGY

Immunohistochemical studies (HÖKFELT et al., 1973)

Dopamine-β-hydroxylase. Using antibodies against dopamine-β-hydroxylase (DβH) it has been possible to discover DβH containing, and thus probably noradrenergic, terminals in certain parts of the lateral external layer of the median eminence. These terminals appeared as a thin dotted zone of specific immunofluorescence immediately adjacent to the primary capillary plexus. Scattered DβH containing terminals were also found in the medial external part which is rich in capillary loops. These latter DβH containing terminals seem to correspond to those CA terminals discovered with the Falck-Hillarp technique, to have a very slow disappearance of fluorescence following tyrosine-hydroxylase inhibition. In agreement with previous pharmacological work (see FUXE and HÖKFELT, 1969; JONSSON et al., 1972) the DβH containing nerve terminals in the subependymal and fibre layer of the median eminence appeared to have a similar distribution and frequency as the CA nerve terminals in these areas.

The immunohistochemical evidence therefore underline the view that the majority of the CA nerve terminals in the subependymal zone and the fibre layer, where they appear as large droplets, contain noradrenaline (NA), whereas the majority of the

CA nerve terminals of the external layer of the median eminence contain DA (see HÖKFELT and FUXE, 1972; JONSSON et al., 1972; BJÖRKLUND et al., 1973). On the other hand, it should be underlined that using DβH immunofluorescence technique, a new NA system has been discovered in the external lateral layer close to the primary capillary plexus. These terminals, although few compared to the DA terminals, could play an important functional role. It has to be considered that these NA terminals can mediate the LRF releasing activity of intraventricularly administered DA (SCHNEIDER and McCANN, 1970) and adrenaline (RUBINSTEIN and SAWYER, 1971).

The lack of positive phenylethanolamine-N-methyl-transferase immunofluorescence in these terminals indicate that they do contain NA and not adrenaline. The findings imply that NA nerve terminals in a certain restricted part of the median eminence can exert an axo-axonic influence on peptidergic nerve terminals containing releasing or inhibitory factors.

Studies on dopa-decarboxylase. The results support the view given above from the experiments with DβH. With our present antibodies and technique, the amounts of dopa-decarboxylase (DDC) in the NA nerve terminals are too low to be demonstrated, and in agreement positive DDC immunofluorescence was only found in the external layer, especially in the lateral part and in cell bodies in the arcuate nucleus.

Hypothalamic islands (JONSSON et al., 1972)

These studies demonstrated that the DA nerve terminals in the external layer mainly derived from DA cell bodies in the arcuate nucleus and the anterior periventricular nucleus, ventral part. Thus, the DA nerve terminals remained unchanged in complete hypothalamic islands, in which all other CA cell body groups except A12 were excluded from the island including groups A11 and A13 and CA cell bodies in the most anterior part of the periventricular hypothalamus. Recent results obtained by BJÖRKLUND et al. (1973) are in agreement with this view. The NA nerve terminals in the internal layer including the subependymal layer probably derive from ascending NA tracts from the pons and the medulla oblongata, since they are not present in hypothalamic islands (JONSSON et al., 1972; see also BJÖRKLUND et al., 1973).

Pharmacological analysis

In vitro studies using the neurotoxic compound 6-hydroxy-dopamine (6-OH-DA) (JONSSON et al., 1972) revealed that a resistance of the CA nerve terminals in the external layer and a disappearance of the CA nerve terminals in the internal layer is correlated with an intact DA synthesis and blockade of the NA synthesis, respectively, which is in good agreement with previous pharmacological work (see HÖKFELT and FUXE, 1972; JONSSON et al., 1972).

Recent microfluorimetric studies with FLA 63 in our laboratory (LÖFSTRÖM, unpublished data) demonstrate a certain reduction of the fluorescence in the lateral part of the external layer but not in the neostriatum suggesting in agreement with the immunohistochemical work, that a minor part of these terminals may contain NA. The depletion was, however, greater in the internal and subependymal zone in agreement with previous work (CORRODI et al., 1970).

It should be underlined that although the dopaminergic mechanism in the median eminence probably represent the major one in neuroendocrine control, there are DA nerve terminals in various limbic structures, which could represent important extrahypothalamic sites for dopaminergic control of hypothalamic function. It is known that e.g. fornix transection strongly depresses pituitary-gonadal activity (KAWAKAMI et al., 1972). This idea is also favoured by the fact that our group has recently been able to confirm the presence of new types of DA nerve terminals (THIERRY et al., 1973) e.g. in the limbic cortex especially in the gyrus cinguli. These studies are based on a special pharmacological model utilizing the occurrence of selective reserpine resistant binding of CA in the DA neurons (LIDBRINK et al., 1973; HÖKFELT et al., 1973) in combination with the Vibratome sectioning technique (HÖKFELT and LJUNGDAHL, 1972; LINDVALL et al., 1973).

Although many findings support the idea of an action of CA mainly at the median eminence level, an action at the pituitary level in addition should not be overlooked. Thus, several workers (MACLEOD et al., 1970; KOCH et al., 1970) have found that CA produce a dose-dependent reduction of prolactin release in vitro from anterior pituitaries. These results have to be taken into account as well when trying to explain neuroendocrine effects of monoaminergic drugs. E.g. it is also known that dopa can be taken up and decarboxylated by PAS positive cells in the anterior pituitary (DAHLSTRÖM and FUXE, 1966; HÅKANSON et al., 1972).

FUNCTION

During a number of years we have studied the turnover changes that occur in the DA nerve terminals in the external layer of the median eminence in various endocrine states and after treatment with hormones and psychoactive drugs. Changes in turnover have been evaluated by studying changes in disappearance of fluorescence after treatment with the tyrosine-hydroxylase inhibitor, α-methyl-tyrosine methylester. The pattern of fluorescence disappearance is shown in Fig. 1. It can be seen that there is good agreement between semiquantitative subjective estimations and the microfluorimetric measurements (LÖFSTRÖM, JONSSON and FUXE, unpublished data). The initial rapid decline should be noted, the degree of which may give an estimation of the turnover in a small functionally very active pool of amines. The partial restoration of fluorescence observed 30 min after H 44/68 (see GLOWINSKI, 1972) may possibly be explained by the fact that there is a continuous transport of amine granules from the cell bodies to the terminals and that the amine containing granules in the fibres are not depleted to any extent by H 44/68 treatment before they have reached the nerve terminals and can be released by the nervous impulse flow. From Fig. 1 it can be calculated that $T_{1/2}$ of the main amine pool is in the order of 135 min. In agreement with previous findings (FUXE et al., 1969a) it can be seen that the turnover of the main amine pool is markedly increased in lactation and the calculated $T_{1/2}$ is in the order of 85 min. Also the DA levels appear to be somewhat reduced in lactation compared with normal male rats. In castrated female rats, on the other hand, $T_{1/2}$ of the main amine pool appears to be similar to that in normal male rats.

These fluorescence disappearance curves also clearly indicate that there seems to be a linear fluorescence–concentration relationship in the DA terminals, in contrast to the situation in other CA neuron systems (see JONSSON, 1971). In the future, it will be of interest to study particularly this initial decline of fluorescence, since the

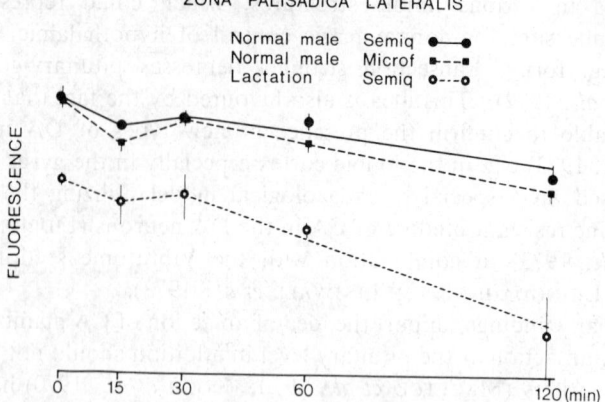

FIG. 1.—The time-curve of fluorescence disappearing after H 44/68 is illustrated in normal male rats and lactating female rats. In normal male rats both micro-fluorori-metric and semiquantitative estimations of fluorescence intensity have been made. Means ± S.E.M. are given.

the small functional active pool may be particularly sensitive to changes in nervous impulse activity (GLOWINSKI, 1972).

In our work (see FUXE and HÖKFELT, 1969; HÖKFELT and FUXE, 1972) we have so far mainly studied the turnover in the main pool in the lateral part of the external layer of the median eminence, and the 2-hr interval after H 44/68 has been found suitable for this purpose (see Fig. 1). The results can be summarised in the following way. In states with low FSH–LH secretion and blockade of ovulation there is an increase in DA turnover in the TIDA neurons, whereas during periods of high LH–FSH secretion and prolactin secretion such as in the "critical period" of adult female rats and of immature rats treated with PMS and in castrated animals, the turnover in the TIDC neurons is low (FUXE et al., 1969a, b; AHRÉN et al., 1971; FUXE et al., 1972). It has also been discovered that estrogen, testosterone, clomiphene, stilbestrol and antifertility steroids of the estrogen type or the nortestosterone type under conditions in which they exert negative inhibitory feedback on gonadotrophin secretion, markedly increase turnover in the TIDA neurons (FUXE et al., 1971a). A similar marked increase in turnover has also been observed after injection of prolactin into hypophysectomised rats. In view of these findings we have postulated that the TIDA neurons may act by inhibiting LRF–FRF secretion and enhancing PIF secretion.

Most of our previous work has mainly involved studies on the lateral part of the external layer. Therefore we have made a study to compare the effects of prolactin on the lateral and medial part of the external layer in hypophysectomised rats. The results are summarised in Fig. 2 (JONSSON, FUXE and LÖFSTRÖM, unpublished data). In the upper part of the figure it can be seen that there is a slight reduction of the fluorescence intensity after hypophysectomy (4 weeks) both in the medial and lateral part of the external layer. In the lower part of the figure the percentage disappearance of fluorescence is given at the 2 hr interval following H 44/68. A dramatic acceleration of fluorescence disappearance is observed after the prolactin injection (5 mg/kg i.v., 24 hr before killing). This dramatic effect occurs in both the

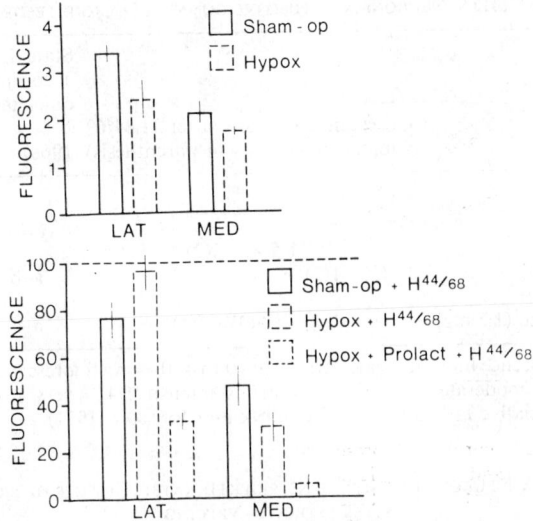

Fig. 2.—Effect of prolactin on the disappearance of fluorescence in hypophysectom-ised rats after H 44/68. Prolactin was given i.v. in a dose of 5 mg/kg 24 hr before killing. Hypophysectomy had been performed 4-weeks earlier. H 44/68 was given i.p. in a dose of 250 mg/kg 2 hr before killing. Microspectrofluorimetric measurements have been performed. The fluorescence in the upper part of the figure is given in fluorescence units and in the lower it is given as percentage of normal and hypophy-sectomised rats respectively. Means ± s.e.m.

medial and lateral part of the external layer. It should be noted, however, that there is a reduction of DA turnover in the lateral part after hypophysectomy, whereas the DA turnover may be increased in the medial part. These interesting observations, however, require further study but points out the importance of looking for possible regional changes in the various DA terminal systems in the median eminence.

Our hypothesis on the role of the TIDA neurons in gonadotrophin control has recently obtained support by studies on the turnover of the TIDA neurons under conditions of positive gonadal steroidal feedback and by pharmacological studies. Thus, it is know that in castrated estrogen primed female rats, progesterone will after certain time interval induce a peak of FSH, LH and prolactin (Caligaris *et al.*, 1971; KALRA *et al.*, 1972). It has now been possible to demonstrate that estrogen alone causes a significant increase of DA turnover in the palisade zone, which is not seen in combination with progesterone (LÖFSTRÖM, FUXE and HÖKFELT, unpublished data; see Table 1). Furthermore, the latter response pattern is not seen in androgen sterilised animals (see Table 2), which are known to have no cyclic gonadotrophin release (Barraclough, 1967). Recent studies with DA receptor stimulating agents, DA precursors and releasing agents such as ET 495, CB 154, dopa and amphetamine, demonstrate that marked increases in DA receptor activity will result in inhibition of ovulation in immature female rats (FUXE *et al.*, 1971b; HÖKFELT and FUXE, 1972; FUXE, HÖKFELT and LÖFSTRÖM, unpublished data). After treatment of normal castrated estrogen primed female rats with progesterone, turnover changes also occur in other monoamine neuron systems. Thus, using H 44/68 in combination with a histochemical and biochemical analysis of NA it has been found that progesterone in-crease NA turnover both in the hypothalamus and in the cortex cerebri (FUXE,

TABLE 1. TIDA NEURONS AND POSITIVE GONADAL STEROID FEEDBACK

Treatment	Fluorescence intensity after H 44/68 (Number of animals in parenthesis)	Statistical significance according to Tukey's quick test (Tukey, 1959 Neave and Granger, 1968)
Castration		
McCann model		
Oil	1·5(4) 2(2) 2·5(1)[a]	
Estrogen (5 µg)	1·5(3) 1(2)[b]	a–b: $P < 0.01$
Estrogen (5 µg)		
+ progesterone (1·5 mg)	1(1) 1·5(3) 2(1)[c]	a–c: N.S.

The fluorescence intensity scale varies from 0 to 4. 0 = no fluorescence; 1 = weak intensity; 2 = moderate intensity; 3 = strong intensity; 4 = very strong intensity. For semiquantiative estimation, see Lidbrink and Jonsson (1971).

TABLE 2. TIDA NEURONS IN ANDROGENSTERILISED RATS: EFFECTS OF PROGESTERONE AFTER ESTROGEN PRIMING

Treatment	Fluorescence intensity after H 44/68 (Number of animals in parenthesis)	Statistical significance according to Tukey's quick test (Tukey, 1959; Neave and Granger, 1968)
Androgensterilisation (1·25 mg testosteron-proprionate day 3) + castration		
McCann model		
Oil	1(2) 1·5(6) 2(2)[a]	
Estrogen (5 µg)	1(3) 1·5(2)[b]	a–b: N.S
Estrogen (5 µg)		
+ progesterone (1·5 mg)	1(5) 1·5(2)[c]	a–c: N.S.

For further details, see text to Table 1.

Hökfelt, Jonsson and Löfstrom, unpublished data). These results support the view view that NA neurons in the hypothalamic and preoptic area as well as in the median eminence could facilitate the release of LRF and possibly FRF (Kalra et al., 1972). Interestingly enough the 5-HT turnover changes as revealed by biochemical analysis of 5-HT and by the use of α-propyldopacetamide, a tryptophane hydroxylase inhibitor, are in the opposite direction. Thus, the 5-HT turnover increases after estrogen priming alone, whereas the NA turnover was unchanged by this treatment (Fuxe, Hökfelt, Jonsson and Löfström, unpublished data). Furthermore, after progesterone treatment of the estrogen primed animals, the 5-HT turnover was reduced back to normal. These results are in good agreement with the view of Kordon et al. (1972) that there exist an inhibitory 5-HT pathway controlling LH secretion and also suggest that this 5-HT system could be involved in the inhibitory feedback action of estrogen on gonadotrophin secretion. In view of the above findings it can be speculated that the *TIDA neurons and the hypothalamic 5-HT neurons have to be turned off, whereas the hypothalamic NA neurons have to be turned on in the "critical period" in order for the LH, FSH and prolactin surge to occur resulting in ovulation.*

At the present time it cannot be excluded that the TIDA neurons are also involved in the control of other releasing and inhibitory factors beside those controlling gonadotrophin secretion, since certain turnover changes have been observed after stress (retardation; LIDBRINK *et al.*, 1972), thyroidectomy (acceleration; FUXE, LÖFSTRÖM and TSUCHIYA, unpublished data) and growth hormone in high doses (HÖKFELT, HALL and FUXE, unpublished data). At present, studies exploring a possible regional differentiation in these states are in progress. However, these effects are much less dramatic than the effects after interference with the pituitary-gonadal axis and could be interpreted to be related to changes in prolactin secretion. It should also be remembered that there probably exist dopaminergic mechanisms in other areas particularly in the limbic system that may control neuroendocrine events.

Acknowledgements—This study has been supported by grants from the Swedish Medical Research Council (04X-715; 04X-2887; 04X-2295) and by grants from the Population Council (M73.73) and from Svenska livförsäkringsbolags nämnd för medicinsk forskning. For skilful technical assistance we thank Mrs. K. Andreasson, Mrs. M. Baidins, Mrs. A. Eliason and Mrs. B. Hagman.

REFERENCES

AHRÉN K., FUXE K., HAMBERGER L. and HÖKFELT T. (1971) *Endocrinology* **88**, 1415.
BARRACLOUGH C. A. (1967) In *Neuroendocrinology* (MARTINI and GANGONG, Eds.), Vol. 2, pp. 61–99. Academic Press, New York.
BJÖRKLUND A., MOORE R. Y., NOBIN A and STENEVI U. (1973) *Brain Res.* **51**, 171.
CALIGARIS L., ASTRADA J. J. and TALEISNIK S. (1971) *Endocrinology* **89**, 331.
CORRODI H., FUXE K., HAMBERGER B. and LJUNGDAHL Å. (1970) *Europ. J. Pharmacol.* **12**, 145.
DAHLSTRÖM A. and FUXE K. (1966) *Acta Endocrinol.* **51**, 310.
FUXE K. and HÖKFELT T. (1969) In *Frontiers in Neuroendocrinology* (GANGONG and MARTINI, Eds.), Oxford University Press, New York, pp. 47–96.
FUXE K., HÖKFELT T. and JONSSON G. (1970) In *Neorochemical Aspects of Hypothalamic Function* (MARTINI and MEITES, Eds.), Academic Press, New York, pp. 61–83.
FUXE K., HÖKFELT T. and JONSSON G. (1971a) In *Hormonal Steroids*, Excerpta Med. Intern. Congr. Ser. No. 219, pp. 805–813.
FUXE K., HÖKFELT T. and NILSSON O. (1967) *Life Sci.* **6**, 2057.
FUXE K., HÖKFELT T. and NILSSON O. (1969a) *Neuroendocrinology* **5**, 527.
FUXE K., HÖKFELT T. and NILSSON O. (1969b) *Neuroendocrinology* **5**, 107.
FUXE K., HÖKFELT T. and SUNDSTEDT C. D. (1971b) In *Fedn. Proc.* **30**, No. 2.
FUXE K., HÖFKLET T., SUNDSTEDT C. D., AHRÉN K. and HAMBERGER L. (1972) *Neuroendocrinology* **10**, 282.
GLOWINSKI J. (1972) In *Perspectives in Neuropharmacology* (SNYDER, Ed.), Oxford University Press, New York, pp. 349–404.
HÅKANSON R., LARSSON L.-I., NOBIN A. and SUNDLER F. (1972) *J. Histochem. Cytochem.* **20**, 908.
HÖKFELT T. and FUXE K. (1972) In *Brain Endocrine Interaction. Median Eminence: Structure and Function*, Karger, Basel, pp. 181–223.
HÖKFELT T., FUXE K. and GOLDSTEIN M. (1973a) *Brain Res.*, **53**, 175.
HÖKFELT T. and LJUNGDAHL Å. (1972) *Histochemie* **29**, 325.
HÖKFELT T., LJUNGDAHL Å., JOHANSSON O. and FUXE K. (1973b) *Brain Res.*, in press.
JONSSON G. (1971) *Progr. Histochem. Cytochem.* **2**, 299.
JONSSON G., FUXE K. and HÖKFELT T. (1972) *Brain Res.* **40**, 271.
KALRA P. S., KRULICH L., FAWCETT C. P. and MCCANN S. M. (1972) *Endocrinology* **90**, 1168.
KAMBERI I. A., MICAL R. S. and PORTER J. C. (1970) *Endocrinology* **87**, 1.
KAWAKAMI M., SETO K., KIMURA F. and YANASE M. (1972) *Endocrinol. Japan* **19**, 429.
KOCH Y., LU K. H. and MEITES J. (1970) *Endocrinology* **87**, 673
KORDON C., GOGAN F., HERY M. and ROTSZTEJN W. H. (1971/72) *Gynecol. Invest.* **2**, 116.
LIDBRINK P. and JONSSON G. (1971) *J. Histochem. Cytochem.* **19**, 747.
LIDBRINK P., CORRODI H., FUXE K. and OLSON L. (1972) *Brain Res.* **45**, 507.
LIDBRINK P., JONSSON G. and FUXE K. (1973) *Brain Res.* in press.
LINDVALL O., BJÖRKLUND A., HÖKFELT T. and LJUNGDAHL Å. (1973) *Histochemie* **35**, 31.
MACLEOD R. M., FONTHAM E. H. and LEHMEYER J. E. (1970) *Neuroendocrinology* **6**, 283
MEITES J., LU K. H., WUTTKE W., WELSCH C. W., NAGASAWA H. and QUADRI S. K. (1972) In *Recent Progress in Hormone Research* (ASTWOOD, Ed.) Vol 28,   Academic Press, New York, pp. 471–516.

NEAVE H. R. and GRANGER W. J. (1968) *Technometrics* **10**, 509.
RUBINSTEIN L. and SAWYER C. H. (1970) *Endocrinology* **86**, 988.
SCHNEIDER H. P. G. and McCANN S. M. (1970) *Endocrinology* **87**, 249.
THIERRY A. M., STINUS L., BLANC G. and GLOWINSKI J. (1973) *Brain Res.* **50**, 230
TUKEY J. W. (1959) *Technometrics* **1**, 31.

Frontiers in Catecholamine Research 1973, pp. 795 to 801. Pergamon Press. Printed in Great Britain.

EFFECT OF PREGNANCY AND SEX HORMONES ON THE TRANSMITTER LEVEL IN UTERINE SHORT ADRENERGIC NEURONS

CH. OWMAN and N.-O. SJÖBERG

Departments of Histology, and Obstetrics and Gynaecology, University of Lund,
Lund, Sweden

COMBINED fluorescence histochemistry and chemical determinations have shown that NE* in the mammalian uterus is stored exclusively in neurons (SJÖBERG, 1967). DA is found in small amounts only in the cervix region (SWEDIN and BRUNDIN, 1968) where it probably is contained in chromaffin cells (OWMAN et al., 1973b). The presence of E has for a long time been controversial. There is reason to believe that the uterine E reported by some investigators is not neuronal, but derives either from the blood or from a population of chromaffin cells in or near the utero-vaginal ganglion (OWMAN et al., 1973b). Thus, among uterine catecholamines, only NE is located in neurons, which supply both myometrial smooth muscle cells and uterine blood vessels. Denervation experiments have established that the rich myometrial adrenergic innervation originates in peripheral sympathetic ganglia located in the utero-vaginal junction, thus belonging to the system of "short adrenergic neurons" (OWMAN et al., 1973c). The vasomotor sympathetic nerves, on the other hand, arise from the paravertebral sympathetic chain. Besides constituting a special anatomical entity, the short adrenergic neurons are unique also in a number of functional respects: they are largely unaffected by postnatal administration of nerve growth factor antiserum, they are resistant to the effect of α-methyltyrosine, reserpine, 6-OH-DA and certain sympathetic ganglion stimulants, and they are difficult to deplete by nerve stimulation (for references, see OWMAN et al., 1973c). The special features of the neurotransmission mechanisms of short adrenergic neurons have been investigated in detail by SWEDIN (1971). One of the functional aspects that make the short adrenergic neurons different from classical sympathetic nerves is their susceptibility to the influence of sex hormones, which can be elucidated by, for example, measurement of the total uterine NE content under various experimental conditions (see SJÖBERG, 1967; OWMAN et al., 1973c).

A hormonal dependence can be demonstrated already in connection with the critical period of postnatal differentiation of the reproductive organs. Experiments to show this were based on the ability of a single postnatal injection of testosterone propionate to produce masculinisation of the female reproductive tract, probably by a primary effect on the hypothalamus (see JACOBSOHN, 1965). Similarly, castration of male animals shortly after birth induces a female pattern of the gonadotrophin control mechanisms (see HARRIS, 1964). These abnormalities are persistent and modify the further growth and development of the reproductive organs. For example, the cyclic events are abolished in these organs of the females which become sterile,

---

* Abbreviations used: DA, dopamine; E, epinephrine; HCG, human chorionic gonadotrophin; NE, norepinephrine.

and the normal androgenisation of the male is prevented. The early hormonal influence on the short adrenergic neurons (Broberg, Nybell, Owman, Rosengren and Sjöberg, unpublished observations) was shown by castration of male rats immediately after birth, which markedly reduces the NE content in vas deferens (innervated by short adrenergic neurons). This reduction is more pronounced than if castration is performed on adult animals. Early androgenisation of female rats does not retard the normal increase in weight of the uterus, but the total organ content of NE is significantly reduced compared to untreated controls, both when measured in 9-week- and 13-week-old animals (Fig. 1). Thus, a single dose of testosterone propionate or postnatal castration appeared to permanently lower NE in short adrenergic neurons of uterus and vas deferens, respectively. The neurons hence seem to form a separate target system for those humoral factors which, via the early differentiation of the hypothalamus, determine the pattern of development of the reproductive tract. Hypothalamic NE does not appear to be affected by neonatal adrogenisation (or castration), whereas a significant reduction of the amine occurs in brain cortex after early androgenisation of female rats (Hyyppä and Rinne, 1971).

The short adrenergic neurons exhibit pronounced changes in their transmitter content also during pregnancy and after administration of female sex hormones. Such changes do not take place in organs, such as heart and ovary, supplied with the classical type of long adrenergic neurons. In a recent study, fluctuations in uterine NE were followed fluorometrically throughout pregnancy (Owman et al., 1973a) and during the post partum period (Gårdmark et al., 1971) in guinea-pigs. These animals were chosen because they often have unilateral pregnancies, which means that one of the uterine horns is not affected by the mechanical strain from the growing conceptus. In all uterine horns containing foetuses, the variation in the NE level (determined as total amine concent per uterine horn by which changes in uterine weight can be disregarded) during the gestation period were similar irrespective of whether pregnancy was bilateral or unilateral. Thus, within the first 10 days the NE content was almost doubled (Fig. 2). From the 15th day onward, the amine level showed a continuous decrease until a near-zero content was reached just before parturition, which occurs at approximately 65 days post coitum. A similar NE increase was found in the uterine horn devoid of foetuses from animals with unilateral pregnancy. However, this enhanced amine level remained constant until the 50th day of pregnancy, i.e. about 2 weeks before term. During these last 2 weeks, NE in the horn fell to the same minimum amounts as in the foetus-containing horns. At term, histochemically visible adrenergic nerves were present only in the cervix. Immediately after parturition (Gårdmark et al., 1971), the NE level in the uterine horns that had contained foetuses increased transiently to have a value that was, however, still less than half of the level in nonpregnant animals. It is possible that this reflected a shortlasting rise in the level of circulating NE, resulting in an uptake into the adrenergic nerves. Uterine NE then returned to lower values, as seen one week post partum, after which NE started to normalise progressively. Thus, two weeks after birth, the NE content was slightly but significantly higher than the values from pregnant animals. Even one month after pregnancy, uterine NE was still about 40 per cent lower than in nonpregnant animals. The original nonpregnant level of NE in the organ was not restored until within a further 5 months. Uterine NE increased progressively during the post partum period also in the horn that had been devoid of conceptus, though the time-course

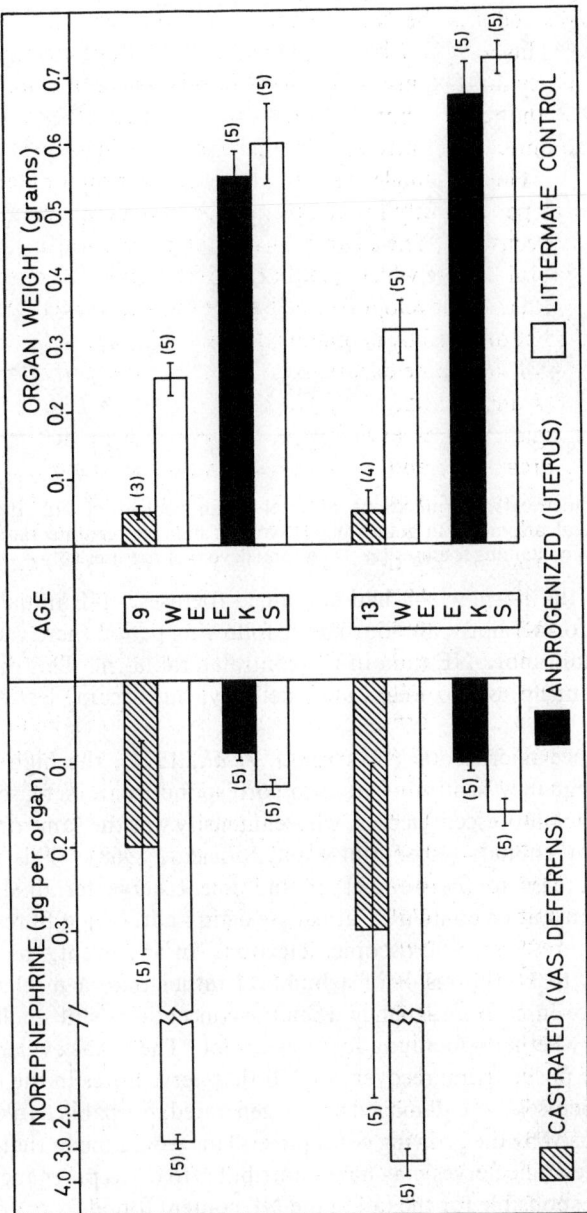

Fig. 1.—Changes in total organ content of NE and organ weight of uterus and vas deferens of 9- and 13-weeks-old rats. The females received 1·25 μg testosterone propionate 5 days after birth, the males were castrated 24 hr *post partum*. Mean ± S.E.M., number of determinations within parentheses.

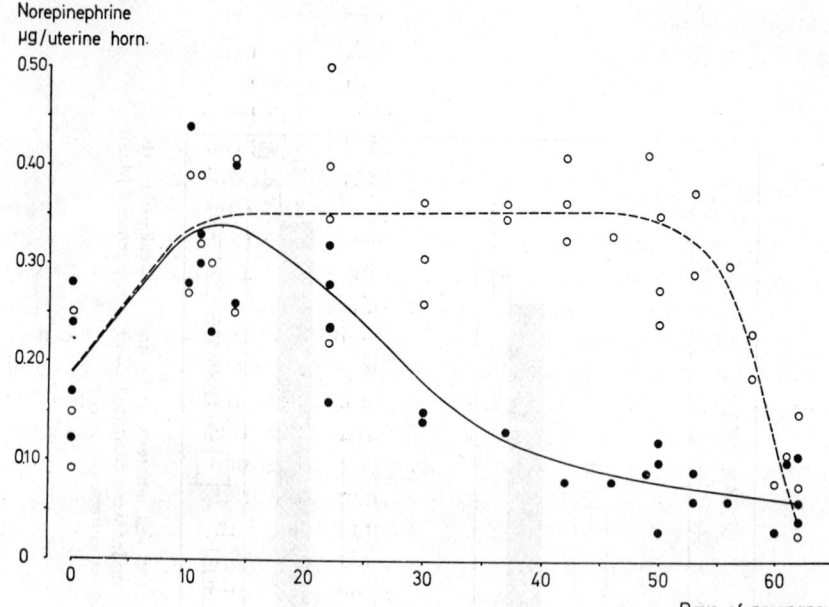

Fig. 2.—Fluorometrical determinations of total organ content of NE at various stages of unilateral pregnancy in guinea-pigs. Symbols indicate separate values from uterine horn containing foetuses (●——●) and devoid of foetuses (○- - -○).

was different from the horn which had contained foetuses: NE initially recovered more rapidly in the former horn, and during the following period there was a tendency of this horn to contain more NE than in the contralateral horn. This difference was statistically significant up to two weeks after delivery, but became less evident from 3 weeks onward.

In the fluorescence microscope (Gårdmark *et al.*, 1971), the high NE level at the beginning of pregnancy has been found to correspond to an increase in the total number of myometrial fluorescent nerves, whose intensity was the same or even higher than in nonpregnant animals (Rosengren and Sjöberg, 1968). This picture may represent a true increase in the number of fluorescent nerve terminals and/or an increment in the transmitter content of already existing adrenergic fibres with a NE level too low for fluorescence microscopic detection. In any event, the elevation in uterine NE appears to be the result of a humoral rather than a mechanical factor because it took place in early pregnancy when the conceptus is still small, and it was equal in both horns whether containing foetuses or not. There was evidence from the measurements of the post partum recovery of NE that nerve fibres in the uterine horn which has carried foetuses were damaged and degenerated, probably as a consequence of the distension caused by the growing conceptus. This would mean that a reduction in the number of adrenergic nerves may have contributed to or, as pregnancy advanced, even been entirely responsible for the fall in the NE content found from about 10 days onwards. In the uterine horn devoid of foetuses and therefore not affected by mechanical trauma, the NE accordingly remained constantly high. The inconspicuous enlargement of this horn also at the end of pregnancy indicates that mechanical trauma is not responsible for the rapid decline of NE in the horn during the last 2 weeks before parturition. Probably, the decline is instead the result of a hormonal influence.

The assumption that the NE changes in the uterine horn lacking foetuses reflects a humoral effect on the short adrenergic neurons with little contribution from mechanical factors, is favoured by the observations that the pattern of NE changes during pregnancy was very similar in the cervix region. In the uterine horn that had been devoid of foetuses, the return of the NE was more rapid probably since it involved only a "refilling" of NE into intact adrenergic nerve fibres, whereas in the horns that had contained foetuses the recovery was slower, particularly during the first 1–3 weeks *post partum*, because it included also regenerative changes in damaged neurons.

An increase in uterine NE of the same magnitude as that seen at the beginning of pregnancy could be induced by daily treatment of normal or oophorectomised animals with 0·5 μg/kg of 17-β-estradiol benzoate. The high NE content was reached already within a week's estrogen administration (SJÖBERG, 1967; FALCK *et al.*, 1969a and b). A very characteristic finding at the microscopic level was that the total number of fluorescent myometrial adrenergic nerves had increased, and their fluorescence intensity was distinctly higher than in the controls, which sometimes gave them "clumsy" appearance. No effect was found either histochemically or chemically in ovarian or cardiac NE, whereas the catecholamine content changed in a manner resembling that of the uterus also in the vagina and oviduct (SJÖBERG, 1967; FALCK *et al.*, 1969a and b), which are supplied by short adrenergic neurons (OWMAN *et al.*, 1966). The estrogen-induced increment in uterine NE remained unaltered for more than 2 weeks following cessation of the treatment. However, when estrogen administration

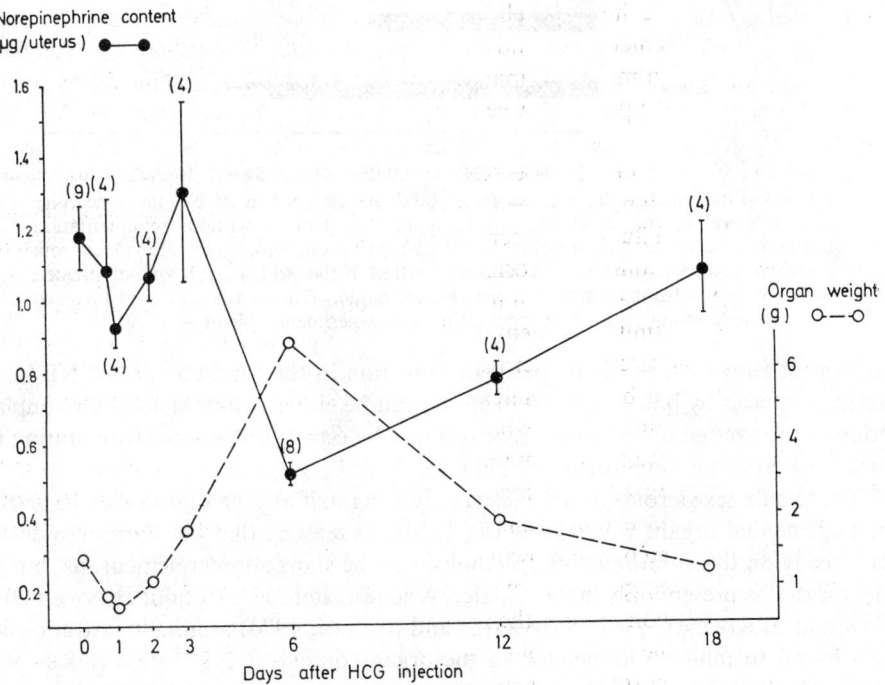

FIG. 3.—Effect of pseudopregnancy in rabbits. Total uterine content of NE (mean ± S.E.M., number of determinations within parentheses; one uterus per determination) and mean uterine weight at various time-periods after a single i.v. injection of 1500 i.u. of HCG.

was either combined with, or followed by, daily injections with 2 mg/kg of progesterone the organ content of NE returned to normal, or even subnormal, values within a week (Falck *et al.*, 1969b). Results from two subsequent experiments offer further support for the view that uterine short adrenergic neurons are under hormonal control, and they indicate that the ovary is involved in the control mechanism. One is that HCG-induced pseudopregnancy within 3 days after the injection causes an abrupt fall in the NE content of the uterus, a mimimum amount being reached in the course of a further 3 days (Fig. 3). The amine then progressively returns to normal values, which are attained within the life-span of the progesterone-producing corpus luteum. A second

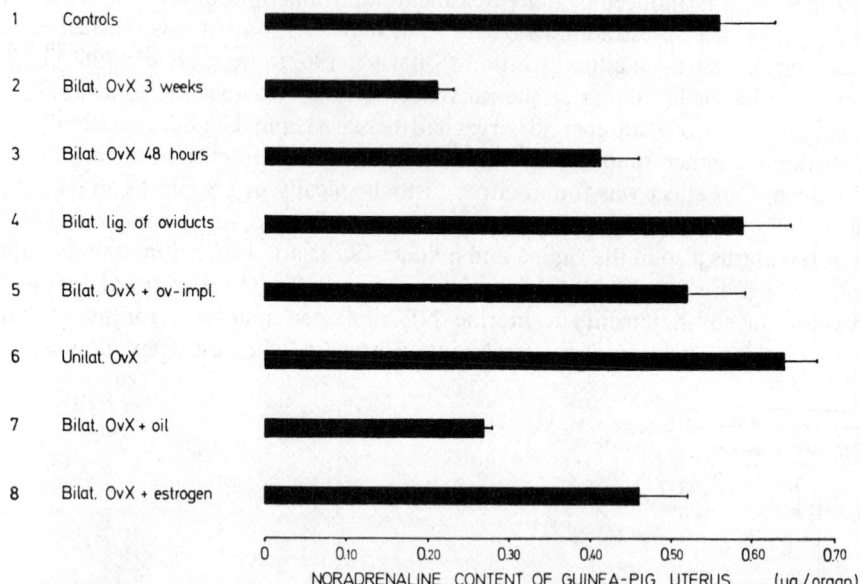

Fig. 4.—Fluorometric determinations of total uterine content of NE in guinea-pigs after bilateral or unilateral oophorectomy (OvX), with or without reimplantation of ovarian tissue, or daily injection of 0·5 μg/kg 17-β-estradiolbenzoate. Values from untreated controls, animals with bilateral ligation of the oviducts and animals treated with estrogen solvent (peanut oil) are also indicated. One uterus was used for each determination on 5–30 animals in each experiment. Mean + s.e.m.

experiment showed that the 60 per cent reduction in the total content of NE in the uterus produced by bilateral oophorectomy can be either counteracted by re-implantation of the excised ovarian tissue, or restored by estrogen administration during the third week after the oophorectomy (Fig. 4).

The female sex steroids tested caused only a slight, if any, change in the NE level of the male genital organs (Owman *et al.*, 1970). It may be that the hormones do not act directly on the noradrenaline metabolism of the short adrenergic neurons, but via a factor that is present only in the female. Also testosterone is without any overt effect (Ryd and Sjöstrand, 1967; Sjöstrand and Swedin, 1970) although castration has been found to inhibit the increase in the organ content of NE normally seen with advancing age.

CONCLUSION

It is suggested that the hormonal influence on the level of neuronal NE in the anatomically and functionally unique system of short adrenergic neurons in the reproductive tract constitutes a peripheral neuro-endocrine mechanism involved in the motor function of the oviducts and uterus.

Acknowledgement—Supported by Ford Foundation grant No. 68–383.

REFERENCES

FALCK B., OWMAN CH., ROSENGREN E. and SJÖBERG N.-O. (1969a) *Acta Endocrinol. (Kbh.)* **62,** 77–81.
FALCK B., OWMAN CH., ROSENGREN E. and SJÖBERG N.-O. (1969b) *Endocrinology* **84,** 958–959.
GÅRDMARK S., OWMAN CH. and SJÖBERG N.-O. (1971) *Am. J. Obstet. Gynecol.* **109,** 997–1002.
HARRIS G. W. (1965) *Endocrinology* **75,** 627–648.
HYYPPÄ M. and RINNE U. K. (1971) *Acta Endocrinol. (Kbh.)* **66,** 317–324.
JACOBSOHN D. (1965) *Acta Univ. Lund II* **17,** 1–19.
OWMAN CH., ROSENGREN E. and SJÖBERG N.-O. (1966) *Life Sci.* **5,** 1389–1396.
OWMAN CH., ROSENGREN E. and SJÖBERG N.-O. (1973a) *Am. J. Obstet. Gynecol.*, in press.
OWMAN CH., ROSENGREN E., SJÖBERG N.-O. and SWEDIN G. (1973b) *Acta physiol. scand.*, in press.
OWMAN CH., SJÖBERG N.-O. and SJÖSTRAND N. O. (1973c) In: *Amine Fluorescence Histochemistry.* (FUJIWARA, M., ed.). Igaku-Shoin, Tokyo. In press.
OWMAN CH., SJÖBERG N.-O., SJÖSTRAND N. O. and SWEDIN G. (1970) *Acta Endocrinol. (Kbh.)* **64,** 459–465.
ROSENGREN E. and SJÖBERG N.-O. (1968) *Acta physiol. scand.* **72,** 412–424.
RYD G. and SJÖSTRAND N. O. (1967) *Experientia* **23,** 816.
SJÖBERG N.-O. (1967) *Acta physiol. scand. Suppl.* **305,** 1–32.
SWEDIN G. (1971) *Acta physiol. scand. Suppl.* **369,** 1–34.
SWEDIN G. and BRUNDIN J. O. (1968) *Experientia* **24,** 1015–1016.

Frontiers in Catecholamine Research 1973, pp. 803 to 809. Pergamon Press. Printed in Great Britain.

CATECHOLAMINES IN SEXUAL HORMONE REGULATION: FOREBRAIN INFLUENCE ON TUBERO-INFUNDIBULAR DOPAMINE NEURONS AND INTERACTION WITH CHOLINERGIC SYSTEMS

W. LICHTENSTEIGER

Department of Pharmacology, University of Zürich, Zürich, Switzerland

NEUROENDOCRINE systems are capable to produce a variety of fast responses, a fact that may be especially important with regard to the coordination of behavioural and hormonal processes. We have been interested in the question whether monoamine systems and in particular the tubero-infundibular dopamine (DA) neurons might come into play in such short-term adjustments.

STIMULATION-INDUCED CHANGES IN CELLULAR FLUORESCENCE INTENSITY AND THEIR POSSIBLE BIOCHEMICAL BACKGROUND

As a tool to detect rapid responses in the tubero-infundibular DA neurons, we used a characteristic short-term change in the intensity of the catecholamine fluorescence of their cell bodies. The latter was measured by a microfluorimetric technique based on the histochemical fluorescence method of Falck and Hillarp (LICHTENSTEIGER, 1969a, 1970, 1971). Various populations of central DA neurons of mice and rats were found to exhibit this acute change in intensity upon a number of treatments such as local electrical or transsynaptic stimulation, acute exposure to cold, morphine or physostigmine (Fig. 1; LICHTENSTEIGER, 1969b; 1971; HEINRICH et al., 1971; LIENHART and LICHTENSTEIGER, 1973). The response is prevented by tyrosine hydroxylase inhibition which indicates that it is linked with an enhancement of DA synthesis. However, it appears that it is not due exclusively to the formation of the amine: Determinations of DA in extracts from substantia nigra-pieces of mice, carried out by means of a fluorimetric micromethod (SCHLUMPF, 1973), yielded an initial intensity change that was opposed to the one observed by microfluorimetry. Since the intensities of DA and DOPA are inversely related in the two procedures, we thought of a possible contribution of DOPA to the final fluorescence intensity. Although DOPA was not detected in extracts of normal whole brain (KEHR et al., 1972), we have recently observed a band corresponding to the position of DOPA in thin layer chromatograms of extracts from mouse midbrain, where amines and DOPA were visualised by reaction with formaldehyde vapour. As a working hypothesis, we would suggest, therefore, that neuronal activation induces a transient shift in the proportion of DOPA vs. DA in the cell bodies which, together with the formation of DA, may account for the intensity changes.

INFLUENCE OF VARIOUS BRAIN REGIONS ON THE TUBERAL DA NEURONS AND INTERACTION WITH CHOLINERGIC SYSTEMS

With regard to the questions put forward in the introduction, it would seem to be important to have some information on the integration of the tuberal DA neurons

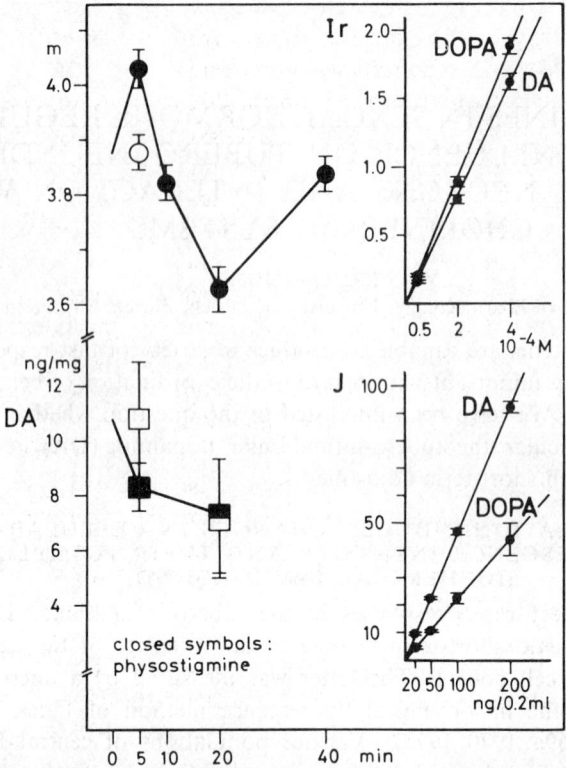

FIG. 1.—Acute response of substantia nigra DA neurons of mice to physoistgmine
(0·25 mg/kg s.c.). *Upper half, left:* Biphasic intensity response in DA nerve cells
as detected by microfluorimetry. Ordinate: means with 99% confidence limits (cell
counts 763–803 per experimental group); abscissa   time in min. *Upper half, right:*
Relative intensities of DA and DOPA in 7 μ sections of gelatin standards, measured
against a NA standard, after treatment according to the histochemical fluorescence
method. Ordinate: means with 99% confidence limits (100 measurements per point);
abscissa: DA and DOPA concentrations in the original 2% gelatin solution. *Lower
half, left:* DA concentrations as determined by a fluorimetric micromethod in extracts
of substantia nigra-pieces of mice subjected to the same experimental conditions.
Ordinate: means with 99% confidence limits (9 assays per point each performed on the
two blocks of substantia nigra of both sides from one animal); abscissa: time as above.
Lower half, right: Intensities of DA and DOPA as found in the extraction method.
Ordinate: means with 99% confidence limits (8 determinations per point) in absolute
instrument values; abscissa: concentrations of DA and DOPA. The changes in
fluorescence intensity observed with the two methods after the first 5 min as well as the
relations of the fluorescence intensities of DA and DOPA are opposed to each other.

into the neuroendocrine organisation. In a search for regions capable of influencing
the DA neurons, we used the initial increase in fluorescence intensity induced by
stimulation as an evoked response. The experiments were mainly performed on
ovariectomised rats pretreated for one day with estrogen and progesterone. A
clearcut intensity response was elicited by intermittent electrical stimulation (10 min)
of the medial preoptic area, nucleus of diagonal tract, ventrolateral part of the bed
nucleus of stria terminalis, medial amygdaloid nucleus and ventromedial tegmental
area (VMT) of the midbrain. Certain effects were also noted after stimulation of
the ventral hippocampus (LICHTENSTEIGER, 1971, 1973 and in preparation). It

appears thus that the tubero-infundibular DA neurons (1) are capable of short-term responses and (2) may serve to transmit signals from higher-order neuroendocrine 'centers', *limbic structures* and *ascending brainstem systems.*

In most cases, the effect exerted on the tuberal DA neurons appeared to depend upon the activity of cholinergic systems (LICHTENSTEIGER, 1973): Atropine administered 15 min before the onset of electrical stimulation markedly reduced the response to stimulation in the medial preoptic area, nuc. of diagonal tract, bed nucleus of stria terminalis and midbrain VMT. The drug also exerted a moderate effect on the reaction to amygdaloid stimulation. The effect of atropine was most probably due to a specific action at some cholinergic synapse(s), since (1) local electrical stimulation in the arcuate nucleus, where most of the DA cell bodies are situated, was effective despite atropine treatment, (2) the reduction of the intensity response to electrical stimulation of the medial preoptic area was dose-dependent (range 0·4–10 mg/kg, s.c.) and (3) methylatropine administered s.c. in a dose that was equimolar to the highest dose of atropine used, was almost ineffective. The fact that the effect of the drug rather did not appear to be linked with a special site of stimulation, and also its complex interaction with hormone secretion, could mean that the cholinergic synapse(s) may not belong to a neuron of a specifically neuroendocrine pathway but rather, to a cholinergic projection exerting some facilitatory influence on the transmission of the stimulatory effect.

RELATIONSHIP TO LUTEINISING HORMONE (*LH*) AND PROLACTIN SECRETION: HOMOGENEITY OR HETEROGENEITY OF THE TUBERAL DA NEURON GROUP?

Whenever responses of tuberal DA neurons and hormonal changes are compared, one should take into consideration that this neuron group does not only project to the external layer of the median eminence but also to intermediary and probably neural lobes (cf. BJÖRKLUND *et al.*, 1973). For safe conclusions, it would be necessary to investigate simultaneously the various hormone axes. Our own information is limited to serum *LH* and prolactin which were determined by radioimmunoassay (LICHTENSTEIGER and KELLER, in preparation).

Despite these limitations, certain indications for functional differentiation within the DA neuron group became evident in our material: Thus, the magnitude of the intensity response to electrical stimulation varied through the antero-posterior extension of the arcuate nucleus (Fig. 2). Moreover, the extent to which the response was inhibited by atropine, also appeared to differ somewhat in the various parts of the nucleus. Topographical differences were further noted when intensity profiles of groups of stimulated rats with different *LH* concentrations ranges were compared. In view of such differences and in consideration of earlier findings (LICHTENSTEIGER, 1969b), the tuberal DA neuron group was divided into two parts (levels 1–7 and 8–15) and fluorescence intensities and hormone levels of individual rats were correlated separately for the two parts.

From these data, it appears that *two types of responses* may tentatively be considered. They are represented by the results obtained with stimulation of the medial preoptic area and of the medial amygdaloid nucleus (Table 1) : (1) *Medial preoptic stimulation* yielded a significant positive correlation between the increased fluorescence intensity of the *anterior* part of the DA cell group and an increase in *LH*

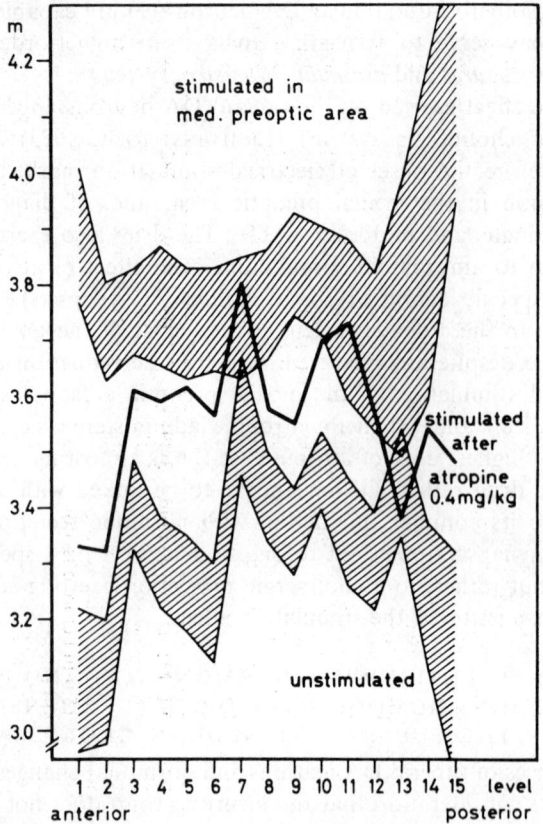

FIG. 2.—Profiles of fluorescence intensity of DA nerve cells through the arcuate nucleus of ovariectomised rats pretreated with estrogen and progesterone (5 rats per group). Abscissa: investigated levels, distance between levels $\sim 84\,\mu$. Ordinate: intensity of the catecholamine fluorescence in natural logarithms. The shaded areas indicate the 99% confidence limits of sham-operated controls (lower half) and animals stimulated for 10 min in the medial preoptic area (upper half). The magnitude of the increase in intensity differs at different levels, with most intense responses in the anterior and the posteriormost parts. The solid line connects the means of a group stimulated preoptically after atropine: It seems that the reduction of the stimulation-induced response is most marked at levels with greater intensity responses. All values are based on $\sim 40–100$ cells per level and group, except for levels 1,2,14,15 with lower cell counts.

concentration. No significant correlation was found with the posterior part, although the intensity increased there, too. This relation is basically in agreement with our earlier findings (KELLER and LICHTENSTEIGER, 1971). Atropine typically reduced both parameters, the correlation remaining thus positive. In contrast, *prolactin* remained largely unchanged after preoptic stimulation in the absence of atropine and accordingly, no significant correlation between intensity and hormone level was seen. Yet, atropine, while reducing the intensity response, allowed an increase in prolactin concentrations to occur upon preoptic stimulation. This led to a positive correlation between intensity and prolactin level (significant in both parts of the region). (2) After *medial amygdaloid stimulation*, the relation between *LH* and

TABLE 1. FLUORESCENCE INTENSITY OF THE ANTERIOR PART OF OF THE TUBERAL DA NEURON GROUP.

(Mean (m), variance (s^2), cell count (n) of logarithmically transformed frequency distributions of relative fluorescence intensity (natural logarithms) and serum concentrations of luteinising hormone (LH) and prolactin (determined by radioimmunoassay) of individual rats.

Electrode site and stimulation parameters	m	s^2	n	LH (ng/ml)	Prolactin (ng/ml)
I. Medial preoptic area 10 min, no stimulation	3·230	0·1563	147	80	500
	3·300	0·1651	75	290	100
	3·182	0·1381	81	170	420
	3·540	0·1549	82	195	—
	3·193	0·2454	132	180	280
II. Medial preoptic area 10 min, 100 μA, 0·5 msec 100 cs, 15 sec on/off	3·853	0·1291	93	245	50
	3·602	0·1405	91	380	280
	3·757	0·1329	76	360	160
	3·719	0·1362	59	230	280
	3·830	0·1854	73	310	460
III. Medial amygdaloid nucleus, 10 min stimulation, same parameters as II.	3·727	0·1458	92	140	560
	3·521	0·1167	101	185	500
	3·481	0·1576	75	140	460
	3·659	0·1254	81	225	500

Electrode site and stimulation parameters	m	s^2	n	LH (ng/ml)	Prolactin (ng/ml)
IV. Medial preoptic area, 10 min, no stimulation, atropine 10 mg/kg s.c. 15 min before electrode placement	3·399	0·1887	105	160	460
	3·411	0·2568	162	250	420
	3·107	0·1898	115	200	420
	3·207	0·1515	148	120	280
	2·974	0·1926	98	130	220
V. Medial preoptic area, 10 min stimulation with same parameters as II., atropine 10 mg/kg s.c. 15 min before onset of stimulation	3·561	0·1262	98	250	—
	3·278	0·2883	141	220	420
	3·346	0·1819	94	250	350
	3·267	0·1551	91	220	460
	3·409	0·1409	93	250	560
VI. Medial amygdaloid nucleus, 10 min stimulation, same parameters as II. atropine 10 mg/kg 15 min before onset of stimulation	3·491	0·1703	92	280	—
	3·378	0·1753	99	250	560
	3·610	0·1405	106	—	560

Correlation coefficients (a = r for anterior part, p = r for posterior part (intensities not shown); * = significant for $P < 0.05$): *Correlation of mean intensities with LH levels*: I and II a = 0·61*, p = 0·54; IV and V a = 0·67*, p = 0·38; II and V a = 0·47, p = 0·46; I and III a = 0·07, p = −0·07; IV and VI a = 0·71*, p = 0·58; III and VI a = −0·45, p = −0·47. *Correlation with prolactin levels*: I and II a = −0·33, p = −0·25; IV and V a = 0·71*, p = 0·62*; II and V a = −0·63*, p = −0·52; I and III a = 0·56, p = 0·50; IV and VI a = 0·82*, p = 0·63; III and VI a = 0·18, p = −0·03.

intensity response of the DA neurons resembled that observed after preoptic stimulation with regard to prolactin: No significant correlation in the absence of atropine,
appearance of a positive correlation between *LH* and intensity in atropine-treated
rats. This time, it was the *LH* level that rose in atropine-treated stimulated animals
On the other hand, fluorescence intensity and *prolactin* levels showed a similar
positive relationship with or without atropine. The type of response observed after
preoptic stimulation was relatively isolated, as the changes induced by stimulation
of the nuc. of diagonal tract, bed nucleus of stria terminalis and midbrain VMT
rather resembled the amygdala-type. A certain analogy to the contrasting effects of
atropine may be found in the action of nicotine which was recently reported to
reduce *LH* as well as prolactin surges, the effect on the latter hormone depending upon
the procedure used to elicit the surge (BLAKE *et al.*, 1972; BLAKE and SAWYER, 1972).

The rather complex results do not allow to design a generally applicable scheme
with regard to the response of the DA cell group and hormonal changes. There
may be several reasons for that: (1) It may well be that the DA neuron population is
inhomogeneous with regard to function (cf. LICHTENSTEIGER, 1969b; BJÖRKLUND,
et al., 1973). Our results suggest an activation of DA neurons located predominantly
in the anterior part in connection with *LH* release (prolactin inhibition?) on one
hand and a relation between increase in prolactin levels and activation of DA neurons
with a more uniform distribution throughout the region, on the other hand. The
latter phenomenon did probably not result from a direct feedback action of prolactin
such as has been described for different conditions (HÖKFELT and FUXE, 1972),
since we did not obtain a general positive correlation between intensities and the
levels of this hormone in all experimental groups. (2) Stimulation in different sites
most probably elicits *different additional effects* on releasing-factor release, either
through synapses at the releasing factor neurons or through effects at the level of the
median eminence. In this context, noradrenergic and serotoninergic projections
may be considered (KALRA and McCANN, 1972; KORDON, 1969) but also eventually
stimulation of releasing factor neurons, even in the preoptic area (cf KELLER and
LICHTENSTEIGER, 1971). (3) It is possible that hormone levels sometimes did not
change because the magnitude of the response of the DA neurons did not reach the
necessary *threshold*. However, this cannot be the only reason for the observed
discrepancies because intensity changes of similar magnitude were accompanied
by different effects on hormone levels.

In conclusion, it appears that influences from a variety of extrahypothalamic
sites, notably from *limbic structures* and *ascending brainstem systems*, reach the
tubero-infundibular DA neurons. The transmission of such influences seems to
depend in part upon the activity of *cholinergic systems*. The indications for a rather
complex response pattern and a possible inhomogeneity of the DA neuron group
may eventually help to reconcile the divergent conclusions that have been reached
with regard to facilitation or inhibition of *LH* and prolactin secretion in various
laboratories, including our own, especially if in addition, the possible existence of
NA neurons at the level of the external layer of the median eminence (BJÖRKLUND
et al., 1970) is considered (cf. AHRÉN *et al.*, 1971; DONOSO *et al.*, 1971; FUXE *et al.*,
1967, 1969, 1972; KAMBERI *et al.*, 1969, 1971; KORDON and GLOWINSKI, 1969;
KORDON, 1971; SCHNEIDER and McCANN, 1970; Van MAANEN and SMELIK, 1968,
WUTTKE *et al.*, 1971).

Acknowledgements—This research was supported by SNSF grant 3.691.71, the Hartmann-Müller Stiftung, the Barell Stiftung and the Jubiläumsspende of the University of Zürich.

REFERENCES

AHRÉN K., FUXE K., HAMBERGER L. and HÖKFELT T. (1971) *Endocrinology* **88**, 1415–1424.

BJÖRKLUND A., HROMEK F., OWMAN C. and WEST K. A. (1970) *Brain Res.* **17**, 1–23.

BJÖRKLUND A., MOORE R. Y., NOBIN A. and STENEVI U. (1973) *Brain Res.* **51**, 171–191.

BLAKE C. A. and SAWYER C. H. (1972) *Science* **177**, 619–621.

BLAKE C. A., SCARAMUZZI R. J., REID L. N., KANEMATSU S. and SAWYER C. H. (1972) *Endocrinology* **91**, 1253–1258.

DONOSO A. O., BISHOP W., FAWCETT C. P., KRULICH L. and MCCANN S. M. (1971) *Endocrinology* **89**, 774–784.

FUXE K., HÖKFELT T. and NILSSON O. (1967) *Life Sci.* **6**, 2057–2061.

FUXE K., HÖKFELT T. and NILSSON O. (1969) *Neuroendocrinology* **5**, 107–120.

FUXE K., HÖKFELT T., SUNDSTEDT C.-D., AHRÉN K. and HAMBERGER L. (1972) *Neuroendocrinology* **10**, 282–300.

HEINRICH U., LICHTENSTEIGER W. and LANGEMANN H. (1971) *J. Pharmacol. Exp. Ther.* **179**, 259–267.

HÖKFELT T. and FUXE K. (1972) *Neuroendocrinology* **9**, 100–122.

KALRA S. P. and MCCANN S. M. (1972) *IV Int. Congr. Endocrinology*, Abstract no. 508, Excerpta Medica Intern. Congr. Series no. 256.

KAMBERI I. A., MICAL R. S. and PORTER J. C. (1969) *Science* **166**, 388–389.

KAMBERI I. A., MICAL R. S. and PORTER J. C. (1971) *Endocrinology* **88**, 1012–1020.

KEHR W., CARLSSON A. and LINDQVIST M. (1972) *Arch. Pharmacol.* **274**, 273–280.

KELLER P. J. and LICHTENSTEIGER W. (1971) *J. Physiol.* **219**, 385–401.

KORDON C. (1969) *Neuroendocrinology* **4**, 129–138.

KORDON C. (1971) *Neuroendocrinology* **7**, 202–209.

KORDON C. and GLOWINSKI J. (1969) *Endocrinology* **85**, 924–931.

LICHTENSTEIGER W. (1969a) *J. Pharmacol. Exp. Ther.* **165**, 204–215.

LICHTENSTEIGER W. (1969b) *J. Physiol.* **203**, 675–687.

LICHTENSTEIGER W. (1970) *Prog. Histochem. Cytochem.* **1**, 185–276.

LICHTENSTEIGER W. (1971) *J. Physiol.* **218**, 63–84.

LICHTENSTEIGER W. (1973) *Endocrinology, Proc. IV Intern. Congress*, Excerpta Medica, Amsterdam, in press.

LIENHART R. and LICHTENSTEIGER W. (1973) *Arch. Pharmacol.* **277**, (suppl), R43.

SCHLUMPF M. (1973) Analytische Mikromethode zur fluorimetrischen Bestimmung von Monoaminen. Thesis, Eidg. Technische Hochschule, Zürich.

SCHNEIDER H. P. G. and MCCANN S. M. (1970) *Endocrinology* **87**, 249–253.

VAN MAANEN J. H. and SMELIK P. G. (1968) *Neuroendocrinology* **3**, 177–186.

WUTTKE W., CASSELL E. and MEITES J. (1971) *Endocrinology* **88**, 737–741.

Frontiers in Catecholamine Research 1973, pp. 811 to 813. Pergamon Press. Printed in Great Britain.

EFFECTS OF CATECHOLAMINES ON SUPRA-OPTIC NUCLEUS NEURONES IN ORGAN CULTURE*

Bernard H. Marks, Kakuichi K. Sakai, Jack M. George and
Adalbert Koestner,
Departments of Pharmacology, Medicine and Veterinary Pathobiology,
The Ohio State University, Columbus, Ohio 43210, U.S.A.

Studying the effects of drugs upon neuro-endocrine systems has been a highly inexact science, largely frustrated by problems of experimental design. Thus for the most part, with the techniques currently available, it has not been possible to know the effective drug concentrations at the presumed site of drug action, to be certain whether a given drug is acting directly upon a neuro-secretory cell or at some upstream site, or even to know whether applied drugs are acting presynaptically or post-synaptically upon neuro-endocrine units. In order to avoid these problems, and permit more direct study of drug effects upon receptor mechanisms associated with neuro-endocrine functions, we have developed an isolated organ-cultured neuro-endocrine test system, in which the neural units are devoid of efferent connections. The system is the organ-cultured supra-optic nucleus derived from the brain of newborn puppies.

The method consists in dissecting, mincing and explanting on to coverslips the supra-optic nuclei from young pups (Koestner et al., 1972). The explants are cultured in Leighton tubes with small volumes of a medium containing balanced salts, bovine fetal serum, bovine adult serum ultrafiltrate and glucose. The medium is changed every 3 days. Under these conditions, death and degeneration of damaged cells is complete in 7–10 days, the remaining supra-optic neurones remaining viable and growing out new axons. Study of the adrenergic receptor mechanisms of these cells were carried out after 2–3 weeks of culture.

Cultured cells were studied after transferring each coverslip bearing an explant to a small warmed (37°C) chamber, volume 0·5 ml, filled with Kreb's–Henseleit medium and gassed with 95% O_2-5% CO_2. By controlled gravity flow and suction, oxygenated K–H medium or drugs dissolved in K–H medium was perfused through the chamber. Under a dissecting microscope, individual neurones in each explant could be penetrated with conventional glass micro-electrodes in order to record resting and action potentials.

Using this method, a resting membrane potential of $-36 \pm 0\cdot69$ mV was recorded from 112 cells. The majority of these cells were quiescent, with only rare spontaneous action potentials. Continuous nerve activity could be generated by perfusion either with 10^{-6} M glutamate or nicotine. Spikes generated by such treatment had an approximate amplitude of 45 mV, and, as illustrated in Fig. 1A, showed a spike duration of approximately 6 msec with a characteristic negative afterpotential that had a very much longer duration.

When either norepinephrine, isoproterenol or phenylephrine was perfused through

* Supported by research grant NS AM 10655, National Institutes of Health.

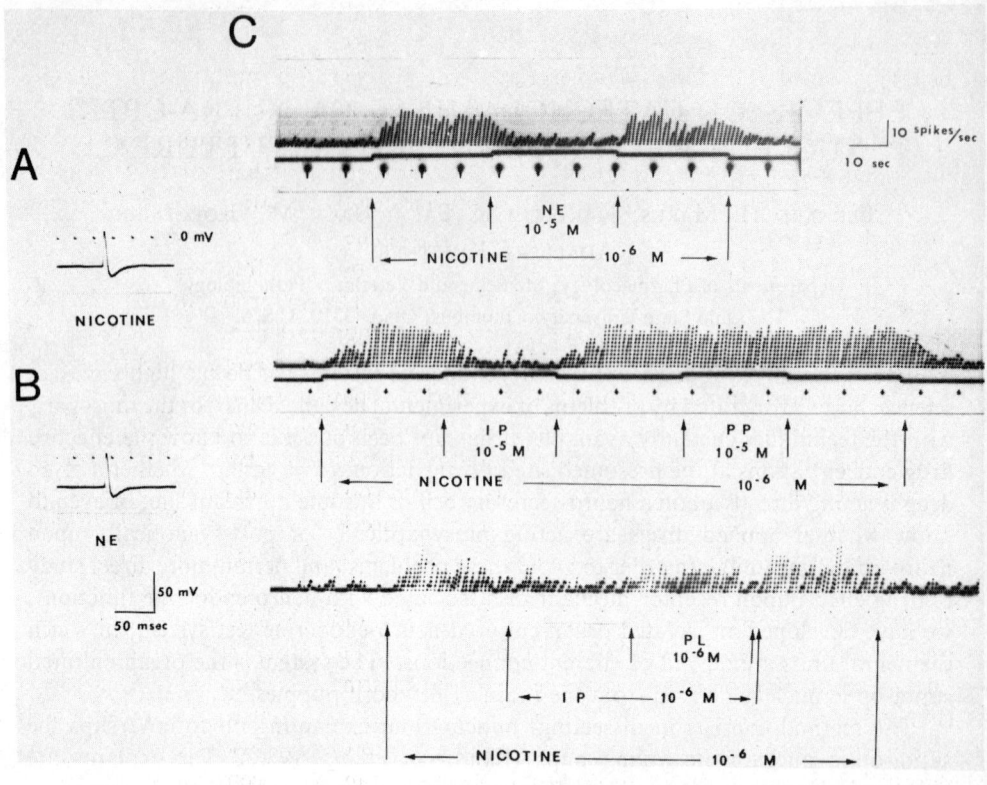

Fig. 1.—Effects of catecholamines on action potential configuration and on spike frequency recorded from supra-optic nucleus neurones in organ culture. A, spike configuration during exposure to nicotine 10^{-6} M. B, spike configuration during exposure to nicotine plus norepinephrine (NE) 10^{-6} M. C, inhibition of nicotine-induced supra-optic neurone spiking activity by NE (upper tracing), by isoproterenol (IP-middle tracing) and absence of effect of phenylephrine (PP-middle tracing); reversal of the effect of isoproterenol by propranolol (PL-lower tracing).

the explants, no change could be seen either in resting potential or in the frequency of the occasional spontaneous action potentials. Therefore, the effect of catecholamines was examined on cultures in which a high spike frequency was generated by continuous exposure either to glutamate or to nicotine 10^{-6} M. Under these conditions concentration-dependent inhibition of spike frequency is produced by concentrations of norepinephrine between 10^{-7} to 10^{-4} M. The upper tracing of Fig. 1C illustrates the inhibition of nicotine spiking produced by exposure to norepinephrine. Exactly the same norepinephrine response is observed if glutamate is used to initiate membrane activity. It is noteworthy that, during the period in which norepinephrine is inhibiting spiking activity, the configuration of those action potentials which do occur is not significantly different from the action potentials that develop during the control period with nicotine (Fig. 1B) or with glutamate. Only the frequency of spike generation is altered. This is in contrast to the effect of gamma-amino butyric acid (GABA), which also reduces the spike frequency with a concentration dependence much like that of norepinephrine. However, the individual spikes recorded during exposure to GABA are found to have a more negative resting potential and the duration of the negative afterpotential is seen to be markedly prolonged by GABA. Thus, one might infer that the ionic basis for the inhibition of nerve spiking activity is different for norepinephrine and for GABA, though the final effect on spike frequency is the same.

In cells in which high frequency spiking is induced by nicotine or by glutamate, it is possible to identify the nature of the catecholamine receptor mechanism mediating the inhibitory effect of norepinephrine. The middle tracing of Fig. 1C illustrates one such approach, showing that the inhibitory effect of norepinephrine is also readily demonstrated by perfusion with isoproterenol, but no effect on spike frequency is observed when cultured cells are exposed to phenylephrine. The lower tracing of Fig. 1C demonstrates that the inhibitory effect of isoproterenol on spike frequency can be reversed by exposure of the cultures to propanolol. Thus, the receptor mechanism in the organ-cultured supra-optic nucleus neurones is apparently of the beta type.

These experiments suggest that it may be feasible to use the isolated organ-cultured supra-optic nucleus as a test system to study the relationship between cell membrane receptor activation and the subsequent processes involved in the function of typical neuro-endocrine units. These explants, after culture for weeks, appear to behave exactly as they do *in situ*, insofar as their response to cholinergic and adrenergic agonists and antagonists are concerned (BARKER *et al.*, 1971). These responses are consistent with the effects of nicotine, acetylcholine and norepinephrine upon vasopressin secretion (BURN *et al.*, 1945; ABRAHAMS and PICKFORD, 1956), and with the known occurrence of acetylcholine and norepinephrine in the supra-optic nucleus (FELDBERG and VOGT, 1948; CARLSSON *et al.*, 1962). Thus, it would appear to be valid to use this isolated neural system to inquire into the processes that regulate the sensitivity of these membrane receptors, and that translate receptor activation into neurosecretion—the fundamental process that transduces neural into endocrine information.

REFERENCES

ABRAHAMS V. C. and PICKFORD M. (1956) *J. Physiol.* **131**, 712–718.
BARKER J. L., CRAYTON J. W. and NICOLL R. A. (1971) *J. Physiol.* **218**, 19–32.
BURN J. H., TRUELOVE L. H. and BURN I. (1945) *Brit. Med. J.* **1**, 403–406.
CARLSSON A., FALCK B. and HILLARP N. A. (1962) *Acta Physiol. Scand.* **56**, suppl. 196, 1–28.
FELDBERG W. and VOGT M. (1948) *J. Physiol.* **107**, 372–381.
KOESTNER A., GEORGE J. M. and LONG J. F. (1972) *J. Neuro-path. exp. Neurol.* **31**, 194.

Frontiers in Catecholamine Research 1973, pp. 815 to 817. Pergamon Press. Printed in Great Britain.

CATECHOLAMINES IN NON-NEURAL CELLS OF THE CNS: A POSSIBILITY

WILLIAM J. SHOEMAKER, PH.D.

Laboratory of Neuropharmacology, St. Elizabeth's Hospital-NIMH,
Washington, D.C. 20032, U.S.A.

CATECHOLAMINES have many functions in biological systems; they transmit messages between cells over long distances as hormones and over short distances as transmitter substances for neurons in the sympathetic nervous system and in certain areas of the central nervous system. However, in the brain, cells other than neurons may contain catecholamines. Such cells are most likely neurosecretory, and the catecholamine compounds contained in these cells may function other than as a neurotransmitter. The non-neural nature of some catecholamine-containing cells in the brain may contribute to the problems encountered in determining the physiological role of these compounds and to the controversy regarding dopamine's role in gonadotrophin secretion (SCHNEIDER and McCANN, 1970; FUXE and HÖKFELT, 1970).

I will focus here on the dopamine-containing cells of the arcuate nucleus-median eminence area and their role in the regulation of anterior pituitary secretion. Firstly, the small dopamine cells of the arcuate nucleus (AN) are at best atypical neurons in that they do not form synapses (AJIKA and HÖKFELT, 1973). That is, using either the standard gluteraldehyde-OsO_4 procedure or the ethanolic-phosphotungstic acid stain, no synaptic profiles were observed in median eminence either between axon and axon, or between axon and dendrite. Rather, the efferent cell processes of AN appear to terminate adjacent to perivascular spaces (STOECKART et al., 1972). The situation in the AN can be contrasted with the substantia nigra-corpus striatum dopamine-containing neurons which do form morphological synapses (AJIKA and HÖKFELT, 1973).

Secondly, there is the extreme resistance of AN dopamine to the depleting effect of 6-hydroxy-dopamine (6-OHDA), (CUELLO, SHOEMAKER and GANONG, 1973). In experiments involving chronic third ventricle cannulation in rats, high doses of 6-OHDA delivered directly into the third ventricle reduced hypothalamic norepinephrine levels to less than 10 per cent of normal. The dopamine levels, however, were unaffected (Table 1). Dopamine levels in the brain are usually more resistant to the depleting action of 6-OHDA (URETSKY and IVERSEN, 1970; LYTLE et al., 1972) but even at doses of 6-OHDA (with Pargyline and DMI pretreatment) that will deplete caudate dopamine, hypothalamic dopamine is not affected (Table 2). Clearly, the two dopamine-containing systems differ from each other in rather important characteristics.

The notion that some catecholamine-containing cells in the central nervous system may be neurosecretory rather than true neurons is not a new one (BERN and KNOWLES, 1966; KNOWLES, 1967). Many hormone producing cells in a variety of peripheral tissues contain large amounts of catecholamines. The frequent association between polypeptide hormone production and the ability to take up and form catecholamines and 5-HT by decarboxylation has led PEARSE (1968) to term these cells members of the APUD series (an acronym for Amine and Precursor Uptake and Decarboxylation). Interestingly,

TABLE 1. Hypothalamic catecholamines 15 days after
injecting 6-OHDA into third ventricle

Treatment	Norepinephrine (μg/g)	Dopamine (μg/g)	Corticosterone (μg/100 ml)
Control (NaBr)	1·13 ± 0·09	0·41 ± 0·03	5·5 ± 0·8
6-OHDA (200 μg ×2)	0·19 ± 0·03	0·42 ± 0·03	8·3 ± 1·5
Significance	$P < 0.001$	N.S.	N.S.

Animals were stereotaxically implanted with a stainless steel
cannula in the midline of the brain directly into the 3rd ventricle
several days prior to drug administration. Each rat received 2
injections (10 μl) 48 hr apart of either 200 μg of 6-OHDA (the
hydrobromide salt dissolved in distilled water) or NaBr equiosmolar
to the drug. Animals were decapitated and bled 15 days after the
second injection in a resting condition, 3 hours after light onset.
(Cuello, Shoemaker and Ganong, 1973).

TABLE 2. Comparison of dopamine depletion in hypothalamus and
'corpus striatum' using 6-OHDA (Pargyline & DMI)

Treatment	Hypothalamus (μg/g)	Corpus striatum (μg/g)	LH (ng/ml)
Control (NaBr ×2)	0·60 ± 0·03	1·48 ± 0·08	47·8 ± 5
6-OHDA (200 μg; Pargyline, DMI, pretreat ×2)	0·60 ± 0·02 N.S.	1·10 ± 0·23 $P < 0.05$	61·0 ± 11 N.S.

Adult male albino rats were twice injected intracisternally 24 hr apart with
either NaBr (10 μl) or 6-OHDA (200 μg). The animals receiving 6-OHDA
were given pargyline (50 mg/kg) i.p. and Desmethylimipramine (DMI)
(25 mg/kg) i.p. 30 min previously. Ether anesthesia was administered before
intracisternal injection; Penicillin-streptomycin (i.m.) after. All animals
were sacrificed 2 weeks post-injection at 3 hr after light-onset: blood
samples were taken at that time for radio-immunoassay of LH using the
N.I.H. standard.

these peripheral amine-containing endocrine cells appear to be derived embryonically
from the neural crest. Although these cells have the enzymatic capacity to decarboxylate amines, there is no evidence that the amines are synthesised in the cells, nor is
the specific function of such amines known.

Owman (1973) has hypothesised that amine ares normal constituents of all polypep-
tide-hormone producing endocrine cells, and that the intracellular amines function
in the formation, storage and/or release of the hormone.

What is needed is an intensive morphological investigation of these catecholamine-
containing cells to provide us with information regarding their secretory capability
and the precise nature and termination of their processes. A combination of electron
microscopy, autoradiography, and fluorescence histology (see Bloom, 1972; Bloom and
Crayton, 1972) could be utilised to yield considerable information. One study using
a combination of techniques has been published on the periventricular cell group in
the toad hypothalamus (McKenna and Rosenbluth, 1971). These workers find the
catecholamine-containing cells do not form synapses, and are more analogous to
peripheral chromaffin cells than to CNS neurons.

Another type of study that could enhance our understanding of catecholamine function in neuroendocrine regulation is turnover studies using radioactive tracers. The use of catecholamine synthesis inhibitors would not be recommended for these studies because the amines may not be synthesised in the AN cells. Although there are certain limitations associated with turnover studies (see discussion by WURTMAN, 1971), these methods could correlate the turnover rate of amines and/or releasing factors with different endocrine states, even if, as RODRIGUEZ (1972) has suggested, amines and hypothalamic releasing factors are contained in the same cell.

The initial descriptions of the catecholamine-containing cells of the hypothalamus (CARLSSON, FALCK and HILLARP, 1962; FUXE, 1964) carefully pointed out that no direct evidence identified such cells as neurons. By design and interpretation, however, virtually all studies of the role played by the dopamine cells of the hypothalamus in pituitary control assume the cells are neurons. A plethora of experiments has been done using synthesis inhibitors, electrical stimulation, receptor blockers, depleting and repleting agents, agonists and antagonists, alone or in combinations, to determine the function of the dopamine 'neurons' in gonadotrophin secretion. The possibility that dopamine may reside in brain cells that do not function as neurons, that it may not be synthesised in these cells, and that it may function in the binding and release of polypeptide hormones is testable. It is a hypothesis which must be explored before further conceptual progress in catecholamine-neuroendocrine regulatory function can occur.

REFERENCES

AJIKA K. and HÖKFELT T. (1973) *Brain Res.* **57,** 97–117.
BERN H. A. and KNOWLES F. G. W. (1966) In: *Neuroendocrinology* (MARTINI L. and GANONG W. F., Eds.) Vol. 1, pp. 139–186, Academic Press, N.Y.
BLOOM F. (1972) In: *Neurotransmitters*, Res. Publ. Assoc. Res. Nerv. Ment. Dis., pp. 25–57, Williams and Wilkins, Baltimore.
BLOOM F. E. and CRAYTON J. W. (1972) In: *Methods in Investigative and Diagnostic Endocrinology* (BERSON S., YALOW R., DORFMAN R., RALL E. and KOPIN I., Eds.) North-Holland, Amsterdam.
CARLSSON A., FALCK B. and HILLARP N.-A. (1962) *Acta Physiol. Scand.* (Suppl. 196), pp. 1–28.
CUELLO A. C., SHOEMAKER W. J. and GANONG W. F. (1973) Submitted to *Brain Res.*
FUXE K. (1964) *Z. Zellforsch.* **61,** 710–724.
FUXE K. (1970) In: *Aspects of Neuroendocrinology* (BARGMANN W. and SCHARRER B., Eds.) pp. 192–205, Springer-Verlag, N.Y.
KNOWLES F. (1967) In: *Neurosecretion* (STUTINSKY F., Ed.) pp. 8–19, Springer-Verlag, N.Y.
LYTLE L. D., SHOEMAKER W. J., COTTMAN K. and WURTMAN R. J. (1972) *J. Pharmacol. Exp. Ther.* **183,** 56–64.
McKENNA O. C. and ROSENBLUTH J. (1971) *J. Cell Biol.* **48,** 650–672.
OWMAN C., HAKANSON R. and SUNDLER F. (1973) *Fedn. Proc.* **32,** 1785–1791.
PEARSE A. G. E. (1968) *Proc. Roy. Soc. B.* **170,** 71–80.
RODRIGUEZ E. M. (1972) In: *Brain Endocrine Interaction. Median Eminance: Structure and Function* (KNIGGE K. M., SCOTT D. E. and WEINDL A., Eds.) pp. 319–334, Karger, Basle.
SCHNEIDER H. P. G. and McCANN S. M. (1970) In: *Aspects of Neuroendocrinology* (BARGMANN W. and SCHARRER B., Eds.) pp. 177–191, Springer-Verlag, N.Y.
STOECKART R., JANSEN H. G. and KREIKE A. J. (1972) *Z. Zellforsch.* **131,** 99–107.
URETSKY N. J. and IVERSEN L. L. (1970) *J. Neurochem.* **17,** 269–278.
WURTMAN R. J. (1971) *Neurosci. Res. Progr. Bull.* 9/2.

Frontiers in Catecholamine Research 1973, pp. 819 to 824. Pergamon Press. Printed in Great Britain.

CATECHOLAMINES AND THE SECRETION OF RENIN, ACTH AND GROWTH HORMONE*

WILLIAM F. GANONG

Department of Physiology, University of California, San Francisco, California 94143, U.S.A.

AN INTERESTING theme in modern neuroendocrine research is the ubiquitous occurrence of adrenergic neurons as regulators of endocrine secretion. This regulation occurs both in glands such as the juxtaglomerular cells and the pancreatic islets, which are innervated by postganglionic sympathetic neurons, and in the hypothalamus, where the cells that secrete the releasing and inhibiting factors that regulate anterior pituitary secretion are apparently innervated by norepinephrine-secreting and dopamine-secreting neurons in the central nervous system. The adrenergic neurons generally do not constitute the sole regulation of individual endocrine glands; rather, there are regulatory factors unique to the individual gland with an additional adrenergic modulation or adjustment.

The renin-secreting juxtaglomerular cells of the kidneys are a good example (see ASSAYKEEN and GANONG, 1971). The secretion of these cells is believed to be regulated partly via an intrarenal baroreceptor mechanism that increases renin secretion when the pressure in the renal artery and arterioles is low. Renin secretion also appears to be increased when the amount of sodium crossing the macula densa in the epithelium of the renal tubule is decreased. In addition, renin secretion is increased by increased sympathetic discharge. For example, renin secretion is increased by hypoglycemia, stimulation of the pressor region of the medulla oblongata, and stimulation of the renal nerves. The response to hypoglycemia is apparently mediated primarily by circulating epinephrine, since it is blocked by adrenal denervation. On the other hand, the response to stimulation of the medulla oblongata is mediated primarily via the renal nerves, since the response is blocked by renal denervation.

These effects of sympathetic stimulation on renin secretion could be mediated via the baroreceptor or macula densa mechanisms, since injected catecholamines and renal nerve stimulation can lower the pressure in the renal arterioles and reduce sodium delivery to the distal tubule, with a resultant decrease in sodium transport across the macula densa. However, α-adrenergic blocking drugs do not inhibit the rise in renin produced by sympathetic stimulation, and most if not all of the hemodynamic effects of catecholamines on the kidney are mediated via α-adrenergic receptors (GUMP et al., 1968). Indeed, in a number of situations, α-adrenergic blockade potentiates renin secretion. Conversely, β-adrenergic blockade with propranolol blocks the increase in renin secretion produced by hypoglycemia (ASSAYKEEN et al., 1970),

* This paper includes the results of experiments supported by USPHS Grant AM06704 and the Kroc and Skaggs Foundations. Previously unpublished work in the paper was carried out in collaboration with Hector Nolly, Ian Reid, Robert Lovinger, Umberto Scapagnini, Selna Kaplan, Melvin Grumbach, Angela Boryczka and Roy Shackelford.

stimulation of the pressor region of the medulla oblongata (PASSO et al., 1971) and stimulation of the renal nerves (LOEFFLER et al., 1972). Furthermore, rapid administration of α-adrenergic blocking agents increases renin secretion by itself, and this increase is prevented by β-adrenergic blockade (LOEFFLER et al., 1972). Additional evidence that sympathetic effects on renin secretion are mediated via β-adrenergic receptors is the finding that in dogs, theophylline increases renin secretion and potentiates the renin response to circulating catecholamines (REID et al., 1972).

The fact that the stimulatory effects of sympathetic activity on renin secretion are β- rather than α-mediated raises the possibility that epinephrine and norepinephrine act directly on the juxtaglomerular cells themselves. There is a dense sympathetic plexus in the arteriolar wall around these cells, and there appear to be "en passant" adrenergic endings on the juxtaglomerular cells (BARAJAS, 1964).

To explore further the mechanism by which catecholamines increase renin secretion, Nolly, Reid, and I (unpublished observations) have studied the release of renin from kidney slices in vitro. These slices were cut from the cortex of the kidneys of normal adult male and female rats, and were incubated in Krebs–Ringer phosphate glucose dextran solution at pH 7·45 at 37° C for 1 hr. Renin in the medium and in the tissue was measured by radioimmunoassay of angiotensin I.

In confirmation of MICHELAKIS et al. (1969), we found that norepinephrine increased renin release in vitro (Table 1). There appears to have been net synthesis of renin, or at least activation of renin in the preparation, because norepinephrine also increased the amount of renin in the incubated kidney tissue. Evidence that the response is mediated by cyclic AMP, and hence presumably by β-receptors, is the observation that theophylline potentiated the renin response to norepinephrine while having no significant effect by itself (Table 1). Potentiation was present at several different concentrations of norepinephrine.

TABLE 1. EFFECT OF NOREPINEPHRINE, THEOPHYLLINE AND PHENOXYBENZAMINE ON RENIN RELEASE in vitro

Treatment	Initial molar concentration in medium	Renin release (ng AI/mg kidney tissue/hr)	P, compared to control
Control	—	5·36 ± 0·84	
Norepinephrine	5 × 10⁻⁶	6·54 ± 0·96	
Norepinephrine	10⁻⁵	7·43 ± 1·05	<0·02
Norepinephrine	2 × 10⁻⁵	10·92 ± 1·61	<0·005
Control	—	7·43 ± 0·24	
Theophylline	10⁻³	7·26 ± 0·41	
Norepinephrine	10⁻⁶	8·22 ± 0·19	<0·005
Norepinephrine and Theophylline	10⁻⁶ and 10⁻³	12·77 ± 0·57	<0·001;*
Control	—	7·02 ± 0·45	
Phenoxybenzamine	10⁻⁴	8·10 ± 0·57	
Norepinephrine	10⁻⁵	10·75 ± 1·09	<0·01
Norepinephrine and Phenoxybenzamine	10⁻⁵ and 10⁻⁴	14·44 ± 2·22	<0·005;†

Values are means ± S.E.
* $P < 0·005$ compared to norepinephrine alone.
† $P < 0·01$ compared to norepinephrine alone.

The effect of the α-adrenergic blocking agent phenoxybenzamine on the renin response to norepinephrine *in vitro* is also shown in Table 1. Phenoxybenzamine had little if any effect by itself, but it produced significant potentiation of the stimulatory effect of norepinephrine. Similar results were obtained with phentolamine *in vitro*. This observation is consistent with our *in vivo* finding that α-adrenergic blockade potentiated the renin response to several sympathetic stimuli. We have also conducted experiments with the β-adrenergic blocking agent propranolol *in vitro*. The preliminary results available to date indicate that the increase in renin produced by norepinephrine is blocked by L-propranolol and unaffected by D-propranolol.

It is true, of course, that the slices of renal cortex incubated in these experiments contained a number of different types of cells. However, it is difficult to see how norepinephrine could act except by stimulating the juxtaglomerular cells directly. There is precedent for this type of effect in another endocrine gland; it appears that the pancreatic islets are innervated by postganglionic sympathetic neurons, and there is considerable evidence that catecholamines act directly on the beta cells, producing increased insulin secretion by a β-mediated effect and decreased insulin secretion by an α-mediated effect (IVERSON, 1971; FRANKEL *et al.*, 1973).

The secretion of glucagon by the alpha cells of the pancreas is regulated in part by catecholamines (IVERSON, 1971; FRANKEL *et al.*, 1973). In addition, the biosynthesis of melatonin in the pineal body is facilitated by norepinephrine secreted by the postganglionic sympathetic neurons innervating the pineal, and this effect is blocked by propranolol (KLEIN, 1973).

The secretion of the anterior pituitary gland also is affected by catecholamines. Anterior pituitary secretion is regulated by a family of approximately eight releasing and inhibiting factors secreted by cells in the hypothalamus and the secretion of these releasing and inhibiting factors is in turn regulated to a large degree by norepinephrine- and dopamine-secreting neurons in the diencephalon (see GANONG, 1973). The secretion of the anterior pituitary hormones is also influenced in a feedback fashion by the thyroid, adrenocortical and gonadal hormones secreted from the endocrine glands that are the target organs of the anterior pituitary hormones, and some of the steroidal effects may be exerted on the hypothalamus as well as the pituitary.

For some years, my associates and I have been studying the relation of brain aminergic neural systems to the regulation of ACTH secretion in the dog and the rat. Our current evidence fits best with the hypothesis that a central neural system inhibits ACTH secretion, that norepinephrine is the mediator in this system, and that its effects are mediated by an α-adrenergic mechanism. The inhibitory effect is probably exerted directly on the cells that secrete the ACTH-regulating hypothalamic factor, CRF (see GANONG, 1972, 1973).

One interesting feature of the changes in ACTH secretion produced by this system is that they are apparently transient, rather than prolonged. For example, reserpine produces a prompt increase in ACTH secretion, but with continued reserpine treatment, ACTH secretion returns to normal even though brain catecholamines remain depleted (MCKINNEY *et al.*, 1971). Similarly, α-methyl-*p*-tyrosine produces an acute increase in circulating corticosterone, but its chronic administration does not lead to adrenal hypertrophy (VAN LOON *et al.*, 1971; WEINER and GANONG, 1972). Six-hydroxydopamine also produces a transient increase in ACTH secretion. In experiments carried out by Cuello, Shoemaker, and Scapagnini in my laboratory,

6-hydroxydopamine was injected directly into the third ventricle in rats with chronically implanted cannulas. Twenty-four hours after the second of two intra-ventric-
ular injections, plasma corticosterone concentration was elevated and there was
marked depletion of hypothalamic norepinephrine. Fifteen days after the second
injection into the ventricle, hypothalamic norepinephrine was still markedly depleted,
but plasma corticosterone had returned to normal. The explanation of the return
of ACTH secretion to normal despite continued hypothalamic norepinephrine
depletion is presently unknown.

Six-hydroxydopamine crosses the blood–brain barrier to a limited degree, and the
median eminence region of the hypothalamus is "outside the blood–brain barrier".
Since norepinephrine-containing neurons end in the median eminence region and in
more dorsal portions of the hypothalamus, we have compared the effects of systemically administered 6-hydroxydopamine to those of intraventricular 6-hydroxy-
dopamine in order to gain some insight into the site at which the noradrenergic neurons
in the diencephalon influence ACTH secretion. Plasma corticosterone was found
to be moderately reduced one day after intraperitoneal administration of 6-hydroxy-
dopamine (Table 2). This suggests that the site of action of norepinephrine is

TABLE 2. EFFECTS OF 6-HYDROXYDOPAMINE AND 6-HYDROXYDOPA (TWO
INTRAPERITONEAL DOSES, EACH 100 mg/kg, 24 hr APART).

Day after treatment	Treatment	N	Plasma corticosterone (μg/100 ml)	Norepinephrine concentration (μg/g)	
				Hypothalamus	Brain Stem
1	6-OH dopamine	12	19·2 ± 3·8*	1·72 ± 0·09†	0·91 ± 0·11
5	6-OH dopamine	15	6·8 ± 2·4	2·01 ± 0·08	0·88 ± 0·01
15	6-OH dopamine	12	5·2 ± 3·2	1·92 ± 0·10	0·89 ± 0·10
1	6-OH dopa	18	28·4 ± 4·2*	1·39 ± 0·09*	0·70 ± 0·09†
5	6-OH dopa	18	10·2 ± 2·2*	1·47 ± 0·12*	0·69 ± 0·10
15	6-OH dopa	15	9·2 ± 3·2	1·50 ± 0·07*	0·72 ± 0·80
1	Saline	9	7·9 ± 2·2	1·93 ± 0·08	0·92 ± 0·08
5	Saline	9	6·4 ± 1·8	1·90 ± 0·10	0·90 ± 0·08
15	Saline	9	5·9 ± 2·8	2·05 ± 0·11	0·88 ± 0·12

Values are means ± S.E.
* $P < 0.01$ vs. saline on corresponding day.
† $P < 0.05$ vs. saline on corresponding day.

in the median eminence, outside the blood–brain barrier. However, 6-hydroxydopa,
a compound which crosses the blood–brain barrier and is converted to 6-hydroxy-
dopamine in adrenergic neurons in all parts of the brain, was also injected. It produced a greater elevation in plasma corticosterone (Table 2). Other indirect evidence
(see GANONG, 1973) suggests that norepinephrine acts above the median eminence.
The 6-hydroxydopamine data plus the other observations therefore suggest that
norepinephrine acts in part in the median eminence region and in part in other
portions of the hypothalamus "inside the blood–brain barrier".

The secretion of growth hormone is stimulated, rather than inhibited, by adrenergic discharge. In humans L-dopa stimulates growth hormone secretion (BOYD
et al., 1970), and the increase in growth hormone secretion produced by insulin hypoglycemia is reduced by α-adrenergic blockade and potentiated by β-adrenergic

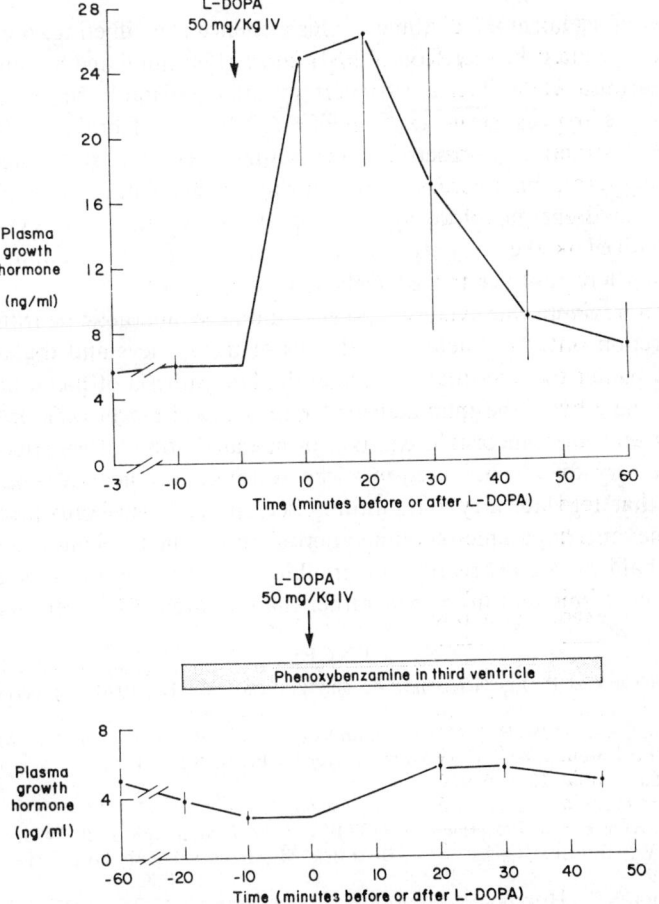

FIG. 1.—Effect of L-DOPA on plasma growth hormone in pentobarbital-anesthetized dogs. In the experiment illustrated in the lower portion of the diagram, phenoxybenzamine was administered as a single dose of 0·04 mg/kg followed by a constant infusion of 0·01 mg/kg over 90 min.

blockade (BLACKARD and HEIDINGSFELDER, 1968). In baboons, injection of norepinephrine directly into the ventromedial nucleus of the hypothalamus increases growth hormone secretion, while the injection of dopamine inhibits growth hormone secretion (TOIVOLA and GALE, 1973). Injection of phentolamine into the third ventricle lowers plasma growth hormone (TOIVOLA et al., 1972). LOVINGER, KAPLAN, GRUMBACH, BORYCZKA and I (unpublished observations) have developed an immunoassay for canine growth hormone, and have shown that the secretion of this hormone is sharply increased by L-dopa (Fig. 1). The increase is prevented by administration of phenoxybenzamine directly into the third ventricle in the same dose that inhibits the effect of L-dopa on ACTH secretion (see GANONG, 1973). We also found that the growth hormone response to L-dopa was blocked by the tetradecapeptide growth hormone inhibiting factor recently synthesized by BRAZEAU et al. (1973).

There is considerable evidence from other laboratories that norepinephrine- and dopamine-secreting neurons are involved in the regulation of the secretion of the other anterior pituitary hormones. Some of this evidence is conflicting, but the bulk of it supports the view that the secretion of FSH and LH is stimulated by norepinephrine-secreting neurons, while the secretion of prolactin is inhibited by dopaminergic neurons (for references, see GANONG, 1973). The role of brain amines in the regulation of TSH secretion is unsettled. Prolactin secretion is stimulated by the dopamine-blocking drug haloperidol and chlorpromazine, but aside from the observation that α-adrenergic blocking drugs prevent ovulation (SAWYER, 1964), the receptors involved in the regulation of the secretion of the remaining anterior pituitary hormones have not been extensively studied.

In summary, norepinephrine-secreting postganglionic sympathetic neurons appear to affect the secretion of the juxtaglomerular cells of the kidneys and the alpha and beta cells of the pancreatic islets, and to affect the biosynthesis of melatonin in the pineal body. In the case of the juxtaglomerular cells and the beta cells of the pancreas, it appears that norepinephrine exerts a β-mediated stimulatory effect and an α-mediated inhibitory effect on secretion. The secretion of the releasing and inhibiting factors that regulate anterior pituitary secretion is also controlled in part by norepinephrine- and dopamine-secreting neurons in the diencephalon. It appears that via this mechanism, ACTH secretion is inhibited and growth hormone secretion stimulated by mechanisms that involve α- rather than β-adrenergic receptors.

REFERENCES

ASSAYKEEN T. A., CLAYTON P. L., GOLDFIEN A. and GANONG W. F. (1970) *Endocrinology* **87**, 1318–1322.

ASSAYKEEN T. A. and GANONG W. F. (1971) In *Frontiers of Neuroendocrinology*, 1971. (MARTINI L. and GANONG W. F., Eds.) pp. 67–102. Oxford University Press, New York.

BARAJAS L. (1964) *Lab. Invest.* **13**, 916–929.

BLACKARD W. F. and HEIDINGSFELDER S. A. (1968) *J. Clin. Invest.* **47**, 1407–1414.

BOYD A. E. III, LEBOVITZ E. and PFEIFFER J. B. (1970) *N. Engl. J. Med.* **283**, 1425–1429.

BRAZEAU P., VALE W., BURGUS R., LING N., BUTCHER M., RIVIER J. and GUILLEMIN R. (1973) *Science* **179**, 77–79.

FRANKEL B. J., GERICH J. E., HUGURA R., FANSKA R. E. and GRODSKY G. M. (1973) *Clin. Res.* **21**, 273 (abstract).

GANONG W. F. (1972) In *Brain-Endocrine Interaction*. (KNIGGE K. M., SCOTT D. E. and WEINDL A., Eds.) pp. 254–266, Karger, Basel.

GANONG W. F. (1973) In *The Neurosciences: Third Study Program*. (WORDEN F., Ed.) The MIT Press, Cambridge, Mass. (in press).

GUMP F. E., MAGILL T., THAL A. P. and KINNEY J. M. (1968) *Surg. Gynec. Obst.* **127**, 319–326.

IVERSON J. (1971) *Diabetologia* **7**, 485 (abstract).

KLEIN D. C. (1973) In *The Neurosciences: Third Study Program*. (WORDEN F., Ed.) The MIT Press, Cambridge, Mass. (in press).

LOEFFLER J. R., STOCKIGT J. R. and GANONG W. F. (1972) *Neuroendocrinology* **10**, 129–138.

MCKINNEY W. T., PRANGE A. J., MAJCHOWICZ E. and SCHLESINGER K. (1971) *Dis. Nerv. Syst.* **32**, 308–313.

MICHELAKIS A. M., CAUDLE J. and LIDDLE G. W. (1969) *Proc. Soc. Exp. Biol. Med.* **130**, 748–753.

PASSO S. S., ASSAYKEEN T. A., GOLDFIEN A. and GANONG W. F. (1971) *Neuroendocrinology* **7**, 97–104.

REID I. A., STOCKIGT J. R., GOLDFIEN A. and GANONG W. F. (1972) *European J. Pharmacol.* **17**, 325–332.

SAWYER C. H. (1963) In *Advances in Neuroendocrinology*. (NALBANDOV A. V., Ed.) pp. 444–457, University of Illinois Press, Urbana, Illinois.

TOIVOLA P. T. K. and GALE C. C. (1973) *Fedn. Proc.* **32**, 265 (abstract).

TOIVOLA P. T. K., GALE C. C., GOODNER C. J. and WERRBACH J. H. (1972) *Hormones* **3**, 193–213.

VAN LOON G. R., SCAPAGNINI U., MOBERG G. P. and GANONG W. F. (1971) *Endocrinology* **89**, 1464–1469.

WEINER R. I. and GANONG W. F. (1972) *Neuroendocrinology* **9**, 65–71.

Frontiers in Catecholamine Research 1973, pp. 825 to 829. Pergamon Press. Printed in Great Britain.

L-DOPA AND PANCREATIC SECRETION

KOROKU HASHIMOTO, YASUHIKO FURUTA and KAZUHIKO IWATSUKI
Department of Pharmacology Tohoku University School of Medicine, Sendai, Japan

GREENGARD, ROBACK and IVY (1942) first observed that epinine (N-methyldopamine), dopamine (dihydroxyphenylethylamine) and few phenylethylamine derivatives given intravenously stimulated the canine pancreatic secretion, while the majority of sympathomimetic amines tested inhibited it. Although this observation appeared strikingly important, no particular attention had been given until ALM, EHINGER and FALCK (1967) when FALCK and his collaborators reported on the L-dopa turnover to dopamine in the mouse pancreas using fluorescense method. "After L-dopa i.v. first the whole of the pancreatic acinar cells displayed a specific diffuse fluorescence within few minutes and then most of the fluorescence was situated in zymogen granules as the second pattern of fluorescence after 40–60 min". Thus, they assumed that L-dopa was decarboxylated to dopamine in the cytoplasm and then incorporated in granules. On the other hand, HOLZ, CREDNER and STRÜBING (1942) demonstrated the existence of dopa decarboxylase in the pancreas, and SCHÜMANN and HELLER (1959) detected a relatively large amount of dopamine in the pancreas of the ox and the sheep.

THE SPECIFICITY OF THE DOPAMINE-INDUCED PANCREATIC SECRETION

In the course of comparative studies on the vascular and the functional responses of the isolated blood-perfused organs to various pharmacological agents, we observed that dopamine stimulated strikingly the exocrine pancreas (TAKEUCHI et al., 1971). Such striking response to dopamine was not observed in any other organ. Since the canine pancreas had its spontaneous secretion even in the fasting state, either stimulation or inhibition could be readily observed when substances were injected intra-arterially. The specificity of dopamine as a secretagogue of the exocrine pancreas was confirmed when effects of various biogenic and foreign substances were compared on the pancreatic secretion in the resting state (TAKEUCHI et al., 1973). Among sympathomimetic amines tested, i.e., dopamine, noradrenaline, adrenaline, isoprenaline, phenylephrine, methoxamine, ephedrine, tyramine, epinine, α-methyldopamine and 6-hydroxydopamine, only dopamine (1–10 μg) and 6-hydroxydopamine (100 μg), one of metabolites of dopamine (SENOH et al., 1959) stimulated the secretion, but the latter was far less active than the former. Others had either no effect, or rather an inhibitory effect, on the secretion. We could not find any stimulation of the pancreatic secretion with epinine given intra-arterially, although GREENGARD et al. described a striking stimulation by an intravenous injection. The contradictory results may be due to impurity of epinine used in their experiments. The intra-arterial administration has the merit of excluding an indirect effect. Among amino acids tested, L-dopa induced a profuse and long-lasting secretion by the intra-arterial infusion at a rate of 100 μg/min for 10 min with a delay up to 5 min for the induction of secretion. Tyrosine and phenylalanine had no stimulatory effect on the pancreatic secretion. Among endogenous substances tested, acetylcholine, histamine, gastrin and glucagon were effective to induce the secretion, and foreign substances such as dibutyryl cyclic AMP, cholinergic drugs, ganglion stimulants, methylxanthines, nitroglycerin, apomorphine and fusaric acid were found to be effective. The stimulation caused by these substances, however, were far less than that induced by L-dopa or dopamine. It was interesting that bradykinin, kallikrein and vasopressin inhibited the spontaneous secretion while angiotensin had no effect. Nucleotides, nucleosides, 5-HT, various vasodilators and cocaine had no effect. The secretory effects of these drugs did not relate with their vascular ones.

MECHANISM OF THE STIMULATION OF THE PANCREATIC SECRETION INDUCED BY L-DOPA OR DOPAMINE

The mode of action of dopamine intra-arterially was quite similar to that of secretin, causing a larger volume of juice with higher concentrations of bicarbonate and protein (Furuta et al., 1972). The secretion was promptly induced and its volume increased in a dose-related manner without any tachyphylaxis. The higher dose was given, the higher concentrations of bicarbonate and protein were measured in a larger volume of juice. The potency of dopamine (10 μg) intra-arterially was approximately equal to that of secretin (Boots) (0·3 units) or secretin (Jorpes) (0·03 units) (Hashimoto et al., 1971). The pancreatic juice stimulated by the infusion of L-dopa contained naturally a high concentration of bicarbonate ions.

The increased secretion induced by the intra-arterial infusion of L-dopa (100 μg/min) was antagonised by Ro 4-4602 (300 μg), a dopa decarboxylase inhibitor, while the secretagogue effect of dopamine (1–10 μg) intra-arterially was not affected by Ro 4-4602. The effect of dopamine was enhanced by the infusion of fusaric acid (100 μg/min), a dopamine-β-hydroxylase inhibitor and also enhanced by treatment with nialamide (100 mg/kg), a monoamine oxidase inhibitor (Furuta et al., 1973). The effect of α-methyl dopa was further investigated, which showed an effective antagonism between L-dopa and α-methyldopa but not between dopamine and the latter, suggesting the inhibition of dopa decarboxylase induced by α-methyldopa (Fig. 1). α-Methyldopa might be converted to α-methyldopamine which did not

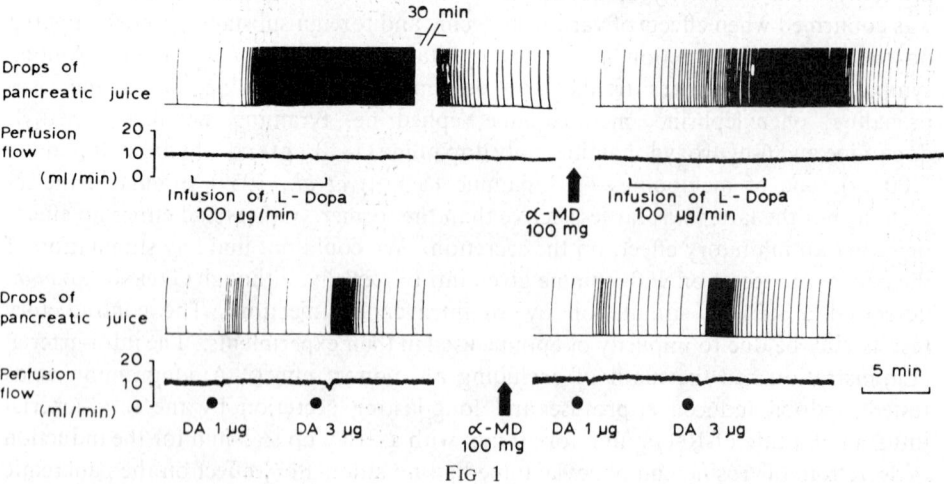

FIG 1

stimulate the exocrine pancreas. We assume that L-dopa will not be a secretagogue but it is converted to dopamine in the exocrine pancreas cells which induced the dopaminergic effect of the stimulation of the secretion. The intracellular level of dopamine may be controlled by enzymatic equilibrium.

CONTENT OF CATECHOLAMINES IN THE PANCREATIC TISSUE

The canine pancreas was quickly removed under anesthesia with sodium pentobarbitone and immediately frozen. The duodenal portion was used for the assay (SHELLENBERGER and GORDON, 1971). Reserpine (0·1 mg/kg s.c. for 5 days) led to selective depletion of noradrenaline while the L-dopa infusion increased selectively the dopamine content in the pancreatic tissue. It is worthy to note that various treatments produced an apparent differentiation between the noradrenaline and the dopamine contents in the pancreatic tissue (Table 1).

TABLE 1. THE DOPAMINE AND THE NORADRENALINE CONTENTS IN THE CANINE PANCREAS

Treatment	Number of animals	Dopamine (ng/g)	Noradrenaline (ng/g)
None	13	139 ± 6	375 ± 40
L-DOPA (5 mg/kg i.v.)	5	818 ± 204*	563 ± 109
Reserpine (0·1 mg/kg s.c., 5 days)	5	135 ± 8	14 ± 5*

* $P < 0.01$

EFFECT OF HALOPERIDOL AND APOMORPHINE

The dopamine-induced secretion was not modified by atropine, phentolamine, propranolol, guanethidine or tetrodotoxin (HASHIMOTO et al., 1971) while haloperidol (1 mg) intra-arterially attenuated the dopamine-induced pancreatic secretion (FURUTA et al., 1973). The vasoconstriction induced by dopamine was completely blocked by phentolamine and that by perivascular stimulation was blocked by tetrodotoxin. The vasodilation induced by dopamine after α-adrenoceptor blockade was also completely blocked by haloperidol. Apomorphine (1 mg) intra-arterially stimulated markedly the exocrine pancreas. Its effect was long-lasting. After the stimulated secretion returned to the initial level, the dopamine-induced stimulation was significantly blocked but the secretin-induced one was not blocked. This characteristic effect may suggest some similarity to the depression of the vomiting center by subsequent doses of apomorphine (GOODMAN and GILMAN, 1970). Such specific antagonism was previously observed by GOLDBERG and MUSGRAVE (1971) in the renal circulation.

EFFECTS OF COCAINE AND DESMETHYLIMIPRAMINE ON THE DOPAMINE-INDUCED SECRETION OF THE EXOCRINE PANCREAS

Cocaine 1 mg intra-arterially enhanced the L-dopa- or the dopamine-induced pancreatic secretion but not the secretin-induced one (Fig. 2). Desmethylimipramine had a similar effect. These phenomena can be understood as follows: The transport of dopamine across the membrane of zymogen granules is blocked with cocaine and

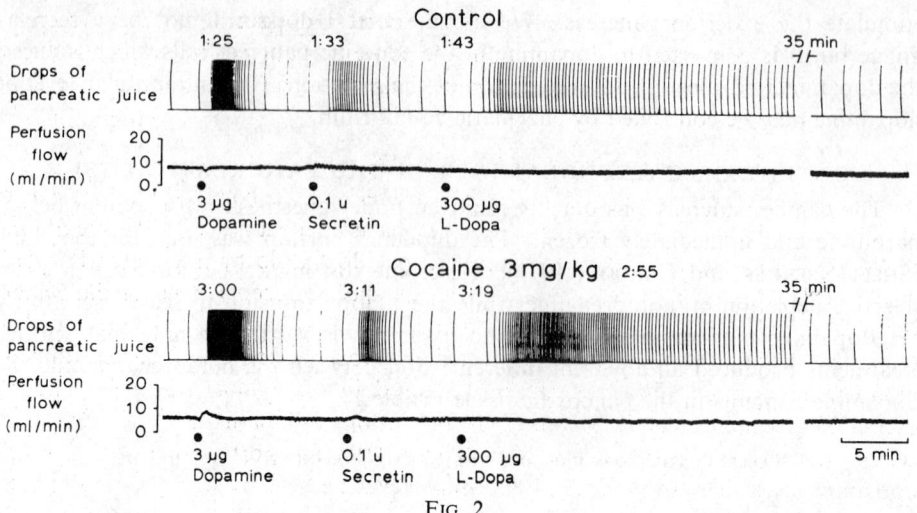

FIG. 2

consequently dopamine acts longer at the receptor sites. Previously ALM and his collaborators observed first diffuse cytoplasmic and later incorporated fluorescence in zymogen granules after L-dopa i.v. The similar phenomenon was observed also with dopamine injection but only a weak fluorescence even in larger doses. Thus they suggested the highly efficient mechanism for the uptake of L-dopa presumably through some active mechanism in the cell membrane (ALM et al., 1967). Their assumption will be natural, because the active transport of amino acids prevails in the pancreas for the protein synthesis.

ABSENCE OF INTERACTION BETWEEN DOPAMINE AND SECRETIN

The stimulation of exocrine pancreas by secretin and that by dopamine were additive and they neither inhibited nor potentiated each other, although both substances had a quite similar mode of action to increase the volume of pancreatic juice and the excretion of bicarbonate ions and protein. By lowering Ca ions concentration in the circulating blood by GEDTA infusion, the stimulation of secretion induced by either secretin or dopamine was profoundly reduced (IWATSUKI et al., 1973b). On the other hand, the difference between secretin and dopamine existed at the receptor site: (1) Prostaglandin $F_{2\alpha}$ (100 μg) inhibited the secretin-induced secretion but it did not modify the dopamine-induced one (IWATSUKI et al., 1973a). (2) On the other hand haloperidol (FURUTA et al., 1973) and apomorphine antagonised the dopamine-induced secretion but not the secretin-induced one. (3) Either cocaine or desmethylimipramine enhanced the dopamine-induced secretion but not the secretin-induced one.

THE CONTROL MECHANISM OF THE PANCREATIC SECRETION INDUCED BY L-DOPA OR DOPAMINE AT THE CELLULAR LEVEL

The incorporation of dopamine in zymogen granules may be probably understood as one of inactivation process, because dopamine will be quickly degradated in the alkaline medium of the pancreatic juice when granules are extruded into the pancreatic ducts. Furthermore dopamine in the cytoplasm is broken down with monoamine oxidase. Kallikrein inhibited the spontaneous secretion of the pancreas and also the dopamine-induced secretion. Thus we can not neglect some possibility of control mechanism of spontaneous pancreatic secretion by natural constituents in the pancreatic tissue.

CONCLUSION

Although the results obtained in the organ level must be confirmed in the cellular and the subcellular levels, we assume that L-dopa is taken up across the membrane of the exocrine pancreas cell and is converted to dopamine by dopa decarboxylase in the cytoplasm. Dopamine interacts with the specific receptor sites of the acinar, centroacinar and ductule cells and stimulates the pancreatic secretion, meanwhile the excess of dopamine is incorporated in the zymogen granules or broken down by monoamine oxidase. When dopamine is stored in the zymogen granules and no excess of dopamine remains in the cytoplasm, the L-dopa-induced secretion will cease. Recently, NG, COLBURN and KOPIN (1972) reported that incorporated dopamine in particles of rat brain homogenates was released by addition of dopamine in the suspending medium. The incorporation in zymogen granules could be understood as an exchangable storage of dopamine. ALM, EHINGER and FALCK pointed out that "L-dopa readily accumulates in the pancreas but is not taken up to significant extent into the adrenergic nerves or into the other mono-aminergic cell systems as the endocrine pancreas". It may be given as a conclusion of these studies that the dopaminergic system exists in the pancreas independently of the adrenergic system.

Acknowledgement—This study was performed partly by grants of Funai's Bounty Inc., Pharmacological Research Foundation, Inc. and Squibb Institute for Medical Research. The authors express many thanks to Dr. K. Saameli, Biological and Medical Research, Sandoz for supplying prostaglandins and epinine, to Dr. Langemann, Hoffmann-La Roche for supplying α-methyl-dopamine and Dr. C. A. Stone, Merck Sharp and Dohme for supplying (+) and (−) α-methyl-dopamine and to Dr. Jorpes, Karolinska Institute for supplying pure secretin.

REFERENCES

ALM P., EHINGER B. and FALCK B. (1967) *Life Sci.* 6, 913–917.
FURUTA Y., IWATSUKI K., TAKEUCHI O. and HASHIMOTO K. (1972) *Tohoku J. exp. Med.* 108, 353–360.
FURUTA Y., HASHIMOTO K., IWATSUKI K. and TAKEUCHI O. (1973) *Br. J. Pharmac.* 47, 77–84.
GOLDBERG L. I. and MUSGRAVE G. E. (1971) *Pharmacologist* 13, 227.
GOODMAN L. S. and GILMAN A. (1970) *The Pharmacological Basis of Therapeutics*, Edition 4, p. 252.
GREENGARD H., ROBACK R. A. and IVY A. C. (1942) *J. Pharmac. exp. Ther.* 74, 309–318.
HASHIMOTO K., SATOH S. and TAKEUCHI O. (1971) *Br. J. Pharmac.* 43, 739–746
HOLZ P., CREDNER K. and STRÜBING C. (1942) *Archs. exp. Path. Pharmak.* 199, 145–152.
IWATSUKI K., FURUTA Y. and HASHIMOTO K. (1973a) *Experientia* 29, 319.
IWATSUKI K., FURUTA Y., HASHIMOTO K. and ISHII Y. (1973b) *Jap. J. Pharmac.* 23, Suppl. 54.
NG L. K. Y., COLBURN K. W. and KOPIN I. J. (1972) *J. Pharmac. exp. Therap.* 183, 316–325.
SCHÜMANN H. J. and HELLER I. (1959) *Archs exp. Path. Pharmak.* 236, 474–482.
SHELLENBERGER M. K. and GORDON J. H. (1971) *Anal. Biochem.* 39, 356–372.
SENOH S., CREVELING C. R., UDENFRIEND S. and WITKOP B. (1959) *J. Amer. Chem. Soc.* 81, 6236–6240.
TAKEUCHI O., SATOH S. and HASHIMOTO K. (1971) *Tohoku J. exp. Med.* 104, 203–204.
TAKEUCHI O., SATOH S. and HASHIMOTO K. (1973) *Jap. J. Pharmac.* in press.

Frontiers in Catecholamine Research 1973, pp. 831 to 833. Pergamon Press. Printed in Great Britain.

CATECHOLAMINE BIOSYNTHETIC ENZYMES IN PANCREATIC ISLET CELLS

Harold E. Lebovitz and Robert W. Downs, Jr.

Division of Endocrinology, Departments of Medicine and Physiology,
Duke University Medical Center, Durham, North Carolina 27710, U.S.A.

Histochemical studies using fluorescent techniques show that dopamine or serotonin or both are present in the cytoplasm of pancreatic islet cells of a number of animals, such as guinea pig, cat, dog, pig and in human fetus (Cegrell 1968). The islet cells of other species such as albino rabbit, albino mouse and golden hamster do not ordinarily contain histochemically stainable monoamines. However, after administration of the monoamine precursors L-dopa or 5-hydroxytryptophan, to these animals, their islet cells show intense histochemical monoamine fluorescence (Cegrell, 1970). Adrenergic nerves containing norepinephrine are present in abundance in the islets of the golden hamster but are only occasionally found in the islets of other species, such as the rabbit, mouse or man (Cegrell, 1968).

Dopamine and serotonin significantly inhibit stimulated insulin secretion from pancreatic islets both *in vivo* and *in vitro* in many animal species (Quickel *et al.*, 1971a) Serotonin antagonists potentiate stimulated insulin secretion *in vitro* from pieces of pancreas from golden hamster and rabbits (Lebovitz and Feldman,1973; Feldman *et al.*, 1972) and *in vivo* in humans with adult onset diabetes mellitus (Quickel *et al.*, 1971b).

From these data it is appealing to hypothesize an important regulatory role for pancreatic β cell monoamines as tonic inhibitors of insulin secretion. In this schema, insulin secretion could be viewed as the net result of external stimulatory agents, such as substrates and hormones, and intracellular inhibitory agents, such as dopamine or serotonin, or both (Lebovitz and Feldman, 1973).

Enigmatic in the above schema are the questions of the origin and regulation of the concentration of the pancreatic β cell monoamines and the significance of the species variation in both the occurrence and nature of the pancreatic β cell monoamines. The present communication is a preliminary progress report of our studies on the biosynthesis of monoamines in pancreatic islet cells.

Pancreatic islet tissue from several sources were examined for tyrosine hydroxylase and L-aromatic amino acid decarboxylase activity (EC 4.1.1.26). Isolated pancreatic islets were obtained by collagenase treatment of pancreas from albino rabbits, pigmented and albino guinea pigs and golden hamsters (Lacy *et al.*, 1972). The β cells of the albino rabbit ordinarily contain no histochemically stainable monoamines; those of the pigmented and albino guinea pig contain large quantities of dopamine and serotonin; and the β cells of the golden hamster contain no monoamines, but their islets have an extensive adrenergic nerve innervation (Cegrell, 1968; Jaim-Etcheverry and Zieher, 1968). A transplantable golden hamster islet cell carcinoma whose cells contain large quantities of cytoplasmic dopamine and serotonin (Cegrell *et al.*, 1969a) and normal hamster adrenals were also assayed for the catecholamine biosynthetic enzymes. Tyrosine hydroxylase was determined by the generation of

$^{14}CO_2$ from 1-^{14}C-L-tyrosine by coupled decarboxylation of the L-dopa formed (WAYMIRE et al., 1971). L-Aromatic amino acid decarboxylase was measured by the conversion of 2-^{14}C-D,L-dopa to ^{14}C-dopamine (CREVELING and DALY, 1971). The supernatant protein in each homogenate was determined as Lowry protein.

TABLE 1. TYROSINE HYDROXYLASE AND L-AROMATIC AMINO ACID DECARBOXYLASE
ACTIVITY IN PANCREATIC ISLETS
Data are expressed as mean ± s.e. The number of determinations are given in the parentheses.

Tissue	Tyrosine hydroxylase (n moles/hr/mg protein)	L-aromatic amino acid decarboxylase (μ moles/hr/mg protein)
Hamster adrenal	10·356	1·937 ± 0·454 (5)
Hamster trans- plantable islet cell carcinoma	1·590 ± 0·441 (7)	16·413 ± 1·776 (5)
Isolated hamster islets	0	5·058 ± 0·070 (2)
Isolated rabbit islets	0	0·280 ± 0·073 (3)
Isolated guinea pig islets	0	2·976 ± 0·049 (2)

Table 1 depicts the results of these studies. All of the isolated pancreatic islets contained measureable quantities of L-aromatic amino acid decarboxylase. The difference in activity is striking. The transplantable hamster islet cell carcinoma has an extremely high activity with isolated hamster islets being about 1/3 as active, guinea pig islets about 1/6 and rabbit islets about 1/50. Tyrosine hydroxylase activity could not be detected in any of the isolated islets. The transplantable islet cell carcinoma contained significant tyrosine hydroxylase activity.

The presence of significant L-aromatic amino acid decarboxylase activity in the islets of all the species suggest that most, or all pancreatic islet tissue can readily convert L-dopa or 5-hydroxytryptophan to their appropriate monoamines and thereby increase intracellular pools of these amines when exposed to sufficient precursor.

The absence of measureable tyrosine hydroxylase activity in isolated hamster and rabbit islets might be expected, since these islets ordinarily do not show intracellular monoamines by histochemical fluorescent studies. In contrast the failure to demonstrate tyrosine hydroxylase activity in isolated guinea pig islets was unexpected, since these islets have been shown to contain considerable dopamine (0·594 ± 0·077 μg/g wet wt) and serotonin (JAIM-ETCHEVERRY and ZIEHER, 1968). The demonstration of tyrosine hydroxylase and L-aromatic amino acid decarboxylase activities in the hamster islet cell carcinoma confirm previous reports using somewhat different techniques (CEGRELL et al., 1969b; AXELSSON et al., 1970). The presence of these enzymes can account for the high concentrations of monoamines (dopamine 2·75 ± 0·35 μg/g wet wt; 5-hydroxytryptamine 0·87 ± 0·09 μg/g) found in these tumor cells (CEGRELL et al., 1969a).

The data fail to provide a clear understanding of the origin and mechanism of species variation in pancreatic islet cell monoamines. The failure to demonstrate tyrosine hydroxylase in the guinea pig islets suggest that the dopamine normally present in the β cells of some species may be (1) the result of tyrosine hydroxylase activity that is below the level of sensitivity of current assays, (2) the consequence of conversion of circulating L-dopa that is taken up by the β cells, (3) the uptake and storage of circulating dopamine by the β cell, or, (4) the synthesis of dopamine by some as yet unidentified biosynthetic pathway. We are unable to comment about the origin of β cell serotonin, as measurements of tryptophan hydroxylase activity in these islets have not been completed.

Further definition of the physiologic role of pancreatic β cell monoamines in the regulation of insulin secretion must await the elucidation of the origin and control mechanisms regulating the type and quantity of these amines.

Acknowledgements—Supported by grants from the National Institute of Arthritis, Metabolic and Digestive Diseases (AM 01324, 5TI AM 5074 and K3 AM 17,954)

REFERENCES

AXELSSON S., CEGRELL L. and ROSENGREN A. M. (1970) *Experientia* **25**, 998–999.
CEGRELL L. (1968) *Acta Physiol. Scand. (Suppl.)* **314**, 1–60.
CEGRELL L., FALCK B. and ROSENGREN A. M. (1969a) *Acta Physiol. Scand.* **77**, 23–27.
CEGRELL L., FALCK B. and ROSENGREN A. M. (1969b) *Experientia* **25**, 969–970.
CEGRELL L. (1970) In: *The Structure and Metabolism of the Pancreatic Islets.* (FALKMER S., HELLMAN B. and TALJEDAL I. B., Eds.) pp. 131–150. Pergamon Press, Oxford.
CREVELING C. R. and DALY J. W. (1971) In: *Methods of Biochemical Analysis: Analysis of Biogenic Amines and Their Related Enzymes.* (GLICK D. ed.) pp. 153–182. Interscience Publishers, New York.
FELDMAN J. M., QUICKEL K. E., JR. and LEBOVITZ H. E. (1972) *Diabetes* **21**, 779–788.
JAIM-ETCHEVERRY G. and ZIEHER L. M. (1968) *Endocrinology* **83**, 917–923.
LACY P. E., WALKER M. M. and FINK C. J. (1972) *Diabetes* **21**, 987–998.
LEBOVITZ H. E. and FELDMAN J. M. (1973) *Fed. Proc.* **32**, 1797–1802.
QUICKEL K. E. JR., FELDMAN J. M. and LEBOVITZ H. E. (1971a) *Endocrinology* **89**, 1295–1302.
QUICKEL K. E. JR., FELDMAN J. M. and LEBOVITZ H. E. (1971b) *J. Clin. Endocrinol. Metab.* **33**, 877–881.
WAYMIRE J. C., BJUR R. and WEINER N. (1971) *Anal. Biochem.* **43**, 588–600.

Frontiers in Catecholamine Research 1973, pp. 835 to 841. Pergamon Press. Printed in Great Britain.

BRAIN MONOAMINES PARTICIPATION IN THE CONTROL OF GROWTH HORMONE SECRETION IN DIFFERENT ANIMAL SPECIES

EUGENIO E. MÜLLER

Second Chair Department of Pharmacology, University of Milan, Italy

A LARGE body of evidence has recently been accumulated for the involvement of brain monoamines in the control of growth hormone (GH) secretion. While there seemed to be sufficient proof of the existence of a dual neurohormonal control of the secretion of this hormone from the anterior pituitary exerted by specific GH-releasing (GRF) and inhibiting (GIF) factors (1), information was still scanty regarding the possible neural influences impinging on the neurosecretory structures responsible for the elaboration of the releasing factors. The introduction of a fluorescence method for the histochemical localization of the neurotransmitters (2) was a turning point in the understanding of their physiological role in the release of anterior pituitary hormones.

This article is an attempt to consisely review the present status of knowledge on the participation of central nervous system (CNS) dopamine (DA), norepinephrine (NE) and serotonin (5-HT) in the control of GH secretion in different animal species.

STUDIES IN RODENTS

The first neuro-pharmacological studies performed in the rat showed that drugs endowed with anti-adrenergic effects (reserpine, chlorpromazine, tetrabenazine, α-methyl-*m*-tyrosine, α-methyl-dopa) were capable of blocking the release of GH induced by insulin hypoglycemia (tibial plate bioassay) (3,4). Administration of NE and DA into the lateral ventricles of normal and hypophysectomized rats to circumvent the blood brain barrier to catecholamines (CA) (5) was followed by decreased GH levels in the anterior pituitary and by depletion of hypothalamic GRF.

Central injection of 5-HT was ineffective (6,7). When given intraventricularly (IVT), NE induced a depletion of pituitary GH activity at doses at which DA was ineffective; this finding suggested a role for NE as the synaptic transmitter for the release of GRF (7). Subsequent development of radioimmunological (RIA) techniques for the measurement of rat GH in the plasma (8), revealed discrepancies between RIA and bioassay (BA) data (9) and affected conclusions in neuropharmacological studies of GH regulation in the rat. In this connection, COLLU et al. (10) reported a reduction of plasma GH levels after IVT administration of DA and a highly significant rise following central injection of 5-HT. KATO et al. (11) observed a significant increase in plasma GH after systemic administration of either phentolamine, an α-adrenergic blocker, or isoproterenol, an adrenergic β-stimulant drug. No increases were detected in response to L-dopa, the immediate biological precursor of both DA and NE (12).

A recent re-investigation of the role of brain monoamines in the control of GH secretion in the rat by using the RIA technique has been undertaken. (COCCHI *et al.*, paper in preparation). These studies assigned a negative role for DA but no confirmation could be obtained for the stimulatory action of 5-HT. This role seems, instead, to be played by NE. Thus, administration of FLA-63, a selective blocker of NE biosynthesis (13), lowered plasma RIA-GH, while central sympathectomy by 6-hydroxydopamine (6-OHDA) or systemic administration of α-methyl-*p*-tyrosine (α-MT), procedures which affect both the noradrenergic and the dopaminergic function (14,15), did not modify plasma RIA-GH. The previous findings on the stimulatory role of DA on GH release (BA method) might be accounted for the high rate of conversion of IVT-injected DA to NE present at hypothalamic level (16).

In mouse, a species which appears particularly refractory to GH-releasing stimuli (17), SINHA *et al.* (18) observed a dramatic decrease in the serum GH levels following systemic administration of either DA or L-dopa and reduced levels were also present after intraperitoneal injection of epinephrine (E). Conversely, central administration of 6-hydroxydopamine markedly increased circulating GH levels (MÜLLER *et al.*, unpublished data).

PIG, SHEEP AND DOG

Data on the involvement of brain amines in the regulation of GH secretion in the pig are rather scanty. In this species, in which the response to GH-releasing stimuli appears to be blunted, no rise in plasma GH has been reported following subcutaneous injection of 1 mg of epinephrine; equally ineffective in this context, was peripheral administration of NE (19). In the sheep, plasma levels of GH were constantly low during an intravenous infusion of E and the plasma GH response to arginine infusion was abolished. Plasma non-esterified fatty acid (NEFA) levels were greatly increased during the 3 hr period of epinephrine infusion (20). Intra-carotid infusion of phenoxybenzamine, an α-adrenergic blocker, resulted in a late increase in plasma levels of GH in mature cyclic ewes and produced additive effects when combined with an arginine stimulus. Infusion of L-dopa by itself had no effect on plasma levels of GH, but suppressed the expected increase of GH due to arginine (21). In the dog, intravenous L-dopa administration induced instead, a brisk increase in growth hormone levels, which was inhibited by phenoxybenzamine given intraventricularly (22).

PRIMATES

Monkey

Single injections of large doses of epinephrine provoked a marked and prompt increase in the plasma GH levels of female rhesus monkeys (23).

Studies conducted in conscious fasted baboons showed that α-adrenergic blockade, resulting from intravenous infusion of phentolamine, significantly depressed GH secretion, while β-adrenergic blockade with propranolol and ganglionic blockade with trimetaphan were associated with a significant prompt rise in GH. The observed GH changes seemed not to be associated with alterations in plasma glucose or NEFA levels, although, lowered plasma NEFA and glucose levels were present following propranolol administration (24). Infusion of much smaller doses of phentolamine directly into the third ventricle or anterior hypothalamus evoked a similar fall in

TABLE 1.

Drug	Animal species	Route of administration	←Dosage←	Effect	References
Epinephrine	Monkey	i.v.	5–40 µg/kg	↑	23
	Human	s.c.	10 µg/kg	→	51
		i.v.	0·14 µg/kg/min	→	35
Norepinephrine	Monkey	IVT	1 µg/min/30 min	→	27
		microinjected into VMN	10 µg	↑	26
	Human	i.v.	0·05–0·2 µg/kg/min	→	36
Methyl-amphetamine	Human	i.v.	15 mg/5 min	↑	46
Dopamine	Monkey	IVT	5 µg	↓	27
L-Dopa	Human	p.o.	500–1000 mg	↑	30
Apomorphine	Human	s.c.	0·75–1·5 mg	↑	38
Phentolamine	Monkey	i.v.	0·013 mg/kg/min	↓	24
		IVT	1–5 µg/min/30 min	↓	25
	Human	i.v.	0·5 mg/min/75 min	inhibits L-dopa	32
		i.v.	0·5 mg/min/90 min	inhibits insulin hypoglycemia	33
		i.v.	0·5 mg/min/90 min	inhibits vasopressin	52
		i.v.	3·5 mg/50 min	inhibits propranolol + E	36
		i.v.	0·5 mg/min/120 min	no effect on GH sleep peak	53
		i.v.	0·5 mg/min/120 min	inhibits arginine	54
Propranolol	Monkey	i.v.	0·011 mg/kg/min	↑	24
	Human	i.v.	0·21 mg/kg/10 min	→	35
		i.v.	0·20 mg/kg/120 min	↑	34
		i.v.	0·15 mg/kg/few min	potentiates amphetamine	46
		i.v.	0·15 mg/kg/few min	potentiates insulin hypoglycemia	33
Propranolol + E	Human	i.v.	0·21 mg/kg/10 min 0·07–0·14 µg/kg/min/ 30 min	↑	35
Timoxamine	Human	i.v.	0·1 mg/kg	potentiates amphetamine	46
Haloperidol	Human	p.o.	2 mg	attenuates insulin hypoglycemia	42
Chlorpromazine	Human	p.o.	100 mg	↓	43
Serotonin	Monkey	microinjected into VMN	10 µg	→	26
	Human*			↑	44
5-hydroxytryptophan	Human	p.o.	150 mg	↑	45
Tryptophan	Human	p.o.	70 mg/kg	slight increase	Author's unpublished data

→ means no effect
↓ means stimulation

↑ means inhibition

* Patients with carcinoid syndrome

GH (25). To exclude the possibility that centrally-administered phentolamine was acting systemically to lower GH, the blocking agent was infused into the abdominal inferior vena cava. It was necessary to infuse a dose 2 to 4 times greater to lower GH significantly (25). These data suggested the existence in the CNS of the baboon of α-adrenergic receptors regulating GH secretion. It was gratifying to observe that NE microinjected into the ventromedial nucleus of the hypothalamus consistently elevated GH in conscious baboons. Micro injections of 5-HT in the same area did not elevate GH (26). In contrast to central administration of NE, intrahypothalamic infusion of DA lowered plasma GH, while NE and isoproterenol given systemically induced no change in the hormone level (27).

Human

In man the infusion of epinephrine in doses sufficient to produce significant hyperglycemia did not provoke GH secretion, nor blunt arginine-stimulated GH release (28,29). The poor penetrability of the blood–brain barrier by CA was circumvented by administering L-dopa, a drug that easily crosses the blood–brain barrier and increases brain levels of both DA and NE (12). The results obtained showed that moderate doses of L-dopa caused a significant rise in plasma GH levels in patients with parkinsonism, requiring gram amounts of the substance (30) as well in normal subjects (31). The stimulatory effect on GH secretion by L-dopa was not blocked by either oral or intravenous glucose (30), but was reduced or potentiated respectively, by the concomitant infusion of phentolamine (32) or propranolol (MASSARA and CAMANNI, unpublished results), stressing the involvement of adrenergic α and β-receptors in the regulation of GH secretion. In line with these findings, previous studies had shown that blockade of α-adrenergic receptors by phentolamine in man prevented the increase in plasma GH that follows insulin-induced hypoglycemia in normal subjects. In contrast, β-adrenergic receptors blockade by propranolol augmented the levels of GH in hypoglycemic subjects, although this was associated with increased hypoglycemia and a decrement in NEFA (33).

However, in this study neither α nor β-adrenergic blockade had a detectable effect in the absence of hypoglycemia, although a stimulant effect was reported for propranolol alone in Japanese subjects (34). Systemic infusions of propranolol combined with epinephrine resulted in increased GH levels (35). In this instance the stimulatory effect of epinephrine on plasma HGH in the presence of propranolol was the result of an unopposed α-receptor activity, since β-stimulation would be expected to inhibit GH secretion. Accordingly, the epinephrine-propranolol stimulation was blocked by phentolamine (36). In addition to L-dopa, apomorphine a direct stimulant of DA receptors (37) induced a rise of HGH (38), an effect which is compatible with a dopaminergic mechanism in the release of HGH. In this line are the observations that diethyldithiocarbamate, a blocker of NE biosynthesis (39) did not prevent the HGH rise due to L-dopa (GIORDANO, MINUTO, MARUGO, BARRECA, FOPPIANI, personal communication) and that haloperidol or chlorpromazine, two neuroleptic drugs mainly antagonistic to DA (40,41) blunted the insulin-induced HGH rise (42,43).

With regard to 5-HT, although increased GH levels have been found in plasma of patients with excessive 5-HT secretion due to the carcinoid syndrome (44) or following administration of the biological precursor, 5-hydroxytriptophan (45), much more information is needed before its role in the GH-releasing mechanism(s) can be definitely assessed. In our hands, the stimulatory effects of large amounts of triptophan on HGH secretion appeared to be slight and rather erratic (MÜLLER, BRAMBILLA, CAVAGNINI, PERACCHI, PANERAI, unpublished results).

Table 1 deals with the effects of drugs altering brain monoamines on GH secretion in primates.

CONCLUDING REMARKS

The studies mentioned are without doubt compatible with the view of brain monoamines intervention in the process of GH secretion; however, some aspects of this neurohumoral control are still unclear and warrant further investigations. In the rat it seems probable that the two adrenergic neurotransmitters NE and DA exert a dual effect on the secretion of RIA-GH, the former having a stimulatory influence, being inhibitory the latter. This concept is in line with the demonstration of both GRF and GIF factors in the rat hypothalamus. In the mouse the central DA tone seems to be predominant : in fact, concomitant reduction of both NE and DA due to 6-OHDA, resulted in increased GH levels in this species. No major role seems to be played by 5-HT, although data are discordant in this regard.

In the pig and the sheep the available evidence is rather scanty and the results available, if ever, favour an inhibitory role for NE (phenoxybenzamine administration in the sheep). The proof presented for an inhibitory role of the adrenergic system in these species is questionable since it is mainly based on the effect of systemic administration of high doses of epinephrine, a drug which does not cross appreciably the blood–brain barrier and might be capable to induce the observed effects through the feedback action of increased NEFA levels. This same observation also applies to some of the results obtained in rodents or the human following systemic administration of this adrenergic compound. In the dog the few available data point to a stimulatory role of the adrenergic system in the control of GH.

The same dual role for NE and DA present in the rat seems to extend to the monkey, a species in which the stimulant action of NE on GH release is particularly well documented as well as the inhibitory effect of α-adrenergic blockers. However, the inhibitory effect of DA in this species rests on a single experiment and confirmation of this point would be desirable.

In the human, the evidence available is compatible with the hypothesis that both NE and DA serve as neurotransmitters controlling GH release. The fact that L-dopa administration causes a GH rise is not "per se" proof of a stimulant role of DA, since L-dopa could act to increase NE levels in the hypothalamus or limbic system and this effect would mediate GH release. Quite interestingly in fact, both phentolamine and propranolol interacted with the effect of L-dopa. However, suggestions for a direct stimulation of DA receptors in the release of GH come from the reported stimulant action of apomorphine, the persistence of L-dopa activity following blockade of NE synthesis and the suppressive effects of antidopaminergic drugs. The likelihood of a dopaminergic mechanism in GRF control is also supported by the effectiveness of amphetamine derivatives to stimulate HGH secretion, and action which is not suppressed by the α blocker, thymoxamine (46). It is known, in fact, that some of the central effects of the sympathomimetic amines result by an enhancement of DA receptor activity (47).

The possibility that the stimulatory effect of L-dopa on GH secretion in the human might be due to a final activation of a serotoninergic receptor, for a displacement of the brain indoleamine from vesicular stores (48), is appealing. It appears, however, rather unlikely on considering the very high doses of L-dopa necessary to elicit this effect in the laboratory animal (49) and the limited action of serotonin as GH releaser. Stricly connected with the nature of the transmitter(s) and still unresolved is the problem of its (their) site of action in the brain, a problem, which for reasons of economy cannot be discussed here. Several loci have been proposed at which monoamines might participate in the control of secretion from anterior pituitary (50).

REFERENCES

1. McCann S. M. and Porter J. C. (1969) *Physiol. Rev.* **49**, 240–284.
2. Falck B., Hillarp N. A., Thieme G. and Torp A. (1962) *J. Histochem. Cytochem.* **10**, 348–354.
3. Müller E. E., Saito T., Arimura A. and Schally A. V. (1967) *Endocrinology* **80**, 109–117.
4. Müller E. E., Sawano S., Arimura A. and Schally A. V. (1967) *Endocrinology* **80**, 471–476.
5. Axelrod J. (1965) *Recent Progr. Hormone Res.* **21**, 597–622.
6. Müller E. E., Dal Pra' P. and Pecile A. (1968) *Endocrinology* **83**, 893–896.
7. Müller E. E., Pecile A., Felici M. and Cocchi D. (1970) *Endocrinology* **86**, 1376–1381.
8. Schalch D. S. and Reichlin S. (1966) *Endocrinology* **79**, 275–280.
9. Schalch D. S. and Reichlin S. (1968) In: *Growth Hormone* (Pecile A. and Müller E. E. Eds) pp. 211–225, Excerpta Medica, Amsterdam.
10. Collu R., Fraschini F., Visconti P. and Martini L. (1972) *Endocrinology* **90**, 1231–1237.
11. Kato Y., Dupre' J. and Beck J. C. (1973) *Endocrinology* **93**, 135–146.
12. Andén N. E., Dahlstrom A., Fuxe K. and Larsson K. (1966) *Acta Pharmacol. Toxic.* **24**, 263–274.
13. Corrodi H., Fuxe K., Hamberger B. and Ljungdahl A. (1970) *Europ. J. Pharmac.* **12**, 145–155.
14. Uretsky N. J. and Iversen L. L. (1970) *J. Neurochem.* **17**, 269–273.
15. Spector S., Sjoerdsma A. and Udenfriend S. (1965) *J. Pharm. Exp. Ther.* **147**, 86–95.
16. Glowinski J. and Iversen L. L. (1966) *J. Neurochem.* **13**, 655–669.
17. Müller E. E., Miedico D., Giustina G. and Cocchi D. (1971) *Endocrinology* **88**, 345–350.
18. Sinha Y. N., Selby F. W., Lewis U. J. and Vanderlaan W. P. (1972) *Endocrinology* **91**, 784–792.
19. Machlin L. J., Takahashi Y., Horino M., Hertelendy F., Gordon R. S. and Kipnis D. M. (1968) In: *Growth Hormone* (Pecile A. and Müller E. E., Eds) pp. 292–305. Excerpta Medica, Amsterdam.
20. Hertelendy F., Machlin L. and Kipnis D. M. (1969) *Endocrinology* **84**, 192–199.
21. Davis S. L. and Borger M. L. (1973) *Endocrinology* **92**, 303–309.
22. Lovinger R., Connors M., Boryzcka A., Kaplan S. L. and Grumbach M. M. (1973) *Fifty Fifth Meet. Endocr. Soc.* Chicago.
23. Meyer V. and Knobil E. (1967) *Endocrinology* **80**, 163–171.
24. Werrbach J. H., Gale C. C., Goodner C. J. and Conway M. J. (1970) *Endocrinology* **86**, 77–82.
25. Toivola P. T. K., Gale C. C., Goodner C. J. and Werrback J. H. (1972) *Hormones* **3**, 193–213.
26. Toivola P. T. K. and Gale C. C. (1972) *Endocrinology* **91**, 895–902.
27. Toivola P. T. K. and Gale C. C. (1970) *Neuroendocrinology* **6**, 210–219.
28. Roth J., Glick S. M., Yalow R. S. and Berson S. A. (1963) *Science* **140**, 987–988.
29. Rabinowitz D., Merimee T. J., Nelson J. K., Schultz R. B. and Burgess J. A. (1968) In: *Growth Hormone* (Pecile A. and Müller E. E. eds) pp. 105–115, Excerpta Medica, Amsterdam.
30. Boyd A. E., Lebovitz H. E. and Pfeiffer J. B. (1970) *New Engl. J. Med.* **238**, 1425–1429.
31. Eddy R. L., Lloyd Jones A., Chakmajkan Z. H. and Silverthone M. C. (1971) *Fifty third Meet. Endocr. Soc.* A-210.
32. Kansal P. C., Buse J., Talbert O. R. and Buse M. G. (1972) *J. Clin. Endocr. Metab.* **34**, 99–105.
33. Blackard W. G. and Heidingsfelder S. A. (1968) *J. Clin. Invest.* **47**, 1400–1414.
34. Yawata M. and Fukase M. (1968) *J. Clin. Endocr. Metab.* **28**, 1079–1081.
35. Massara F. and Strumia E. (1970) *J. Endocr.* **47**, 95–100.
36. Massara F. and Camanni F. (1971) *J. Endocr.* **54**, 195–206.
37. Anden N. E., Rubenson A. A., Fuxe K., Hokfelt T. (1967) *J. Pharm. Pharmac.* **19**, 629.
38. Lal S., De La Vega C. E., Sourkes T. L. and Friesen H. G. (1972) *Lancet* **i**, 661.
39. Goldstein M., Anagnoste B., Lauber E. and McKeregan M. R. (1964) *Life Sci.* **3**, 763–767

40. Janssen P. A. J., Niemergeers C. J., Schellekens K. H. L., Lenaerts F. M. Verbruggen F. J., Van Neuten J. M. and Schaper W. K. A. (1970) *Europ. J. Pharm.* **11,** 139–154.
41. Neff N. H. and Costa E. (1967) In *Antidepressant drugs* pp. 28–34, Excerpta Medica, Amsterdam.
42. Kim S., Sherman L., Kolodny H. D., Benjamin F. and Singh A. (1971) *Clin. Res.* **15,** 718.
43. Sherman L., Kim S., Benjamin F. and Kolodny H. D. (1971) *New Engl. J. Med.* **284,** 72–74.
44. Feldman J. M. and Lebovitz H. E. (1972) In: *Abstracts 4th Int. Congr. Endocr.* p. 35, Excerpta Medica, Amsterdam.
45. Imura H., Nakai I., Yoshimi T. (1973) *J. Clin. Endocr. Metab.* **36,** 204–206.
46. Rees L., Butler P. W. P., Gosling C., Besser G. M. (1970) *Nature, Lond.* **228,** 565–566.
47. Carlsson A. (1970) In: *Amphetamine and related compounds* (Costa E. and Garattini S., Eds) pp. 289–300 Raven Press, New York.
48. Ng K. Y., Chase T. N., Colburn R. W. and Kopin I. J. (1970) *Science* **170,** 76–78.
49. Algeri S. and Cerletti C. (1972) In: *Abstracts V° Int. Congr. Pharmac.* p. 4 S. Francisco
50. Wurtman R. J. (1971) In: *Brain monoamines and endocrine function*, Neur. Res. Progr. Bull. **9,** 214–217.
51. Schalch D. S. (1967) *J. Lab. Clin. Med.* **69,** 256–269
52. Heidingsfelder S. A. and Blackard W. G. (1968) *Metabolism* **17,** 1019–1024.
53. Lucke C. and Glick S. M. (1971) *J. Clin. Endocr. Metab.* **32,** 729–736.
54. Buckler J. M. H., Bold A. M., Taberner M. and London D. R. (1969) *Brit. Med. J.* **3,** 153–154.

Frontiers in Catecholamine Research 1973, pp. 843 to 847. Pergamon Press. Printed in Great Britain.

EFFECTS OF L-DOPA ON PROLACTIN SECRETION IN HUMANS

Andrew G. Frantz, Han K. Suh and Gordon L. Noel

Department of Medicine, Columbia University College of Physicians and Surgeons and the Presbyterian Hospital, New York, N.Y., U.S.A.

Although prolactin was identified as a constituent of animal pituitaries more than 40 years ago, it is only within the last few years that the regulation of this hormone has come under intensive investigation, largely as a result of the development of sensitive radioimmunoassays in several species. Even more recent has been the demonstration of a human prolactin, separate from growth hormone, which circulates in human blood (Frantz and Kleinberg, 1970). The isolation of this hormone from monkey and human pituitaries (Lewis et al., 1971; Hwang et al., 1972) has permitted the development of homologous human radioimmunoassays (Hwang et al., 1971; Sinha et al., 1973). Heterologous radioimmunoassay systems have also been developed by which the human hormone can be studied (Jacobs et al., 1972; L'Hermite et al., 1972). Many physiologic and pharmacologic factors have been found to affect plasma prolactin in humans, including nursing or breast stimulation, sleep, stress, pregnancy, estrogens, sexual intercourse, neuroleptic drugs, thyrotropin releasing hormone (TRH), and L-dopa; these and other factors in the regulation of human prolactin have recently been reviewed (Frantz, 1973).

The studies to be presented here concern some effects of L-dopa on prolactin in humans. A prominent role of dopamine in the regulation of prolactin had been indicated by animal experiments, and we as well as others had found that L-dopa given orally to humans could produce transient depression of both pathologically elevated and normal plasma prolactin concentrations (Kleinberg et al., 1971; Malarkey et al., 1971; Friesen et al., 1972; Frantz et al., 1973). We had also found that L-dopa pre-treatment could inhibit the rise in prolactin produced by chlorpromasine (Kleinberg et al., 1971), a drug known to act at the hypothalamic level by blocking release of prolactin inhibiting factor (PIF). The present studies explore the effects of L-dopa on TRH-stimulated prolactin release, as well as its use in conjunction with the L-dopa decarboxylase inhibitor L-alpha-methyldopa hydrasine (MK-486).

MATERIALS AND METHODS

Subjects were healthy normal volunteers aged 21–27, studied in the morning after an overnight fast. Blood samples were collected through an antecubital vein at the start of the experiment. L-dopa and MK-486 were administered orally. TRH, either 100 μg or 500 μg, was given intravenously over 30 sec. Prolactin was measured by homologous human radioimmunoassay as previously described (Frantz et al., 1972). Growth hormone was also measured by radioimmunoassay (Frantz and Rabkin, 1964).

RESULTS

Response to TRH alone and with L-dopa pre-treatment

TRH given alone to 7 normal women and 7 normal men caused the expected acute rises in plasma prolactin shown by the solid lines in Figs. la and 1b. These results are

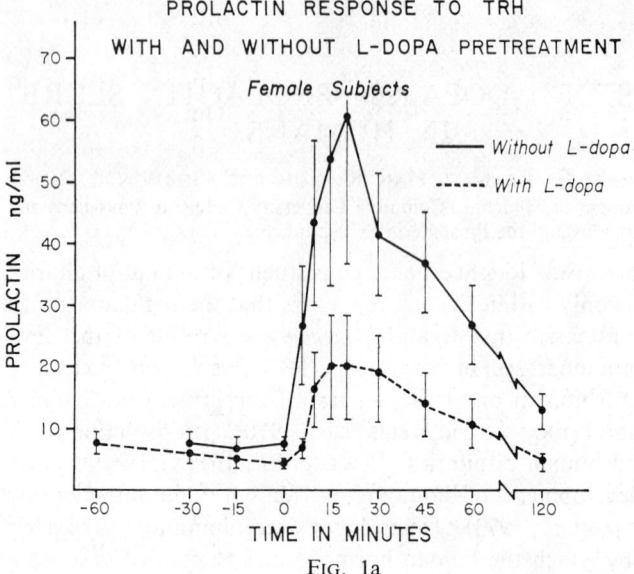

FIG. 1a

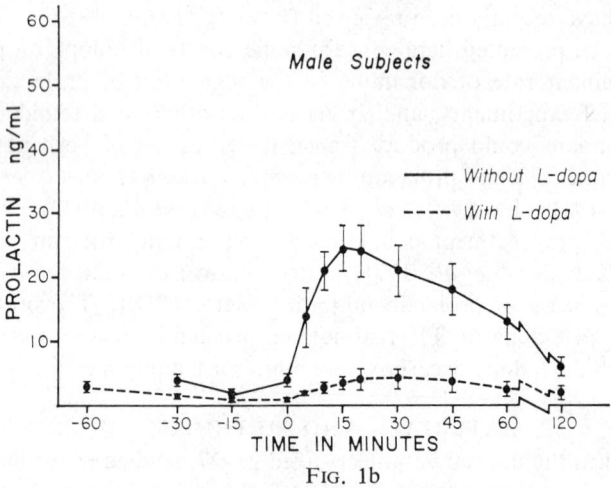

FIG. 1b

FIG. 1.—Prolactin response to intravenous TRH at 0 min without and with pretreatment with L-dopa. 1a (top): Response in 7 normal women. 1b (bottom):
Response in 7 normal men. Vertical bars denote standard error of the mean.

similar to what has been reported by others (BOWERS *et al.*, 1971; JACOBS *et al.*, 1971).
Women as a group were more responsive than men to TRH; a greater responsiveness
in women has also been noted to other kinds of prolactin-releasing stimuli (NOEL
et al., 1972). Several days after the studies with TRH alone, the same subjects were
given the same dose of TRH but were pretreated with 500 mg of L-dopa an hour
beforehand; 11 of the 14 subjects received an additional 250 mg of L-dopa 30 min

before the TRH. The responses are shown in the dotted lines of Figs. 1a and 1b. Except for one woman whose TRH response was unchanged by L-dopa, a clear suppressive effect was evident in all subjects. Mean prolactins after L-dopa were significantly lower at all time points from +10 to +60 min in the group of women as a whole ($P < 0.02$), as well as in the group of men ($P < 0.005$). No difference in peak prolactin response in either sex was evident after 100 μg of TRH as compared with 500 μg, either with or without L-dopa pre-treatment. Those who received 750 mg of L-dopa did not show more or less suppression than those who received 500 mg.

L-*dopa compared with* L-*dopa plus MK-486 in normals*

The effect of a single dose of 500 mg of L-dopa was compared with that of 100 mg of L-dopa given together with 50 mg of MK-486 on a subsequent occasion in each of 6 normal subjects (3 men, 3 women). On a third occasion a placebo was administered and the subjects were followed as before. The effects on prolactin are shown in Fig. 2. A pronounced and statistically significant drop in prolactin, maximal at 2 hr, is evident in subjects receiving either L-dopa alone or L-dopa with inhibitor. An apparent rebound to greater than normal levels occurs at 6 hr, though the differences at this point are not statistically significant because of wide scatter. There were no statistical differences between the prolactin responses to L-dopa alone and to L-dopa with MK-486. Growth hormone stimulation, previously noted by others after L-dopa administration (BOYD *et al.*, 1970) occurred to an equal degree with both L-dopa regimens (Fig. 3). Mean plasma L-dopa concentrations* showed no statistical difference at any time from 20 min to 24 hr when the two regimens were compared.

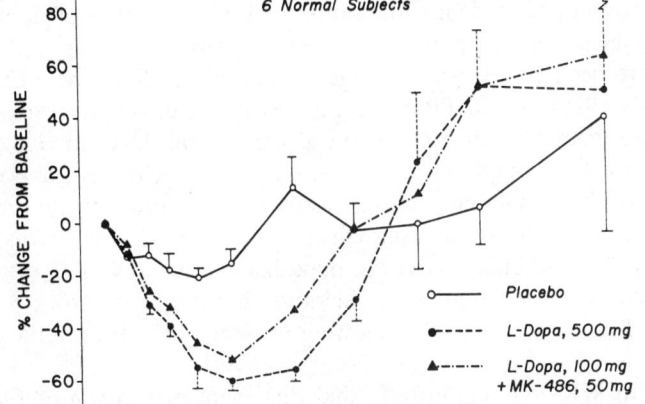

PLASMA PROLACTIN AFTER L-DOPA,
L-DOPA + MK-486, AND PLACEBO

FIG. 2.—Plasma prolactin after L-dopa, 500 mg; after L-dopa, 100 mg, plus MK-486, 50 mg; and after placebo. All drugs were given orally at 0 time on separate occasions several days apart. Vertical bars denote standard error of the mean.

---

* Performed by Dr. George Breault of the Merck Sharp & Dohme Research Laboratories.

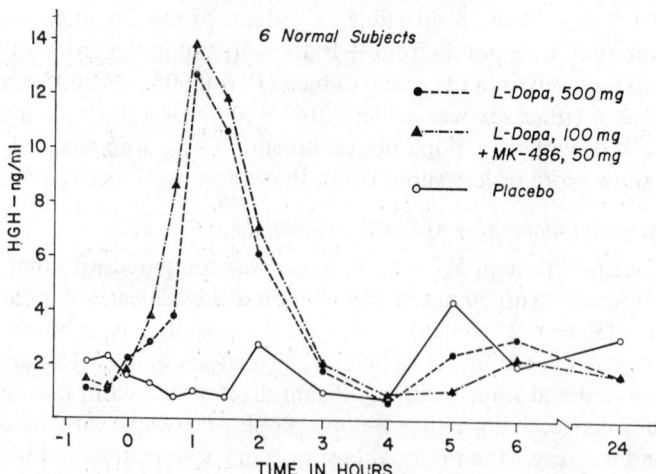

Fig. 3.—Plasma growth hormone concentrations in the same samples as those depicted in Fig. 2.

DISCUSSION

The inhibitory effect of L-dopa on prolactin has been assumed to be mediated via stimulation of hypothalamic PIF, largely on the basis of work by Kamberi et al. (1970; 1971) showing that dopamine infused into the third ventricle raised PIF in portal blood and decreased prolactin release, whereas direct infusions of dopamine into the anterior pituitary did not affect prolactin levels in peripheral blood. Three other groups have obtained evidence for a direct inhibitory action of dopamine or other catecholamines on pituitary prolactin release in vitro, however, although the doses required have been comparatively large (Birge et al., 1970; MacLeod et al. 1970; Koch et al., 1970). Since TRH is presumed to act directly on the pituitary, the present studies suggest that dopamine may also act at this locus in vivo to antagonise TRH; they are also compatible with the more widely held view, however, that dopamine acts via the hypothalamus; in this case the PIF liberated would presumably be sufficient to overcome the stimulating effect of TRH on the pituitary. A third possibility, namely that TRH may act on the hypothalamus as well as on the pituitary to promote prolactin release, deserves consideration in view of the work of Prange et al. (1972) and Kastin et al. (1972) showing evidence of extra-pituitary actions of TRH.

The studies with MK-486 demonstrate that this agent may prove useful in conjunction with L-dopa when the latter is used for its effect on the pituitary in promoting growth hormone release or suppressing prolactin. The latter possibility has particular relevance in view of the recent findings by ourselves and others that remission of disease may occur in some individuals with metastatic breast cancer treated chronically with L-dopa (Frantz et al., 1973).

Acknowledgements—The cooperation of Dr. Marvin E. Jaffe and the Merck Sharp and Dohme Reseach Laboratories in providing MK–486 and assisting with these studies is gratefully acknowledged. Dr. Henry Friesen kindly supplied human prolactin standards and anti–prolactin antibody. Mr. Robert Sundeen and Mrs. Irene Conwell provided expert technical assistance. This work was supported by U.S. Public Health Service grants AM–11294, TIAM–5397, CA–11704, CA–13696, and American Heart Association grant 68–111.

REFERENCES

BIRGE C. A., JACOBS L. S., HAMMER C. T. and DAUGHADAY W. H. (1970) *Endocrinology* **86,** 120–130.

BOWERS C. Y., FRIESEN H. G., HWANG P., GUYDA H. J. and FOLKERS K. (1971) *Biochem. Biophys. Res. Comm.* **45,** 1033–1041.

BOYD A. E., III, LEBOVITZ H. E. and PFEIFFER J. B. (1970) *New Engl. J. Med.* **283,** 1425–1429.

FRANTZ A. G. (1973) In: *Frontiers in Neuroendocrinology* 1973. (GANONG W. F. and MARTINI L., eds.) pp. 337–374. Oxford University Press, New York.

FRANTZ A. G. and KLEINBERG D. L. (1970) *Science* **170,** 745–749.

FRANTZ A. G. and RABKIN M. T. (1964) *New Engl. Med. J.* **271,** 1375–1381.

FRANTZ A. G., KLEINBERG D. L. and NOEL G. L. (1972) *Recent Progr. Hormone Res.* **28,** 527–573.

FRANTZ A. G., HABIF D. V., HYMAN G. A., SUH H. K., SASSIN J. F., ZIMMERMAN E. A., NOEL G. L. and KLEINBERG D. L. (1973) In: *International Symposium on Human Prolactin.* (ROBYN C., ed.) Excerpta Medica, Netherlands (in press).

FRIESEN H., GUYDA H., HWANG P., TYSON J. E. and BARBEAU A. (1972) *J. Clin. Invest.* **51,** 706–709.

HWANG P., GUYDA H. and FRIESEN H. (1971) *Proc. Nat. Acad. Sci. U.S.A.* **68,** 1902–1906..

HWANG P., GUYDA H. and FRIESEN H. (1972) *J. Biol. Chem.* **247,** 1955–1958.

JACOBS L. S., SNYDER P. J., WILBER J. F., UTIGER R. D. and DAUGHADAY W. H. (1971) *J. Clin. Endocr.* **33,** 996–998.

KAMBERI I. A., MICAL R. S. and PORTER J. C. (1970) *Experientia* **26,** 1150–1151.

KAMBERI I. A., MICAL R. S. and PORTER J. C. (1971) *Endocrinology* **88,** 1288–1293.

KASTIN A. B., EHRENSING R. H., SCHALCH D. S. and ANDERSON M. S. (1972) *Lancet* **2,** 740–742.

KLEINBERG D. L., NOEL G. L. and FRANTZ A. G. (1971) *J. Clin. Endocr.* **33,** 873–876.

KOCH Y., LU K. H. and MEITES J. (1970) *Endocrinology* **87,** 673–675.

LEWIS U. J., SINGH R. N. P., SINHA Y. N. and VANDER LAAN W. P. (1971) *J. Clin. Endocr.* **33,** 153–156.

L'HERMITE L., DELVOYE P., NOKIN J., VEKEMANS M. and ROBYN C. (1972) In: *Prolactin and Carcinogenesis.* (BOYNS A. R. and GRIFFITHS K., eds.) pp. 81–97. Alpha Omega Alpha Publishing, Cardiff.

MACLEOD R. M., FONTHAM E. H. and LEHMEYER J. E. (1970) *Neuroendocrinology* **6,** 283–294.

MALARKEY W. B., JACOBS L. S. and DAUGHADAY W. H. (1971) *New Engl. J. Med.* **285,** 1160–1163.

NOEL G. L., SUH H. K., STONE G. and FRANTZ A. G. (1972) *J. Clin. Endocr.* **35,** 840–851.

PRANGE A. J., JR., WILSON I. C., LARA P. P., ALLTOP L. B. and BREESE G. R. (1972) *Lancet* **2,** 999–1002.

SINHA Y. N., SELBY F. W., LEWIS U. J. and VANDER LAAN W. P. (1973) *J. Clin. Endocr.* **36,** 509–516.

Frontiers in Catecholamine Research 1973, pp. 849 to 852. Pergamon Press. Printed in Great Britain.

HYPOTHALAMIC CATECHOLAMINES (CA) AND THE SECRETION OF GONADOTROPINS AND GONADOTROPIN RELEASING HORMONES

I. A. KAMBERI

Department of Obstetrics and Gynecology, UCLA School of Medicine,
Harbor General Hospital Campus, Division of Reproductive Biology,
Torrance, California, U.S.A.

IT IS now well established that the function of the pituitary is modulated by the hypo-thalamus (Ht) (HARRIS, 1970). In this paper we shall discuss the role of catechol-aminergic mechanism on the alteration of the release of luteinizing hormone releasing hormone (LH–RH), follicle-stimulating hormone releasing hormone (FSH–RH), and prolactin release inhibiting hormone (PR–IH), which in turn regulates the release of luteinizing hormone (LH), follicle-stimulating hormone (FSH) and prolactin. *In vitro* and *in vivo* experiments have been applied to the problem. Hormones were measured by radioimmunoassays.

In the initial experiments, we measured monoamine oxidase activity (MAOA) in the entire Ht and different Ht regions, in the amygdala and cerebral cortex of male and 4 day cyclic female rats, using radioisotopic methods (KAMBERI and KOBAYASHI, 1970).

As shown in Fig. 1, measurements of MAOA in the Ht and amygdala demon-strated cyclic changes during the estrous cycle. Extracts from the whole Ht show a peak of activity at P1 of the day of proestrus. A dramatic fall in MAOA occurred in the afternoon of the day of proestrus (P3). Subsequently, there was a smaller peak of MOAO on the day of estrus. Lowest levels occurred at P3, D1 and M2. The amyg-dala, also showed cyclic changes with an increase at P1 and peaking at P2 on the day of proestrus. The cerebral cortex possessed much lower levels of MAOA and showed no cyclic changes (Fig. 1). Cyclic changes were found when various Ht zones were dis-sected and measured individually. The median eminence showed the highest activity throughout the cycle, mirroring the levels of MAOA of the whole Ht. MAOA in other Ht-zones (anterior, lateral, posterior) was considerably less. No cyclic changes in MAOA have been found in the Ht of male rats (KAMBERI and KOBAYASHI, 1970).

Number of studies recently have shown that 17β-estradiol is a major ovarian hormone responsible for preovulatory release of gonadotropins. Recently we have found that α-methylthyrosine, an inhibitor of CA synthesis, when injected to the 4 day cycling rats, in the morning of the day of diestrus (D2) prior to the day of proestrus, affectively inhibited the proestrus rise in 17β-estradiol, abolished cyclic changes in MAOA in the Ht and amygdala during the day of proestrus (Fig. 1), and prevented proestrus surge of gonadotropins. This result suggests that CA are involved in reg-ulation of secretion of gonadotropins, but do not specify which of the CA are in question.

In our further work, we decided that these results would have to be supplemented by further experiments, before the role of CA as synaptic transmitters could be con-sidered established. For this purpose different CA compounds and CA-depleting or

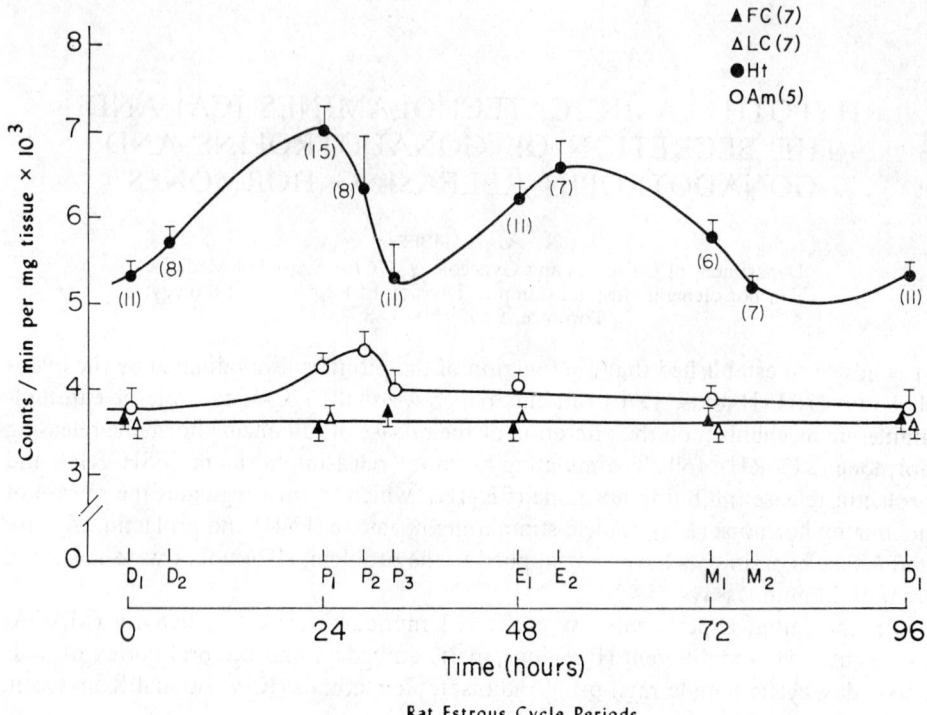

FIG. 1.—Brain monoamine oxidase activity during the estrous cycle. The number of rats analysed is in brackets; each point represents the mean ± SEM. Abbreviations: FC, frontal cortex; LC, lateral cortex; Ht, hypothalamus; Am, amygdala; D, diestrus; P, proestrus; E, estrus; M, metestrus; D1, P1, E1, M1 = 10 a.m.; D2, P2, E2, M2 = 3 p.m.; P3 = 6–7 p.m. (From KAMBERI and KOBAYASHI, 1970.)

adrenergic blocking agents were applied to the problem under *in vitro* and *in vivo* conditions (KAMBERI, 1973a, b). It was found that, of the CA, epinephrine and norepinephrine in relatively higher dose levels (5–100 μg) and dopamine in lower doses ($\sim$1 μg) had a stimulatory effect on release of LH, FSH, and an inhibitory effect on the release of prolactin (KAMBERI *et al.*, 1970a, b; 1971a, b). The effect of CA on LH, FSH, and prolactin release were observed only when anterior pituitary halves have been co-incubated in the presence of ventral hypothalamic fragments or CA were administered intraventricularly. The CA precursor, L-3,4-dihydroxyphenylalanine (L-dopa), injected systemically, also stimulated the release of gonadotropin and inhibited the release of prolactin (KAMBERI, 1973a, b). The effects of CA tested were actually on the discharge of hypothalamic hypophysiotropic hormones, since we have demonstrated that LH–RH, FSH–RH, and PR–IH activities in hypophysial portal blood (KAMBERI *et al.*, 1969; 1970c, d; 1971c) were elevated after intraventricular injection or co-incubation of pituitary halves in presence of hypothalamic fragments (KAMBERI *et al.*, 1970b). Conversely, no effect was observed when CA agents were infused directly into anterior pituitary *via* a cannulated hypophysial portal vessel or were incubated in the presence of anterior pituitary halves alone (KAMBERI *et al.*, 1970a, b; 1971a, b, c). It was found that this effect of CA on release of LH–RH, FSH–RH and PR–IH, could be potentiated, when clonidine-HCl, a noradrenergic stimulating agent, is simultaneously administered with the CA. It is tempting to postulate that CA can enter the intracellular pool of the hypothalamus, be taken up by adrenergic nerve terminals, and subsequently re-released as norepinephrine or some other active monoamine metabolite, through which the discharge of LH–RH, FSH–RH and PR–IH is effected. Furthermore, this stimulatory effect of CA on the release of LH–RH, FSH–RH and PR–IH is linked to the α-adrenergic receptors of the hypothalamus. This is supported by our finding that administration of phenoxybenzamine, an α-adrenergic blocking agent along with CA, prevented the response seen with CA alone, whereas administration of pronethalol, a β-adrenergic blocking agent, with CA failed to do so (KAMBERI *et al.*, 1970b, e).

SUMMARY AND COMMENTS

From these experiments, it is clear that catecholaminergic mechanism is involved in the control of gonadotropin and prolactin secretion. In both *in vivo* and *in vitro* studies, relatively small doses of dopamine and large doses of norepinephrine or epinephrine, induced the release of FSH and LH and inhibited prolactin. The release of these hormones is mediated through the hypothalamus, secondary to a discharge of FSH–RH, LH–RH and PR–IH, and this effect seems to be linked to the α-adrenergic receptors of the hypothalamus.

However, other investigators are not in uniform agreement with the above action of catecholamines on the release of FSH–RH, LH–RH and PR–IH. FUXE and HÖKFELT (1969) on the basis of their fluorescence microscopic data suggest that dopamine has an inhibitory effect on LH–RH release and a stimulatory effect on PR–IH. The recent results of Hale and SYMINGTON (1972) may help to explain these divergent findings. They found that dopamine stimulates the release of FSH from pituitaries incubated in the presence of whole hypothalami, whereas dopamine suppressed gonadotropin release from pituitary tissue incubated in the presence of stalk median eminence only. These findings suggest that two different mechanisms concerning the gonadotropin release may exist. One which involves the perikaryons of the Ht-hypophysiotropic hormone (HHH) secreting neurons upon which dopamine or catecholamines have a stimulatory action, the other may involve the nerve terminals of the HHH secreting cells upon which dopamine or catecholamines have an inhibitory effect. If so, discrepancies in results obtained by us using physiological-biochemical techniques, with those reported by Fuxe and his associates, using histofluorescence microscopy, can be easily explained. Namely, intraventricular injection of dopamine in our experiments causes an increase of dopamine in the perikaryons of HHH secreting neurons, which could be reflected in a decrease of dopamine in nerve terminals, as observed by Fuxe and his associate. It is interesting to note, however, that more recent results of LICHTENSTEIGER (1973) similarly using histofluorescence microscopy, are not in agreement with the data of FUXE and HÖKFELT (1969).

Acknowledgements—The author wishes to thank Miss Lynette Meyers, Miss Elisa Bacleon, and Mrs. Maria A. Kamberi for their generous assistance in the preparation of this manuscript. This work was supported by General Research Support Grant 5SO1RRO555 from N.I.H.

REFERENCES

FUXE K. and HÖKFELT T. (1969) In: *Frontiers in Neuroendocrinology*. (GANONG W. F. and MARTINI L., eds.) pp. 47–96. Oxford University Press, New York.
HALE D. H. and SYMINGTON R. B. (1972) *S. Afr. Med. J.* **46**, 787–791.

HARRIS G. W. (1970) In: *Hypophysiotropic Hormones of the Hypothalamus: Assay and Chemistry.* (MEITES J., ed.) pp. 1–20. Williams and Wilkins, Baltimore.
KAMBERI I. A. (1973a) In: *Proceedings of the IV International Congress of Endocrinology.* (SCOW R. O., ed.) pp. 485–492. Excerpta Medica Fndn., Amsterdam.
KAMBERI I. A. (1973b) *Prog. Brain Res.* **39,** in press.
KAMBERI I. A. and KOBAYASHI Y. (1970) *J. Neurochem.* **17,** 261–268.
KAMBERI I. A., MICAL R. S. and PORTER J. C. (1970a) *Endocrinology* **87,** 1–12.
KAMBERI I. A., SCHNEIDER H. P. G. and MCCANN S. M. (1970b) *Endocrinology* **86,** 728–284.
KAMBERI I. A., MICAL R. S. and PORTER J. C. (1971a) *Endocrinology* **88,** 1003–1011.
KAMBERI I. A., MICAL R. S. and PORTER J. C. (1971b) *Endocrinology* **88,** 1012–1020.
KAMBERI I. A., MICAL R. S. and PORTER J. C. (1969) *Science* **166,** 388–390.
KAMBERI I. A., MICAL R. S. and PORTER J. C. (1970c) *Nature (Lond.)* **227,** 714–715.
KAMBERI I. A., MICAL R. S. and PORTER J. C. (1970d) *Experientia* **26,** 1150–1151.
KAMBERI I. A., MICAL R. S. and PORTER J. C. (1971c) *Endocrinology* **89,** 1042–1046.
KAMBERI I. A., MICAL R. S. and PORTER J. C. (1970e) *Physiologist* **13,** 239.
LICHTENSTEIGER W. (1973) In: *Proc. IV Int. Congr. Endocrinology.* (SCOW R. O., ed.) Excerpta Medica Fndn., Amsterdam, in press.

Frontiers in Catecholamine Research 1973, pp. 853 to 857. Pergamon Press. Printed in Great Britain.

THE MECHANISM OF ACTION OF CLONIDINE

R. Bloch, P. Bousquet, J. Feldman and J. Schwartz

Institut de Pharmacologie et de Médecine Expérimentale. Faculté de Médecine
Strasbourg, France

Numerous experiments have documented the central hypotensive action of clonidine. As a matter of fact perfusion of the cerebral cavities from the lateral ventricle to the cisterna magna (Bousquet et al., unpublished), intracisternal injections (Schmitt et al., 1973) and perfusion of clonidine through the vertebral artery (Sattler and Van Zwietten, 1967), always produce a fall in arterial pressure together with a slight bradycardia.

(a) Nevertheless, the mechanism of the hypotensive action of clonidine is not well known; that mechanism could be of an adrenergic type (Schmitt, 1971) although noradrenaline, when injected intraventricularly or intracisternally, has not, as a rule, produced a hypotensive effect.

The structure of the clonidine molecule has been established from an analysis of its i.r., u.v. and NMR spectra (Wermuth et al., 1973). In the protonated form which predominates in vivo (pK_a of clonidine is 8·2) the positive charge is delocalized on the carbon atom and the three nitrogen atoms in the guanidine function. The imidazoline ring rotates around the ArN-C bond. Clonidine has the characteristics of an α-mimetic drug as described by Coubeils et al., (1972) when the rings of the molecule lie at right angles to each other. The distance between the center of the aromatic nucleus and one of the imidazoline nitrogen atoms of clonidine or the amine function of a sympathomimetic agent is identical, i.e. 5·1Å. Similarities between clonidine and noradrenaline are evident and even more so between clonidine and dopamine.

(b) As for the mechanism of action of clonidine, its site of action also has not been well established. However, in a recent study, Bousquet and Guertzenstein (in press) have demonstrated that the action of clonidine is localized on the ventral surface of the brain stem of the cat; these data are consistent with the work of Feldberg and Guertzenstein (1972) which had enabled localization of the vaso-depressive action of pentobarbitone sodium in a region caudally to the trapezoid bodies and laterally to the pyramids. In this region a topical application of clonidine produced a fall in arterial pressure.

The anatomic description of this area has been made by Petrovicky (1968): it is a very precisely delimited region where, the glia marginalis being very thin, the neurons are found immediately under the pia mater whether cell bodies or dendrites. These neurons form a small nucleus which is 2-mm long, 1-mm wide and 1-mm deep. This nucleus is made up of two kinds of cells and its rostral part is in contact with the nucleus paragiganto-cellularis.

METHODS

The experiments were performed on cats weighing 1·7–3 kg. The cats were anaesthetized by intraperitoneal injection of pentobarbitone sodium (30 mg/kg). In a few experiments, anaesthesia was induced with ether.

For artificial respiration (Logic 03 respirator) the trachea was cannulated. The left femoral vein was catheterized for intravenous injections. The left femoral artery was also cannulated and connected to a physiological pressure transducer (Statham P23 Db). The arterial pressure was recorded by a Minipolygraph Gilson M 5 P (module CHCBPP). In some experiments arterial pressure was recorded by means of a potentimetric module (SE 21 servomodule Gilson) and the mean heart rate by the means of a cardiotachymeter (module CT 27 Gilson) connected to the module CHCBPP.

The method using the topical application of drugs on the ventral surface of the brain stem was realized in the same way as originally described by Feldberg and Guertzenstein (1972) and Guertzenstien (1973) by means of two plastic rings. The drugs were applied in each ring in a volume of 10 µl. The rings were placed just under the pons on each side of the midline on the surface of the medulla oblongata under light pressure; thus the fluid cannot flow out of the rings. After each application of the drug, the surface was washed out with artificial CSF (Merlis, 1940). The drugs were also dissolved in artificial CSF.

RESULTS

(1) Figure 1 shows the fall in arterial pressure which is regularly caused by the topical application of 10 µl of a solution of clonidine during 5 min on each side of the ventral surface of the brain stem.

Clonidine proves to be active in very low concentrations: 100 µg/ml (which is equivalent to the application of a dose of 2µg/kg) induces a fall in arterial pressure of

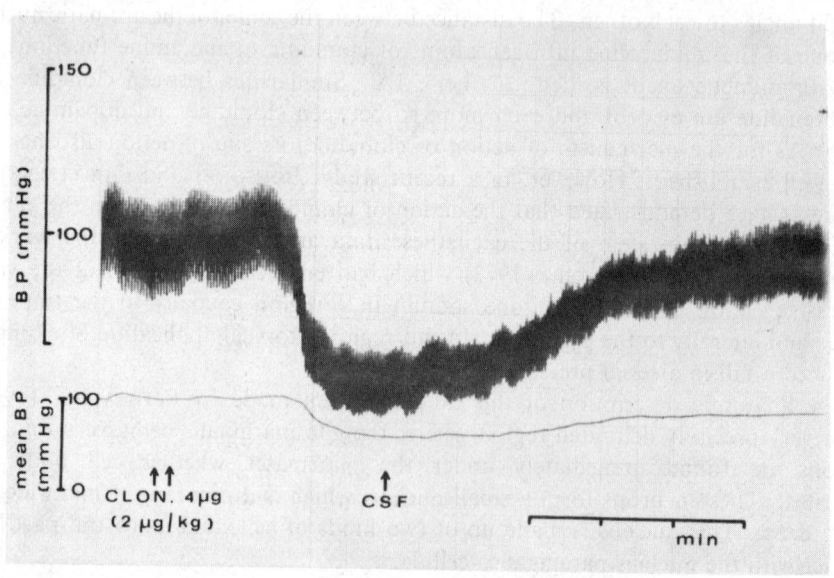

Fig. 1.—Effect of clonidine on arterial blood pressure when topically applied on the ventral surface of the cat's medulla.
↑↑ left then right application of clonidine
↑ washout with artificial CSF.

30–40 mm Hg. The application on one side only is not as effective even when the concentrations are higher, but, as a rule, hypotension appears as soon as clonidine comes into contact with the second side. The duration of the hypotensive effect is variable; the arterial pressure generally returns to the initial level after 20–60 min, sometimes even after several hours, and this delay is all the more longer that the hypotensive effect is greater. In many experiments hypotension is accompanied by a slight bradycardia.

It is impossible to state that this area of the ventral surface of the brain stem where the application of low doses of clonidine causes a fall in arterial pressure is its only site of hypotensive action; nevertheless it is a hypothesis which may be assumed. In fact, the intraventricular perfusion of clonidine does not cause any fall in arterial pressure when the drug is not in contact with the ventral face of the brain stem. Furthermore BOUSQUET and GUERTZENSTEIN (in press) have demonstrated that clonidine injected into the lateral ventricle of the cat proves to be hypertensive when the drug flows through a cannula placed in the aqueduct of Sylvius. Consequently the site of hypotensive action of clonidine can only be localized in the brain stem. Some current experiments enable us to assert that this action does not take place in the floor of the fourth ventricule, that is to say, on the dorsal surface of the brain stem.

(2) The analysis of the clonidine molecule shows that it has the same structure as noradrenaline and dopamine.

(a) Therefore we applied noradrenaline in concentration of 0·5–10 mg/ml to the site where clonidine is active. At low doses, less than 2 mg/ml, noradrenaline does not modify arterial pressure; at higher concentrations of 5–10 mg/ml. a hypertension is produced which lasts for several minutes before recovery takes place; there is never any hypotensive phase (Fig. 2). Besides, the hypertensive effect which occurs when high concentrations of noradrenaline are applied may be due to a fraction of the product passing into the blood stream (the total dose applied under these circumstances corresponds to amounts of 100–200 μg).

(b) On the other hand, dopamine in concentrations of 12·5–100 mg/ml (that is to say application of 250 μg–2 mg) shows a hypotensive effect as a rule, and there is a certain linear relationship between the dose and the hypotensive effect. Like clonidine, dopamine induces an important, immediate and long lasting hypotension (providing that this amine is applied on both sides) (Fig. 2); it is accompanied by a slowing of the heart. At high doses the effect may sometimes be diphasic: an initial hypertension which lasts for a few minutes followed by hypotension which remains for more than an hour even though the product is applied for not more than 5 min.

We have been able also to show in the dog that arterial hypotension can be caused by infusion of dopamine into the lateral ventricle with an outflow cannula in the cisterna magna, in doses of 10–50 μg/kg, in a total volume of 1 ml during 5 min. Same results are obtained with clonidine in doses of 1–10 μg/kg. On the contrary, under these circumstances, noradrenaline never causes hypotension and shows hypertensive effects at higher doses.

Besides, when injected intravenously, amounts of dopamine similar to these applied in cats or infused in dogs, always induce an important rise in arterial pressure.

856   R. BLOCH, P. BOUSQUET, J. FELDMAN and J. SCHWARTZ

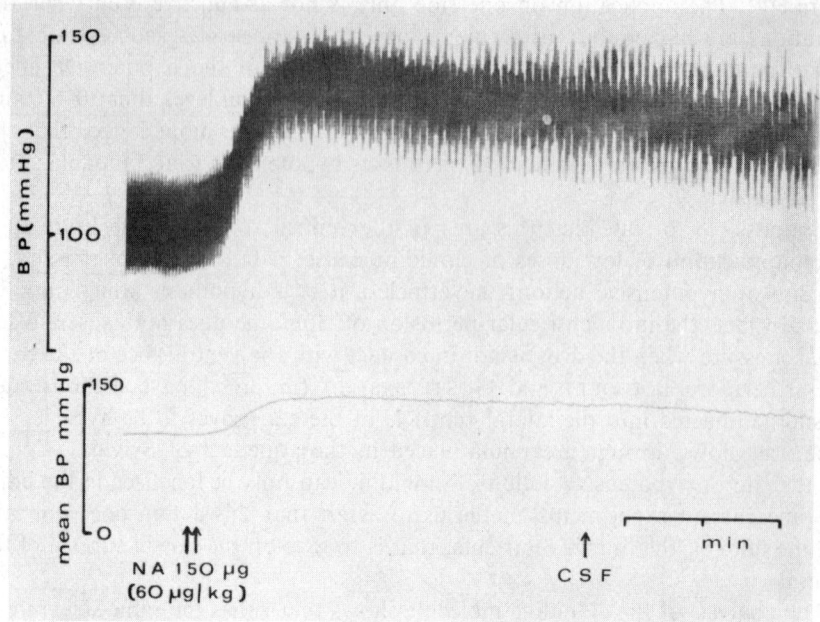

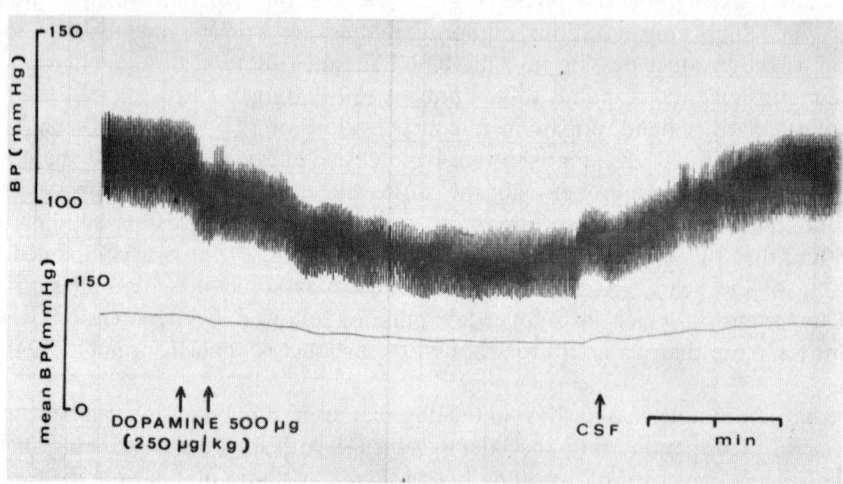

FIG. 2.—Effect of noradrenaline (above) and dopamine (below) on arterial blood
pressure when applied on the ventral surface of the cat's medulla.
↑↑ left then right topical application of the drug solutions
↑ washout with artificial CSF

DISCUSSION

Many authors have studied the structure of the ventral surface of the medulla oblongata and earlier works took into account a chemosensitive area at this level. It has indeed been demonstrated that pledgets of filter paper soaked in artificial cerebrospinal fluid with high pCO_2 or H^+ concentrations as well as when containing nicotine or acetylcholine produced a hyperpnea, (LOESCHCKE et al., 1958; MITCHELL et al., 1963) when applied to a zone corresponding closely to that described by PETROVICKY (1968). On the contrary procaine induced a depression of respiration and a fall in arterial pressure (LOESCHCKE and KOEPCHEN, 1958a, b). Localized cooling in this area of the ventral surface of the medulla resulted in a reduction of the tidal volume and a drop in arterial pressure (SCHLAEFKE and LOESCHCKE, 1967).

Subsequently, FELDBERG and GUERTZENSTEIN (1972) have shown that pentobarbitone sodium has a vasodepressive effect in this area. Moreover, GUERTZENSTEIN (1973) demonstrated that in the same area cholinomimetic substances (physostigmine and carbachol) as well as glycine and GABA induce a fall in arterial pressure antagonized by atropine. With strychnine, leptazol and tubocurarine, an increase in arterial pressure can be observed. It is in that same area that BOUSQUET and GUERTZENSTEIN (in press) have localized the hypotensive action of clonidine. And in that area also experiments we report here demonstrate a specific vasodepressive activity of dopamine, structurally related to clonidine, while noradrenaline, also analogous to clonidine, proves to be vasopressive. Nevertheless clonidine is obviously more potent (a total dose of 4–10 μg of clonidine causes an evident hypotension which requires a dose of 250 μg of dopamine). Our experiments show that clonidine acts in an analogous fashion to dopamine but not to noradrenaline.

At the level of the structures we studied, dopamine may have a direct mediatory action, or may play the role of a modulator in underlying cholinergic mechanisms (BERTLER and ROSENGREN, 1966). Based on the results of experiments, which demonstrate hypotensive actions of both cholinomimetic agents and dopamine on the ventral surface of the brain stem, the second of these hypothesis seems more plausible.

REFERENCES

BERTLER A. and ROSENGREN E. (1966) Pharmacol. Rev. 18, (1) 769–773.
BOUSQUET P., BLOCH R., FELDMAN J. and SCHWARTZ J. (unpublished observations)
BOUSQUET P. and GUERTZENSTEIN P. G. (in press) Br. J. Pharmac.
COUBEILS J. L., COURRIERE Ph. and PULLMAN B. (1972) J. med. Chem. 15, 453–455.
FELDBERG W. and GUERTZENSTEIN P. G. (1972) J. Physiol. 224, 83–103.
GUERTZENSTEIN P. G. (1973) J. Physiol. 229, 395–408.
LOESCHCKE H. H., KOEPCHEN H. P. and GERTZ K. H. (1958) Pflügers Arch. ges. Physiol. 266, 569–585.
LOESCHCKE H. H. and KOEPCHEN H. P. (1958a) Pflügers Arch. ges. Physiol. 266, 611–627.
LOESCHCKE H. H. and KOEPCHEN H. P. (1958b) Pflügers Arch. ges. Physiol. 266, 628–641.
MERLIS J. K. (1940) Am. J. Physiol. 131, 67–72.
MITCHELL R. A., LOESCHCKE H. H., SEVERINGHAUS J. W., RICHARDSON B. W. and MASSION W. H. (1963) Ann. N.Y. Acad. Sci. 109, 661–681.
PETROVICKY P. (1968) Z. Anat. Entw. Gesch. 127, 221–231.
SATTLER R. W. and VAN ZWIETEN P. A. (1967) Europ. J. Pharmac. 2, 9–13.
SCHLAEFKE M. and LOESCHCKE H. H. (1967) Pflügers Arch. ges. Physiol. 297, 201–220.
SCHMITT H. (1971) Actualités Pharmacologiques 24, 93–131.
SCHMITT H., SCHMITT H. and FENARD S. (1973) Arzneim. Forsch. 23, 40–45.
WERMUTH C., SCHWARTZ J., LECLERC G., GARNIER J. P. and ROUOT B. (1973) Chim. Thér. 1, 115–116.

Frontiers in Catecholamine Research 1973, pp. 859 to 864. Pergamon Press. Printed in Great Britain.

ROLE OF SYMPATHETIC FIBRES AND OF ADRENAL MEDULLA IN THE MAINTENANCE OF CARDIO-VASCULAR HOMEOSTASIS IN NORMOTENSIVE AND HYPERTENSIVE RATS

JACQUES DE CHAMPLAIN* and MARIE-REINE VAN AMERINGEN

Centre de Recherches en Sciences Neurologiques, Département de Physiologie,
Faculté de Médecine, Université de Montréal, Montréal, Canada

THE CONCEPT of modulation of cardiovascular system by the autonomic nervous system has been accepted for a long time. However, the efferent sympathetic component of the autonomic nervous system is more dynamically related to the peripheral cardiovascular system by a dense plexus of excitatory fibres distributed to the heart and to all vascular beds. Small arterioles which are crucial to the homeostasis of arterial blood pressure were found to be the most densely innervated segments of the vascular tree (NORBERG, 1967). In addition to this network of excitatory fibres, the sympathetic nervous system is also responsible for the control of catecholamine secretion by the adrenal medulla which can also influence the heart and blood vessels by acting on specific receptors to catecholamines localized in these structures.

Little is known about the relative contribution of each component of the sympathetic system in the maintenance of normal cardiovascular functions. In numerous previous studies, the effects of sympathectomy were studied after surgical denervation, after treatment with nerve growth factor antiserum or after administration of catecholamine depleting drugs such as reserpine. Most of these techniques however were either non specific to the peripheral fibres or, resulted in an incomplete sympathectomy so that the conclusions reached with these procedures may be questionable.

The discovery that 6-hydroxydopamine (6-OH-DA) could specifically and selectively destroy adrenergic nerve fibres (TRANZER and THOENEN, 1967) has given a more specific and more useful tool for the study of the role of the sympathetic nervous fibres. In periphery, 6-OH-DA when given in sufficient amount, causes the degeneration of the great majority of sympathetic fibres without destroying the ganglion cell bodies or the adrenal medulla.

NORMOTENSIVE RATS

In normotensive rats and dogs, the intravenous administration of 6-OH-DA is immediately followed by a marked and prolonged sympathomimetic effect characterized by an increase of blood pressure and heart rate due to the massive release of norepinephrine from the nerve endings occurring during the acute phase of degeneration (DE CHAMPLAIN and VAN AMERINGEN, 1972; GAUTHIER et al., 1972). In the following hours and days, the blood pressure stabilizes about 30 mm Hg lower and the heart rate is slightly reduced (Fig. 1, Table 1). On the other hand, bilateral adrenalectomy alone causes only a slight insignificant decrease in blood pressure of

---

* Member of the Medical Research Council Group in Neurological Sciences at the Université de Montréal.

about 10 mm Hg and no change in the heart rate. It therefore appears from these results that the sympathetic nervous fibres are more dynamically related to the maintenance of blood pressure and heart rate than the adrenal medulla. Nevertheless, after the removal of either component of the sympathetic system, the animals appeared relatively normal and could easily survive. The combination of both procedures produced more dramatic changes than expected from the addition of either individual effects. After chemical sympathectomy, bilateral adrenalectomy caused a rapid fall in blood pressure and heart rate of more than 40 mm Hg and 100 beats/min respectively (Fig. 1, Table 1). In the hours following adrenalectomy the blood pressure stabilized around 50 mm Hg and the animals remained in a state of shock until death which occurred in all cases within 2–3 hr.

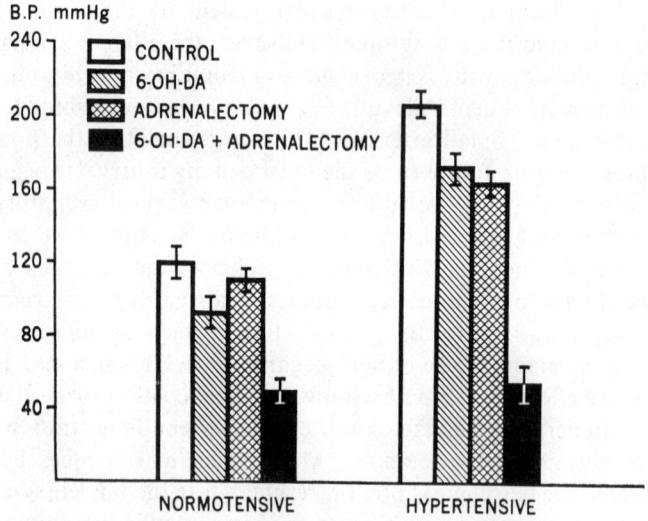

FIG. 1.—Systolic arterial pressure (mm Hg) in unanaesthetized normotensive and (DOCA and saline) hypertensive rats 8 days after one intravenous injection of 6-OH-DA (100 mg/kg) or bilateral adrenalectomy. The black bars represent the blood pressure values 60 min after adrenalectomy in animals previously sympathectomised with 6-OH-DA. These latter values were recorded under anaesthesia. Each bar represents the mean ± S.E. of 4–8 animals.

TABLE 1. EFFECT OF 6-OH-DA AND/OR BILATERAL
ADRENALECTOMY ON THE HEART RATE (BEATS/MIN) OF
ANAESTHETIZED NORMOTENSIVE AND (DOCA AND SALINE)
HYPERTENSIVE RATS

	Normotensive	Hypertensive
Control	283 ± 22	245 ± 12
6-OH-DA	255 ± 29	228 ± 44
Adrenalectomy	285 ± 32	238 ± 25
6-OH-DA + adrenalectomy	184 ± 11*	182 ± 28

The values represented in this table are those observed 18 hr after one intravenous injection of 100 mg/kg of 6-OH-DA or 60 min after adrenalectomy. Each number is the mean ± S.E. of 4–6 rats.
    * $P < 0.01$ vs control values.

These findings strongly suggest that after the removal of sympathetic fibres, the adrenal medulla has the capacity to increase its activity and substitute partially the function of sympathetic nervous fibres in maintaining blood pressure and heart rate. The acute and marked effects of adrenal clamping or adrenalectomy in sympathectomised animals are most probably the result of the interruption of catecholamine secretion by the adrenal medulla. The effects observed occur too rapidly to be attributed to the adrenocortical steroids and moreover, similar effects were also observed in animals which were given cortisone replacement therapy. In addition, identical fall in blood pressure could also be observed in sympathectomised rats after treatment with an alpha blocker. The compensatory role of the adrenal medulla is also supported by the study of MUELLER et al. (1969), who reported that the synthesis of catecholamine by the adrenal medulla was doubled in sympathectomised rats forty hours after the injection of 6-hydroxydopamine.

It appears that the sympathetic fibres may also have the capacity to compensate in the absence of the adrenal medulla. In the hours following bilateral adrenalectomy, a marked increase in the norepinephrine turnover could be observed in peripheral sympathetic nerve fibres (DE CHAMPLAIN and VAN AMERINGEN, 1972; LANDSBERG and AXELROD, 1968). It is therefore possible to postulate that a functional balance exists between the activity of either component of the sympathetic system in the regulation of the heart rate and blood pressure in normotensive animals. This functional relationship is probably determined partly through baroreceptor reflexes and partly through circulating catecholamine levels.

The present observations were made under acute conditions and it is not excluded that other more chronic compensatory mechanisms could develop and eventually contribute to the maintenance of cardiovascular functions following the chemical sympathectomy. It is likely that the development of a supersensitivity to catecholamines, the change in blood volume and the activation of the renin-angiotensin system could contribute to the cardiovascular homeostasis in chronically sympathectomised animals.

HYPERTENSIVE ANIMALS

It is known that the regulation of blood pressure involves the interaction of various systems and factors. An elevation of blood pressure could result from a variety of dysfunctions occurring at any point in this regulatory mechanism. Several systems and factors have been suspected to participate in the physiopathology of hypertension and it seems difficult, at this point, to consider exclusively only one given factor or system as the basic mechanism of hypertension. Hypertensive disorders result most probably from a variety of dysfunctions many of which we are still unable to recognize. For these reasons, it is risky to recognize only one experimental model for human hypertension. Several models have been developed in various animal species and it is likely that each one illustrates different facets of the human disease. We have more specifically studied the role of sympathetic fibres and adrenal medulla in two experimental models of hypertension using an approach similar to that used in normotensive animals.

In previous studies made in collaboration with Krakoff, Mueller and Axelrod on one experimental model produced by the administration of desoxycorticosterone (DOCA 10 mg/week) and saline (1 % drinking solution), we were able to demonstrate that the turnover of norepinephrine was markedly increased in the sympathetic fibres of various peripheral vascular organs (DE CHAMPLAIN, 1972). Studies on the subcellular distribution of norepinephrine and on the pattern of metabolites suggested that there are greater quantities of norepinephrine available at the receptor sites in this condition. In addition, it was also found that the blood pressure levels varied in relation to the sodium intake in parallel with the variations in peripheral sympathetic activity thus suggesting that sodium or other related ions might be a determinant factor in the modulation of sympathetic activity (DE CHAMPLAIN et al., 1969).

Chemical sympathectomy and adrenalectomy alone produced greater changes in blood pressure in DOCA and sodium hypertensive animals that in normotensive animals (Fig. 1). After either one of these procedures the blood pressure was lowered by about 40 mm Hg thus suggesting a more active contribution of both components of the sympathetic system in the maintenance of blood pressure in this group of animals than in normotensive rats. Nevertheless, after the removal of either component of the sympathetic system, the blood pressure remained at hypertensive levels. When the adrenal glands were removed in 6-OH-DA treated hypertensive rats, the blood pressure fell rapidly and markedly by more than 70 mm Hg thus indicating that the adrenal medulla had the capacity to compensate the loss of sympathetic fibres in these animals as well. It is also interesting to note that after adrenalectomy and sympathectomy, the basal blood pressure and heart rate reached in DOCA and sodium hypertensive animals were identical to those found in normotensive animals after the same procedures (Fig. 1, Table 1). This observation strongly suggests that the most likely factor which could account for an elevated blood pressure in rats treated with DOCA and sodium is a synergic hyperactivity of both the sympathetic fibres and the adrenal medulla.

Since both components of the sympathetic system appear to be hyperactive in this form of hypertension it is possible that this might be the result of a dysfunction at the site of the pressure regulatory centres. In support of that hypothesis, a marked decrease in the norepinephrine turnover rate was found in the brain stem of DOCA hypertensive animals whereas the turnover rates were found unchanged in the telediencephalon and spinal cord (VAN AMERINGEN, M. R., and DE CHAMPLAIN, J., unpublished observations). This decrease in the norepinephrine turnover does not seem to be a consequence of the elevation of blood pressure since it was still present in DOCA and saline treated rats in which blood pressure was restored to normotensive levels following a cervical spinal cord section.

The same protocol was also used to investigate the role of the sympathetic system in another model of experimental hypertension induced by stenosis of one renal artery and contralateral nephrectomy. Although, an hyperactivity of the renin-angiotensin system has been associated with this form of hypertension (GROSS et al., 1965), there are several indications that the sympathetic system might be involved at some stages in the physiopathology of hypertension in that model (DE CHAMPLAIN, 1972). Although adrenalectomy did not produce any significant change, chemical sympathectomy by 6-OH-DA produced a greater fall in blood pressure and heart rate than in normotensive and in DOCA hypertensive rats (Fig. 2). However, after the removal of both components of the sympathetic fibres, the residual blood pressure was considerably higher than normotensive and DOCA hypertensive rats thus suggesting an additional pressor mechanism in that form of hypertension. The removal of the clamped kidney in sympathectomised and adrenalectomised animals lowered the blood pressure rapidly to the same residual blood pressure than in normotensive and DOCA hypertensive animals after removal of both components of the sympathetic system. Thus, both the sympathetic nervous fibres and the renal pressor system appear to participate in the maintenance of an elevated blood pressure in renal hypertension.

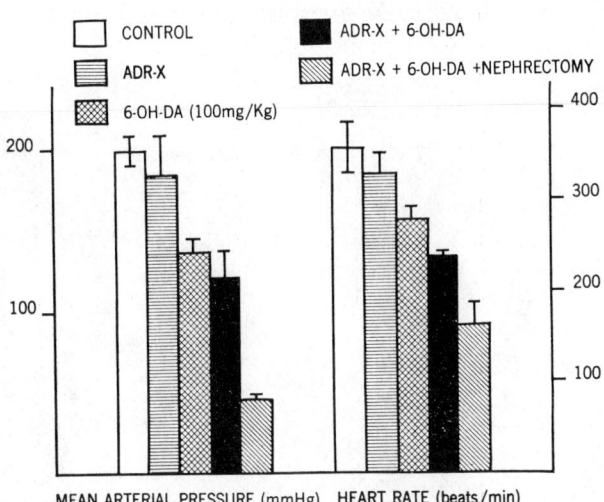

FIG. 2.—Cardiovascular effects of chemical sympathectomy (6-OH-DA 100 mg/kg), bilateral adrenalectomy (ADR-X), or both, with and without nephrectomy in rats made hypertensive by renal artery stenosis and contralateral nephrectomy 45 days previously. The values were recorded under anaesthesia and observed 18 hr after one intravenous injection of 6-OH-DA (100 mg/kg), 60 min after adrenalectomy and 120 min after total nephrectomy. Each bar is the mean ± S.E. of 3 animals.

CONCLUSION

In normotensive animals it seems that the sympathetic nervous system (the sympathetic fibres and the adrenal medulla) is essential for the maintenance of cardiovascular functions. It is difficult at this point to determine whether the sympathetic nervous system plays a primary role in the pathogenesis of certain forms of human and experimental hypertension or whether it contributes only to certain stages of the hypertensive disease. It is likely that both possibilities exist. In the two models of experimental hypertension which were studied, an hyperactivity of the sympathetic system could be observed. In DOCA and sodium hypertensive animals it appears that an activation of the whole sympathetic system is the main factor responsible for an elevation of blood pressure, whereas, in renal hypertensive rats, an activation of the renin-angiotensin system is associated with an hyperactivity of the sympathetic fibres. It is hoped that the development of more sensitive and sophisticated means of investigation in coming years will permit a better understanding of the functioning of the sympathetic nervous system in the physiopathology of various human hypertensive diseases so that a more rational therapy could be developed for these diseases.

REFERENCES

DE CHAMPLAIN J. (1972) In: *Perspectives in Neuropharmacology: A tribute to Julius Axelrod.* (SNYDER S. H., ed.) pp. 215–265. Oxford University Press.

DE CHAMPLAIN J., KRAKOFF L. R. and AXELROD J. (1969) *Circulat. Res.* **24,** (Suppl. 1), 75–92.
DE CHAMPLAIN J. and VAN AMERINGEN M. R. (1972) *Circulat. Res.* **31,** 617–628.
FALCK B., HILLARP N. A., THIEME G. and TORP A. (1962) *J. Histochem. Cytochem.* **10,** 348–354.
GAUTHIER P., NADEAU R. and DE CHAMPLAIN J. (1972) *Circulat. Res.* **31,** 207–217.
GROSS F., BRUNNER H. and ZIEGLER M. (1965) *Recent Prog. Hormone Res.* **21,** 119–177.
LANDSBERG L. and AXELROD J. (1968) *Circulat. Res.* **22,** 559–571.
MUELLER R. A., THOENEN H. and AXELROD J. (1969) *Science* **163,** 468–469.
NORBERG K. A. (1967) *Brain Res.* **5,** 125–170.
TRANZER J. P. and THOENEN H. (1967) *Naunyn-Schmiedeberg's Arch. exp. Pathol. Pharmak.* **257,** 343–344.

Frontiers in Catecholamine Research 1973, pp. 865 to 866. Pergamon Press. Printed in Great Britain.

SODIUM AND CATECHOLAMINE EXCRETION

Yehuda Gutman, Giora Feuerstein and Punya Boonyaviroj
Department of Pharmacology, The Hebrew University-Hadassah Medical School
Jerusalem, Israel

The development of hypertension has been correlated with increased sympathetic activity and/or with increased vascular response to sympathetic stimulation or to catecholamine (= CA) administration. Hypertension also develops or is aggravated by loading with sodium and administration of mineralocorticoids (DOCA). Hypertension induced by Na-DOCA has been reported to be accompanied by increased turnover rate of CA in heart and reduced retention of CA (De Champlain et al., 1969). However, this phenomenon was abolished by ganglion-blocking agents in spite of continued administration of Na-DOCA (De Champlain et al., 1969). On the other hand, it is also unclear whether the kinetics of development of abnormal CA storage fit that of the increased blood pressure. We, therefore, set to study the rate of catecholamine secretion to clarify whether it was related to hypertension or to sodium administration.

The experiments were carried out in rats and in cats. Intragastric loading of 50 ml/kg of 0·15 M NaCl caused increased CA excretion in the urine (4·12 ± 0·49 μg/kg × 8 hr compared to 2·61 ± 0·30 in non-loaded rats and 2·51 ± 0·26 in rats loaded with an equivalent volume of water; $P < 0.01$). Since intragastric loading involved handling of the animals and stress of the introduction of a gastric tube, in the next experiment rats were adapted to metabolic cages with food and water ad lib. and, after 10 days the water was switched for a solution of NaCl. Urinary CA excretion on the day of presentation of NaCl increased from 0·528 ± 0·043 μg/rat × 24 hr to 0·950 ± 0·060; $N = 10$, $P < 0.001$, paired analysis). The possibility that passage of sodium through the gastrointestinal tract was the cause of increased CA excretion was eliminated because intra-peritoneal sodium-loading also resulted in a rise of CA excretion (from 2·61 ± 0·30 to 3·59 ± 0·30 μg/kg × 8 hr, $N = 10$, $P < 0.05$).

The increased CA excretion caused by sodium-loading was abolished after treatment of the rats with a ganglion-blocking agent (Pentapyrrolidinium), thus suggesting that the effect of sodium was mediated through increased sympathetic activity rather than through a direct effect at adrenergic nerve terminals or on the storage mechanism in these nerve endings.

Although the effect of sodium loading on catecholamine excretion was evident within a short time (hours) the type of experiment described did not exclude the possibility that inactivation of CA was reduced rather than that secretion of CA was enhanced by the sodium load. Therefore, the effect of sodium loading was also studied in the cat with the technique of collection of adrenal vein blood. Intravenous infusion of NaCl, as shown in Fig. 1, resulted in immediate (within 10 min) increase of CA secretion from adrenal medulla. The secretion of CA following i.v. sodium loading increased by 7·51 ± 2·31 ng/kg × min from each adrenal at the peak.

In cats with bilateral cervical vagotomy infusion of 0·15 M NaCl caused an even greater increase of CA secretion from the adrenal gland (10·22 ± 3·86 ng/kg × min)

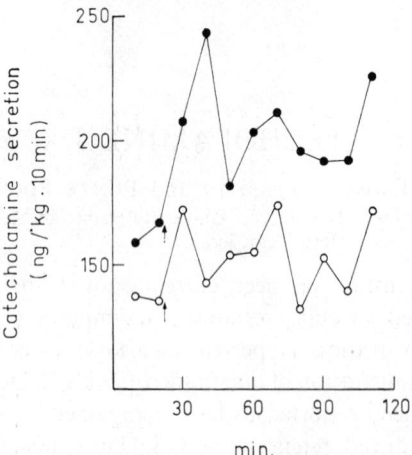

Fig. 1.—Effect of sodium loading on catecholamine secretion by the cat adrenal gland *in vivo*. Sodium infusion started at arrow and continued at the rate of 0·15 mEq/kg × min. ○———○ 0·15 M NaCl (11 experiments); ●———● 1·5 M NaCl (6 experiments).

and the fall in the secretion rate noted after the initial rise (Fig. 1) was abolished following vagotomy. This fall was, therefore, apparently due to hypervolemia.

The various experiments described indicate that sodium-loading can induce an immediate increase of CA secretion. These findings seem to raise doubt about the correlation of sodium and hypertension on the one hand, and sodium and catecholamines on the other hand, since hypertension follows only after long-term loading with sodium while the effect on catecholamines in our experiments seems to be immediate.

To further investigate this point a long-term experiment was performed with rats. The rats were placed in individual cages on food *ad lib.* and either water or 1·4% NaCl as drinking fluid. Catecholamine excretion in the urine was followed for two months. An immediate increase in CA excretion was observed on the day of switching from water to NaCl solution as drinking fluid (from 0·811 ± 0·102 to 1·285 ± 0·159 μg/rat × 24 hr, $P < 0.05$). However, while the CA excretion in the control group (on H_2O) was constant throughout the 2 months, it increased gradually in the experimental group (on NaCl) to 3·914 ± 0·436 μg/rat × 24 hr. During the first two weeks the increase of CA excretion was mainly in norepinephrine but from one month onwards epinephrine excretion increased even more than norepinephrine. At the end of two months on NaCl no significant change in blood pressure was observed between the two groups. Tyrosine hydroxylase activity was increased fourfold in the adrenals of the NaCl-treated rats and cardiac endogenous catecholamines were significantly reduced (0·447 ± 0·028 compared to 0·534 ± 0·019 μg/g in hearts of the control group, $P < 0.02$). These latter findings corroborate those reported for rats with Na-DOCA hypertension (De Champlain *et al.*, 1969). However, in our experiment no significant change of blood pressure accompanied these alterations of catecholamine storage and synthesis. Thus, again, there is an indication of a dissociation between the effect of sodium on catecholamines and its effect on blood pressure. Whether the effect of sodium on catecholamines is through a central mechanism or through other, as yet unidentified, mediators deserves further study.

REFERENCE

De Champlain J., Krakoff L. and Axelrod J. (1969) *Circulat. Res.* **XXIV & XXV** (Suppl. I), 75–92.

Frontiers in Catecholamine Research 1973, pp. 867 to 873. Pergamon Press. Printed in Great Britain.

HYPERTENSION, ANTIHYPERTENSIVE DRUGS, AND NOREPINEPHRINE IN BLOOD VESSELS

Barry Berkowitz, Trajko Trajkov, James Tarver and Sydney Spector

Pharmacology Section, Roche Institute of Molecular Biology, Nutley, New Jersey U.S.A.

It is controversial whether hypertension is associated with an abnormal activity of the sympathetic nervous system (De Champlain et al., 1968; De Champlain et al., 1969; De Champlain et al., 1972; Hickler et al., 1970; Louis et al., 1973; Louis et al., 1969; Molinoff et al., 1972; Nagatsu et al., 1964; Spector et al., 1972; Tarver et al., 1971; Yamori et al., 1970). One approach to this question has been to utilize animal models of hypertension (De Champlain et al., 1968; De Champlain et al., 1969; De Champlain and Van Ameringen, 1972; Louis et al., 1973; Louis et al., 1969; Okamoto, 1969) and examine the dynamics of catecholamine metabolism in the heart, adrenal glands and brain. Our studies have focused on the disposition of norepinephrine in blood vessels (Berkowitz et al., 1972; Berkowitz et al., 1971; Spector et al., 1972; Tarver et al., 1971;) and in this review we will summarise our findings in two hypertensive model systems: in the genetic spontaneously hypertensive rat (SHR) (Okamoto, 1969) and in the uninephrectomised–deoxcorticosterone acetate (DOCA)-salt treated rat (De Champlain et al., 1968; De Champlain et al., 1969; De Champlain and Van Ameringen, 1972). We also will present studies on the effects of anti-hypertensive drugs on vascular norepinephrine disposition and metabolism in normal and hypertensive animals.

HYPERTENSION

SHR

It is clear that blood vessels synthesise catecholamines and possess both anabolic and catabolic enzymes (Berkowitz et al., 1972; Berkowitz et al., 1971; Gillis and Roth, 1970; Tarver et al., 1971; Trajkov et al., 1973; Verity et al., 1972). In order to better assess the role of the adrenergic neuro-transmitter in hypertension we have examined in the vasculature those enzymes involved in its metabolism.

The activity of tyrosine hydroxylase in the mesenteric artery was markedly reduced in the SHR compared to normotensive Wistar rats (Fig. 1). We also utilized a "back-crossed SHR" with an intermediate degree of hypertension which was obtained by mating the SHR with normotensive Wistar rats. These rats had an intermediate fall in tyrosine hydroxylase activity in the mesenteric artery. There thus appeared to be an inverse relationship between blood pressure and tyrosine hydroxylase activity.

Aromatic L-amino acid decarbocylase activity may also be decreased in the vasculature of the SHR as Tanaka has reported a decreased activity of this enzyme in cerebral blood vessels (Tanaka, 1972)

Another of the enzymes of catecholamine metabolism that has been the subject of increasing attention, particularly as a possible marker of sympathetic activity, is dopamine-β-hydroxylase (DBH). Only a few studies pertaining to vascular DBH have been presented. Based on indirect evidence, Weinshilboum and Axelrod (Weinshilboum and Axelrod, 1973) suggested that the bulk of circulating DBH may originate

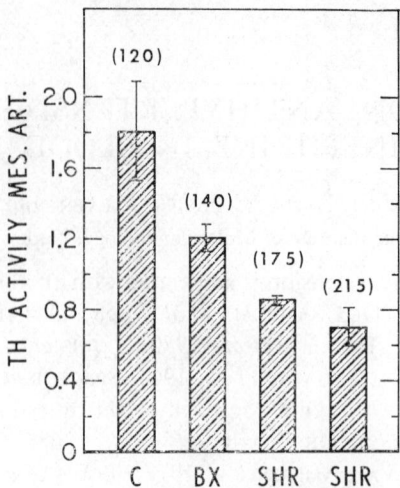

FIG. 1.—Tyrosine hydroxylase (TH) activity of the mesenteric artery in hypertensive rats. C = control normotensive Wistar; BX = Backcrossed SHR-Wistar, SHR = spontaneously hypertensive. Parentheses above bars indicate systolic blood pressure.

from blood vessels. ROFFMAN et al. (1973), found a direct relationship between alterations in vascular DBH and circulating DBH and HARTMAN and UDENFRIEND (1972) reported a high concentration of DBH in kidney blood vessels using a histo-fluorescent method. Utilising a spectrophotometric method (NAGATSU and UDEN-FRIEND, 1972) we measured the activity of DBH in the mesenteric artery of the SHR and found that the activity of this enzyme was reduced by 50 per cent in these hypertensive rats (TRAJKOV et al., 1973).

The regulation of catecholamine metabolism has been primarily associated with an alteration in synthesis. The role of degradation has been less extensively studied. VERITY, SU and BEVAN (1972) measured catechol-O-methyl transferase (COMT) activity in the aorta of normal rabbits and found the highest enzyme activity in the medial layer. In the hypertensive rat, Crevling reported that COMT activity is increased in the heart, liver and kidney CREVLING et al., 1969, but did not examine blood vessels. We have found that the COMT activity is elevated in the aorta ($+50\%$), mesenteric artery ($+60\%$) and heart ($+40\%$) of the SHR with the highest activity seen in those rats having the highest blood pressure. In contrast, monoamine oxidase activity was unaltered in the heart and mesenteric artery of the SHR.

DOCA-salt-uninephrectomised hypertensive rats

The interrelationship between hypertension and vascular catecholamines in another model of hypertension, the DOCA-salt-uninephrectomised rat (De CHAMPLAIN et al., 1968; DE CHAMPLAIN et al., 1969; DE CHAMPLAIN and VAN AMERINGEN, 1972) was examined. Tyrosine hydroxylase activity was altered in a biphasic fashion in this hypertensive specie as shown in Table 1. As hypertension developed, tyrosine hydroxylase was initially increased in the mesenteric artery. However, as the pressure continued to rise, the tyrosine hydroxylase activity began to fall below normal values. Dopamine-β-hydroxylase activity may also be diminished in the vasculature of these rats. With respect to degradative enzymes, their activity was increased in the

TABLE 1. TYROSINE HYDROXYLASE (TH), MONOAMINE OXIDASE (MAO) AND CATECHOL-*O*-METHYL TRANSFERASE (COMT) ACTIVITY IN THE MESENTERIC ARTERY OF THE DOCA-SALT-UNINEPHRECTO-MISED HYPERTENSIVE RAT

B.P. (mmHg)	Time (days)	TH†	MAO† activity	COMT†
115–120	0	100	100	100
130–135	15	152*	95	—
135–140	21	130*	—	—
160–170	32	76*	169*	—
170–180	50	77*	—	150
>210	62	47*	176*	—

* $P < 0.05$ compared to control rats.

† Enzyme activities are as percentage of values obtained from control rats with B.P. 115–120 mmHg.

mesenteric artery. Monoamine oxidase and COMT activities rose by 50–70 per cent (Table 1). In DOCA-salt hypertension it is likely that tissue hypertrophy is a factor in the alterations of the metabolic enzymes.

Our findings in both of these models of hypertension show that the activity of synthetic catecholamine enzymes in the mesenteric artery tend to decrease whereas the activity of the degradative enzymes appear to increase. We propose two hypotheses: (1) that both the enzymes of synthesis and degradation may be altered simultaneously to regulate neurotransmitter activity and (2) that in these hypertensive rats, the enzymatic changes represent a compensatory response by the sympathetic nervous system to diminish the vasoconstrictor stimulus of norepinephrine at the vascular level. Although the synthesis of norepinephrine is clearly diminished in the heart of the SHR (LOUIS et al., 1969) it should be recognised that it remains to be established whether the turnover or release of catecholamines by vascular beds in vivo are altered by hypertension. Moreover the enzymatic adaptations noted in the aorta and mesenteric arteries of hypertensive rats need not necessarily mean that identical enzymatic alterations occur in veins or in all arterial beds.

ANTIHYPERTENSIVE DRUGS

Many of the drugs which are used to treat hypertension interfere with the function of the sympathetic nervous system. Perhaps that antihypertensive drug for which there is the best evidence for its mechanism of action involving norepinephrine and the sympathetic nervous system is reserpine. We have previously shown that reserpine depletes vascular and cardiac norepinephrine stores in the normotensive rat and that mesenteric artery tyrosine hydroxylase activity was increased whereas monoamine oxidase activity was decreased (BERKOWITZ et al., 1971). Since hypertension can alter vascular catecholamine enzymes (TARVER et al., 1971) it was not surprising that reserpine-induced hypotension also modified these enzymes. However, recent evidence indicates that alterations in amine concentration may also be a determinant of catecholamine enzyme levels (DAIRMAN, 1972; MOLINOFF, 1972). We therefore administered reserpine and other hypotensive drugs to the SHR in doses which reduced blood pressure towards normal but did not result in a fall in blood pressure below that of a normotensive rat and measured the activity in the mesenteric artery of the synthetic catecholamine enzymes, dopamine-β-hydroxylase and tyrosine hydroxylase.

Reserpine administration resulted in a fall in systolic blood pressure from 190 mm to 135 mm Hg and a 60 per cent increase in vascular dopamine-β-hydroxylase activity (Table 2). Moreover tyrosine hydroxylase activity was also increased (C. Kohler, unpublished observation). When pressure was reduced in the SHR with phenoxybenzamine, an α-receptor blocker, dopamine-p-hydroxylase activity was again increased (Fig. 2).

In order to lower blood pressure without also depleting catecholamines, L-dopa was administered daily for 3 weeks. In contrast to the two previous drugs, no change

Table 2. Influence of reserpine† on the dopamine-β-hydroxylase (DBH) and tyrosine hydroxylase (TH) activity in the mesenteric artery of the SHR

Treatment	B.P.‡	DBH‡	TH
SHR	193/148	50 ± 0.4	0.98 ± 0.08
SHR + Reserpine	136/100	$72 \pm 0.6*$	$2.15 \pm 0.20*$

* Significantly differ from control SHR, $P < 0.05$.
† Reserpine was given in a dose of 0.25 mg/kg i.p. once a day for 4 days and rats killed on day 5.
‡ DBH activity is as nmol octopamine/hr/mg protein. The results are the average $\pm$ standard error of the mean from 8–13 animals for each group. Blood pressures are the average systolic/diastolic pressure. Tyrosine hydroxylase activity is expressed as nmoles Dopa/15 min/mg/protein.

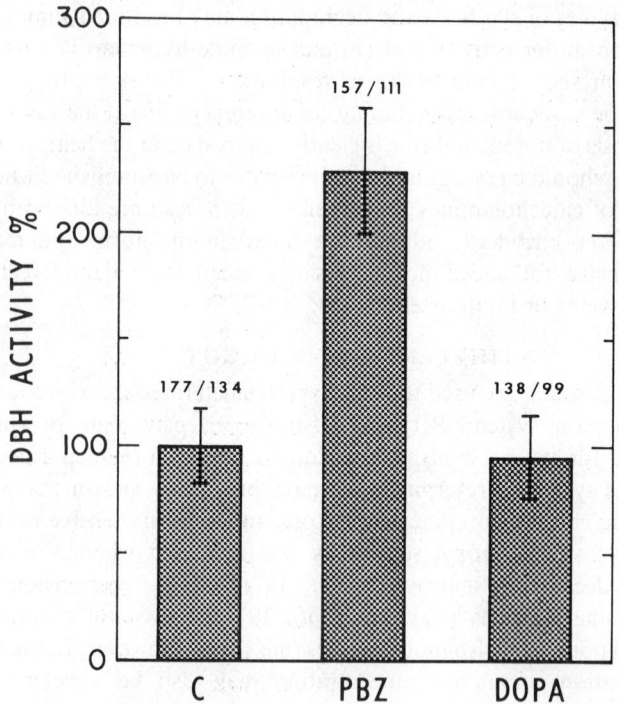

Fig. 2.—Effect of reduction of blood pressure by phenoxybenzamine or L-dopa on the dopamine-β-hydroxylase (DBH) activity in the mesenteric artery of the SHR. Results are expressed as percentage control DBH activity. Phenoxybenzamine was given in a dose of 10 mg/kg once a day for 2 days and rats killed on the third day. L-dopa was given once daily s.c. for 3 weeks, 300 mg/kg, week one, 600 mg/kg, week two and 1000 mg/kg, week three. Systolic/diastolic blood pressure is shown above bars. Results are the average of 5–8 determinations.

in dopamine-β-hydroxylase accompanied the fall in blood pressure (Fig. 2). Similar results have been obtained utilizing the monoamine oxidase inhibitor, pargyline (unpublished observation). It appears that both blood pressure and amine concentration are important in determining whether vascular catecholamine enzyme activity is altered. Moreover, a drug which lowers blood pressure but increases the activity of catecholamine synthetic enzymes in blood vessels may initiate mechanism which tend to limit its action or result in tolerance to its antihypertensive effects.

TABLE 3. FEEDBACK CONTROL OF NOREPINEPHRINE SYNTHESIS

Tissue (N)†	Drug	Time	μg/g (%Δ)	NE† dis/min per g (%Δ)	dis/min per μg (%Δ)
Heart (8)	Control	—	1·83 ± 0·08	2233 ± 321	1205 ± 392
Heart (8)	MAOI	24	3·18 ± 0·01 (+73)	699 ± 137*(−68)	214 ± 38* (−82)
Aorta (3)	Control	—	1·66 ± 0·71	649 ± 80	488 ± 122
Aorta (3)	MAOI	24	1·55 ± 0·32 (−7)	399 ± 58* (−39)	273 ± 53 (−44)
Mesenteric Art. (3)	Control	—	2·44 ± 0·16	2971 ± 390	1205 ± 88
Mesenteric Art. (3)	MAOI	24	6·20 ± 0·30*(+154)	3233 ± 524 (+9)	533 ± 118*(−56)
Mesenteric Vein (2)	Control	—	1·03–1·47	641–793	539–622
Mesenteric Vein (2)	MAOI	24	2·43–3·14 (+122)	320–481 (−44)	102–197 (−74)

Effect of monoamine oxidase inhibition on the concentration and synthesis of ^{14}C-norepinephrine in the guinea pig cardiovascular system. Pargyline hydrochloride 75 mg/kg was administered i.p. 25 hr prior to sacrifice; 30 μCi of ^{14}C-tyrosine was administered i.v. 1 hr prior to sacrifice.

* Statistically significant $P < 0.05$ compared to control tissue.

† (N) indicates the number of individual hearts or groups of vessels assayed. Each group of vessels was pooled from 4 to 5 animals for aorta and mesenteric arteries and the range from two groups of 8 mesenteric veins.

‡ Norepinephrine is dis/min per g tissue or μg norepinephrine ± S.E.M., (%Δ) indicates percentage difference from control tissues.

Among the most enigmatic of the antihypertensive drugs, are those classed as monoamine oxidase inhibitors. The paradox as to why drugs which can increase the tissue content of norepinephrine, a vasoconstrictor, may also lower blood pressure has not been resolved (NICKERSON, 1970). Drugs which elevate norepinephrine levels could diminish catecholamine synthesis in the vasculature by feedback inhibition (NAGATSU et al., 1964; WEINER et al., 1972) and conceivably diminish the amount of amine reaching the medial smooth muscle. We have studied the end product control of norepinephrine formation in blood vessels and heart of the guinea pig in vitro and in vivo.

As shown in Table 2, the susceptibility to decreased norepinephrine synthesis after pargyline was heart > mesenteric vein > aorta > mesenteric artery. A 73 per cent elevation in heart norepinephrine content caused an 82 per cent fall in cardiac norepinephrine specific activity (dis/min per μg) whereas a 154 per cent increase in mesenteric arterial norepinephrine levels, twice that of the heart, resulted in only a 56 per cent decline in norepinephrine specific activity in the artery.

Another index of norepinephrine synthesis is the amount of radioactive norepinephrine formed from tyrosine-^{14}C per amount of tissue (dis/min per g tissue). Using this measure of conversion the heart catecholamine synthesis was reduced to a far greater extent than any vascular bed with no reduction in labelled norepinephrine detected in the mesenteric artery.

Because vascular tyrosine hydroxylase is susceptible to catecholamine inhibition in *vitro* we believe it is likely that this mechanism is also operative *in vivo*. However, in some vascular beds such as the mesenteric arteries the high tyrosine hydroxylase levels (SPECTOR *et al.*, 1972; TARVER *et al.*, 1971) may require a greater concentration of catecholamines for further inhibition of the enzyme than can easily be achieved *in vivo*. Therefore after a drug, like a monoamine oxidase inhibitor, end product regulation of norepinephrine synthesis may vary in different parts of the cardiovascular system. The present findings suggest that the heart and veins may be more readily subject to decreases in norepinephrine synthesis by feed-back inhibition than are the arteries. If applicable to man this data may explain the clinical evidence for monoamine oxidase inhibitors decreasing cardiac output and producing postural hypotension but only minimally altering recumbent blood pressure (NICKERSON *et al.*, 1970).

CONCLUSIONS

(1) In the mesenteric arteries and the aorta of hypertensive rats the enzymes of norepinephrine metabolism are altered. Synthetic enzyme activities are reduced and degradative enzyme activities are increased. We believe that this is a compensatory adaptation to hypertension at the vascular level.

(2) Some anti-hypertensive drugs increased the activity of vascular tyrosine hydroxylase and dopamine-β-hydroxylase as they reduced blood pressure in the SHR whereas other drugs lowered blood pressure without altering enzyme activities. Both blood pressure and catecholamine concentrations are apparently regulators of enzyme levels in blood vessels.

(3) Monoamine oxidase inhibitors raised the content of norepinephrine in the heart and blood vessels of guinea pigs and diminished norepinephrine synthesis by end product inhibition at the tyrosine hydroxylase step. However, some arterial beds were more resistant to end product regulation of norepinephrine than were the veins or heart.

REFERENCES

BERKOWITZ B. A., SPECTOR S. and TARVER J. H. (1972). *Brit. J. Pharm.* **44**, 10–16.
BERKOWITZ B. A., TARVER J. H. and SPECTOR S. (1971). *J. Pharm Exp. Therap.* **177**, 119–126.
CREVLING C., DALGARD N., and NIKODIJEVIC B. (1969). *Fedn Proc.* **28**, 416.
DAIRMAN W. (1972). *Molecular Pharmacology* **8**, 293–299.
DE CHAMPLAIN J., KRAKOFF L. R. and AXELROD J. (1968). *Circulat. Res.* **28**, 479–491.
DE CHAMPLAIN J., MUELLER R. A. and AXELROD J. (1969). *Circulat Res.* **25**, 285–291.
DE CHAMPLAIN J. and VAN AMERINGEN M. R. (1972). *Circulat Res.* **4**, 617–628.
GILLIS C. M. and ROTH R. H. (1970). *Experientia* **26**, 960–961.
HARTMAN B. and UDENFRIEND S. (1972). *J. Pharm. Exp. Ther.* **24**, 311.
HICKLER R. B. and VANDAM L. (1970). *Anesthesiology* **33**, 214–228.
LOUIS W., DOYLE A. E., ANOVEKAR S. (1973). *N. Eng. J. Med.* **288**, 599–601.
LOUIS W. J., SPECTOR S., TABEI R. and SJOERDSMA A. (1969). *Circulat. Res.* **24**, 85–91.
MOLINOFF P. B., BRIMIJOIN S. and AXELROD J. (1972). *J. Pharm. Exp. Ther.* **182**, 116–124.
NAGATSU T., LEVITT M. G. and UDENFRIEND S. (1964). *J. Biol. Chem.* **238**, 2910–2917.
NAGATSU T. and UDENFRIEND S. (1972). *Clinical Chem.* **18**, 980–983.
NAKAMURA K., GEROLD M. and THOENEN H. (1972). In *Spontaneous Hypertension* (OKAMOTO, K., Ed). pp. 51–58, Igaku Shoin, Tokyo.
NICKERSON M. (1970). *The pharmacologic Basis of Therapeutics*, (GOODMAN L. S. and GILMAN, A. Eds.) pp. 732–733, Macmillan, New York.
OKAMOTO K. (1969). *Int. Rev. Exp. Path.* **7**, 227–270.
ROFFMAN M. and GOLDSTEIN M. (1973). *Life Sciences* In press.
SPECTOR S., TARVER J. and BERKOWITZ B. A. (1972). *Pharm. Rev.* **24**, 191–202.
TANAKA C. (1972). In *Spontaneous Hypertension* (Okamoto, K., Ed). pp. 60–62 Igaku-Shoin, Tokyo.
TARVER J., BERKOWITZ B. and SPECTOR S. (1971). *Nature, Lond.* **231**, 252–253.

TRAJKOV T., BERKOWITZ B. A. and SPECTOR S. (1973). *Eur. J. Pharm.* (submitted).
VERITY A., SU C. and BEVAN J. (1972). *Biochem. Pharm.* **21**, 193.
WEINER M., CLOUTIER G. BJUR R. and PFEFFER R. J. (1972). *Pharm. Rev.* **24**, 203–221.
WEINSHILBOUM R. and AXELROD J. (1971). *Science* **173**, 931–934.
YAMORI Y., LOVENBERG W. and SJOERDSMA A. (1970). *Science* **170**, 544–546.

Frontiers in Catecholamine Research 1973, pp. 875 to 877. Pergamon Press. Printed in Great Britain.

INTRACISTERNAL 6-HYDROXYDOPAMINE (6-OHDA) AND 5,6 DIHYDROXYTRYPTAMINE (5,6-DHT) IN EXPERIMENTAL HYPERTENSION

J. P. CHALMERS, J. L. REID and L. M. H. WING

Department of Medicine, University of Sydney, Sydney, N.S.W. 2006, Australia

THE experiments reported here were designed to examine the role of central nora-drenergic and serotonergic nerves in the pathogenesis of experimental renal hyper-tension and experimental neurogenic hypertension.

Small doses of 6-OHDA (600 $\mu g/kg$) or of 5,6-DHT (300 $\mu g/kg$) were injected intracisternally (i.c.) into New Zealand white rabbits to produce selective ablation of either catecholaminergic (THOENEN and TRANZER, 1968; URETSKY and IVERSEN, 1970; CHALMERS and REID, 1972) or serotonergic (BAUMGARTEN and LACHENMAYER, 1972; BAUMGARTEN et al., 1971) nerves in the brain and spinal cord. Small doses of these drugs given in this way do not have any permanent effects on peripheral autonomic nerves (CHALMERS and REID, 1972).

Neurogenic hypertension was produced by section of the carotid sinus and aortic nerves causing an immediate increase in mean arterial pressure which persisted throughout a two-week observation period in control animals (Fig. 1A). In animals pretreated with 6-OHDA, sinoaortic denervation only produced a transient increase in arterial pressure lasting two days and thereafter pressure returned to pre-denerva-tion control levels, (Fig. 1A). When 6-OHDA was given to rabbits with sustained neurogenic hypertension produced by sinoaortic denervation, it caused an immediate and persistent return of pressure to pre-denervation levels (Fig. 1B). Central nore-pinephrine concentrations were reduced in all brain and cord areas in the 6-OHDA treated animals, especially in the spinal cord where the levels fell to $<10\%$ of the values seen in control rabbits (CHALMERS and REID, 1972).

Intracisternal 6-OHDA was then used in rabbits with experimental renal hyper-tension produced by bilateral wrapping of the kidneys with cellophane. Pretreatment with 6-OHDA markedly reduced the rise in arterial pressure following renal wrapping (Fig. 1C). In rabbits with sustained hypertension following renal wrapping, intra-cisternal 6-OHDA caused the pressure to return towards control levels whether given six weeks (Fig. 1D, black circles) or 18 weeks (Fig. 1D, black squares) after the operation.

5,6-DHT given intracisternally caused a reduction in endogenous serotonin concentration, most marked in the spinal cord, where the levels fell to approximately 25 per cent of control. There were no significant changes in endogenous catechol-amine concentrations. Pretreatment with 5,6-DHT caused a small reduction in mean arterial pressure in normal animals and completely prevented the hypertension that usually follows sinoaortic denervation (Fig. 1E). When given to animals with sus-tained neurogenic hypertension produced by section of the buffer nerves, 5,6-DHT caused a significant ($P < 0.05$) fall in arterial pressure, though the pressure remained above initial normotensive control levels (Fig. 1F).

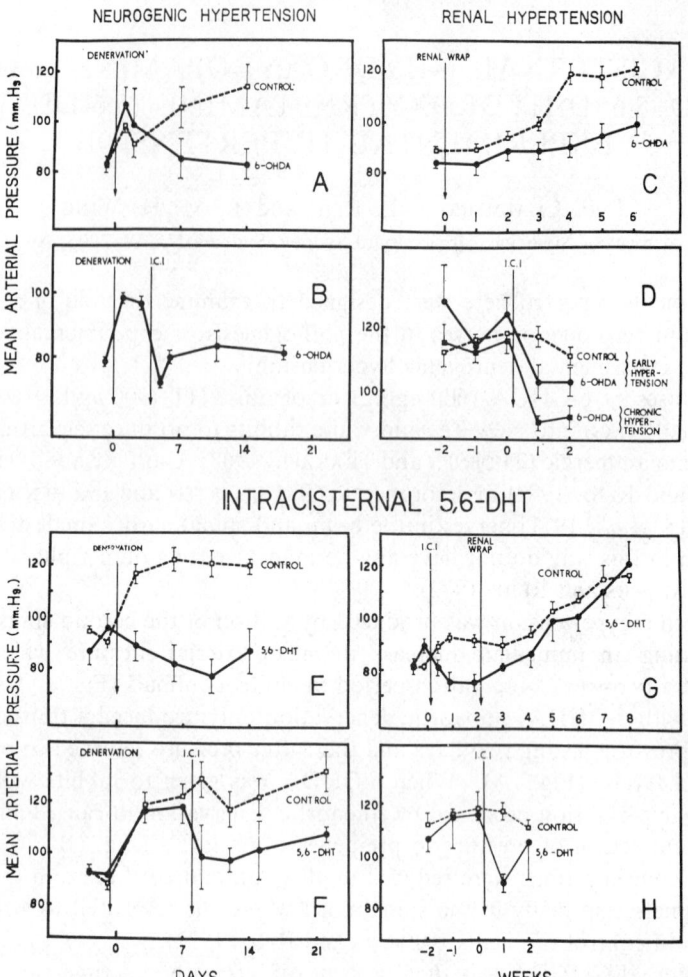

FIG. 1.—Mean arterial pressure changes in animals subjected to sinoaortic denervation (4 left panels) or bilateral renal wrapping (4 right panels). "ICI" indicates intra-cisternal injection of vehicle solution in control animals and of 6-OHDA or 5,6-DHT in test animals as labelled. Values represent means ± s.e. of the mean.

5,6-DHT given intracisternally did not in any way modify the development of renal hypertension produced by bilateral wrapping of the kidneys with cellophane; in animals pretreated with 5,6-DHT, renal wrapping caused the pressure to rise at the same rate and to the same extent as it did in the control rabbits (Fig. 1G). When the 5,6-DHT was given intracisternally six weeks after the development of renal hypertension, it caused only a transient and minor lowering of arterial pressure (Fig. 1H).

These experiments suggest that central noradrenergic and serotonergic nerves both participate in the regulation of systemic arterial pressure, though in different ways. Both types of neurones appear to play a role in neurogenic hypertension but only the noradrenergic nerves seem to be important in renal hypertension.

REFERENCES

BAUMGARTEN H. G., BJORKLUMD A., LACHENMAYER L., NOBIN A. and STENEVI U. (1971) *Acta Physiol. Scanda. Suppl.* **373**, 1–15.
BAUMGARTEN H. G. and LACHENMAYER, L. (1972) *Brain Res.* **38**, 228–232.
CHALMERS J. P. and REID J. L. (1972) *Circulat. Res.* **31**, 789–804.
THOENEN H. and TRANZER J. P. (1968) *Naunyn Schmiedebergs Arch. Pharmakol.* **261**, 271–288.
URETSKY N. J. and IVERSEN L. L. (1970) *J. Neurochem.* **17**, 269–278.

Frontiers in Catecholamine Research 1973, pp. 879 to 881. Pergamon Press. Printed in Great Britain.

CENTRAL ADRENERGIC NEURONS AND THE CONTROL OF BLOOD PRESSURE

G. HAEUSLER

Department of Experimental Medicine of F. Hoffmann-La Roche & Co., Ltd.,
Basel, Switzerland

EVIDENCE that central adrenergic neurons are involved in the cardiovascular control stems mainly from two types of observations. Firstly, some hypotensive drugs, in particular α-methyldopa and clonidine reduce peripheral sympathetic tone and lower blood pressure predominantly by an interaction with central adrenergic structures (SCHMITT and SCHMITT, 1970; RUBENSON, 1971; SCHMITT et al., 1971; HEISE and KRONEBERG, 1972; FINCH and HAEUSLER, 1973; HAEUSLER, 1973a). Secondly, in experimental hypertension changes in the turnover of hypothalamic, bulbar or spinal norepinephrine seem to occur (CHALMERS and WURTMAN, 1971; NAKAMURA et al., 1971a,b) and destruction of central adrenergic neurons interferes with the initiation of experimental hypertension (HAEUSLER et al., 1972). In all probability central adrenergic neurons are involved in the peripheral sympathetic activation which follows electrical stimulation of the posterior hypothalamus (PRZUNTEK et al., 1971; PRZUNTEK and PHILIPPU, 1973). The present report deals with the possible existence of another central adrenergic system which seems to be intimately related to the baroreceptor reflex arc.

In cats anaesthetised with urethane electrical stimulation of the posterior hypothalamus causes an increase in peripheral sympathetic nerve activity—recorded from the preganglionic splanchnic and a postganglionic renal sympathetic nerve—and an elevation of blood pressure and heart rate (Fig. 1, panel 1). The original recording from the oscilloscope screen (Fig. 1, panel 1) shows that the stimulation-induced augmentation of sympathetic nerve activity is characterised by a strong initial burst, followed by a phase of inhibition and a final stabilisation of the discharges at a reduced level.

Previous experiments (HAEUSLER, 1973a) have shown that the phase of inhibition in the stimulation-induced discharge pattern is due to the counterregulation by the depressor baroreceptor reflex which is activated by the rapid rise in blood pressure during hypothalamic stimulation. For instance, the phase of inhibition disappears after an artificial lowering of the blood pressure by bleeding and it is readily re-introduced after re-infusion of the blood and re-elevation of the blood pressure. Furthermore, cutting the buffer nerves (both vagus and sinus nerves) irreversibly converts the normal stimulation-induced discharge pattern containing an inhibitory phase into a continuous firing.

If hypothalamic stimulation meets an already activated depressor baroreceptor reflex, the phase of inhibition is also absent. Panel 2 of Fig. 1 shows that electrical stimulation of both sinus nerves virtually abolished spontaneous sympathetic nerve activity; additional hypothalamic stimulation resulted in a continuous low amplitude firing without an inhibitory phase (Fig. 1, panel 3). Therefore, the phase of inhibition in the discharge pattern during hypothalamic stimulation can be considered as a

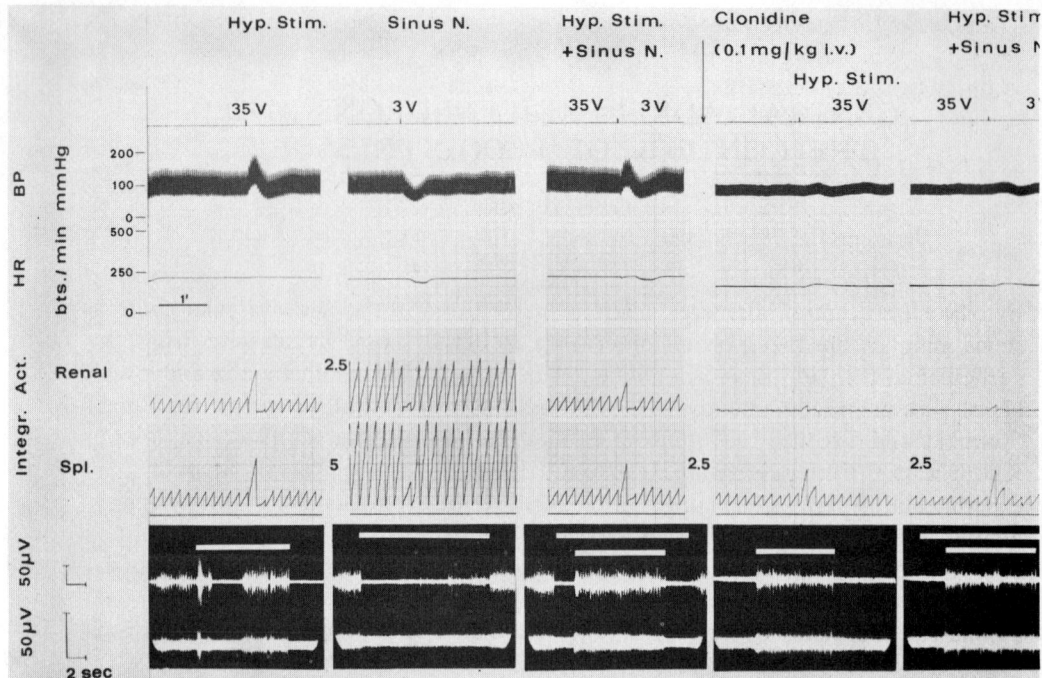

FIG. 1.—Cat, urethane anaesthesia (0·9 g/kg i.p.). Similarity between the effect of clonidine (0·1 mg/kg i.v.) and that of bilateral sinus nerve stimulation. The four channels of
the record paper show from top to bottom blood pressure, heart rate, integrated electrical
activity of a postganglionic renal sympathetic nerve and the right sympathetic preganglionic splanchnic nerve in arbitrary units. Below each panel the original oscilloscope recordings show the effect of the electrical stimulation of the posterior hypothalamus on the discharges in the renal (upper trace) and splanchnic (lower trace) sympathetic
nerves. The upper white horizontal bar in the oscilloscope recordings indicates the
duration (14 sec) of bilateral sinus nerve stimulation, the lower bar the duration (10 sec)
of hypothalamic stimulation. The voltage and rate of stimulation of the posterior hypothalamus were 35 V and 60 Hz, those for stimulation of the sinus nerves were 3 V and
32 Hz. The sensitivity of the integrators was increased by a factor of 2·5 or 5 in the panels
2, 4 and 5 as shown by the numbers on the left-hand side of these panels. The intervals
between the panels were 5–6 min each, except the interval between the panels 3 and 4
which was 30 min. From: HAEUSLER, G.: Activation of the Central Pathway of the
Baroreceptor Reflex, a Possible Mechanism of the Hypotensive Action of Clonidine.
Naunyn-Schmiedeberg's Arch. Pharmacol. **278,** 231–246 (1973). Berlin-Heidelberg-New
York: Springer.

reliable criterion for an inhibition of peripheral sympathetic nerve activity by the
counterregulating depressor baroreceptor reflex. This holds also true for the peripheral sympathetic activation which is achieved by electrical stimulation of the anterior
part of the fastigial nucleus of the cerebellum (HAEUSLER, 1973a).

Intravenous administration of clonidine to cats is followed by a decrease in
spontaneous sympathetic nerve activity, blood pressure and heart rate. Hypothalamic stimulation after clonidine causes a similar discharge pattern in the peripheral sympathetic nerves as does combined stimulation of the sinus nerves and the
posterior hypothalamus (compare panels 3 and 4 of Fig. 1). This discharge pattern
is not further modified by additional stimulation of the sinus nerves (Fig. 1, panel 5).

Thus, clonidine induced a state which closely resembles that of an activation of the depressor baroreceptor reflex. In view of the well-known property of clonidine to stimulate α-adrenoceptors the possibility was considered that the central part of the baroreceptor reflex arc is under the influence of adrenergic neurons.

Support for this hypothesis was provided by experiments in rats. The fall in blood pressure due to the stimulation of the baroreceptor fibres of the carotid sinus was inhibited in a dose-dependent manner by the intraventricular injection of the α-adrenoceptor blocking agents phentolamine, phenoxybenzamine and piperoxan (HAEUSLER, 1973b).

Thus, the intraventricular injection of three α-adrenoceptor blocking agents from different chemical classes prevents the activation of the depressor baroreceptor reflex. Such an activation is, however, induced by the α-stimulating agent clonidine. Both observations support each other and make it highly probable that the central part of the baroreceptor reflex arc either contains adrenergic neurons or is profoundly modulated by such neurons.

REFERENCES

CHALMERS J. P. and WURTMAN R. J. (1971). *Circul. Res.* **28**, 480–491.
FINCH L. and HAEUSLER G. (1973). *Br. J. Pharmacol.* **47**, 217–288.
HAEUSLER G. (1973a) *Naunyn-Schmiedeberg's Arch. Pharmacol.* **278**, 131–246.
HAEUSLER G. (1973b) *Naunyn-Schmiedeberg's Arch. Pharmacol.* (Suppl. to Vol. 277), R 27.
HAEUSLER G., FINCH L. and THOENEN H. (1972) *Experientia (Basel)* **28**, 1200–1203.
HEISE A. and KRONEBERG G. (1972) *Europ. J. Pharmacol.* **17**, 315–317.
NAKAMURA K., GEROLD M. and THOENEN H. (1971a) *Naunyn-Schmiedeberg's Arch. Pharmak.* **268**, 125–139.
NAKAMURA K., GEROLD M. and THOENEN H. (1971b) *Naunyn-Schmiedeberg's Arch. Pharmak.* **271**, 157–169.
PRZUNTEK H., GUIMARAES S. and PHILIPPU A. (1971) *Naunyn-Schmiedeberg's Arch. Pharmak.* **271**, 311–319.
PRZUNTEK H. and PHILIPPU A. (1973) *Naunyn-Schmiedeberg's Arch. Pharmacol.* **276**, 119–122.
RUBENSON A. (1971) *J. Pharm. Pharmac.* **23**, 228–230.
SCHMITT H. and SCHMITT H. (1970) *Europ. J. Pharmacol.* **9**, 7–13.
SCHMITT H., SCHMITT H. and FENARD S. (1971) *Europ. J. Pharmacol.* **14**, 98–100.

Frontiers in Catecholamine Research 1973, pp. 883 to 889. Pergamon Press. Printed in Great Britain.

THE CENTRAL NEURAL REGULATION BY BARORECEPTORS OF PERIPHERAL CATECHOLAMINERGIC MECHANISMS

Donald J. Reis, Nobutaka Doba[1] and M. Samir Amer[2]

[1] Laboratory of Neurobiology, Department of Neurology, Cornell University Medical College, New York, New York 10021, U.S.A. and [2] Mead Johnson Research Center, Evansville, Indiana, U.S.A.

The reflexes from baroreceptors play an important role in governing the discharge of sympathetic neurons to blood vessels. As such, they serve two principal functions. First they are of importance in the reflex regulation of the circulation. Second, since the activity of the sympathetic innervation of the vasculature is a major source of catecholamines (Spector, 1972) and since the rate of synthesis, the activities and possibly quantities of synthetic enzymes, and the rate of release of catecholamines varies directly with the intensity of nerve impulse activity (Pletscher, 1972), the baroreceptors probably help regulate much of the catecholamine metabolism in the body.

The baroreceptors, under normal circumstances, are tonically active (Korner, 1971) thereby exerting a continuous inhibition on the discharge of sympathetic nerves. Withdrawal of baroreceptor input, either by reduced stretch of the vascular receptors [most commonly initiated by assumption of an upright posture (Gauer and Thron, 1965)] or by transection of afferent nerves, results in a differentiated release of preganglionic sympathetic fibres from inhibition. The major cardio-vascular response to peripheral deafferentation of baroreceptors is arterial hypertension (Korner, 1965; Krieger, 1967; De Quattro et al., 1969).

The magnitude of the total inhibition of sympathetic nerve activity exerted by baroreceptors has previously never been determined. Denervation of baroreceptors of the carotid sinus and aortic arch region by transection of the carotid sinus and aortic depressor nerves (sino-aortic denervation) is not sufficient to interrupt afferent activity from baroreceptors of other major arteries or of stretch receptors of the heart and the lesser circulation which are carried in the vagus (Paintal, 1969). The hypertension produced by sino-aortic denervation is only moderate (Korner, 1965; Krieger, 1967; De Quattro et al., 1969).

We have recently sought to determine in a new way just how much inhibition is exerted by baroreceptors on the sympathetic neural drive to the circulation. We have done this by denervating baroreceptors *centrally* rather than peripherally. This has been accomplished by placing small electrolytic lesions bilaterally in a major site of termination of baroreceptors in the brain, the area of the *nucleus tractus solitarii* (NTS) located at the obex (Miura and Reis, 1969; Miura and Reis, 1972). Lesions at this site will destroy not only primary afferent fibres but also first order neurons and will abolish all depressor responses elicited from vascular stretch receptors in arteries and the heart (Lee et al., 1972; Miura and Reis, 1972).

The effects of such lesions in the unanaesthetised rat are profound (Doba and

Reis, 1973). Following the placement of such lesions and immediately upon awakening from anaesthesia the animals develop an elevation of systolic, diastolic, and mean arterial blood pressures to about 160 per cent of control without changes in heart rate. The elevation of blood pressure is entirely attributable to the 2·5-fold increase in the total peripheral resistance. The elevated resistance produces an overload of the left ventricle with consequent reduction of stroke volume, and an increase in end-diastolic pressure resulting in a fall of cardiac output to 60 per cent of normal. This ultimately leads to failure of the left ventricle and by 4–6 hr after placing the lesion the animals develop acute pulmonary oedema and die.

The hypertension is not due to changes in blood gases nor the release of pressor substances from kidneys or adrenal glands since prior adreno-nephrectomy does not reduce the magnitude of the hypertension. However, the hypertension appears to depend on the integrity of as yet unknown regions of the upper brainstem and the forebrain since the hypertension, once established, is abolished by mid-collicular decerebration. It probably is mediated by ascending polysynaptic pathways (Miura and Reis, 1969) since it is extremely sensitive to anaesthetics.

All present evidence suggests that the hypertension is neurogenic, is due to differentiated activation of sympathetic neurons, and is mediated by alpha-adrenergic receptors. First, the hypertension can be abolished by ganglionic blocking agents or by the alpha-adrenergic blocking agent, phentolamine. Second, treatment of rats with 6-hydroxydopamine (6-OHDA) will also block the experimental hypertension produced by NTS lesions but only after the adrenal glands have been removed (Fig. 1). This finding, paralleling the recent observations by de Champlain and Van Ameringen (1972), in animals with DOCA-salt hypertension indicates that adreno-medullary secretion is increased following a lesion of the NTS, and in the absence of sympathetic fibers the amount of adrenal catecholamines released is sufficient to maintain blood pressure. The observation also implies that adreno-medullary secretion is tonically inhibited by baroreceptors. Indeed, that the release of adrenal catecholamines may be suppressed by baroreceptor activity is also indirectly suggested by the experiments of De Quattro et al., (1969), who have demonstrated that chronic denervation of sinoaortic nerves in rabbit produces a trans-synaptically mediated increase in the activities of adrenal tyrosine hydroxylase and PNMT.

Some recent findings suggest that the profound and sustained increase of the release of catecholamines from sympathetic neurons resulting from central baro-receptor deafferentation may have some effects on receptor mechanisms within the blood vessels themselves. Amer (1973), has found that the levels of cyclic AMP in the aorta of spontaneously hypertensive rats and rats made hypertensive by prolonged stress may be significantly reduced. The reduction is due to an increase in degradation of cyclic AMP by enhanced activity of the high affinity phosphodiesterase (PDE II). The reduction of cyclic AMP has been considered as a possible mechanism whereby arterial resistance is increased since a reduction of the cyclic nucleotide in smooth muscle is generally associated with increased contractility (Sobel and Mayer, 1973). In the artery this may also reflect a decreased sensitivity of the beta receptor.

A similar reduction in cyclic AMP in aorta and also in the heart can be found in rats made hypertensive by NTS lesions within 90 min after placement of NTS lesions

EFFECTS OF SYSTEMIC 6-OHDA ON NEUROGENIC
HYPERTENSION

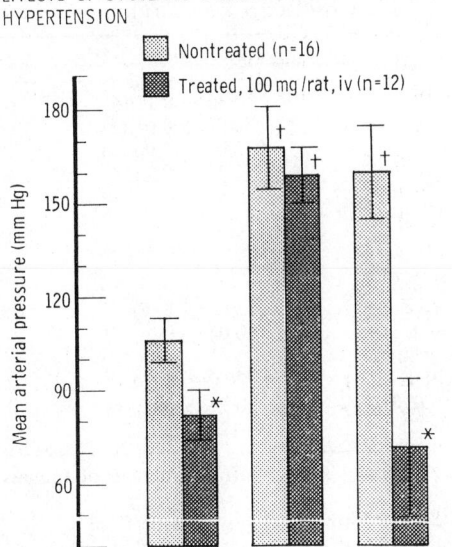

† Differs from pre-lesion, p<.001
* Differs from nontreated group, p<.001

Fig. 1.—Effects of systemic 6-hydroxydopamine (6-OHDA) on neurogenic hypertension in rat. 6-OHDA was administered i.v. (100 mg/rat) in 0·5 ml of ascorbic acid vehicle 24 hr prior to surgery. Under halothane anaesthesia a cannula was inserted into the tail artery in treated and untreated groups. Anaesthesia was discontinued and basal blood pressure (BP) measured. Animals were reanaesthetised, lesions placed in NTS (Doba and Reis, 1973) and anaesthesia discounted. Ninety minutes later BP was measured, the animals reanaesthetized and adrenals removed. BP was again measured after recovery from anaesthesia. In separate experiments this dose of 6-OHDA produced a 70–80 per cent fall ($P < ·001$) in NE in heart and spleen but not in brainstem.

(Table 1). As in other models for hypertension the fall of cyclic AMP is associated with an increase in the activity of the high affinity phosphodiesterase system without a change in the basal activity of adenylate cyclase. Paralleling the fall of cyclic AMP there is an elevation of cyclic GMP in the aorta. This may reflect in some manner activation of alpha-adrenergic receptors by norepinephrine.

These changes in the metabolism of cyclic nucleotides in acute neurogenic hypertension taken with the previous studies (Amer, 1973) suggest that intense activation of sympathetic neurons to blood vessels can lead to an activation of the phosphodiesterase system resulting in a reduction of the cyclic nucleotide AMP, an increase in GMP, a reduced availability of beta-adrenergic mechanisms, and a resultant vasoconstriction. Whether this may represent a mechanism of presynaptic control of postsynaptic vascular reactivity by the baroreceptors is a fascinating question for future research.

While baroreceptor reflex mechanisms in the brain appear to play a role in modulating peripheral catecholamine release, metabolism and possibly the catecholamine receptors themselves, the excitability of the baroreceptor reflex is itself controlled by other brain regions including forebrain, hypothalamus and cerebellum

TABLE 1. CHANGES IN CYCLIC NUCLEOTIDES AND ASSOCIATED ENZYMES IN AORTAS OF
RATS WITH ACUTE NEUROGENIC HYPERTENSION

	Sham	Lesion
Cyclic AMP[a]	0·76 ± 0·11 (4)	0·43 ± 0·01 (5)†
Cyclic GMP[a]	0·07 ± 0·01 (5)	0·12 ± 0·02 (5)*
Adenylyl cyclase[b]	24·7 ± 3·00 (5)	19·2 ± 5·10 (5)‡
Guanylyl cylase[c]	3·78 ± 0·76 (4)	4·15 ± 0·45 (4)‡
Phosphodiesterase (PDE) Activity[d]		
Cyclic AMP as substrate		
High Substrate (PDE I)	20·0 ± 2 (7)	20·0 ± 2 (7)
Low Substrate (PDE II)	0·04 ± 0·01 (7)	0·06 ± 0·004 (6)†
Cyclic GMP as substrate		
High Substrate (PDE I)	20·0 ± 4 (6)	20·0 ± 2 (6)‡
Low Substrate (PDE II)	0·14 ± 0·02 (6)	0·33 ± 0·02 (7)†
Blood pressure (mm Hg)	122 ± 2·1 (7)	173 ± 6·9 (7)†

Each value represents mean ± S.E.M.   (n) = number of animals
  (a) pm/mg wet tissue
  (b) nmoles cyclic AMP formed/10 mg wet tissue/10 min
  (c) nmoles cyclic GMP formed/10 mg wet tissue/20 min
  (d) nmoles cyclic nucleotide hydrolysed/5 mg wet tissue/10 min at 30°C.
  * $P < 0.05$
  † $P < 0.01$
  ‡ not significant.

    Bilateral lesions were placed in NTS with animals anaesthetised with halothane
(3%). Blood pressure was recorded from the tail artery. Immediately after lesions
the anaesthesia was discontinued. The animals were killed 90 min later and aortas
removed and frozen in liquid N_2. Sham controls had electrodes inserted into NTS
but no lesions were made. The results obtained in sham animals did not differ from
unoperated controls. The method for assay of cyclic AMP, adenylyl cyclase and
phosphodiesterase were as described by AMER (1973). Phosphodiesterase activity
was determined at two substrate concentrations (0·8 × 10⁶ and 0·5 × 10⁻³ M, cyclic
AMP; 1 × 10⁻⁶ and 0·5 × 10⁻³ M, cyclic GMP). Cyclic GMP was isolated in 4N
formic acid fraction eluted from Dowex-1-formate columns as described by MURAD
er al. (1971), and assayed by the method of MURAD and GILMAN (1971). Guanylyl
cyclase was assayed by a method similar to that used for adenylyl cyclase except
that GTP was substituted for ATP and Mn^{2+} was substituted for Mg^{2+}.

(MORUZZI, 1940; REIS and CUENOD, 1965; GEBBER and SNYDER, 1970). Moreover
some of the control may be by those neurons which synthesise, store and release the
neurotransmitter NE.

    The evidence for this is two-fold. First CHALMERS and REID (1972), have recently
shown that the hypertension produced in rabbit by sino-aortic denervation can be
aborted or abolished by the intra-cisternal injection of the adrenolytic agent 6-
OHDA in doses which do not alter the content of catecholamine peripherally, but
will significantly reduce them within the CNS. Second, we have recently found
that the hypertension produced in rats by NTS lesions may similarly be affected by
intra-cisternal administration of 6-OHDA. The intra-cisternal injection of 600
μg/rat of 6-OHDA, while not affecting the mean blood pressure, will significantly
attenuate the hypertension produced by NTS lesions (Fig. 2) confirming Chalmers'
and Reid's observations in rabbit. This effect is dose-related. In rat the effects of
intra-cisternal 6-OHDA on the hypertension produced by NTS lesions can only be
attributed to effects on central catecholamines since it does not alter the concentrations

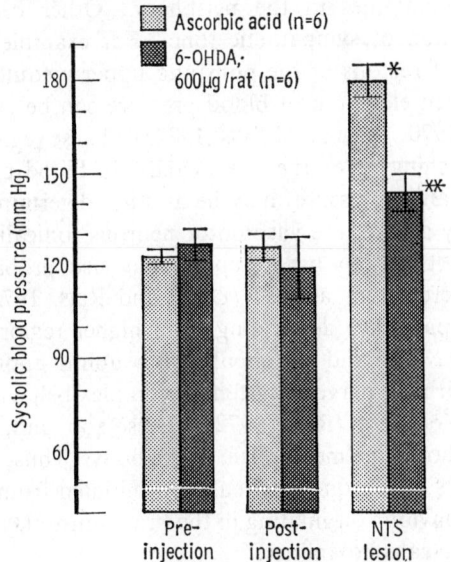

EFFECTS OF INTRACISTERNAL 6-OHDA ON NEURO-
GENIC HYPERTENSION

* Differs from pre-lesion, p<.001
** Differs from ascorbic acid, p<.01

FIG. 2.—Effects of intracisternal 6-OHDA on neurogenic hypertension in rat. 600 μg of 6-OHDA in 6 μl of ascorbic acid vehicle or vehicle alone was administered intracisternally and blood pressure measured daily by a tail cuff method. Three days later animals were anaesthetised with halothane, lesions placed in NTS (DOBA and REIS, 1973) and blood pressure measured 90 min later. In separate experiments such treatment resulted in a 35 per cent ($P < .0.01$) fall of NE in spinal cord but not in brainstem, hypothalamus or cerebellum.

of catecholamines in peripheral adrenergic endings nor will this dose when administered intravenously in the animal with adrenal glands intact in any way attenuate the hypertension. On the assumption that the hypertension produced by NTS lesions is a consequence of central deafferentation of baroreceptors, the evidence strongly suggests, in agreement with CHALMERS and REID (1972), that central catecholaminergic neurons are involved in mediating the hypertension resulting from withdrowal of the inhibition of preganglionic sympathetic neurons by baroreceptors. Moreover they are inhibitory, directly or indirectly, to the discharge of preganglionic neurons in the spinal cord.

The site at which the adrenergic terminals act and the location of the cell bodies giving rise to these terminals is not known. It does not seem likely that all central catecholamine systems are involved. For example, the local injection of 6-hydroxydopamine into the lateral hypothalamus in doses sufficient to impair catecholamine levels within the hypothalamus and forebrain (SMITH et al., 1972) does not attenuate the hypertension. However, when 6-OHDA was administered intracisternally in a dose sufficient to abort NTS hypertension (Fig. 2) the only area showing a reduction in NE was the spinal cord. We would therefore agree with CHALMERS and REID (1972), that it is more likely that the effects of NE are mediated via a descending NE system which in some manner is inhibitory for the expression of sympathetic activity at the level of the spinal cord.

This brief review has only touched on one neural system involved in the mediation of reflex drive to catecholamines in the periphery. Other central systems also participate in the regulation of sympathetic tone. For example, we have recently discovered the presence of regions of the pons and upper medulla distinct from the NTS from which a marked elevation of blood pressure can be evoked by local distortion (Hoff and Reis, 1970; Doba and Reis, 1972a). These regions which probably mediate the so-called Cushing reflex (i.e. the increase of blood pressure in response to an increase of intracranial pressure) may be another determinant of sympathetic nerve activity possibly by driving the vasomotor neurons tonically. The orthostatic reflexes are modulated not only by baroreceptors but also probably from the vestibular apparatus and cerebellum as well (Doba and Reis, 1972b, c, d). Finally, there are the important pathways descending from higher regions which appear to be the link between behaviour and the circulation coupling appropriate changes of the activity of the sympathetic nervous system to complex behavioural patterns such as fighting, feeding and exercise (Reis, 1972). These too undoubtedly contribute to the regulation of catecholamine metabolism in the body. Thus, the central nervous system integrating reflexes acting on the circulation initiated from the periphery and those associated with behaviour originating in the brain must play an essential role in the regulation of systemic catecholamines.

SUMMARY

The activity of preganglionic sympathetic neurons to blood vessels are tonically inhibited by baroreceptors. Removal of this inhibition by central deafferentation produced by lesions at a site of baroreceptor termination in the brainstem results in a marked and differentiated increase in sympathetic nerve activity resulting in neurogenic hypertension, alteration of cyclic nucleotide metabolism in the aorta and secretion of adrenal medullary catecholamines. Central baroreceptor mechanisms, themselves modulated by noradrenergic neurons, serve to regulate the release and metabolism of peripheral catecholamines and possibly their receptors.

Acknowledgements—This research was supported by grants from NIH (NS 04876, NS 3346) and NASA (33–010–179).

REFERENCES

Amer M. S. (1973) *Science* **179**, 807–809.
Chalmers J. P. and Reid J. L. (1972) *Circ. Res.* **31**, 789–804.
de Champlain J. and van Ameringen M. R. (1972) *Circ. Res.* **31**, 617–628.
de Quattro V., Nagatsu T., Maronde R. and Alexander N. (1969) *Circ. Res.* **24**, 545–555.
Doba N. and Reis D. J. (1972a) *Brain Res.* **47**, 487–491.
Doba N. and Reis D. J. (1972b) *Fedn. Proc.* **31**, 814.
Doba N. and Reis D. J. (1972c) *Brain Res.* **39**, 495–500.
Doba N. and Reis D. J. (1972d) *J. Physiol. (Lond.)* **227**, 729–747.
Doba N. and Reis D. J. (1973) *Circ. Res.* (in press)
Gauer O. H. and Thron H. C. (1965) In: *Handbook of Physiology*, section 2, Circulation, vol. III, (Hamilton W. F. and Daw P., Eds.) American Physiological Society., Washington, D.C.
Gebber G. L. and Snyder D. W. (1970) *Am. J. Physiol.* **218**, 124–131.
Hoff J. T. and Reis D. J. (1970) *Archs Neurol.* **23**, 228–240.
Korner P. I. (1965) *J. Physiol. (Lond.)* **180**, 266–278.
Korner P. I. (1971) *Physiol. Rev.* **51**, 312–367.
Krieger E. M. (1967) *Am. J. Physiol.* **213**, 139–142.
Lee T. M., Kuo J. S. and Chai C. Y. (1972) *Am. J. Physiol.* **222**, 713–720.
Miura M. and Reis D. J. (1969) *Am. J. Physiol.* **217**, 142–153.
Miura M. and Reis D. J. (1972) *J. Physiol. (Lond.)* **223**, 525–548.

Moruzzi G. (1940) *J. Neurophysiol.* **3**, 20–32.
Murad F. and Gilman A. G. (1971) *Biochem Biophys. Acta* **252**, 397–400.
Murad F., Manganiello V. and Vaughn M. (1970) *Proc. Natl. Acad. Sci.* **68**, 736–739.
Paintal A. S. (1969) *J. Physiol. (Lond.)* **203**, 511–532.
Pletscher A. (1972) *Pharmacol. Rev.* **24**, 225–232.
Reis D. J. and Cuenod M. (1965) *Am. J. Physiol.* **209**, 1267–1279.
Reis D. J. (1972) In: *Neural and Psychological Mechanisms in Cardiovascular Disease* (Zanchetti, A., Ed.) Casa Editrice, Il-Ponte, Milano.
Smith G. P., Strohmayer A. J. and Reis D. J. (1972) *Nature (New Biol.)* **235**, 27–29.
Sobel B. E. and Mayer S. E. (1973) *Circ. Res.* **32**, 407–414.
Spector S., Tarver J. and Berkowitz B. (1972) *Pharmacol. Rev.* **24**, 191–202.

Frontiers in Catecholamine Research 1973, pp. 891 to 895. Pergamon Press. Printed in Great Britain.

GENETIC VARIATION OF THE CATECHOLAMINE BIOSYNTHETIC ENZYME ACTIVITIES IN VARIOUS STRAINS OF RATS INCLUDING THE SPONTA- NEOUSLY HYPERTENSIVE RAT

WALTER LOVENBERG, HIROHIKO YAMABE, WYBREN DE JONG* and
CARL T. HANSEN
Section on Biochemical Pharmacology, Experimental Therapeutics Branch,
National Heart and Lung Institute,

and

Veterinary Resources Branch, Division of Research Services, National Institutes of
Health, Bethesda, Maryland 20014, U.S.A.

NOREPINEPHRINE, the neurotransmitter of the sympathetic nervous system, is synthesised by a well established three step enzymic pathway. These enzymes have been characterised and some of the mechanisms which control their intracellular activity are known. Since the sympathetic system plays a role in regulating the tone of the vasculature and central noradrenergic systems appear to be involved in blood pressure regulation, many investigators have attempted to relate catecholamine metabolism to hypertension. No gross changes in the easily measured parameters of catecholamine metabolism are evident in any type of human hypertension except that resulting from pheochromocytoma; several recent reports, however, indicate that plasma norepinephrine is slightly elevated in certain patients with essential hypertension (ENGELMAN et al., 1970; DEQUATTRO and CHAN, 1972; LOUIS et al., 1973).

The catecholamine metabolism of the spontaneously hypertensive rat (SHR) (OKAMOTO and AOKI, 1963) has also been investigated intensively by our laboratory (LOUIS et al., 1969, 1970; YAMORI et al., 1970, 1972a,b; YAMABE et al., 1973) and by other laboratories (OZAKI, 1966; NAGATSU et al., 1971; NAKAMURA et al., 1971; SPECTOR et al., 1972). No unambiguous relationship of catecholamine metabolism to the pathogenesis of hypertension in these rats has been established.

There is an alternate approach to evaluating cause and effect relationships in hypertension. The results of studies with different species clearly indicate that there is a genetic component in hypertension. For example HANSEN (1972) found in a survey of the blood pressures of inbred rats that considerable interstrain variation exists, although within-strain variation is relatively small. Comparison of factors which are suspected to participate in the development of hypertension in several strains may provide insight into physiologic-biochemical relationships.

In the current study we have measured tissue norepinephrine content and the activities of the three norepinephrine biosynthetic enzymes (tyrosine hydroxylase, aromatic L amino acid decarboxylase, and dopamine β hydroxylase,) in 9 different inbred strains of rats with a wide range of "normal" blood pressures.

---

* Present address: Rudolf Magnus Institute for Pharmacology, University of Utrecht-Medical Faculty, Vondellaan 6, Utrecht, The Netherlands

MATERIALS AND METHODS

The animals used in these studies were maintained in inbred colonies in the Animal Production Section of the National Institutes of Health (NIH). The animals were matched as closely as possible for age, sex and size in all the experiments. Deviations from complete matching are noted in the various experiments. The strains of rats used and their abbreviations are as follows: The Kyoto spontaneously hypertensive rat maintained as an inbred strain at the NIH, SHR; Wistar/NIH, W; Wistar/Kyoto, W/Ky; Roman High Avoidance, RHA; Osborne-Mendel, OM; Albany, ALB; ACI, PETH, and M-520. The details of all the physiologic and analytical methods used in this study have been presented previously (Yamabe et al., 1973). Table 1 gives the mean systolic blood pressure for each of these strains.

Table 1. Systolic blood pressure of 10-week-old inbred
rat strains of the NIH*

	Mean blood pressure (mm Hg)		Mean blood pressure (mm Hg)
SHR	182 ± 5	PETH	128 ± 7
OM	160 ± 6	W/K	128 ± 6
RHA	158 ± 11	W	128 ± 6
ALB	143 ± 14	M520	124 ± 9
		ACI	116 ± 6

* Taken from Hansen, C. T. (1972). Values given are mean systolic pressures ± s.d.

RESULTS AND DISCUSSION

The various parameters of catecholamine metabolism were first examined in brainstem. The level of norepinephrine was relatively constant within any one strain but ranged from a low 455 ng/g in the W/Ky to a high of 739 ng/g in W. All other strains examined were approximately midway between the two extremes.

Table 2 gives the brainstem activity for each of the norepinephrine biosynthetic enzymes. There are several observations of interest in these data. First, the activity of tyrosine hydroxylase varies very little among the different strains. The absence of strain differences is not surprising since this is the rate limiting enzyme in norepinephrine synthesis and since a certain minimal level of norepinephrine is necessary for normal neuronal functions. Conversely, it should be noted that the variation in norepinephrine levels is somewhat greater than that in tyrosine hydroxylase levels and therefore factors other than total tyrosine hydroxylase activity must contribute to amine level. This phenomenon has been discussed elsewhere (Yamabe et al., 1973).

Second, dopamine β hydroxylase which has about 10 times the specific activity of tyrosine hydroxylase shows more variability between strains. There appears to be a partial relationship between dopamine β hydroxylase and norepinephrine content, although insufficient data have been accumulated to determine statistical or physiological significance of such a relationship.

Third, aromatic L amino acid decarboxylase activity which is about 100 times higher than tyrosine hydroxylase shows marked variation between strains. This enzyme is thought to be present in neuronal cells in substantial excess. In this limited study there appeared to be two strain-types. The SHR, RHA and W/Ky being a

TABLE 2. NOREPINEPHRINE BIOSYNTHETIC ENZYME ACTIVITY IN THE
BRAINSTEM OF INBRED RAT STRAINS

Strain	Enzyme activity* (nmole/hr/mg protein)		
	Tyrosine hydroxylase	Aromatic L amino acid decarboxylase	Dopamine-β-hydroxylase
SHR	0·55 ± 0·01	38 ± 1	5·7 ± 0·3
OM	0·52 ± 0·02	68 ± 2	6·7 ± 0·1
RHA	0·49 ± 0·02	45 ± 1	7·0 ± 0·2
ALB	0·51 ± 0·06	83 ± 1	5·9 ± 0·1
PETH		64 ± 1	5·4 ± 0·2
W/K	0·47 ± 0·02	41 ± 1	4·5 ± 0·1
W	0·55 ± 0·03	71 ± 3	6·4 ± 0·3
M520	0·50 ± 0·02	66 ± 2	5·2 ± 0·3
ACI	0·54 ± 0·02	62 ± 1	5·4 ± 0·1

* The values given are the mean ± S.E.M. for at least four individual animals analysed in duplicate. Tyrosine hydroxylase and aromatic L amino acid decarboxylase were measured in 5-week-old animals. The age of the animals for dopamine-β-hydroxylase was 5–7 weeks.

low activity type and the other six strains a high activity type. When additional strains are examined, however, the types may merge into a continuous spectrum.

The activities of the norepinephrine biosynthetic enzymes have also been measured in the adrenal glands of each of these inbred strains (Table 3). Since there are significant strain differences in adrenal weights and amounts of cortical tissue, the activities are reported per pair of glands. As in the brainstem, strain differences are evident in all three enzymes. The variation in tyrosine hydroxylase is more marked whereas aromatic L amino acid decarboxylase activities fall into a much tighter range. It is also notable that the relative proportions of the various enzymes are quite different when the brainstem is compared to adrenal tissue. There appears to be no relationship between any of the enzymes and blood pressure.

The levels of each of the norepinephrine biosynthetic enzymes appear to be under independent genetic regulation (Tables 2 and 3). This is apparent in both the central nervous system and adrenal gland. Furthermore the relationships that occur in one organ are not present in the other organ. Using a completely different experimental approach THOENEN (1972) also showed that these enzymes are independently regulated.

BARCHAS and coworkers (CIARANELLO et al., 1972; KESSLER et al., 1972) have measured tyrosine hydroxylase in the adrenal gland and brainstem of a number of inbred mouse strains. They found greater variation in tyrosine hydroxylase than we observed among the rat strains, however, their conclusions were similar to ours; i.e., there are significant genetic variations in the catecholamine biosynthetic enzymes.

In the current study, the nine inbred rat strains with a wide range of normal blood pressures showed marked variations in norepinephrine biosynthetic enzymes in the brainstem and adrenal glands. No correlation between any of the enzymes and blood pressure was apparent. This divergence was exemplified by a comparison of aromatic L amino acid decarboxylase and blood pressure in W, W/Ky and SHR (Table 4). All three of these strains are of the Wistar type, however the SHR is genetically similar to the W/Ky, being separated from that strain about 30 generations previously.

TABLE 3. Norepinephrine biosynthetic enzyme activities in the
adrenal gland of inbred rat strains

Strain	Enzyme activity* (nmole/hr/gland)		
	Tyrosine hydroxylase	Aromatic L amino acid decarboxylase	Dopamine-β-hydroxylase
SHR	11·7 ± 0·5	319 ± 20	210 ± 5
OM	8·1 ± 0·2	310 ± 29	396 ± 10†
RHA	7·7 ± 1·0	455 ± 20	313 ± 17
ALB	12·3 ± 0·7		573 ± 52†
PETH		363 ± 14	303 ± 30†
W/Ky	12·3 ± 0·4	405 ± 37	295 ± 7†
W	9·9 ± 0·3	336 ± 25	275 ± 26
M520	8·1 ± 0·9	321 ± 18	551 ± 25
ACI	9·0 ± 0·8	329 ± 16	355 ± 22

    * The values given are the mean ± s.e.m. for at least four individual animals. Tyrosine hydroxylase and aromatic L amino acid decarboxylase in 5-week-old animals and dopamine-β-hydroxylase in 5–7 week old animals.
    † 7-weeks-old.

The decarboxylase activity of brainstem and heart is similar in the two Kyoto strains, but is 2 to 3 times higher in the Wistar from the NIH. In addition to these organs, other peripheral tissues show significant differences in many parameters of catecholamine metabolism when the W is compared with the two strains developed in Kyoto, Japan (Yamabe and Lovenberg, to be published).

TABLE 4. Aromatic L amino acid decarboxylase and blood pressure
in 3 rat strains*

Rat strain	Adult systolic blood pressure (mm Hg)	Decarboxylase (nmole/hr/mg protein)	
		Heart	Brainstem
Wistar/NIH	135 ± 2	5·9 ± 0·1	71 ± 3
Wistar/Kyoto–NIH	132 ± 1	1·9 ± 0·1	41 ± 1
SHR/NIH	187 ± 1	1·9 ± 0·2	38 ± 1

    * Data is taken from male animals 10–12 weeks of age for blood pressure and 5 weeks of age for the enzyme studies. Decarboxylase activity is measured as described by Yamabe et al. (1973).

CONCLUSIONS

    (1) There is significant genetic variation in the level of catecholamine biosynthetic enzymes in various inbred rat strains.

    (2) The genetic variation of catecholamine synthesis in the brainstem and adrenal glands does not appear to relate to the "normal" blood pressure of individual inbred strains.

    (3) Each of the three norepinephrine biosynthetic enzymes appear to be under independent genetic regulation.

REFERENCES

CIARANELLO R. D., BARCHAS R., KESSLER S. and BARCHAS J. D. (1972) *Life Sciences* **11,** 565–572.

DEQUATTRO V. and CHAN S. (1972) *Lancet* **1,** 806–809.

ENGELMAN K., PORTNOY B. and SJOERDSMA A. (1970) *Circ. Res.* **27,** (Suppl. 1) 141–146.

HANSEN C. T. (1972). In: *Spontaneous Hypertension* (OKAMOTO, K., Ed). pp. 13–17, Igaku Shoin, Tokyo.

KESSLER S., CIARANELLO R. D., SHIRE J. G. M. and BARCHAS J. D. (1972) *Proc. Nat. Acad. Sci.* **69,** 2448–2450.

LOUIS W. J., DOYLE A. E. and ANAVEKAR S. (1973) *New Eng. J. Med.* **288,** 599–601.

LOUIS W. J., SPECTOR S., TABEI R. and SJOERDSMA A. (1969) *Circ. Res.* **24,** 85–91.

LOUIS W. J., KRAUSS K. R., KOPIN I. J. and SJOERDSMA A. (1970) *Circ. Res.* **27,** 589–594.

NAGATSU I., NAGATSU T., MIZUTANI K., UMEZAWA H., MATSUZAKI M. and TAKEUCHI T. (1971) *Nature, Lond.* **230,** 381–382.

NAKAMURA K., GEROLD M. and THOENEN H. (1971) *Arch. Pharmak.* **271,** 157–169.

OKAMOTO K. and AOKI K. (1963) *Jap. Circ. J.* **27,** 282–293.

OZAKI M. (1966) *Jap. J. Pharm.* **16,** 257–263.

SPECTOR S., TARVER J. and BERKOWITZ B. (1972) In: *Spontaneous Hypertension.* (OKAMOTO K., Ed.) pp. 41–45, Igaku Shoin, Tokyo.

THOENEN H. (1972) *Pharm. Rev.* **24,** 255–267.

YAMABE H., DEJONG W. and LOVENBERG W. (1973) *Europ. J. Pharm.* **22,** 91–98.

YAMORI Y., LOVENBERG W. and SJOERDSMA A. (1970) *Science* **170,** 544–546.

YAMORI Y., YAMABE H., DEJONG W., LOVENBERG W. and SJOERDSMA A. (1972) *Europ. J. Pharm.* **17,** 135–140.

YAMORI Y., DEJONG W., YAMABE H., LOVENBERG W. and SJOERDSMA A. (1972) *J. Pharm. Pharmacol.* **24,** 690–695.

Frontiers in Catecholamine Research 1973, pp. 897 to 899. Pergamon Press. Printed in Great Britain.

NOREPINEPHRINE SYNTHESIS AND TURNOVER IN THE BRAIN: ACCELERATION BY PHYSOSTIGMINE

Tomislav Kažić*†

Laboratory of Clinical Science, National Institute of Mental Health,
Bethesda, Maryland 20014, U.S.A.

Cholinergic-adrenergic interactions have been shown to exist in peripheral organs supplied with dual autonomic innervation (Kažić, 1971; Muscholl, 1970; Paton and Vizi, 1969). Both cholinergic and adrenergic neurons are well identified in the central nervous system (CNS), and acetylcholine (ACh) and norepinephrine (NE) are generally regarded as transmitter substances in the brain. Their close spatial distribution and a considerable body of experimental evidence indicate that an active interrelationship exists between the two parts of the autonomic nervous system in the brain.

After systemic administration, physostigmine, a reversible cholinesterase inhibitor, produces a hypertensive response in the rat, as do some other cholinesterase inhibitors which penetrate into the brain (Varagić, 1955, 1966). Because physostigmine has been shown to produce glycogenolysis in the brain and liver (Mršulja, 1968; Varagić et al., 1967), increased neuronal activity in the preganglionic fibres of the cervical sympathetic chain (Stamenović, 1955) and hypothermia (Myers and Yaksh, 1968; Varagić et al., 1971), it is believed that the drug produces a general noradrenergic activation. The sympathetic activation produced by physostigmine is of central origin (Mršulja et al., 1968). Inhibition of cholinesterase enhances cholinergic activity which appears to trigger the central adrenergic mechanism responsible for the activation of the peripheral sympathetic nervous system (Varagić et al., 1968). The present experiments were performed in order to examine the role of brain catecholamines on the sympathetic activating effect of physostigmine.

METHODS

Experiments have been carried out on male Sprague–Dawley rats (200–350 g body weight). Physostigmine (200 μg/kg, i.v.) was administered 5 min before a single rapid injection of L-tyrosine-^{14}C (44 μCi, i.v.; spec. activity 476 mCi/mmole), and 10 min later the rats were decapitated. All tissues were rapidly removed, brains carefully dissected into: brainstem, hypothalamus and forebrain (rest of brain), and all structures including spinal cords, hearts and salivary glands were quickly frozen in liquid nitrogen and stored at $-70°$C. Radioactive catecholamines were determined according to Kopin (1972), while the endogenous NE was assayed using the technique of Haggendal (1963).

6-Hydroxydopamine (6-OH-DA) was administered intracisternally (2×200 μg at 48 hr interval), and seven days after the second injection blood pressure responses

* International Postdoctoral Fellow, U.S. Public Health Service, Fellowship number 1 FO5 TW01827-01.

† Home Institute, Department of Pharmacology, Faculty of Medicine, 11.000 Belgrade, Yugoslavia.

to physostigmine were recorded under urethane anesthesia. A separate group of rats treated with 6-OH-DA intracisternally was used for determination of endogenous NE content in the brain, heart and salivary gland.

Atropine (2 mg/kg, i.p.), phenoxybenzamine (10 mg/kg, i.p.) and propranolol (10 mg/kg, i.p.) were administered 20–40 min before physostigmine (200 μg/kg, i.v.) in order to study their effects on the endogenous NE in the brain, and their inter- action with the physostigmine-induced changes.

Tyrosine hydroxylase (TH) activity in the brain areas was determined after a prolonged administration of physostigmine (2 × 200 μg/kg, i.v., during 7 days) according to the technique of Coyle (1972).

RESULTS AND DISCUSSION

In rats under urethane anesthesia, the hypertensive response to physostigmine was shown to depend on the integrity of catecholaminergic neurons in the brain. After intracisternal pretreatment with 6-OH-DA this response was diminished by 40 per cent. Endogenous NE content in the brain was severely reduced (15–36 per cent of the control level). In conscious rats, treatment with physostigmine produced a rapid decrease in the endogenous NE content of the brainstem and hypothalamus, but not of the forebrain. A linear dose–response relationship was obtained with doses of physostigmine (100, 200 and 300 μg/kg, i.v.) and the reduction in the endog- enous NE in the brain areas. Pharmacological analysis of this catecholamine- depleting action of physostigmine showed that it can be completely prevented with both atropine and propranolol. Atropine was also found not only to prevent the effect of physostigmine, but also to increase the endogenous NE content of the hypothalamus when given alone. Propranolol produced no change in the endogenous NE, however, pretreatment with this drug completely prevented the NE depleting action of physostigmine. Results obtained with phenoxybenzamine are inconclusive since this substance was found to produce a significant depletion of the endogenous NE stores in the brain, by itself.

Determination of the synthesis of radioactive NE from the labeled precursor, L-tyrosine-^{14}C, provided evidence that physostigmine, simultaneously with the ele- vation of the blood pressure in the periphery, produced an increase in the synthesis of the radioactive transmitter in the CNS. Synthesis of NE-^{14}C was enhanced in all brain areas, including those in which the endogenous NE levels had not been changed. Specific activity of the NE-^{14}C, considered as a measure of turnover, was also significantly increased.

TH assay of the enzyme activity in the brain areas performed after prolonged administration of physostigmine has shown that this substance is capable to produce a significant activation of the TH, particularly in the brainstem and hypothalamus, areas which seem to be main targets of the action of physostigmine. In addition, it should be pointed out that the repeated administration of physostigmine produced no change in the blood pressure levels of the treated animals, as compared to the saline-treated control group.

The present experiments support the view that the peripheral sympathetic acti- vation, seen after systemic administration of physostigmine, involves activation of a central adrenergic mechanism. Administration of physostigmine results in increased synthesis and turnover of NE both in the CNS and in the peripheral sympathetic nervous system. Increased activity of TH, rate-limiting enzyme in the catecholamine synthesis, has also been found in the hypothalamus and brainstem of the animals treated with physostigmine. Destruction of the brain catecholaminergic neurons with 6-OH-DA, blockade of the central muscarinic receptors with atropine and administration of a beta-adrenergic blocking agent, propranolol, were found to diminish or completely prevent the activating action of physostigmine on the adrenergic neurons in the brain.

It can be concluded, that the present investigation provides direct experimental evidence that an increase in synthesis, turnover and release of NE in the brain are the underlying mechanisms for the general adrenergic activation produced by physostigmine.

REFERENCES

COYLE J. T. (1972) *Biochem. Pharmacol.* **21,** 1935–1944.
HAGGENDAL J. (1963) *Acta physiol. Scand.* **57,** 242–254.
KAŽIĆ T. (1971) *Europ. J. Pharmacol.* **16,** 367–373.
KOPIN I. J. (1972) In: *The Thyroid and Biogenic Amines* (RALL and KOPIN I. J., Eds.) pp. 489–496, North-Holland, Amsterdam.
MRŠULJA B. B., TERZIĆ M. and VARAGIĆ V. M. (1968) *J. Neurochem.* **15,** 1329–1333.
MUSCHOLL E. (1970) In: *New Aspects of Storage and Release Mechanisms of Catecholamines.* (SCHÜMANN H. J. and KRONEBERG G., Eds.) pp. 168–186, Springer–Verlag, Berlin.
MYERS R. D. and YAKSH T. L. (1968) *Physiol. Behav.* **3,** 917–928.
PATON W. D. M. and VIZI E. S. (1969) *Br. J. Pharmac.* **35,** 10–28.
STAMENOVIĆ B. A. and VARAGIĆ V. M. (1970) *Neuropharmacology* **9,** 561–566.
VARAGIĆ V. M. (1955) *Br. J. Pharmac. Chemother.* **10,** 349–353.
VARAGIĆ V. M. and KRSTIĆ M. (1966) *Pharmacol. Rev.* **18,** 799–800.
VARAGIĆ V. M., KAŽIĆ T. and ROSIĆ N. (1968) *Yugoslav. Physiol. Pharmacol. Acta* **4,** (Suppl. 1), 113–120.
VARAGIĆ V. M., TERZIĆ M. and MRŠULJA B. B. (1967) *Arch. Pharmak. Exp. Path.* **258,** 229–237.
VARAGIĆ V. M., ŽUGIĆ M. and KAŽIĆ T. (1971) *Arch. Pharmak. Exp. Path.* **270,** 407–418.

Frontiers in Catecholamine Research 1973, pp. 901 to 903. Pergamon Press. Printed in Great Britain.

AMINOACYL DERIVATIVES OF DOPAMINE AS ORALLY EFFECTIVE RENAL VASODILATORS

J. H. BIEL[1], P. SOMANI[1], P. H. JONES[1], F. N. MINARD[1] and L. I. GOLDBERG[2]

[1]Division of Pharmacology and Medicinal Chemistry, Abbott Laboratories, North Chicago Illinois, 60064, U.S.A. and [2]Department of Clinical Pharmacology, Emory University School of Medicine, Atlanta, Georgia, 30303, U.S.A.

THE FINDING by Goldberg and associates [J. L. McNAY, R. H. McDONALD and L. I. GOLDBERG, Circ. Res. 16, 510 (1965)] that dopamine produced a selective dilatation of the renal vascular bed prompted the synthesis and pharmacological investigation of a large variety of dopamine derivatives, in an effort to find a clinically useful agent which would be orally effective and exert its effect over a relatively prolonged period of time (4–6 hr). An agent of this type would be an important adjunct in the drug treatment of congestive heart failure, hypertension, and acute renal shutdown. Many of the currently available antihypertensive drugs reduce renal blood flow in hypertensive patients. Increased renin secretion is believed to be due to reduced renal blood flow (RBF) and may result in the overproduction of angiotensin, thereby perpetuating the hypertensive process. An effective and selectively acting renal vasodilator would be useful both in incipient hypertension and in overcoming the deficiencies of the present drugs used in the treatment of moderate to severe hypertension. Reduced RBF also prevails in congestive heart failure and conventional diuretic therapy could be rendered significantly more effective in the presence of decreased renal vascular resistance.

While dopamine produces a pronounced increase in RBF in both animals and man which is accompanied by substantial diuresis and natriuresis, it has to be infused continuously because of its poor absorption from the G.I. tract and short duration of action. The instability of dopamine in the gut is primarily due to the presence of monoamine oxidase (MAO). On the other hand, the gut could act as a useful repository for the gradual absorption of a "protected" or latentiated dopamine which would be absorbed intact and then converted by appropriate enzymes to the active dopamine, preferably at the target site, i.e., the kidney.

Several hundred derivatives of dopamine were prepared protecting the molecule at its metabolically most vulnerable sites, the amino and phenolic hydroxyl groups. The protecting groups had to be of such a nature, as to be cleaved by the body's enzyme systems at an optimal rate to afford the release of significant quantities of dopamine over a protracted period of time, thereby producing a significant rise in renal blood flow of several hours duration without greatly compromising other hemodynamic parameters.

The present report deals with one such compound, ABBOTT-41596(N-L-isoleucyl-dopamine):

$$HO-\langle\bigcirc\rangle-CH_2HC_2NHCO-CH(NH_2)CH(CH_3)C_2H_5$$

HO

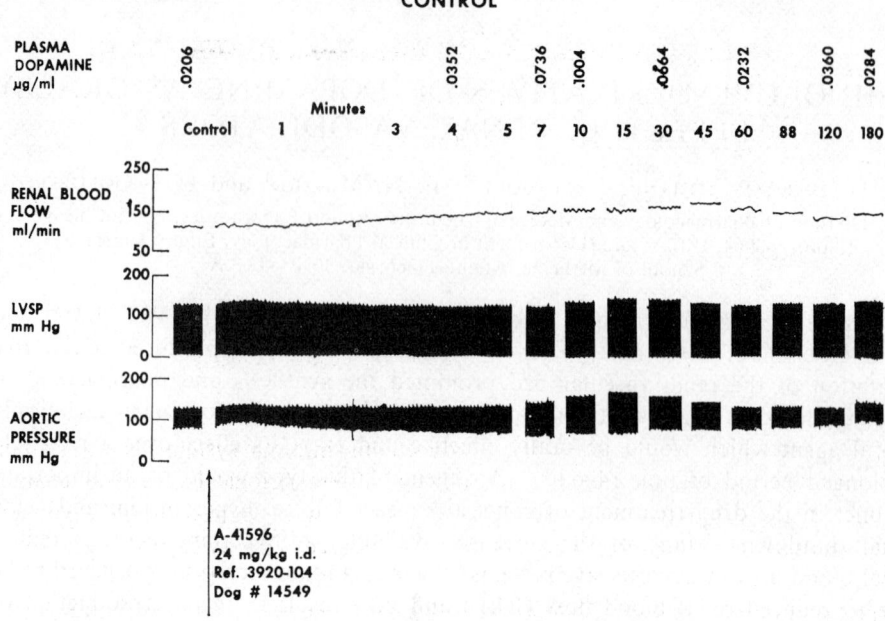

FIG. 1.—Effect of intraduodenal injection of A-41596 in an anesthetised dog. In each panel, data shown represent from top to bottom: plasma dopamine, renal blood flow, left ventricular systolic pressure (LVSP) and aortic blood pressure. The drug was given at the arrow. Tracings obtained at various time intervals after the drug administration are included to show a prolonged increase in plasma dopamine and renal blood flow.

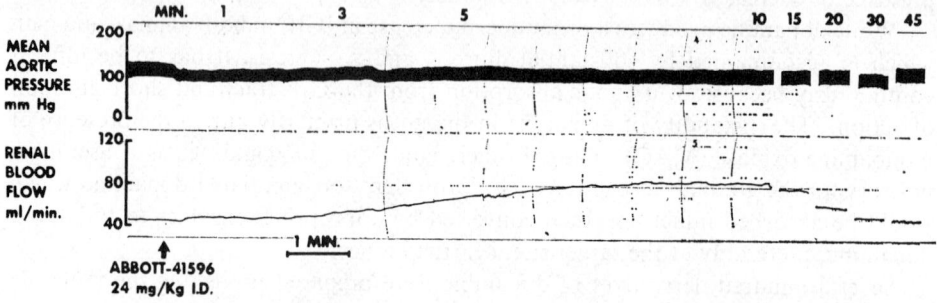

FIG. 2.—Effect of intraduodenal injection of A-41596 in an anesthetised monkey. The drug was administered at the arrow in a dose of 24 mg/kg, which is equimolar to 15 mg/ kg of dopamine. Note the selective increase in renal blood flow which remained elevated for 38 min in this animal.

which fulfills some of these requirements. This amide of dopamine is cleaved at a rather slow rate by the enzyme, aminoacylarylarylamidase, which is particularly abundant in renal tissue.

The intraduodenal administration of 12 or 24 mg/kg of ABBOTT-41596 to anesthetised dogs produced a significant increase in RBF for 155 and 172 min, with a peak increase of 31 and 44 per cent respectively. RBF was measured by an electromagnetic flow probe around the renal artery. Plasma samples taken from these

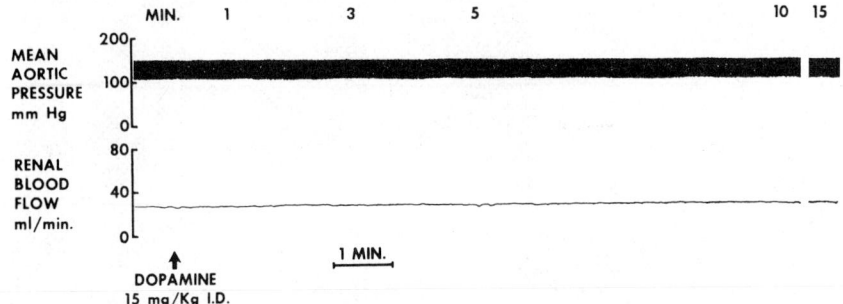

Fig. 3.—Effect of intraduodenal injection of dopamine in an anesthetised rhesus monkey. In each panel, aortic blood pressure and renal blood flow are shown. Dopamine was given at the arrow in a dose of 15 mg/kg.

animals demonstrated the presence of both intact amide and markedly increased concentrations of dopamine. The drug produced only minimal systemic hemodynamic effects at the lower dose; at the higher dose, pharmacological responses characteristic of both α and β-adrenoreceptor stimulation were seen also. Pretreatment with both phenoxybenzamine and propranolol increased RBF by 55–75% at doses of 12–24 mg/kg/i.d., suggesting that the renal vasodilator effect is independent of α or β-adrenoreceptor stimulation. In Rhesus monkeys, 12 and 24 mg/kg/i.d. of ABBOTT-41596 increased RBF by an average of 25 and 37 per cent respectively, with an onset time of 3–5 min and a duration greater than 30 min. No changes in heart rate or systemic blood pressure were observed in these animals.

ABBOTT-41596 is the first in a series of latentiated derivatives of dopamine which produces significant increases in RBF in both dog and monkey over a protracted period of time, when administered by the i.d. route. The increase in RBF can be accomplished at doses which give only minimal systemic hemodynamic changes. Slow but consistent hydrolysis by the enzyme, aminoacylarylamidase, of ABBOTT-41596 appears to be responsible for the prolonged presentation of physiologically active quantities of free dopamine to the renal vasculature, resulting in a substantial decrease in renal vascular resistance.

Frontiers in Catecholamine Research 1973, pp. 905 to 906. Pergamon Press. Printed in Great Britain.

REPORTER'S NOTES—FIRST SESSION

Barry Hoffer

THE OPEN discussion for this session focused on the complexities which surround the extrapolation of catecholamine function into the realms of electrophysiology and behaviour. These complexities could be divided into three groups: first, the problem of relating a specific catecholamine to a specific electrophysiological or behavioural event; second, the need for selective lesions or pathway stimulation; and third, the nature of the functional compensation after destruction of central catecholamine pathways.

The first issue arose from the diversity of catecholamine actions on neurons of the mammalian central nervous system. Ascher took this problem one step further by demonstrating that single neurons themselves may have both excitatory and inhibitory receptors for the same catecholamine. The existence of such neurons in the central nervous system may provide for a unique transynaptic influence of catecholamines.

In the realm of behaviour, the discussion focused on the need for multisystem analysis of even simple responses. Although turning of rats has been linked to dopamine in the caudate by several of the participants, Bartholini emphasised the mutual interaction of dopaminergic and cholinergic mechanisms in this structure. Thus behavioural changes may result from imbalances of antagonistic systems, rather than simple lesions of catecholamine terminals. Similar conclusions were reached in the analysis of the role of catecholamines in sleep. Rather than equate activity in the norepinephrine and serotonin systems to the two phases of sleep, it was clearly indicated that manipulations of the ascending norepinephrine pathways, either at the level of the locus coeruleus or dorsal bundle, alters serotonin turnover and, conversely, lesions of the serotonin containing neurons of the raphé changes norepinephrine turnover. Thus sleep may also involve balance of antagonistic systems.

The need for selective lesions or activation of catecholamine systems was also stressed; yet this is especially difficult in view of the close juxtaposition of adrenergic and nonadrenergic systems in the brain. Some controversy arose concerning the selectivity of 6-hydroxydopamine and differences between 6-hydroxydopamine and electrolytic lesions. While differences in the specificity of 6-hydroxydopamine for catecholamine containing neurons (ranging from very specific in the hands of Ungerstedt and Aghajanian to nonspecific in the hands of Butcher) appear to be largely methodological, differences in the effects of drug vs electrolytic lesions were readily apparent. Fibiger indicated that defects after 6-hydroxydopamine in the substantia negra were quantitatively *smaller* than after electrolytic lesions in the lateral hypothalamus. Zigmond extended this by pointing out that defects in water intake and body temperature, prominent in the 'lateral hypothalamic syndrome' are not seen after 6-hydroxydopamine. Reis pointed out that electrolytic lesions probably destroy more nonadrenergic tissue than 6-hydroxydopamine and this may be an important determinant in the degree of deficit seen.

The third and most important issue, is the role of functional compensation in the determination of 'permanent' deficits after lesion of catecholamine pathways.

This is not a problem unique to catecholamine studies, since compensation has been known for many years to occur after lesions of nonadrenergic brain loci. At the simplest level, minimal deficits may be produced if the lesion involves two antagonistic systems. The studies of Leibwitz and Slangen suggest that, in some cases, both antagonistic systems may be adrenergic and, hence, destroyed by a single drug such as 6-hydroxy-dopamine.

Many of the discussants presented evidence that the compensation that follows catecholamine lesions possess both pre and postsynaptic components. Evidence for presynaptic function i.e. an increased influence of the few residual catecholamine fibres that survive was presented by Zigmond, Schoenfeld and Breese, who showed decreased conditioned performance in rats given alpha methyltyrosine, even if the animals were pretreated with 6-hydroxydopamine.

Functional compensation also involves postsynaptic mechanisms. Ungerstedt, in particular, emphasised denervated postsynaptic receptors develop suprasensitivity. Several discussants showed that low doses of apormorphine, which stimulates dopamine receptors, are able to elicit behaviour in lesioned animals at levels that have no effect in normal animals.

The role of non-catecholamine systems, although doubtlessly of great importance was not evaluated. Do cholinergic, serotonergic or other pathways possess sufficient plasticity to participate in the compensation? Answers to these questions are not available at the present time.

This session emphasised the complexities of evaluating the functional role of catecholamines in the realm of electrophysiology and behaviour. The problems of multiple receptors, functional compensation and plasticity within and between afferent systems, and the mutual interaction of catecholamine and noncatecholamine inputs must be resolved. The pieces of the puzzle were laid out today, but their assembly will challenge us all for many years to come.

Frontiers in Catecholamine Research 1973, p. 907. Pergamon Press. Printed in Great Britain.

REPORTER'S NOTES

M. SANDLER

IN THE open discussion, it was observed that it is well known from fluorescence histochemical analysis that the external layer of the median eminence contains high levels of dopamine (DA). Recent work by the group in Cambridge (CUELLO, HORN, MACKAY and IVERSEN) has been concerned with estimating the levels of DA in this organ and examining the possibility of the existence of an active uptake for this amine, as well as investigating recent pharmacological claims of a possible noradrenergic innervation to this area. Using specific radiochemical enzymatic techniques high levels of DA have been found to be present, $10 \cdot 3 \pm 0 \cdot 98 \ \mu g/g$ as well as large amounts of noradrenaline (NA), $4 \cdot 1 \pm 0 \cdot 53 \ \mu g/g$. Studies on DA uptake were performed on synaptosome rich homogenates of the median eminence. It was found that about 40 per cent of the catecholamine uptake areas were sensitive to inhibition by a concentration of desipramine (5×10^{-7} M) that did not significantly affect DA uptake, thus suggesting the existence of noradrenergic uptake sites. In order to specifically study the DA, rather than NA transport system in this organ, all uptake studies were therefore carried out in the presence of 5×10^{-7} M desipramine. The uptake of DA was found to be similar to the process occurring in the corpus striatum, i.e., it was inhibited by lowering the incubation temperature and the sodium content of the medium. It was also found to have an IC_{50} of $9 \cdot 0 \times 10^{-8}$ M for inhibition by benztropine. It has been shown, therefore, that there probably is a major noradrenergic input to the median eminence and also that an uptake process for DA occurs in this brain area.

Amphetamines and Other Drugs of Abuse
(May 24, 1973; 9:00 A.M.–6:30 P.M.)

CHAIRMEN: William Bunney and J. R. Boissie

COUNCILLOR: Arnold Mandell

REPORTER: Susan Iversen

Frontiers in Catecholamine Research 1973, pp. 911 to 915. Pergamon Press. Printed in Great Britain.

CLINICAL PHARMACOLOGICAL STUDY AND EEG CHANGES OF DELTA-9 TETRAHYDRO-CANNABINOL EFFECTS IN HUMAN VOLUNTEERS*

J. R. BOISSIER

Unité de Recherches de Neuropsychopharmacologie de l'INSERM, 2, rue d'Alésia, 75014, Paris, France

(1) VIGILANCE CHANGES FOLLOWING THC ADMINISTRATION

(a) *Experimental setting and protocol*

SIX MEMBERS of our psychiatric staff volunteered for the first group-study of Delta 9 THC effects in human subjects. After a 20 min control period of EEG recording prior to drug administration, they received a total 10 mg oral-dose of Delta 9 THC in a sesame-oil vehicle. A second EEG recording session of 90 min was followed by successive 15 min recording sessions every hour and for 5 or 6 hr following THC. Blood sample was collected 3 hr after THC and urine collections 6 and 24 hr later. A light lunch followed the blood sample collection.

A neurological and psychophysiological checking was taken every hour, independently of the EEG's recorded on magnetic tape from five different leads (right and left fronto-parietal, right and left parieto-occipital, bilateral occipital) and further submitted to computer's statistical spectral analysis. One-hour pre-control and post-control EEG recording sessions were conducted under the same conditions, the subjects were lying on a bed with their eyes closed most of the time in a dim-light and sound-attenuated room. Horizontal ocular movements, EKG and respiration were also recorded.

(b) *Mean time-course changes of EEG*

The EEG polygraphic recordings taken over 5 hr following THC administration have been scored according to AGNEW and WEBB (1972) procedures. A distinction was made between slight drowsiness (IA state, with interrupted and diffused alpha rhythm) and sedation (IB state, with low amplitude EEG suppressed alpha and occurring theta rhythm) distinguished from light sleep (II state), moderate and deep sleep (III and IV) and REM-state. Percentages of wakefulness (O-state) and IA, IB, II, III, IV, REM states were computed from each successive hour EEG samples and for each subject. These values were plotted for each subject and averaged between subjects for visual 1 hr sequential analysis of vigilance changes following THC. The mean percentages time-course reveals a dominant arousal state during the first 2 hr and then a sedated state increased towards moderate sleep in the 5th hr following THC.

---

* This paper is part of an article to be published in *Therapie:* by P. DENIKER, J. R. BOISSIER, P. ETEVENON, D. GINESTET, P. PERON-MAGNAN and G. VERDEAUX; Etude de Pharmacologie clinique du delta-9 tétrahydrocannabinol chez des sujets volontaires sains avec contrôle polygraphique.

(c) *Vigilance changes and individual variability*

In fact, these mean percentages values are not representative of the six subjects, a large variability between subjects being noted. Three subjects presented a tendency towards moderate sleep stages and one of them for instance (Fig. 1) oscillated during the first 4 hr between IA and IB sedated phases, ending with a dominant III state of moderate sleep during the 5th hr after THC. However, the three other subjects were more aroused; one of them presented after the initial arousal a IA state of light drowsiness from the 2nd to the 5th hr after THC (Fig. 2). This last polygraphic

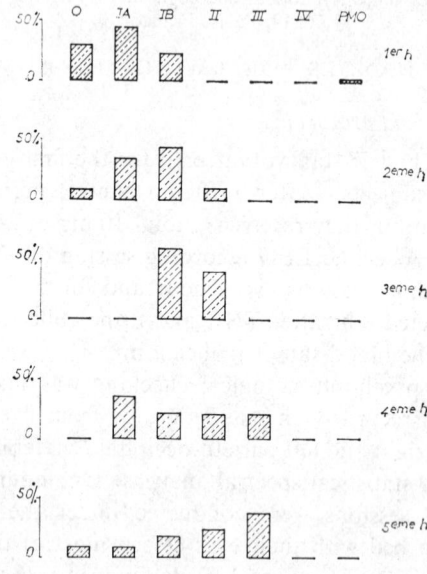

FIG. 1.—One-hour percentage changes of vigilance states following THC administration in one subject presenting a sleep trend through time.

recording presented a time-course hypovariability with an alpha rhythm of high amplitude diffusing over anterior areas. The comparison between the time-courses of 1-hr dominant vigilance phases of each one of the six subjects, indicates a great dispersion between subjects. A trend towards increased deepening of sleep appears at the 5th and 6th hr following THC, which may also be related with the habituation and the monotony of the experimental conditions.

(2) NEUROLOGICAL AND PSYCHOPHYSIOLOGICAL CHANGES

The well-described initial tachychardia (10–20 beats increase in a few minutes), appears when the subject reports the first THC effects, 20–90 min following the oral administration. Some vertiginous sensations have been reported by some subjects during the first 2 hr, whereas conjunctival hyperhemia was observed in all subjects. It appears that the 'eyes-open situation', needed during the clinical and psychophysiological examination is very different from the inner subjective state of consciousness of the 'eyes-closed situation' when the subject is left alone and recorded lying on the bed in order to minimise vertiginous or hypotensive effects of THC. During the

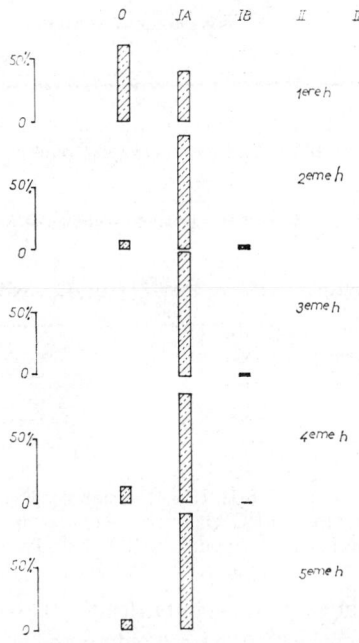

FIG. 2.—One-hour percentage changes of vigilance states following THC administration
in another subject presenting of IA-state hypovariability.

successive interviews and clinical testings the subjects have described, after the recording sessions, the intense and vivid visual imagery that occurs in the eyes closed
situation. This enhancement of visual scenery was experienced in a moderate euphoric
or dysphoric state, depending of the subject, with a more-or-less refusal of answering to
the clinical examiner. As many authors have already reported, the subjective reports
of our subjects have described difficulties of rational thinking, relaxation and body
image changes with oral preoccupations. Three subjects have considered that their total
experience provided insight in favour of what can be lived as a primary schizophrenic
experience which is in agreement with the model already proposed by JONES (1972).

These dissociated states of 'floating' consciousness, oscillating from hypomanic
or oniroïd or confused crepuscular states of vigilance can be related with the observed
EEG changes. The problem appears if we have to distinguish or not between these
THC-induced subjective dreamy-states and the normal hypnagogic imagery occurring
during the on-going of sleep, the REM periods and the hypnopompic imagery sometimes described after awakenings.

(3) VISUAL ANALYSIS OF EEG RECORDINGS FOLLOWING THC

In 4 of our 6 subjects the first 2 hr following THC presented greater arousal time
than sedation. Such an arousal recording presents short desynchronised periods
which appear similar to hyperaroused phases which occur under an intense mental
task or following analeptic treatment.

The peculiar time-course hypervariability of the EEG, following THC administration may be illustrated by the rapid succession of different states of vigilance (Fig. 3).
After a deep sedation (IB) followed by some slow-wave sleep patterns (II-III) a mild
drowsiness phase occurs (IA) with a high amplitude alpha rhythm specially under

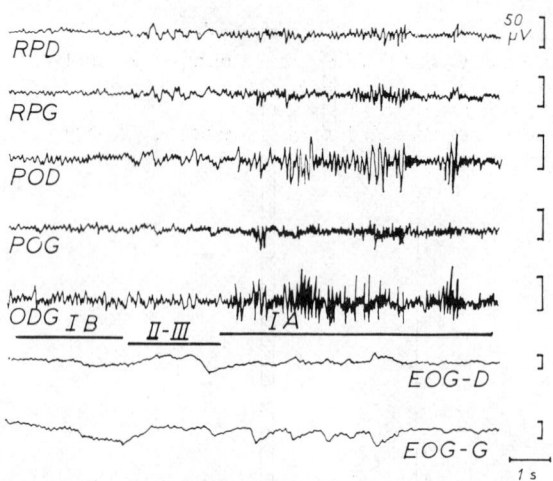

FIG. 3.—Rapid shifts between IB-(II, III)—IA states of vigilance following THC administration in one subject. EEG channels: Right Rolando-Parietal: RPD, left Rol. Par.: RPG, Right Parieto-Occipital: POD, Left Par. Occ.: POG, Occipital Transverse: ODG.

right-left bilateral and right parieto-occipital leads. This time-course variability of vigilance states may to some extent be related to the floating impression and the extreme mobility of the inner subjective impressions and the visual fantasies described by the subjects under THC effects. A REM episode was obtained occurring at the end of an afternoon session, during a control experiment of a subject recorded one month after the THC experiment. The subject being awakened after this recording, was able to describe the content of his dream. This REM episode followed a deep sleep episode and was compared with a recording obtained in the same subject during the THC previous experience and where rapid eye movements occurred between two phases of deep sedation (IB-state). Of course, without chin-muscle recordings or other eye-movements channels, it will be hazardous to assimilate this observed phase of ocular movements with the previous REM episode. However, the Dement and Kleitman hypothesis considering phase I as a 'descending stage one of sleep' and REM as 'emergent stage one of sleep', may be retained at first for those tracings observed in different subjects following THC administration.

In conclusion, from the visual inspection of polygraphic recordings following THC administration, differences in EEG patterns and changes in vigilance time-courses appear, which cannot be wholly explained by considering normal day-sleep recordings more or less facilitated by THC. It is of course difficult to speak of a specific THC-drug induced state, especially when the THC dose is low. The subjective reports have shown an inner visualisation, a highly variable state which appears less inhibited and submitted to the cartesian reasoning and the analytical thinking. A physiological explanation will be founded if a normal functional hemispheric lateralisation was modified following THC administration. Computer's analysis in progress should be able to test this hypothesis. Besides, HEATH (1972–1973) has found EEG changes in deep cerebral structures following THC administration of much greater amplitude than the observed scalp potentials changes and that is another argument in favour of computer's EEG analysis for assessing the slight changes between EEG channels after Delta-9 tetrahydrocannabinol treatment.

REFERENCES

AGNEW H. W. Jr. and WEBB W. B. (1972) Sleep state scoring American Psychol. Ass., May 1972.
HEATH R. G. (1972) *Archs. Ged. Psychiat.*, **26,** 577–584.
HEATH R. G. (1973) *Neuropharmac.* **12,** 1–14.
JONES R. T. (1972) American Psychopathol. Meeting, Session I, New-York.

Frontiers in Catecholamine Research 1973, pp. 917 to 925. Pergamon Press. Printed in Great Britain.

AMPHETAMINE AND COCAINE ON AMINE TURNOVER*

A. Groppetti, F. Zambotti, A. Biazzi and P. Mantegazza

2nd Chair, Department of Pharmacology, University of Milan, Italy

Although it is generally accepted that brain catecholamines are implicated in the central action of d-amphetamine (Glowinski and Axelrod, 1965, 1966; Stein and Wise, 1969; Carr and Moore, 1969; Hanson, 1966; Randrup and Munkvad, 1966; McLean and McCartney, 1961; Sanan and Vogt, 1962; Moore, 1963) conflicting reports exist whether norepinephrine (NE) or dopamine (DM) or both are involved in modulating single behavioural changes induced by this drug.

Recently it has been suggested that stereotyped behaviour elicited by d-amphetamine depends on the availability of DM, whereas NE appears to be required for locomotor stimulation (Randrup and Scheel-Kruger, 1966; Coyle and Snyder, 1969). On the other hand Carlsson (Carlsson, 1970) has proposed that in mice, hypermotility caused by d-amphetamine is mediated through a release of brain DM. This hypothesis is consistent with studies by Van Rossum et al., 1962 and Smith, 1963 in disputing that d-amphetamine effect on motor activity involves a release of brain NE.

The role of brain catecholamine in the regulation of food intake is also controversial.

It has been reported that the noradrenergic terminals in the lateral hypothalamus are associated with satiety (Margules, 1970). The anorexia produced by amphetamine, with particular reference to the indirect action on the central noradrenergic system of this drug, is brought forward in support of this theory (Margules, 1969).

However in apparent contradiction to the noradrenergic satiety theory, other investigators have shown that direct application of NE in the lateral area of the hypothalamus produces eating in satiated animals (Grossman, 1960, 1962, 1968). This effect can be suppressed by contemporary administration of α-blocking agents and potentiated by desmethylimipramine (Leibowitz, 1970; Berger et al., 1971). According to these investigators amphetamine anorexia does not depend on the release of brain NE but derives from a direct interaction of this drug with brain receptors.

Recently Ungersted (1971) has given evidences that the nigrostriatum dopaminergic system plays an important role in regulating food intake, suggesting that amphetamine anorexia may be mediated by brain DM.

The purpose of the present investigation was to further explore the individual roles of NE and DM in amphetamine induced anorexia, hypermotility and hyperthermia.

In these studies changes of turnover rates of brain norepinephrine and dopamine have been chosen as indices of functional status of the two neuronal systems at the presynaptic site.

In order to establish whether the experimental data concerning behavioural and

---

* Supported by C.N.R. grant 71.00849.04.115.2380

neurochemical parameters were related to each other and, if so, whether their relationship was unique, the effects of cocaine, fenfluramine and *l*-amphetamine on brain catecholamine turnover rate and animal behaviour were also investigated.

The experimental evidence resulting from this study seems to suggest:

(1) an indirect action on the noradrenergic neuronal system in the hypothalamus and telediencephalon does not appear to be required in eliciting anorexia or hypermotility.

(2) *d*-amphetamine and cocaine may elicit psychomotor stimulation by releasing dopamine from striatum nerve terminals.

(3) brain DM may play some role in regulating food intake in *d*-amphetamine treated rats.

BEHAVIOURAL EFFECTS

Table 1 reports the structures and the behavioural effects of *d*-amphetamine, *l*-amphetamine, fenfluramine and cocaine. The dose for each drug has been chosen in order to have comparable and as selective as possible effects on food intake.

It is known that high doses of *d*-amphetamine decrease food intake and increase body temperature and motor activity. However, at smaller doses, behavioural effects of *d*-amphetamine are more selective. In fact doses of 0·3 mg/kg i.v. do not change body temperature while they still increase motor activity and give anorexia (Table 1).

The optical isomer, *l*-amphetamine, has much less central stimulant effect than *d*-amphetamine (Prinzmetal and Alles, 1940; Stein, 1964). In non striatal brain areas *l*-amphetamine is 10 fold less potent than *d*-isomer in inhibiting NE uptake by synaptosomes, whereas in the corpus striatum the two isomers are almost equally active (Coyle and Snyder, 1969).

Table 1. Behavioural effects of *d*-amphetamine, *l*-amphetamine, fenfluramine and cocaine in rat

Drug	Structure	Dose (mg/kg i.v.)	Food* intake	Motor† activity	Body‡ temperature
Saline		—	2·3 ± 0·3 (5)	103 ± 15 (14)	36·2 ± 0·2 (5)
d-Amphetamine	—CH₂—CH—NH₂ , CH₃	0·3	1·2 ± 0·3§ (5)	335 ± 35§ (4)	36·5 ± 0·2 (5)
l-Amphetamine		1·0	1·3 ± 0·6§ (5)	107 ± 19 (12)	38·1 ± 0·1 (5)
Fenfluramine	F₃C— —CH₂—CH—N(H)(C₂H₅) , CH₃	1·0	0·76 ± 0·2§ (5)	132 ± 30 (4)	36·0 ± 0·2 (5)
Cocaine	H₃C—O·OC / CH—CH—CH₂ / —CO·O—CH N—CH₃ / CH₂—CH—CH₂	3·0	0·80 ± 0·1§ (5)	515 ± 130§ (4)	37·4 ± 0·1§ (5)

* Food intake: g of food eaten/100 g of body weight ± s.e. during the first 30 min after drug injection. The rats were trained for two weeks to eat for 4 hr/day. Food was presented immediately after drug injection.

† Motor activity: events ± s.e. during the first 30 min after drug injection. Rats were singularly housed in compartments of an I.R. Electronic Motility Meter. Locomotor activity was measured by infrared photocells equally spaced in the floor of the cages.

‡ Body temperature: °C ± s.e. measured rectally 30 min after drug injection.

§ $P < 0.01$ Number of animals in parentheses

The administration of 1 mg/kg i.v. of *l*-amphetamine to rats, decreases food intake without affecting motor activity (Table 1). Body temperature was also increased.

Fenfluramine, a compound chemically related to *d*-amphetamine (Table 1), also causes anorexia but, unlike amphetamine, fails to change motor activity and body temperature in the rat (ALPHIN *et al.*, 1964). Only at high doses this drug decreases motor activity (COLMORE and MOORE, 1966). Fenfluramine also reduces brain 5-HT and NE concentrations (DUCE and GESSA, 1967; OPITZ, 1967).

Cocaine, although having an entirely different structure than amphetamine, shows similar behavioural effects. This drug, given at 3 mg/kg i.v., increases body temperature and motor activity and decreases food intake (Table 1).

NEUROCHEMICAL EFFECTS

The turnover rates of the two catecholamines were estimated "*in vivo*" from the conversion of labeled tyrosine into NE and DM respectively. In these experiments the following procedure was adopted: At 0 time $3,5^3$H-tyrosine (S.A. 30 Ci/mM) was injected (1 mCi/kg i.v.). After 10 min drugs or saline were given intravenously. 25 min after tyrosine injections the rats were sacrificed and the brains dissected. Concentrations of labeled as well as endogenous tyrosine, NE and DM were determined by a method previously described by NEFF *et al.* (1971).

As a model to calculate the incorporation rate of tyrosine into NE or DM, a two compartments closed system has been selected. In this system, first proposed by SEDVALL *et al.* (1968), the amount of catecholamine formed can be calculated from the ratio between the radioactivity incorporated in the amines and the specific activity of the aminoacid precursor, in this case tyrosine. The quotient of this equation is called "conversion index" to indicate that this value is not an absolute measurement of incorporation rate. This model can be used for comparative estimation of incorporation rate of aminoacids into amines providing:

(1) aminoacid and amine concentrations have reached steady state; (2) specific activity of amino acid precursor are higher than specific activity of tissue amines.

The data listed in Table 2, some of which have been previously reported (COSTA *et al.*, 1972), indicate that conversion index of tyrosine into striatal DM was significantly higher in rats receiving *d*-amphetamine and cocaine. Since neither steady state concentrations of tyrosine and DM nor the specific activity of striatum tyrosine were changed by these two drugs, it follows that the turnover time of striatal DM in *d*-amphetamine and cocaine treated rats is faster than in rats injected with saline.

These results are consistent with "*in vivo*" studies by MACKENZIE and SZERB (1968) indicating that amphetamine applied directly into the caudate nucleus enhances the release of DM from dopaminergic terminals. JORI and BERNARDI (1969) have also observed an increase of brain homovanillic acid in amphetamine treated rats.

This effect is restricted to these two drugs. In fact both fenfluramine and *l*-amphetamine fail to change the incorporation rate of tyrosine into striatal dopamine. Table 2 also includes data showing the effects of *d*- and *l*-amphetamine, cocaine and fenfluramine on the conversion index of tyrosine into telediencephalic NE.

None of these drugs seems to modify the steady state or the incorporation rate of tyrosine into NE.

Previous reports (COSTA *et al.*, 1971; GROPPETTI *et al.*, 1972) have indicated that fenfluramine may increase the turnover rate of both NE and DM in the rat brain.

TABLE 2. EFFECT OF d-AMPHETAMINE, l-AMPHETAMINE, COCAINE AND FENFLURAMINE ON CONVERSION OF $3,5^3$H-TYROSINE INTO TELDIENCEPHALIC NOREPINEPHRINE (NE) AND STRIATAL DOPAMINE (DM).*

Treatment	Dose (mg/kg i.v.)	Teldiencephalon			Striatum	
		Tyrosine (dpm/mμmol ± SE × 10³)	NE mμmol/g ± SE	Conversion index mμmol NE/g/25 min. ± SE	DM mμmol/g ± SE	Conversion index mμmol DM/g/25 min ± SE
Saline	—	4·1 ± 0·1 (11)	2·2 ± 0·1	0·38 ± 0·05	71 ± 13 (6)	15 ± 2
			—	—		—
d-Amphetamine	0·3	3·9 ± 0·3 (6)	2·5 ± 0·1	0·41 ± 0·04	83 ± 3 (6)	27 ± 2†
			—	—		—
Saline	—		—		53 ± 5 (11)	17 ± 1
						—
l-Amphetamine	1·0	3·8 ± 0·2 (11)	2·8 ± 0·2	0·39 ± 0·04	54 ± 2 (8)	21 ± 2
			—	—		—
Saline	—		—		58 ± 6 (9)	15 ± 1
						—
Cocaine	3·0	3·7 ± 0·1 (10)	2·6 ± 0·1	0·43 ± 0·04	62 ± 4 (6)	24 ± 2†
			—	—		—
Saline	—	2·9 ± 0·2 (5)	2·8 ± 0·1	0·91 ± 0·09	81 ± 5 (6)	24 ± 1
Fenfluramine	1·0	3·4 ± 0·3 (5)	3·3 ± 0·1	0·89 ± 0·08	87 ± 6 (5)	28 ± 3

* Part of the data reported in this table were taken from Costa et al., Brit. J. Pharmacol. (1972) **44**, 742
† $P < 0.01$ Number of animals in parentheses

However the apparent discrepancy may be due to the higher doses used in those experiments.

The present findings suggest that the receptors that control feeding behaviour, motor activity and body temperature may be influenced without affecting the biochemistry of presynaptic noradrenergic neurons in the rat telediencephalon.

However it is not possible to exclude that in some more discrete areas of the brain the turnover rate of NE is changed and this change may be masked in our experiments by the size of tissue sample.

Since the hypothalamus is generally considered as the brain area where the processes regulating food intake mostly occur (ANAND et al., 1955), a study of how anorexic doses of d- and l-amphetamine, fenfluramine and cocaine affect NE turnover rate in this brain area was thought more suitable.

In order to increase the sensitivity of the method, otherwise insufficient, 3,5³H tyrosine had to be injected intraventricularly instead of intravenously. Two permanent cannulas of poliethylene were placed one for each lateral brain ventricle of rats (ROBINSON et al., 1969) 48 hr before injecting labeled tyrosine.

On the day of the experiment 7 μCi of 3,5³H-tyrosine (S,A. 30 Ci/mM) were injected in each ventricle. Interferences with uptake and distribution of labeled tyrosine were minimised by injecting the drugs, intravenously, 12 min after labeling. Twenty-five min after the injection of 3,5³H-tyrosine the rats were sacrificed and the hypothalamus separated from the rest of the brain. Chemical determinations and conversion index have been calculated as indicated above.

The method offers the following advantages:
(1) minimal stress to the animal at the time of injection,
(2) avoidance of anesthesia,
(3) accumulation of sufficient radioactivity in the hypothalamus to permit measurements of specific activity of tyrosine and norepinephrine in individual animals.

Figure 1 shows the changes, as a function of time, of the specific activity of tyrosine and NE in brain and hypothalamus of rats treated with 3,5³H-tyrosine, intravenously (1·0 mCi/kg) or intraventricularly (7 μCi/ventricle) respectively.

K_m values were calculated according to the equation proposed by NEFF et al., (1971). (Fig. 1)

For this calculation each graph was divided into consecutive 5 min intervals. As shown in Fig. 1 while the individual K_m values calculated in the brain after intravenous injection of labeled tyrosine were uniform during the entire 40 min. period of investigation, after intraventricular injection 10–15 min must elapse before K_m values in the hypothalamus become relatively constant. These results suggest that even after intraventricular injection the equilibrium between the injected labeled and the endogenous tyrosine is rapidly reached but that it takes at least 10–15 min before being complete.

The data listed in Table 3 show that of d- and l-amphetamine, fenfluramine and cocaine change neither the endogenous levels of NE nor the S.A. of tyrosine nor the incorporation rate of this aminoacid into NE in rat hypothalamus.

These results are consistent with the findings indicating that when given in moderate pharmacological doses d-amphetamine does not significantly increase the release of stored H³–NE into the perfusate from the hypothalamus in vivo (STRADA and SULSER 1970).

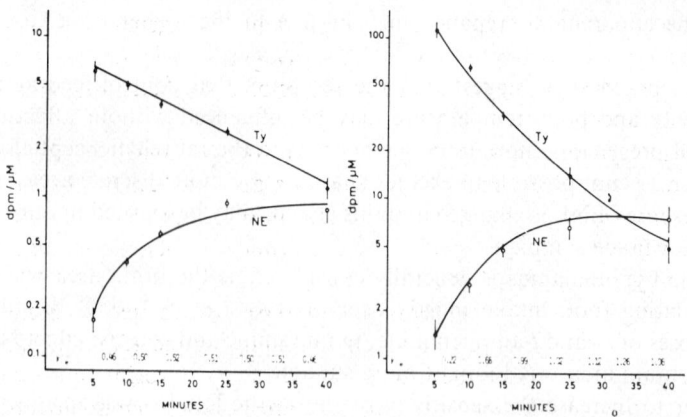

Fig. 1.—Specific activity of brain (left) and hypothalmic (right) tyrosine and norepinephrine after intravenous (1 mCi/kg) or intraventricular (7 μCi/ventricle)
injection of 3,5-³H-tyrosine in the rat.

K_m values were calculated from the formula

$$K_m \approx \cfrac{\dfrac{M_{t_2} - M_{t_1}}{t_2 - t_1}}{\dfrac{[A - M]t_2 - [A - M]t_1}{2}}$$

Where: M = specific activity of the amine
A = specific activity of the aminoacid

RELATIONSHIP BETWEEN BEHAVIOURAL AND NEUROCHEMICAL EFFECTS

The data reported in Tables 1, 2 and 3 tend to exclude the possibility that anorexia
as well as hyperthermia and hypermotility induced by d-amphetamine, l-amphetamine,
fenfluramine and cocaine occur together with an increase of the extraneuronal release
of telediencephalic or hypothalamic NE.

TABLE 3. EFFECT OF d-AMPHETAMINE, l-AMPHETAMINE, COCAINE AND FENFLURAMINE ON CON-
VERSION INDEX OF TYROSINE INTO NOREPINEPHRINE (NE) IN RAT HYPOTHALAMUS AFTER INTRA-
VENTRICULAR INJECTION OF 3,5³H-TYROSINE.*

Treatment	Dose mg/kg i.v.	Tyrosine dpm/mμmol ± SE × 10³	NE mμmol/g ± SE	Conversion index mμmol NE/g ± SE
Saline	—	14·43 ± 1·27 (6)	11·48 ± 0·62 (6)	5·15 ± 0·60 (6)
d-Amphetamine	0·3	16·90 ± 1·03 (7)	10·61 ± 0·53 (7)	5·58 ± 0·77 (7)
l-Amphetamine	1·0	18·09 ± 1·14 (7)	9·35 ± 0·53 (7)	3·84 ± 0·26 (7)
Saline	—	16·64 ± 1·65 (5)	8·3 ± 0·91 (5)	2·26 ± 0·03 (5)
Cocaine	3·0	21·42 ± 4·97 (6)	9·3 ± 0·28 (7)	2·76 ± 0·50 (6)
Fenfluramine	1·0	22·22 ± 1·58 (6)	8·1 ± 0·47 (7)	2·05 ± 0·30 (6)

* 3,5³H-tyrosine (7 μCi/ventricle)
Number of rats in parentheses

These results support and extend to amphetamine congeners the proposal made by other investigators (LEIBOWITZ, 1970; BERGER et al.,1971; GROPPETTI et al.,1972) that the central action of minimal effective doses of d-amphetamine does not involve a release of brain NE.

Changes in DM turnover rates do not seem to be related to hyperthermia that follows some of these drug treatments. So that a dose of d-amphetamine (0·3 mg/kg i.v.) that increases incorporation rate of tyrosine into striatal DM does not affect body temperature,while on the other hand an hyperthermic dose of l-amphetamine (1 mg/kg i.v.) does not change striatal DM turnover.

The increase of DM turnover seems instead to be more related to changes in motor activity. Both, d-amphetamine (0·3 mg/kg i.v.) and cocaine (3mg/kg i.v.) stimulate motor activity and accelerate the incorporation rate of tyrosine into striatum DM. On the other hand fenfluramine and l-amphetamine at doses that do not change motor activity failed to affect DM turnover rate. COSTA et al. (1971) have already suggested that dopaminergic axons could be involved in the motor stimulation elicited by aminorex and p-chloroamphetamine.

The role of brain DM in regulating food intake is still unresolved. While d-amphetamine and cocaine accelerate DM turnover and also produce anorexia, suggesting that the increase of striatal DM turnover may be related to the anorexic effect of these drugs, on the other hand it is possible to reduce food intake without affecting striatal DM conversion index by administration of l-amphetamine and fenfluramine.

WEISSMAN et al. (1966) have reported that α-methyltyrosine (α-MT), an inhibitor of catecholamine synthesis (SPECTOR et al., 1965), antagonizes the anorexic effect of amphetamine. This observation has been generally interpreted as evidence that amphetamine anorexia is due to an indirect effect of this drug on NE neurons. However brain DM as well as other non specific mechanisms entirely unrelated to the inhibition of catecholamine synthesis may also be involved, but the data reported in Table 4 seem to negate the possibility that α-MT may block the d-amphetamine induced anorexia by these non specific mechanisms.

TABLE 4. EFFECT OF α-METHYLTYROSINE (α-MT) AND PIMOZIDE (PZ) PRETREATMENT ON d-AMPHETAMINE, FENFLURAMINE AND APOMORPHINE INDUCED ANOREXIA IN RAT.

Treatment*	Food intake†		
	Saline	α-MT‡	PZ§
Saline	3·0 ± 0·28 (5)	3·71 ± 0·45 (5)	3·14 ± 0·52 (5)
d-amphetamine 1·5 i.p.	0·90 ± 0·20 (5)	2·1 ± 0·15 (5)	2·1 ± 0·40 (5)
Fenfluramine 3 mg/kg i.p.	0·96 ± 0·27 (5)	0·36 ± 0·10 (5)	0·34 ± 0·16 (5)
Apomorphine 1 mg/kg s.c.	0·56 ± 0·16 (5)	0·75 ± 0·12 (5)	1·44 ± 0·44 (5)

  * Drugs were given 15 minutes before presentation
  † g of food/100 g of Body Weight ± SE measured 1 hr after food presentation
  ‡ α-MT 50 mg/kg i.p. 2 hrs. before food presentation
  § PZ 0·25 mg/kg i.p. 4 hrs. before food presentation

In fact α-MT selectively antagonizes *d*-amphetamine anorexia while it does not block but eventually potentiates the fenfluramine effect on food intake.

On the other hand the fact that dopaminergic system may play some role in amphetamine anorexia is suggested by the findings that pimozide, a drug indicated to block dopaminergic receptors (Andén *et al.*, 1970), antagonizes the depressive effect of amphetamine on food intake. This antagonism appears to be selective because fenfluramine anorexia was not inhibited but potentiated by pimozide pretreatment. (Table 4).

In this context it is interesting to note that apomorphine, an agonist of dopaminergic receptors (Ernst and Smelik, 1966) also causes anorexia and that this effect is antagonized by pimozide and not by α-MT. (Table 4)

This observation gives further support to the hypothesis that activation of the brain dopaminergic system may be responsible, at least in part, for the reduction of food intake by *d*-amphetamine. This correlation seems to be restricted to this drug because fenfluramine anorexia is not antagonized either by α-MT or pimozide pretreatment.

REFERENCES

Alphin R. S., Funderburk W. H. and Ward J. W. (1964) *Toxicol. Appl. Pharmac.* **6**, 340.
Anand B. K., Dua S. and Schoenberg K. (1955) *J. Physiol.* **127**, 143.
Andén N. E., Butcher S. G., Corrodi H., Fuxe K. and Ungerstedt U. (1970) *Europ. J. Pharmacol.* **11**, 303.
Berger B. D., Weise C. D. and Stein L. (1971) *Science*, **172**, 281.
Carlsson A. (1970) In: *Amphetamine and Related Compounds* (Costa E. and Garattini S. Eds.) pp. 289 Raven Press, New York.
Carr L. A. and Moore K. E. (1969) *Science*, **164**, 322.
Colmore J. P. and Moore J. D. (1966) *J. New Drugs* **6**, 123.
Costa E., Groppetti A. and Revuelta A. (1971) *Brit. J. Pharmacol.*, **41**, 57.
Costa E., Groppetti A. and Naimzada M. K. (1972) *Brit. J. Pharmacol.* **44**, 742.
Costa E., Naimzada M. K. and Revuelta A. (1971) *Brit. J. Pharmacol.* **43**, 570.
Coyle J. T. and Snyder S. H. (1969) *J. Pharmacol. Exp. Ther.* **170**, 221.
Duce M. and Gessa G. L. (1967) *Boll. Soc. It. Biol. Sper.* **42**, 1631.
Ernst A. M. and Smelik P. G. (1966) *Experientia*, **22**, 837.
Glowinski J. and Axelrod J. (1965) *J. Pharmacol. Exp. Ther.* **149**, 43.
Glowinski J. and Axelrod J. (1966) *J. Pharmacol. Exp. Ther.* **153**, 30.
Groppetti A., Mischer A., Naimzada M. K., Revuelta A. and Costa E. (1972) *J. Pharmacol. Exp. Ther.* **182**, 464.
Grossman S. P. (1960) *Science*, **132**, 301.
Grossman S. P. (1962) *Am. J. Physiol.* **202**, 872.
Grossman S. P. (1968) *Fedn. Proc.* **27**, 1349.
Hanson L. C. (1966) *Psychopharmacologia* **9**, 78.
Jori A. and Bernardi D. (1969) *J. Pharm. Pharmacol.* **21**, 694.
Leibowitz S. F. (1970) *Nature, Lond.* **226**, 963.
McLean J. R. and McCartney M. (1961) *Proc. Soc. Exp. Biol.* **107**, 77.
MacKenzie G. M. and Szerb J. C. (1968) *J. Pharmacol. Exp. Ther.* **162**, 302.
Margules D. L. (1969) *Life Sciences* **8**, 693.
Margules D. L. (1970) *J. Comp. Physiol. Psychol.* **73**, 1.
Moore K. E. (1963) *J. Pharmacol. Exp. Ther.* **142**, 6.
Neff N. H., Spano P. F., Groppetti A., Wang C. T. and Costa E. (1971) *J. Pharmacol. Exp. Ther.* **176**, 701.
Opitz K. (1967) *Arch. Pharmakol.* **259**, 56.
Prinzmetal M. and Alles G. A. (1940) *Am. J. Med. Sci.* **200**, 665.
Randrup A. and Scheel-Kruger J. (1966) *J. Pharm. Pharmacol.* **18**, 752.
Randrup A. and Munkvad J. (1966) *Nature, Lond.* **211**, 540.
Robinson C. A., Hengeveld G. A. and De Balbian Vester F. (1969) *Physiol. Behav.*, **4**, 123.
Sedvall G. C., Weise V. K. and Kopin I. J. (1968) *J. Pharmacol. Exp. Ther.* **159**, 274.
Sanan S. and Vogt M. (1962) *Brit. J. Pharmacol.* **18**, 109.

Smith C. B. (1963) *J. Pharmacol. Exp. Ther.* **142**, 343.
Smith C. B. (1965) *J. Pharmacol. Exp. Ther.* **147**, 96.
Spector S., Sjoerdsma A., Udenfriend S. (1965) *J. Pharmacol. Exp. Ther.* **147**, 86.
Stein L. and Wise C. D. (1969) *J. Comp. Physiol. Psychol.* **67**, 189.
Stein L. (1964) *Fedn. Proc.* **23**, 836.
Strada S. J. and Sulser F. (1970) *Fedn. Proc.* **29**, 963.
Ungerstedt U. (1971) *Acta Physiol. Scand., Supp.* **367**, 95.
Van Rossum J. M., Van Der Schoot J. B. and Hurkmans J. A. (1962) *Experientia* **18**, 229.
Weissman A., Koe B. K., Tenen S. S. (1966) *J. Pharmacol. Exp. Ther.*, **151**, 339.

Frontiers in Catecholamine Research 1973, pp. 927 to 932. Pergamon Press. Printed in Great Britain.

COMPARATIVE METABOLISM OF SOME AMPHETAMINES IN VARIOUS SPECIES

R. T. WILLIAMS, J. CALDWELL and L. G. DRING

Department of Biochemistry, St. Mary's Hospital Medical School, London, W2 1PG U.K.

FEW DETAILED metabolic studies have been carried out on drugs based upon amphetamine (No. 1 in Table 1). Amphetamine, methamphetamine (No. 2), norephedrine (No. 5) and 4-hydroxyamphetamine have been examined in this laboratory. Other workers have examined, in varying detail, ephedrine (BRALET et al., 1968) (No. 6), ethylamphetamine (BECKETT et al., 1969) (No. 3), Pondinil (LONG, 1970) (No. 4), mephentermine (WALKENSTEIN et al., 1955) (No. 8) and fenfluramine (BRUCE and MAYNARD, 1968) (No. 7).

Metabolism of these drugs in different species may be correlated with a) their physical properties and b) their biological properties such as pharmacological activity and toxicity. At pH 7·4 these drugs occur almost entirely as cations (see Table 1), being over 99 per cent ionised except fenfluramine which is 98 per cent ionised. The unionised drugs, however, differ considerably in lipid solubilities which range over 3000 if one excludes fenfluramine.

The structures of these drugs (see Table 1) suggests several routes of biotransformation,

(a) *aromatic hydroxylation* introducing a phenolic group particularly at the *para*-position of the benzene ring.
(b) N-*dealkylation* in the case of compounds Nos. 2, 3, 4, 6, 7 and 8 of Table 1.
(c) *Oxidative deamination* (following N-dealkylation where this occurs) to a ketone which may then be reduced to other compounds or oxidised to benzoic acid.
(d) *Aliphatic hydroxylation* of the carbon atoms in the side chain especially the one in the methylene group next to the benzene ring, except in compounds Nos. 5 and 6 (Table 1.)
(e) N-*Oxidation* to form hydroxylamines or N-oxides.

· Apart from the above reactions, the drug may be excreted unchanged in the urine to varying extents. There is actually little faecal excretion of these drugs in any of the species examined.

All the above reactions, except possibly (e), have been shown to occur *in vivo*, but the extent to which they occur as measured by the metabolites found in the urine shows a remarkable species variation.

AROMATIC HYDROXYLATION

The *p*-hydroxylation of the benzene ring of the amphetamines is a reaction which occurs extensively and consistently in the rat but not to any great extent in the other species listed in Table 2 except in one or two cases such as Pondinil in man and the N-alkylated drugs, methamphetamine, Pondinil and mephentermine, in the dog. There is no evidence of hydroxylation in the *o*- or *m*-positions. In the rat, the extent

TABLE 1. THE STRUCTURE AND SOME PHYSICOCHEMICAL PROPERTIES OF THE AMPHETAMINES
(Data from Vree et al., 1969)

No.	Drug	R_1	R_2	R_3	R_4	pK_a	cation at pH 7·4%	Relative lipid solubility*
1	Amphetamine	H	H	H	H	9·90	99·6	1·00
2	Methamphetamine	H	Me	H	H	10·11	99·8	2·31
3	Ethylamphetamine	H	Et	H	H	10·23	99·8	5·56
4	Pondinil	H	C_3H_6Cl	H	H	—	—	—
5	Norephedrine	OH	H	H	H	9·55	99·3	0·002
6	Ephedrine	OH	Me	H	H	9·60	99·4	0·031
7	Fenfluramine	H	Et	H	CF_3	9·10	98·0	67·40
8	Mephentermine	H	Me	Me	H	10·25	99·8	2·54

* Amphetamine taken as 1·00; calculated from the apparent partition coefficients at pH 7·4 between $CHCl_3$ and H_2O.

of aromatic hydroxylation seems correlated with lipid solubility since the more lipid soluble drugs, amphetamine and methamphetamine are more extensively hydroxylated than the less lipid soluble drugs, ephedrine and norephedrine. Much of the latter drugs are excreted unchanged by the rat (see Table 6). In the rabbit and guinea pig, aromatic hydroxylation is a minor reaction, and may not occur in the guinea pig with amphetamine and methamphetamine. In the dog, some 20–30 per cent of those drugs containing an N-alkyl group, except ephedrine, are hydroxylated. In man, aromatic hydroxylation is a minor reaction except with Pondinil and methamphetamine, and the hydroxylation of amphetamine is approximately the same (2–3 per cent of the dose) in normal and tolerant subjects.

TABLE 2. THE EXTENT OF AROMATIC HYDROXYLATION OF VARIOUS AMPHETAMINES IN DIFFERENT SPECIES
(Values are taken from various sources in the literature)

	Rat	Rabbit	Guinea pig	Dog	Man	Tolerant Man
Amphetamine	60	6	0	6	2	3
Methamphetamine	53	—	0	30	18	—
Norephedrine	28	3	—	—	0–1	—
Ephedrine	14	11	1	1	0	—
Pondinil	58	6	15	22	37	—
Mephentermine	32	5	—	28	—	—
Fenfluramine	—	—	—	0	0	—

Since in the rat the extent of aromatic hydroxylation of these drugs seems to be correlated with their lipid solubility and as the aromatic hydroxylation of many compounds occurs in the lipid containing liver microsomes, then this reaction would be expected to occur in rat liver microsomes. Although amphetamine is hydroxylated in the liver, it is not hydroxylated by liver microsomes (DINGELL and BASS, 1969).

N-DEALKYLATION

Three drugs have been examined for this reaction in various species, namely methamphetamine, ephedrine and mephentermine, which contain an N-methyl group. From the data available it occurs significantly (20–90 per cent) in the rat, rabbit, guinea pig and dog (see Table 3) but not extensively in man (about 10 per cent). On the limited data the rabbit and guinea pig appear to be the species likely to carry out this reaction to the greatest extent. Unlike aromatic hydroxylation, this reaction occurs in the liver microsomes (AXELROD, 1955).

TABLE 3. EXTENT OF N-DEALKYLATION OF CERTAIN AMPHETAMINES IN DIFFERENT SPECIES
(Values are taken from various sources in the literature)

	Rat	Rabbit	Guinea pig	Dog	Man
Methamphetamine	28	—	79	45	14
Ephedrine	20	93	39	58	10
Mephentermine	45	27	—	44	—

DEAMINATION

Data for the extent of deamination of four of these drugs are summarised in Table 4. Deamination is a very minor reaction in the rat, an extensive one in the rabbit and guinea pig and a relatively minor one in man except for amphetamine. The possible mechanisms of this reaction are discussed later.

Only in the metabolism of amphetamine can the rabbit and guinea pig be compared. After deamination, considerable differences occur between these two species in the subsequent fate of the ketone produced. In the guinea pig, the major urinary metabolites of both amphetamine and methamphetamine are benzoic acid and its conjugates (CALDWELL et al., 1972a; DRING et al., 1970). In the guinea pig, the ketone produced is mainly oxidised to benzoic acid whereas in the rabbit it also undergoes reduction and/or some form of conjugation. A major metabolite of amphetamine in the rabbit is a conjugate of benzyl methyl ketone together with appreciable amounts of conjugated phenylpropan-2-ol (DRING et al., 1970), and after norephedrine considerable amounts of conjugated 1,2-dihydroxy-1-phenyl-propane and 1-hydroxy-2-oxo-1-phenylpropane are excreted (SINSHEIMER et al., 1973).

TABLE 4. EXTENT OF DEAMINATION OF VARIOUS AMPHETAMINES IN DIFFERENT SPECIES
(Values are taken from various sources in the literature)

	Rat	Rabbit	Guinea pig	Dog	Man	Tolerant Man
Amphetamine	3	54	62	30	24	23
Methamphetamine	4	—	74	—	6	—
Norephedrine	1	76	—	—	3	—
Ephedrine	3	91	—	—	10	—

ALIPHATIC C-HYDROXYLATION

Amphetamines contain a straight side-chain of three carbon atoms, each of which could be hydroxylated. There is no evidence for an attack on the terminal C atom (C-3) which might yield phenylalanine (DRING et al., 1970).

Hydroxylation of the C-2 atom giving an unstable carbinolamine (BRODIE et al., 1958) has indirect support from recent studies (HUCKER et al., 1971; PARLI et al., 1971) and has been invoked in one mechanism for the deamination of amphetamines.

The hydroxylation of the 1-carbon atom, a reaction of importance in explaining tolerance to amphetamines has been definitely proved. Table 5 shows the extent of this reaction in four species as measured by the urinary output of norephedrine and 4-hydroxynorephedrine. This table suggests that β-hydroxylation is a relatively minor metabolic reaction of the amphetamines. In the guinea pig and rat, β-hydroxylation occurs more extensively with methamphetamine than with amphetamine and paredrine. The values of 16 per cent of the dose in the rat and 19 per cent in the guinea pig are for doses of methamphetamine of 45 mg/kg, whereas the 1 per cent quoted for the guinea pig was obtained with 10 mg/kg. This might suggest that in the guinea pig the production of norephedrine is dose dependent. In normal humans β-hydroxylation is about 3–5 per cent but in the amphetamine tolerant human, β-hydroxylation is higher (5 per cent) than in normals (3 per cent) (CALDWELL et al., 1972b; SEVER et al., 1973). If these values are acecpted as being different, the probable explanation is that the normal humans had been given ($\pm$)-amphetamine and the tolerant subjects (+)-amphetamine, whilst only (+)-amphetamine is a substrate for dopamine β-hydroxylase (GOLDSTEIN and ANAGNOSTE, 1965).

TABLE 5. THE URINARY EXCRETION OF NOREPHEDRINE AND 4-HYDROXYNOREPHEDRINE
AFTER CERTAIN AMPHETAMINES
(Values are taken from various sources in the literature)

Drug	Rat		Guinea pig		Rabbit		Normal Human		Tolerant Human	
	N	HN	N	HN	N	HN	N	HN	N	HN
Amphetamine	0·3	0·3	—	—	—	—	2·4	0·4	4·1	1·3
Methamphetamine	0	16	19	0	—	—	2	1·7	—	—
4-Hydroxyamphetamine	—	4	—	3	—	—	—	4·7	—	—
Norephedrine	48	28	—	—	8	3	86	0–1	—	—

Norephedrine and especially 4-hydroxynorephedrine are regarded as false neurotransmitters. Several authors have suggested that some aspects of the development of tolerance to amphetamines may be related to the metabolic production of 4-hydroxynorephedrine (BRODIE et al., 1970; LEWANDER, 1971). In man, it would appear that 4-hydroxynorephedrine is produced from 4-hydroxyamphetamine but hardly at all from norephedrine. The formation of 4-hydroxynorephedrine from amphetamine or methamphetamine requires both an aliphatic and an aromatic hydroxylation. It is possible in man that the intermediate in the production of 4-hydroxynorephedrine from these drugs is 4-hydroxyamphetamine since more 4-hydroxynorephedrine is excreted after methamphetamine than after amphetamine (Table 5) and there is considerably more aromatic hydroxylation of methamphetamine than amphetamine by man (see Table 2).

It is also of interest to compare production of norephedrines from methamphetamine in the rat and guinea pig. Only 4-hydroxynorephedrine is excreted by the rat whilst only norephedrine by the guinea pig. If current theories are correct, this should mean that the guinea pig would not become tolerant to those effects of the amphetamines in which 4-hydroxynorephedrine apparently plays a role. However, no work which compares the two species with respect to tolerance development is recorded.

N-OXIDATION

The biological oxidation of nitrogen occurring in aliphatic or aromatic amines, in amides or as an atom in a heterocyclic system is well-known (BRIDGES et al.

1972). Several amphetamines are reported to yield hydroxylamines in liver micro-somal preparations of rat, rabbit and guinea pig (BECKETT, 1971) but their formation *in vivo* has not yet been satisfactorily shown. Such metabolites, however, could be excreted as the more stable *N-O*-glucuronides or even sulphates (IRVING, 1971). These hydroxylamines have been suggested as intermediates in deamination.

EXCRETION OF AMPHETAMINES UNCHANGED

In Table 6, the literature data on the extent to which different amphetamines are excreted unchanged in the urine are shown. The amphetamines are extensively metabolised by rabbits and guinea pigs and least metabolised by man.

TABLE 6. EXTENT TO WHICH SOME AMPHETAMINES ARE EXCRETED UNCHANGED IN
VARIOUS SPECIES
(The values given are from various sources in the literature and are intended to indicate the approxi-mate extent to which the drugs are excreted unchanged)

Drug	Rat	Rabbit	Guinea pig	Dog	Man
Amphetamine	12	4	19	30	34 [tolerant 44]
Methamphetamine	11 (14)	2 (2)	1–3 (4–16)	20 (35)	23 (26)
Ethylamphetamine	—	—	—	—	17 (24)
Pondinil	1	18	3	17	1
Norephedrine	48	8	—	—	86
Ephedrine	42 (45)	0·5 (2·5)	2 (41)	6 (64)	61 (73)
Fenfluramine	—	—	—	—	7 (10)
Mephentermine	0 (13)	0 (22)	—	0 (16)	—

Values in brackets are the sum of *N*-alkylamine and free amine.

In man, the least lipid soluble drugs (see Table 1), ephedrine and norephedrine, are the most extensively excreted unchanged (70–80 per cent of the dose). The more lipid soluble drugs, amphetamine, methamphetamine and ethylamphetamine (in order of increasing lipid solubility) the excretion unchanged (Table 6) diminishes as lipid solubility (Table 1) increases, but the excretion of primary amine plus alkyl amine for each of these compounds is about 30–40 per cent. The excretion of total amine could be greater than this since this depends upon urinary pH (BECKETT and ROWLAND, 1965). The only drugs almost completely metabolised in man are the highly lipid-soluble Pondinil (only 1 per cent excreted unchanged) and Fen-fluramine (7 per cent, or 10 per cent as total amine). Metabolism of these drugs in man is thus correlated with the lipid solubility of the un-ionised forms.

The rat is similar to man except that all the drugs are metabolised to a greater extent than in man. Thus, some 40–50 per cent of the least soluble drugs, ephedrine and norephedrine, are excreted unchanged by the rat compared with over 70 per cent in man, and about 12 per cent of the more soluble drugs, amphetamine and meth-amphetamine, in the rat compared with some 30 per cent in man. The dog appears to be between man and rat in this respect.

In the rabbit and guinea pig, there is no correlation between the extent of metabo-lism and lipid solubility since they metabolise some 90 per cent of the dose of all these drugs. In the rabbit these drugs, despite the fact that at pH 7·4 they occur almost entirely as cations, are metabolised by the liver microsomes which, according to accepted views on drug metabolism, are mainly concerned with the biotrans-formation of non-polar lipid-soluble compounds. Several explanations are possible, one being that rabbit liver microsomes are different from those of the rat, in allowing cations to penetrate them or in being able to metabolise the small amount of un-ionised drug present (<1 per cent) very rapidly so that all the ionised form is quickly converted to un-ionised drug as required by the laws of ionisation equilibria.

REFERENCES

AXELROD J. (1955) *J. Biol. Chem.* **214**, 753–763.
BECKETT A. H. (1971) *Xenobiotica* **1**, 53–72.
BECKETT A. H., BROOKES L. G. and SHENOY E. V. B. (1969) *J. Pharm. Pharmacol.* **21**, 151S–156S.
BECKETT A. H. and ROWLAND M. (1965) *J. Pharm. Pharmacol.* **17**, 109S–114S.
BRALET J., COHEN Y. and VALLETTE G. (1968) *Biochem. Pharmacol.* **17**, 2319–2331.
BRIDGES J. W., GORROD J. W. and PARKE D. V. (1972) *Biological Oxidation of Nitrogen in Organic Molecules*, Taylor and Francis, London.
BRODIE B. B., CHO A. K. and GESSA G. L. (1970) In: *Amphetamines and Related Compounds*, (COSTA E. and GARRATINI S., Eds), pp. 217–230, Raven Press, New York.
BRODIE B. B., GILLETTE J. R. and LA DU B. N. (1958) *Ann. Rev. Biochem.* **27**, 427–454.
BRUCE R. B. and MAYNARD W. R. (1968) *J. Pharm. Sci.* **57**, 1173–1176.
CALDWELL J., DRING L. G. and WILLIAMS R. T. (1972a) *Biochem. J.* **129**, 11–22.
CALDWELL J., DRING L. G. and WILLIAMS R. T. (1972b) *Biochem. J.* **129**, 23–24.
DINGELL J. V. and BASS A. D. (1969) *Biochem. Pharmacol.* **18**, 1535–1538.
DRING L. G., SMITH R. L. and WILLIAMS R. T. (1970) *Biochem. J.* **116**, 425–435.
GOLDSTEIN M. and ANAGNOSTE B. (1965) *Biochim. Biophys. Acta* **107**, 168–170.
HUCKER H. B., MICHNIEWICZ B. M. and RHODES R. E. (1971) *Biochem. Pharmacol.* **20**, 2123–2128.
IRVING C. C. (1971) *Xenobiotica* **1**, 75–86.
LEWANDER T. (1971) *Psychopharmacologia* (*Berlin*) **21**, 17–31.
LONG R. F. (1970) quoted by SMITH R. L. and DRING L. G. (1970) In: *Amphetamines and Related Compounds* (COSTA E. and GARRATINI S., Eds), pp. 121–139, Raven Press, New York.
PARLI C. J., WANG N. and McMAHON R. E. (1971) *Biochem. Biophys. Res. Commun.* **44**, 1204–1210.
SEVER P. S., CALDWELL J., DRING L. G. and WILLIAMS R. T. (1973) *Eur. J. Clin. Pharmacol.*, in press.
SINSHEIMER J. E., DRING L. G. and WILLIAMS R. T. (1973) *Trans. Biochem. Soc.* in press.
VREE T. B., MUSKENS A. T. and VAN ROSSUM J. M. (1969) *J. Pharm. Pharmacol.* **21**, 774–775.
WALKENSTEIN S. S., CHUMAKOV N. and SEIFTER J. (1955) *J. Pharmacol. Exp. Ther.* **115**, 16–20.

Frontiers in Catecholamine Research 1973, pp. 933 to 937. Pergamon Press. Printed in Great Britain.

BEHAVIOURAL PHARMACOLOGY OF D-AMPHETAMINE: SOME METABOLIC AND PHARMACOLOGICAL CONSIDERATIONS

H. C. FIBIGER

Division of Neurological Sciences University of British Columbia
Vancouver, British Columbia, Canada

FOOD deprivation and amphetamine are both known to increase behavioural arousal in the rat. Amphetamine is generally believed to induce psychomotor excitation by increasing the synaptic release of catecholamines (CA) and by blocking their reuptake from the synapse.[7,13,30] Food deprivation increases electrophysiological measures of arousal in the mesencephalic reticular formation and other brain regions[26] but the neurochemical substrates of this response are presently not known. DELL[10] postulated that the behavioural arousal induced by starvation resulted from the release of adrenal catecholamines and their subsequent action on excitatory mechanisms in the brain stem. This hypothesis has not been supported, however, by the finding that rats continue to show starvation-induced arousal after adrenal demedullation.[14] Recently, the question of how starvation-induced and amphetamine-induced increases in behavioural arousal might interact was investigated.[6] The stimulant effects of amphetamine were found to interact synergistically with starvation-induced arousal. Thus, food-deprived rats were significantly more responsive to the stimulant effects of amphetamine than were controls. In addition, the response to a given dose of amphetamine (1 mg/kg) increased with increasing days of food deprivation. A more detailed analysis of the results indicated that both the peak effect and the duration of drug action were significantly increased by food deprivation.[6]

During starvation liver metabolism of certain psychoactive drugs is altered.[11,23] ANGEL,[1] for example, gave intraperitoneal injections of cocaine and found higher levels of the drug in the brains of starved rats as compared with controls. These findings suggest that the increased stimulant effects of amphetamine observed during food deprivation may depend on impaired metabolism of the drug. To test this hypothesis H^3-amphetamine was injected intraperitoneally (i.p.) into animals food-deprived for 4 days and controls. One hour later, when the peak behavioural effects occurred, the rats were sacrificed and the whole brain levels of H^3-amphetamine and its metabolites were measured. No significant difference between control and test groups were obtained, indicating that impaired metabolism of amphetamine cannot account for its increased psychomotor stimulant effects during food deprivation (Table 1).

Food deprivation is a highly stressful situation for the rat, and when stressed, increased turnover of brain norepinephrine (NE) has been reported by a number of investigators. For example, foot shock,[2,40] extremes in temperature,[8,36] immobilisation,[9,46] and extreme muscular exercise[22] have all been shown to increase the turnover of brain NE. Dopamine (DA) turnover can also be altered by stress.[2,3,28] The increased responsiveness to amphetamine during food deprivation may therefore reflect changes in the interaction between amphetamine and brain CA turnover.

TABLE 1. EFFECT OF FOOD DEPRIVATION ON WHOLE BRAIN LEVELS
OF d-AMPHETAMINE

	Whole supernatant ($\times 10^4$ dis/min per g)	Extracted amphetamine ($\times 10^4$ dis/min per g)
2 days food deprivation		
Control	$3 \cdot 90 \pm 0 \cdot 32$	$3 \cdot 79 \pm 0 \cdot 57$
Food deprived	$3 \cdot 29 \pm 0 \cdot 25$	$2 \cdot 85 \pm 0 \cdot 25$
4 days food deprivation		
Control	$3 \cdot 39 \pm 0 \cdot 19$	$3 \cdot 26 \pm 0 \cdot 25$
Food deprived	$3 \cdot 21 \pm 0 \cdot 30$	$3 \cdot 17 \pm 0 \cdot 30$

Rats were injected intraperitoneally with a solution of unlabelled d-amphetamine sulfate (1·0 mg/kg) containing ^{3}H-d-amphetamine sulfate ($8 \cdot 56 \times 10^7$ dis/min per g.). Animals were sacrificed 1 hr later and the whole brain (including cerebellum) was homogenised in acid. An aliquot of the supernatant was counted (whole supernatant) and amphetamine was extracted from the remaining supernatant by the method of GLOWINSKI, AXELROD and IVERSEN.[20] Data represent means ($\pm$ s.e.m.) of 4 animals in each group. No statistically significant differences were obtained between the control and food deprived groups.

The fact that stress significantly increases amphetamine toxicity is consistent with such a hypothesis.[21,44] To examine this possibility, animals were maintained on *ad libitum* food and water or they were food deprived for 4 days and then injected i.p. with α-methyl-p-tyrosine (α-MPT). Food deprivation did not significantly alter the rate of depletion of whole brain NE or DA after α-MPT (Table 2). These results are consistent with the observation that 4 days of food deprivation do not alter tyrosine hydroxylase [EC.1.14.3a] activity in the mid-brain, hypothalamus, hippocampus or caudate (FIBIGER and McGEER, unpublished). It appears unlikely therefore that the increased stimulant effects of amphetamine during food deprivation are the result of an altered interaction between amphetamine and CA turnover, and at present the neurochemical basis of this response remains to be elucidated.

TABLE 2. EFFECT OF FOOD DEPRIVATION (4 DAYS) ON α-METHYL-p-TYROSINE
INDUCED DEPLETION OF BRAIN NORADRENALINE AND DOPAMINE

	0 hr	0·5 hr	2·0 hr
Noradrenaline (μg/g)			
Controls	$0 \cdot 338 \pm 0 \cdot 002$ (4)	$0 \cdot 249 \pm 0 \cdot 015$ (9)	$0 \cdot 137 \pm 0 \cdot 019$ (6)
Food deprived	$0 \cdot 377 \pm 0 \cdot 015$ (4)	$0 \cdot 226 \pm 0 \cdot 015$ (9)	$0 \cdot 147 \pm 0 \cdot 018$ (6)
Dopamine (μg/g)			
Controls	$0 \cdot 466 \pm 0 \cdot 036$ (4)	$0 \cdot 415 \pm 0 \cdot 009$ (9)	$0 \cdot 259 \pm 0 \cdot 014$ (8)
Food deprived	$0 \cdot 495 \pm 0 \cdot 028$ (4)	$0 \cdot 427 \pm 0 \cdot 017$ (9)	$0 \cdot 260 \pm 0 \cdot 009$ (9)

Rats were food deprived for 4 days or maintained on *ad libitum* food and water. L-α-methyl-p-tyrosine (Regis Co.) was injected intraperitoneally (200 mg/kg) and the animals were sacrificed at various intervals thereafter. The whole brain (including cerebellum) was homogenised in acid and brain noradrenaline and dopamine levels were determined fluorometrically by the method of McGEER and McGEER.[29] Numbers in parentheses indicate number of animals in each group. Data represent means ($\pm$ s.e.m.). No statistically significant differences were obtained between the control and food deprived groups.

There are several possibilities which require further investigation however. In the above experiments only the whole brain levels of amphetamine and its metabolites were measured. Food deprivation may produce changes in either the regional accumulation of amphetamine in brain or in the subcellular distribution of the drug. Similarly, although whole brain DA and NE turnover do not appear to be affected by food deprivation, discrete and local changes may occur in the turnover of these amines which are obscured by whole brain measurements. Strong support for the latter possibility has recently been provided by FRIEDMAN, STARR and GERSHON[19] who found that 22 hr of food deprivation significantly increased α-MPT induced depletion of NE in the hypothalamus but not in the remainder of the brain. These workers also observed that whole brain DA depletion was slightly increased in food-deprived rats 4 hrs after α-MPT, but that finding was not supported by the present experiments.

Whatever mechanisms underlie the increased amphetamine response during food deprivation, it is obvious that the nutritive state of the animal can be of considerable importance in determining the magnitude and the nature of the drug response. An example of the difficulties which can be encountered when this factor is not controlled for is found in the reports of increased sensitivity to amphetamine and increased spontaneous locomotor activity after chronic reserpine treatment in the rat.[33,34,37] Such treatment produces *subsensitivity* to indirectly acting sympathomimetics such as amphetamine in the peripheral nervous system.[41] The resolution of this inconsistency may lie in the fact that when rats are treated chronically (10–14 days) with small doses of reserpine, many of the animals become hypophagic and hypodipsic and show considerable losses in body weight.[15,37] The importance of this factor in the potentiation of amphetamine stimulation after chronic reserpine treatment has been evaluated.[15] In accordance with the previous reports, groups of rats given chronic reserpine treatment showed significantly increased spontaneous locomotor activity and an enhanced response to D-amphetamine as compared with controls. It is noteworthy however, that the chronic reserpine treatment produced variable effects on final body weight and the increased spontaneous activity and the enhanced responsiveness to amphetamine were observed only in those individual animals which suffered marked weight loss. In a second experiment, the food intake of the control group was restricted so that it was similar to the *ad libitum* intake of the rats treated with reserpine. In this case, where both the saline and the reserpine-treated groups suffered similar weight losses over 10 days, the chronic reserpine group was in no sense more responsive to amphetamine than were the controls. On the contrary, on one measure (lowest amphetamine dosage which significantly increased activity) the control group showed slightly greater responsiveness to amphetamine than did the chronic reserpine group. Taken together, these experiments suggest that the increased stimulant effect of D-amphetamine and the increase in spontaneous activity which develops during chronic reserpine administration is a result of the severe hypophagia observed during the drug treatment.[15] These results again point to the importance of controlling for the nutritive state of the organism in behavioural pharmacological research.

The fact that chronic reserpine administration produces hypophagia and hypodipsia supports the view that brain biogenic amines may be of critical importance in regulating food and water intake in the rat.[24,31] This hypothesis has recently received strong experimental support in the finding that 6-hydroxydopamine (6-OHDA), an agent which can selectively destroy CA neurons in the brain, can when injected intraventricularly into monoamine oxidase (MAO) inhibited animals or directly into the substantia nigra, produce profound aphagia and adipsia.[16,17,43] The syndrome produced shows a remarkable resemblance to the well-known "lateral hypothalamic syndrome" in which animals with bilateral electrolytic lesions of the lateral hypothalamus gradually recover from aphagia and adipsia but continue to show more subtle but permanent deficits.[17,39,47] These more permanent deficits are also found in the 6-OHDA treated animals.[17,47] To date, these experiments suggest that the DA nigrostriatal projection may have a major role in the control of food and water intake,[17,43] since bilateral injections of 6-OHDA into the substantia nigra produce a syndrome which most closely resembles the lateral hypothalamic syndrome. A contribution of the mesolimbic system, the other major DA projection,[42] to this syndrome cannot at present be ruled out however. MALER and FIBIGER (unpublished observations) using the Fink Heimer technique, have found that when 6-OHDA is injected into the substantia nigra in the doses necessary to destroy the nigro-striatal projection (8 μg), extensive damage is also observed in the mesolimbic DA system.

TABLE 3. EFFECTS OF 6-HYDROXYDOPAMINE ON LOCOMOTOR STIMULATION, STEREOTYPY AND ANOREXIA INDUCED BY AMPHETAMINE SULFATE

	Locomotor stimulation	Stereotypy	Anorexia	
			Baseline intake	Amphetamine intake
Amphetamine dosage (expressed as the salt)	1 mg/kg	5 mg/kg		1·5 mg/kg
Controls	491 ± 72	3·20 ± 0·10	9·07 ± 0·77	1·97 ± 0·30
6-OHDA	34 ± 10*	0·59 ± 0·14*	8·24 ± 0·38	6·75 ± 0·54*

6-hydroxydopamine treated animals received intraventricular injections (250 μg) into the lateral ventricle 30 min after tranylcypromine sulfate (5 mg/kg i.p.). Controls received intraventricular injections of the vehicle. Behavioural tests were conducted after recovery of food and water intake. In the locomotor stimulation test, animals were adapted to a photocell cage for 1 hr, then injected i.p. with d-amphetamine and the resulting activity (number of photobeams interrupted) was recorded for 1 hr. In the stereotypy test, stereotypy was measured by the method of FIBIGER, FIBIGER and ZIS[42] for 3 hr after i.p. amphetamine administration. In the anorexia test food intake (in g) during 75 min (after 24 hr of food deprivation) was measured after saline (baseline) or amphetamine injections. Data represent means (± S.E.M.) of 10 animals in each group. *Significantly different from controls. $P < 0.01$.

The use of 6-OHDA has also permitted further investigations of the neuro-chemical mechanisms underlying the behavioural effects of amphetamine. Because considerable evidence indicates that amphetamine is an indirectly acting sympath-omimetic,[25,38,45] it is somewhat surprising that large doses of 6-OHDA have failed to attenuate amphetamine-induced motor-stimulation despite a reduction of brain catecholamines by 75–80 per cent.[12] This finding is in accord with other reports which have described a general lack of long-lasting behavioural changes (except for increased irritability) after intraventricular 6-OHDA treatment.[5,27,32,35] In these reports 6-OHDA was administered to animals in which MAO had not been inhibited, and the depletion of brain DA was consequently less complete than that of brain NE. Since MAO inhibition increases the destructive capacity of 6-OHDA, particularly on DA neurons,[4,16] some of these earlier experiments were repeated in animals in which MAO was inhibited before the intraventricular injection of 6-OHDA.[18] This procedure produced a 90 per cent depletion in both brain NE and brain DA. After the animals had recovered the ability to regulate food and water intake, they were tested for amphetamine-induced motor stimulation, stereotyped behaviour, and amphetamine anorexia. In these animals all of the above behavioural effects of amphetamine were drastically reduced[17,18] (Table 3), suggesting that the earlier failures to significantly alter the behavioural effects of amphetamine by 6-OHDA treatment were due to an incomplete destruction of CA neurons.

These observations are consistent with the view that amphetamine exerts its behavioural effects indirectly through its action on brain catecholamines, but suggest that the CA systems which subserve these behaviours are present in large excess of the requirements for the maintenance of normal behavioural function. Since only slight effects on behaviour are observed when 70–80 per cent of CA systems are destroyed by 6-OHDA and marked effects are sometimes not observed until this destruction reaches 90 per cent or more, the relationship between brain CA levels and behaviour is not linear.

REFERENCES

1. ANGEL C. (1969) *Dis. Nerv. Sys.* **30**, 94.
2. BLISS E. L., AILION J. and ZWANZIGER J. (1968) *J. Pharmacol. exp. Ther.* **164**, 122–134.
3. BLISS E. L. and AILION J. (1971) *Life Sci.* **10**, 1161–1169.
4. BREESE G. R. and TRAYLOR T. D. (1971) *Brit. J. Pharmac.* **42**, 88–89.
5. BURKHARD W. P., JALFRE M. and BLUM J. (1969) *Experientia* **25**, 1295.
6. CAMPBELL B. A. and FIBIGER H. C. (1971) *Nature Lond.* **233**, 424–425.
7. CARR L. A. and MOORE K. E. (1970) *Biochem. Pharmacol.* **19**, 2361–2374.
8. CORRODI H., FUXE K. and HÖKFELT T. (1967) *Acta physiol. scand.* **71**, 224–232.
9. CORRODI H., FUXE K. and HÖKFELT T. (1968) *Life Sci.* **7**, 107–112.
10. DELL P. C. (1958) In: *Neurological Basis of Behaviour* (WOLSTENHOLME G. E. W. and O'CONNER J., Eds.). Churchill, London.
11. DIXON R. L., SHUNTTICE R. W. and FOUTS J. R. (1960) *Fedn Proc.* **103**, 333.
12. EVETTS K. D., URETSKY N. J., IVERSEN L. L. and IVERSEN S. D. (1970) *Nature, Lond.* **225**, 961–962.
13. FARNEBO L.-O. (1971) *Acta physiol. scand. Suppl.* **371**, 45–52.
14. FIBIGER H. C. and CAMPBELL B. A. (1971) *Physiol. & Behav.* **6**, 403–405.
15. FIBIGER H. C., TRIMBACH C. and CAMPBELL B. A. (1972a) *Neuropharmacology* **11**, 57–67.
16. FIBIGER H. C., LONSBURY B., COOPER H. P. and LYTLE L. D. (1972b) *Nature, Lond.* **236**, 209–211.
17. FIBIGER H. C., ZIS A. P. and MCGEER E. G. (1973a) *Brain Res.* **55**, 135–148.
18. FIBIGER H. C., FIBIGER H. P. and ZIS A. P. (1973b) *Brit. J. Pharmac.* **47**, 683–692.
19. FRIEDMAN E., STARR N. and GERSHON S. (1973) *Life. Sci.* **12**, 317–326.
20. GLOWINSKI J., AXELROD J. and IVERSEN L. L. (1966) *J. Pharmac. exp. Ther.* **153**, 30–41.
21. GOLDBERG M. E. and SALAMA A. I. (1969) *Toxicol. Appl. Pharmacol.* **14**, 447.
22. GORDON R., SPECTOR S., SJÖERDSMA A. and UDENFRIEND S. (1966) *J. Pharmacol. exp. Ther.* **153**, 440–447.
23. GRAM T. E., GUARINO A. M., SCHROEDER D. H., DAVIS D. C., REAGAN R. L. and GILLETTE J. R. (1970) *J. Pharmac. exp. Ther.* **176**, 12–21.
24. GROSSMAN S. P. (1968) *Fedn Proc.* **27**, 1349–1360.
25. HANSON L. C. F. (1967) *Psychopharmacologia (Berl.)* **10**, 289–297.
26. HOCKMAN C. H. (1964) *EEG clin. Neurophysiol.* **17**, 420.
27. LAVERTY R. and TAYLOR K. M. (1970) *Brit. J. Pharmac.* **40**, 836–846.
28. LIDBRINK P., CORRODI H., FUXE K. and OLSON L. (1972) *Brain Res.* **45**, 507–524.
29. MCGEER E. G and MCGEER P. L. (1962) *Can. J. Biochem. Physiol.* **40**, 1141–1151.
30. MCKENZIE G. M. and SZERB C. (1968) *J. Pharmac. exp. Ther.* **162**, 302–308.
31. MILLER N. E. (1965) *Science* **148**, 328–338.
32. NAKAMURA K. and THOENEN H. (1972) *Psychopharmacologia (Berl.)* **24**, 359–372.
33. PIRCH J. H. and RECH R. H. (1968) *Psychopharmacologia* **12**, 115–122.
34. PIRCH J. H. (1969) *Psychopharmacologia* **16**, 253–260.
35. SHOENFELD R. I. and ZIGMOND M. J. (1970) *Pharmacologist* **12**, 227.
36. SIMMONDS M. A. (1969) *J. Physiol. (Lond.)* **203**, 199–210.
37. STOLK J. M. and RECH R. H. (1968) *J. Pharmac. exp. Ther.* **163**, 75–83.
38. STOLK J. M. and RECH R. H. (1970) *Neuropharmacology* **9**, 249–263.
39. TEITELBAUM P. and EPSTEIN A. N. (1962) *Psychol. Rev.* **69**, 74–90.
40. THIERRY A. M., BLANC G. and GLOWINSKI J. (1971) *J. Neurochem.* **18**, 449–461.
41. TRENDELENBURG U. (1966) *Pharmac. Rev.* **18**, 629–640.
42. UNGERSTEDT U. (1971a) *Acta physiol. scand. Suppl.* **367**, 1–48.
43. UNGERSTEDT U. (1971b) *Acta physiol. scand. Suppl.* **367**, 95–122.
44. WEISS B., LATIES V. G. and BLANTON F. L. (1961) *J. Pharmacol. exp. Ther.* **132**, 366.
45. WEISSMAN A. and KOE B. K. (1965) *Life Sci.* **4**, 1037–1048.
46. WELCH B. L. and WELCH A. S. (1968) *Nature, Lond.* **218**, 575–577.
47. ZIGMOND M. J. and STRICHER E. M. (1972) *Science* **177**, 1211–1214.

Frontiers in Catecholamine Research 1973, pp. 939 to 941. Pergamon Press. Printed in the Great Britain.

CATECHOLAMINE METABOLISM AND AMPHETAMINE EFFECTS ON SENSITIVE AND INSENSITIVE MICE

A. JORI and S. GARATTINI

Istituto di Ricerche Farmacologiche "Mario Negri" Via Eritrea, 62-
20157 Milano, Italy

IT HAS been found that amphetamine elicits in mice a strain-dependent symptomatology (WEAVER and KERLEY, 1962; BROWN, 1965). C_3H mice have been previously reported to be considerably less sensitive than other strains to the stimulating activity of amphetamine (DOLFINI et al., 1969a, 1969b; DOLFINI et al., 1970). The data presented summarise some differences and similarities between C_3H and NMRI mice in the action of d-amphetamine (Table 1).

DIFFERENCES IN THE ACTION OF d-AMPHETAMINE IN THE TWO STRAINS

(1) d-Amphetamine neither increases the spontaneous motility, nor, does it induce the stereotyped behaviour and grouping effect in C_3H mice as do the doses active in NMRI mice.

(2) d-Amphetamine does not significantly increase the body temperature of C_3H mice, while it elicits a dose-dependent hyperthermia in NMRI mice (3·75–30 mg/kg i.p.). On the contrary, at low doses, it decreases the body temperature in C_3H mice (CACCIA et al., 1973).

(3) d-Amphetamine decreases noradrenaline (NE) in the brain-stem and increases homovanillic acid (HVA) in the striatum, but it only slightly affects noradrenaline and does not modify HVA concentrations of C_3H mice (CACCIA et al., 1973).

SIMILARITIES IN THE ACTION OF d-AMPHETAMINE IN THE TWO STRAINS

(1) The distribution of amphetamine is similar in C_3H and NMRI mice. No differences were observed between the two strains in the concentration of amphetamine in whole brain and in specific areas such as striatum and brain-stem at various times after administration of the drug.

(2) The anorexigenic effect of d-amphetamine seems to be similar in C_3H and in NMRI mice.

(3) d-Amphetamine increases plasma FFA to a similar extent in C_3H and in NMRI mice.

(4) d-Amphetamine elicits a clear dose-dependent hyperthermic effect in reserpinised mice of both strains. These data are in sharp contrast with the lack of hyperthermic activity in normal untreated C_3H mice.

The thermic response to amphetamine in normal untreated mice is probably a polygenically-inherited trait. This hypothesis is supported by the results of pharmacogenetic experiments conducted in our Institute involving the cross mating of C_3H and NMRI mice (Fig. 1); (JORI and PRICE-EVANS, unpublished results). In an attempt to understand this genetically determined different reactivity to amphetamine, the basal brain concentrations of the biogenic amines and of their metabolites were compared

A. JORI and S. GARATTINI

TABLE 1

No. determinations	Effects of amphetamine	Strain	
		C$_3$H	NMRI
72	Hypermotility	NO	YES
72	Stereotyped behaviour	NO	YES
18	Hyperthermia (°C)	-0.7 ± 0.1	$+2.5 \pm 0.1\ddagger$
8	NE decrease (ng/g)	-60 ± 15	$-150 \pm 20\ddagger$
4	HVA increase (ng/g)	-14 ± 20	$+224 \pm 15\ddagger$
5	Brain Amphetamine (µg/g)	5.2 ± 0.1	5.5 ± 0.2
3	Food intake * (g/2 hr/6 mice)	$0 \pm 1\ddagger$	$2 \pm 1\ddagger$
5	Plasma-FFA increase (mequiv/1.)	$+715\ddagger \pm 35$	$+727\ddagger \pm 39$
18	Hyperthermia in reserpinised mice* (°C)	$+3.9\ddagger \pm 0.4$	$+3.5\ddagger \pm 0.2$

$\ddagger$ $P < 0.01$ vs untreated mice. Mice were caged in groups of 6 and were given d-amphetamine-sulphate at a dose of 7·5 mg/kg i.p. or* 5 mg/kg i.p. Temperature and biochemical determinations were performed 30 or 60 min after treatment. Food intake of control mice was 11 g $\pm$ 0.5

in the two strains (Table 2). No differences were noted for the brain serotonin (5HT) and 5-hydroxyindolacetic acid (5HIAA), for NE in the brain-stem, and for

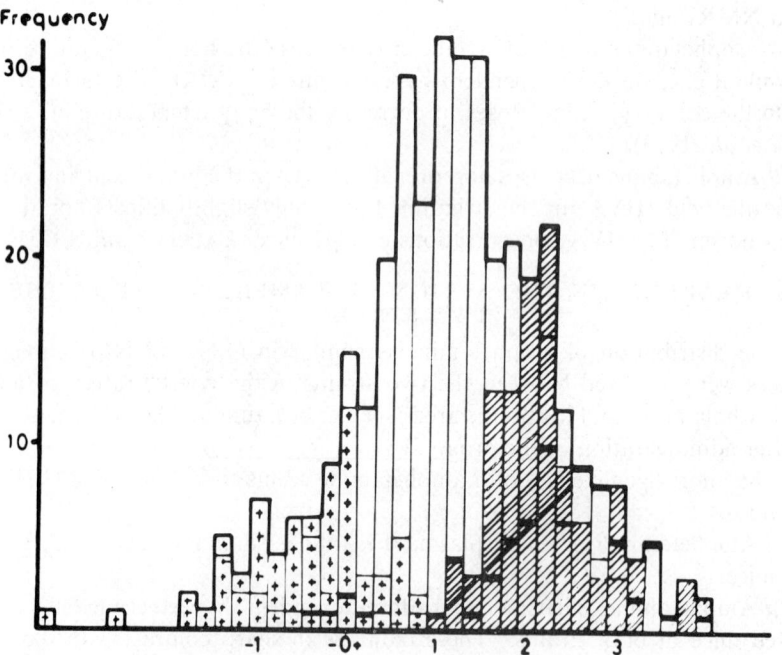

FIG. 1.—Frequency distribution of temperature changes 30 min after d-amphetamine sulphate (7·5 mg/kg/i.p.), in the various groups of mice. (312 mice were used.)

NMRI mice

C$_3$H mice

F$_1$ mice

F$_2$ mice

dopamine (DA) and HVA in the striatum   However a basic difference in the striatum dopamine metabolism was suggested by determination of DA disappearance after blocking catecholamine synthesis with α-methyltyrosine and of the HVA accumulation after blocking its active transport with probenecid. These experiments demonstrated

TABLE 2

No. determinations	Biochemical parameters	Strain	
		C_3H	NMRI
		(ng/g $\pm$ s.e.)	
8	Brain 5HT	285 ± 25	250 ± 20
8	Brain 5HIAA	490 ± 10	480 ± 10
8	Brain Stem NE	570 ± 20	540 ± 20
6	Striatum DA	4280 ± 160	4260 ± 200
4	Striatum HVA	279 ± 15	240 ± 15
4	Striatum HVA after Probenecid*	$389 \pm 8\ddagger$	630 ± 36
	Rate constant of DA loss after αMTyrosine K (hr^{-1})†	$0.136 \pm 0.017\ddagger$	0.188 ± 0.021

$\ddagger$ $P < 0.01$ versus NMRI mice. * Probenecid was given at a dose of 200 mg/kg i.p. HVA was determined 90 min after Probenecid. † K represents the slope × 2·3 of the curve obtained by plotting the log-concentrations (μg/g) of dopamine at 0, 1, 2 and 3 hr after the administration of α-methyltyrosine (αMT) (500 mg/kg/ i.p.).

a lower DA turnover rate in C_3H than in NMRI mice. These last data are preliminary and other experiments are in progress. We consider that the comparative use of genetically insensitive and sensitive strains, can provide a useful model for the understanding of the mechanism of the action of amphetamine and the factors responsible for its tolerance and its abuse.

Acknowledgment—These studies have been supported by a grant of the G. and L. Pfeiffer Foundation, N.Y., U.S.A.

REFERENCES

BROWN A. M. (1965) *Lab. Anim. Care* **15**, 111–118.
CACCIA S., CECCHETTI G., GARATTINI S. and JORI A. (1973) *Br. J. Pharmacol.* (in press).
DOLFINI E., GARATTINI S. and VALZELLI L. (1969a) *Europ. J. Pharmacol.* **7**, 220–223.
DOLFINI E., GARATTINI S. and VALZELLI L. (1969b) *J. Pharm. Pharmac.* **21**, 871–872.
DOLFINI E., RAMIREZ DEL ANGEL A. and GARATTINI S. (1970) *Europ. J. Pharmacol.* **9**, 333–336.
WEAVER L. C. and KERLEY T. L. (1962) *J. Pharmac. Exp. Ther.* **135**, 240–244.

Frontiers in Catecholamine Research 1973, p. 943. Pergamon Press. Printed in Great Britain.

EFFECTS OF d-AMPHETAMINE ON BODY TEMPERATURE AND ON BEHAVIORAL THERMOREGULATION IN RATS

Shlomo Yehuda and Richard J. Wurtman

Laboratory of Neuroendocrine Regulation, Department of Nutrition and Food Science
Massachusetts Institute of Technology, Cambridge, Massachusetts 02139, U.S.A.

We have previously shown that the administration of d-amphetamine (7·5–15·0 mg/kg i.p.) to rats (100–150 g body wt) kept in a cold ambient temperature (4°C–15°C) causes a significant decrease in colonic temperature (from 37·0°C to 34·6°C after 60 min) (Yehuda and Wurtman 1972a). This effect is apparently mediated by dopaminergic brain neurons inasmuch as it is blocked by pretreatment with halo-peridol (2.0 mg/kg), pimozide (25·0 mg/kg) or ditran (4·0 mg/kg) but not by propranolol (12·0 mg/kg) or phenoxybenzamine (20·0 mg/kg). Similar effects on body temperature is produced by L-dopa (100·0 mg/kg) apomorphine (10·0 mg/kg) and clonidine (2·0 mg/kg). The hypothermic effect of d-amphetamine is blocked by lesions that destroy the dopaminergic system in the limbic forebrain (i.e., the olfactory tubercules and adjacent tissues) (Yehuda and Wurtman, 1972b).

Present studies show that d-amphetamine also causes paradoxical thermoregulatory behavior. If rats are kept at a cold ambient temperature (4°C) and allowed to locate themselves at various distances from the beam of a heating lamp the animals normally will place themselves under the lamp (about 80% of the time). d-Amphetamine administration causes rats to choose to avoid the lamp. This effect is also blocked by pimozide (5·0 mg/kg), haloperidol (2·0 mg/kg) and spiroperidol (2·0 mg/kg) and to some extent by phenoxybenzamine (10·0–20·0 mg/kg). Animals placed in a warm environment (20°C–30°C) will normally avoid the beam of the heat lamp; d-amphet-amine administration causes them to choose to place themselves under the lamp. The effect is also blocked by pimozide, haloperidol and sprioperidol, but not by phenoxybenzamine.

These results indicate that among those effects of d-amphetamine probably mediated by central dopaminergic neurons (e.g., stereotyped behavior and rotational behavior) are its effects on body temperature thermoregulatory responses. The hypothermic effect may lend itself to use in screening dopamine receptor stimulants, and dopamine-receptor blocking agents, (including antipsychotic agents).

Acknowledgements—These studies have been supported in part by the United States Public Health Service.

REFERENCES

Yehuda S. and Wurtman, R. J. (1972a) *Life Sciences (I)*, **11** (18): 851–859.
Yehuda S. and Wurtman, R. J. (1972b) *Nature, Lond.* **240** (5382) 477–478.

Frontiers in Catecholamine Research 1973, pp. 945 to 950. Pergamon Press. Printed in Great Britain.

THE METABOLISM AND EXCRETION OF 'AMPHETAMINES' IN MAN

J. W. GORROD

Department of Pharmacy, Chelsea College, University of London,
Manresa Road, London, S.W.3.

THE 'amphetamines' are a group of drugs, based on β-phenylisopropylamine, which possess a diverse variety of pharmacological properties. They include a broad range of central stimulants, hallucinogens, psychotomimetic and anorexic agents and are widely used in clinical medicine, amphetamine itself being still the therapy of choice in the treatment of narcolepsy.[1]

The structure-activity relationships of 'amphetamines' and some related compounds have been discussed by BIEL[2] and some of these compounds in general use are shown in Table 1. It is the purpose of this communication to present an overview of the current situation regarding the excretion and metabolism of certain 'amphetamines' in man.

The metabolism and excretion of amphetamine itself in man had long been a subject of study for many groups of workers[3-10] prior to the development of specific assay procedures based on gas–liquid chromatography,[11-14] and it was early realised that, as expected from the work of MILNE et al.[15] and WEINER and MUDGE[16], the excretion was greatly influenced by the urinary pH of the subject[17-20] and unless this factor was controlled, comparison between individuals or the effect of genetic variation, drugs, diet or disease etc. on the excretion and metabolism of 'amphetamines' may only be a reflection of urinary pH.[21-23] Under uncontrolled urinary pH conditions the excretion of unchanged amphetamine averaged 14·5 per cent with a range of 5–30 per cent; concomitant acidification of the urine by administration of ammonium chloride greatly increased the urinary excretion to about 55 per cent of the dose and a much narrower range of intersubject variation was observed.[18] An alkaline urine virtually prevented the excretion of unchanged amphetamine in man.[18] The clinical significance of these findings have been discussed by ROWLAND and BECKETT.[21,43]

Parallel studies on the metabolism of ^{14}C-amphetamine were undertaken by Professor R. T. WILLIAMS, group,[24-26] and these showed that in man, under uncontrolled urinary pH conditions, 4-hydroxylation was a minor route of metabolism accounting for less than 4 per cent of the dose, whereas deamination accounted for 24 per cent.

These results are in good agreement with ELLISON et al.[27] and BALTES et al.[31], the latter authors using tritiated amphetamine. That changes in urinary pH could actually change the extent by which the amphetamine was metabolised through a given route in man was first shown by BECKETT et al.[32]. In this report the deamination route was shown to fall from 17 per cent under fluctuating urine conditions to 5 per cent under controlled acidic conditions. These results were confirmed by DAVIS et al.[20] who also showed that 4-hydroxylation of amphetamine was not however greatly influenced by urinary pH variation.

TABLE 1. THE STRUCTURES OF SOME 'AMPHETAMINES' USED IN CLINICAL PRACTICE

Drug	[ring]—CH— $\overset{\underset{\displaystyle CH_3}{\displaystyle\mid}}{C}$ —NH				Use
Amphetamine	—	H	H	H	Stimulant, Narcolepsy
Methamphetamine	—	H	H	CH_3	Stimulant
Paradrine	4-OH	H	H	H	Nasal decongestant
Ephedrine	—	OH	H	CH_3	Bronchodilator
Mephentermine	—	H	CH_3	CH_3	Nasal decongestant
Chlorphentermine	4-Cl	H	CH_3	H	Anorexic
Fenfluramine	3-CF_3	H	H	C_2H_5	Anorexic
Methoxyphenamine	2-OCH_3	H	H	CH_3	Bronchodilator
Prenylamine	—	H	H	$(C_6H_5)_2CHCH_2CH_2$	Angina Pectoris
Fenetylline	—	H	H	Aminoethyl theophylline	Psychostimulant, Antidepressant

The work of BECKETT et al.[32] is also particularly important in that it showed that under acidic, but not under fluctuating, urinary conditions, the rate of urinary excretion of amphetamine is directly proportional to its plasma concentration.

Other minor metabolites of amphetamine which have been reported to occur in man are formed by β-hydroxylation of either amphetamine or 4-hydroxyamphetamine to yield norephedrine[20,28] or 4-hydroxynorephedrine.[20,28,29] The β-hydroxylation of 4-hydroxyamphetamine had previously been observed in man.[30] 4-Hydroxy-nor-ephedrine is known to have the properties of a false transmitter[29,62] and these β-hydroxylated metabolites may play some role in the development of amphetamine psychosis.[33]

Studies on the excretion of individual stereo-isomers by BECKETT and ROWLAND[34] indicated that stereoselective metabolism occurred, as less of the (+)-isomer was recovered in the subjects' urine. This has been confirmed by GUNNE[35] using a G.L.C. technique which allowed separation of the stereo-isomers. The data obtained from ^{14}C-amphetamine experiments would suggest that stereo-selective deamination may be one of the determining factors[26], although in vitro stereo-specific β-hydroxylation is known to occur.[63]

Beckett's group have carried out detailed studies on the influence of molecular modification on the pharmacokinetic characteristics of N-substituted amphetamine in man.[34,36-43] In each case an influence of urinary pH on the excretion of the un-changed drug and its dealkylated metabolite has been observed.

The results of some of these studies are presented in Table 2. Stereo-selective metabolism is emphasised with increasing N-alkyl chain length of up to a C_4 chain, but on increasing the substituent to a n-butyl or more bulky group not only is the stereo-selectivity lost but the total drug and metabolite recovered reduced to a much lower level. To some extent these results reflect the partition coefficients of the unionised form of this series of drugs determined using heptane as the organic phase,[54] although far better predictions as to the pharmacokinetic characteristics of these compounds were made possible by the introduction by BECKETT et al.[52-54] of the buccal absorption test.

TABLE 2. THE URINARY RECOVERIES (%) OF ENANTIOMORPHS OF N-ALKYLAMPHETAMINES TOGETHER WITH THE RECOVERY (%) OF AMPHETAMINE PRODUCED BY METABOLIC DEALKYLATION.

Drug administered	Recovery of compound administered	Recovery of amphetamine	Total recovery	Unaccounted for
(+)-amphetamine	66	←	66	34
(−)-amphetamine	78	←	78	22
(+)-N-methylamphetamine	46	7	53	47
(−)-N-methylamphetamine	65	3	68	32
(+)-N-ethylamphetamine	24	16	40	60
(−)-N-ethylamphetamine	73	6	79	21
(+)-N-n-propylamphetamine	16	11	27	73
(−)-N-n-propylamphetamine	50	15	65	35
(+)-N-n-butylamphetamine	2	8	10	90
(−)-N-n-butylamphetamine	3	6	9	91
(+)-N-benzylamphetamine	traces	3	3+	97
(−)-N-benzylamphetamine	traces	4	4+	96
(+)-N-o-chlorbenzylamphetamine	traces	3	3+	97
(−)-N-o-chlorbenzylamphetamine	traces	4	4+	96

Subjects maintained under acidic urinary conditions. The data represents the mean urinary recovery. Compiled from references 34, 38, 36, 40, 41, 42.

This test allows compounds to be classified according to their absorption characteristics across the buccal membrane at various pH values, compounds which show the greatest change in buccal absorption with pH being those most sensitive to changes in urinary pH and renal reabsorption. Generally, compounds which are well absorbed at all pH values are well metabolised.[55]

The influence of certain ring substituents on the excretion and metabolism of amphetamine and N-ethylamphetamine is indicated in Table 3. Whilst exact comparison is not possible due to lack of pH control, certain trends emerge; thus the introduction of a lipophilic group tends to lower the urinary recovery of the unchanged drug, results which parallel the characteristics of these compounds in the buccal absorption test. When the compound is capable of undergoing N-dealkylation

TABLE 3. THE INFLUENCE OF RING AND N-SUBSTITUTION ON THE EXCRETION AND METABOLISM OF SOME 'AMPHETAMINES'

Drug	Recovery of drug		dealkylated metabolite		Reference
	Acid urine	Uncontrolled	Acid urine	Uncontrolled	
(+)-amphetamine	37–61	5–29	—	—	18
(+)-p-chloramphetamine	8–23	12	—	—	37
*p-methoxyamphetamine	—	0·3–15	—	—	44
(±)-m-trifluormethyl-amphetamine	24–38	—	—	—	45
*p-hydroxyamphetamine	—	40–52	—	—	30
(±)-N-ethylamphetamine	42–46	12–24	7–13	4–10	36
(±)-N-cyanoethyl-amphetamine	5–9	3	34–56	30–50	51
(±)-N-ethyl-m-trifluormethyl-amphetamine	20–30	5·3	12–23	3·6	45

Figures indicate range of percentage recovery found in urine under urinary conditions indicated after an oral dose.
* Optical isomer not stated.

as in the case of N-ethylamphetamine, then the effect of the ring substitution is to increase the importance of this route of metabolism. In the case of 4-hydroxy-amphetamine metabolism prior to conjugation is unnecessary and high levels of unchanged drug are excreted as conjugates. Recently, using ^{14}C-4-hydroxyamphetamine Sever et al.[46] have shown that up to 90 per cent of the drug is excreted as a sulphate conjugate in man.

In the case of some N-alkylamphetamines studied, only low recoveries of drug and metabolite have been found, which poses the problem as to their metabolic fate. Ring and β-hydroxylation routes have been recognised in man and the extent of these routes in various drugs is indicated in Table 4. N-Substitution with a relatively small group, e.g. methyl or chlorpropyl, greatly increases the amount of drug hydroxylated. In the case of drugs with larger substituents the amount of drug metabolised via this route seems to be small, although even in these cases some 4-hydroxyamphetamine is formed.

TABLE 4. THE INFLUENCE OF STRUCTURE ON THE EXCRETION OF RING AND β-HYDROXYLATED DERIVA-
TIVES OF SOME 'AMPHETAMINES' BY MAN

Drug	4-hydroxylation of		nor-ephedrine	p-hydroxy-norephedrine	References
	Drug	Amphetamine			
amphetamine	—	3–7	2–3	0·3–0·4	20, 28, 29
N-methylamphetamine	15	1	2–3	1–2	47
N-chlorpropyl-amphetamine	28–33	2–5	—	0·5–5	48
p-hydroxyamphetamine	—	—	—	3–9	30 46
N-(3, 3-diphenylpropyl)-amphetamine	?	+	+	+	49
Fenetylline†	—	5–9	—	—	50

Figures represent the range of percentage recovery from an oral dose.
+ indicates present but not quantified
† N-(aminoethyltheophylline)amphetamine.

It is of interest that β-hydroxylation of an N-alkylamphetamine has not been reported; thus whilst methylamphetamine is converted into both norephedrine and 4-hydroxynorephedrine, it is not converted to ephedrine.[47] This suggests that dealkylation is a prerequisite for this metabolic route.

The metabolic oxidation of nitrogen in organic molecules has recently attracted considerable attention[64] and N,N-dimethylamphetamine has been shown to be extensively converted to the N-oxide in man.[65] The extent to which this reaction occurs with amphetamine or its N-monosubstituted derivatives remains a matter for conjecture.

The precise mechanism of metabolic oxidation of 'amphetamines' is still unclear. Whilst one would have thought that ring hydroxylation was a typical microsomal reaction, results reported by Dingell and Bass[56] indicate that this reaction did not proceed in a microsomal system which was able to oxidise other substrates.

Similarly whilst the process of in vitro deamination has been known for a long time[57] controversy still exists regarding the mechanism by which this process is mediated.[58–60] Recently a general metabolic scheme for the metabolism of 'amphetamines' has been proposed[61] in which hydroxylation of α-carbon atoms of substituents leads to cleavage of the substituent. In the case of secondary 'amphetamines' this would lead to two separate 'dealkylated' products due to the presence of two α-carbon atoms in the original molecule. This phenomenon has been recognised in man in the case of the drug Fenetylline.[50]

The alternative N-oxidative pathway would lead to the formation of oximes[58-60] by oxidative breakdown of the hydroxylamine. Traces of phenylacetone oxime have been found to be formed in the urine of man receiving amphetamine[42] on treatment of the urine with alkali.

From the foregoing it can be seen that considerable advances have been made in our knowledge regarding the metabolism and excretion of 'amphetamines' in man. However, the gaps which remain must continue to act as a stimulus, as until these are filled the picture inevitably remains fragmentary and extrapolation of experience from one compound to another must be metered with caution.

REFERENCES

1. PARKE D. and FENTON C. W. (1973) *J. Neurol. Neurosurg and Psych.* In press.
2. BIEL J. H. (1970) In: *Amphetamines and Related Compounds.* (Ed. COSTA E. and GARATTINI S.) Publ. Raven Press. N.Y. pp. 3–19.
3. RICHTER D. (1938) *Biochem. J.* **32,** 1763–1769.
4. JACOBSEN E. and GAD I. (1940) *Arch Exp. Path. Pharmak.* **196,** 280–289.
5. BEZER K. H. and SKINNER J. T. (1940). *J. Pharmacol. and Exptl. Therap.* **68,** 419–432.
6. HARRIS S. C., SEARLE L. M. and IVY A. C. (1947) *J. Pharmacol. and Expt. Therap.* **89,** 92–96.
7. KELLER R. E. and ELLENBOGEN W. C. (1952) *J. Pharmacol. and Expt. Therap.* **106,** 77–82.
8. CAMPBELL J. A., NELSON E. and CHAPMAN D. G. (1959) *Canad. Med. Assn. J.* **81,** 15–20.
9. CHAPMAN D. G., SHENNOY K. G. and CAMPBELL J. A. (1959) *Canad. Med. Assn. J.* **81,** 470–477.
10. ALLES G. A. and WISEGARVER B. B. (1961) *Toxicol and Appl. Pharmacol.* **3,** 678–688.
11. FALES H. M. and PISANO J. J. (1962) *Anal. Biochem.* **3,** 337–342.
12. CARTONI G. P. and DE STEFANO F. (1963) *Ital. J. Biochem.* **12,** 296–309.
13. BECKETT A. H. and ROWLAND M. (1964) *J. Pharm. Pharmac.* **16,** 27T–31T.
14. BECKETT A. H. and ROWLAND M. (1965) *J. Pharm. Pharmac.* **17,** 60–428.
15. MILNE M. D., SCRIBNER B. H. and CRAWFORD M. A. (1958) *Amer. J. Med.* **24,** 709–729.
16. WEINER I. M. and MUDGE G. H. (1964) *Amer. J. Med.* **36,** 743–762.
17. BECKETT A. H. and ROWLAND M. (1964) *Nature* **204,** 1203–1204.
18. BECKETT A. H., ROWLAND M. and TURNER P. (1965) *Lancet.* **1,** 303.
19. ASATOOR A. M., GALMAN B. R., JOHNSON J. R. and MILNE M. D. (1965) *Brit. J. Pharmacol.* **24,** 293–300.
20. DAVIS J. M., KOPIN I. J., LEMBERGER L. L. and AXELROD J. (1971) *Ann. N.Y. Acad. Sci.* **179,** 493–501.
21. ROWLAND M. and BECKETT A. H. (1966) *Arzneimittel Forschung.* **16,** 1369–1373.
22. GORROD J. W. and BECKETT A. H. (1969) *Carworth-Europe Collected Papers.* **3,** 79–99.
23. WESLEY–HADZIYA B. (1969) *J. Pharm. Pharmac.* **21,** 196–197.
24. DRING L. G., SMITH R. L. and WILLIAMS R. T. (1966) *J. Pharm. Pharmac.* **18,** 402–495.
25. SMITH R., and DRING L. G. (1970) In: *Amphetamines and Related Compounds,* (ed. COSTA E. and GARATTINI S.), pp. 121–139. Raven Press, N.Y.
26. DRING L. G., SMITH R. L. and WILLIAMS R. T. (1970) *Biochem. J.* **116,** 425–435.
27. ELLISON T., GUTZAIT L. and VAN LOON E. J. (1965) *Fedn. Proc.* **24,** 688.
28. CALDWELL J., DRING L. G. and WILLIAMS R. T. (1972) *Biochem. J.* **129,** 23–24.
29. CAVANAUGH J. H., GRIFFITH J. D. and OATES J. A. (1970) *Clin. Pharmacol. Ther.* **11,** 656–664.
30. SJOERDSMA A. and VON STUDNITZ W. (1963) *Brit. J. Pharmacol.* **20,** 278–284.
31. BALTES B. J., ELLISON T., LEVY L. and OKUN R. (1966) *Pharmacologist* **8,** 220.
32. BECKETT A. H., SALMON J. A. and MITCHARD M. M. (1969) *J. Pharm. Pharmac.* **21,** 251–258.
33. ANGGARD E., JONSSON L. E. and HOGMARK A-L. (1973) *J. Clin. Pharmacol.* (in press).
34. BECKETT A. H. and ROWLAND M. (1965) *J. Pharm. Phamac.* **17,** 628–639.
35. GUNNE L-M. (1967) *Biochem. Pharmacol.* **16,** 863–869.
36. BECKETT A. H., BROOKES L. G. and SHENOY E. V. B. (1969) *J. Pharm. Pharmac.* **21,** 151S–156S.
37. SALMON J. A. (1971) Ph.D. Thesis. (Univ. of London).
38. BECKETT A. H. and ROWLAND M. (1965) *J. Pharm. Pharmac.* **17,** 109S–114S.
39. BECKETT A. H. and ROWLAND M. (1965) *Nature., Lond.* **206,** 1260–1261.
40. BECKETT A. H. and BROOKES L. G. (1970) In: *Amphetamines and Related Compounds,* (ed. COSTA E. and GARATTINI, S.) pp. 109–120. Raven Press, N.Y.

41. SHENOY E. V. B. (1971) Ph.D. Thesis. (Univ. of London).
42. VAN DYK J. M. (1972) Ph.D. Thesis. (Univ. of London).
43. BECKETT A. H. (1966) *Dansk. Tid. Farm.* **40,** 197–223.
44. SCHWEITZER J. W., FRIEDLOFF A. J., ANGRIST B. M. and GERSHONS S. (1971) *Nature, Lond.* **229,** 133–134.
45. BECKETT A. H., and BROOKES L. G. (1967) *J. Pharm. Pharmac.* **19,** 42S–49S.
46. SEVER P. S., DRING L. G. and WILLIAMS R. T. (1973) *Trans. Biochem. Soc.* (in press).
47. CALDWELL J., DRING L. G. and WILLIAMS R. T. (1972) *Biochem. J.* **129,** 11–22.
48. ZIEGLER W., LONG R. F. and BREIDERER H. (1968) *Abst. 2nd Intern. Symp. Pharm. Chem.* (Munster). 48.
49. PALM D., FENGLER H. and GROBECKER H. (1969) *Life Sci.* **8,** 247–257.
50. ELLISON J., LEVY L., BOLGER J. W. and OKUN R. (1970) *Europ. J. Pharmacol.* **13,** 123–128.
51. BECKETT A. H., SHENOY E. V. B. and SALMON J. A. (1972) *J. Pharm. Pharmac.* **24,** 194–202.
52. BECKETT A. H. and TRIGGS E. J. (1967) *J. Pharm. Pharmac.* **19,** 31S–41S.
53. BECKETT A. H., BOYES R. N. and TRIGGS E. J. (1968) *J. Pharm. Pharmac.* **20,** 92–97.
54. BECKETT A. H. and MOFFAT A. C. (1969) *J. Pharm. Pharmac.* **21,** 144S–150S.
55. BECKETT A. H. and BROOKES L. G. (1971) *J. Pharm. Pharmac.* **23,** 288–294.
56. DINGLE J. F. and BASS A. D. (1969) *Biochem. Pharmacol.* **18,** 1535.
57. AXELROD J. (1955) *J. Biol. Chem.* **214,** 753–763.
58. HUCKER H. B., MICHNIEWICZ B. M. and RHODES R. E. (1971) *Biochem. Pharmac.* **20,** 2123–2128.
59. PARLI C. J., WANG N. and MCMAHON R. E. (1971) *Biochem. biophys. Res. Commun.* **43,** 1204–1208.
60. BECKETT A. H., VAN DYK J. M., CHISSICK H. H. and GORROD J. W. (1971) *J. Pharm. Pharmac.* **23,** 560.
61. BECKETT A. H., VAN DYK J. M., CHISSICK H. H. and GORROD J. W. (1971) *J. Pharm. Pharmac.* **23,** 809–812.
62. FISCHER J. E., HORST W. D. and KOPIN I. J. (1965) *Brit. J. Pharmacol.* **24,** 477–484.
63. GOLDSTEIN M., MCKEREGHAN M. R. and LANBER E. (1964) *Biochim biophys. Acta* **89,** 191–193.
64. BRIDGES J. W., GORROD J. W. and PARKE D. V. (Eds) (1972) *The Biological Oxidation of Nitrogen in Organic Molecules.* Taylor and Francis.
65. BECKETT A. H., GORROD J. W. and WATSON C. L. unpublished observation.

Frontiers in Catecholamine Research 1973, pp. 951 to 956. Pergamon Press. Printed in Great Britain.

THE POSTSYNAPTIC EFFECT OF AMPHETAMINE ON STRIATAL DOPAMINE-SENSITIVE NEURONES

PAUL FELTZ* and JACQUES DE CHAMPLAIN

Department of Research in Anaesthesia, McGill University and Centre de Recherche en Sciences Neurologiques, Université de Montréal, Montreal, Canada

IT IS generally agreed that most of the behavioural effects of amphetamine involving the nigro-striatal dopaminergic pathway are related to a drug action at the presynaptic level (CARLSSON, 1970; UNGERSTEDT, 1971a; IVERSEN, 1971; TAYLOR and SNYDER, 1971; BOULU et al., 1972). Amphetamine is known to cause release of dopamine from the striatal nerve terminals of dopamine-containing neurones of the substantia nigra (MCKENZIE and SZERB, 1968; CARR and MOORE, 1970; BESSON et al., 1971). This same drug is also known to act on the inactivation of dopaminergic neurotransmission by inhibiting the reuptake of striatal dopamine (e.g. COYLE and SNYDER, 1969). However, these presynaptic effects of amphetamine, along with those ascribed to an inhibition of monoamine-oxidase (GLOWINSKI et al., 1966; RUTLEDGE et al., 1970) are not specific to dopaminergic pathways since noradrenergic systems are known to be similarly affected. Therefore many correlative behavioural and biochemical studies of amphetamine have been directed more towards an assessment of the relative involvement of dopamine and noradrenaline in a given behavioural pattern (CHRISTIE and CROW, 1971; COSTA et al., 1972; CREESE and IVERSEN, 1972) rather than towards an elucidation of some possible direct effect of amphetamine on striatal neurones receiving a dopaminergic innervation.

Our interest in possible postsynaptic effects of amphetamine arose from electrophysiological experiments on dopamine sensitive neurones in the cat caudate nucleus. Amphetamine was chosen as an index of complete pharmacological disruption of the nigro-striatal dopaminergic pathway since the biochemical evidence clearly indicates that its effects in vivo are essentially presynaptic (e.g. JAVOY et al., 1970; BUNNEY et al., 1972). However we found and report herein that amphetamine mimics the effect of dopamine on neurones in the caudate nucleus even after the dopaminergic nigro-striatal pathway has been destroyed or functionally impaired by two different pharmacological techniques.

The dopaminergic nigro-striatal system was substantially destroyed in 17 cats by treatment with a series of bilateral intraventricular injections of 6-hydroxydopamine (6-OH-DA) given in 7 increasing doses over a period of 5 days (total dose 8–10 mg). Electrophysiological experiments lasting 12–24 hr were performed 2–25 days later and the extent of destruction of the dopaminergic system was assessed subsequent to the acute experiments using biochemical and histochemical techniques (FELTZ and DE CHAMPLAIN, 1972a).

In a second group of 9 cats with intact dopaminergic pathways, the endogenous storage pool of dopamine was severely depleted by the injection of reserpine (5 mg/kg, i.p.) 20 hr prior to the acute experiment. These experiments lasted for about 10 hr.

---

* Present address: Laboratoire de Physiologie des Centres Nerveux 4, avenue Gordon-Bennett 75016 Paris (France).

During the acute experiments, 3 or 4 systemic injections of D-L-α-methyl-*p*-tyrosine methylester (α-M-*p*-T; 150–200 mg/kg) were given at intervals of 2–3 hr; this was assumed to inactivate the pool of newly synthetised dopamine.

A third group of 13 untreated cats served as controls.

The acute electrophysiological experiments were performed under general anaesthesia with diallylbarbituric acid (60 mg/kg) and urethane (240 mg/kg). The activity of caudate neurones was recorded with five-barreled micropipettes for simultaneous recordings and microiontophoretic injections of D-amphetamine (from 0·3 M solutions of the sulphate salt at pH 5·5–6·3, Sigma), L-amphetamine sulphate (0·3 M, pH 5·5–6, K and K Laboratories) and the following other substances: Na-glutamate and K-aspartate, as excitatory amino-acids; acetylcholine and γ-aminobutyric acid, for changing neuronal excitability; dopamine, noradrenaline and serotonine for full scanning of monoaminergic sensitivities; and papaverine and dimethoxyphenylethylamine (Gonzales Vegas, 1972) as possible antagonists of dopaminergic effects. Before pharmacological testing, neurones were characterised as either anti- or orthodromically excited on stimulation of the substantia nigra (Feltz and De Champlain, 1972a).

In the 13 control cats which had a mean endogenous dopamine content in the caudate nucleus of 11·6 ± 2·3 μg/g (SD), both isomers of amphetamine strongly depressed the neuronal firing evoked by excitatory amino-acids even when this firing was further enhanced by acetylcholine (Fig. 1A). This depressant action was seen on the 136 neurones tested but was effective in blocking the synaptically evoked responses of these neurones in 63 cases only. These results are consistent with those of previous experiments in which irrigation of the ventricular surface of the caudate nucleus with D-amphetamine (10^{-6} M solutions in Ringer) inhibited firing of caudate neurones (Feltz, 1970).

It was found that amphetamine retained full blocking potency in the cats treated with 6-OH-DA (Fig 1B). This was true for cells in areas of the caudate nucleus both partially and totally depleted of dopamine. At least 78 neurones tested for their sensitivity to amphetamine and dopamine were found to have been in the main zone of the caudate nucleus where the green fluorescence of dopamine-containing fibres was always shown to have totally disappeared (one third of the total volume of the nucleus). Another 157 neurones were recorded in sites obviously out of the region of maximal effectiveness of the 6-OH-DA treatment. These neurones were located within a neuronal network characterised by a marked reduction in the number and intensity of dopamine-containing terminals and by numerous swollen degenerating dopaminergic fibres. Total dopamine content in the caudate nucleus was reduced by between 42% and 87% (mean concentration and S.D. for the 17 cats: 4·2 ± 1·9 μg/g) with no systematic relation to the number of days of survival after the last 6-OH-DA injection. Although a substantial relation was found between biochemical estimates of dopamine depletion and the histochemical profile of the caudate nucleus from its ventricular surface down to its base, no clear correlation was found between the extent of 6-OH-DA induced damage and the doses of iontophoretic amphetamine (10–140 nA) required to block the firing. This was also true with respect to amphetamine induced blockage of nigro-caudate orthodromic and antidromic responses (26 cases out of 65 and 24 out of 32 respectively).

Although there was clearly a direct postsynaptic effect of amphetamine on the 78 caudate neurones studied in zones of total destruction of the dopaminergic terminals,

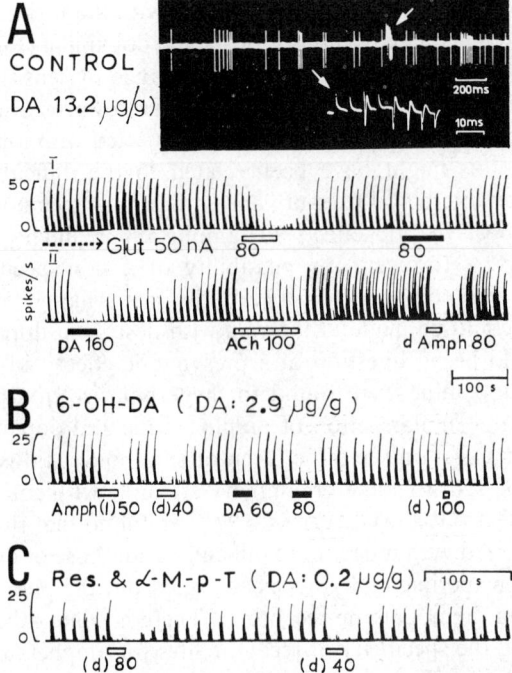

FIG. 1.—Responses to iontophoretically applied compounds of typical caudate neurones for the three groups of cats described in the text. Dopamine (DA) content in the caudate nucleus is given in each case. A: Normal cat. Photographic traces show excitatory response of caudate neuron to 4 shocks (arrows) delivered in substantia nigra. Penwriter recordings show periodic firing evoked by 50 nA of glutamate which was given in regular pulses throughout the recording period. I, responses to glutamate were blocked by D-amphetamine and dopamine (white and black bars respectively—numbers indicate iontophoretic current doses in nA; bars indicate periods of passage of current); II, enhanced firing following acetylcholine (Ach) is also depressed by amphetamine. B: 6-OH-DA treated cat; blocking effect seen with D- and L-amphetamine and dopamine. C: Reserpine and α-M-p-T treated cat; note prolonged effect of the highest dose of amphetamine. For comparisons of amphetamine and dopamine doses note that equal currents release at least 5 times more dopamine owing to different transport numbers.

at least one other explanation must be considered for the effects of amphetamine on the 157 neurones in zones of the caudate nucleus where some dopaminergic terminals remained. It is conceivable that the lack of change in the amphetamine induced depression of firing of these cells was the result of a release of dopamine from remaining undamaged terminals in conjunction with some form of real or apparent supersensitivity of postsynaptic neurones to extraneuronal dopamine (UNGERSTEDT, 1971b; FELTZ and DE CHAMPLAIN, 1972b). Although this implies a maintained significant contribution of presynaptic effects of amphetamine in spite of greatly reduced presynaptic innervation, it is difficult to imagine how it could account for the total absence of a physiological gradient of amphetamine sensitivity corresponding to the histochemical gradient observed within the caudate nucleus after treatment with 6-OH-DA. For this reason we suspected that amphetamine also has a major postsynaptic effect on cells in both the normal caudate nucleus and in that with a partially destroyed dopaminergic innervation.

The conclusion that amphetamine has a clear cut post-synaptic effect was further supported by the results obtained in the 9 cats with subtotal dopamine depletion (by 98 per cent) and inactivated synthesis of striatal dopamine (Fig. 1C). None of the 47 neurones tested in these cats showed a real change in sensitivity to either D- or L-amphetamine. In two cats we observed some trend towards a slight decrease in the number of neurones responding to amphetamine injected with iontophoretic currents below 80 nA, but this might have been related to the difficulties encountered in maintaining a reasonable blood pressure and satisfactory metabolic conditions.

The observed postsynaptic effects of amphetamine might well be specific to dopamine receptors. However, the possibility of a non-specific drug effect was investigated. The best comparison between the postsynaptic effects of dopamine and amphetamine could be made in the cats with destroyed dopaminergic terminals in which there could be no question of a presynaptic effect; in these animals both amphetamine and dopamine were found to 'hyperpolarize' the post-synaptic membrane (Fig. 2). If the similar action of amphetamine and dopamine can in fact be accounted for by an effect on the same receptor site, both actions should be blocked by the same antagonists. In some preliminary attempts with compounds free of any local anaesthetic effect (Gonzales Vegas, 1971) we found that the effects of amphetamine were antagonized with much more difficulty than those of dopamine. However the data from these experiments by no means ruled out the possibility that amphetamine was acting on the dopamine receptors. There is some other indirect evidence available concerning the specificity of receptor sites for amphetamine.

It is possible to identify, using physiological criteria, several different kinds of neurones in the caudate nucleus. We have already referred to the two groups of caudate neurones which are antidromically and orthodromically excited by nigral

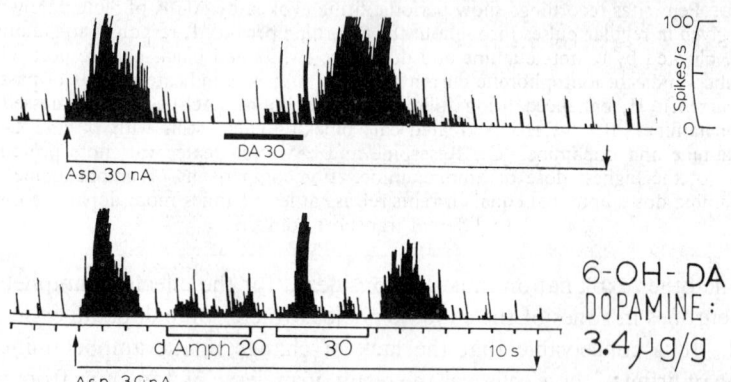

Fig. 2.—Evidence for a hyperpolarizing effect of both dopamine (DA) and amphetamine (Amph). The prolonged application of aspartate (between arrows) induced firing and then excessive depolarization manifested by the drop to zero of the firing frequency 30–40 s after beginning aspartate injection; firing was restored by either 30 nA of dopamine (above) or by 20 or 30 nA of amphetamine (below); note tendency of amphetamine to cause subsequent reduction of firing frequency when 30 nA application was maintained (antagonistic effect of depolarizing amino-acid and of potent hyperpolarizing action of amphetamine). In all tests, substantia nigra was stimulated by a train of 4 stimuli every 10 s to control further the amino-acid induced excessive depolarization characterised by a progressive decrease in spike amplitude and finally a disappearance of action potentials even in response to the nigral stimulus.

stimulation and inhibited by amphetamine; there is, for example, another group of caudate neurones, receiving a γ-aminobutyric mediated nigro-caudate inhibition without preceding excitation (FELTZ, 1972) which are similarly inhibited by amphetamine. The fact that these three and indeed other groups of caudate neurones are sensitive to amphetamine raises the question of the specificity of its action. However, it appears that the distribution of dopamine receptors (as shown in animals treated with 6-OH-DA) may be even wider within the caudate neuronal network than is reflected in estimates performed with an intact uptake mechanism (FELTZ and DE CHAMPLAIN, 1972b). Thus it appears not unlikely that the postsynaptic effects of amphetamine on striatal neurones reflect a direct interaction with dopaminergic receptor sites.

This raises the question of the appropriate interpretation of our observation that both D- and L-isomers proved to be equipotent. TAYLOR and SNYDER (1971) made a comparable observation on typical behaviours related to dopaminergic pathways (but see FERRIS et al., 1972). Our data might well suggest a large contribution of post-synaptic effects of amphetamine when a given behaviour is equally affected by both isomers. Furthermore, this could well be an important feature of dopaminergic as opposed to noradrenergic mechanisms.

Such an assumption could explain the discrepancy between our findings of a post-synaptic effect of amphetamine in a specific dopaminergic system like the striatum and the demonstration by BOAKES et al. (1972) that the action of amphetamine in the noradrenergic system of the brain stem was entirely pre- rather than postsynaptic (but see HOFFER et al., 1971). This may well be of some relevance for an understanding of some complex behavioural effects of amphetamine (COOLS, 1971) dependent on the assumption of a "dopamine-receptor-stimulating" drug action.

Acknowledgements—This work was supported in part by a Fellowship, awarded to P. FELTZ at McGill University from the Foundation's Fund for Research in Psychiatry and in part by a grant from the Medical Council of Canada to the MRC Group in Neurological Sciences at the Université de Montréal.
    The authors wish to acknowledge their debt to Dr. K. KRNJEVIC for advice and encouragement and to Miss L. FARLEY for her assistance in biochemical analysis.

REFERENCES

BESSON M. J., CHERAMY A., FELTZ P. and GLOWINSKI J. (1971) *Brain Res.* **32**, 407–424.
BOAKES R. J., BRADLEY P. B. and CANDY J. M. (1972) *Br. J. Pharmac.* **45**, 391–403.
BOULU R., RAPIN J. R., LEBAS M. and JACQUOT C. (1972) *Psychopharmacologia (Berl.)* **26**, 54–61.
BUNNEY B. S., AGHAJANIAN G. K., ROTH R. H. and WALTERS S. (1972) *Soc. for Neurosciences, Second Annual Meeting Abstracts*, p. 174.
CARLSSON A. (1970) In: *Amphetamine and related compounds* (COSTA, E. and GARATTINI, S., Eds.) pp. 289–300, Raven Press, New York.
CARR L. A. and MOORE K. E. (1970) *Biochem. Pharmac.* **19**, 2361–2374.
CHRISTIE J. E. and CROW T. J. (1971) *Br. J. Pharmac.* **43**, 658–667.
COOLS A. R. (1971) *Archs. Int. Pharmacodyn.* **194**, 259–269.
COSTA E., GROPPETTI A. and NAIMZADA M. K. (1972) *Br. J. Pharmac.* **44**, 742–751.
COYLE J. T. and SNYDER S. M. (1969) *J. Pharmac. exp. Ther.* **170**, 221–231.
CREESE I. and IVERSEN S. D. (1972) *Nature, New Biology* **238**, 247–248.
FELTZ P. (1970) *J. Physiol. (Paris)*, **62**, 374–375.
FELTZ P. (1972) *Proc. Can. Fed. Biol. Soc.* **15**, 420.
FELTZ P. and DE CHAMPLAIN J. (1972a) *Brain Res.* **43**, 595–600.
FELTZ P. and DE CHAMPLAIN J. (1972b) *Brain Res.* **43**, 601–605.
GONZALES-VEGAS J. A. (1971) *Brain Res.* **35**, 264–267.
HOFFER B. J., SIGGINS G. R. and BLOOM F. E. (1971) *Brain Res.* **25**, 523–534.
IVERSEN S. D. (1971) *Brain Res.* **31**, 295–311.

Javoy F., Hamon M. and Glowinski J. (1970) *Eur. J. Pharmacol.* **10**, 178–188.
McKenzie G. M. and Szerb J. C. (1968) *J. Pharmac. exp. Ther.* **162**, 302–308.
Rutledge C. O. (1970) *J. Pharmac. exp. Ther.* **171**, 188–195.
Taylor K. M. and Snyder S. (1971) *Brain Res.* **28**, 295–309.
Ungerstedt U. (1971a) *Acta Physiol. Scand. (suppl.)* **367**, 49–68.
Ungerstedt U. (1971b) *Acta Physiol. Scand. (suppl.)* **367**, 69–93.

Frontiers in Catecholamine Research 1973, pp. 957 to 962. Pergamon Press. Printed in Great Britain.

ELECTROPHYSIOLOGICAL EFFECTS OF AMPHETAMINE ON DOPAMINERGIC NEURONS

BENJAMIN S. BUNNEY and GEORGE K. AGHAJANIAN

Departments of Psychiatry and Pharmacology, Yale University School of Medicine, and the Connecticut Mental Health Center New Haven, Conn. 06508, U.S.A.

INTRODUCTION

D-AMPHETAMINE (d-AMP) has many interesting CNS effects, including the ability to induce paranoid psychosis in man and stereotyped behaviour in animals (SNYDER, 1972). Over the years a large body of indirect biochemical and behavioural evidence (WEISSMAN and KOE, 1965; ERNST, 1967; HANSON, 1967) has accumulated suggesting that d-AMP exerts its central action through its effect on the catecholamine (CA) systems of the brain. In low doses d-AMP has been shown to increase dopamine (DA) turnover and preferentially release newly synthesized DA from dopaminergic terminals in the neostriatum (BESSON et al., 1969a, b). d-AMP also blocks DA uptake (SNYDER and COYLE, 1969). CORRODI et al. (1967) have hypothesized that the release of CA by d-AMP results in increased postsynaptic CA concentrations, leading to an inhibition of catecholaminergic cell activity via a neuronal feedback circuit.

This paper presents direct evidence supporting this hypothesis based on extracellular recordings from single DA neurons in the substantia nigra zona compacta (A9) and adjacent ventral tegmental areas (A10). DA containing cells in these areas were first demonstrated anatomically by Dahlström and Fuxe using fluorescence histochemical techniques (DAHLSTRÖM and FUXE, 1964). Recently we have reported (BUNNEY et al., 1973b), based on combined neurophysiological and fluorescence histochemical methods, that the firing pattern of DA neurons is distinctive and easily recognizable. This identification of DA neurons on the basis of histochemical and neurophysiological characteristics has made it possible to determine the effect of drugs on the firing rate of these units.

EFFECT OF AMPHETAMINE AND ANTAGONISTS ON ACTIVITY OF DOPAMINERGIC NEURONS

In previous single unit recording studies (BUNNEY et al., 1973b), using albino rats (Charles River) we have shown that intravenously administered d-AMP stops or markedly slows A9 and A10 cell activity in low doses (Fig. 1A). This depressant action was highly specific in that non-DA cells in the midbrain did not decrease their rate. In doses as low as 0·25 mg/kg d-AMP stopped 20 per cent of DA cells. The mean dose of d-AMP for inhibition of DA cells to 50 per cent of baseline rate was 1·6 mg/kg (± 0·35 S.E.M.). All DA cells ceased firing by the time a dose of 6·4 mg/kg had been reached. This sensitivity of DA cells to d-AMP is in marked contrast to their response to l-AMP. Recently (BUNNEY, KUHAR and AGHAJANIAN, unpublished data) we have found that over 50 per cent of DA cells are relatively unresponsive to l-AMP in that even when intravenous doses as high as 25 mg/kg are given 50 per cent inhibition is not achieved. For those DA units that are responsive it appears to take 6–10 times as much l-AMP as d-AMP to produce 50 per cent inhibition.

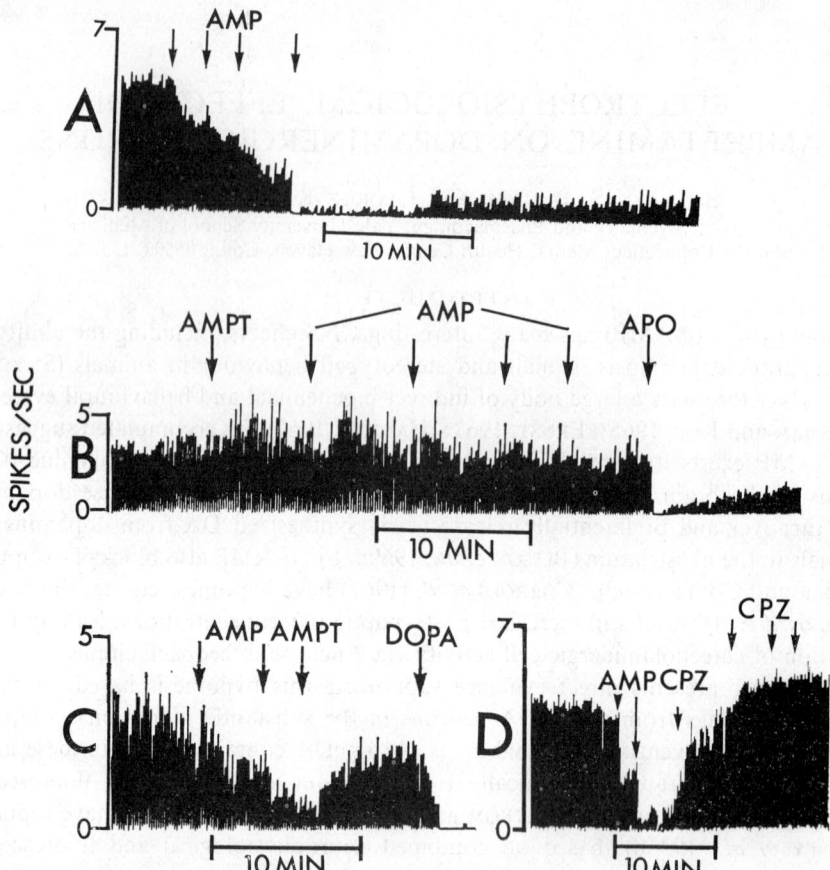

Fig. 1.—A. Typical effect of d-amphetamine (d-AMP) on the firing rate of a dopaminergic cell. Serial injections of d-AMP (0·25, 0·25, 0·5, 1·0 mg/kg) progressively depressed unit activity until firing ceased. Recovery was slow—25 per cent in ½ hr.

B. Prevention by α-methyl-p-tyrosine (AMPT) of d-AMP induced slowing of dopaminergic cell activity. AMPT (50 mg/kg) completely prevented any significant decrease in firing rate in ZC and VT cells. In the example given, d-AMP was administered in total dose of 3·5 mg/kg (0·50 mg/kg followed by 1·0 mg/kg × 3). AMPT, however, had no effect on the usual depressant response of these cells to apomorphine (APO)—0·10 mg/kg (Bunney et al., 1973a).

C. AMPT reversal of d-AMP induced slowing of dopaminergic cell activity with a subsequent decrease and cessation of activity after administration of l-dihydroxyphenylalanine (DOPA). AMP in consecutive doses of 0·5 mg/kg each, slowed this DA cell markedly. AMPT (50 mg/kg) rapidly reversed the d-AMP effect in this and all other DA cells tested. Subsequent l-DOPA (50 mg/kg) overcame the blockade of d-AMP depression by AMPT and inhibited the cell (Bunney et al., 1973a).

D. Reversal of d-AMP induced depression of a dopaminergic cell by chlorpromazine (CPZ). Two injections of d-AMP (0·25, 0·50 mg/kg) resulted in a 75 per cent decrease in the activity of a DA containing cell. CPZ (0·25 mg/kg) rapidly reversed the d-AMP induced depression and returned DA unit activity almost to baseline rate. CPZ (0·5 mg/kg) increased firing rate beyond baseline rate. Two further doses of CPZ (0·5 mg/kg each) had little additional effect on unit activity.

All drugs were administered intravenously. Methods as described in Bunney et al. (1973a, b).

Antipsychotic phenothiazines and haloperidol (drugs thought to be DA receptor blockers) were shown to reverse (Fig. 1D) and block the depressant effect of d-AMP (BUNNEY et al., 1973b). In addition, we found that blockade of DA synthesis by the intraperitoneal administration of 50 mg/kg of the tyrosine hydroxylase inhibitor, α-methyl-p-tyrosine (AMPT) (SPECTOR et al., 1965) prevented and reversed the slowing of DA cells induced by intravenous d-AMP (BUNNEY et al., 1973a) (Fig. 1B, 1C). When the blockade of DA synthesis by AMPT was bypassed by the intravenous administration of the immediate precursor of DA*l* dihydroxyphenylalanine (*l*-dopa), the cell firing rate again decreased (Fig. 1C). Taken together, these results suggest that ongoing DA synthesis is necessary for d-AMP to exert its effect and that DA receptors may be involved in d-AMP's ability to depress DA neurons. Since AMPT prevented the d-AMP depression of DA unit activity a significant *direct* action on postsynaptic neurons by d-AMP seems unlikely. How then is d-AMP affecting the firing of DA cells? Neuronal feedback inhibition remains a possibility. Although as yet unidentified anatomically, two forms of this pathway are conceivable: a postsynaptic striatal-nigral pathway or a direct recurrent collateral system consisting of dopaminergic terminals with synapses on DA cell bodies. In the latter case iontophoresis of d-AMP into the vicinity of the DA cell body should cause a depression of firing rate.

INHIBITION OF DOPAMINERGIC NEURONS BY AMPHETAMINE: EVIDENCE FOR A NEURONAL FEEDBACK PATHWAY

To test this hypothesis d-AMP was iontophoresed onto DA cells in the midbrain of rats through multibarreled micropipettes. d-AMP was found either to have no effect on DA unit activity (Fig. 2A) or to cause a minimal slowing the first time ejected. This effect did not increase significantly with increasing ejection currents (Fig. 2B). Multiple application rapidly led to an attenuation of the depressant response (Fig. 2B). However, d-AMP administered intravenously in a dose of 0·5 mg/kg had its usual marked inhibitory effect (Fig. 2A). In many cases d-AMP produced a local anesthetic action as indicated by a decreasing spike amplitude in association with DA cell slowing. DA iontophoresed onto DA cells had a greater depressant effect than d-AMP but demonstrated "tachyphylaxis" to the same degree (AGHAJANIAN and BUNNEY, this volume). No local anesthetic effect was observed with DA. These iontophoretic results suggest that a DA receptor may be present on the dendrites and/or cell bodies of DA neurons. However, it is doubtful that the depressant effect seen with iontophoretic d-AMP has any physiological significance as the response was small even at concentrations many times that which would reach these cells when small doses of d-AMP are administered intravenously. In addition, the attenuation of response seen after repeated iontophoretic administrations of d-AMP is never seen when d-AMP is given systemically, suggesting further that the small direct effect of iontophoretically applied d-AMP on DA cell activity has no physiological significance.

In contrast to the minimal response of DA cells to iontophoretic d-AMP we found postsynaptic cells in the caudate nucleus, accumbens nucleus and olfactory tubercles that were markedly depressed by low ejection currents of d-AMP and DA (Fig. 2C). No attenuation of response could be demonstrated. In cells responsive to both DA and d-AMP local anesthetic effects of d-AMP were seen only at higher ejection currents (> 20 nA). Interestingly, adenosine 3', 5'-monophosphate applied iontophoretically also markedly depressed many of these cells (Fig. 2C).

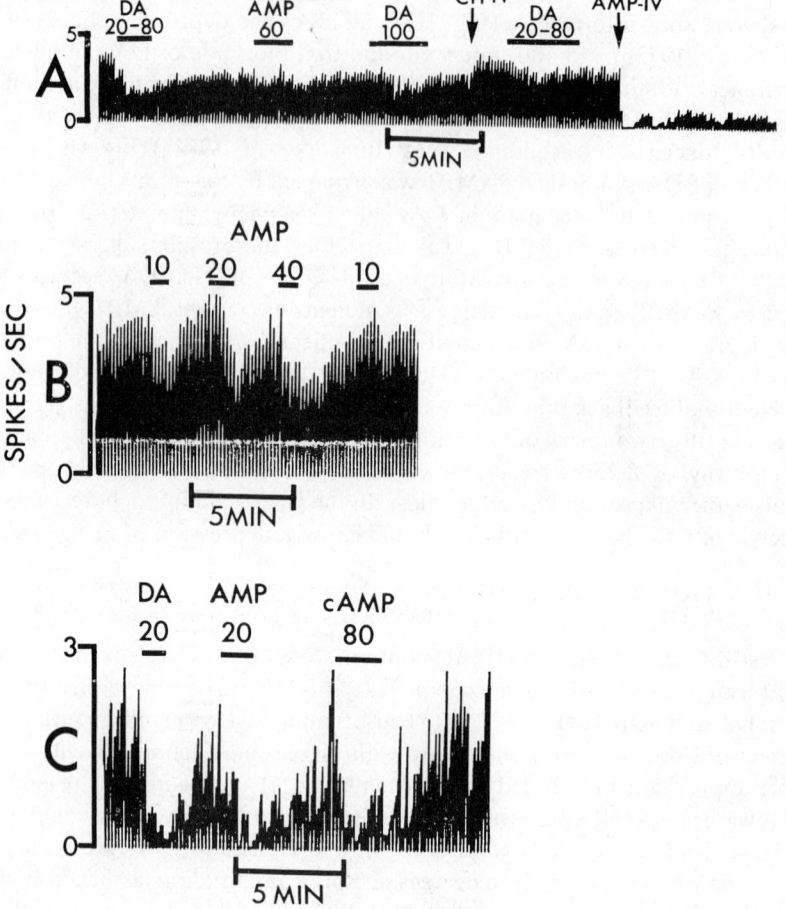

FIG. 2.—A. Effects of microiontophoretic and intravenous *d*-AMP on the firing rate of a dopaminergic neuron. Using a five barrel micropipette DA (0·2 M, pH 4) and *d*-AMP (0·2 M, pH 4) were ejected onto this DA cell in the substantia nigra zona compacta ($n = 16$). DA ejected with currents ranging from 20 to 80 nA had only a minimal depressant effect. Subsequently when applied to the same cell DA had less and less of an effect. *d*-AMP (60 nA) had no significant effect on this cell's firing rate. However, when given intravenously *d*-AMP (0·5 mg/kg) produced the usual marked inhibition of neuronal activity. Also shown in this figure is the typical increase in baseline rate of DA neurons seen after intravenous chloral hydrate (CH-I.V.) (40 mg/kg).

B. Effect of serial applications of microiontophoretically ejected *d*-AMP on the firing rate of a dopaminergic neuron. *d*-AMP ejected with a 10 nA current produced a moderate depressant effect on cell activity. Subsequent ejections of *d*-AMP at higher currents (20, 40 nA) failed to significantly further depress cell activity. A repeat application of *d*-AMP ejected at 10 nA produced no change in firing rate. This figure demonstrates the typical diminution in DA cell response seen after repeated iontophoretic applications of *d*-AMP.

C. Effect of DA, *d*-AMP and adenosine 3′, 5′-monophosphate (cAMP) on the firing rate of a cell in the accumbens nucleus. DA and *d*-AMP, in contrast to their effect on DA cells (see A), caused this postsynaptic cell to markedly decrease in rate. Typical of the postsynaptic cells tested this cell showed no diminution of response after repeated ejections of DA and *d*-AMP (not shown in this figure; $n = 20$). cAMP (0·2 M, pH 8) also depressed the firing rate of this cell.

Each tracing shows the activity of a single cell. Each vertical line represents the integrated rate of firing expressed as spikes/sec. The duration of each microiontophoretic drug application is depicted by length of bar above tracing. The number associated with each bar indicates the intensity of the ejection current (in nA). Methods as described in HAIGLER and AGHAJANIAN (1973).

Thus it would appear from the relative lack of response of DA cells to microion-tophoretically ejected d-AMP that recurrent collaterals are not involved in the marked slowing of these cells by low doses of systemically administered d-AMP. However, these results do not provide *positive* proof for a postsynaptic non-dopaminergic feedback pathway. If such a pathway exists one should be able to abolish the in-hibitory effect of systemically administered d-AMP on DA cells by destroying the pathway with a lesion. Accordingly, under chloral hydrate anesthesia, using a stereotoxically placed fine retractable knife the rat brain was transected at a level just in front of the most anterior edge of the substantia nigra zona compact (A2970-A3290μ, KÖNIG and KLIPPEL, 1970) ($n = 8$). Following this lesion almost all DA cells were firing at abnormally high rates ($\sim$10/sec). Intravenous administration of d-AMP in doses up to 14 mg/kg either had no effect or caused only a slight decrease in activity below baseline rate. Thus doses of d-AMP twice that needed to stop all DA cells in unlesioned animals produced only a minimal slowing.

DISCUSSION

Our results show that d-AMP in small systemic doses has a marked depressant effect on the activity of dopaminergic neurons. However, DA cells appear to be rela-tively insensitive to l-AMP. This finding directly parallels the report by COSTA *et al.* (1972) that 0·3 mg/kg of d-AMP increases striatal DA turnover while 1·0 mg/kg of l-AMP has no effect. Our results are also in accord with the recent report that d-AMP administered intravenously is 3 to 4 times more potent than l-AMP in increasing striatal ^{3}H–DA release (VON VOIGTLANDER and MOORE, 1973). On the other hand, in studies of DA uptake in synaptosomes little difference was found between the blocking effects of d- and l-AMP (COYLE and SNYDER, 1969). It is not clear how the latter finding relates to our own due to the difference in the preparation used.

Inhibition of DA synthesis by AMPT blocked the slowing of DA cells induced by d-AMP. Until recently this finding would have been interpreted as conclusive proof that ongoing synthesis is necessary for d-AMP to exert its effect. However, ENNA *et al.* (1973) have reported that in addition to inhibition of tyrosine hydroxylase AMPT inhibits d-AMP induced efflux of exogenous ^{3}H-norepinephrine (^{3}H–NE) from brain slices. They suggest that AMPT interferes with movement of NE to an AMP releas-able site. We do not know yet whether AMPT has a similar action in the DA system or whether this action is present *in vivo* as well as *in vitro*. In addition, the concentra-tion of d-AMP used by Enna *et al.* is higher than could have been achieved intracere-brally when low doses of d-AMP are administered intravenously. A hypothesis which might explain these various results is that release of NE by d-AMP is synthesis coupled, i.e. NE release may be dependent upon the availability of newly synthesized NE to replace it. It this were true then synthesis inhibition by AMPT would lead to the decreased efflux of ^{3}H–NE from brain slices reported by Enna *et al.* Final resolution of this question, however, will have to await further studies.

The discovery of this new action of AMPT calls into question the validity of using AMPT inhibition of the depressant effect of d-AMP on DA cell activity as the only evidence that ongoing DA synthesis is necessary for d-AMP to exert its action. We have therefore inhibited DA synthesis at the next step in its synthetic pathway and again examined the effect of d-AMP on DA cell activity. N^1-(DL-seryl)-N^2-(2,3,4-trihydroxybenzyl) hydrazine (R04-4602) completely blocked the depressant effect of d-AMP on these cells when given intraperitoneally in a dose of 800 mg/kg. At this high dose RO4-4602 has been shown to be an effective dopa-decarboxylase inhibitor in the CNS (WEISS *et al.*, 1972).

SUMMARY

(1) d-AMP administered intravenously in low doses markedly depresses the firing rate of dopaminergic neurons. This effect of d-AMP appears dependent upon the presence of ongoing DA synthesis as it is abolished by AMPT (a tyrosine hydroxylase inhibitor) and by high doses of RO4-4602 (a decarboxylase inhibitor).

(2) We have confirmed the previously proposed hypothesis (CORRODI *et al.*, 1967) that d-AMP depresses DA cell firing rate indirectly via a postsynaptic non-dopaminergic feedback pathway. This conclusion is based on the following findings:

(a) As predicted, d-AMP selectively depresses the firing rate of dopaminergic neurons in the substantia nigra zona compacta (A9) and adjacent ventral tegmental (A10).

(b) The direct effect of d-AMP on DA cell activity when applied iontophoretically, even with high ejection currents, is weak and transient as compared to its marked depressant effect when applied with low currents in the vicinity of postsynaptic neurons (i.e. neurons receiving a dopaminergic input).

(c) Transection between the substantia nigra and striatum abolishes the depressant effect on DA cell firing rate of systemically administered d-AMP.

The latter result strongly suggests the existence of a neuronal feedback pathway although its precise anatomical location has yet to be determined.

Acknowledgement—This research was supported by NIMH Grant MH-17871, USPHS Research Scientist Development Award MH-14459 (to G.K.A.), and the State of Connecticut.

REFERENCES

BESSON M. J., CHERAMY A., FELTZ P. and GLOWINSKI J. (1969a). *Proc. Nat. Acad. Sci.* **62**, 741–748.
BESSON M. J., CHERAMY A. and GLOWINSKI J. (1969b). *Europ. J. Pharmacol.* **7**, 111–114.
BUNNEY B. S., AGHAJANIAN G. K. and ROTH R. H. (1973a). *Nature*, in press.
BUNNEY B. S., WALTERS J. R., ROTH R. H. and AGHAJANIAN G. K. (1973b). *J. Pharmacol. Exp. Ther.*, 185, 560–571.
CORRODI H., FUXE K. and HÖKFELT T. (1967). *Europ. J. Pharmacol.* **1**, 363–368.
COSTA E., GROPPETTI A. and NAIMZADA (1972). *Brit. J. Pharmacol.* **44**, 742–751.
DAHLSTRÖM A. and FUXE K. (1964). *Acta Physiol. Scand.* **62**, Suppl. 232, 1–55.
ENNA S. J., DORRIS R. L. and SHORE P. A. (1973). *J. Pharmacol. Exp. Ther.* **184**, 576–582.
ERNST A. M. (1967). *Psychopharmacologia* **10**, 316–323.
HAIGLER H. J. and AGHAJANIAN G. K. (1973). *Europ. J. Pharmacol.* **21**, 53–60.
HANSEN L. C. F. (1967). *Psychopharmacologia* **10**, 289–297.
KÖNIG J. F. R. and KLIPPEL R. A. (1970). *The Rat Brain: A Stereotaxic Atlas.* Robert E. Krieger.
SNYDER S. H. (1972). *Archs Gen. Psychiat.* **27**, 169–179.
SNYDER S. H. and COYLE J. T. (1969). *J. Pharmacol. Exp. Ther.* **165**, 78–86.
SPECTOR S., SJOERDSMA A. and UDENFRIEND S. (1965). *J. Pharmacol. Exp. Ther.* **147**, 86–95.
VON VOIGTLANDER P. F. and MOORE K. E. (1973). *J. Pharmacol. Exp. Ther.* **184**, 542–552.
WEISS B. F., MUNROE L. A., ORDONEZ R. J. and WURTMAN R. J. (1972). *Science* **177**, 613–616.
WEISSMAN A. and KOE B. K. (1965). *Life Sci.* **4**, 1037–1048.

Frontiers in Catecholamine Research 1973, pp. 963 to 968. Pergamon Press. Printed in Great Britain.

AMPHETAMINE AND NORADRENERGIC REWARD PATHWAYS

LARRY STEIN and C. DAVID WISE

Wyeth Laboratories, Philadelphia, Pa., U.S.A.

IN HIGHER animals, brain mechanisms have evolved for the selective facilitation or reinforcement of successful responses (i.e., behaviour that leads to reward or to the avoidance of punishment). Analysis of these central reward mechanisms has been accelerated by studies of amphetamine. At low doses, this agent causes a remarkable facilitation of all reinforced responses (STEIN, 1964a). The facilitation does not depend on the nature of the goal that motivates the behaviour: amphetamine facilitates both positively-reinforced behaviours maintained by reward and negatively-reinforced behaviours maintained by the avoidance of punishment. Furthermore, if small doses of amphetamine are given regularly after a response, as in self-administration experiments in animals, the response that delivers the drug is strongly reinforced (PICKENS and HARRIS, 1968); a similar demonstration of the powerful rewarding action of amphetamine in man is regularly provided by the amphetamine addict. These behavioural observations suggest that amphetamine acts on, or in intimate relation to, the reward system in the brain. Understanding of the biochemical nature of the facilitatory amphetamine response could therefore be expected to increase understanding of the biochemical nature of the reward system.

AMPHETAMINE AND BRAIN NOREPINEPHRINE

In early studies, it was found that the facilitating action of amphetamine on behaviour was indirect and probably dependent on the release of a catecholamine (STEIN, 1964b). In these experiments, the self-stimulation method of OLDS and MILNER (1954) was used to measure the behaviour enhancing effect of amphetamine. Depletion of catecholamine stores by reserpine decreased the facilitating action of amphetamine and preservation of catecholamines by monoamine oxidase inhibitors increased the facilitating action of amphetamine. Furthermore, in rats pretreated with monoamine oxidase inhibitors, a powerful enhancement of self-stimulation was obtained with phenethylamine, the chemical structure common to both amphetamine and the catecholamines. Subsequently, by the use of different behavioural methods and more specific inhibitors of catecholamine synthesis, other workers confirmed the conclusion that facilitatory action of amphetamine is mediated by a catecholamine (WEISSMAN, KOE and TENEN, 1966; HANSON, 1967; RECH and STOLK, 1970).

More recent work suggests that the relevant catecholamine is norepinephrine (WISE and STEIN, 1970). Selective blockade of norepinephrine biosynthesis by inhibition of dopamine-β-hydroxylase (E.C. 1.14.2.1), the enzyme responsible for the conversion of dopamine to norepinephrine, eliminated the facilitating action of amphetamine. Intraventricular infusion of norepinephrine after dopamine-β-hydroxylase inhibition reinstated the action of amphetamine. The noradrenergic receptor involved in behavioural reinforcement appears to be of the α-type (WISE, BERGER and STEIN, 1973). Intraventricular administration of the α-noradrenergic antagonist phentolamine, but not the β-antagonist propranolol, reduced the rate of self-stimulation and blocked the facilitatory effect of amphetamine. In other experiments, rewarding brain stimulation or moderate doses of amphetamine administered to freely-moving rats with permanently-indwelling cannulas released norepinephrine and its metabolites into brain perfusates (Stein and Wise, 1969). Both treatments caused shifts in the pattern of metabolites towards O-methylated products (see also, Glowinski and Axelrod, 1965).

These pharmacological observations fit nicely with the results of self-stimulation mapping studies on the one hand (Olds, 1962) and histochemical maps of norepinephrine pathways on the other (Fuxe, 1965; Fuxe, Hökfelt and Ungerstedt, 1968). Self stimulation sites and noradrenergic areas overlap to a surprising degree: the most intensely rewarding points in the brain fall precisely along noradrenergic fibre bundles and in rich noradrenergic terminal areas (Stein, 1968). Although most noradrenergic cell groups have not yet been systematically studied, high rates of self-stimulation are obtained in the almost exclusively noradrenergic cell concentrations of the locus coeruleus (Crow, Spear and Arbuthnott, 1972; Ritter and Stein, 1972). Self-stimulation of this region is quite sensitive to facilitation by amphetamine and suppression by chlorpromazine, but it is largely unaffected by the dopamine-antagonist pimozide (Ritter and Stein, in press). These pharmacological observations support the inference from two independent mapping studies that self-stimulation can be localised in a relatively homogeneous noradrenergic site (the locus coeruleus). Contrary to the view that noradrenergic neurons merely activate behaviour by a nonspecific increase in arousal (Roll, 1970; Antelman, Lippa and Fisher, 1972), this demonstration of 'pure' noradrenergic self-stimulation is presumptive evidence that at least some noradrenergic neurons specifically mediate rewarding effects.

AMPHETAMINE AND BRAIN DOPAMINE

While it is evident that norepinephrine is involved in the behavioural-facilitating action of amphetamine, the role of dopamine is still unclear. At high doses, amphetamine induces a stereotyped behaviour pattern that appears to be mediated by the release of dopamine (Randrup and Munkvad, 1967, 1970); however, at these high doses of amphetamine, goal-directed behaviour is usually suppressed rather than facilitated (Fig. 1). Similarly, activation of central dopamine receptors by apomorphine over

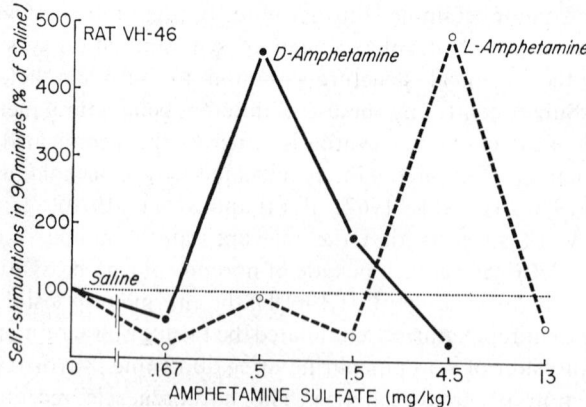

Fig. 1.—Facilitation and suppression of internal capsule self-stimulation by various doses of d- and l-amphetamine. Note 9-fold difference between the two isomers in potency for peak facilitation. Rat VH-46.

a wide dose range suppresses self-stimulation and other rewarded behaviours (DE OLIVEIRA and GRAEFF, 1972). It is conceivable that dopamine-mediated stereotyped behaviour can compete with and disrupt norepinephrine-mediated goal-directed behaviour. This idea implies that it might be possible to reverse the suppressive effects of dopaminergic agents on goal-directed behaviour by administration of noradrenergic agents. Such seems to be the case. The suppressive effects of apomorphine on self-stimulation are reversed by d-amphetamine, but only poorly by l-amphetamine (KOJIMA, RITTER, WISE and STEIN, in preparation) (Fig. 2). The different activities of the two isomers is consistent with the suggestion that d-amphetamine is many times more potent than l-amphetamine as a potentiator of central noradrenergic activity (TAYLOR and SNYDER, 1971).

The effects of norepinephrine and dopamine on self-stimulation were directly examined by injecting these agents in the lateral ventricle via permanently-indwelling cannulas. l-Norepinephrine facilitated medial forebrain bundle self-stimulation over a wide range of doses (WISE, BERGER and STEIN, 1973) (Fig. 3). Similar doses of dopamine are much less effective or may even suppress self-stimulation. The mild facilitating effects of dopamine are sometimes observed after a delay of several minutes, and thus may reflect conversion of the dopamine to norepinephrine. If this step is

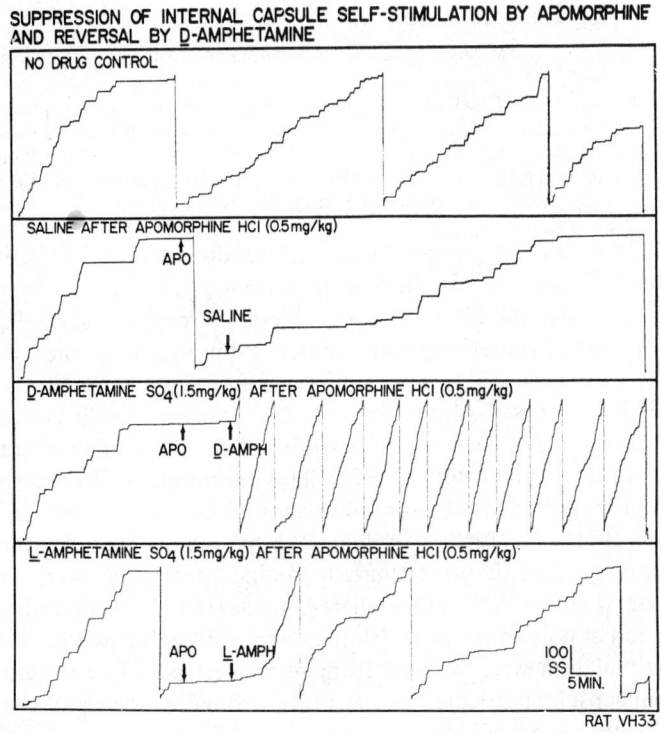

FIG. 2.—Suppression of self-stimulation by apomorphine (APO) and reversal by d-amphetamine. The same dose of l-amphetamine has only a small effect. Pen cumulates self-stimulation responses (SS) over time and resets automatically after 500 responses (see Key). Intraperitoneal injections marked by arrows. Rat VH-33: internal capsule.

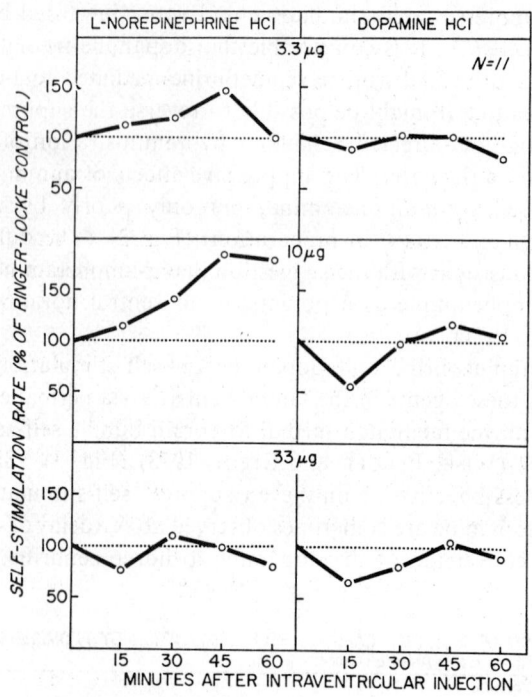

FIG. 3.—Facilitation of self-stimulation by norepinephrine and suppression by dopamine. The catecholamines were dissolved in 10 μl of Ringer–Locke solution and injected in the lateral ventricle one-half hour after the start of the 90-min test. Averaged data of 11 rats with electrodes in the medial forebrain bundle (level of posterior hypothalamus).

blocked by inhibitors of dopamine-β-hydroxylase, dopamine is ineffective (WISE and STEIN, 1969). In contrast, the facilitating action of exogenous norepinephrine is especially evident after inhibition of dopamine-β-hydroxylase, since depletion of the endogenous stores of norepinephrine causes suppression of the self-stimulation baseline.

Recent findings of self-stimulation in the dopamine cell concentrations of the substantia nigra have been taken as evidence of a dopaminergic reward system (CROW, 1972; PHILLIPS and FIBIGER, 1973). This interpretation is rendered hazardous, however, by the presence of noradrenergic fibres of passage in this region (UNGERSTEDT, 1971). To further evaluate the hypothesis that dopamine mediates rewarding effects, rostally-projecting dopamine pathways were mapped for self-stimulation at points in the brain where UNGERSTEDT (1971) describes a somewhat better separation of noradrenergic and dopaminergic systems than that in the substantia nigra. Self-stimulation was obtained from some electrodes in the dopamine tracts of the internal capsule, particularly from sites surrounding the tip of the crus cerebri, but maximal rates were only about 20 per cent of the maximal rates obtained from medial forebrain bundle electrodes (KOJIMA, RITTER, WISE and STEIN, in preparation). The most reinforcing internal capsule placements were located either in medial sites that border on the noradrenergic fibre system in the medial forebrain bundle, or in ventrolateral sites just dorsal to the noradrenergic projection into the amygdala.

Internal capsule self-stimulation was facilitated by *d-* and *l-*amphetamine, but *d-*amphetamine was about 9 times more potent (Fig. 1). As already noted, such a result has been taken to reflect a noradrenergically-mediated behavioural process (TAYLOR and SNYDER, 1971).

Although the above observations may be interpreted in a number of ways, the overall pattern of results does not support the idea that dopamine is an important transmitter in the reward system. Indeed, while we cannot at this point rule out the possibility that some dopamine pathways may facilitate behaviour, our evidence generally favours the idea that at least some dopamine systems can suppress goal-directed behaviour.

AMPHETAMINE AND BEHAVIOURAL PUNISHMENT: POSSIBLE ROLE OF BRAIN SEROTONIN

Like other higher functions, the regulation of goal-directed behaviour appears to be mediated by antagonistic mechanisms — the behaviourally-facilitatory reward system mentioned above, and a behaviourally-suppressant punishment system (MAGOUN, 1958). The reward and punishment mechanisms seem to be neurochemically distinct. As already suggested, the reward system may be mainly noradrenergic, whereas the punishment system appears to contain both cholinergic and serotonergic components (MARGULES and STEIN, 1967, 1969; WISE, BERGER and STEIN, 1970, 1973; STEIN, WISE and BERGER, 1973).

Amphetamine has long been known to intensify the suppressive effects of punishment on goal-directed behaviour (GELLER and SEIFTER, 1960; KELLEHER and MORSE, 1964). This effect seems paradoxical, since amphetamine generally facilitates goal-directed behaviour. Furthermore, because intraventricular administration of norepinephrine markedly reduces the suppressant effects of punishment (WISE, BERGER and STEIN, 1973), it might be expected that a strong potentiator of central noradrenergic activity like amphetamine would lessen rather than increase punishment effects. In attempts to resolve this paradox, the possible involvement of other monoamine transmitters should be considered. Recent evidence suggests that the suppressant effects of punishment may be mediated at least in part by serotonergic neurons (WISE, BERGER and STEIN, 1970, 1973). Since amphetamine releases serotonin as well as catecholamines from central neurons (FUXE and UNGERSTEDT, 1970; RUTLEDGE, AZZARO and ZIANCE, 1972), it is logical to speculate that the punishment-enhancing effects of amphetamine may be mediated via the release of serotonin (STEIN, WISE and BERGER, 1973). In the amphetamine addict, enhancement of the effects of punishment could lead to apprehension and suspiciousness, and eventually to paranoia.

Whether in any given case amphetamine predominantly facilitates the release of norepinephrine, or whether it mainly facilitates the release of serotonin, probably depends to a large extent on the nature of the test situation. Norepinephrine release may be predominantly facilitated in tests that depend on the activation of the reward system (e.g., self-stimulation), whereas serotonin release may be selectively facilitated in tests that depend on the activation of the punishment system (e.g., conflict test of GELLER and SEIFTER, 1960).

In summary, amphetamine enhances both the behaviourally-facilitatory effects of reward and the behaviourally-suppressant effects of punishment; at high doses, a species-typical pattern of stereotyped behaviour is observed. All of these effects of amphetamine may be mediated via the release of different brain monoamines; reward (and addiction) by norepinephrine, punishment (and paranoia) by serotonin, and stereotyped behaviour by dopamine.

Acknowledgements—We thank Drs. Sue Ritter and Hideki Kojima for stimulating discussions, and A. T. Shropshire and W. C. Carmint for expert technical assistance.

REFERENCES

ANTELMAN S. M., LIPPA A. S. and FISHER A. (1972) *Science* **175,** 919–920.
CROW T. J. (1972) *Brain Res.* **36,** 265–274.
CROW T. J., SPEAR P. J. and ARBUTHNOTT G. W. (1972) *Brain Res.* **36,** 275–288.
DE OLIVEIRA L. and GRAEFF F. G. (1972) *Europ. J. Pharmacol.* **18,** 159–165.
FUXE K. (1965) *Acta Physiol. Scand.* (Suppl. 247) **64,** 37–102.
FUXE K., HÖKFELT T. and UNGERSTEDT U. (1968) In: *Metabolism of Amines in the Brain* (HOOPER G., Ed.) pp. 10–22. Macmillan, London.
FUXE K. and UNGERSTEDT U. (1970) In: *Amphetamines and Related Compounds* (COSTA E. and GARATTINI S. Eds.) pp. 257–288. Raven Press, New York
GELLER I. and SEIFTER J. (1960). *Psychopharmacologia (Berl.)* **1,** 482–492.
GLOWINSKI J. and AXELROD J. (1965) *J. Pharmacol. Exptl. Therap.* **149,** 43–49.
HANSON L. C. F. (1967) *Psychopharmacologia (Berl.)* **10,** 289–297.
KELLEHER R. T. and MORSE W. H. (1964) *Fedn. Proc.* **23,** 808–835.
KOJIMA H., RITTER S., WISE C. D. and STEIN L. (In preparation).
MAGOUN H. W. (1958) *The Waking Brain*, Charles C. Thomas, Springfield,
MARGULES D. L. and STEIN (1967) In: *Neuropsychopharmacology* (BRILL H., COLE J. O., DENIKER P., HIPPIUS H. and BRADLEY P. B., Eds.) pp. 108–120. Exerpta Medica Foundation, New York.
MARGULES D. L. and STEIN L. (1969) *Am. J. Physiol.* **217,** 475–480.
OLDS J. and MILNER P. (1954) *J. Comp. Physiol. Psychol.* **47,** 419–427.
OLDS J. (1962) *Physiol. Rev.* **42,** 554–604.
PHILLIPS A. G. and FIBIGER H. C. (1973) *Science* **179,** 575–577.
PICKENS R. and HARRIS W. C. (1968) *Psychopharmacologia (Berl.)* **12,** 158–163.
RANDRUP A. and MUNKVARD I. (1967) *Psychopharmacologia (Berl.)* **1,** 300–310.
RANDRUP A. and MUNKVAD I. (1970) In: *Amphetamine and Related Compounds* (COSTA E. and GARATTINI S., Eds.) pp. 695–713. Raven Press, New York.
RECH R. H. and STOLK J. M. (1970) In: *Amphetamines and Related Compounds* (COSTA E. and GARATTINI S., Eds.), pp. 385–414. Raven Press, New York.
RITTER S. and STEIN L. (1972) *Fedn. Proc.* **31,** 820.
RITTER S. and STEIN L. (in press). *J. Comp. Physiol. Psychol.*
ROLL S. K. (1970) *Science* **168,** 1370–1372.
RUTLEDGE C. O., AZZARO A. J. and ZIANCE R. J. (1972) In: *Monoamine Oxidases-New Vistas* (COSTA E. and SANDLER M., Eds.) pp. 379–392. Raven Press, New York.
STEIN L. (1964a) In: *Ciba Foundation Symposium on Animal Behaviour and Drug Action* (STEINBERG H., DEREUCK A. V. S. and KNIGHT J., Eds.) pp. 91–113. Churchill, London.
STEIN L. (1964b) *Fedn. Proc.* **23,** 836–850.
STEIN L. (1968) In: *Psychopharmacology: A Review of Progress*, 1957–1967 (EFRON D. H., Ed.) pp. 105–123. U.S. Government Printing Office, Washington, D.C.
STEIN L. and WISE C. D. (1969) *J. Comp. Physiol. Psychol.* **67,** 189–198.
STEIN L., WISE C. D. and BERGER B. D. (1973) In: *The Benzodiazepines* (RANDELL L. and GARATTINI S., Eds.) pp. 299–326. Raven Press, New York.
TAYLOR K. M. and SNYDER S. H. (1971) *Brain Res.* **28,** 295–309.
UNGERSTEDT U. (1971) *Acta Physiol. Scand. Suppl.* **367,** 1–48.
WEISSMAN A., KOE B. K. and TENEN S. S. (1966) *J. Pharmacol. Exptl. Therap.* **151,** 339–352.
WISE C. D. and STEIN L. (1969) *Science* **163,** 299–301.
WISE C. D., BERGER B. D. and STEIN L. (1970) *Dis. Nerv. Syst.* (Suppl.) **31,** No. 11, 34–37.
WISE C. D. and STEIN L. (1970) In: *Amphetamine and Related Compounds* (COSTA E. and GARATTINI S., Eds.) pp. 463–485. Raven Press, New York.
WISE C. D., BERGER B. D. and STEIN L. (1973) *Biol. Psychiat.* **6,** 3–21.

Frontiers in Catecholamine Research 1973, pp. 969 to 972. Pergamon Press. Printed in Great Britain.

OPERANT BEHAVIOURAL DEMONSTRATION OF QUALITATIVE DIFFERENCES BETWEEN THE d- AND l-ISOMERS OF AMPHETAMINE*

S. B. Sparber and D. W. Peterson

Department of Pharmacology and Psychiatry Research Unit of the Department of
Psychiatry, University of Minnesota, Minneapolis, Minnesota 55455, U.S.A.

Most of the the literature on the pharmacology of amphetamine indicates that the d and l isomers are about equiactive in their effects upon measures reflective of responses by peripheral organs (Alles, 1939; Van Liere, et al., 1951; Day, 1965; Rudzik and Eble, 1967; Clay, et al., 1971; Schneider, 1972), while the best estimate of their potencies upon behavioural variables or reversal of sedation or depression induced by other drugs indicates that d-amphetamine is at least 2–4 times more active than l-amphetamine (Alles, 1939; Prinzmetal and Alles, 1940; Schulte, et al., 1941; Dewhurst and Marley, 1965; Maj, et al., 1972; Wallach and Gershon, 1972). Although implicit in the above statement is their effects upon the central nervous system, there is evidence that peripheral tissue, alone or in combination with a central component, may also show a greater potency for the d- over the l-isomer. (Blaschko and Stromblad, 1960; Bhargava, et al., 1963; Burgen and Iverson, 1965; Daly, et al., 1966; Maling, et al., 1972; Schneider, 1972).

Although the neurochemical studies purporting to show regional differences in the degree to which the two isomers inhibit the uptake or facilitate the release of norepinephrine (NE) and dopamine (DM) are equivocal, (Taylor and Snyder, 1970; Ferris, et al., 1972), the differences appear not whether or not a potency ratio different from 1 exists between d and l-amphetamine but where in the CNS this difference emerges. In addition, Svensson (1971) reported that along with an increase in locomotor activity produced by both isomers, l-amphetamine increased DM levels and decreased NE levels while d-amphetamine decreased both catecholamines in mouse brain.

One of the major problems associated with studying the behavioural and/or neurochemical consequences of pharmacological agents is that no one unitary hypothesis can describe the mechanism by which drugs act. For example, to assume a common mechanism because patterns or levels of unconditioned or conditioned behaviour are 'identically' affected by two different drugs, would be highly questionable. Why then must we assume that isomers of the same drug act through the same mechanism but are only quantitatively different, if more of one isomer is necessary to produce the 'same effect' as the other isomer, especially in light of neurochemical evidence to the contrary.

---

* Supported in part by grants from the U.S.P.H.S. MH 08565, DA 00532 and GM 01117.

We have been addressing ourselves to methodological as well as theoretical problems of this type during the past several years and some of our data support the possibility that *d* and *l*-amphetamine, in addition to or aside from being quantitatively different, are also qualitatively different regarding the mechanism by which they affect certain classes of operant behaviour.

We have gone about this in several ways and I would like to discuss the advantages and limitations of our methods. In choosing a more controllable behaviour measure we have focussed upon operant techniques. However, like other measures of behaviour, the schedule of reinforcement and history of the organism can generate fairly resistant behaviour or labile behaviour which is more easily disrupted or otherwise altered by drug administration (KELLEHER and MORSE, 1969). Fixed ratio behaviour is quite stable and requires relatively greater doses of drug to disrupt. On the other hand, schedules which generate lower or moderate rates of responding appear to be more sensitive for measuring drug effects, often demonstrating rate-increasing effects prior to disruption. We have attempted to use these principles as well as those of tolerance and cross-tolerance to assess isomeric differences.

When a dose of 1 mg *d*-amphetamine sulfate/kg is administered (i.p.) to rats trained to bar-press for food pellets on a fixed ratio 30 (FR-30) schedule, the behaviour is altered so that the subjects are responding at about 40% of control rates. A dose of 4 mg l-amphetamine sulfate/kg likewise brought FR-30 responding down to about 40% of control. However, when tolerance to the disruptive effect of either isomer (at equieffective doses) developed, a challenge by the other isomer resulted in a demonstration of no cross-tolerance, suggesting differential mechanisms (TILSON and SPARBER, in press). If a more sensitive measure of operant behaviour is used, we see a differential effect of *d* and *l*-amphetamine on components of fixed-interval (FI-75 seconds) responding. Whereas low doses of *d*-amphetamine (0·16 and 0·50 mg/kg) increase responding early in the interval as well as late in the interval, resulting in an overall increase in FI behaviour, *l*-amphetamine increases responding late in the interval at 0·5 and 1·0 mg/kg. Overall FI behaviour is likewise increased slightly (beyond 2 SD of control values) but the pattern or distribution of responding among the segments of the FI is not the same. If one had to choose equieffective rate-increasing doses of *d* and *l*-amphetamine, the best approximation would be 0·16 mg/kg of the *d*-isomer and 0·50 mg/kg of the *l*-isomer, again the 1:3 ratio. However, after tolerance to the rate-increasing effects of either isomer, challenge the next session by the opposite isomer resulted in a drug effect or lack of cross tolerance (TILSON and SPARBER, in press).

Injections of 6-hydroxydopamine into the lateral ventricle of rats, at dosages that have little or no effect upon DM in various regions of brain but significantly lower NE concentrations, increase overall FR-responding significantly as long as 155 days after 6-OHDM. However, while the dose-response curves for *d*-amphetamine are virtually identical for both treated and control groups, (PETERSON and SPARBER, unpublished) the curves for *l*-amphetamine are not, the 6-OHDM treated rats being more resistant to the disruptive action of *l*-amphetamine (Fig. 1).

One interpretation of these last data might be a demonstration of the importance of DM or lack of relative importance of NE for at least one effect of *d*-amphetamine. Alternatively, it might indicate a relatively greater importance of NE for the action of *l*-amphetamine. In any event, the general conclusion we have come to is that in

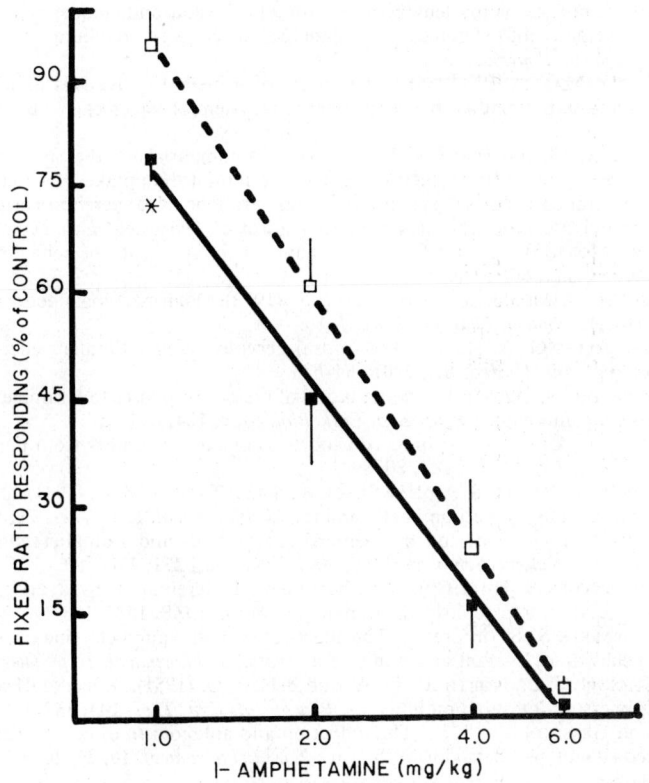

Fig. 1.—Log dose response curves demonstrating significant ($P < ·025$) differences between intraventricularly 6-OHDM (□) treated rats and vehicle controls (■) in their responsiveness to the FR-30 disruptive action of L-amphetamine. Animals were injected via their right lateral ventricles with either 200 μg 6-OHDM/15 μl of 0·1% ascorbic acid in saline or vehicle. L-Amphetamine's effect was determined, along with that of *d-*amphetamine, between 30 and 72 days after treatment. Data is derived for nine rats in each group, using saline injection as the control one day before each dose of *l-*amphetamine. Vertical lines at each point show 1 S.E.M. Data was analyzed according to analysis of variance for bioassay curves and Student's t-test.

addition to differential neurochemical actions described *in vitro*, with appropriate tests, behavioural differences of a qualitative nature likewise emerge.

REFERENCES

ALLES G. A. (1939). Comparative actions of optically isomeric phenisopropylamines. *Am. J. Physiol.* **126**, P 420.

BHARGAVA K. P., KAR K. and PARMAR S. S. (1963). Effects of enzyme inhibitors on the adrenergic mechanisms in isolated vas deferens preparation. *Life Sci.* **2**, 989–992.

BLASCHKO H. and STROMBLAD B. C. R. (1960). The inhibition of human amine oxidase by the two isomers of amphetamine. *Arzneim. Forsch.* **10**, 327.

BURGEN A. S. V. and IVERSON L. L. (1965). The inhibition of noradrenaline uptake by sympathomimetic amines in the rat isolated heart. *Br. J. Pharmacol.* **25**, 34–49.

CLAY G. A., CHO A. K. and ROBERFROID M. (1971). Effect of diethylaminoethyl diphenylpropylacetate hydrochloride (SKF-525A) on the norepinephrine-depleting actions of *d-*amphetamine. *Biochem. Pharmacol.* **20**, 1821–1831.

DALY J. W., CREVELING C. R. and WITKOP B. (1966). The chemorelease of norepinephrine from mouse hearts. Structure-activity relationships. I. Sympathomimetic and related amines. *J. Med. Chem.* **9**, 273–284.

Day M. D. (1965). Influence of the length of the stimulus period and frequency of sympathetic stimulation on the response of the guinea-pig isolated vas deferens to bretylium, guanethidine and amphetamine. *J. Pharm. Pharmacol.* **17**, 619–627.

Dewhurst W. G. and Marley E. (1965). The effects of α-methyl derivatives of noradrenaline, phenylethylamine and tryptamine on the central nervous system of the chicken. *Br. J. Pharmacol.* **25**, 682–704.

Ferris R. M., Tang F. L. M. and Maxwell R. A. (1972). A comparison of the capacities of isomers of amphetamine, deoxypipradrol and methylphenidate to inhibit the uptake of tritiated catecholamines into rat cerebral cortex slices, synaptosomal preparations of rat cerebral cortex, hypothalamus and striatum and into adrenergic nerves of rabbit aorta. *J. Pharmacol. Exp. Ther.* **181**, 407–416.

Kelleher R. T. and Morse W. H. (1968). Determinants of the specificity of behavioral effects of drugs. *Ergbn. der. Physiol.* **60**, 1–56.

Maj J., Grabowski M., Gajda L. and Michaluk, J. (1972). Pharmacologic action of d and l-amphetamine. *Dissert. Pharm. Pharmacol.* **24**, 7–16.

Prinzmetal M. and Alles G. A. (1940). The central nervous system stimulant effects of dextroamphetamine sulfate. *Am. J. Med. Sci.* **200**, 665–673.

Rudzik A. D. and Eble J. N. (1967). The potentiation of pressor responses to tyramine by a number of amphetamine-like compounds. *Proc. Soc. Eptl. Biol. Med.* **124**, 655–657.

Schneider F. H. (1972). Amphetamine-induced exocytosis of catecholamines from the cow adrenal medulla. *J. Pharmacol. Exp. Ther.*, **183**, 80–89.

Schulte J. W., Reif E. C., Bacher J. A., Lawrence W. S. and Tainter M. L. (1941). Further study of central stimulation from sympathomimetic amines. *J. Pharmacol. Exp. Ther.* **71**, 62–74.

Svensson T. H. (1971). Functional and biochemical effects of d- and 1-amphetamine on central catecholamine neurons, *Naunyn-Schmiederbergs Arch. Pharmak.* **271**, 170–180.

Taylor K. M. and Snyder S. H. (1970). Amphetamine: Differentation by d and 1-isomers of behavior involving brain norepinephrine or dopamine. *Science* **168**, 1487–1489.

Tilson H. A. and Sparber S. B. (in press). The effects of *d*- and *l*-amphetamine on fixed-interval and fixed-ratio behavior in tolerant and non-tolerant rats. *J. Pharmacol. Exp. Ther.*

Van Liere E. J., Stickney J. C., Northrup D. W. and Bell R. O. (1951). Effect of dl-amphetamine sulfate and its isomers on intestinal mobility. *J. Pharamcol. Exp. Ther.* **103**, 187–189.

Wallach M. B. and Gershon S. (1972). The induction and antagonism of central nervous system stimulant-induced stereotyped behavior in the cat. *Eur. J. Pharmacol.* **18**, 22–26.

Frontiers in Catecholamine Research 1973, pp. 973 to 975. Pergamon Press. Printed in Great Britain.

DISSOCIATION OF AMPHETAMINE-INDUCED RELEASE OF NOREPINEPHRINE FROM INHIBITION OF NEURONAL UPTAKE IN ISOLATED BRAIN TISSUE*

CHARLES O. RUTLEDGE, ALBERT J. AZZARO[†] and RONALD J. ZIANCE[‡]

Department of Pharmacology, University of Colorado School of Medicine,
Denver, Colorado 80220, U.S.A.

AMPHETAMINE and amphetamine analogues increase the efflux of norepinephrine from neurons in the central nervous system. The mechanism by which this effect is produced has not been satisfactorily resolved. Amphetamine has been observed to inhibit neuronal uptake of biogenic amines (ROSS and RENYI, 1964; GLOWINSKI and AXELROD, 1965; HÄGGENDAL and HAMBERGER, 1967), and thus the efflux of norepinephrine could be explained by inhibition of neuronal uptake of spontaneously released norepinephrine. Indirect measurements of uptake and release observed *in vivo* suggest that inhibition of neuronal uptake is probably not the major mechanism by which amphetamine increases the efflux of norepinephrine since the tricyclic antidepressants appear to inhibit neuronal uptake without releasing norepinephrine (CARLSSON and WALDECK, 1966; FUXE and UNGERSTEDT, 1968). Amphetamine also has been observed to release ^{3}H-norepinephrine from presumed intraneuronal binding sites (GLOWINSKI and AXELROD, 1965; STEIN and WISE, 1969; ZIANCE et al., 1972), and the efflux of norepinephrine from central nervous system neurons could be largely the result of this effect. In the present report, the effects of amphetamine and other drugs on release and neuronal uptake are studied in *in vitro* systems where these two processes can be measured quantitatively. The aim is to establish conditions in which the actions of drugs on these two processes can be distinguished.

Release of ^{3}H-norepinephrine from isolated brain tissue was studied by methods previously described (ZIANCE and RUTLEDGE, 1972; ZIANCE et al., 1972). The method involves incubating chopped rat brain tissue with 10^{-6} M ^{3}H-norepinephrine, allowing the ^{3}H-amine to be taken up into the neurons, washing the unbound and non-specifically bound ^{3}H-amine from the tissue and then measuring the effect of drugs on the amount of ^{3}H-norepinephrine in the incubation medium and tissue. Since a portion of the ^{3}H-norepinephrine is deaminated under the conditions of this experiment, ^{3}H-norepinephrine was separated from deaminated metabolites by cation exchange chromatography with Dowex 50, Na$^+$. O-methylation plays a minor role in the metabolism of norepinephrine in this system (ZIANCE et al., 1972) and thus norepinephrine was not separated from its O-methylated metabolite, normetanephrine. Release is expressed as ^{3}H-norepinephrine in the medium as a percentage of

* This study was supported by Grant NS90951 from the U.S.P.H.S.

† Former U.S. Public Health Service Postdoctoral Fellow (Fellowship No. NS 46683). Present address: Department of Neurology, West Virginia University Medical Center, Morgantown, West Virginia 26505.

‡ Present address: Department of Pharmacology, School of Pharmacy, University of Georgia, Athens, Georgia 30601.

^{3}H-norepinephrine in the medium + tissue. Previous studies (ZIANCE *et al.*, 1972; AZZARO and RUTLEDGE, 1973) have shown that release of norepinephrine from brain tissue by amphetamine as measured by this procedure is: (1) sensitive, the threshold for release of ^{3}H-norepinephrine from cerebral cortex by amphetamine is 10^{-7} M; (2) tissue selective, the ED_{50} for release of ^{3}H-norepinephrine from cerebral cortex is less than the ED_{50}'s for release from medulla oblongata or corpus striatum; (3) substrate specific, the concentration–effect curve for release of ^{3}H-norepinephrine by amphetamine is markedly different from the curves for release of ^{3}H-dopamine or ^{3}H-5-hydroxytryptamine; (4) temperature dependent, maximal release occurs at 37°C and there is no release at 0°C; (5) stereoselective, *d*-amphetamine is more potent than the *l*-isomer.

Neuronal uptake of ^{3}H-norepinephrine into chopped rat brain tissue was measured by a method involving a 10 min incubation of chopped rat brain tissue with 10^{-7} M ^{3}H-norepinephrine and measuring total radioactivity in the incubation medium and tissue (ZIANCE and RUTLEDGE, 1972). Drugs were added to the incubation medium 10 min before and during the incubation with ^{3}H-norepinephrine. Experimental samples were incubated at 37°C and compared to control samples incubated at 0°C. Since there is relatively little metabolism of ^{3}H-norepinephrine in this 10 min exposure, ^{3}H-norepinephrine was not separated from the deaminated metabolites. Results were expressed as tissue to medium ratios: (dis/min per g of tissue)/(dis/min per ml of medium). This ratio approaches 1·0 at 0°C which indicates a lack of specific uptake at this temperature. The ratio was 7·1 ± 0·4 at 37°C in the absence of drugs, and this reflects approximately a seven-fold increase in uptake and accumulation of ^{3}H-norepinephrine into the tissue. From these ratios the percentage inhibition of neuronal uptake produced by drugs was determined.

TABLE 1. EFFECT OF AMPHETAMINE, COCAINE AND DESIPRAMINE ON RELEASE AND INHIBITION OF NEURONAL UPTAKE OF ^{3}H-NOREPINEPHRINE IN ISOLATED CEREBRAL CORTEX TISSUE

Drug	Inhibition of uptake* (%)	Release† (%)
Amphetamine 10^{-5} M	85·0 ± 1·3 (8)‡	30·4 ± 1·3 (14)
Cocaine 10^{-5} M	83·6 ± 2·4 (6)	10·2 ± 1·3 (3)
Desipramine 10^{-5} M	78·9 ± 3·5 (3)	0·7 ± 0·7 (4)

 * All samples were incubated for 10 min with 10^{-7} M ^{3}H-NE. Each value represents the mean ± S.E.M.

 † Calculated as $\dfrac{\text{dis/min } ^3\text{H-NE in medium (100)}}{\text{dis/min } ^3\text{H-NE in medium + dis/min } ^3\text{H-NE in tissue.}}$ Release in the absence of drug was 11·7 ± 1·1 and was subtracted from each of the drug values. Each value is the mean ± S.E.M.

 ‡ Number within the parenthesis represents the number of experiments.

The effect of 10^{-5} M concentrations of amphetamine, cocaine and desipramine upon neuronal uptake and release of ^{3}H-norepinephrine from chopped cerebral cortex tissue can be seen in Table 1. Inhibition of neuronal uptake by the three drugs was approximately the same while release of ^{3}H-norepinephrine was markedly different. Release of ^{3}H-norepinephrine by amphetamine was much greater than that produced by cocaine and ^{3}H-norepinephrine was not released by desipramine.

The effects of various concentrations of amphetamine (10^{-7}–10^{-3} M) on release and neuronal uptake of ^{3}H-norepinephrine were measured and it was observed that amphetamine released ^{3}H-norepinephrine from cerebral cortex with concentrations which were only slightly less than those required for inhibition of neuronal uptake. The ID_{50} for inhibition of neuronal uptake by amphetamine was $7 \cdot 1 \times 10^{-7}$ M while the ED_{50} for release was $1 \cdot 2 \times 10^{-6}$ M. Although the ED_{50} for cocaine induced release of ^{3}H-norepinephrine was approximately the same as that of amphetamine, the maximal release ($11 \cdot 2 \pm 2 \cdot 1 \%$) produced by cocaine ($10^{-4}$ M) was much less than that of amphetamine (10^{-3} M, $38 \cdot 1 \pm 1 \cdot 6 \%$). Desipramine, on the other hand, was much less potent than the other drugs in releasing ^{3}H-norepinephrine; the lowest concentration of desipramine which released ^{3}H-norepinephrine was 10^{-4} M. It is not likely that the efflux of ^{3}H-norepinephrine produced by amphetamine is the result of inhibition of reuptake of spontaneously released ^{3}H-norepinephrine since cocaine and desipramine were equipotent with amphetamine in inhibiting uptake of ^{3}H-norepinephrine but were much less efficacious in releasing the ^{3}H-amine.

Results have also been obtained (RUTLEDGE et al., 1972) which indicate that when neuronal uptake is blocked with cocaine or desipramine, amphetamine is still capable of releasing ^{3}H-norepinephrine even though the concentration effect curve is shifted to the right. Studies on the uptake of ^{3}H-amphetamine into isolated synaptosomes from rat cerebral cortex suggest that low concentrations of amphetamine (10^{-7}–10^{-6} M) enter the neuron by the neuronal uptake mechanism, since the accumulation of ^{3}H-amphetamine is inhibited by cocaine and desipramine (RUTLEDGE et al., 1972). Thus, these drugs probably inhibit the release of ^{3}H-norepinephrine which is observed with low concentrations of amphetamine by inhibiting the uptake of amphetamine into the neuron. Amphetamine in higher concentrations (10^{-5}–10^{-3} M) appears to enter the neuron by a nonspecific process which is not inhibited by cocaine or desipramine and thus the release produced by high concentrations of amphetamine is not prevented by blockade of neuronal uptake.

In summary, inhibition of neuronal uptake is probably not the primary mechanism by which the efflux of ^{3}H-norepinephrine is enhanced by amphetamine. However, inhibition of neuronal uptake may play a role in blocking recapture of the released amine.

Acknowledgements—The skillful technical assistance of Mrs. Elisabeth Dreyer is gratefully acknowledged. The authors are grateful to Smith, Kline and French Laboratories for the contribution of *d*-amphetamine sulfate and to Geigy Pharmaceuticals for the contribution of desipramine hydrochloride.

REFERENCES

AZZARO A. J. and RUTLEDGE C. O. (1973) *Biochem. Pharmacol.* In press.
CARLSSON A. and WALDECK B. (1966) *J. Pharm. Pharmacol.* **18**, 252–253.
FUXE K. and UNGERSTEDT U. (1968) *Europ. J. Pharmacol.* **4**, 135–144.
GLOWINSKI J. and AXELROD J. (1965) *J. Pharmacol. Exp. Ther.* **149**, 43–49.
HÄGGENDAL J. and HAMBERGER B. (1967) *Acta physiol. scand.* **70**, 277–280.
ROSS S. B. and RENYI A. L. (1964) *Acta pharmacol. et toxicol.* **21**, 226–239.
RUTLEDGE C. O., AZZARO A. J. and ZIANCE, R. J. (1972) *Fedn. Proc.* **31**, 601.
STEIN L. and WISE C. D. (1969) *J. Comp. Physiol. Psychol.* **67**, 189–198.
ZIANCE R. J. and RUTLEDGE C. O. (1972) *J. Pharmacol. Exp. Ther.* **180**, 118–126.
ZIANCE R. J., AZZARO A. J. and RUTLEDGE C. O. (1972) *J. Pharmacol. Exp. Ther.* **182**, 284–294.

Frontiers in Catecholamine Research 1973, pp. 977 to 981. Pergamon Press. Printed in Great Britain.

AMPHETAMINE AND METHYLPHENIDATE PSYCHOSIS

John M. Davis and David S. Janowsky

Illinois State Psychiatric Institute-Dept. of Psychiatry University of Chicago,
Chicago, Illinois,
and
Tennessee Neuropsychiatric Institute and Dept. of Psychiatry, Vanderbilt
University, Nashville, Tennessee.

Amphetamine and other similar psychomotor stimulants such as methylphenidate (Ritalin), when taken in large amounts, can produce a psychosis, essentially indistinguishable from paranoid schizophrenia. Many subjects who develop amphetamine psychosis on the street are not overtly or latently schizophrenic prior to taking amphetamine; this suggests that the psychosis is a drug produced psychosis. Recently amphetamine has been given experimentally to subjects who were not overtly schizophrenic. It did produce an "amphetamine" psychosis (Griffith et al., 1972).

Since amphetamine psychosis is clinically indistinguishable from an endogenous paranoid psychosis, it may be the best drug-induced model of schizophrenia. Drugs which block stereotyped behaviour produced by amphetamine in animals also have antipsychotic properties in schizophrenic patients. There is a significant amount of indirect evidence linking changes in central dopamine to schizophrenia. d-Amphetamine is approximately two times as potent as l-amphetamine in producing stereotyped behaviour in rats pretreated with monoamine oxidase inhibitor. In addition d-amphetamine is ten times as potent as l-amphetamine in producing locomotor stimulation. This evidence would be consistent with the suggestion that stereotyped behaviour is produced by dopamine and locomotor activity by norepinephrine (Taylor and Snyder, 1970). Several authors have shown that cholinomimetic drugs can block stereotyped behaviour produced by the psychomoto stimulants, indicating that stereotyped behaviour may be controlled by a balance of neurotransmitters, such as dopamine vs. acetylcholine (Janowsky et al., 1972). If one assumes that in man, d-amphetamine is ten times more potent than l-amphetamine in producing psychomotor stimulation, and that this is under noradrenergic control and two times as potent in producing amphetamine psychosis than l-amphetamine, which is under dopaminergic control, then the isomers may be valuable tools to investigate amphetamine psychosis in man. Angrist et al. (1971), has produced amphetamine psychosis experimentally in patients who abuse psychomotor stimulants, using both d and l-amphetamine. d-Amphetamine is 1·3 times as potent as l-amphetamine in producing the typical amphetamine psychotic picture.

It is important to note that there are two types of psychomotor stimulant produced psychosis. The psychosis produced by large oral doses of amphetamine administered every hour or so for several days is a paranoid psychosis. Recently Janowsky et al. (1972a, 1973), reported that small dose intravenous administration of methylphenidate 0·5 mg/kg can produce a marked worsening of pre-existing psychosis in patients with active schzophrenic illness. It does not produce psychotic symptoms in normal patients or patients who are in remission. The phenomena of markedly worsening a pre-existing psychosis may be a different one than that producing a typical paranoid psychosis in the non-schizophrenic subject. A patient's psychosis worsens both qualitatively and quantitatively in the direction of their pre-existing psychosis. Thus, catatonic schizophrenics become more catatomic, without showing paranoid symptoms. Thus, psychosis worsening after the acute intravenous administration produces an exacerbation of the schizophrenic symptoms rather than a uniform paranoid psychosis.

METHOD

A total of 17 actively ill schizophrenic patients, who were in good health without cardiovascular or other physical illness were administered active drug by a single injection, preceded and followed by placebo injections every 5 min. Equimolar doses of d-amphetamine sulphate solution, l-amphetamine succinate solution, or methylphenidate hydrochloride solution in random order on different days was administered. A blind rater noted each patient for changes in a number of variables every 10 min using a 5 point rating scale. The items rated included: Psychosis, conceptual disorganization, unusual thoughts, anger, irritable interaction, talkativeness, as well as "activation" and "inhibition" as defined previously (Janowsky et al., 1972).

In each experiment, the rater and patient were blind to when in an IV injection sequence active drug was substituted for placebo and which active drug was given. Blood pressures and pulses were monitored every 5 min. Data was analysed by comparing average scores and change scores between the baseline–placebo phases and the 10–20, 20–30 and 30–40 min periods after active drug injection for each of the experimental drugs. Thus, each patient served as his own control. Patients received d and l-amphetamine, 20 mg (0·11 mM.) and 28 (0·11 mM) respectively, methylphenidate 29 mg (0·11 mM) and d-amphetamine 10 mg (0·55 mM) and d-amphetamine 4 mg (0·022 mM).

For ethical reasons, it was sometimes not possible to give every subject all the injections in the series. All comparisons between drugs were made in subjects who received both injections.

In addition, in order to evaluate the ability of acetylcholine to antagonise psychostimulant effects, a total of 24 schizophrenic patients received a series of placebo injections every 5 min followed by injection of 0·5 mg/kg methylphenidate over a 30 sec. period of time. In nine patients two placebo injections at 5 min intervals were followed by a series of either: (1) every 5 min placebo injections, or (2) 0·5 mg physostigmine injections every 5 min until methylphenidate antagonism occurred or 2·5 mg had been given, or (3) neostigmine 0·25 mg injection given every 5 min until 1·25 mg had been given. After the above series had been given, placeboes were injected every 5 min for 30 min.

Also, the ability of methylphenidate to reverse the physostigmine induced "inhibitory state" was evaluated using the same design as above except that the physostigmine or neostigmine was given first, followed by methylphenidate.

RESULTS

In actively schizophrenic patients, methylphenidate produced a marked worsening of the schizophrenic symptoms causing a doubling of the psychosis scores (Janowsky et al., 1972 and Table 1). d-Amphetamine worsened psychotic symptoms somewhat, but was less potent than methylphenidate. l-Amphetamine was the least potent of the 3 drugs. d-Amphetamine 10 mg and d-amphetamine 4 mg were less potent than l-amphetamine 28 mg. The results from each of several experiments were combined

TABLE 1. CHANGE SCORES REPRESENTING DIFFERENCES BETWEEN THE AVERAGE BASELINE-PLACEBO PHASE RATINGS AND THE AVERAGE OF RATINGS DONE AT 10 AND 20 MIN AFTER INTRAVENOUS PSYCHOSTIMULANT ADMINISTRATION

	Methylphenidate† 29 mg (N = 10)	d-amp. 20 mg (N = 18)	l-amp. 28 mg (N = 14)	d-amp. 10 mg (N = 6)	d-amp. 4 mg (N = 8)
Psychosis (global)	0·48 ± 0·16**	0·36 ± 0·15*	0·22 ± 0·11*	0·20 ± 0·09*	0·14 ± 0·08*
Conceptual disorganisation	0·88 ± 0·27**	0·53 ± 0·17**	0·22 ± 0·16	0·0	0·04 ± 0·09
Unusual thoughts	0·72 ± 0·23**	0·52 ± 0·18**	0·25 ± 0·11*	0·08 ± 0·09	0·06 ± 0·08
Combined psychosis score	2·08 ± 0·60**	1·42 ± 0·44**	0·69 ± 0·32*	0·28 ± 0·14*	0·24 ± 0·17
Anger	0·56 ± 0·35	0·09 ± 0·22	0·35 ± 0·19*	0·0	0·26 ± 0·20
Irritable	0·26 ± 0·26	−0·08 ± 0·12	0·23 ± 0·18	0·22 ± 0·22	0·14 ± 0·17
Activation (interaction and talkativeness)	1·80 ± 0·26***	1·22 ± 24***	0·56 ± 0·21*	0·45 ± 0·14*	0·63 ± 0·27*

$* = P < 0.05$, $** = P < 0.01$, $*** = P < 0.001$ = Statistical significance of change scores.
† = d-amphetamine (20 mg), l-amphetamine (28 mg), and methylphenidate (29 mg) are equimolar.

to arrive at drug potencies relative to the psychosis worsening effects of l-amphetamine. Methylphenidate was also more potent in activating the patients by increasing their talkativeness and interactions. d-Amphetamine was less potent than methylphenidate but more potent than l-amphetamine. d-Amphetamine-4 mg and 10 mg were of comparable potency to l-amphetamine 28 mg.

The experiment concerning dopaminergic-cholinergic balance can be done in one of two ways. One may produce an increase in psychosis and activation by methylphenidate and block this by the administration of physostigmine, or produce an inhibitory state by physostigmine and then reverse this by methylphenidate.

In the experiment in which methylphenidate was given followed by physostigmine, methylphenidate produced a 50 per cent increase in psychosis ratings. When physostigmine was administered after methylphenidate there was essentially no net increase in psychosis, so that physostigmine effectively reversed the psychosis worsening property of methylphenidate (Table 2). Methylphenidate also increased interactions, and this effect was blocked by physostigmine. When a psychomotor retardation syndrome was produced by physostigmine, this can be reversed by methylphenidate. Physostigmine failed to alter the baseline psychosis when given alone.

DISCUSSION

All of the psychomotor stimulants can markedly worsen psychosis and can also increase activation as manifested by talkativeness and increased interactions. In order of potency, methylphenidate is more potent than d-amphetamine, which itself is more potent than l-amphetamine. Lower doses of d-amphetamine (10 mg, 4 mg) are essentially of equal potency to l-amphetamine-28 mg in producing activation. If one focuses on the relationship of d and l-amphetamine, d-amphetamine is slightly more potent than l-amphetamine in activating psychosis and increasing activation. Furthermore, the degree of its greater potency in these two behaviours is approximately equal. Thus, a different relationship between the differential potency of

TABLE 2A. Effects of methylphenidate in antagonising physostigmine induced state ($N = 7$)

	Baseline	P	Physostigimine	P	Physostigmine Methylphenidate
Inhibition	7.7 ± 2.8	0.02	13.9 ± 3.5	0.02	8.9 ± 2.0
Activation	4.9 ± 1.1	0.07	2.9 ± 0.9	NS	3.8 ± 0.6
Psychoses	1.4 ± 0.6	NS	1.5 ± 0.5	NS	2.0 ± 0.3

B. Effects of physostigmine in antagonising methylphenidate induced psychosis activation and increased interactions in schizophrenics

Methylphenidate + placebo	Baseline score	P	15 min. post methylphenidate score	P	45 min post methylphenidate score
Psychosis	1.96 ± 0.24	<0.0002	3.00 ± 0.27	<NS	2.89 ± 0.27
Interaction	1.67 ± 0.17	<0.0003	2.76 ± 0.20	<0.03	2.54 ± 0.20
Methylphenidate + physostigmine**					Physostigmine
Psychosis	1.89 ± 0.46	<0.003	2.93 ± 0.53	<0.004	1.96 ± 0.48
Interaction	1.96 ± 0.34	<0.002	3.04 ± 0.24	<0.006	1.81 ± 0.37

*$N = 24$, **$N = 9$

P values represent the comparison between two adjacent columns.

the two isomers of amphetamine on these two behaviours exist in man than in the rat, if one assumes that psychoses corresponds to stereotyped behaviour and activation corresponds to locomotor activity. In man, both behaviours occur with a potency ratio consistent with a dopaminergic theory of schizophrenia. Space does not permit a detailed review here, and reference is made to our previous exposition (Janowsky et al., 1972).

If brain monoamine oxidase is decreased in schizophrenic brain, it is tempting to speculate that the methylphenidate-induced psychosis activation is related to the monoamine oxidase deficit, particularly in that central catecholamines released by methylphenidate could be expected to be released in active form, since they would not be effectively metabolized intraneuronally (Murphy and Wyatt, 1972). It is relevant to note that methylphenidate releases preferentially from the monoamine stores, a finding which would be consistent with the greater potency of methylphenidate, relative to amphetamine in worsening psychosis (Scheel-Kruger, 1971).

There is a substantial body of evidence indicating that stereotyped behaviour is controlled by a balance between the dopaminergic and the cholinergic system. It is relevant to examine whether increasing brain acetylcholine with physostigmine can block the psychosis-activating properties of methylphenidate. When physostigmine was administered prior to methylphenidate, it prevented the psychosis worsening. This indicates that the worsening of the psychosis produced by methylphenidate and presumably mediated by dopamine is controlled by a dopaminergic-cholinergic balance, as is stereotyped behavior. Is the underlying psychosis also controlled by an endogenous dopaminergic factor which is balanced by a cholinergic system? The observation that physostigmine does not reduce psychosis would suggest that the underlying psychotic process is not as easily amenable to the effects of altering cholinergic tone as is the worsening of the psychosis produced by methylphenidate. It is relevant to note that antipsychotic drugs do not always immediately block psychosis, the therapeutic improvement occurring over weeks rather than hours in many instances.

The activating effects of methylphenidate are blocked by physostigmine, and the inhibition syndrome produced by physostigmine is blocked by methylphenidate: indicating the activation is under control of adrenergic-cholinergic balance.

It is appropriate to record the obvious caution in interpreting experiments such as this due to the inexactitude of clinical studies, the speculative nature of the dopaminergic theory of schizophrenia, as well as the reliance in this theorizing on the assumption that data derived from rats by Snyder is generalisable to (1) a greater variety of situations involving the relationship between d and l-amphetamine, stereotyped behavior, dopamine and norepinephrine release and uptake and (2) to man.

Acknowledgements—This research is supported by Grant MH-11468 and NIH 15431 from the National Institute of Mental Health, and the State of Tennessee Department of Mental Health.

REFERENCES

ANGRIST B. M., SHOPSIN B. and GERSHON S. (1971) *Nature, Lond.* **234**, 152–153.
GRIFFITH J. J. *et al.*, (1972) *Archs Gen. Psychiat.* **26**, 97–100.
JANOWSKY D. S., EL-YOUSEF M. K., DAVIS J. M. and SEKERKE H. J. (1973) *Archs Gen. Psychiat.* **28**, 185–191.
JANOWSKY D. S., EL-YOUSEF M. K. and DAVIS J. M. (1972) *Comp. Psychiat.* **13**, 83.
JANOWSKY D. S., EL-YOUSEF M. K., DAVIS J. M. and SEKERKE H. J. (1973) *Archs Gen. Psychiat.* **28**, 542–547.
JANOWSKY D. S., EL-YOUSEF M. K., DAVIS J. N. and SEKERKE H. J. (1972) *Psychopharmacologia* **27**, 295–303.
MURPHY D. L., WYATT R. J. (1972) *Nature, Lond.* **238**, 225–226.
SCHEEL-KRUGER J. (1971) *Europ. J. Pharmacol.* **14**, 47–59.
TAYLOR K. M. and SNYDER S. H. (1970) *Science* **168**, 1487–1489.

Frontiers in Catecholamine Research 1973, pp. 983 to 985. Pergamon Press. Printed in Great Britain.

AMPHETAMINE METABOLISM IN AMPHETAMINE-INDUCED PSYCHOSIS

LARS-M. GUNNE, M.D. and ERIK ÄNGGÅRD, M.D.

Psychiatric Research Center, Ulleråker Hospital, University of Uppsala,
S-750 17 Uppsala, Sweden

and

Department of Pharmacology, Karolinska Institutet, S- 104 01 Stockholm 60,
Sweden

ELEVEN intravenous abusers of amphetamine were admitted to our clinic in a state of paranoid psychosis, with delusions of persecution, visual and auditory hallucinations and varying degrees of anxiety and motor unrest. All had a history of periodic intravenous abuse of high doses of amphetamine and the presence of amphetamine in their urine was verified in all cases. After admission 6 were given ammonium chloride in order to enhance the urinary elimination of amphetamine, whereas 5 received sodium bicarbonate, by which the excretion of amphetamine is retarded (BECKETT and ROWLAND, 1965). During the first 1–2 days after admission 50 mg of amphetamine were given 3 times daily and during this initial stabilisation period 500 μCi of ^{3}H-dl-amphetamine was administered at 8 a.m. Plasma and urine samples were collected during the period of psychosis and the psychotic symptoms were rated 4 times daily (ÄNGGÅRD et al., 1970; JÖNSSON,1972). The radioactive compounds of the urine were separated into four different fractions (ÄNGGÅRD et al., 1973): (1) non-polar bases (mostly amphetamine), (2) polar bases (hydroxylated metabolites), (3) acidic and neutral compounds (deaminated metabolites), (4) watersoluble residue.

It was found that administration of bicarbonate, which retarded the elimination of basic compounds, tended to prolong the course of the psychosis and to intensify the symptoms compared with the group which had an acidic urine. There was an inverse relationship between the urinary output of amphetamine and both basic and deaminated metabolites. Thus, when the pH of urine was high (and the output of amphetamine low) both basic and acidic metabolites increased considerably.

When the ratings of psychotic symptoms were plotted against the plasma levels of amphetamine at 8 a.m. on the second day after admission no correlation was obtained [Fig. 1(a)]. Figure 1(b) shows the relationship between the relative excretion of labelled basic hydroxylated metabolites and psychotic symptoms. Both during conditions of acidic urine (3 cases) and alkaline urine (4 cases) there appeared to be a correlation between the intensity of the psychotic manifestations and the urinary output of hydroxylated metabolites.

The identity of these compounds was established by a series of extraction and ion exchange column procedures in combination with gas chromatography with

Supported by grant B73-04X-2566-05B from the Swedish Medical Research Council and grant 150 from the Swedish Bank Tricentennial Fund.

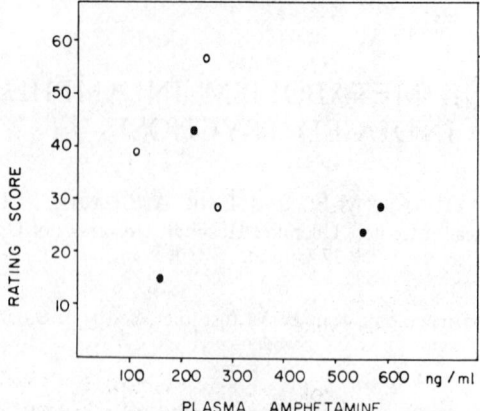

Fig. 1(a).—Plasma amphetamine levels and mean ratings of psychosis during first day after admission.

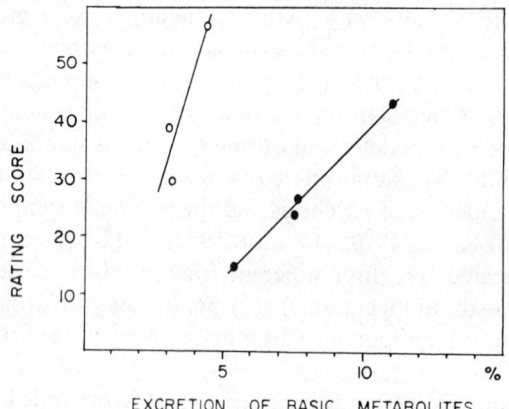

Fig. 1(b).—Relationship between psychosis scores and urinary basic hydroxylated metabolites of amphetamine (expressed as percent of administered ^{3}H-amphetamine). Patients with acidic urine; open circles; alkaline urine: filled circles.

radioactivity monitoring and also by mass spectrometry after conversion into the trifluoroacetyl derivatives. Radioactive peaks were obtained with retention times corresponding to p-hydroxyamphetamine, norephedrine and p-hydroxynorephedrine. All these compounds give a dominant peak at m/e 140 due to a common fragment resulting from cleavage β to the ring. When the mass spectrometer was focused on this fragment, peaks were observed with the retention times of these three metabolites (Änggård et al., 1973). In accordance with Schweitzer and Friedhoff (1972), no p-methoxyamphetamine could be detected.

A correlation between the intensity of the symptoms of psychosis and the urinary excretion of hydroxylated metabolites may indicate a role for one or more of these metabolites in the production of the psychotic symptoms. The plasma levels of amphetamine on the other hand did not indicate a direct effect of amphetamine on the psychotic manifestations.

REFERENCES

ÄNGGÅRD E., GUNNE L.-M., JÖNSSON L.-E. and NIKLASSON F. (1970) *Europ. J. clin. Pharmacol.* **3,** 3–11.

ÄNGGÅRD E., JÖNSSON L.-E., HOGMARK A.-L. and GUNNE L.-M. (1973) *J. Clinical Pharmacol.* (In press).

BECKETT A. H. and ROWLAND M. (1965) *J. Pharm. Pharmacol.* **17,** 628–639.

JÖNSSON L.-E. (1972) *Acta Universitatis Upsaliensis.*

SCHWEITZER J. W. and FRIEDHOFF A. J. (1972) *Proc. Int. Conf. Drug Abuse* (Ed. C. J. D. ZARAFONETIS) Philadelphia 1972.

Frontiers in Catecholamine Research 1973, pp. 987 to 989. Pergamon Press. Printed in Great Britain.

ON AMPHETAMINE TOLERANCE AND ABSTINENCE IN RATS

TOMMY LEWANDER, GULL MOLIIS and INGER BRUS
Psychiatric Research Center, University of Uppsala, Ulleråker Hospital,
S 750 17 Uppsala, Sweden

INTRODUCTION

DEVELOPMENT of tolerance to a number of pharmacological effects of amphetamine on chronic administration of the drug has been described in animals and man (see reviews by KOSMAN and UNNA, 1968, LEWANDER, 1970, 1972, KALANT et al., 1971). In rats it has been concluded that amphetamine tolerance seem to be functional rather than dispositional (LEWANDER, 1968, 1971b). The depletion of noradrenaline and incorporation of p-hydroxynorephedrine, a metabolite of amphetamine in rats, as a false transmittor into central and peripheral noradrenaline neurons, seem to be one mechanism of tolerance to amphetamine in rats (BRODIE et al., 1969; LEWANDER, 1971b).

The present communication concerns a preliminary account of a study of some aspects of amphetamine tolerance, such as its speed of production, its extent and duration, the occurrence of carry-over of tolerance, which do not seem to have been investigated previously. In addition, the development of tolerance to the inhibitory action of amphetamine on hoarding behaviour (BLUNDELL, 1971) and the unexpected finding of an increased hoarding activity on withdrawal of chronic amphetamine administration is demonstrated.

METHODS

Male Sprague–Dawley rats, approximately 300 g body wt., were used. The rats were kept in individual cages in a temperature constant room (21°C).

In the first experiment body (colonic) temperature was measured by an electric thermometer (Ellab, Copenhagen) at hourly intervals after the morning injection for 7 hr a day for 45 days. d-Amphetamine, 20 mg/kg i.p., or saline was administered twice daily to two groups of 15 rats each for 15 days (day 1–15, Fig. 1a). During the following 15 days (day 16–30) all rats received saline except on day 21, 25 and 30, when two subgroups of five rats each from the chronically amphetamine and saline treated groups respectively, were tested with amphetamine. d-Amphetamine or saline was again given chronically for a second 15 days period (day 31–45). Food and water was freely accessible between 4.30 p.m. and 8.30 a.m.

The temperature response to amphetamine was calculated each day as the mean area limited by the time–response curves between 0 and 6 hr for amphetamine and saline respectively. For convenient presentation of the results (Fig. 1a) the daily areas were expressed as percentages of the day 1 area, which was set to 100 per cent.

Hoarding behaviour was measured as the number of food pellets (2, 6 g, Anticimex no. 214) hoarded during a 15-min trial (BLUNDELL, 1971). Twelve rats received dl-amphetamine sulphate, 16 mg/kg i.p. twice daily for 14 days and 12 control rats received saline. At day 18 (Fig. 1b) all rats were injected with amphetamine and from day 19 on all rats were given saline. The hoarding trials were run daily at 1, 3 and 5 hr after the morning injections; food was then freely available for 3 hr. The mean hoarding score for the 3 trials were calculated for each rat and the daily median of the hoarding scores for each group are given in Fig. 1b. During the chronic amphetamine treatment period (from day 6 through day 18), however, the hoarding trials for each time point are given separately for the amphetamine treated rats. Filled circles in Fig. 1B indicates statistically significant differences ($p < 0.05$) between the experimental and control groups (Mann–Whitney U-test; SIEGEL, 1956).

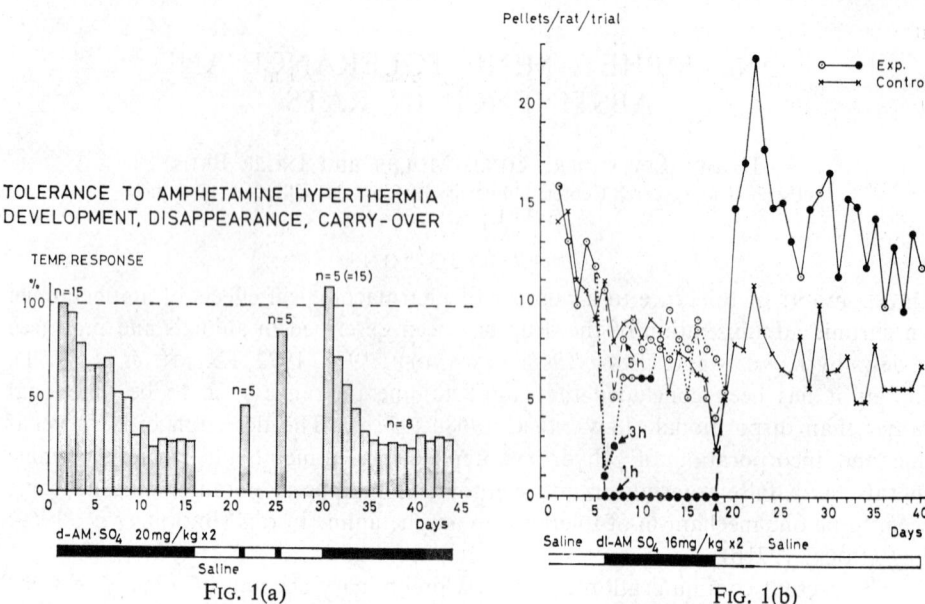

FIG. 1(a)

FIG. 1(b)

RESULTS AND DISCUSSION

It is concluded from the present experiment, Fig. 1a, that maximal tolerance, about 30 per cent of the initial response, to the hyperthermic effect of a high dose of amphetamine developed after 9 days of chronic treatment. The duration of tolerance was estimated to 10–15 days, when initial temperature response was reattained. The presence of the phenomenon of carry-over of amphetamine tolerance could be demonstrated, since maximal tolerance was reached already after 6 days during a second period of chronic amphetamine administration. Tolerance to the hyperthermic action of amphetamine has been related to the biochemical effects of amphetamine on noradrenaline neurons (Lewander, 1971b). The present results prompts a study of central and peripheral catecholamines at various time points in a similarly designed experiment in order to find explanations for the present observations.

In the second experiment, Fig. 1b, the development of tolerance to the inhibitory effect of amphetamine on hoarding behaviour was demonstrated, which was dependent on the time-point after the injection of amphetamine. At 1 hr after the amphetamine injection, the hoarding score was practically zero throughout the treatment period and at 5 hr hoarding was not significantly affected by amphetamine. At 3 hr after injection, however, the hoarding scores were initially decreased but after 7 days did not differ significantly from control scores of chronic treatment. Hoarding activity increased unexpectedly after withdrawal of the chronic amphetamine treatment. This sign of amphetamine abstinence appeared to have an initial phase for about 7 days with a maximal hoarding score on the fourth day, and a second protracted disappearance phase for at least 2 additional weeks. To the knowledge of the authors only slight sedation (Lewander, 1968) and a small rebound increase in food intake (Tormey and Lasagna, 1960) after withdrawal of amphetamine in tolerant rats have been noticed previously. In humans an increase in rapid eye movement sleep been documented during amphetamine withdrawal in amphetamine dependent subjects (OSWALD and THACORE, 1963).

Acknowledgements—The study was supported by the Swedish Medical Research Council (project No B73-04X-1017-09, TL).

REFERENCES

BLUNDELL J. E. (1971) *Psychopharmacologia (Berl.)* **22**, 224–229.
BRODIE B. B., CHO A. K., STEFANO F. J. E. and GESSA G. L. (1969) In: *Advances in Biochemical Psychopharmacology* Vol. **1**, (COSTA E. and GREENGARD P., Eds.) pp. 219–238.
KALANT H., LEBLANC A. E. and GIBBINS R. J. (1971) *Pharmacological Reviews* **23**, 135–191.
LEWANDER T. (1968) *Psychopharmacologia (Berl.)* **13**, 394–407.
LEWANDER T. (1970) *Acta Universitatis Upsaliensis* **96**, 1–24.
LEWANDER T. (1971a) *Naunyn Schmiedebergs Arch. Pharmak.* **271**, 211–233.
LEWANDER T. (1971b) *Psychopharmacologia (Berl.)* **21**, 17–31.
LEWANDER T. (1972) In *Biochemical and Pharmacological Aspects of Dependence and Reports on Marihuana Research.* (VAN PRAAG H. M., ed.) pp. 69–84. De ERVEN F. BOHN N. V., HAARLEM.
KOSMAN M. E. and UNNA H. R. (1968) *Clin. Pharmacol. Ther.* **9**, 240–254.
OSWALD I. and THACORE W. A. (1963) *Brit. Med. J.* **2**, 427–431.
SIEGEL S. (1956) *Nonparametric Statistics.* (MORGAN O. T. ed.) McGraw-Hill, New York.
TORMEY J. and LASAGNA L. (1960) *J. Pharmacol. exp. Ther.* **128**, 201–209.

Frontiers in Catecholamine Research 1973, pp. 991 to 994. Pergamon Press. Printed in Great Britain.

BEHAVIOURAL AND BIOCHEMICAL EFFECTS OF L-DOPA IN PSYCHIATRIC PATIENTS*

BURTON M. ANGRIST, M.D.[1], GREGORY SATHANANTHAN, M.D.[1],
SHERWIN WILK, Ph.D.[2] and SAMUEL GERSHON, M.D.[1]

Neuropsychopharmacology Research Unit, Department of Psychiatry, New York University
Medical Center, 550 First Avenue, New York, N.Y. 10016, U.S.A. and Department of
Pharmacology, Mt. Sinai School of Medicine, 100th Street and Fifth Avenue, New York,
N.Y. 10029, U.S.A.

PRIOR studies in our laboratories have suggested that dopamine may play an important role in the etiology of amphetamine psychoses (ANGRIST et al., 1971): this was consistent with an increasing body of evidence linking central dopaminergic hyperactivity with animal stereotypy and some forms of human psychosis. This evidence is derived from several sources. The extensive data indicating that animal stereotyped behaviour is dopaminergically mediated has been reviewed by RANDRUP and MUNKVAD (1970). Many of the drugs that induce this behaviour have been shown to be capable of causing psychotic states (GRIFFITH et al., 1970; RYLANDER, 1969; SPENSLEY and ROCKWELL, 1972) which bear a striking resemblance to some forms of schizophrenia (CONNELL, 1958; ANGRIST and GERSHON, 1970). On the other hand, drugs which block central dopamine receptors both block stereotyped behaviour and frequently prove to be clinically effective as neuroleptics (ANDEN et al., 1970; JANSSEN et al., 1965; VAN ROSSUM, 1966; SNYDER, 1973).

Administration of the dopamine precursor, L-dopa to Parkinsonian patients has frequently precipitated psychotic reactions of various types (BROGDEN et al., 1972; GOODWIN, 1972), a finding consistent with this hypothesis. Accordingly, it was felt desirable to attempt a controlled assessment of the behavioural effects of L-dopa in psychiatric patients whose pre-drug behavioural status was well documented and who were free of neurologic disease.

METHODS

The methodology for this study is reported in detail elsewhere (ANGRIST et al., 1973). Briefly, patients who consented to participate were maintained on matched placebo for 5–10 days before L-dopa was administered. During this time no other drugs were given with the exception of chloral hydrate or sodium amytal as required for agitation or insomnia. L-Dopa was initiated at a dosage of 750 mg/day and then gradually increased to the maximum daily dosage that was tolerated. A clinical laboratory work-up and EKG were done at baseline and weekly. Behaviour was assessed both by daily clinical notes and by a weekly BPRS and CGI. In those patients who consented to lumbar puncture, this was done at baseline and at the maximum dose. Cerebrospinal fluid was analysed for MHPG (WILK et al., 1971) and HVA (GERBODE and BOWERS, 1968). Results were as follows.

RESULTS

Schizophrenic patients

Several patients showed deterioration during the placebo period and were dropped from the study and treated with standard neuroleptics. Ten patients received L-dopa

* This work was supported by USPHS Grant MH 04669.

for a mean of 21·2 days (range 11–29 days). The mean maximum daily dose attained was 5·050 g (range 3–6 g). This resulted in significant deterioration on the BPRS with respect to conceptual disorganization, emotional withdrawal and agitation-excitement (analyses of variance repeated measures model for both raters).

While all patients showed behavioural deterioration, a qualitative distinction into two response patterns could be made. Three patients showed stimulant effects with psychomotor activation, loquaciousness leading to freer expression of pathology, intrusiveness and irritability. No changes, however, were noted in their original psychotic symptomatology.

A second response pattern was shown by 7 of 10 patients. This consisted of both the stimulant effects noted in the first group and a dose-related increase in the original symptom pattern present at baseline. Deterioration was particularly prominent with regard to thought disorder, emotional lability, bizarre behaviour, agitation and hallucinations. Two patients showed the emergence of *de novo* symptoms while on L-dopa, auditory hallucinations in one case and mannerisms in the second.

Non-schizophrenic patients

These patients tolerated larger doses of L-dopa. The mean maximum daily dose was 8·830 g (range 6–10 g). The mean duration of administration was 22 days (range 19–28 days). The responses of this group (6 patients) were variable. The first patient showed hypersexuality and compulsive masturbation without other signs suggestive of hypomania such as euphoria or hyperactivity. The second showed no objective behavioural effects and tolerated dosages of 10 g L-dopa/day. A third patient abruptly developed a toxic confusional state at 9 g L-dopa/day. Two other patients showed a combination of nausea, diaphoresis, a sense of dysphoric stimulation and anxiety. One of these developed ideas of reference. The remaining patient, who had no history of psychosis or psychiatric signs suggestive of schizophrenia developed a paranoid schizophreniform psychosis. This 39-year-old male had been a successful stock broker but had been drinking heavily for 2 years. At a dose of 9 g L-dopa/day, he became intrusive and would not permit conversations to be terminated. He compulsively dwelt on past injustices done to him by a brother-in-law who, he claimed, had unjustly accused him of "interstate narcotics sales" and had dealings with "the mafia". He claimed his own lawyer had joined the conspiracy and tried to have him sent to jail because of vaguely described "big business interests". The connection between the lawyer and brother-in-law could not be clearly explained but he alleged that both were part of a "web" created to ruin him. Changes in formal aspects of speech were striking. He was compulsively circumstantial, unable to give a focused account and constantly introduced irrelevant details so that his line of thought became so rambling that it could not be followed. His speech showed occasional blocking and his affect appeared incongruously blunted when compared to the content of his speech. These signs resolved completely within 16 hr after L-dopa was discontinued and 5 mg haloperidol was administered intramuscularly. Thus, 1 of these 6 showed a clear psychotic reaction to L-dopa administration.

Biochemical data

CSF levels of MHPG and HVA before and after L-dopa administration are given in Table 1.

TABLE 1. CEREBROSPINAL FLUID LEVELS OF MHPG AND HVA
(ng/ml) BEFORE AND AFTER L-dopa ADMINISTRATION

Patient No.	MKPG Pre/Post	HVA Pre/Post	L-dopa dosage (mg/day)
1	17/7	0/110	4·000
2	11/12	0/123	3·000
3	9/13	55/78	6·000
4	7/15	19/202	3·000
5	15/8	0/26	3·000
6	14/12	0/82	3·000
7	5/10	0/271	9·000
8	5/11	0/574	9·000
9	7/9	17/235	10·000
10	6/8	30/129	6·000

DISCUSSION

Since L-dopa induces CNS stimulation one might question whether the behavioural deterioration of the schizophrenic patients was due to increased dopaminergic activity or represents an inability of this group to tolerate non-specific CNS stimulation of any sort. We attempted to clarify this by administering caffeine, which does not induce stereotypy in animals (WILLNER et al., 1970), to these subjects. All showed tremor, anxiety and increased heart rate but none showed exacerbation of psychotic symptoms. This suggests that if the two can be separated, the behavioural worsening observed was more likely secondary to dopaminergic events than to non-specific stimulation.

The precipitation of a schizophreniform psychosis by administration of L-dopa in a patient documented to be non-schizophrenic prior to receiving the drug also supports the hypothesised relationship between psychosis and central dopaminergic activity that has been proposed by several investigators. However, the variability of response in the non-schizophrenics, the differing response patterns in the schizophrenics, and the differences in sensitivity of the two patient groups to the drug all suggest the concept of an individually variable threshold of dopaminergic activity that can be tolerated without psychosis. Such a concept would be consistent with recently reported observations by JANOWSKI et al. (1973) of striking variation in response to intravenous methylphenidate in psychiatric patients both across diagnostic categories and in the same patient in differing phases of his disease. This concept would also be consistent with the marked variability in sensitivity to the psychotogenic effects of amphetamine that have been noted when this drug has been given in large doses (GRIFFITH et al., 1970; ANGRIST and GERSHON, 1970).

REFERENCES

ANDEN N. E., BUTCHER S. G., CORRODI H., FUXE K. and UNGERSTEDT U. (1970) Europ. J. Pharmacol. 11, 303–314.
ANGRIST B. M. and GERSHON S. (1970) Biol. Psychiat. 2, 95–107.
ANGRIST B. M., SATHANANTHAN G. and GERSHON S. (1973) Psychopharmacologia (in press).
ANGRIST B. M., SHOPSIN B. and GERSHON S. (1971) Nature, Lond. 234, 152–153.
BROGDEN R. N., SPEIGHT T. M. and AVERY G. S. (1971) Drugs 2, 257–408.
CONNELL P. H. (1958) In: Amphetamine Psychosis. Maudsley Monographs No. 5, Oxford University Press.

Gerbode F. and Bowers M. B., JR (1968) *J. Neurochem.* **15,** 105–1055.

Goodwin F. K. (1962) In: *Psychiatric Complications of Medical Drugs.* (Shader R. C., Ed.) pp. 149–174. Raven Press, New York.

Griffith J. J., Cavanaugh J. and Oates J. (1970) In: *Psychotomimetic Drugs.* (Efron D. H., Ed.) pp. 287–294. Raven Press New York.

Janowsky D. S., El-Yousef K., Davis J. M. and Sckerke H. J. (1973) *Archs. Gen. Psychiat.* **28,** 185–191.

Janssen P. A. J., Niemegeers J. E. and Schellekens K. H. L. (1965) *Arzneim. Forsch.* **15,** 104–117.

Randrup A. and Munkvad I. (1970) In: *Amphetamines and Related Compounds* (Costa E. and Garattini S., Eds) pp. 695–713.

Rylander G. (1969) In: *Abuse of Central Stimulants* (Sjoqvist F. and Tattie M., Eds.) pp. 251–273. Raven Press, New York.

Snyder S. H. (1973) *Am. J. Psychiat.* **130,** 61–67.

Spensley J. and Rockwell D. A. (1972) *New Eng. J. Med.* **286,** 880–881.

Van Rossum J. M. (1966) *Archs. Int. Pharmacodyn. Ther.* **160,** 492–494.

Wilk S., Davis K. L., Thacker S. B. (1971) *An. Biochem.* **39,** 498.

Willner J. H., Samach M., Angrist B. M., Wallach M. B. and Gershon S. (1970) *Comm. Behav. Biol.* **5,** 135–149.

Frontiers in Catecholamine Research 1973, pp. 995 to 1001. Pergamon Press. Printed in Great Britain.

HALOGEN SUBSTITUTION OF AMPHETAMINE BIOCHEMICAL AND PHARMACOLOGICAL CONSEQUENCES

F. SULSER and E. SANDERS-BUSH

Department of Pharmacology, Vanderbilt University School of Medicine,
and
Tennessee Neuropsychiatric Institute, Nashville, Tennessee, U.S.A.

METABOLIC CONSEQUENCES OF HALOGEN SUBSTITUTION IN p-POSITION OF AMPHETAMINE

THE MAJOR pathways for the metabolism of amphetamine involve p-hydroxylation of the aromatic ring of amphetamine to p-hydroxyamphetamine (POH) followed by β-hydroxylation of the d-isomer to p-hydroxynorephedrine or oxidative deamination of the side chain. The relative extent of these two routes of metabolism varies markedly with the species (AXELROD, 1954; ELLISON et al., 1966; DRING et al., 1970). In the rat, where p-hydroxylation is the predominant pathway, one would expect that p-substituted derivatives of amphetamine such as p-chloroamphetamine (PCA) should be metabolised much more slowly than amphetamine (FULLER and HINES, 1967). MILLER et al. (1971) have demonstrated that the biological half-life of PCA in the rat was about seven times that of amphetamine, while in mice, where p-hydroxylation is less important, the rate of metabolism of PCA closely approximated that of amphetamine. Moreover, iprindole, an inhibitor of the aromatic hydroxylation of amphetamine (FREEMAN and SULSER, 1972) markedly enhances and prolongs the action of amphetamine, causes increased levels of the drug in brain and prolongs the half-life of amphetamine from 45 to 190 min. As expected, however, the pharmacological action and the half-life of PCA are not influenced by iprindole in the rat. Halogen substitution in p- or m-position changes the subcellular distribution of the amphetamine derivatives. Thus, PCA and m-chloroamphetamine (in DMI pretreated rats) are present mainly in the particulate fraction of brain homogenates whereas amphetamine is mainly localised in the supernatant fraction (FULLER et al., 1972). Whether the localisation of PCA in the particulate fraction is due to uptake into synaptic vesicles has not been established but WONG et al. (1972) have recently demonstrated that PCA is associated with the synaptosomal fraction of brain homogenates. β,β-Difluoro substitution of PCA shortens the half-life of PCA and leads to a marked accumulation of the drug in fat tissue while in all other tissues, the difluoro compound is present in concentrations lower than those of PCA (FULLER et al., 1973b). Interestingly, desmethylimipramine which enhances the levels of amphetamine in rat brain through inhibition of its metabolism by para-hydroxylation (SULSER et al., 1966; CONSOLO et al., 1967; LEWANDER, 1969) does not affect the levels of β,β-difluoroamphetamine (FULLER et al., 1973a), suggesting that the metabolism of the β,β-difluoro derivative occurs by a pathway other than para-hydroxylation. PARLI and LEE (1972) have recently demonstrated the oxime of difluorophenylacetone in free and conjugated form in the urine of rats given β,β-difluoroamphetamine thus indicating oxidative deamination. The differences in the half-life of PCA in rats and mice (Miller et al., 1971) and between PCA and β,β-difluoro-PCA in rats (Fuller et al., 1973b) are reflected in differences in the duration of the effects of the drugs on brain 5-hydroxyindoles.

It is noteworthy that PCA like POH is a substrate for β-hydroxylase and that the rate of formation of p-chloronorephedrine from PCA approximates that of p-hydroxynorephedrine from POH (personal communication by Dr. F. C. Brown). Since it has been demonstrated that d-POH but not l-POH is a substrate for β-hydroxylase (Goldstein and Anagnoste, 1965), it will be of interest to investigate whether or not the conversion of PCA to p-chloronorephedrine is also stereoselective for the d-isomer and to study the contributions, if any, of p-chloronorephedrine to the action of the parent drug on adrenergic and serotonergic mechanisms.

EFFECT OF HALOGEN SUBSTITUTION OF AMPHETAMINE ON ADRENERGIC MECHANISMS

p-Chlorinated derivatives of amphetamine exert pharmacological effects which are similar to those of the parent compounds (Nielsen et al., 1967; Frey and Magnussen, 1968). In contrast to amphetamine, PCA does not alter the concentration of either norepinephrine or dopamine in brain but markedly changes the metabolism of cerebral serotonin (Pletscher et al., 1964; 1966; Fuller et al., 1965). The initial central excitatory action of PCA is, like that of amphetamine, related to its effect on the metabolism of cerebral catecholamines (Strada et al., 1970). Like amphetamine, PCA causes a large increase in the concentration of normetanephrine and a reduction in the levels of the deaminated metabolites of ^{3}H-norepinephrine. Accordingly, similar mechanisms might be responsible for the effects: release of catecholamines, blockade of their reuptake and possibly either direct inhibition of MAO (Glowinski and Axelrod, 1965; Glowinski et al., 1966; Fuller, 1966) or indirect inhibition of the enzyme resulting from the blockade of neuronal uptake of catecholamines (Rutledge, 1970). The effects elicited by PCA on the metabolism of ^{3}H-norepinephrine are, however, more pronounced and longer lasting in accordance with its longer biological half-life (Strada et al., 1970; Carr and Moore, 1970).

Studies with tyrosine hydroxylase inhibitors have lead to the view that the central action of amphetamine depends on an uninterrupted synthesis of catecholamines (Weissman et al., 1966; Hansen, 1967; Dingell et al., 1967; Sulser et al., 1968; Scheel-Krüger, 1971) whereas that of PCA appears to be mediated through the release of stored catecholamines (Strada and Sulser, 1971). Recent results from our laboratory strengthen this view. Amphetamine and PCA were studied in animals whose noradrenergic terminals in brain had been destroyed by intraventricular 6-hydroxydopamine. This procedure reduces the level of norepinephrine to about 16 per cent of its control value while that of dopamine in the striatum is decreased only by about 40 per cent. The activity of tyrosine hydroxylase, measured by the coupled assay of Waymire et al. (1972) and expressed as nCi^{14}CO$_2$/30 min/20mg tissue, decreases in the striatum from $22\cdot8 \pm 1\cdot1$ to $7\cdot7 \pm 0\cdot9$ and in the diencephalon from $3\cdot5 \pm 0\cdot1$ to $2\cdot1 \pm 0\cdot1$. It is of interest that such a procedure does not alter the central action of PCA and only slightly reduces that of amphetamine. Reserpinisation of animals (depletion of remaining stores) whose noradrenergic neurons were destroyed by 6-hydroxydopamine, completely blocks the action of PCA and either enhances or does not change that of amphetamine (Fig. 1). These data provide more

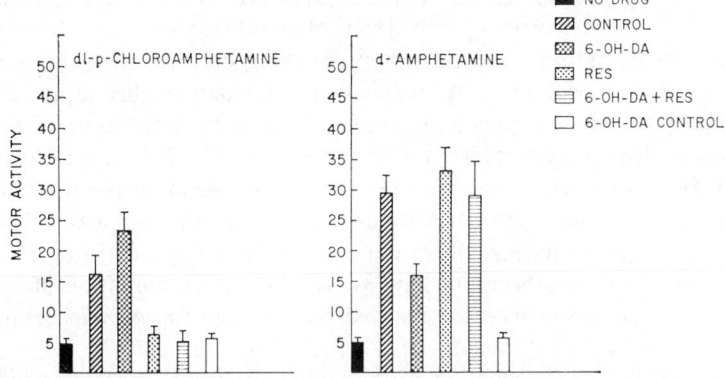

FIG. 1.—Modification by 6-hydroxydopamine (6-OH-DA) and reserpine (RES) of the psychomotor stimulation elicited by *d*-amphetamine (3mg/kg i.p.) and *dl-p*-chloroamphetamine (PCA; 5mg/kg i.p.). Two daily doses of 6-OH-DA (250μg in 10μl) were injected intraventricularly through a polyethylene cannula. The animals were used one week after the last dose of 6-OH-DA. Reserpine was administered intraperitoneally (5mg/kg) 4 hr prior to the injection of amphetamine or PCA. Psychomotor stimulation was measured in Williamson activity cages. The data are expressed as integrated counts $\pm$ S.E.M.

direct evidence that the central stimulatory action of amphetamine is mediated predominantly through newly synthesised catecholamines, whereas the action of PCA depends on the store of catecholamines, predominantly dopamine. These results are also compatible with data showing that motor activity of rats can be increased by doses of PCA which increase the turnover rate of dopamine but not necessarily that of norepinephrine in various parts of the brain (COSTA *et al.*, 1971). Since the steady state concentration of dopamine remains unchanged after PCA, the data might suggest that PCA alters the synthesis of catecholamines, particularly dopamine in nerve terminals and cell bodies in brain. However, PCA (5 mg/kg i.p.) does not change the activity of tyrosine hydroxylase in the striatum or diencephalon (Table 1). Moreover, the *in vitro* addition of PCA (10^{-9}–10^{-5} M) does not alter the activity of tyrosine hydroxylase in preparations from striatum or diencephalon. Since we measured total enzyme activity, we cannot rule out that the drug may cause a shift in the activity of tyrosine hydroxylase from the soluble to the particulate 'synaptosomal' fraction as has been reported to occur with methamphetamine (MANDELL *et al.*, 1972).

TABLE 1. EFFECT OF *p*-CHLOROAMPHETAMINE (PCA) ON THE ACTIVITY OF TYROSINE HYDROXYLASE

Time after PCA* (hr)	Tyrosine hydroxylase activity nCi $^{14}CO_2$/30 min/20mg tissue			
	Striatum		Diencephalon	
	Control	PCA	Control	PCA
1	21·90 $\pm$ 0·78	19·28 $\pm$ 1·15	3·78 $\pm$ 0·20	3·48 $\pm$ 0·10
4	17·80 $\pm$ 1·31	17·89 $\pm$ 1·85	3·71 $\pm$ 0·20	3·69 $\pm$ 0·14

dl-PCA was administered intraperitoneally (5mg/kg).
Tyrosine hydroxylase activity was measured according to WAYMIRE *et al.* (1972).

EFFECT OF HALOGEN SUBSTITUTION OF AMPHETAMINE ON SEROTONERGIC MECHANISMS

With regard to the serotonergic system, striking differences in the biochemical effects of amphetamine and PCA are evident. Unlike amphetamine, the chloro derivative causes a simultaneous and prolonged decrease in the brain levels of 5HT and its principal metabolite 5HIAA (PLETSCHER et al., 1964). A decrease in the turnover of 5HT also occurs as demonstrated by a decrease in the rise of brain 5HIAA after the administration of probenecid (SANDERS-BUSH and SULSER, 1970). COSTA and REVUELTA (1972a) have confirmed these results using an isotopic technique. Recent studies from our laboratory have suggested that an inhibition of brain tryptophan hydroxylase can explain the marked reduction in turnover of 5HT after the administration of PCA (SANDERS-BUSH et al., 1972a).

The mechanism of the reduction of the activity of tryptophan hydroxylase after the administration of PCA is not yet understood. One day after the in vivo administration of PCA, a dose related reduction in tryptophan hydroxylase was found (Table 2). However, in agreement with data of PLETSCHER et al. (1970), PCA did not modify the activity of tryptophan hydroxylase when added in vitro. Kinetic studies of enzymes isolated from control rats and from rats treated with PCA showed that the drug did not change the apparent K_m for either tryptophan (3.0×10^{-4} M) or DMPH$_4$ (1.5×10^{-4} M). Moreover, the reduction of tryptophan hydroxylase in preparations from animals treated with PCA is not reversed by dialysis. These results suggest that the administration of PCA may reduce the amount of active enzyme without altering its properties. Therefore, it was important to examine the time course of the reduction in the activity of tryptophan hydroxylase and brain 5HT by PCA and to compare it to that caused by p-chlorophenylalanine, an irreversible inhibitor of tryptophan hydroxylase. Both drugs cause a marked reduction in the activity of tryptophan hydroxylase and the levels of 5HT in brain after one day. However, two weeks after injection, the effect of p-chlorophenylalanine has disappeared while both enzyme activity and levels of 5HT are still maximally reduced in rats treated with PCA (SANDERS-BUSH et al., 1972b). Even 4 months after a single dose of PCA, although some recovery has occurred, the levels of 5HT and 5HIAA and the activity of tryptophan hydroxylase are still significantly reduced (Table 3). The parent compound

TABLE 2. EFFECT OF THE INTRAPERITONEAL ADMINISTRATION OF p-CHLOROAMPHETAMINE ON THE ACTIVITY OF TRYPTOPHAN HYDROXYLASE

Dose (mg/kg*)	Tryptophan hydroxylase activity	
	nCi^{14}5C-HT formed/g/ hr ± s.e.	Per cent inhibition
0 (5)	31·4 ± 2·9	0
2 (4)	25·2 ± 0·4	20
5 (7)	21·0 ± 1·2	33
7·5 (7)	18·0 ± 2·2	43
10 (7)	12·5 ± 1·2	60

* p-Chloroamphetamine was administered, i.p., 16 hr prior to sacrifice. The number of animals is indicated in parentheses.
Tryptophan hydroxylase activity was measured according to the procedure described by SANDERS-BUSH et al. (1972a).

TABLE 3. CEREBRAL LEVELS OF 5HT AND 5HIAA AND ACTIVITY OF TRYPTOPHAN
HYDROXYLASE AT VARIOUS TIMES AFTER A SINGLE DOSE OF *p*-CHLOROAMPHETAMINE

Time after injection	Percent of control $\pm$ s.e.m.		
	5HT	5HIAA	Tryptophan hydroxylase
16 hr	39·0 $\pm$ 3·6(4)*	39·6 $\pm$ 3·0(4)*	57·6 $\pm$ 1·3(4)*
4 days	39·6 $\pm$ 5·1(5)*	31·5 $\pm$ 2·7(8)*	52·3 $\pm$ 7·9(5)†
10 days	38·2 $\pm$ 3·3(5)*	—	49·6 $\pm$ 1·5(5)*
2 weeks	47·9 $\pm$ 3·2(9)*	25·8 $\pm$ 2·1(6)*	41·4 $\pm$ 7·6(9)*
4 weeks	62·6 $\pm$ 4·5(11)†	35·3 $\pm$ 5·7(5)*	57·7 $\pm$ 6·6(11)*
6 weeks	73·1 $\pm$ 2·3(4)*	49·2 $\pm$ 3·7(4)*	72·3 $\pm$ 7·7(4)†
2 months	73·2 $\pm$ 2·7(5)†	—	—
4 months	79·5 $\pm$ 8·5(4)‡	60·1 $\pm$ 3·9(7)*	60·5 $\pm$ 3·0(4)†

Rats were injected i.p. with a single dose of either saline or 10 mg/kg of *p*-chloroamphetamine. The animals were sacrificed at various times after the injection. Results are mean values and are expressed as percentage of the respective control values. The number of animals is shown in parentheses. Mean values for all control animals were: 5HT, 0·69 $\pm$ 0·02 μg/g ($n = 43$): 5HIAA, 0·29 $\pm$ 0·01 μg/g ($n = 33$); tryptophan hydroxylase, 78·1 $\pm$ 3·9 nCi ^{14}C-5HT formed/g/hr ($n = 39$). (From SANDERS-BUSH *et al.*, 1972b)
*$P < 0.001$. †$P < 0.01$. ‡$P < 0.05$.

amphetamine does not reduce the activity of tryptophan hydroxylase after a single acute dose or after chronic treatment with the drug.

Recently we have found another important difference between the action of *p*-chlorophenylalanine and PCA. Unlike *p*-chlorophenylalanine, the amphetamine derivative does not reduce the synthesis of 5HT in peripheral organs such as intestine. Thus, the mechanism of the decrease in the activity of tryptophan hydroxylase after these two drugs must certainly be different.

Other halogen substituted phenylethylamine derivatives which decrease brain 5HT have been described. In rats the *meta*-chloro derivative of amphetamine only lowers 5HT in animals treated with desmethylimipramine, which increases the brain levels of the drug presumably by inhibiting its *para*-hydroxylation (FULLER *et al.*, 1972). The ortho-derivative does not lower brain 5HT even after treatment with desmethylimipramine. Other derivatives of amphetamine with marked effects on brain 5HT are fenfluramine (DUHAULT and VERDAVAINNE, 1967) and norfenfluramine (MORGAN *et al.*, 1972). The mechanism for the decrease in 5HT after PCA and norfenfluramine is apparently different. Although both drugs cause a decrease of 5HT and 5HIAA in brain, PCA decreases the turnover of brain 5HT (SANDERS-BUSH and SULSER, 1970; COSTA and REVUELTA, 1972a) while norfenfluramine has been reported to increase it (COSTA and REVUELTA, 1972b).

OTHER DIFFERENTIAL EFFECTS BETWEEN AMPHETAMINE AND ITS *p*-CHLORINATED DERIVATIVES

The bulk of evidence indicates that the CNS stimulation elicited by both amphetamine and PCA is mediated through catecholamines. Since norepinephrine has been shown to increase the level of cyclic AMP in rat brain slices *in vitro* (KAKIUCHI and RALL, 1968; PALMER *et al.*, 1972) and a dopamine sensitive adenyl cyclase has been found in the cerebral cortex (MCCUNE *et al.*, 1971) and in the caudate nucleus (KEBABIAN *et al.*, 1972), an increase in the level of endogenous cyclic AMP would be expected following the administration of amphetamine or PCA. Studies from our laboratory have shown, however, that amphetamine causes no detectable effects in the concentrations of cyclic AMP in any brain area at any time even though marked behavioural activation and increased sympathetic activity are evident (SCHMIDT *et al.*, 1972). Moreover, *dl*-PCA actually causes a decrease in the concentration of cyclic AMP in brain (PALMER *et al.*, 1972). These unexpected and puzzling findings remain a challenge for further investigation.

The clinical profile of *p*-chlorinated derivatives of amphetamine also differs from that of the parent drugs. Thus, while amphetamine has no value in the treatment of depressive illness, *p*-chloromethamphetamine has been reported to be a true antidepressant without causing central motor stimulation and insomnia (VAN PRAAG *et al.*, 1969; 1971).

Acknowledgements—The original investigations reported in this paper have been supported by United States Public Health Service Grant MH–11468.

REFERENCES

AXELROD J. (1954) *J. Pharmacol. Exp. Therap.* **110**, 315–326.
CARR L. A. and MOORE K. E. (1970) *Biochem. Pharmacol.* **19**, 2671–2675.
CONSOLO S., DOLFINI E., GARATTINI S. and VALZELLI L. (1967) *J. Pharm. Pharmacol.* **19**, 253–256.
COSTA E., NAIMZADA K. M. and REVUELTA A. (1971) *Br. J. Pharmacol.* **43**, 570–579.
COSTA E. and REVUELTA A. (1972a) *Neuropharmacology* **11**, 291–295.
COSTA E. and REVUELTA A. (1972b) *Biochem. Pharmacol.* **21**, 2385–2393.
DINGELL J. V., OWENS M. L., NORVICH M. R. and SULSER F. (1967) *Life Sci.* **6**, 1155–1162.
DRING L. G., SMITH R. L. and WILLIAMS R. T. (1970) *Biochem. J.* **116**, 425–435.
DUHAULT J. and VERDAVAINNE C. (1967) *Archs. Int. Pharmacodyn.* **170**, 276–286.
ELLISON T., GUTZATI L. and VAN LOON E. J. (1966) *J. Pharmacol. Exp. Ther.* **152**, 383–387.
FREEMAN J. J. and SULSER F. (1972) *J. Pharmacol. Exp. Ther.* **183**, 307–315.
FREY H. H. and MAGNUSSEN M. P. (1968) *Biochem. Pharmacol.* **17**, 1299–1307.
FULLER R. W. (1966) *Life Sci.* **5**, 2247–2252.
FULLER R. W., HINES C. W. and MILLS J. (1965) *Biochem. Pharmacol.* **14**, 483–488.
FULLER R. W. and HINES C. W. (1967) *J. Pharm. Sci.* **56**, 302–303.
FULLER R. W., SCHAFFER R. J., ROUSH B. W. and MOLLOY B. B. (1972) *Biochem. Pharmacol.* **21**, 1413–1417.
FULLER R. W., MOLLOY B. B. and PARLI C. J. (1973a) In: *Advances in Neuropsychopharmacology*, Avicencun Press, Praha, in press.
FULLER R. W., SNODDY H. D. and MOLLOY B. B. (1973b) *J. Pharmacol. Exp. Ther.* **184**, 278–284.
GLOWINSKI J. and AXELROD J. (1965) *J. Pharmacol. Exp. Ther.* **149**, 43–49.
GLOWINSKI J., AXELROD J. and IVERSEN L. L. (1966) *J. Pharmacol. Exp. Ther.* **153**, 30–41.
GOLDSTEIN M. and ANAGNOSTE B. (1965) *Biochem. Biophys. Acta.* **107**, 166–168.
HANSEN L. C. F. (1967) *Psychopharmacologia* **10**, 289–297.
KAKIUCHI S. and RALL T. W. (1968) *Molec. Pharmacol.* **4**, 367–378.
KEBABIAN J. W., PETZOLD G. L. and GREENGARD P. (1972) *Proc. Nat. Acad. Sci.* **69**, 2145–2149.
LEWANDER T. (1969) *Europ. J. Pharmacol.* **6**, 38–44.
MANDELL A. J., KNAPP S., KUCZENSKI R. T. and SEGAL D. S. (1972) *Biochem. Pharmacol.* **21**, 2737–2750.
McCUNE R. W., GILL T. H., VON HUNGER K. and ROBERTS S. (1971) *Life Sci.* **10**, Part II, 443–450.
MILLER K. W., SANDERS-BUSH E. and DINGELL J. V. (1971) *Biochem. Pharmacol.* **20**, 500–503.
MORGAN D., LÖFSTRANDH S. and COSTA E. (1972) *Life Sci.* **11**, 83–93.
NIELSEN C. K., MAGNUSSEN M. P., KAMPMANN E. and FREY H. H. (1967) *Archs. Internat. Pharmacodyn.* **170**, 428–444.
PALMER G. C., ROBISON G. A., MANIAN A. and SULSER F. (1972) *Psychopharmacologia* **23**, 201–211.
PARLI C. J. and LEE N. W. (1972) In: *Abstracts, Fifth Int. Congr. Pharmacology*, San Francisco, p. 176.
PLETSCHER A., BARTHOLINI G., BRUDERER H., BURKARD W. P. and GEY K. F. (1964) *J. Pharmacol. Exp. Ther.* **145**, 344–350.
PLETSCHER A., DA PRADA M., BURKARD W. P., BARTHOLINI G., STEINER F. A., BRUDERER H. and BIGLER F. (1966) *J. Pharmacol. Exp. Ther.* **154**, 64–72.
PLETSCHER A., DA PRADA M. and BURKARD W. P. (1970) In: *Amphetamines and Related Compounds*, (Eds. COSTA E. and GARATTINI S.) Raven Press, N.Y., 331–342.
PRAAG VAN, H. M. KITS T. P., SCHUT T. and DIJKSTRA P. (1969) *Behav. Neuropsychiat.* **1**, 17–24.

PRAAG VAN, H. M., SCHUT T., BOSMA E. and BERGH VANDEN R. (1971) *Psychopharmacologia* **20**, 66–76.

RUTLEDGE C. O. (1970) *J. Pharmacol. Exp. Ther.* **171**, 188–195.

SANDERS-BUSH E. and SULSER F. (1970) *J. Pharmacol. Exp. Ther.* **175**, 419–426.

SANDERS-BUSH E., BUSHING J. A. and SULSER F. (1972a) *Biochem. Pharmacol.* **21**, 1501–1510.

SANDERS-BUSH E., BUSHING J. A. and SULSER F. (1972b) *Europ. J. Pharmacol.* **20**, 385–388.

SCHEEL-KRÜGER J. (1971) *Europ. J. Pharmacol.* **14**, 47–59.

SCHMIDT M. J., HOPKINS J. T., SCHMIDT D. E. and ROBISON G. A. (1972) *Brain Res.* **42**, 465–477.

STRADA S. J., SANDERS-BUSH E. and SULSER F. (1970) *Biochem. Pharmacol.* **19**, 2621–2629.

STRADA S. J. and SULSER F. (1971) *Europ. J. Pharmacol.* **15**, 45–51.

SULSER F., OWENS M. L. and DINGELL J. V. (1966) *Life Sci.* **5**, 2005–2010.

SULSER F., OWENS M. J., NORVICH M. R. and DINGELL J. V. (1968) *Psychopharmacologia* **12**, 322–332.

WAYMIRE J. C., BJUR R. and WEINER N. (1971) *Anal. Biochem.* **43**, 588–600.

WEISSMAN A., KOE K. B. and TENEN S. St. (1966) *J. Pharmacol. Exp. Therap.* **151**, 339–352.

WONG D. T., HORUG J. S., VAN FRANK R. M. and FULLER R. W. (1972) *J. Pharm. Pharmacol.* **24**, 171–173.

Frontiers in Catecholamine Research 1973, pp. 1003 to 1010. Pergamon Press. Printed in Great Britain.

NARCOTIC ANALGESICS AND THE REGULATION OF NEURONAL CATECHOLAMINE STORES

E. Costa, A. Carenzi, A. Guidotti and A. Revuelta

Laboratory of Preclinical Pharmacology, National Institute of Mental Health,
St. Elizabeth's Hospital, Washington, D.C. 20032, U.S.A.

INTRODUCTION

Several putative central neurotransmitters (Maynert, 1967; Way *et al.*, 1968; Clouet, 1971) including noradrenaline and dopamine (Clouet and Ratner, 1970; Sesame *et al.*, 1972; Smith *et al.*, 1972; Loh *et al.*, 1973) have been implicated in the pharmacological actions of morphine. Clouet and Ratner (1970) measured radioactive 3,4-dihydroxyphenylalanine (DOPA) norepinephrine (NE) and dopamine (DA) in various brain parts of rats receiving acutely (1 or 2 hr before labeling) or chronically (1 dose per day for 5 days) 210 μmoles/kg of morphine. They found that morphine increases the conversion of ^{14}C tyrosine into catecholamines and that tolerance to this morphine effect fails to develop. In contrast a report by Smith *et al.* (1972) shows that tolerance to the analgesic effects of morphine is associated with tolerance to the increase of catecholamine synthesis elicited by morphine. Loh *et al.* (1973) also performed a study of radioactive tyrosine conversion into NE and DA in the whole brain of mice. They used two doses of morphine (13 and 52 μmoles/kg) and estimated the turnover rate of brain catecholamines using the accumulation index which takes in account the specific activity of the precursor tyrosine. They found that although morphine (52 μmoles/kg s.c.) increases the conversion of radioactive tyrosine into brain DA and NE it does not significantly change the accumulation index. Friedler *et al.* (1972) have reported that 6-hydroxydopamine (6OHDA) which causes a selective degeneration of the adrenergic neurons in the brain, reduces morphine analgesia, but it does not prevent the development of morphine tolerance. Moreover, the pretreatment with 6OHDA exacerbates the signs of morphine withdrawal.

ESTIMATION OF CATECHOLAMINE TURNOVER RATE: EFFECT OF ACUTE MORPHINE, CHRONIC MORPHINE, VIMINOL AND NALOXONE

We have estimated the turnover rate of cerebellar NE, spinal cord NE and striatal DM by injecting intravenously 1 mCi/kg of L-tyrosine 3,5-^{3}H (1 mCi/33 nmoles) and by measuring the specific radioactivity of tyrosine, NE or DA. We selected these brain areas, because each area contains only dopaminergic or noradrenergic axons. Drugs were injected i.p. 10 min before the radioactive tyrosine and the animals were killed 10 min after the label. Specific radioactivity of tyrosine, NE and DA and the turnover rate of these amines were determined as reported earlier (Neff *et al.*, 1971; Costa *et al.*, 1972). Animals were rendered tolerant to and physically dependent on morphine by pellet implantation (Ho *et al.*, 1972). Physical dependence was assessed by the naloxone test (Cheney *et al.*, 1972). The data reported in Table 1 show that morphine (52 μmoles/kg, i.p.) increases the turnover rate of DA in striatum,

1003

E. Costa, A. Carenzi, A. Guidotti and A. Revuelta

TABLE 1. Turnover rate of NE and DA in brain parts of rats
receiving various doses of morphine

Morphine (μmoles/kg, i.p.)	Turnover rate (nmoles/g/hr ± s.e.)		
	Striatal DM	Cerebellar NE	Spinal cord NE
None	15 ± 1·7	0·45 ± 0·054	0·72 ± 0·074
13	22 ± 6·6	0·34 ± 0·053	0·81 ± 0·16
26	25 ± 6·4	0·47 ± 0·014	0·86 ± 0·12
52	24 ± 2·6*	0·50 ± 0·068	0·96 ± 0·093
Pellets (1950 s.c.)	18 ± 4·2	0·52 ± 0·067	0·64 ± 0·11

Each value represents the mean of four experiments. Pellets were implanted
72 hr before measuring the catecholamine turnover rate.
* $P < 0.02$

but it does not change the turnover rate of spinal cord and cerebellar NE. Moreover, in rats implanted with morphine pellets the turnover rate of striatal DA is normal. To study the tolerance to the increase of striatal DA turnover elicited by morphine we implanted rats with morphine pellets for 72 hr. Ten minutes before injecting the label, they received an additional dose of morphine intraperitoneally. The results of these experiments are shown in Table 2. An intraperitoneal injection of 52 μmoles/kg of morphine to rats implanted with morphine pellets no longer increases the turnover of striatal DA. The data reported in Table 2 also show that high doses of morphine injected to rats implanted with morphine pellets increase the striatal DA turnover rate but these doses increase the turnover rate of neither cerebellar nor spinal cord NE.

Using the tail flick test we found that the morphine dose that increases the turnover rate of striatal DA is an AD_{80} for analgesia in 20 min. We found that like the analgesia also the increase of striatal DA turnover rate exhibits tolerance. The relationship between morphine analgesia and DA turnover is corroborated by the data reported in Table 3. They concern some stereoisomers if viminol (1[α(N-0-chlorobenzyl) pyrryl] 2-di-sec butylamine ethanol) a central analgesic which exhibits cross tolerance to morphine (Della Bella et al., 1973). The data listed in Table 3 show that the stereoisomer R_2 can elicit central analgesia and a dose AD_{80} for analgesia can

TABLE 2. Turnover rate of NE and DA in brain parts of rats
implanted with morphine pellets and receiving various doses
of morphine

Morphine (μ moles /kg, i.p.)	Turnover rate (nmoles/g/hr ± s.e.)		
	Striatal DM	Cerebellar NE	Spinal cord NE
Pellets (1950 s.c.) + saline	18 ± 4·2	0·52 ± 0·063	0·64 ± 0·11
Pellets + 52	28 ± 2·7	0·69 ± 0·093	0·75 ± 0·068
Pellets + 208	32 ± 2·9*	0·52 ± 0·011	0·61 ± 0·072
Pellets + 416	48 ± 5·0†	0·68 ± 0·11	0·93 ± 0·16

Each value represents the mean of four experiments. Pellets were implanted
72 hr before injecting morphine
* $P < 0.05$
† $P < 0.005$

TABLE 3. TURNOVER RATE OF STRAITAL DA IN RATS RECEIVING
VARIOUS DOSES OF VIMINOL**

Type of stereoisomer	μmoles/kg, i.p.	Analgesia AD	Turnover rate (nmoles/g/hr $\pm$ S.E.)
None	—	—	$25 \pm 1\cdot8$
R_2	$1\cdot7$	20	$22 \pm 3\cdot2$
R_2	$3\cdot5$	60	$24 \pm 2\cdot9$
R_2	$7\cdot0$	80	$48 \pm 3\cdot6*$
R_2	$14\cdot0$	90	$54 \pm 3\cdot8*$
S_2	$56\cdot0$	0	$26 \pm 2\cdot4$
MESO	28	0	$29 \pm 3\cdot0$

Each value represents the mean of at least 4 experiments
* $P < 0\cdot001$

R_2 is $C_1 = R$; $C_2 = R$; $C_3 = ([\alpha]_D 20°\Delta\ 19\cdot46)$
S_2 is $C_1 = S$; $C_2 = S$; $C_3 = -([\alpha]_D\ 20 + 1\cdot10)$

increase the turnover rate of striatal DA. In contrast the stereoisomer S_2 was injected in doses four times greater than those of R_2, but it neither increased the turnover rate of striatal DA nor did it cause analgesia. Similarly the Meso isomers, which are weak analgesics, did not increase the turnover rate of striatal DA when injected in doses devoid of analgesic activity (Table 3).

However, the relationship between increase of striatal DA turnover rate and analgesia failed to be completely supported when it was tested using naloxone. This pure morphine antagonist which is devoid of analgesic activity was injected intraperitoneally in various doses, it increased the turnover rate of striatal DA and cerebellar NE when given in high doses (Table 4).

TABLE 4. TURNOVER RATE OF NE AND DA IN BRAIN PARTS
OF RATS INJECTED WITH VARIOUS DOSES OF NALOXONE

Naloxone (μmoles/kg, i.p.)	Turnover rate (nmoles/g/hr $\pm$ S.E.)	
	Striatal DN	Cerebellar NE
None	$17 \pm 2\cdot7$	$0\cdot65 \pm 0\cdot043$
22	$22 \pm 2\cdot9$	$0\cdot62 \pm 0\cdot052$
44	$20 \pm 2\cdot8$	$0\cdot69 \pm 0\cdot014$
88	$29 \pm 2\cdot4*$	$0\cdot83 \pm 0\cdot091†$

Each value represents the mean of four experiments.
* $P < 0\cdot01$
† $P < 0\cdot05$

INTERPRETATION OF THE MORPHINE ACTION ON STRIATAL DA
TURNOVER RATE

Current understanding of the relationship between drug effects on striatal DA turnover rate and their ability to cause an indirect stimulation of dopaminergic receptors is far from perfect. In fact, amphetamine which indirectly stimulates dopaminergic receptors (Costa et al., 1972) and chlorpromazine which blocks DA receptors (Neff and Costa, 1966) increase the turnover rate of striatal DA. A morphine dose twice that increasing the turnover rate of striatal DA causes catalepsy, a postural plasticity with increased muscle tone, which resembles parkinsonian rigidity. This similarity would support the theory that morphine blocks doapminergic receptors, although, muscular rigidity can be elicited pharmacologically without involving directly or indirectly dopaminergic neuronal functions.

Evidence is now available indicating that striatum contains a dopamine sensitive adenylate cyclase (Kebabian et al., 1972). This information suggested to us to investigate whether the dose of (+) amphetamine that selectively increases the turnover rate of striatal dopamine also increases the concentrations of cyclic 3′,5′-adenosine monophosphate (cAMP) in striatum. These studies have become feasible with the availability of microwave sources that can be focused to allow inactivation of brain enzymes in 2 sec. Previously such studies were plagued by continuous post mortem changes of cAMP concentrations in brain (Uzunov and Weiss, 1972).

The data of Table 5 show a dose dependent correlation between the increase of DA turnover rate, and of cAMP concentrations in striatum and the increase of motor activity elicited by (+) amphetamine. The data reported in Fig. 1 show that 104 and 260 μmoles/kg, i.p. of morphine increased the concentrations of cAMP but 52 μmoles/kg did not. The latter dose was threshold for increasing the turnover rate of striatal DA. We have analysed the content of cAMP at 30 min after morphine injection (104 μmoles/kg) in cerebellum and hypothalamus and found that the cAMP content of these brain areas was not increased. The data of Fig. 1 also show that 52 μmoles/kg, i.p. of morphine increase the concentrations of cAMP in pituitary, adrenal cortex and medulla. As shown in Fig. 2 (+) amphetamine increases cAMP concentrations in striatum, but fails to change the concentrations of cAMP in pituitary, adrenal cortex and medulla. The increase of cAMP in striatum, pituitary, adrenal medulla and cortex elicited by morphine lasts for several hours (Fig. 3). This long lasting increase of the cAMP is quite unusual, the cAMP increase elicited

TABLE 5. EFFECTS OF VARIOUS DOSES OF (+) AMPHETAMINE ON MOTOR ACTIVITY, TURNOVER RATE
OF TELENCEPHALIC NE, STRIATAL DA AND STRIATAL CONCENTRATIONS OF cAMP

(+) amphet-amine (μmoles/ kg, i.p.)	Motor activity (events/min $\pm$ SE)	Telencephalic NE (nmoles/g/hr $\pm$ SE)	Striatal DA (nmoles/g/hr $\pm$ SE)	Striatal cAMP (pmoles/mg protein)
NO	4·5 $\pm$ 1·7	2·1 $\pm$ 0·31	28 $\pm$ 3·2	2·8 $\pm$ 0·20
0·4	2·5 $\pm$ 0·71	1·9 $\pm$ 0·27	24 $\pm$ 2·9	4·2 $\pm$ 0·74
3·2	25 $\pm$ 7·4*	2·3 $\pm$ 0·41	67 $\pm$ 7·2*	9·2 $\pm$ 1·8*

* $P < 0.01$

Each value is the average of at least four measurements. Motor activity was measured as described by Costa et al. (1972).

Striatal cAMP concentrations, motor activity and monamine turnover rate were measured at 15 min post injection.

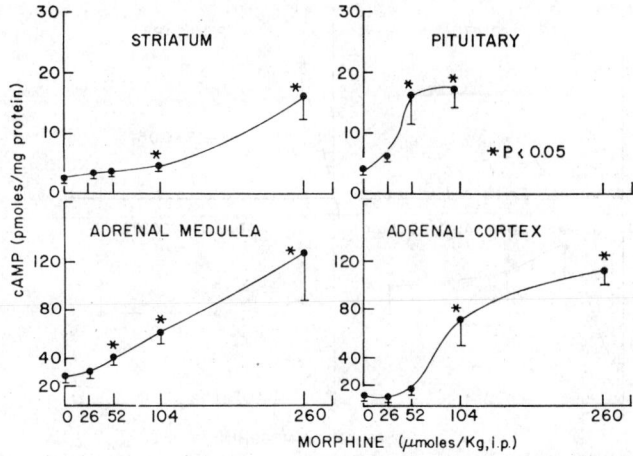

FIG. 1.—Concentrations of cAMP in striatum, pituitary adrenal cortex and medulla thirty minutes after increasing doses of morphine. cAMP concentrations were measured with a modification of the method reported by EBADI *et al.* (1971).

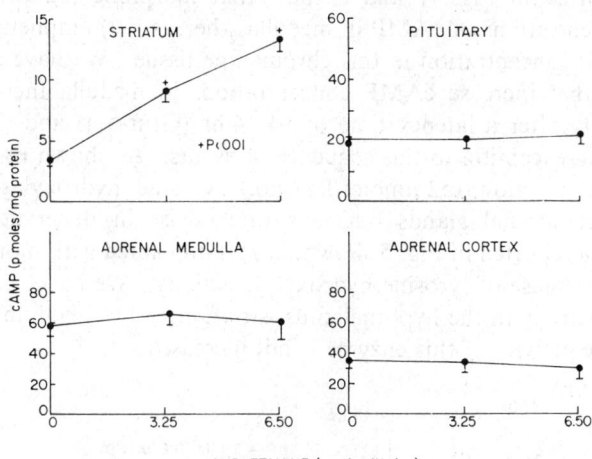

FIG. 2.—Concentrations of cAMP in striatum, pituitary, adrenal cortex and medulla fifteen minutes after increasing doses of (+) amphetamine.

by various drugs and by exposure of the rat to 4°C (GUIDOTTI and COSTA, 1973; GUIDOTTI *et al.*, 1973) does not last longer than 2 hr. Preliminary data suggest that the increase of striatal cAMP elicited by (+) amphetamine does not last longer than 1 hr.

EFFECTS OF MORPHINE ON cAMP CONCENTRATIONS AND TYROSINE HYDROXYLASE ACTIVITY OF ADRENAL MEDULLA

The data of Fig. 4 show that the cAMP concentration in adrenal cortex and medulla of rats implanted with morphine pellets is equal to that of rats sham operated. Moreover, when sham operated or morphine pellets implanted rats are exposed to 4°C for 30 min the increase of cAMP is greater in morphine pellets implanted rats than in rats receiving saline. This greater increase of cAMP concentration can be seen in both cortex and adrenal medulla.

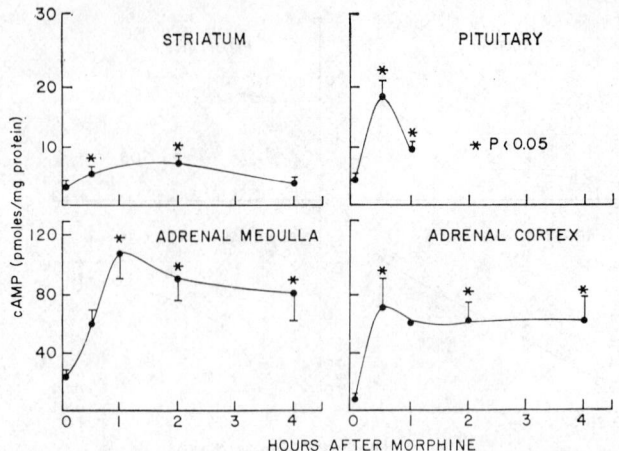

Fig. 3.—Time course of the increase of cAMP concentrations elicited by 104 μmoles/kg, i.p. of morphine in striatum, pituitary adrenal medulla and cortex.

The data reported in Figs. 1 and 2 show that morphine (52 μmoles/kg, i.p.) increases the concentrations of cAMP in medulla whereas (+) amphetamine fails to increase the cAMP concentration in this chromaffine tissue. We have shown that a number of drugs that increase cAMP concentrations in medulla increase tyrosine hydroxylase activity after a latency time of 10–14 hr (Guidotti and Costa, 1973). Morphine is not an exception to this sequence of events. As shown in Fig. 5, 24 hr after a morphine injection (52 μmoles/kg, i.p.) tyrosine hydroxylase activity is increased in intact adrenal glands but fails to increase in denervated adrenals. Moreover, the data reported in Fig. 5 show that rats implanted with morphine pellets fail to exhibit an increase of tyrosine hydroxylase activity. We have measured tyrosine hydroxylase activity in the hypothalamus, striatum and cerebellum of these rats and found that the activity of this enzyme is not increased.

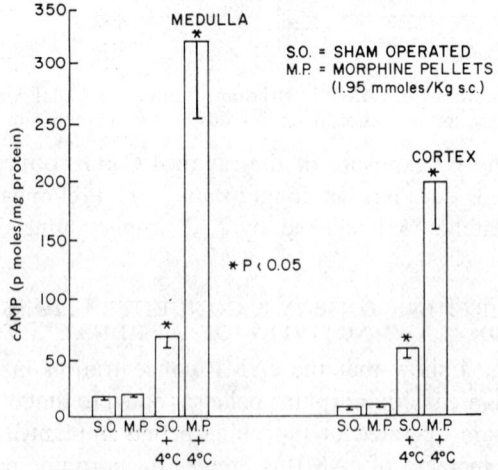

Fig. 4.—cAMP concentrations in adrenal cortex of rats implanted with morphine pellets (1·95 mmoles/kg, s.c.) or sham implanted. Both groups of rats were either kept at 24°C or at 4°C for 30 min.

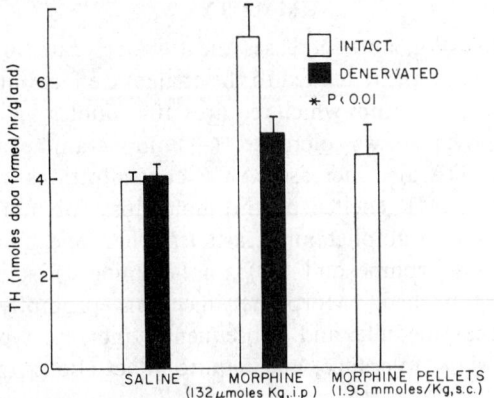

F<small>IG</small>. 5.—Tyrosine hydroxylase activity in intact and denervated (5 days before) adrenal glands of rats injected with either 132 μmoles/kg, i.p. of morphine, or morphine pellets (1·95 mmoles/kg, s.c.). Enzyme activity was measured according to W<small>AYMIRE</small> *et al.* (1971).

CONCLUSION

The data presented suggest the following:

(1) Doses of morphine and viminol which are very close to the AD_{80} for analgesia, increase the turnover rate of striatal DA. The turnover rate of NE in spinal cord and cerebellum of these animals is not increased.

(2) Like analgesia, the increase of DA turnover elicited by morphine shows tolerance.

(3) Naloxone, a pure morphine antagonist, increases turnover rate of DA and cerebellar NE when injected in very high doses (88 μmoles/kg, i.p.).

(4) (+) amphetamine like morphine, selectively stimulates the turnover rate of striatal DA. Both drugs increase the concentrations of cAMP in striatum. However, the effect of morphine persists longer than that of amphetamine, in addition, unlike amphetamine the action of morphine on cAMP does not strictly correlate to its action on DA turnover rate.

(5) If the increase of cAMP in striatum reflects a stimulation of postsynaptic dopaminergic receptors, judging from the persistence of this response, it might be inferred that morphine stimulates persistently, these receptors. The location of the morphine action on cAMP concentrations and DA turnover rate to the striatum is of interest in view of reports that this brain structure contains the highest concentration of opiate receptors among various brain areas studied (P<small>ERT</small> and S<small>NYDER</small>, 1973).

(6) Morphine injections promptly increase cAMP concentration in adrenal medulla and stimulate tyrosine hydroxylase activity of this tissue after a time delay of almost 24 hr. This increase of enzyme activity, like the increase of cAMP, requires the presence of cholinergic nerves suggesting that morphine does not increase cAMP because it blocks phosphodiesterase (like aminophyllin C<small>OSTA</small> and G<small>UIDOTTI</small>, 1973) or because it stimulates medullary cholinergic receptors (like carbamylcholine C<small>OSTA</small> and G<small>UIDOTTI</small>, 1973).

(7) The action of morphine on cAMP concentrations in adrenal medulla and cortex shows tolerance. However, rats tolerant to the increase of cAMP elicited by morphine are still responsive to the increase of cAMP elicited by cold exposure (4°C).

SUMMARY

Morphine (52 μmoles/kg, i.p.) increases the turnover rate of DA in striatum. Quantitatively, this action appears related to the analgesic effect but not to an increase of cAMP concentration of striatum which requires 104 μmoles/kg, i.p. The action of morphine on DA turnover shows tolerance. (+) amphetamine doses that enhance the turnover of striatal DA also increase cAMP concentrations in striatum. While the increase of striatal cAMP elicited by threshold doses of morphine lasts longer than two hours that of (+) amphetamine lasts less than one hour. An identity of the mode of action of morphine and (+) amphetamine on striatal dopaminergic functions was not established. Morphine injections promptly increase cAMP concentration in adrenal medulla and subsequently increase tyrosine hydroxylase activity. Both effects show tolerance, and require intact cholinergic innervation.

REFERENCES

CHENEY D. L. and GOLDSTEIN A. (1971) *Nature, Lond.* 232, 477–478.
CLOUET D. H., (Ed.) *Narcotic Drugs—Biochem. Pharmacology*, Plenum Press, New York, 1971.
CLOUET C. H. and RATNER M. (1970) *Science* 168, 845–855.
COSTA E. and NEFF N. H. (1966) In: *Biochemistry and Pharmacology of the Basal Ganglia* (Eds. E. COSTA, COTE', L., YAHR M. D.) Raven Press, New York, pp. 141–156.
COSTA E., GROPPETTI A., NAIMZADA M. K. (1972) *Br. J. Pharmacol.* 44, 742–751.
DELLA BELLA D., FERRA V., FRIGENI V. and LUALDI P. (1973) *Nature, New Biology* 241, 282–283.
EBADI M. S., WEISS B. and COSTA E. (1971) *J. Neurochem.* 18, 183–192.
FRIEDLER G., BHARGAVA H. N., QUOCK R. and WAY E. L. (1972) *J. Pharmacol. exp. Ther.* 183, 49–55.
GUIDOTTI A. and COSTA E. (1973) *Science* 179, 902–904.
GUIDOTTI A., ZIVKOVIC B., PFEIFFER R. and COSTA E. (1973) *Naunyn-Schmiedeberg's Arch Pharmacol.* 278, 195–206.
HO I. K., LU S. E., STOLMAN S., LOH H. H. and WAY E. L. (1972) *J. Pharmacol. exp. Ther.* 182, 155–165.
KEBABIAN J. W., PETZOLD G. L. and GREENGARD P. (1972) *Proc. Natn. Acad. Sci.* 69, 2145–2149.
LOH H., HITZEMANN R. J. and WAY E. L. (1973) *Life Sci.* 12, 33–41.
MAYNERT E. W. (1967) *Archs. Biol. Med. Exp.* 4, 36–41.
NEFF N. H. and COSTA E. (1966) In: (S. GARATTINI and M. N. C. DUKES Eds.) *Proc. International Symposium on Antidepressant Drugs*, New York, 122, 28.
NEFF N. H., SPANO P. F., GROPPETTI A., WANG C. T. and COSTA E. (1971) *J. Pharmacol. exp. Ther* 176, 701–709.
PERT C. B. and SNYDER S. H. (1973) *Science* 179, 1011–1014.
SESAME H. A., PEREZ-CRUET J., DI CHIARA G., TAGLIAMONTE A., TAGLIAMONTE P., and GESSA G. L. (1963) *J. Neurochem.* 19, 1953–1957.
SMITH C. B., SHELDON M. I., BENDARCZYK J. H. and VILLARREAL J. E. (1972) *J. Pharmacol. exp. Ther.* 180, 547–557.
UZUNOV P. and WEISS B. (1972) *Adv. Cyclic Neucleotides Res.* 1, 435–453.
WAU E. L., LOH H. H. and SHEN F. H. (1968) *Science* 162, 1290–1292.
WAYMIRE J. C., BJUR R. and WEINER N. (1971) *Anal. Biochem.* 43, 588–600.

Frontiers in Catecholamine Research 1973, pp. 1011 to 1014. Pergamon Press. Printed in Great Britain.

EFFECT OF METHADONE ON BRAIN DOPAMINE METABOLISM

G. L. Gessa, L. Vargiu, G. Biggio and A. Tagliamonte

University of Cagliari, Cagliari, Italy

Previous studies from our laboratory (Sasame et al., 1972) have shown that methadone shares with phenothiazine and butyrophenone neuroleptics several pharmacological and biochemical actions. Thus, methadone causes catalepsy and hypothermia in rats, and the latter effect is more pronounced in animals kept in a cold environment. It also blocks apomorphine-induced gnawing, increases the brain HVA level, but not that of 5-HIAA, and stimulates the synthesis of brain DA. Because the cataleptic action of methadone is potentiated by α-methyl-tyrosine (α-MT), and reversed by apomorphine we suggested that this effect is secondary to an inhibition by methadone of dopaminergic receptors: that is the increase in DA synthesis found after methadone administration may be a compensatory feedback response to the blockade of dopaminergic receptors. Indeed a similar theory has been proposed for neuroleptics. This hypothesis was supported by the finding that apomorphine prevented methadone-induced accumulation of HVA levels.

From a point of view of the structure–activity relationship it is possible to recognise in the molecule of methadone the chetone group and the tertiary amine separated by a three carbon chain as in the haloperidol molecule.

On the other hand, Kuschyinsky and Hornykiewicz (1972) have suggested that the catalepsy induced by morphine and methadone might originate from an increased intraneuronal breakdown of the newly formed DA with a resultant decrease of the amine at the receptor sites. According to either theory the increase in DA turnover by methadone or morphine would be secondary to the functional deficiency of DA at receptor sites.

However, several considerations suggest that methadone might exhert an amphetamine like effect at the DA nerve terminals in addition to its inhibitory action at DA receptor sites. Thus, the increase in HVA levels and DA synthesis produced by methadone lasts much longer that the cataleptic response the drug. Moreover, the present results show that methadone stimulates DA turnover in mice, in which it does not produce catalepsy but increased motor activity; this effect, as in the case of d-amphetamine is blocked by α-MT. Moreover, DL-methadone, similarly to d-amphetamine inhibits monoamine-oxidase (MAO) both in vitro and in vivo.

Finally, the compound appears to meet the steric requirements of a sympathomimetic agent since in its molecule a phenylalkylamino structure is present, which in the flexible molecule of methadone may approximate that of amphetamine.

MATERIAL AND METHODS

Male Wistar rats of 200–230 g and male albino mice of 20–25 g were used. The animals were kept under the conditions previously described (Sasame et al., 1972).

Brain 5-HT, 5-HIAA, DA and HVA were assayed fluorometrically as previously described (Sasame et al., 1972).

The conversion of labelled tyrosine to labelled DA was measured according to

Costa and Groppetti (1970). Monoamine-oxidase (MAO) activity was measured with the method of Krajl (1965).

RESULTS

In agreement with previous results (Sasame et al., 1971, 1972) D,L-methadone significantly increased brain HVA, but did not influence 5-HIAA levels in rats. In addition, Table 1 shows that methadone increased both HVA and 5-HIAA levels in the mouse brain. In this species, the compound did not produce catalepsy but increased motor activity and produced the Straub phenomenon. Moreover, a dose of Haloperidol, which produced only a moderate catalepsy, increased brain HVA both in rats and mice by a much greater extent than did the maximal dose of methadone used.

On the other hand, d-amphetamine, which has been shown to increase DA turnover (Costa and Groppetti, 1970) did not increase brain HVA but significantly increased 5-HIAA in the rat brain.

These results prompted us to investigate whether methadone might interfere with the formation of HVA by inhibiting MAO.

Table 2 shows that DL-Methadone concentrations of 10^{-4}, 10^{-5} and 10^{-6} M inhibited the MAO activity of rat brain homogenates by 83, 56 and 12 per cent, respectively. Similar degree of inhibition was exerted by d-amphetamine. In contrast, haloperidol inhibited MAO activity by about 50 per cent at the highest concentration used.

The finding that methadone is a potent monoamine oxidase inhibitor *in vitro*, raised the question of whether this compound might effectively inhibit the enzyme *in vivo*. We tried to clarify this problem by studying the effect of methadone on the disposition of brain DA and serotonin released by reserpine.

As Table 3 shows, the administration of DL-methadone prevented the accumulation of brain HVA and 5-HIAA produced by reserpine. Interestingly, the combination of methadone and reserpine resulted in a smaller accumulation of these metabolites than did the administration of methadone given alone. It is possible that the drug combination caused an increased concentration of free DA within the nerve endings, which inhibits tyrosine hydroxylase activity.

As these results show, HVA changes do not always reflect parallel and proportional changes in DA turnover. Therefore we studied the effect of DL-methadone on the conversion of (^{3}H) tyrosine to DA in the mouse brain. In Table 4, are reported

TABLE 1. EFFECT OF DL-METHADONE ON THE LEVELS OF BRAIN HVA AND 5-HIAA IN RATS AND MICE.

Treatment	Dose (mg/kg i.p.)	Rats		Mice	
		HVA (ng/g)	5-HIAA (μg/g)	HVA (ng/g)	5-HIAA (μg/g)
None	—	70 ± 8	0·54 ± 0·02	180 ± 11	0·85 ± 0·02
DL-methadone	10	138 ± 15	0·52 ± 0·03	260 ± 11	0·95 ± 0·03
	20	150 ± 11	0·65 ± 0·03	360 ± 15	1·40 ± 0·02
Haloperidol	5	210 ± 15		525 ± 20	
d-Amphetamine	10	81 ± 13	0·77 ± 0·02	180 ± 8	0·85 ± 0·04

Drugs given 1 hr before death

TABLE 2. EFFECT OF DL-METHADONE, HALOPERIDOL AND *d*-AMPHETA-
MINE ON THE MAO ACTIVITY (*) OF RAT BRAIN HOMOGENATES.

Compound	% Inhibition at the molar concentration		
	10^{-4}	10^{-5}	10^{-6}
DL-methadone	83·4	46·6	12·2
Haloperidol	56·3	14·2	0
d-Amphetamine	100	42·8	10·6

(*) Calculated as % of kynuramine metabolised in 30 min.

TABLE 3. EFFECT OF RESERPINE ON BRAIN HVA AND 5-HIAA IN MICE TREATED
WITH DL-METHADONE.

Treatment	Dose (mg/kg, i.p.)	HVA (ng/g)	5-HIAA (μg/g)
None	—	180 ± 11	0·85 ± 0·02
Reserpine	5	350 ± 21	1·60 ± 0·03
DL-methadone + reserpine	20 + 5	270 ± 12*	0·90 ± 0·03*

Reserpine was given 60 min before sacrifice, methadone 15 min prior to reserpine.
* $P < 0.001$ in respect to reserpine treated group.

TABLE 4. EFFECT OF DL-METHADONE ON THE CONVERSION OF ^{3}H-TYROSINE
TO ^{3}H-DA IN THE MOUSE BRAIN.

	S.A. 20 min after the i.p. injection of H^3-tyrosine	
	Tyrosine	DA
Controls	625 ± 12	54 ± 3
DL-methadone	432 ± 21	87 ± 11

DL-methadone was given 30 min prior to ^{3}H-tyrosine

the specific activities (S.A.) of tyrosine and DA 20 min after the intraperitoneal injection of a pulse dose of (^{3}H) tyrosine to control mice and mice treated with DL-methadone. The S.A. of brain tyrosine was 30 per cent lower than normal in mice treated with methadone. On the other hand, the specific activity of brain DA was 40 per cent higher in methadone-treated mice than in control ones.

These results indicate that also in mice, as in rats, DL-methadone accelerates the conversion of ^{3}H tyrosine to brain DA. The decrease in brain tyrosine S.A. may be the result of its decreased transport into the brain or its decreased absorption from peritoneal cavity.

DISCUSSION

The results of the experiments reported and several contrasting effects of methadone might be interpreted by assuming that the drug has two actions on the dopaminergic system: an inhibitory one at DA receptor sites (weaker and more easily surmountable than that of haloperidol), and an amphetamine-like action at DA nerve terminals.

The two actions would produce opposite effects on behaviour and the resultant effect would depend on the dose, the animal species and the experimental condition. As a species, the amphetamine-like effect would prevail in mice and cats, which are usually wildly excited by methadone. The stimulant effect of methadone, similar to that of *d*-amphetamine, is antagonised by α-methyl tyrosine (α-MT), by haloperidol and chlorpromazine.

In rats, the prevailing effect of methadone during the first hour would be that of a blockade of DA receptors with a resultant catalepsy. As the level of the drug wears off, the amphetamine-like effect would counteract the cataleptic one.

In humans, the usual effect of methadone is one of sedation, but some subjects became very excited by the drug. The difference in response might depend on differences in DA receptors in different animal species or on differences in DA turnover rates.

It is likely that the analgesic effect of these compounds is more related to the amphetamine-like effect than to the blockade of DA receptors. In fact, α-MT antagonises the analgesic effect of morphine, but potentiates catalepsy (see Kushinsky and Hornkiewicz, 1972). Moreover, haloperidol possesses no analgesic activity. On the other hand, the inhibitory effect of methadone on DA receptors might play a role on its suppressant effect on the withdrawal syndrome in heroin addicts.

REFERENCES

Costa E. and Groppetti A. (Eds) (1970) In: *Amphetamines and related compounds:* Proc. Mario Negri Inst. for Pharmacological Research, Milan, Italy p. 231, Raven Press, New York.
Krajl M. (1965) *Biochem. Pharmacol.* **14,** 1683–1685.
Kuschinsky K. and Hornykiewicz O. (1972) *Europ. J. Pharmacol.* **19,** 119–122.
Sasame H. A., Perez-Cruet J., Dichiara G., Tagliamonte A., Tagliamonte P., and Gessa G. L. (1972) *J. Neurochem.* **19,** 1953–1957.

Frontiers in Catecholamine Research 1973, pp. 1015 to 1020. Pergamon Press. Printed in Great Britain.

EFFECTS OF SOME TETRAHYDROCANNABINOLS ON THE BIOSYNTHESIS AND UTILISATION OF CATECHOLAMINES IN THE RAT BRAIN

L. Maitre, P. C. Waldmeier and P. A. Baumann

Research Department, Pharmaceuticals Division, CIBA-GEIGY Limited, Basle, Switzerland

Δ^9-TETRAHYDROCANNABINOL (Δ^9-THC) is the active constituent of marijuana. It has variously been reported to accelerate (Maitre et al., 1970a; Schildkraut and Efron, 1971), retard (Truitt and Andersson, 1971) or have no influence (Leonard, 1971) on the turnover and metabolism of catecholamines in the rat brain. Other biochemical and histological investigations have indicated that Δ^9-THC and Δ^8-THC increase noradrenaline turnover in several brain regions and decrease the turnover of dopamine in the neostriatum (Fuxe and Jonsson, 1971).

We examined the effects of Δ^9-THC, Δ^8-THC and $\Delta^{3,4}$-dimethylheptyltetrahydrocannabinol (DMHP) on the accumulation and disappearance of ^{3}H-catecholamines formed from ^{3}H-tyrosine. Determinations of the endogenous content of homovanillic acid (HVA) and dihydroxyphenylacetic acid (DOPAC) in the corpus striatum were made, to gain a better insight into the metabolism of dopamine in this region.

MATERIALS AND METHODS

The experiments were performed on male albino rats, weighing 180–230 g, which had been acclimatised for 2–3 weeks in an animal room kept at a constant temperature of 22–23°C under controlled lighting conditions consisting of 14 hr light followed by 10 hr darkness. Δ^9-THC, Δ^8-THC and DMHP were synthesised in our Chemistry Department by Dr. W. Bencze.

The cannabinols were suspended in Tween 80 with the aid of glass homogenisers. Physiological saline was added in small portions to give a final suspension containing 15 mg/ml of the cannabinols and 2% of Tween 80. In the experiments reported here the cannabinols were injected intraperitoneally in a dose of 30 mg/kg. Control rats received the vehicle alone. All the experiments were started between 7·00 and 8·00 a.m.

A detailed account of the estimation of the accumulation and disappearance of ^{3}H-catecholamines following the injection of ^{3}H-tyrosine has already been published (Maitre et al., 1970b and 1972). Briefly in the accumulation experiments the rats received the cannabinols 1 hr before and in the disappearance experiments 1 hr after 3,5-^{3}H-L-tyrosine (the Radiochemical Centre, Amersham, England). The radioactive tyrosine was diluted with cold L-tyrosine, so that the rats received 1 mCi and 200 μg L-tyrosine/ml/kg body weight.

In the accumulation experiments, the brains were removed 1 hr after the injection of ^{3}H-tyrosine (i.e. 2 hr after treatment with the cannabinols). In the disappearance experiments, the control brains were removed 1 and 3 hr after the ^{3}H-tyrosine injection. The 1-hr group showed the amounts of ^{3}H-catecholamines present in the brain at the time of treatment with the cannabinols. The brains of the drug-treated rats were removed 3 hr after the injection of ^{3}H-tyrosine (i.e. 2 hr after treatment with the cannabinols).

In both types of experiment, the amounts of ^{3}H-catecholamines found 1 hr after the ^{3}H-tyrosine injection were taken as a measure of biosynthesis. The disappearance of ^{3}H-catecholamines between 1 and 3 hr after the tyrosine injection, i.e. between the cannabinol injection and the removal of the brains was taken as a measure of utilisation.

^{3}H-noradrenaline and ^{3}H-dopamine were extracted from the tissues and separated from ^{3}H-tyrosine or its other metabolites by adsorption onto alumina and subsequent paper chromatography, or by passage through Dowex 50WX4 columns as described in the paper mentioned above.

The determinations were made both in whole brain and in individual brain regions dissected out as described by Glowinski and Iversen (1966). HVA and DOPAC were isolated according to the technique used by Murphy et al. (1969). HVA was determined fluorometrically by an automated procedure based on the method described by Anden et al. (1963). DOPAC was estimated as described by Sharman (1971). These determinations were only carried out in the corpus striatum.

RESULTS

1. *Experiments on the whole brain*

The effects of Δ^9-THC, Δ^8-THC and DMHP on the accumulation and release of ^{3}H-catecholamines formed from intravenously injected ^{3}H-tyrosine are shown in Fig. 1. All three cannabinols had the same qualitative effect on catecholamine accumulation and release. The accumulation of ^{3}H-noradrenaline and of ^{3}H-dopamine was increased. Their effect on ^{3}H-noradrenaline was very marked whereas their effect on ^{3}H-dopamine was much less pronounced. The increases produced by Δ^9-THC and by Δ^8-THC were of the same order of magnitude and slightly smaller than those caused by DMHP.

The effects of the cannabinols on the disappearance of ^{3}H-catecholamines paralleled their effects on its accumulation. ^{3}H-noradrenaline utilisation was markedly accelerated to a similar extent by all three cannabinols. The disappearance rate of ^{3}H-dopamine, however, was not significantly altered by Δ^9-THC and Δ^8-THC. The mean values showed a trend towards a greater disappearance rate, but increased utilisation was only observed after treatment with DMHP ($P < 0.05$).

2. *Experiments on different brain regions*

The accumulation of ^{3}H-noradrenaline was estimated in the hypothalamus, the pons-medulla and the remaining brain tissue, that of ^{3}H-dopamine in the same parts of the brain and in the corpus striatum. Δ^9-THC was not included in this series of experiments. The accumulation of ^{3}H-noradrenaline increased after treatment with Δ^8-THC by $82 \pm 10\%$ in the hypothalamus, by $41 \pm 9\%$ in the pons-medulla and by $52 \pm 7\%$ in the remaining tissue. The corresponding figures for DMHP were $165 \pm 30\%$, $39 \pm 12\%$ and $57 \pm 12\%$ respectively. The accumulation of ^{3}H-dopamine increased much less than that of ^{3}H-noradrenaline and only in the corpus striatum and the hypothalamus. In the corpus striatum, the increases averaged

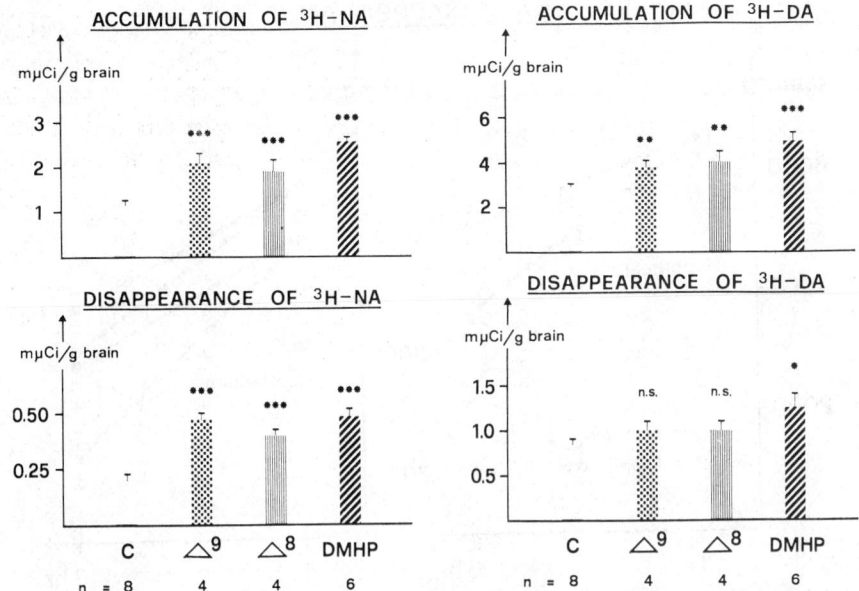

FIG. 1.—Effect of cannabinols on accumulation and disappearance of ³H-noradrenaline (³H-NA) and ³H-dopamine (³H-DA) formed from intravenously injected ³H-L-
tyrosine in the rat brain.

In *accumulation* experiments ³H-tyrosine was injected 1 hr after the drugs. The brains
were removed 1 hr after ³H-tyrosine injection.

In *disappearance* experiments, ³H-tyrosine was injected 1 hr before the drugs. ³H-NA
and ³H-DA were measured 1 hr and 3 hr after ³H-tyrosine injection (controls) or
2 hr after drug treatment. The columns represent the amounts of ³H-catecholamines
which disappeared from the brain between 1 and 3 hr after ³H-tyrosine injection.

$42 \pm 6\%$ after Δ^8-THC and $34 \pm 10\%$ after DMHP. In the hypothalamus, they
were $26 \pm 9\%$ and $57 \pm 14\%$ respectively. In the other regions, no significant
changes were observed. These data were published recently (MAITRE *et al.*, 1972).

The disappearance of ³H-catecholamines from different brain regions was estimated in rats treated with DMHP. That of ³H-noradrenaline was accelerated in all
regions of the brain, but particularly in the hypothalamus and cerebellum (Fig. 2).
In the other regions, the effects were slight and of the same order of magnitude. As
an example, disappearance in the cortex is illustrated in Fig. 2.

The utilisation of ³H-dopamine was slightly increased in some brain regions, but
this accelerating effect of DMHP never reached statistical significance at a probability
level of $P = 0.05$.

The action of DMHP on striatal dopamine was further evaluated by measuring
the endogenous content of its major metabolites, HVA and DOPAC. The values
obtained 2 hr after a single injection are shown in Table 1. The content of HVA was
not altered markedly, whereas that of DOPAC was increased. The latter finding
is in keeping with the recent observation that treatment of rats with Δ^9-THC or
Δ^8-THC roughly doubled the amount of ³H-DOPAC found in the whole brain 1 hr
after an intravenous injection of ³H-tyrosine, while DMHP was even more active
(MAITRE *et al.*, 1972).

The preferential increase in DOPAC has been confirmed by studying the time-
course of the effect of DMHP on striatal dopamine metabolism. The maximum

NA – DISAPPEARANCE

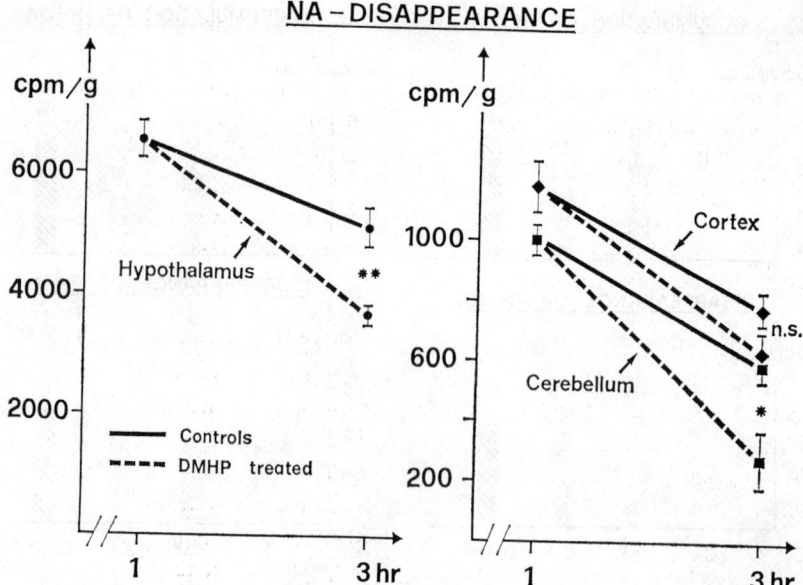

Fig. 2.—Effect of DMHP on noradrenaline (NA) disappearance from rat hypothalamus, cortex and cerebellum.
³H-L-tyrosine was injected intravenously 1 hr before the administration of DMHP or its vehicle. ³H-noradrenaline was estimated at the time of DMHP (or vehicle) injection (= 1 hr after ³H-tyrosine) and 2 hr later.

increase in DOPAC was seen after 2 hr ($P < 0.05$) and normalisation occurred between 4 and 16 hr. No concomitant alteration in HVA was detected.

The lack of effect of DMHP on HVA levels and the small increase in DOPAC levels in the corpus striatum contrast with the huge increases seen after treatment with neuroleptic drugs such as chlorpromazine or haloperidol (Table 1), which also increase the accumulation of ³H-dopamine formed from ³H-tyrosine under the experimental conditions described here.

DISCUSSION

The cannabinols influence catecholamine metabolism in the rat brain. Judging from the data available so far, Δ^9-THC, Δ^8-THC and DMHP appear to produce qualitatively similar changes. For convenience, these changes will be discussed

TABLE 1. Effect of DMHP on HVA and DOPAC levels in the rat corpus striatum. Comparison with chlorpromazine and haloperidol

Drug	Dose (mg/kg, route)	HVA % of control values	DOPAC
DMHP	30 i.p.	106 ± 13 n.s.	126 ± 6*
Chlorpromazine	10 p.o.	338 ± 39†	246 ± 15‡
Haloperidol	30 p.o.	650 ± 56‡	313 ± 36‡

    * $P < 0.05$ † $P < 0.01$ ‡ $P < 0.001$ $n = 4$
Results are expressed as per cent of the control experiments carried out on the same day. P values were also calculated against the controls of the same day.
Absolute values were 369 ± 20 ng/g for HVA ($n = 24$) and 559 ± 27 ng/g for DOPAC ($n = 19$)

without distinction between the cannabinols, on the assumption that they are in fact similar.

Both the biosynthesis and the utilisation of noradrenaline are increased markedly. This is the most obvious effect, being easily recognisable in each of the brain regions studied as well as in the whole brain. The greatest increases were observed in the hypothalamus. Recent studies (MAITRE et al., 1972) had already shown that Δ^8-THC and DMHP have a great affinity for hypothalamic sites; it was found that the accumulation of ^{3}H-noradrenaline from ^{3}H-tyrosine was still marked 5 days after a single dose of the drug. In investigations based on histochemical techniques, MIRAS (1971) and CONSTANTINIDIS and MIRAS (1971) also reported an increase in the intensity of noradrenaline-induced fluorescence in the terminal varicosities of some hypothalamic nuclei after treatment of the rat with a standardised hashish-smoke sublimate. This seems to indicate that Δ^9-THC—the major active constituent of this extract— also exhibits a particular affinity for the hypothalamus. The biosynthesis and utilisation of dopamine are affected to a much smaller extent than those of noradrenaline. Interestingly enough, the accumulation of dopamine was more markedly influenced than its utilisation. This might reflect the fact that newly synthesised dopamine disappears from a pool with a relatively small specific radioactivity. Another possibility which could lead to erroneous disappearance values is that in the accumulation experiments the cannabinols were shown to enhance the amounts of ^{3}H-catecholamines in the tissue. This increase might play a role in the utilisation experiments (when the cannabinols are given after tyrosine), as the ^{3}H-tyrosine remaining in the blood and tissues at the time of the cannabinol injections may be more readily incorporated into ^{3}H-catecholamine in the drug-treated rats than in the controls. The disappearance rate might therefore have been greater than that actually measured. If such a process did occur, it must have taken place to an even greater extent in respect of noradrenaline, and the figures obtained for utilisation would consequently also represent minimum values. The increase in DOPAC levels without alteration in HVA in the corpus striatum is in striking contrast to the effects of typical neuroleptics on these dopamine metabolites. First, the effectiveness of the cannabinols is marginal, as compared with that of chlorpromazine or haloperidol, and secondly, the increase in HVA is greater than that of DOPAC after treatment with the neuroleptics. These results indicate that the cannabinols mainly increase intraneuronal dopamine metabolism.

SUMMARY

Δ^9-THC, Δ^8-THC and DMHP increased catecholamine biosynthesis and utilisation in the rat brain. Their effects on noradrenaline were marked, particularly in the hypothalamus. The effects on dopamine biosynthesis were more clearly demonstrable than those on dopamine utilisation. Endogenous levels of DOPAC were increased without alteration of HVA levels, indicating a preferential intraneuronal pathway of dopamine metabolism. No qualitative difference between the three cannabinols has been observed so far.

REFERENCES

ANDÉN N. -E., ROOS B. -E. and WERDINIUS B. (1963) Life Sci. 7, 448–458.
CONSTANTINIDIS J. and MIRAS C. J. (1971) Psychopharmacologia (Berl.), 22, 80–90.
FUXE K. and JONSSON G. (1971) Acta pharmaceutica Suecica 8, 695.

GLOWINSKI J. and IVERSEN L. L. (1966) *J. Neurochem.* **13,** 655–669.
LEONARD B. E. (1971) *Pharmacol. Res. Commun.* **3,** 139–145.
MAITRE L., BAUMANN P. A. and DELINI-STULA A. (1972) In: *Cannabis and its derivatives,* (PATON W. D. M. and CROWN J. Eds.) pp. 101–117, Oxford University Press, London.
MAITRE L., STAEHELIN M. and BEIN H. J. (1970a) *Agents and Actions,* **1,** 136–143.
MAITRE L., STAEHELIN M. and BEIN H. J. (1970b) *Biochem. Pharmacol.* **19,** 2875–2892.
MIRAS C. J. (1971). *Acta pharmaceutica Suecica* **8,** 694–695.
MURPHY G. F., ROBINSON D. and SHARMAN D. F. (1969) *Brit. J. Pharmacol.* **36,** 107–115.
SCHILDKRAUT J. J. and EFRON D. H. (1971) *Psychopharmacologia (Berl.)* **20,** 191–196.
SHARMAN D. F. (1971) In *Methods of Neurochemistry* (FRIED R. Ed.) Vol. **1,** pp. 83–127, Dekker, New York.
TRUITT E. B. JR. and ANDERSON S. M. (1971) *Ann. N.Y. Acad. Sci.* **191,** 68–73.

Frontiers in Catecholamine Research 1973, pp. 1021 to 1026. Pergamon Press. Printed in Great Britain.

ALCOHOL AND CATECHOLAMINE DISPOSITION:

A ROLE FOR TETRAHYDROISOQUINOLINE ALKALOIDS

GERALD COHEN

College of Physicians & Surgeons of Columbia University, 640 West 168th Street,
New York, New York 10032, U.S.A.

INTRODUCTION

THE PLANT alkaloids comprise a broad grouping of compounds (ROBINSON, 1968), many of which exhibit drug-like actions. A subclassification is the tetrahydroiso-quinoline (TIQ) group (SHAMMA, 1972), which contains alkaloids that appear to be derived from dopamine (DA). Some of the more complex alkaloids such as morphine and codeine originate from DA and pass through a benzyl-TIQ as a biosynthetic intermediate (KIRBY, 1967). Plant alkaloids related to norepinephrine (NE) or epinephrine (E) are unknown, presumably because plants do not β-hydroxylate the DA side chain. Recently, laboratory synthesis of an NE-derived TIQ was achieved by COLLINS and KERNOZEK (1972).

The main thesis of this paper is that a group of TIQ alkaloids can be biosynthe-sised in people when they drink alcoholic beverages. These substances might interfere with normal adrenergic neurotransmission in the brain and in the periphery and, by so doing, they might be responsible for alterations in mood, mentation and behaviour.

PATHWAY OF BIOSYNTHESIS OF TIQ ALKALOIDS

The work of SCHÖPF and BAYERLE (1934) showed that a spontaneous condensation occurred with DA and acetaldehyde under conditions amenable to plant life (neutral pH, ambient temperature). The presence of an activating hydroxyl group opposite (para) to the point of ring closure facilitates the formation of the TIQ ring system at neutral pH (ZENKER, 1966). All catecholamines can undergo this reaction and a variety of aldehydes can be used. For example, the condensation of dopa with pyridoxal phosphate (vitamin B_6, an enzyme cofactor) causes substrate inhibition of dopa decarboxylase (3,4-dihydroxy-L-phenylalanine carboxy-lyase, E.C. 4.1.1.26) (SCHOTT and CLARK, 1952).

A pathway for the biosynthesis of TIQ alkaloids in people becomes evident. During ethanol intoxication, acetaldehyde circulates in the bloodstream (MAJCHROWICZ and MENDELSON, 1970). It seems likely that this acetaldehyde would condense with DA, NE and E in their storage sites in the brain and in the periphery to form TIQ derivatives. SANDLER and coworkers (1973) recently reported that administration of ethanol to Parkinson patients who were being treated with large doses of L-dopa, provoked urinary excretion of the TIQ condensation product of DA with acetaldehyde.

SYNTHESIS OF TIQ ALKALOIDS AND RELEASE FROM THE ADRENAL MEDULLA BY ACETYLCHOLINE

Retrograde perfusion of isolated, fresh cow adrenal glands with solutions of acetaldehyde or formaldehyde results in the synthesis of TIQ derivatives of E and NE

(COHEN and COLLINS, 1970). These TIQs are bound, in part, to the chromaffin granules (GREENBERG and COHEN, 1972). With 33 mM formaldehyde (1 mg/ml), a total conversion of catecholamines to TIQs is achieved within one hour (COHEN, 1971a); such large conversions have not been observed with acetaldehyde. With 23 μM acetaldehyde (1 μg/ml), which is in the concentration range reported in the blood of persons ingesting alcoholic beverages, some formation of TIQs has been detected (COHEN, 1971b). There is evidence for synthesis of formaldehyde-derived TIQs in the adrenals of rats receiving i.p. injections of methanol (COHEN and BARRETT, 1969).

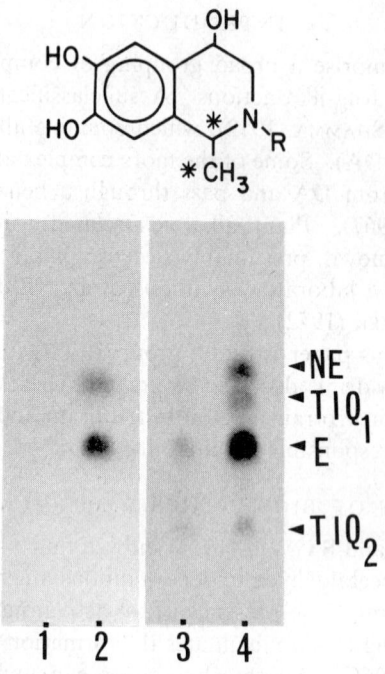

FIG. 1.—Thin-layer chromatographic assays of Al(OH)₃-extracts of perfusates of cow adrenal glands. Shown are paired glands prior to stimulation (1, control; 3, acetaldehyde-perfused) and during stimulation with acetylcholine (2, control; 4, acetaldehyde-perfused). TIQ₁ = NE derivative (R = H); TIQ₂ = E derivative (R = CH₃). The acetaldehyde portion of the TIQ structure is shown by asterisks.

Release of acetaldehyde-derived TIQs from cow adrenal glands was recently studied (GREENBERG and COHEN, 1973). Paired glands were perfused for 1 hr with either Tyrode's solution (controls) or with 23 mM acetaldehyde (1 mg/ml) in Tyrode's solution (experimental glands). The relatively high concentration of acetaldehyde was used to obtain a reasonable yield of TIQs in order to facilitate analyses by thin-layer chromatography. The glands were rinsed by perfusion with fresh solution without acetaldehyde for an additional hour. Subsequently, stimulation by perfusion with Tyrode's solution containing 0·16 mM acetylcholine for 2 min, resulted in release of catecholamines from the control gland, and catecholamines plus TIQs from the experimental gland (Fig.1). Similar observations were made with carbachol (carbamylcholine) as the secretagogue.

The secretion process for TIQs, like that for catecholamines, was dependent upon calcium ions. When glands were stimulated with carbachol in the absence of Ca^{2+}, neither the catecholamines nor the TIQs were found in the perfusates. Secretory responses were restored by replenishing Ca^{2+} to the gland. No secretion of catecholamines or TIQs was evident in the presence of 0·1 mM tetracaine, an agent which prevents the inflow of Ca^{2+} into stimulated glands. Since TIQs are bound to the chromaffin granules and since they are released along with catecholamines in a Ca^{2+}-dependent process, it appears that the secretion of TIQs and catecholamines takes place by the same process.

UPTAKE AND RELEASE OF TIQS BY PERIPHERAL SYMPATHETIC NERVES

The formaldehyde-derived TIQs are intermediates in the well-known method (reviewed by CORRODI and JONSSON, 1967) for the visualisation of catecholamines in tissues by means of fluorescence microscopy. In this method, tissues are heated with formaldehyde gas under carefully defined conditions of humidity and temperature. The catecholamines first condense with formaldehyde to form TIQs, which become further transformed to fluorescent 3,4-dihydroisoquinolines. Since TIQs are reaction intermediates, it follows that TIQs in tissues can be visualised by the same procedure. Fluorescence microscopy was used to study the uptake and release of these alkaloids by peripheral adrenergic nerves. However, in order to avoid interference from fluorescence of endogenous NE, the animals were first treated with reserpine or α-methyl-p-tyrosine methyl ester to deplete the catecholamines.

In *in vitro* studies (COHEN, MYTILINEOU and BARRETT, 1972), irides from reserpinised or α-methyl-p-tyrosinised rats were incubated for 30 min at 37°C in isotonic buffer containing NE, DA or 6,7-dihydroxy-TIQ (1–10 µg/ml). Fluorescence microscopy revealed that the TIQ was taken up into the adrenergic plexus of the iris and that it was particularly well accumulated in the varicosities (nerve terminals), even in reserpinised preparations. TIQ accumulation was better than that for DA, but about 1/10th that for NE, as judged by fluorescence microscopy. Uptake was completely blocked by 10^{-5} M desmethylimipramine.

In recent studies (MYTILINEOU, COHEN and BARRETT, 1973), rats were treated with α-methyl-p-tyrosine methyl ester (500 mg/kg). These animals showed very little evidence of an adrenergic nerve plexus in the iris (Fig. 2a) due to depletion of endogenous NE. Under urethane anaesthesia (2 g/kg), 6,7 dihydroxy-TIQ (10 mg/kg) was injected into the femoral vein. Both cervical sympathetic trunks were cut and then one trunk was stimulated at parameters which were supramaximal for a normal animal (6 V amplitude, 2 msec duration, 15 biphasic pulses per sec). After 30 min of stimulation, the irides were removed and examined by fluorescence microscopy. The control iris (unstimulated) exhibited a rich adrenergic nerve plexus due to the presence of 6,7-dihydroxy-TIQ (Fig. 2b). There were prominent varicosities and the overall appearance was similar to a normal iris filled with NE. In contrast, the iris that had been subjected to preganglionic stimulation showed diminution in fluorescence intensity, with smaller varicosities and a smoother overall appearance (Fig. 2c). The depletion was even greater when desmethylimipramine was used to prevent reuptake of released TIQ.

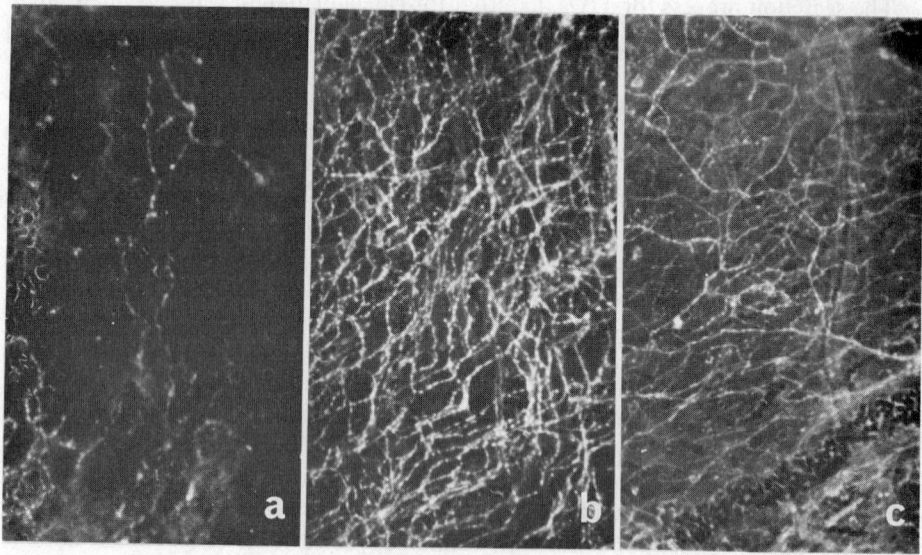

FIG. 2.—Formaldehyde vapour-treated stretch preparations of irides from NE-depleted rats. Control (a) and injected with 6,7-dihydroxy-TIQ (b, c). The unstimulated (b) and stimulated (c) irides are from the same animal.

During stimulated release of 6,7-dihydroxy-TIQ, there was marked retraction of the eyelid and protrusion of the eyeball. Dilatation of the pupil was also evident. These responses were similar to those seen in normal (NE-containing) animals. These responses were absent just prior to the injection of the TIQ.

CONCLUSIONS

Catecholamines in tissues can condense with acetaldehyde (ethanol metabolite) or formaldehyde (methanol metabolite) to form a series of hydroxylated TIQ alkaloids. The TIQs possess the following properties:

(1) They are bound to vesicles in the adrenal medulla (GREENBERG and COHEN, 1972) and in peripheral adrenergic nerves (TENNYSON et al., 1973).

(2) They can be taken up into peripheral adrenergic nerve terminals both in vitro (COHEN, MYTILINEOU and BARRETT, 1972) and in vivo (LOCKE, DEMBIEC and COHEN, 1973) by means of a cocaine-sensitive and desmethylimipramine-sensitive process.

(3) They can be taken up in vitro into nerve endings in the brain and they block the uptake of catecholamines (HEIKKILA, COHEN and DEMBIEC, 1971). When injected into the brain, they release catecholamines and cause hypothermia (BREZENOFF and COHEN, 1973).

(4) They can be released from tissues by exposure directly to acetylcholine GREENBERG and COHEN, 1973) or by preganglionic stimulation.

(5) When released by preganglionic stimulation, 6,7-dihydroxy-TIQ can interact with adrenergic receptors in the eye. It was recently reported (LEE et al., 1973) that 1-methyl-6,7-dihydroxy-TIQ (salsolinol) activates adrenergic receptors in rat adipose tissue (cf. also HJORT et al., 1942).

These properties indicate that catecholamine-derived TIQ alkaloids can be classified as potential neurotransmitter agents in adrenergic systems. The criteria for classifying a compound as a false transmitter include uptake, storage and release of the substance (KOPIN, 1968). The TIQs possess these properties and, additionally, they can react with certain adrenergic receptors. The properties of the TIQs raise some interesting speculations:

Speculation No. 1.

Is it possible that some of the well-known actions of alcoholic beverages are dependent, in part, upon alterations in adrenergic neurotransmission produced by the presence of TIQ alkaloids in nerve terminals? Although present in small amount, these alkaloids can achieve a precision of action by their discharge from nerve terminals directly onto adrenergic receptor areas. Released TIQs might activate some receptors, block others, or interfere with a variety of enzymes that act on or interact with the catecholamines.

Speculation No. 2.

The TIQs possess a property that is attractive for a consideration of post-intoxication states. The TIQs are bound in vesicles and, therefore, they can be retained, perhaps for prolonged periods of time. Furthermore, they are not attacked by monoamine oxidase (monoamine:oxygen oxidoreductase, E.C. 1.4.3.4.) (unpublished observation) and they can be taken back up after discharge and "used" again. Many bodily changes (JAFFE, 1970) such as psychomotor agitation or "hangover" occur as the general sedative or depressant actions of alcohol are wearing off. Still others, such as tremulousness, hyperexcitability, hallucinosis and seizure, are seen after particularly heavy or long-term drinking, and they occur when blood alcohol levels have either declined considerably or are absent. Are some of these changes due, in part, to the persistent actions of TIQs remaining in nerve terminals and at receptor areas? Are TIQ actions masked during alcohol intoxication, but made more evident in the absence of sedative?

SUMMARY

The current work was undertaken to explore the possibility that TIQs may play a role during alcohol intoxication and/or in post-intoxication states. At the present time, the available data show only that the catecholamine-derived TIQs are interesting pharmacologic agents for manipulating aspects of catecholaminergic function. Much remains to be done to explore the potential actions of the TIQs and to delineate their role(s) in the bodily responses to alcohol.

Acknowledgement—Portions of this work were supported by Grants AA-00243 and NS-05184 from the U.S. Public Health Service. I am indebted to the following friends and associates for their contributions to this work: Dr. Robert Barrett, Dr. Henry Brezenoff, Dr. Michael Collins, Ms. Dorothy Dembiec, Dr. Robert Greenberg, Dr. Richard Heikkila, Dr. Steven Locke, Dr. Catherine Mytilineou, Dr. Virginia Tennyson.

REFERENCES

BREZENOFF H. E. and COHEN G. (1973) *Neuropharm.* In press.
COHEN G. (1971a) In: *Advances in Mental Science III. Biologic Aspects of Alcohol* (ROACH M. K., MCISAAC W. M. and CREAVEN P. J., Eds.) pp. 267–284, University of Texas Press, Austin.
COHEN G. (1971b) *Biochem. Pharmacol.* **20,** 1757–1761.

COHEN G. and BARRETT R. (1969) *Fed. Proc.* **28,** 288 Abs.
COHEN G. and COLLINS M. A. (1970) *Science* (*Wash.*) **167,** 1749–1751.
COHEN G., MYTILINEOU C. and BARRETT R. E. (1972) *Science* (*Wash.*) **175,** 1269–1272.
COLLINS M. A. and KERNOZEK F. J. (1972) *J. Heterocyclic Chem.* **9,** 1437–1440.
CORRODI H. and JONSSON G. (1967) *J. Histochem. Cytochem.* **15,** 65–78.
GREENBERG R. S. and COHEN G. (1972) *Eur. J. Pharmacol.* **18,** 291–294.
GREENBERG R. S. and COHEN G. (1973) *J. Pharmacol. Exp. Ther.* **184,** 119–128.
HEIKKILA R., COHEN G. and DEMBIEC D. (1971) *J. Pharmacol. Exp. Ther.* **179,** 250–258.
HJORT A. M., DE BEER E. J., BUCK J. S. and RANDALL L. O. (1942) *J. Pharmacol. Exp. Ther.* **76,** 263–269.
JAFFE J. H. (1970) In: *The Pharmacological Basis of Therapeutics* (GOODMAN, L. S. and GILMAN A., Eds.) 4th ed., pp. 291–293, Macmillan, New York.
KIRBY G. W. (1967) *Science* **155,** 170–173
KOPIN I. J. (1968) *Ann. Rev. Pharmacol.* **8,** 377–394.
LEE O. S., MEARS J. E., BARDIN J. J., MILLER D. D. and FELLER D. R. (1973) *Fedn. Proc.* **32,** 723 Abs.
LOCKE S., DEMBIEC D. and COHEN G. (1973) *Fedn. Proc.* **32,** 526 Abs.
MYTILINEOU C., COHEN G. and BARRETT R. E. (1973) *Fedn. Proc.* **32,** 739 Abs.
ROBINSON T. (1968) *The Biochemistry of Alkaloids.* Springer-Verlag, New York.
SANDLER M., CARTER S. B., HUNTER K. R. and STERN G. M. (1973) *Nature, Lond.* **241,** 439–443.
SCHÖPF C. and BAYERLE H. (1934) *Ann. Chem.* **513,** 190–202.
SCHOTT H. F. and CLARK W. G. (1952) *J. Biol. Chem.* **196,** 449–462.
SHAMMA M. (1972) *Organic Chemistry* 25, *The Isoquinoline Alkaloids: Chemistry and Pharmacology.* Academic Press, New York.
TENNYSON V. M., COHEN G., MYTILINEOU C. and HEIKKILA R. (1973) *Brain Res.* **51,** 161–169.
ZENKER N. (1966) *J. Med. Chem.* **9,** 826–828.

Frontiers in Catecholamine Research 1973, pp. 1027 to 1029. Pergamon Press. Printed in Great Britain.

ON THE POSSIBLE INTERRELATIONSHIP IN MECHANISM OF ACTION BETWEEN MORPHINE, AMPHETAMINE AND NEUROLEPTIC DRUGS

JØRGEN SCHEEL-KRÜGER

Central laboratory, Department E, Sct. Hans Hospital,
DK-4000 Roskilde, Denmark

MORPHINE interacts with the brain catecholamines, dopamine and noradrenaline (CLOUET and RATNER, 1970; GUNNE et al., 1969; LOH et al., 1973; SMITH et al., 1972), but neither the biochemical nor the pharmacological mechanism of action of morphine is presently understood.

In our laboratory the interest on morphine arose from reported studies in mice and rats which on one side indicate that morphine produces central stimulant effects which apparently may be related to amphetamine and on the other side points to similarities between morphine and neuroleptic drugs. These findings seem apparently paradoxical since it is well-established that amphetamine and neuroleptic drugs show mutual antagonism and that this effect is a highly characteristic property of neuroleptica (FOG, 1972; JANSSEN et al., 1960; VAN ROSSUM, 1966). However, our comparative studies in rats between amphetamine sulphate (0·25–10 mg/kg) and small acute doses of morphine chloride (0·5–2 mg/kg) revealed significant differences in mechanism of action: The morphine hyperactivity induced by 1–2 mg/kg appeared as a general stimulation of locomotion, rearing, grooming, eating and drinking in the rats. The morphine induced locomotion and rearing was very characteristic and seen in phases as bursts of activity followed by phases of sedation. Morphine in doses above 5 mg/kg induced sedation and catalepsy. For further details see FOG, 1970; AYHAN and RANDRUP, 1973a,b.

In contrast amphetamine induced a more selective and continuous stimulation of locomotion and rearing which at increasing doses became more and more stereotyped. At no dose levels stimulation of grooming, eating and drinking have been observed. High doses of amphetamine (5–10 mg/kg) led, after a prephase of locomotion and rearing, to an extremely stereotyped stimulation consisting only of sniffing, licking or gnawing of the cage wires (SCHEEL-KRÜGER, 1971).

As further differences should be underlined that amphetamine very strongly abolish social activity in rats (and monkeys) (SCHIØRRING and RANDRUP, 1971) whereas morphine did not abolish any social activity (SCHIØRRING and HECHT, in preparation).

Recently α-methyltyrosine, an inhibitor of the catecholamine biosynthesis, has been shown to antagonise the locomotion of small acute doses of morphine in rats (DAVIS et al., 1972) and mice (CARROL et al., 1972; STRUBELT et al., 1970). This finding was confirmed in the rat and furthermore that inhibition of noradrenaline biosynthesis with FLA-63 or blockade of noradrenaline or dopamine receptors in the brain also antagonised the morphine-induced locomotion, rearing and grooming. (AYHAN and RANDRUP, 1973a).

These findings thus strongly indicate a significance of the catecholamines in the morphine stimulation. A similar relationship in the amphetamine stimulation is already well-established (SCHEEL-KRÜGER, 1971), but the biochemical and pharmacological mechanisms of actions must be different since it was found that these two drugs showed mutual antagonism. (FOG, 1970; AYHAN and RANDRUP, 1973b).

The simultaneous injection of morphine (2 mg/kg) and amphetamine led to the inhibition of morphine type excitation: bursts of locomotion and rearing, increased grooming. 0·25 mg/kg of amphetamine produced a significant inhibition and a further increase of amphetamine dose (0·5 and 1·0 mg/kg) caused almost complete inhibition. The mutal antagonism was evident since morphine (2 mg/kg) antagonised the stimulation of locomotion and rearing after higher doses of amphetamine (AYHAN and RANDRUP, 1973b). Stimulation of central dopamine mechanism with apomorphine (0·5 mg/kg) or L-dopa (100 mg/kg) given after a peripheral decarboxylase inhibitor (Ro-4-4602) produced also a significant inhibition of the morphine-type excitation.

The mechanism of the morphine stimulation must thus be different from that of amphetamine since apomorphine or L-dopa in the above mentioned doses potentiated the amphetamine stimulation (AYHAN and RANDRUP, 1973b).

It is well-known that the behavioural sedation and catalepsy after a single high dose of morphine to rats changes to behavioural excitation (EIDELBERG and SCHWARTZ, 1970) and typical amphetamine-like stereotyped, licking gnawing activity (FOG, 1970; AYHAN and RANDRUP, 1972) during chronically injection of morphine. Pretreatment with drugs which deplete the catecholamines, reserpine, α-methyltyrosine, FLA-63, diethyldithiocarbamate or antagonised dopamine or noradrenaline receptors antagonise the excitation after chronic morphine. (EIDELBERG and SCHWARTZ, 1970; AYHAN and RANDRUP, 1972). Comparison between morphine and amphetamine stereotypy indicated that brain noradrenaline plays a more important role than dopamine in morphine stereotypy (AYHAN and RANDRUP, 1972) in contrast to amphetamine stereotypy where dopamine is the most important amine (SCHEEL-KRÜGER, 1971).

The simultaneous injection of amphetamine (20 mg/kg) with the daily injection of morphine (100 mg/kg) to chronically morphinized rats antagonised the stereotyped licking/biting activity which otherwise would have been present. Mutual antagonism between amphetamine and morphine is thus even present at dose levels of these drugs which given separately induce stereotypy (FOG, 1970). Morphine in a single high dose shows antagonism of apomorphine (JANSSEN et al., 1960) and amphetamine stereotypy (FOG, 1970). Other characteristic neuroleptic properties in rats are development of catalepsy which biochemically seems correlated with the increase of homovanillic acid a major dopamine metabolite (KUSCHINSKY and HORNYKIEWICS, 1972; STILLE, 1971; STILLE and LAUENER, 1971).

In conclusion of the present discussion it is hoped that the presented evidence has underlined the clear-cut key position between amphetamine and neuroleptic drugs which morphine occupy and that further investigation along these lines might provide a very fruitful understanding of the central mechanism underlying these types of drugs.

REFERENCES

AYHAN I. H. and RANDRUP A. (1972) *Psychopharmacologia* **27**, 203–212.
AYHAN I. H. and RANDRUP A. (1973a) *Psychopharmacologia* **29**, 317–328.
AYHAN I. H. and RANDRUP A. (1973b) *Archs. int. Pharmacodyn.* in press.
CARROL B. J. and SHARP P. T. (1972) *Brit. J. Pharmacol.* **46**, 124–139.
CLOUET D. H. and RATNER M. (1970) *Science* **168**, 854–856.
DAVIS W. M., BABBINI M. and KHALSA I. H. (1972) *Research Com. Chem. Path. Pharmacol.* **4**, 267–278.
EIDELBERG E. and SCHWARTZ A. S. (1970) *Nature., Lond.* **225**, 1152–1153.
FOG R. (1970) *Psychopharmacologia* **16**, 305–312.
FOG R. (1972) *Acta. Neurol. Scand.* **48**, (suppl. 50), 1–66.
GUNNE L. M., JÖNSSON J. and FUXE K. (1969) *Europ. J. Pharmacol.* **5**, 338–342.
JANSSEN P. A. J., NIEMEGEERS C. J. C. and JAGENEAU A. H. M. (1960) *Arzneimittel-Forsch.* **12**, 1003–1005.
KUSCHINSKY K. and HORNYKIEWICS O. (1972) *Europ. J. Pharmacol.* **19**, 119–122.
LOH H. H., HITZEMANN R. J. and WAY, E. L. (1973) *Life Sci.* **12**, 33–41.
SCHEEL-KRÜGER J. (1971) *Europ. J. Pharmacol.* **14**, 47–59.
SCHIØRRING E. and RANDRUP A. (1971) *Pharmakopsychiat. Neuro-Psychopharm.* **4**, 1–12.
SMITH C. B., SHELDON M. I., BEDNARCZYK J. H. and VILLARREAL J. E. (1972) *J. Pharm. Exptl. Ther.* **180**, 547–557.
STRUBELT O., VOGELBERG W. and ZETLER G. (1970) *Arzneimittel-Forsch.* **20**, 32–37.
STILLE G. (1971) *Arzneimittel-Forsch.* **21**, 386–392.
STILLE G. and LAUENER H. (1971) *Arzneimittel-Forsch.* **21**, 252–255.
VAN ROSSUM J. (1966) *Archs. int. Pharmacodyn.* **160**, 492–494.

Frontiers in Catecholamine Research 1973, pp. 1031 to 1034. Pergamon Press. Printed in Great Britain.

IMPORTANCE OF BRAIN DOPAMINE FOR THE STIMULANT ACTIONS OF AMPHETAMINE*

K. E. MOORE and J. E. THORNBURG

Department of Pharmacology, Michigan State University, East Lansing, Michigan 48823, U.S.A.

IT HAS been suggested (TAYLOR and SNYDER, 1971) that the central stimulant actions of amphetamine are mediated by brain norepinephrine (NE) whereas the stereotype behaviours that follow high doses of this drug are mediated by dopamine (D). We have investigated the behavioural effects of a number of drugs that alter the dynamics of brain catecholamines and have concluded that many pharmacological effects previously attributed to modifications of NE actions result instead from alterations of dopaminergic systems. Experiments on how drugs which block catecholamine synthesis influence the actions of d-amphetamine will serve as one example of the studies which have led to this conclusion.

The antiamphetamine actions of α-methyltyrosine (αMT), an inhibitor of tyrosine hydroxylase (EC 1.14.3.1), are well documented (e.g. WEISSMAN et al., 1966; DOMINIC and MOORE, 1969; MOORE and DOMINIC, 1971) and suggest that the central stimulant actions of d-amphetamine are mediated by a newly synthesised pool of brain catecholamines. Since αMT blocks the synthesis of both NE and D, it is not possible to relate the antiamphetamine properties of this drug to a disruption of synthesis exclusive to either of these two catecholamines in the brain. If the antiamphetamine effects of αMT result from inhibition of NE synthesis, then inhibition of dopamine-β-hydroxylase (EC 1.14.2.1, DBH) should duplicate the actions of αMT. On the other hand, if DBH inhibitors do not block the stimulant actions of amphetamine, then the αMT effect may result from inhibition of D synthesis.

It has been reported that DBH inhibitors produce behavioural depression and block the central stimulation produced by amphetamines and related drugs (SVENSSON and WALDEK, 1969; MAJ et al., 1968). Nevertheless, the behavioural depressant and NE depleting actions of these inhibitors do not appear to be casually related (MOORE, 1969). Intraperitoneal administration of various DBH inhibitors [disulfiram; FLA-63, bis (4-methyl-l-homopiperazinyl-thiocarbonyl disulfide); U-14,624, 1-phenyl-3-(2-thiazolyl)-2-thiourea] reduces locomotor activity in rodents and markedly elevates plasma corticosterone and blood glucose concentrations (THORNBURG and MOORE, 1971). These effects may have resulted from peritoneal irritation produced by the intraperitoneal administration of the insoluble inhibitors. When administered in the diet DBH inhibitors reduced brain NE concentrations, blocked NE synthesis, but did not reduce spontaneous locomotor activity (MOORE, 1969; VON VOIGTLANDER and MOORE, 1970).

The effects of adding αMT, U-14,624, and FLA-63 to the diet of mice that have been accommodated to consuming their daily food in 4 hr are compared in Table 1.

---

* Supported by USPHS Grants NSO9174 and MH13174.

Table 1. Effects of 4 hr diets containing 0·4% α-methyltyrosine, 0·4% U-14,624 or 0·05% FLA-63 on brain contents of (^{14}C) and of endogenous tyrosine and catecholamines and on locomotor activity in mice. One hr after the 4 hr test diet mice were sacrificed and their brains analyzed for endogenous compounds as described by Carr and Moore (1968). Other mice were injected i.v. with 10 μCi L-(^{14}C) tyrosine, sacrificed 30 min later, and their brains analysed for radioactive compounds as described by Bhatnagar and Moore (1972). Locomotor activity was determined in the following manner. One hour after the 4 hr test diet pairs of mice were placed in Woodard actophotometers and activity recorded 10–20 min later. The mice were then injected i.p. with saline or d-amphetamine sulfate (2 mg/kg), returned to actophotometers and activity recorded 20–60 min later. Amphetamine-stimulated activity represents the activity of animals injected with amphetamine less the activity of animals injected with saline. Numbers represent means ± S.E.M. and underlined values are statistically different from controls ($P < 0·01$); values in parentheses represent the number of determinations.

Diet	Control	αMT	U-14,624	FLA-63
Endogenous compounds (6–18)				
Tyrosine (μg/g)	13·2 ± 1·5	14·3 ± 3·7	10·4 ± 0·4	13·1 ± 1·8
Norepinephrine (μg/g)	0·35 ± 0·01	0·22 ± 0·02	0·17 ± 0·01	0·15 ± 0·01
Dopamine (μg/g)	0·75 ± 0·04	0·31 ± 0·03	0·79 ± 0·04	0·90 ± 0·13
^{14}C-compounds (3–5)				
^{14}C-Tyrosine (dis/min × 10²/brain)	270 ± 12	287 ± 41	300 ± 35	324 ± 22
^{14}C-Norepinephrine (dis/min/brain)	72 ± 5	24 ± 3	31 ± 4	24 ± 3
^{14}C-Dopamine (dis/min/brain)	176 ± 5	78 ± 15	372 ± 109	308 ± 22
Locomotor activity (12–42)				
Exploratory activity	647 ± 25	450 ± 31	695 ± 26	582 ± 39
Amphetamine-stimulated activity	2927 ± 229	1268 ± 78	2724 ± 439	2541 ± 384

When administered in this manner none of the drugs altered plasma corticosterone concentrations. All drugs, however, reduced the brain contents of NE, but only αMT reduced the brain contents of D. All three drugs significantly reduced the accumulation of (^{14}C)NE 30 min after the i.v. administration of L-(^{14}C) tyrosine. Only αMT reduced the brain content of (^{14}C)D, whereas the DBH inhibitors increased the brain content of this radioactive amine. None of the diets altered the endogenous or (^{14}C) tyrosine contents of the brain. Only mice on the αMT diet exhibited significantly reduced spontaneous and amphetamine-stimulated activity. Similar results have been obtained with mice placed on 24 hr diets of the inhibitors of catecholamine synthesis (Thornburg, 1972). These results suggest, therefore, that the locomotor stimulant effects of amphetamine involve a dopaminergic mechanism.

This proposal is supported by the results of the experiment illustrated in Fig. 1. The activities of groups of 4 mice were recorded in their home cages during the 12 hr dark–light cycle. By using this procedure animal activity could be recorded with minimal disturbance to ongoing behaviour and without possible nonspecific disturbing effects that might accompany parenteral drug administration. In the top panel the first two pairs of bars represent activity on two consecutive days of control diet; this activity was set at 100 per cent. The next day the mice received a diet containing 0·4 per cent αMT and there was a slight reduction in activity. The next day 0·05 per cent amphetamine was added to the αMT diet and little effect was noted. Animals then received two days of control diet and finally a diet containing 0·05 per cent amphetamine which caused a marked increase in activity. Thus, the antiamphetamine property of αMT was clearly demonstrated.

In the bottom panel the same experimental design was utilised except that U-14,624 was added to the diet in place of αMT. U-14,624 alone had little effect on

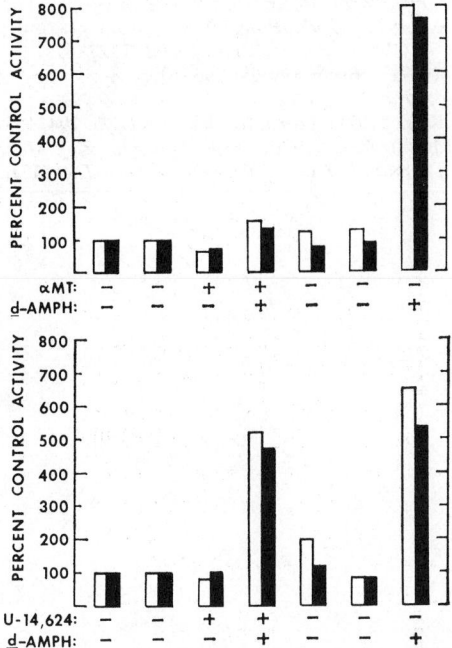

FIG. 1.—Effects of amphetamine on the activity of mice in their home cages. Activity was recorded and food was presented only during the 12 hr dark phase of a 24 hr dark–light cycle. Activity is expressed as a percentage of control activity. Open bars represent results obtained with an instrument that utilises electromagnetic proximity sensors (Selective Activity Meter, Columbus Instruments) and solid bars represent results obtained with an instrument that utilises 40 infrared-sensitive photocells (Electronic Motility Meter, Motron Products). DL-α-methyltyrosine (0·4%), U-14,624 (0·4%) and d-amphetamine sulfate (0·05%) were added to the diet as indicated.

activity, but when amphetamine was added to the U-14,624 diet a marked stimulation was observed that was approximately equivalent to that seen two days later when amphetamine alone was added to the diet. Thus, blockade of DBH with U-14,624 did not influence amphetamine-induced stimulation of activity. Since disruption of both NE and D synthesis by αMT blocks d-amphetamine-stimulated activity while disruption of only NE synthesis does not, it appears that the stimulation produced by amphetamine is primarily dependent upon a dopaminergic mechanism. This conclusion, however, is based upon the premise that antiamphetamine properties of αMT are related to the ability of this drug to inhibit tyrosine hydroxylase. It has recently been reported (ENNA et al., 1973) that αMT blocks amphetamine-induced release of amines from brain slices. Using an in situ cerebroventricular perfusing technique, an inhibitory effect of αMT on amphetamine-induced release of (^{3}H)D from the caudate nucleus of the cat could not be demonstrated (CHIUEH and MOORE, 1973).

REFERENCES

BHATNAGAR R. K. and MOORE K. E. (1972). J. Pharmacol. Exp. Ther. 180, 265–276.
CARR L. A. and MOORE K. E. (1968). Neuroendocrinology 3, 285–302.
CHIUEH C. C. and MOORE K. E. (1973). The Pharmacologist 15, in press.
DOMINIC J. A. and MOORE K. E. (1969). Arch. int. Pharmacodyn. 178, 166–176.
ENNA S. J., DORRIS R. L. and SHORE P. A. (1973). J. Pharmacol. Exp. Ther. 184, 576–582.
MAJ J., PRZEGALINSKI E. and WIELOSZ M. (1968). J. Pharm. Pharmacol. 20, 247–248.

MOORE K. E. (1969). *Biochem. Pharmacol.* **18,** 1627–1634.
MOORE K. E. and DOMINIC J. A. (1971). *Fedn. Proc.* **30,** 859–870.
SVENSSON T. H. and WALDECK B. (1969). *Europ. J. Pharmacol.* **7,** 278–282.
TAYLOR K. M. and SNYDER S. (1971). *Brain Res.* **28,** 295–309.
THORNBURG J. E. (1972). *Fedn. Proc.* **31,** 530 Abs.
THORNBURG J. E. and MOORE K. E. (1971). *Arch. int. Pharmacodyn.* **194,** 158–167.
VON VOIGTLANDER P. F. and MOORE K. E. (1970). *Proc. Soc. exp. Biol. Med.* **133,** 817–820.
WEISSMAN A., KOE B. K. and TENEN S. (1966). *J. Pharmacol. Exp. Ther.* **151,** 339–352.

Frontiers in Catecholamine Research 1973, pp. 1035 to 1037. Pergamon Press. Printed in Great Britain.

p-HYDROXYNOREPHEDRINE: ITS SELECTIVE DISTRIBUTION IN DIFFERENT RAT BRAIN AREAS

F. Cattabeni, G. Racagni and A. Groppetti[*]

Institute of Pharmacology and Pharmacognosy, School of Pharmacy,
University of Milan, 20129 Milano, Italy

THE MECHANISMS proposed for the explanation of the various effects elicited by *d*-Amphetamine (*d*-A) invoke either the direct action of *d*-A on norepinephrine (NE) or dopamine (DA) stores, or the accumulation of one of its metabolites: *p*-Hydroxynorephedrine (*p*OHNE) (Brodie *et al.*, 1970; Costa and Groppetti, 1970; Sulser and Sanders-Bush, 1971; Boakes, 1972). In fact the depletion of heart and brain NE outlasts for more than 24 hr the *d*-A disappearance from these tissues, whereas *p*OHNE accumulates in heart and brain of rats and its concentrations have a half life of about 20 hr (Costa and Groppetti, 1970).

The elucidation of the central role of *p*OHNE seems therefore to be of importance, because if more is known about its formation, localization and disappearance from different brain areas, more light would be shed on the mechanisms underlying effects such as a tachyphylaxis and psychosis in man after *d*-A treatment.

It has been suggested, on the basis of results obtained in peripheral nervous system, that also in brain *p*OHNE may be synthetised from *p*-Hydroxyamphetamine (*p*OHA) (Goldstein and Anagnoste, 1965) and selectively stored in noradrenergic nerve endings, acting as a false neurotransmitter (Fisher *et al.*, 1965; Thoenen *et al.*, 1966; Groppetti and Costa, 1968).

Since only circumstantial evidence has been offered in support of this hypothesis, mainly because the analytical methods available did not have sufficient specificity and sensitivity, we have developed a mass fragmentographic assay for *p*OHNE at picomole levels (Cattabeni *et al.* (1973) in press).

Utilising this technique, we have measured *p*OHNE levels after 6-hydroxydopamine (6-OHDA) treatment. 6-OHDA has shown to be a useful tool in the study of the noradrenergic system function. Evidences from morphological studies suggest that this compound causes an ultra-structural modification of the axons and nerve terminals of the sympathetic nervous system. When injected intracisternally or into the lateral ventricles of rat brain, 6-OHDA produces a selective and long lasting depletion of brain NE while dopaminergic and serotonergic nerve terminals are only slightly affected (Tranzer and Thoenen, 1968; Ungerstedt, 1968; Uretsky and Iversen, 1969; Bloom *et al.*, 1969). In our experiment, the animals were given intraventricularly 200 μg per rat of 6-OHDA and after one week *d*-A (7 mg/kg i.p.) and sacrificed 5 hr later.

In treated animals *p*OHNE levels were 60 per cent lower when compared with controls and, since brain NE concentrations were also decreased by a similar extent, while 80 per cent of DA was still present, it seems appropriate to infer that also in brain *p*OHNE is associated with the noradrenergic system. To further substantiate

---

* Institute of Pharmacology, School of Medicine, 2nd Chair, University of Milan, 20129 Milano, Italy

the relationship between the noradrenergic system and *p*OHNE formation and/or storage, we have studied the distribution of this compound in different brain areas after high doses of *d*-A.

As expected (see Table 1) only traces of *p*OHNE are present in the caudate nucleus, whereas consistent levels of *p*OHNE are observed in the hypothalamus and

TABLE 1. CORRELATION BETWEEN THE DISTRIBUTION OF *p*-OH-NOREPHEDRINE (*p*OHNE) IN BRAIN AREAS OF AMPHETAMINE TREATED RATS* AND NOREPINEPHRINE (NE) LEVELS IN CONTROL ANIMALS

Brain area	*p*OHNE μg/g $\pm$ s.e.	NE† μg/g $\pm$ s.e.	*p*OHNE/NE
Cerebellum	0·192 ± 0·017	0·18 ± 0·01	1:1
Tel-diencephalon	0·175 ± 0·044	0·46 ± 0·01	1:2·5
Brain stem	0·172 ± 0·029	0·71 ± 0·03	1:4
Hypothalamus	0·420 ± 0·072	2·10 ± 0·13	1:5
Caudate N.	0·040 ± 0·004	n.d.	

* *d*-Amphetamine 7 mg/kg i.p., 5 hr before sacrifice.
† NE was measured spectrophotofluorimetrically according to C. C. Chang, *J. Neuropharm.* **3**, 643 (1964).
n.d.: not detectable.
The values are a mean ± s.e. of 3 experiments.

other brain areas rich in NE. Considering the *p*OHNE to NE ratio, as indicated in the Table, it is possible to infer that *p*OHNE synthesis and/or storage occur preferentially at nerve endings rather than in cell bodies. In fact the ratio is higher in brain structures containing predominantly nerve terminals, as in the cerebellum and tel-diencephalon. Obviously, it would be more appropriate to relate *p*OHNE levels to the functional activity of noradrenergic nerve endings in different brain areas rather than to NE levels. Unfortunately indications on turnover rate in all these brain areas are still incomplete.

We are now investigating whether this preferential accumulation of *p*OHNE is time dependent and if the rate of disappearance follows the same time course in different brain areas, in the hope to shed more light on the possible role played by *p*OHNE in the behavioural effects that follow *d*-A treatment.

Preliminary data have been obtained also for the levels of *p*OHA, the other hydroxylated metabolite of *d*-A in several species, including man. Goldstein and Anagnoste (1965) have demonstrated that this compound can be in turn utilised as a substrate for dopamine-β-hydroxylase to form *p*OHNE.

We have observed that after *d*-A administration the highest concentrations of *p*OHA are present in caudate nucleus. In brain stem, cerebellum and tel-diencephalon the levels of this metabolite are only 50 per cent of those found in the caudate.

Carlsson (1970) and Costa *et al.*, (1972) have suggested that an action on dopaminergic neurones in caudate nucleus may be entertained as a possible indirect mechanism involved in hypermotility elicited by *d*-A in rats. Therefore it is tempting to speculate that *p*OHA could play a role in this behavioural effect.

In conclusion, the selective distribution of *p*OHNE in different rat brain areas indicates that also in the central nervous system this *d*-A metabolite is associated with noradrenergic nerve endings.

Preliminary results indicate that also *p*OHA has a preferential distribution in rat brain areas, being localised maximally in the caudate nucleus. Experiments are in progress to ascertain if these compounds can be involved in the mechanism of some of the behavioural and biochemical effects due to *d*-A treatment.

REFERENCES

BLOOM F. E., ALGERI S., GROPPETTI A., REVUELTA A. and COSTA E. (1969) *Science* **166**, 1284–1286.
BOAKES R. J., BRADLEY P. B. and CANDY J. M. (1972) *Br. J. Pharmacol.* **45**, 391–403.
BRODIE B. B., CHO A. K. and GESSA G. L. (1970) In: *International Symposium on Amphetamines and Related Compounds* (E. COSTA and S. GARATTINI, Eds.), pp. 217–230, Raven Press, New York.
CARLSSON A. (1970) In: *International Symposium on Amphetamines and Related Compounds* (E. COSTA and S. GARATTINI, Eds.), pp. 289–300, Raven Press, New York.
CATTABENI F., RACAGNI G. and GROPPETTI A. (1973). (in press) In: *Advances in Biochemical Psychopharmacology* Vol. 7 (E. COSTA and B. HOLMSTEDT, Eds), Raven Press, New York.
COSTA E. and GROPPETTI A. (1970) In: *International Symposium on Amphetamines and Related Compounds* (E. COSTA and S. GARATTINI, Eds), pp. 231–255, Raven Press, New York.
COSTA E., GROPPETTI A. and NAIMZADA K. (1972) *Br. J. Pharmacol.* **44**, 742–751.
FISHER J. E., HORST W. D. and KOPIN I. J. (1965) *Br. J. Pharmacol.* **24**, 477–484.
GOLDSTEIN M. and ANAGNOSTE B. (1965) *Biochim. Biophys. Acta* **107**, 166–168.
GROPPETTI A. and COSTA E. (1968) *Atti Accademia Medica Lombarda* **23**, 1105–1107.
SULSER F. and SANDERS-BUSH E. (1971) *Ann. Rev. Pharmacol.* **11**, 209–230.
THOENEN H., HURLIMAN K. F., GEY K. F. and HAEFELY W. (1966) *Life Sci.* **5**, 1715–1722.
TRANZER J. P. and THOENEN H. (1968) *Experientia* **24**, 155–156.
UNGERSTEDT U. (1968) *Europ. J. Pharmacol.* **5**, 107–110.
URETSKY N. J. and IVERSEN L. L. (1969) *Nature, Lond.* **221**, 557–559.

Frontiers in Catecholamine Research 1973, pp. 1039 to 1042. Pergamon Press. Printed in Great Britain.

THE EFFECT OF MORPHINE ON RAT BRAIN CATECHOLAMINES: TURNOVER *IN VIVO* AND UPTAKE IN ISOLATED SYNAPTOSOMES*

D. H. CLOUET, J. C. JOHNSON, M. RATNER,
N. WILLIAMS and G. J. GOLD

New York State Narcotic Addition Control Commission, Testing and Research
Laboratory, 80 Hanson Place, Brooklyn, New York 11217, U.S.A.

THE ACUTE administration of morphine and other narcotic analgesic drugs to animals produces transient changes in the levels of biogenic amines in the central nervous system (WAY and SHEN, 1971). Among the most striking acute responses are the depletion of dopamine from the striatum (GUNNE et al., 1969) and norepinephrine from the hypothalamus (VOGT, 1954; REIS et al., 1969). As measured by the accumulation of ^{14}C-dopamine and -norepinephrine from ^{14}C-tyrosine *in vivo*, the biosynthesis of dopamine is transiently increased after a single injection of morphine, especially in the hypothalamus and striatum (CLOUET and RATNER, 1970). A more pronounced increase in the rates of dopamine biosynthesis was found in the same brain regions of morphine-tolerant rats.

The present report is the results of the exploration of the relationship between brain biogenic amines and opiates in three aspects: (1) biosynthesis of catecholamines; (2) tyrosine hydroxylase levels; and (3) synaptosomal uptake of catecholamines.

MATERIALS AND METHODS

Experimental animals

Male Wistar rats were injected subcutaneously with morphine sulphate in doses of 5, 20 or 60 mg/kg, which produces analgesic and hypothermic responses in this strain for 63, 138 and 192 min, respectively. After sacrifice, the brains were removed and dissected into six regions (GLOWINSKI and IVERSEN, 1966). Chronic treatment with morphine was by daily injection of 60 mg/kg/day or by pellet implantation.

Catecholamine biosynthesis

^{14}C-tyrosine was administered intracisternally under light ether anaesthesia for a 10 min period. Norepinephrine (NE) and dopamine (DA) levels were measured by the method of ANTON and SAYRE (1962) after separation on Dowex-50 (NYBACK and SEDVALL, 1968), and the fractions examined for radioactivity by scintillation spectrometry. Protein was measured by the Folin method of LOWRY et al. (1951).

Tyrosine hydroxylase

The sum of the accumulations of ^{14}C radioactivity in dopa, DA and NE in a 10 min period *in vivo* was taken as an estimation of tyrosine hydroxylase *activity* during the period. The enzyme *levels* were assayed in brain areas in independent experiments using a 40 per cent saturated ammonium sulphate precipitate, as described by MUSACCHIO et al. (1969).

Catecholamine uptake

Pooled samples of striatum (for DA) and hypothalamus (for NE) were homogenised and centrifugally fractionated to sediment either a crude mitochondrial fraction containing the synaptosomes or a purer synaptosomal fraction isolated by centrifuging the crude mitochondrial fraction through a sucrose density gradient (HAGA, 1971). The synaptosomes were incubated in Krebs-Henseleit medium with 2×10^{-6} M amine and the MAO inhibitor, nialamide, for 5 min at $37°$ in the presence or absence of opiates. The reaction was stopped by separating the particles and medium by Millipore® filtration, and each fraction was examined for radioactivity.

RESULTS AND DISCUSSION

The levels of NE were increased in the hypothalamus after a single injection of morphine (Table 1). DA levels were too low to measure, except in the striatum, in which the levels were significantly lower at 30 min, and above control at 60 min. The rates of biosynthesis of norepinephrine were relatively unaffected by morphine. Dopamine biosynthesis was increased in most areas after a single injection of 60 mg/kg of morphine, and in all areas in tolerant animals (Table 1). After 5 or 20 mg/kg, the responses were in the same direction but less, except for dopamine biosynthesis which was even faster after 20 mg/kg than after 60 mg/kg.

The results suggest that the responses of the dopaminergic system are predominant after morphine administration.

TABLE 1. CHANGES EFFECTED BY MORPHINE ADMINISTRATION IN CATECHOLAMINE LEVELS AND RATES OF BIOSYNTHESIS

	Cerebellum	Medulla	Hypothalmus	Striatum	Midbrain	Cortex
NE Levels (μg/g)	163 ± 2	587 ± 37	1144 ± 53	221 ± 25	459 ± 25	210 ± 12
Acute morphine	ns	ns	1943 ± 221	ns	ns	297 ± 26
Chronic morphine pellets	ns	429 ± 34	1599 ± 192	ns	344 ± 41	250 ± 34
DA Levels (μg/g)				2067 ± 44		
Acute morphine 30 min				1005 ± 112		
60 min				2427 ± 189		
Chronic morphine pellets				2439 ± 90		
NE Synthesis (nmoles/hr/ mg protein)	0.8 ± 0.0	5.1 ± 0.2	21.0 ± 2.0	174 ± 24	23 ± 1.0	0.23 ± 0.0
DA Synthesis (nmoles/hr/mg)	8.9 ± 0.7	13.7 ± 0.6	72.4 ± 8.0	24.1 ± 3.0	10.3 ± 0.5	26.2 ± 3.0
Acute morphine 30 min	ns	ns	90.9 ± 1.0	ns	12.2 ± 0.2	ns
60 min	ns	ns	119.7 ± 23	33.8 ± 0.3	ns	ns
Chronic morphine pellets	13.8 ± 2.0	23.1 ± 0.6	118.9 ± 5.0	38.5 ± 3.0	18.1 ± 1.0	49.3 ± 5.0

The levels and turnover values are averages of at least three experiments.
Only values different from control values at $P < 0.01$ are listed.
ns = not significant.

TABLE 2. THE EFFECT OF MORPHINE ADMINISTRATION ON TYROSINE HYDROXYLASE IN BRAIN
Tryosine hydroxylase (nmoles/hr/g wet weight)

	Cerebellum	Medulla	Hypothalamus	Striatum	Midbrain	Cortex
Controls						
in vivo	0·43	0·99	4·14	1·34	1·40	1·77
in vitro	2·12	4·71	9·98	12·48	12·29	3·82
Acute						
in vivo	0·41	0·89	4·61	2·57*	1·53	1·50
in vitro	1·86	5·27	9·85	10·39	12·81	4·07
Chronic-injected						
in vivo	0·59*	1·39*	5·61*	3·79*	1·84*	2·20*
in vitro	1·19	5·27	10·37	12·97	10·62	3·87
Chronic-pellets						
in vivo	0·46	0·83	3·32*	2·14*	1·20	2·06
in vitro	1·66	5·48	13·17*	17·40*	10·73	3·89

The starred values are different from controls at $P < 0.01$.

The in vivo accumulation of ^{14}C in dopa, DA and NE, and the in vitro assay for tyrosine hydroxylase are described in the methods section. The acute treated rats were injected with 60 mg/kg morphine and killed 1 hr later. The chronically injected animals were injected with 60 mg/kg morphine each day for four (in vivo) or five (in vitro) days and sacrificed 2 hr after the last dose.

Because the rates of synthesis of the individual catecholamines varied widely during morphine treatment, the sums of ^{14}C-catecholamines synthesised from tyrosine also varied widely, with significant differences in each area at some point after morphine injection. The most pronounced increase in tyrosine hydroxylase activity was found in the brains of rats killed 2 hr after the fourth daily injection of morphine (Table 2). The actual *levels* of tyrosine hydroxylase, however, were remarkably constant during morphine treatment, with significant increases in isolated enzyme activity only in the striatum and hypothalamus of long-time morphine-treated rats (Table 2). These increases may be taken as an indication that tyrosine hydroxylase was induced during chronic morphine treatment. It is possible that induction at lower levels occurred at earlier times (or in other tissue areas) in amounts too low to be detectable.

The nature of the opiate-catecholamine interaction was not defined by the results of the previous studies. In order to explore the possibility that narcotic analgesics have a direct action on the transport of catecholamines across the neuronal membranes, the effect of morphine on the uptake and release of the amines in isolated synaptosomes was examined. The rates of uptake of ^{14}C-NE into hypothalamic synaptosomes and ^{14}C-DA into striatal synaptosomes were inhibited in the presence of morphine-HCl, but only at concentrations of 10^{-4} M or higher. When this inhibition was examined in relation to DA concentration, it was found that only the low affinity uptake, representing diffusion, was inhibited by morphine. These results suggest that one effect of morphine in vivo is an interference with dopamine re-uptake following after an excessive release of the amine triggered by a prior event in the central nervous system. The rates at which the biogenic amines were released after preloading with labelled amines in vitro was not affected by morphine.

SUMMARY

(1) Acute morphine administration produced an initial depletion of DA in the striatum and an increased rate of DA biosynthesis in the same area. In tolerant animals, the biosynthesis of DA was increased in all areas of rat brain.

(2) While the *activity* of tyrosine hydroxylase varied widely during morphine treatment, the *levels* were increased in some brain areas only after long-time morphine-treatment.

(3) The uptake of NE and DA into isolated synaptosomes was inhibited by morphine at concentrations of 10^{-4} M or higher. A kinetic analysis of DA uptake in the presence of morphine suggested that DA re-uptake by diffusion was the mode of transport inhibited by the opiate.

Acknowledgements—Supported in part by NIMH Grant DA 00087.

REFERENCES

ANTON A. H. and SAYRE D. F. (1962) *J. Pharmacol. Exp. Therap.* **138**, 360–375.
CLOUET D. H. and RATNER M. (1970) *Science* **168**, 854–856.
GLOWINSKI J. and IVERSEN L. L. (1966) *J. Neurochem.* **13**, 655–669.
GUNNE L. M., JOHNSON J. and FUXE K. (1969) *Europ. J. Pharmacol.* **9**, 338–342.
HAGA T. (1971) *J. Neurochem.* **18**, 781–790.
LOWRY O. H., ROSEBROUGH N. J., FARR A. L. and RANDALL R. J. (1951) *J. Biol. Chem.* **193**, 265–275.
MUSACCHIO J. M., JULOU L., KETY S. S. and GLOWINSKI J. (1969) *Proc. Nat. Acad. Sci.* **63**, 1117–1119.
NYBACK H. and SEDVALL G. (1968) *J. Pharmacol. Exp. Therap.* **162**, 294–304.
REIS D. J., RIFKIN M. and CORVELLI A. (1969) *Europ. J. Pharmacol.* **9**, 149–152.
VOGT M. (1954) *J. Physiol.* **123**, 451–481.
WAY E. L. and SHEN F.-H. (1971) In: *Narcotic Drugs. Biochemical Pharmacology.* (CLOUET D. H., ed.) pp. 229–253. Plenum Press, New York.

Frontiers in Catecholamine Research 1973, pp. 1043 to 1044. Pergamon Press. Printed in Great Britain.

PSYCHOPHARMACOLOGICAL PROFILE OF COCAINE

Pierre Simon

Unité de Recherche de NeuroPsychoPharmacologie et Département de Pharmacologie
de la Faculté de Médecine Pitié-Salpétrière, 91, Bd. de l'Hôpital, 75643 Paris,
Cedex 13, France.

THE REVIEW of literature (see Simon et al., 1972) concerning the psychopharmacological effects of cocaine can be summarised under the following headings:

—in humans, stimulant, "antifatigue" and psychodysleptic effects of cocaine are
reported. The route of administration could influence the quality of the observed
effects.

—in animals, the published studies are quite fragmentary: they always concern
a particular property of cocaine, often in comparison with amphetamine or impira-
mine-like antidepressants.

—from a biochemical viewpoint, cocaine is considered, especially at the peripheral
sympathetic system level, as the typical representative of catecholamine reuptake
inhibitors.

From these different facts, one can establish correlations which differ from one
research hypothesis to another.

To determine the effects of cocaine in animals, we conducted a systematic study
of its psychopharmacological profile. The obtained spectrum of activity can be compared with that of amphetamine and imipramine-like drugs. Table 1 summarises our
findings (for technical details, see Simon et al., 1972).

Table 1

	Amphetamine	Cocaine	Imipramine
Increased motor activity (M)	+++	++	0
Stereotyped behaviour (R)	+++	++	0
Greater toxicity in aggregated mice	+++	+	0
Antagonism of hypnotics (M)	+	+	0
"Antifatigue activity" (R)	+++	++	0
Antagonism of neuroleptic-induced catalepsy (R)	+	++	+
Antagonism of reserpine (M)	+	++	+

(M = mice; R = rats)

Cocaine produces amphetamine-like stimulant effects: increase of locomotor
activity, stereotyped behaviour, greater toxicity in aggregated animals than in isolated
ones, reappearance of an avoidance reaction in exhausted animals (considered as
"antifatigue" activity (Boissier and Simon, 1968), antagonism of the effect of hypnotics.
Other effects of the same type have been reported: increase of autostimulation
(Benesova, 1969; Stein and Wise, 1970), similar EEG patterns (Monnier, 1957),
anorexigenic effect (Monnier, 1957), potentiation of the effects on operant behaviour
by imipramine (Scheckel and Boff, 1964).

Generally, with these tests, the maximum effects are weaker than those of amphetamine. On the other hand, two effects are more pronounced with cocaine: antagonism of neuroleptic-induced catalepsy and antagonism of reserpine effects. It must be noticed that these two effects also exist with imipramine-like drugs.

From a pharmacological viewpoint, cocaine simultaneously acts as a moderate amphetamine and tricyclic antidepressant. It would be interesting, but difficult from ethical considerations, to study this possible antidepressant effect in man.

The study of cocaine shows how careful one has to be in order to establish cause and effect relations between biochemical and pharmacological data or between biochemical and clinical data. Two examples can be given:

—the pharmacological effects of cocaine and imipramine-like drugs are not identical, although their biochemical effects are very similar.

TABLE 2

	Amphetamine	Cocaine
Hyperactivity in mice		
after α MpT	suppressed	unchanged
after *p*CPA	decreased	unchanged
Stereotyped behaviour in rats		
after α MpT	suppressed	decreased
after *p*CPA	decreased	unchanged

—the stimulant effect of cocaine in animals appears to be of the same type as that of amphetamine. However, pretreatment of animals, with α-methylparatyrosine, a tyrosinehydroxylase inhibitor, or with parachlorophenylalanine, a tryptophane-hydroxylase inhibitor, does not modify in the same way the effects of cocaine and those of amphetamine (Table 2).

REFERENCES

BENESOVA O. (1969) In: *The present status of psychotropic drugs* (CERLETTI A. and BOVE J. F., Eds.) pp. 247–249. Excerpta Med. Foundn, Amsterdam.
BOISSIER J. R. and SIMON P. (1968) *Thérapie*, **23**, 1267–1276.
MONNIER M. (1957) In: *Psychotropic drugs* (GARATTINI S. et GETTI V. eds.) Elseviers, Amsterdam.
SCHECKEL C. L. and BOFF E. (1964). *Psychopharmacologia (Berlin)* **5**, 198–208.
SIMON P., SULTAN Z., CHERMAT R. and BOISSIER J. R. (1972) *J. Pharmacol (Paris)* **3**, 129–142.
STEIN L. and WISE C. D. (1970) In: *Psychotomimetic drugs* (EFRON D. H., Ed.) pp. 123–145, Raven Press, New York.

Frontiers in Catecholamine Research 1973, pp. 1045 to 1049. Pergamon Press. Printed in Great Britain.

REPORTERS NOTES

SUSAN D. IVERSEN

IT IS easy to understand why the amphetamines continue to attract a great deal of neuropharmacological interest. They are a group of drugs with varied and pronounced behavioural effects both in animals and in man, and have become commonly used drugs of abuse. For several years it has been known that amphetamines interact with amine transmitters in the brain and although it is premature to conclude that these are the only neuropharmacological effects of amphetamines, there is accumulating evidence that at least some of their specific behavioural effects can be explained in such terms.

The presented papers illustrated the achievements of this field but perhaps more importantly, the methodological innovations which will guide the research on amphetamine and related compounds during the next few years.

STUDIES OF THE METABOLIC FATE OF AMPHETAMINE

Williams *et al.*, reviewed comparative work on the metabolism of amphetamine and related compounds which is important for identifying the species with the metabolic routes most similar to man. Gorrod described in detail methods for studying the metabolic pathways in man and in particular the value of manipulating urinary pH to accelerate or slow the rate of degradation of amphetamines. Acid conditions resulted in 60 per cent of the drug being excreted unchanged, while alkali conditions increased the percentage of drug metabolised to 90 per cent. It is significant, as Gunne pointed out, that such manipulations of pH shorten or lengthen the duration of psychosis in man, which suggests that the metabolite levels may correlate more closely with psychosis than do levels of unchanged drug.

Aromatic hydroxylation and deamination are the principal degradation routes for amphetamine. In the rat, for example, hydroxylation is the major metabolic pathway while in the rabbit, guinea pig, mouse and man deamination plays the more important role. Interestingly, however, although *d*-amphetamine is not hydroxylated to any extent in man, methamphetamine is, by some 20 per cent. The possible significance of this was emphasised by Gunne, who pointed out that methamphetamine is the most potent psychosis-inducing amphetamine in man. He found that the intensity of psychosis in methamphetamine addicts was not correlated with plasma levels of the unchanged drug but with levels of the hydroxylated metabolite. In discussion, Iversen pointed out that the metabolic difference between rat and mouse could sometimes explain discordant behavioural results obtained with amphetamine. In the rat, tricyclic antidepressants, through an effect on liver microsomal enzymes retard hydroxylation of amphetamine, increase its level in the brain and potentiate its stimulatory effects on locomotion. However, in the mouse, where liver hydroxylation plays only a minor role, imipramine, does not increase brain amphetamine levels or potentiate its behavioural effects. In contrast, MAO inhibitors do increase brain amphetamine levels in the mouse, presumably by interfering with liver metabolising enzymes, and do potentiate amphetamine induced locomotion (LEW, IVERSEN and IVERSEN, 1971).

METHODS OF ANALYSING THE MULTIPLICITY OF BEHAVIOURIAL AND NEUROCHEMICAL EFFECTS OF AMPHETAMINE

Amphetamine simultaneously induces stimulation of behaviour, depresses appetite and increases body temperature. When efforts are made to correlate these changes with endogenous neurochemical substrates, it would clearly be an advantage to be able to study one effect at a time. Several speakers illustrated different approaches to this problem. Groppetti noted that both small doses of amphetamine and cocaine decrease food intake and stimulate locomotor activity. Both increase the conversion of tyrosine to dopamine (DA) in the striatum in the absence of any effect on noradrenaline (NA) metabolism. The suggestion that DA rather than NA mediates these two behavioural effects is strengthened by the observation that the DA blocking agent, pimozide, antagonised amphetamine-induced anorexia and the DA receptor stimulant apomorphine induced anorexia.

Jorri advocated the genetic approach for dissociating these behavioural effects, and described two strains of mice, C_3H and NMRI, showing different patterns of amphetamine sensitivity. Despite a similar anorexigenic response in both strains, C_3H did not show increased motility and stereotypy at doses active in NMRI, neither did this strain show a hyperthermic response. The brain levels and distribution of amphetamine were identical in the two species. A difference in striatal DA mechanisms which may explain these effects is suggested by the observation of higher striatal DA turnover and of larger amphetamine-induced increases in homovanillic acid levels in the striatum in the NMRI strain. Sulser illustrated the value of manipulating the chemical structure of the drug molecule. Chlorination of the *para* position of the aromatic ring of amphetamine, reduces the efficiency of the aromatic hydroxylation enzymes and greatly retards the breakdown of the drug. This chemical modification also profoundly alters the neurochemical effects of the drug. Although both amphetamine and *p*-chloroamphetamine increase the turnover of catecholamines, only the latter compound inhibits synaptosomal uptake of serotonin and inhibits its rate of turnover by inhibiting the enzyme tryptophan hydroxylase.

CORRELATION OF NA AND DA WITH AMPHETAMINE INDUCED BEHAVIOUR

The controversy as to whether NA or DA mediates the locomotor and anorexic responses to amphetamine continues. Groppetti, for example, suggested that DA mediates both of these effects. Intracranial self stimulation is another behaviour which has been associated with NA mechanisms. Stein reviewed his studies which suggest that in the rat NA mediates ICS behaviour maintained by electrodes placed in the medial forebrain bundle. Blocking of amine receptors and inhibition of amine synthesis abolish ICS and in the latter experiment *d*-NA reinstates the behaviour whereas *l*-NA and DA do not. *d*-Amphetamine stimulates ICS behaviour, presumably because of its ability to release endogenous NA. More recently Stein has investigated the evidence of CROW (1972) and PHILLIPS and FIBIGER (1973) that ICS can be sustained not only by a noradrenergic system which projects to the forebrain via the lateral hypothalamus, but also by a DA system, the nigro-striatal pathway. Stein reported that he had also found some sites in the internal capsule, presumably at the site where the fibres from the substantia nigra enter the striatum, which would sustain ICS albeit at much lower rates than those obtained in classical hypothalamic sites. However, when he compared the ability of d- and l-amphetamine to stimulate ICS rates at these sites, he found that the d-isomer was 10 times more potent than the l-isomer; a ratio which has been suggested to indicate that a NA rather than a DA system is operating.

In discussion Crow defended his position that the nigro-striatal system sustains a form of ICS behavioural which is dintinct from that obtained on NA pathways. The evidence that DA, as well as NA neurones are involved in the electrical self-stimulation phenomenon is as follows:

(1) Self-stimulation responding is abolished by administration of α-methyl-p-tyrosine but, at least with lateral hypothalamic electrodes, is not abolished by administration of dopamine-β-hydroxylase inhibitors (ROLL, 1970)

(2) electrical self-stimulation sites in the ventral mesencephalon correspond closely to the location of the A9 and A10 (DA-containing) cell body groups (DRESSE, 1966; CROW, 1972).

(3) electrical self-stimulation through ventral mesencephalic electrodes is associated with the contraversive turning behaviour (GRASTYAN et al., 1969; ARBUTHNOTT et al., 1970; ANLEZARK et al., 1971) that has been attributed to unilateral activation of the nigrostriatal DA system, as well as with elements of the sniffing, licking, and gnawing syndrome. These latter behaviours are almost invariably concomitants of electrical self-stimulation through ventral mesencephalic electrodes.

Fibiger suggested that the different current levels used in ICS studies may explain the apparently contradictory results. If threshold level currents were used the DA system sustains a form of ICS behaviour, which is blocked by the specific DA receptor blocking agent pimozide and which is equally stimulated by d- and l-amphetamine. He thought that the higher current levels used by Stein although acceptable to the NA system would make it more difficult to identify the DA system, which even under ideal conditions sustained relatively low rates of ICS behaviour.

Turning to human behaviour, Davis reviewed the accumulating evidence that DA systems are involved in schizophrenia and that amphetamine psychosis, which in the clinic is difficult to distinguish from schizophrenia, can also be accounted for in terms of an interaction of amphetamine with DA systems. In support of this position Davis reported some of his recent clinical experiments in which normal and schizophrenic subjects were given small doses of methylphenidate and d- or l-amphetamine. Control subjects showed mild euphoria to the doses used with methylphenidate being the most potent stimulant and d-amphetamine twice as effective as the l-isomer. When the same doses were given to schizophrenic patients who were no longer reporting hallucinations and delusions, a dramatic intensification of these symptoms was observed, with the stimulants again ranged in the same order of potency. Davis noted that on the basis of rat studies a 2:1 ratio of potency in favour of the d-isomer is thought to indicate involvement of DA, but he and others commented on the danger of applying this dictum across species and indeed across widely different behaviours within a species.

In the rat it has been shown that there is a balance between DA and ACh in the striatum and interestingly Davis has shown that the cholinomimetic physostigmine, if given before methylphenidate, markedly reduced the worsening of symptomology in schizophrenics.

ELECTROPHYSIOLOGICAL CORRELATES OF AMPHETAMINE ACTION

Having characterised the neuropharmacology and behavioural effects of amphetamine the ultimate aim is to localise the effects to a particular region of the brain. Electrophysiological studies of amine pathways provide a crucial link on this path, and the study of Bunney and Aghajanian on the nigro-striatal pathway will, no doubt, come to be seen as a landmark. Recording from identified DA-containing neurones in the substantia nigra, they found that systemic injections of amphetamine depressed the firing rate and pretreatment with DA blocking agents like haliperidol and chlorpromazine abolished this effect of d-amphetamine. The necessity of endogenous presynaptic DA for this effect is suggested because α-methyl tyrosine blocked the effect and yet after α-methyltyrosine, DA receptor stimulation with apomorphine still depressed firing in the substantia nigra. The striatum is seen as the crucial site for the amphetamine action, because if amphetamine was applied iontophretically to the substantia nigra it had no effect. When applied in the striatum postsynaptically, spontaneous activity was depressed and it is concluded that striatal neurones projecting to the SN constitute the necessary feedback loop for this effect of amphetamine. If the afferent fibres to the substantia nigra, from the striatum are severed, then, as predicted, systemic amphetamine no longer depressed neuronal activity in the substantia nigra. The challenge now is to identify which, if any, of the overt behavioural effects of amphetamine are correlated with this electrophysiological event.

NEUROPHARMACOLOGICAL STUDIES OF AMPHETAMINE AND OTHER EUPHORIA PRODUCING DRUGS

The biochemical effects of amphetamine on *in vivo* and *in vitro* neurotransmitter systems have been well studied in the last ten years. Although the general picture is established there is clearly still need for methological refinements which permit a more precise evaluation of effects on amine uptake, release and turnover. The use of a pulse labelling technique with ^{3}H-tyrosine to measure the rates of catecholamine turnover *in vivo* was described by Costa and his colleagues. Rutledge described new techniques for measuring the releasing action of amphetamine on catecholamine stores in *in vitro* brain preparations prelabelled by exposure to radioactive catecholamine. However, while the details of amphetamine action are unravelled, attention is turning to other mood changing drugs which are also abused by man. This is an area which will undoubtedly see development before the Fourth International Catecholamine Symposium. We learned from Costa that morphine in analgesic doses, like amphetamine, increases the turnover of amines, particularly of DA in the striatum. He also presented evidence that amphetamine increased cyclic AMP levels in the striatum, and that morphine had a similar effect in the adrenal medulla and anterior pituitary. While the significance of cyclic AMP changes to euphoria must await further clarification, an involvement of presynaptic amines (particularly DA) in analgesia is further supported by the work of Gessa who reported that methadone also increased the conversion of tyrosine to DA and increased homovanilic acid levels in the striatum on a time scale which exceeds the cataleptic response. Turning to cannabis, Maitre presented evidence that the active Δ^9 and Δ^8 isomers increased the rate of formation of NA and DA from ^{3}H-tyrosine in the brain, and the rate of disappearance of these amines from brain. Finally Cohen presented evidence that alcohol, under certain conditions may react with endogenous amine transmitters to form isoquinolines which are taken up and released by the storage mechanisms of the adrenergic nerve terminals and can apparently act as false transmitters. He suggested that this reaction could explain the psychological changes associated with alcohol intake.

The session as a whole provided breadth and detail and fully justified the dedication of a session to recent work on amphetamine and related mood changing drugs. Few areas of research in neuropharmacology have yielded advances on such a broad front.

REFERENCES

ANLEZARK G. M., ARBUTHNOTT G. W., CHRISTIE J. E. and CROW T. J. (1971) Br. J. Pharmac. 41, 405P.
ARBUTHNOTT G. W., CROW T. J., FUXE K. and UNGERSTEDT U. (1970) J. Physiol. 210, 61–62P.
CROW T. J. (1972) Brain Res. 36, 265–273.
DRESSE A. (1966) Life Sci. 5, 1003–1014.
GRASTYAN E., SZABO I., MOLNAR P. and KOLTA P. (1969) Comm. Behav. Biol. 2, 235–266.
LEW C., IVERSEN S. D. and IVERSEN L. L. (1971) Europ. J. Pharmac. 14, 351–359.
PHILLIPS A. G. and FIBIGER H. C. (1973) Science 179, 575–577.
ROLL S. K. (1970) Science 168, 1370–1372.

Catecholamines in Man
(May 25, 1973; 9:00 A.M.–5:30 P.M.)

CHAIRMEN: Seymour S. Kety and Irwin J. Kopin
COUNCILLOR: Dennis Murphy
REPORTER: Morris Lipton

Frontiers in Catecholamine Research 1973, pp. 1053 to 1054. Pergamon Press. Printed in Great Britain.

CATECHOLAMINE METABOLITES IN CSF

BJÖRN-ERIK ROOS

Department of Pharmacology, University of Göteborg, Göteborg, Sweden

THE ONLY metabolites of monoamines in cerebrospinal fluid (CSF) known to be of practical value today are homovanillic acid (HVA), methoxyphenylethylglycol (MOPEG)-sulphate and 5-hydroxyindoleacetic acid (5-HIAA).

The efflux of these from CSF to blood has obviously three main components, which are due to diffusion, bulkflow of CSF and active transport. The third phenomenon is at least due to one, but possibly two or three, active transport mechanisms. How these functions act on the efflux of HVA and 5-HIAA is rather obscure, and very little is known about the outflow of MOPEG-sulphate.

In dogs has been shown that while the value of CSF-5-HIAA rises to the same level in the hydrocephalic dog, where the outflow of 5-HIAA is almost completely blocked, and in the normal dog pretreated with probenecid and examined when reaching a steady state level of this drug in blood, the HVA concentration in the normal dog under same conditions reaches only half of the hydrocephalic level (ANDERSSON, VON ESSEN and ROOS, 1973). This phenomenon could be due to the higher lipid solubility of HVA. A relatively larger part of HVA than of 5-HIAA might be eliminated from the CSF by diffusion which is suggested to take place mainly on the convexities of the brain. In experimentally induced hydrocephalus, where the normal flow from the ventricular system is blocked, the acids cannot reach the diffusion sites which has different consequences for the elimination of HVA than for 5-HIAA.

Recently VON ESSEN in our laboratory has shown a significant increase in cerebral blood flow after dopamine intravenously to dogs and, which is more important in the clinic, a decrease after haloperidol and pimozide. The increase and the decrease is about 25 per cent of the normal flow (VON ESSEN, 1972).

The bulk flow carrying both large and small molecules by the same route is, of course, of some importance also for 5-HIAA and HVA, but the difference in the type of efflux between the acids cannot be explained in this way. Drugs reducing the intracranial pressure might possibly change the rate of disappearance of the acids via the bulk flow.

Probenecid blocks the active outtransport of 5-HIAA and HVA from CSF to blood. This has been used in clinical studies both in Parkinsonism (OLSSON and ROOS, 1968; for rev. see KORF, 1971) and manic-depressive psychosis (ROOS and SJÖSTRÖM, 1969; SJÖSTRÖM and ROOS, 1972; for rev. see SJÖSTRÖM, 1973). Also the basic levels of both 5-HIAA and HVA have been studied in many different conditions and I would like to review some of our recent findings.

Results of screening of material from 154 neurological patients (JOHANSSON and ROOS, 1973, in preparation) show that low basic values of both HVA and 5-HIAA can be seen in parkinsonism, in multiple sclerosis and in amyotrophic laterial sclerosis. (ROOS and STÖRTEBECKER, 1973, in preparation) low values for HVA were also seen in two cases with spinal cord tumor.

In material of depressed and manic patients we matched 15 depressed and 17 manic pairs (one male and one female in every pair) and compared them to 11 pairs of control patients and 11 pairs of healthy volunteers (SJÖSTRÖM and ROOS, 1972). There was a significant difference in the levels of 5-HIAA between depressed and manic male and female patients. Females had higher level of 5-HIAA in CSF than males. In material of young schizophrenic patients SEDVALL (personal communications) and his coworkers have recently seen changes in CSF-HVA in the similar direction. Females had significantly higher levels of the acid metabolite in lumbar CSF (SEDVALL, 1973, in preparation). In a material of 63 Parkinsonian patients, however, no such sex differences could be seen (GRANERUS, MAGNUSSON, ROOS and SVANBORG, 1973). The same negative finding is seen in our control material, (GOTTFRIES, GOTTFRIES, JOHANSSON, OLSSON, PERSSON, ROOS and SJÖSTRÖM, 1971) in which 100 patients are investigated according to 5-HIAA and 84 according to HVA in lumbar CSF. The results are thus divergent, but a recent study with autopsy material (GOTTFRIES, ROOS and WINBLAD 1973, in preparation) might throw some light upon the problem and give new information. There seems to be a sex difference of distribution of 5-HIAA in some parts of the human brain but not in all. In cortex lobus occipitalis and in cortex lobus hippocampi as in hypothalamus there are significantly higher levels of 5-HIAA in female than in male. In all the other parts investigated we could not find any difference. Sharman (personal communication) has seen the same difference in 5-HIAA in mice during the oestrus cycle. In a small series, where we studied HVA in caudate nucleus and putamen, we could see a tendency towards higher levels of HVA in these two nuclei in females. Our material is now bigger and a new statistical evaluation seems to show that the difference might be even more significant than in the first preliminary study. Our finding will possibly give new data to the discussion about the biological background to the psychological differences between males and females. We could tentatively speculate about what we know about disturbances in the balance between catecholamines and serotonin-release as a possible cause of aggressive behaviour. Maybe we should, until we know more, say only 'vive la différence'.

REFERENCES

ANDERSSON H., VON ESSEN C. and ROOS B.-E. (1973) *Acta Pharmacol. et toxicol* **32**, 139–147.
ESSEN C. VON (1972) *J. Pharm. Pharmac.* 24, 668.
GRANERUS A. K., MAGNUSSON T., ROOS B.-E. and SVANBORG A. (1973) *Europ. J. clin. Pharmacol.* In press.
GOTTFRIES C. G., GOTTFRIES L., JOHANSSON B., OLSSON R., PERSSON T., ROOS B.-E. and SJÖSTRÖM R. (1971) *Neuropharmac.* **10**, 665–672.
KORF J. (1971) Amine metabolism in the human brain: evaluation of the probenecid test. Thesis, Groningen.
OLSSON R. and ROOS B.-E. (1968) *Nature, Lond.,* **219**, 502–503.
ROOS B.-E. and SJÖSTRÖM R. (1969) *Pharmacologia Clinica* **1**, 153–155.
SJÖSTRÖM R. (1973) *Acta Univ. Uppsaliensis* **154**,
SJÖSTRÖM R. and ROOS B.-E. (1972) *Europ. J. clin. Pharmacol.* **4**, 170–176.

Frontiers in Catecholamine Research 1973, pp. 1055 to 1061. Pergamon Press. Printed in Great Britain.

METABOLITES OF CATECHOLAMINES IN THE CEREBROSPINAL FLUID

D. F. SHARMAN

Agricultural Research Council, Institute of Animal Physiology,
Babraham, Cambridge, CB2 4AT

THE MAJOR ACIDIC METABOLITES OF THE CATECHOLAMINES IN CEREBROSPINAL FLUID

FIFTEEN years ago, G.W. Ashcroft and I were working together under the tutelage of T. B. B. Crawford in the Department of Pharmacology in the University of Edinburgh. We had found that a substance, behaving like 5-hydroxyindol-3-ylacetic acid (5-HIAA) on paper chromatography and showing chemical reactions similar to this metabolite of 5-hydroxytryptamine (5-HT), was present in brain tissue (SHARMAN, 1960). We then examined human cerebrospinal fluid for the presence of 5-hydroxy-indoles and found that we could detect a substance or substances having similar fluorescence characteristics to those exhibited by 5-hydroxyindole compounds. In order to try to identify this material we obtained relatively large volumes of cerebrospinal fluid from patients suffering from hydrocephalus and who were under-going ventricular drainage. These samples of cerebrospinal fluid were found to contain a concentration of 5-hydroxyindoles which was higher than normal (ASH-CROFT and SHARMAN, 1960). The work of Armstrong and his colleagues on the phenolic acids of human urine (ARMSTRONG, SHAW and WALL, 1956) led us to apply similar analytical techniques to cerebrospinal fluid. In the first samples we usually found four substances which gave a clear colour reaction with diazotised p-nitro-aniline. One of these was found to be salicylic acid. Paper chromatography of an extract of cerebrospinal fluid from a patient who was *not* receiving acetylsalicylic acid indicated the presence of three phenolic substances. The R_f values, colour and fluorescence reactions which were applied to these substances showed that Spot 1 was 5-hydroxyindol-3-ylacetic acid. It was concluded that Spot 2 was probably 4-hydroxy-3-methoxyphenylacetic acid (homovanillic acid; HVA) a metabolite of 3,4-dihydroxyphenylethylamine (dopamine) but the further examination of this substance was not carried out until later. Spot 3 appeared to be isographic on paper chromatograms with 4-hydroxy-3-methoxymandelic acid (vanillyl mandelic acid; VMA) a metabolite of noradrenaline. However, there was one reaction which distinguished Spot 3 from this metabolite. When VMA is reacted with 2,6-dichlo-roquinonechloroimide (Indophenol reaction; GIBBS, 1927) an intense blue colour is developed, but Spot 3 gave no reaction with this reagent. It is now thought that this substance is p-hydroxyphenyllactic acid SJÖQUIST and ÄNGGÅRD, 1972). This could be a product of transamination of tyrosine.

In 1963, fluorimetric methods for the estimation of HVA were published (ANDÉN, ROOS and WERDINIUS, 1963a; SHARMAN, 1963) and the former authors clearly demonstrated the presence of HVA in cerebrospinal fluid. These reactions are based on the oxidation of HVA to form a fluorescent dimer, 2,2¹-dihydroxy-3,3¹-dimethoxy-biphenyl-5,5¹-diacetic acid (CORRODI and WERDINIUS, 1965). The procedure of ANDÉN *et al.*, (1963a) has formed the basis for most of the methods now in use for the fluorimetric analysis of HVA in the cerebrospinal fluid. HVA is extracted from a salt saturated, deproteinised extract of cerebrospinal fluid at an acid pH, into an organic solvent such as di-ethylether, ethylacetate or *n*-butylacetate. The HVA is then back extracted into an alkaline buffer solution. ANDÉN *et al.* (1963a) used 0·05M-tris (hydroxymethyl) amino methane, pH 8·5, GERBODE and BOWERS (1968) used 0·5M-sodium phosphate buffer pH 8·5 and borate buffer may also be used (PULLAR, personal communication). The choice of buffer is important since the conditions for the development of the fluorescence are critical. Small amounts of other ions in the buffer solution can greatly reduce the intensity of the fluorescence produced or completely prevent its development. KORF, VAN PRAAG and SEBENS (1971) have tried to improve the extraction procedure by using a column of Sephadex G10 to extract HVA from cerebrospinal fluid, a procedure which allows the esti-mation of 4-hydroxy-3-methoxyphenylethyleneglycol (MHPG), a metabolite of noradrenaline, to be made on the same sample. The HVA is eluted from the Seph-adex G10 with 0·5M-sodium phosphate buffer. The fluorescent dimer is formed by oxidation in a very dilute (1–50 μg/ml) solution of potassium ferricyanide. The oxidation reaction is terminated and the fluorescent product stabilised with a reducing agent, usually cysteine. The fluorescence shows a maximum activation wavelength of about 320 nm and a maximum fluorescence wavelength of about 430 nm. (The reported values for these wavelengths are uncorrected instrumental values.) The fluorescence developed from the apparent HVA in cerebrospinal fluid shows similar maxima.

The specificity of the reaction is reasonable. GJESSING, VELLAN, WERDINIUS and CORRODI (1967) have shown that several substances having the vanyl (i.e. 4-hydroxy-3-methoxyphenyl) group in their structure will give rise to a similar fluorescence. Among these vanyllactic acid must be considered since it is found in the urine of patients treated with L-3,4-dihydroxyphenylalanine (L-dopa) and would be extracted, if present in the cerebrospinal fluid of such patients. When other phenolic acids including *p*-hydroxyphenylacetic acid are oxidised with hydrogen peroxide in the presence of peroxidase, a fluorescent derivative is produced (GUILBAULT, BRIGNAC and JUNEAU, 1968) but this acid and related compounds gave little or no fluorescence when oxidised with ferric chloride (SHARMAN, 1963) and SATO (1965) has shown that *p*-hydroxyphenylacetic acid yields less than 2 per cent of the fluorescence seen when HVA is treated by the method of ANDÉN *et al.*, (1963a).

Recently an isomer of HVA, 3-hydroxy-4-methoxyphenylacetic acid (homo-isovanillic acid; iso-HVA) has been demonstrated to be present in cerebrospinal fluid, a finding which may have far reaching implications. It is known that, *in vitro*, the enzyme catechol-*O*-methyl transferase, which is responsible for the introduction of the methyl group into position 3 of the phenyl group of catecholamines and their catechol metabolites can also methylate the phenolic hydroxyl group at the 4 position. About 10 per cent of the substrate is methylated at position 4 by the enzyme ex-tracted from rat liver (AXELROD, 1966). MATHIEU, REVOL and TROUILLAS (1972) and MATHIEU, CHARVET, CHAZOT and TROUILLAS (1972) have separated HVA and iso-HVA by paper chromatography of the products formed by coupling these two acids with diazotised *p*-nitroaniline, since it is difficult to separate the parent compounds by the usual analytical procedures. However, iso-HVA does not give rise to a fluo-rescence if subjected to the method of ANDÉN *et al*. (1963) for the fluorimetric assay of HVA (KIRSCHBERG, COTE, LOWE and GINSBURG (1972) and MATHIEU, CHARVET, CHAZOT and TROUILLAS (1972) have shown that the fluorimetric assay of HVA in cerebrospinal fluid gives results which are comparable with the values obtained by their colorimetric procedure after separation of the two isomers. Iso-HVA can depress the fluorescence yield from HVA as can other acids which might be present in cerebrospinal fluid (KIRSCHBERG *et al.*, 1972). In practice, the other acids normally present in extracts appear not to affect the development of the fluorescence to any great extent, but this should be checked particularly if samples are taken from subjects which have been treated with large amounts of catecholamine precursors or metab-olites. There has been little use made of gas–liquid chromatography for the routine estimation of HVA in cerebrospinal fluid. SJÖQUIST and ÄNGGÅRD (1972) have described a method for the estimation of HVA in human cerebrospinal fluid using heptafluorobutyryl derivative of the methyl ester of HVA. This substance was detected after gas chromatography by electron capture or by the multiple ion detector of a mass spectrometer. When the latter method of detection was used, the deu-terated-methyl ester of HVA could be used as a standard. The observations of these authors confirm the presence of HVA in human cerebrospinal fluid. The derivative described by DZIEDZIC, BERTANI, CLARKE and GITLOW (1972) has possibilities in this respect. These authors have shown that the hexafluoro-isopropyl ester of 4-tri-fluoroacetyl-HVA has excellent electron capture properties. At Babraham, we have used this derivative in a method for the estimation of HVA in nervous tissues and have shown that it can be applied to cisternal cerebrospinal fluid from the pig.

In our early experiments we did not detect any 3,4-dihydroxyphenylacetic acid (DOPAC), another known metabolite of dopamine, on the paper chromatograms, a result similar to that obtained by ANDÉN, ROOS and WERDINIUS (1963b). However, ASHCROFT, CRAWFORD, DOW and GULDBERG (1968) showed that DOPAC was present in extracts of dog cerebrospinal fluid and used the fluorescence which is developed when DOPAC is condensed with 1,2-diaminoethane to estimate its concentration. In fluid taken from the lateral ventricle of the dog, the concentration of DOPAC was 237 ng/ml and that of HVA 2298 ng/ml. In normal human lumbar cerebrospinal fluid the concentration of DOPAC is near the limit of the sensitivity of the fluori-metric method of assay but WATERBURY and PIERCE (1972) have demonstrated the presence of this metabolite in human lumbar cerebrospinal fluid by a gas chromato-graphic method involving the formation of the trimethylsilyl derivative of its methyl ester combined with mass spectrometry. The gas chromatographic method for HVA in which the trifluoroacetyl-hexafluoro-iso-propyl derivative is made can also be used for DOPAC. We have found a small peak at the appropriate relative retention time for the derivative of DOPAC on our records of extracts from pig cerebrospinal fluid but we have not yet confirmed that this peak represents DOPAC only.

The presence of either of two possible acidic metabolites of noradrenaline 3,4-dihydroxymandelic acid (DOMA) and VMA in cerebrospinal fluid has not been unequivocally demonstrated.

ALCOHOL METABOLITES OF CATECHOLAMINES IN CEREBROSPINAL FLUID

During the past ten years evidence has accrued to show that the major metabolite of noradrenaline in the brain is 1-(4-hydroxy-3-methoxyphenyl) ethane-1,2-diol

(4-hydroxy-3methoxyphenylethyleneglycol, MHPG, MOPEG). The formation of this metabolite was demonstrated in cat brain *in vivo* by MANNARINO, KIRSHNER and NASHOLD (1963) and in rabbit brain *in vitro* by RUTLEDGE and JONASON (1967). In 1968, SCHANBERG, SCHILDKRAUT, BREESE and KOPIN found that in the rat brain, the major metabolite formed from radioactively labelled 1-(4-hydroxy-3-methoxy-phenyl)-2-aminoethan-1-ol (normetanephrine, NM) was the ethereal sulphate conjugate of MHPG. The presence of MHPG-sulphate in the brains of some other mammalian species in addition to the free compound was demonstrated by SCHAN-BERG, BREESE, SCHILDKRAUT, GORDON and KOPIN (1968) who also showed that MHPG and its sulphate conjugate was present in human cerebrospinal fluid.

A method for the estimation of MHPG in urine, using gas chromatography and electron capture of the trifluoroacetyl derivative of MHPG was described by WILK, GITLOW, CLARKE and PALEY (1968) and this method was applied by SCHANBERG, BREESE, SCHILDKRAUT, GORDON and KOPIN (1968) to human cerebrospinal fluid. In this method, the conjugated MHPG is hydrolysed by incubation with a sulphatase preparation and the MHPG is absorbed onto a column of Biorad AG 1 × 4 exchange resin in the chloride form. The MHPG is eluted with water and then extracted into ethylacetate. After evaporating to dryness the extract is treated with trifluoroacetic anhydride to yield the tri-trifluoroacetyl derivative of MHPG. More recently the absorption and elution step has been omitted by some workers for the analysis of cerebrospinal fluid (WILK, DAVIS and THACKER, 1970; GORDON and OLIVER, 1971). Another derivative of MHPG, acetyl-di-trifluoroacetyl-MHPG, has been used, by BOND (1972) to estimate MHPG in urine and cerebrospinal fluid. This author ran into difficulties when using the method of WILK *et al.* (1967) and found that acetylation of MHPG before reacting with trifluoracetic anhydride gave satisfactory results. Acetylation of MHPG and of 1-β,4-dihydroxyphenyl)ethane-1,2-diol (DHPG, DOPEG) in aqueous solution followed by extraction and the formation of di-heptafluorobutyryl derivatives has been used to measure the concentrations of these two metabolites in brain tissue (SHARMAN, 1969; CEASAR and SHARMAN, 1972). The acetylation procedure imparts a greater stability to the compounds and enables more rigorous separation procedures to be used. The responses of the gas chromatograph, perhaps because of the exquisite sensitivity of such methods, are just as prone to be misinterpreted as those of the fluorimeter, particularly when preceded only by a simple solvent extraction. However, when the amounts of the substances to be estimated are very small, it may only be possible to use such a simple extraction procedure in order to avoid excessive losses. The use of an ion exchange resin in the methods for the estimation of MHPG (WILK *et al.*, 1967; SCHANBERG, BREESE, SCHILDKRAUT, GORDON and KOPIN, 1968; BOND, 1972) improves the specificity of the method and PULLAR (personal communication) has found that a small column of the anion exchange resin in the hydroxide form can be used to great advantage for the extraction of MHPG in his method which also uses gas chromatography and electron capture detection of the acetyl-di-trifluoroacetyl derivative. WATERBURY and PIERCE (1972) have been able to identify MHPG in human cerebrospinal fluid by gas chromatography of its trimethylsilyl derivative.

Fluorimetric methods for the estimation of MHPG have also been used. KORF, VAN PRAAG and SEBENS (1971) have isolated MHPG by means of absorption on to and elution from a column of Sephadex G10 in their method which also permits the estimation of 5-HIAA and HVA in the same sample of cerebrospinal fluid. Potassium ferricyanide solution and a mixture of 1,2 diaminoethane, ammonium chloride and ammonium hydroxide is added to the eluate and the mixture is heated for 15 min after which a solution of cysteine is added. This procedure develops a fluorophore from MHPG, the fluorescence of which is measured, with activating light set at 400 nm wavelength, at a wavelength of 500 nm. A fluorimetric method for the estimation of the sulphate conjugate of MHPG has been described by MEEK and NEFF (1972) but there have been no reports of its application to cerebrospinal fluid. The above methods clearly demonstrate that MHPG is a normal constituent of cerebrospinal fluid and if used correctly give accurate estimates of its concentration. The possibility that 1-(3-hydroxy-4-methoxyphenyl) ethan-1,2-diol(ISO-MHPG) might be present in cerebrospinal fluid has been considered by MATHIEU et al. (1972) but as yet this substance has not been shown to be present.

The presence in or absence from cerebrospinal fluid of alcohol metabolites of dopamine, i.e. 2-(3,4-dihydroxyphenyl)ethan-1-ol (DHPE, DOPOL) and 2-(4-hydroxy-3-methoxyphenyl)ethan-1-ol (MHPE, MOPOL) is at present a subject for discussion. WATERBURY and PIERCE (1972) have reported the presence of DHPE and MHPE in human cerebrospinal fluid on the basis of the behaviour of their trimethylsilyl derivatives on gas chromatography and in the case of MHPE confirmed this by mass spectral data. However, WILK (1971) was unable to detect MHPE in human cerebrospinal fluid. GOLDSTEIN, FRIEDHOFF, POMERANTZ and CONTRERA (1961) showed that DHPE and MHPE were metabolites of dopamine in the rat. BRAESTRUP (1972, 1973) has presented gas chromatographic evidence for the presence of conjugated MHPE in rat brain but has suggested that this is derived from the peripheral metabolism of dopamine. Dr. P. Ceasar and I have been unable to detect free MHPE in the stratum of the mouse even though we were able to measure the small amount of MHPG which is present in this tissue. The problem of the alcohol metabolites of dopamine in the cerebrospinal fluid requires further investigation.

THE REFLECTION OF NEURONAL ACTIVITY IN THE CONCENTRATION OF THE METABOLITES OF CATECHOLAMINES IN THE CEREBROSPINAL FLUID

The presence of metabolites of catecholamines in the brain and the cerebrospinal fluid poses the question whether the concentrations of such metabolites are related to the activity of the noradrenergic and dopaminergic neurons in the brain. With regard to the acidic metabolites it is unlikely that these are derived from the blood since the intravenous injection of HVA or 5HIAA, the acidic metabolite of 5-HT does not result in an increased concentration of these acids in the cerebrospinal fluid (ASHCROFT, DOW and MOIR, 1968; BARTHOLINI, PLETSCHER and TISSOT, 1966). The concentration of an acidic metabolite in cerebrospinal fluid appears to reflect the concentration of the metabolite in regions of the brain adjacent to the site of sampling when the two concentrations are allowed to come to equilibrium. MOIR, ASHCROFT, CRAWFORD, ECCLESTON and GULDBERG (1970) in reviewing this problem have pointed out that when the concentrations of HVA and 5HIAA in the caudate nucleus of the dog are increased after the administration of chlorpromazine the ratios of the concentrations of each of the acids in the caudate nucleus and the lateral ventricular cerebrospinal fluid remain constant (GULDBERG and YATES, 1969). The concentration of 5HIAA in the cisternal fluid appears to reflect predominantly the concentration of 5HIAA in the brain stem region. Bulat and Zivkovic (1971) have concluded that the 5-HIAA in the lumbar cerebrospinal fluid has its origin in the spinal cord. A similar origin for the MHPG in lumbar cerebrospinal fluid might be supposed since the spinal cord contains both 5-HT and noradrenaline. Although shown to be present, there is but very little dopamine in the rat spinal cord (Atack, personal communication) and we have found only traces of HVA in rabbit spinal cord. It is thus unlikely that the greater part of the HVA in the lumbar cerebrospinal fluid is derived from this source. Bulat and Zivkovic (1971) have shown that 5-HIAA, injected into the cisterna magna does not appear in the lumbar fluid although it disappears from the cisternal fluid. However, many samples of lumbar cerebrospinal fluid are taken from subjects which have had their life time to reach equilibrium and usually when more than one sample is taken a period of several days is allowed to elapse before a second sample is obtained. However, the observations of Korf, Van Praag and Sebens (1971) and of Gordon, Oliver, Goodwin, Chase and Post (1973) show that when active transport of acids out of the cerebrospinal fluid is inhibited by probenecid then the concentrations of HVA and 5HIAA on the lumbar cerebrospinal fluid start to increase after four hours whereas the concentration of MHPG-sulphate is increased only after 18 hr. One further point, germane to this problem, is that, in the lateral ventricle, equilibrium between the acid metabolites in the tissue and the cerebrospinal fluid appears to be reached within one hour provided that small samples (0·5 ml) only are removed from the ventricle (Ashcroft, Dow and Moir, 1968).

The time course of a change in the activity of mono-aminergic neurons in the brain must be greatly attenuated in its reflection in a change in the concentration of the corresponding acidic metabolite in the cerebrospinal fluid. Portig and Vogt (1969) have shown that when the substantia nigra of the cat was stimulated electrically for four minutes there was an increased release of HVA into a perfusate of the lateral ventricle which could persist for more than one hour. This suggests that metabolites formed deep in the brain substance take a long time to reach the fluid in the ventricular system. There might be even greater difficulty in interpreting any changes which occur in concentrations of metabolites in the cisternal and lumbar cerebrospinal fluids since the acidic metabolites are known to be removed from the cerebrospinal fluid by a probenecid-sensitive, active transport system. This is thought to be situated in the choroid plexuses (Ashcroft, Dow and Moir, 1968; Cserr and Van Dyke, 1971; Pullar, 1971; Forn, 1972). The deciphering of the message contained on the concentrations of the metabolites of catecholamines in the cerebrospinal fluid is far from complete but the observations which have so far been made are such as to require that further research into this problem be carried out.

REFERENCES

Andén N.-E., Roos B.-E. and Werdinius B. (1963a) Life Sci. 2, 448–458.
Andén N.-E. Roos B.-E. and Werdinius B. (1963b) Experientia, 19, 359–360.
Armstrong M. D., Shaw K. N. F. and Wall P. E. (1956) J. biol. Chem. 218, 293–303.
Ashcroft G. W., Crawford T. B. B., Dow R. C. and Guldberg H. C. (1968) Br. J. Pharmac. Chemother., 33, 441–456.
Ashcroft G. W., Dow R. C. and Moir A. T. B. (1968) J. Physiol., Lond., 199, 397–425.
Ashcroft G. W. and Sharman D. F. (1960). Nature, Lond., 186, 1050–1051.
Axelrod J. (1966). Pharmac. Rev., 18, 95–113.

BARTHOLINI G., PLETSCHER A. and TISSOT R. (1966) *Experientia*, **22**, 609–610.
BOND P. A. (1972). *Biochem. Med.*, **6**, 36–45.
BRAESTRUP C. (1972). *Biochem. Pharmac.*, **21**, 1775–1776.
BRAESTRUP C. (1973). *J. Neurochem.*, **20**, 519–527.
BULAT M. and ŽIVKOVIĆ B. (1971) *Science*, **173**, 738–740.
CEASAR P. M. and SHARMAN D. F. (1972) *Br. J. Pharmac.*, **44**, 340–341P.
CORRODI H. and WERDINIUS B. (1965). *Acta chem. scand.*, **19**, 1854–1858.
CSERR H. F. and VAN DYKE D. H. (1971) *Amer. J. Physiol.*, **220**, 718–723.
DZIEDZIC S. W., BERTANI L. M., CLARKE D. D. and GITLOW S. E. (1972). *Analyt. Biochem.*, **47**, 592–600.
FORN J. (1972). *Biochem. Pharmac.*, **21**, 619–624.
GERBODE F. A. and BOWERS M. B. (1968) *J. Neurochem.*, **15**, 1053–1055.
GIBBS M. D. (1927) *J. biol. Chem.* **72**, 649–664.
GJESSING L. R., VELLAN E. J., WERDINIUS B. and CORRODI H. (1967). *Acta chem. scand.*, **21**, 820–821.
GOLDSTEIN M., FRIEDHOFF A. J., POMERANTZ S. and CONTRERA J. F. (1961) *J. Biol. Chem.*, **236**, 1816–1812.
GORDON E. K. and OLIVER J. (1971) *Clin. chim. Acta*, **35**, 145–150.
GORDON E. K., OLIVER J., GOODWIN F. K., CHASE T. N. and POST R. M. (1973). *Neuropharmacology*, **12**, 391–396.
GUILBAULT G. G., BRIGNAC P. J. Jr. and JUNEAU M. (1968). *Analyt. Chem.*, **40**, 1256–1263.
GULDBERG H. C. and YATES C. M. (1969) *Br. J. Pharmac.*, **36**, 535–548.
KIRSCHBERG C. J., COTE L. M., LOWE Y. H. and GINSBURG S. (1972) *J. Neurochem.*, **19**, 2873–2876.
KORF J., VAN PRAAG H. M. and SEBENS J. B. (1971) *Biochem. Pharmac.*, **20**, 659–668.
MANNARINO E., KIRSHNER N. and NASHOLD B. S. Jr. (1963) *J. Neurochem.*, **10**, 373–379.
MATHIEU P., CHARVET J. C., CHAZOT G. and TROUILLAS P. (1972) *Clin. Chim. Acta*, **41**, 5–17.
MATHIEU P., REVOL L. and TROUILLAS P. (1972) *J. Neurochem.*, **19**, 81–86.
MEEK J. L. and NEFF N. H. (1972). *Br. J. Pharmac.*, **45**, 435–441.
MOIR A. T. B., ASHCROFT G. W., CRAWFORD T. B. B., ECCLESTON D. and GULDBERG H. C. (1970) *Brain*, **93**, 357–368.
PORTIG P. J. and VOGT M. (1969). *J. Physiol., Lond.*, **204**, 687–715.
PULLAR I. A. (1971). *J. Physiol., Lond.*, **216**, 201–211.
RUTLEDGE C. O. and JONASON J. (1967) *J. Pharmac. exp. Ther.*, **157**, 493–502.
SATO T. L. (1965) *J. lab. clin. Med.*, **66**, 517–525.
SCHANBERG S. M., BREESE G. R., SCHILDKRAUT J. J., GORDON E. K. and KOPIN I. J. (1968) *Biochem. Pharmac.*, **17**, 2006–2008.
SCHANBERG S. M., SCHILDKRAUT J. J., BREESE G. R. and KOPIN I. J. (1968). *Biochem. Pharmac.*, **17**, 247–254.
SHARMAN D. F. (1960). The significance of pharmacologically active amines in animal tissues and body fluids. Ph.D. Thesis, University of Edinburgh.
SHARMAN D. F. (1963) *Br. J. Pharmac. Chemother.*, **20**, 204–213.
SHARMAN D. F. (1969) *Br. J. Pharmac. Chemother.*, **36**, 523–534.
SJÖQUIST B. and ÄNGGÅRD E. (1972) *Analyt. Chem.*, **44**, 2297–2301.
WATERBURY L. D. and PEARCE L. A. (1972) *Clin. Chem.* **18**, 258–262.
WILK S. (1971) *Biochem. Pharmac.*, **20**, 2095–2096.
WILK S., DAVIS K. L. and THACKER S. B. (1971). *Analyt. Biochem.*, **39**, 498–504.
WILK S., GITLOW S. E., CLARKE D. D. and PALEY D. H. (1967). *Clin. chim. Acta*, **16**, 403–408.

Frontiers in Catecholamine Research 1973. pp. 1063 to 1065. Pergamon Press. Printed in Great Britain.

SELECTIVE DEPLETION OF BRAIN REGIONAL NORADRENALINE BY SYSTEMIC 6-HYDROXYDOPAMINE IN NEWBORN RATS*

R. LAVERTY, M-C. LIEW and K. M. TAYLOR

Department of Pharmacology, University of Otago Medical School, Dunedin, New Zealand

WHEREAS the therapeutic use of amphetamine is declining, its use as a pharmacological tool in increasing. However, before the results of an increasing number of behavioural and neurochemical studies with amphetamine can be applied to any clinical situation two questions must be answered: (i) which behavioural effect of amphetamine in animals corresponds to its behavioural effects in humans and (ii) which neurochemical effect of amphetamine is involved in the mediation of these behavioural effects.

An ideal answer would be that amphetamine acted on a specific neuronal pathway involving only one neurotransmitter which resulted in a specific behavioural effect in animals that corresponded to a given disorder of the human nervous system. However, current research indicates that the situation is more complex than this. Catecholamines are involved in the mediation of the behavioural effects of amphetamines but the important problem is not so much the site of action on catecholamine neurones e.g. release, uptake or enzyme inhibition or direct action, but rather which catecholamine is involved in the behavioural effects of amphetamine. We have developed a technique which may help to resolve this problem.

Administration of 6-hydroxydopamine systemically to newborn rats causes not only a permanent destruction of the peripheral adrenergic nervous system (CLARK, LAVERTY and PHELAN, 1972) but also a prolonged depletion of noradrenaline from various regions of the central nervous system (TAYLOR, CLARK, LAVERTY and PHELAN, 1972). In subsequent experiments (LIEW and TAYLOR, 1972) we found that by restricting the period of intraperitoneal administration of 6-hydroxydopamine to one or two days during the first 12 days after birth it was possible to deplete the noradrenaline in various brain regions selectively.

6-Hydroxydopamine (100 mg kg^{-1}) was injected intraperitoneally into newborn male white rats daily on each of two consecutive days between days 1 and 20 after birth. Litter mate controls were similarly injected with the saline ascorbic acid (0·5 mg ml^{-1}) solvent. From age one month to age 3 months various physiological and behavioural measures were studied including rectal temperature, fluid intake, overnight cage activity, Y-runway exploration (STEINBERG, RUSHTON and TINSON, 1961), shock-induced aggression (ULRICH and AZRIN, 1962) and sleeping time after chloral hydrate (FASTIER, SPEDEN and WAAL, 1957). At age 3 months the rats were killed and the brains removed, dissected and analysed fluorimetrically for noradrenaline content (LAVERTY and TAYLOR, 1968).

Injection on days 1 and 2 after birth caused a depletion of noradrenaline in the cerebral cortex, hippocampus and spinal cord whereas injection on days 9 and 10 after birth caused predominantly a depletion of cerebellar noradrenaline (Fig. 1). Noradrenaline levels in the thalamus and hypothalamus and dopamine levels in the

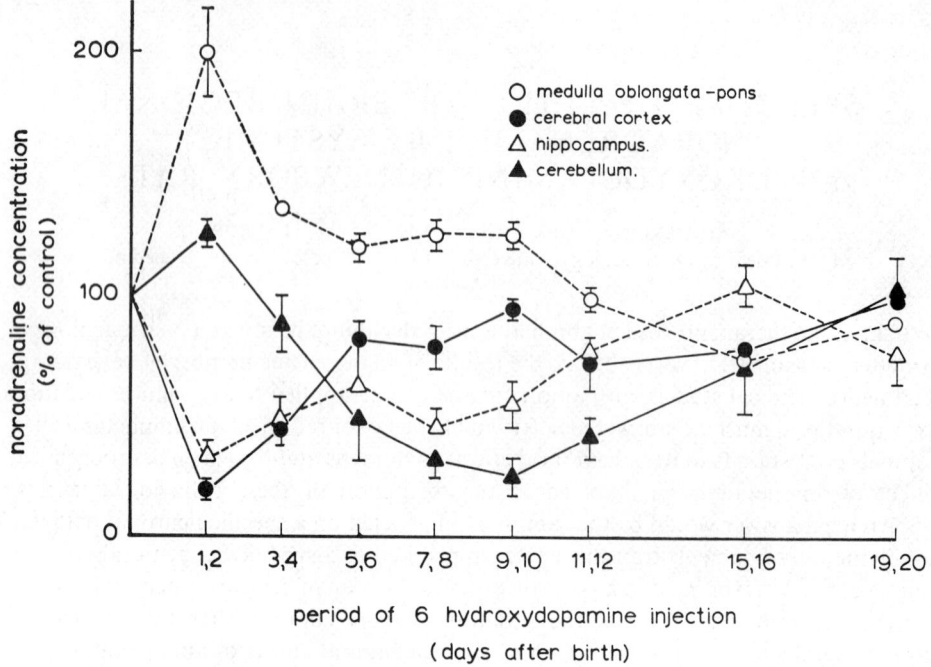

Fig. 1.—The effect of 6-hydroxydopamine (100 mg kg^{-1} intraperitoneally on 2 consecutive days) on the noradrenaline concentration of brain regions of rats aged 3 months. Each point represents the mean $\pm$ s.e.m. from 6 to 8 animals. The control group consisted of 20 animals.

striate regions were not affected; noradrenaline levels in the pons-medulla were increased, particularly following treatment on days 1 and 2.

Behavioural tests on these animals showed that animals treated on days 1 and 2 were lighter in weight, but had similar temperature regulation, food and fluid consumption to controls or animals treated on 9 and 10 days. There was no sign of the fluid appetite changes reported in rats given repeated intraventricular injections of 6-hydroxydopamine (Sorenson, Ellison and Masuoka, 1972).

Only rats treated on days 1 and 2 showed a prolonged sleeping time after chloral hydrate or increased shock-induced aggression (Table 1). This suggests that these effects may be due to loss of cortical or hippocampal noradrenaline. That shock-induced aggression is affected by brain noradrenaline levels has already been suggested (Thoa, Eichelman, Richardson and Jacobowitz, 1972). Activity as measured by entries (Table 1) or by rearing in a Y-runway or by overnight cage activity (unpublished observations) was not markedly changed by these treatments with 6-hydroxydopamine.

Rats treated on 1 and 2 days had an increased response to a low dose of amphetamine (Table). At higher doses of amphetamine no difference in Y-runway activity between treatments was observed; all groups showed stereotyped behaviour.

At this stage it is not possible to state whether this is an action of amphetamine causing stimulation of supersensitive activating adrenoceptors in cortical neurones or due to the removal of a cortical noradrenergic inhibitory system. Effects on sleeping time may be explained by the removal of an adrenergic activating system whereas

TABLE 1.

	Control	6-Hydroxydopamine Days 1 & 2	Day 9 & 10
No. of rats	8	9	9
Body weight (g)	189·5 ± 6·6	122·6 ± 12·3*	160·5 ± 4·3
Rectal temperature (°C)			
at room temperature (20°)	36·8 ± 0·1	36·8 ± 0·1	36·3 ± 0·1
after 1 hr at 4°C	36·9 ± 0·1	36·1 ± 0·1	35·2 ± 0·2
after 3 hr at 4°C	36·6 ± 0·1	36·2 ± 0·2	35·8 ± 0·2
Fluid intake (ml day^{-1} rat^{-1})			
Water	36·8 ± 4·4	37·3 ± 2·1	33·5 ± 1·4
Sucrose (2%)	57·4 ± 4·5	59·1 ± 3·7	67·1 ± 3·0
Quinine (0·02%)	25·4 ± 2·3	24·2 ± 0·8	25·4 ± 3·3
Duration of hypnosis after chloral hydrate (min)	42·8 ± 2·8	65·7 ± 3·3*	40·6 ± 4·3
Shock-induced aggression (responses in 20 trials)	2·6 ± 1·7	7·5 ± 1·7*	1·4 ± 0·6
Activity in Y-runway (entries in 3 min)			
Baseline	5·2 ± 0·4	5·9 ± 0·5	4·0 ± 0·4
After amphetamine (0·5 mg kg^{-1} i.p.)	4·8 ± 1·0	10·4 ± 1·5*‡	4·4 ± 0·7
Noradrenaline content (ng g^{-1})			
Cortex	146 ± 10	34 ± 4*	120 ± 10
Hippocampus	339 ± 30	50 ± 10*	177 ± 20
Spinal Cord	233 ± 20	112 ± 20*	293 ± 40
Cerebellum	120 ± 10	173 ± 20	34 ± 10*
Hypothalamus	869 ± 60	802 ± 90	986 ± 70

Results are expressed as means ± S.E.M.
* Significantly different ($P < 0.05$) from controls
‡ Significantly different ($P < 0.05$) from baseline.

increased shock-induced aggression could be due to removal of an inhibitory system. These preliminary experiments show the application of these techniques of selective regional depletion of brain noradrenaline to the problems of site and mode of action of psychotropic drugs such as amphetamine.

Acknowledgements—This work was supported by the Medical Research Council of New Zealand.

REFERENCES

CLARK D. W. J., LAVERTY R. and PHELAN E. L. (1972) *Br. J. Pharmac.* **44**, 233–243.
FASTIER F. N., SPEDEN R. N. and WAAL H. (1957) *Br. J. Pharmac.* **12**, 251–256.
LAVERTY R. and TAYLOR K. M. (1968) *Analyt. Biochem.* **22**, 269–279.
LIEW M. C. and TAYLOR K. M. (1972) *Proc. Univ. Otago med. Sch.* **50**, 58–59.
SORENSON C. A., ELLISON G. D. and MASUOKA D. (1972) *Nature, New Biol.* **237**, 279–281.
STEINBERG H., RUSHTON R. and TINSON C. (1961) *Nature, Lond.* **192**, 533–535.
TAYLOR K. M., CLARK D. W. J., LAVERTY R. and PHELAN E. L. (1972) *Nature, New Biol.* **239**, 247–248.
THOA N. B., EICHELMAN B., RICHARDSON J. S. and JACOBOWITZ D. (1972) *Science* **178**, 75–77.
ULRICH R. E. and AZRIN N. H. (1962) *J. exp. Analysis Behav.* **5**, 511–520.

Frontiers in Catecholamine Research 1973, pp. 1067 to 1069. Pergamon Press. Printed in Great Britain.

VMA IN SPINAL FLUID: EVALUATION OF THE PATHWAYS OF CEREBRAL CATECHOLAMINE METABOLISM IN MAN

SHERWIN WILK and ERIC WATSON

Department of Pharmacology, Mount Sinai School of Medicine of The City University of New York, Fifth Avenue and 100th Street, New York, N.Y. 10029, U.S.A.

THE PATHWAYS of cerebral catecholamine metabolism have been evaluated in animals by application of radioisotopic techniques. These studies involved incubation of brain slices with labelled norepinephrine (NE), dopamine (DA) and their precursors or intraventricular or intracisternal administration of these compounds. These studies have indicated that dopamine is preferentially metabolised to the acids, homovanillic acid (HVA) and dihydroxyphenylacetic acid (DOPAC), whereas norepinephrine is preferentially metabolised to the alcohols, 3-methoxy-4-hydroxy-phenylethylene glycol (MHPG) and dihydroxyphenylglycol (MANNARINO et al., 1963; RUTLEDGE and JONASON, 1967; BREESE et al., 1969; JONASON, 1969); Because of the inaccessibility of the human brain to direct study, similar information for man is lacking. Evaluation of cerebral catecholamine metabolism in man is best achieved by study of the metabolites of the catecholamines in cerebrospinal fluid (CSF) (MOIR et al., 1970). The occurrence of HVA and MHPG in CSF has been documented and these metabolites can be accurately quantitated (ANDEN et al., 1963; WILK et al., 1971). In man the major NE metabolite excreted in the urine is vanillylmandelic acid (VMA) (ARMSTRONG et al., 1957). The presence of this compound in brain and CSF of man or other species has not been documented (SHARMAN, 1971). To evaluate the metabolic significance of VMA in human brain, a highly sensitive method was developed which is of general use for the quantitation of acidic metabolites of biogenic amines in brain and CSF (WILK and WATSON, 1973). This procedure utilises gas–liquid chromatography and electron capture detection (GLC-ECD).

The esterification of acids with trichloroethanol using trifluoroacetic anhydride as catalyst was reported in 1971 (SMITH and TSAI, 1971). Utilising the principles described in this publication, we developed a new reagent, 20% pentafluoropropanol in pentafluoropropionic anhydride, for the preparation of derivatives of phenolic acids which possess excellent electron capture properties. In this reaction, the anhydride catalyses esterification of the carboxyl group with pentafluoropropanol and itself reacts with the hydroxyl groups. Derivatives of VMA, HVA, DOPAC, 5-hydroxy indoleacetic acid (5-HIAA), and p-hydroxy phenylacetic acid were prepared. Excellent separation of these compounds was obtained on a 3% OV-17 column. The derivative of 5-HIAA was chromatographed at 145° whereas the other acid derivatives were chromatographed at 115°.

Although as little as 15 pg of these derivatives can be detected, the true sensitivity of this method is dependent upon the fraction of sample that can be injected onto the column. This in turn is governed by the sample background. The anticipated low levels of VMA in CSF led to the consideration of a system that could offer the highest degree of sample purification. This was achieved by thin layer chromatography (TLC).

The procedure developed for the quantitation of VMA and HVA in CSF involves acidification of the sample and extraction of the acids into ethyl acetate. After evaporation of the solvent, the residue is applied to a cellulose coated TLC plate and chromatographed in a benzene–acetic acid–water (100–50–2) system. Zones corresponding to the acids are eluted, the eluate acidified and the acids reextracted into ethyl acetate. The solvent is removed under nitrogen and the residue treated with the anhydride–alcohol mixture for 15 min at 75°. The derivatives are chromatographed on the 3% OV-17 column. Recoveries have averaged 60 per cent. The purity of these samples permits one-tenth of the total sample to be applied to the column.

All spinal fluid samples examined had very low levels of VMA (< 2ng/ml). Accurate quantitation of a 10 ml CSF sample gave a value of 0·50 ng VMA/ml (Fig. 1). Because of these low levels the identity of the peak attributed to VMA could not be confirmed by auxillary techniques such as mass spectrometry. However, these studies indicate that VMA in CSF is present at levels less than 2 ng/ml. In contrast, HVA can be easily quantitated. HVA levels in CSF have averaged 30 ng/ml.

The alcohol metabolites of NE and DA can be measured by GLC-ECD after formation of trifluoroacetyl derivatives. Levels of MHPG in CSF average 16 ng/ml (WILK et al., 1971). The alcohol metabolite of dopamine 3-methoxy-4-hydroxyphenyl ethanol, was not detected in CSF (WILK, 1971). If present levels of this compound are less than 1·5 ng/ml.

The MHPG/VMA ratio in CSF (∼16) is in sharp contrast to this ratio in urine (∼0·5) (WILK et al., 1965; WILK et al., 1967). To eliminate the possibility that VMA is rapidly transported out of CSF, preliminary studies were carried out on patients who were treated with large doses of probenecid (100 mg/kg) (GOODWIN et al., 1973). Measurement of CSF probenecid levels in such studies is of importance since it was shown

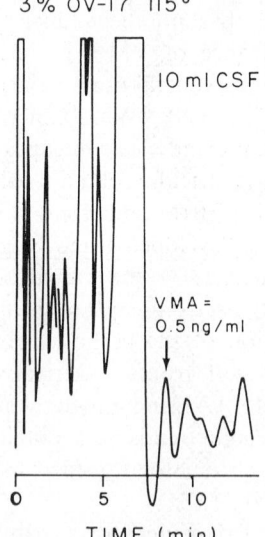

FIG. 1.—The demonstration of VMA in CSF. 10 ml of CSF was purified by thin layer chromatography where R_f values were: VMA = 0·25, 5-HIAA = 0·25, DOPAC = 0·25, p-hydroxyphenylacetic acid = 0·70, HVA = 0·80. Retention time of derivatives on 3% OV-17 at 115° relative to VMA: p-hydroxyphenylacetic acid = 0·58, DOPAC = 0·82, HVA = 1·90. The 5-HIAA derivative is chromatographed at 145°.

that the accumulation of acidic metabolites in CSF is linearly related to the CSF probenecid concentration (KORF and VAN PRAAG, 1971; SJOSTROM, 1972). The pentafluoropropanol–pentafluoropropionic anyhdride reagent can derivatise probenecid and measurement of CSF probenecid levels can be achieved using as little as 50 microliters of CSF. In these studies HVA accumulated to several hundred ng/ml whereas after probenecid the level of VMA in CSF averaged 2 ng/ml. Therefore rapid transport of VMA out of CSF cannot account for the finding of low levels of this metabolite.

The metabolism of catecholamines in human brain has been evaluated by measurement of the endogenous levels of metabolites in CSF. This approach avoids the difficulties inherent in the use of radioisotopes in which one assumes mixing of endogenous and exogenous material. The assumption made in the studies reported here is that metabolites in the CSF accurately reflect metabolism in brain. While there is good evidence that HVA in CSF has its origin in brain (PAPESCHI et al., 1971; CURZON et al., 1971) the relative contribution of spinal cord NE to the levels of the NE metabolites in CSF is still unknown.

On the basis of these gas chromatographic determinations one may conclude that within the central nervous system of man NE is preferentially metabolised to the alcohol, MHPG, whereas DA is preferentially metabolised to the acid HVA. These results are in agreement with a growing body of literature from animal studies and provide the first direct evidence that within the central nervous system of man the predominant pathway of NE metabolism is reductive to MHPG.

Acknowledgements—Supported by a grant from the NIMH: MH-21638-02 and a Research Career Development Award to S. W. #1K4-GM-40 793-01.

REFERENCES

ANDEN N.-E., ROOS B.-E. and WERDINIUS B. (1963) *Life Sci.* **2**, 448–458.
ARMSTRONG M. D., McMILLAN A. and SHAW K. N. F. (1957) *Biochim. Biophys. Acta* **25**, 422–423.
BREESE G. R., CHASE T. N. and KOPIN I. J. (1969) *J. Pharmacol. Exp. Ther.* **165**, 9–13.
CURZON G., GUMPERT E. J. W. and SHARPE D. M. (1971) *Nature, New Biol.* **231**, 189–191.
GOODWIN F. K., POST R. M., DUNNER D. L. and GORDON E. K. (1973) *Am. J. Psychiat.* **130**, 73–79.
JONASON J. (1969) *Acta Physiol. Scand. Suppl.* **320**, 1–50.
KORF J. and VAN PRAAG H. M. (1971) *Brain Res.* **35**, 221–230.
MANNARINO E., KIRSHNER N. and NASHOLD B. (1963) *J. Neurochem.* **10**, 373–379.
MOIR A. T. B., ASHCROFT G. W., CRAWFORD T. B. B., ECCLESTON D. and GULDBERG H. C. (1970) *Brain* **93**, 357–368.
PAPESCHI R., SOURKES T. L., POIRIER L. J. and BOUCHER R. (1971) *Brain Res.* **28**, 527–533.
RUTLEDGE C. O. and JONASON J. (1967) *J. Pharmacol. Exp. Ther.* **157**, 493–502.
SHARMAN D. F. (1971) in *Methods of Neurochemistry* (FRIED R., ed.) Vol. 1, pp. 83–127. Marcel Dekker New York.
SJOSTROM R. (1972) *Psychopharmacologia (Berl.)* **25**, 96–100.
SMITH R. V. and TSAI S. L. (1971) *J. Chromatogr.* **61**, 29–34.
WILK S. (1971) *Biochem. Pharmacol.* **20**, 2095–2096.
WILK S., DAVIS K. L. and THACKER S. B. (1971) *Anal. Biochem.* **39**, 498–504.
WILK S., GITLOW S. E., CLARKE D. D. and PALEY D. H. (1967) *Clin. Chim. Acta* **16**, 403–408.
WILK S., GITLOW S. E., MENDLOWITZ M., FRANKLIN M. J., CARR H. E. and CLARKE D. D. (1965) *Anal. Biochem.* **13**, 544–551.
WILK S. and WATSON E. (1973) *Fedn. Proc.* **32**, 798.

Frontiers in Catecholamine Research 1973, pp. 1071 to 1075. Pergamon Press. Printed in Great Britain.

OXYGEN-18 IN MEASUREMENT OF DOPAMINE TURNOVER IN RAT BRAIN

GÖRAN SEDVALL AVRAHAM MAYEVSKY, DAVID SAMUEL
and CLAES-GÖRAN FRI
Department of Pharmacology, Karolinska Institutet, Stockholm, Sweden, and the
Isotope Department, The Weizmann Institute of Science, Rehovot, Israel

INTEREST for dopamine (DA) in brain function was stimulated by the discovery that levels of this compound in brain of parkinsonian patients were markedly reduced (HORNYKIEWICZ, 1966). The demonstration that *amphetamine* (COSTA and GROP-PETTI, 1970) and antipsychotic drugs markedly affect brain DA metabolism (NYBÄCK and SEDVALL, 1968) indicated a role for this transmitter substance also in psychotic states like *schizophrenia* (KLAWANS et al., 1972).

For studies on brain DA metabolism in man, determination of base line levels or *probenecid* induced elevations of homovanillic acid (HVA), the major DA metabolite, in lumbar cerebrospinal fluid (CSF) seem to be the best procedures elaborated so far (PAPESCHI, 1972; SJÖSTRÖM, 1973). However, since mechanisms for HVA transport from brain to lumbar cerebrospinal fluid are not readily controlled and probenecid in tolerable doses only gives a sub-maximal blockade of HVA transport from CSF, the search for alternative methods is necessary. In animals the use of catecholamine synthesis inhibitors or precursors labelled with radioactive isotopes seem to be the most valid procedures developed so far for quantitative determination of brain catecholamine turnover (COSTA, 1972).

The development of mass spectrometric methods for determination of catecholamines and their metabolites (ÄNGGÅRD and SEDVALL, 1969; KOSLOW, CATTABENI and COSTA, 1972, FRI et al., 1973, in preparation) has recently made possible the use of stable isotopes to label brain catecholamines *in vivo* for turnover determinations. Thereby the risk with radioactive isotopes can be avoided. Since catechol- and indoleamines are formed in rate limiting reactions by hydroxylases using molecular oxygen, the possibility to label brain monoamines *in vivo* with stable oxygen isotopes is of considerable interest. Molecular oxygen can easily be administered by inhalation and by a single exposure it is possible to label a number of compounds of biological interest.

In a recent series of experiments we could demonstrate the use of *stable oxygen isotopes* for labelling of HVA in rat brain. Following *in vivo* exposure of rats to atmospheres highly enriched with oxygen-18, mass spectrometric evidence was obtained for the incorporation of at least one oxygen isotope in brain HVA (SEDVALL et al., 1973, MAYEVSKY et al., 1973). Figure 1 demonstrates the high mass range in the mass spectra of the methyl ester heptafluorobutyryl derivative of authentic HVA and apparent HVA from brain extracts. HVA has the molecular ion at m/e 392 which is also the base peak. An abundant fragment is also present at m/e 333. An almost identical spectrum was obtained from brain extracts of animals exposed to an $^{16}O_2$ containing atmosphere. On the other hand, in animals exposed to $^{18}O_2$ gas, abundant

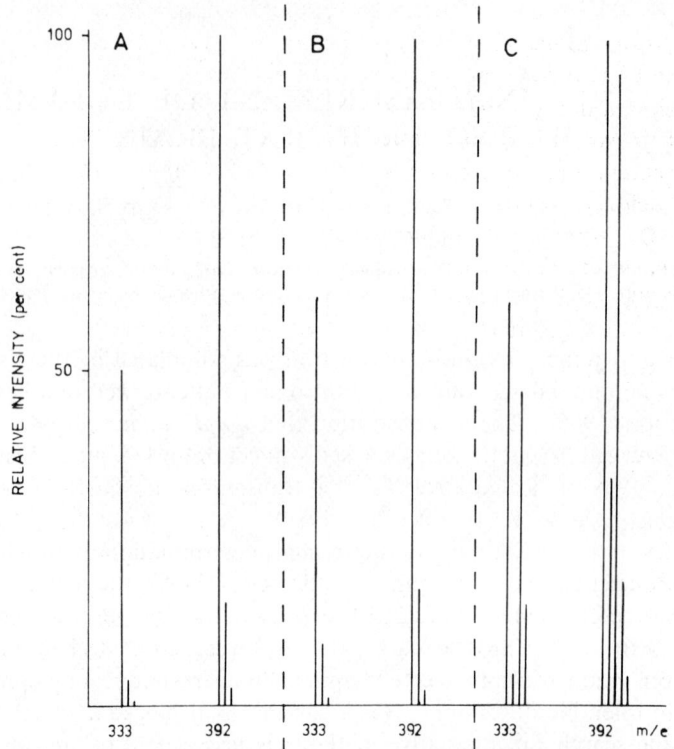

FIG. 1.—High mass range of HVA in the form of Me-HFB derivatives. A represents authentic HVA. B and C are obtained from brain of rats exposed to $^{16}O_2$ and $^{18}O_2$ respectively. Spectra were obtained by subtracting background intensities at retention time 4·15 min from values recorded at 4·20 min—the retention time of the HVA derivatives on the 1% OV-17 column used.

fragments at m/e 334 and 335 were also present, demonstrating the presence of HVA containing one ^{18}O atom per molecule.

The atmospheric origin of the ^{18}O atom incorporated was proved by injecting ^{18}O labelled water into the animals which was not followed by incorporation of significant amounts of ^{18}O in brain HVA (Mayevsky et al., 1973). The usefulness of the technique for studies on brain DA turnover was demonstrated by studying the effect of chlorpromazine treatment on ^{18}O, incorporation in brain HVA. Chlorpromazine treatment was followed by an approximately three-fold increase in the rate of formation of ^{18}O labelled HVA in the rat brain (Sedvall et al., 1973). For the quantitative determination of DA turnover an analysis of precursor–product relationships for the incorporation of ^{18}O in precursor amino acids, DA and HVA is required. For that purpose mass fragmentometric methods for the determination of phenylalanine, tyrosine, DA and HVA in the same rat brain sample have been developed (Fri et al., 1973, in preparation). In the present communication mass fragmentometric evidence for the incorporation of ^{18}O in DA of rat brain during *in vivo* exposure to ^{18}O containing atmospheres is presented.

METHODS

Male Sprague–Dawley rats weighing about 60 g were introduced into a system (Mayevsky et al., 1973), which contained normal air at the beginning of the experiment. As the system was closed and oxygen was consumed by the animals, $^{18}O_2$ or $^{16}O_2$ gas containing about 95% ^{18}O or 100% ^{16}O was introduced by a pressure regulated valve. Humidity and carbon dioxide formed within the system were removed by circulating the atmosphere over silica gel and a CO_2 trap. Immediately after the end of exposure, which lasted 3 hr, the animals were removed from the system, decapitated and the corpora striata were dissected out and homogenised in 0.1 M formic acid containing 0.5 μM ascorbic acid. Aliquots of the extracts were analysed for DA and HVA by mass fragmentographic procedures described by KOSLOW, CATTABENI and COSTA (1972) and FRI et al., (1973). For dopamine determination α-methyldopamine (α-MDA) was used as internal standard, for HVA determination a deuterated HVA molecule was used. A schematical outline of the chemical procedures used is depicted in Fig. 2. For further details see MAYEVSKY

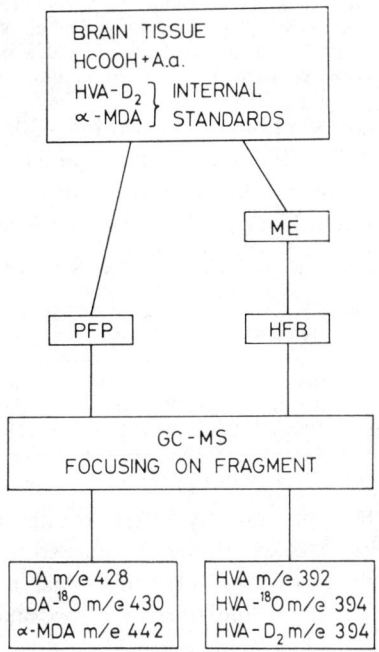

FIG. 2.—Flow sheet for extraction, derivatisation and mass fragmentometric analysis of DA and HVA in rat brain tissue.

et al. (1973). Recordings were made using an LKB 9000 gas chromatograph-mass spectrometer.

RESULTS

Mass spectra of pentafluoropropionic (PFP) acid derivates of DA and α-MDA were shown by KOSLOW et al. (1972) to exhibit major fragments at m/e values 428 and 442 respectively. The latter authors also reported the mass fragmentometric identification of DA in rat striatum. Figure 3 demonstrates a mass fragmentogram from authentic DA and DA extracted from striata of rats exposed for 3 hr to ^{16}O and ^{18}O containing atmospheres respectively. The figure illustrates how in control rats significant intensities are obtained at m/e 428 but not 430 as in authentic DA. In ^{18}O exposed animals however, there is a relative diminution of the intensities at m/e 428,

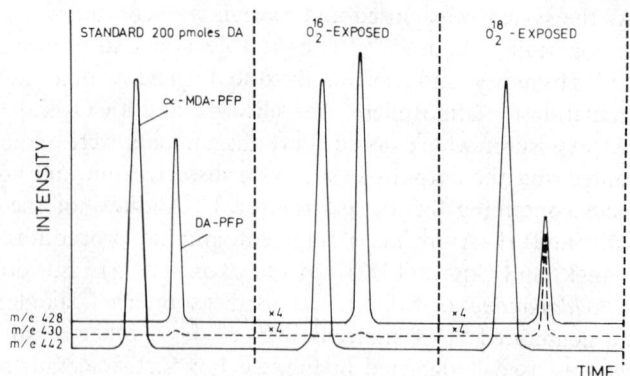

FIG. 3.—Mass fragmentograms of pentafluoropropionyl derivates of DA extracted from brain of (B) $^{16}O_2$ and (C) $^{18}O_2$ exposed rats. The curve for m/e 442 represents the derivative of the internal standard α-MDA. M/e 428 represents authentic DA, whereas 430 represents DA labelled with one ^{18}O atom.

whereas significant intensities are obtained also at m/e 430. By mass fragmentography and the use of α-MDA and HVA-D_2 as internal standards the absolute amounts of ^{16}O and ^{18}O labelled DA and HVA were determined in brains from ^{16}O and ^{18}O exposed animals. It is evident from Table 1 that following exposure to an ^{18}O con-

TABLE 1. OXYGEN ISOTOPIC LEVELS (pmole/striata) OF DA AND HVA IN RAT BRAIN AFTER EXPOSURE TO $^{16}O_2$ OR $^{18}O_2$

Exposure	DA	DA-^{18}O	HVA	HVA-^{18}O	% ^{18}O in air
$^{16}O_2$	2860 ± 310	not detect.	422 ± 31	not detect.	0·2
$^{18}O_2$	1130 ± 400	103 ± 130	222 ± 16	186 ± 18	81·5

Data represent mean ± s.e. from 5 to 6 rats.

taining atmosphere for 3 hr approximately 50% of the amount of DA as well as HVA are labelled with ^{18}O. The mass spectrometer was also focussed on m/e 432, which should represent DA labelled with two ^{18}O atoms. Also here intensities significantly above back-ground were obtained, but they were too small to allow exact quantification.

DISCUSSION

We have previously demonstrated the incorporation of one ^{18}O atom into HVA of rat brain following *in vivo* exposure to ^{18}O containing atmospheres. The present experiments give strong evidence for the labelling also of DA with ^{18}O. This is a direct *in vivo* confirmation of previous *in vitro* data demonstrating that hydroxylation of tyrosine to DOPA involves the incorporation of molecular oxygen (NAGATSU, LEVITT and UDENFRIEND 1964, DALY *et al.*, 1968). Since mass fragmentometric methods for the determination of phenylalanine, tyrosine, DA and HVA in single rat striata were recently developed (FRI *et al.*, 1973, in preparation), the prerequisite for determination of precursor-product relationships for ^{18}O incorporation in the DA pathway are now available.

Inhalation of ^{18}O containing atmospheres for several months in rats was not followed by any signs of toxicity (SAMUEL, 1973, in preparation). The availability of techniques for the mass fragmentometric determination of HVA levels in lumbar CSF, serum and urine of man makes it possible therefore to initiate studies with ^{18}O for determination of brain DA turnover also in humans.

For clinical studies ^{18}O gas could easily be administered by a short term inhalation from a closed system. Labelling of brain monoamines by such a procedure and subsequent determination of changes in specific activity of amine metabolites in CSF should allow calculation of brain DA turnover rates. By measuring specific activities in serum or urine following blockade of peripheral HVA synthesis at optimal intervals following ^{18}O exposure, selective determination of brain DA synthesis could also be obtained. Future studies have to elucidate the practicability of such methods which might be of value for studies on brain DA synthesis during neuropsychiatric disease states. Such procedures might also be of value to control specific drug therapies directed to alteration of brain DA metabolism. Besides for DA, the described technique may be preferable also to study the metabolism of *noradrenaline* and *serotonin* in the central nervous system, since the latter amines too are formed by hydroxylases using molecular oxygen.

Acknowledgements—The present study was supported by the Swedish Medical Research Council (40X-3560) and the Weizmann Institute of Science, Israel. The skilful technical assistance of Mrs. Varda Greenberger is gratefully acknowledged.

REFERENCES

COSTA E. (1972) *Adv. Biochem. Psychopharmacol.* **4,** 171–193.
COSTA E. and GROPPETTI A. (1970) In: *International Symposium on Amphetamines and Related Compounds.* (COSTA E. and GARATTINI S., Eds.) pp. 231–255. Raven Press, New York.
DALY J., LEVITT M., GUROFF G. and UDENFRIEND S. (1968) *Archs. Biochem. Biophys.* **126,** 593–598.
HORNYKIEWICZ O. (1966) *Pharmacol. Rev.* **18,** 925–964.
KLAWANS H. L., GOETZ B. A. and WESTHEIMER R. (1972) *Dis. Nerv. Syst.* **33,** 711–719.
KOSLOW S. H., CATTABENI F. and COSTA E. (1972) *Science* **176,** 177–180.
MAYEVSKY A., SJÖQUIST B., FRI C. -G., SAMUEL D. and SEDVALL G. (1973) *Biochem. Biophys. Res. Comm.* **51,** 746–756.
NAGATSU T., LEVITT M. and UDENFRIEND S. (1964) *J. Biol. Chem.* **239,** 2910–2917.
NYBÄCK H. and SEDVALL G. (1968) *J. Pharmacol. Exp. Ther.* **162,** 294–301.
PAPESCHI R. (1972) *Psychiat. Neurol. Neurochir.* **75,** 13–48.
SEDVALL G., MAYEVSKY A., FRI C. -G., SJÖQUIST B. and SAMUEL D. (1973) *Adv. Biochem. Psychopharmacol.* **7,** in press.
SJÖSTRÖM R. (1973) *Acta Universitatis Upsaliensis* **154,** 5–39.
ÄNGGÅRD E. and SEDVALL G. (1969) *Anal. Chem.* **41,** 1250–1256.

Frontiers in Catecholamine Research 1973, pp. 1077 to 1082. Pergamon Press. Printed in Great Britain.

TECHNICAL STRATEGIES FOR THE STUDY OF CATECHOLAMINES IN MAN

DENNIS L. MURPHY

Laboratory of Clinical Science, NIMH, Bethesda, Maryland, U.S.A.

BECAUSE of necessary limitations in obtaining tissue and in administering certain drugs and isotopes, studies of catecholamine metabolism in man often depend on indirect procedures or utilise specialised methods not required in animal investigations. Most human studies have been directed towards the identification of alterations in catecholamine metabolism produced by drugs or by physiologic changes (e.g. exercise) or pathologic states (e.g. cardiovascular, neurologic and psychiatric disorders). The major tactics utilised in the study of biogenic amines in man are outlined in Table 1. Only a small amount of material subsumed under this title can be mentioned here, but some major strategies for catecholamine studies in man (e.g., the probenecid approach to CSF amine metabolite formation) are reviewed in detail elsewhere in this symposium, and some reviews directed toward the study of amine synthesis, turnover and metabolic pathways of biogenic amines in man are listed in the bibliography (SCHILDKRAUT and KETY, 1967; MANDELL and SPOONER, 1968; FRANZEN and EYSELL, 1969; KOPIN, 1972; BARCHAS and USDIN, 1973). This paper will more narrowly focus on attempts to study catecholamines, catecholamine-related enzymes and catecholamine transport, storage and release mechanisms at the cellular level in man. Examples from current studies in this laboratory will be used for illustration.

STUDY OF BIOGENIC AMINE-RELATED ENZYMES IN MAN

Increasing attention is being devoted to the study of enzymes from biogenic amine pathways found in plasma, blood cells and cultured fibroblasts. Many genetically-based disorders of metabolism in other fields have been identified on the basis of enzyme alterations in leukocytes (HSIA, 1972), and in erythrocytes, platelets and plasma (HARRIS, 1971).

Monoamine oxidase (EC 1.4.3.4, MAO)

MAO has been studied in human platelets and plasma. The platelet enzyme is located in mitochondria, and has many substrate and inhibitor characteristics in common with mitochondrial MAO's of the B type found in other tissues (PAASONEN and SOLATUNTURI, 1965; ROBINSON, LOVENBERG, KEISER and SJOERDMA, 1968; COLLINS and SANDLER, 1971; NEFF and GORIDIS, 1972, MURPHY and WEISS, 1972). The purified platelet enzyme is electrophoretically homogenous, with a molecular weight of 235,000 (COLLINS and SANDLER, 1971). The soluble plasma enzyme is a pyridoxal-containing enzyme which has different substrate specificities compared to platelet and other tissue MAO's and is inhibited by carbonyl reagents but is relatively insensitive to tissue MAO inhibitors such as pargyline, tranylcypromine and iproniazid (ROBINSON *et al.*, 1968; MCEWEN, 1972).

Treatment with MAO inhibiting drugs (MAOI) under ordinary clinical conditions leads to marked inhibition of the human platelet (ROBINSON *et al.*, 1968) and brain

1078   D. L. Murphy

TABLE 1. METHODOLOGIC APPROACHES TO THE STUDY OF CATECHOLAMINES IN MAN

I. Direct measurement of amines, amine metabolites and amine-related enzymes in blood, urine, cerebrospinal fluid and biochemical or histofluoresence analysis in tissues (e.g. blood cells, surgically-obtained specimens and autopsy material).
II. Use of radioactively-labeled amine precursors and amines to study pathways of amine synthesis and metabolism.
III. Use of metabolic loads to stress pathways of amine synthesis and degradation.
IV. Use of drugs with relatively specfic effects to alter amine synthesis, degradation, turnover, transport, storage and release, with measurement of effects on amines and amine metabolites as above.

enzymes (Ganrot, Rosengren and Gottfries, 1962), with MAO activity reduced 70–90 per cent after several days of treatment. Inhibition persists for up to 2–3 weeks after discontinuation of the MAOI. Platelet MAO activity appears little affected by drugs other than MAOI or by environmental changes and menstrual cycle changes, although MAO activity in platelets appears to increase with age, particularly over age 50 (Robinson, Davis J. M., Nies, Colburn, Davis J. N., Bourne, Bunney, Shaw and Coppen, 1972; Murphy D. L., unpublished); females tend to have 10–20 per cent higher MAO levels, a difference which again is especially evident over age 50 and also prior to puberty (Robinson et al., 1972; Murphy, unpublished).

Studies comparing MAO activity in normal and psychiatrically-ill monozygotic and dizygotic twins, sibling pairs and random unrelated pairs suggest a high order of genetic influence on platelet MAO activity (Table 2). Moderately reduced platelet MAO activity has been described in bipolar manic–depressive patients and in individuals with acute schizophreniform psychoses, while chronic schizophrenic individuals have more markedly reduced levels (Murphy and Weiss, 1972; Murphy and Wyatt, 1972). In contrast, slightly elevated (Murphy and Weiss, 1972) or moderately elevated (Nies, Robinson, Ravaris and Davis, 1971) levels are found in other depressed patients. In monozygotic twins discordant for schizophrenia, the twins with and without schizophrenia have MAO values which are reduced in comparison to normal controls, suggesting that the reduction in schizophrenic individuals does not represent an effect of the disorder or its treatment, but rather may provide a genetically-related measure reflecting vulnerability to schizophrenia (Wyatt, Murphy, Belmaker, Cohen, Donnelly and Pollin, 1973).

TABLE 2. INTRACLASS CORRELATION COEFFICIENTS FOR PLATELET MONOAMINE OXIDASE ACTIVITY MEASURED IN NORMAL, SCHIZOPHRENIC AND BIPOLAR TWINS AND IN SIB AND RANDOM PAIRS

	Intraclass correlation
Normals	
Monozygotic twins ($N = 9$)	0·88
Dizygotic twins ($N = 10$)	0·45
Sib pairs ($N = 37$)	0·28
Random pairs matched for age and sex ($N = 37$)	0·12
Monozygotic twins discordant for schizophrenia ($N = 13$)	0·65
Monozygotic twins concordant for bipolar manic-depressive illness ($N = 3$)	0·83

Catechol-O-methyl transferase (EC 2.1.16, COMT)

This enzyme has been identified in human erythrocytes (AXELROD and COHN, 1971) and has similar properties and substrate specificities as COMT in other animal tissues (COHN, 1970). The enzyme appears to be a stable characteristic of the individual and is not affected by tricyclic antidepressants, phenothiazines, lithium carbonate, α-methyl-*para*-tyrosine or *para*-chlorophenylalanine (COHN, 1970; DUNNER, 1971), although chronic L-dopa administration has been reported to increase red cell COMT activity in Parkinsonian patients (WEISS, COHN and CHASE, 1971). Age and sex are said not to influence activity, although reduced COMT levels were observed in women but not in men hospitalised for depression (COHN, 1970). Females with unipolar depression had a significantly greater reduction in COMT activity than did female bipolar patients, while schizophrenic females and men with antisocial personalities were no different from normal controls. Recovery from depression or the development of mania was not associated with any change in enzyme activity (DUNNER, 1971).

Dopamine β-hydroxylase (DBH) and other catecholamine-related enzymes

Plasma DBH also appears to be a relatively stable characteristic of the individual (WEINSHILBOUM and AXELROD, 1971), with twin studies suggesting a marked genetic influence on measured activity (LAMPRECHT *et al.*, 1973). The characteristics of this enzyme in man and alterations found in some clinical populations, including such genetically-based disorders as torsion dystonia (WOOTEN, ELDRIDGE, AXELROD and STERN, 1973) and dysautonomia (WEINSHILBOUM and AXELROD, 1971), are reviewed elsewhere in this symposium. Other catecholamine-related enzymes including tyrosine hydroxylase, *l*-amino acid decarboxylase and phenylethanolamine *N*-methyl transferase have been little studied in man.

STUDY OF BIOGENIC AMINE CELLULAR TRANSPORT AND
STORAGE IN MAN

The platelet model

Platelets accumulate biogenic amines and store them in vesicles, reaching concentration gradients over 100:1 for serotonin and dopamine and 5–13:1 for norepinephrine and epinephrine (PLETSCHER, 1968; ABRAMS and SOLOMON, 1969; BORN and SMITH, 1970; BOULLIN and O'BRIEN, 1970; PAASONEN, AHTEE and SOLATUNTURI, 1971; MURPHY and KOPIN, 1972). The transport K_m for serotonin in platelets is 3×10^{-7} M, which is very similar to K_m for serotonin uptake in brain slices (SNYDER, KUHAR, GREEN, COYLE and SHASKAN, 1970). Although dopamine, norepinephrine and epinephrine competitively inhibit serotonin uptake in platelets, a saturable uptake mechanism for catecholamines has only been demonstrated for dopamine (BOULLIN, 1970). Since isolated platelet vesicles also bind catecholamines less avidly than serotonin (DAPRADA and PLETSCHER, 1969), it would appear that the platelet provides a better model system for serotonergic than catecholaminergic neurons. However, the similarity of effects of many drugs such as reserpine, MAO inhibitors and tricyclic antidepressants on both catecholaminergic and serotonergic cell processes permits utilisation of the platelet model system in monitoring drug effects in individuals (ROBINSON *et al.*, 1968; MURPHY, COLBURN, DAVIS and BUNNEY, 1970).

However, this apparent specificity as a serotonergic cell provides an opportunity to investigate the accumulation and fate of other amines not ordinarily found in serotonergic cells; for example, the accumulation of dopamine or a dopamine metabolite during treatment with L-dopa (BOULLIN, 1970; MURPHY *et al.*, 1970), and the accumulation of octopamine during monoamine oxidase inhibitor treatment (CAHAN, MURPHY and MOLINOFF, 1973). *In vitro*, serotonin stores appear to be displaced by dopamine, norepinephrine, epinephrine, tyramine and octopamine. *In vivo*, L-dopa administration leads to a reduction in platelet serotonin content (MURPHY, *et al.*, 1970). Whether this alteration results from competition by catecholamines for uptake of serotonin at the cell membrane level or from serotonin displacement from vesicles, this evidence suggests the possibility that 'false-transmitter' type effects demonstrated in animals may occur in man. Furthermore, the demonstration of several-fold higher levels of octopamine in platelets isolated from individuals receiving monoamine-oxidase inhibiting drugs compared to normal controls strengthens the suggestions made on the basis of animal studies that octopamine accumulation in cells might contribute to some effects of MAO-inhibiting drugs in man such as hypotension (KOPIN, FISCHER, MUSACCHIO and HORST, 1964).

Other cell systems

While human erythrocytes do not contain amine storage vesicles and do not achieve concentration gradients for biogenic amines greater than 2–3:1, some studies of amine distribution across the erythrocyte cell membrane and of catecholamine metabolism by these cells have been accomplished (MURPHY and KOPIN, 1972; DANON and SAPIRA, 1973). Some biochemical (BOURNE, BUNNEY, COLBURN, DAVIS J. M., DAVIS J. N., SHAW and COPPEN, 1968; BEVAN-JONES, PARE, NICHOLSON, PRICE and STACEY, 1972) and histofluorescence studies (DE LA TORRE, 1972) of catecholamines in human brain tissue have been made. Axonal transport of dopamine β-hydroxylase has been studied in biopsy specimens of human sural nerves (BRIMIJOIN and DYCK, in press).

STUDY OF BIOGENIC AMINE RECEPTOR FUNCTIONS IN MAN

Platelets and leukocytes possess an adenyl cyclase system which is responsive to biogenic amines and biogenic amine-affecting drugs (MOSKOWITZ, HARWOOD, REID and KRISHNA, 1971). Platelets can be stimulated by prostaglandin E_1 to produce a 10-fold increase in cell cyclic AMP content; in intact cells, norepinephrine reduces this cyclic AMP formation elicited by prostaglandin E_1, while the norepinephrine effects are antagonised by phentolamine but not by propranolol. This apparent adrenergic α-receptor function in platelets is affected by some psychoactive drugs *in vitro* and by at least one such drug, lithium carbonate, *in vivo* in man (MURPHY, DONNELLY and MOSKOWITZ, in press). Lithium has similar antagonistic effects on catecholamine and other hormone-induced adenyl cyclase activity in other tissues. Norepinephrine and epinephrine also produce platelet aggregation, a cell response which can be prevented by phentolamine and phenoxybenzamine, while propranolol is much less potent than the α-receptor antagonists (MUSTARD and PACKHAM, 1970).

CONCLUSION

These investigations of biogenic amine enzymes, transport and metabolic functions, and receptor effects of amines in human cells suggest a potential for the use of such approaches in confirming amine metabolic changes studied by other techniques, and in determining individual differences in response to amine-affecting drugs. While cells such as platelets do not synthesise biogenic amines, their enzymes and their transport and storage functions appear to be affected by various drugs in ways similar to those observed in nerve tissue and brain. The possibility of studying genetically-based differences in these membrane-based processes is suggested by the evidence that genetic factors are major determinants of the activity of such enzymes as mono-amine oxidase in these cells.

REFERENCES

ABRAMS W. B. and SOLOMON H. M. (1969) *Clin. Pharmacol. Ther.* **10**, 702–709.

AXELROD J. and COHN C. K. (1971) *J. Pharmacol. Exp. Ther.* **176**, 650–654.

BARCHAS J. and USDIN E. (Eds.) (1973) *Serotonin and Behaviour.* Academic Press, New York.

BEVAN-JONES A. B., PARE C. M. B., NICHOLSON W. J., PRICE K. and STACEY R. S. (1972) *Bri. Med. J.* **1**, 17–19.

BORN G. V. R. and SMITH J. B. (1970) *Bri. J. Pharmacol.* **39**, 765–778.

BOULLIN D. J. and O'BRIEN R. A. (1970) *Bri. J. Pharmacol.* **39**, 779–788.

BOURNE H. R., BUNNEY W. E. JR., COLBURN R. W., DAVIS J. M., DAVIS J. N., SHAW D. M. and COPPEN A. J. (1968) *Lancet* **ii**, 805–808.

BRIMIJOIN S. and DYCK P. J. (1973) *Science*, in press.

CAHAN D., MURPHY D. L. and MOLINOFF P. (1973) *Fedn. Proc.* **32**, 708.

COHN C. K., DUNNER D. L. and AXELROD J. (1970) *Science* **170**, 1323–1324.

COLLINS G. G. S. and SANDLER M. (1971) *Biochem. Pharmacol.* **20**, 289–296.

DANON A. and SAPIRA J. D. (1973) *Clin. Pharmacol. Exp. Ther.* **13**, 916–922.

DAPRADA M. and PLETSCHER A. (1969) *Life Sci.* **8**, 65–72.

DE LA TORRE J. C. (1972) In: *Dynamics of Brain Monoamines* pp. 33–37. Plenum Press, New York.

DUNNER D. L., COHN C. K., GERSHON E. S. and GOODWIN F. K. (1971) *Archs. Gen. Psychiat.* **25**, 348–353.

FRANZEN F. and EYSELL K. (1969) *Biologically Active Amines Found in Man.* Pergamon Press, Oxford.

GANROT P. O., ROSENGREN E. and GOTTFRIES C. G. (1962) *Experientia* **18**, 260–261.

HARRIS H. (1971) *The Principles of Human Biochemical Genetics.* Elsevier, New York.

HSIA D. Y. Y. (1972) *Enzyme* **13**, 161–168.

KOPIN I. J. (ed.) (1972) In: *Methods in Investigative and Diagnostic Endocrinology* (BERSON, S. O., Ed.), pp. 309–669, Elsevier, New York.

KOPIN I. J., FISCHER J. E., MUSACCHIO J. and HORST W. D. (1964) *Biochem.* **52**, 716–721.

LAMPRECHT F., *et al.* (This Symposium).

MANDELL A. J. and SPOONER C. E. (1968) *Science* **162**, 1442–1453.

MCEWEN C. M. JR. (1972) In: *Advances in Biochemical Psychopharmacology.* pp. 151–165, Raven Press, New York.

MOSKOWITZ J., HARWOOD J. P., REID W. D. and KRISHNA G. (1971) *Biochim. Biophys. Acta* **230**, 279–285.

MURPHY D. L. (1972) *Am. J. Psychiat.* **129**, 141–148.

MURPHY D. L., COLBURN R. W., DAVIS J. M. and BUNNEY W. E., JR. (1970) *Amr. J. Psychiat.* **127**, 339–345.

MURPHY D. L., DONNELLY C. H., MOSKOWITZ J. (1973) *Clin. Pharmacol. Exp. Ther.*, in press.

MURPHY D. L. and KOPIN I. J. (1972) In: *Metabolic Transport* (HOKIN, L. E., ed.) Academic Press, New York.

MURPHY D. L. and WEISS R. (1972) *Amr. J. Psychiat.* **128**, 1351–1357.

MURPHY D. L. and WYATT R. J. (1972) *Nature, Lond.* **238**, 225–226.

MUSTARD J. F. and PACKHAN M. A. (1970) *Pharmacol. Rev.* **22**, 97–188.

NEFF N. H. and GORIDIS C. (1972) In: *Advances in Biochemical Psychopharmacology* pp. 307–323. Raven Press, New York.

NIES A., ROBINSON D. S., RAVARIS C. L. and DAVIS J. M. (1971) *Psychosom. Med.* **33**, 440.

PAASONEN M. K., AHTEE L. and SOLATUNTURI E. (1971) In: *Progress in Brain Research* (ERANKO E., ed.) Vol. **34**, pp. 269–279, Elsevier, Amsterdam.

PAASONEN M. K. and SOLATUNTURI E. (1965) *Annals Med. Exp. Biol. Fenn.* **43**, 98–105.

PLETSCHER A. (1968) *Brit. J. Pharmacol.* **32**, 1–16.

ROBINSON D. S., DAVIS J. M., NIES A., COLBURN R. W., DAVIS J. N., BOURNE H. R., BUNNEY W. E. SHAW D. M. and COPPEN A. J. (1972) *Lancet* **i**, 290–291.

ROBINSON D. S., LOVENBERG W., KEISER H. and SJOERDSMA J. (1968) *Biochem. Pharmacol.* **17,** 109–119.
SCHILDKRAUT J. J. and KETY S. S. (1967) *Science* **156,** 21–30.
SNYDER S. H., KUHAR M. J., GREEN A. I., COYLE J. T. and SHASKAN E. G. (1970) *Int. Rev. Neurobiol.*
      13, 127–159.
WEINSHILBOUM R. and AXELROD J. (1971) *Cir. Res.* **28,** 307–315.
WEINSHILBOUM R. and AXELROD J. (1971) *New Eng. J. Med.* **285,** 958–942.
WEISS J. L., COHN C. K. and CHASE T. N. (1971) *Nature, Lond.* **234,** 218–219.
WOOTEN G. F., ELDRIDGE E., AXELROD J. and STERN R. S. (1973) *New Engl. J. Med.* **228,**
      284–287.
WYATT R. J., MURPHY D. L., BELMAKER R., COHEN S. DONNELLY C. H. and POLLIN W. (1973)
      Science **173,** 916–918.

Frontiers in Catecholamine Research 1973, pp. 1083 to 1084. Pergamon Press. Printed in Great Britain.

CYCLIC AMP AND A POSSIBLE ANIMAL MODEL OF RECEPTOR SUPERSENSITIVITY

D. ECCLESTON

MRC Brain Metabolism Unit, 1 George Square, Edinburgh, Scotland.

CONVENTIONAL techniques for investigating aminergic function in man: namely the estimation of amine metabolites in C.S.F. have suggested that factors other than transmitter output should be considered in affective illness. Levels of 5HIAA have been found to be low in unipolar depressed patients but normal in bipolar depression (MRC BRAIN METABOLISM UNIT, 1972). These workers also suggested that clinical observations of changes in the facial dyskinesias with the phase of manic depressive illness may be related to changes in post synaptic receptor supersensitivity. The demonstration that receptor supersensitivity in brain can be linked to adenyl cyclase activity is, then, important. In the animal model described this is indicated.

The locus coeruleus (LC) is a bilateral noradrenergic nucleus supplying cortex almost ipsilaterally (UNGERSTEDT, 1971). The production of supersensitivity in dopaminergic receptors has been established by lesions of the cell bodies in the substantia nigra (UNGERSTEDT, 1971). Similar supersensitivity in the noradrenergic system should follow lesions of one LC and this should be unilateral. In these experiments electrolytic lesions are made in one LC and the animals allowed to recover. Subsequently cortical slices are prepared and cyclic AMP production measured by the

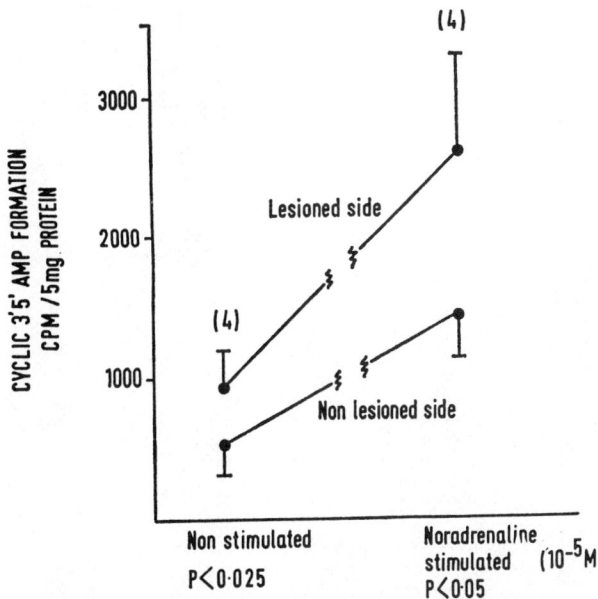

FIG. 1.—Formation of ^{14}C cyclic AMP from ^{14}C adenine in animals with unilateral lesions of locus coeruleus when no NA added to incubation medium or when incubated with 10^{-5} M NA. (paired t-test).

method of SHIMIZU *et al.* (1969). In these experiments two points arise: firstly there is a significantly increased cyclic production on the side of the lesion both in unstimulated slices and in response to NA stimulation compared with slices prepared from the unlesioned side of the animal (Fig. 1), and secondly this increased response to NA develops over at least 4 weeks.

Perhaps the efforts now being directed to determine adenyl cyclase function in man—more specifically in brain, are important. This study which shows that the development of changes is over some weeks may give us a clue to the paradoxically slow response to the cyclic antidepressants and perhaps also to the long term phasic nature of affective illness.

REFERENCES

MRC BRAIN METABOLISM UNIT (1972) *Lancet* **ii**, 573–577.
SHIMIZU H., CREVELING C. R. and DALY J. (1970) *Proc. Natl Acad. Sci.* **65,** 1033–1040.
UNGERSTEDT U. (1971) *Acta Physiol. Scand.* suppl. 367.

Frontiers in Catecholamine Research 1973, pp. 1085 to 1090. Pergamon Press. Printed in Great Britain.

APPLICATION OF MASS FRAGMENTOGRAPHY TO THE QUANTITATION OF ENDOGENOUS CATECHOLAMINES

STEPHEN H. KOSLOW

Laboratory of Preclinical Pharmacology, National Institute of Mental Health,
Saint Elizabeths Hospital, Washington, D.C. 20032, U.S.A.

THE USE of quantitative gas chromatography-mass spectrometry (GC-MS) for the measurement of punative neurotransmitters (KOSLOW et al., 1972; CATTABENI et al., 1972) offers many advantages over the presently used methods. It can be applied to the simultaneous measurement of endogenous norepinephrine (NE), epinephrine (E) and dopamine (DA) over a wide range of tissue samples. The preparation of the samples is fast and efficient and the analysis offers a high degree of specificity with sensitivity greater than 10^{-15} mole. With this high sensitivity as will be described in this paper, it is possible to measure the catecholamine content of 30 μm sections of neuronal tissue.

The GC serves to resolve the various neurotransmitters from each other, while the MS is used for quantitation by measuring the ion density of specific characteristic fragments (mass-to-charge, m/e) of each compound. Prior to analysis it is necessary to prepare catecholamine derivatives with the appropriate vapour pressure for GC. This is done by reacting the catecholamines with pentafluoropropionic anhydride (PFPA) resulting in the acylation of the phenoxy and β-hydroxy and primary and secondary amino groups. Using an LKB 9000 GC-MS the structures of the derivitised (PFP) catecholamines were confirmed by mass spectral analysis. Figure 1 shows a partial mass spectra of the three catecholamine-PFP derivatives (complete spectra have been published, KOSLOW et al., 1972).

For quantitation, the ion density of the most abundant fragment (base peak, 100 per cent relative intensity) is recorded at the compound's retention time (Fig. 2, Table 1). α-Methylnorepinephrine (αMNE) and α-methyldopamine (αMDA) are included as internal standards and fulfill all the requirements of an internal standard (KOSLOW et al., 1972); αMNE is the internal standard for NE and E, while αMDA is the internal standard for DA. For analysis tissues samples are homogenised in 0·1 M formic acid (50 mM ascorbate) and an aliquot of the supernatant is dried under nitrogen with an exact concentration of the internal standard. Concomitantly, authentic standards (over the appropriate concentration range) containing the same concentration of internal standards are prepared. After reacting the dried samples with 100 μl PFPA, 20 μl ethylacetate (30 min), the sample is dried under nitrogen and reconstituted in 10 μl ethylacetate. Usually, 2 μg of this solution is injected into the GC port of the GC-MS. A typical recording is shown in Fig. 2.

The ratio formed by dividing the peak height (or area) of the catecholamine by the peak height (or area) of the appropriate internal standard is plotted against the known concentration of catecholamine reacted. In this way the linear response obtained is used to quantitate endogenous catecholamine levels.

1086   STEPHEN H. KOSLOW

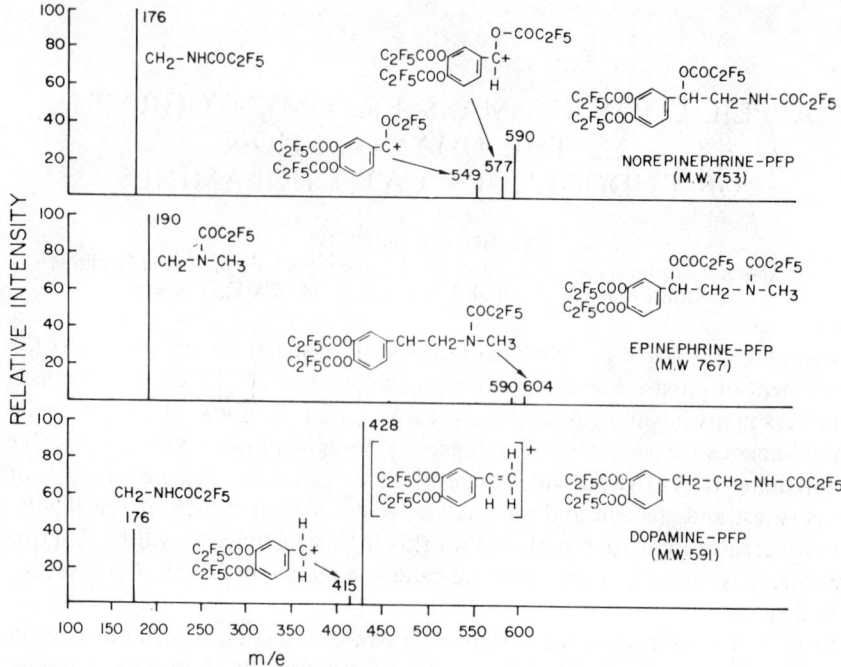

FIG. 1.—Partial mass spectra of catecholamines acylated with pentafluoropropionic anhydride. The base peaks (100 per cent relative abundance) are monitored for quantitation and the other fragments whose theoretical structures are shown are monitored for multiple ion detection (MID) for specificity. The fragment at m/e 590 cannot be monitored for MID since it is present for both norepinephrine and epinephrine and both of these compounds have the same retention time.

TABLE 1. GAS CHROMATOGRAPHIC (GC) RETENTION TIME AND POSITIVE IONS MONITORED
FOR QUANTITATION AND MULTIPLE ION DETECTION (MID) ANALYSIS OF ACYLATED
CATECHOLAMINE DERIVATIVES

PFP amine derivatives	GC retention times (min)	Quantitation m/e	Multiple ion detection m/e (%)*	Fragment ratio
α-Methyl-nor-epinephrine	2·08	190		
Norepinephrine	2·83	190	577 (12)	
			549 (7)	1·7
Epinephrine	2·83	176	190 (100)	
			604 (4)	25
α-Methyl dopamine	4·36	442		
Dopamine	5·83	428	428 (100)	
			415 (5)	20

* (%) refers to the relative intensity of the fragment compared to the most abundant (base) peak which is 100%. MID is not done on internal standards, tissue samples are however, processed without internal standards to insure the absence of any peak at this GC time and m/e due to "biological background".

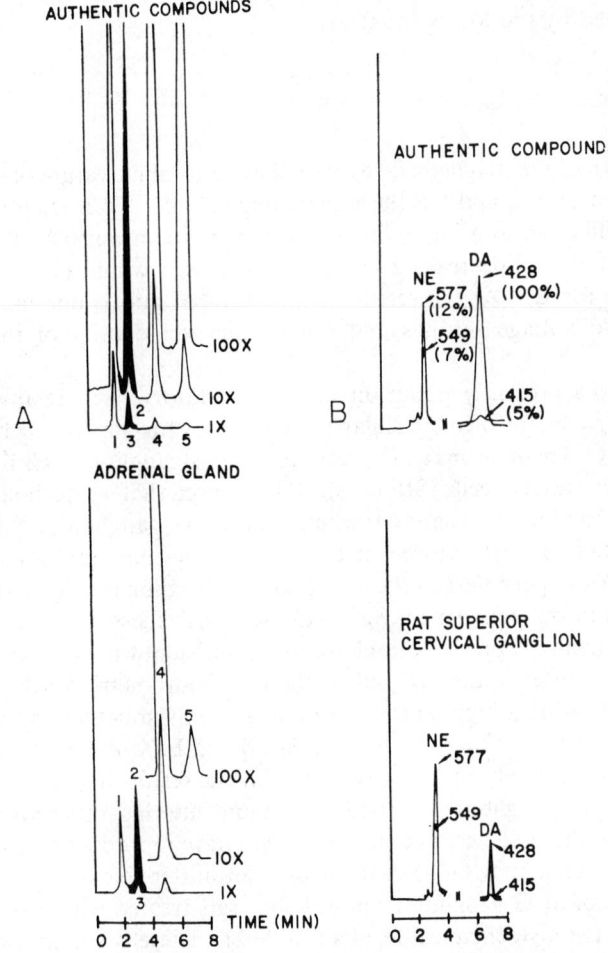

FIG. 2A.—Simultaneous analyses of norepinephrine (NE, 3); epinephrine (E, 2), and dopamine (DA, 5). Internal standards 1 α-methylnorepinephrine and 4, α-methyl-dopamine. For fragments recorded, see Fig. 1 or Table 1. Top record (A) authentic standards E 30 pm; NE 21 pm; DA, 47 pm. Bottom record (A) adrenal gland extract 1 adrenal gland is homogenised in 500 μl. 1 MFA (50 mM ascorbate) and 1 μl sampled for analysis E, 99·2 pm; NE, 50·6 pm; DA 3·32 pm measured. Concentration (nmoles/adrenal pair) E, 81; NE, 53; and DA, 3·4.

FIG. 2B.—Multiple ion detection. Number indicates m/e recorded. Numbers in parentheses indicate relative intensity of each fragment. Top record (B), authentic compounds. Bottom record (B) rat superior cervical ganglion extract. The ratio at m/e 577 and 549 for NE = 1·7 and the ratio at m/e 425 and 415 for DA = 20. The agreement in these ratios to those obtained from authentic compound substantiates the specificity of the measurement.

It is obvious from the GC retention times listed, that NE and E are not chromatographically resolved. It is, however, possible to resolve these compounds mass spectrometrically. NE has its base peak at m/e 176; E does not have any fragment at this m/e. The base peak for E is at m/e 190, and NE lacks any fragment at this m/e. Thus, NE is measured by recording the fragment at m/e 176, and E the fragment at 190. This can be done in a single analysis by alternating the accelerating voltage (V).

The m/e recorded by the MS is equal to:

$$\frac{H^2 r^2}{2V}$$

H^2 is the strength of the magnetic field; r^2 is the radius of curvature of the path of the ions which is a constant; and V is the accelerating voltage. To measure NE (m/e 176) the magnetic field is set to bring in focus m/e 176, r is constant and the accelerating voltage is V initial. By decreasing V, the m/e focussed would be increased (in this case to m/e 190 for E). At the retention time for NE and E one merely alternates between the two voltage settings and obtains the ion density of these fragments (Fig. 2A).

In experiments presently going on in our laboratory, we are investigating the neurotransmitter content of the cellular population of the superior cervical ganglion (SCG) of the rat. The neurons of interest are the postganglionic cell bodies, and the small intensity fluorescent cells (SIF cells). It is believed that a catecholamine released by the SIF cells modulates synaptic transmission in this ganglion (Costa et al.,1961; Eccles and Libet, 1961). More current experiments have shown that the local application of DA hyperpolarises the membrane of ganglion cells (Libet and Tosaka, 1969, 1970) and in this manner may alter the responsiveness of these cells to acetylcholine released from the presynaptic elements. Consistent with the view that DA may be the neurotransmitter in the SIF cells is the histofluorometric work of Bjorklund et al. (1970), indicating a high concentration of DA in sympathetic ganglion cells.

Recent histochemical experiments by Eränkö and Eränkö (1971) have indicated that the treatment of newborn rats with guanethidene results in the loss of all fluorescence from the postganglionic cell bodies, without altering the fluorescence of the SIF cells and that these SFI cells contain DA. We have reproduced these experiments in an attempt to identify by GC-MS the neurotransmitter remaining in the SIF cells. GC-MS analysis of SCG ganglion from newborn rats treated with guanethidine does not substantiate the histochemical results (Table 2). There is still at least 50 per cent of the NE remaining, and the decrease in DA is not statistically significant; this latter point does not, however, refute the theory that DA is contained in the SIF cells. As a control, histochemistry was done on the contralateral SCG of the normal and guanethidine treated rats and the results were the same as those described by Eränkö and Eränkö (1971).

To be sure of the specificity of the GC-MS measurement, multiple ion detection (MID) was done on the ganglion extract. MID takes advantage of the fact that the fragmentation pattern of any one compound is completely reproducible in terms of the

TABLE 2. Effect of postnatal guanethidine treatment on the rat superior cervical ganglion cathecholamine content

	Norepinephrine	Dopamine
	(pmoles/mg protein)	
Control	746 ± 28	275 ± 57
Guanethidine	473 ± 42*	193 ± 84

New born animals were treated days 1–8 with guanethidine (20 mg/kg i.p.). Animals were sacrificed at 90 days of age. Number of animals in each group is 6. Value is the mean ± s.e.
* $P < 0.005$.

relative intensity of each of the fragments to each other. MID varification is done by simultaneously monitoring two characteristic fragments and comparing the relative intensities of these fragments to the fragment ratios obtained from authentic compounds. Only when the ratio values are in agreement is identification considered true. MID analysis for SCG extracted catecholamine showed the NE ratio to be 1·7 and DA 20 (Fig. 2B) thereby giving an extremely high degree of specificity to the measurement.

The most direct approach for determination of the neurotransmitter content of the SIF cells would be to dissect out the SIF cells and measure the content directly. Although we believe we have the sensitivity to do this measurement, we cannot as yet dissect pure SIF cells from the ganglion. The identification of the catecholamines stored in SIF cells was, therefore, approached indirectly by making the assumption that the SIF cells contain only DA and no NE. If this is true, the ratio of DA/NE in total ganglia should be greater than the DA/NE ratio in the postganglionic cell bodies. By cutting serial 10 μm transverse sections through the ganglion, it is possible by fluorescence microscopy to find many areas where three or more successive sections contain only postganglionic cells. In order to measure the NE and DA in sections free of SIF cells, several (10 μm) sections were made and alternate sections taken for fluorescence microscopy or for possible GC-MS analysis. When two fluorescent sections were found to be free of SIF cells, then the middle section saved for GC-MS analysis was taken and the NE and DA content determined (Table 3). The average DA as a percentage of NE in ganglion sections without SIF cells is 23 per cent as compared to 37 per cent in the whole ganglion. It would appear that 41 per cent or 112 pmoles/mg protein DA is in excess and is most probably of SIF cell origin. This calculation is tested against the results with guanethidine treated animals (Table 3) and appears to hold.

TABLE 3. INDIRECT CALCULATION OF THE DOPAMINE CONCENTRATION IN SMALL INTENSELY FLUORESCENT CELLS (SIF CELLS) OF THE RAT SUPERIOR CERVICAL GANGLION

10 μm sections without SIF cells	NE	DA	$\frac{DA}{NE} \times 100$
	(picomole/sample)		
1	1·56	0·399	25
2	2·02	0·376	19
3	0·816	0·212	26

A = Average DA as percentage NE in ganglion sections not containing SIF cells
23%
B = Average DA as percentage NE in whole ganglion with SIF cells
37%

A/B—Non SIF cell DA = 62%
    SIF cell DA   = 38%—pmoles/mg protein

Test of hypothesis:
    Consider non SIF cell DA = 62%
    With guanethidine, NE decrease = 37% ×
    Then guanethidine DA should = 23%
Control DA-23% = 211 pmoles/mg protein
Measured DA in guanethidine treatment = 193 pmoles/mg protein.

Alternate sections were taken for fluorescence microscopy and GC-MS analysis. Sections or parts of sections without SIF cells were pooled and analysed for amine content. No more than three 10 μm sections were analysed per sample. See text for complete discussion.

In summary we have described a method for the simultaneous measurement of NE, E and DA by gas chromatography-mass spectrometry. The sensitivity ($> 10^{-15}$ mole) and the specificity is extremely high. Since the time required to run each analysis is 5–6 min, it is extremely easy to analyse 50 samples a day, thereby generating 150 values if all three catecholamines are being measured. Using this technique, the catecholamine content of the postganglionic elements of the SCG of the rat has been measured. In comparing these values to the values obtained from whole ganglia, the DA content is apparently excessive and possibly of SIF cell origin. This conclusion is tentative and must be substantiated by direct measurement of the catecholamines in the SIF cells.

REFERENCES

Bjorklund A., Cegvell L., Falck B., Ritzen M. and Rosengreen E. (1970) *Acta Physiol. Scand.* **78**, 334–338.

Cattabeni F., Koslow S. H. and Costa E. (1972) *Science* **178**, 166–168.

Costa E., Revzin A. M., Kuntzman R., Specter S. and Brodie B. B. (1961) *Science* **133**, 1822–1823.

Eccles R. M. and Libet B. (1961) *J. of Physiol.* **157**, 484–503.

Eränkö L. and Eränkö O. (1971) *Acta Pharmacologica (Kbh)* **30**, 403–416.

Koslow S. H., Cattabeni F. and Costa E. (1972) *Science* **176**, 177–180.

Libet B. and Tosaka T. (1969) *J. Neurophysiol.* **321**, 43–50.

Libet B. and Tosaka T. (1970) *Proc. Natn. Acad. Sci.* **67**, 667–673.

Frontiers in Catecholamine Research 1973, pp. 1091 to 1096. Pergamon Press. Printed in Great Britain.

THE IDENTIFICATION OF DEPRESSED PATIENTS WHO HAVE A DISORDER OF NE METABOLISM AND/OR DISPOSITION

JAMES W. MASS[1], HAROUTUNE DEKIRMENJIAN[2], and FRANK JONES[2]

[1]Department of Psychiatry, Yale University School of Medicine, 333 Cedar Street, New Haven, Connecticut 06510, and [2]Illinois State Psychiatric Institute, 1601 West Taylor Street, Chicago, Illinois 60612, U.S.A.

INTRODUCTION

3-METHOXY-4-hydroxyphenethyleneglycol (MHPG) is a naturally occurring catecholamine metabolite which was initially discovered by AXELROD et al. (1959). A number of subsequent reports have shown that MHPG is the major metabolite of brain norepinephrine (NE) (MANNARINO et al., 1963; GLOWINSKI et al., 1965; RUTLEDGE and JONASON, 1967; MAAS and LANDIS, 1968; SCHANBERG et al., 1968a, b; SHARMAN, 1969). In addition it has been shown that either direct stimulation of the locus coeruleus or stress produces an increased turnover of NE and an increase in the sulphate conjugate of MHPG in the rat cerebrum, and that these effects are abolished by ablation of the locus coeruleus suggesting that MHPG in brain may reflect functional activity of central noradrenergic neurons (KORF et al., 1973a, b, c). Finally, there is some evidence which suggests that a significant fraction of urinary MHPG has its origins in the metabolism of NE within brain, whereas urinary NE, normetanephrine (NM), and perhaps 3-methoxy-4-hydroxymandelic acid (VMA) originate in pools of catecholamines outside the central nervous system (GLOWINSKI et al., 1965; MAAS and LANDIS, 1965, 1968, 1971; MAAS et al., 1972b, 1973). It should be noted, however, that while there is general agreement that MHPG is the major metabolite of brain NE, definitive information is not available as to the exact amount of MHPG excreted in urine which is derived from brain (BREESE et al., 1972). Despite this uncertainty there is the possibility that urinary MHPG may reflect directly or indirectly NE metabolism in brain, and for this reason it has seemed that assaying urinary MHPG along with other catecholamine metabolites which have their origins outside the central nervous system might be a reasonable strategy for beginning clinical investigations of the catecholamine hypothesis of the affective disorders (SCHILDKRAUT, 1965; SCHILDKRAUT and KETY, 1967; BUNNEY and DAVIS, 1965). In this paper a brief summary of the results of some clinical studies of depressed patients in which this approach has been used will be noted. In addition, attention will be focussed upon pharmacological and clinical methods by which a subgroup of depressed patients who are thought to have a disorder of NE metabolism and/or disposition may be identified. Finally, reports bearing upon the role which state variables, particularly activity, may have in affecting the excretion of NE and its metabolites into urine by normal subjects and depressed patients will be noted.

DIAGNOSIS AND CATECHOLAMINE METABOLITE EXCRETION

In an initial pilot study of MHPG excretion by depressed patients it was found that depressed patients as a group excreted significantly less MHPG than did normal subjects, whereas the urinary NM and metanephrine (M) levels for the two groups were the same (Maas et al., 1968). Subsequently, Greenspan et al. (1970), Bond et al. (1972) and Jones et al. (1973c) published data indicating that during periods of depression patients excreted significantly less MHPG than they did during episodes of euthymia or mania. It is also of interest that it has been reported that MHPG in the CSF of depressed patients is significantly less than that of control subjects (Gordon and Oliver, 1971; Post et al., 1973). However, another group did not find decrements in the CSF MHPG of depressed patients, although they did note a tendency for CSF MHPG levels to be elevated during mania (Wilks et al., 1972; Shopsin et al., 1973). More recently, in a study containing more subjects, 19 healthy male subjects, 20 seriously depressed male patients, 21 healthy female subjects, and 48 seriously depressed female patients were compared as to the urinary excretion of NM, M, VMA and MHPG. All groups were age matched, were on a VMA exclusion diet, and were free of medication for at least three weeks prior to the collection of two or more separate 24-hr urine specimens which were assayed for the noted catecholamine metabolites. The value for a given metabolite for each subject was taken as the average of the separate determinations. It was found that while male and female patients and controls did not differ as to urinary NM, M, or VMA, there were statistically significant decrements in MHPG excretion by the patient group. These data are presented in Table 1 (Dekirmenjian et al., 1973). There thus seems to be some agreement among different investigators that some patients do in fact have decrements in MHPG excretion during periods of depression.

TABLE 1.

	Males			Females		
	patients N = 20	controls N = 19	P	patients N = 48	controls N = 21	P
M	114 ± 22	86 ± 9	NS	97 ± 6	83 ± 9	NS
NM	172 ± 17	196 ± 17	NS	214 ± 17	223 ± 33	NS
VMA	4601 ± 318	4165 ± 656	NS	4378 ± 257	4612 ± 408	NS
MHPG	1394 ± 89	1674 ± 117	<·05	1155 ± 58	1348 ± 65	<·05

Values are expressed as $\mu g/24$ hr.

Inspection of the data presented in the initial report (Maas et al., 1968), as well as subsequent findings (Dekirmenjian et al., 1973; Prange et al., 1971; Schildkraut et al., in press) and see Table 1), indicate that not every depressed patient excretes less than normal quantities of urinary MHPG. It has been found, however, that those patients who excrete less than normal quantities of MHPG can be identified pharmacologically and clinically. This data may be briefly summarised as follows. In two separate studies (N = 12 patients, first study; and N = 16 patients, second study), it was found that a low *pretreatment* urinary MHPG predicted a favorable response to treatment with desipramine or imipramine, whereas pretreatment values for NM, M, or VMA were not significantly related to treatment response.* Further, patients

---

* Patients were not given other specific antidepressant treatments such EST or monoamine oxidase inhibitors, and therefore the relationship of pretreatment MHPG to improvment with these types of therapies is unknown.

who excreted less than normal quantities of MHPG responded with a mood elevation when given d-amphetamine, whereas those patients who excreted normal or greater than normal quantities of MHPG had either no response or a worsening of mood after d-amphetamine administration. With either d-amphetamine, desipramine, or imipramine induced mood elevations, there were modest increments in MHPG excretion, whereas those patients who did not respond had decrements in MHPG excretion (FAWCETT et al., 1972; MAAS et al., 1972a). In contrast, it has been reported that depressed patients who excrete normal or greater than normal quantities of MHPG respond well to amitryptilline (SCHILDKRAUT et al., 1971). These findings are interpreted as being generally consistent with and supportive of the possibility that there is a functional deficiency of central nervous system NE in a subgroup of depressed patients and that these patients may be identified by pretreatment urinary MHPG levels.

While these pharmacological-behavioural methods of identifying patients who excrete less than normal amounts of MHPG are of particular interest in terms of the biochemical genesis of depressive illness, they are time consuming, do not lend themselves readily to use by other investigators, and make large N studies difficult if not impossible. For these reasons, investigations as to the possibility that those patients who excrete less than normal amounts of MHPG might be easily and quickly identified by explicit clinical criteria are of particular interest. The results of studies from our group dealing with this issue of the relationship between clinical classifications of depressed patients and catecholamine metabolite excretion have been presented in detail elsewhere (JONES et al., 1973b) and are summarised in that which follows. The clinical classifications chosen for study were (1) severe agitated and retarded depressions (2) unipolar (single episode and recurrent types) or bipolar illness (types 1 and 11) (3) psychotic or neurotic depressions (with psychosis being defined by the presence of a thought disorder), and (4) the method used by the Washington University group, i.e., primary affective disorder, secondary affective disorder, or affective disorder undiagnosed type (ROBBINS and GUZE, 1971; FEIGHNER et al., 1972; ROBBINS et al., 1972; BAKER et al., 1971). The sample studied consisted of 32 female patients having depressions of sufficient severity to warrant hospitalisation and 21 outpatient, healthy women of approximately the same age who were clinically judged to be free of gross psychopathology and significant depression. All subjects were off medication for at least three weeks prior to study and were maintained on a VMA exclusion diet. A minimum of two separate 24-hr urines were collected and assayed for M, NM, VMA and MHPG without knowledge of diagnosis. It was found that (i) urinary levels of M, NM, or VMA were not significantly different between patients and controls, nor were there significant differences for these metabolites between any of the subgroups. (ii) There were no significant differences in urinary MHPG levels between agitated depressives or retarded depressives in the total sample, and normal subjects. (iii) The presence or absence of a thought disorder per se, i.e., psychosis, does not distinguish patients from controls in terms of MHPG excretion. (iv) The patient group diagnosed as having unipolar illness ($N = 27$) did not differ significantly from control subjects in urinary MHPG. (v) The mean MHPG excretion by the bipolar group ($N = 5$) was 916 ± 43 μg MHPG/24 hr, and that for the control subjects was 1328 ± 85 μg/24 hr, and the difference between these groups is statistically significant ($P < 0.025$). This finding is generally consistent with other data which indicates that bipolar type depressed patients excrete significantly less MHPG than do patients with depressions of a characterological type (Schildkraut *et al.*, in press). (vi) As noted previously, the *total* patient group excreted less MHPG than did the healthy subjects, but the level of statistical significance was weak ($P < 0.05$). In contrast, those depressed patients who had been classified as having a primary affective disorder ($N = 21$) excreted 1032 ± 63 μg MHPG/24 hr, whereas the control subjects ($N = 21$) excreted 1348 ± 65 μg/24 hr, and here the difference was highly significant ($P = 0.0005$). When those subjects who were diagnosed as having primary affective disorders were subdivided into bipolar, unipolar recurrent, unipolar single episode types and were compared with healthy subjects by ANOVA, a significant *F* was obtained, and the use of the Duncan's Multiple Range Test indicates that none of the subtypes of the primary affective disorder group differ from each other, whereas all of them are significantly different from the control group. This data is presented in Table 2.

Table 2.

	Primary affective disorders			Comparison group
	Bipolar ($N = 5$)	Unipolar recurrent ($N = 9$)	Unipolar single epsiode ($N = 7$)	($N = 21$)
MHPG (μg/24 hr)	916 ± 152	1066 ± 86	1073 ± 140	1348 ± 65

No statistically significant differences between the primary affective disorder group or any of its subtypes and control subjects for VMA, NM, or M were found. The primary affective disorder group did not differ significantly from the control subjects in age [patients 48 ± 3 (range 24–62) and controls 42 ± 2 (range 21–59)]. Within this primary affective disorder depressed group there were 7 patients who were markedly agitated and 10 patients who were retarded. The former group excreted 1030 ± 108 μg MHPG/24 hr, and the latter 986 ± 65 μg MHPG/24 hr. An ANOVA in which these two groups and healthy subjects were compared indicated statistically significant differences, with the agitated and retarded groups both being significantly different from the controls but not from each other. Finally, those depressed patients who as a group were diagnosed as having affective disorders, undiagnosed type ($N = 11$) excreted 1322 ± 174 μg MHPG/24 hr. The variance for the undiagnosed group is quite large, and their mean is not significantly different from the means of either the primary or control groups. As measured by Nurse's depression ratings, Hamilton scores, or an item on the BPRS, there were no differences in the severity of depression between the primary and undiagnosed affective disorder groups.

In brief, those patients who were classified as having primary affective disorders, depressed type, had levels of urinary MHPG which were clearly and consistently less than normal. This relationship between the diagnosis of primary affective disorders and low urinary MHPG has implications for both improved treatment and biochemical studies of depression.

THE PROBLEM OF STATE VARIABLES AND ARTIFACTUAL RESULTS

Data presented elsewhere indicate that age, smoking, 24-hr urine volume, height, weight, body surface area, and creatinine are not significantly related to urinary MHPG (24-hr specimens) in patients or controls (DEKIRMENJIAN et al., 1973). There are sex differences in MHPG excretion, as well as pharmacological effects of tricyclic drugs (DEKIRMENJIAN et al., 1973; PRANGE et al., 1971; MAAS et al., 1972a; SCHILDKRAUT et al., 1965). All patients and control subjects in the study reported above were free of medication for three weeks, however, and in these summarised clinical classification studies (JONES et al., 1973b) all subjects were females, and for this reason it seems unlikely that the above variables are operative in producing the cited results.

It has been reported that activity may alter urinary MHPG levels (EBERT et al., 1972), but our data which indicates that this variable cannot alone account for the cited relationship between urinary MHPG and certain types of depression may be summarised as follows. (i) Five healthy, young male subjects were exercised for 2 hr on two of three consecutive days using standardised isometric and isotonic exercse procedures. The exercise protocols were moderately severe in that marked fatigue was produced in all subjects. Urinary levels of NE, NM, M and MHPG and plasma MHPG were measured for the two hour periods before, during and after exercise. A 3- to 5-fold increase in urinary NE was present after both exercise procedures. No increase in urinary MHPG or M was observed during or 2 hr after exercise. Slight increases of plasma MHPG and urinary NM were observed during one of the two exercise periods compared to the 2 hr preceding exercise (GOODE et al., in press) (ii) As noted above, agitated and retarded depressed patients were chosen for study as to MHPG excretion. Only retarded patients who were clinically judged to be free of an agitation component to their illness were selected. One patient was classified as having an agitated stupor and one a retarded stupor. No significant differences in urinary quantities of MHPG between these two types of patients within the group as a whole or within the primary affective disorder group were found (JONES et al., 1973a, b). (iii) A patient who shifted from depression into mania on two occasions and from mania into depression once has been followed longitudinally in terms of 24-hr urinary MHPG or NM levels. It was found that the increments in MHPG and possibly NM preceded by as much as four days the shifts from a retarded depression into a hyperactive manic state (JONES et al., 1973c).

SUMMARY

The data given here lead us to conclude that in a subgroup of depressed patients who may be identified clinically (primary affective disorders), biochemically (low MHPG), and pharmacologically (good response to desipramine, imipramine and amphetamine), there is a significant alteration in the metabolism and/or disposition of NE which is integral to the depressive illness *per se*.

REFERENCES

AXELROD J., KOPIN I. J., and MANN J. D. (1959) *Biochem. Biophys. Acta* **36,** 576.
BAKER M., DORZAB J., WINOKUR G. and CADORET R. J. (1971) *Compreh. Psychiat.* **12,** 354–365.
BOND P. A., JENNER F. A., SAMPSON G. A. (1972) *Psychol. Med.* **2,** 81.
BREESE G. R., PRANGE A. J., HOWARD J. L., LIPTON M. A., McKINNEY W. T., BOWMAN R. E. and BUSHNELL P. (1972) *Nature New Biol.* **240,** 286–287.
BUNNEY W. E. JR. and DAVIS J. M. (1965) *Archs. Gen. Psychiat.* **13,** 483–494.

Dekirmenjian H., Maas J. and Fawcett J. (1973) Presented at Am. Psychiat. Assoc. Ann. Meeting, Hawaii, May, 1973.

Ebert M. H., Post R. M. and Goodman F. K. (1972) *Lancet* **ii,** 706.

Fawcett J., Maas J. W. and Dekirmenjian H. (1972) *Archs. Gen. Psychiat.* **26,** 246–251.

Feighner J. P., Robbins E., Guze S. B., Woodruff R. A., Winkour G. and Munoz R. (1972) *Archs. Gen. Psychiat.* **26,** 57–63.

Glowinski J., Kopin I. J., Axelrod J. (1965) *J. Neurochem.* **12,** 25–30.

Goode D., Dekirmenjian H., Meltzer H. and Maas J. (1973) *Archs. Gen. Psychiat.* In press.

Gordon E. K. and Oliver J. (1971) *J. Clin. Chim. Acta.* **35,** 145–150.

Greenspan J., Schildkraut J. J., Gordon E. K., Baer L., Arnoff M. S., and Durell J. (1970) *J. Psychiat. Res.* **7,** 171.

Jones F. D., Maas J. W. and Dekirmenjian H. (1973) Presented at Annual Psychosomatic Soc. meetings, Denver, Colorado, April 1973. (1973a)

Jones F., Maas J., Dekirmenjian H. and Sanchez J. (1973b) Depressive syndromes and catecholamine metabolites. Presented at Amr. Psychiat. Ass. Ann. meeting, Hawaii, May, 1973.

Jones F. D., Maas J. W., Dekirmenjian H. and Fawcett J. A. (1973c) *Science* **179,** 200–202.

Korf J., Aghajanian G. K. and Roth R. H. (1973a) *Neuropharmacology.* In press.

Korf J., Roth R. H., Aghajanian G. K. (1973a) *Europ. J. Pharmacol.* In press.

Korf J., Aghajanian G. K. and Roth R. H. (1973) *Europ. J. Pharmacol.* In press.

Maas J. W. and Landis D. H. (1965) *Psychosom. Med.* **26,** 339.

Maas J. W. and Landis D. H. (1968) *J. Pharmacol. Exp. Ther.* **163,** 147–162.

Maas J. W. and Landis D. H. (1971) *J. Pharmacol. Exp. Ther.* **177,** 600.

Maas J. W., Fawcett J. A. and Dekirmenjian H. (1968) *Archs. Gen. Psychiat.* **19,** 129–134.

Maas J. W., Fawcett J. and Dekirmenjian H. (1972a) *Archs. Gen. Psychiat.* **26,** 252–262.

Maas J. W., Dekirmenjian H., Garver D., Redmond D. E. Jr. and Landis D. H. (1972b) *Brain Res.* **41,** 507–511.

Mass J. W., Dekirmenjian H., Garver D., Redmond D. E. Jr. and Landis D. H. (1973) *Europ. J. Pharmacol.,* In press.

Mannarino E., Kirshner N. and Nashold B. S. (1963) *J. Neurochem.* **10,** 373.

Prange A. J., Wilson I. C., Knox A. E., McClane T. K., Breese G. R., Martin B. R., Altop L. B. and Lipton M. A. (1971) *Brain Chemistry and Mental Disease,* Plenum Press, New York.

Post R. M., Gordon E. K., Goodwin F. K. and Bunney W. E. Jr. (1973) *Science* **179,** 1002–1003.

Robbins E. and Guze S. B. (1971) In: *Recent Adbances in Psychobiology of the Depressive Illnesses.* (Proceedings of a Workshop sponsored by NIMH. Eds. Thomas A. Williams, M. M. Katz, and J. A. Shielf, Jr.) U.S. Govt. Printing Off. pp. 283–293.

Robbins E., Munoz R. A., Marten S., and Gentry K. A. (1972) In: *Disorders of Mood.* Zubin J. and Freyhan F. A. Eds.) pp. 33–45. The Johns Hopkins Press, Baltimore.

Rutledge C. O. and Jonason J. (1967) *J. Pharmacol. Exp. Ther.* **157,** 493.

Schanberg S. M., Breese G. R., Schildkraut J. J., Gordon E. K. and Kopin I. J. (1968b) *Biochem. Pharmacol.* **17,** 2006.

Schanberg S. M., Schilkraut J. J., Breeze G. R. and Kopin I. J. (1968a) *Biochem. Pharmacol.* **17,** 247–254.

Schildkraut J. J. (1965) *Am. J. Psychiat.* **122,** 509–522.

Schildkraut J. J. and Kety S. S. (1967) *Science* **156,** 21–30.

Schildkraut J. J., Gordon E. K. and Durell J. (1965) *J. Psychiat. Res.* **3,** 213–228.

Schildkraut J. J., Draskoczy P. R., Gershon E. S. Reich P. and Grab E. L. (1971) *Brain Chemistry and Mental Disease,* pp. 215–236. Plenum Press, New York.

Schildkraut J. J., Keeler B. A., Hartmann E. and Papousek M. *Science* In press.

Sharman D. F. (1969) *Br. J. Pharmacol.* **36,** 523.

Shopsin B., Wilk S., Gershon S., Davis K. and Suhl M. (1973) *Archs. Gen. Psych.* **28,** 230–233.

Wilk S., Shopsin B., Gershon S., Suhl M. (1972) *Nature, Lond.* **235,** 440–441.

Frontiers in Catecholamine Research 1973, pp. 1097 to 1100. Pergamon Press. Printed in Great Britain.

MONOAMINE NEURON SYSTEMS IN THE NORMAL AND SCHIZOPHRENIC HUMAN BRAIN: FLUORESCENCE HISTOCHEMISTRY OF FETAL, NEUROSURGICAL AND POST MORTEM MATERIAL*

LARS OLSON, BO NYSTRÖM and ÅKE SEIGER

Department of Histology, Karolinska Institutet, Stockholm, Sweden

THREE sources of material are presently being used to study the distribution of mono-amine neuron systems in the human brain using Falck–Hillarp fluorescence histo-chemistry (FALCK et al., 1962). The possibility of using fetal brain material from abortions was shown by OLSON and UNGERSTEDT (1970) and has recently resulted in detailed mapping studies (NOBIN and BJÖRKLUND, 1973, OLSON et al., 1973a). Neuro-surgery material has demonstrated the presence of catecholamine (CA) nerve terminals in the cerebral and cerebellar cortices (NYSTRÖM et al., 1972). Using post mortem material, the presence of CA terminals in selected areas was shown by CONSTANTINIDIS et al. (1969) and by DE LA TORRE (1972). Our recent studies have shown that the post mortem time limit for histochemical analysis given by DE LA TORRE, 45 min, can be extended by several hours, especially when using in vitro incubations in amine solutions. Thus aspects on the distribution of CA cell bodies and of CA and 5-hydroxytryptamine (5-HT) nerve terminal areas have been described (OLSON et al., 1973b).

While fetal material is excellent for mapping studies, due to the well-known increased histochemical detectability of non-terminal monoamine axons during de-velopment, the possibility of using post mortem brain material no doubt provides the most important source of material in order to gain insights into the possible involve-ment of the monoamine neurons in various diseases. The need for fluorescence histochemistry of the schizophrenic brain has been directly expressed by PLUM (1972). It provides a possibility to test the hypothesis of Stein and Wise (see STEIN, WISE and BERGER, 1972) of a destruction of noradrenergic nerves by endogenous 6-OH-dopamine (6-OH-DA) formation in schizophrenia. In the following we will briefly describe methodology and results mainly of the post mortem analyses.

METHODOLOGY

For distribution of analysed cases of schizophrenia, see Table 1. The histo-chemistry of two of the cases (F9 + F15) have been briefly described elsewhere, together with our control material (see OLSON et al., 1973). The procedure followed was a so called "partial autopsy". Following death, as diagnosed by an independent physician, and after permission of the relatives, dissection was commenced as soon as possible. This procedure has been approved upon by the staff of the Ulleråker Hospital.

* Supported by the Swedish Medical Research Council (04X-3185), "Magnus Bergvalls Stiftelse" and "Karolinska Institutets Forskningsfonder". The post mortem brains were obtained through the kind help of Dr. L. Wetterberg, Psychiatric Research Center, Ulleråker Hospital, Uppsala.

We use smears (Olson and Ungerstedt, 1970b) and freeze-dried material (Olson and Ungerstedt, 1970a). Fluorescence microscopy of the adult human brain is complicated by the heavy accumulations of lipofuscin in neurons and glia having a strong yellow autofluorescence. The specific paraformaldehyde-induced green and yellow neuronal fluorescence is, however, easily recognized in the microscope due to color, morphology, diffusion and fading characteristics, especially using a narrow band excitation filter (TAL405) and a Zeiss 47 barrier filter. *In vitro* incubations of thin slices of tissue with α-methyl-NA (10^{-6} or 10^{-5} M) sometimes following preincubation in 6-OH-DA (10^{-4} M) or DMI (10^{-4} M) were carried out according to Hamberger (1967). Nerve densities and fluorescence intensities were estimated semiquantitatively. For the reliability of such estimations as correlated to biochemistry see Olson *et al.* (1968), Olson and Malmfors (1970), Jonsson (1971) and Lidbrink and Jonsson (1971).

RESULTS AND DISCUSSION

We conclude from our three sources of material (about 40 fetal brains, 75 neurosurgery operations and 20 *post mortem* brains) that the basic cytoarchitecture of the monoamine neurons present in the rat (see Fuxe *et al.*, 1973) is present also in man. Thus, there are descending bulbospinal noradrenaline (NA) and 5-HT systems, as well as ascending NA and 5-HT systems reaching almost all areas of the brain, having their cell bodies in the lower brain stem. Furthermore, the predicted nigro-neostriatal dopamine (DA) system with cell bodies in the substantia nigra area and a dense arrangement of terminals in the nuc. caudatus and putamen has been visualized.

Endogenous CA fluorescence located to varicose nerve terminals was a constant finding in subcortical areas of all the *post mortem* brains, the largest *post mortem* time interval being 7·3h. Like in the rat, there was a mixture of thin fibers with small varicosities and thick fibers with large, strongly fluorescent varicosities. In addition, scattered fibers showed extremely large "varicosities" with an intense yellow-green fluorescence not seen in the rodent brain. Especially dense innervation patterns were found in hypothalamic areas, e.g. periventricularly, in the tuber cinereum and in nuc. supraopticus, but moderate numbers of CA nerve terminals were found throughout the central nervous system from the filum terminale, and the spinal cord, through the lower brain stem, the hypothalamic areas, amygdala and in the olfactory bulb, to give some examples.

The pineal glands were yellow fluorescent, the habenulae richly provided by CA and 5-HT fibers.

The nuc. caudatus and putamen were always diffusely green fluorescent. Following incubation in α-m-NA an extremely dense innervation by fine CA varicosities were disclosed in these areas, identical in appearance to the DA innervation of the neostriatum as seen in rats.

Untreated cerebral and cerebellar cortex showed no or very few CA nerve terminals *post mortem*. A marked concentration dependent uptake of α-m-NA could always be demonstrated. Following incubation in α-m-NA the typical (as compared to rats) thin varicose fibers were found in all layers of the cerebral cortex and in the molecular layer of the cerebellar cortex. This uptake was inhibited by 6-OH-DA and by DMI. 5-HT nerve terminals were likewise found in the cerebral cortex.

TABLE. 1. DISTRIBUTION OF ANALYZED CASES.
(All four patients were repeatedly diagnosed by different psychiatrists as having typical schizophrenia with a relatively early onset and a long duration.)

Case	Age	Sex	Duration of disease (years)	Analyzed material	Post mortem time (min)
F9	85	female	45	post mortem brain	100
F15	63	female	38	post mortem brain	65
F19	69	male	45	post mortem brain	70
Op.*	53	female	32	cortex cerebri re-sected at tumour neurosurgery	—

* This patient had also undergone a frontal lobotomy.

In case "op." (Table 1), where fresh cerebral cortex from a schizophrenic was obtained at neurosurgery, both CA and 5-HT nerve terminals showed endogenous fluorescence. Work in progress shows that these terminals are able to take up and accumulate labelled NA and 5-HT, respectively, and to release their labelled amines upon field stimulation *in vitro* (FARNEBO, NYSTRÖM, OLSON and SEIGER, in preparation).

Green fluorescent cell bodies, simultaneously heavily loaded by neuromelanin were found in the locus coeruleus and substantia nigra.

A sympathetic adrenergic innervation of larger intracranial blood vessels was constantly found.

In view of the Stein and Wise theory (cited above) it is important to note that **we** found *no differences between the schizophrenic and non-schizophrenic brains* (Fig. 1).

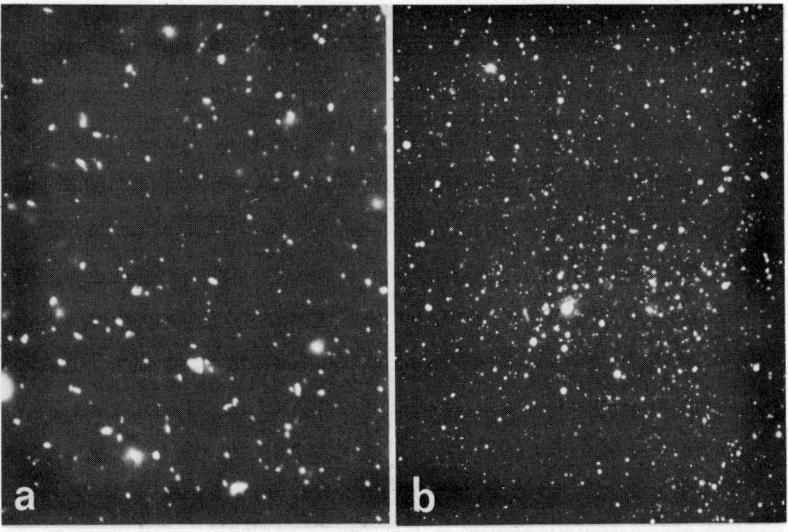

FIG. 1.—Fluorescence micrographs of smears from a schizophrenic brain (case F19). (a) A moderate number of green fluorescent varicosities is seen in the cerebral cortex after incubation in α-m-NA (10^{-5} M) $\times 300$. (b) A high number of varicosities showing endogenous CA fluorescence seen in the hypothalamus close to the ventricular surface. Very small, sharply outlined dots represent unspecific lipofuscin fluorescence in both pictures $\times 120$.

Thus, although our data are only preliminary, not fully quantitative, and does not discriminate between DA and NA, they clearly demonstrate an abundant presence of NA-like nerve terminals in all investigated areas of the schizophrenic brain. The schizophrenic brain likewise seem to contain an amount of DA and 5-HT nerve terminals similar to that of the non-schizophrenic and to have a well developed sympathetic adrenergic blood vessel innervation.

REFERENCES

CONSTANTINIDIS J., TISSOT R., DE LA TORRE J. C. and GEISSBUHLER F. (1969) *Path-Biol.* **17**, 361–363.
DE LA TORRE J. C. (1972) *Acta Neuropathol. (Berl.)* **21**, 165–168.
FALCK B., HILLARP N.-Å., THIEME G. and TORP A. (1962) *Histochem. Cytochem.* **10**, 348–354.
FUXE K., HÖKFELT T., OLSON L. and UNGERSTEDT U. (1973) In *The Pharmacology of the extrapy-ramidal system.* (HORNYKIEWICZ O., Ed.). *Int. Encyclopedia of Pharmacol. and Therapeut.* Pergamon Press, Oxford.
HAMBERGER B. (1967) *Acta physiol. scand. (Suppl.* **295**) 1–56.
JONSSON G. (1971) *Progr. Histochem. Cytochem.* **2**, 299–334.
LIDBRINK P. and JONSSON G. (1971) *J. Histochem. Cytochem.* **19**, 747–757.
NOBIN A. and BJÖRKLUND A. (1973) *Acta physiol. scand.* (Suppl. **388**) 1–40.
NYSTRÖM B., OLSON L. and UNGERSTEDT U. (1972) *Science* **176**, 924–926.
OLSON L., BORÉUS L. O. and SEIGER Å. (1973) *Z. Anat. Entwickl.-Gesch* **139**, 259–282.
OLSON L., HAMBERGER B., JONSSON G. and MALMFORS T. (1968) *Histochemie* **15**, 38–45.
OLSON L., NYSTRÖM B. and SEIGER Å. (1973) *Brain Res.* (in press).
OLSON L. and UNGERSTEDT U. (1970a) *Histochemie* **22**, 8–19.
OLSON L. and UNGERSTEDT U. (1970b) *Brain Res.* **17**, 343–347.
PLUM F. (1972) In *Prospects for Research on Schizophrenia* (KETY S. S. and MATTYSSE S., Eds.) *Neurosciences Research Program Bulletin* **10**, 384–388.
STEIN L., WISE C. D. and BERGER B. D. (1972) In *The Chemistry of Mood, Motivation and Memory* (MCGAUGH J. L., Ed.) pp. 81–103. Plenum, New York.

Frontiers in Catecholamine Research 1973, pp. 1101 to 1107. Pergamon Press. Printed in Great Britain.

METABOLISM OF DOPAMINE AND L-DOPA IN HUMAN BRAIN

OLEH HORNYKIEWICZ

Department of Psychopharmacology, Clarke Institute of Psychiatry and
Department of Pharmacology, University of Toronto, Toronto, Canada

INTRODUCTION

STUDIES concerned directly with the metabolism of dopamine (DA) in the human brain have been limited largely to the use of post mortem material. Despite this limitation, the results leave little doubt that in the human brain the formation and degradation of DA (and other biogenic amines) follows routes well established in laboratory animals. Detailed studies in human brain material obtained at autopsy demonstrated that the distribution pattern of DA and other monoamines, as well as their metabolites, was quite similar to the pattern found in freshly obtained animal brains (cf. HORNYKIEWICZ, 1966, 1972a,b, 1973). Differences between human and animal material in respect to the actual concentrations of the compounds detected are well within the range of interspecies variations observed in fresh animal brain material. An obvious point of caution are certain enzymes whose activity may decline sharply after death. However, animal studies show that the enzymes concerned with the synthesis and catabolism of catecholamines (CA) are comparatively stable in this respect. Therefore, there is at present no reason to doubt that human post mortem brain material represents a valuable and valid source of information on the behaviour and functional role of biogenic amines.

METABOLISM OF DA IN THE NORMAL HUMAN BRAIN

As in all mammalian species studied, in man the nuclei of the basal ganglia complex, notably the caudate nucleus and putamen (these constitute the corpus striatum or striatum), the substantia nigra and the globus pallidus, contain a major portion (80 per cent or more) of the total brain DA (SANO et al., 1959; EHRINGER and HORNYKIEWICZ, 1960; BERTLER, 1961) and its metabolic end product homovanillic acid (HVA) (BERNHEIMER, 1964) (Table 1). The above nuclei are also among those brain regions containing the highest activites of the CA-synthesising enzymes L-tyrosine hydroxylase (TH) (VOGEL et al., 1969; LLOYD, 1972) and aromatic L-amino acid (L-DOPA) decarboxylase (DOPA D) (LLOYD and HORNYKIEWICZ, 1972) (Table 1) as well as the catabolising enzymes monoamine oxidase (MAO) (BIRKHÄUSER, 1940) and catechol-O-methyl transferase (COMT) (VOGEL et al., 1969; LLOYD, 1972). In the caudate nucleus and putamen the concentration of HVA is in the same range as that of DA; in contrast, in the substantia nigra and the globus pallidus, the levels of HVA are 5–10 fold higher than those of the parent compound (cf. Table 1). The functional significance of these inter-regional differences in the metabolic rate of DA is not yet established, but it is logical to assume that they reflect the fact that in the striatum DA is contained exclusively in nerve terminals whereas in the substantia nigra this amine is found in neuronal pericarya only. This suggests the possibility

TABLE 1. NEUROCHEMISTRY OF BASAL GANGLIA DISORDERS: PARKINSON'S
DISEASE (PD) AND HUNTINGTON'S CHOREA

		Caudate nucleus	Putamen	Substantia nigra	Globus pallidus
DA	Controls	2·64 ± 0·30 (28)	3·44 ± 0·29 (28)	0·49 ± 0·09 (5)	0·42 ± 0·08 (4)
	PD	0·43 ± 0·09 (13)	0·04 ± 0·01 (13)	0·07 ± 0·01 (10)	0·10 (3)
	mild akinesia	0·58 ± 0·12 (13)	0·44 ± 0·21 (13)	—	—
	marked akinesia	0·22 ± 0·08 (9)	0·05 ± 0·02 (9)	—	—
	HemiP right side	1·25	0·93	<0·01	—
	left side	0·59	0·13	<0·01	—
	Huntington's Chorea	1·56 ± 0·20 (10)	2·71 ± 0·35 (10)	0·51 ± 0·15 (5)	—
HVA	Controls	3·23 ± 0·27 (8)	4·29 ± 0·68 (8)	1·79 ± 0·18 (5)	2·12 ± 0·27 (8)
	PD	1·05 ± 0·16 (13)	0·89 ± 0·12 (13)	0·41 ± 0·08 (9)	0·77 ± 0·12 (12)
	mild tremor	1·68 ± 0·35 (7)	2·03 ± 0·55 (7)	—	1·71 ± 0·36 (7)
	marked tremor	1·26 ± 0·20 (11)	1·07 ± 0·22 (11)	—	0·70 ± 0·12 (10)
	Huntington's Chorea	2·02 ± 0·48 (8)	3·88 ± 0·63 (8)	2·09 ± 0·25 (3)	2·02 ± 0·27 (6)
TH	Controls	11·0 ± 0·7 (16)	10·2 ± 0·4 (17)	12·0 ± 2·3 (9)	7·7 ± 1·0 (10)
	PD	7·8 ± 1·1 (11)	6·5 ± 0·8 (12)	7·9 ± 0·5 (3)	4·8 (2)
DOPA D	Controls	364 ± 95 (19)	432 ± 109 (18)	549 ± 294 (15)	22 ± 3 (9)
	PD	54 ± 14 (13)	32 ± 7 (13)	21 ± 6 (10)	18 ± 3 (12)

All values are means ± S.E.M. (number of cases in brackets); dopamine (DA) and homovanillic acid (HVA) in μg/g, L-tyrosine hydroxylase (TH) and L-DOPA decarboxylase (DOPA D) in nmol/ 100 mg protein/2 hr. HemiP = Hemiparkinsonism (symptoms on right side of the body). The data are taken from: EHRINGER and HORNYKIEWICZ (1960); BERNHEIMER and HORNKYIEWICZ (1965); BAROLIN et al. (1964); BERNHEIMER et al. (1965; 1973), LLOYD (1972); LLOYD and HORNYKIEWICZ (1972); LLOYD et al. (1973).

that cell bodies are less effecient than the nerve endings in storing the newly synthesised, and recapturing the released, amine.

METABOLISM OF BRAIN DA IN BASAL GANGLIA DISORDERS

Parkinson's disease

Parkinson's disease is a disorder of the basal ganglia, characterised morphologically by degeneration of the melanin-containing nerve cells in the substantia nigra: the same nerve cells that give rise to the nigrostriatal DA pathway. The main extrapyramidal symptoms of Parkinson's disease are: akinesia, rigidity and tremor.

Neurochemically, Parkinson's disease is characterised by a marked decrease in the concentrations of DA (EHRINGER and HORNYKIEWICZ, 1960) and HVA (BERNHEIMER and HORNYKIEWICZ, 1965) as well as TH (LLOYD, 1972) and DOPA D (LLOYD and HORNYKIEWICZ, 1970) in the nigro-striato-pallidal system (Table 1). These neurochemical changes are characteristic of all Parkinsonian syndromes (EHRINGER and HORNYKIEWICZ, 1960; BERNHEIMER et al., 1965; 1973) regardless of etiological factors involved (including the reversible condition produced by neuroleptic drugs such as reserpine, phenothiazines and butyrophenones). Therefore, Parkinsonism can be regarded, from a neurochemical point of view, as a "Striatal Dopamine Deficiency Syndrome" (HORNYKIEWICZ, 1972c). In Parkinson's disease proper there exists a significant correlation between the degree of cell loss in the substantia nigra and the degree of DA and HVA deficiency in the striatum (BERNHEIMER et al., (1965; 1973).

The severity of the main symptoms also correlates significantly with the degree of striatal DA deficiency. This is demonstrated by the observations (Table 1) that (1) in a case with Hemiparkinsonism, the DA deficiency was markedly more severe in the striatum contra-lateral to the side of the symptoms (BAROLIN et al., 1964); (2) the degree of akinesia was significantly correlated ($P < 0.05$) with the degree of DA deficiency in the caudate nucleus (with a similar trend in putamen and globus pallidus) (BERNHEIMER et al., 1973); and (3) the degree of tremor was significantly correlated ($P < 0.01$) with the degree of HVA decrease in the globus pallidus (and possibly putamen, but not caudate nucleus) (BERNHEIMER et al., 1973).

As could be expected, the degeneration of the nigrostriatal DA pathway produces a denervation supersensitivity of the striatum to DA. This is shown by the observation that in Parkinsonian patients the sensitivity of the akinesia to L-DOPA, DA's immediate precursor substance, was inversely related to the DA content of the striatum: cases with mild akinesia (i.e. significantly milder degree of striatal DA deficiency) responded less promptly to an i.v. test dose of L-DOPA than more severe cases (BERNHEIMER et al., 1973). The presence of a (denervation) supersensitivity of the Parkinsonian striatum to DA explains the high susceptibility of the extrapyramidal symptoms to doses of L-DOPA that have hardly any significant effect on the basal ganglia in normal subjects.

Huntington's chorea

The extrapyramidal symptomatology of Huntington's chorea is dominated by the presence of a hyperkinetic-hypotonic state, that is abnormal involuntary movements associated with decreased muscle tone. This disturbance of the basal ganglia function can be related, on the morphological level, to the severe neuronal degeneration (especially loss of the small neurons) regularly observed in the caudate nucleus and putamen.

In Huntington's chorea the only abnormality of DA metabolism within the basal ganglia so far detected (Table 1) was a mild, but statisically significant ($P < 0.05$), decrease (by approximately 40%) of DA and HVA in the caudate nucleus (BERN-HEIMER and HORNYKIEWICZ, 1973; BERNHEIMER et al., 1973). The levels of DA and HVA in the putamen and the other DA-containing nuclei of the basal ganglia remained essentially unchanged. Thus, it can be postulated that in Huntington's chorea there is a significant shifting of the dopaminergic balance "caudate-putamen" in favour of the putamen. In view of the clinical observations showing that antidopaminergic drugs (reserpine, phenothiazines, butyrophenones) depress the abnormal involuntary movements in the patients with Huntington's chorea and L-DOPA exacerbates them, it is tempting to speculate that the dopaminergic pre-dominance of the putamen over the caudate nucleus may be responsible for the choreatic hyperkinesias.

Dopaminergic balance "caudate–putamen"

The possibility of an unbalanced dopaminergic relationship between the caudate nucleus and putamen as the cause of choreatic hyperkinesias, although speculative at the present time, is not lacking in indirect experimental support. Neurophysiological evidence shows that DA may have opposite actions in the caudate nucleus and putamen with the activity of caudate units inhibited (BLOOM et al., 1965; CONNOR, 1970), and those in the putamen facilitated (YORK, 1970) by this amine. Thus from the point of view of striatal DA, there might well exist a functional differentiation between the caudate nucleus and putamen, with caudate nucleus subserving mainly inhibitory, and putamen facilitatory functions for the dopaminergic control of motor functions.

METABOLISM OF L-DOPA IN THE BRAINS OF PATIENTS WITH PARKINSON'S DISEASE

Major metabolites of L-DOPA—DA replacement

Chemical analyses in brains of Parkinsonian patients treated chronically with high oral doses (2–6g daily) of L-DOPA until death disclosed (DAVIDSON et al., 1971) that, in principle, the Parkinsonian brain metabolises L-DOPA along the same chemical pathways as the brain of untreated, normal laboratory animals. Thus, small amounts of DOPA and 5–10 fold higher levels of 3-O-methyl-DOPA were found throughout the brain (Table 2). The quantitative differences between the concentrations of DOPA and 3-O-methyl-DOPA reflect the fact that 3-O-methyl-DOPA has a much longer half-life (12–14 hr) as compared to the short half-life of L-DOPA (30 min) (BARTHOLINI and PLETSCHER, 1968). In contrast to the diffuse occurrence of these compounds in L-DOPA treated patients, DA accumulated exclusively, and HVA predominantly, in the striatal areas (Table 2). The magnitude of DA's increase in the caudate nucleus and putamen was strictly dependent on the size of the last dose of L-DOPA, and even more so on the time this dose was given before death; the latter fact is in agreement with the comparatively short half-life (2 hr) of the striatal DA (COSTA and NEFF, 1966.)

TABLE 2. NEUROCHEMISTRY OF L-DOPA IN PARKINSON'S DISEASE (PD)

		Caudate nucleus	Putamen	Substantia nigra	Globus pallidus
DA	PD untreated	0.43 ± 0.09 (13)	0.04 ± 0.01 (13)	0.07 ± 0.01 (10)	0·10 (3)
	PD + L-DOPA	1.76 ± 1.10 (4)	2.06 ± 0.72 (4)	—	—
	poor responders	0·22 (3)	0·05 (3)	—	—
	good responders	1·94 (3)	1·04 (3)	—	—
HVA	PD untreated	1.05 ± 0.16 (3)	0.89 ± 0.12 (13)	0.41 ± 0.08 (9)	0.77 ± 0.12 (12)
	PD + L-DOPA	9.14 ± 2.59 (4)	2.69 ± 3.84 (4)	7·98 (1)	10·93 (1)
	poor responde	2·69 (3)	4·18 (3)	—	—
	good responders	6·61 (3)	7·22 (3)	—	—
DOPA	PD + L-DOPA	$0.53 + 0.09$ (5)	0.58 ± 0.14 (5)	—	—
Me-DOPA	PD + L-DOPA	2.76 ± 1.14 (5)	3.58 ± 1.78 (5)	—	—
GAD	PD untreated	641 (2)	583 (2)	526 (2)	776 (2)
	PD + L-DOPA <8 months	339 ± 53 (4)	249 ± 24 (4)	300 ± 102 (4)	504 ± 50 (4)
	PD + L-DOPA >12 months	1172 ± 173 (5)	887 ± 95 (5)	1210 ± 110 (4)	1210 ± 109 (5)

All values are means $\pm$ S.E.M. (number of cases in brackets); dopamine (DA) and homovanillic acid (HVA), DOPA and 3-O-methyl-DOPA (Me-DOPA) in $\mu g/g$, L-glutamic acid decarboxylase (GAD) in nmol/100 mg protein/2 hr. Except for GAD, the data for the L-DOPA treated patients with Parkinson's disease (PD + L-DOPA) are from a group of patients who received the last dose of L-DOPA (0·5–1·5 g orally) no more than $2\frac{1}{4}$ to 9 hr before death. Compiled from: LLOYD (1972); LLOYD and HORNYKIEWICZ (1973); LLOYD et al. (1973).

Although no DA could be detected, in the L-DOPA treated cases, in any of the other brain regions analysed, substantial concentrations of HVA accumulated also in extrastriatal brain areas. This shows that (1) in analogy to normal laboratory animals the Parkinsonian patient metabolises L-DOPA to DA essentially in all brain regions containing significant levels of DOPA D activity, and (2) only the basal ganglia appear to possess any significant storing capacity for DA, and, although heavily damaged, this DA-storing capicity is not completely lost in Parkinson's disease.

In conclusion, from the point of view of its main therapeutic activity, L-DOPA treatment in Parkinson's disease can be regarded as a DA replacement therapy. This conclusion is strongly supported by the observation that good response to L-DOPA could be related to the accumulation of markedly higher amounts of DA in the striatum than poor response to the drug (cf. Table 2, "good responders" and "poor responders") (LLOYD et al., 1973).

L-DOPA and the dopaminergic balance "caudate-putamen"

There is evidence to show that the DA formed from L-DOPA is used up in the putamen of the treated cases significantly faster than in the caudate nucleus. This is probably due to the more severe damage of the DA storage and inactivation (re-uptake) mechanisms in the former region. Thus, in contrast to what is seen shortly after the administration of the drug, (cf. Table 2) in cases receiving the last dose of L-DOPA 10–24 hr before death, the concentration of DA in the putamen declined to very low (pre-DOPA) values whereas in the caudate nucleus the amine level was still elevated (approximately 1 $\mu g/g$); this was accompanied by an opposite behaviour of HVA (LLOYD and HORNYKIEWICZ, in preparation). This suggests the possibility that in Parkinsonian patients chronic L-DOPA treatment may result in a dopaminergic predominance of the putamen (cf. faster "utilisation" of DA) over the caudate nucleus. Thus, it is possible that L-DOPA's major side effect of abnormal involuntary (chorei-form) movements may be due to such a putamenal predominance. It will be remembered that this factor has been considered as a possible cause of the hyperkinetic behaviour in patients with Huntington's chorea (see above).

"Minor" metabolites of L-DOPA—possible "false" transmitters or DA antagonists

In the animal organism, L-DOPA can give rise, via DA, to small quantities of complex condensation products (with aldehydes) such as tetra-hydroisoquinoline and tetrahydropapaveroline derivatives (SANDERS et al., 1973). In the light of this possibil-ity it is important to note that such compounds, if accumulated in high enough amounts might interfere with DA's effectiveness at the level of the striatal receptors, either by virtue of their being "false transmitters" within the CA terminals (cf. COHEN et al., 1972) or partial agonists of DA at the receptors. The latter possibility is clearly borne out by observations that apomorphine, which may serve as the prototype of some of these complex L-DOPA metabolites and which has direct DA receptor simulating properties, is a potential DA antagonist (i.e. partial agonist), both in the periphery (GOLDBERG and MUSGRAVE, 1971; SIMON and VAN MAANEN, 1971; FERRINI and MIRAGOLI, 1972) and the CNS (BIEGER et al., 1972). Thus, the possibility should be kept in mind that accumulation of pharmacologically effective concentrations of these complex L-DOPA metabolites might be involved in some puzzling side effects of

L-DOPA therapy, for example the "on-off" phenomenon; this side-effect is characterised by a sudden and transient loss of L-DOPA's therapeutic activity during the course
of an otherwise successful chronic treatment with the drug. At present there is no
pharmacological evidence that any of these complex L-DOPA condensation products
possess significant direct antiakinesia and antirigidity activity of their own.

L-GLUTAMIC ACID DECARBOXYLASE (GAD) IN PARKINSON'S DISEASE— A POSSIBLE DA-GABA LINK IN THE BASAL GANGLIA

Subnormal GAD *levels in Parkinson's disease—effect of* L-DOPA

In parkinson's disease, the activity of GAD is significantly decreased in the nuclei
of the nigro-striato-pallidal complex (BERNHEIMER and HORNYKIEWICZ, 1962;
LLOYD and HORNYKIEWICZ, 1973) (Table 2). Recent observations disclosed that
prolonged treatment of the patients with high doses of L-DOPA had a significant
influence on the levels of GAD in the basal ganglia (LLOYD and HORNYKIEWICZ, 1973).
Thus, in patients receiving L-DOPA for 12 months or longer the GAD levels in the
caudate nucleus, putamen, globus pallidus and substantia nigra (but not in other brain
areas) were well within the range of control values, being significantly higher than in
patients who received L-DOPA for shorter periods of time (8 months or less) (Table 2).
An attractive possibility to explain these findings is that the decrease in GAD activity
in the basal ganglia in Parkinson's disease may in fact be secondary to the severe
degeneration of the dopaminergic neurons in this brain region. One may speculate
that the striatal and pallidal GAD (GABA)-containing neurons might be under a
continuous "trophic" influence of the dopaminergic system. Degeneration of the
latter system might then result in a biochemical atrophy of these GAD-containing
neurons. This possibility, though hypothetical at present, would help to explain
the decrease of GAD in the basal ganglia in Parkinson's disease and the positive effect
of prolonged L-DOPA administration on the activity of this enzyme.

GABA *neurons in the basal ganglia—possible relation to tremor in Parkinson's disease*

From a clinical point of view, the possibility of a DA-GABA link in the basal
ganglia may have a significant bearing on the etiology and neurochemistry of the
Parkinsonian tremor. It is known that tremor responds rather late in the course of
chronic L-DOPA treatment, thus paralleling to some extent the course of L-DOPA's
effect on GAD; in addition, the antitremor effect of L-DOPA requires particularly
high doses of the drug. Thus, it is possible that a deficiency of a GABA-containing
neuronal system in the basal ganglia, via changes originating in the dopaminergic
system, may be directly involved in the etiology of Parkinsonian tremor; this possibility would furnish, for the first time, a neurochemical basis for this symptom. It may
be relevant in this context that recently a GABA-like compound (3,4,5-trimethoxy-
benzoyl-1,4-aminobutyrate) has been reported to exert a prompt effect on tremor
in patients with Parkinson's disease (CURCI and PRANDI, 1972).

REFERENCES

BAROLIN G. S., BERNHEIMER H. and HORNYKIEWICZ O. (1964) *Schweiz. Arch. Neurol. Psychiat.*
    94, 241–248.
BARTHOLINI G. and PLETSCHER A. (1968) *J. Pharmacol.* **161**, 14–20.
BERNHEIMER H. (1964) *Nature (Lond.)* **204**, 587–588.
BERNHEIMER H. and HORNYKIEWICZ O. (1962) *Arch. exp. Path. Pharmakol.* **243**, 295.
BERNHEIMER H. and HORNKYKIEWICZ O. (1965) *Klin. Wschr.* **43**, 711–715.

BERNHEIMER H., BIRKMAYER W., HORNYKIEWICZ O., JELLINGER K. and SEITELBERGER F. (1965)
Proc. 8th Internat. Congr. Neurol. Vol. IV, pp. 145–148,Wiener Medizinische Akademie, Vienna.
BERNHEIMER, H., BIRKMAYER W., HORNYKIEWICZ O., JELLINGER K. and SEITELBERGER, F. (1973)
J. neurol. Sci. In Press.
BERTLER A. (1961) Acta physiol. scand. 51, 97–107.
BIEGER D., LAROCHELLE L. and HORNYKIEWICZ O. (1972) Europ. J. Pharmacol. 18, 128–136.
BIRKHÄUSER H. (1940) Helv. chim. acta. 23, 1071–1086.
BLOOM F. E., COSTA E. and SALMOIRAGHI G. (1965) J. Pharmacol. 150, 244–252.
COHEN G., MYTILINEOU C. and BARRETT R. E. (1972) Science 175, 1269–1272.
CONNOR J. D. (1970) J. Physiol. (Lond.) 208, 691–703.
COSTA E. and NEFF N. H. (1966) Biochemistry and Pharmacology of the Basal Ganglia (COSTA E.,
CÔTÉ, L. J. and YAHR, M. D., Eds.), pp. 141–155, Raven Press, Hewlett, New York.
CURCI P. and PRANDI G. (1972) Rev. Farmacol. Terap. 111, 197–203.
DAVIDSON L., LLOYD K., DANKOVA J. and HORNYKIEWICZ O. (1917) Experientia 27, 1048–1049.
EHRINGER H. and HORNYKIEWICZ O. (1960) Klin. Wschr. 38, 1236–1239.
FERRINI R. and MIRAGOLI G. (1972) Pharmacol. Res. Commun. 4, 347–352.
GOLDBERG L. I. and MUSGRAVE G. (1971) Pharmacologist 13, 227.
HORNYKIEWICZ O. (1966) Pharmacol. Rev. 18, 925–964.
HORNYKIEWICZ O. (1972a) Handbook of Neurochemistry (LAJTHA A., Ed.) Vol 7, pp. 465–501
Plenum Press. New York.
HORNYKIEWICZ O. (1972b) In The Structure and Function of the Nervous Tissue (BOURNE G. H., Ed.).
Vol. VI. pp. 367–415, Academic Press, New York.
HORNYKIEWICZ O. (1972c) In Neurotransmitters, Res. Publ. Ass. Res. Nerv. Ment. Dis. Vol. 50,
(KOPIN I. J., Ed.) pp. 390–412. Williams & Wilkins, Baltimore.
HORNYKIEWICZ O. (1973) Fedn. Proc. 32, 183–190.
LLOYD K. G. (1972) Ph. D. Thesis, University of Toronto.
LLOYD K. and HORNYKIEWICZ O. (1970) Science 170, 1212–1213.
LLOYD K. and HORNYKIEWICZ O. (1972) J. Neurochem. 19, 1549–1559.
LLOYD K. G., DAVIDSON L. and HORNYKIEWICZ O. (1973) In Advances in Neurology, Vol. 3, Treat-
ment of Parkinsonism (CALNE D. B., Ed.) Raven Press, New York. In Press.
LLOYD K. G. and HORNYKIEWICZ O. (1973) Nature (Lond.) 243, 521–523.
SANDLER M., CARTER S. B., HUNTER K. R. and STERN G. M. (1973) Nature (Lond.) 241, 439-43.
SANO I., GAMO T., KAKIMOTO Y., TANIGUCHI K., TAKESADA M. and NISHINUMA K. (1959) Biochim.
Biophys. Acta. 32, 586–587.
SIMON A. and VAN MAANEN E. F. (1971) Fedn. Proc. 30, 624.
VOGEL W. H., ORFEI V. and CENTURI G. (1969) J. Pharmacol. 165, 196–203.
YORK D. H. (1970) Brain Res. 20, 233–249.

Frontiers in Catecholamine Research 1973, pp. 1109 to 1114. Pergamon Press. Printed in Great Britain.

CHANGES IN HUMAN SERUM DOPAMINE-β-HYDROXYLASE ACTIVITY IN VARIOUS PHYSIOLOGICAL AND PATHOLOGICAL STATES

Lewis S. Freedman, Mark Roffman
and Menek Goldstein

Department of Psychiatry, Neurochemistry Laboratories, New York University
Medical Center, New York, N.Y. 10016, U.S.A.

INTRODUCTION

Dopamine-β-hydroxylase (DβH) (EC 1.14.2.1) catalyses the β-hydroxylation of dopamine (DA) to norepinephrine (NE) (Friedman and Kaufman, 1965). DβH is localised in the chromaffin granules of the adrenal medulla (Kirshner, 1957) and storage vesicles of sympathetic nerves (Potter and Axelrod, 1963). Recent evidence indicates that DβH is released concomitantly with catecholamines from sympathetic nerve terminals and the adrenal medulla (Schneider et al., 1967; De Potter et al., 1969; Weinshilboum et al., 1971a). DβH has been found in the serum of man and animals (Weinshilboum and Axelrod, 1971; Goldstein et al., 1972). In animals activation of the sympathetic nervous system by immobilisation or swim stress, results in increases in circulatory DβH (Weinshilboum et al., 1971; Roffman et al., 1973). Thus experimental evidence suggests that serum DβH might reflect peripheral sympathetic nerve activity. In order to determine whether DβH could also serve as an index of human sympathetic nervous system activity, serum DβH levels were monitored in patients with sympathetic nervous system dysfunction and aberrant catecholamine metabolism.

RESULTS

Developmental aspects of human serum DβH

Serum samples were analysed from 141 normal subjects of varying age groups. Serum DβH activity was measured by a previously described sensitive coupled enzymatic reaction (Goldstein et al., 1971; Weinshilboum and Axelrod, 1971). Serum DβH levels vary widely in the normal population, but are maintained at a relatively constant level in each individual (Freedman et al., 1971). DβH activity exhibits a marked developmental rise with age (Table 1). Low levels of activity are a striking feature of the first year of life and thereafter enzyme activity progressively increases attaining adult values in the 16–20 year age group. These findings emphasised the necessity of the use of proper age controls in any clinical study.

Neuroblastoma

Neuroblastoma is the most common solid malignant tumor in infants and children Abnormal and variable urinary catecholamine secretion is a common clinical finding. Some patients excrete mainly DA and its major metabolite homovanillic acid (HVA), others secrete both DA, NE and their respective metabolites HVA and vanillymandelic acid (VMA). Serum DβH activity and urinary catecholamine levels were

TABLE 1. SERUM DβH ACTIVITY: VARIATION WITH
AGE

Age (yr)	Serum DβH activity (units*)
0–1	7·5 ± 1·50 (12)
1–5	22·5 ± 5·01 (16)
6–10	50·7 ± 8·25 (18)
11–15	43·2 ± 6·78 (20)
16–20	105·0 ± 13·04 (15)
21–40	92·1 ± 8·46 (38)
21–40	92·1 ± 8·46 (38)
41–60	101·7 ± 8·76 (23)

* Units, nm^{14}C product/hr/ml; results are means ± S.E.M.; number of individuals in parenthesis. Control individuals (141) were male and females from NYU pediatric and adult clinics, and volunteers. None of these patients had neurological or psychiatric disorders. Values of nonclinic individuals were slightly higher than other patients.

monitored in 20 children with active neuroblastoma and 11 patients with inactive or "cured" disease (GOLDSTEIN et al., 1972; FREEDMAN, ROFFMAN, GOLDSTEIN and HELSON, unpublished data). Serum DβH activity was elevated* in 9 of 20 active cases. In these patients urinary catecholamine excretion was characterised by abnormal levels of NE and VMA. No correlation of serum DβH activity and DA or HVA was apparent. Thus, serum DβH levels paralleled the catecholamine secretory processes of the active tumor. However, high serum DβH activity was also observed in 6 of 10 "cured" patients. Long term studies are underway to ascertain whether serum DβH activity may be predictive of tumor reoccurrence in these patients. Thus, the monitoring of serum DβH in conjunction with other diagnostic tests might be of significant usefulness in the diagnosis and prognosis of neuroblastoma.

Familial dysautonomia

Familial dysautonomia (F.D.) is a rare syndrome prevalent in Ashkenazie Jews and clinically characterised by autonomic nervous system dysfunction. Abnormally low serum DβH activity has previously been reported in F.D. (WEINSHILBOUM and AXELROD, 1971; FREEDMAN et al., 1972). Further studies from our laboratory have extended these findings. It is evident (Fig. 1) that serum DβH activity varies over a wide range of activity in these patients. A significant number of patients had low enzyme activity whereas others exhibited high enzyme activity. Reduced serum DβH activity could reflect reduced peripheral sympathetic activity. Recent histologic studies of sural nerve biopsies have suggested morphologic neuronal degeneration in F.D. (AGUAYO et al., 1971). Elevated activity could reflect compensatory release mechanisms in the peripheral sympathetic systems. Therefore, in patients with F.D. serum DβH activity must be cautiously interpreted. It has not yet been established what processes serum DβH activity might reflect in the etiology and/or development of symptomatology of F.D.

---

* Activity greater than two standard deviations above control mean value.

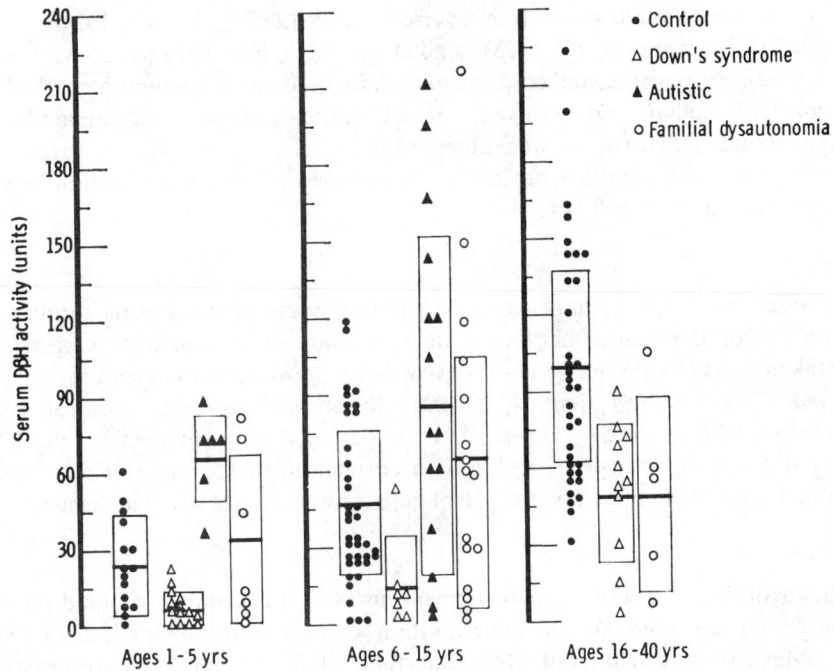

FIG. 1.—Human serum dopamine-β-hydroxylase (DβH) activity. DβH activity units = nmoles ^{14}C-product/hr/ml serum

Down's syndrome

Down's syndrome (D.S.) is characterised by trisomy-21 and mental and growth retardation. Since low blood serotonin levels have been reported (BAZELON *et al.*, 1967; ROSNER *et al.*, 1965) it has been suggested that biogenic amine metabolism is abnormal in D.S. In order to assess the involvement of the peripheral nervous system serum DβH activity was monitored in 32 children with D.S. We and others have observed reduced serum DβH activity in D.S. (WETTERBERG *et al.*, 1972a; COLEMAN, 1973). The results (Fig. 1) indicate that enzyme levels are reduced in all age groups studied (Fig. 1). Serum DBH activity was analysed in non-mongoloid mentally retarded (autistic) children. In this sample of autistic children serum DβH activity was slightly higher than in controls. Thus, in comparison to autistic patients, D.S. serum enzyme activity reduction is more pronounced. Studies are now in progress with radioimmunoassay techniques to determine the levels of circulating DβH protein in these patients.

Parkinson's disease and Huntington's chorea

Parkinson's disease (P.D.) and Huntington's chorea (H.C.) are disorders of the extrapyramidal system. Since central monoamine pathways seem to play a significant role in these diseases, it was interesting to determine whether serum DβH activity could serve as a useful index of clinical status of these diseases.

Longitudinal studies of 28 patients with H.C. and 49 patients with P.D. have been carried out (LIEBERMAN *et al.*, 1972; LIEBERMAN *et al.*, 1973). P.D. patients on L-dopa therapy ($N = 34$) exhibited a wide range of serum enzyme activity with a mean value comparable to that of controls ($\bar{X} = 122\cdot7 \pm 11\cdot70$ vs $102\cdot6 \pm 6\cdot15$). Untreated P.D. patients ($N = 15$) tended to have low enzyme activity ($\bar{X} = 55\cdot2 \pm 9\cdot12$). In patients studied prior to and during L-dopa therapy a marked rise in serum DβH activity was observed. In H.C. patients ($N = 28$), a preponderance of high serum DβH values was observed ($\bar{X} = 128\cdot7 \pm 13\cdot84$). A preliminary study of several H.C. families has not yet established a familial correlation between serum enzyme activity and H.C.

Coma

In order to further attempt to assess the usefulness of monitoring serum DβH in disorders of the central nervous system, a study of 24 comatose patients was undertaken (LIEBERMAN *et al.*, 1972). A wide range of serum enzyme activity was observed. Mean values were significantly lower than controls (control $102 \pm 6\cdot15$ vs coma $64\cdot8 \pm 10\cdot25$). However, in 4 patients who survived, serum DβH activity did not significantly rise. Thus, a correlation of serum DβH activity and functional state of central nervous system activity was not clearly established.

Hypertension

The involvement of the sympathetic nervous system in control of blood pressure has long been suggested. We, in collaboration with Dr. Serrano (Institute of Cardiology, Mexico) have instituted a clinical study of 50 patients with hypertension. A wide range of serum DβH activity was observed. No clear relationship existed between serum DβH activity and blood pressure. In some patients with prolonged history of hypertension, serum DβH levels tend to lower levels of control population.

Psychiatric disorders

Aberrant catecholamine metabolism has been suggested in psychiatric disorders. (BUNNEY and DAVIS, 1965; SCHILDKRAUT, 1965). Serum DβH activity was studied in 56 patients with various psychiatric disorders: manic depressives, schizophrenics and character disorders (SHOPSIN *et al.*, 1972). Serum DβH levels in these patients did not differ from controls.

DISCUSSION

The past 3 years have been a period of active research in many laboratories to evaluate the usefulness of monitoring serum DβH as an index of human sympathetic nervous system dysfunction. At this time a critical evaluation of the results seem warranted. In general, these studies seem to indicate a relative lack of correlation between serum DβH activity and sympathetic nervous system function. However, certain factors should be considered before this issue is finalised. Although serum DβH activity is relatively stable over long periods of time in normal individuals, this stability might not be characteristic of certain pathological conditions. Therefore, longitudinal studies are necessary to more fully evaluate the relationship of serum DβH values to the clinical course of the disease. Since control values vary widely, it would be desirable to use the individual as his own control. This is especially important during prolonged drug treatment. If groups of patients are to be compared to normals, proper age controls are required. These factors are crucial if the monitoring of serum DβH is to be of clinical significance. On the other hand, it is possible that in man the circulatory DβH values may not reliably reflect sympathetic activity. In contrast to data obtained from animal studies, stress induced changes in human serum DβH activity are small in magnitude and do not occur in all cases (WOOTEN and CARDON, 1972; FREEDMAN and GOLDSTEIN, unpublished data). Also, changes in circulatory levels of NE are not reflected by parallel changes in serum DβH (MUELLER et al., 1972).

The circulating levels of DβH probably reflect not only the rate of release but also the rate of degradation of the enzyme (GOLDSTEIN et al., 1972). DβH, like other plasma proteins may be degraded and removed from the circulation by the reticuloendothelial system. In some preliminary studies we have observed abnormal DβH levels in disorders not associated with dysfunction of the sympathetic nervous system (GOLDSTEIN et al., 1973). It must be considered that these abnormally high values in various neoplastic conditions (leukemia, hepatoma, lymphosarcoma) might reflect a defect in the rate of enzyme degradation. An intensive study of the factors involved in control of human serum DβH activity levels is required. Why do normal individuals exhibit such a wide range of values? Are DβH and NE proportionally released into the human circulation? Do genetic factors play an important role in the individual levels of circulatory DβH?

Also, since serum DβH activity probably reflects enzyme released from peripheral sources, the monitoring of DβH activity in cerebrospinal fluid (CSF) might provide a more direct index of central sympathetic nervous activity. We have recently developed an assay for DβH in the CSF (EBSTEIN, FREEDMAN and GOLDSTEIN, unpublished data) and are presently studying DβH in the CSF of normal and neurological patients.

Acknowledgements—Without the help of many clinicians these studies could not have been undertaken. We especially thank Drs. J. Dancis, F. Axelrod, A. Lieberman, I. Fish and B. Shopsin (NYU), Dr. L. Helson (Sloan Kettering), Dr. Mary Coleman (Washington, D.C.) Dr. P. Serrano (Mexico), and Dr. C. Bohuon (Institute Gustave-Roussy, Paris).

This work was supported by USPHS grant MH-02717.

REFERENCES

AGUAYO A. J., NAIR C. P. V. and BRAY G. M. (1971) *Archs Neurol.* **24**, 106–116.
BAZELON M., PAINE R. S., COWIE V. A., HUNT P., HOUCK J. C. and NAHANAND D. (1967) *Lancet* **1**, 1191–1193.
BUNNEY W. E. and DAVIS J. M. (1965) *Archs Gen. Psychiat.* **13**, 483–486.
FREEDMAN L. S., OHUCHI T., LIEBERMAN A. N., ANAGNOSTE B. and GOLDSTEIN M. (1971) *Trans. Am. Soc. Neurochem.* **2**, 70.
FREEDMAN L. S., OHUCHI T., GOLDSTEIN M., AXELROD F., FISH I. and DANCIS J. (1972a) *Nature, Lond.* **236**, 310–311.
FREEDMAN L. S., GOLDSTEIN M., ORES C. N. BOHUON A. C. and GUERINOT F. (1972b) *Proc. Am. Assn. Canc. Res.* **13**, 75.
FRIEDMAN S. and KAUFMAN S. (1965) *J. Biol. Chem.* **240**, 4763–4765.
GOLDSTEIN M., FREEDMAN L. S., and BONNAY M. (1971) *Experientia* **27**, 632–633.
GOLDSTEIN M., FREEDMAN L. S., BOHUON A. C. and GUERINOT F. (1972) *New Eng. J. Med.* **286**, 1123–1125.
GOLDSTEIN M., FREEDMAN L. S., ROFFMAN M. and HELSON L. (1973) *Europ. J. Cancer.* **9**, 233–234.
KIRSHNER N. (1957) *J. Biol. Chem.* **226**, 821–825.
LIEBERMAN A. N., FREEDMAN L. S. and GOLDSTEIN M. (1972) *Lancet* **1**, 153–154.
LIEBERMAN A. N., FREEDMAN L. S., ROFFMAN M., FEIGENSON J. and GOLDSTEIN M. (1973) In: *Advances in Neurology.* (BARBEAU A., CHASE T. N. and PAULSON G. W., Eds.) Vol. **1**, pp. 559–568. Raven Press, New York.
MUELLER R. A., FISHBURNE J. I., BRENNER W. E., BRAAKSMA J. T., STAUROVSKY L. G., HOFFER J. L. and HENDRICKS C. (1972) *Prostaglandins* **2**, 219–226.

POTTER L. T. and AXELROD J. (1963) *J. Pharmacol. Exp. Therap.* **142**, 299–305.

ROFFMAN M., FREEDMAN L. S. and GOLDSTEIN M. (1973) *Life Sci.* **12**, 369–376.

ROSNER F., ONG B. H., PAINE R. S. and MAHANAND D. (1965) *Lancet*, **1**, 1191–1193.

SCHILDKRAUT J. J. (1965) *Am. J. Psychiat.* **122**, 509–514.

SCHNEIDER F. H., SMITH A. D. and WINKLER H. (1967) *Br. J. Pharmacol.* **31**, 94–99.

SHOPSIN B., FREEDMAN L. S., GOLDSTEIN M. and GERSHON S. (1972) *Psychopharmacologia* **27**, 11–16.

WEINSHILBOUM R. M. and AXELROD J. (1971a) *Circulat. Res.* **28**, 307–315.

WEINSHILBOUM R. M., KVETNANSKY R., AXELROD J. and KOPIN I. J. (1971b) *Nature, New Biol.* **230**, 287–288.

WEINSHILBOUM R. M. and AXELROD J. (1971c) *New Eng. J. Med.* **285**, 938–942.

WEINSHILBOUM R. M., THOA N. B., JOHNSON D. G., KOPIN I. J. and AXELROD J. (1971d) *Science* **174**, 1349–1351.

WETTERBERG L., ABERG H., ROSS S. B. and FRODEN O. (1972a) *Scand. J. Clin. Lab. Invest.* **30**, 283–289.

WETTERBERG L., GUSTAVSON K.-H., BACKSTROM M., ROSS S. B. and FRODEN O. (1972b) *Clinical Genetics* **3**, 152–157.

WOOTEN G. F. and CARDON P. V. (1973) *Archs Neurol.* **28**, 103–106.

Frontiers in Catecholamine Research 1973, pp. 1115 to 1121. Pergamon Press. Printed in Great Britain.

DOPAMINE-β-HYDROXYLASE ACTIVITY IN SERUM

RICHARD M. WEINSHILBOUM, FREDRICK A. RAYMOND, LILA R. ELVEBACK, and
WILLIAM H. WEIDMAN

The Mayo Clinic, Rochester, MN, U.S.A.

INTRODUCTION

CLINICAL research into the role of the sympathetic nervous system in human disease has been hampered by the lack of a convenient and sensitive measure of the level of function of the sympathetic nervous system in man. Blood is among the most easily sampled of human tissues. In the course of evolution processes have evolved at the synaptic terminal which limit access of neuro-transmitter substances to the peripheral circulation. This fact increases the technical problems involved in the determination of levels of neurotransmitters in blood and complicates the interpretation of these values even when it is possible to determine them accurately. The recent discovery that the release of catecholamines from the adrenal medulla and from sympathetic nerves is accompanied by the release of proteins found within catecholamine-containing vesicles (BANKS and HELLE, 1965; GEFFEN et al., 1969; DE POTTER et al., 1969) has raised the possibility that the determination of circulating levels of these releasable vesicular proteins might serve as a measure of the level of function of the sympathetic nervous system, the adrenal medulla, or both.

The development of a sensitive enzymatic radiochemical assay procedure for the determination of the activity of the enzyme dopamine-β-hydroxylase (E.C. 1.14.2.1, DBH) (MOLINOFF et al., 1971; GOLDSTEIN et al., 1971), one of the releasable vesicular proteins found in sympathetic nerve terminals and the adrenal medulla (GEFFEN et al., 1969; DE POTTER et al., 1969; VIVEROS et al., 1968), led to the observation that DBH activity is present in the blood of man and other animals (WEINSHILBOUM and AXELROD, 1970; WEINSHILBOUM and AXELROD, 1971a). In order to interpret measurements of this circulating enzyme activity in a clinical setting, however, it was necessary to determine (a) the biochemical identity of the circulating enzyme activity with that found in tissues, (b) the source of the circulating enzyme activity, and (c) the factors which are important in the regulation of the level of this enzyme activity in the blood.

BIOCHEMICAL DATA

The DBH activity found in blood is biochemically similar to DBH activity in the adrenal medulla and sympathetic nerves with regard to co-factor requirements, requirement for oxygen, Michaelis–Menten constant for substrate, and response to the addition of cupric ion to inhibit endogenous inhibitors of the enzyme (WEINSHILBOUM and AXELROD, 1971a). Starch block electrophoresis of serum DBH activity and the enzyme activity found in sympathetic nerves and the adrenal medulla has demonstrated that there are differences in the electrophoretic mobilities of the DBH activity among various species, but that within a given species the enzymatic activity found in the blood has the same electrophoretic mobility as that found in other tissues (ROSS et al., 1972). Immunochemical studies have shown that there is immunochemical cross reactivity within a given species of serum DBH with antibody formed in response to the purified adrenal enzyme from that same species (Rush and Geffen, 1971; Goldstein *et al.*, 1972). All of these data are compatible with the hypothesis that the circulating DBH activity is the same as that found in other tissue in the same species.

SOURCE OF CIRCULATING ENZYME

Circulating DBH activity in the rat is unchanged after adrenalectomy and adrenal demedullation (Weinshilboum *et al.*, 1971a). Partial chemical sympathectomy performed by the use of intravenous 6-hydroxydopamine results in a significant decrease in serum DBH activity in this animal (Weinshilboum and Axelrod, 1971b). When rats are subjected to forced immobilization, a procedure which is known to increase urinary catecholamine excretion, serum DBH activity increases acutely, and the magnitude of the increase is the same in animals with or without adrenal glands (Weinshilboum *et al.*, 1971a). These data suggest that at least a portion of the serum DBH activity originates from sympathetic nerve terminals, and that the adrenal gland is not necessary for the maintenance of serum DBH activity in the rat.

The release of DBH and norepinephrine are directly proportional when the nerves to the isolated perfused spleen or the isolated vas deferens are stimulated (Smith *et al.*, 1970; Weinshilboum *et al.*, 1971b). This observation is significant with regard to attempts to use circulating DBH activity as a measure of the level of function of the sympathetic nervous system.

FACTORS WHICH AFFECT HUMAN SERUM DBH ACTIVITY

Age

Serum DBH activity in man is age dependent (Weinshilboum and Axelrod, 1971c; Freedman *et al.*, 1972). The activity increases with increasing age for the first 4–5 years of life, and there is little change thereafter at least into the mid-thirties or forties. The greatest increase in DBH activity in blood occurs during the first 2–3 years of life. It has been suggested that the age related increase of serum DBH activity in man may represent the functional or anatomical development of the sympathetic nervous system (Weinshilboum and Axelrod, 1971c; Freedman *et al.*, 1972). Until the factors that are involved in the clearance of DBH activity from blood are better understood, however, it can not be assumed that this change in circulating levels of the enzyme represents either a change in the rate of release or the quantity of enzyme released from sympathetic nerve terminals.

Environmental factors

In man, as in the rat, it has been demonstrated that stress such as the cold pressor test and exercise can lead to transient elevations in serum DBH activity (Wooten and Cardon, 1973). The magnitude of the elevations of serum DBH activity in normal humans in response to such stress is variable, and is small when compared to the total range of enzymatic activity in a control population. Changes in circulating DBH activity of greater than 100 per cent have been demonstrated in quadraplegic patients during hypertensive crises (Naftchi *et al.*, 1973).

Genetic factors

A wide range of values of enzymatic serum DBH activity in control subjects has been reported in all studies performed thus far. Because changes in the enzyme activity in the blood of control subjects in response to relatively minor stress are small when compared to the total range of enzyme activity in the entire population, the possibilty arises that genetic factors may play an important role in the determination of baseline values in serum DBH activity in control populations. In a recent study we have determined serum DBH activity in blood samples obtained from 433 children aged 6–12 and from 227 adult control subjects (WEINSHILBOUM et al., submitted for publication). Samples from children were obtained in the morning at school after an overnight fast, and samples from adult subjects were obtained from blood donors receiving no medications.

DBH activity was measured by a sensitive radiochemical enzymatic assay as described elsewhere (MOLINOFF et al., 1971; WEINSHILBOUM and AXELROD, 1971c) except that acetate buffer, 1·0 M, pH 4·9, was used instead of Tris HCl buffer, 1·0 M, pH 6 in the DBH reaction. This change resulted in a final incubation pH of 5·2, the optimal pH for the determination of DBH activity in human serum in this assay system. One unit of DBH activity represented the formation of 1 nmole of β-phenyl-β-ethanolamine from β-phenylethylamine per ml of serum per hr.

The percentage frequency distribution of serum DBH values in 433 control children (233 boys, 200 girls) is shown in Fig. 1. 4·6 per cent of the children (9 boys, 11 girls) had a very low serum DBH activity (less than 50 units). This group of children included 3 out of the 4 children tested in one family and 2 sets of 2 siblings each. Serum DBH values in this population are skewed to the right. The skewness in distribution could be corrected by expressing the values as the square root of DBH activity. To eliminate the possibility that the frequency distribution had been biased by the inclusion of data from siblings, values in one sibling from each set of siblings were chosen randomly and the frequency distribution for serum DBH activity in these 280 unrelated children is shown in Fig. 1. The percentage frequency distribution of serum DBH activity in blood samples obtained from 227 unrelated adult blood donors is also plotted in Fig. 1. The adult control population included 134 men and 93 women. The median age of this population was 32·1 years. 3·1 per cent of these adults had a very low enzyme activity (less than 50 units). The distribution of values for serum enzyme activity in this population was also skewed to the right. Data from the first 317 consecutive control children were analysed, and in this group serum enzyme activity in girls did not change from age 6 through 12 while that in boys increased only approximately 50 units. In the adult population, no change in serum activity with age was detected in either women or men.

A highly significant correlation, r = 0·57 ($p < 0·001$), was found between serum DBH activity in the blood of 94 sibling-sibling pairs included in the first 317 consecutive children examined. This correlation was established in terms of age and sex specific relative deviates about the sex specific regression of the square root of DBH activity on age to correct for the small increase in enzyme activity with age in boys and for the lack of a normal distribution (Fig. 2). There were no differences in the degree of correlation between brother–brother, brother–sister, or sister–sister pairings. When random pairs of single children with no siblings drawn from the same population were generated by using tables of random numbers, no correlation of serum DBH activity was found between members of non-sibling pairs.

To test for the possibility that the differences in serum DBH activity found in different members of a control population were due to differences in circulating

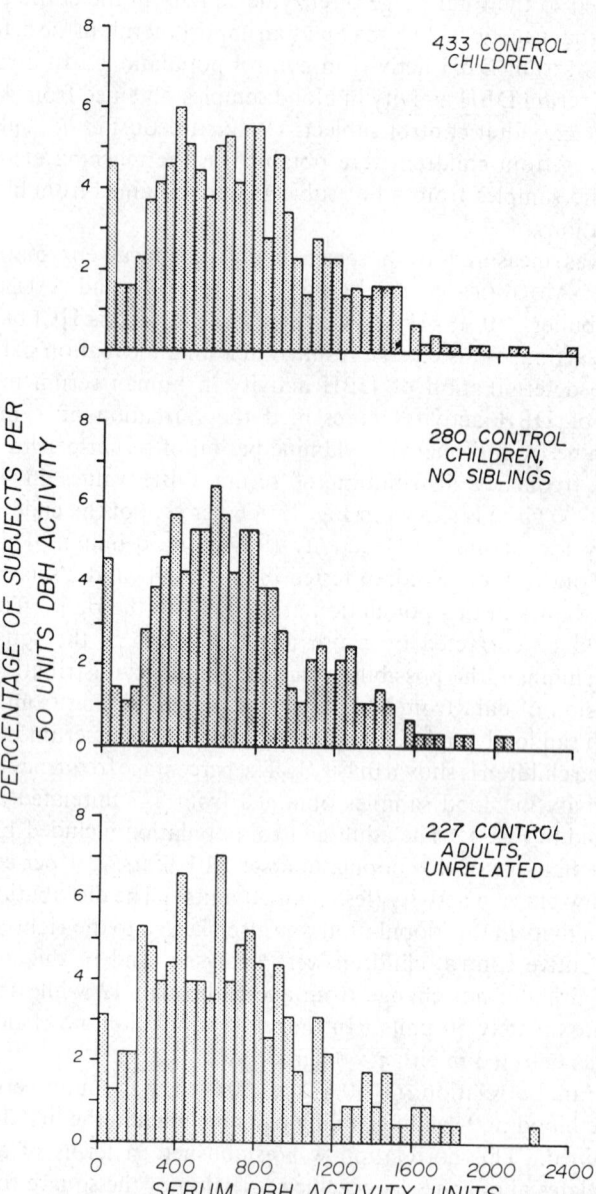

Fig. 1.—Percentage frequency distributions of serum DBH activity. The percentage of subjects with serum DBH activity per 50 unit increments are shown for: top, 433 control children; middle, 280 of the same children with only one child from each family represented; and bottom, 227 control adult subjects.

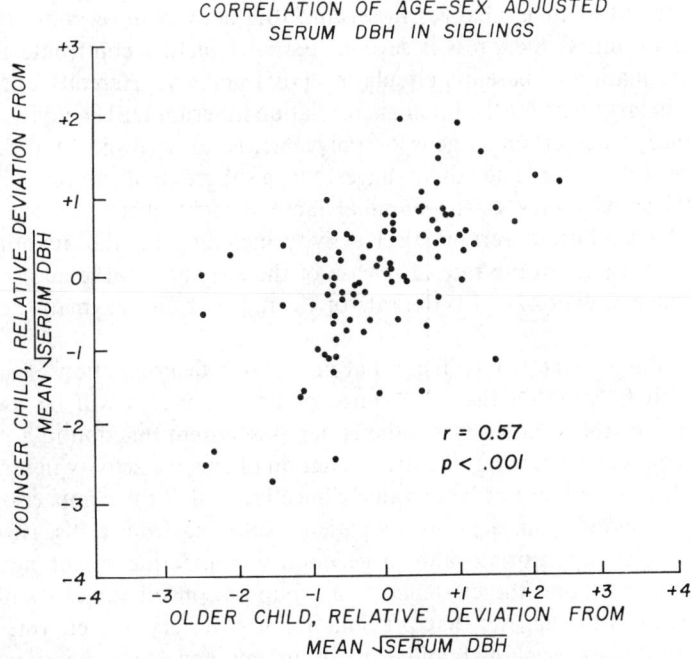

FIG. 2.—Sibling-sibling correlation of serum DBH activity. The correlation of specific
relative deviates of the square root of serum DBH activity in sibling pairs is shown.
$r = 0.57$, $p < 0.001$.

inhibitors of the enzyme, purified bovine adrenal DBH activity was added to samples
of blood assayed under the standard assay procedure. In multiple samples from
different individuals with either very low activity (less than 50 units), low normal
activity (200–350 units), or high normal activity (850–1000 units), there was no
demonstrable inhibition of exogenously added purified bovine adrenal DBH activity.
The wide range in serum DBH activity found in control human subjects could not
be accounted for on the basis of different levels of circulating inhibitor(s) of the
enzyme.

These data demonstrated a familial correlation of enzymatic DBH activity in the
blood of control human subjects. It was not possible to determine how much of this
correlation was due to the effects of shared environment and how much was due to
factors of heredity. Ross and his co-workers in Sweden have studied serum DBH
activity in monozygotic and dizygotic twins (Ross et al., 1973). They found a correlation coefficient for monozygotic twins of 0·96 and for dizygotic twins of 0·75.
Both of these populations showed significant correlation between members of twin
pairs, but a significantly higher degree of correlation was found in enzyme activity
in monozygotic than in dizygotic twins. Their interpretation of these results was
that heredity was the major factor which contributed to the familial correlation of
serum DBH activity.

CONCLUSION

There is a wide range of serum DBH activity in control human populations.
Circulating enzyme activity levels change dramatically during the first years of life.

After that period of time, changes in serum DBH activity in response to stress are relatively small unless the stress is severe. Familial factors contribute significantly to the determination of baseline circulating DBH activity. Heredity appears to be responsible in large part for the familial correlation in serum DBH activity. The mode of inheritance, whether single gene or polygenic, is not known. Studies of a large group of control children and adults suggest that a subgroup of subjects with very low serum DBH activity may exist. Familial factors might affect any one of several variables which influence serum DBH activity including (a) the quantity of DBH available for release, (b) the rate of release of the enzyme, (c) the access of enzyme protein to the circulation, or (d) the rate of clearance of the enzymatic activity from the blood.

These findings make it less likely that an isolated determination of serum DBH activity which falls within the wide range of normal values will be useful in the determination of the levels of sympathetic nervous system function in a given individual. This does not mean that the determination of enzyme activity in serial samples from an individual might not be of value clinically, or that the comparison of values determined in blood from a group of patients suffering from a disease with those determined in an appropriate and large control population might not be useful clinically. Furthermore, the existence of a group of control subjects with very low serum DBH activity suggests that previous reports of very low enzyme activity in patients with diseases such as familial dysautonomia and Down's syndrome (Weinshilboum and Axelrod, 1971c; Wetterberg et al., 1972) must be interpreted with caution.

It is not possible to predict the ultimate clinical usefulness of either the enzymatic measure of serum DBH activity or the immunochemical radioimmunoassay of DBH protein as measures of the level of function of the sympathetic nervous system in man. Until the factors which control both the enzymatic activity and immunoreactive DBH are better understood, clinical studies in which these measures of the rate of release of transmitter from sympathetic nerve are assayed will have to be interpreted with caution. Whether the measurement of serum DBH activity and serum immunoreactive DBH protein in man will eventually prove to be unrelated to sympathetic nervous system function, will prove to be merely a different way of obtaining the same information provided by the accurate determination of blood catecholamines, or will provide us with new and different insights into neural function remain to be investigated. Today the clinician dealing with a patient with hepatic disease measures not only the bilirubin levels in the blood but also the levels of glutamic oxaloacetic transaminase activity and alkaline phosphatase activity. The time may be approaching when the clinician will have available several procedures by which the function of the sympathetic nervous system may be monitored.

Acknowledgements—R. M. W. and W. H. W. were supported in part by a Faculty Development Award in Clinical Pharmacology sponsored by the Pharmaceutical Manufacturers Association Foundation, Inc. (R. M. W.), by PHS grant 5 SOl RR 05530-10 (R. M. W.), and NHLI grant HE 14196E (W. H. W.).

REFERENCES

Banks P. and Helle K. (1965) *Biochem. J.* **97**, 40–41c.
Freedman L. S., Ohuchi T., Goldstein M., Axelrod F., Fish I. and Dancis J. (1972) *Nature, Lond.* **236**, 310–311.

GEFFEN L. B., LIVETT B. G. and RUSH R. A. (1969) *J. Physiol.* **204**, 58–59P.
GOLDSTEIN M., FREEDMAN L. S. and BONNAY M. (1971) *Experientia* **27**, 632-633.
GOLDSTEIN M. FUXE K. and HÖFELT T. (1972) *Pharmacol. Rev.* **24**, 293–309.
MOLINOFF P. B., WEINSHILBOUM R. and AXELROD J. (1971) *J. Pharmacol. exp. Ther.* **178**, 425–431.
NAFTCHI N. E., WOOTEN G. F., LOWMAN E. W. and AXELROD J. (1973) *Fedn. Proc.* **32**, 708 Abs.
DE POTTER W. P., DE SCHAEPDRYVER A. F., MOREMAN E. J. and SMITH A. D. (1969) *J. Physiol.* **204**, 102–104P.
ROSS S. B., WEINSHILBOUM R. M., MOLINOFF P. B., VESELL E. S. and AXELROD J. (1972) *Mol. Pharmacol* **8**, 50–58.
ROSS S. B., WETTERBERG L. and MYRHED M. (1973) *Life Sci.* in press.
RUSH R. A. and GEFFEN L. B. (1972) *Circulat. Res.* **31**, 444–452.
SMITH A. D., DE POTTER W. P., MOERMAN E. J. and DE SCHAEPDRYVER A. F. (1970) *Tissue and Cell* **2**, 547–568.
VIVEROS O. H., ARQUEROS L. and KIRSHNER N. (1968) *Life Sci.* **7**, 609–618.
WEINSHILBOUM R. and AXELROD J. (1970) *The Pharmacologist* **12**, 214.
WEINSHILBOUM R. and AXELROD J. (1971a) *Circulat. Res.* **28**, 307–315.
WEINSHILBOUM R. and AXELROD J. (1971b) *Science* **173**, 931–934.
WEINSHILBOUM R. M. and AXELROD J. (1971c) *New Eng. J. Med.* **285**, 938–942.
WEINSHILBOUM R. M., KVETNANSKY R., AXELROD J. and KOPIN I. J. (1971a) *Nature, New Biol.* **230**, 287–288.
WEINSHILBOUM R. M., THOA N. G., JOHNSON D. G., KOPIN I. J. and AXELROD J. (1971b) *Science* **174**, 1349–1351.
WETTERBERG L., GUSTAVSON K. H., BÄCKSTRÖM M., ROSS S. B. and FRÖDEN O. (1972) *Clin. Genetics* **3**, 152–153.
WOOTEN G. F. and CARDON P. V. (1973) *Arch. Neurol.* **28**, 103–106.

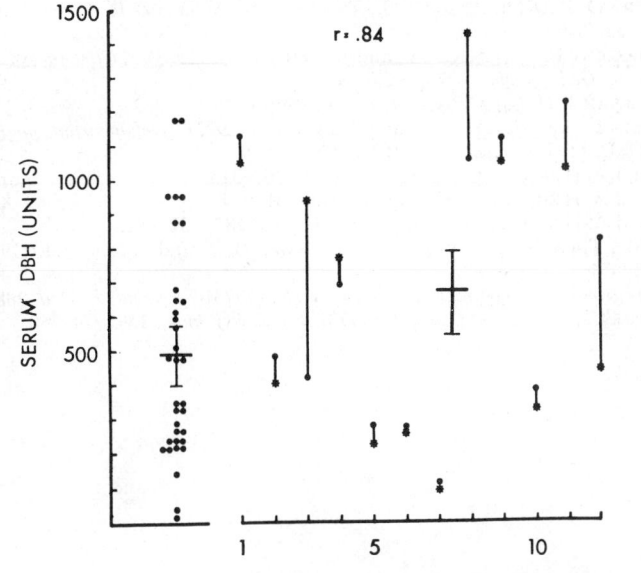

SEVERITY OF DISEASE (SCHIZOPHRENIA)

FIG. 1.—Serum dopamine-beta-hydroxylase (DBH) in 34 normal controls compared with 12 monozygotic twins discordant for schizophrenia. The schizophrenic twins are indicated by an asterisk (*), and the normal cotwins' values are indicated by a connecting line (see also Legend to Table).

Since blood samples were taken at different times and cotwins of nine of the twin pairs investigated lived in different cities, these results strongly suggest that environmental factors play a minor role in influencing DBH activity and that plasma DBH activity may be genetically determined. Family studies of individuals with torsion dystonia (WOOTEN *et al.*, 1973) and dysautonomia (WEINSHILBOUM and AXELROD, 1971c) have also suggested that genetic factors are important determinants of DBH activity in blood. No pathophysiological relationship between plasma DBH activity and schizophrenia has been demonstrated. Catecholamine-related enzymes can be increased in the central nervous system without concommitant peripheral changes (LAMPRECHT *et al.*, 1972) and an almost complete destruction of central noradrenergic neurons by intraventricularly administered 6-hydroxydopamine had no effect on serum DBH levels (LAMPRECHT *et al.*, 1973). Thus, it does not necessarily follow from the data presented that brain DBH is unaltered in schizophrenia since plasma DBH is mainly derived from peripheral sympathetic nerves (WEINSHILBOUM and AXELROD, 1971b) and separate genetic loci may control plasma and central nervous system DBH.

REFERENCES

ANGST V. J. (1965) *Schweiz. Med. Wschr.* **86,** 1304–1306.
DUNNER D. L., COHN C. K., WEINSHILBOUM R. M. and WYATT R. J. (1973) *Biol. Psychiat.,* In press.
GOLDSTEIN M., ANAGNOSTE B., LAUBER E. and McKEREGHAN M. R. (1964) *Life Sci.* **3,** 763–767.
HEATH R. G., NESSELHOF W., BISHOP M. P. and BYERS L. W. (1965) *Dis. Nerv. System* **26,** 99–105.
HOFFER A. and POLLIN W. (1970) *Arch. Gen. Psychiat.* **23,** 469–477.
KAUFMAN S. and FRIEDMAN S. (1965) *Pharmacol. Rev.* **17,** 71–100.
LAMPRECHT F. (1973) In: *Seelische Störungen bei Anfallskranken.* (PENIN H., ed.) Schattauer Verlag, Stuttgart (Germany), In press.
LAMPRECHT F., EICHELMAN B., THOA N. B., WILLIAMS, R. B. and KOPIN I. J. (1972) *Science* **177,** 1214–1215.

Lamprecht F., Henry D. P., Richardson J. S., Thomas J. A., Williams R. B. and Bartter F. C. (1973) *Fedr. Proc.* **32,** 763.

Miller W. L., Lamprecht F., Cardon P. V. and Bartter F. C. (1973) *J. Clin. Invest.*, **52,** 57a.

Pollin W. (1972) *Arch. Gen. Psychiat.* **27,** 29–37.

Pollin W. and Stabenau J. (1968) *J. Psychiat. Res.* **6,** (Suppl. 1) 317–332.

Shopsin B., Freedman L. S., Goldstein M. and Gershon S. (1972) *Psychopharmacologia* **27,** 11–16.

Stein L. and Wise C. D. (1971) *Science* **171,** 1032–1036.

Weinshilboum R. and Axelrod J. (1971a) *Circ. Res.* **28,** 307–315.

Weinshilboum R. and Axelrod J. (1971b) *Science* **173,** 931–934.

Weinshilboum R. and Axelrod J. (1971c) *New Eng. J. Med.* **285,** 938–942.

Weinshilboum R. M., Thoa N. B., Johnson D. G., Kopin I. J. and Axelrod J. (1971) *Science* **174,** 1349–1351.

Wooten G. F., Eldridge R., Axelrod J. and Stern R. S. (1973) *New Eng. J. Med.* **288,** 284–287.

Wyatt R. J., Saavedra J. M. and Axelrod J. (1973) *Am. J. Psychiat.*, **130,** 754–760.

Frontiers in Catecholamine Research 1973, pp. 1127 to 1132. Pergamon Press. Printed in Great Britain.

CATECHOLAMINE METABOLISM AND NEUROLOGIC DISEASE

THOMAS N. CHASE

Neurology Unit, National Institute of Mental Health, Bethesda, Maryland, U.S.A.

ALTERATIONS in central catecholamine metabolism have been found to attend a number of neurologic disorders. These changes have often been assumed to reflect altered nerve impulse activity in neural systems containing these amines. In some instances the metabolic abnormality appears directly attributable to the degeneration of a specific catecholaminergic pathway. In other disorders it is conceivable that altered catecholamine metabolism may be a functional response to a drug or disease induced change in the sensitivity of dopaminergic receptors or in the activity of some related neuronal system.

DOPAMINE METABOLISM

Early attempts to study catecholamine metabolism in the central nervous system of the living patient made extensive use of steady state levels of the principal metabolites of dopamine and norepinephrine in cerebrospinal fluid (CSF). The rationale for this approach derives from observations in the experimental animal as well as in man suggesting that CSF homovanillic acid (HVA), a major metabolite of dopamine, and 3-methoxy-4-hydroxyphenylglycol (MHPG), a major product of norepinephrine degradation, arises largely from central rather than peripheral metabolism and that CSF levels of these metabolites tend to follow those in the cerebral tissues (MOIR *et al.*, 1970; CHASE *et al.*, 1973). Substantial reductions in the CSF content of HVA have been found in patients with Parkinson's disease and related extrapyramidal disorders (Table 1). The validity of these observations in Parkinson's disease or Huntington's chorea has been confirmed by direct biochemical assay of brain tissues at autopsy (HORNYKIEWICZ, 1966, 1973). On the other hand, diminished HVA levels have also been found in disorders such as amyotrophic lateral sclerosis where lesions are not known to involve dopamine-containing neuronal systems. Such observations cast suspicion upon the presumed relationship between reduced HVA levels in lumbar CSF and functional alterations in cerebral dopaminergic systems.

More recent studies of central dopamine metabolism have employed the probenecid loading technique. Since the acute administration of probenecid inhibits HVA transport from the spinal fluid compartment, without substantially altering cerebral dopamine concentrations, the HVA rise in CSF should provide an index to the rate of metabolite formation and thus to the central turnover of the parent amine (WERDINIUS, 1967; KORF and VAN PRAAG, 1971; BOWERS, 1972). The probenecid-induced accumulation of HVA is substantially diminished in parkinsonian patients (Table 2). Less marked, yet statistically significant, reductions in HVA turnover have also been found in patients with Huntington's chorea and amyotrophic lateral sclerosis (Table 2). It remains to be determined, however, whether orally administered probenecid affects HVA transport to the same extent in patient and control groups.

TABLE 1. HOMOVANILLIC ACID (HVA) AND 3-METHOXY-4-HYDROXYPHENYLETHY-
LENE GLYCOL (MHPG) LEVELS IN LUMBAR SPINAL FLUID*

	HVA	MHPG
Controls	36 ± 4·0 (25)	15 ± 2·3 (11)
Parkinsonism	10 ± 1·2 (30)‡	15 ± 2·5 (13)
Huntington's chorea	16 ± 3·3 (14)‡	16 ± 0·7 (7)
Dystonia musculorum deformans	20 ± 3·6 (15)†	14 ± 2·1 (6)
Down's syndrome	56 ± 11 (9)	14 ± 3·5 (6)
Parkinsonism-dementia	9 ± 2·7 (16)‡	—
Progressive supranuclear palsy	8 ± 2·6 (3)‡	—
Jakob–Creutzfeldt disease	10 ± 3·7 (6)‡	—
Tardive dyskinesia	31 ± 4·4 (10)	—
Amyotrophic lateral sclerosis	13 ± 2·4 (21)‡	—
Spinocerebeller degenerations	25 ± 4·7 (10)	—

* Values are the means ± S.E.M. for the number of untreated patients given
in parentheses, expressed in ng/ml. Spinal fluid collection and assay methods are
as previously described (CHASE and NG, 1972).
    † $P < 0.01$.
    ‡ $P < 0.001$.

NOREPINEPHRINE METABOLISM

Studies of central noradrenergic mechanisms based on CSF levels of MHPG have
yet to yield much useful information. Steady state MHPG values are not significantly altered in any of the neurologic disorders studied (Table 1), including
Parkinson's disease, where biochemical measurements at autopsy have shown a
depression in brain norepinephrine (HORNYKIEWICZ, 1966). Moreover, estimates of
central norepinephrine turnover based on the probenecid-induced accumulation of
MHPG have not been possible, since the oral probenecid loading technique currently
employed does not significantly influence lumbar CSF concentrations of either the
free or conjugated form of this metabolite (CHASE et al., 1973). Mechanisms in man
subserving the transfer of MHPG from the central tissues to CSF or from CSF to
blood would thus appear to differ from those mediating the efflux of HVA and
5-hydroxyindoleacetic acid (5-HIAA). This contention is supported by the lack of
any difference between ventricular and lumbar CSF levels of free or conjugated
MHPG in contrast to the steep concentration gradients for both HVA and 5-HIAA

TABLE 2. EFFECT OF PROBENECID ON HOMOVANILLIC ACID LEVELS IN LUMBARSPINAL FLUID*

	Number of patients	Baseline	Treatment	Difference
Controls	8	28 ± 6·8	194 ± 24	166 ± 24
Parkinsonism	20	12 ± 1·7	73 ± 12	62 ± 11‡
Huntington's chorea	12	14 ± 5·0	115 ± 14	100 ± 14†
Tardive dyskinesia	8	29 ± 5·0	146 ± 24	116 ± 22
Dystonia musculorum deformans	8	14 ± 4·9	168 ± 22	154 ± 19
Down's syndrome	6	48 ± 12	225 ± 35	178 ± 25
Amyotrophic lateral sclerosis	7	23 ± 7·7	128 ± 22	104 ± 16†

* Values are the means ± S.E.M., expressed in ng/ml, for untreated patients. Probenecid (2 g)
was administered orally immediately after obtaining the baseline spinal fluid sample and again 3
and 6 hr later (CHASE and NG, 1972). The second lumbar puncture was performed 9 hr after the
initial one.
    † $P \leq 0.05$.
    ‡ $P < 0.001$.

(CHASE *et al.*, 1973). The absence of altered MHPG values in the lumbar CSF of patients with brain diseases such as parkinsonism might thus reflect the special nature of efflux mechanisms for MHPG or possibly be a consequence of rapid MHPG formation by noradrenergic terminals in the spinal cord.

NATURALLY-OCCURRING EXTRAPYRAMIDAL DISORDERS

Studies based on the probenecid loading technique suggest a close relationship between central dopaminergic mechanisms and overall parkinsonian severity (Fig. 1). Separate analysis of the 3 cardinal parkinsonian signs indicates, however, that this relationship holds for rigidity and akinesia but not for tremor (CHASE and NG, 1972). This observation as well as results from certain animal models of Parkinson's disease (BATTISTA *et al.*, 1969; GUMULKA *et al.*, 1970) support the contention that neuro-humoral mechanisms subserving tremor may differ from those relating to rigidity or akinesia. It is probable that the amount of reduction in central dopamine turnover in untreated parkinsonian patients reflects the degree of degeneration of dopaminergic neurons originating in the substantia nigra.

Assays of post-mortem tissues (HORNYKIEWICZ, 1973) or lumbar CSF (CHASE, 1973a) suggest that a reduction in central dopamine metabolism also occurs in patients

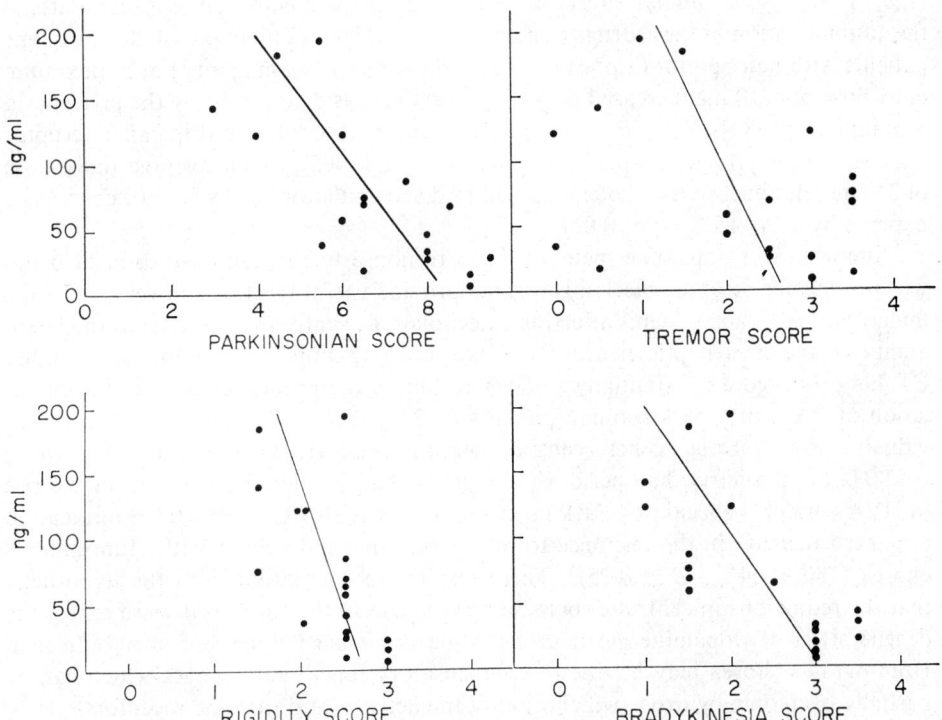

FIG. 1.—Effect of probenecid on homovanillic acid (HVA) levels in lumbar CSF of untreated parkinsonian patients. Rigidity, bradykinesia and tremor were each rated on a score of 0 (absent) to 4 (very severe) and combined to give an overall index of parkinsonian severity. Significant inverse correlations were found between the probenecid-induced accumulation of HVA and overall parkinsonian severity ($P < 0.001$), rigidity ($P < 0.01$) and bradykinesia ($P < 0.01$), but not resting tremor ($P > 0.05$).

with Huntington's disease. Although there is a tendency for an inverse correlation between choreatic severity and the probenecid-induced accumulation of HVA, this relationship is not statistically significant in the 11 patients studied to date ($r = 0.590$; $P > 0.05$). The characteristic histopathologic changes in Huntington's disease do not involve neurons known to contain dopamine. In further contrast to Parkinson's disease, drugs which elevate brain dopamine (for example, L-dopa) exacerbate choreatic movements in patients with Huntington's chorea, while those which presumably diminish dopaminergic function (for example, α-methylparatyrosine, tetrabenazine, or haloperidol) tend to ameliorate the extrapyramidal signs of this disorder (Chase, 1973a). Altered dopamine metabolism in Huntington's chorea might thus be a secondary, functional response of neurons within the inhibitory nigrostriatal dopaminergic pathway to the degeneration of striatal neurons upon which they make synaptic contact.

Pharmacologic observations in the experimental animal suggest that a close relationship may exist between the activity of postsynaptic dopaminergic receptors and dopamine turnover in the presynaptic neuron. Drugs which appear to block dopamine receptors such as the psychotropic phenothiazine and butyrophenone derivatives tend to increase dopamine turnover, while those which are believed to act mainly as dopamine receptor agonists exert the opposite metabolic effect (Corrodi et al., 1967, 1972). Similar effects on central dopamine metabolism appear to attend the administration of these drugs to man (Chase, 1973b). Treatment of 20 neurologic patients with haloperidol (a presumed dopamine receptor antagonist) at a maximum daily dose of 8–10 mg increased dopamine turnover, as determined by the probenecid technique, by $98 \pm 32\%$ ($P < 0.01$). Conversely, the putative dopamine receptor agonist, 1-(2″-pyrimidyl)-4-piperonylpiperazine (ET 495), at an average daily dose of 230 mg, diminished the probenecid-induced accumulation of HVA in 10 neurologic patients by $51 \pm 15\%$ ($P < 0.01$).

Alterations in dopamine metabolism in response to the administration of drugs which influence receptor mechanisms may provide useful information relative to the integrity of feedback systems affecting catecholamine synthesis as well as to the latent ability of the presynaptic neuron to synthesize dopamine. In preliminary studies ET 495 produced a substantially smaller decline in the probenecid-induced accumulation of HVA in 6 parkinsonian patients ($-23 \pm 13\%$; $P > 0.05$) than in 4 individuals with various other central nervous system disorders ($-91 \pm 14\%$; $P < 0.001$). Similarly, haloperidol (10 mg/day) had no significant effect on the rise in HVA with probenecid in 4 Parkinson's patients ($21 \pm 19\%$; $P > 0.05$) but lead to a marked increase in the response to probenecid in 5 individuals with Huntington's chorea ($189 \pm 64\%$; $P < 0.05$). These results are compatible with the hypothesis that the reductions in central dopamine metabolism in Parkinson's disease reflect the degeneration of dopaminergic neurons while diminished dopamine metabolism in Huntington's chorea may be due to compensatory functional changes which can be partially overcome by drugs which modify the activity of dopamine receptors.

DRUG-INDUCED EXTRAPYRAMIDAL DISORDERS

Studies of dopamine metabolism in patients with drug-induced parkinsonism or tardive dyskinesia suggest that the former state may in part be the clinical expression of diminished activity of nigrostriatal dopaminergic neurons while the latter condition may reflect hyperfunction of this neural system. Dopamine turnover, as estimated by steady state HVA values in CSF, appears diminished in patients who develop parkinsonism while receiving α-methylparatyrosine, an inhibitor of catecholamine synthesis. In contrast to these results, as well as those in patients with the naturally-occurring disorder (Table 2), the probenecid-induced accumulation of HVA was normal or slightly increased in 10 patients who manifested parkinsonian signs during treatment with haloperidol (137 ± 21% of levels in 8 untreated control subjects). Inhibition of dopaminergic transmission due to receptor blockade rather than enhanced transsynaptic activity owing to increased amine synthesis would appear to have been the preponderant functional effect of haloperidol in these patients.

Preliminary results suggest that dopamine turnover may also be depressed in patients who manifest tardive dyskinesias long after neuroleptic withdrawal (Table 2). Such metabolic changes might reflect structural alterations in dopamine-containing neurons or be a functional response to a primary change in the sensitivity of dopaminergic receptors. The haloperidol-induced increase in dopamine turnover, as estimated by the probenecid test, appeared to be less in 6 patients with tardive dyskinesia (79 ± 18 ng/ml) than in 5 individuals with Huntington's chorea (185 ± 93 ng/ml). In view of evidence suggesting that morphologic changes occur in the substantia nigra and other brain stem areas of patients with tardive dyskinesias (CHRISTENSEN et al., 1970) these biochemical findings support the view that some dopamine-containing neurons may be damaged in patients receiving long-term neuroleptic treatment. In such individuals dyskinesias might be the result of denervation supersensitivity of postsynaptic dopamine receptors.

CONCLUSIONS

Although methods for studying catecholamine metabolism in the central nervous system of man have advanced considerably during the past decade, they remain indirect and of uncertain validity. Moreover, the relation between the metabolic changes which these techniques attempt to measure and the state of neural transmission across synapses containing these amines has yet to be established. Furthermore, pathogenetic mechanisms in the human brain are undoubtedly more complex than suggested by the simplistic models currently advanced to explain available results. Neurohumoral mechanisms, nevertheless, appear to be sensitive indicators of central nervous dysfunction and of the effects of centrally active drugs. Efforts to improve our ability to examine catecholamines as well as other putative central neurotransmitters should prove eminently worthwhile.

REFERENCES

BATTISTA A. F., GOLDSTEIN M., NAKATANI S. and ANAGNOSTE B. (1969) *Confin. Neurol.* **31**, 135–144.
BERNHEIMER H. and HORNYKIEWICZ O. (1973) In: *Advances in Neurology* (BARBEAU A., CHASE T. N. and PAULSEN G. W., Eds.), Vol. 1, pp. 525–531, Raven Press, New York.
BOWERS M. B. (1972) *Neuropharmacology* **11**, 101–111.
CHASE T. N. (1973a) In: *Advances in Neurology* (BARBEAU A., CHASE T. N. and PAULSEN G. W., Eds.), Vol. 1, pp. 533–542, Raven Press, New York.
CHASE T. N. (1973b) *Arch. Neurol.* In press.
CHASE T. N., GORDON E. K. and NG L. K. Y. (1973) *J. Neurochem.* In press.
CHASE T. N. and NG L. K. Y. (1972) *Arch. Neurol.* **27**, 486–491.
CHRISTENSEN E., MOLLER J. E. and FAURBYE A. (1970) *Acta Psychiat. Scand.* **46**, 14–23.
CORRODI H., FARNEBO L. -O., FUXE K., HAMBERGER B. and UNGERSTEDT U. (1972) *Europ. J. Pharmacol* **20**, 195–204.
CORRODI H., FUXE K. and HOKFELT T. (1967) *Life Sci.* **6**, 767–774
GUMULKA W., ANGEL A. R. D., SAMANIN R. and VALZELLI L. (1970) *Europ. J. Pharmacol.* **10**, 79–82.

HORNYKIEWICZ O. (1966) *Pharmac. Rev.* **18,** 925–964.
KORF J. and VAN PRAAG H. M. (1971) *Brain Res.* **35,** 221–230.
MOIR A. T. B., ASHCROFT G. W., CRAWFORD T. B. B., ECCLESTON D. and GULDBERG H. C. (1970)
    Brain **93,** 357–368.
WERDINIUS B. (1967) *Acta Pharmacol.* **25,** 18–23.

Frontiers in Catecholamine Research 1973, pp. 1133 to 1138. Pergamon Press. Printed in Great Britain.

AMINES IN SCHIZOPHRENIA

SOLOMON H. SNYDER and SHAILESH P. BANERJEE

Departments of Pharmacology and Experimental Therapeutics,

and

Psychiatry and the Behavioural Sciences, The Johns Hopkins University School of Medicine, Baltimore, Maryland 21205, U.S.A.

THIS essay is not confined to catecholamines in schizophrenia, because to do so would convey the impression that other amines are irrelevant to the disease. Some drug actions suggest a role for dopamine; others implicate the indoleamines. The 'dopamine story' derives from the influences of antischizophrenic phenothiazines and butyrophenones upon the synaptic activities of dopamine and studies of amphetamine psychosis as a model schizophrenia mediated via brain dopamine. Aspects of indole-amines relevant to schizophrenia are: (a) the psychotomimetic actions of indole-amine-related psychedelic drugs, and (b) the existence of enzymes capable of methyl-ating indoleamines to form psychedelic drugs in the human body.

DOPAMINE

Phenothiazine–catecholamine interactions

Effects of phenothiazines upon brain catecholamines would be relevant to schizophrenia only if the phenothiazine drugs exerted a selective antischizophrenic action so that their primary site of action might be assumed to play a prominent role in brain dysfunction in schizophrenia. This is a very strong assertion which can only be proposed tentatively. However, several lines of investigation do favour the notion that phenothiazines exert a unique antischizophrenic therapeutic action. In very extensive and well controlled studies comparing phenothiazines with sedatives such as barbiturates and the antianxiety agents, a greater efficacy for the phenothiazines has consistently been demonstrated. Moreover, nonsedating phenothiazine drugs are just as effective in treating schizophrenics as the 'sedating' ones. In addition, phenothiazines exert just as much therapeutic benefit in withdrawn as in hyperactive patients. Careful analysis of various symptoms of schizophrenic patients shows that the so called 'fundamental' symptoms of schizophrenia respond selectively to phenothiazine drugs, while accessory and non-schizophrenic symptoms are not as dramatically affected (KLEIN and DAVIS, 1969). These sort of data suggest strongly that the phenothiazine and butyrophenone drugs do exert a very specific therapeutic action in schizophrenia. Of course this action need not necessarily be at 'the schizo-phrenia receptor' in the brain.

How do the phenothiazines exert their antischizophrenic action? Because they are highly reactive chemicals, phenothiazine effects have been demonstrated upon almost every biochemical system studied. However, most of these actions do not correlate at all with clinical efficacy. Certain phenothiazines, such as the antihistamine prometh-azine are essentially devoid of antischizophrenic activity, yet are equally active as chlorpromazine in mediating many biochemical effects. The best correlation with clin-ical efficacy has emerged from studies of the actions of these drugs upon dopamine and norepinephrine disposition in the brain. CARLSSON and LINDQVIST (1963) first advanced the concept that phenothiazines block catecholamine receptors and that a feedback mechanism causes the catecholamine neurons to increase their firing rate. An increased firing rate should be reflected in enhanced turnover of the catecholamines, which has been well documented (ANDÉN, CARLSSON and HÁGGENDAL, 1969). The acceleration of catecholamine turnover after phenothiazine or butyrophenone drug treatment is more striking for dopamine than norepinephrine, and changes in turnover rate of dopamine correlate better with clinical efficacy than do changes in norepinephrine turnover. Thus, if phenothiazine action involves catecholamine receptor blockade, dopamine is the best candidate. Dopamine receptor blockade in the caudate by phenothiazines has been directly demonstrated by measurements of caudate adenylate cyclase (KEBABIAN, PETZOLD and GREENGARD, 1972) and more recently by direct recordings from olfactory tubercle cells which receive dopamine terminals and upon which the inhibitory effect of iontophoresed dopamine is blocked by phenothiazine or butyrophenone administration (AGHAJANIAN, BUNNEY and KUHAR, 1973).

A molecular mechanism whereby phenothiazines can fit into dopamine receptors is suggested by the close similarity between the optimal conformation of chlorpromazine determined by X-ray crystallography and the optimal conformation of catecholamines (Fig. 1). Indeed the ability of phenothiazines to assume the conformation which mimics dopamine very closely parallels their clinical efficacy. For instance, compounds lacking A ring substituents would be less capable of mimicking the dopamine conformation and, in the case of mepazine and promazine, have been shown to be less effective than other phenothiazines in alleviating schizophrenic symptoms (HORN and SNYDER, 1972). Moreover phenothiazines with a shorter side chain than chlorpromazine, e.g. promethazine, would also mimic the dopamine conformation less well and are less effective in the treatment of schizophrenia.

Amphetamines and catecholamines

Amphetamines link catecholamines and schizophrenia in a fashion opposite to the phenothiazines. While the phenothiazines antagonise schizophrenic symptoms and block dopamine receptors, amphetamines exascerbate schizophrenic symptoms and can elicit a schizophrenia-like psychosis in non-schizophrenic individuals, while enhancing the synaptic actions of dopamine. The amphetamine psychosis produced by large doses of the drug is an acute paranoid psychosis which is frequently mistaken by experienced psychiatrists for acute paranoid schizophrenia (CONNELL, 1958; BELL, 1965). Of all drug psychoses, amphetamine psychosis best mimics schizophrenia, since it is the only drug psychosis which frequently deceives skilled clinicians into a diagnosis of schizophrenia.

Some psychiatrists have speculated that amphetamine psychosis is merely a drug precipitation of latent schizophrenia, an effect of sleep deprivation or of overstimulation. However experimental studies in human subjects with no demonstrable schizoid tendencies have shown that amphetamine can reproducibly elicit psychosis in almost all subjects (GRIFFITH, CAVANAUGH, HELD and OATES, 1972). Since some subjects become psychotic within twenty four hours and since after the first 12 hr most subjects are not overtly stimulated, it is unlikely that the psychosis is related to sleep deprivation or hyperstimulation.

It has been argued that the psychosis does not faithfully mimic schizophrenia, because a typically schizophrenic thought disorder and disturbance of affect cannot be

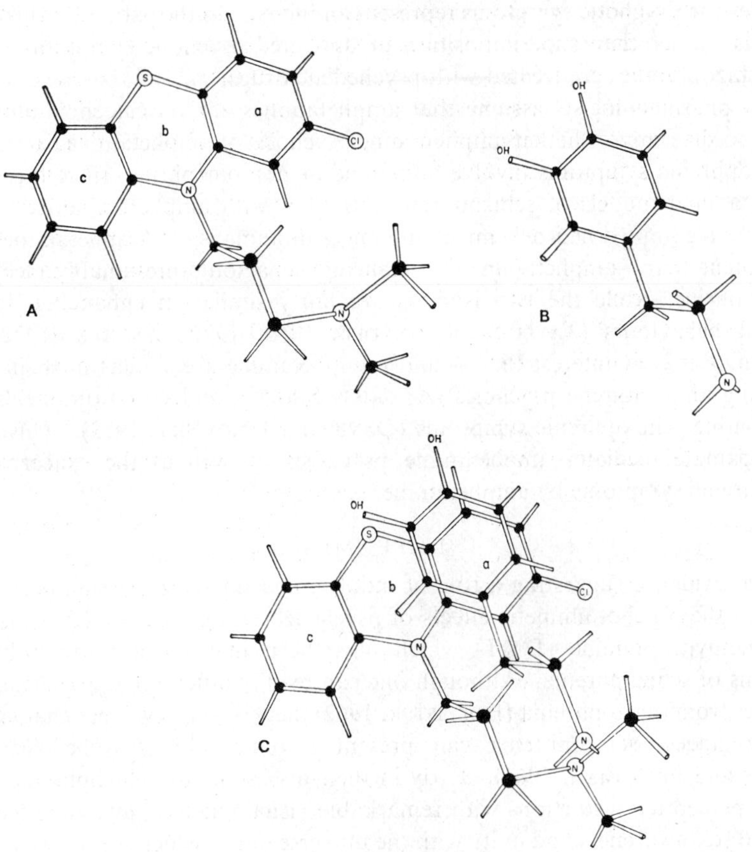

FIG. 1

proven in these individuals. However, in some experimental studies of amphetamine psychosis, the subjects did show thought disorder and affect disturbance (ANGRIST and GERSHON, 1970, 1971).

It is possible that a 'contaminating' alerting action of amphetamine transforms a schizophrenia-mimicking effect of the drug into a paranoid picture (SNYDER, 1972, 1973). In an acute paranoid process, clear-cut thought and affective disorders would not be as readily observed as in a more slowly developing schizophrenia. In addition the concept of a dual action of amphetamine, namely an alerting effect transforming a purer schizophrenia-like state into a paranoid process would explain another objection to the concept of amphetamine psychosis as a model schizophrenia, namely its failure to mimic undifferentiated schizophrenia. Thus, because amphetamine psychosis is almost always manifested as a paranoid psychosis, it is too restricted to be a 'model schizophrenia', unless one postulates extraneous pharmacological actions of the drug transforming the 'schizophreniform psychosis' into a paranoid psychosis.

Another link between amphetamine and schizophrenia is the ability of very small doses of amphetamine or related agents such as methylphenidate, to produce a florid exascerbation of schizophrenic symptoms (JANOWSKY, EL-YOUSEL, DAVIS and SEKERKE, 1973). These drugs do not exascerbate depressive or manic psychosis.

Moreover the psychotic symptoms represent an increase in the patient's schizophrenic psychosis, rather than superimposition of unrelated psychotic symptoms as occurs when schizophrenics are treated with psychedelic drugs.

Most pharmacologists assume that amphetamines act via catecholamines. It is difficult to determine whether amphetamine psychosis or amphetamine exasercbation of schizophrenic symptoms involve dopamine or norepinephrine, since there are no suitable animal models of schizophrenia. Studies with amphetamine isomers have made possible limited neurochemical inferences in humans. *d*-Amphetamine is much more potent than *l*-amphetamine in enhancing behaviours presumably mediated by norepinephrine, while the two isomers are fairly similar in enhancing dopamine mediated behaviours (TAYLOR and SNYDER, 1970, 1971; SNYDER, 1972, 1973). Accordingly it is of interest that *d*- and *l*-amphetamine are similar in their potency in eliciting amphetamine psychosis (ANGRIST, SHOPSIN and GERSHON, 1971) and in exasercbating schizophrenic symptoms (DAVIS and JANOWSKY, 1973). This suggests that dopamine mediates amphetamine psychosis as well as the exacerbation of schizophrenic symptoms by amphetamine.

INDOLEAMINES

Initial evidence suggesting a role of indoleamines in schizophrenia derived from studies of the psychotomimetic effects of psychedelic drugs such as LSD, psilocybin and dimethyltryptamine (DMT) which have been interpreted as resembling the symptoms of schizophrenia. Although one can readily differentiate psychedelic drug psychoses from schizophrenia (HOLLISTER, 1962) there is some evidence that, at least in its early stages, schizophrenia can present a remarkably 'psychedelic' picture (BOWERS and FREEDMAN, 1966). Early in their disease many schizophrenic patients describe perceptual distortions with remarkable visual changes, awesome feelings of enhanced self-awareness and unity with the universe, all of which seem very much like the symptoms evoked by psychedelic drugs.

The existence of psychedelic drug psychosis and the fact that drugs such as mescaline and dimethyltryptamine resemble catecholamines and indoleamines respectively prompted the speculation that the brains of schizophrenics might methylate normally occurring amines to form psychotomimetic agents which account for schizophrenic symptoms. *O*-dimethylation of catecholamines might form dimethoxy-dopamine, which resembles mescaline, but is not itself psychotomimetic (FRIEDHOFF and VAN WINKLE, 1967). Alternatively one might postulate the *N*-methylation of indoleamines (POLLIN, CARDON and KETY, 1961; MANDELL and SPOONER, 1968). Enzymes which can *N*-methylate a variety of biogenic amines have been reported in rabbit lung (AXELROD, 1962), chick and human brain (MORGAN and MANDELL, 1969; MANDELL and MORGAN, 1971) and in several mammalian tissues (SAAVEDRA and AXELROD, 1972; SAAVEDRA, COYLE and AXELROD, 1973). Unfortunately, these are all relatively feeble enzymatic activities. While all of these *N*-methylations utilise *S*-adenosylmethionine (AMe) as a methyl donor, recently LADURON (1972) showed that methyltetrahydrofolic acid (MTHF) can serve as the methyl donor in the methylation of dopamine to epinine. We have found that MTHF can serve as a methyl donor in the methylation of a variety of indoleamines as well as phenethylamines (BANERJEE and SNYDER, 1973). With MTHF as methyl donor, this enzymatic activity is much more vigourous than with AMe suggesting a more important biological role for this reaction than was evident from earlier studies with AMe.

We found amine methylating activity in a variety of mammalian tissues. The enzyme from rabbit lung (AXELROD, 1962) was quite different from that in other tissues, since it was the only one which preferred AMe as methyl donor. In all of the tissues MTHF was considerably more active than AMe as methyl donor (Table 1). Both indoleamines and phenothylamines were methylated. Strikingly, no methylation of serotonin could be demonstrated with AMe, while with MTHF serotonin was the best amine substrate.

TABLE 1. SPECIES AND TISSUE DISTRIBUTION OF METHYLTRANSFERASE ACTIVITY

Tissues were homogenised in 10 volumes of 5 μM Na-phosphate buffer, pH 7·9, and enzyme activity was assayed in the 100,000 g supernatant fraction after dialysis with serotonin (5 mM) or tyramine (5 mM) as substrates and S-adenosylmethionine (AMe) (1 μM) or 5-methyltetrahydrofolic acid (MTHF) (1 μM) as methyl donors. Data are the mean of 3 experiments whose results varied less than 20 per cent).

| Tissue | Methyltransferase activity* | | | | | | Ratio of enzyme activity with AMe to activity with MTHF | |
| | AMe as methyl donor | | | MTHF as methyl donor | | | | |
	Serotonin	Tyramine	Ratio Serotonin/Tyramine	Serotonin	Tyramine	Ratio Serotonin/Tyramine	Serotonin	Tyramine
Rabbit lung	32·0	24·0	1·33	8·0	4·0	2·0	4·0	6·0
Brain	0	0·05	0	2·6	1·4	1·9	0	0·04
Liver	0	0·2	0	2·1	1·2	1·7	0	
Rat brain	0	0·4	0	3·0	1·3	2·3	0	0·31
Liver	0	1·2	0	4·3	1·4	3·1	0	0·90
Lung	0	0·35	0	4·0	1·3	3·1	0	0·27
Heart	0	0·4	0	8·0	3·0	2·7	0	0·14
Chick brain	0	1·0	0	6·0	4·0	1·5	0	0·25
Heart	0	2·0	0	26·0	11·0	2·3	0	0·18

* nmoles/mg protein/hr.

Purification of the enzyme 20-fold from rat brain by ammonium sulphate fractionation and negative adsorption on alumina-C-γ gel (BANERJEE and SNYDER, 1973) produced a preparation which methylated vigourously with MTHF but was completely inactive with AMe, suggesting the existence of an enzyme which uses MTHF exclusively in methylating amines, although the possibility of a change in methyl donor properties during purification cannot be ruled out. The purified enzyme as well as the crude supernatant fraction was most active towards serotonin (Table 2). The K_m of the partially purified enzyme for MTHF was 1 μM and its K_m values for serotonin and tyramine were 0·1 mM and 1 mM respectively.

Thin layer chromatographic analysis in several systems showed that, while phenethylamines, tryptamine, and 5-methoxytryptamine are methylated on the amine nitrogen, serotonin is predominantly methylated on the 5-hydroxyl group to form 5-methoxytryptamine. Confirmation of the O-methylation of indoleamines was obtained by showing that bufotenin, in which the amine nitrogen is already dimethylated, is vigourously methylated to form the potent hallucinogen, 5-methoxy-N,N-dimethyltryptamine. The relative methylation of bufotenin and 5-methoxytryptamine is constant in 4 different ammonium sulphate fractions of the enzyme purified from rat liver, suggesting that the same enzyme mediates O- and N-methylation. The failure of serotonin to function as a substrate with AMe may indicate that O-methylation requires MTHF, while both MTHF and AMe can serve as donors for N-methylation.

In considering what role this enzyme might play in the patho-physiology of schizophrenia or other psychoses, it is of interest that serotonin is its best substrate.

Table 2. Substrate specificity of partially purified rat brain methyltransferase with MTHF as methyl donor

The enzyme preparation used was the supernatant fraction after 75%
ammonium sulphate precepitation Enzymatic activity is expressed as
nmoles/mg protein/hr. The following specific extraction procedures were
used for the optimal isolation of a given product: (1) extracted with a mixture
of toluene and isoamyl alcohol (3:2) and dried overnight at 80°C in a chromatography oven: (2) extracted with a toluene isoamyl alcohol mixture
(97:3) and dried in oven; (3) extracted with isomyl alcohol and dried in
oven; (4) extracted as in (2) but counted with no previous drying procedure
because of the volatility of the product.)

Substrate	Specific activity	Extraction procedure
Tyramine	4·4	1
Tryptamine	4·3	2
Serotonin	8·8	3
β-phenylethylamine	2·0	4
Octopamine	2·7	1
N-methyltryptamine	3·4	2
N-methylserotonin	7·2	3
Desmethylimipramine	0·0	3

However, it is important to bear in mind that brain was one of the least active of
tissues examined.

Acknowledgements—Supported by USPHS Grants MH-18501, NS-07275, DA-00266, grants of
the John A. Hartford Foundation and the Scottish Rite foundation, Research Scientist Development
Award MH-33128 to S. H. S. and a fellowship of the Medical Research Council of Canada to S. P. B.

REFERENCES

Aghajanian G. K., Bunney B. S. and Kuhar M. J. (1973) in *New Concepts in Neurotransmitter
Regulation* (Mandell A. J., ed.), Plenum Press, New York, pps. 115–134.
Anden N. E., Carlsson A. and Haggendal J. (1969) *Ann. Rev. Pharmacol.* 9, 119–134.
Angrist B. and Gershon S. (1970) *Amer. J. Psychiat.* 126, 95–107.
Angrist B. M., Shopsin B. and Gershon S. (1971) *Nature* 234, 152–154.
Axelrod J. (1962) *J. Pharmacol. Exp. Ther.* 138, 28–33.
Banerjee S. P. and Snyder S. H. (1973) *Science*, 1973, in press.
Bell D. S. (1965) *Brit. J. Psychiat.* 111, 701–707.
Bowers M. J., Jr. and Freedman D. X. (1966) *Arch. Gen. Psychiat.* 15, 240–248.
Carlsson A. and Lindqvist (1963) *Acta Pharmacol. Toxicol.* 20, 140–144.
Connell P. H. (1958) *Amphetamine Psychosis.* Chapman and Hall, London.
Davis J. M. and Janowsky D. S. (1973) *This Volume*, in press.
Friedhoff A. and Van Winkle E. (1967) in *Amines and Schizophrenia* (Himwich H. E., Kety S. S.
and Smythies Jr., eds.) Pergamon Press, Oxford, pps. 19–22.
Griffith J. D., Cavanaugh J., Held J. and Oates J. A. (1972) *Arch. Gen. Psychiat.* 26, 97–100.
Hollister L. E. (1962) *Ann. N.Y. Acad. Sci.* 96, 80–88.
Horn A. S. and Snyder S. H. (1971) *Proc. Natl. Acad. Sci. U.S.A.* 68, 2325–2328.
Janowsky D. S., El-Yousel M. K., Davis J. M. and Sekerke H. J. (1973) *Arch. Gen. Psychiat.*,
28, 185–191.
Kebabian J., Petzold G. L. and Greengard P. (1972) *Proc. Natl. Acad. U.S.A.* 69, 2145–2149.
Klein D. F. and Davis J. M. (1969) *Diagnosis and Drug Treatment in Psychiatry.* Williams and
Wilkins, Baltimore, pps. 52–138.
Laduron P. (1972) *Nature New Biology* 238, 212–213.
Mandell A. J. and Morgan M. (1971) *Nature New Biology* 230, 85–87.
Mandell A. J. and Spooner C. E. (1968) *Science* 162, 1442–1453.
Morgan M. and Mandell A. J. (1969) *Science* 165, 492–493.
Pollin W., Cardon P. V., Jr. and Kety S. S. (1961) *Science* 133, 104–105.
Saavedra J. M. and Axelrod J. (1972) *Science* 172, 1365–1367.
Saavedra J. M., Coyle J. T. and Axelrod J. (1973) *J. Neurochem.* 20, 743–752.
Snyder S. H. (1972) *Arch. Gen. Psychiat.* 27, 169–179.
Snyder S. H. (1973) *Amer. J. Psychiat.* 130, 61–67.
Taylor K. M. and Snyder S. H. (1970) *Science* 168, 1487–1489.
Taylor K. M. and Snyder S. H. (1971) *Brain Res.* 28, 295–309.

Frontiers in Catecholamine Research 1973, pp. 1139 to 1142. Pergamon Press. Printed in Great Britain.

IMPLICATIONS OF FEEDBACK CONTROL IN CATECHOLAMINE NEURONAL SYSTEMS

STEVEN MATTHYSSE

Massachusetts General Hospital, Boston, Mass. 02114, U.S.A.

THE THEORY of feedback control of catecholamine neuronal systems leads one to expect certain peculiarities in their response to drugs which block or potentiate transmission at catecholamine synapses, peculiarities which may be relevant both to the interpretation of experiments with the drugs and to their use in therapy. These ideas were first suggested, I believe, by the neurologist Dr. Janice Stevens, and Dr. Kety made use of them in his talk at the conference on 'Catecholamines and their Enzymes in the Neuropathology of Schizophrenia'. This presentation adds only a slightly more complete theoretical development.

It is very well documented that the level of activity in the nigrostriatal dopamine tract is under negative feedback control; agents which block transmission, such as neuroleptics, increase the rate of neuronal activity in the substantia nigra and also the release of dopamine, as measured by its metabolite homovanillic acid. Conversely, as was reported by Dr. Aghajanian at this Symposium, amphetamine decreases the rate of firing in the substantia nigra. From the data obtained by his group, there is reason to think that feedback control is also present in the noradrenaline system emanating from the locus coeruleus, since amphetamine decreases activity in the locus coeruleus and this decrease is blocked by chlorpromazine.

If the feedback loop characteristic of these systems is a one-to-one inverse projection, that is, if the cell innervated by a given catecholamine neuron projects downward, without branching, directly to the cell of origin, the consequence of negative feedback will simply be to attenuate responses to blocking and potentiating drugs. On the other hand, it may be important to consider the case where the identity of the ascending neuron is lost through branching, so that the cell of origin may receive its feedback innervation from quite another neuron than the one to which it sends its ascending signals. There may even be such thorough mixing that the feedback compensation becomes only a statistical average applied diffusely to the cells of origin. Let us call this the 'mixing hypothesis'. When the mixing hypothesis is valid, the application of drugs may give rise to some unexpected effects.

Suppose, for example, that some neurons arising in the locus coeruleus make alpha-adrenergic connections in the cortex, while others are of the beta type. Just for discussion, let us suppose that most neurons originating in the locus coeruleus are alpha, and only a few are beta. Now imagine treatment with an alpha blocker. Because of the loss of negative feedback from the principal alpha pathway, the level of activity in the nucleus goes up substantially. Since by the mixing hypothesis the feedback is applied diffusely, the beta neurons will increase in firing rate as well as the alpha. They, however, are not synaptically affected by the alpha blocker, so the net effect is potentiation of the beta-adrenergic pathway to the cortex, along with inhibition of the alpha-adrenergic pathway. It is easily seen that the mixing condition

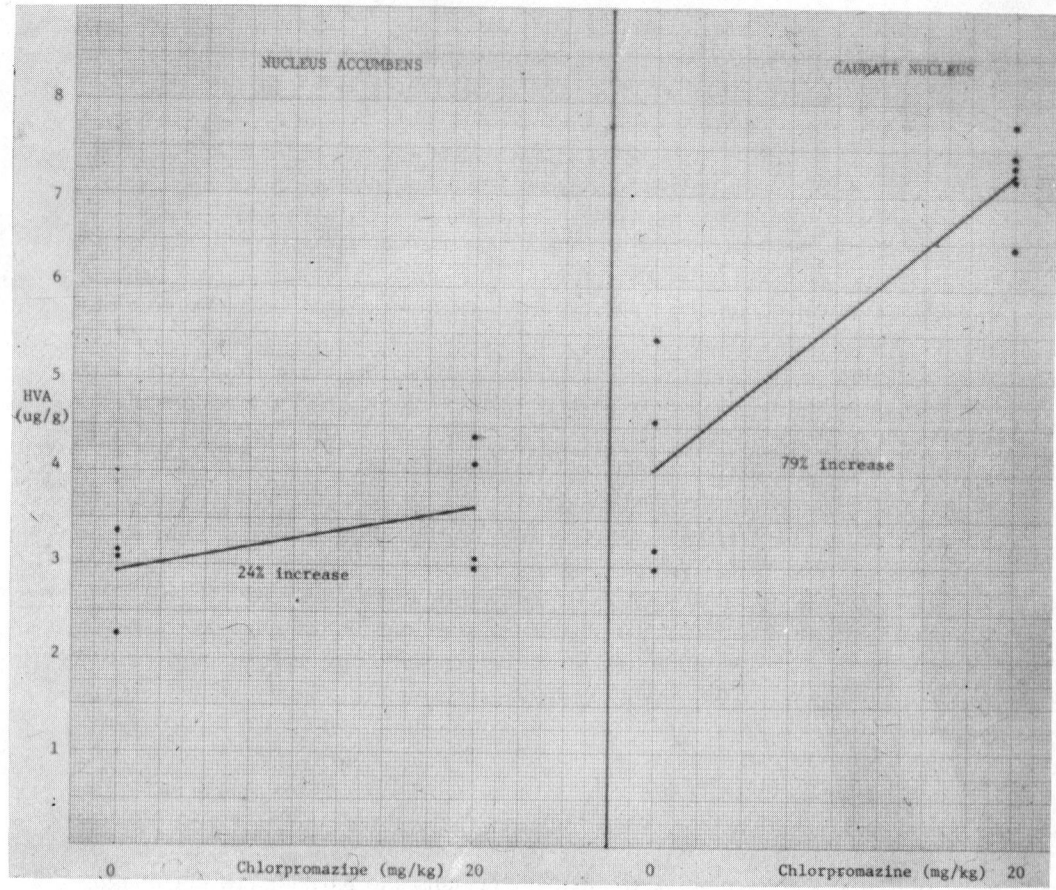

always amplifies the divergence between the magnitude of a drug's action on different classes of synapses. Mixing will be of no consequence if a particular drug has no differential effect on efferent branches from a nucleus. On the other hand, differential effects should be a relatively common occurrence. In addition to the alpha- and beta-receptor case just mentioned, Bradley observed that chlorpromazine blocked excitatory noradrenaline synapses in the brain stem, but did not block inhibitory noradrenaline responses. The magnitude of the amplification of differential drug effects will depend on the level of feedback control and the degree of asymmetry of the fibre bundles emanating from the nucleus in which mixing takes place. If feedback control were complete, blocking drugs would have no effect at all on major pathways, and effects exactly opposite to what is expected on minor pathways where they are less effective. With less complete feedback control, the major pathway would be blocked to some degree, and the minor pathway would either be facilitated or inhibited, depending on the quantitative relationships. The greater the discrepancy in size between the major and minor fibre bundles, the greater the paradoxical effects to be expected in the minor pathway.

There are several ways in which the validity of the mixing hypothesis for a particular catecholamine cell group can be tested; one of them is suggested by the experiments of Mr. Allen Nineberg in our laboratory, which are illustrated in the accompanying figure. Cats are injected with chlorpromazine or saline and sacrificed after four hours. The caudate and nucleus accumbens septi are dissected separately in the frozen tissue, and assayed for homovanillic acid. Because of the small size of the nucleus accumbens, it is necessary to pool tissue from two animals for each data point. The rise in homovanillic acid after chlorpromazine is much larger in the caudate than in the nucleus accumbens. If the mixing hypothesis were true for the system whose terminals are represented by these two structures, the percent rise should have been the same in both regions. According to data obtained by Andén and by Lloyd, after haloperidol the percentage rise in homovanillic acid in caudate and limbic system is approximately the same. The mixing hypothesis requires, however, that the percent rise be the same in all regions innervated by the cell group in which mixing takes place, not merely for some, but for all drugs, at least insofar as they act by blockade or potentiation of catecholamine synapses. The most likely explanation of the difference between chlorpromazine and haloperidol is that haloperidol blocks caudate and accumbens dopamine synapses equally, whereas chlorpromazine is less effective in the nucleus accumbens; but this inequality of effect would not manifest itself in differential homovanillic acid rise if the mixing hypothesis were valid for the cell group of origin. Of course it is no surprise that it is invalid, since the A9 region which innervates the caudate, and the A10 region which innervates the accumbens, are morphologically distinct cell groups. I think it is nevertheless reasonable to conjecture that within each of the catecholamine cell clusters, the mixing hypothesis applies. A diffuse downward projection is likely in view of the highly branched and ramified terminal plexi of the neurons of these groups.

The possibility of feedback mixing introduces some complexities in the interpretation of experiments. The most obvious is uncertainty whether a behavioural effect results from activation of a major pathway or inhibition of a minor pathway (or vice-versa). Furthermore, we should not be suprised to discover paradoxical effects of intraventricular drug administration. For example, endogenous norepinephrine may be involved in behavioural alerting through its effects on the cerebral cortex, but when introduced intraventricularly it could cause sedation by feedback inhibition of noradrenaline neurons, resulting from activation of cells postsynaptic to noradrenergic fibres in the periventricular zone of the hypothalamus. We might also expect qualitative rather than merely quantitative changes in drug effects as dose increases, that is, the emergence of new phenomena at high doses. At low doses, when the level of feedback control is high, a drug effect on a minor branch may not be observed because of feedback mixing from a more strongly affected pathway. In the case of synaptic blocking agents, the system may partially escape from feedback control at high doses, because facilitation of firing in the nucleus of origin can never exceed the maximum rate of which the neurons are capable. In the case of synaptic facilitating agents, the more strongly affected pathway will eventually reach a point of maximum potentiation, after which the degree of asymmetry in drug effect on the major and minor branch will decrease steadily as dose increases, permitting the minor branch effect to emerge. Examples of emergence of new phenomena at high doses include amphetamine stereotypy, which is not continuous with the behavioural

activation observed at low doses, and neuroleptic-induced catalepsy, which is unlike
the sedative effects observed with lower doses of the drug.

Certain behavioural antagonisms may also find explanation through the mixing
hypothesis. One is tempted to apply the principle to Dr. Leibowitz's finding that
alpha receptors stimulate feeding and inhibit drinking, while beta receptors do the
reverse. The mutual antagonism of these two behaviours may be caused by reciprocal
inhibitory innervation of the hypothalamic receptors, but it could also be accounted
for by the mixing hypothesis. If the feeding and drinking centers receive their
adrenergic innervation from the same norepinephrine cell group, an alpha blocker
might potentiate the beta branch through feedback facilitation, and vice versa.
Similar reasoning could be applied to other behavioural antagonisms. Dr. Randrup
has pointed out that during apomorphine or amphetamine-induced stereotypy there
is not only facilitation of the compulsive activities, but also inhibition of ordinary
goal-seeking motivated behaviour. Conceivably this antagonism could be caused by
differential effects of the drugs on different branches of the A10 system.

There may be some potential applications of the mixing hypothesis to paradoxical
drug responses in man. Why does pimozide, a dopamine blocker, increase arousal
and motivation in chronic, withdrawn schizophrenics? Why do acute dystonic
reactions which resemble dopamine-induced hyperkinesias sometimes occur during
treatment with drugs expected to block dopamine synapses? Why does amphetamine
have a quieting effect on hyperactive children? The application of feedback facilitation to the first of these phenomena was suggested by Dr. Kety; similar reasoning
might apply to the others. The existence of mixing in the catecholamine feedback
control systems, and its significance for behaviour in response to drugs, remain to be
demonstrated; but I do not think they lie outside the range of experimentation.

Frontiers in Catecholamine Research 1973, pp. 1143 to 1147. Pergamon Press. Printed in Great Britain.

INCREASED SERUM DOPAMINE-β-HYDROXYLASE ACTIVITY DURING NEUROGENIC HYPERTENSION IN QUADRIPLEGIA

N. Eric Naftchi[1],* G. Frederick Wooten[2],
Edward W. Lowman[1]* and
Julius Axelrod[2]

[1] From the Laboratory of Biochemical Pharmacology, Institute of Rehabilitation Medicine, New York University Medical Center, New York, N.Y. 10016, U.S.A.

and

[2] From the Laboratory of Clinical Science, National Institute of Mental Health, Bethesda, Maryland, 20014, U.S.A.

Patients with high level spinal cord injury, above the sympathetic outflow at the level of thoracic six dermatome very often develop spontaneous hypertensive crises due to any noxious stimuli (Guttman and Whitteridge, 1947). These stimuli usually arise from the urinary bladder due to cystitis or kidney stone formation, or from the rectum because of rectal impaction.

Synthesis of norepinephrine (NE) is catalysed by dopamine-β-hydroxylase enzyme (DBH) from the precursor 3,4-dihydroxyphenylalanine (dopamine) (Kaufman and Friedman, 1965). DBH has been found in the catecholamine containing granules in the heart (Potter and Axelrod, 1963), synaptosomes of the brain (Coyle and Axelrod, 1942), splenic nerves (Stjarne and Lishajko, 1967) and localised in chromaffin granules of the adrenal medulla (Molinoff et al., 1970). It is also present in the serum of a variety of mammalian species (Weinshilboum and Axelrod, 1971b). It has been shown that DBH is proportionally released together with neurotransmitter NE by the process of exocytosis (Weinshilboum et al., 1971a). Furthermore, adrenalectomy in rat was found not to alter the base line levels of serum DBH (Weinshilboum et al., 1971b).

An increase in HVA output (Smith and Dancis, 1967) and a decrease in serum DBH activity has been demonstrated in familial dysautonomia (Freedman et al., 1972; Weinshilboum and Axelrod, 1971a). It has also been shown that catecholamine metabolites were significantly enhanced during hypertension in chronic quadriplegia (Naftchi et al., 1971; Sell et al., 1972). Serum DBH activity, therefore, was analysed in chronic quadriplegic subjects as an index of sympathetic activity.

METHODS

A group of eight quadriplegic patients were self-compared before and after expansion of the urinary bladder by means of water intake. All patients were during the chronic phase, 6 months or longer after the onset of the injury, and had suffered a complete physiologic transverse lesion at the level of the 5th and 7th cervical dermatomes (C_5–C_7). Brachial blood pressure was measured by auscultatory technique and digital blood flow was measured calorimetrically in the fourth finger.

* Supported by The Edmond A. Guggenheim Clinical Research Endowment and in part by S.R.S., Department of H.E.W.

TABLE 1. SERUM DOPAMINE-β-HYDROXYLASE ACTIVITY AND URINARY CATECHOL-
AMINE METABOLITES BEFORE AND DURING HYPERTENSIVE CRISES IN SPINAL MAN

No. of subjects	Brachial blood pressure (mmHg)		VMA*		HVA*		HMPG*		DβH†	
	Before	During	Before	During	Before	During	Before	During	Before	During
1	108/68	230/130	1·5	5	2	10	2	6	718	1652
2	144/102	210/126	3	7	6	16	4	8	1155	1916
3	128/80	176/122	3	5	4	7	2	3	594	646
4	136/98	172/110	2·5	5	6	12	3	5·5	1394	1617
5	112/78	150/90	1	2·5	2	5	2	4	307	484
Mean	126/85	188/116	2·2	4·9	4·0	10	2·6	5·3	833	1263
± S.D.	—	—	0·91	1·6	2·0	4·3	0·89	1·9	437	650
P				>0·001		>0·001		>0·001		>0·05

* μg/mg creatinine.
† nmoles phenylethanolamine/ml serum/hr.

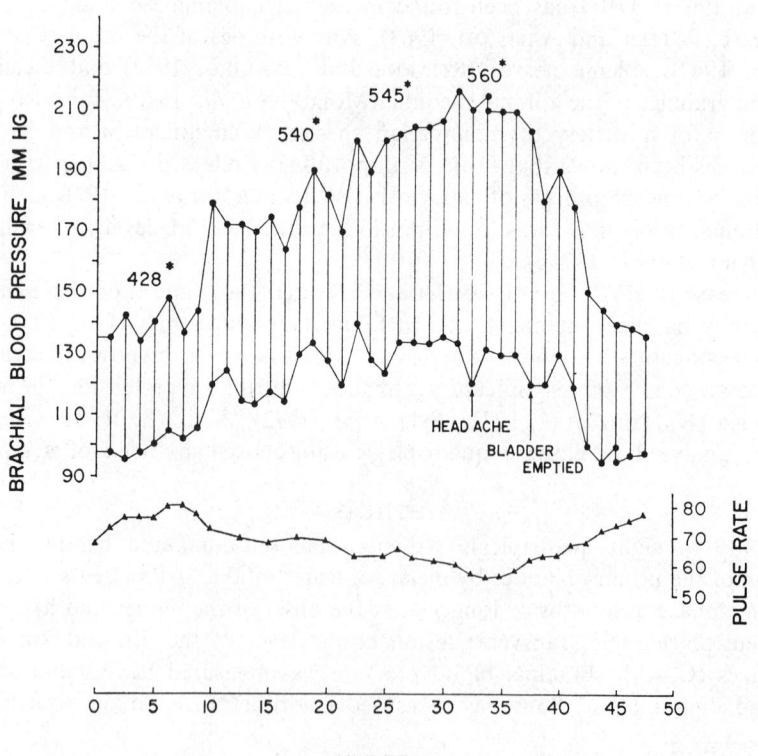

CORRELATION OF CHANGES IN ARTERIAL BLOOD
PRESSURE WITH DβH ACTIVITY.

FIG. 1.—Numbers represent DβH (nmoles phenylethanolamine formed/ml serum/hr)

Serum DBH activity was analysed before and during the height of hypertension. Serum DBH activity was also determined in five patients during spontaneous hypertensive crises and at the time when the patients' blood pressure was at resting control levels. Serum DBH activity was determined by a modification (WEINSHILBOUM and AXELROD, 1971a) of the sensitive isotopic method of MOLINOFF *et al.* (1970). In addition, urine specimens from the five quadriplegic subjects were collected just before, during or immediately after a spontaneous hypertensive crisis.

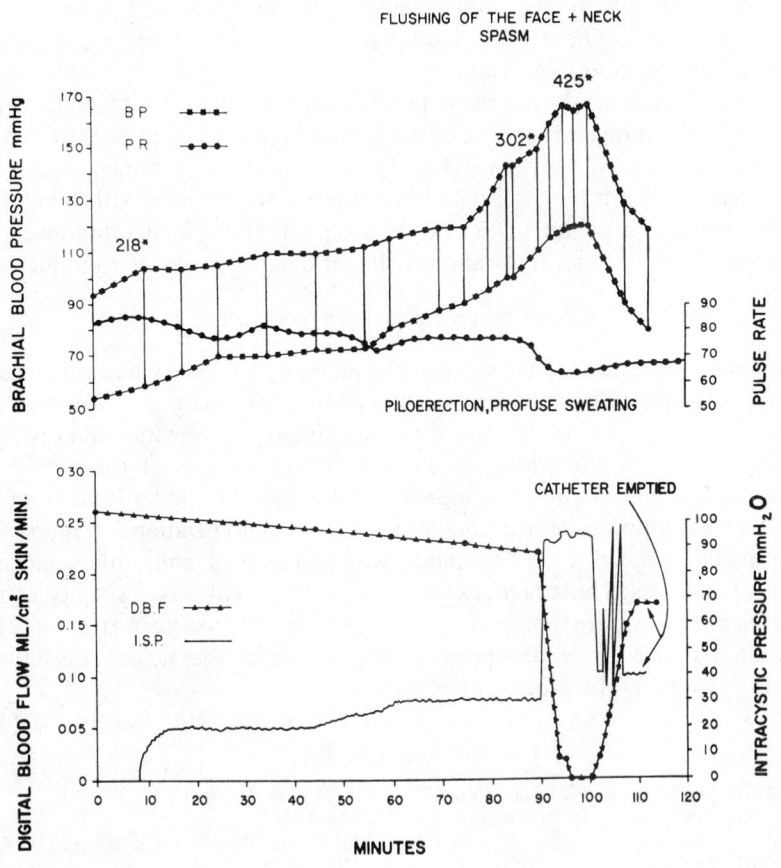

EFFECTS OF BLADDER FILLING ON AUTONOMIC
RESPONSE AND DβH ACTIVITY

*NUMBERS REPRESENT DβH (nMOLES PHENYLETHANOLAMINE FORMED/ml SERUM/hr)

FIG. 2.—Crystometric studies were performed in a C_7 quadreplegic man. Note that the height of the DBH activity coincides with that of brachial and intracystic pressure, at which time the digital blood flow has dropped to immeasurable amounts and pulse rate is at its lowest level.

BP = Brachial blood pressure   PR = Pulse rate
DBF = Digital blood flow   ISP = Intracystic pressure

Urine samples were analysed for their content of catecholamine metabolites, 4-hydroxy-3-methoxymandelic acid, vanillmandelic acid (VMA), 4-hydroxy-3-methoxy-phenylacetic acid, homovanillic acid (HVA), and 4-hydroxy-3-methoxyphenylethylene glycol (HMPG). VMA and HVA were analysed by bidimensional paper chromatography (Armstrong et al., 1956) and HMPG was analysed by gas–liquid chromatography (Wilk et al., 1967).

RESULTS

In five C_7-C_8 quadriplegic subjects who developed spontaneous hypertensive crises (Table 1) the mean brachial blood pressure rose from a resting value of 126/85 mmHg to 188/116 mmHg. The mean values for each CM metabolite, expressed in micrograms per milligram of creatinine in urine, increased significantly during hypertensive crisis ($P < 0.001$). Such high values are found only in patients with pheochromocytoma or neuroblastoma. Similarly, serum DBH was significantly enhanced during hypertension (Table 1).

In one C_5-C_6 subject, during the expansion of the urinary bladder, the gradual rise in arterial blood pressure is accompanied by progressive increase in serum DBH activity and a decrease in pulse rate (Fig. 1). In another C_7 quadriplegic subject it is demonstrated that the height of arterial blood pressure coincides with that of DBH activity and intracystic pressure (Fig. 2). At this point of maximal autonomic activity the pulse rate decreases markedly and the digital blood flow drops to immeasurable amounts.

CONCLUSIONS AND SUMMARY

Autonomic response in quadriplegia is manifested by headache, piloerection, a red flush, and profuse diaphoresis above the level of transection. In contrast to the vasodilatation cephalad to the lesion, there is a marked vasoconstriction of peripheral blood vessels (Fig. 2) and pallor of the skin below the level of transection. The hypertension, brought about spontaneously or by means of water intake, is caused by increased elaboration of catecholamines. The concentration of homovanillic acid, the major metabolite of dopamine, was enhanced significantly concomitant with that of other catecholamine metabolites and with enhanced activity of serum DBH. In agreement with previous *in vitro* studies (Weinshilboum et al., 1971), these results indicate an *in vivo* proportional release of the neurotransmitter NE with DBH possibly by the process of exocytosis.

REFERENCES

Armstrong M. D., Shaw K. N. F. and Wall P. E. (1956) *J. Biol. Chem.* **218**, 293–303.
Coyle J. T. and Axelrod J. (1972) *J. Neurochem.* **19**, 449–459.
Freedman L. S., Ohuchi T., Goldstein M., Axelrod J., Fish I. and Dancis J. (1972) *Nature, (Lond.)* **236**, 310–311.
Guttman L. and Whitteridge D. (1947) *Brain* **70**, 361–404.
Kaufman S. and Friedman S. (1965) *Pharmacol. Rev.* **17**, 71–100.
Molinoff P. B., Brimijoin W. S., Weinshilboum R. M. and Axelrod J. (1970) *Proc. Nat. Acad. Sci. U.S.A.* **66**, 453–458.
Naftchi N. E., Lowman E. W., Sell H. and Rusk H. (1971) *Fedn. Proc.* **30**, 678.
Potter L. T. and Axelrod J. (1963) *J. Pharmacol. Exp. Ther.* **142**, 299–305.
Sell G. H., Natchi N. E., Lowman E. W. and Rusk H. (1972) *Arch. Phys. Med. Rehab.* **53**, 415–417.
Smith A. A. and Dancis J. (1967) *New Eng. J. Med.* **227**, 61–64.
Stjarne L. and Lishajko F. (1967) *Biochem. Pharmacol.* **16**, 1719–1728.

WEINSHILBOUM R. M. and AXELROD J. (1971a) *New Eng. J. Med.* **285,** 938–942.
WEINSHILBOUM R. M. and AXELROD J. (1971b) *Circ. Res.* **28,** 307–315.
WEINSHILBOUM R. M., THOA N. B., JOHNSON D. G., KOPIN I. J. and AXELROD J. (1971a) *Science* **174,** 1349–1351.
WEINSHILBOUM R. M., KVETNANSKY R., AXELROD J. and KOPIN I. J. (1971b) *Nature (New Biol.)* **230,** 287–288.
WILK S., GITLOW S. E., CLARKE D. D. and PALEY D. H. (1967) *Clin. Chim. Acta* **16,** 403–408.

Frontiers in Catecholamine Research 1973, pp. 1149 to 1155. Pergamon Press. Printed in Great Britain.

HYPOTHALAMIC RELEASING HORMONES AND CATECHOLAMINES: A NEW INTERFACE

ARTHUR J. PRANGE, JR.[1] IAN C. WILSON[2], GEORGE R. BREESE[1],

NICHOLAS P. PLOTNIKOFF[3], PATRICIO P. LARA[2] and MORRIS A. LIPTON[1]

[1]Department of Psychiatry, Medical School and the Biological Sciences Research Center of the Child Development Institute, University of North Carolina, Chapel Hill, North Carolina 27514, U.S.A.

[2]Division of Research, North Carolina Mental Health Department, Raleigh, North Carolina 27611, U.S.A.

and

[3]Abbott Laboratories, North Chicago, Illinois 60064, U.S.A.

SEVERAL years ago we demonstrated that small quantities of triiodothyronine (T3) added to imipramine in the treatment of depression accelerated the recovery process (PRANGE et al., 1970). Additional investigations as to the mechanism of interaction between the hormone and the drug led to the demonstration that in hypothyroidism there was increased synthesis and turnover of catecholamines (LIPTON et al., 1968), while in hyperthyroidism the converse was true (PRANGE et al., 1970a). These and other studies led us to suggest (PRANGE et al., 1972a) that thyroid hormones affect receptor sensitivity, and that while the concentration of catecholamines alone is insufficient to explain pathological mood states, the concentration of amines in relation to receptor sensitivity might be sufficient to do so.

Continuing our studies on the role of thyroid hormone in the neurobiology of the mood disorders, we moved up the thyroid axis and were able to demonstrate that thyrotropin (thyroid stimulating hormone, TSH) had effects similar to those of T3 (PRANGE et al., 1970b). Somewhat later synthetic thyrotropin releasing hormone (TRH) became available, and we began studies with this substance.

The hypothalamus elaborates a series of substances whose only generally accepted function is to migrate via the portal system to the anterior lobe of the pituitary gland and there to prompt or inhibit the release of tropic hormones. Many if not all these substances are polypeptides (SCHALLY et al., 1973). Our recent evidence as well as that of others suggests that three of these hypothalamic hormones, or 'factors,' exert behavioural effects, that these effects may be independent of pituitary actions, and that they involve at least in part catecholaminergic mechanisms.

THYROTROPIN RELEASING HORMONE (TRH)

TRH, pyroglutamyl-histadyl-proline amide, is a potent releaser of pituitary prolactin (BOWERS et al., 1971) and TSH (FLEISCHER et al., (1970).

Behavioural effects in humans

(a) *Depression*. The logic which led to our trials of TRH in clinical depression has already been described. This was reinforced by the finding that in animals TRH is active in the pargyline-DOPA mouse activation test (PLOTNIKOFF et al., 1972), which has been useful in identifying antidepressant substances (EVERETT, 1966).

In our clinical trial ten women, ages 25–45, with primary, unipolar depression but in good physical health were treated in a double-blind, placebo-controlled, cross-over design. Randomly assigned, they received a single i.v. injection of TRH (600 μg) followed by saline injection one week later or received the injections in reverse order. On injection days mental status examination and the 100-mm line test (a subjective assessment of depression) (AITKEN, 1969; ZEALLEY and AITKEN, 1969) were performed seven times. They were also performed daily throughout the study. The Hamilton Rating Scale for Depression (HRS) (HAMILTON, 1960) and the Taylor Manifest Anxiety Scale (TAS) (TAYLOR, 1955) were used daily, including injection days. Data pertaining to the ten patients under the TRH condition were compared to data from the same patients under the saline condition.

Results from the four clinical measures were compared as they pertained to time periods (after injections) of three durations. TRH conferred significantly more benefit than saline in five of twelve instances; saline was never better than TRH. Most patients were markedly improved on TRH days, often within three hours. Later they tended to relapse somewhat, though significant benefits persisted as long as the fourth day (HRS and TAS, $P < 0.01$) (Fig. 1) (PRANGE and WILSON, 1972; PRANGE *et al.*, 1972b).

The surprising results of this experiment, which we wish to emphasise, were that TRH acted alone rather than in combination with imipramine, and that a single injection acted very rapidly though transiently. In this experiment as well as in the experiments with normal and schizophrenic women to be reported below, psychological and behavioural changes were demonstrated in an environment characterised by pleasant stability and the absence of environmental 'noise' such as may be introduced by intensive psychotherapeutic intervention.

The antidepressant action of TRH was immediately confirmed by ITIL (personal

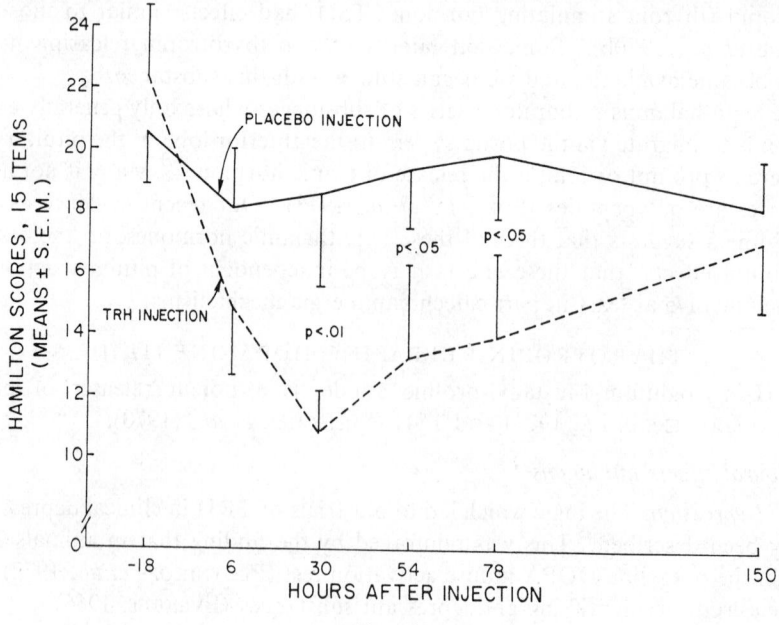

FIG. 1

communication) and quickly confirmed by KASTIN *et al.* (1972) and by VAN DER VIS-MELSEN and WIENER (1972).

(b) *Normal subjects.* Believing that TRH might exert psychotropic effects that would be masked in a depressed population, we next turned to a study of normal women, age-matched to our depressed patients. In a double-blind, placebo-controlled, cross-over design we gave an i.v. injection of TRH, 500 μg, on one occasion and saline on another. Injections were given 1–3 weeks apart and the sequence was randomised. Since measures of depression were inappropriate in this population we relied mainly on a line analog to indicate whether, at each hour after injection, subjects felt happier or sadder. After 9 a.m. saline injection, subjects showed no change whatever. After TRH injection, they rated themselves as becoming steadily happier, the effect reaching its maximum at 4 p.m. At this time they had rated themselves on the average 20 mm happier (maximum possible score, 50 mm) (Fig. 2).

Each subject was "blindly" interviewed hourly after injection and interview content was taped, later transcribed and rated "blindly" by another psychiatrist. These data supported the line test findings and in addition showed that in the first few hours after TRH, subjects experienced relaxation without sedation. Mild euphoria occurred later in the day.

Intensity of side-effects and magnitude of endocrine changes (see below) bore no reliable relationships to intensity of euphoric response (WILSON *et al.*, 1973).

(c) *Schizophrenia.* Encouraged by these findings and the report of Tiwary et al. (see below), we performed a preliminary, single-blind, uncontrolled study of four schizophrenic women. Patients were in the third decade of life. We viewed their illnesses as early process schizophrenia since all had had previous hospitalisations for schizophrenia, all had two or more absences from hospital on medication for extended periods of time with reasonable social adjustments and all had relapsed again requiring rehospitalisation. Their current diagnoses were as follows: chronic

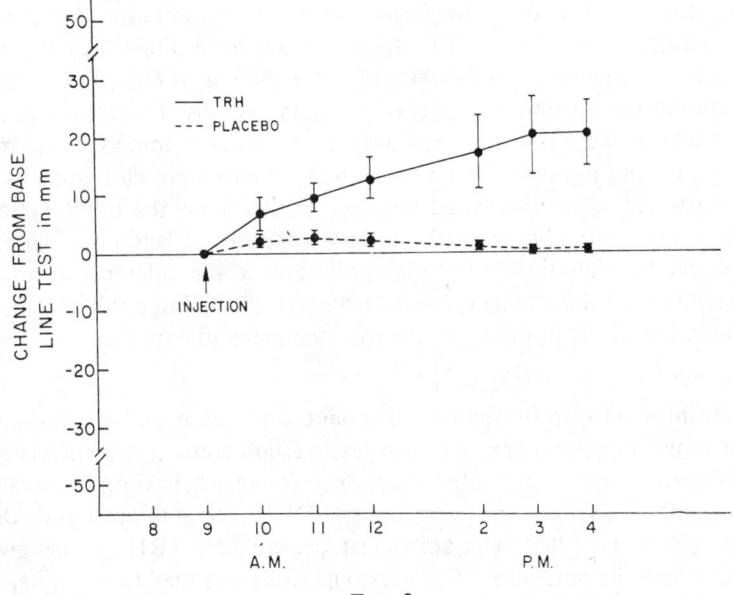

FIG. 2

undifferentiated schizophrenia; schizo-affective disorder, depressed type; catatonic schizophrenia, mute type; paranoid schizophrenia. All had been in hospital for at least two weeks before entering our study. All were drug free for at least one week prior to hormone administration and all were on oral placebos during the washout period and throughout the study.

Approaching these patients with caution, we injected TRH, 200 μg i.v. on Day 1 and 400 μg on Day 8. They were rated on the Brief Psychiatric Rating Scale (BPRS) (OVERALL and GORHAM, 1962) daily and twice on injection days. Within a few hours after each injection all patients were substantially improved, and improvement was generally maintained for 3–4 days. It was not limited to affective change but occurred as well in the areas of hallunicatory behaviour and mutism, for example.

Definitive studies are in progress to assess the apparent beneficial action of TRH in schizophrenia. It is of interest to note that TIWARY (1972) and his colleagues had previously reported that a single dose of TRH given for diagnostic purposes had a striking beneficial effect on the disturbed behaviour of a 16-year-old boy with cerebral gigantism. The effect was demonstratable within a few hours and lasted about one day.

Pituitary-thyroid effects

As stated, TRH causes release from the pituitary of prolactin and TSH. After 600 μg TRH our ten unipolar depressed women showed what appeared to be a sub-normal rise in TSH and showed no change whatever in serum T3 or thyroxine (T4). KASTIN (1972) and his colleagues also found a blunted TSH response in five de-pressed patients.

Our ten normal female subjects, after 500 μg TRH, showed a rise in TSH, which peaked in 30 min, and showed a distinct rise in T3 and a more delayed and lesser rise in T4 (PRANGE *et al.*, 1973). These latter findings confirmed the work of HOLLANDER's (1972) group.

When the TSH responses of our depressed women were compared to those of our normal women it was found that the former were, in fact, diminished but the difference was not statistically significant. Two factors could have diminished the difference: first, normals were given a smaller dose of TRH (500 μg vs 600), though both doses may be supramaximal from the aspect of pituitary response (ANDERSON *et al.*, 1971); second, injections were given more slowly to normals (2 min vs. 1 plus min), thus probably producing lower blood levels. Since patients were euthyroid by all usual criteria, hyperthyroidism (increased negative feedback on the pituitary) cannot be invoked to explain probably blunted pituitary responses. Clearly the matter requires additional investigation both to test the significance of the difference in response and to understand it. At the moment, the data suggest, but fail to establish, a diminished tonic stimulation of the pituitary by the hypothalamus in depression.

Animal studies

Relationships between the activity of rodents and the mental aberrations of man are, not astonishingly, obscure. Nevertheless, certain connections have been empiri-cally established. For example, substances which enhance behavioural activation when given to mice that have received pargyline and DOPA, tend to ameliorate depression in the clinic (EVERETT, 1966). One such substance is TRH. TRH whether given orally or i.p. is active in the pargyline-DOPA test and dose is related to response. Further-more, TRH loses no potency in this model when animals have been hypophysectomised

(PLOTNIKOFF *et al.*, 1972). That is, the only generally accepted target for TRH, the pituitary, need not be present for the hormone to increase running, squeaking, salivation, piloerection, etc. when pargyline-DOPA have been administered. Later work by PLOTNIKOFF *et al.* (1973, in press) and by BREESE *et al.* (1973, in press) suggests that when the test is adapted to rats, thyroidectomy in this species similarly does not block the activity of TRH. The adaptation of the test to rats, however, is subject to a number of qualifications which limit its convenience and generality.

These data suggest a central action of TRH, an action that is independent of its pituitary–thyroid effects. What this mechanism is cannot yet be stated. For example, it has not been excluded that TRH might somehow alter brain uptake of administered DOPA. Nor can a direct action of TRH on specific receptors be excluded. The importance of catecholamines in the regulation of affect (PRANGE, 1964; SCHILDKRAUT, 1965; BUNNEY and DAVIS, 1965) and the potency of TRH in a test that involves a catecholamine precursor, DOPA, prompts an investigation of the possible influence of TRH on catecholamine metabolism. Positive results are sparse. In a series of detailed investigations BREESE (1973, in press) and his colleagues have shown that a variety of doses of systemically administered TRH have no effect in rat brain on levels or turnover rates of norepinephrine, dopamine (DA) or serotonin. On the other hand, if mice submitted to the pargyline-DOPA test are killed 20 min after DOPA, when behavioural activity is maximum, one finds that brain DA is 50 per cent higher in TRH-treated than in control-injected mice (BREESE *et al.*, 1973, in press).

Recently SEGAL and MANDELL (1973, in press) have applied their intraventricular infusion technique (1970) to the study of hypothalamic polypeptides. Substances are slowly introduced into the lateral ventricle of the free-roving rat and his spontaneous activity is automatically counted. When TRH is infused his activity over time is about twice as great as when control solution is infused. This experiment tells us nothing about TRH-catecholamine interplay. It stands, however, as the first demonstration that a polypeptide indigenous to the hypothalamus exerts a behavioural effect when applied directly to brain.

MELANOCYTE STIMULATING HORMONE RELEASE INHIBITING FACTOR (MIF)

MIF, prolyl-leucyl-glycine amide, inhibits the release from the pituitary gland of melanocyte stimulating hormone (SCHALLY *et al.*, 1973).

Behavioural effects in humans

KASTIN and BARBEAU (1972) have conducted a study that suggests that MIF has anti-Parkinsonian value. This finding, if corroborated, is, of course, consistent with the notion that MIF possesses prodopaminergic activity as animal studies suggest.

Animal studies

PLOTNIKOFF *et al.* (1971) have shown that MIF, like TRH, is active in the pargyline-DOPA test. MIF, like TRH, loses no potency when mice are hypophysectomised. On the other hand, MIF, unlike TRH, lacks activity in an animal test designed to identify serotonergic facilitation (PLOTNIKOFF, personal communication). If one entertains what we have termed the biogenic amine permissive hypothesis of affective disorders (PRANGE, 1973), then one must suppose that MIF would have less clinical antidepressant activity than TRH. This comparison has not yet been performed.

SOMATOTROPIN-RELEASE INHIBITING FACTOR (SRIF)

SRIF is a tetradecapeptide which appears to inhibit the secretion of growth hormone and to block the secretion of growth hormone stimulated by other substances (Brazeau *et al.*, 1973; Vale *et al.*, 1973).

Animal studies

Segal and Mandell (1973, in press) have shown that the intraventricular infusion of SRIF grossly inhibits the activity of the free-roving rat. That is, it has an opposite effect to that of TRH in the same preparation.

SUMMARY AND CONCLUSIONS

Three hypothalamic polypeptides appear to influence the behavior of animals or man. In the case of TRH and MIF this action may be independent of the pituitary gland. Catecholamines and dopaminergic mechanisms may be involved in the control of synthesis and release of these hormones from the hypothalamus, but evidence is far from complete on this. Even more sparse is information pertinent to the question of whether the behavioural effects of the hypothalamic hormones are mediated via their effects on biogenic amine synthesis, turnover or release. Similarly nothing is known about direct effects of these hypothalamic hormones upon amine receptors. The possibility that they may act as transmitters or transmission modulators at localised sites must be entertained. It is even conceivable that alterations in catecholamines exert behavioural effects by producing changes in hypothalamic hormones.

The mechanism by which these hormones exert their behavioural effects is thus far from clear. Actions through the pituitary to peripheral endocrine glands are clearly established, but new data raise the question of whether these are the only actions or whether more direct central nervous system actions may exist (Plotnikoff *et al.*, 1972; Segal and Mandell, 1973, in press; Plotnikoff *et al.*, 1971).

That behavioural effects occur in humans, at least with TRH seems certain. These effects have been shown to occur in normal and depressed women and are suggested by a single-blind study of schizophrenic women. The effects of a single injection are rapid, occurring within a few hours, and lasting not more than a few days. They are best demonstrated in a pleasant, stable and relatively "noise" free environment. The effects are subtle, and it seems likely that intrinsic hormonal effects may be amplified or dampened by the environment. In normal women the hormone effect has been relaxing and mildly euphorient. In depressed women it has reduced anxiety and depression. In early process schizophrenics it has improved both affect and thought and diminished secondary symptoms such as hallucinations.

No claims are made for therapeutic utility at this time because of the limited numbers of patients, the limited number of trials and the transient nature of the responses. Rather we see the hypothalamic hormones as opening an exciting vista in both basic and clinical neurobiology. It is striking that hormones with putative central actions have psychological effects which cut across diagnostic boundaries. However, in well established clinical entities psychological symptoms frequently overlap as well.

Acknowledgements—This work was supported in part by N.I.M.H. grant # MH-15631, N.I.M.H. Career Scientist Award (A. J. P.) # MH-22536, N.I.C.H.D. grant # HD-03110, N.I.C.H.D. Career Development Award (G. R. B.) # HD-24585, and a contract from Abbott Laboratories, who generously furnished TRH.

REFERENCES

AITKEN R. C. B. (1969) *Proc. Roy. Soc. Med.* **62,** 989–993.

ANDERSON M. S., BOWERS C. Y., KASTIN A. J., SCHALCH D. S., SCHALLY A. V., SNYDER P. J., UTIGER R. D., WILBER J. F. and WISE A. J. (1971) *N. Engl. J. Med.* **285,** 1279–1283.

BOWERS C. Y., FRIESEN H. G., HWANG P., GUYDA H. J. and FOLKERS K. (1971) *Biochem. Biophys. Res. Commun.* **45,** 1033–1041.

BRAZEAU P., VALE W., BURGUS R., LING N., BUTCHER M., RIVIER J. and GUILLEMIN R. (1973) *Science* **179,** 77–79.

BREESE G. R., PLOTNIKOFF N. P., PRANGE A. J., Jr., WILSON I. C. and LIPTON M. A. (1973) In: *The Thyroid Axis, Drugs, and Behavior* (PRANGE A. J., Jr., Ed.), Raven Press, New York, in press.

BUNNEY W. E., Jr. and DAVIS J. M. (1965) *Archs. Gen. Psychiat.* **13,** 483–494.

EVERETT G. M. (1966) *Excerpta. Med. Int. Congr. Ser.* **122,** 164–167.

FLEISCHER N., BURGUS R., VALE W., DUNN T. and GUILLEMIN R. (1970) *J. Clin. Endocrinol. Metab.* **31,** 109–112.

HAMILTON M. (1960) *J. Neurol. Neurosurg. Psychiat.* **23,** 56–62.

HOLLANDER C. S., MITSUMA T., SHENKMAN L., WOOLF P. and GERSHENGORN M. C. (1972) *Science* **175,** 209–210.

KASTIN A. J. and BARBEAU A. (1972) *Can. Med. Assoc. J.* **107,** 1079–1081.

KASTIN A. J., EHRENSING R. H., SCHALCH D. S. and ANDERSON M. S. (1972) *Lancet* **ii,** 740–742.

LIPTON M. A., PRANGE A. J., Jr., DAIRMAN W. and UDENFRIEND S. (1968) *Fedn. Proc., Fedn. Am. Socs. Exp. Biol.* **27,** 399.

OVERALL J. E. and GORHAM D. R. (1962) *Psychol. Rep.* **10,** 799–812.

PLOTNIKOFF N. P., KASTIN A. J., ANDERSON M. S. and SCHALLY A. V. (1971) *Life Sci.* **10,** 1279–1283.

PLOTNIKOFF N. P., PRANGE A. J., Jr., BREESE G. R., ANDERSON M. S. and WILSON I. C. (1972) *Science* **178,** 417–418.

PLOTNIKOFF N. P., PRANGE A. J., Jr., BREESE G. R., ANDERSON M. S. and WILSON I. C. (1973) In: *The Thyroid Axis, Drugs, and Behavior* (PRANGE A. J., Jr., Ed.). Raven Press, New York, in press.

PRANGE A. J., Jr. (1964) *Dis. Nerv. Sys.* **25,** 217–221.

PRANGE A. J., Jr. (1973) *Psychiat. Ann.* **3,** 56–74.

PRANGE A. J., Jr. and WILSON I. C. (1972) *Psychopharmacol. Suppl.* **26,** 82.

PRANGE A. J., Jr., WILSON I. C., RABON A. M. and LIPTON M. A. (1969) *Am. J. Psychiat.* **126,** 457–469.

PRANGE A. J., Jr., MEEK J. L. and LIPTON M. A. (1970a) *Life Sci.* **9,** 901–907.

PRANGE A. J., Jr., WILSON I., C., KNOX A. E., McCLANE T. K. and LIPTON M. A. (1970b) *Am. J. Psychiat.* **127,** 101–199.

PRANGE A. J., Jr., WILSON I. C., KNOX A. E., McCLANE T. K., BREESE G. R., MARTIN B. R., ALLTOP L. B. and LIPTON M. A. (1972a) *J. Psychiat. Res.* **9,** 187–205.

PRANGE A. J., Jr., WILSON I. C., LARA P. P., ALLTOP L. B. and BREESE G. R. (1972b) *Lancet* **ii,** 999–1002.

PRANGE A. J., Jr., WILSON I. C., LARA P. P., WILBER J. F., BREESE G. R., ALLTOP L. B. and LIPTON M.A. (1973) *Archs. Gen. Psychiat.* **29,** 28–32.

SCHALLY A. V., ARIMURA A. and KASTIN A. J. (1973) *Science* **179,** 341–350.

SCHILDKRAUT J. J. (1965) *Am. J. Psychiat.* **122,** 509–522.

SEGAL D. S. and MANDELL A. J. (1970) *Proc. Natn. Acad. Sci. U.S.A.* **66,** 289–293.

SEGAL D. S. and MANDELL A. J. (1973) In: *The Thyroid Axis, Drugs, and Behavior* (PRANGE A. J., Jr., Ed.), Raven Press, New York, in press.

TAYLOR J. A. (1955) *J. Abnorm Soc. Psychol.* **48,** 285–292.

TIWARY C. M., FRAIS J. L. and ROSENBLOOM A. L. (1972) *Lancet.* **ii,** 1086.

VALE W., BRAZEAU P., RIVIER C., RIVIER J., GRANT G., BURGUS R. and GUILLEMIN R. (1973) *Fedn. Proc., Fedn. Am. Soc. Exp. Biol.* **32,** 1.

VAN DER VIS-MELSEN M. J. E. and WEINER J. D. (1972) *Lancet* **ii,** 1415.

WILSON I. C., PRANGE A. J., Jr., McCLANE T. K., RABON A. M. and LIPTON M. A. (1970) *N. Engl. J. Med.* **282,** 1063–1067.

WILSON I. C., PRANGE A. J., Jr., LARA P. P., ALLTOP L. B., STIKELEATHER R. A. and LIPTON M. A. (1973) *Archs. Gen. Psychiat.* **29,** 15–21.

ZEALLY A. K. and AITKEN R. C. B. (1969) *Proc. Roy. Soc. Med.* **62,** 993–996.

Frontiers in Catecholamine Research 1973, pp. 1157 to 1164. Pergamon Press. Printed in Great Britain.

AFFECTIVE DISORDERS: THE CATECHOLAMINE HYPOTHESIS REVISITED

FREDERICK K. GOODWIN and ROBERT L. SACK

Laboratory of Clinical Science, NIMH, Bethesda, Maryland, U.S.A.

THE CATECHOLAMINE hypothesis of affective disorders states that depression is associated with a functional deficit of one or more neurotransmitter catecholamines at critical synapses in the central nervous system, and conversely mania is associated with a functional excess of these amines. The hypothesis was generated largely from observations of the clinical effects of drugs known to alter central amine function in animals. Thus, reserpine, a depletor of central catecholamines was associated with the onset of depression in some individuals, while monoamine oxidase (MAO) inhibitors, which increase levels of catecholamines in the brains of animals, were reported to have some antidepressant activity. Further, amphetamine, a stimulant in man, was found to increase catecholamine function at the synapse through several mechanisms. Finally, the tricyclic antidepressants were shown to block the presynaptic reuptake of catecholamines, thereby increasing their availability at the receptor site; conversely, lithium, an effective antimanic agent, was found to alter catecholamines in ways that should result in decreases in functionally active transmitter at the receptor. This extensive body of clinical-pharmacological data has been reviewed in detail elsewhere (BUNNEY and DAVIS, 1965; SCHILDKRAUT, 1965).

The purpose of this paper is to re-examine this important hypothesis, particularly in light of some recent clinical data concerning affective illness. Some refocusing of our thinking concerning the possible relationships between mood disorders and brain catecholamines will be suggested. The data to be reviewed can be included under four general headings: (a) a brief re-evaluation of the clinical-pharmacological data focusing on those observations which do not appear to 'fit' so easily into the straightforward version of the catecholamine hypothesis; (b) the relevance of biologically meaningful subgroups of affective illness, specifically the unipolar–bipolar dichotomy; (c) evidence for similarities between the clinical states of depression and mania; (d) recent data on the direct assessment of catecholamine function in patients with affective illness, as reflected in cerebrospinal fluid (CSF) and urinary amine metabolites; in this regard, state dependent variables affecting the level of these metabolites will be discussed.

THE CLINICAL-PHARMACOLOGICAL DATA

Table 1 lists the major mood altering drugs which have been hypothesised to exert their clinical effects through alterations of brain catecholamines; the diversity of the clinical data illustrates some of the difficulties involved when one attempts to correlate mood with the functional state of brain catecholamines. Although the neurochemical effects of these drugs are complex, and some controversy remains concerning their specific mechanisms of action, the effects on central catecholamines as illustrated in simplified form in Table 1, are those generally reported as the

TABLE 1. DRUG-CATECHOLAMINE RELATIONSHIPS

Drug	Effect on catecholamines at receptor	Behavioural effects in man			
		Normals	Predisposed to affective illness	Depressed patients	Manic patients
MAOI	↑	No effect or mild sedation	Can precipitate mania	Some antidepressant activity	?
Tricyclics	↑	No effect or mild sedation	Can precipitate mania; Prevent recurrences of depression (?)	Antidepressant	Antimanic (?)
Amphetamine	↑	Stimulation	Can precipitate mania (?)	Poor antidepressant	?
Cocaine	↑	Stimulation	Can precipitate mania (?)	Poor antidepressant	?
L-dopa	↑	No effect	Can precipitate hypomania	Activation without antidepressant effect	?
Reserpine	↓	Sedation	Can precipitate depression	?	Sedation and/or tranquilisation
Lithium	↓	No effect or mild sedation	Prevent recurrences of mania & depression	Moderate antidepressant effect in some	Antimanic
AMPT	↓	Sedation	?	Sedation	Antimanic (?)

acute effects of these drugs in most animal species. First it can be noted that drugs which are stimulants in normal individuals are not generally found to be therapeutic in patients suffering from major depressive illness (KLEIN and DAVIS, 1969). Conversely, those drugs which do have antidepressant activity are not stimulants in normals (OSWALD, BREZINOVA and DUNLEAVY, 1972). This data on differential response to drugs support the concept that depression may not simply represent a quantitative extension of a normal mood state, but may reflect a qualitatively different psychobiological substrate.

The weight of the available clinical evidence suggests that the MAO inhibitors are not very effective antidepressants in patients with more severe, 'primary' or 'endogenous' depressions—that is, the group that has been the major focus of studies concerned with catecholamine dysfunction (reviewed in WECHSLER, GROSSER, and GREENBLATT 1965). Rather these drugs are most efficacious in outpatients with mixed anxiety–depression syndromes or in 'depressives' with atypical features such as anxiety, fatigue, phobia or somatic complaints (WEST and DALLY, 1959; KELLY et al., 1970; KELLNER, 1970, POLLIT and YOUNG, 1971). Thus, their spectrum of usefulness does not coincide with the tricyclics which are more likely to benefit hospitalised depressed patients with classical 'endogenous' symptoms.

Amphetamine, a stimulant in normals, is generally conceded not to be an effective antidepressant agent, although the occurrence of transient improvement in some depressed patients may have predictive value for subsequent response to tricyclics (FAWCETT and SIOMOPOULOS, 1971).

Cocaine, a powerful stimulant in normals, is a potent inhibitor of catecholamine reuptake at the synapse, an effect similar to that of the tricyclic antidepressants. Our recent studies with this drug in hospitalised depressed patients (POST, KOTIN and GOODWIN, Annual Meeting, Amer. Psychiat. Ass. 1973) suggest that its actions are not primarily antidepressant, but rather non-specific arousal and activation.

Early clinical trials with L-dihydroxyphenylalanine (L-dopa—the amino acid precursor of the catecholamines) in depressed patients represented the first effort to provide a more direct evaluation of the catecholamine hypothesis. Since the generally negative results of these trials were difficult to interpret because of the small doses employed, we undertook a trial of L-dopa in 26 hospitalised depressed patients using high oral doses (averaging 100 mg/kg) administered over periods ranging from 18 to 45 days with and without a peripheral decarboxylase inhibitor (GOODWIN et al., 1970). Eighty per cent of the patients failed to show any improvement in depression on L-dopa, although a small subgroup of five patients evidenced clear improvement which appeared related to the drug. It is of interest that this small subgroup of responders were all patients with prominent psychomotor retardation. In the majority who did not have an antidepressant response there was nevertheless consistent evidence of some activation with increases in anger and psychosis ratings in some patients and hypomanic episodes superimposed on the depression in 8 out of the 9 patients with a prior history of mania (bipolar) (MURPHY et al., 1970).

These predominantly negative results with L-dopa in depression suggest that catecholamine depletion as defined by the reserpine model (which is reversible by L-dopa) is not a sufficient explanation for the pathophysiology of the majority of depressions.

Since our CSF data suggest a substantial increase in brain dopamine in the depressed patients on L-dopa (Goodwin *et al.*, 1971) the negative clinical results suggest that most depressions probably do not involve a depletion of central dopamine as the critical pathophysiological mechanism.

The question of the effect of L-dopa on norepinephrine in brain is more complex; although in normal animals L-dopa does not increase brain norepinephrine (NE) (Iversen, 1967; Butcher and Engel, 1969; Everett and Borcherding 1970) it does increase NE in animals previously *depleted* of this catecholamine (Seiden and Peterson, 1968) which is, of course, the situation hypothesised to exist in depression.

In relation to the clinical effects of drugs which *decrease* functional catecholamines in brain we have recently completed a critical review of the original reports of reserpine induced depressions (Goodwin, Ebert and Bunney, 1971) and noted that the incidence of patients who experienced major depressive symptoms (analogous to 'endogenous' depressions) across all studies (approximately six per cent) was almost identical to the per cent incidence of individuals with prior histories of depression. From this revised data, it appears more likely that reserpine is capable of *precipitating* depression in susceptible individuals rather than inducing it *de novo*; this is an important distinction, since depressions can be precipitated by a variety of agents or conditions not directly related to amine function.

Lithium, another drug which can presumably decrease functional amines (Corrodi *et al.*, 1967; Katz, Chase and Kopin, 1968; Colburn *et al.*, 1968) should increase depressive symptoms in patients according to the catecholamine hypothesis. However, to the contrary, lithium has been shown in controlled studies to have moderate antidepressant properties in some depressed patients (Goodwin, Murphy and Bunney, 1969; Mendels, Secunda and Dyson, 1972) and to effectively prevent recurrences of depression when used prophylactically (Baastrup and Schou, 1967). Thus, drug–catecholamine relationships in depressed patients contain a number of findings which are discrepant in relation to the catecholamine hypothesis of depression.

In the case of manic or hypomanic reactions the behaviour–catecholamine relationships appears more consistent (see Table 1). Thus the five drugs which increase functional catecholamines can all precipitate manic or hypomanic reactions in susceptible individuals, whereas the three drugs which decreases functional brain catecholamines all have some beneficial effect on mania.

THE UNIPOLAR–BIPOLAR DICHOTOMY

Recent evidence from clinical, genetic, pharmacologic and biological studies suggest that 'endogenous' depressions can be meaningfully subdivided into bipolar and unipolar groups on the basis of the presence or absence of a prior history of mania. Comparing the clinical features of patients with unipolar and bipolar affective illness, the unipolar patients have a later age of onset and their depressions have mixed features of agitation and retardation accompanied by a significantly higher frequency of symptoms of anger, anxiety and physical complaints (Beigel and Murphy, 1971). Family history data indicates a significantly higher frequency of mania in the first degree relatives of bipolar patients compared to unipolar patients (Winokur, Clayton and Reich, 1969). In a series of biological studies in hospitalised depressed patients we have noted significant differences between unipolar and bipolar groups (Table 2).

TABLE 2.

Biological variable	Bipolar	Unipolar
Cortical evoked potential (BUCHSBAUM et al., 1971)	'Augmenter'	'Reducer'
Plasma Mg²⁺ following lithium (GOODWIN, MURPHY and BUNNEY, 1968)	Elevated	Unchanged
Urinary 17-OHCS (DUNNER et al., 1972)	Reduced	Elevated
Platelet MAO activity (MURPHY and WEISS, 1972)	Reduced	Normal
Urinary dopamine following L-dopa (MURPHY et al., 1973)	Higher	Lower
Red cell COMT (DUNNER et al., 1971)	Reduced	Normal
CSF HVA accumulation on Probenecid (GOODWIN et al., 1973)	Higher	Lower

We have also observed some unipolar–bipolar differentiation in the antidepressant response to tricyclics and to lithium: compared to patients with unipolar depression, bipolar patients tend to respond less well to imipramine (BUNNEY et al., 1970); the reverse appears to obtain with the antidepressant effects of lithium—that is a higher frequency of antidepressant responses in the bipolar group compared to the unipolar group (GOODWIN et al., 1972).

THE RELATIONSHIP BETWEEN DEPRESSION AND MANIA

Clinical observations and independent behavioural ratings of both spontaneous and drug-induced manic episodes suggest that in many respects the clinical states of depression and mania do not represent phenomenologically 'opposite' poles of the same continuum (KOTIN and GOODWIN, 1972). In a comparison of bipolar depression and mania the aspect most clearly 'opposite' is the level of psychomotor activity. In addition, the evidence that lithium has some therapeutic efficacy in *both* the manic and depressive phases of bipolar affective illness is difficult to reconcile with a concept of these states as simply representing opposite poles of a continuum.

URINARY 3 METHOXY 4 HYDROXYPHENYLGLYCOL (MHPG): THE ROLE OF ACTIVITY

Evidence from animal studies suggests that MHPG is the urinary catecholamine metabolite which best reflects brain NE metabolism (MAAS and LANDIS, 1968; SCHANBERG et al., 1968). Thus the reports of MAAS et al. (1968) and of GREENSPAN et al. (1970) that urinary MHPG is low in depressed patients and elevated in manics is intriguing.

Since the level of activity is strikingly different in depressed and manic patients we have been evaluating the effect of altered activity on the excretion of urinary MHPG in depressed patients. Preliminary data (EBERT, POST and GOODWIN, 1972) suggests that moderate increases in activity (without observable changes in mood) are associated with marked increases in urinary MHPG in 10 of the 11 depressed patients studied (mean increase 104 per cent, $P = <0.01$). In 6 of the 11 patients, urinary NE and epinephrine (E) were also determined for baseline and activity periods (Fig. 1). There was no significant change in urinary NE, while the increase in E was roughly comparable to that observed for MHPG, suggesting a possible stress factor contributing to these changes. Although the design of these studies does not allow us to differentiate the contributions of activity and stress, it nevertheless suggests that substantial alterations in urinary MHPG can be produced in depressed patients secondary to non-specific changes.

1162   F. K. GOODWIN and R. L. SACK

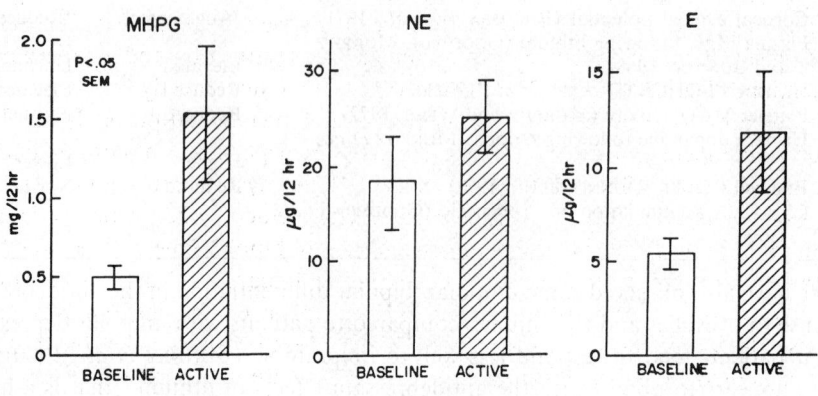

FIG. 1

CSF catecholamine metabolites

Both MHPG and Homovanilic Acid (HVA) are low in our depressed patients
and tend to be normal or elevated in manics (POST *et al.*, 1973; GOODWIN *et al.*, 1973),
see Table 3.

TABLE 3.

	Controls	Depressed patients	Manic patients
HVA	22·4 ± 2·4 (28)	15·2 ± 2·1 (55)*	25·7 ± 4·3 (16)
MHPG	15·1 ± 3·6 (44)	9·2 ± 2·4 (30)*	15·4 ± 5·5 (9)

(Number of subjects in brackets); *Indicates significant difference from
normals.

To evaluate the possible contribution of activity differences to these data, a group
of 10 moderately depressed patients were asked to simulate the hyperactivity of
mania for four hours preceding a lumbar puncture (POST *et al.*, 1972). The following
changes were noted (Table 4).

TABLE 4.

	Baseline (Mod. Dep.)	Activity
HVA	16·9 ± 2·8	42·1 ± 5·7 ($P = <0·01$)
MHPG	7·5 ± 2·2	14·2 ± 3·4 ($P = <0·1$)

Although these data suggest a possible activity artifact in CSF catecholamine
metabolite data, we were not able to demonstrate any relationship between MHPG
levels and ratings of agitation-retardation in the depressed patients; there was a
trend for HVA levels to be higher in the less retarded and more agitated depressed
patients. Analysis of the frequency distribution of MHPG values in the CSF of 45
depressed patients suggests a bimodal pattern. Although CSF MHPG values do not
significantly correlate with any single subdivision of the clinical population, there
are trends for the MHPG values to be lower in patients with a combination of more
endogenous features, bipolar histories and more severe depressive symptoms.

It is of considerable interest that the low MHPG did not show a tendency to
increase with recovery from the depression.

CONCLUSIONS

These data suggests that unipolar and bipolar affective illness may represent biologically distinct entities with different clinical responses to drugs affecting brain catecholamines. In regard to bipolar affective illness we would suggest a broadening of biological theories to include both an underlying dysfunction and a superimposed dysfunction. The underlying dysfunction (which may not directly involve catecholamines) is present in both the depressive and manic phases and probably involves a genetic component constituting the biological predisposition to both manic and depressive episodes. The existence of an underlying dysfunction common to both phases of bipolar affective illness is consistent with the acute and prophylactic effects of lithium in both the manic and depressive phases of bipolar illness. We would suggest that the underlying dysfunction common to both manic and depressive patients, may involve an abnormality of regulatory mechanism controlling brain aminergic responses to stress and that this possibility may best be approached by studies of individuals predisposed to affective illness when they are in a normal phase.

REFERENCES

BEIGEL A. and MURPHY D. L. (1971) *Archs. Gen. Psychiat.* **24**, 215–220.
BUCHSBAUM M., GOODWIN F. K., MURPHY D. L. and BORGE G. (1971) *Am. J. Psychiat.* **128**, 19–25.
BUNNEY W. E., JR. and DAVIS J. M. (1965) *Archs. Gen. Psychiat.* **13**, 483–494.
BUNNEY W. E., JR., BRODIE H. K. H., MURPHY D. L. and GOODWIN F. K. (1970) *Proc. 78th Ann. Conv. Am. Psychol. Ass.*, 829–830.
BUNNEY W. E., JR., GOODWIN F. K. and MURPHY D. L. (1972) *Archs. Gen. Psychiat.* **27**, 312–317.
BUTCHER L. L. and ENGEL J. (1969) *Brain Res.* **15**, 233–242.
COLBURN R. W., GOODWIN F. K., MURPHY D. L., BUNNEY W. E., JR. and DAVIS J. M. (1968) *Biochem. Pharmacol.* **17**, 957–964.
CORRODI H., FUXE K., HOKFELT T. and SCHOU M. (1967) *Psychopharmacol.* **11**, 345–353.
DUNNER D. L., COHN C. K., GERSHON E. S. and GOODWIN F. K. (1971) *Archs. Gen. Psychiat.* **25**, 348–353.
DUNNER D. L., GOODWIN F. K., GERSHON E. S., MURPHY D. L. and BUNNEY W. E., JR. (1972) *Archs. Gen. Psychiat.* **26**, 360–363.
EBERT M. H., POST R. M. and GOODWIN F. K. (1972) *Lancet* **ii**, 766.
EVERETT G. M. and BORCHERDING J. W. (1970) *Science* **168**, 849–850.
FAWCETT J. and SIOMOPOULOS V. (1971) *Archs. Gen. Psychiat.* **25**, 247–255.
GOODWIN F. K., MURPHY D. L. and BUNNEY W. E., JR. (1968) *Sci. Proc. Am. Psychiat. Ass.* **121**, 233–234.
GOODWIN F. K., MURPHY D. L. and BUNNEY W. E., JR. (1969) *Archs. Gen. Psychiat.* **21**, 486–496.
GOODWIN F. K., BRODIE H. K. H., MURPHY D. L. and BUNNEY W. E., JR. (1970) *Biol. Psychiat.* **2**, 341–366.
GOODWIN F. K., EBERT M. and BUNNEY W. E., JR. (1971) In: *Psychiatric Complications of Medical Drugs* (SHADER R. I., Ed.), Raven Press, New York.
GOODWIN F. K., MURPHY D. L., DUNNER D. L. and BUNNEY W. E., JR. (1972) *Am. J. Psychiat.* **129**, 44–47.
GOODWIN F. K., POST R. M. DUNNER D. L. and GORDON E. K. (1973) *Am. J. Psychiat.* **130**, 73–79.
GREENSPAN K., SCHILDKRAUT J. J., GORDON E., BAER L., ARONOFF M. S. and DURELL J. (1970) *J. Psychiat. Res.* **7**, 171–183.
IVERSEN L. L. (1967) *Nature, Lond.* **214**, 8–14.
KATZ R. I., CHASE R. M. and KOPIN I. J. (1968) *Science* **162**, 466–467.
KELLNER R. (1970) *J. Nerv. Ment. Dis.* **151**, 85–96.
KELLY D., GUIRGUIS W., FROMMER E., MITCHELL-HEGGS N., and SARGANT W. (1970) *Br. J. Psychiat.* **116**, 387–398.
KLEIN D. and DAVIS J. M. (1969) *Diagnosis and Drug Treatment of Psychiatric Disorders.* William & Wilkins, Baltimore.
KOTIN J. and GOODWIN F. K. (1972) *Am. J. Psychiat.* **129**, 679–686.
MAAS J. W., FAWCETT J. and DEKIRMENJIAN H. (1968) *Archs. Gen. Psychiat.* **19**, 129–134.
MAAS J. W. and LANDIS D. H. (1968) *J. Pharmac. Exp. Ther.* **163**, 147–152.

MENDELS J., SECUNDA S. K. and DYSON W. C. (1972) *Archs. Gen. Psychiat.* **26,** 154–157.
MURPHY D. L., BRODIE H. K. H., GOODWIN F. K. and BUNNEY W. E., JR. (1971) *Nature, Lond.* **229,** 135–136.
MURPHY D. L., GOODWIN F. K., BRODIE H. K. H. and BUNNEY W. E., JR. (1973) *Am. J. Psychiat.* **130,** 79–82.
MURPHY D. L. and WEISS R. (1972) *Am. J. Psychiat.* **128,** 1351–1357.
OSWALD I., BREZINOVA V. and DUNLEAVY D. L. F. (1972) *Br. J. Psychiat.* **120,** 673–677.
POLLIT J. and YOUNG J. (1971) *Br. J. Psychiat.* **119,** 143–149.
POST R. M., KOTIN J., GOODWIN F. K. and GORDON E. K. (1973) *Am. J. Psychiat.* **130,** 67–72.
POST R. M., GORDON E. K., GOODWIN F. K. and BUNNEY W. E., JR. (1973) *Science* **179,** 1002–1003.
SCHANBERG S. M., SCHILDKRAUT J. J., BREESE G. R. and KOPIN I. J. (1968) *Biochem. Pharmac.* **17,** 247–254.
SCHILDKRAUT J. J. (1965) *Am. J. Psychiat.* **122,** 509–522.
SEIDEN L. S. and PETERSON D. D. (1968) *J. Pharmacol. Exp. Ther.* **163,** 84–90.
WECHSLER S., GROSSER G. H. and GREENBLATT M. (1965) *J. Nerv. Ment. Dis.* **141,** 231–239.
WEST E. D. and DALLY P. J. (1959) *Br. Med. J.* **1,** 1491–1494.
WINOKUR G., CLAYTON P. J. and REICH T. (1969) *Manic Depressive Illness.* C. V. Mosby Company, St. Louis.

Frontiers in Catecholamine Research 1973, pp. 1165 to 1171. Pergamon Press. Printed in Great Britain.

CATECHOLAMINE METABOLISM AND AFFECTIVE DISORDERS: STUDIES OF MHPG EXCRETION

JOSEPH J. SCHILDKRAUT

Neuropsychopharmacology Laboratory, Massachusetts Mental Health Center,
Department of Psychiatry, Harvard Medical School,
Boston, Massachusetts 02115, U.S.A.

STUDIES of catecholamine metabolism in affective disorders (depressions and manias) have been one of the more productive areas of investigation in biological psychiatry during the past decade (SCHILDKRAUT, 1965, 1970; BUNNEY and DAVIS, 1965). Since the literature in this field has been reviewed recently (SCHILDKRAUT, 1973a), and because of space limitations, this paper will not provide another comprehensive review. Instead, I shall summarise aspects of our ongoing research in this area, focusing on recent studies of the urinary excretion of 3-methoxy-4-hydroxyphenyl-glycol (MHPG), a metabolite of norepinephrine which may provide some index of the synthesis and metabolism of norepinephrine in the brain (SCHANBERG et al., 1968a, b; MAAS and LANDIS, 1968; MAAS et al., 1972a).*

In one aspect of our research we have examined the changes in MHPG excretion which were associated with changes in affective state. The findings from our longitudinal studies of individual patients with naturally occurring or amphetamine-induced manic-depressive episodes indicate that levels of urinary MHPG are relatively lower during depressions and higher during manic or hypomanic episodes than after clinical remissions (GREENSPAN et al., 1970; SCHILDKRAUT et al., 1971, 1972b), and these findings have been confirmed by other investigators (BOND et al., 1972; JONES et al., 1973b) but not by all (BUNNEY et al., 1972). The relationship between MHPG excretion and clinical state that we have observed in manic-depressive patients, is illustrated in Table 1, which summarises the changes in MHPG excretion in a manic-depressive patient studied longitudinally through 5 successive periods which were defined by differences in clinical state.

However, all depressed patients do not excrete comparably low levels of MHPG, and recent preliminary findings have suggested that MHPG excretion may provide a biological criterion for classifying the depressive disorders and for predicting the responses to specific forms of antidepressant pharmacotherapy (MAAS et al., 1968, 1972b; SCHILDKRAUT et al., 1972a; SCHILDKRAUT, 1973b). For example, in a preliminary study of a small group of depressed patients, we observed favorable responses to treatment with amitriptyline in depressed patients with relatively higher levels of urinary MHPG but not in patients with lower levels of MHPG (SCHILDKRAUT, 1973b).

In order to explore further the relationship between MHPG excretion and the clinical classification of depressive disorders, we recently compared the urinary excretion of MHPG in a small group of patients (6 men and 6 women) with various

---

* While urinary MHPG may also derive, in part, from the peripheral sympathetic nervous system as well as the brain, recent studies in non-human primates suggest that an appreciable fraction of urinary MHPG may derive from norepinephrine originating in the brain (MAAS et al., 1972a).

TABLE 1. MHPG EXCRETION AND AFFECTIVE STATE IN A MANIC-DEPRESSIVE
PATIENT

Clincial state*	Treatment	N†	MHPG (µg/day)	P‡	MHPG (µg/g creat.)	P‡
Mildly depressed	None	5	1284 ± 76		966 ± 72	
				<0·05		<0·05
Very depressed	None	4	817 ± 176		609 ± 105	
						<0·05
Decreasing depression	ECT	12	999 ± 78		897 ± 63	
				<0·01		<0·001
Mildly hypomanic	None	4	1500 ± 70		1386 ± 59	
						<0·01
Mildly depressed	None	5	1384 ± 58		1117 ± 38	

* Sequentially occurring periods defined by clinical state.
† N = number of urine samples analysed in each period.
‡ P values are given for adjacent periods.

MHPG EXCRETION IN DEPRESSIVE DISORDERS

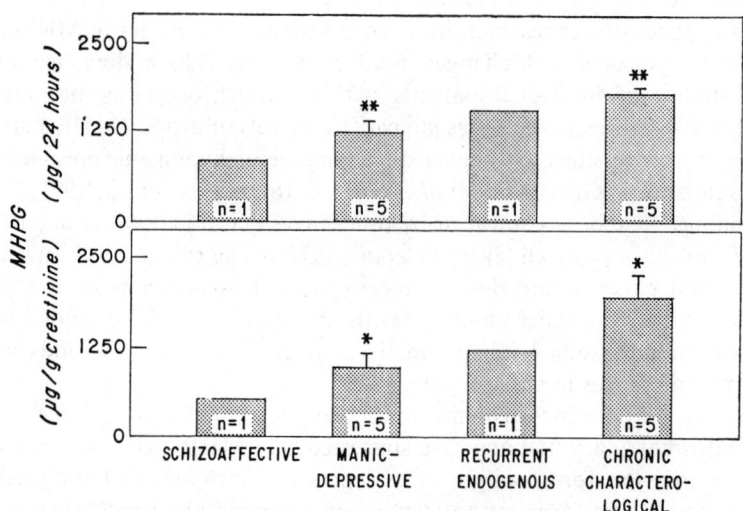

FIG. 1.—MHPG excretion in depressive disorders. MHPG was determined in 3–10 separate 24-hr urine samples from each patient. These individual values were averaged to obtain an overall value for each subject; this value was then used in computing the means and standard errors of the means for the two groups (manic depressive and chronic characterological depressions). According to our diagnostic criteria SCHILDKRAUT, 1970 three of the patients with chronic characterological depressions and one of the patients with manic-depressive depressions had signs and symptoms suggesting possible involutional depressive syndromes—e.g. agitation, somatic preoccupation, guilt, and paranoid ideation.
$*P < 0.05$, $**P < 0.02$ for difference between manic-depressive and chronic characterological depressions.
(Reproduced with permission from SCHILDKRAUT J. J. KEELER B. A., GRAB E. L., KANTROWICH J., HARTMANN E. MHPG Excretion and Clinical Classification in Depressive Disorders, Lancet, 1973).

clinically defined subtypes of depressive disorders* studied in hospital prior to treatment during drug free periods, and we examined our findings for possible relationships between urinary MHPG excretion and clinical measures of motor activity and anxiety (SCHILDKRAUT *et al.*, 1973a, b). As shown in Fig. 1, MHPG excretion was significantly lower in 5 patients with manic-depressive depressions (i.e., bipolar disorders with histories of hypomanias or manias) than in 5 patients with chronic characterological depressions (i.e. dysphoric depressive syndromes with no history of hypomania or mania). In a patient with a schizoaffective depression, MHPG excretion was lower than the mean for the manic-depressive depressions, while in a patient with a recurrent endogenous depression (but no history of hypomania or mania) MHPG excretion fell between the means for manic-depressive and chronic characterological depressions. These differences in MHPG excretion may be related to the recent observations of other investigators that platelet monoamine oxidase activity was decreased in patients with bipolar affective disorders as well as schizo-phrenias but relatively increased in other types of depressive disorders (MURPHY and WYATT, 1972; MURPHY and WEISS, 1972; NIES *et al.*, 1971).

We also measured the urinary excretion of norepinephrine, normetanephrine, epinephrine, metanephrine and 3-methoxy-4-hydroxymandelic acid (VMA) in these patients. As indicated in Table 2, the urinary excretion of norepinephrine tended to be lower in manic-depressive than in chronic characterological depressions, but this difference was not statistically significant. There were no differences in norm-etanephrine, epinephrine, metanephrine or VMA excretion between these two groups. Thus, of the urinary catecholamines and metabolites that we measured, only MHPG showed significant differences when values in the manic-depressive and chronic characterological depressions were compared (Table 2).

A correlation (of borderline statistical significance) between MHPG excretion and age was observed in the total group of 12 depressed patients. However, this may be related to the fact that the manic-depressives tended to be younger than the patients with chronic characterological depressions†, since other investigators found no correlation between MHPG excretion and age in a series of control subjects (DEKIRMENJIAN and MAAS, 1973).

TABLE 2. EXCRETION OF CATECHOLAMINES AND METABOLITES IN MANIC-DEPRESSIVE AND CHRONIC CHARACTEROLOGICAL DEPRESSIONS

	Manic-dep.	Chron. char. dep.	P
Norepinephrine*	21 ± 5	36 ± 11	NS
Normetanephrine*	180 ± 30	200 ± 50	NS
Epinephrine*	10 ± 2	9 ± 3	NS
Metanephrine*	190 ± 30	160 ± 30	NS
VMA*	3700 ± 310	3590 ± 230	NS
MHPG*	1240 ± 160	1800 ± 90	$< 0·02$

See caption to Fig. 1.
* Expressed as means $\pm$ S.E.M. in μg per day.

___

* The clinical criteria used in the classification of the depressive disorders have been described previously (SCHILDKRAUT, 1970).

† A similar difference in age has been noted in other studies, and may be explained by the fact that the age of onset is lower for manic-depressive disorders (bipolar) than for other types of depressive disorders (unipolar) (WINOKUR *et al.*, 1969).

Because it has been suggested recently that differences in physical activity or stress could conceivably account for the differences in MHPG excretion observed in patients with affective disorders (Ebert *et al.*, 1972; Maas *et al.*, 1971; Rubin *et al.*, 1972), we examined our data in several different ways for possible relationships between MHPG excretion and clinical assessments of motor activity and anxiety as reflected by scores on a modified Hamilton Depression Rating Scale (Hamilton, 1960). As shown in Table 3, there were no significant differences in scores for retardation, agitation and psychic or somatic anxiety between the manic-depressive and chronic characterological depressions. Moreover, when the entire group of 12 depressed patients were subdivided into the 4 with lowest, the 4 with intermediate and the 4 with highest scores on each of these items, no corresponding differences in MHPG excretion emerged (Tables 4 and 5). There were also no meaningful correlations between MHPG excretion and scores for retardation, agitation and psychic

TABLE 3. Hamilton depression rating scale items in manic-depressive and chronic characterological depressions

Hamilton item	Manic-dep.	Chron. char. dep.	P
Retardation	0·75 ± 0·28	0·59 ± 0·20	N.S.
Agitation	1·01 ± 0·17	1·20 ± 0·06	N.S.
Anxiety (Psychic)	2·50 ± 0·15	2·44 ± 0·18	N.S.
Anxiety (Somatic)	0·80 ± 0·34	1·08 ± 0·42	N.S.

TABLE 4. MHPG excretion in depressive disorders with low and high retardation or agitation

A. Retardation	H.D.R.S.* Score	MHPG (μg/day)
Low (N = 4)	0·21 ± 0·09	1510 ± 280
Middle (N = 4)	0·49 ± 0·03	1290 ± 180
High (N = 4)	1·24 ± 0·10	1620 ± 160

B. Agitation	H.D.R.S.* Score	MHPG (μg/day)
Low (N = 4)	0·77 ± 0·10	1510 ± 150
Middle (N = 4)	1·16 ± 0·01	1510 ± 70
High (N = 4)	1·44 ± 0·06	1390 ± 360

* H.D.R.S. = Hamilton Depression Rating Scale

TABLE 5. MHPG excretion in depressive disorders with low and high anxiety

A. Anxiety (Psychic)	H.D.R.S.* Score	MHPG (μg/day)
Low (N = 4)	2·11 ± 0·01	1510 ± 140
Middle (N = 4)	2·40 ± 0·12	1400 ± 210
High (N = 4)	2·83 ± 0·07	1500 ± 290

B. Anxiety (Somatic)	H.D.R.S.* Score	MHPG (μg/day)
Low (N = 4)	0·22 ± 0·08	1440 ± 120
Middle (N = 4)	0·83 ± 0·15	1460 ± 200
High (N = 4)	1·83 ± 0·17	1510 ± 320

* H.D.R.S. = Hamilton Depression Rating Scale

TABLE 6. CORRELATIONS BETWEEN HAMILTON DEPRESSION RATING SCALE ITEMS AND MHPG EXCRETION

Hamilton item	Correlation coefficient	P
Retardation	0·24	N.S.
Agitation	−0·19	N.S.
Anxiety (Psychic)	−0·19	N.S.
Anxiety (Somatic)	0·11	N.S.

or somatic anxiety (Table 6). Other investigators have recently made similar observations concerning the absence of an association between MHPG excretion and marked retardation or agitation in depressed patients (JONES et al., 1973a). These findings thus provide no support for the suggestion that differences in MHPG excretion in patients with affective disorders reflect only differences in activity or stress.

In interpreting these data, one cannot exclude the possibility that these Hamilton Depression Rating Scale items were not specific or sensitive enough to detect differences in retardation, agitation or anxiety which may have been related to MHPG excretion. However, this possibility is weakened by the fact that we did observe meaningful relationships between several of these Hamilton Depression Rating Scale items and certain other biochemical variables. For example, patients with high scores on retardation excreted significantly less epinephrine and tended to excrete less metanephrine than patients with low retardation scores, whereas patients with high scores on somatic anxiety excreted significantly more metanephrine than did patients with low scores on this item.

In collaboration with Dr. Ernest Hartmann, we also examined the relationship of MHPG excretion, in these depressed patients, to certain aspects of central nervous system activity as reflected by all-night electroencephalographic sleep recordings. We were particularly interested in the relationship between MHPG excretion and the amount of time spent in desynchronised—i.e., REM—sleep (D-time), since pharmacological studies in animals and man suggest that there may be an inverse relationship between central catecholaminergic activity and D-time (HARTMANN, 1970; HARTMANN et al., 1971a, b; WYATT, 1972; KUPFER and BOWERS, 1972), although this has not been confirmed in all studies (JOUVET, 1968; KING, 1971).

As shown in Table 7, when the data from the total group of 12 depressed patients were examined, there were no statistically significant correlations between MHPG excretion and total sleep time or slow-wave sleep time. However, a statistically significant inverse correlation was observed between MHPG excretion and D-time. When the subgroups of depressive disorders were examined separately the inverse correlation between MHPG excretion and D-time was high and significant in the

TABLE 7. CORRELATIONS BETWEEN MHPG EXCRETION AND SLEEP MEASURES IN DEPRESSIVE DISORDERS

Sleep measure	Correlation coefficient	P
Total sleep	−0·35	N.S.
Slow wave sleep	−0·15	N.S.
D-Time	−0·66	<0·02

patients with manic-depressive depressions. In addition, MHPG excretion was higher and D-time was lower in two patients studied during hypomanic episodes than in any of the patients with manic-depressive disorders studied during depressive episodes (SCHILDKRAUT *et al.*, 1973b). Taken in conjunction with pharmacological observations that have suggested an inverse relationship between D-time and central catecholaminergic activity, these findings appear to support the view that MHPG excretion may reflect central noradrenergic activity, particularly in patients with manic-depressive disorders.

In summary, during several longitudinal studies of individual patients we have observed that MHPG excretion was higher during hypomanias or manias, inter-mediate during well intervals and lower during depressions in patients with naturally occurring or amphetamine-induced manic-depressive disorders. Moreover, in a recent cross-sectional comparison of a small group of patients with various clinically defined subtypes of depressive disorders examined prior to treatment with anti-depressant drugs or electroconvulsive therapy, we observed that MHPG excretion was significantly lower in patients with manic-depressive depressions than in patients with chronic characterological depressions. MHPG excretion did not appear to be related to the degree of retardation, agitation or anxiety in these depressed patients; but MHPG excretion was inversely related to the time spent in desynchronised—i.e. REM—sleep (D-time), particularly in the patients with manic-depressive disorders. In the aggregate, these findings provide further evidence that alterations in central norepinephrine metabolism may be of importance in the underlying pathophysiology of at least some types of depressive disorders. Moreover, these findings suggest that MHPG excretion may provide a clinically useful biochemical criterion for classifying the depressive disorders and possibly also for predicting responses to specific forms of antidepressant pharmacotherapy.

Acknowledgements—This work was supported in part by USPHS Grant No. MH 15,413 from the National Institute of Mental Health.

REFERENCES

BOND P. A., JENNER F. A. and SAMPSON G. A. (1972) *Psychological Med.* **2,** 81–85.

BUNNEY W. E., JR. and DAVIS J. M. (1965) *Archs. Gen. Psychiat.* **13,** 483–494.

BUNNEY W. E., JR., GOODWIN F. K., MURPHY D. L., HOUSE K. M. and GORDON E. K. (1972) *Archs. Gen. Psychiat.* **27,** 304–309.

DEKIRMENJIAN H. and MAAS J. W. (1973) presented at the Annual Meeting, American Psychiatric Association, Honolulu, May, 1973.

EBERT M. H., POST R. M. and GOODWIN F. K. (1972) *Lancet* II, 766.

GREENSPAN K., SCHILDKRAUT J. J., GORDON E. K., BAER L., ARONOFF M. S. and DURELL J. (1970. *J. Psychiat. Res.* **7,** 171–183.

HAMILTON M. (1960) *J. Neurol. Neurosurg. Psychiat.* **23,** 56–62.

HARTMANN E. (1970) In: *Sleep and Dreaming.* (E. HARTMANN, Ed.) pp. 308–328, Little, Brown & Co., Boston.

HARTMANN E., BRIDWELL T. J. and SCHILDKRAUT J. J. (1971a) *Psychopharmacologia* (Berl.) **21,** 157–164.

HARTMANN E., CHUNG R., DRASKOCZY P. R. and SCHILDKRAUT J. J. (1971b) *Nature, Lond.* **233,** 425–427.

JONES F. D., DEKIRMENJIAN H. and MAAS J. W. (1973a) presented at the annual meeting of the American Psychosomatic Society, Denver, Colorado, April, 1973.

JONES F. D., MAAS J. W., DEKIRMENJIAN H. and FAWCETT J. A. (1973b) *Science* **179,** 300–302.

JOUVET M. (1968) *Science* **163,** 32–41.

KING C. D. (1971) *Adv. Pharmacol. & Chemother.* **9,** 1–91.

KUPFER D. J. and BOWERS M. B., JR. (1972) *Psychopharmacologia (Berl.)* **27,** 183–240.

MAAS J. W. and LANDIS D. H. (1968) *J. Pharmacol. Exp. Ther.* **163,** 147–162.

MAAS J. W , FAWCETT J. A., DEKIRMENJIAN H. (1968) *Archs. Gen. Psychiat.* **19**, 129–134.

MAAS J. W., DEKIRMENJIAN H. and FAWCETT J. (1971) *Nature, Lond.* **230**, 330–331.

MAAS J. W., DEKIRMENJIAN H., GARVER D., REDMOND D. E,. JR. and LANDIS H. D. (1972a) *Brain Res.* **41**, 507–511.

MAAS J. W., FAWCETT J. A. and DEKIRMENJIAN H. (1972b) *Archs. Gen. Psychiat.* **26**, 252–262.

MURPHY D. L. and WEISS R. (1972) *Am. J. Psychiat.* **128**, 1351–1357.

MURPHY D. L. and WYATT R. J. (1972) *Nature, Lond.* **238**, 225–226.

NIES A., ROBINSON D. S., RAVARIS C. L. and DAVIS J. M. (1971) *Psychosomatic Med.* **33**, 470.

RUBIN R. T., MILLER R. G., CLARK B. R., POLAND R. E. and ARTHUR R. J. (1972) *Psychosomatic Med.* **32**, 589–597.

SCHANBERG S. M., SCHILDKRAUT J. J., BREESE G. R. and KOPIN I. J. (1968a) *Biochem. Pharmacol.* **17**, 247–254.

SCHANBERG S. M., BREESE G. R., SCHILDKRAUT J. J., GORDON E. K. and KOPIN I. J. (1968b) *Biochem. Pharmacol.* **17**, 2006–2008.

SCHILDKRAUT J. J. (1965) *Am. J. Psychiat.* **122**, 509–522.

SCHILDKRAUT J. J. (1970) *Neuropsychopharmacology and the Affective Disorders* Little, Brown & Co., Boston.

SCHILDKRAUT J. J. (1973a) *Ann. Rev. Pharm.* **13**, 427–454.

SCHILDKRAUT J. J. (1973b) *Am J. Psychiat.* **130**, 695–699.

SCHILDKRAUT J. J., WATSON R., DRASKOCZY P. R. and HARTMANN E. (1971) *Lancet* ü, 485–486.

SCHILDKRAUT J. J., KEELER B. A., ROGERS M. P. and DRASKOCZY P. R. (1972b) *Psychosomatic Med.* **34**, 470.

SCHILDKRAUT J. J., DRASKOCZY P. R., GERSHON E. S., REICH P. and GRAB E. L. (1972a) *J. Psychiat. Res.* **9**, 173–175.

SCHILDKRAUT J. J., KEELER B. A., GRAB E. L., KANTROWICH J. and HARTMANN E. (1973a) *Lancet* I, 1251–1252.

SCHILDKRAUT J. J., KEELER B. A., PAPOUSEK M. and HARTMANN E. (1973b) *Science* **181**, 762–764.

WINOKUR G., CLAYTON P. J. and REICH T. (1969) *Manic-Depressive Illness* C. V. Mosby, St. Louis.

WYATT R. J. (1972) *Biol. Psychiat.* **5**, 33–64.

Frontiers in Catecholamine Research 1973, pp. 1173 to 1179. Pergamon Press. Printed in Great Britain.

COLLABORATIVE PSYCHOPHARMACOLOGIC STUDIES EXPLORING CATECHOLAMINE METABOLISM IN PSYCHIATRIC DISORDERS

Baron Shopsin,[1] Sherwin Wilk,[2] Samuel Gershon,[1]
Mark Roffman,[3] and Menek Goldstein[3]

[1] Department of Psychiatry, Neuropsychopharmacology Research Section, New York University School of Medicine, New York, N.Y.; [2] Department of Pharmacology, Mount Sinai School of Medicine of the City of New York, New York, N.Y.; [3] Neurochemistry Laboratories, Department of Psychiatry New York University School of Medicine, New York, N.Y., U.S.A.

INTRODUCTION

Our departments have closely collaborated in designing and carrying out various clinical-biochemical studies over the past several years in an attempt to elucidate the possible role of catecholamines and particularly norepinephrine, in the etiopathogenesis underlying the affective psychiatric illnesses (Angrist et al., 1972 Shopsin et al., 1971, 1972; Wilk et al., 1971b, 1972). The present paper represents a report of progress and includes updated findings as well as data from new experimental approaches to the problem which were initiated during the past year.

SPECIFIC STUDIES

Cerebrospinal fluid MHPG

Studies of amines and their metabolites in cerebrospinal fluid (CSF) likely represent the most direct available means for evaluating amine metabolism in the human subject (Moir et al., 1970). The occurrence of 3-methoxy-4-hydroxyphenyl glycol (MHPG) in CSF has been documented (Schanberg et al., 1968). It has been our belief, based on the interpretation of available data, that measuring CSF-MHPG in man likely yields the most direct feasible information on the relationship between catecholamines and affective disorder. With several modifications (Wilk et al., 1971a) of the gas–liquid chromatographic technique developed for measuring MHPG in urine (Wilk et al., 1967), we have analysed CSF-MPHG levels in carefully selected patient groups before and after drug therapy (Shopsin et al., 1971; Wilk et al., 1971b, 1972). Our studies collectively indicate that the vast majority of psychiatric patients investigated to date, regardless of diagnosis, show CSF-MHPG levels falling within the range of values seen in normal control subjects. All depressed patients, whether they be endogenous or reactive, bipolar or unipolar and irrespective of the degree of psychomotor agitation or retardation show normal MHPG levels in CSF. In a previous study (Wilk et al., 1972), we showed that some manic and schizophrenic patients have CSF-MHPG levels elevated beyond the range of our normal controls; for the manic and schizophrenic groups taken as a whole, statistical significance was attained for each group, with greatest significance in the manics ($P < 0.01$). There was no statistically significant difference between the manic and schizophrenic groups. Those patients in each group showing abnormal values do not show clinical or chemical characteristics distinguishing them from other patients having the same diagnosis nor do they differ as regards age, severity, or duration of illness, or the degree of motor activity.

With the addition of more patients, we currently have 15 manic individuals, 10 endogenous depressives, 2 reactive depressives and 26 schizophrenics who have been explored with regard to CSF-MHPG values (Fig. 1). The greater "n" has eliminated any significance between the manic or schizophrenic populations compared to controls.

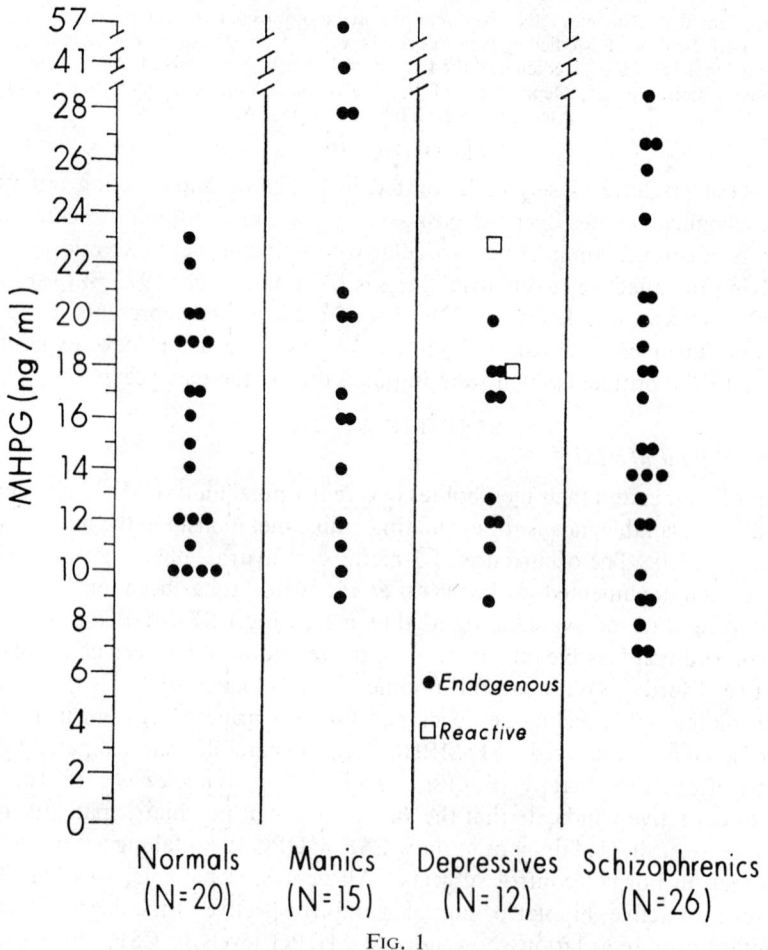

Lumbar CSF MHPG Levels in Hospitalized Psychiatric Patients and Normals (5/8/73)

Fig. 1

Again, the additional endogenously depressed patients show normal values of CSF-MHPG and for the depressed population as a whole there is no statistical difference compared to the mean of the control group.

Coppen has recently found that MHPG in the CSF of his endogenously depressed individuals is within the normal range (personal communication); the NIMH group finds CSF-MHPG lower in some depressed individuals (Post *et al.*, 1973). Because there are no apparent clinically discernible features in those patients with high or low values from others within the same diagnostic category showing normal CSF-MHPG levels, implications are that the "abnormal" values seen in some manic, schizophrenic and depressed patients across different studies relate to secondary factors. It is less likely that these elevations represent a primary biochemical defect common to these illnesses.

It is noteworthy that clinical change in some of our bipolar manic-depressives is clearly associated with a marked shift in MHPG levels corresponding to the direction of mood swing (SHOPSIN et al., 1971). While this cannot be duplicated in all bipolar subjects, this would be consistent with previous data by the NIMH group suggesting that catecholamines may only be involved in the "switch process" rather than the steady state, i.e., that only the onset and cessation of the manic state may be associated with abnormal catecholamine metabolism (GOODWIN et al., 1970).

CSF-VMA

Recent data question whether conjugated MHPG is the major metabolite of brain NE (STOLK and BARCHAS, 1970). A study by CHASE et al. (1971) suggests that the major metabolites of cerebral NE, at least in the dog, may be vanylmandelic acid (VMA) and not MHPG. Quantitation of VMA in CSF would, therefore, represent an important tool with which to explore noradrenergic function in the affective illnesses. However, CSF levels of VMA were reported by Dr. Wilk at this meeting to be less than 2ng/ml. VMA represents a minor metabolite of NE in CSF, the levels of which are too low to be of significant value for the assessment of abnormalities in catecholanime metabolism.

Longitudinal studies of MHPG in urine

A longitudinal study was carried out in 3 depressed females and one male manic patient in an attempt to monitor changes in urinary MHPG with change in clinical state during drug treatment. Urinary MHPG was measured using a modification (DEKIRMENJIAN and MAAS, 1970) of the method of WILK et al. (1967). In the manic individual, MHPG levels were measured both in mg/24 hr and by μg/mg creatinine for comparison.

The results in the manic indicate that MHPG levels when plotted as mg/24 hr or μg/mg creatinine gave similar patterns.

There was a wide variability of MHPG levels in each patient during drug treatment irrespective of diagnoses. Clearly, the daily values of this NE metabolite in urine do not correspond to change in mood-affect-behaviour or state of motor activity in these patients; both high and low MHPG excretion rates are obtained during any documented shift in clinical affective state.

In the 3 depressed females, and despite the marked variability in daily urinary MHPG, all values were within the normal range throughout the study. Two of the depressives showed psychomotor retardation, the third was agitated; all 3 individuals showed symptom improvement on imipramine (Tofranil) with total remission on 300 mg by the third week of study. The MHPG excretion was generally low in the manic patient during 3 weeks of treatment with chlorpromazine: he was severely manic throughout.

Correlation of MHPG in CSF and urine

An attempt was made to correlate baseline lumbar CSF and urinary levels of MHPG in 9 hospitalized psychiatric patients including 3 unipolar depressed females,

2 manics, 3 schizophrenics and 1 alcoholic. Computerised linear regression analysis failed to reveal any significant correlation between MHPG values in these two biological fluids (Fig. 2).

We have previously reported (Angrist *et al.*, 1972) that urinary MHPG excretion was widely variable showing no consistent change during or after large dose amphetamine administration to healthy, non-psychiatric volunteer subjects. Likewise, there was no consistent change in CSF MHPG levels; change in CSF MHPG, when apparent, could not be correlated with mood alteration or change in urinary MHPG excretion.

MHPG: Urinary Values (μg/mg Creatinine)
Plotted Against CSF Values (ng/ml)

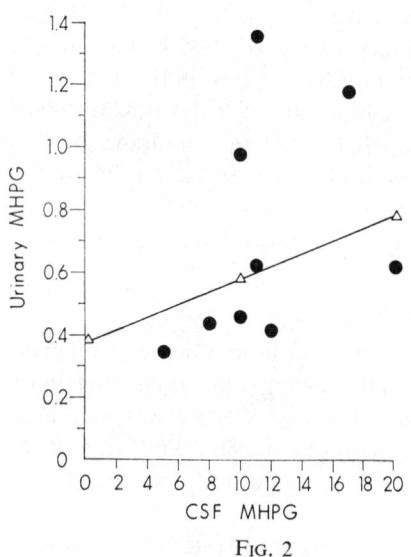

FIG. 2

Erythrocyte, COMT (EC2.11.6) activity in psychiatric patients

In efforts to discriminate between the kinetic properties of catechol-O-methyltransferase (COMT) in psychiatric patients compared to normal control subjects, we have measured the activity of this enzyme by varying the substrate concentration. From the initial velocity plots, the apparent K_m and V_{max} values were calculated.

Bloods were drawn on hospitalized psychiatric patients after a minimum washout period of 1 week to 10 days; in a large number of schizophrenics, bloods were sampled for COMT only after showing a "Forrest-negative reaction".* Several patients had never previously ingested psychoactive drugs. The method employed for COMT determination was that of McCaman (1965) with minor modifications. Enzyme source was lysed red blood cells in an aqueous solution containing 5×10^{-4} M dithiothreotol. Incubations were carried out for 10 min at 37·5°C; this reaction was stopped by adding 0·1 ml of 3 N HCl. ^{14}C product was extracted in tolueneisoamylalcohol (7:3). An aliquot of the organic phase was counted by liquid scintillation spectrophotometry.

The results reveal no apparent differences in either the K_m for the substrate or

---

* Colour reaction in urine for the presence of chlorpromazine and its metabolites.

V_{max} in any of the patient groups or normal controls. However, if the schizophrenic group is subdivided into schizo-affectives, chronic undifrererentiated types and paranoids, an increase in K_m emerges for the paranoid schizophrenics compared to either schizo-affectives or chronic undifferentiated types, these latter groups are both similar to normals. When paranoid schizophrenics are compared to "all other groups" (including normals, manics, depressives and other schizophrenics) the K_m is biometrically greater (Fig. 3); no difference is noted in V_{max}. However, the difference between the apparent K_m values for these two groups is $\pm$ 70 per cent; the statistical significance therefore, requires validation on a larger population of patients.

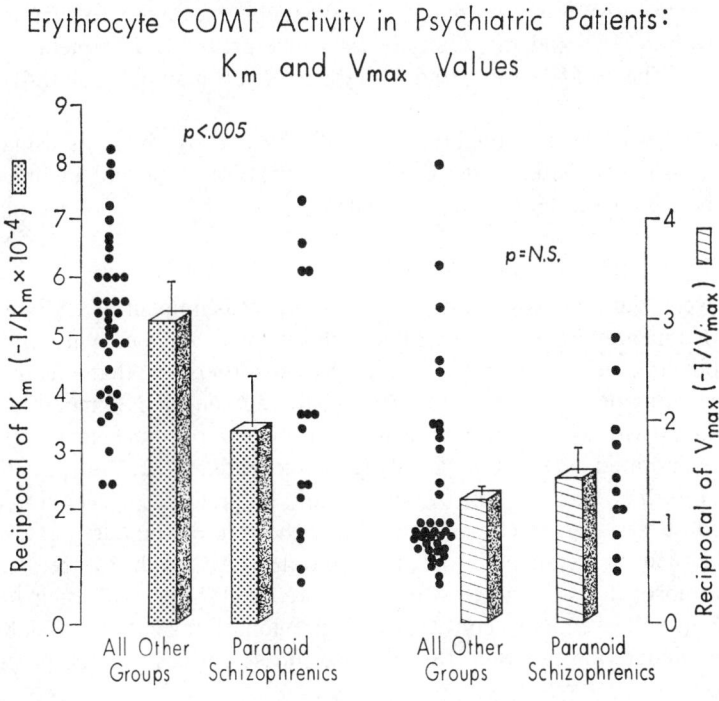

Erythrocyte COMT Activity in Psychiatric Patients: K_m and V_{max} Values

FIG. 3

Steroids such as estrogen may act as substrate for COMT (KNUPPEN *et al.*, 1969). Our data is presently being analysed to isolate the influence of age and sex including data comparison between post-menopausal women and those having ovulatory phases.

These findings await duplication and critical evaluation; at present they should be regarded as preliminary and tentative. The actual meaning of different apparent K_m values for the substrate of human erythrocyte COMT is questionable. Even *assuming* that changes in a peripheral enzyme reflect similar central events, the K_m values in such a heterogenous system as human erythrocytes could be influenced by various factors, e.g., non-competitive endogenous inhibitors of this enzyme, albeit these variables should hold across the different hospitalized psychiatric populations explored. If indeed the apparent K_m values represent a different isozyme of COMT, it is then conceivable that inactivation of catecholamines occurs at a different (slower) rate. A study dealing with the purification characteristics of COMT obtained from paranoid schizophrenics and other psychiatric populations appears warranted.

Case report

Alpha-methylparatyrosine(α-MPT) lowers brain catecholamines (Guldberg, 1969; Sjostrom, 1971; Bunney *et al.*, 1971). In an attempt to appraise the catecholamine hypothesis of depression, a female depressive (recurrent unipolar) was placed on imipramine (Tofranil) and increased to 375 mg/day by the third week of treatment at which time clinical response was assessed at 3+ (of a possible 4). At this point α-MPT was *added* (500 mg/day) and gradually increased to 3 g/day over the next 3 weeks (imipramine was continued). The patient continued to respond during treatment with α-MPT and she was rated as 4+ improved at treatment termination with α-MPT. When α-MPT was withdrawn the patient (on imipramine still) continued to do well.

The patient was rated throughout the study by two psychiatrists using different objective assessments including the Hamilton Depression Scale and a Clinical Global Rating. These data will be recorded elsewhere.

COMMENT

Collectively, our studies suggest that either the measurement of CSF and urinary MHPG in manic and depressed patients will not provide a clinically useful biological tool in helping to elucidate the role of NE in these disorders, or, that affective illnesses may not be explained solely on the basis of altered central NE metabolism. Our present findings with α-MPT appear relevant in this regard. Such new experimental findings as elaborated here and previously (Angrist *et al.*, 1972; Shopsin *et al.*, 1971, 1972; Wilk *et al.*, 1971b, 1972) by our group, suggest that other, more sensitive studies need be designed in efforts to elucidate the role of the adrenergic system as well as other neuronal systems in the affective states. Although, by necessity, the experimental approaches and systems to be explored will differ, we anticipate that the resulting data will eventually converge with previous information to elaborate the mechanisms underlying not only the affective illness entities, but other psychiatric disorders as well.

Acknowledgements—This work was supported by USPHS Grant Nos MH-17436[1] and MH-21638-02[2].

REFERENCES

Angrist B., Shopsin B., Gershon S. and Wilk S. (1972) *Psychopharmacologia* **26**, 1–10.
Bunney W. E. Jr., Brodie K. E., Murphy D. L. and Goodwin F. K. (1971) *Am. J. Psychiat.* **127**, 872–881.
Chase T. N., Breese G. R., Gordon E. K. and Kopin I. J. (1971) *J. Neurochem.* **18**, 135–140.
Dekirmenjian H. and Maas J. W. (1970) *Anal. Biochem.* **35**, 113–122.
Goodwin F. K., Murphy D. L., Brodie H. K. H. and Bunney W. E. Jr. (1970) *Biol. Psychiat.* **2**, 341–346.
Guldberg H. C. (1969) In: *Metabolism of Amines in the Brain* (Hooper G., Ed.) Macmillan, London, pp. 110–131.
Knuppen A. R., Holler M., Tilmann D. and Breuer H. (1969) *J. Physiol. Chem.* **350**, 1301–1309.
McCaman R. E. (1965) *Life Sci.* **4**, 2353–2359.
Moir A. T. B., Ashcroft G. W., Crawford T. B. B., Eccleston D. and Guldberg H. C. (1970) *Brain* **93**, 357–368.
Post R. M., Gordon E. K., Goodwin F. K. and Bunney W. E. Jr. (1973) *Science* **170**, 1002–1003.

SCHANBERG S. M., SCHILDKRAUT J. J., BREESE G. R. and KOPIN I. J. (1968) *Biochem. Pharmacol.* **17**, 247–254.

SHOPSIN B., WILK S., GERSHON S., DAVIS K. and SUHL M. (1971) *Arch. Gen. Psychiat.* **28**, 230–238.

SHOPSIN B., FREEDMAN L. S., GOLDSTEIN M. and GERSHON S. (1972) *Psychopharmacologia* **27**, 11–16.

SJOSTROM R. (1971) *Psychopharmacologia* **22**, 214–216.

STOLK J. M. and BARCHAS J. D. (1970) *Pharmacologist* **12**, 268.

WILK S., GITLOW S. E., CLARKE D. D. and PALEY D. H. (1967) *Clin. Chim. Acta* **16**, 403–408.

WILK S., DAVIS K. L. and THACKER S. B. (1971a) *Anal. Biochem.* **39**, 498–504.

WILK S., SHOPSIN B., GERSHON S., GREEN J. P. and DAVIS K. L. (1971b) *Pharmacologist* **13**, 352.

WILK S., SHOPSIN B., GERSHON S. and SUHL M. (1972) *Naturel, Lond.* **234**, 440–441.

Frontiers in Catecholamine Research 1973, pp. 1181 to 1185. Pergamon Press. Printed in Great Britain.

REPORTER'S NOTES

M. A. LIPTON

To PARODY a phrase common to enzymologists studying catecholamine metabolism, methods may be considered the rate limiting step in the synthesis of new information. The portion of the International Catecholamine Symposium devoted to the study of catecholamines in man dramatically demonstrates the advance in methods over the past decade. From these have come new strategies for investigation, new data and new concepts regarding the pathogenesis of and the therapeutic mechanisms involved in pharmacotherapy for several mental and neurological diseases. Fifteen papers and six invited discussions were presented. Each of them utilised methods unavailable a decade ago. Since new methods often compete with others as tools for the solution of problems and frequently have imperfections, it is not surprising that open discussion from the floor was often challenging, animated and limited only by the time available.

Well established methods such as fluorometry have not been discarded, but have been supplemented by the common use of gas liquid chromatography with and without simultaneous mass fragmentography. Radioimmune assays have been developed recently by Spector and others for small molecules in the biogenic amine series. These are coupled with larger molecules to render them antigenic so that antibodies may be produced in animals. Antibodies have been produced for norepinephrine and serotonin which can now be measured in picogram quantities. These methods have thus far been used relatively little in the study of catecholamine metabolism in man, but their specificity and versatility is such that they will undoubtedly be increasingly employed for the study of not only metabolites but enzymes as well. More commonly used are the elegant microenzymatic assays developed by Axelrod and others employing isotopically labelled substrates Using known quantities of the appropriate substrate they can be used for assays of enzyme activity. Alternatively using specific enzymes in excess they can be used for the measurement of metabolites of interest. Improved methods now exist for the measurement of tyrosine and tryptophan hydroxylase, decarboxylases, dopamine beta hydroxylase, catechol methyl transferase, N-methyl transferase and monoamine oxidase. Combined with appropriate chromatographic and electrophoretic techniques for separation of proteins they have already shown that isoenzymes of differing activities may exist. Enzymatic or gas chromatographic methods now exist that permit the measurement in nanogram and even picogram quantities of phenylalanine, tyrosine, norepinephrine, dopamine, serotonin, tryptamine, octopamine, methylated tryptamines, homovanillic acid, MHPG, dopac, 5-hydroxy indole acetic, acetylcholine GABA and others. These methods and results obtained with them were presented by all of the speakers and discussants. Sedvall presented a new method for studying *in vivo* turnover of dopamine in which an atmosphere containing O^{18} is inhaled and then incorporated into appropriate substrates. Though still limited to animal studies he clearly demonstrated that tyrosine is the predominant precursor of brain dopamine and that turnover of dopamine is markedly altered by chlorpromazine.

With the new methods and from animal studies using them have come new strategies for determining the chemical events of the human brain when this tissue is not directly available. Available fluids and tissues are obviously used but these have expanded in scope and sophistication. Urinary studies carried out by Maas, Schilkdraut and Goodwin are now almost invariably longitudinal and are investigated while the patient is ill, during treatment and after recovery. In such studies greater attention is given to control of physical activity and stress. Correlations with mood and sleep patterns are also made. Cerebrospinal fluid may also be studied longitudinally though clearly not as frequently. Results of studies with this fluid were reported by Drs. Roos, Sharman, Laverty, Wilk, Maas, Chase, Goodwin, Schildkraut, Shopsin, Gershon and Angrist. The administration of probenecid which diminishes transport of important metabolites out of the CSF, though not devoid of technical and interpretive difficulties probably permits the judgment that levels of catecholamine metabolites found in CSF after probenecid more accurately reflects synthesis, release and turnover in the brain than did earlier methods.

Whole blood, red cells, serum and platelets are also useful tissues for human analysis. Not only have they been employed for measurement of neurohumors and their metabolites, but even more interestingly, for the detection and quantitation of enzymes involved in the synthesis and degradation of the biogenic amines. The finding of low red cell COMT in depressed women which persists after the depression has lifted may offer a clue about the genetics of this illness. Freedman, Roffman and Goldstein feel that plasma DBH reflects peripheral sympathetic activity and offered interesting data regarding changes in this serum enzyme in neuroblastoma, familial dysautonomia, Huntington's Chorea, Down's syndrome and Parkinson's disease. They found no changes in affective disorders, but do find changes in neoplastic disorders which do not involve the sympathetic nervous system. This may reflect changes in the degradation rate of the serum enzyme. Weinshilboum offered biochemical evidence that serum DBH is biochemically and immunologically identical to that found in sympathetic nerves and the adrenal medulla. He also reported very low figures in neonates with a large increase in the first few years of life and a very interesting sibling-sibling corelation of serum DBH substantiating some genetic control. Lamprecht in his invited discussion offered new data relating intact pituitary function to serum DBH and also reported lowered DBH in hypertension.

Platelets are especially interesting because they contain amine storage vesicles and ordinarily store and release serotonin. However, they can also transport and store epinephrine, norepinephrine, dopamine and octopamine by displacing the stored serotonin. While platelets do not contain tyrosine hydroxylase nor tryptophan hydroxylase and therefore cannot synthesise the biogenic amines, they can degrade them with monoamine oxidase (MAO). Murphy reported that platelet MAO has substrate and inhibitor characteristics which resemble neural MAO more than plasma MAO. They may therefore also offer a small window to the events of the brain. In this regard Murphy's finding that levels of platelet MAO are genetically determined and that monozygotic twins discordant for schizophrenia both have low platelet MAO is of the greatest interest.

Still another strategy involves the study of the brains of patients with psychiatric or neurologic diseases who die from suicide or other causes not directly related to the disease under scrutiny. This strategy which has been considerably improved by the development of micromethods has been ingeniously employed by Hornykiewicz to obtain further insights into the nature of the chemical-pathological lesion of the basal ganglia as well as into mechanism of the therapeutic action of Levodopa. He sees the Parkinsonian Syndrome which occurs in the endogenous disease or which can be induced reversibly by reserpine or the phenothiazines as a "striatal dopamine deficiency disease". Probably because of redundancy in the CNS and then super-sensitivity of receptors, the striatum can functionally compensate for the loss of some DA. Later, decompensation occurs. Hornykiewicz was also able to demonstrate at autopsy that patients treated with high doses of dopa had higher levels of DA in the caudate and putamen than did untreated patients and that good responders achieved higher levels than did poor responders. Parkinson's disease is however not exclusively a dopamine deficiency disease. Hornykiewicz also finds reduction of glutamic acid decarboxylase suggesting an abnormality in the GABA system of the striatal system. He also finds evidence in animals that an intact noradrenergic system is required for full effectiveness. Finally, the serotonergic system may also be involved since Prange and others found that the administration of pyridoxine and tryptophan in large doses increased Parkinsonian symptoms.

Still another effective strategy is that derived from experimental therapeutics. Following the finding that triiodothyronine enhanced the therapeutic efficacy of imipramine in affective disorders, Prange and coworkers tested the recently available thyrotropin releasing hormone (TRH) in depression and reported that alone it caused a rapid though transient diminution of depressive symptoms. Hypothalamic hor-mones have usually been considered to directly effect only the pituitary. However, Breese and others have found that TRH has CNS effects in hypophysectomized animals. This raises the interesting possibility that the effects of TRH in elevating mood may be through some direct influence on the CNS. Although the role of biogenic amines acting on the hypothalamus to cause synthesis and discharge of the releasing factors is gradually being unravelled, the converse is not true. Thus TRH has no known effects on synthesis or turnover of biogenic amines. The only positive effect thus far noted is an elevation of dopamine in an MAO inhibited mouse given dopa which is higher in the animal receiving TRH than in controls.

The nature of receptors and their variation with stress, drugs and adaptation remains an area of extreme interest, but still suffers from inadequate methods for study.

Indirect evidence was presented by Hornykiewicz, Prange and Snyder that alterations in receptor function may occur in Parkinsonism, affective disorders and schizophrenia. Snyder in particular presented evidence implicating dopamine in schizophrenia. This evidence is based upon the mimicking of schizophrenic symptoms by amphetamines, X-ray crystallographic evidence showing a structural resemblance between antipsychotic phenothiazines and dopamine and the accumulating evidence that antipsychotic drugs act as receptor blockers. Since biogenic amines do not act in vacuo but rather on specific receptors the biological defect in schizophrenia may as well reside at the receptor level as at the transmitter level. Eccleston, an invited discussant supported the view that receptor sensitivity is the basic problem in depres-sion and offered animal evidence that following lesions of the locus coeruleus more cyclic AMP is made on the lesioned side.

Questions and discussion from invited discussants and from the floor ranged in scope from disagreements with methods and results to interpretations of data. For example a statement by one of the speakers regarding the difficulties in separating isomers by gas chromatography was partially answered by Wilk who pointed out that Dziedzic had recently reported a method for separating HVA from isoHVA by gas chromatography. Wilk who in his presentation reported a new method for the measurement of VMA in cerebrospinal fluid which separates VMA, HVA, 3,4-dihydroxy phenylacetic and parahydroxy phenylacetic acids and who with this method finds only 2 ng of VMA and 33 ng of HVA was asked why dopamine is not metabolized to a glycol but goes only to the acid, while NE goes mainly to the glycol with very little going to VMA. Sedvall was questioned about the adequacy of the O^{18} method by Kaufman who reported that Udenfriend had demonstrated that only 60 per cent of O^{18} was incorporated with pure enzyme *in vitro*. Roos, who reported that in the CSF HVA was low in both depression and mania, was questioned by Kety as to why this should be so. The validity of the serum DBH method was questioned by Geffen who stated that by immunoassay most serum DBH was inactive and that a new immuno assay for DBH using human antibodies was being developed by Dr. Udenfriend. Dr. Youdim questioned Murphy's model of platelet MAO as resembling brain MAO, stating that substrate specificity of platelet MAO differed from brain and liver MAO and that electrophoresis of partially purified platelet MAO on polyacrylamide gel gave a single band in contrast to brain and liver MAO. The dopa decarboxylase assay was also questioned by Dr. B. L. Goodwin who offered evidence that the assay based on the formation of CO_2 was susceptible to errors such as the autooxidation of dopa to melanin and suggested that dopamine must be measured to accurately assay dopa decarboxylase. A discussant from the floor pointed out an apparent discrepancy between the results presented by Drs. Hornykiewicz and Chase. The former found increased DA turnover, the latter decreased in Parkinson's disease. In response to the question, Dr. Chase pointed out that his CSF method reflected total brain activity, while Dr. Hornykiewicz was examining only basal ganglia. He suggested that both might be correct.

The subject of the meaning of urinary MHPG in relation to the affective disorders was discussed to the point of debate by speakers and discussants. Maas who works mainly with urinary MHPG found that many depressed patients have a normal level of MHPG excretion, but he identifies a subgroup with low urinary MHPG. Such patients suffer from primary affective disorders, show an elevation of mood with amphetamines and respond well to imipramine. Unfortunately he offered no data on MHPG excretion following drug treatment and clinical recovery. Maas finds no changes in urinary MHPG with exercise. Goodwin finds substantial changes. Shopsin and his group found normal levels of MHPG in the CSF of depressed patients. Furthermore, they found no correlation between CSF, MHPG and urinary MHPG. Schildkraut who generally finds low urinary MHPG in depression and elevation in mania also finds exceptions. Patients with relatively high urinary MHPG excretion apparently respond better to amitriptyline than to imipramine. If the data were truly clear cut, urinary MHPG measurements might be of value in determining the drug of choice. Low MHPG, says Maas, means a good therapeutic response to imipramine; high, says Schildkraut, means a good response to amitriptyline. Clearly disagreements and overlap are of such magnitude that much more must be done before chemical classifications can be done and choices made.

This reporter's impression is that the fields of biological psychiatry and neurology involving the catecholamines are in an exciting and tumultuous state. New methods are quickly developed and used in clinical studies before they are fully standardized and acceptable. Ingenious but perhaps simplistic hypotheses are offered and found to be only partially adequate. Significant gaps in information remain. One of these is in the area of understanding the receptors and their alterations. Another is in the almost certain interaction of the various aminergic systems with each other and with the GABA and acetylcholine systems. The hypothalamic polypeptides have been introduced into both neurologic and psychiatric treatment, but their complete physiologic role requires much more understanding. Compared to 10 years ago the field is much more technically sophisticated and somewhat more conceptually so. It is mind boggling to speculate on what the next decade will bring.

Frontiers in Catecholamine Research 1973, pp. 1187 to 1188. Pergamon Press. Printed in Great Britain.

CLOSING REMARKS

MERTON SANDLER

Bernhard Baron Memorial Research Laboratories and Institute of Obstetrics and
Gynaecology, Queen Charlotte's Maternity Hospital,
London W6 OXG, U.K.

THE SHEER size and scope of this meeting testify to the vast expansion which has occurred in the catecholamine field since the Milan symposium in 1965. It must be obvious even to the casual observer that such progress has largely been made possible by technical advances. If the interval between the first meeting at Bethesda in 1959, and that in Milan was dominated by the availability of radioactive catecholamines of high specific activity, the last intercongress era was the one in which the formaldehyde fluorescence technique reached its full flowering. Whilst it would be rash to try to extrapolate forward, we may certainly take note that two important new techniques have arrived, each of which seems destined to make an overriding contribution to this research area; I refer to immunofluorescence localisation procedures and gas chromatographic-mass spectrometric identification and assay (GC-MS).

The present meeting has been characterised by a massive virtuoso display of new information by the acknowledged leaders in the field. The data, for the most part, were impeccable—but do we lay too great a stress, perhaps, on the same, tired old enzymes? It is true that we are still a long way from understanding their mechanisms. However, with certain honourable exceptions, scant attention was paid to metabolic transformations such as transamination, or m-dehydroxylation which fashion alone has dubbed 'minor'; or to enzymes of great potential importance such as phenol sulphotransferase or aldehyde reductase. Above all, the very fact of our having a meeting on this theme means that we separate and encapsulate the catecholamines artificially out of their biological context.

As I have indicated elsewhere (SANDLER, 1972a), we measure catecholamines largely because we are *able* to measure them, because they stick so obligingly on to alumina. Similarly, 5-hydroxytryptamine possesses a relatively specific fluorescence spectrum in 3 N HCl which makes its assay a simple one. And so we have meetings about them, whereas relatively undistinguished molecules such as phenylethylamine, phenylethanolamine, m- and p-tyramine and octopamine, tryptamine and their N-methylated derivatives tend to be ignored. Never mind! They are starting to emerge into the literature, albeit timidly, thanks to a battery of new enzymological techniques (see AXELROD and SAAVEDRA, 1973) and, perhaps, of greater long-term importance, GC-MS. Indeed, at this present meeting, one heard for the first time the word 'octopaminergic' whispered shyly.

It does not need much insight to reach the conclusion that all these substances must work in concert, not only with each other but with diamines and polyamines, peptides, prostaglandins, amino acids, and other putative transmitters still to be identified. Future meetings of this type may well concentrate on such interrelationships rather than study a single group of compounds in isolation, away from its biological milieu.

If the philosophy of meetings such as this is somewhat suspect, there was much that was new and startling in the present example; however, any item selected for special mention must perforce be highly subjective and arbitrary. There were three topics at this meeting which excited me personally more than any other:

(1) The whole question of an extrapituitary action of the hypothalamic releasing factors, as raised by Dr. Lipton (p. 1181). Parenteral administration in the rat of large amounts of two of them, thyrotropin releasing hormone and melanocyte stimulating hormone release inhibiting factor, seems to have no gross effect, however, on catecholamine metabolism (SANDLER et al., 1973b).

(2) The demonstration by Dr. Fellman of trihydroxyphenolic acid formation in patients on L-dopa treatment, according to earlier prediction (CALNE et al. 1969), and its therapeutic implications, with special reference to the general role of L-dopa transamination (SANDLER, 1972b).

(3) The finding by Dr. Garattini and his colleagues of tetrahydropapaveroline, the Schiff base condensation product of dopamine with the aldehyde generated from its own oxidative deamination, in the brains of rats chronically treated with L-dopa. This observation complements our earlier detection of these compounds in L-dopa treated parkinsonian patients (SANDLER et al., 1973a), the first conclusive demonstration of their in vivo formation in the animal kingdom.

It should be noted that only the first of these topics was the subject of a formal presentation; the second emerged during general discussion; the third was a personal communication during one of the coffee breaks and was not raised at the actual conference sessions. This point underlines one of the most important functions of a meeting like this; it provides a forum for personal interchange of ideas, and intellectual cross-fertilisation. Thus, scientists mulling over the same problem can view it afresh from a slightly different angle, so that the eventual solution can be attained more speedily. Exchange of ideas by printed word alone has its limitations. To get an insight into what a man really is thinking, you have to meet him face to face. Perhaps it would not be such a bad thing after all if we set the wheels in motion for a Fourth International Catecholamine Symposium.

REFERENCES

AXELROD J. and SAAVEDRA J. M. (1973) Octopamine, phenylethanolamine, phenylethylamine and tryptamine in the brain. In: Aromatic Amino Acids in the Brain, (Eds. G. E. W. WOLSENHOLME and D. FITZSIMONS) Associated Scientific Publishers, Amsterdam, in press.
CALNE D. B., KAROUM F., RUTHVEN C. R. J. and SANDLER M. (1969) Brit. J. Pharmac. 37, 57–68.
SANDLER M. (1972) Proc. R. Soc. Med. 65, 584–585.
SANDLER M. (1972) Catecholamine synthesis and metabolism in man (with special reference to parkinsonism). In: Handbook of Experimental Pharmacology, Vol. 33, (Ed. H. BLASCHKO and E. MUSCHOLL) Springer, Berlin, pp. 845–899.
SANDLER M., BONHAM CARTER S., HUNTER K. R. and STERN G. M. (1973) Nature, Lond. 241, 439–443.
SANDLER M., GOODWIN B. L., LEASK B. G. S. and RUTHVEN C. R. J. (1973) Lancet i, 612.

INDEX

Printed in Great Britain by A. Wheaton & Co., Exeter

F